Handbook of Corrosion Engineering

Introduction to
Corrosion
Engineering

Handbook of Corrosion Engineering

Pierre R. Roberge

McGraw-Hill

New York   San Francisco   Washington, D.C.   Auckland   Bogotá
Caracas   Lisbon   London   Madrid   Mexico City   Milan
Montreal   New Delhi   San Juan   Singapore
Sydney   Tokyo   Toronto

Library of Congress Cataloging-in-Publication Data

Roberge, Pierre R.
   Handbook of Corrosion Engineering / Pierre R. Roberge.
     p.   cm.
   Includes bibliographical references.
   ISBN 0-07-076516-2 (alk. paper)
   1. Corrosion and anti-corrosives. I. Title.
TA418.74.R63   1999
620.1'1223—dc21   99-35898
                                          CIP

McGraw-Hill
*A Division of The **McGraw·Hill** Companies*

Copyright © 2000 by The McGraw-Hill Companies, Inc. All rights reserved. Printed in the United States of America. Except as permitted under the United States Copyright Act of 1976, no part of this publication may be reproduced or distributed in any form or by any means, or stored in a data base or retrieval system, without the prior written permission of the publisher.

2 3 4 5 6 7 8 9 DOC/DOC 0 4 3 2 1 0

ISBN 0-07-076516-2

The sponsoring editor of this book was Robert Esposito. The editing supervisor was David E. Fogarty, and the production supervisor was Sherri Souffrance. This book was set in New Century Schoolbook by Joanne Morbit and Paul Scozzari of McGraw-Hill's Professional Book Group in Hightstown, N.J.

 This book was printed on recycled, acid-free paper containing a minimum of 50% recycled, de-inked fiber.

McGraw-Hill books are available at special quantity discounts to use as premiums and sales promotions, or for use in corporate training programs. For more information, please write to the Director of Special Sales, McGraw-Hill, Professional Publishing, Two Penn Plaza, New York, NY 10121-2298. Or contact your local bookstore.

Information contained in this work has been obtained by The McGraw-Hill Companies, Inc. ("McGraw-Hill") from sources believed to be reliable. However, neither McGraw-Hill nor its authors guarantee the accuracy or completeness of any information published herein and neither McGraw-Hill nor its authors shall be responsible for any errors, omissions, or damages arising out of use of this information. This work is published with the understanding that McGraw-Hill and its authors are supplying information but are not attempting to render engineering or other professional services. If such services are required, the assistance of an appropriate professional should be sought.

To be returned on or before
the date below.

Avril Robarts LRC

Liverpool John Moores University

Contents

Preface ix
Acknowledgments xi

Introduction 1

1.1 The Cost of Corrosion 1
1.2 Examples of Catastrophic Corrosion Damage 3
1.3 The Influence of People 5
References 12

Chapter 1. Aqueous Corrosion 13

1.1 Introduction 13
1.2 Applications of Potential-pH Diagrams 16
1.3 Kinetic Principles 32
References 54

Chapter 2. Environments 55

2.1 Atmospheric Corrosion 58
2.2 Natural Waters 85
2.3 Seawater 129
2.4 Corrosion in Soils 142
2.5 Reinforced Concrete 154
2.6 Microbes and Biofouling 187
References 216

Chapter 3. High-Temperature Corrosion 221

3.1 Thermodynamic Principles 222
3.2 Kinetic Principles 229
3.3 Practical High-Temperature Corrosion Problems 237
References 265

Chapter 4. Modeling, Life Prediction and Computer Applications   267

- 4.1 Introduction   267
- 4.2 Modeling and Life Prediction   268
- 4.3 Applications of Artificial Intelligence   303
- 4.4 Computer-Based Training or Learning   322
- 4.5 Internet and the Web   324
- References   327

Chapter 5. Corrosion Failures   331

- 5.1 Introduction   332
- 5.2 Mechanisms, Forms, and Modes of Corrosion Failures   332
- 5.3 Guidelines for Investigating Corrosion Failures   359
- 5.4 Prevention of Corrosion Damage   360
- 5.5 Case Histories in Corrosion Failure Analysis   368
- References   369

Chapter 6. Corrosion Maintenance Through Inspection And Monitoring   371

- 6.1 Introduction   372
- 6.2 Inspection   374
- 6.3 The Maintenance Revolution   583
- 6.4 Monitoring and Managing Corrosion Damage   406
- 6.5 Smart Sensing of Corrosion with Fiber Optics   448
- 6.6 Non-destructive Evaluation (NDE)   461
- References   481

Chapter 7. Acceleration and Amplification of Corrosion Damage   485

- 7.1 Introduction   486
- 7.2 Corrosion Testing   488
- 7.3 Surface Characterization   562
- References   574

Chapter 8. Materials Selection   577

- 8.1 Introduction   578
- 8.2 Aluminum Alloys   584
- 8.3 Cast Irons   612
- 8.4 Copper Alloys   622
- 8.5 High-Performance Alloys   664
- 8.6 Refractory Metals   692
- 8.7 Stainless Steels   710
- 8.8 Steels   736
- 8.9 Titanium   748
- 8.10 Zirconium   769
- References   777

Chapter 9. Protective Coatings   781

- 9.1 Introduction   781
- 9.2 Coatings and Coating Processes   782

9.3 Supplementary Protection Systems 829
9.4 Surface Preparation 831
References 831

Chapter 10. Corrosion Inhibitors 833

10.1 Introduction 833
10.2 Classification of Inhibitors 834
10.3 Corrosion Inhibition Mechanism 838
10.4 Selection of an Inhibitor System 860
References 861

Chapter 11. Cathodic Protection 863

11.1 Introduction 863
11.2 Sacrificial Anode CP Systems 871
11.3 Impressed Current Systems 878
11.4 Current Distribution and Interference Issues 886
11.5 Monitoring the Performance of CP Systems for Buried Pipelines 904
References 919

Chapter 12. Anodic Protection 921

12.1 Introduction 921
12.2 Passivity of Metals 923
12.3 Equipment Required for Anodic Protection 927
12.4 Design Concerns 930
12.5 Applications 932
12.6 Practical Example: Anodic Protection in the Pulp and Paper Industry 933
References 938

Appendix A. SI Units 939

Appendix B. Glossary 947

Appendix C. Corrosion Economics 1001

C.1 Introduction 1001
C.2 Cash Flows and Capital Budgeting Techniques 1002
C.3 Generalized Equation for Straight Line Depreciation . . . 1004
C.4 Examples 1006
C.5 Summary 1009
References 1009

Appendix D. Electrochemistry Basics 1011

D.1 Principles of Electrochemistry 1011
D.2 Chemical Thermodynamics 1029
D.3 Kinetic Principles 1047

Appendix E. Chemical Compositions of Engineering Alloys   1061

Appendix F. Thermodynamic Data and E-pH Diagrams   1101

Appendix G. Densities and Melting Points of Metals   1125

Index   1129

Preface

The design and production of the *Handbook of Corrosion Engineering* are drastically different than other handbooks dealing with the same subject. While other corrosion handbooks have been generally the results of collective efforts of many authors, the *Handbook of Corrosion Engineering* is the result of an extensive survey of state-of-the-art information on corrosion engineering by a principal author. Although only one author appears on the cover, this Handbook is indeed the result of cumulative efforts of many generations of scientists and engineers in understanding and preventing the effects of corrosion, one of the most constant foes of human endeavors. The design and construction of this Handbook were made for the new millennium with the most modern information-processing techniques presently available. Many references are made to sources of information readily accessible on the World Wide Web and to software systems that can simplify the most difficult situation. It also provides elements of information management and tools for managing corrosion problems that are particularly valuable to practicing engineers. Many examples, for example, describe how various industries and agencies have addressed corrosion problems. The systems selected as supportive examples have been chosen from a wide range of applications across various industries, from aerospace structures to energy carriers and producers.

This Handbook is aimed at the practicing engineer, as a comprehensive guide and reference source for solving material selection problems and resolving design issues where corrosion is possibly a factor. During the past decades, progress in the development of materials capable of resisting corrosion and high temperatures has been significant. There have been substantial developments in newer stainless steels, high-strength low-alloy steels, superalloys, and in protective coatings. This Handbook should prove to be a key information source concerning numerous facets of corrosion damage, from detection and monitoring to prevention and control.

The Handbook is divided into three main sections and is followed by supporting material in seven appendixes. Each section and its chapters are relatively independent and can be consulted without having to go through previous chapters. The first main section (Introduction and

Chapters 1 to 3) contains fundamental principles governing aqueous corrosion and high-temperature corrosion and covers the main environments causing corrosion such as atmospheric, natural waters, seawater, soils, concrete, as well as microbial and biofouling environments.

The second section (Chapters 4 to 7) addresses techniques for the prediction and assessment of corrosion damage such as modeling, life prediction, computer applications, inspection and monitoring and testing through acceleration and amplification of corrosion damage. The second section also contains a detailed description of the various types of corrosion failures with examples and ways to prevent them. The third section (Chapters 8 to 12) covers general considerations of corrosion prevention and control with a focus on materials selection. This chapter is particularly valuable for its detailed descriptions of the performance and maintenance considerations for the main families of engineering alloys based on aluminum, copper, nickel, chrome, refractory metals, titanium and zirconium, as well as cast irons, stainless steels and other steels. This section also provides elements for understanding protective coatings, corrosion inhibitors, cathodic protection and anodic protection.

The first appendix contains a table of appropriate SI units making references to most other types of units. This table will hopefully compensate for the systematic usage of SI units made in the book. Another appendix is an extensive glossary of terms often used in the context of corrosion engineering. A third appendix summarizes corrosion economics with examples detailing calculations based on straight value depreciation. The fourth appendix provides a detailed introduction to basic electrochemical principles. Many examples of E-pH (Pourbaix) diagrams are provided in a subsequent appendix. The designations and compositions of engineering alloys is the subject of a fifth appendix.

Pierre R. Roberge

Acknowledgments

The *Handbook of Corrosion Engineering* was designed entirely in collaboration with Martin Tullmin. In fact, Martin is the sole author of many sections of the book (corrosion in concrete, soil corrosion and cathodic protection) as well as an important contributor to many others. My acknowledgments also go to Robert Klassen who contributed to the atmospheric corrosion section as well as for his study of the fiber optic sensors for corrosion monitoring.

As I mentioned in the Preface, this book tries to summarize the present state of our knowledge of the corrosion phenomena and their impact on our societies. Many of the opinions expressed in the Handbook have come either from my work with collaborators or, more often, from my study of the work of other corrosion engineers and scientists. Of the first kind I am particularly indebted to Ken Trethewey with whom I have had many enlightening discussions that sometimes resulted in published articles. I also have to thank the congenial experts I interacted with in corrosion standard writing committees (ISO TC 156 and ASTM G01) for their expert advice and the rigor that is required in the development of new procedures and test methods.

Of the second kind I have to recognize the science and engineering pillars responsible for the present state of our knowledge in corrosion. The names of some of these giants have been mentioned throughout the book with a particular recognition made in the Introduction in Table I.4. In this respect, my personal gratitude goes to Professor Roger Staehle for his pragmatic vision of the quantification of corrosion damage. I have been greatly inspired by the work of this great man.

I would also like to take this occasion to express my love to those close to me, and particularly to Diane whose endurance of my working habits is phenomenal.

Introduction

I.1	The Cost of Corrosion	1
I.2	Examples of Catastrophic Corrosion Damage	3
I.2.1	Sewer explosion, Mexico	3
I.2.2	Loss of USAF F16 fighter aircraft	3
I.2.3	The Aloha aircraft incident	3
I.2.4	The MV KIRKI	4
I.2.5	Corrosion of the infrastructure	4
I.3	The Influence of People	5

Corrosion is the destructive attack of a material by reaction with its environment. The serious consequences of the corrosion process have become a problem of worldwide significance. In addition to our everyday encounters with this form of degradation, corrosion causes plant shutdowns, waste of valuable resources, loss or contamination of product, reduction in efficiency, costly maintenance, and expensive overdesign; it also jeopardizes safety and inhibits technological progress. The multidisciplinary aspect of corrosion problems combined with the distributed responsibilities associated with such problems only increase the complexity of the subject. Corrosion control is achieved by recognizing and understanding corrosion mechanisms, by using corrosion-resistant materials and designs, and by using protective systems, devices, and treatments. Major corporations, industries, and government agencies have established groups and committees to look after corrosion-related issues, but in many cases the responsibilities are spread between the manufacturers or producers of systems and their users. Such a situation can easily breed negligence and be quite costly in terms of dollars and human lives.

I.1 The Cost of Corrosion

Although the costs attributed to corrosion damages of all kinds have been estimated to be of the order of 3 to 5 percent of industrialized countries' gross national product (GNP), the responsibilities associated with these problems are sometimes quite diffuse. Since the first significant report by Uhlig[1] in 1949 that the cost of corrosion to nations is indeed great, the conclusion of all subsequent studies has been that corrosion represents a constant charge to a nation's GNP.[2] One conclusion of the 1971 UK government-sponsored report chaired by Hoar[3] was that a good fraction of corrosion failures were avoidable and that improved education was a good way of tackling corrosion avoidance.

Corrosion of metals cost the U.S. economy almost $300 billion per year at 1995 prices.[4] Broader application of corrosion-resistant materials and the application of the best corrosion-related technical practices could reduce approximately one-third of these costs. These estimates result from a recent update by Battelle scientists of an earlier study reported in 1978.[5] The initial work, based upon an elaborate model of more than 130 economic sectors, had revealed that metallic corrosion cost the United States $82 billion in 1975, or 4.9 percent of its GNP. It was also found that 60 percent of that cost was unavoidable. The remaining $33 billion (40 percent) was said to be "avoidable" and incurred by failure to use the best practices then known.

In the original Battelle study, almost 40 percent of 1975 metallic corrosion costs were attributed to the production, use, and maintenance of motor vehicles. No other sector accounted for as much as 4 percent of the total, and most sectors contributed less than 1 percent. The 1995 Battelle study indicated that the motor vehicles sector probably had made the greatest anticorrosion effort of any single industry. Advances have been made in the use of stainless steels, coated metals, and more protective finishes. Moreover, several substitutions of materials made primarily for reasons of weight reduction have also reduced corrosion. Also, the panel estimated that 15 percent of previously unavoidable corrosion costs can be reclassified as avoidable. The industry is estimated to have eliminated some 35 percent of its "avoidable" corrosion by its improved practices. Table I.1 summarizes the costs attributed to metallic corrosion in the United States in these two studies.

TABLE I.1 Costs Attributed to Metallic Corrosion in the United States

	1975	1995
All industries		
Total (billions of 1995 dollars)	$82.5	$296.0
Avoidable	$33.0	$104.0
Avoidable	40%	35%
Motor vehicles		
Total	$31.4	$94.0
Avoidable	$23.1	$65.0
Avoidable	73%	69%
Aircraft		
Total	$3.0	$13.0
Avoidable	$0.6	$3.0
Avoidable	20%	23%
Other industries		
Total	$47.6	$189.0
Avoidable	$9.3	$36.0
Avoidable	19%	19%

I.2 Examples of Catastrophic Corrosion Damage

I.2.1 Sewer explosion, Mexico

An example of corrosion damages with shared responsibilities was the sewer explosion that killed over 200 people in Guadalajara, Mexico, in April 1992.[6] Besides the fatalities, the series of blasts damaged 1600 buildings and injured 1500 people. Damage costs were estimated at 75 million U.S. dollars. The sewer explosion was traced to the installation of a water pipe by a contractor several years before the explosion that leaked water on a gasoline line laying underneath. The subsequent corrosion of the gasoline pipeline, in turn, caused leakage of gasoline into the sewers. The Mexican attorney general sought negligent homicide charges against four officials of Pemex, the government-owned oil company. Also cited were three representatives of the regional sewer system and the city's mayor.

I.2.2 Loss of USAF F16 fighter aircraft

This example illustrates a case that has recently created problems in the fleet of USAF F16 fighter aircraft. Graphite-containing grease is a very common lubricant because graphite is readily available from steel industries. The alternative, a formulation containing molybdenum disulphide, is much more expensive. Unfortunately, graphite grease is well known to cause galvanically induced corrosion in bimetallic couples. In a fleet of over 3000 F16 USAF single-engine fighter aircraft, graphite grease was used by a contractor despite a general order from the Air Force banning its use in aircraft.[7] As the flaps were operated, lubricant was extruded into a part of the aircraft where control of the fuel line shutoff valve was by means of electrical connectors made from a combination of gold- and tin-plated steel pins. In many instances corrosion occurred between these metals and caused loss of control of the valve, which shut off fuel to the engine in midflight. At least seven aircraft are believed to have been lost in this way, besides a multitude of other near accidents and enormous additional maintenance.

I.2.3 The Aloha aircraft incident

The structural failure on April 28, 1988, of a 19-year-old Boeing 737, operated by Aloha airlines, was a defining event in creating awareness of aging aircraft in both the public domain and in the aviation community. This aircraft lost a major portion of the upper fuselage near the front of the plane in full flight at 24,000 ft.[8] Miraculously, the pilot managed to land the plane on the island of Maui, Hawaii. One flight attendant was swept to her death. Multiple fatigue cracks were detected in the remaining aircraft structure, in the holes of the upper row of rivets in several fuselage skin lap joints. Lap joints join large panels of skin together and run longitudinally along the fuselage. Fatigue cracking was not anticipated to be a problem, provided the overlapping panels remained strongly bonded together. Inspection of other similar aircraft revealed disbonding, corrosion, and cracking problems in the lap joints. Corrosion processes and the subsequent buildup of voluminous corrosion products inside the lap joints, lead to "pillowing," whereby the faying surfaces are separated. Special instrumentation has been developed to detect this dangerous condition. The aging aircraft problem will not go away, even if airlines were to order unprecedented numbers of new aircraft. Older planes are seldom scrapped, and the older planes that are replaced by some operators will probably end up in service with another operator. Therefore, safety issues regarding aging aircraft need to be well understood, and safety programs need to be applied on a consistent and rigorous basis.

I.2.4 The MV KIRKI

Another example of major losses to corrosion that could have been prevented and that was brought to public attention on numerous occasions since the 1960s is related to the design, construction, and operating practices of bulk carriers. In 1991 over 44 large bulk carriers were either lost or critically damaged and over 120 seamen lost their lives.[9] A highly visible case was the MV KIRKI, built in Spain in 1969 to Danish designs. In 1990, while operating off the coast of Australia, the complete bow section became detached from the vessel. Miraculously, no lives were lost, there was little pollution, and the vessel was salvaged. Throughout this period it seems to have been common practice to use neither coatings nor cathodic protection inside ballast tanks. Not surprisingly therefore, evidence was produced that serious corrosion had greatly reduced the thickness of the plate and that this, combined with poor design to fatigue loading, were the primary cause of the failure. The case led to an Australian Government report called "Ships of Shame." MV KIRKI is not an isolated case. There have been many others involving large catastrophic failures, although in many cases there is little or no hard evidence when the ships go to the bottom.

I.2.5 Corrosion of the infrastructure

One of the most evident modern corrosion disasters is the present state of degradation of the North American infrastructure, particularly in the snow belt where the use of road deicing salts rose from 0.6M ton in 1950 to 10.5M tons in 1988. The structural integrity of thousands of bridges, roadbeds, overpasses, and other concrete structures has been impaired by corrosion, urgently requiring expensive repairs to ensure public safety. A report by the New York Department of Transport has stated that, by 2010, 95 percent of all New York bridges would be deficient if maintenance remained at the same level as it was in 1981. Rehabilitation of such bridges has become an important engineering practice.[10] But the problems of corroding reinforced concrete extend much beyond the transportation infrastructure. A survey of collapsed buildings during the 1974 to 1978 period in England showed that the immediate cause of failure of at least eight structures, which were 12 to 40 years old, was corrosion of reinforcing or prestressing steel. Deterioration of parking garages has become a major concern in Canada. Of the 215 garages surveyed recently, almost all suffered varying degrees of deterioration due to reinforcement corrosion, which was a result of design and construction practices that fell short of those required by the environment. It is also stated that almost all garages in Canada built until very recently by conventional methods will require rehabilitation at a cost to exceed $3 billion. The problem surely extends to the northern United States. In New York, for example, the seriousness of the corrosion problem of parking garages was revealed dramatically during the investigation that followed the bomb attack on the underground parking garage of the World Trade Center.[11]

I.3 The Influence of People

The effects of corrosion failures on the performance maintenance of materials would often be minimized if life monitoring and control of the environmental and human factors supplemented efficient designs. When an engineering system functions according to specification, a three-way interaction is established with complex and variable inputs from people (p), materials (m), and environments (e).[12] An attempt to translate this concept into a fault tree has produced the simple tree presented in Fig. I.1 where the consequence, or top event, a corrosion failure, can be represented by combining the three previous contributing elements. In this representation, the top event probability (P_{sf}) can be evaluated with boolean algebra, which leads to Eq. (I.1) where P_m and P_e are, respectively, the probability of failure caused by materials and by the environment, and Factor$_p$ describes the influence of people on the lifetime of a system. In Eq. (I.1), Factor$_p$ can be either inhibiting (Factor$_p$ < 1) or aggravating (Factor$_p$ > 1):

$$P_{sf} = P_m P_e \text{Factor}_p \qquad (I.1)$$

The justification for including the people element as an inhibit gate or conditional event in the corrosion tree should be obvious (i.e., corrosion

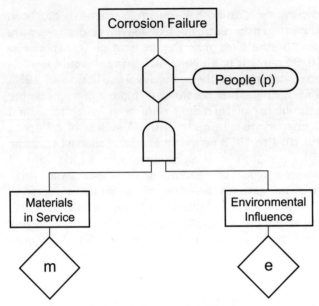

Figure I.1 Basic fault tree of a corrosion failure.

is a natural process that does not need human intervention to occur). What might be defined as purely mechanical failures occur when P_m is high and P_e is low. Most well-designed engineering systems in which P_e is approximately 0 achieve good levels of reliability. The most successful systems are usually those in which the environmental influence is very small and continues to be so throughout the service lifetime. When P_e becomes a significant influence on an increasing P_{sf}, the incidence of corrosion failures normally also increases.

Minimizing P_{sf} only through design is difficult to achieve in practice because of the number of ways in which P_m, P_e, and Factor$_p$ can vary during the system lifetime. The types of people that can affect the life and performance of engineering systems have been regrouped in six categories (Table I.2).[13] Table I.2 also contains a brief description of the main contributions that each category of people can make to the success or premature failure of a system. Table I.3 gives an outline of methods of corrosion control[14] with an indication of the associated responsibility.

However, the influence of people in a failure is extremely difficult to predict, being subject to the high variability level in human decision making. Most well-designed engineering systems perform according to specification, largely because the interactions of people with these systems are tightly controlled and managed throughout the life of the systems. Figure I.2 breaks down the causes responsible for failures

TABLE I.2   Positions and Their Relative Responsibilities in System Management

Procurer

What is the main system being specified?
What is the function of the main system?
Did the budget introduce compromise into the design?
How was a subsystem embodied into the main system?
Does the envelope of the subsystems fit that for the main system?

Designer

What is the subsystem being specified for?
What is the function of the subsystem?
What is the optimum materials selection?
Has the correct definition of the operating environment been applied?
By what means will the component be manufactured?
What is the best geometrical design?
Have finishing operations, protective coatings, or corrosion control techniques been specified?
Have the correct operating conditions been specified?
Has the best maintenance schedule been specified?
Does the design embody features that enable the correct maintenance procedures to be followed?

Manufacturer

Were the same materials used as were originally specified?
Did the purchased starting materials conform to the specification in the order?
Has the manufacturing process been carried out correctly?
Has the design been reproduced accurately and has the materials specification been precisely followed?
Have the correct techniques been used?
Have the most suitable joining techniques been employed?
Have the specified conditions/coatings necessary for optimum performance been implemented?
Did the component conform to the appropriate quality control standards?
Was the scheme for correct assembly of the subsystem implemented correctly so that the installation can be made correctly?

Installer

Has the system been installed according to specification?
Has the correct setting-to-work procedure been followed?
Have any new features in the environment been identified that are likely to exert an influence and were not foreseen by the design process?

Maintainer

Has the correct maintenance schedule been followed?
Have the correct spares been used in repairs?
Have the correct maintenance procedures been carried out?
Has the condition of the system been correctly monitored?

User

Has the system been used within the specified conditions?
Is there a history of similar failures or is this an isolated occurrence?
Do aggravating conditions exist when the system is not in use?
Is there any evidence that the system has been abused by unauthorized personnel?

TABLE I.3 Outline of Methods of Corrosion Control

Method	Responsibility	
	Direct	Managerial
Selection of Materials		
Select metal or alloy (on nonmetallic material) for the particular environmental conditions prevailing (composition, temperature, velocity, etc.), taking into account mechanical and physical properties, availability, method of fabrication and overall cost of structure	Designer	Procurer (for user)
Decide whether or not an expensive corrosion-resistant alloy is more economical than a cheaper metal that requires protection and periodic maintenance	Designer	Procurer (for user)
Design		
If the metal has to be protected, make provision in the design for applying metallic or nonmetallic coatings or applying anodic or cathodic protection	Designer	Designer

Avoid geometrical configurations that facilitate corrosive conditions such as
  Features that trap dust, moisture, and water
  Crevices (or else fill them in) and situations where deposits can form on the metal surface
  Designs that lead to erosion corrosion or to cavitation damage
  Designs that result in inaccessible areas that cannot be reprotected (e.g., by maintenance painting)
  Designs that lead *to* heterogeneities in the metal (differences in thermal treatment) or in the environment (differences in temperature, velocity)
	Designer	Designer

Contact with other materials		

Avoid metal-metal or metal-nonmetallic contacting materials that facilitate corrosion such as
  Bimetallic couples in which a large area of a more positive metal (e.g., Cu) is in contact with a small area of a less noble metal (e.g., Fe, Zn, or Al)
  Metals in contact with absorbent materials that maintain constantly wet conditions or, in the case of passive metals, that exclude oxygen
  Contact (or enclosure in a confined space) with substances that give off corrosive vapors (e.g., certain woods and plastics)
	Designer, user	Designer, user

Mechanical factors		
Avoid stresses (magnitude and type) and environmental conditions that lead to stress-corrosion cracking, corrosion fatigue, or fretting corrosion:	Designer, user	Designer, user

TABLE I.3 Outline of Methods of Corrosion Control (*Continued*)

Method	Responsibility	
Selection of Materials	Direct	Managerial
For stress corrosion cracking, avoid the use of alloys that are susceptible in the environment under consideration, or if this is not possible, ensure that the external and internal stresses are kept to a minimum.		
For a metal subjected to fatigue conditions in a corrosive environment ensure that the metal is adequately protected by a corrosion-resistant coating.		
Processes that induce compressive stresses into the surface of the metal such as shot-peening, carburizing, and nitriding are frequently beneficial in preventing corrosion fatigue and fretting corrosion.		
Coatings		

If the metal has a poor resistance to corrosion in the environment under consideration, make provision in the design for applying an appropriate protective coating such as
  Metal reaction products (e.g., anodic oxide films on Al), phosphate coatings on steel (for subsequent painting or impregnation with grease), chromate films on light metals and alloys (Zn, Al, cd, Mg)
  Metallic coatings that form protective barriers (Ni, Cr) and also protect the substrate by sacrificial action (Zn, Al, or cd on steel)
  Inorganic coatings (e.g., enamels, glasses, ceramics)
  Organic coatings (e.g., paints, plastics, greases)
	Designer	Designer

Environment		

Make environment less aggressive by removing constituents that facilitate corrosion; decrease temperatures decrease velocity; where possible prevent access of water and moisture.
  For atmospheric corrosion dehumidify the air, remove solid particles, add volatile corrosion inhibitors (for steel).
  For aqueous corrosion remove dissolved O_2, increase the pH (for steels), add inhibitors.
	Designer, user	Designer, user

Interfacial potential		
Protect metal cathodically by making the interfacial potential sufficiently negative by (1) sacrificial anodes or (2) impressed current.		
Protect metal by making the interfacial potential sufficiently positive to cause passivation (confined to metals that passivate in the environment under consideration).		

TABLE I.3 Outline of Methods of Corrosion Control (*Continued*)

Method	Responsibility	
Selection of Materials	Direct	Managerial
Corrosion testing and monitoring		
When there is no information on the behavior of a metal or alloy or a fabrication under specific environmental conditions (a newly formulated alloy and/or a new environment), it is essential to carry out corrosion testing.	Designer	Designer, user
Monitor composition of environment, corrosion rate of metal, interfacial potential, and so forth, to ensure that control is effective.	Designer	Designer, user
Supervision and inspection		
Ensure that the application of a protective coating (applied in situ or in a factory) is adequately supervised and inspected in accordance with the specification or code of practice.	Designer, user	User

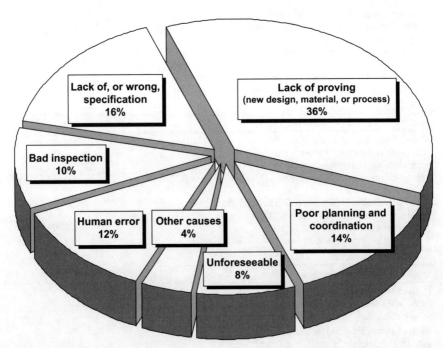

Figure I.2 Pie chart attribution of responsibility for corrosion failures investigated by a large chemical company.

investigated by a large process industry.[15] But the battle against such an insidious foe has been raging for a long time and sometimes with success. Table I.4 presents some historical landmarks of discoveries related to the understanding and management of corrosion. Although the future successes will still relate to improvements in materials and their performance, it can be expected that the main progress in corrosion prevention will be associated with the development of better information-processing strategies and the production of more efficient monitoring tools in support of corrosion control programs.

TABLE I.4 Landmarks of Discoveries Related to the Understanding and Management of Corrosion

Date	Landmark	Source
1675	Mechanical origin of corrosiveness and corrodibility	Boyle
1763	Bimetallic corrosion	HMS Alarm report
1788	Water becomes alkaline during corrosion of iron	Austin
1791	Copper-iron electrolytic galvanic coupling	Galvani
1819	Insight into electrochemical nature of corrosion	Thenard
1824	Cathodic protection of Cu by Zn or Fe	Sir Humphrey Davy
1830	Microstructural aspect of corrosion (Zn)	De la Rive
1834–1840	Relations between chemical action and generation of electric currents	Faraday
1836	Passivity of iron	Faraday, Schoenbein
1904	Hydrogen overvoltage as a function of current	Tafel
1905	Carbonic and other acids are not essential for the corrosion of iron	Dunstan, Jowett, Goulding, Tilden
1907	Oxygen action as cathodic stimulator	Walker, Cederholm
1908–1910	Compilation of corrosion rates in different media	Heyn, Bauer
1910	Inhibitive paint	Cushman, Gardner
1913	Study of high-temperature oxidation kinetics of tungsten	Langmuir
1916	Differential aeration currents	Aston
1920–1923	Season-cracking of brass = intergranular corrosion	Moore, Beckinsale
1923	High-temperature formation of oxides	Pilling, Bedworth
1924	Galvanic corrosion	Whitman, Russell
1930–1931	Subscaling of "internal corrosion"	Smith
1931–1939	Quantitative electrochemical nature of corrosion	Evans
1938	Anodic and cathodic inhibitors	Chyzewski, Evans
1938	E-pH thermodynamic diagrams	Pourbaix
1950	Autocatalytic nature of pitting	Uhlig
1956	Tafel extrapolation for measurement of kinetic parameters	Stern, Geary
1968	Electrochemical noise signature of corrosion	Iverson
1970	Study of corrosion processes with electrochemical impedance spectroscopy (EIS)	Epelboin

References

1. Uhlig, H. H., The Cost of Corrosion in the United States, *Chemical and Engineering News,* **27:**2764 (1949).
2. Cabrillac, C., Leach, J. S. L., Marcus P., et al., The Cost of Corrosion in the EEC, *Metals and Materials,* **3:**533–536 (1987).
3. Hoar, T. P., *Report of the Committee on Corrosion and Protection.* 1971. London, UK, Her Majesty's Stationary Office.
4. Holbrook, D., Corrosion Annually Costs $300 Billion, According to Battelle Study, *http://www.battelle.org/pr/12corrode.html,* 1-1-1996, Battelle Memorial Institute.
5. Bennett, L. H., Kruger, J., Parker, R. L., Passaglia, E., Reimann, C., Ruff, A. W., and Yakowitz, H., *Economic Effects of Metallic Corrosion in the United States: A Report to the Congress,* NBS Special Pub. 511-1. 1-13-1978. Washington, DC, National Bureau of Standards.
6. Up Front, *Materials Performance,* **31:**3 (1992).
7. Vasanth, K., *Minutes of Group Committee T-9 - Military, Aerospace, and Electronics Equipment Corrosion Control,* 3-30-1995. Houston, Tex., NACE International.
8. Miller, D., Corrosion control on aging aircraft: What is being done? *Materials Performance,* **29:**10–11 (1990).
9. Hamer, M., Clampdown on the Rust Buckets, *New Scientist,* **146:**5 (1991).
10. Broomfield, J. P., *Five Years Research on Corrosion of Steel in Concrete: A Summary of the Strategic Highway Research Program Structures Research,* paper no. 318 (Corrosion 93), 1993. Houston, Tex., NACE International.
11. Trethewey, K. R., and Roberge, P. R., Corrosion Management in the Twenty-First Century, *British Corrosion Journal,* **30:**192–197 (1995).
12. Roberge, P. R., Eliciting Corrosion Knowledge through the Fault-Tree Eyeglass, in Trethewey, K. R., and Roberge, P. R. (eds.), *Modelling Aqueous Corrosion: From Individual Pits to Corrosion Management,* The Netherlands, Kluwer Academic Publishers, 1994, pp. 399–416.
13. Trethewey, K. R., and Roberge, P. R., Lifetime Prediction in Engineering Systems: The Influence of People, *Materials and Design,* **15:**275–285 (1994).
14. Shreir, L. L., Jarman, R. A., and Burstein, G. T., *Corrosion Control.* Oxford, UK, Butterworths Heinemann, 1994.
15. Congleton, J., Stress Corrosion Cracking of Stainless Steels, in Shreir, L. L., Jarman, R. A., and Burstein, G. T. (eds), *Corrosion Control.* Oxford, UK, Butterworths Heinemann, 1994, pp. 8:52–8:83.

Chapter 1

Aqueous Corrosion

1.1	Introduction	13
1.2	Applications of Potential-pH Diagrams	16
	1.2.1 Corrosion of steel in water at elevated temperatures	17
	1.2.2 Filiform corrosion	26
	1.2.3 Corrosion of reinforcing steel in concrete	29
1.3	Kinetic Principles	32
	1.3.1 Kinetics at equilibrium: the exchange current concept	32
	1.3.2 Kinetics under polarization	35
	1.3.3 Graphical presentation of kinetic data	42
	References	54

1.1 Introduction

One of the key factors in any corrosion situation is the environment. The definition and characteristics of this variable can be quite complex. One can use thermodynamics, e.g., Pourbaix or E-pH diagrams, to evaluate the theoretical activity of a given metal or alloy provided the chemical makeup of the environment is known. But for practical situations, it is important to realize that the environment is a variable that can change with time and conditions. It is also important to realize that the environment that actually affects a metal corresponds to the microenvironmental conditions that this metal really "sees," i.e., the local environment at the surface of the metal. It is indeed the reactivity of this local environment that will determine the real corrosion damage. Thus, an experiment that investigates only the nominal environmental condition without consideration of local effects such as flow, pH cells, deposits, and galvanic effects is useless for lifetime prediction.

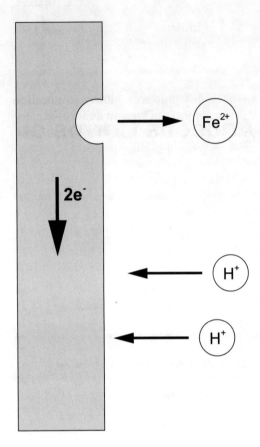

Figure 1.1 Simple model describing the electrochemical nature of corrosion processes.

In our societies, water is used for a wide variety of purposes, from supporting life as potable water to performing a multitude of industrial tasks such as heat exchange and waste transport. The impact of water on the integrity of materials is thus an important aspect of system management. Since steels and other iron-based alloys are the metallic materials most commonly exposed to water, aqueous corrosion will be discussed with a special focus on the reactions of iron (Fe) with water (H_2O). Metal ions go into solution at anodic areas in an amount chemically equivalent to the reaction at cathodic areas (Fig. 1.1). In the cases of iron-based alloys, the following reaction usually takes place at anodic areas:

$$Fe \rightarrow Fe^{2+} + 2e^- \qquad (1.1)$$

This reaction is rapid in most media, as shown by the lack of pronounced polarization when iron is made an anode employing an external current. When iron corrodes, the rate is usually controlled by the

cathodic reaction, which in general is much slower (cathoc
In deaerated solutions, the cathodic reaction is

$$2H^+ + 2e^- \rightarrow H_2$$

This reaction proceeds rapidly in acids, but only slowly in alkaline or neutral aqueous media. The corrosion rate of iron in deaerated neutral water at room temperature, for example, is less than 5 μm/year. The rate of hydrogen evolution at a specific pH depends on the presence or absence of low-hydrogen overvoltage impurities in the metal. For pure iron, the metal surface itself provides sites for H_2 evolution; hence, high-purity iron continues to corrode in acids, but at a measurably lower rate than does commercial iron.

The cathodic reaction can be accelerated by the reduction of dissolved oxygen in accordance with the following reaction, a process called depolarization:

$$4H^+ + O_2 + 4e^- \rightarrow 2H_2O \tag{1.3}$$

Dissolved oxygen reacts with hydrogen atoms adsorbed at random on the iron surface, independent of the presence or absence of impurities in the metal. The oxidation reaction proceeds as rapidly as oxygen reaches the metal surface.

Adding (1.1) and (1.3), making use of the reaction $H_2O \leftrightarrow H^+ + OH^-$, leads to reaction (1.4),

$$2Fe + 2H_2O + O_2 \rightarrow 2Fe(OH)_2 \tag{1.4}$$

Hydrous ferrous oxide ($FeO \cdot nH_2O$) or ferrous hydroxide [$Fe(OH)_2$] composes the diffusion-barrier layer next to the iron surface through which O_2 must diffuse. The pH of a saturated $Fe(OH)_2$ solution is about 9.5, so that the surface of iron corroding in aerated pure water is always alkaline. The color of $Fe(OH)_2$, although white when the substance is pure, is normally green to greenish black because of incipient oxidation by air. At the outer surface of the oxide film, access to dissolved oxygen converts ferrous oxide to hydrous ferric oxide or ferric hydroxide, in accordance with

$$4Fe(OH)_2 + 2H_2O + O_2 \rightarrow 4Fe(OH)_3 \tag{1.5}$$

Hydrous ferric oxide is orange to red-brown in color and makes up most of ordinary rust. It exists as nonmagnetic αFe_2O_3 (hematite) or as magnetic αFe_2O_3, the α form having the greater negative free energy of formation (greater thermodynamic stability). Saturated $Fe(OH)_3$ is nearly neutral in pH. A magnetic hydrous ferrous ferrite, $Fe_3O_4 \cdot nH_2O$, often forms a black intermediate layer between hydrous Fe_2O_3

and FeO. Hence rust films normally consist of three layers of iron oxides in different states of oxidation.

1.2 Applications of Potential-pH Diagrams

E-pH or Pourbaix diagrams are a convenient way of summarizing much thermodynamic data and provide a useful means of summarizing the thermodynamic behavior of a metal and associated species in given environmental conditions. E-pH diagrams are typically plotted for various equilibria on normal cartesian coordinates with potential (E) as the ordinate (y axis) and pH as the abscissa (x axis).[1] For a more complete coverage of the construction of such diagrams, the reader is referred to Appendix D (Sec. D.2.6, Potential-pH Diagrams).

For corrosion in aqueous media, two fundamental variables, namely corrosion potential and pH, are deemed to be particularly important. Changes in other variables, such as the oxygen concentration, tend to be reflected by changes in the corrosion potential. Considering these two fundamental parameters, Staehle introduced the concept of overlapping mode definition and environmental definition diagrams,[2] to determine under what environmental circumstances a given mode/submode of corrosion damage could occur (Fig. 1.2). Further information on corrosion modes and submodes is provided in Chap. 5, Corrosion Failures. It is very important to consider and define the environment on the metal surface, where the corrosion reactions take place. Highly corrosive local environments that differ greatly from the nominal bulk environment can be set up on such surfaces, as illustrated in some examples given in following sections.

In the application of E-pH diagrams to corrosion, thermodynamic data can be used to map out the occurrence of corrosion, passivity, and nobility of a metal as a function of pH and potential. The operating environment can also be specified with the same coordinates, facilitating a thermodynamic prediction of the nature of corrosion damage. A particular environmental diagram showing the thermodynamic stability of different chemical species associated with water can also be derived thermodynamically. This diagram, which can be conveniently superimposed on E-pH diagrams, is shown in Fig. 1.3. While the E-pH diagram provides no kinetic information whatsoever, it defines the thermodynamic boundaries for important corrosion species and reactions. The observed corrosion behavior of a particular metal or alloy can also be superimposed on E-pH diagrams. Such a superposition is presented in Fig. 1.4. The corrosion behavior of steel presented in this figure was characterized by polarization measurements at different potentials in solutions with varying pH levels.[3] It should be noted that the corrosion behavior of steel appears to be defined by thermody-

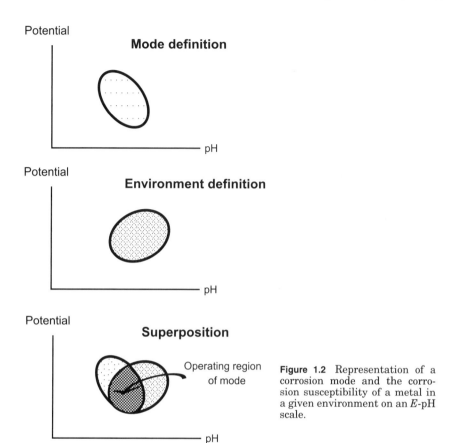

Figure 1.2 Representation of a corrosion mode and the corrosion susceptibility of a metal in a given environment on an E-pH scale.

namic boundaries. Some examples of the application of E-pH diagrams to practical corrosion problems follow.

1.2.1 Corrosion of steel in water at elevated temperatures

Many phenomena associated with corrosion damage to iron-based alloys in water at elevated temperatures can be rationalized on the basis of iron-water E-pH diagrams. Marine boilers on ships and hot-water heating systems for buildings are relevant practical examples.

Marine boilers. The boilers used on commercial and military ships are essentially large reactors in which water is heated and converted to steam. While steam powering of ships' engines or turbines is rapidly drawing to a close at the end of the twentieth century, steam is still required for other miscellaneous purposes. All passenger ships require

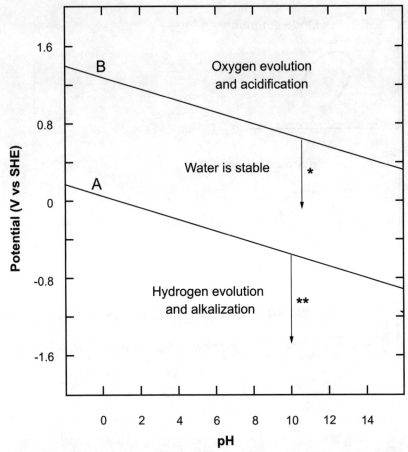

Figure 1.3 Thermodynamic stability of water, oxygen, and hydrogen. (A is the equilibrium line for the reaction: $H_2 = 2H^+ + 2e^-$. B is the equilibrium line for the reaction: $2H_2O = O_2 + 4H^+ + 4e^-$. * indicates increasing thermodynamic driving force for cathodic oxygen reduction, as the potential falls below line B. ** indicates increasing thermodynamic driving force for cathodic hydrogen evolution, as the potential falls below line A.)

steam for heating, cooking, and laundry services. Although not powered by steam, motorized tankers need steam for tank cleaning, pumping, and heating.

Steel is used extensively as a construction material in pressurized boilers and ancillary piping circuits. The boiler and the attached steam/water circuits are safety-critical items on a ship. The sudden explosive release of high-pressure steam/water can have disastrous consequences. The worst boiler explosion in the Royal Navy, on board *HMS Thunderer*, claimed 45 lives in 1876.[4] The subsequent inquiry revealed that the boiler's safety valves had seized as a result of corro-

Aqueous Corrosion   19

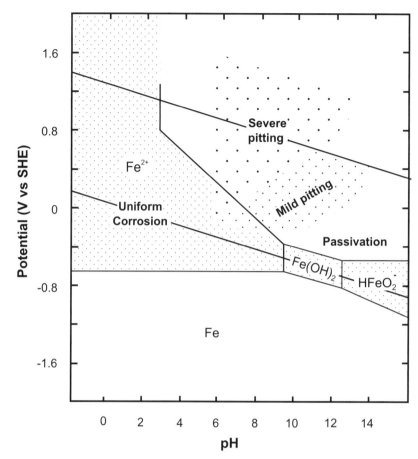

Figure 1.4 Thermodynamic boundaries of the types of corrosion observed on steel.

sion damage. Fortunately, modern marine steam boilers operate at much higher safety levels, but corrosion problems still occur.

Two important variables affecting water-side corrosion of iron-based alloys in marine boilers are the pH and oxygen content of the water. As the oxygen level has a strong influence on the corrosion potential, these two variables exert a direct influence in defining the position on the E-pH diagram. A higher degree of aeration raises the corrosion potential of iron in water, while a lower oxygen content reduces it.

When considering the water-side corrosion of steel in marine boilers, both the elevated-temperature and ambient-temperature cases should be considered, since the latter is important during shutdown periods. Boiler-feedwater treatment is an important element of minimizing corrosion damage. On the maiden voyage of *RMS Titanic,* for

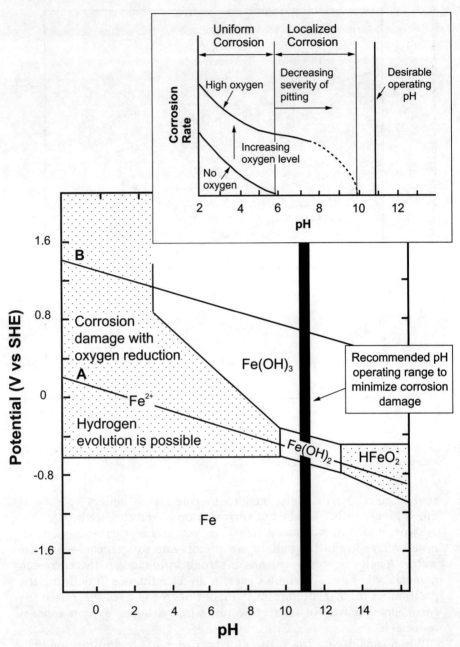

Figure 1.5 E-pH diagram of iron in water at 25°C and its observed corrosion behavior.

example, no fewer than three engineers were managing the boiler room operations, which included responsibility for ensuring that boiler-water-treatment chemicals were correctly administered. A fundamental treatment requirement is maintaining an alkaline pH value, ideally in the range of 10.5 to 11 at room temperature.[5] This precaution takes the active corrosion field on the left-hand side of the E-pH diagrams out of play, as shown in the E-pH diagrams drawn for steel at two temperatures, 25°C (Fig. 1.5) and 210°C (Fig. 1.6). At the recommended pH levels, around 11, the E-pH diagram in Fig. 1.5 indicates the presence of thermodynamically stable oxides above the zone of immunity. It is the presence of these oxides on the surface that protects steel from corrosion damage in boilers.

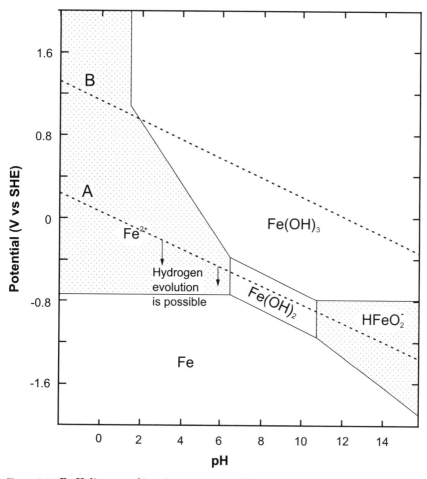

Figure 1.6 E-pH diagram of iron in water at 210°C.

Practical experience related to boiler corrosion kinetics at different feedwater pH levels is included in Fig. 1.5. The kinetic information in Fig. 1.5 indicates that high oxygen contents are generally undesirable. It should also be noted from Figs. 1.5 and 1.6 that active corrosion is possible in acidified untreated boiler water, even in the absence of oxygen. Below the hydrogen evolution line, hydrogen evolution is thermodynamically favored as the cathodic half-cell reaction, as indicated. Undesirable water acidification can result from contamination by sea salts or from residual cleaning agents.

Inspection of the kinetic data presented in Fig. 1.5 reveals a tendency for localized pitting corrosion at feedwater pH levels between 6 and 10. This pH range represents a situation in between complete surface coverage by protective oxide films and the absence of protective films. Localized anodic dissolution is to be expected on a steel surface covered by a discontinuous oxide film, with the oxide film acting as a cathode. Another type of localized corrosion, caustic corrosion, can occur when the pH is raised excessively on a localized scale. The E-pH diagrams in Figs. 1.5 and 1.6 indicate the possibility of corrosion damage at the high end of the pH axis, where the protective oxides are no longer stable. Such undesirable pH excursions tend to occur in high-temperature zones, where boiling has led to a localized caustic concentration. A further corrosion problem, which can arise in highly alkaline environments, is caustic cracking, a form of stress corrosion cracking. Examples in which such microenvironments have been proven include seams, rivets, and boiler tube-to-tube plate joints.

Hydronic heating of buildings. Hydronic (or hot-water) heating is used extensively for central heating systems in buildings. Advantages over hot-air systems include the absence of dust circulation and higher heat efficiency (there are no heat losses from large ducts). In very simple terms, a hydronic system could be described as a large hot-water kettle with pipe attachments to circulate the hot water and radiators to dissipate the heat.

Heating can be accomplished by burning gas or oil or by electricity. The water usually leaves the boiler at temperatures of 80 to 90°C. Hot water leaving the boiler passes through pipes, which carry it to the radiators for heat dissipation. The heated water enters as feed, and the cooled water leaves the radiator. Fins may be attached to the radiator to increase the surface area for efficient heat transfer. Steel radiators, constructed from welded pressed steel sheets, are widely utilized in hydronic heating systems. Previously, much weightier cast iron radiators were used; these are still evident in older buildings. The hot-water piping is usually constructed from thin-walled copper tubing or steel pipes. The circulation system must be able to cope with the water expansion resulting from heating in the boiler. An expansion tank is provided for these purposes. A return pipe carries the cooled water from the radiators back to the boiler. Typically, the temperature of the water in the return pipe is 20°C lower than that of the water leaving the boiler.

An excellent detailed account of corrosion damage to steel in the hot water flowing through the radiators and pipes has been published.[6] Given a pH range for mains water of 6.5 to 8 and the E-pH diagrams in Figs. 1.7 (25°C) and 1.8 (85°C), it is apparent that minimal corrosion damage is to be expected if the corrosion potential remains below -0.65 V (SHE). The position of the oxygen reduction line indicates that the cathodic oxygen reduction reaction is thermodynamically very

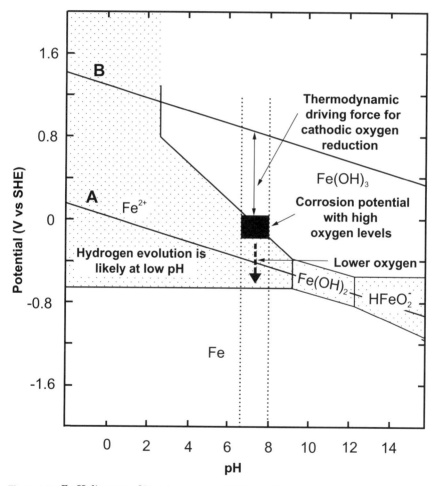

Figure 1.7 E-pH diagram of iron in water at 25°C, highlighting the corrosion processes in the hydronic pH range.

favorable. From kinetic considerations, the oxygen content will be an important factor in determining corrosion rates. The oxygen content of the water is usually minimal, since the solubility of oxygen in water decreases with increasing temperature (Fig. 1.9), and any oxygen remaining in the hot water is consumed over time by the cathodic corrosion reaction. Typically, oxygen concentrations stabilize at very low levels (around 0.3 ppm), where the cathodic oxygen reduction reaction is stifled and further corrosion is negligible.

Higher oxygen levels in the system drastically change the situation, potentially reducing radiator lifetimes by a factor of 15. The undesirable oxygen pickup is possible during repairs, from additions of fresh water to compensate for evaporation, or, importantly, through design

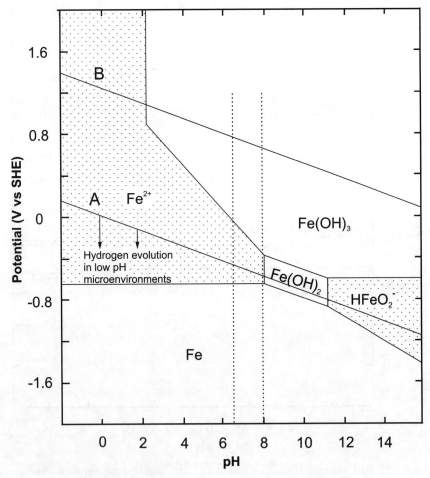

Figure 1.8 E-pH diagram of iron in water at 85°C (hydronic system).

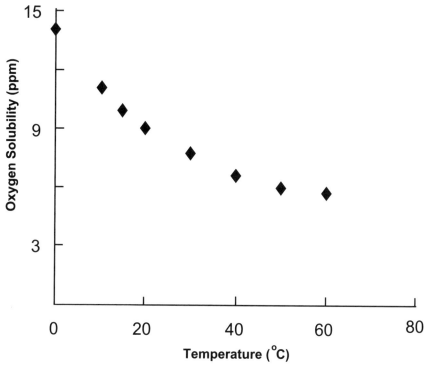

Figure 1.9 Solubility of oxygen in water in equilibrium with air at different temperatures.

faults that lead to continual oxygen pickup from the expansion tank. The higher oxygen concentration shifts the corrosion potential to higher values, as shown in Fig. 1.7. Since the $Fe(OH)_3$ field comes into play at these high potential values, the accumulation of a red-brown sludge in radiators is evidence of oxygen contamination.

From the E-pH diagrams in Figs. 1.7 and 1.8, it is apparent that for a given corrosion potential, the hydrogen production is thermodynamically more favorable at low pH values. The production of hydrogen is, in fact, quite common in microenvironments where the pH can be lowered to very low values, leading to severe corrosion damage even at very low oxygen levels. The corrosive microenvironment prevailing under surface deposits is very different from the bulk solution. In particular, the pH of such microenvironments tends to be very acidic. The formation of acidified microenvironments is related to the hydrolysis of corrosion products and the formation of differential aeration cells between the bulk environment and the region under the deposits (see Crevice Corrosion in Sec. 5.2.1). Surface deposits in radiators can result from corrosion products (iron oxides), scale, the settling of suspended solids, or microbiological activity. The potential range in which the hydrogen reduction reaction can participate in corrosion reactions clearly widens toward the low end of the pH scale. If such deposits are not removed periodically by cleaning, perforations by localized corrosion can be expected.

1.2.2 Filiform corrosion

Filiform corrosion is a localized form of corrosion that occurs under a variety of coatings. Steel, aluminum, and other alloys can be particularly affected by this form of corrosion, which has been of particular concern in the food packaging industry. Readers living in humid coastal areas may have noticed it from time to time on food cans left in storage for long periods. It can also affect various components during shipment and storage, given that many warehouses are located near seaports. This form of corrosion, which has a "wormlike" visual appearance, can be explained on the basis of microenvironmental effects and the relevant E-pH diagrams.

Filiform corrosion is characterized by an advancing head and a tail of corrosion products left behind in the corrosion tracks (or "filaments"), as shown in Fig. 1.10. Active corrosion takes place in the head, which is filled with corrosive solution, while the tail is made up of relatively dry corrosion products and is usually considered to be inactive.

The microenvironments produced by filiform corrosion of steel are illustrated in Fig. 1.11.[7] Essentially, a differential aeration cell is set up under the coating, with the lowest concentration of oxygen at the head

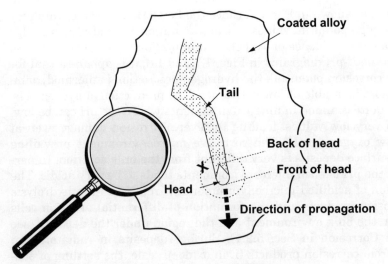

Figure 1.10 Illustration of the filament nature of filiform corrosion.

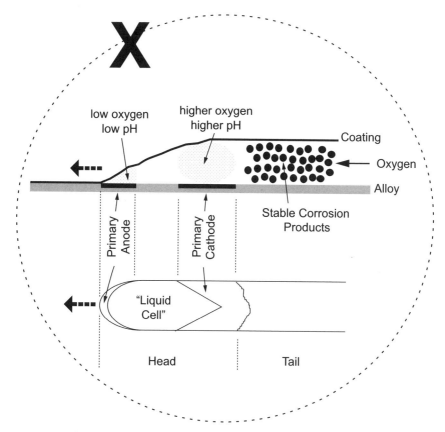

Figure 1.11 Graphical representation of the microenvironments created by filiform corrosion.

of the filament. The oxygen concentration gradient can be rationalized by oxygen diffusion through the porous tail to the head region. A characteristic feature of such a differential aeration cell is the acidification of the electrolyte with low oxygen concentration. This leads to the formation of an anodic metal dissolution site at the front of the head of the corrosion filament (Fig. 1.11). For iron, pH values at the front of the head of 1 to 4 and a potential of close to −0.44 V (SHE) have been reported. In contrast, at the back of the head, where the cathodic reaction dominates, the prevailing pH is around 12. The conditions prevailing at the front and back of the head for steel undergoing filiform corrosion are shown relative to the E-pH diagram in Fig. 1.12. The diagram confirms active corrosion at the front, the buildup of ferric hydroxide at the back of the head, and ferric hydroxide filling the tail.

In filiform corrosion damage to aluminum, an electrochemical potential at the front of the head of −0.73 V (SHE) has been report-

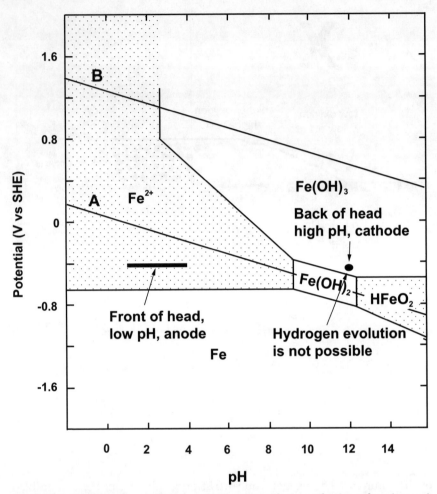

Figure 1.12 E-pH diagram of the iron-water system with an emphasis on the microenvironments produced by filiform corrosion.

ed, together with a 0.09-V difference between the front and the back of the head.[8] Reported acidic pH values close to 1 at the head and higher fluctuating values in excess of 3.5 associated with the tail allow the positions in the E-pH diagram to be determined, as shown in Fig. 1.13. Active corrosion at the front and the buildup of corrosion products toward the tail is predicted on the basis of this diagram. It should be noted that the front and back of the head positions on the E-pH diagram lie below the hydrogen evolution line. It is thus not surprising that hydrogen evolution has been reported in filiform corrosion of aluminum.

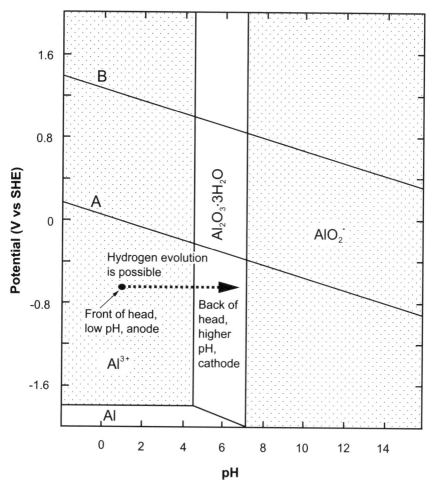

Figure 1.13 E-pH diagram of the aluminum-water system with an emphasis on the microenvironments produced by filiform corrosion.

1.2.3 Corrosion of reinforcing steel in concrete

Concrete is the most widely produced material on earth; its production exceeds that of steel by about a factor of 10 in tonnage. While concrete has a very high compressive strength, its strength in tension is very low (only a few megapascals). The main purpose of reinforcing steel (rebar) in concrete is to improve the tensile strength and toughness of the material. The steel rebars can be considered to be macroscopic fibers in a "fiber-reinforced" composite material. The vast majority of reinforcing steel is of the unprotected carbon steel type. No significant alloying additions or protective coatings for corrosion resistance are associated with this steel.

In simplistic terms, concrete is produced by mixing cement clinker, water, fine aggregate (sand), coarse aggregate (stone), and other chemical additives. When mixed with water, the anhydrous cement clinker compounds hydrate to form cement paste. It is the cement paste that forms the matrix of the composite concrete material and gives it its strength and rigidity, by means of an interconnected network in which the aggregate particles are embedded. The cement paste is porous in nature. An important feature of concrete is that the pores are filled with a highly alkaline solution, with a pH between 12.6 and 13.8 at normal humidity levels. This highly alkaline pore solution arises from by-products of the cement clinker hydration reactions such as NaOH, KOH, and $Ca(OH)_2$. The maintenance of a high pH in the concrete pore solution is a fundamental feature of the corrosion resistance of carbon steel reinforcing bars.

At the high pH levels of the concrete pore solution, without the ingress of corrosive species, reinforcing steel embedded in concrete tends to display completely passive behavior as a result of the formation of a thin protective passive film. The corrosion potential of passive reinforcing steel tends to be more positive than about -0.52 V (SHE) according to ASTM guidelines.[9] The E-pH diagram in Fig. 1.14 confirms the passive nature of steel under these conditions. It also indicates that the oxygen reduction reaction is the cathodic half-cell reaction applicable under these highly alkaline conditions.

One mechanism responsible for severe corrosion damage to reinforcing steel is known as carbonation. In this process, carbon dioxide from the atmosphere reacts with calcium hydroxide (and other hydroxides) in the cement paste following reaction (1.6).

$$Ca(OH)_2 + CO_2 \rightarrow CaCO_3 + H_2O \tag{1.6}$$

The pore solution is effectively neutralized by this reaction. Carbonation damage usually appears as a well-defined "front" parallel to the outside surface. Behind the front, where all the calcium hydroxide has reacted, the pH is reduced to around 8, whereas ahead of the front, the pH remains above 12.6. When the carbonation front reaches the reinforcement, the passive film is no longer stable, and active corrosion is initiated. Figure 1.14 shows that active corrosion is possible at the reduced pH level. Damage to the concrete from carbonation-induced corrosion is manifested in the form of surface spalling, resulting from the buildup of voluminous corrosion products at the concrete-rebar interface (Fig. 1.15).

A methodology known as re-alkalization has been proposed as a remedial measure for carbonation-induced reinforcing steel corro-

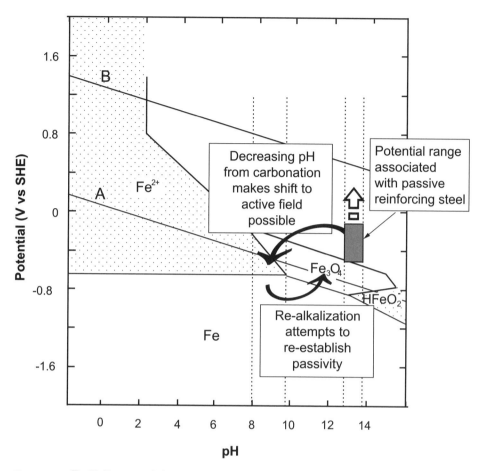

Figure 1.14 E-pH diagram of the iron-water system with an emphasis on the microenvironments produced during corrosion of reinforcing steel in concrete.

sion. The aim of this treatment is to restore alkalinity around the reinforcing bars of previously carbonated concrete. A direct current is applied between the reinforcing steel cathode and external anodes positioned against the external concrete surface and surrounded by electrolyte. Sodium carbonate has been used as the electrolyte in this process, which typically requires several days for effectiveness. Potential disadvantages of the treatment include reduced bond strength, increased risk of alkali-aggregate reaction, microstructural changes in the concrete, and hydrogen embrittlement of the reinforcing steel. It is apparent from Fig. 1.14 that hydrogen reduction can occur on the reinforcing steel cathode if its potential drops to highly negative values.

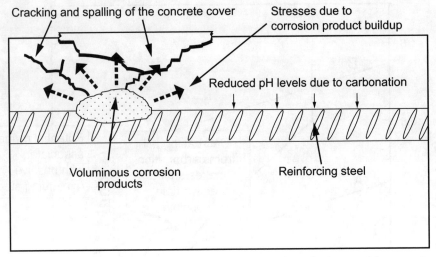

Figure 1.15 Graphical representation of the corrosion of reinforcing steel in concrete leading to cracking and spalling.

1.3 Kinetic Principles

Thermodynamic principles can help explain a corrosion situation in terms of the stability of chemical species and reactions associated with corrosion processes. However, thermodynamic calculations cannot be used to predict corrosion rates. When two metals are put in contact, they can produce a voltage, as in a battery or electrochemical cell (see Galvanic Corrosion in Sec. 5.2.1). The material lower in what has been called the "galvanic series" will tend to become the anode and corrode, while the material higher in the series will tend to support a cathodic reaction. Iron or aluminum, for example, will have a tendency to corrode when connected to graphite or platinum. What the series cannot predict is the rate at which these metals corrode. Electrode kinetic principles have to be used to estimate these rates.

1.3.1 Kinetics at equilibrium: the exchange current concept

The exchange current I_0 is a fundamental characteristic of electrode behavior that can be defined as the rate of oxidation or reduction at an equilibrium electrode expressed in terms of current. The term *exchange current,* in fact, is a misnomer, since there is no net current flow. It is merely a convenient way of representing the rates of oxidation and reduction of a given single electrode at equilibrium, when no loss or gain is experienced by the electrode material. For the corrosion of iron, Eq. (1.1), for example, this would imply that the exchange current is related to the current in each direction of a reversible reaction, i.e., an anodic current I_a representing Eq. (1.7) and a cathodic current I_c representing Eq. (1.8).

$$Fe \rightarrow Fe^{2+} + 2e^- \qquad (1.7)$$

$$Fe \leftarrow Fe^{2+} + 2e^- \qquad (1.8)$$

Since the net current is zero at equilibrium, this implies that the sum of these two currents is zero, as in Eq. (1.9). Since I_a is, by convention, always positive, it follows that, when no external voltage or current is applied to the system, the exchange current is as given by Eq. (1.10).

$$I_a + I_c = 0 \qquad (1.9)$$

$$I_a = -I_c = I_0 \qquad (1.10)$$

There is no theoretical way of accurately determining the exchange current for any given system. This must be determined experimentally. For the characterization of electrochemical processes, it is always preferable to normalize the value of the current by the surface area of the electrode and use the current density, often expressed as a small i, i.e., $i = I/\text{surface area}$. The magnitude of exchange current density is a function of the following main variables:

1. *Electrode composition.* Exchange current density depends upon the composition of the electrode and the solution (Table 1.1). For redox reactions, the exchange current density would depend on the composition of the electrode supporting an equilibrium reaction (Table 1.2).

TABLE 1.1 Exchange Current Density (i_0) for M^{z+}/M Equilibrium in Different Acidified Solutions (1 M)

Electrode	Solution	$\log_{10} i_0$, A/cm^2
Antimony	Chloride	−4.7
Bismuth	Chloride	−1.7
Copper	Sulfate	−4.4; −1.7
Iron	Sulfate	−8.0; −8.5
Lead	Perchlorate	−3.1
Nickel	Sulfate	−8.7; −6.0
Silver	Perchlorate	0.0
Tin	Chloride	−2.7
Titanium	Perchlorate	−3.0
Titanium	Sulfate	−8.7
Zinc	Chloride	−3.5; −0.16
Zinc	Perchlorate	−7.5
Zinc	Sulfate	−4.5

TABLE 1.2 Exchange Current Density (i_0) at 25°C for Some Redox Reactions

System	Electrode Material	Solution	$\log_{10} i_0$, A/cm^2
Cr^{3+}/Cr^{2+}	Mercury	KCl	−6.0
Ce^{4+}/Ce^{3+}	Platinum	H_2SO_4	−4.4
Fe^{3+}/Fe^{2+}	Platinum	H_2SO_4	−2.6
	Rhodium	H_2SO_4	−7.8
	Iridium	H_2SO_4	−2.8
	Palladium	H_2SO_4	−2.2
H^+/H_2	Gold	H_2SO_4	−3.6
	Lead	H_2SO_4	−11.3
	Mercury	H_2SO_4	−12.1
	Nickel	H_2SO_4	−5.2
	Tungsten	H_2SO_4	−5.9
O_2 reduction	Platinum	Perchloric acid	−9.0
	Platinum 10%–Rhodium	Perchloric acid	−9.0
	Rhodium	Perchloric acid	−8.2
	Iridium	Perchloric acid	−10.2

TABLE 1.3 Approximate Exchange Current Density (i_0) for the Hydrogen Oxidation Reaction on Different Metals at 25°C

Metal	$\log_{10} i_0$, A/cm^2
Pb, Hg	−13
Zn	−11
Sn, Al, Be	−10
Ni, Ag, Cu, Cd	−7
Fe, Au, Mo	−6
W, Co, Ta	−5
Pd, Rh	−4
Pt	−2

Table 1.3 contains the approximate exchange current density for the reduction of hydrogen ions on a range of materials. Note that the value for the exchange current density of hydrogen evolution on platinum is approximately 10^{-2} A/cm^2, whereas that on mercury is 10^{-13} A/cm^2.

2. *Surface roughness.* Exchange current density is usually expressed in terms of projected or geometric surface area and depends upon the surface roughness. The higher exchange current density for the H^+/H_2 system equilibrium on platinized platinum (10^{-2} A/cm^2) compared to that on bright platinum (10^{-3} A/cm^2) is a result of the larger specific surface area of the former.

3. *Soluble species concentration.* The exchange current is also a complex function of the concentration of both the reactants and the products involved in the specific reaction described by the exchange current. This function is particularly dependent on the shape of the charge transfer barrier β across the electrochemical interface.

4. *Surface impurities.* Impurities adsorbed on the electrode surface usually affect its exchange current density. Exchange current density for the H^+/H_2 system is markedly reduced by the presence of trace impurities like arsenic, sulfur, and antimony.

1.3.2 Kinetics under polarization

When two complementary processes such as those illustrated in Fig. 1.1 occur over a single metallic surface, the potential of the material will no longer be at an equilibrium value. This deviation from equilibrium potential is called *polarization*. Electrodes can also be polarized by the application of an external voltage or by the spontaneous production of a voltage away from equilibrium. The magnitude of polarization is usually measured in terms of overvoltage η, which is a measure of polarization with respect to the equilibrium potential E_{eq} of an electrode. This polarization is said to be either anodic, when the anodic processes on the electrode are accelerated by changing the specimen potential in the positive (noble) direction, or cathodic, when the cathodic processes are accelerated by moving the potential in the negative (active) direction. There are three distinct types of polarization in any electrochemical cell, the total polarization across an electrochemical cell being the summation of the individual elements as expressed in Eq. (1.11):

$$\eta_{total} = \eta_{act} + \eta_{conc} + iR \qquad (1.11)$$

where η_{act} = activation overpotential, a complex function describing the charge transfer kinetics of the electrochemical processes. η_{act} is predominant at small polarization currents or voltages.

η_{conc} = concentration overpotential, a function describing the mass transport limitations associated with electrochemical processes. η_{conc} is predominant at large polarization currents or voltages.

iR = ohmic drop. iR follows Ohm's law and describes the polarization that occurs when a current passes through an electrolyte or through any other interface, such as surface film, connectors, etc.

Activation polarization. When some steps in a corrosion reaction control the rate of charge or electron flow, the reaction is said to be under activation or charge-transfer control. The kinetics associated with apparently simple processes rarely occur in a single step. The overall anodic reaction expressed in Eq. (1.1) would indicate that metal atoms in the metal lattice are in equilibrium with an aqueous solution containing Fe^{2+} cations. The reality is much more complex, and one would need to use at least two intermediate species to describe this process, i.e.,

$$Fe_{lattice} \rightarrow Fe^{+}_{surface}$$

$$Fe^{+}_{surface} \rightarrow Fe^{2+}_{surface}$$

$$Fe^{2+}_{surface} \rightarrow Fe^{2+}_{solution}$$

In addition, one would have to consider other parallel processes, such as the hydrolysis of the Fe^{2+} cations to produce a precipitate or some other complex form of iron cations. Similarly, the equilibrium between protons and hydrogen gas [Eq. (1.2)] can be explained only by invoking at least three steps, i.e.,

$$H^+ \rightarrow H_{ads}$$

$$H_{ads} + H_{ads} \rightarrow H_{2 \text{ (molecule)}}$$

$$H_{2 \text{ (molecule)}} \rightarrow H_{2 \text{ (gas)}}$$

The anodic and cathodic sides of a reaction can be studied individually by using some well-established electrochemical methods in which the response of a system to an applied polarization, current or voltage, is studied. A general representation of the polarization of an electrode supporting one redox system is given in the Butler-Volmer equation (1.12):

$$i_{reaction} = i_0 \left\{ \exp\left(\beta_{reaction} \frac{nF}{RT} \eta_{reaction}\right) - \exp\left[-(1 - \beta_{reaction}) \frac{nF}{RT} \eta_{reaction}\right] \right\} \quad (1.12)$$

where $i_{reaction}$ = anodic or cathodic current
$\beta_{reaction}$ = charge transfer barrier or symmetry coefficient for the anodic or cathodic reaction, close to 0.5
$\eta_{reaction} = E_{applied} - E_{eq}$, i.e., positive for anodic polarization and negative for cathodic polarization
n = number of participating electrons
R = gas constant
T = absolute temperature
F = Faraday

When η_{reaction} is anodic (i.e., positive), the second term in the Butler-Volmer equation becomes negligible and i_a can be more simply expressed by Eq. (1.13) and its logarithm, Eq. (1.14):

$$i_a = i_0 \left[\exp\left(\beta_a \frac{nF}{RT} \eta_a \right) \right] \tag{1.13}$$

$$\eta_a = b_a \log_{10}\left(\frac{i_a}{i_0}\right) \tag{1.14}$$

where b_a is the Tafel coefficient that can be obtained from the slope of a plot of η against log i, with the intercept yielding a value for i_0.

$$b_a = 2.303 \frac{RT}{\beta nF} \tag{1.15}$$

Similarly, when η_{reaction} is cathodic (i.e., negative), the first term in the Butler-Volmer equation becomes negligible and i_c can be more simply expressed by Eq. (1.16) and its logarithm, Eq. (1.17), with b_c obtained by plotting η versus log i [Eq. (1.18)]:

$$i_c = i_0 \left\{ -\exp\left[-(1-\beta_c)\frac{nF}{RT} \eta_c \right] \right\} \tag{1.16}$$

$$\eta_c = b_c \log_{10}\left(\frac{i_c}{i_0}\right) \tag{1.17}$$

$$b_c = -2.303 \frac{RT}{\beta nF} \tag{1.18}$$

Concentration polarization. When the cathodic reagent at the corroding surface is in short supply, the mass transport of this reagent could become rate controlling. A frequent case of this type of control occurs when the cathodic processes depend on the reduction of dissolved oxygen. Table 1.4 contains some data related to the solubility of oxygen in air-saturated water at different temperatures, and Table 1.5 contains some data on the solubility of oxygen in seawater of different salinity and chlorinity.[10]

Because the rate of the cathodic reaction is proportional to the surface concentration of the reagent, the reaction rate will be limited by a drop in the surface concentration. For a sufficiently fast charge transfer, the surface concentration will fall to zero, and the corrosion process will be totally controlled by mass transport. As indicated in Fig. 1.16, mass transport to a surface is governed by three forces: dif-

TABLE 1.4 Solubility of Oxygen in Air-Saturated Water

Temperature, °C	Volume, cm³*	Concentration, ppm	Concentration (M), μmol/L
0	10.2	14.58	455.5
5	8.9	12.72	397.4
10	7.9	11.29	352.8
15	7.0	10.00	312.6
20	6.4	9.15	285.8
25	5.8	8.29	259.0
30	5.3	7.57	236.7

*cm³ per kg of water at 0°C.

TABLE 1.5 Oxygen Dissolved in Seawater in Equilibrium with a Normal Atmosphere

Chlorinity,* %	0	5	10	15	20
Salinity,† %	0	9.06	18.08	27.11	36.11
Temperature, °C			ppm		
0	14.58	13.70	12.78	11.89	11.00
5	12.79	12.02	11.24	10.49	9.74
10	11.32	10.66	10.01	9.37	8.72
15	10.16	9.67	9.02	8.46	7.92
20	9.19	8.70	8.21	7.77	7.23
25	8.39	7.93	7.48	7.04	6.57
30	7.67	7.25	6.80	6.41	5.37

*Chlorinity refers to the total halogen ion content as titrated by the addition of silver nitrate, expressed in parts per thousand (%).
†Salinity refers to the total proportion of salts in seawater, often estimated empirically as chlorinity × 1.80655, also expressed in parts per thousand (%).

fusion, migration, and convection. In the absence of an electric field, the migration term is negligible, and the convection force disappears in stagnant conditions.

For purely diffusion-controlled mass transport, the flux of a species O to a surface from the bulk is described with Fick's first law (1.19),

$$J_O = -D_O \left(\frac{\delta C_O}{\delta x} \right) \quad (1.19)$$

where J_O = flux of species O, mol · s⁻¹ · cm⁻²
D_O = diffusion coefficient of species O, cm² · s⁻¹
$\frac{\delta C_O}{\delta x}$ = concentration gradient of species O across the interface, mol · cm⁻⁴

The diffusion coefficient of an ionic species at infinite dilution can be estimated with the help of the Nernst-Einstein equation (1.20), which relates D_O to the conductivity of the species (λ_O):

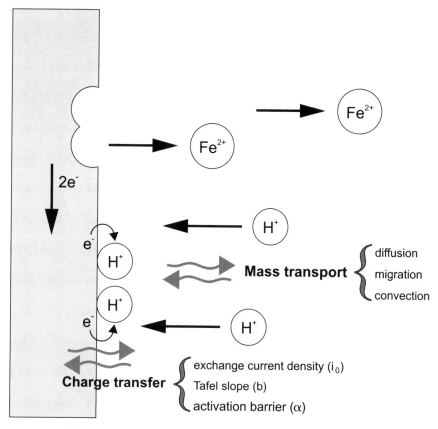

Figure 1.16 Graphical representation of the processes occurring at an electrochemical interface.

$$D_O = \frac{RT\lambda_O}{|z_O|^2 F^2} \qquad (1.20)$$

where z_O = the valency of species O
R = gas constant, i.e., 8.314 J · mol^{-1} · K^{-1}
T = absolute temperature, K
F = Faraday's constant, i.e., 96,487 C · mol^{-1}

Table 1.6 contains values for D_O and λ_O of some common ions. For more practical situations, the diffusion coefficient can be approximated with the help of Eq. (1.21), which relates D_O to the viscosity of the solution μ and absolute temperature:

$$D_O = \frac{TA}{\mu} \qquad (1.21)$$

where A is a constant for the system.

TABLE 1.6 Conductivity and Diffusion Coefficients of Selected Ions at Infinite Dilution in Water at 25°C

Cation	$|z|$	λ, S·cm²·mol⁻¹	$D \times 10^5$, cm²·s⁻¹	Anion	$|z|$	λ, S·cm²·mol⁻¹	$D \times 10^5$, cm²·s⁻¹
H^+	1	349.8	9.30	OH^-	1	197.6	5.25
Li^+	1	38.7	1.03	F^-	1	55.4	1.47
Na^+	1	50.1	1.33	Cl^-	1	76.3	2.03
K^+	1	73.5	1.95	NO_3^-	1	71.4	1.90
Ca^{2+}	2	119.0	0.79	ClO_4^-	1	67.3	1.79
Cu^{2+}	2	107.2	0.71	SO_4^{2-}	2	160.0	1.06
Zn^{2+}	2	105.6	0.70	CO_3^{2-}	2	138.6	0.92
O_2	—	—	2.26	HSO_4^{-1}	1	50.0	1.33
H_2O	—	—	2.44	HCO_3^{-1}	1	41.5	1.11

The region near the metallic surface where the concentration gradient occurs is also called the diffusion layer δ. Since the concentration gradient $\delta C_O/\delta x$ is greatest when the surface concentration of species O is completely depleted at the surface (i.e., $C_O = 0$), it follows that the cathodic current is limited in that condition, as expressed by Eq. (1.22):

$$i_c = i_L = -nFD_O \frac{C_{O,bulk}}{\delta} \qquad (1.22)$$

For intermediate cases, η_{conc} can be evaluated using an expression [Eq. (1.23)] derived from the Nernst equation:

$$\eta_{conc} = \frac{2.303RT}{nF} \log_{10}\left(1 - \frac{i}{i_L}\right) \qquad (1.23)$$

where $2.303RT/F = 0.059$ V when $T = 298.16$ K.

Ohmic drop. The ohmic resistance of a cell can be measured with a milliohmmeter by using a high-frequency signal with a four-point technique. Table 1.7 lists some typical values of water conductivity.[10] While the ohmic drop is an important parameter to consider when designing cathodic and anodic protection systems, it can be minimized, when carrying out electrochemical tests, by bringing the reference electrode into close proximity with the surface being monitored. For naturally occurring corrosion, the ohmic drop will limit the influence of an anodic or a cathodic site on adjacent metal areas to a certain distance depending on the conductivity of the environment. For naturally occurring corrosion, the anodic and cathodic sites often are adjacent grains or microconstituents and the distances involved are very small.

TABLE 1.7 Resistivity of Waters

Water	ρ, $\Omega \cdot$ cm
Pure water	20,000,000
Distilled water	500,000
Rainwater	20,000
Tap water	1000–5000
River water (brackish)	200
Seawater (coastal)	30
Seawater (open sea)	20–25

1.3.3 Graphical presentation of kinetic data

Electrode kinetic data are typically presented in a graphical form called Evans diagrams, polarization diagrams, or mixed-potential diagrams. These diagrams are useful in describing and explaining many corrosion phenomena. According to the mixed-potential theory underlying these diagrams, any electrochemical reaction can be algebraically divided into separate oxidation and reduction reactions with no net accumulation of electric charge. In the absence of an externally applied potential, the oxidation of the metal and the reduction of some species in solution occur simultaneously at the metal/electrolyte interface. Under these circumstances, the net measurable current is zero and the corroding metal is charge-neutral, i.e., all electrons produced by the corrosion of a metal have to be consumed by one or more cathodic processes (e^- produced equal e^- consumed with no net accumulation of charge).

It is also important to realize that most textbooks present corrosion current data as current densities. The main reason for that is simple: Current density is a direct characteristic of interfacial properties. Corrosion current density relates directly to the penetration rate of a metal. If one assumes that a metallic surface plays equivalently the role of an anode and that of a cathode, one can simply balance the current densities and be done with it. In real cases this is not so simple. The assumption that one surface is equivalently available for both processes is indeed too simplistic. The occurrence of localized corrosion is a manifest proof that the anodic surface area can be much smaller than the cathodic. Additionally, the size of the anodic area is often inversely related to the severity of corrosion problems: The smaller the anodic area and the higher the ratio of the cathodic surface S_c to the anodic surface S_a, the more difficult it is to detect the problem.

In order to construct mixed-potential diagrams to model a corrosion situation, one must first gather (1) the information concerning the activation overpotential for each process that is potentially involved and (2) any additional information for processes that could be affected by concentration overpotential. The following examples of increasing complexity will illustrate the principles underlying the construction of mixed-potential diagrams.

The following sections go through the development of detailed equations and present some examples to illustrate how mixed-potential models can be developed from first principles.

1. For simple cases in which corrosion processes are purely activation-controlled
2. For cases in which concentration controls at least one of the corrosion processes

Activation-controlled processes. For purely activation-controlled processes, each reaction can be described by a straight line on an E versus $\log i$ plot, with positive Tafel slopes for anodic processes and negative Tafel slopes for cathodic processes. The corrosion anodic processes are never limited by concentration effects, but they can be limited by the passivation or formation of a protective film.

Note: Since $1\ \mathrm{mA \cdot cm^{-2}}$ corresponds to a penetration rate of 1.2 cm per year, it is meaningless, in corrosion studies, to consider current density values higher than $10\ \mathrm{mA \cdot cm^{-2}}$ or $10^{-2}\ \mathrm{A \cdot cm^{-2}}$.

The currents for anodic and cathodic reactions can be obtained with the help of Eqs. (1.14) and (1.17), respectively, which generally state how the overpotential varies with current, as in the following equation:

$$\eta = b\ \log_{10}(I/I_0) = b\ \log_{10}(I) - b\ \log_{10}(I_0)$$

where $\eta = E - E_{eq}$
$E = E_{\text{applied}}$
E_{eq} = equilibrium or Nernst potential
I_0 = exchange current = $i_0 S$
i_0 = exchange current density
S = surface area

One normally uses the graphical representation, illustrated in cases 1 to 3, to determine E_{corr} and I_{corr}. It is also possible to solve these problems mathematically, as illustrated in the following transformations.

The applied potential is

$$E = E_{eq} + b\ \log_{10}(I) - b\ \log_{10}(I_0)$$

and the applied current can then be written as

$$\log_{10}(I) = \frac{\eta}{b} + \log_{10}(I_0) = \frac{E - E_{eq}}{b} + \log_{10}(I_0)$$

or

$$I = 10^{[(E - E_{eq})/b\ +\ \log_{10}(I_0)]}$$

at E_{corr},

$$I_a = I_c \quad \text{and} \quad E_a = E_c = E_{\text{corr}}$$

and hence

$$\frac{E_{corr} - E_{eq,a}}{b_a} + \log_{10}(I_{0,a}) = \frac{(E_{corr} - E_{eq,c})}{b_c} + \log_{10}(I_{0,c})$$

or

$$b_c(E_{corr} - E_{eq,a}) + b_c b_a \log_{10}(I_{0,a}) = b_a(E_{corr} - E_{eq,c}) + b_c b_a \log_{10}(I_{0,c})$$

and

$$b_c E_{corr} - b_a E_{corr} = b_c E_{eq,a} - b_a E_{eq,c} + b_c b_a [\log_{10}(I_{0,c}) - \log_{10}(I_{0,a})]$$

finally

$$E_{corr} = \frac{b_c E_{eq,a} - b_a E_{eq,c}}{b_c - b_a} + \frac{b_c b_a [\log_{10}(I_{0,c}) - \log_{10}(I_{0,a})]}{b_c - b_a}$$

One can obtain I_{corr} by substituting E_{corr} in one of the previous expressions, i.e.,

$$E_{corr} = E_{eq,a} + b_a \log_{10}(I_{corr}) - b \log_{10}(I_{0,a})$$

or

$$b_a \log_{10}(I_{corr}) = E_{corr} - E_{eq,a} + b \log_{10}(I_{0,a})$$

and

$$\log_{10}(I_{corr}) = \frac{E_{corr} - E_{eq,a} + b \log_{10}(I_{0,a})}{b_a}$$

First case: iron in a deaerated acid solution at 25°C, pH = 0.

Anodic reaction

Surface area = 1 cm²
$Fe \rightarrow Fe^{2+} + 2e^-$
$E^0 = -0.44$ V versus SHE
For a corroding metal, one can assume that $E_{eq} = E^0$.
$i_0 = 10^{-6}$ A · cm^{-2}
$I_0 = 1 \times 10^{-6}$ A
$b_a = 0.120$ V/decade

Cathodic reaction

Surface area = 1 cm²

$2H^+ + 2e^- \rightarrow H_2$

$E^0 = 0.0$ V versus SHE

$E_{eq} = E^0 + 0.059 \log_{10} a_{H^+} = 0.0 + 0 = 0.0$ V versus SHE

$i_0 = 10^{-6}$ A · cm^{-2}

$I_0 = 1 \times 10^{-6}$ A

$b_c = -0.120$ V/decade

The mixed-potential diagram of this system is shown in Fig. 1.17, and the resultant polarization plot of the system is shown in Fig. 1.18.

Second case: zinc in a deaerated acid solution at 25°C, pH = 0.

Anodic reaction

$Zn \rightarrow Zn^{2+} + 2e^-$

$E^0 = -0.763$ V versus SHE

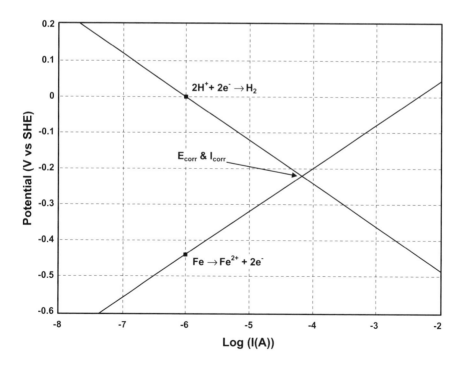

Figure 1.17 The iron mixed-potential diagram at 25°C and pH 0.

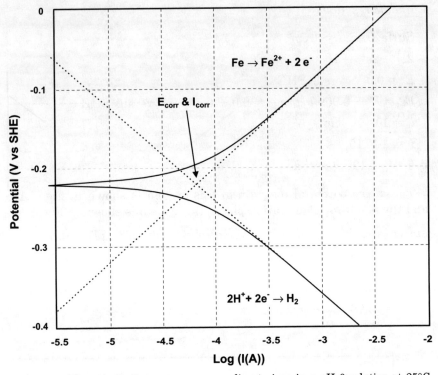

Figure 1.18 The polarization curve corresponding to iron in a pH 0 solution at 25°C (Fig. 1.17).

For a corroding metal, one can assume that $E_{eq} = E^0$.

$i_0 = 10^{-7}$ A · cm^{-2}

$b_a = 0.120$ V/decade

Cathodic reaction

$2H^+ + 2e^- \rightarrow H_2$

$E^0 = 0.0$ V versus SHE

$E_{eq} = E^0 + 0.059 \log a_{H^+} = 0.0 + 0 = 0.0$ V versus SHE

$i_0 = 10^{-10}$ A · cm^{-2}

$b_a = -0.120$ V/decade

The mixed-potential diagram of this system is shown in Fig. 1.19, and the resultant polarization plot of the system is shown in Fig. 1.20.

Third case: iron in a deaerated neutral solution at 25°C, pH = 5.

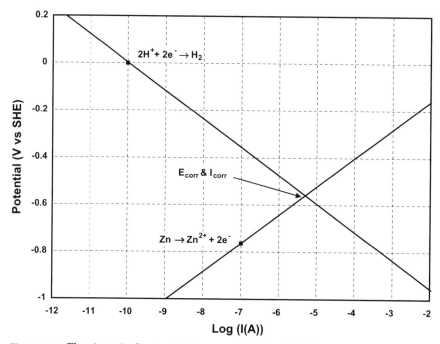

Figure 1.19 The zinc mixed-potential diagram at 25°C and pH 0.

Anodic reaction

Surface area = 1 cm²
Fe → Fe²⁺ + 2e⁻
$E^0 = -0.44$ V versus SHE
For a corroding metal, one can assume that $E_{eq} = E^0$.
$i_0 = 10^{-6}$ A · cm⁻²
$I_0 = 1 \times 10^{-6}$ A
$b_a = 0.120$ V/decade

Cathodic reaction

Surface area = 1 cm²
2H⁺ + 2e⁻ → H₂
$E_{eq} = E^0 + 0.059 \log_{10} a_{H^+} = 0.0 - 0.059 \times (-5) = -0.295$ V versus SHE
$i_0 = 10^{-6}$ A · cm⁻²
$I_0 = 1 \times 10^{-6}$ A
$b_c = -0.120$ V/decade

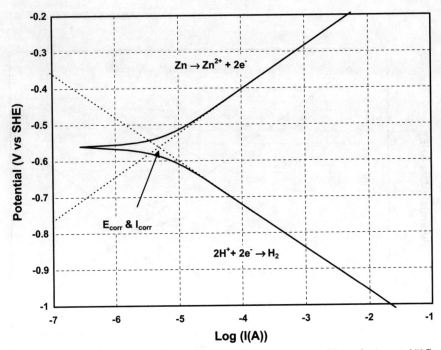

Figure 1.20 The polarization curve corresponding to zinc in a pH 0 solution at 25°C (Fig. 1.19).

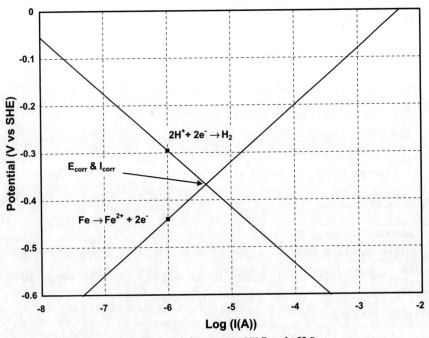

Figure 1.21 The iron mixed-potential diagram at 25°C and pH 5.

48

Aqueous Corrosion 49

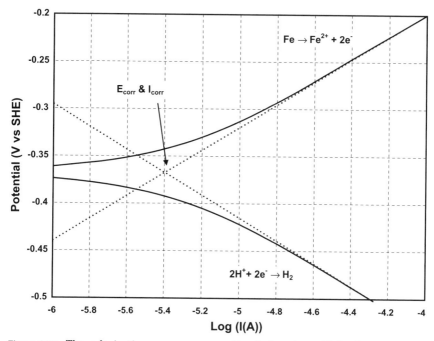

Figure 1.22 The polarization curve corresponding to iron in a pH 5 solution at 25°C (Fig. 1.21).

The mixed-potential diagram of this system is shown in Fig. 1.21, and the resultant polarization plot of the system is shown in Fig. 1.22.

Concentration-controlled processes. When concentration control is added to a process, it simply adds to the polarization, as in the following equation:.

$$\eta_{tot} = \eta_{act} + \eta_{conc}$$

We know that, for purely activation-controlled systems, the current can be derived from the voltage with the following expression:

$$I = 10^{[(E - E_{eq})/b\, +\, \log_{10}(I_0)]}$$

In order to simplify the expression of the current in the presence of concentration effects suppose that

$$A = 10^{[(E - E_{eq})/b\, +\, \log_{10}(I_0)]}$$

$$\eta_{tot} = E - E_{eq} = \eta_{act} + \eta_{conc}$$

and

$$I = I_1 \cdot A/(I_1 + A)$$

where I_1 is the limiting current of the cathodic process.

Fourth case: iron in an aerated neutral solution at 25°C, pH = 5, $I_1 = 10^{-4}$ A.

Anodic reaction

Surface area = 1 cm²
Fe → Fe²⁺ + 2e⁻
For a corroding metal, one can assume that $E_{eq} = E_0$.
$i_0 = 10^{-6}$ A · cm⁻²
$I_0 = 1 \times 10^{-6}$ A
$b_a = 0.120$ V/decade

Cathodic reactions

Surface area = 1 cm²
2H⁺ + 2e⁻ → H₂
$E_{eq} = E^0 + 0.059 \log_{10} a_{H^+} = 0.0 + 0.059 \times (-5) = -0.295$ V versus SHE
$i_0 = 10^{-6}$ A · cm⁻²
$I_0 = 1 \times 10^{-6}$ A
$b_c = -0.120$ V/decade
O₂ + 4H⁺ + 4e⁻ → 2H₂O
$E^0 = 1.229$ V versus SHE
$E_{eq} = E^0 + 0.059 \log_{10} a_{H^+} + (0.059/4) \log_{10}(pO_2)$

Supposing pO₂ = 0.2,

$E_{eq} = 1.229 - 0.059 \times (-5) + 0.0148 \times (-0.699) = 0.9237$ V versus SHE
$i_0 = 10^{-7}$ A · cm⁻²
$I_0 = 1 \times 10^{-7}$ A
$b_c = -0.120$ V/decade
$i_1 = I_1 = 10^{-4}$ A

The mixed-potential diagram of this system is shown in Fig. 1.23, and the resultant polarization plot of the system is shown in Fig. 1.24.

Fifth case: iron in an aerated neutral solution at 25°C, pH = 2, $I_1 = 10^{-4.5}$ A.

Surface area = 1 cm²

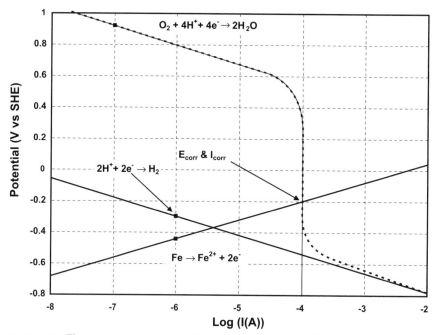

Figure 1.23 The iron mixed-potential diagram at 25°C and pH 5 in an aerated solution with a limiting current of 10^{-4} A for the reduction of oxygen.

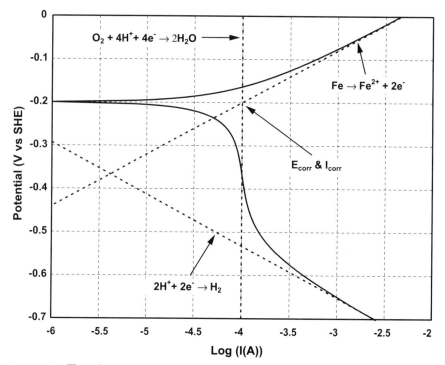

Figure 1.24 The polarization curve corresponding to iron in a pH 5 solution at 25°C in an aerated solution with a limiting current of 10^{-4} A for the reduction of oxygen (Fig. 1.23).

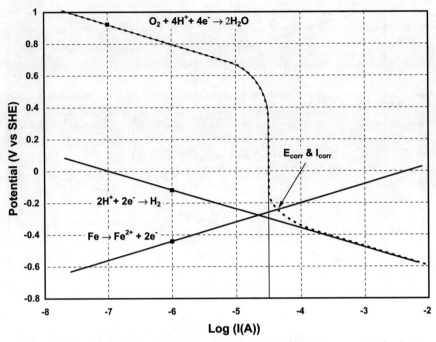

Figure 1.25 The iron mixed-potential diagram at 25°C and pH 2 in an aerated solution with a limiting current of $10^{-4.5}$ A for the reduction of oxygen.

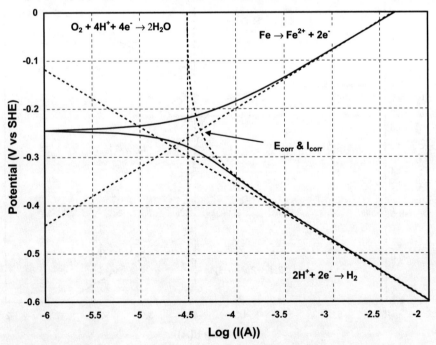

Figure 1.26 The polarization curve corresponding to iron in a pH 2 solution at 25°C in an aerated solution with a limiting current of $10^{-4.5}$ A for the reduction of oxygen (Fig. 1.25).

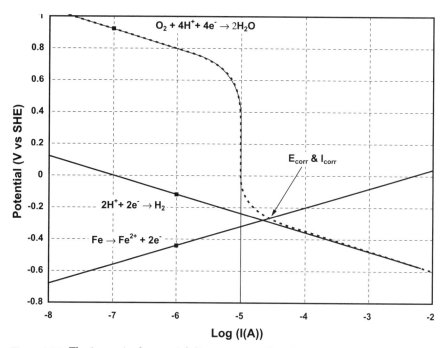

Figure 1.27 The iron mixed-potential diagram at 25°C and pH 2 in an aerated solution with a limiting current of 10^{-5} A for the reduction of oxygen.

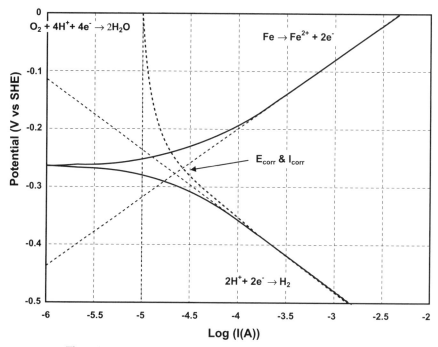

Figure 1.28 The polarization curve corresponding to iron in a pH 2 solution at 25°C in an aerated solution with a limiting current of 10^{-5} A for the reduction of oxygen (Fig. 1.27).

The only differences from the previous case are that (1) the pH has become more acidic and (2) the limiting current of the cathodic reaction has decreased to $10^{-4.5}$ A.

$$2H^+ + 2e^- \rightarrow H_2$$

$$E_{eq} = E^0 + 0.059 \log_{10} a_{H^+} = 0.0 + 0.059 \times (-2) = -0.118 \text{ V versus SHE}$$

The mixed-potential diagram of this system is shown in Fig. 1.25, and the resultant polarization plot of the system is shown in Fig. 1.26.

Sixth case: iron in an aerated neutral solution at 25°C, pH = 2, $I_1 = 10^{-5}$ A.

Surface area = 1 cm²

The only difference from the previous case is that the limiting current of the cathodic reaction has decreased to 10^{-5} A. The mixed-potential diagram of this system is shown in Fig. 1.27, and the resultant polarization plot of the system is shown in Fig. 1.28.

References

1. Pourbaix, M., *Atlas of Electrochemical Equilibria in Aqueous Solutions,* Houston, Tex., NACE International, 1974.
2. Staehle, R. W., Understanding "Situation-Dependent Strength": A Fundamental Objective in Assessing the History of Stress Corrosion Cracking, in *Environment-Induced Cracking of Metals,* Houston, Tex., NACE International, 1989, pp. 561–612.
3. Pourbaix, M. J. N., *Lectures on Electrochemical Corrosion,* New York, Plenum Press, 1973.
4. Guthrie, J., *A History of Marine Engineering,* London, Hutchinson of London, 1971.
5. Flanagan, G. T. H., *Feed Water Systems and Treatment,* London, Stanford Maritime London, 1978.
6. Jones, D. R. H., *Materials Failure Analysis: Case Studies and Design Implications,* Headington Hill Hall, U.K., Pergamon Press, 1993.
7. Ruggeri, R. T., and Beck, T. R., An Analysis of Mass Transfer in Filiform Corrosion, *Corrosion* **39:**452–465 (1983).
8. Slabaugh, W. H., DeJager, W., Hoover, S. E., et al., Filiform Corrosion of Aluminum, *Journal of Paint Technology* **44:**76–83 (1972).
9. ASTM, Standard Test Method for Half-Cell Potentials of Uncoated Reinforcing Steel in Concrete, in *Annual Book of ASTM Standards,* Philadelphia, American Society for Testing and Materials, 1997.
10. Shreir, L. L., Jarman R. A., and Burstein, G. T., *Corrosion Control,* Oxford, U.K., Butterworth Heinemann, 1994.

Chapter 2

Environments

2.1 Atmospheric Corrosion	58
2.1.1 Types of atmospheres and environments	58
2.1.2 Theory of atmospheric corrosion	61
The cathodic process	62
The anodic process	63
Important practical variables in atmospheric corrosion	66
2.1.3 Atmospheric corrosivity and corrosion rates	69
The ISO methodology	69
Corrosivity classification according to PACER LIME algorithm	78
Direct measurement of atmospheric corrosion and corrosivity	81
2.1.4 Atmospheric corrosion rates as a function of time	84
2.2 Natural Waters	85
2.2.1 Water constituents and pollutants	87
Carbon dioxide and calcium carbonate	92
Dissolved mineral salts	93
Hardness	94
pH of water	96
Organic matter	96
Priority pollutants	97
2.2.2 Essentials of ion exchange	99
Synthesis	100
Physical and chemical structure of resins	101
Selectivity of resins	103
Kinetics	103
Types of ion-exchange resins	104
2.2.3 Saturation and scaling indices	105
The Langelier saturation index	106
Ryznar stability index	108
Puckorius scaling index	108
Larson-Skold index	109

Stiff-Davis index	110
Oddo-Tomson index	110
Momentary excess (precipitation to equilibrium)	110
Interpreting the indices	111
2.2.4 Ion association model	112
Optimizing storage conditions for low-level nuclear waste	114
Limiting halite deposition in a wet high-temperature gas well	115
Identifying acceptable operating range for ozonated cooling systems	117
Optimizing calcium phosphate scale inhibitor dosage in a high-TDS cooling system	122
2.2.5 Software systems	123
Scaling of cooling water	124
Scaling of deep well water	126
2.3 Seawater	129
2.3.1 Introduction	129
Salinity	129
Other ions	131
Precipitation of inorganic compounds from seawater	131
Oxygen	133
Organic compounds	135
Polluted seawater	136
Brackish coastal water	137
2.3.2 Corrosion resistance of materials in seawater	138
Carbon steel	139
Stainless steels	140
Nickel-based alloys	140
Copper-based alloys	140
Effect of flow velocity	140
Effect of temperature	141
2.4 Corrosion in Soils	142
2.4.1 Introduction	142
2.4.2 Soil classification systems	142
2.4.3 Soil parameters affecting corrosivity	143
Water	143
Degree of aeration	143
pH	143
Soil resistivity	146
Redox potential	146
Chlorides	146
Sulfates	147
Microbiologically influenced corrosion	147
2.4.4 Soil corrosivity classifications	148
2.4.5 Corrosion characteristics of selected metals and alloys	151
Ferrous alloys	151
Nonferrous metals and alloys	151
Reinforced concrete	153
2.4.6 Summary	154

2.5 Reinforced Concrete	154
2.5.1 Introduction	154
2.5.2 Concrete as a structural material	155
2.5.3 Corrosion damage in reinforced concrete	156
Mehta's holistic model of concrete degradation	156
Corrosion mechanisms	159
Chloride-induced rebar corrosion	159
Carbonation-induced corrosion	165
2.5.4 Remedial measures	166
Alternative deicing methods	166
Cathodic protection	168
Electrochemical chloride extraction	170
Re-alkalization	171
Repair techniques	173
Epoxy-coated reinforcing steel	175
Stainless steel rebar	175
Galvanized rebars	177
Corrosion inhibitors	178
Concrete cover and mix design	178
2.5.5 Condition assessment of reinforced concrete structures	180
Electrochemical corrosion measurements	182
Chloride content	183
Petrographic examination	184
Permeability tests	184
2.5.6 Life prediction for corroding reinforced concrete structures	184
2.5.7 Other forms of concrete degradation	186
Alkali-aggregate reaction	186
Freeze-thaw damage	187
Sulfate attack	187
2.6 Microbes and Biofouling	187
2.6.1 Basics of microbiology and MIC	187
Classification of microorganisms	190
Bacteria commonly associated with MIC	191
Effect of operating conditions on MIC	195
Identification of microbial problems	197
2.6.2 Biofouling	200
Nature of biofilm	201
Biofilm formation	202
Marine biofouling	205
Problems associated with biofilms	206
2.6.3 Biofilm control	208
Introduction	
A practical example: ozone treatment for cooling towers	215
References	216

2.1 Atmospheric Corrosion

Atmospheric corrosion can be defined as the corrosion of materials exposed to air and its pollutants, rather than immersed in a liquid. Atmospheric corrosion can further be classified into dry, damp, and wet categories. This chapter deals only with the damp and wet cases, which are respectively associated with corrosion in the presence of microscopic electrolyte (or "moisture") films and visible electrolyte layers on the surface. The damp moisture films are created at a certain critical humidity level (largely by the adsorption of water molecules), while the wet films are associated with dew, ocean spray, rainwater, and other forms of water splashing.

By its very nature, atmospheric corrosion has been reported to account for more failures in terms of cost and tonnage than any other factor. A case study of costly atmospheric corrosion damage on the Statue of Liberty is presented in Galvanic Corrosion in Sec. 5.2.1. Atmospheric corrosion damage involving aircraft is presently receiving much attention. An example of the serious consequences of aircraft corrosion damage is also described in Chap. 5, in Crevice Corrosion in Sec. 5.2.1. The risk and costs of corrosion are particularly high in aging aircraft. In one of the few detailed aircraft corrosion cost analyses that have been performed, it has been estimated that the direct costs alone of corrosion in U.S. Air Force aircraft exceeded $0.7 billion (FY 1990 dollars), with the oldest aircraft types accounting for approximately half the cost.[1] Similar figures are expected for U.S. Navy aircraft. The total annual costs in the U.S. aircraft industry have been estimated at around $4 billion. It is no longer uncommon for aircraft corrosion maintenance hours to be greater than flight hours.

2.1.1 Types of atmospheres and environments

The severity of atmospheric corrosion tends to vary significantly among different locations, and, historically, it has been customary to classify environments as rural, urban, industrial, marine, or combinations of these. These types of atmosphere have been described as follows:[2]

- *Rural.* This type of atmosphere is generally the least corrosive and normally does not contain chemical pollutants, but does contain organic and inorganic particulates. The principal corrodents are moisture, oxygen, and carbon dioxide. Arid and tropical types are special extreme cases in the rural category.

- *Urban.* This type of atmosphere is similar to the rural type in that there is little industrial activity. Additional contaminants are of the SO_x and NO_x variety, from motor vehicle and domestic fuel emissions.

- *Industrial.* These atmospheres are associated with heavy industrial processing facilities and can contain concentrations of sulfur dioxide, chlorides, phosphates, and nitrates.
- *Marine.* Fine windswept chloride particles that get deposited on surfaces characterize this type of atmosphere. Marine atmospheres are usually highly corrosive, and the corrosivity tends to be significantly dependent on wind direction, wind speed, and distance from the coast. It should be noted that an equivalently corrosive environment is created by the use of deicing salts on the roads of many cold regions of the planet.

Maps have been produced for numerous geographic regions, illustrating the macroscopic variations in atmospheric corrosivity. Such a map of North America is presented in Fig. 2.1, based on the corrosion of automobile bodies.[3] A similar map of South Africa is shown in Fig. 2.2, schematically representing 20 years of atmospheric exposure testing.[4] The coastal regions, extending some 4 to 5 km inland, tend to have the most corrosive atmospheres because of the effect of windswept chlorides. High humidity levels tend to exacerbate the detrimental effects of such chlorides. The effects of rainfall tend to be more ambiguous. Arguably, rain provides the moisture necessary for corrosion reactions, but on the other hand it tends to have a cleansing effect by washing away or diluting corrosive surface species.

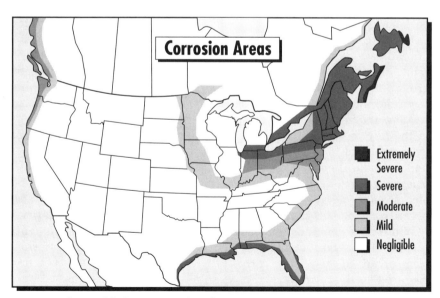

Figure 2.1 Geographical representation of car corrosion severity in North America.

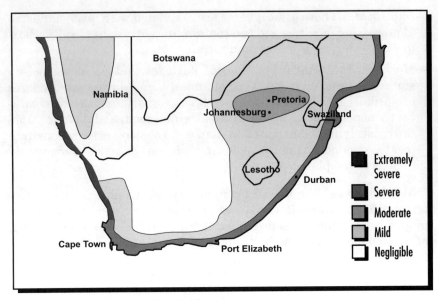

Figure 2.2 Corrosivity map of South Africa.

The high corrosion rates along the Gulf Coast and in Florida in Fig. 2.1 can be attributed to the corrosive marine environment. In the northeastern regions, deicing salts applied to road surfaces in winter are primarily responsible for the high corrosion rates. While accelerated laboratory testing can be satisfactory for evaluating the corrosion resistance of new materials and coatings, the automobile proving grounds are definitively the primary means for testing completed systems. Proving grounds are, in effect, large laboratories. But the proving ground test contents and procedures can differ sharply among manufacturers. Because each test is expressly different, each brings different results, and in this type of test, proper interpretation of the test results is the key to successful testing. For many years, bare steel coupons were attached to different vehicles in the northeastern United States and Canada, then periodically removed and measured for metal loss. The data from these coupons were used to target the corrosion test objectives to metal loss and to determine the localities with the most severe corrosion for captive fleet testing and future survey evaluations.

While it is generally important to rank macro-level environments according to a normalized corrosivity classification, specific information about atmospheric corrosivity and corrosion rates is often required on the micro level. For example, a corrosion risk assessment may be required for a military aircraft operating out of a specific air base environment. One such requirement resulted in a report of the

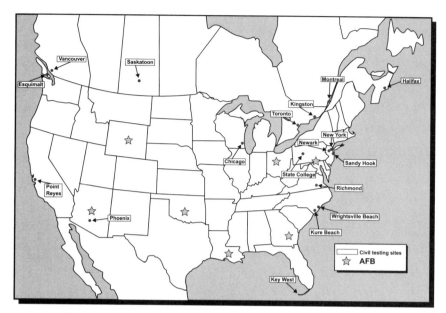

Figure 2.3 Locations of atmospheric corrosion testing sites in North America.

corrosion rates of several aluminum alloys after long-term exposure to different types of outdoor environments, shown in Fig. 2.3, ranging from relatively benign rural to aggressive industrial and marine environments.[5] For the sake of completeness, the results obtained from another valid source of information were added to Fig. 2.3. These results were compiled by the International Standards Organization (ISO) Technical Committee (TC) 156, Corrosion.[6]

2.1.2 Theory of atmospheric corrosion

A fundamental requirement for electrochemical corrosion processes is the presence of an electrolyte. Thin-film "invisible" electrolytes tend to form on metallic surfaces under atmospheric exposure conditions after a certain critical humidity level is reached. It has been shown that for iron, the critical humidity is 60 percent in an atmosphere free of sulfur dioxide. The critical humidity level is not constant and depends on the corroding material, the tendency of corrosion products and surface deposits to absorb moisture, and the presence of atmospheric pollutants.

In the presence of thin-film electrolytes, atmospheric corrosion proceeds by balancing anodic and cathodic reactions. The anodic oxidation reaction involves the dissolution of the metal, while the cathodic reaction is often assumed to be the oxygen reduction reaction. For iron,

Anode Reaction: $2Fe \rightarrow 2Fe^{2+} + 4e^-$

Cathode Reaction: $O_2 + 2H_2O + 4e^- \rightarrow 4OH^-$

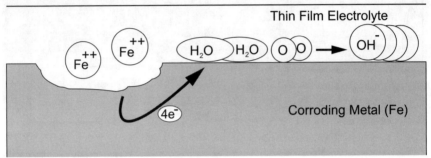

Figure 2.4 Atmospheric corrosion of iron.

these reactions are illustrated schematically in Fig. 2.4. It should be noted that corrosive contaminant concentrations can reach relatively high values in the thin electrolyte films, especially under conditions of alternate wetting and drying. Oxygen from the atmosphere is also readily supplied to the electrolyte under thin-film corrosion conditions.

The cathodic process. If it is assumed that the surface electrolyte in extremely thin layers is neutral or even slightly acidic, then the hydrogen production reaction [Eq. (2.1)] can be ignored for atmospheric corrosion of most metals and alloys.

$$2H^+ + 2e^- \rightarrow H_2 \tag{2.1}$$

Exceptions to this assumption would include corrosive attack under coatings, when the production of hydrogen can cause blistering of the coating, and other crevice corrosion conditions. The reduction of atmospheric oxygen is one of the most important reactions in which electrons are consumed. In the presence of gaseous air pollutants, other reduction reactions involving ozone and sulfur and nitrogen species have to be considered.[7] For atmospheric corrosion in near-neutral electrolyte solution, the oxygen reduction reaction is applicable [Eq. (2.2)].

$$O_2 + 2H_2O + 4e^- \rightarrow 4OH^- \tag{2.2}$$

Two reaction steps may actually be involved, with hydrogen peroxide as an intermediate, in accordance with Eqs. (2.3) and (2.4).

$$O_2 + 2H_2O + 2e^- \rightarrow H_2O_2 + 2OH^- \quad (2.3)$$

$$H_2O_2 + 2e^- \rightarrow 2OH^- \quad (2.4)$$

If oxygen from the atmosphere diffuses through the electrolyte film to the metal surface, a diffusion-limited current density should apply. It has been shown that a diffusion transport mechanism for oxygen is applicable only to an electrolyte-layer thickness of approximately 30 μm and under strictly isothermal conditions.[8] The predicted theoretical limiting current density of oxygen reduction in an electrolyte-layer thickness of 30 μm significantly exceeds practical observations of atmospheric corrosion rates. It can be argued, therefore, that the overall rates of atmospheric corrosion are likely to be controlled not by the cathodic oxygen reduction process, but rather by the anodic reaction(s).

The anodic process. Equation (2.5) represents the generalized anodic reaction that corresponds to the rate-determining step of atmospheric corrosion.

$$M \rightarrow M^{n+} + ne^- \quad (2.5)$$

The formation of corrosion products, the solubility of corrosion products in the surface electrolyte, and the formation of passive films affect the overall rate of the anodic metal dissolution process and cause deviations from simple rate equations. Passive films distinguish themselves from corrosion products, in the sense that these films tend to be more tightly adherent, are of lower thickness, and provide a higher degree of protection from corrosive attack. Atmospheric corrosive attack on a surface protected by a passive film tends to be of a localized nature. Surface pitting and stress corrosion cracking in aluminum and stainless alloys are examples of such attack.

Relatively complex reaction sequences have been proposed for the corrosion product formation and breakdown processes to explain observed atmospheric corrosion rates for different classes of metals. Fundamentally, kinetic modeling rather than equilibrium assessments appears to be appropriate for the dynamic conditions of alternate wetting and drying of surfaces corroding in the atmosphere. A framework for treating atmospheric corrosion phenomena on a theoretical basis, based on six different regimes, has been presented by Graedel[9] (Fig. 2.5). The regimes in this so-called GILDES-type model are the gaseous region (G), the gas-to-liquid interface (I), the surface liquid (L), the deposition layer (D), the electrodic layer (E), and the corroding solid (S).

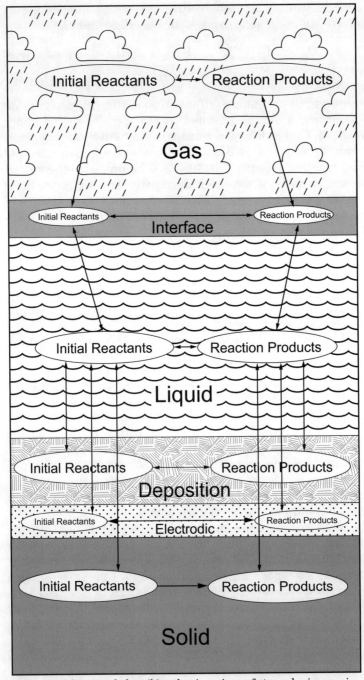

Figure 2.5 A framework describing the six regimes of atmospheric corrosion.

For the gaseous-layer effects, such as entrainment and detrainment of species across the liquid interface, chemical transformations in the gas phase, the effects of solar radiation on photosensitive atmospheric reactions, and temperature effects on the gas phase, reaction kinetics are important. In the interface regime, the transfer of molecules into the liquid layer prior to their chemical interaction in the liquid layer is studied. Not only does the liquid regime "receive" species from the gas phase, but species from the liquid are also volatilized into the gas phase. Important variables in the liquid regime include the aqueous film thickness and its effect on the concentration of species, chemical transformations in the liquid, and reactions involving metal ions originating from the electrochemical corrosion reactions.

In the deposition zone, corrosion products will accumulate, following their nucleation on the substrate. The corrosion products formed under thin-film atmospheric conditions are closely related to the formation of naturally occurring minerals. Over long periods of time, the most thermodynamically stable species will tend to dominate. The nature of corrosion products found on different metals exposed to the atmosphere is shown in Fig. 2.6. The solution known as the "inner electrolyte" can be trapped inside or under the corrosion products formed. The deposited corrosion product layers can thus be viewed as membranes, with varying degrees of resistance to ionic transport. Passivating films tend to represent strong barriers to ionic transport.

	Common Species	Rarer Species
Al	$Al(OH)_3$ $Al_2O_3, Al_2O_3 \cdot 3H_2O$	$AlOOH, Al_x(OH)_y(SO_4)_z,$ $AlCl(OH)_2 \cdot 4H_2O$
Fe	$Fe_2O_3, FeOOH,$ $FeSO_4 \cdot 4H_2O$	$Fe_x(OH)_yCl_z, FeCO_3$
Cu	$Cu_2O, Cu_4SO_4(OH)_6,$ $Cu_4SO_4(OH)_6 \cdot 2H_2O,$ $Cu_3SO_4(OH)_4$	$Cu_2Cl(OH)_3, Cu_2CO_3(OH)_2,$ $Cu_2NO_3(OH)_3$
Zn	$ZnO, Zn_5(OH)_6(CO_3)_2,$ $ZnCO_3$	$Zn(OH)_2, ZnSO_4,$ $Zn_5Cl_2(OH)_8 \cdot H_2O$

Figure 2.6 Nature of corrosion products formed on four metals.

Any corroding surface has a complex charge distribution, producing in the adjacent electrolyte a microscopic layer with chemical and physical properties that differ from those of the nominal electrolyte. This electrodic regime influences the overall reaction kinetics in atmospheric corrosion processes. In the solid regime, the detailed mechanistic steps (sequences) in the dissolution of the solid and their kinetic characteristics are relevant.

Specialized knowledge from different scientific fields is required in order to formulate mathematically the transition and transformation processes in these regimes:[9]

- *Gaseous layer.* Atmospheric chemistry
- *Interface layer.* Mass transport engineering and interface science
- *Liquid layer.* Freshwater, marine, and brine chemistry
- *Deposition layer.* Colloid chemistry and mineralogy
- *Electrodic layer.* Electrochemistry
- *Solid layer.* Solid-state chemistry

Important practical variables in atmospheric corrosion

Time of wetness. From the above theory, it should be apparent that the time of wetness (presence of electrolyte on the corroding surface) is a key parameter, directly determining the duration of the electrochemical corrosion processes. This variable is a complex one, since all the means of formation and evaporation of an electrolytic solution on a metal surface must be considered.

The time of wetness is obviously strongly dependent on the critical relative humidity. Apart from the primary critical humidity, associated with clean surfaces, secondary and even tertiary critical humidity levels may be created by hygroscopic corrosion products and capillary condensation of moisture in corrosion products, respectively. A capillary condensation mechanism may also account for electrolyte formation in microscopic surface cracks and the metal surface–dust particle interface. Other sources of surface electrolyte include chemical condensation (by chlorides, sulfates, and carbonates), adsorbed molecular water layers, and direct moisture precipitation (ocean spray, dew, rain). The effects of rain on atmospheric corrosion damage are somewhat ambiguous. While providing electrolyte for corrosion reactions, rain can act in a beneficial manner by washing away or diluting harmful corrosive surface species.

Sulfur dioxide. Sulfur dioxide, a product of the combustion of sulfur-containing fossil fuels, plays an important role in atmospheric corrosion in urban and industrial atmospheres. It is adsorbed on metal surfaces, has a high solubility in water, and tends to form sulfuric acid in the presence of surface moisture films. Sulfate ions are formed in the surface moisture layer by the oxidation of sulfur dioxide in accordance with Eq. (2.6).

$$SO_2 + O_2 + 2e^- \rightarrow SO_4^{2-} \qquad (2.6)$$

The required electrons are thought to originate from the anodic dissolution reaction and from the oxidation of ferrous to ferric ions. It is the formation of sulfate ions that is considered to be the main corrosion-accelerating effect from sulfur dioxide. For iron and steel, the presence of these sulfate ions ultimately leads to the formation of iron sulfate ($FeSO_4$). Iron sulfate is known to be a corrosion product component in industrial atmospheres and is mainly found in layers at the metal surface. The iron sulfate is hydrolyzed by the reaction expressed by Eq. (2.7).

$$FeSO_4 + 2H_2O \rightarrow FeOOH + SO_4^{2-} + 3H^+ + e^- \qquad (2.7)$$

The corrosion-stimulating sulfate ions are liberated by this reaction, leading to an autocatalytic type of attack on iron.[8–10] The acidification of the electrolyte could arguably also lead to accelerated corrosion rates, but this effect is likely to be of secondary importance because of the buffering effects of hydroxide and oxide corrosion products. In nonferrous materials such as zinc, sulfate ions also stimulate corrosion, but the autocatalytic corrosion mechanism is not easily established. Corroding zinc tends to be covered by stable zinc oxides and hydroxides, and this protective covering is only gradually destroyed at its interface with the atmosphere. In moderately corrosive atmospheres, sulfates present in zinc corrosion products tend to be bound relatively strongly, with limited water solubility. At very high levels of sulfur dioxide, dissolution of protective layers and the formation of more soluble corrosion products is associated with higher corrosion rates.

Chlorides. Atmospheric salinity distinctly increases atmospheric corrosion rates. Apart from the enhanced surface electrolyte formation by hygroscopic salts such as NaCl and $MgCl_2$, direct participation of chloride ions in the electrochemical corrosion reactions is also likely. In ferrous metals, chloride anions are known to compete with hydroxyl ions to combine with ferrous cations produced in the anodic reaction. In the case of hydroxyl ions, stable passivating species tend to be produced. In contrast, iron-chloride complexes tend to be unstable (soluble), resulting in further stimulation of corrosive attack. On this basis, metals such as zinc and copper, whose chloride salts tend to be less soluble than those of iron, should be less prone to chloride-induced corrosion damage,[8] and this is consistent with practical experience.

Other atmospheric contaminants. Hydrogen sulfide, hydrogen chloride, and chlorine present in the atmosphere can intensify atmospheric corrosion damage, but they represent special cases of atmospheric corrosion that are invariably related to industrial emissions in specific microclimates. Hydrogen sulfide is known to be extremely corrosive to most metals/alloys, and the corrosive effects of gaseous chlorine and hydrogen chloride in the presence of moisture tend to be stronger than those of "chloride salt" anions because of the acidic character of the former species.[8]

Nitrogen compounds, in the form of NO_x, also tend to accelerate atmospheric attack. NO_x emission, largely from combustion processes, has been reported to have increased relative to SO_2 levels. However, measured deposition rates of these nitrogen compounds have been significantly lower than those for SO_2, which probably accounts for the generally lower importance assigned to these.

Until recently, the effects of ozone (O_3) had been largely neglected in atmospheric corrosion research. It has been reported that the presence of ozone in the atmosphere may lead to an increase in the sulfur dioxide deposition rate. While the accelerating effect of ozone on zinc corrosion appears to be very limited, both aluminum and copper have been noted to undergo distinctly accelerated attack in its presence.[7]

The deposition of solid matter from the atmosphere can have a significant effect on atmospheric corrosion rates, particularly in the initial stages. Such deposits can stimulate atmospheric attack by three mechanisms:

- Reduction in the critical humidity levels by hygroscopic action
- The provision of anions, stimulating metal dissolution
- Microgalvanic effects by deposits more noble than the corroding metal; carbonaceous deposits deserve special mention in this context.

Temperature. The effect of temperature on atmospheric corrosion rates is also quite complex. An increase in temperature will tend to stimulate corrosive attack by increasing the rate of electrochemical reactions and diffusion processes. For a constant humidity level, an increase in temperature would lead to a higher corrosion rate. Raising the temperature will, however, generally lead to a decrease in relative humidity and more rapid evaporation of surface electrolyte. When the time of wetness is reduced in this manner, the overall corrosion rate tends to diminish.

For closed air spaces, such as indoor atmospheres, it has been pointed out that the increase in relative humidity associated with a drop in temperature has an overriding effect on corrosion rate.[11] This implies that simple air conditioning that decreases the temperature without additional dehumidification will accelerate atmospheric corrosion damage. At temperatures below freezing, where the electrolyte film solidifies, electrochemical corrosion activity will drop to negligible levels. The very low atmospheric corrosion rates reported in extremely cold climates are consistent with this effect.

2.1.3 Atmospheric corrosivity and corrosion rates

The nature and rate of atmospheric corrosive attack are dependent on the composition and properties of the thin-film surface electrolyte. Time of wetness and the type and concentration of gaseous and particulate pollutants in the atmosphere largely affect these in turn. The classification of atmospheric corrosivity is important for specifying suitable materials and corrosion protection measures at the design stage and for asset maintenance management to ensure adequate service life. Two fundamental approaches to classifying atmospheric corrosivity have been followed, as shown in Fig. 2.7. These two approaches to environmental classification can be used in a complementary manner to derive relationships between atmospheric corrosion rates and the dominant atmospheric variables. Ultimately, the value of atmospheric corrosivity classifications is enhanced if they are linked to estimates of actual corrosion rates of different metals or alloys.

The ISO methodology. A comprehensive corrosivity classification system has been developed by the International Standards Organization (ISO). The applicable ISO standards are listed in Table 2.1. Verification and evolution of this system is ongoing through the largest exposure program ever, undertaken on a worldwide basis.[12]

Procedure and limitations. The ISO corrosivity classification from atmospheric parameters is based on the simplifying assumption that the time of wetness (TOW) and the levels of corrosive impurities determine the corrosivity. Only two types of corrosive impurities are considered, namely, sulfur dioxide and chloride. Practical definitions for all the variables involved in calculating an ISO corrosivity index follow.

Time of wetness. Units: hours per year (h·year^{-1}) when relative humidity (RH) > 80 percent and $t > 0°C$

TOW ≤ 10	T_1
10 < TOW ≤ 250	T_2
250 < TOW ≤ 2500	T_3
2500 < TOW ≤ 5500	T_4
5500 < TOW	T_5

Airborne salinity. Units: chloride deposition rate (mg·m^{-2}·day^{-1})

$S \leq 60$	S_1
$60 < S \leq 300$	S_2
$300 < S$	S_3

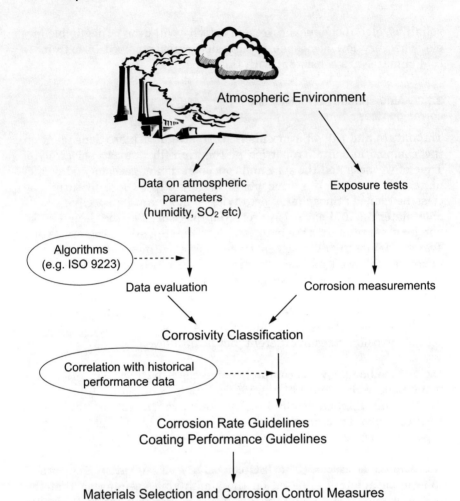

Figure 2.7 Two fundamental approaches to classifying atmospheric corrosivity.

TABLE 2.1 List of ISO Standards Related to Atmospheric Corrosion

ISO standard	Title
ISO 9223	Classification of the Corrosivity of Atmospheres
ISO 9224	Guiding Values for the Corrosivity Categories of Atmospheres
ISO 9225	Aggressivity of Atmospheres—Methods of Measurement of Pollution Data
ISO 9226	Corrosivity of Atmospheres—Methods of Determination of Corrosion Rates of Standard Specimens for the Evaluation of Corrosivity

Industrial pollution by SO_2. Two types of units are used:

Concentration (μg·m^{-3}), P_C

$P_C \leq 40$	P_1
$40 < P_C \leq 90$	P_2
$90 < P_C$	P_3

Deposition rate (mg·m^{-2}·day^{-1}), P_D

$P_D \leq 35$	P_1
$35 < P_D \leq 80$	P_2
$80 < P_D$	P_3

Corrosion rate categories. Two types of corrosion rates are predicted:

Category	Short-term, g·m^{-2}·year^{-1}	Long-term, μm·year^{-1}
C_1	CR ≤ 10	CR ≤ 0.1
C_2	10 < CR ≤ 200	0.1 < CR ≤ 1.5
C_3	200 < CR ≤ 400	1.5 < CR ≤ 6
C_4	400 < CR ≤ 650	6 < CR ≤ 20
C_5	650 < CR	20 < CR

The TOW categorization is presented in Table 2.2, and the sulfur dioxide and chloride classifications are presented in Table 2.3. TOW values can be measured directly with sensors, or the ISO definition of TOW as the number of hours that the relative humidity exceeds 80 percent and the temperature exceeds 0°C can be used. The methods for determining atmospheric sulfur dioxide and chloride deposition rates are described more fully in the relevant standards (Table 2.1).

Following the categorization of the three key variables, the applicable ISO corrosivity category can be determined using the appropriate ISO chart (Table 2.4). Different corrosivity categories apply to different types of metal. As the final step in the ISO procedure, the rate of atmospheric corrosion can be estimated for the determined corrosivity category. Table 2.5 shows a listing of 12-month corrosion rates for different

TABLE 2.2 ISO 9223 Classification of Time of Wetness

Wetness category	Time of wetness, %	Time of wetness, hours per year	Examples of environments
T_1	<0.1	<10	Indoor with climatic control
T_2	0.1–3	10–250	Indoor without climatic control
T_3	3–30	250–2500	Outdoor in dry, cold climates
T_4	30–60	2500–5500	Outdoor in other climates
T_5	>60	>5500	Damp climates

TABLE 2.3 ISO 9223 Classification of Sulfur Dioxide and Chloride "Pollution" Levels

Sulfur dioxide category	Sulfur dioxide deposition rate, mg/m^2·day	Chloride category	Chloride deposition rate, mg/m^2·day
P_0	≤10	S_0	≤3
P_1	11–35	S_1	4–60
P_2	36–80	S_2	61–300
P_3	81–200	S_3	301–1500

TABLE 2.4 ISO 9223 Corrosivity Categories of Atmosphere

TOW	Cl$^-$	SO$_2$	Steel	Cu and Zn	Al
T_1	S_0 or S_1	P_1	1	1	1
		P_2	1	1	1
		P_3	1–2	1	1
	S_2	P_1	1	1	2
		P_2	1	1	2
		P_3	1–2	1–2	2–3
	S_3	P_1	1–2	1	2
		P_2	1–2	1–2	2–3
		P_3	2	2	3
T_2	S_0 or S_1	P_1	1	1	1
		P_2	1–2	1–2	1–2
		P_3	2	2	3–4
	S_2	P_1	2	1–2	2–3
		P_2	2–3	2	3–4
		P_3	3	3	4
	S_3	P_1	3–4	3	4
		P_2	3–4	3	4
		P_3	4	3–4	4
T_3	S_0 or S_1	P_1	2–3	3	3
		P_2	3–4	3	3
		P_3	4	3	3–4
	S_2	P_1	3–4	3	3–4
		P_2	3–4	3–4	4
		P_3	4–5	3–4	4–5
	S_3	P_1	4	3–4	4
		P_2	4–5	4	4–5
		P_3	5	4	5
T_4	S_0 or S_1	P_1	3	3	3
		P_2	4	3–4	3–4
		P_3	5	4–5	4–5
	S_2	P_1	4	4	3–4
		P_2	4	4	4
		P_3	5	5	5
	S_3	P_1	5	5	5
		P_2	5	5	5
		P_3	5	5	5
T_5	S_0 or S_1	P_1	3–4	3–4	4
		P_2	4–5	4–5	4–5
		P_3	5	5	5
	S_2	P_1	5	5	5
		P_2	5	5	5
		P_3	5	5	5
	S_3	P_1	5	5	5
		P_2	5	5	5
		P_3	5	5	5

TABLE 2.5 ISO 9223 Corrosion Rates after One Year of Exposure Predicted for Different Corrosivity Classes

Corrosion category	Steel, g/m^2·year	Copper, g/m^2·year	Aluminum, g/m^2·year	Zinc, g/m^2·year
C_1	≤10	≤0.9	Negligible	≤0.7
C_2	11–200	0.9–5	≤0.6	0.7–5
C_3	201–400	5–12	0.6–2	5–15
C_4	401–650	12–25	2–5	15–30
C_5	651–1500	25–50	5–10	30–60

metals for different corrosivity categories. The establishment of corrosion rates is complicated by the fact that these rates are not linear with time. For this reason, initial rates after 1 year and stabilized longer-term rates have been included for the different metals in the ISO methodology.

In situations in which TOW and pollution levels cannot be determined conveniently, another approach based on the exposure of standardized coupons over a 1-year period is available for classifying the atmospheric corrosivity. Simple weight loss measurements are used for determining the corrosivity categories. The nature of the specimens used is discussed more fully in a later section of this chapter.

Although the ISO methodology represents a rational approach to corrosivity classification, it has several inherent limitations. The atmospheric parameters determining the corrosivity classification do not include the effects of potentially important corrosive pollutants or impurities such as NO_x, sulfides, chlorine gas, acid rain and fumes, deicing salts, etc., which could be present in the general atmosphere or be associated with microclimates. Temperature is also not included as a variable, although it could be a major contributing factor to the high corrosion rates in tropical marine atmospheres. Only four standardized pure metals have been used in the ISO testing program. The methodology does not provide for localized corrosion mechanisms such as pitting, crevice corrosion, stress corrosion cracking, or intergranular corrosion. The effects of variables such as exposure angle and sheltering cannot be predicted, and the effects of corrosive microenvironments and geometrical conditions in actual structures are not accounted for.

Dean[13] has reported on a U.S. verification study of the ISO methodology. This study was conducted over a 4-year time period at five exposure sites and with four materials (steel, copper, zinc, and aluminum). Environmental data were used to obtain the ISO corrosivity classes, and these estimates were then compared to the corrosion classes obtained by direct coupon measurement. Overall, agreement was found in 58 percent of the cases studied. In 22 percent of the cases the estimated corrosion class was lower than the measured, and in 20 percent of the cases it was higher. It was also noted that the selected atmospheric variables (TOW, temperature, chloride deposition, sulfur dioxide deposition, and exposure time) accounted for a major portion of the variation in the corrosion data, with the exception of the data gathered for the corrosion of aluminum. Further refinements in the ISO procedures are anticipated as the worldwide database is developed.

ISO corrosivity analysis at two air bases. Use of the ISO methodology can be illustrated by applying it to a corrosivity assessment performed for two contrasting air bases: a maritime base in Nova Scotia and an inland base in Ontario (Fig. 2.8). The motivation for determining atmospheric corrosivity at these locations can be viewed in the context of the idealized corrosion surveillance strategy shown in Fig. 2.9. Essentially this scheme revolves around predicting where and when the risk of corrosion damage is greatest and tailoring corrosion control efforts accordingly. The principle and importance of linking selected maintenance and inspection schedules to the prevailing atmospheric corrosivity has been described in detail elsewhere.[14] An underlying consideration in these recommendations is that military aircraft spend the vast majority of their lifetime on the ground, and most corrosion damage occurs at ground level.

The ISO TOW parameter could be derived directly from relative humidity and temperature measurements performed hourly at the bases. The average daily TOW at the maritime base is shown in Fig. 2.10, together with the corresponding ISO TOW categories, as determined by the criteria of Table 2.2. The overall TOW profile for the inland base was remarkably similar.

In the case of the air bases, no directly measured data were available for the chloride and sulfur dioxide deposition rates. However, data pertaining to atmospheric sulfur dioxide levels and chloride levels in precipitation had been recorded at sites in relatively close proximity. On the basis of these data, the likely ISO chloride and sulfur dioxide categories for the maritime base were S_3 and P_0–P_1, respectively. Under these assumptions, the applicable ISO corrosivity ratings are at the high to very high levels (C_4 to C_5) for aluminum. Using ISO chloride and sulfur dioxide categories of S_0 and P_0–P_1, respectively, for the inland air base, the corrosivity rating for aluminum is at the C_3 level. The step-by-step procedure for determining these categories and the different corrosion rates predicted for aluminum at the two bases are shown in Fig. 2.11.

The main implications of the analysis of atmospheric corrosivity at the maritime air base are that aircraft are at considerable risk of corrosion damage in view of the high corrosivity categories and that the

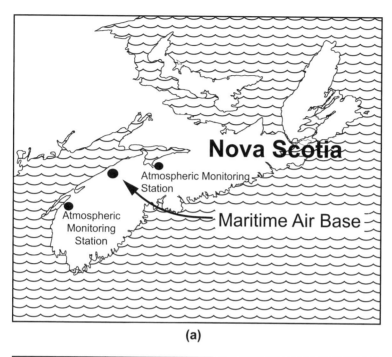

(a)

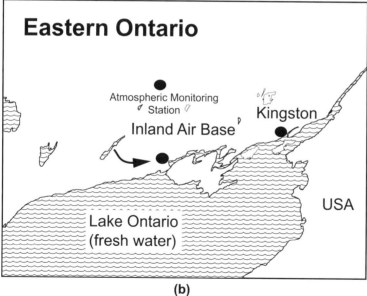

(b)

Figure 2.8 Geographical location of two Canadian air bases: (*a*) a maritime air base on the Bay of Fundy; (*b*) an inland air base on the shore of Lake Ontario.

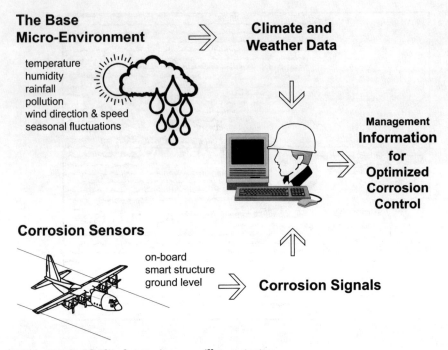

Figure 2.9 An idealized corrosion surveillance strategy.

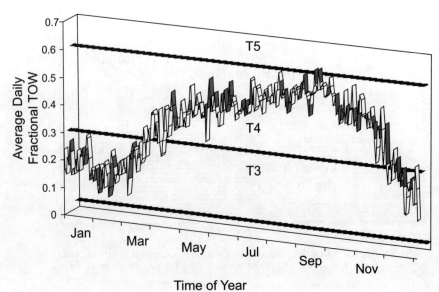

Figure 2.10 Average time of wetness (TOW) at a maritime air base.

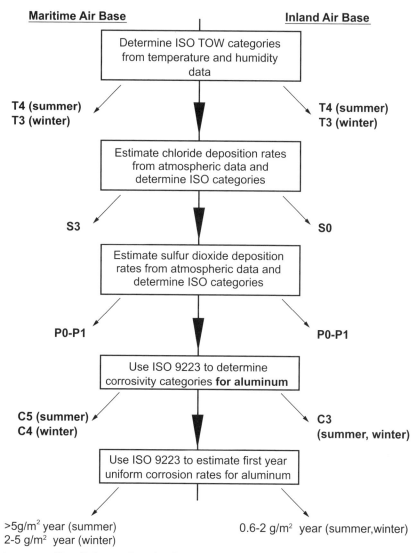

Figure 2.11 Detailed procedure for determining the ISO corrosivity categories.

fluctuations in corrosivity with time deserve special attention. Present "routine" maintenance and inspection schedules and corrosion control efforts do not take such variations into account.

As a simple example of how corrosion control could be improved by taking such variations into account, the effects of aircraft dehumidification can be considered. It is assumed that dehumidification would be applied only on a seasonal basis, when the T_4 TOW category is reached on a monthly average (refer to Fig. 2.10). It is further assumed that the time of wetness can be reduced to an average T_3 level in these critical months by the application of dehumidification systems. The emphasis in dehumidification should be placed on the nighttime, on the basis of Fig. 2.12. The projected cumulative corrosion rates of aluminum with and without this simple measure, based on ISO predictions, are shown in Fig. 2.13. The S_3 chloride and P_1 sulfur dioxide categories were utilized in this example, together with the most conservative 12-month corrosion rates of the applicable ISO corrosivity ratings. The potential benefits of dehumidification, even when it is applied only in selected time frames, are readily apparent from this analysis. Aircraft dehumidification is a relatively simple, practical procedure utilized for aircraft corrosion control in some countries. Dehumidified air can be circulated through the interior of the aircraft, or the entire aircraft can be positioned inside a dehumidified hangar. It should be noted that the numeric values for uniform corrosion rates of aluminum predicted by the ISO analysis are not directly applicable to actual aircraft, which are usually subject to localized corrosion damage under coatings or some other form of corrosion prevention measures.

Corrosivity classification according to PACER LIME algorithm. An environmental corrosivity scale based on atmospheric parameters has been developed by Summitt and Fink.[15] This classification scheme was developed for the USAF for maintenance management of structural aircraft systems, but wider applications are possible. A corrosion damage algorithm (CDA) was proposed as a guide for anticipating the extent of corrosion damage and for planning the personnel complement and time required to complete aircraft repairs. This classification was developed primarily for uncoated aluminum, steel, titanium, and magnesium aircraft alloys exposed to the external atmosphere at ground level.

The section of the CDA algorithm presented in Fig. 2.14 considers distance to salt water, leading either to the very severe AA rating or a consideration of moisture factors. Following the moisture factors, pollutant concentrations are compared with values of Working Environmental Corrosion Standards (WECS). The WECS values were adopted from the 50th percentile median of a study aimed at determining ranges of environmental parameters in the United States and represent "averages of averages." For example, if any of the three pollutants sulfur dioxide, total suspended particles, or ozone level exceeds the WECS values, in combination with a high moisture factor, the severe A rating is obtained. An algorithm for aircraft washing based on similar corrosivity considerations is presented in Fig. 2.15.

Environments 79

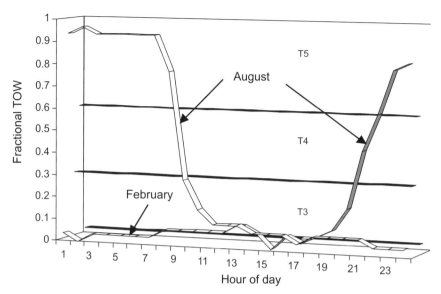

Figure 2.12 Relative TOW as a function of time of day for a dry month (February) and a humid month (August) at a maritime air base.

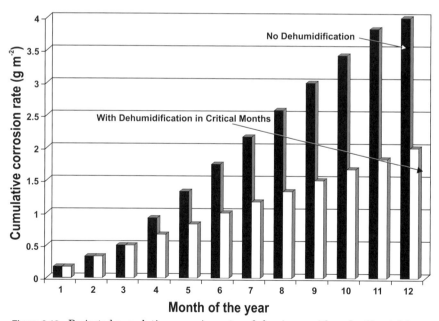

Figure 2.13 Projected cumulative corrosion rates of aluminum with and without dehumidification.

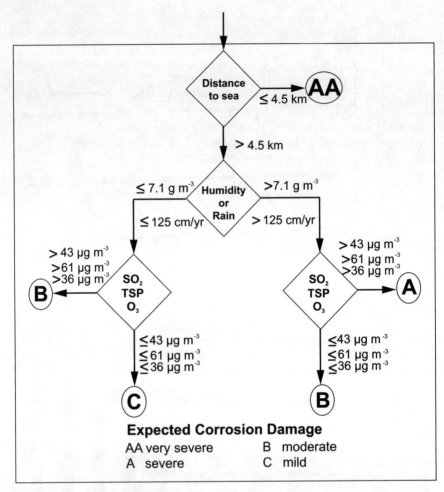

Figure 2.14 Section of the corrosion damage algorithm that considers distance to salt water, leading to either the very severe AA rating or a consideration of moisture factors.

The environmental corrosivity, predicted from the CDA algorithm, of six sea patrolling aircraft bases has been compared to the actual corrosion maintenance effort expended on the aircraft at each base. Considering the simplicity of the algorithms and simplifying assumptions in obtaining relevant environmental and maintenance data, the correlation obtained can be considered to be reasonable.

Further validation of the CDA algorithm approach was sought by comparison of the predicted corrosivity data to actual coupon exposure results. Despite various experimental difficulties in the exposure program involving various bases, good agreement was reported between the algorithm rankings and available experimental data.[15]

Environments 81

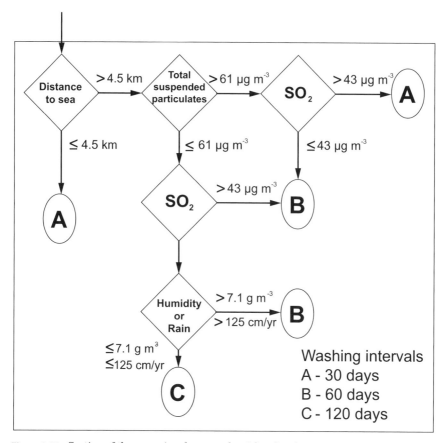

Figure 2.15 Section of the corrosion damage algorithm for planning a washing schedule.

Direct measurement of atmospheric corrosion and corrosivity. Atmospheric corrosion damage has to be assessed by direct measurement if no preexisting correlation between atmospheric corrosion rates and atmospheric parameters is available. Such a correlation and even data on basic atmospheric parameters rarely exist for specific microenvironments, necessitating direct measurement of the atmospheric corrosivity and corrosion rates.

Corrosion coupons. The simplest form of direct measurement of atmospheric corrosion is by coupon exposure. Subsequent to their exposure, the coupons can be subjected to weight loss measurements, pit density and depth measurements, and other types of examination. Flat panels exposed on exposure racks are a common coupon-type device for atmospheric corrosivity measurements. Various other specimen configurations have been used, including stressed U-bend or C-ring specimens for SCC studies. The main drawback associated with conventional coupon measurements is that extremely long exposure times are usually required to obtain meaningful data, even on a relative scale. It is not uncommon for such programs to run for 20 years or longer.

Two variations of the basic coupon specimens that can facilitate more rapid material/corrosivity evaluations deserve a special mention. The first is the use of a helical coil of material, as adopted in the ISO 9226 methodology. The high surface area/weight ratio in the helix configuration gives higher sensitivity than that with a panel coupon. The use of bimetallic specimens in which a helical wire is wrapped around a coarsely threaded bolt can provide additional sensitivity and forms the basis of the CLIMAT test. For aluminum wires, it was established that copper and steel bolts provide the highest sensitivity in industrial and marine environments, respectively.[16] Exposure times for atmospheric corrosivity classification can be conveniently reduced to 3 months with the CLIMAT specimen configuration. In the CLIMAT tests, atmospheric corrosivity indexes are determined as the percentage mass loss of the aluminum wires, and a subjective severity classification has been assigned for industrial and marine atmospheres, as shown in Table 2.6.

The ability of the CLIMAT devices to detect corrosivity fluctuations on a microenvironmental scale is apparent from the results presented in Fig. 2.16. These CLIMAT data were obtained from an exposure program on the grounds of the Royal Military College of Canada (RMC). The distinctly higher corrosivity in winter, associated with proximity to a road treated with deicing salts, should be noted. Furthermore, with the CLIMAT devices, it has been possible to detect significant seasonal corrosivity fluctuations which would not have been detected with other, less sensitive, coupon-type testing. For example, in the summer months (in the absence of deicing salts), the corrosivity at the RMC test point near the road decreased substantially.

Instrumented corrosion sensors. Electrochemical sensors are based on the principle of electrochemical current and/or potential measurements and facilitate the measurement of atmospheric corrosion damage in real time in a highly sensitive manner. There are special requirements for the construction of atmospheric corrosion sensors. For the measurement of corrosion currents and potentials, electrically isolated sensor elements are required. Fundamentally, the metallic sensor elements must be extremely closely spaced under the thin-film electrolyte conditions, in which ionic current flow is restricted. Electrochemical techniques utilized to measure atmospheric corrosion processes include zero resistance ammetry (ZRA), electrochemical noise (EN),

TABLE 2.6 Severity Classification for CLIMAT Testing

Industrial corrosion index (ICI)	Classification	Examples
0–1	Negligible	Rural and suburban areas
1–2	Moderate	Urban residential areas
2–4	Moderately severe	Urban industrialized areas
4–7	Severe	Industrialized areas
>7	Very severe	Heavily industrialized areas

Marine corrosion index (MCI)	Classification	Examples
0–2	Negligible	Average habitable area
2–5	Moderate	Seaside
5–10	Moderately severe	Seaside and exposed
10–20	Severe	Very exposed
>20	Very severe	Very exposed, windswept and sandswept

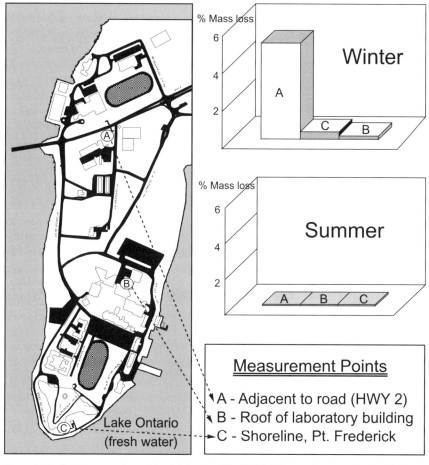

Figure 2.16 Positions and results obtained with CLIMAT corrosion monitoring devices at three locations on the Royal Military College campus.

linear polarization resistance (LPR), and electrochemical impedance spectroscopy (EIS).

The quartz crystal microbalance (QCM) is an example of a piezoelectric crystal whose frequency response to mass changes can be used for atmospheric corrosion measurements. In this technique, a metallic corrosion sensor element is bonded to the quartz sample. Mass gains associated with corrosion product buildup induce a decrease in resonance frequency. A characteristic feature of the QCM is exceptional sensitivity to mass changes, with a mass resolution of around 10 ng/cm^2. The classification of indoor corrosivity, based on the approach of the Instrument Society of America (ISA) S71.01-1985 standard and the use of a copper sensing element and QCM technology, is presented in Table 2.7.

Other technologies that have been used for atmospheric corrosion sensing include electrical resistance (ER) sensors and more recently fiber-optic sensing systems. Additional information may be found on this topic in Chap. 6, Corrosion Maintenance Through Inspection and Monitoring.

2.1.4 Atmospheric corrosion rates as a function of time

As already pointed out, atmospheric corrosion penetration usually is not linear with time. The buildup of corrosion products often tends to reduce the corrosion rate over time. Pourbaix[17] utilized the so-called linear bilogarithmic law for atmospheric corrosion, to describe atmospheric corrosion damage as a function of time on a mathematical basis. This law was shown to be applicable to different types of atmospheres (rural, marine, industrial) and for a variety of alloys, such as carbon steels, weathering steels, galvanized steels, and aluminized steels. This mathematical model has also been applied more recently

TABLE 2.7 Environmental Corrosivity Classification Based on ISA S71.01-1985

Copper oxide film thickness, angstroms*	ISA classification	Severity	Effects
0–300	G1	Mild	Corrosion is not a factor in equipment reliability
300–1000	G2	Moderate	Corrosion may be a factor in equipment reliability
1000–2000	G3	Harsh	High probability of corrosive attack
2000+	GX	Severe	Only specially designed and packaged equipment is expected to survive

*Based on a 30-day exposure period.

in a comprehensive exposure program.[13] It should be noted, however, that not all alloy/environment combinations would follow this law.

According to the linear bilogarithmic law expressed in Eq. (2.8),

$$p = At^B \quad \text{or} \quad \log_{10} p = A' + B \log_{10} t \tag{2.8}$$

where p is the corrosion penetration and t is the exposure time. It follows that the mean corrosion rate can be expressed by Eq. (2.9),

$$p/t = At^{B-1} \quad \text{or} \quad \log_{10}(p/t) = A' + (B-1)\log_{10} t \tag{2.9}$$

and the instantaneous corrosion rate by Eq. (2.10),

$$dp/dt = ABt^{B-1} \quad \text{or} \quad \log_{10}(dp/dt) = A' + B' + (B-1)\log_{10} t \tag{2.10}$$

According to the linear bilogarithmic law, the atmospheric behavior of a specific material at a specific location can be defined by the two parameters A and B. The initial corrosion rate, observed during the first year of exposure, is described by A, while B is a measure of the long-term decrease in corrosion rate. When B equals 0.5, the law of corrosion penetration increase is parabolic, with diffusion through the corrosion product layers as the rate-controlling step. At B values appreciably smaller than 0.5, the corrosion products show protective, passivating characteristics. Higher B values, greater than 0.5, are indicative of nonprotective corrosion products. Loosely adherent, flaky rust layers are an example of this case.

An important aspect of the linear bilogarithmic law is that it facilitates the prediction of long-term corrosion damage from short exposure tests. According to Pourbaix,[17] this extrapolation is valid for up to 20 to 30 years. A caveat of long-term tests is that changes in the environment may affect the corrosion rates more significantly than a fundamental deviation from the linear bilogarithmic law.

2.2 Natural Waters

Abundant supplies of fresh water are essential to industrial development. Enormous quantities are required for cooling of products and equipment, for process needs, for boiler feed, and for sanitary and potable water. It was estimated in 1980 that the water requirements for industry in the United States approximated 525 billion liters per day. A substantial quantity of this water was reused. The intake of "new" water was estimated to be about 140 billion liters daily.[18] If this water were pure and contained no contaminants, there would be little need for water conditioning or water treatment.

Water possesses several unique properties, one being its ability to dissolve to some degree every substance occurring on the earth's crust and in the atmosphere. Because of this solvent property, water typically contains a variety of impurities. These impurities are a source of potential trouble through deposition of the impurities in water lines, in boiler tubes, and on products which are contacted by the water. Dissolved oxygen, the principal gas present in water, is responsible for the need for costly replacement of piping and equipment as a result of its corrosive attack on metals with which it comes in contact.

The origin of all water supply is moisture that has evaporated from land masses and oceans and has subsequently been precipitated from the atmosphere. Depending on weather conditions, this may fall in the form of rain, snow, sleet, or hail. As it falls, this precipitation contacts the gases that make up the atmosphere and suspended particulates in the form of dust, industrial smoke and fumes, and volcanic dust and gases. It, therefore, contains the dissolved gases of the atmosphere and mineral matter that has been dissolved from the suspended atmospheric impurities.

The two most important sources of fresh water are surface water and groundwater. A portion of the rain or melting snow and ice at the earth's surface soaks into the ground, and part of it collects in ponds and lakes or runs off into creeks and rivers. This latter portion is termed surface water. As the water flows across the land surface, minerals are solubilized and the force of the flowing water carries along finely divided particles and organic matter in suspension. The character of the terrain and the nature of the geological composition of the area will influence the kind and quantity of the impurities found in the surface waters of a given geographic area.

That portion of water which percolates into the earth's crust and collects in subterranean pools and underground rivers is groundwater. This is the source of well and spring water. Underground supplies of fresh water differ from surface supplies in three important respects, two of which are advantageous for industrial use. These are a relatively constant temperature and the general absence of suspended matter. Groundwater, like surface water, is subject to variations in the nature of dissolved impurities; that is, the geological structure of the aquifer from which the supply is drawn will greatly influence the predominant mineral constituents. Groundwater is often higher in mineral content than surface supplies in the same geographic area because of the added solubilizing influence of dissolved carbon dioxide. The higher carbon dioxide content of groundwater as compared with surface water stems from the decay of organic matter in the surface soil.

In many areas, the availability of new intake water is limited. Thus, in those industries that require large amounts of cooling water, it is necessary to conserve available supplies by recirculating the water over cooling towers. The primary metals, petrochemical, and papermaking industries are good examples of industries requiring large volumes of water in the manufacturing process that condition a portion of the wastewater for reuse. Use of purified effluent streams from sewage treatment plants is another example of water reuse and conservation.

When purification and water-conditioning techniques are practiced in order to produce water that is acceptable for industrial use, certain analytical tests must be performed to ensure that the objectives of treatment are being achieved. Table 2.8 is a listing of the analytical determinations made in the examination of most natural waters. Described in the list are the general categories of substances, the difficulties commonly encountered as a result of the presence of each substance, and the usual means of treatment to alleviate the difficulties. In Table 2.9 the methods of water treatment are presented, which can be divided into two major groups:

1. Chemical procedures, which are based on material modifications as a result of chemical reactions. These can be monitored by analyzing the water before and after the treatment (softening, respective demineralization).

2. Physical treatments that can alter the crystal structure of the deposits.

The criteria for a successful water treatment are

- Capability of meeting the target process
- Protection of the construction materials against corrosion
- Preservation of the specific water characteristics (quality)

There is no generally valid solution with regard to water treatments. The specific conditions of water supplies can be vastly different, even when the supplies are separated by only a few meters. The basis for all evaluation of water quality must be a specific chemical water analysis.

2.2.1 Water constituents and pollutants

The concentrations of various substances in water in dissolved, colloidal, or suspended form are typically low but vary considerably. A hardness value of up to 400 ppm of $CaCO_3$, for example, is sometimes tolerated in public supplies, whereas 1 ppm of dissolved iron would be unacceptable. In treated water for high-pressure boilers or where radiation effects are important, as in some nuclear reactors, impurities are measured in very small units, such as parts per billion (ppb). Water

TABLE 2.8 Difficulties and Means of Treatment for Common Impurities Found in Fresh Water

Constituent	Chemical formula	Difficulties caused	Means of treatment
Turbidity	None—expressed in analysis as units	Imparts unsightly appearance to water. Deposits in water lines, process equipment, etc. Interferes with most process uses	Coagulation, settling, and filtration
Hardness	Calcium and magnesium salts expressed as $CaCO_3$	Chief source of scale in heat-exchange equipment, boilers, pipelines, etc. Forms curds with soap, interferes with dyeing, etc.	Softening; demineralization; internal boiler water treatment; surface active agents
Alkalinity	Bicarbonate (HCO_3^-), carbonate (CO_3^{2-}), expressed as $CaCO_3$	Foaming and carryover of solids with steam. Embrittlement of boiler steel. Bicarbonate and carbonate produce CO_2 in steam, a source of corrosion in condensate lines	Lime and lime soda softening; acid treatment; hydrogen zeolite softening; demineralization; dealkalization by anion exchange
Free mineral acid	H_2SO_4, HCl. expressed as $CaCO_3$	Corrosion	Neutralization with alkalies
Carbon dioxide	CO_2	Corrosion in water lines and particularly steam and condensate lines	Aeration; deaeration; neutralization with alkalies
pH	(H^+)	pH varies according to acidic or alkaline solids in water. Most natural waters have a pH of 6.0–8.0	pH can be increased by alkalies and decreased by acids
Sulfate	(SO_4^{2-})	Adds to solids content of water, but in itself is not usually significant. Combines with calcium to form calcium sulfate scale	Demineralization
Chloride	Cl^-	Adds to solids content and increases corrosive character of water	Demineralization

Nitrate	$(NO_3)^-$	Adds to solids content, but is not usually significant industrially. High concentrations cause methemoglobinemia in infants. Useful for control of boiler metal embrittlement	Demineralization
Fluoride	F^-	Cause of mottled enamel in teeth. Also used for control of dental decay. Not usually significant industrially	Adsorption with magnesium hydroxide, calcium phosphate, or bone black; alum coagulation
Sodium	Na^+	Adds to solid content of water. When combined with OH^-, causes corrosion in boilers under certain conditions	Demineralization
Silica	SiO_2	Scale in boilers and cooling-water systems. Insoluble turbine blade deposits due to silica vaporization	Hot process removal with magnesium salts; adsorption by highly basic anion exchange resins, in conjunction with demineralization
Iron	Fe^{2+} (ferrous) and Fe^{3+} (ferric)	Discolors water on precipitation. Source of deposits in water lines, boilers, etc. Interferes with dyeing, tanning, papermaking, etc.	Aeration; coagulation and filtration; lime softening; cation exchange; contact filtration; surface-active agents for iron retention
Manganese	Mn^{2+}	Same as iron	Same as iron
Aluminum	Al^{3+}	Usually present as a result of floc carryover from clarifier. Can cause deposits in cooling systems and contribute to complex boiler scales	Improved clarifier and filter operation
Oxygen	O_2	Corrosion of water lines, heat-exchange equipment, boilers, return lines, etc.	Deaeration; sodium sulfite; corrosion inhibitors

TABLE 2.8 Difficulties and Means of Treatment for Common Impurities Found in Fresh Water *(Continued)*

Constituent	Chemical formula	Difficulties caused	Means of treatment
Hydrogen sulfide	H_2S	Cause of "rotten egg" odor. Corrosion	Aeration; chlorination; highly basic anion exchange
Ammonia	NH_3	Corrosion of copper and zinc alloys by formation of complex soluble ion	Cation exchange with hydrogen zeolite; chlorination, deaeration
Dissolved solids	None	A measure of total amount of dissolved matter, determined by evaporation. High concentrations of dissolved solids are objectionable because of process interference and as a cause of foaming in boilers	Various softening processes, such as lime softening and cation exchange by hydrogen zeolite, will reduce dissolved solids; demineralization
Suspended solids	None	A measure of undissolved matter, determined gravimetrically. Suspended solids cause deposits in heat-exchange equipment, boilers, water lines, etc.	Subsidence; filtration, usually preceded by coagulation and settling
Total solids	None	The sum of dissolved and suspended solids, determined gravimetrically	See "dissolved solids" and "suspended solids"

TABLE 2.9 Methods of Water Treatment

	Chemical Procedures
Pretreatment	Methods for clarifying: • Coagulation • Flocculation • Sedimentation to clear floating and grey particles
In operation	Softening methods: • Lime milk/soda principle • Cation exchange (full softening) • Acid dosage (partly softening) Demineralization method: • Cation and anion exchanges (presently the most effective and economical method) Hardness stabilization: • Inhibitor dosage, also as dispersion and corrosion protection agents
Posttreatment	Acid and caustic solution for cleaning of polluted thermal systems, including the neutralization of applied chemical detergents
	Physical Procedures
Pretreatment	Filtration of the subsoil water, predominantly using sand as filtering medium, in pressure and gravity filters
In operation	Reverse osmosis for demineralization by use of diaphragms Transformation of the crystal structures of the hardening-causing substances: • Magnetic field method by means of electrical alternating or permanent magnet • Electrostatic method by applied active anodes
Posttreatment	Automatic cleaning of heat-exchanger tubes by sponge rubber balls or brushes without interruption of plant operation

analysis for drinking-water supplies is concerned mainly with pollution and bacteriological tests. For industrial supplies, a mineral analysis is of more interest. Table 2.10 includes a typical selection and gives some indication of the wide concentration range that can be found. The important constituents can be classified as follows:

1. Dissolved gases (oxygen, nitrogen, carbon dioxide, ammonia, sulfurous gases)
2. Mineral constituents, including hardness salts, sodium salts (chloride, sulfate, nitrate, bicarbonate, etc.), salts of heavy metals, and silica
3. Organic matter, including that of both animal and vegetable origin, oil, trade waste (including agricultural) constituents, and synthetic detergents
4. Microbiological forms, including various types of algae and slime-forming bacteria. This topic is covered in Sec. 2.6.

TABLE 2.10 Typical Water Analyses (Results in ppm)

	A	B	C	D	E	F	G
pH	6.3	6.8	7.4	7.5	7.1	8.3	7.1
Alkalinity	2	38	90	180	250	278	470
Total hardness	10	53	120	230	340	70	559
Calcium hardness	5	36	85	210	298	40	451
Sulfate	6	20	39	50	17	109	463
Chloride	5	11	24	21	4	94	149
Silica	Trace	0.3	3	4	7	12	6
Dissolved solids	33	88	185	332	400	620	1670

A = very soft lake water
B = moderately soft surface water
C = slightly hard river water
D = moderately hard river water
E = hard borehole water
F = slightly hard borehole water containing bicarbonate ions
G = very hard groundwater

Of the dissolved gases occurring in water, oxygen occupies a special position, as it stimulates corrosion reactions. In surface waters, the oxygen concentration approximates saturation, but in the presence of green algae, supersaturation may occur. Please refer to Tables 1.4 and 1.5 in Chap. 1, Aqueous Corrosion, for data on the solubility of oxygen in water. Underground waters are more variable in oxygen content, and some waters containing ferrous bicarbonate are oxygen-free. The solubility is slightly less in the presence of dissolved solids, but this effect is not very significant in natural waters containing less than 1000 ppm dissolved solids. Hydrogen sulfide and sulfur dioxide are also usually the result of pollution or of bacterial activity. Both gases may initiate or significantly accelerate corrosion of most metals.

For some applications, notably feedwater treatment for high-pressure boilers, removal of oxygen is essential. For most industrial purposes, however, deaeration is not applicable, since the water used is in continuous contact with air, from which it would rapidly take up more oxygen. Attention must therefore be given to creating conditions under which oxygen will stifle rather than stimulate corrosion. It has been shown that pure distilled water is least corrosive when fully aerated and that some inhibitors function better in the presence of oxygen. In these cases, oxygen acts as a passivator of the anodic areas of the corrosion cells.

Carbon dioxide and calcium carbonate. The effect of carbon dioxide is closely linked with the bicarbonate content. Normal carbonates are rarely found in natural waters, but sodium bicarbonate is found in some underground supplies. Calcium bicarbonate is the most important of the bicarbonates, but magnesium bicarbonate may be present in smaller quantities. In general, it may be regarded as having properties similar to those of the calcium compound, except that upon decomposition by heat it deposits magnesium hydroxide, whereas calcium bicarbonate precipitates as carbonate. The concentrations of carbon dioxide in water can be classified as follows:

1. The amount required to produce carbonate
2. The amount required to convert carbonate to bicarbonate
3. The amount required to keep the calcium bicarbonate in solution
4. Any excess over that accounted for in types 1, 2, and 3

With less carbon dioxide than required for type 3 (let alone type 4), the water will be supersaturated with calcium carbonate, and a slight increase in pH (at the local cathodes) will tend to cause its precipitation or scaling. If the deposit is continuous and adherent, the metal surface may become isolated from the water and hence protected from corrosion. If type 4 carbon dioxide is present, there can be no deposition of calcium carbonate and existing deposits will be dissolved; there cannot therefore be any protection by calcium carbonate scale. Please refer to Sec. 2.2.3 for detailed coverage of the indices and equilibria-associated precipitation and scaling associated with common chemicals found in natural waters.

Dissolved mineral salts. The principal ions found in water are calcium, magnesium, sodium, bicarbonate, sulfate, chloride, and nitrate. A few parts per million of iron or manganese may sometimes be present, and there may be traces of potassium salts, whose behavior is very similar to that of sodium salts. From the corrosion point of view, the small quantities of other acid radicals present, e.g., nitrite, phosphate, iodide, bromide, and fluoride, generally have little significance. Larger concentrations of some of these ions, notably nitrite and phosphate, may act as corrosion inhibitors, but the small quantities present in natural waters will usually have little effect.

Chlorides have probably received the most study in relation to their effect on corrosion. Like other ions, they increase the electrical conductivity of the water, so that the flow of corrosion currents will be facilitated. They also reduce the effectiveness of natural protective films, which may be permeable to small ions. Nitrate is very similar to chloride in its effects but is usually present in much smaller concentrations. Sulfate in general appears to behave very similarly, at least on carbon steel materials. In practice, high-sulfate waters may attack concrete, and the performance of some inhibitors appears to be adversely affected by the presence of sulfate. Sulfates have also a special role in bacterial corrosion under anaerobic conditions.

Another mineral constituent of water is silica, present both as a colloidal suspension and dissolved in the form of silicates. The concentration varies very widely, and, as silicates are sometimes applied as corrosion inhibitors, it might be thought that the silica content would affect the corrosive properties of a water. In general, the effect appears to be trivial; the fact that silicate inhibitors are used in waters with a high initial silica content suggests that the form in which silica is present is important.

Hardness. The hardness of a water supply is determined by the content of calcium and magnesium salts. Calcium and magnesium can combine with bicarbonates, sulfates, chlorides, and nitrates to precipitate as solids. Mineral salts make water more basic and lead to more aggressive corrosion on many materials. The presence of salts in lime scale deposits is one of the most common causes of corrosion; these deposits cause damage in water pipelines and boilers. Table 2.11 presents a comparison of the various hardness units used in Europe and North America. Any descriptive or numerical classification of hardness of water is rather arbitrary. A water that is termed hard in some areas may be considered soft in other areas. The U.S. Geological Survey uses the following classification:[19]

Soft.   Less than 60 ppm (as $CaCO_3$)

Moderately hard.   60 to 120 ppm

Hard.   120 to 180 ppm

Very hard.   Above 180 ppm

There are basically two types of hardness:

1. Temporary hardness caused by Ca and Mg bicarbonates (precipitate minerals upon heating)
2. Permanent hardness due to Ca and Mg sulfates or chlorides (dissolve with sodium)

Temporary hardness salts

1. *Calcium carbonate ($CaCO_3$).*   Also called calcite or limestone, rare in water supplies. Causes alkalinity in water.
2. *Calcium bicarbonate [$Ca(HCO_3)_2$].*   Forms when water containing CO_2 comes in contact with calcite. Also causes alkalinity in water. When it is heated, CO_2 is released and the calcium bicarbonate reverts back to calcium carbonate, thus forming scale.
3. *Magnesium carbonate ($MgCO_3$).*   Also called magnesite, it has properties similar to those of calcium carbonate.

TABLE 2.11 Comparison of Hardness Units

Hardness units per liter water	mval/L 50 mg $CaCO_3$	°dH (German) 10 mg CaO	°fH (French) 10 mg $CaCO_3$	°eH (British) 14.3 mg $CaCO_3$*	ppm (American) 1 mg $CaCO_3$	mmol/L (international)[†] 100 mg $CaCO_3$
1 mval/L	1	2.8	5	3.51	50	0.5
1 °dH	0.357	1	1.786	1.25	17.86	0.1786
1 °fH	0.2	0.5599	1	0.7	10	0.1
1 °eH	0.285	0.7999	1.429	1	14.29	0.1429
1 ppm	0.02	0.056	0.1	0.07	1	0.01
1 mmol/L	2	5.6	10	7	100	1

*One grain $CaCO_3$ per gallon.
[†]The international hardness scale (mmol/L) is to be preferred to the national hardness scales.

...n bicarbonate [$Mg(HCO_3)_2$]. Similar to calcium bicar-
...ts properties.

...rdness salts

1. *Calcium sulfate ($CaSO_4$)*. Also known as gypsum, used to make plaster of paris. Gypsum will precipitate and form scale in boilers when concentrated.
2. *Calcium chloride ($CaCl_2$)*. This salt hydrolyzes in boiler water to produce a low pH as follows: $CaCl_2 + 2H_2O \rightarrow Ca(OH)_2 + 2HCl$.
3. *Magnesium sulfate ($MgSO_4$)*. Commonly known as epsom salts; may have laxative effect if it is concentrated enough.
4. *Magnesium chloride ($MgCl_2$)*. This salt has properties similar to those of calcium chloride.
5. *Sodium salts*. Also found in household water supplies, but considered harmless as long as they do not exist in large quantities.

pH of water. The pH of natural waters is, in fact, rarely outside the fairly narrow range of 4.5 to 8.5. High values, at which corrosion of steel may be suppressed, and low values, at which gaseous hydrogen evolution occurs, are not often found in natural waters. Copper is affected to a marked extent by pH value. In acidic waters, slight corrosion occurs, and the small amount of copper in solution causes green staining of fabrics and sanitary ware. In addition, redeposition of copper on aluminum or galvanized surfaces sets up corrosion cells, resulting in severe pitting of the metals. The use of these different materials in a water system should thus be avoided. In most waters the critical pH value is about 7.0, but in soft water containing organic acids it may be higher. Chapter 1, Aqueous Corrosion, contains detailed coverage of the effects of pH and temperature on the corrosion of metals.

Organic matter. The types of organic matter in water supplies are very diverse, and organic matter may be present in suspension or in colloidal or true solution. It is largely decaying vegetable matter, but there are many other possible sources, including runoff from fields and domestic and industrial wastes.

Biochemical oxygen demand. Biochemical oxygen demand (BOD or BOD5) is an indirect measure of biodegradable organic compounds in water, and is determined by measuring the decrease in dissolved oxygen in a controlled water sample over a 5-day period. During this 5-day period, aerobic (oxygen-consuming) bacteria decompose organic matter in the sample and consume dissolved oxygen in proportion to the amount of organic material that is present. In general, a high BOD reflects high concentrations of substances that can be biologically degraded, thereby consuming oxygen and potentially resulting in low dissolved oxygen in the receiving water. The BOD test was developed for samples dominated by oxygen-demanding pollutants like sewage. While its merit as a pollution parameter continues to be debated, BOD has the advantage of a long period of record and a large database of results.[20]

Nutrients. Nutrients are chemical elements or compounds essential for plant and animal growth. Nutrient parameters include ammonia, organic nitrogen, Kjeldahl nitrogen, nitrate nitrogen (for water only), and total phosphorus. High amounts of nutrients have been associated with eutrophication, or overfertilization of a water body, while low levels of nutrients can reduce plant growth and (for example) starve higher-level organisms that consume phytoplankton.

Organic carbon. Most organic carbon in water occurs as partly degraded plant and animal materials, some of which are resistant to microbial degradation. Organic carbon is important in the estuarine food web and is incorporated into the ecosystem by photosynthesis of green plants, which are then consumed as carbohydrates and other organic compounds by higher animals. In another process, formerly living tissue containing carbon is decomposed as detritus by bacteria and other microbes. Total organic carbon (TOC) bears a direct relationship to biological and chemical oxygen demand; high levels of TOC can result from human sources, the high oxygen demand being the main concern.

Oil and grease. Oil and grease is one of the most common parameters for quantifying organics from human sources and, to a lesser extent, biogenic sources (e.g., algae and fish). Some examples of oil and grease loadings are leaks from automobile crankcases, illegal dumping into storm sewers, motorboats, oil spills, and discharge from oil production platforms in the bay. Oil and grease is a generic term for material which actually contains numerous and variable chemical compounds, some of which are typically toxic.

Priority pollutants. Priority pollutants refers to a list of 126 specific pollutants, which include heavy metals and specific organic chemicals. The priority pollutants are a subset of "toxic pollutants" as defined in the Clean Water Act (United States). These 126 pollutants were assigned a high priority for development of water quality criteria and effluent limitation guidelines because they are frequently found in wastewater. Many of the heavy metals, pesticides, and other chemicals listed below are on the priority pollutant list.[20]

Heavy metals (total and dissolved). In the water treatment field, "heavy metal" refers to heavy, dense, metallic elements that occur only at trace levels in water, but are very toxic and tend to accumulate. Some key metals of concern and their primary sources are listed below.[21]

- Arsenic from fossil fuel combustion and industrial discharge
- Cadmium from corrosion of alloys and plated surfaces, electroplating wastes, and industrial discharges
- Chromium from corrosion of alloys and plated surfaces, electroplating wastes, exterior paints and stains, and industrial discharges
- Copper from corrosion of copper plumbing, antifouling paints, and electroplating wastes
- Lead from leaded gasoline, batteries, and exterior paints and stains
- Mercury from natural erosion and industrial discharges
- Zinc from tires, galvanized metal, and exterior paints and stains

High levels of mercury, copper, and cadmium have been proven to cause serious environmental and human health problems. Some of the sources listed above, such as lead in gasoline and heavy metals in some paints, are now being phased out as a result of environmental regulations issued in the past 10 years. Most heavy metals are too rarely found in water to justify government regulation at all, but a few have been given maximum contaminant limits (MCLs) and MCL goals by the Environmental Protection Agency (EPA). These include the following:[22]

- Cadmium occurs mostly in association with zinc and gets into water from corrosion of zinc-coated ("galvanized") pipes and fittings.
- Antimony occurs mostly in association with lead, where it is used as a hardening agent. It gets into water from corrosion of lead pipes and fittings, but even then it is rarely detectable. More antimony is found in food than in water.
- Barium is chemically similar to calcium and magnesium and is usually found in conjunction with them. It is not very toxic and is only rarely found at toxic levels. However, it is common at low levels in hard-water areas.
- Mercury is notorious as an environmental toxin, but it is generally not a big problem in water supplies, as it is found only at very low levels in water. Certain bacteria are able to transform it into methyl mercury, which is concentrated in the food chain and can cause malformations.
- Thallium is as toxic as lead or mercury, but is extremely rare and is not often a problem in water.
- Lead is the most significant of the heavy metals because it is both very toxic and very common. It gets into water from corrosion of

plumbing materials, where lead has been used freely since Roman times. In addition, lead can be found in the solder used to join copper pipes and in fittings and faucets made from brass.

- Copper and lead are regulated together because both are commonly used in plumbing materials and because they are corrodible, even though copper is not very toxic. A few tenths of a ppm of copper is common and not a problem, but if as much as 1.3 ppm copper or 15 ppb lead are detected in tap water, the waterworks is required to modify the water chemistry to make it less corrosive toward lead.

Pesticides. Pesticides are a large class of compounds of concern. Typical pesticides and herbicides include DDT, aldrin, chlordane, endosulfan, endrin, heptachlor, and Diazinon. Surprisingly, concentrations of pesticides in urban runoff may be equal to or greater than the concentrations in agricultural runoff. Besides toxicity, persistence in the environment is a key concern. Some of the more persistent compounds, including DDT and dioxin (not a pesticide), are subject to stringent regulation, including outright bans.

Polycyclic aromatic hydrocarbons (PAHs). Polycyclic aromatic hydrocarbons are a family of semivolatile organic pollutants such as naphthalene, anthracene, pyrene, and benzo(a)pyrene. There are typically two main sources of PAHs: spilled or released petroleum products (from oil spills or discharge of oil production brines) and combustion products that are found in urban runoff. Specifically, phenanthrene, pyrene, and fluoranthene are products of the incomplete combustion of fossil fuels. Naphthalene is found in asphalt and creosote. PAHs from combustion products have been identified as carcinogenic.

Polychlorinated biphenyls (PCBs). Polychlorinated biphenyls are organic chemicals that formerly had widespread use in electrical transformers and hydraulic equipment. This class of chemicals is extremely persistent in the environment and has been proven to bioconcentrate in the food chain, thereby leading to environmental and human health concerns in areas such as the Great Lakes. Because of their potential to accumulate in the food chain, PCBs were intensely regulated and subsequently prohibited from manufacture by the Toxic Substances Control Act (TSCA) of 1976. Disposal of PCBs is tightly restricted by the TSCA.[20]

2.2.2 Essentials of ion exchange

Ion-exchange resins are particularly well suited for the removal of ionic impurities for several reasons:[23]

- The resins have high capacities for ions that are found in low concentrations.
- The resins are stable and readily regenerated.
- Temperature effects are for the most part negligible.
- The process is excellent for both large and small installations, from home water softeners to huge utility installations.

Synthesis. Most ion-exchange bead materials are manufactured by a suspension polymerization process using styrene and divinylbenzene (DVB). The styrene and DVB, both liquids at the start, are put into a chemical reactor with roughly the same amount of water. A surfactant is also present to keep everything dispersed. The chemical reactor has an agitator which begins to mix the water–organic chemical solution. The styrene and DVB begin to form large globules of material, and as the speed of agitation increases, the globules break up into smaller droplets until they reach the size of about a millimeter. At this point, the polymerization reaction is initiated by the addition of benzoyl peroxide, which causes the styrene and DVB molecules to form small plastic beads. The DVB is a cross-linking agent that gives the beads their physical strength, and without which the styrene would be water-soluble.

The polystyrene-DVB beads need to be chemically activated in order to perform as an ion-exchange material. Active groups are attached to provide chemical functionality to the beads. Each active group has a fixed electric charge which is balanced by an equivalent number of oppositely charged ions which are free to exchange with other ions of the same charge.[23]

Strong acid cation resins are formed by treating the beads with concentrated sulfuric acid (a process called sulfonation) to form permanent, negatively charged sulfonic acid groups throughout the beads. Important here is the fact that the exchange sites thus formed are located throughout the bead. The ion-exchange process is not a surface phenomenon; more than 99 percent of the capacity of an ion-exchange material is found in the interior of the bead.

Strong-base anion resins are activated in a two-step process that consists of chloromethylation followed by amination. The two-step process begins with the same styrene-DVB material as is used for cation resins. The only difference is that the amount of DVB used is less, to allow for a more porous bead. The first reaction step is the attachment of a chloromethyl group to each of the benzene rings in the bead structure. This intermediate chloromethylated plastic material needs to be reacted with an amine in a process called amination. The type of amine used determines the functionality of the resin. A common amine used is trimethylamine (TMA), which creates a type 1 strongly basic anion exchanger. Using dimethylthanolamine (DMEA) will make a type 2 anion resin. Table 2.12 resumes the main advantages and disadvantages of the common resin types used for purifying and softening water.

Physical and chemical structure of resins. The basic material requirements for ion-exchange beads are insolubility, bead size, and resistance to fracture. The resin must be insoluble under normal operating conditions. The beads must be in the form of spheres of uniform dimension; normal size range is between 16 and 50 U.S. Mesh. The swelling and contraction of the resin bead during exhaustion and regeneration must not cause the beads to burst. Also, an important property of ion-exchange resins is that the active site is permanently

TABLE 2.12 Advantages and Disadvantages of the Main Ion-Exchange Resins

Resin	Advantages	Disadvantages
Strong-acid cation	Useful on all waters Complete cation removal Variable capacity, quality Good physical stabilities Good oxidation stabilities Low initial cost	Operating efficiency
Weak-acid cation	Very high capacity Very high operating efficiency	Only partial cation removal Useful only on specific waters Fixed operating capacity Poor physical stability High initial cost Poor kinetics
Strong-base anion	Complete anion removal (including silica and CO_2) Lower initial cost Variable efficiency and quality Excellent kinetics Shorter rinses	Less organic fouling resistance Limited life Thermodynamically unstable Efficiency vs. quality
Weak-base anion	High operating capacity removal High regeneration efficiency Excellent organic fouling resistance Good thermal stability Good oxidation stability Can be regenerated with leftover caustic from strong-base resin, alkaline by-products, ammonia, soda ash, and other weak bases and waste streams	Only partial anion removal Does not remove silica or CO_2 High initial cost Long rinses Poor kinetics

attached to the bead. Ion-exchange resins can be manufactured into one of two physical structures, gel or macroporous.[23]

- Gel resins are homogeneous cross-linked polymers and are the most common resins available. They have exchange sites distributed evenly throughout the bead. The amount of DVB cross-linking used in the synthesis of a bead determines the relative strength of the bead. Standard strong-acid cation resin used for softening, which is the most common ion-exchange medium, is almost always an 8 percent DVB gelular material. The amount of DVB that this resin contains has proved to be the most economical in terms of resin price and expected operating life. Resins are available today with a DVB content from 2 to 20 percent and higher. Higher DVB content gives the bead additional strength, but the additional cross-linking can hinder kinetics by making the bead too resistant to the shrinking and swelling necessary during normal operation.

- Macroporous resins were introduced commercially in 1959 and are made with large pores that permit access to interior exchange sites. They are also referred to as macroreticular or fixed-pore resins. Macroporous resins are manufactured by a process that leaves a network of pathways throughout the bead. This spongelike structure allows the active portion of the bead to contain a high level of DVB cross-linking without affecting the exchange kinetics. Unfortunately, it also means that the resin has a lower capacity because the beads contain fewer exchange sites. The "pores" can take up to 10 to 30 percent of the polymer. This reduces the ion-exchange capacity proportionately.

Gel resins usually have higher operating efficiencies and cost less. A macropore gives better physical stability, primarily because of its spongelike structure, which gives more stress relief. It also eliminates some of the breakage that may occur from osmotic stress. The higher surface area in a macroporous anion resin gives better organic fouling resistance. In a cation resin, the higher cross-linking level gives better oxidation resistance.

There are two basic types of chemical structures, styrene and acrylic. The styrene-based materials described above are aromatic hydrocarbons. Acrylic resins are straight-chained hydrocarbons based on polyacrylate and polymethacrylate. DVB is still used as a cross-linker in these resins, but the acrylics differ from the styrenics in that the active exchange site is part of the physical structure. This means that their physical and chemical stabilities are intertwined. When an acrylic resin chemically degrades, it is usually at the exchange site, which is the weak link. This destroys the physical structure. As an acrylic resin oxidizes, it will swell and become mushy. Another disadvantage of the acrylic materials is that they are not fully FDA approved. Therefore, they are usually limited to industrial applications. The acrylics are advantageous in applications where organics are present because they do not foul nearly as much as a styrene-based product.[23]

Selectivity of resins. The selectivity or affinity of ion-exchange resins is influenced by the properties of the bead, the ions being exchanged, and the solution in which the ions are present. Water is an essential component of ion-exchange resins. For example, strong-acid cation resins contain about 50 percent moisture. The amount of cross-linking of the bead has an impact on the moisture content of the bead, and the moisture content, in turn, has an impact on the selectivity. A bead with a high moisture content has a high porosity, and the active groups are spaced further from one another. Ion-exchange resins generally have greater selectivities for ions with increasing valence or charge. Among ions with the same charge, higher affinities are seen for ions with a higher atomic number.

These affinity relationships are reversed in concentrated solutions. This is what makes regeneration of exhausted resins possible. An exhausted cation resin used for softening is predominantly in the calcium and magnesium form, both divalent ions. The resin is restored to its regenerated condition, the sodium form, by the introduction of 10 percent sodium chloride. This sodium chloride solution is concentrated enough (10,000 ppm) to reverse the selectivity. The driving force of the monovalent sodium ion then converts the resin back to the sodium form.

Kinetics. The rate of exchange, or kinetics, of ion-exchange reactions is governed by several factors. The solution being treated has an effect; higher solution concentrations can speed up the rate of reaction. The amount of DVB cross-linking of the bead determines the porosity of the bead and, in turn, the ionic mobility within the bead. The size of the ions being exchanged also influences the kinetic rate and is somewhat dependent on the size of the pores in the resin structure. The size of the bead also has an effect; smaller beads present a shorter diffusion path to active sites in the interior of the beads.

Resin has a greater affinity for ions with higher valences, so a predominance of high-valence ions can cause a relatively higher rate of reaction. Other influences include temperature, the ionic form of the exchange sites, and the strength of the exchange sites. Increasing temperature can speed up chemical reactions. The exchange reaction is a diffusion process, so the diffusion rate of the ion on the exchange site has some effect. Also, the strength of the exchange site—whether it is strongly or weakly acidic or basic—affects the reaction rate.

Types of ion-exchange resins

Strong-acid cation resins. Strongly acidic cation resins derive their functionality from the sulfonic acid groups. These strong-acid exchangers operate at any pH, split all salts, and require substantial amounts of regenerant. This is the resin of choice for almost all softening applications and as the first unit in a two-bed demineralizer or the cation component of a mixed bed.

Weak-acid cation resins. The weakly acidic cation resins have carboxylic groups as the exchange site. These resins are highly efficient, for they are regenerated with a nearly 100 percent stoichiometric amount of acid, as compared to the 200 to 300 percent required for strong-acid cations. The weak-acid resins are subject to reduced capacity from increasing flow rate, low temperatures, and a hardness-to-alkalinity ratio below 1.0. They are used very effectively in conjunction with a strong-acid cation resin operating in the hydrogen form, in either a separate-bed or a stratified-bed configuration. In both cases, the influent water first contacts the weak-acid resin, where the cations associated with alkalinity are removed. The remaining cations are removed by the strong-acid cation resin. The weak-acid cation resin is regenerated with the waste acid from the strong-acid unit, making for a very economical arrangement.[23]

Strong-base anion resins. Strongly basic anion resins derive their functionality from quaternary ammonium exchange sites. The two main groups of strong-base anion resins are type 1 and type 2, depending on the type of amine used during the chemical activation process. Chemically, the two types differ in the species of quaternary ammonium exchange sites they exhibit: Type 1 sites have three methyl groups, whereas in type 2, an ethanol group replaces one of the methyl groups.

Type 1 resins are suitable for total anion removal on all waters. They are more difficult to regenerate, and they swell more from the chloride form to the hydroxide form than type 2. They are more resistant to high temperatures and should be used on high-alkalinity and high-silica waters.

Type 2 resins also feature removal of all anions, but they can be less effective in removing silica and carbon dioxide from waters where these weak acids constitute more than 30 percent of the total anions. Type 2 anions give best results on waters that predominantly contain free mineral acids, such as chlorides and sulfates, as in the effluent from a cation unit followed by a decarbonator. Type 2 anion resins operating in the chloride form are typically used in dealkalizers.

Weak-base anion resins. Weakly basic anion resins contain the polyamine functional group, which acts as an acid adsorbed, removing strong acids (free mineral acidity) from the cation effluent stream. This weakly ionized resin is regenerated efficiently by nearly stoichiometric amounts of base, such as sodium hydroxide, which restore the exchange sites to the free base form. The regeneration step is essentially a neutralization of the strong acids that are collected on the resin, and it can use waste caustic from a strong-base anion unit to enhance economics. Weak-base anion resins should be used on waters with high levels of sulfates or chlorides, or where removal of alkalinity and silica is not required.

2.2.3 Saturation and scaling indices

The saturation of water refers to the solubility product K_{sp} of a compound. By definition, the ion-activity product (IAP) of reactants—i.e., $a_{Ca^{2+}}$ and $a_{CO_3^{2-}}$ when $CaCO_3$ is the scalant—is, at equilibrium, equal to K_{sp}:

$$K_{sp} = \text{IAP} = a_{Ca^{2+}} \, a_{CO_3^{2-}}$$

The saturation level (SL) of water is defined as the ratio of the ion-activity product to K_{sp}, as in the following:

$$\text{SL} = \frac{a_{Ca^{2+}} \, a_{CO_3^{2-}}}{K_{sp}} = \frac{\text{IAP}}{K_{sp}}$$

In this example, water is said to be saturated with calcium carbonate when it will neither dissolve nor precipitate calcium carbonate scale. This equilibrium condition is based upon an undisturbed water at constant temperature which is allowed to remain undisturbed for an infinite period of time. Water is said to be undersaturated if it can still dissolve calcium carbonate. Supersaturated water will precipitate calcium carbonate if allowed to rest. If water is undersaturated with respect to calcium carbonate, the SL value will be less than 1.0. When water is at equilibrium, SL will be 1.0 by definition. Water which is supersaturated with calcium carbonate will have a saturation level greater than 1.0. As the saturation level increases beyond 1.0, the driving force for calcium carbonate crystal formation or crystal growth increases.

The SL definition can be simplified if the activity coefficients are incorporated into the solubility product in order to use a more practical concentration unit. The conditional solubility product K_{spc} incorporates the activity coefficients into the solubility product.

$$K_{spc} = \frac{K_{sp}}{\gamma_{Ca^2} + \gamma_{CO_3^{2-}}} = [\ ]_{Ca^{2+}} [\ ]_{CO_3^{2-}}$$

at equilibrium since $a_{ion} = [\ ]_{ion}\gamma_{ion}$, with $[\ ]_{ion}$ being the readily measurable molar concentration.

A distinction must be made between a thick layer of deposit, whether calcium carbonate or any other material, and a protective layer. The ideal protection in fact consists of layers of negligible thickness which do not impede water or heat flow and which are self-healing. This is difficult to achieve with natural waters. A water which is exactly in equilibrium with respect to calcium carbonate is normally corrosive to steel because it has no power to form a calcium carbonate deposit. Supersaturated waters, on the other hand, unless suitably treated, will form a substantial scale, but whether this inhibits corrosion or not depends on its adherence to the metal and its porosity.

Saturation levels, no matter how refined, are an equilibrium-based index. They provide a measure of the thermodynamic driving force that leads a scalant to form, but they do not incorporate the capacity of the water for continued scaling. A water can have a high saturation level with no visible scale formation. The driving force might be present, but there is insufficient mass for gross precipitation. Saturation levels should be viewed as another tool for developing an overall picture of a water's scale potential. They can point out what scales will not form under the conditions evaluated, but they cannot predict whether deposits of any significant quantity will form.

The Langelier saturation index. The Langelier saturation index (LSI) is an equilibrium model derived from the theoretical concept of saturation and provides an indicator of the degree of saturation of water with respect to calcium carbonate. It can be shown that the Langelier saturation index approximates the base 10 logarithm of the calcite saturation level. The Langelier saturation level approaches the concept of saturation using pH as a main variable. The LSI can be interpreted as the pH change required to bring water to equilibrium. Water with a Langelier saturation index of 1.0 is one pH unit above saturation. Reducing the pH by 1 unit will bring the water into equilibrium. This occurs because the portion of total alkalinity present as CO_3^{2-} decreases as the pH decreases, according to the equilibria describing the dissociation of carbonic acid [Eqs. (2.11) and (2.12)].

$$H_2CO_3 \leftrightarrows HCO_3^- + H^+ \qquad (2.11)$$

$$HCO_3^- \leftrightarrows CO_3^{2-} + H^+ \qquad (2.12)$$

The LSI is probably the most widely used indicator of cooling-water scale potential. It is purely an equilibrium index and deals only with the thermodynamic driving force for calcium carbonate scale formation and growth. It provides no indication of how much scale ($CaCO_3$)

will actually precipitate to bring water to equilibrium. It simply indicates the driving force for scale formation and growth in terms of pH as a master variable. LSI is defined as

$$\text{LSI} = \text{pH} - \text{pH}_s$$

where pH = measured water pH
pH_s = pH at saturation in calcite ($CaCO_3$)

In the cooling-water pH range of 6.5 to 9.5, the pH_s calculation simplifies to

$$\text{pH}_s = (pK_2 - pK_s) + pCa + pAlk$$

where pK_2 = negative $\log_{10}$ of the second dissociation constant for carbonic acid [Eq. (2.12)]
pK_s = negative $\log_{10}$ of the solubility product for calcite
pCa = negative $\log_{10}$ of calcium measured in the water
$pAlk$ = negative $\log_{10}$ of the total alkalinity measured for the water being evaluated

A pH decrease of 1 unit will decrease the CO_3^{2-} concentration of the water about tenfold. This affects the saturation level directly by also decreasing the IAP tenfold. So a 1 pH unit decrease will decrease the Langelier saturation index by 1 unit. A 1 pH unit decrease will also decrease the saturation level (IAP/K_{sp}) tenfold. A negative Langelier saturation index indicates that water is undersaturated with respect to calcium carbonate (calcite). If the LSI is -1.0, raising the pH of the water 1 unit will increase the calcium carbonate saturation level to equilibrium. The 1 pH unit increase does this by increasing the CO_3^{2-} portion of the carbonate alkalinity present tenfold. The calcite saturation level increases accordingly (ten times).

Although saturation-level-based indices are very useful, a second factor must be considered in interpreting them. Saturation-level-based indices indicate the potential for scale formation if water is unperturbed for an infinite period of time. Most cooling-water systems have a substantially shorter holding time index. The LSI was not intended as an indicator of corrosivity toward mild steel or other metals of construction. The LSI describes only the stability of an existing calcium carbonate scale or other calcium carbonate–bearing structure. The LSI does describe the tendency of water to dissolve calcite scale. It has been postulated that supersaturated water will form an eggshell-like film of calcium carbonate scale that will act as an inhibitor for corrosion of mild steel. This can occur in highly buffered waters. The LSI and other saturation-based indices do not guarantee this inhibitory behavior. Calcium carbonate film formation is typically observed in highly buffered waters.

It has been shown that water supersaturated with calcium carbonate often develops tubercular deposits which do not inhibit corrosion on mild steel. This behavior is typically associated with water of low buffer capacity. Puckorius also warned against using saturation-level-derived indices as the basis for predicting corrosion problems in cooling systems.

Ryznar stability index. The Ryznar stability index (RSI) attempts to correlate an empirical database of scale thickness observed in municipal water systems to the water chemistry. Like the LSI, the RSI has its basis in the concept of saturation level. Ryznar attempted to quantify the relationship between calcium carbonate saturation state and scale formation. The Ryznar index takes the form

$$RSI = 2\,(pH_s) - pH$$

The empirical correlation of the Ryznar stability index can be summarized as follows:

RSI < 6   The scale tendency increases as the index decreases.
RSI > 7   The calcium carbonate formation probably does not lead to a protective corrosion inhibitor film.
RSI > 8   Mild steel corrosion becomes an increasing problem.

Puckorius scaling index. The previously discussed indices account for only the driving force for calcium carbonate scale formation. They do not account for two other critical parameters: the buffering capacity of the water and the maximum quantity of precipitate that can form in bringing water to equilibrium. The Puckorius (or Practical) scaling index (PSI) attempts to further quantify the relationship between saturation state and scale formation by incorporating an estimate of the buffering capacity of the water into the index.

Water that is high in calcium but low in alkalinity and buffering capacity can have a high calcite saturation level. The high calcium level increases the ion-activity product. A plot of ion-activity product versus precipitate for the water would show a rapid decrease in pH as calcium precipitated because of the low buffering capacity. Even minuscule decreases in carbonate concentration in the water would drastically decrease the ion-activity product because of the small quantity present prior to the initiation of precipitation. Such water might have a high tendency to form scale as a result of the driving force, but the quantity of scale formed might be so small as to be unobservable. The water has the driving force but no capacity and no ability to maintain pH as precipitate forms.

The PSI is calculated in a manner similar to the Ryznar stability index. Puckorius uses an equilibrium pH rather than the actual system pH to account for the buffering effects:

$$\text{PSI} = 2\,(\text{pH}_{eq}) - \text{pH}_s$$

where $\text{pH}_{eq} = 1.465 \times \log_{10}[\text{Alkalinity}] + 4.54$
$[\text{Alkalinity}] = [\text{HCO}_3^-] + 2[\text{CO}_3^{2-}] + [\text{OH}^-]$

Larson-Skold index. The Larson-Skold index describes the corrosivity of water toward mild steel. The index is based upon evaluation of in situ corrosion of mild steel lines transporting Great Lakes waters. The index is the ratio of equivalents per million (epm) of sulfate (SO_4^{2-}) and chloride (Cl^-) to the epm of alkalinity in the form bicarbonate plus carbonate ($HCO_3^- + CO_3^{2-}$).

$$\text{Larson-Skold index} = \frac{\text{epm } Cl^- + \text{epm } SO_4^{2-}}{\text{epm } HCO_3^- + \text{epm } CO_3^{2-}}$$

As outlined in the original paper, the Larson-Skold index correlated closely to observed corrosion rates and to the type of attack in the Great Lakes water study. It should be noted that the waters studied in the development of the relationship were not deficient in alkalinity or buffering capacity and were capable of forming an inhibitory calcium carbonate film, if no interference was present. Extrapolation to other waters, such as those of low alkalinity or extreme alkalinity, goes beyond the range of the original data.

The index has proved to be a useful tool in predicting the aggressiveness of once-through cooling waters. It is particularly interesting because of the preponderance of waters with a composition similar to that of the Great Lakes waters and because of its usefulness as an indicator of aggressiveness in reviewing the applicability of corrosion inhibition treatment programs that rely on the natural alkalinity and film-forming capabilities of a cooling water. The Larson-Skold index might be interpreted by the following guidelines:

Index < 0.8	Chlorides and sulfate probably will not interfere with natural film formation.
0.8 < index < 1.2	Chlorides and sulfates may interfere with natural film formation. Higher than desired corrosion rates might be anticipated.
Index > 1.2	The tendency toward high corrosion rates of a local type should be expected as the index increases.

Stiff-Davis index. The Stiff-Davis index attempts to overcome the shortcomings of the Langelier index with respect to waters with high total dissolved solids and the impact of "common ion" effects on the driving force for scale formation. Like the LSI, the Stiff-Davis index has its basis in the concept of saturation level. The solubility product used to predict the pH at saturation (pH_s) for a water is empirically modified in the Stiff-Davis index. The Stiff-Davis index will predict that a water is less scale forming than the LSI calculated for the same water chemistry and conditions. The deviation between the indices increases with ionic strength. Interpretation of the index is by the same scale as for the Langelier saturation index.

Oddo-Tomson index. The Oddo-Tomson index accounts for the impact of pressure and partial pressure of CO_2 on the pH of water and on the solubility of calcium carbonate. This empirical model also incorporates corrections for the presence of two or three phases (water, gas, and oil). Interpretation of the index is by the same scale as for the LSI and Stiff-Davis indices.

Momentary excess (precipitation to equilibrium). The momentary excess index describes the quantity of scalant that would have to precipitate instantaneously to bring water to equilibrium. In the case of calcium carbonate,

$$K_{spc} = [Ca^{2+}][CO_3^{2-}]$$

If water is supersaturated, then

$$[Ca^{2+}][CO_3^{2-}] \gg K_{spc}$$

Precipitation to equilibrium assumes that one mole of calcium ions will precipitate for every mole of carbonate ions that precipitates. On this basis, the quantity of precipitate required to restore water to equilibrium can be estimated with the following equation:

$$[Ca^{2+} - X][CO_3^{2-} - X] = K_{spc}$$

where X is the quantity of precipitate required to reach equilibrium.

X will be a small value when either calcium is high and carbonate low, or carbonate is high and calcium low. It will increase to a maximum when equal parts of calcium and carbonate are present. As a result, these calculations will provide vastly different values for waters with the same saturation level. Although the original momentary excess index was applied only to calcium carbonate scale, the index can be extended to other scale-forming species. In the case of sulfate, momentary excess is calculated by solving for X in the relationship

$$[Ca^{2+} - X][SO_4^{2-} - X] = K_{spc}$$

The solution becomes more complex for tricalcium phosphate:

$$[Ca^{2+} - 3X]^3[PO_4^{3-} - 2X]^2 = K_{spc}$$

While this index provides a quantitative indicator of scale potential and has been used to correlate scale formation in a kinetic model, the index does not account for two critical factors: First, the pH can often change as precipitates form, and second, the index does not account for changes in driving force as the reactant levels decrease because of precipitation. The index is simply an indicator of the capacity of water to scale, and can be compared to the buffer capacity of a water.

Interpreting the indices. Most of the indices discussed previously describe the tendency of a water to form or dissolve a particular scale. These indices are derived from the concept of saturation. For example, saturation level for any of the scalants discussed is described as the ratio of a compound's observed ion-activity product to the ion-activity product expected if the water were at equilibrium K_{sp}. The following general guidelines can be applied to interpreting the degree of supersaturation:

1. If the saturation level is less than 1.0, a water is undersaturated with respect to the scalant under study. The water will tend to dissolve, rather than form, scale of the type for which the index was calculated. As the saturation level decreases and approaches 0.0, the probability of forming this scale in a finite period of time also approaches 0.

2. A water in contact with a solid form of the scale will tend to dissolve or precipitate the compound until an IAP/K_{sp} ratio of 1.0 is achieved. This will occur if the water is left undisturbed for an infinite period of time under the same conditions. A water with a saturation level of 1.0 is at equilibrium with the solid phase. It will not tend to dissolve or precipitate the scale.

3. As the saturation level (IAP/K_{sp}) increases above 1.0, the tendency to precipitate the compound increases. Most waters can carry a moderate level of supersaturation before precipitation occurs, and most cooling systems can carry a small degree of supersaturation. The degree of supersaturation acceptable for a system varies with parameters such as residence time, the order of the scale reaction, and the amount of solid phase (scale) present in the system.

2.2.4 Ion association model

The saturation indices discussed previously can be calculated based upon total analytical values for all possible reactants. Ions in water, however, do not tend to exist totally as free ions.[24] Calcium, for example, may be paired with sulfate, bicarbonate, carbonate, phosphate, and other species. Bound ions are not readily available for scale formation. This binding, or reduced availability of the reactants, decreases the effective ion-activity product for a saturation-level calculation. Early indices such as the LSI are based upon total analytical values rather than free species primarily because of the intense calculation requirements for determining the distribution of species in a water. Speciation of a water requires numerous computer iterations for the following:[25]

- The verification of electroneutrality via a cation-anion balance, and balancing with an appropriate ion (e.g., sodium or potassium for cation-deficient waters; sulfate, chloride, or nitrate for anion-deficient waters).
- Estimating ionic strength; calculating and correcting activity coefficients and dissociation constants for temperature; correcting alkalinity for noncarbonate alkalinity.
- Iteratively calculating the distribution of species in the water from dissociation constants. A partial listing of these ion pairs is given in Table 2.13.
- Verification of mass balance and adjustment of ion concentrations to agree with analytical values.
- Repeating the process until corrections are insignificant.
- Calculating saturation levels based upon the free concentrations of ions estimated using the ion association model (ion pairing).

The ion association model has been used by major water treatment companies since the early 1970s. The use of ion pairing to estimate the concentrations of free species overcomes several of the major shortcomings of traditional indices. While indices such as the LSI can correct activity coefficients for ionic strength based upon the total dissolved solids, they typically do not account for common ion effects. Common ion effects increase the apparent solubility of a compound by reducing the concentration of available reactants. A common example is sulfate reducing the available calcium in a water and increasing the apparent solubility of calcium carbonate. The use of indices which do not account for ion pairing can be misleading when comparing waters in which the TDS is composed of ions which pair with the reactants and of ions which have less interaction with them.

TABLE 2.13 Examples of Ion Pairs Used to Estimate Free Ion Concentrations

Aluminum
[Aluminum] = [Al^{3+}] + [Al(OH)$^{2+}$] + [Al(OH)$_2^+$] + [Al(OH)$_4^-$] + [AlF^{2+}] + [AlF$_2^+$] + [AlF$_3$] + [AlF$_4^-$] + [AlSO$_4^+$] + [Al(SO$_4$)$_2^-$]

Barium
[Barium] = [Ba^{2+}] + [BaSO$_4$] + [BaHCO$_3^+$] + [BaCO$_3$] + [Ba(OH)$^+$]

Calcium
[Calcium] = [Ca^{2+}] + [CaSO$_4$] + [CaHCO$_3^+$] + [CaCO$_3$] + [Ca(OH)$^+$] + [CaHPO$_4$] + [CaPO$_4^-$] + [CaH$_2$PO$_4^+$]

Iron
[Iron] = [Fe^{2+}] + [Fe^{3+}] + [Fe(OH)$^+$] + [Fe(OH)$^{2+}$] + [Fe(OH)$_3^-$] + [FeHPO$_4^+$] + [FeHPO$_4$] + [FeCl^{2+}] + [FeCl$_2^+$] + [FeCl$_3$] + [FeSO$_4$] + [FeSO$_4^+$] + [FeH$_2$PO$_4^+$] + [Fe(OH)$_2^+$] + [Fe(OH)$_3$] + [Fe(OH)$_4^-$] + [Fe(OH)$_2$] + [FeH$_2$PO$_4^{2+}$]

Magnesium
[Magnesium] = [Mg^{2+}] + [MgSO$_4$] + [MgHCO$_3^+$] + [MgCO$_3$] + [Mg(OH)$^+$] + [MgHPO$_4$] + [MgPO$_4^-$] + [MgH$_2$PO$_4^+$] + [MgF$^+$]

Potassium
[Potassium] = [K$^+$] + [KSO$_4^-$] + [KHPO$_4^-$] + [KCl]

Sodium
[Sodium] = [Na$^+$] + [NaSO$_4^-$] + [Na$_2$SO$_4$] + [NaHCO$_3$] + [NaCO$_3^-$] + [Na$_2$CO$_3$] + [NaCl] + [NaHPO$_4^-$]

Strontium
[Strontium] = [Sr^{2+}] 1 [SrSO$_4$] 1 [SrHCO$_3^+$] + [SrCO$_3$] + [Sr(OH)$^+$]

The ion association model provides a rigorous calculation of the free ion concentrations based upon the solution of the simultaneous nonlinear equations generated by the relevant equilibria.[26] A simplified method for estimating the effect of ion interaction and ion pairing is sometimes used instead of the more rigorous and direct solution of the equilibria.[27] Pitzer coefficients estimate the impact of ion association upon free ion concentrations using an empirical force fit of laboratory data.[28] This method has the advantage of providing a much less calculation-intensive direct solution. It has the disadvantages of being based upon typical water compositions and ion ratios, and of unpredictability when extrapolated beyond the range of the original data. The use of Pitzer coefficients is not recommended when a full ion association model is available.

When indices are used to establish operating limits such as maximum concentration ratio or maximum pH, the differences between indices calculated using ion pairing can have some serious economic significance. For example, experience on a system with high-TDS water may be translated to a system operating with a lower-TDS water. The high indices that were found acceptable in the high-TDS water may be unrealistic when translated to a water where ion pairing is less significant in reducing the apparent driving force for scale formation. Table 2.14 summarizes the impact of TDS upon LSI when it is calculated using total analytical values for calcium and alkalinity, and when it is calculated using the free calcium and carbonate concentrations determined with an ion association model.

Indices based upon ion association models provide a common denominator for comparing results between systems. For example, calcite saturation level calculated using free calcium and carbonate concentrations has been used successfully as the basis for developing models which describe the minimum effective scale inhibitor dosage that will maintain clean heat-transfer surfaces.[29] The following cases illustrate some practical usage of the ion association model.

Optimizing storage conditions for low-level nuclear waste. Storage costs for low-level nuclear wastes are based upon volume. Storage is therefore most cost-effective when the aqueous-based wastes are concentrated to occupy the minimum volume. Precipitation is not desirable because it can turn a low-level waste into a high-level waste, which is much more costly to store. Precipitation can also foul heat-transfer equipment used in the concentration process. The ion association model approach has been used at the Oak Ridge National Laboratory to predict the optimum conditions for long-term storage.[30] Optimum conditions involve the parameters of maximum concentration, pH, and temperature. Figures 2.17 and 2.18, respectively, depict a profile of the degree of supersaturation for silica and for magnesium hydroxide as a function of pH and temperature. It can be seen that amorphous silica deposition may present a problem when the pH falls below approximately 10, and that magnesium hydroxide or brucite deposition is predicted when the pH rises above approximately 11. Based upon this preliminary run, a pH range of 10 to 11 was recommended for storage and concentration. Other potential precipitants can be screened using the ion association model to provide an overall evaluation of a wastewater prior to concentration.

TABLE 2.14 Impact of Ion Pairing on the Langelier Scaling Index (LSI)

Water	LSI		TDS impact on LSI
	Low TDS	High TDS	
High chloride			
No pairing	2.25	1.89	−0.36
With pairing	1.98	1.58	−0.40
High sulfate			
No pairing	2.24	1.81	−0.43
With pairing	1.93	1.07	−0.86

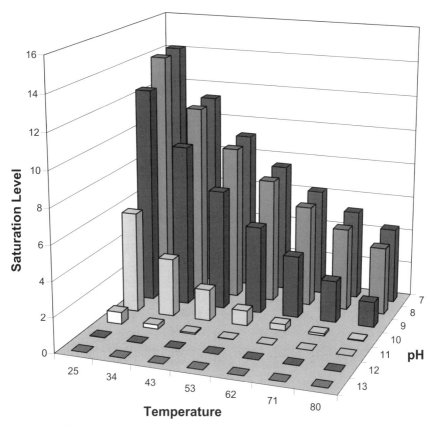

Figure 2.17 Amorphous silica saturation in low-level nuclear wastewater as a function of pH and temperature (WaterCycle).

Limiting halite deposition in a wet high-temperature gas well. There are several fields in the Netherlands that produce hydrocarbon gas associated with very high TDS connate waters. Classical oilfield scale problems (e.g., calcium carbonate, barium sulfate, and calcium sulfate) are minimal in these fields. Halite (NaCl), however, can be precipitated to such an extent that production is lost in hours. As a result, a bottom-hole fluid sample is retrieved from all new wells. Unstable components are "fixed" immediately after sampling, and pH is determined under pressure. A full ionic and physical analysis is also carried out in the laboratory.

The analyses were run through an ion association model computer program to determine the susceptibility of the brine to halite (and other scale) precipitation. If a halite precipitation problem was predicted, the ion association model was run in a "mixing" mode to determine if mixing the connate water with boiler feedwater would prevent the problem. This

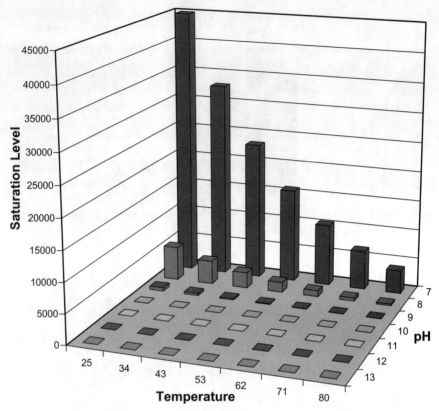

Figure 2.18 Brucite saturation in low-level nuclear wastewater as a function of pH and temperature.

approach has been used successfully to control salt deposition in the well with the composition outlined in Table 2.15. The ion association model evaluation of the bottom-hole chemistry indicated that the water was slightly supersaturated with sodium chloride under the bottom-hole conditions of pressure and temperature. As the fluids cooled in the well bore, the production of copious amounts of halite was predicted.

The ion association model predicted that the connate water would require a minimum dilution with boiler feedwater of 15 percent to prevent halite precipitation (Fig. 2.19). The model also predicted that overinjection of dilution water would promote barite (barium sulfate) formation (Fig. 2.20). Although the well produced H_2S at a concentration of 50 mg/L, the program did not predict the formation of iron sulfide because of the combination of low pH and high temperature. Boiler feedwater was injected into the bottom of the well using the downhole injection valve normally used for corrosion inhibitor injection. Injection of dilution water at a rate of 25 to 30 percent has allowed the well to produce successfully since start-up. Barite and iron sulfide precipitation have not been observed, and plugging with salt has not occurred.

Identifying acceptable operating range for ozonated cooling systems. It has been well established that ozone is an efficient microbiological control agent in open recirculating cooling-water systems (cooling towers). It has also been reported that commonly encountered scales have not been observed in ozonated cooling systems under conditions where scale would otherwise be expected. The water chemistry of 13 ozonated cooling systems was evaluated using an ion association model. Each system was treated solely with ozone on a continuous basis at the rate of 0.05 to 0.2 mg/L based upon recirculating water flow rates.[31]

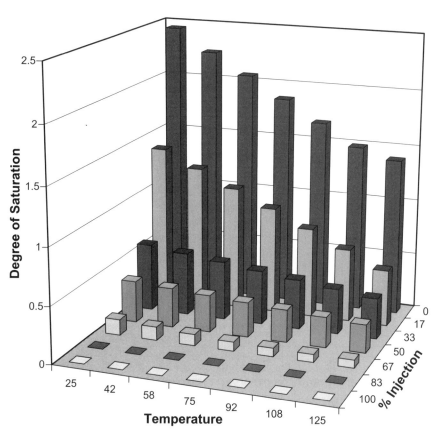

Figure 2.19 Degree of saturation of halite in a hot gas well as a function of temperature and reinjected boiler water (DownHole SAT).

TABLE 2.15 Hot Gas Well Water Analysis

	Bottom hole connate	Boiler feedwater
Temperature, °C	121	70
Pressure, bars	350	1
pH, site	4.26	9.10
Density, kg/m^3	1.300	1.000
TDS, mg·L^{-1}	369,960	<20
Dissolved CO_2, mg·L^{-1}	223	<1
H_2S (gas phase), mg·L^{-1}	50	0
H_2S (aqueous phase), mg·L^{-1}	<0.5	0
Bicarbonate, mg·L^{-1}	16	5.0
Chloride, mg·L^{-1}	228,485	0
Sulfate, mg·L^{-1}	320	0
Phosphate, mg·L^{-1}	<1	0
Borate, mg·L^{-1}	175	0
Organic acids <C_6, mg·L^{-1}	12	<5
Sodium, mg·L^{-1}	104,780	<1
Potassium, mg·L^{-1}	1,600	<1
Calcium, mg·L^{-1}	30,853	<1
Magnesium, mg·L^{-1}	2,910	<1
Barium, mg·L^{-1}	120	<1
Strontium, mg·L^{-1}	1,164	<1
Total iron, mg·L^{-1}	38.0	<0.01
Lead, mg·L^{-1}	5.1	<0.01
Zinc, mg·L^{-1}	3.6	<0.01

The saturation levels for common cooling-water scales were calculated, including calcium carbonate, calcium sulfate, amorphous silica, and magnesium hydroxide. Brucite saturation levels were included because of the potential for magnesium silicate formation as a result of the adsorption of silica upon precipitating magnesium hydroxide. Each system was evaluated by[31]

- Estimating the concentration ratio of the systems by comparing recirculating water chemistry to makeup water chemistry.
- Calculating the theoretical concentration of recirculating water chemistry based upon makeup water analysis and the apparent, calculated concentration ratio from step 1.
- Comparing the theoretical and observed ion concentrations to determine precipitation of major species.
- Calculating the saturation level for major species based upon both the theoretical and the observed recirculating water chemistry.
- Comparing differences between the theoretical and actual chemistry to the observed cleanliness of the cooling systems and heat exchangers with respect to heat transfer surface scale buildup, scale formation in valves and on non–heat-transfer surfaces, and precipitate buildup in the tower fill and basin.

Three categories of systems were encountered:[31]

- *Category 1.* The theoretical chemistry of the concentrated water was not scale-forming (i.e., undersaturated).
- *Category 2.* The concentrated recirculating water would have a moderate to high calcium carbonate scale–forming tendency. Water chemistry observed in these systems is similar to that in systems run successfully using traditional scale inhibitors such as phosphonates.
- *Category 3.* These systems demonstrated an extraordinarily high scale potential for at least calcium carbonate and brucite. These systems operated with a recirculating water chemistry similar to that of a softener rather than of a cooling system. The Category 3 water chemistry was above the maximum saturation level for calcium carbonate where traditional inhibitors such as phosphonates are able to inhibit scale formation.

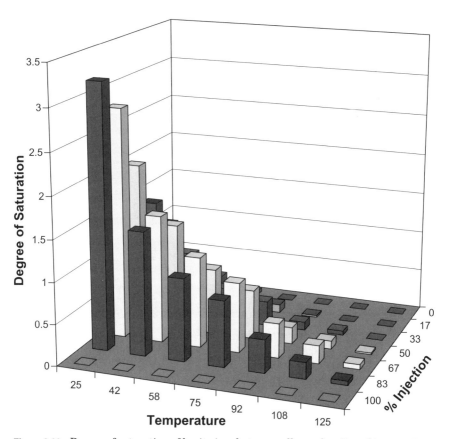

Figure 2.20 Degree of saturation of barite in a hot gas well as a function of temperature and reinjected boiler water.

TABLE 2.16 Theoretical vs. Actual Recirculating Water Chemistry

System (Category)	Calcium			Magnesium			Silica			System cleanliness
	T*	A†	Δ‡	T*	A†	Δ‡	T*	A†	Δ‡	
1 (1)	56	43	13	28	36	−8	40	52	−12	No scale observed
2 (2)	80	60	20	88	38	50	24	20	4	Basin buildup
3 (2)	238	288	−50	483	168	315	38	31	7	Heavy scale
4 (2)	288	180	108	216	223	−7	66	48	18	Valve scale
5 (3)	392	245	147	238	320	−82	112	101	11	Condenser tube scale
6 (3)	803	163	640	495	607	−112	162	143	19	No scale observed
7 (3)	1464	200	1264	549	135	414	112	101	11	No scale observed
8 (3)	800	168	632	480	78	402	280	78	202	No scale observed
9 (3)	775	95	680	496	78	418	186	60	126	No scale observed
10 (3)	3904	270	3634	3172	508	2664	3050	95	2995	Slight valve scale
11 (3)	4170	188	3982	308	303	5	126	126	0	No scale observed
12 (3)	3660	800	2860	2623	2972	−349	6100	138	5962	No scale observed
13 (3)	7930	68	7862	610	20	590	1952	85	1867	No scale observed

*T = theoretical (ppm).
†A = actual (ppm).
‡Δ = difference (ppm).

TABLE 2.17 Theoretical vs. Actual Recirculating Water Saturation Level

System (Category)	Calcite T*	Calcite A†	Brucite T*	Brucite A†	Silica T*	Silica A†	Observation
1 (1)	0.03	0.02	<0.001	<0.001	0.20	0.25	No scale observed
2 (2)	49	5.4	0.82	0.02	0.06	0.09	Basin buildup
3 (2)	89	611	2.4	0.12	0.10	0.12	Heavy scale
4 (2)	106	50	1.3	0.55	0.13	0.16	Valve scale
5 (3)	240	72	3.0	0.46	0.21	0.35	Condenser tube scale
6 (3)	540	51	5.3	0.73	0.35	0.49	No scale observed
7 (3)	598	28	10	0.17	0.40	0.52	No scale observed
8 (3)	794	26	53	0.06	0.10	0.33	No scale observed
9 (3)	809	6.5	10	<0.01	0.22	0.27	No scale observed
10 (3)	1198	62	7.4	0.36	0.31	0.35	Slight valve scale
11 (3)	1670	74	4.6	0.36	0.22	0.44	No scale observed
12 (3)	3420	37	254	0.59	1.31	0.55	No scale observed
13 (3)	7634	65	7.6	0.14	1.74	0.10	No scale observed

*T = theoretical (ppm).
†A = actual (ppm).

Table 2.16 outlines the theoretical versus actual water chemistry for the 13 systems evaluated. Saturation levels for the theoretical and actual recirculating water chemistries are presented in Table 2.17. A comparison of the predicted chemistries to observed system cleanliness revealed the following:[31]

- *Category 1* (recirculating water chemistry undersaturated). The systems did not show any scale formation.
- *Category 2* (conventional alkaline cooling system control range). Scale formation was observed in eight of the nine Category 2 systems evaluated.
- *Category 3* (cooling tower as a softener). Deposit formation on heat-transfer surfaces was not observed in most of these systems.

The study revealed that calcium carbonate (calcite) scale formed most readily on heat-transfer surfaces in systems operating in a calcite saturation level range of 20 to 150, the typical range for chemically treated cooling water. At much higher saturation levels, in excess of 1000, calcite precipitated in the bulk water. Because of the overwhelming high surface area of the precipitating crystals relative to the metal surface in the system, continuing precipitation leads to growth on crystals in the bulk water rather than on heat-transfer surfaces. The presence of ozone in cooling systems does not appear to influence calcite precipitation and/or scale formation.[31]

Optimizing calcium phosphate scale inhibitor dosage in a high-TDS cooling system. A major manufacturer of polymers for calcium phosphate scale control in cooling systems has developed laboratory data on the minimum effective scale inhibitor (copolymer) dosage required to prevent calcium phosphate deposition over a broad range of calcium and phosphate concentrations, and a range of pH and temperatures. The data were developed using static tests, but have been observed to correlate well with the dosage requirements for the copolymer in operating cooling systems. The data were developed using test waters with relatively low levels of dissolved solids. Recommendations from the data were typically made as a function of calcium concentration, phosphate concentration, and pH. This database was used to project the treatment requirements for a utility cooling system that used geothermal brine for makeup water. An extremely high dosage (30 to 35 mg/L) was recommended based upon the laboratory data.[25]

It was believed that much lower dosages would be required in the actual cooling system because of the reduced availability of calcium anticipated in the high-TDS recirculating water. As a result, it was believed that a model based upon dosage as a function of the ion association model saturation level for tricalcium phosphate would be more appropriate, and accurate, than a simple lookup table of dosage versus pH and analytical values for calcium and phosphate. Tricalcium phosphate saturation levels were calculated for each of the laboratory data points. Regression analysis was used to develop a model for dosage as a function of saturation level and temperature.

The model was used to predict the minimum effective dosage for the system with the makeup and recirculating water chemistry found in Table 2.18. A dosage in the range of 10 to 11 mg/L was predicted, rather than the 30 ppm derived from the lookup tables. A dosage minimization study was conducted to determine the minimum effective dosage. The system was initially treated with the copolymer at a dosage of 30 mg/L in the recirculating water. The dosage was decreased until deposition was observed. Failure was noted when the recirculating water concentration dropped below 10 mg/L, validating the ion association–based dosage model.

2.2.5 Software Systems

Some software systems are available for water treatment personnel. The products combine the calculation sophistication of university-based mainframe programs with a practical, commonsense engineering approach to evaluating and solving water treatment problems. Color-coded graphics in combination with 3-D representation can be quite useful in visualizing water treatment problems over a user-defined probable dynamic operating range. Graphics reduce advanced physical chemistry concepts and profiles to a level where even laypeople can understand the impact of changing parameters such as pH,

TABLE 2.18 Calcium Phosphate Inhibitor Dosage Optimization Example

Water analysis at 6.2 cycles		Deposition potential indicators	
Cations		Saturation level	
Calcium (as $CaCO_3$)	1339	Calcite	38.8
Magnesium (as $CaCO_3$)	496	Aragonite	32.9
Sodium (as Na)	1240	Silica	0.4
Anions		Tricalcium phosphate	1074
Chloride (as Cl)	620	Anhydrite	1.3
Sulfate (as SO_4)	3384	Gypsum	1.7
Bicarbonate (as HCO_3)	294	Fluorite	0.0
Carbonate (as CO_3)	36	Brucite	<0.1
Silica (as SiO_2)	62	Simple indices	
Parameters		Langelier	1.99
pH	8.40	Ryznar	4.41
Temperature, °C	36.7	Practical	4.20
Half-life, h	72	Larson-Skold	0.39
Recommended Treatment			
100% active copolymer, mg/L	10.53		

temperature, or concentration. These products serve niche water treatment markets, including the cooling-water and oilfield markets. An ion association model engine forms the basis for the sophisticated predictions of scale, corrosion, and inhibitor optimization provided by these software systems.

Scaling of cooling water. Watercycle is a computer-based system that allows a water treatment chemist to evaluate the scale potential for common scalants over the range of water chemistry, temperature, and pH anticipated in an operational cooling system.[32] This computer system, which was developed to allow water treaters to readily evaluate the scale potential for common scalants over the broadest of operating ranges without the necessity for tedious manual calculations, has been used to generate the analyses presented in this section.

Even when scaling indices can be calculated, they often offer conflicting results that can easily cloud the interpretation of what they are foretelling. The program can be applied to long- as well as short-residence-time systems. The computer system uses the mean salt activities for estimating ion-activity coefficients based upon temperature and ionic strength.[24] The use of ion pairing expands the usefulness of calculated saturation levels. The system can assist the cooling tower operator or water treaters in establishing control limits based on concentration ratio (cycles of concentration), pH, and temperature profiles. The program can be used to

- Develop an overall profile of scale potential for common cooling-system scalants over the entire range of critical operating parameters anticipated.
- Evaluate the scale potential of an open recirculating cooling system versus concentration ratio as an aid in establishing control limits.
- Evaluate the benefits of pH control with respect to scale potential and to estimate acid requirements.
- Review these indicators as water quality changes or environmental constraints force operation with reduced water quality and increased scale potential.
- Learn about the interaction of water chemistry and operating conditions (pH, temperature) by using the program as a system simulator.

Many cooling-water evaluations assume that the cooling system is static. Indices for scale potential are calculated at the "harshest" conditions for the foulant under study. What-if scenario modeling provides one of the greatest benefits from using Watercycle. The "what-if scenario" modules allow one to

- Visualize what will happen to the scale potential and corrosivity of a cooling water as operating parameters and water chemistry change.
- Evaluate the current cooling water over the entire range of operating parameters.
- Predict water scaling behavior for use in evaluating new cooling systems, and as an aid in establishing control ranges and operating parameters.

In the case of calcium carbonate scale, indices are typically calculated at the highest expected temperature and highest expected pH—the conditions under which calcium carbonate is least soluble. In the case of silica, the opposite conditions are used. Amorphous silica has its lowest solubility at the lowest temperature and lowest pH encountered. Indices calculated under these conditions would be acceptable in many cases. Unfortunately, cooling systems are not static. The foulants silica and tricalcium phosphate are used as examples to demonstrate the use of operating range profiles in developing an in-depth evaluation of scale potential and the impact of loss of control.

Silica. Guidelines for the upper silica operating limits have been well defined in water treatment practice, and have evolved with the treatment programs. In the days of acid chromate cooling-system treatment, an upper limit of 150 ppm silica as SiO_2 was common. The limit increased to 180 ppm with the advent of alkaline treatments and pH control limits up to 9.0. Silica control levels approaching or exceeding 200 ppm as SiO_2 have been reported for the current high-pH, high-alkalinity all-organic treatment programs where pH is allowed to equilibrate at 9.0 or higher.[26]

The evolution of silica control limits can be readily understood by reviewing the silica solubility profile. As depicted in Fig. 2.21, solubility of amorphous silica increases with increasing pH. Silica solubility also increases with increasing temperature. In the pH range of 6.0 to 8.0 and temperature range of 20 to 30°C, cooling water will be saturated with amorphous silica when the concentration reaches 100 ppm (20°C) or 135 ppm (30°C) as SiO_2. These concentrations correspond to a saturation level of 1.0. The traditional silica limit for this pH range has been 150 ppm as SiO_2. As outlined in Table 2.19, a limit of 150 ppm would correspond roughly to a saturation level of 1.4 at 20°C and 1.1 at 30°C.

At the upper end of the cooling-water pH range (9.0), silica solubility increases to 115 ppm (20°C) and 140 ppm (30°C). A control limit of 180 ppm would correspond to saturation levels of 1.5 and 1.3, respectively. In systems where concentration ratio is limited by silica solubility, it is recommended that the concentration ratio limit be reestablished seasonally based on amorphous silica saturation level or whenever significant temperature changes occur.[26]

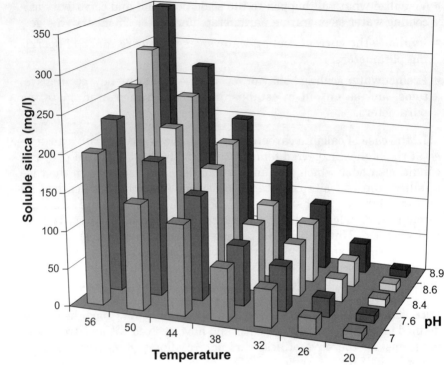

Figure 2.21 Solubility of amorphous silica as a function of temperature and pH.

TABLE 2.19 Silica Limits for Three Treatment Schemes

	Low pH (6.0)		Moderate pH (7.6)		High pH (8.9)	
Temperature (°C)	20	30	20	30	20	30
Silica level (ppm)	130	150	150	150	>180	>180
Saturation level limit	1.2	1.1	1.4	1.1	1.5	1.3

Calcium phosphate. Neutral phosphate programs can benefit from saturation-level profiles for tricalcium phosphate. Treatment programs using orthophosphate as a corrosion inhibitor must operate in a narrow pH range in order to achieve satisfactory corrosion inhibition without catastrophic calcium phosphate deposition occurring. Operating-range profiles for tricalcium phosphate can assist the water treatment chemist in establishing limits for pH, concentration ratio, and orthophosphate in the recirculating water. Such profiles are also useful in showing operators the impact of loss of pH control, chemical overfeed, or overconcentration.

Scaling of deep well water. DownHole SAT is another specialized computer program that allows a water treatment specialist to evaluate the scale potential for common scalants over a broad range of water chemistry parameters, such as temperature, pressure, pH, and pCO_2.[33] As with the previous computer system, "what-if scenario" modules provide an easy way to visualize what could happen to the scale potential and corrosivity of a water as environmental parameters and water chemistry change. The what-if scenarios also allow evaluating the impact of bringing a water to the surface, or finding the safe ratios for mixing waters under varying conditions. The scenarios can provide a predictor for use in anticipating problems in new or proposed wells. The following indices and the scaling behavior of the solid species shown in Table 2.20 are all calculated by DownHole SAT.

- Stiff-Davis
- Oddo-Tomson
- Ryznar
- Puckorius
- Larson-Skold

Convenience groups. Three "convenience groups" have been programmed into the computer system to allow multiple graph selection for common groups:

- The common foulants group includes calcite, barite, witherite, and anhydrite saturation levels.
- The common indices group includes the Langelier, Stiff-Davis, Oddo-Tomson, and Ryznar indices.

TABLE 2.20 Scales Modeled by DownHole SAT

Scale	Formula
Calcite	$CaCO_3$
Aragonite	$CaCO_3$
Witherite	$BaCO_3$
Magnesite	$MgCO_3$
Siderite	$FeCO_3$
Barite	$BaSO_4$
Anhydrite	$CaSO_4$
Gypsum	$CaSO_4 \cdot 2H_2O$
Celestite	$SrSO_4$
Fluorite	CaF_2
Amorphous iron	$Fe(OH)_3$
Amorphous silica	SiO_2
Brucite	$Mg(OH)_2$
Strengite	$FePO_4 \cdot 2H_2O$
Tricalcium phosphate	$Ca_3(PO_4)_2$
Hydroxyapatite	$Ca_5(PO_4)_3(OH)$
Thenardite	Na_2SO_4
Halite	$NaCl$
Iron sulfide	FeS

- The calcium carbonate group includes calcite saturation level, the Langelier saturation index, the Stiff-Davis index, and the Oddo-Tomson index.

Dosages for scale inhibitors should be applied as a function of a driving force for scale formation and growth (e.g., calcite saturation level), temperature as it affects reaction rates, pH as it affects the dissociation state of the inhibitor, and time. A version of the computer program allows the development of mathematical models for the minimum effective scale inhibitor dosage as a function of these parameters: driving force, temperature, pH, and time.

Mathematical models. Mathematical models for an inhibitor are developed by the program using multiple regression. The goodness of fit for the data can be presented in table and graphical format. Models are discussed by parameter. The basic parameter to which scale inhibitor dosages have been correlated historically is the driving force for crystal formation and crystal growth. Early models attempted to develop models based upon the Langelier saturation index or the Ryznar stability index. Most water treaters are in agreement that dosage requirements increase with the driving force for scale formation. Calcite saturation level provides an excellent driving force for calcium carbonate scale inhibitor models, gypsum saturation level for calcium sulfate in the cooling-water temperature range, and tricalcium phosphate saturation level for calcium phosphate scale prevention. The momentary excess indices can also be used effectively to model dosage requirements.

A second critical factor in determining an effective dosage or developing a model for an inhibitor is time. Time is the residence time of scale-forming species in the system you wish to treat. The time factor for scale inhibition can be as short as 4 to 10 s in a utility condenser system, or extend into days for cooling towers. In high-saturation-level systems, the induction period can be very short. In systems where water is barely supersaturated, the induction time can approach infinity. Scale inhibitors have been observed to extend the induction time before scale formation or growth on existing scale substrate occurs.[34]

Inhibitors extend the time before scale will form in a system by interfering with the kinetics of crystal formation and growth. Rate decreases as inhibitor dosages increase. Additional parameters include temperature, as it affects the rate of crystal formation and/or growth. Dosage changes with temperature can be modeled with a simple Arrhenius relationship. pH is an important parameter to include in these models when an inhibitor can exist in two or more forms within the pH range of use, and one of the forms is much more active as a scale inhibitor than the other(s). pH can also affect the type of scale that forms (e.g., tricalcium phosphate versus hydroxylapatite).

2.3 Seawater

2.3.1 Introduction

Seawater systems are used by many industries, such as shipping, offshore oil and gas production, power plants, and coastal industrial plants. The main use of seawater is for cooling purposes, but it is also used for firefighting, oilfield water injection, and desalination plants. The corrosion problems in these systems have been well studied over many years, but despite published information on materials behavior in seawater, failures still occur. Most of the elements that can be found on earth are present in seawater, at least in trace amounts. However, 11 of the constituents account for 99.95 percent of the total solutes, as indicated in Table 2.21, with chloride ions being by far the largest constituent.

The concentration of dissolved materials in the sea varies greatly with location and time because rivers dilute seawater, rain, or melting ice, and seawater can be concentrated by evaporation. The most important properties of seawater are[35]

- Remarkably constant ratios of the concentrations of the major constituents worldwide
- High salt concentration, mainly sodium chloride
- High electrical conductivity
- Relatively high and constant pH
- Buffering capacity
- Solubility for gases, of which oxygen and carbon dioxide in particular are of importance in the context of corrosion
- The presence of a myriad of organic compounds
- The existence of biological life, to be further distinguished as microfouling (e.g., bacteria, slime) and macrofouling (e.g., seaweed, mussels, barnacles, and many kinds of animals or fish)

Some of these factors are interrelated and depend on physical, chemical, and biological variables, such as depth, temperature, intensity of light, and the availability of nutrients. The main numerical specification of seawater is its salinity.

Salinity. Salinity was defined, in 1902, as the total amount of solid material (in grams) contained in one kilogram of seawater when all halides have been replaced by the equivalent of chloride, when all the carbonate

TABLE 2.21 Average Concentration of the 11 Most Abundant Ions and Molecules in Clean Seawater (35.00 ‰ Salinity, Density of 1.023 g·cm^{-3} at 25°C)

Species	Concentration	
	mmol^{-1}·kg^{-1}	g·kg^{-1}
Na$^+$	468.5	10.77
K$^+$	10.21	0.399
Mg^{2+}	53.08	1.290
Ca^{2+}	10.28	0.4121
Sr^{2+}	0.090	0.0079
Cl$^-$	545.9	19.354
Br$^-$	0.842	0.0673
F$^-$	0.068	0.0013
HCO$_3^-$	2.30	0.140
SO$_4^{2-}$	28.23	2.712
B(OH)$_3$	0.416	0.0257

is converted to oxide, and when all organic matter is completely oxidized. The definition of 1902 was translated into Eq. (2.13), where the salinity (S) and chlorinity (Cl) are expressed in parts per thousand (‰).

$$S\ (\text{‰}) = 0.03 + 1.805\text{Cl}\ (\text{‰}) \tag{2.13}$$

The fact that the equation of 1902 gives a salinity of 0.03 ‰ for zero chlorinity was a cause for concern, and a program led by the famous United Nations Scientific, Education and Cultural Organization (UNESCO) helped to determine a more precise relation between chlorinity and salinity. The definition of 1969 produced by that study is given in Eq. (2.14):

$$S\ (\text{‰}) = 1.80655\text{Cl}\ (\text{‰}) \tag{2.14}$$

The definitions of 1902 and 1969 give identical results at a salinity of 35 ‰ and do not differ significantly for most applications. The definition of salinity was reviewed again when techniques to determine salinity from measurements of conductivity, temperature, and pressure were developed. Since 1978, the Practical Salinity Scale defines salinity in terms of a conductivity ratio:

> The practical salinity, symbol S, of a sample of sea water, is defined in terms of the ratio K of the electrical conductivity of a sea water sample of 15°C and the pressure of one standard atmosphere, to that of a potassium chloride (KCl) solution, in which the mass fraction of KCl is 0.0324356, at the same temperature and pressure. The K value exactly equal to one corresponds, by definition, to a practical salinity equal to 35.

The corresponding formula is given in Eq. (2.15).[36]

$$S = 0.0080 - 0.1692K^{0.5} + 25.3853K + 14.0941K^{1.5} \\ - 7.0261K^2 + 2.7081K^{2.5} \tag{2.15}$$

Note that in this definition, (‰) is no longer used, but an old value of 35‰ corresponds to a new value of 35. Since the introduction of this practical definition, salinity of seawater is usually determined by measuring its electrical conductivity and generally falls within the range 32 to 35 ‰.[35]

Other ions. A large part of the dissolved components of seawater is present as ion pairs or in complexes, rather than as simple ions. While the major cations are largely uncomplexed, the anions other than chloride are to varying degrees present in the form of complexes. About 13 percent of the magnesium and 9 percent of the calcium in ocean waters exist as magnesium sulfate and calcium sulfate, respectively. More than 90 percent of the carbonate, 50 percent of the sulfate, and 30 percent of the bicarbonate exist as complexes. Many minor or trace components occur primarily as complexed ions at the pH and the redox potential of seawater. Boron, silicon, vanadium, germanium, and iron form hydroxide complexes. Gold, mercury, and silver, and probably calcium and lead, form chloride complexes. Magnesium produces complexes with fluorides to a limited extent.

Surface seawater characteristically has pH values higher than 8 owing to the combined effects of air-sea exchange and photosynthesis. The carbonate ion concentration is consequently relatively high in surface waters. In fact, surface waters are almost always supersaturated with respect to the calcium carbonate phases, calcite and aragonite. The introduction of molecular carbon dioxide into subsurface waters during the decomposition of organic matter decreases the saturation state with respect to carbonates. While most surface waters are strongly supersaturated with respect to the carbonate species, the opposite is true of deeper waters, which are often undersaturated in carbonates.

Precipitation of inorganic compounds from seawater. The value of calcareous deposits in the effective and efficient operation of marine cathodic protection systems is generally recognized by corrosion engineers. The calcareous films are known to form on cathodic metal surfaces in seawater, thereby enhancing oxygen concentration polarization and reducing the current density needed to maintain a prescribed cathodic potential. For most cathodic surfaces in aerated waters, the principal reduction reaction is described by Eq. (2.16):

$$O_2 + 2H_2O + 4e^- \rightarrow 4OH^- \tag{2.16}$$

In cases where the potential is more negative than the reversible hydrogen electrode potential, the production of hydrogen as described in Eq. (2.17) becomes possible:

$$2H_2O + 2e^- \rightarrow H_2 + 2OH^- \tag{2.17}$$

In either case, the production of hydroxyl ions results in an increase in pH for the electrolyte adjacent to the metal surface. In other terms, an increase in OH^- is equivalent to a corresponding reduction in acidity or H^+ ion concentration. This situation causes the production of a pH profile in the diffuse layer, where the equilibrium reactions can be quite different from those in the bulk seawater conditions. Temperature, relative electrolyte velocity, and electrolyte composition will all influence this pH profile. There is both analytical and experimental evidence that such a pH increase exists as a consequence of the application of a cathodic current. In seawater, pH is controlled by the carbon dioxide system described in Eqs. (2.18) through (2.20):

$$CO_2 + H_2O \rightarrow H_2CO_3 \quad (2.18)$$

$$H_2CO_3 \rightarrow H^+ + HCO_3^- \quad (2.19)$$

$$HCO_3^- \rightarrow H^+ + CO_3^{2-} \quad (2.20)$$

If OH^- is added to the system as a consequence of one of the above cathodic processes [Eqs. (2.16) and (2.17)], then the reactions described in Eqs. (2.21) and (2.22) become possible, with Eq. (2.23) describing the precipitation of a calcareous deposit.

$$CO_2 + OH^- \rightarrow HCO_3^- \quad (2.21)$$

$$OH^- + HCO_3^- \rightarrow H_2O + CO_3^{2-} \quad (2.22)$$

$$CO_3^{2-} + Ca^{2+} \rightarrow CaCO_{3(s)} \quad (2.23)$$

The equilibria represented by Eqs. (2.18) through (2.23) further indicate that as OH^- is introduced, then Eqs. (2.19) and (2.20) are displaced to the right, resulting in proton production. This opposes any rise in pH and accounts for the buffering capacity of seawater. Irrespective of this, however, Eqs. (2.18) through (2.23) indicate that this buffering action is accompanied by the formation of calcareous deposits on cathodic surfaces exposed to seawater.

Magnesium compounds, $Mg(OH)_2$ in particular, could also contribute to the protective character of calcareous deposits. However, calcium carbonate is thermodynamically stable in surface seawater, where it is supersaturated, whereas magnesium hydroxide is unsaturated and less stable. In fact, $Mg(OH)_2$ would precipitate only if the pH of seawater were to exceed approximately 9.5. This is the main reason why the behavior of $CaCO_3$ in seawater has been so extensively studied, since calcium carbonate sediments are prevalent and widespread in the oceans.[37]

It has been demonstrated that calcium carbonate occurs in the oceans in two crystalline forms, i.e., calcite and aragonite. Partly because calcite and magnesium carbonate have similar structures, these compounds form solid solutions, the Ca:Mg ratio of which depends on the ratio of these ions in seawater. Theoretical calculations suggest that calcite in equilibrium with seawater should contain between 2 and 7 mol% $MgCO_3$. But although low magnesium calcite is the most stable carbonate phase in seawater, its precipitation and crystal growth are strongly inhibited by dissolved magnesium. Consequently, aragonite is the phase that actually precipitates when seawater is made basic by the addition of sodium carbonate. The degree of saturation for aragonite is described in Eq. (2.24),

$$K_{sp, \text{aragonite}} = (Ca^{2+})(CO_3^{2-}) \qquad (2.24)$$

where (Ca^{2+}) and (CO_3^{2-}) are the molalities of the Ca^{2+} and CO_3^{2-} ions, respectively, and $K_{sp, \text{aragonite}}$ is the solubility product of aragonite (at 25°C, $K_{sp, \text{aragonite}} = 6.7 \times 10^{-7}$).

In order to understand the buildup of carbonate ions at a metallic surface under cathodic protection (CP), one can combine Eqs. (2.17), (2.21), and (2.22) to obtain an expression describing the electrochemical production of carbonate ions [Eq. (2.25)]:

$$H_2O + CO_2 + 2e^- \rightarrow H_2 + CO_3^{2-} \qquad (2.25)$$

By referring to Chap. 1, Aqueous Corrosion, one can also develop an expression for the limiting current corresponding to this reaction [Eq. (2.26)]:

$$i_L = nFD_{CO_3^{2-}} \frac{C_{CO_3^{2-}, \text{surface}} - C_{CO_3^{2-}, \text{bulk}}}{\delta} \qquad (2.26)$$

where, at neutral bulk pH, the concentration of carbonate ions in seawater is basically zero, and the expression of i_L is correctly described by Eq. (2.27):

$$i_L = nFD_{CO_3^{2-}} \frac{C_{CO_3^{2-}, \text{surface}}}{\delta} \qquad (2.27)$$

Oxygen. The oxygen content depends primarily on factors such as salinity and temperature. Relationships have been derived from which the equilibrium concentration of dissolved oxygen can be calculated if the absolute temperature T (K) and salinity S (‰) are known:[35]

$$\ln [O_2] \text{ (mL} \cdot \text{L}^{-1}) = A_1 + A_2 (100/T) + A_3 \ln (T/100) + A_4 (T/100) + S[B_1 + B_2 (T/100) + B_3 (T/100)^2]$$

where $A_1 = -173.4292$
$A_2 = 249.6339$
$A_3 = 143.3483$
$A_4 = -21.8492$
$B_1 = -0.033096$
$B_2 = 0.014259$
$B_3 = -0.0017000$

The primary source of the dissolution of oxygen is the air-sea exchange with oxygen in the atmosphere, leading to near saturation (within 5 percent). However, mainly because of biological processes, deviations may occur with the seasons; e.g., in spring, when significant photosynthesis develops, supersaturation levels up to 200 percent may be found. Another action that can cause supersaturation of oxygen is the entrainment of air bubbles as a result of wave action, resulting in supersaturation values up to 10 percent.

The normal profile of corrosion of unprotected steel, as in the case of pilings or the supporting legs for offshore oil-drilling structures, is shown in Fig. 2.22 based on the measurements of the distribution of corrosion of test pilings exposed in a partially enclosed basin at Kure Beach, North Carolina.[38]

The reverse of the process is the biochemical oxidation of organic matter, leading to oxygen consumption and undersaturation coupled with carbon dioxide production and acidification. The rate and occurrence of such processes are strongly dependent on the availability of nutrients and dissolved oxygen. It is for this reason that very low oxygen concentrations can be found below the zone of surface mixing, as is the case in some locations in the Pacific Ocean.[39] At still greater depths the oxygen level can increase again as a result of the supply of oxygen-rich cold water by deep oceanographic currents. However, such situations are strongly related to local conditions and can also depend on the season. Examples are known where in winter the mixed zone extends to the bottom because of the action of storms, whereas in summer the same water may become stratified, as in parts of the North Sea.

At any location there are seasonal variations in salinity, temperature, and other parameters. There are also variations with the depth of water, as illustrated in Fig. 2.23, representing data collected during studies at U.S. Naval Engineering test sites in the Pacific Ocean. It should not be assumed that the variations found in these studies can be extrapolated to other oceanographic sites. For example, observations within the same depth range in the Atlantic Ocean showed a much higher concentration of dissolved oxygen to the bottom, even approaching the concentration found at the surface. The effects of depth on corrosion will thus vary from location to location, depending principally on the variations in concentration of dissolved oxygen and bacterial activity, for which the information is slowly developing.

Organic compounds. Seawater contains a wide variety of dissolved organic compounds. The total amount is low (~2 ppm), but their composition is very complex. Some of the organic compounds are resistant to decomposition and are relatively old. However, most are biologically active and are constantly being modified. The organic content of the oceans is very important to biological life processes, and the effects are much greater than might be assumed from the amount of material present. A large number of soluble compounds have been identified in seawater, including amino and organic acids and carbohydrates.

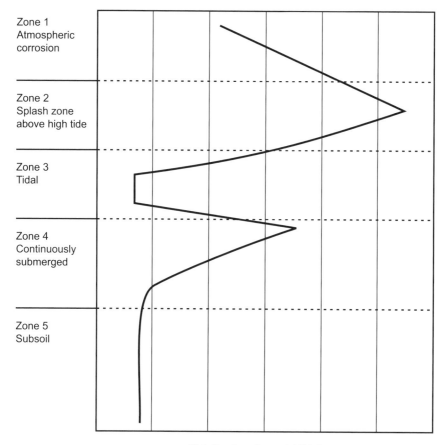

Figure 2.22 Corrosion profile of steel piling in seawater.

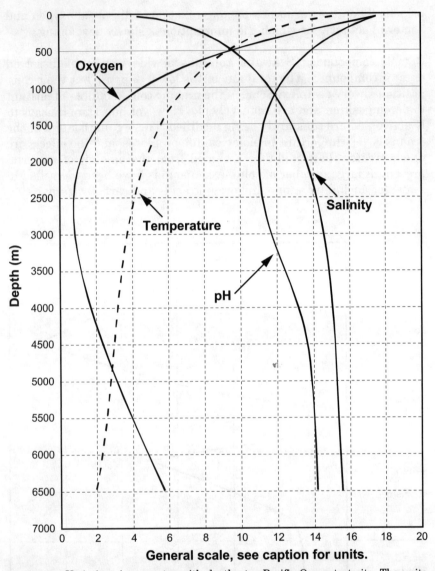

Figure 2.23 Variations in seawater with depth at a Pacific Ocean test site. The units have to be estimated with the following conversion: temperature, scale × 1 (°C); oxygen, scale × 0.333 (ppm); pH, 6.4 + scale × 0.1 (pH unit); salinity, 33.0 + scale × 0.1 (‰).

Polluted seawater. The main effect of polluted seawater arises from a combination of low oxygen content and generally decreased pH, together with the presence of sulfide ions and/or ammonia. It may be that, depending on the design of a cooling-water system, there is the risk that the water velocity will be below the design value in some areas. Organic matter entering such a system can be deposited in layers in some areas rather than being either filtered off and rejected or allowed to pass through the system. Such layers become anaerobic and yield significant amounts of sulfides, which are introduced into the cooling water and so become available for reaction with adjacent metal surfaces. In particular, a number of copper alloys will be affected by these high sulfide levels and become more susceptible to pitting.

Pollution can also occur when biofouling present in systems decays to produce sulfur-containing compounds. This form of pollution is a major problem in situations where the use of hypochlorite and other biocides is prohibited or restricted. In a large cooling system this can result in significant amounts of biological matter being generated in the form of thick layers of barnacles, mussels, and shellfish. During stagnant or low-flow water conditions, the system is likely to become anaerobic, resulting in death of the organisms followed by their gradual decomposition.

Brackish coastal water. Brackish water is defined as natural ocean water diluted to a certain extent with fresh water. The ionic concentration will diminish, depending upon the dilution factor, as will the electrical conductivity. However, under normal circumstances, even with a dilution to a salinity of 5 to 10 ‰, the chlorinity ratios of the major ions will not change. In contrast, the concentrations of the minor constituents can be changed by several orders of magnitude. Brackish water differs from open seawater in certain other respects. The biological activity, for example, can be significantly modified by higher concentrations of nutrients. Fouling is also likely to be more severe as a consequence of the greater availability of nutrients. An additional factor can be a significant increase in the proportion of suspended solids in brackish water, which can be as much as two orders of magnitude greater than in open seawater.[35] The main differences between seawater and brackish coastal water are:

1. Oxygen content may change owing to decreased salt concentration, generally increased temperature, and pollution.

2. Chloride content decreases owing to increased dilution.

3. Specific conductivity decreases owing to increased dilution.

4. The concentration and the diversity of the organic compounds will generally increase.

5. The increased amount of fouling often arising in brackish waters will lead to increased shielding, and thus a decrease in the general corrosion rate, as a result of oxygen reduction.

6. The increase in the level of suspended solids, often associated with brackish waters, is likely to have a marked effect on corrosion processes, often in association with water velocity effects.

Within harbors, bays, and other estuaries, marked differences in the amount and type of fouling can exist. The main environmental factors responsible, singly or in combination, for these differences are the salinity, the degree of pollution, and the prevalence of silt. Moreover, the influence of these factors can be very specific to the type of organism involved. Apart from differences that can develop between different parts of the same estuary, there can also be differences between fouling in enclosed waters and on the open coast. In this respect, the extent of offshore coastal fouling is strongly determined by the accessibility to a natural source of infection. Local currents, average temperature, seasonal effects, depth, and penetration of light are operative factors. The pollution can also be quite important in coastal areas. Two main sources of pollution have been identified:

- Waste products of industrial, farming, or domestic origin: heavy metal ions, nutrients such as phosphates and nitrates, dissolved organic material, etc.
- Products arising from bacteriological and biological processes in the seawater itself

There are many examples of the detrimental effects of decaying organic material in cooling systems, for instance, seaweed, barnacles, mussels, and shellfish accumulated in heat-exchanger systems. For unpolluted seawater, it normally suffices to measure the salinity or chlorinity, the pH, and perhaps the oxygen content. However, in the case of polluted seawater, it is often necessary to obtain additional data. These can include the concentrations of heavy metal ions, sulfide, and ammonia as well as the chemical oxygen demand (COD) and total organic carbon (TOC) values.

2.3.2 Corrosion resistance of materials in seawater

Table 2.22 lists the materials commonly used in seawater systems as a function of whether they pertain to a low-cost (high-maintenance) or low-maintenance (high-cost) category. For low-initial-cost systems, materials such as mild steel and cast iron with and without coatings can often be used. In marine engineering, upgrading from steel has traditionally meant a change to copper-based alloys, and this trend is also occurring for offshore oil and land-based plants where high reliability is required. However, in recent years, systems based on high-performance stainless steels such as the 6% Mo superaustenitic and the super duplexes have been used by the offshore industry.[40]

TABLE 2.22 Materials Used in Seawater Systems

Component	Low-cost system	Low-maintenance system
Pipe	Galvanized steel	90/10 cupronickel
Flanges	Steel	Cast or forged 90/10 cupronickel Steel welded overlayed with cupronickel Gunmetal 6% Mo austenitic high duplex
Tube plates	60/40 brass/naval brass	Nickel aluminum bronze 90/10 cupronickel 6% Mo austenitic high duplex
Tubes	Aluminum brass	70/30 cupronickel (particularly 2% Fe + 2% Mn) 90/10 cupronickel
Pump casing	Cast iron or leaded gunmetal	Cast cupronickel Nickel aluminum bronze Admiralty gunmetal Ni-resist type D2
Pump impeller	Gunmetal	Monel alloy 410 Alloy 20 (CN7M) Stainless steel (CF3 and CF8) Nickel aluminum bronze
Pump shaft	Naval brass	Monel alloy 400 or 500 Nickel aluminum bronze Ni-resist iron type D2 Nickel aluminum 6% Mo austenitic high duplex UNS 31600 stainless steel
Strainer body	Cast iron	Cast cupronickel Gunmetal 6% Mo austenitic high duplex
Strainer	Galvanized iron	Monel alloy 400 6% Mo austenitic high duplex
Plate	Munz metal	6% Mo austenitic high duplex

Carbon steel. Corrosion of carbon steel in seawater is controlled by the availability of oxygen to the metal surface. Thus, under static conditions, carbon steel corrodes at between 100 and 200 μm/year, reflecting the oxygen level and temperature variations in different locations. As velocity causes a mass flow of oxygen to the surface, corrosion is very dependent on flow rate and can increase by a factor of 100 in moving from static or zero velocity to velocity as high as 40 m·s^{-1}. Galvanizing confers only limited benefit under flow conditions, as corrosion of zinc also increases with velocity. For the thickness normally used in seawater piping, it will extend the life of the pipe for about 6 months.

s. Stainless steels are not subject to impingement prone to pitting and crevice corrosion under low-velocity d this must be taken into consideration when these l in seawater. Attempts to build seawater systems from standard grades of stainless steel, such as Type 316, have proved unsuccessful. In recent years, grades of stainless steel with high resistance to pitting and crevice corrosion have been developed.

The first successful major use of stainless steel for seawater systems was in the Gullfaks oilfield in the Norwegian offshore sector where Avesta 254SMO (21% Cr, 18% Ni, 6% Mo, 0.2% N) was adopted. The reason for this selection was the need for a material resistant to alternate exposure to seawater and sulfide-containing oil in the storage/ballast spaces in concrete platforms. Several thousand tonnes of superaustenitic stainless steel are now in service, mainly on offshore platforms.[40]

Nickel-based alloys. Nickel-based alloys such as Inconel 625, Hastelloys C-276 and C-22, and titanium are not subject to pitting or crevice corrosion in low-velocity seawater, nor do they suffer impingement attack at high velocity. However, price limits their use to special applications in seawater systems.

Copper-based alloys. The copper-based alloys are velocity-limited, as impingement attack occurs when the hydrodynamic effect caused by seawater flow across the surface of such alloys exceeds the value at which protective films are removed and erosion-corrosion occurs. Thus, if these alloys are to exhibit high corrosion resistance, they must be used at design velocities below this limiting value. A more detailed coverage of the marine usage of two important copper-nickel alloys is presented in the section on copper alloys.

Effect of flow velocity. Velocity is the most important single factor influencing design and corrosion in seawater systems. The design velocity chosen controls the dimensions of many components, such as piping and valves. Velocity also influences the corrosion behavior of the materials, and the design value chosen is often controlled by corrosion considerations. When the corrosion rate is subject to mass transfer control, flow velocity at the metal surface becomes the rate-determining factor. This is also true with active-passive alloys, where flow, and thereby the ample supply of oxygen to the metal surface, provides the oxygen necessary to maintain the metal in the passive state. Stainless steels, for example, can perform satisfactorily provided that the water flow in the system is uninterrupted. However, in the case of zero or low flow, special precautions have to be taken. Low flow may also result in the settling of deposits from the water, with the possible consequence of local corrosion cells being set up, possibly leading to localized corrosion attack.

High flow rates can also have detrimental effects in some cases. They can increase the rate of various corrosion processes and lead to erosion corrosion, impingement attack, enhanced graphite corrosion, etc. Uneven flow over an alloy surface can be undesirable when it leads to differential aeration effects. Table 2.23 provides data on the effect of velocity on some of the materials commonly used in seawater systems.[40] In considering velocity, it is important to note that local velocities may vary considerably from design velocity. This is particularly important where features of the system such as small-radius bends, orifices, partly throttled valves, or misaligned flanges can generate turbulence and accelerate corrosion. It follows that a major consideration during the design and fabrication of a system should be to minimize turbulence raisers.

Effect of temperature. Not much information exists on the effect of temperature within the range normally encountered in seawater systems. It has been noted, at the LaQue Centre, that corrosion of carbon steel increases by approximately 50 percent between the winter (average temperature 7°C) and summer (27 to 29°C) months. Although oxygen solubility tends to fall with a rise in temperature, the higher temperature tends to increase reaction rate. Evidence from work on steel in potable waters suggests that the temperature effect is more important and that corrosion, for steel, will increase with temperature.[41]

For copper alloys, increase in temperature accelerates film formation. While it takes about 1 day to form a protective film at 15°C, it may take a week or more at 2°C. It is important to continue initial circulation of clean seawater long enough for initial film formation for all copper alloys. For stainless steels and other alloys that are prone to pitting and crevice corrosion, an increase in temperature tends to facilitate initiation of these types of attack. However, data on propagation rate suggest

TABLE 2.23 Effect of Velocity on the Corrosion of Metals in Seawater

Alloy	Deepest pit, mm	Average corrosion rate, mm·y^{-1}		
		Quiet seawater	Flowing seawater	
			8.2 m·s^{-1}	35–42 m·s^{-1}
Carbon steel	2.0	0.075	—	4.5
Grey cast iron (graphitized)	4.9	0.55	4.4	13.2
Admiralty gunmetal	0.25	0.027	0.9	1.07
85/5/5/5 Cu/Zn/Pb/Zn	0.32	0.017	1.8	1.32
Ni resist cast iron type 1B	Nil	0.02	0.2	0.97
Ni Al bronze	1.12	0.055	0.22	0.97
70/30 Cu/Ni + Fe	0.25	<0.02	0.12	1.47
Type 316 stainless steel	1.8	0.02	<0.02	<0.01
6% Mo stainless steel	Nil	0.01	<0.02	<0.01
Ni-Cu alloy 40	1.3	0.02	<0.01	0.01

that this declines with rise in temperature. The net effect of these conflicting tendencies is not always predictable. Temperature also influences biological activity, which may, in turn, influence corrosion.[40]

2.4 Corrosion in Soils

2.4.1 Introduction

Soil is an aggregate of minerals, organic matter, water, and gases (mostly air). It is formed by the combined weathering action of wind and water, and also organic decay. The proportions of the basic constituents vary greatly in different soil types. For example, humus has a very high organic matter content, whereas the organic content of beach sand is practically zero. The properties and characteristics of soil obviously vary as a function of depth. A vertical cross section taken through the soil is known as a *soil profile,* and the different layers of soil are known as *soil horizons*. The following soil horizons have been classified:

- A. Surface soil (usually dark in color due to organic matter)
- O. Organic horizon (decaying plant residues)
- E. Eluviation horizon (light color, leached)
- B. Accumulation horizon (rich in certain metal oxides)
- C. Parent material (largely nonweathered bedrock)

Corrosion in soils is a major concern, especially as much of the buried infrastructure is aging. Increasingly stringent environmental protection requirements are also placing a focus on corrosion issues. Topical examples of soil corrosion are related to oil, gas, and water pipelines; buried storage tanks (a vast number are used by gas stations); electrical communication cables and conduits; anchoring systems; and well and shaft casings. Such systems are expected to function reliably and continuously over several decades. Corrosion in soils is a complex phenomenon, with a multitude of variables involved. Chemical reactions involving almost each of the existing elements are known to take place in soils, and many of these are not yet fully understood. The relative importance of variables changes for different materials, making a universal guide to corrosion impossible. Variations in soil properties and characteristics across three dimensions can have a major impact on corrosion of buried structures.

2.4.2 Soil classification systems

Soil texture refers to the size distribution of mineral particles in a soil. Sand (rated from coarse to very fine), silt, and clay refer to textures of decreasing particle coarseness (Table 2.24). Soils with a high proportion of sand have very limited storage capacity for water, whereas clays are excellent in retaining water. One soil identification system has defined eleven soil types on the basis of their respective proportions of clay, silt, and sand. The eleven types are sand, loamy sand, sandy loam, sandy clay loam, clay loam, loam, silty loam, silt, silty clay loam, silt clay, and clay. A further identification scheme has utilized chemical composition, organic content, and history of formation to define types such as gravel, humus, marsh, and peat.

A newer soil classification system has evolved in the United States that can be utilized to classify soils globally, at any location. In this "universal" classification system, soils are considered as individual three-dimensional entities that can be grouped according to similar physical, chemical, and mineralogical properties. The system uses a hierarchical approach, with the amount of information about a soil increasing down the classification ladder. From top to bottom, the hierarchy is structured in the following categories: order, suborder, great groups, subgroups, families, and series. Further details are provided in Table 2.25.

2.4.3 Soil parameters affecting corrosivity

Several important variables have been identified that have an influence on corrosion rates in soil; these include water, degree of aeration, pH, redox potential, resistivity, soluble ionic species (salts), and microbiological activity. The complex nature of selected variables is presented graphically in Fig. 2.24.[42]

Water. Water in liquid form represents the essential electrolyte required for electrochemical corrosion reactions. A distinction is made between saturated and unsaturated water flow in soils. The latter represents movement of water from wet areas toward dry soil areas. The groundwater level is important in this respect. It fluctuates from area to area, with water moving from the water table to higher soil, against the direction of gravity. Saturated water flow is dependent on pore size and distribution, texture, structure, and organic matter.

TABLE 2.24 Particle Sizes in Soil Texture

Category	Diameter (mm)
Sand (very coarse)	1.00–2.00
Sand (coarse)	0.50–1.00
Sand (medium)	0.25–0.50
Sand (fine)	0.10–0.25
Sand (very fine)	0.05–0.10
Silt	0.002–0.05
Clay	<0.002

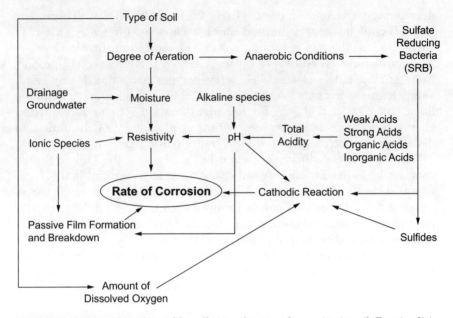

Figure 2.24 Relationship of variables affecting the rate of corrosion in soil. For simplicity, only the MIC effects of sulfate-reducing bacteria are shown.

Water movement in soil can occur by the following mechanisms: gravity, capillary action, osmotic pressure (from dissolved species), and electrostatic interaction with soil particles. The water-holding capacity of a soil is strongly dependent on its texture. Coarse sands retain very little water, while fine clay soils store water to a high degree.

Degree of aeration. The oxygen concentration decreases with increasing depth of soil. In neutral or alkaline soils, the oxygen concentration obviously has an important effect on corrosion rate as a result of its participation in the cathodic reaction. However, in the presence of certain microbes (such as sulfate-reducing bacteria), corrosion rates can be very high, even under anaerobic conditions. Oxygen transport is more rapid in coarse-textured, dry soils than in fine, waterlogged textures.

Excavation can obviously increase the degree of aeration in soil, compared with the undisturbed state. It is generally accepted that corrosion rates in disturbed soil with greater oxygen availability are significantly higher than in undisturbed soil.

pH. Soils usually have a pH range of 5 to 8. In this range, pH is generally not considered to be the dominant variable affecting corrosion rates. More acidic soils obviously represent a serious corrosion risk to common construction materials such as steel, cast iron, and zinc coat-

TABLE 2.25 Soil Classification System using Hierarchical Approach

Category	Basis for classification	Example(s)	Comments
Order	Differences in measurable and visible characteristics of soil horizons	Entisol, Vertisol, Inceptisol, Aridisol, Mollisol, Spodosol, Alfisol, Ultisol, Oxisol, Histosol	Nine orders for mineral soils and one order for all organic soils
Suborder	Differences in development characteristics	Aquod, Udult	Grouping according to accumulation of soluble materials, presence or absence of B horizons, mineralogy, and chemistry
Great group	Presence or absence of certain horizons	Kandihumult	Relative thickness of horizons is important
Subgroup	Typical or dominant concept of the great group	Typic Kandihumult	Coded as either the great group name with the "typic" prefix or a combination of great group names
Family	Differences in textural classes, mineralogy, acidity, and temperature	Clayey oxidic isothermic Typic Kandihumult	Plants generally react in a similar manner to the same soil family
Series	Differences in texture	Paaola	Usually named after the location where the soil was first described

ings. Soil acidity is produced by mineral leaching, decomposition of acidic plants (for example, coniferous tree needles), industrial wastes, acid rain, and certain forms of microbiological activity. Alkaline soils tend to have high sodium, potassium, magnesium, and calcium contents. The latter two elements tend to form calcareous deposits on buried structures, and these have protective properties against corrosion. The pH level can affect the solubility of corrosion products and also the nature of microbiological activity.

Soil resistivity. Resistivity has historically often been used as a broad indicator of soil corrosivity. Since ionic current flow is associated with soil corrosion reactions, high soil resistivity will arguably slow down corrosion reactions. Soil resistivity generally decreases with increasing water content and the concentration of ionic species. Soil resistivity is by no means the only parameter affecting the risk of corrosion damage. A high soil resistivity alone will not guarantee absence of serious corrosion. Variations in soil resistivity along the length of a pipeline are highly undesirable, as this will lead to the formation of macro corrosion cells. Therefore, for structures such as pipelines, the merit of a corrosion risk classification based on an absolute value of soil resistivity is limited.

Soil resistivity can be measured by the so-called Wenner four-pin technique or, more recently, by electromagnetic measurements. The latter allows measurements in a convenient manner and at different soil depths. Another option for soil resistivity measurements is the so-called soil box method, whereby a sample is taken during excavation. Preferably sampling will be in the immediate vicinity of a buried structure (a pipe trench, for example).

Redox potential. The redox potential is essentially a measure of the degree of aeration in a soil. A high redox potential indicates a high oxygen level. Low redox values may provide an indication that conditions are conducive to anaerobic microbiological activity. Sampling of soil will obviously lead to oxygen exposure, and unstable redox potentials are thus likely to be measured in disturbed soil.

Chlorides. Chloride ions are generally harmful, as they participate directly in anodic dissolution reactions of metals. Furthermore, their presence tends to decrease the soil resistivity. They may be found naturally in soils as a result of brackish groundwater and historical geological seabeds (some waters encountered in drilling mine shafts have chloride ion levels comparable to those of seawater) or come from external sources such as deicing salts applied to roadways. The chloride ion concentration in the corrosive aqueous soil electrolyte will vary as soil conditions alternate between wet and dry.

Sulfates. Compared to the corrosive effect of chloride ions, sulfates are generally considered to be more benign in their corrosive action toward metallic materials. However, concrete may be attacked as a result of high sulfate levels. The presence of sulfates does pose a major risk for metallic materials in the sense that sulfates can be converted to highly corrosive sulfides by anaerobic sulfate-reducing bacteria.

Microbiologically influenced corrosion. Microbiologically influenced corrosion (MIC) refers to corrosion that is influenced by the presence and activities of microorganisms and/or their metabolites (the products produced through their metabolism). Bacteria, fungi, and other microorganisms can play a major part in soil corrosion. Spectacularly rapid corrosion failures have been observed in soil as a result of microbial action, and it is becoming increasingly apparent that most metallic alloys are susceptible to some form of MIC. The mechanisms potentially involved in MIC have been summarized as follows:[43]

- Cathodic depolarization, whereby the cathodic rate-limiting step is accelerated by microbiological action.
- Formation of occluded surface cells, whereby microorganisms form "patchy" surface colonies. Sticky polymers attract and aggregate biological and nonbiological species to produce crevices and concentration cells, the basis for accelerated attack.
- Fixing of anodic reaction sites, whereby microbiological surface colonies lead to the formation of corrosion pits, driven by microbial activity and associated with the location of these colonies.
- Underdeposit acid attack, whereby corrosive attack is accelerated by acidic final products of the MIC "community metabolism," principally short-chain fatty acids.

Certain microorganisms thrive under aerobic conditions, whereas others thrive in anaerobic conditions. Anaerobic conditions may be created in the microenvironmental regime even if the bulk conditions are aerobic. The pH conditions and availability of nutrients also play a role in determining what types of microorganisms can thrive in a soil environment. In general, microbial activity is highest in the surface O and A horizons, because of the availability of both organic carbon nutrients and oxygen. Microorganisms associated with corrosion damage in soils include the following:

- Anaerobic bacteria, which produce highly corrosive species as part of their metabolism.
- Aerobic bacteria, which produce corrosive mineral acids.

- Fungi, which may produce corrosive by-products in their metabolism, such as organic acids. Apart from metals and alloys, they can degrade organic coatings and wood.
- Slime formers, which may produce concentration corrosion cells on surfaces.

A summary of the characteristics of bacteria commonly associated with soil corrosion (mostly for iron-based alloys) is provided in Table 2.26.

2.4.4 Soil corrosivity classifications

For design and corrosion risk assessment purposes, it is desirable to estimate the corrosivity of soils, without conducting exhaustive corrosion testing. Corrosion testing in soils is complicated by the fact that long exposure periods may be required (buried structures are usually expected to last for several decades) and that many different soil conditions can be encountered. Considering the complexity of the parameters affecting soil corrosion, it is obvious that the use of relatively simple soil corrosivity models is bound to be inaccurate. These limitations should be considered when applying any of the common aids/methodologies.

One of the simplest classifications is based on a single parameter, soil resistivity. Table 2.27 shows the generally adopted corrosion severity ratings. Sandy soils are high on the resistivity scale and therefore are considered to be the least corrosive. Clay soils, especially those contaminated with saline water, are on the opposite end of the spectrum. The soil resistivity parameter is very widely used in practice and is generally considered to be the dominant variable in the absence of microbial activity.

The American Water Works Association (AWWA) has developed a numerical soil corrosivity scale that is applicable to cast iron alloys. A severity ranking is generated by assigning points for different variables, presented in Table 2.28.[44] When the total points of a soil in the AWWA scale are 10 (or higher), corrosion protective measures (such as cathodic protection) have been recommended for cast iron alloys. It should be appreciated that this rating scale remains a relatively simplistic, subjective procedure for specific alloys. Therefore, it should be viewed as a broad indicator and should not be expected to accurately predict specific cases of corrosion damage.

A worksheet for estimating the probability of corrosion damage to metallic structures in soils has been published, based on European work in this field. The worksheet consists of 12 individual ratings (R1 to R12), listed in Table 2.29.[45] This methodology is very detailed and comprehensive. For example, the effects of vertical and horizontal soil homogeneity are included, as outlined in Table 2.30. Even details such as the presence of coal or coke and other pollutants in the soil are con-

TABLE 2.26 Characteristics of Bacteria Commonly Associated with Corrosion in Soils

Species	Likely soil conditions	Metabolic action	Species produced	Comments
Sulfate-reducing bacteria (SRB)	Anaerobic, close to neutral pH values, presence of sulfate ions. Often associated with waterlogged clay soils	Convert sulfate to sulfide	Iron sulfide, hydrogen sulfide	Very well known for corrosion of iron and steel. *Desulfovibrio* genus very widespread
Iron-oxidizing bacteria (IOB)	Acidic, aerobic	Oxidize ferrous ions to ferric ions	Sulfuric acid, iron sulfate	*Thiobacillus ferrooxidans* is a well-known example
Sulfur-oxidizing bacteria (SOB)	Aerobic, acidic	Oxidize sulfur and sulfide to form sulfuric acid	Sulfuric acid	*Thiobacillus* genus is a common example
Iron bacteria (IB)	Aerobic, close to neutral pH values	Oxidize ferrous ions to ferric ions	Magnetite	*Gallionella* genus is an example. Usually associated with deposit and tubercle formation

TABLE 2.27 Corrosivity Ratings Based on Soil Resistivity

Soil resistivity, Ω·cm	Corrosivity rating
> 20,000	Essentially noncorrosive
10,000–20,000	Mildly corrosive
5000–10,000	Moderately corrosive
3000–5000	Corrosive
1000–3000	Highly corrosive
< 1000	Extremely corrosive

TABLE 2.28 Point System for Predicting Soil Corrosivity According to the AWWA C-105 Standard

Soil parameter	Assigned points
Resistivity, Ω·cm	
< 700	10
700–1000	8
1000–1200	5
1200–1500	2
1500–2000	1
> 2000	0
pH	
0–2	5
2–4	3
4–6.5	0
6.5–7.5	0
7.5–8.5	0
> 8.5	3
Redox potential, mV	
> 100	0
50–100	3.5
0–50	4
< 0	5
Sulfides	
Positive	3.5
Trace	2
Negative	0
Moisture	
Poor drainage, continuously wet	2
Fair drainage, generally moist	1
Good drainage, generally dry	0

sidered. The assessment is directed at ferrous materials (steels, cast irons, and high-alloy stainless steels), hot-dipped galvanized steel, and copper and copper alloys. Summation of the individual ratings produces an overall corrosivity classification into one of the four categories listed in Table 2.31. It has been pointed out that sea or lake beds cannot be assessed using this worksheet.

TABLE 2.29 Variables Considered in Worksheet of Soil Corrosivity

Rating number	Parameter
R1	Soil type
R2	Resistivity
R3	Water content
R4	pH
R5	Buffering capacity
R6	Sulfides
R7	Neutral salts
R8	Sulfates
R9	Groundwater
R10	Horizontal homogeneity
R11	Vertical homogeneity
R12	Electrode potential

TABLE 2.30 R10 and R12 Worksheet Ratings

Resistivity variation between adjacent domains (all positive R2 values are treated as equal)		Rating
R10, Horizontal Soil Homogeneity		
R2 difference <2		0
R2 difference ≥2 and ≤3		−2
R2 difference >3		−4
R11, Vertical Soil Homogeneity		
Adjacent soils with same resistivity	Embedded in soils with same structure or in sand	0
	Embedded in soils with different structure or containing foreign matter	−6
Adjacent soils with different resistivity	R2 difference ≥2 and ≤3	−1
	R2 difference >3	−6

TABLE 2.31 Overall Soil Corrosivity Classification

Summation of R1 to R12 ratings	Soil classification
≥0	Virtually noncorrosive
−1 to −4	Slightly corrosive
−5 to −10	Corrosive
≤10	Highly corrosive

2.4.5 Corrosion characteristics of selected metals and alloys

Ferrous alloys. Steels are widely used in soil, but almost never without additional corrosion protection. It may come as something of a surprise that unprotected steel is very vulnerable to localized corrosion damage (pitting) when buried in soil. Such attack is usually the result of differential aeration cells, contact with different types of soil, MIC, or galvanic cells when coal or cinder particles come into contact with buried steel. Stray current flow in soils can also lead to severe pitting attack. A low degree of soil aeration will not necessarily guarantee low corrosion rates for steel, as certain microorganisms associated with severe MIC damage thrive under anaerobic conditions.

The primary form of corrosion protection for steel buried in soil is the application of coatings. When such coatings represent a physical barrier to the environment, cathodic protection in the form of sacrificial anodes or impressed current systems is usually applied as an additional precaution. This additional measure is required because coating defects and discontinuities will inevitably be present in protective coatings.

Cast iron alloys have been widely used in soil; many gas and water distribution pipes in cities are still in use after decades of service. These have been gradually replaced with steel (coated and cathodically protected) and also with polymeric pipes. While cast irons are generally considered to be more resistant to soil corrosion than steel, they are subject to corrosion damage similar to that described above for steel. Coatings and cathodic protection with sacrificial anodes tend to be used to protect buried cast iron structures.

Stainless steels are rarely used in soil applications, as their corrosion performance in soil is generally poor. Localized corrosion attack is a particularly serious concern. The presence of halide ions and concentration cells developed on the surface of these alloys tends to induce localized corrosion damage. Since pitting tends to be initiated at relatively high corrosion potential values, higher redox potentials increase the localized corrosion risk. Common grades of stainless steel (even the very highly alloyed versions) are certainly not immune to MIC, such as attack induced by sulfate-reducing bacteria.

Nonferrous metals and alloys. In general, copper is considered to have good resistance to corrosion in soils. Corrosion concerns are mainly related to highly acidic soils and the presence of carbonaceous contaminants such as cinder. Chlorides and sulfides also increase the risk of corrosion damage. Contrary to common belief, copper and its alloys are not immune to MIC. Cathodic depolarization, selective leaching, underdeposit corrosion, and differential aeration cells have been cited as MIC mechanisms for copper alloys.[46] Corrosive products produced by microbes include carbon dioxide, hydrogen sulfide and other sulfur compounds, ammonia, and acids (organic and inorganic).

In the case of brasses, consideration must be given to the risk of dezincification, especially at high zinc levels. Soils contaminated with detergent solutions and ammonia also pose a higher corrosion risk for copper and copper alloys. Additional corrosion protection for copper and copper alloys is usually considered only in highly corrosive soil conditions. Cathodic protection, the use of acid-neutralizing backfill (for example, limestone), and protective coatings can be utilized.

The main application of zinc in buried applications is in galvanized steel. Performance is usually satisfactory unless soils are poorly aerated, acidic, or highly contaminated with chlorides, sulfides, and other solutes. Well-drained soils with a coarse texture (the sandy type) provide a high degree of aeration. It should also be borne in mind that zinc corrodes rapidly under highly alkaline conditions. Such conditions can arise on the surface of cathodically overprotected structures. The degree of corrosion protection afforded by galvanizing obviously increases with the thickness of the galvanized coating. Additional protection can be afforded by so-called duplex systems, in which additional paint coatings are applied to galvanized steel.

The corrosion resistance of lead and lead alloys in soils is generally regarded as being in between those of steel and copper. The corrosion resistance of buried lead sheathing for power and communication cables has usually been satisfactory. Caution needs to be exercised in soils containing nitrates and organic acids (such as acetic acid). Excessive corrosion is also found under highly alkaline soil conditions. Silicates, carbonates, and sulfates tend to retard corrosion reactions by their passivating effects on lead. Barrier coatings can be used as additional protection. When cathodic protection is applied, overprotection should be avoided because of the formation of surface alkalinity.

Aluminum alloys are used relatively rarely in buried applications, although some pipelines and underground tanks have been constructed from these alloys. Like stainless steels, these alloys tend to undergo localized corrosion damage in chloride-contaminated soils. Protection by coatings is essential to prevent localized corrosion damage. Cathodic protection criteria for aluminum alloys to minimize the risk of generating undesirable alkalinity are available. Aluminum alloys can undergo accelerated attack under the influence of microbiological effects. Documented mechanisms include attack by organic acid produced by bacteria and fungi and the formation of differential aeration cells.[46] It is difficult to predict the corrosion performance of aluminum and its alloys in soils with any degree of confidence.

Reinforced concrete. Steel-reinforced concrete (SRC) pipes are widely used in buried applications to transport water and sewage, and their use dates back nearly a century. So-called prestressed concrete cylinder pipes (PCCP) were already developed prior to 1940 for designs requiring relatively high operating pressures and large diameters.

PCCP applications include water transmission mains, distribution feeder mains, water intake and discharge lines, low-head penstocks, industrial pressure lines, sewer force mains, gravity sewer lines, subaqueous lines, and spillway conduits.[47]

There are three dominant species in soils that lead to excessive degradation of reinforced concrete piping. Sulfate ions tend to attack the tricalcium aluminate phase in concrete, leading to severe degradation of the concrete/mortar cover and exposure of the reinforcing steel. The mechanism of degradation involves the formation of a voluminous reaction product in the mortar, which leads to internal pressure buildup and subsequent disintegration of the cover. Sulfate levels exceeding about 2 percent (by weight) in soils and groundwater reportedly put concrete pipes at risk. Chloride ions are also harmful, as they tend to diffuse into the concrete and lead to corrosion damage to the reinforcing steel. A common source of chloride ions is soil contamination by deicing salts. This corrosion phenomenon is discussed in detail in Sec. 2.5, Reinforced Concrete. Finally, acidic soils present a corrosion hazard. The protective alkaline environment that passivates the reinforcing steel can be disrupted over time. Carbonic acid and humic acid are examples of acidic soil species.

2.4.6 Summary

Corrosion processes in soil are highly complex phenomena, especially since microbiologically influenced corrosion can play a major role. Soil parameters tend to vary in three dimensions, which has important ramifications for corrosion damage. Such variations tend to set up macrocells, leading to accelerated corrosion at the anodic site(s). The corrosion behavior of metals and alloys in other environments should not be extrapolated to their performance in soil. In general, soils represent highly corrosive environments, often necessitating the use of additional corrosion protection measures for common engineering metals and alloys.

2.5 Reinforced Concrete

2.5.1 Introduction

Concrete is the most widely produced material on earth. The use of cement, a key ingredient of concrete, by Egyptians dates back more than 3500 years. In the construction of the pyramids, an early form of mortar was used as a structural binding agent. The Roman Coliseum is a further example of a historic landmark utilizing cement mortar as a construction material. Worldwide consumption of concrete is close to 9 billion tons and is expected to rise even further.

Contrary to common belief, concrete itself is a complex composite material. It has low strength when loaded in tension, and hence it is common practice to reinforce concrete with steel, for improved tensile mechanical properties. Concrete structures such as bridges, buildings, elevated highways, tunnels, parking garages, offshore oil platforms, piers, and dam walls all contain reinforcing steel (rebar). The principal cause of degradation of steel-reinforced structures is corrosion damage to the rebar embedded in the concrete. The scale of this problem has reached alarming proportions in various parts of the world. In the early 1990s, the costs of rebar corrosion in the United States alone were estimated at $150 to $200 billion per year.[48]

The durability of concrete should not simply be equated to high-strength grades of concrete. There are several methods for controlling rebar corrosion in new structures, and valuable lessons can be learned from previous failures. In existing structures, the choices for correcting rebar corrosion problems are relatively limited. The corrosion mechanisms involved in the repair of existing structures may be fundamentally different from those that affect new constructions. A gamut of inspection methods is available for assessment of the condition of reinforced concrete structures.

2.5.2 Concrete as a structural material

In order to understand corrosion damage in concrete, a basic understanding of the nature of concrete as an engineering material is required. A brief summary follows for this purpose. It is important to distinguish clearly among terms such as *cement, mortar,* and *concrete*. Unfortunately, these tend to be used interchangeably in household use.

The fundamental ingredients required to make concrete are cement clinker, water, fine aggregate, coarse aggregate, and certain special additives. Cement clinker is essentially a mixture of several anhydrous oxides. For example, standard Portland cement consists mainly of the following compounds, in order of decreasing weight percent: $3CaO \cdot SiO_2$, $2CaO \cdot SiO_2$, $3CaO \cdot Al_2O_3$, and $4CaO \cdot Al_2O_3 \cdot Fe_2O_3$. The cement reacts with water to form the so-called cement paste. It is the cement paste that surrounds the coarse and fine aggregate particles and holds the material together. The importance of adequately mixing the concrete constituents should thus be readily apparent. The fine and coarse aggregates are essentially inert constituents. In general, the size of suitable aggregate is reduced as the thickness of the section of a structure decreases.

The reaction of the cement and water to form the cement paste is actually a series of complex hydration reactions, producing a multi-phase cement paste. One example of a specific hydration reaction is the following:

$$2(3\text{CaO} \cdot \text{SiO}_2) + 6\text{H}_2\text{O} \rightarrow 3\text{Ca(OH)}_2 + 3\text{CaO} \cdot 2\text{SiO}_2 \cdot 3\text{H}_2\text{O} \quad (2.28)$$

Following the addition of water, the cement paste develops a fibrous microstructure over time. Importantly for corrosion considerations, the cement paste is not a continuous solid material on a microscopic scale. Rather, the cement paste is classified as a "gel" to describe its limited crystalline character and the water-filled spaces between the solid phases. These microscopic spaces are also known as gel "pores" and, strictly speaking, are filled with an ionic solution rather than "water." Additional pores of larger size are found in the cement paste and between the cement paste and the aggregate particles. The pores that result from excess water in the concrete mix are known as capillary pores. Air voids are also invariably present in concrete. In so-called air-entrained concrete, microscopic air voids are intentionally created through admixtures. This practice is widely used in cold climates to minimize freeze-thaw damage. Clearly then, concrete is a porous material, and it is this porosity that allows the ingress of corrosive species to the embedded reinforcing steel.

A further important feature of the hydration reactions of cement with water is that the resulting pore solution in concrete is highly alkaline [refer to Eq. (2.28) above]. In addition to calcium hydroxide, sodium and potassium hydroxide species are also formed, resulting in a pH of the aqueous phase in concrete that is typically between 12.5 and 13.6. Under such alkaline conditions, reinforcing steel tends to display completely passive behavior, as fundamentally predicted by the Pourbaix diagram for iron. In the absence of corrosive species penetrating into the concrete, ordinary carbon steel reinforcing thus displays excellent corrosion resistance.

From the above discussion, the complex nature of concrete as a particulate-strengthened ceramic-matrix composite material and the difference between the terms *concrete* and *cement* should be apparent. The term *mortar* refers to a concrete mix without the addition of any coarse aggregate.

2.5.3 Corrosion damage in reinforced concrete

Mehta's holistic model of concrete degradation. The large-scale environmental degradation of the reinforced concrete infrastructure in many countries (often prematurely) has indicated that traditional approaches to concrete durability may be in need of revision. Historically, the general approach has been to relate concrete durability directly to the strength of concrete. It is well known that higher water-to-cement ratios in concrete lead to lower strength and increase the degree of porosity in the concrete. A generally accepted argument is that low-strength, more permeable concrete is less durable. However, in real reinforced concrete structures, durability issues are more complex, and consideration of the strength variable alone is inadequate.

The approach adopted by Mehta in his holistic model of concrete degradation was to focus on the soundness of concrete under service conditions as a fundamental measure of concrete durability rather than on the strength of concrete. In simplistic terms, soundness of concrete implies freedom from cracking.[49] Mehta's proposed model of concrete degradation has been adapted in the illustration of environmental damage in Fig. 2.25. According to this model, concrete manufactured to high quality standards is initially considered to be an impermeable structure. This condition exists so long as interior pores and microcracks do not form interconnected paths extending to the exterior surfaces.

Under environmental weathering and loading effects, the permeability of the concrete gradually increases as the network of "defects" becomes more interconnected over time. It is then that water, carbon dioxide, and corrosive ions such as chlorides can enter the concrete and produce detrimental effects at the level of the reinforcing steel. The corrosion mechanisms involved are discussed in more detail in subsequent sections. The buildup of corrosion products leads to a buildup of internal pressure in the reinforced concrete because of the voluminous nature of these products. The volume of oxides and hydroxides associated with rebar corrosion damage relative to steel is shown in Fig. 2.26. In turn, these internal stresses lead to severe cracking and spalling of the concrete covering the reinforcing steel. Extensive surface damage produced in this manner is shown in Figs. 2.27 and 2.28. It is clear that the damage inflicted by formation of corrosion products (and other effects) reduces the soundness of concrete and facilitates further deterioration at an increasing rate.

In the light of the importance that Mehta's model of environmental concrete degradation attaches to defects such as cracks, the reliance on the high strength of concrete alone for satisfactory service life becomes questionable. High strength levels in concrete alone certainly do not guarantee a high degree of soundness; several arguments can be made for high-strength concrete being potentially more prone to cracking.

The importance of concrete cracks in rebar corrosion has also been highlighted by Nürnberger.[50] Both carbonation and chloride ion diffusion, two important processes associated with rebar corrosion, can proceed more rapidly into the concrete along the crack faces, compared with uncracked concrete. Nürnberger argued that corrosion in the vicinity of the crack tip could be accelerated further by crevice corrosion effects and galvanic cell formation. The steel in the crack will tend to be anodic relative to the cathodic (passive) zones in uncracked

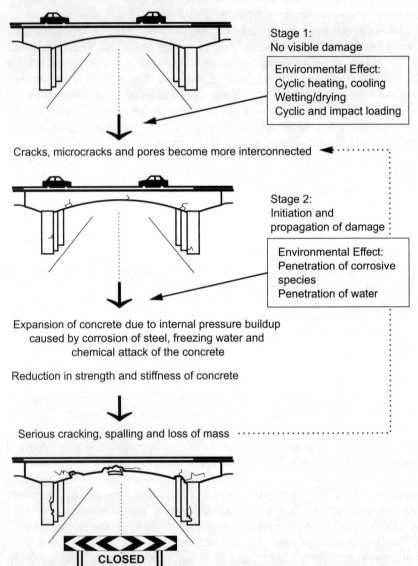

Figure 2.25 Concrete degradation processes resulting from environmental effects.

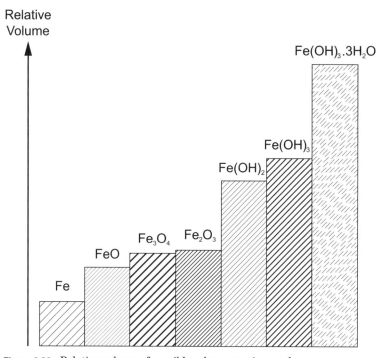

Figure 2.26 Relative volume of possible rebar corrosion products.

concrete. The particularly harmful effects of dried-out cracks (as opposed to those that are water-filled), which allow rapid ingress of corrosive species, were also emphasized. Even casual visual examinations of most reinforced concrete structures invariably reveal the presence of macroscopic cracks in concrete.

Corrosion mechanisms. The two most common mechanisms of reinforcing steel corrosion damage in concrete are (1) localized breakdown of the passive film by chloride ions and (2) carbonation, a decrease in pore solution pH, leading to a general breakdown in passivity. Harmful chloride ions usually originate from deicing salts applied in cold climate regions or from marine environments/atmospheres. Carbonation damage is predominantly induced by a reaction of concrete with carbon dioxide (CO_2) in the atmosphere.

Chloride-induced rebar corrosion. Corrosion damage to reinforcing steel is an electrochemical process with anodic and cathodic half-cell reactions. In the absence of chloride ions, the anodic dissolution reaction of iron,

Figure 2.27 Concrete degradation caused by rebar corrosion damage in a highway structure in downtown Toronto, Ontario. Extensive repair work was underway on this structure at the time the picture was taken. The annual maintenance costs for this structure were recently reported at around $18 million.

Figure 2.28 Concrete degradation caused by rebar corrosion damage near Kingston, Ontario. This bridge underwent extensive rehabilitation shortly after this picture was taken.

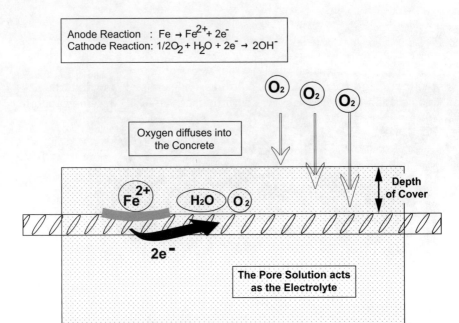

Figure 2.29 Schematic illustration of electrochemical corrosion reactions in concrete.

$$Fe \rightarrow Fe^{2+} + 2e^- \quad (2.29)$$

is balanced by the cathodic oxygen reduction reaction,

$$\tfrac{1}{2}O_2 + H_2O + 2e^- \rightarrow 2OH^- \quad (2.30)$$

Oxygen diffuses to the reinforcing steel surface through the porous concrete, with cracks acting as fast diffusion paths, especially if they are not filled with water. The Fe^{2+} ions produced at the anodes combine with the OH^- ions from the cathodic reaction to ultimately produce a stable passive film. This electrochemical process is illustrated schematically in Fig. 2.29.

Chloride ions in the pore solution, having the same charge as OH^- ions, compete with these anions to combine with the Fe^{2+} cations. The resulting iron chloride complexes are thought to be soluble (unstable); therefore, further metal dissolution is not prevented, and ultimately the buildup of voluminous corrosion products takes place. Chloride ions also tend to be released from the unstable iron chloride complexes, making these harmful ions available for further reaction with the reinforcing steel. As the iron ultimately precipitates out in the form of iron oxide or hydroxide corrosion products, it can be argued that the consumption of hydroxide ions leads to localized pH reduction and therefore enhanced metal dissolution.

Chloride-induced rebar corrosion tends to be a localized corrosion process, with the original passive surface being destroyed locally under the influence of chloride ions. Apart from the internal stresses created by the formation of corrosion products leading to cracking and spalling of the concrete cover, chloride attack ultimately reduces the cross section and significantly compromises the load-carrying capability of steel-reinforced concrete.

Sources of chloride ions and diffusion into concrete. The harmful chloride ions leading to rebar corrosion damage either originate directly from the concrete mix constituents or diffuse into the concrete from the surrounding environment. The use of seawater or aggregate that has been exposed to saline water (such as beach sand) in concrete mixes creates the former case. Calcium chloride has been deliberately added to certain concrete mixes to accelerate hardening at low temperatures, mainly before the harmful corrosion effects were widely known.

An important source of chlorides from the external environment is the widespread use of deicing salts on road surfaces in cold climates. Around 10 million tons of deicing salt is used annually in the United States; the Canadian figure is about 3 million tons. The actual tonnage used each year fluctuates with the severity of the particular winter season. The main purpose of deicing salt application is to keep roadways safe and passable in winter and to minimize the disruption of economic activity. The application of salt to ice and snow results in the formation of brine, which has a lower freezing point.

Salt, primarily in the form of rock salt, is the most widely used deicing agent in North America because of its low cost, general availability, and ease of storage and handling. Rock salt is also known as halite and has the well-known chemical formula NaCl. The rate of salt application to roads varies with traffic and weather conditions. Other chloride compounds in use for deicing purposes are calcium chloride ($CaCl_2$) and magnesium chloride ($MgCl_2$).

Other obvious important sources of corrosive chloride ions are seawater and marine atmospheres. Alternate drying and wetting cycles promote the buildup of chloride ions on surfaces. Hence actual surface concentrations of chlorides can be well in excess of those of the bulk environment.

Clearly the diffusion rate of external chlorides into concrete to the reinforcing steel is very important. While some simplified models such as Fick's second law of diffusion have been used for life prediction purposes in combination with so-called critical chloride levels, the actual processes are much more complex than such simplistic models. Considering the complex nature of concrete as a material on the microstructural scale, this complexity must be anticipated. Chloride diffusion processes are affected by capillary suction and chemical and physical interaction in the concrete. Weather/climatic conditions, the pore structure in concrete, and other microstructural parameters are important variables. If only the capillary suction mechanism is considered, the rate of chloride ingress from exposure to a saline solution will be higher in dry concrete than in water-saturated concrete. Furthermore, the surface concentration of chlorides is obviously time-dependent, particularly in deicing salt applications, adding more complications to diffusion models. The effects of cracks on both the macroscopic and microscopic levels are also important practical considerations, since they function as rapid chloride diffusion paths.

Chlorides in concrete and critical chloride levels. Chlorides in concrete exist in two basic forms, so-called free chlorides and bound chlorides. The former are mobile chlorides dissolved in the pore solution, whereas the latter type represents relatively immobile chloride ions that interact (by chemical binding and/or adsorption) with the cement paste. At first glance, it may appear that only the free chlorides should be considered for corrosion reactions. However, Glass and Buenfeld have recently reviewed the role of both bound and free chlorides in corrosion processes in detail and have concluded that both types may be important.[51] Bound chloride may essentially buffer the chloride ion activity at a high value, and localized acidification at anodic sites may release some bound chloride.

The determination of a critical chloride level, below which serious rebar corrosion damage does not occur, for design, maintenance planning, and life prediction purposes is appealing. Not surprisingly, then, several studies have been directed at defining such a parameter. Unfortunately, the concept of a critical chloride content as a universal parameter is unrealistic. Rather, a critical chloride level should be defined only in combination with a host of other parameters. After all, a threshold chloride level for corrosion damage will be influenced by variables such as

- The pore solution pH
- Moisture content of the concrete
- Temperature
- Age and curing conditions of the concrete
- Water-to-cement ratio
- Pore structure and other "defects"
- Oxygen availability (hence cover and density of concrete)
- Presence of prestressing
- Cement and concrete composition

Considering the above, it is apparent that the specification of critical chloride levels should be treated with extreme caution. Furthermore, it should not be surprising that an analysis of 15 chloride levels reported for the initiation of corrosion of steel produced a range of 0.17 to 2.5 percent, expressed as total chlorides per weight of cement.[51]

Carbonation-induced corrosion. Carbon dioxide present in the atmosphere can reduce the pore solution pH significantly by reacting with calcium hydroxide (and other hydroxides) to produce insoluble carbonate in the concrete as follows:

$$Ca(OH)_2 + CO_2 \rightarrow CaCO_3 + H_2O \tag{2.31}$$

Carbonation is manifested as a reduction in the pH of the pore solution in the outer layers of the concrete and often appears as a well-defined "front" parallel to the external surface. This front can conveniently be made visible by applying a phenolphthalein indicator solution to freshly exposed concrete surfaces. Behind the front, where all the calcium hydroxide has been depleted, the pH is around 8, whereas ahead of the front, the pH remains in excess of 12.5.[52] The passivating ability of the pore solution diminishes with the decrease in pH. Carbonation-induced corrosion tends to proceed in a more uniform manner over the rebar surface than chloride-induced corrosion damage.

The rate of ingress of carbonation damage in concrete decreases with time. Obviously carbon dioxide has to penetrate greater distances into the concrete over time. The precipitation of calcium carbonate and possibly additional cement hydration are also thought to contribute to the reduced rate of ingress.[52]

Several variables affect the rate of carbonation. In general, low-permeability concrete is more resistant. Carbonation tends to proceed most rapidly at relative humidity levels between 50 and 75 percent. At lower humidity levels, carbon dioxide can penetrate into the concrete relatively rapidly, but little calcium hydroxide is available in the dissolved state for reaction with it. At higher humidity levels, the water-filled pore structure is a more effective barrier to the ingress of carbon dioxide. Clearly, environmental cycles of alternate dry and wet conditions will be associated with rapid carbonation damage.

In many practical situations, carbonation- and chloride-induced corrosion can occur in tandem. Research studies have shown that corrosion caused by carbonation was intensified with increasing chloride ion concentration, provided that the carbonation rate itself was not retarded by the presence of chlorides.[52] According to these studies, chloride attack and carbonation can act synergistically (the combined damage being more severe than the sum of its parts) and have been responsible for major corrosion problems in hot coastal areas.

2.5.4 Remedial measures

In principle, a number of fundamental technical measures can be taken to address the problem of reinforcing steel corrosion, such as

- Repairing the damaged concrete
- Modifying the external environment
- Modifying the internal concrete environment
- Creating a barrier between the concrete and the external environment
- Creating a barrier between the rebar steel and the internal concrete environment
- Applying cathodic protection to the rebar
- Using alternative, more corrosion-resistant rebar materials
- Using alternative methods of reinforcement

Alternative solutions to periodic repair of damaged concrete are being sought. After all, this is generally a costly corrective maintenance approach after serious damage has already set in. In view of the overwhelming magnitude of the problem and increasingly limited government budgets, various alternative approaches have come to the forefront over the last two decades. Several of these are still in emerging stages with limited track records. Given that rebar corrosion problems are typically manifested only over many decades, it takes significant time for new technologies to acquire credibility in industrial practice.

An important distinction has to be made in the applicability of remedial measures to new and existing structures. Unfortunately, the options for the most pressing problems in aging existing structures are fairly limited. Obviously even the "best" technologies for new construction are of limited value if education and technology transfer efforts directed at designers and users are not effective. This aspect is particularly challenging in the fragmented construction industry.[52] A further important prerequisite for advancing the cause of effective corrosion control in reinforced concrete structures is acceptance and implementation of life-cycle costing, as opposed to awarding contracts on the basis of the lowest initial capital cost outlay.

Alternative deicing methods. Since chloride-based deicing agents are a major factor in rebar corrosion, one obvious consideration is the possible use of alternative noncorrosive deicing chemicals. Such chemicals are indeed available and are used in selective applications, such as for airport runway deicing and on certain bridges. In addition to the corrosive action on reinforcing steel, the details of the deicing mechanism (temperature ranges, texture of products, etc.) and possible damage to the concrete itself obviously need to be considered for alternative chemicals. Strictly speaking, a distinction is also made between anti-icing and deicing, depending on whether chemical application is done before or after snow and ice accumulation. An excellent summary of highway deicing practices has been published by the Ministry of Transportation, Ontario.[53]

The potential use of calcium magnesium acetate (CMA) has been extensively researched in North America, and field trials have been conducted in several states and provinces. The CMA specification in terms of composition, particle size and shape, color, and density has evolved over time. CMA application rates have generally been higher than those for salt. The majority of trials conducted have indicated effectiveness similar to that of salt at temperatures down to $-5°C$, but slower performance than salt at lower temperatures. Unfortunately, costs are reportedly more than 10 times higher than those of road salt on a mass basis. If a higher application rate of 1.5 times that of salt is assumed, a cost factor increase of 45 has been reported.[53] Cost issues surrounding the use of CMA are complex and include factors such as potential environmental benefits, reduced automobile corrosion, mass production technology, and alternative raw materials.

The use of formate compounds as highway deicers was explored as early as 1965. Lower reaction rates of sodium formate with snow and ice have been reported in Canadian field trials. In the Canadian studies, commercial grades of sodium formate were found to be "contaminated" with chlorides.[53] Concerns related to automobile corrosion and increased costs have been expressed, and little information is available concerning possible adverse effects on the environment.

Urea is widely used as an airport runway deicer, as it is not corrosive to aircraft materials. However, urea is generally not considered to be a viable alternative deicing chemical for highway applications. Reported limitations include higher application rates, longer reaction times, effectiveness only at temperatures above $-10°C$, relatively high cost, and significant adverse effects on the environment.[53]

Verglimit, a patented compound, is often mentioned in the context of alternative deicing compounds. In this product, capsules that contain calcium chloride are incorporated into asphalt paving. With gradual wear and tear of the asphalt surface, the capsules are exposed and broken open, releasing the deicing chemical. This methodology was specifically designed for exposed bridge decks that freeze over more rapidly than adjacent road surfaces. Many North American readers will be familiar with the traffic warning signs, "Caution: Bridge Freezes First."

Abrasives are widely used in Europe and North America to improve skid resistance, and using them exclusively as a means of eliminating deicing salts has been considered. While mixtures of sand and road salt are widely used, elimination of deicing chemicals has not proved feasible in geographic areas such as Ontario. A major problem is that abrasives alone do not assist snowplows in removing ice bonded to pavements. Other problems include the blocking of storm sewers and accumulation in catch basins. Importantly, without mixed-in deicing chemicals, stockpiles of abrasives would tend to freeze in winter, with resultant reduced workability and difficulty in spreading. Such stockpiles invariably contain moisture, causing abrasive particles to freeze together in winter.

The concept of embedding electrical heating elements in concrete to keep road surfaces ice-free has received some attention. Considering the fact that electric power is routinely fed to street lighting, the potential merits of such systems can be appreciated. A Canadian experimental concept of electrically conducting concrete also appears to hold promise for heating purposes. Other innovative experimental approaches that have been explored include noncontact deicing with acoustic or microwave energy.

Alternative deicing methods are largely applicable to new structures; arguably, they may also benefit existing structures, provided that no serious corrosion damage or chloride ion ingress has taken place.

Cathodic protection. Cathodic protection (CP) is one of the few techniques that can be applied to control corrosion on existing structures. Cathodic protection of conventional rebar is well established, with applications dating back well over 20 years. The subject of the applicability of CP to prestressed concrete (pre- and posttensioned systems) is much more controversial, with the main concern being hydrogen embrittlement of the high-strength prestressing steel. To the author's knowledge, CP for prestressed concrete has not progressed beyond initial laboratory tests. The difficult issues surrounding CP and prestressed concrete have been reviewed by Hartt.[54]

The principles and theory of cathodic protection are the subject of Chap. 11. Essentially the concept involves polarizing the rebar to a cathodic potential, where anodic dissolution of the rebar is minimized. A direct current source (rectifier) is usually employed to establish the rebar as the cathode of an electrochemical cell, and a separate anode is required to complete the electric circuit. Three basic methods are available for controlling the output of a rectifier:

- In *constant-current mode,* the rectifier maintains a constant current output. The output voltage will vary with changes in the circuit resistance. The potential of the reinforcing steel can be measured with a reference cell as a function of the applied current, to ensure that certain protection criteria are met.

- In *constant-voltage mode,* a constant output voltage is maintained by the rectifier. The applied current will change with variations in circuit resistance. Low concrete resistance, often associated with increased risk for corrosion damage, will result in increased current output. It should be noted that in this mode, the rebar potential is not necessarily constant. It can again be monitored with a reference cell.

- In *constant rebar potential mode,* the current output is adjusted continuously to provide a constant (preselected) rebar potential. The rebar potential, measured continuously with reference electrodes, is fed back to the rectifier unit. Successful operation in this mode depends on minimizing the *IR* drop error in the rebar potential measurements and on the accuracy and stability of the reference electrodes over time.

An important issue in CP of reinforcing steel is how much current should be impressed between the reinforcing steel and the anode. Too little current will result in inadequate corrosion protection of the rebar, while excessive current can result in problems such as hydrogen embrittlement and concrete degradation. Furthermore, a uniform current distribution is obviously desirable.

Unfortunately, the current requirement cannot be measured directly, and various indirect criteria have been proposed (see Table 2.32). The CP current requirements are often expressed in terms of the potential of the reinforcing steel (or a shift in the potential when the CP system is activated or deactivated) relative to a reference electrode. The reference electrodes can be located externally, in contact with the outside concrete surface, or be embedded in the concrete with the rebar. It is important that potential readings should be free from so-called *IR* drop errors; this fundamental aspect is discussed in more detail in Chap. 11. The current densities involved in meeting commonly used protection criteria are typically around 10 mA per square meter of rebar surface.

Adequate anode lifetime is obviously also an important factor related to the magnitude and uniformity of current flow. A variety of anode systems have evolved for cathodic protection of reinforcing steel, each with certain advantages and limitations. Continuous surface anodes have been based on conductive bituminous overlays and conductive surface coatings. The former are suited only to horizontal surfaces. In general, good current distribution is achievable with such systems. Discrete anodes have been used without overlays and with cementitious overlays. For horizontal surfaces, anodes without overlays can be recessed in the concrete surface. Nonuniform current distribution is a

TABLE 2.32 Cathodic Protection Criteria for Steel in Concrete

Criterion	Details	Comments
Potential shift	100-mV shift of rebar potential in the positive direction when system is depolarized.	Depolarization occurs when CP current is switched off. Time period required for rebar to depolarize is debatable. The potential reading before interrupting the CP current should be IR corrected.
Potential shift	300-mV shift of rebar potential in the negative direction due to application of CP current.	The potential reading with the CP current on should be IR corrected. The method relies on a stable rebar potential before the application of CP current.
$E \log i$ curve	The decrease in corrosion rate due to the application of CP current can be determined provided the relationship between rebar potential E and current i can be measured and modeled. A simple model is Tafel behavior with a linear relationship between E and $\log i$.	This methodology is structure-specific, and the measurements involved are relatively complex and require specialist interpretation. Ideal Tafel behavior is rarely observed for steel in concrete.
Current density	Application of 10 mA/m^2 of rebar surface area.	Empirical approach based on limited experience. Does not consider individual characteristics of structures and environments.

fundamental concern in these systems. Anodes in the form of a titanium mesh, with proprietary surface coatings of precious metals, are commonly used in concrete structures, in conjunction with cementitious overlays. These systems are applicable to both horizontal and vertical surfaces and generally provide uniform current distribution.

Although the underlying principle of cathodic protection is a relatively simple one, considerable attention needs to be directed at details such as sound electrical connections, reliable reference electrodes, durable control cabinets, possible short circuits between the anodes and rebar, and maintenance schedules for the CP hardware.

Electrochemical chloride extraction. A further technique, applicable to existing concrete structures that have been contaminated with chlorides, involves the electrochemical removal of these harmful ions. The hardware involved is similar to that involved in cathodic protection. Electrochemical extraction of chloride ions is achieved by establishing an anode and a caustic electrolyte on the external concrete surface, and impressing a direct current between the anode and the reinforcing steel, which acts as the cathode (Fig. 2.30). Under the application of this electric field, chloride ions migrate away from the negatively charged steel and toward the positively charged external anode.

Chloride extraction has been recommended for structures that do not contain pre- or posttensioned steel and have little damage to the concrete itself. The current densities involved are significantly higher than those used in cathodic protection. The unsuitability of the technique to prestressed concrete is thus not surprising. The risk of hydrogen evolution on the rebar and subsequent hydrogen embrittlement is clearly much greater than in cathodic protection. Further requirements are a high degree of rebar electrical continuity and preferably low concrete resistance. Since the extraction processes require several days or even weeks using suitable current densities, the technique is more applicable to highway substructures than to bridge decks (most readers will agree that long traffic closures are highly unpopular).

In practice, the chloride extraction process does not remove the chloride ions from the concrete completely. Rather, a certain percentage is removed and the balance is redistributed away from the reinforcing bars. Importantly, through the cathodic reaction on the rebar surface, OH^- ions are generated, which have an important effect in counteracting the harmful influence of chloride ions, as explained earlier.

As with cathodic protection, the applied current density has to be controlled. If the current magnitude is excessive, several problems can arise, such as reduction in bond strength, softening of the cement paste around the rebar steel, and cracking of the concrete. Concrete containing alkali-reactive aggregates is not considered a suitable candidate for the process, as the expansive reactions leading to cracking and spalling associated with these aggregates tend to be aggravated.[55]

Electrochemical chloride extraction has been applied industrially for a number of years and can be an effective control method for chloride-induced corrosion of existing structures. Its limitations and drawbacks must be recognized, and it is clear that it is a relatively complex methodology, requiring specialized knowledge.

Re-alkalization. This treatment is applied to existing structures, to restore alkalinity around reinforcing bars in previously carbonated concrete. The electrochemical principle and hardware are similar to those for electrochemical chloride extraction. Direct current is applied between the cathodic rebar and external anodes positioned at the external concrete surface and surrounded by electrolyte (Fig. 2.30). Compared to cathodic protection, the current densities in re-alkalization

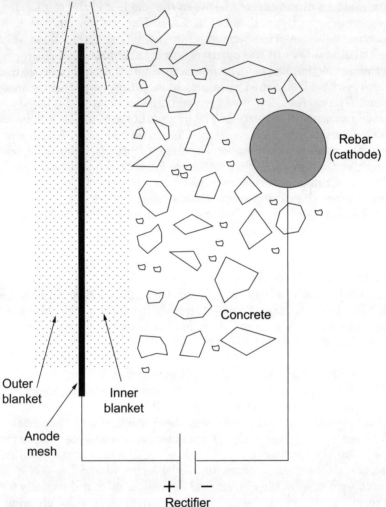

Figure 2.30 Principle of electrochemical chloride extraction and re-alkalization treatments (schematic).

are again significantly higher. Typically, the process is applied for several days to restore alkalinity in carbonated concrete.

The external electrolyte used in re-alkalization is a sodium carbonate solution, with a caustic pH. In addition to the generation of hydroxyl (OH^-) ions at the cathode and their migration away from the rebar under the electric field, other mechanisms can account for the formation of alkaline solution in the concrete. First, simple diffusion effects may arise as a result of concentration gradients in the concrete. Furthermore, "bulk" flow of external solution into the concrete may occur by either direct absorption or electro-osmosis. In dry concrete, absorption effects can occur to a depth of several centimeters in a matter of a day.

The potential disadvantages of re-alkalization are similar to those of chloride extraction, namely, risk of reduced bond strength, hydrogen embrittlement, alkali-aggregate reaction, and other microstructural changes in the concrete. Several practical applications of this technology have been documented in recent years.

Repair techniques. Given the vast scale of concrete infrastructure deterioration by corrosion processes, concrete repair is practiced widely to maintain the functionality of existing structures. Anyone traveling in North America during the road repair season can attest to this. An important fundamental consideration that should be respected in dealing with concrete repairs is that corrosion protection in repaired systems has different requirements from corrosion protection in new

New Structures	Repaired Structures

Service life includes corrosion initiation and propagation phases	Service life involves mostly the propagation phase - corrosion is generally more severe
Reinforcement experiences a relatively uniform internal concrete environment (at least initially)	The internal environment affecting the rebar is very heterogeneous - corrosion macrocells can be set up
Durability requirements are related to design life	Durability requirements are related to minimizing further corrosion
Low permeability concrete generally offers excellent protection	Low permeability concrete in one area can lead to problems in another area
Corrosive species usually penetrate from the outside into the interior, toward the rebar	Transport effects from outside through the protective cover but also from old concrete to new concrete
Good and relatively uniform bond between rebar and concrete	Bond between rebar and concrete often weakened and variable
Protective coatings on rebar can be applied under controlled, off-site conditions	Existing rebar cannot be removed from site, hence surface preparation and coating application is more challenging

Figure 2.31 Differences between corrosion protection in new and repaired structures.

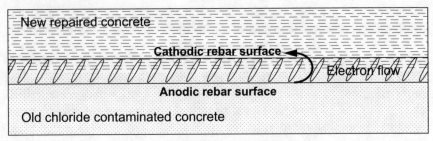

Figure 2.32 Galvanic corrosion cell in concrete repair (schematic).

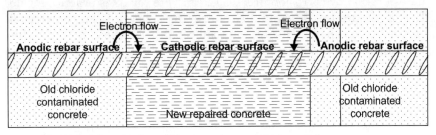

Figure 2.33 Galvanic corrosion cell in concrete repair (schematic).

construction. Some important differences between new and repaired structures are highlighted in Fig. 2.31.

Two basic approaches to concrete repair have been followed. The first repair methodology utilizes concrete or other cementitious materials alone. Essentially, these procedures involve the removal of loose, spalled concrete, followed by further systematic removal of the concrete surrounding the corroded rebar. Finally, the rebar and concrete surfaces are cleaned and primed before the new repair concrete is applied. The repair procedures thus create three different material zones that interact with the reinforcing steel: (1) the old chloride-contaminated/carbonated concrete, (2) the new concrete, and (3) the interface between the old and new concrete. The interface may represent a zone of weakness with respect to further ingress of corrosive species.

Importantly, the existing concrete should be removed to a depth well below the corroded reinforcing bars. Failure to do this can easily produce a detrimental galvanic corrosion cell in the repaired area, as depicted in Fig. 2.32. An undesirable galvanic corrosion cell involving the new and existing concrete can still be created despite this precaution, as shown in Fig. 2.33. To avoid rebar corrosion damage in the existing concrete in this situation, more extensive removal of the old chloride-contaminated concrete is necessary. Cathodic protection of the rebar or chloride extraction could also be considered as part of the repair specifications. The complex nature of electrochemical compatibility in repaired concrete structures is well illustrated by the examples in Figs. 2.32 and 2.33.

The second type of repair methodology involves additional corrosion protection schemes, apart from replacing the damaged concrete. Cathodic protection of the rebar is an obvious candidate for this role. Other approaches include the following:

- Zinc epoxy primer applied to the rebar, with sacrificial corrosion protection for the rebar in the repair zone
- Corrosion-inhibiting admixtures in conjunction with polymer-modified cementitious coatings applied to the rebar
- Other cement-based barrier coatings in combination with corrosion inhibitors
- Migrating corrosion inhibitors

It has been pointed out that reliable information and guidelines for the selection of repair strategies remain scarce and that there is an urgent need to establish suitable (preferably short-term) test methods for evaluating corrosion protection in repair systems.

Epoxy-coated reinforcing steel. Epoxy coatings provide an inert physical barrier that isolates the reinforcing steel from the corrosive environment. In North America, the use of epoxy-coated rebar dates back more than two decades, and at present it represents the most commonly used alternative to standard reinforcing steel. Standards covering these materials include ASTM A 775 and BS 7295. It is important to recognize that different types of epoxy coatings will display different properties. Variables such as surface cleanliness and preparation, coating thickness, coating adhesion to the rebar, coating continuity, and coating thickness have to be considered for optimal corrosion resistance.

While epoxy coatings have reportedly performed satisfactorily in many applications, such as bridge decks, incidents of severe corrosion have been observed in the substructure of four bridges in the Florida Keys (United States) after only 6 to 10 years of exposure.[56] The marine environment concerned was a particularly corrosive one.

Since the epoxy coating functions by providing a corrosion barrier, the coating continuity is obviously very important. While it may be possible to control coating defects (holidays) within tight limits in the manufacturing plant, the risk of coating damage during transportation, off-loading, storage, installation on site, and concrete pouring and vibration is considerably greater. Efforts have been directed at repairing visible damage on site, prior to placement in concrete.

Stainless steel rebar. The use of stainless steel rebar is as yet not widespread and is still a "novelty" in the construction industry. It may thus appear surprising that some industrial applications of stainless rebar date back more than 10 years. A list of selected international applications is presented in Table 2.33. The results from a number of research projects have indicated the superior corrosion performance of stainless alloys compared with carbon steel; reviews of international research findings have been published.[57] Several potential advantages could lead to growing interest in stainless rebars:

- Corrosion resistance is integral to the material (this does not imply that the material is always immune to corrosive attack).
- No coatings are involved that could chip, crack, or degrade.
- They have the capability to withstand shipping, handling, and bending.
- There are no "exposed" ends to cover or coat.
- Common rebar grades have good ductility, strength, and weldability.
- They can be magnetic or nonmagnetic, depending on grade.

TABLE 2.33 Examples of Stainless Steel Rebar Applications

Application	Date	Comments
Bridge deck in I-696 highway near Detroit, Michigan	1995	Type 304 rebars. Exposure to winter deicing salts.
Bridge deck in I-295 highway near Trenton, New Jersey	1985	Carbon steel rebars with type 304 cladding. Exposure to winter deicing salts. If ends of clad products are exposed, these represent a galvanic corrosion risk.
Bridge deck in 407 toll highway, near Toronto, Ontario	1996	Type 316LN rebars. Exposure to winter deicing salts.
Seafront structure restoration, Scarborough, U.K.	mid 1980s	Type 316 for replacement columns and precast beams.
Guild Hall Yard East project, London	1996	Type 304 selected for very long design life, in keeping with the famous historic buildings on the site.
Road slab of underpass, Cradlewell, U.K.	1995	Type 316.
Sydney Opera House forecourt restoration, Australia	~1990	Type 316 in a marine environment.

A range of stainless steels is available for rebar applications; final selection depends on mechanical design requirements, expected corrosivity, and cost considerations. For rebar, the austenitic and duplex (austenitic-ferritic) grades have received the most attention. There are currently two standards dealing with stainless steel rebar, the British BS 6744 (dating back to 1986) and the American ASTM A 955 (first published in 1996). The British Standard specifies austenitic alloys (Types 304, 304L, 316, and 316L, where "L" denotes alloys with lower carbon contents). The ASTM standard covers a wider range of alloys, including the well-known duplex alloy 2205.

Naturally, the initial cost of a structure with stainless steel rebar will be higher than that of a conventional structure. However, the overall construction cost increase may actually be relatively modest. The case for stainless steel rebar can be strengthened when a life-cycle cost approach is followed. This approach helps to focus attention on total costs over the lifetime of a structure, including the frequency and cost of future maintenance and replacement work. In such an analysis performed for a bridge, the cost benefits of austenitic stainless steels over carbon steel were clearly apparent after a time period of 18 to 23 years, at which time major repair costs would be incurred for the conventionally reinforced structure.[58]

When considering the use of stainless rebar, a further type of "hidden" cost is of relevance. Anyone who has been trapped in a traffic jam resulting from concrete repair work (a common experience in North America) can obviously also attest to such costs as lost productivity, wasted fuel, delivery delays, disruption of trade, etc., which are not easily quantified for highways. In the case of toll bridges and tunnels and harbor facilities, such disruptions have a direct impact on revenue.

Galvanized rebars. Arguments advocating corrosion protection of rebars by galvanizing are based on three principles. First, zinc-coated rebar is thought to remain passive in concrete at somewhat lower pH levels than ordinary steel. Second, the zinc coating represents a sacrificial anode that will tend to protect the steel cathodically. The galvanized coating is clearly consumed in the protection of rebar rather than being of the inert type, as epoxy coatings are. Third, it is argued that the usual corrosion product(s) of zinc occupy lower volume than the corrosion products of steel, implying lower expansive stresses.

Despite the above considerations, the performance of galvanized reinforcing steel has had somewhat mixed reviews. One explanation provided for low performance levels has been the formation of a particularly voluminous corrosion product involving expansions greater than those of many iron corrosion products.

Corrosion inhibitors. Corrosion-inhibiting admixtures are essentially designed to improve the performance of good-quality reinforced concrete. It has been emphasized that the function of such admixtures is *not* to "make good concrete out of poor quality concrete."[59] Adherence to this important principle is important in order to avoid unrealistic performance expectations for corrosion inhibitors in concrete.

One of the better-known corrosion-inhibitor admixtures used in attempting to control chloride-induced rebar corrosion is calcium nitrite, $Ca(NO_2)_2$. The mechanism of inhibition involves nitrite ions competing with chloride ions to react with Fe^{2+} ions produced at the anode.[59] Essentially, the nitrite ions limit the formation of unstable iron chloride complexes and promote the formation of stable compounds that passivate the rebar surface. The following reactions have been proposed:

$$2Fe^{2+} + 2OH^- + 2NO_2^- \rightarrow 2NO(g) + Fe_2O_3 + H_2O$$

or

$$Fe^{2+} + OH^- + NO_2^- \rightarrow NO(g) + FeOOH$$

The results of surface analysis of rebar steel exposed to concrete pore solutions inhibited with calcium nitrite have been entirely consistent with this proposed mechanism.[60] Since nitrite ions compete with chloride ions to interact with ferrous ions, the ratio of nitrite to chloride ions is important for adequate corrosion protection.

An important consideration for any corrosion-inhibiting admixture is its effects on the properties of concrete, such as workability, curing time, and strength. Many mechanistic details of commercial rebar corrosion inhibitors have remained relatively obscure because of the proprietary nature of their formulations.

Concrete cover and mix design. Given that chlorides and other corrosive species diffuse to the reinforcing steel, one obvious method of mitigating corrosion damage is to increase the concrete cover. The rule of thumb that a twofold increase in the concrete cover produces a fourfold life extension, that a threefold cover increase results in a ninefold benefit, etc., is often quoted in industry. This relationship is based on the (overly) simplistic model described by Fick's second law of diffusion. The beneficial effects of increased cover are also applicable to cracked concrete, but are less significant in exposure to chloride solutions involving alternative wetting and drying cycles.[50]

The diffusion rate of chlorides into concrete increases distinctly with increasing porosity, which in turn is increased by higher water-to-cement ratios and lower cement content. In the case of Portland cement, the penetration of chlorides by diffusion decreases with increasing content of the $3CaO \cdot Al_2O_3$ (C_3A) cement component. The following recommendations have therefore been made for ameliorating the corrosion risk for Portland cement concrete:[50]

- Water/cement ratio < 0.45
- Cement content > 400 kg/m^3
- C_3A content in cement: > 11% (by weight)

A further important consideration in controlling corrosion by reducing the permeability of concrete is that adequate curing (2 to 4 weeks' hydration) is required for the development of a dense internal texture with low porosity.

An important development in concrete mix design has been the addition of so-called supplementary cementitious materials. Two classifications apply to these compounds: *Pozzolans,* such as fly ash and silica fume, react with the cement hydration products, notably calcium hydroxide; *hydraulic materials,* such as granulated blast furnace slag, undergo direct hydration reactions. As these materials are of lower cost than conventional cement and essentially represent "environmental waste products," there are obvious incentives for blending them in concrete mixes. As pointed out by Hansson,[52] these materials can improve the strength and durability of concrete, with the important proviso that the concrete is cured adequately.

Concrete mixes known as high-performance concrete (HPC) have received considerable attention in recent years. This new generation of concrete has resulted from advances in the fields of admixtures and new cementitious materials. Examples of recent major HPC construction projects in highly corrosive environments include the Canadian Hibernia offshore oil platform and the P.E.I. Fixed Link, a 13-km bridge structure linking Prince Edward Island to New Brunswick. HPC does not refer to any specific mix design and should not simply be equated with high-strength concrete. Rather, it refers to various mixes with enhanced attributes compared with those of traditional concrete. Improvements in mechanical properties, durability, early-age strength, ease of placement and compaction, chemical resistance, and adhesion to hardened concrete all fit into the HPC concrete family.

HPC mixes with particularly low water-to-cement ratios, resulting in high compressive strengths, have been achieved with superplasticizer additives. These additives obviously play a crucial role in ensuring satisfactory workability at the low water contents. High density and low permeability typically characterize such mixes, which can be expected to represent an effective barrier to the ingress of corrosive species, provided the concrete is in the uncracked condition. Certain additives also increase the electrical resistance of the cement paste, thereby arguably retarding the kinetics of ionic corrosion reactions.

2.5.5 Condition assessment of reinforced concrete structures

In view of the large-scale reinforcing steel corrosion problems, the ability to assess the severity of corrosion (and other) damage is assuming increasing importance. Techniques that can provide early warning of imminent corrosion damage are particularly helpful. Once rebar corrosion has proceeded to such an advanced state that its effects are visually apparent on external surfaces, it is usually too late to implement effective corrosion control measures, and high repair or replacement costs are inevitable. Techniques with high sensitivity are required for early warning capability, and also for assessing the effectiveness of remedial measures in short, practical time frames.

Specific codes, guides, and standards related to the assessment of reinforced concrete structures are generally not as well developed as in, say, the mechanical engineering and metallic materials domains. However, the essential purpose of a detailed condition survey provides useful terms of reference. This is usually threefold: First, the extent of deterioration has to be determined; second, the mechanisms and causes of deterioration should be established; and finally, a corrosion control and/or repair strategy has to be specified.

An example of a comprehensive algorithm for reinforced concrete condition assessment is one developed for concrete bridge components under a contract from the U.S. Strategic Highway Research Program (SHRP). This algorithm (SHRP product 2032) is based on 13 conventional, well-established test methods and 7 new methodologies (refer to Table 2.34). An excellent summary of the SHRP methodology has been published by the Canadian Strategic Highway Research Program.[61]

Briefly, in the SHRP algorithm, the evaluation of bridge components is divided into three types of surveys. The initial (baseline) evaluation survey focuses on parameters that undergo relatively little change over time. The test methods recommended for this baseline survey are listed in Table 2.35 and essentially represent tests that should form part of the acceptance testing of new concrete bridge components. The second type of survey in the SHRP scheme is subsequent evaluation. The initial step in these surveys is visual inspection. The nature of subsequent inspection techniques depends on the morphology of damage observed in the visual inspection phase. The emphasis on reinforcing steel corrosion damage is placed under concrete spalling phenomena. The recommended assessment procedures for this form of

TABLE 2.34 Conventional and New Methods for Reinforced Concrete Structures

Property	Test methodology
Existing Techniques in SHRP Bridge Assessment Guide	
Depth of concrete cover	Magnetic flux devices
Concrete strength from test cylinders	ASTM C 39 for test cylinders
Concrete strength from core samples	ASTM C 42
Concrete strength from pullout tests	ASTM C 900
Concrete strength/quality from rebound hammer tests	ASTM C 805
Concrete strength/quality from penetration tests	ASTM C 803
Air void system characterization in hardened concrete	ASTM C 457
Microscopic evaluation of hardened concrete quality	ASTM C 856 petrographic examination
Alkali-silica reactivity	
Delamination detection	ASTM D 4580 sounding
Damage assessment by pulse velocity	ASTM C 597
Cracking damage	ACI 224.1R
Probability of active rebar corrosion	ASTM C 876 based on corrosion potential (note that no corrosion rate is determined)
New Techniques in SHRP Bridge Assessment Guide	
Instantaneous corrosion-rate measurement	Electrochemical measurements, applicable to uncoated steel
Condition assessment of asphalt-covered decks with pulsed radar	
Condition assessment of preformed membranes on decks using pulse velocity	
Evaluating relative effectiveness of penetrating concrete sealers with electrical resistance	
Evaluating penetrating concrete sealers by water absorption	
Evaluating chloride content in concrete by specific ion sensor	
Evaluating relative concrete permeability by surface air flow	

TABLE 2.34 (Continued)

	Other
Chloride ion content by titration	AASHTO T-260
Rebar location	X-ray and radar
pH and depth of carbonation	Phenolphthalein solution or pH electrode in extracted pore solution
Concrete permeability with respect to chloride ions	ASTM C 1202 and ASTM C 642
Delamination, voids, and other hidden defects	Impact echo, infrared thermography, pulse echo, and radar
Material properties	Density (ASTM C 642), moisture content (ASTM C 642), shrinkage (ASTM C 596, C 426), dynamic modulus (ASTM C 215), modulus of elasticity (ASTM C 464)

TABLE 2.35 Initial Evaluation Survey of Reinforced Bridge Components

Assessment procedure	Comments
Air void and petrographic core samples	Applicable to structures up to 15 years old
Alkali-silica reactivity test	Applicable to structures that are between 1 and 15 years old
Concrete strength	Applicable at all ages
Relative permeability	Applicable at all ages
Rebar cover	Applicable at all ages

Note: Test and sampling details may vary, depending on the age of the structure.

damage are presented in Fig. 2.34. The third survey category is special surveys applicable to asphalt-covered decks, pretensioned and post-tensioned concrete members, and rigid deck overlays.

Certain basic methods, such as visual inspection, core sampling for compressive strength tests, and petrographic analysis and chain drag sounding, have formed the basis of "traditional" condition assessments. However, as is apparent from Table 2.34, a host of additional new NDE methods and corrosion-monitoring techniques are available in modern engineering practice. A brief description of selected individual techniques specifically related to reinforcing steel corrosion damage follows.

Electrochemical corrosion measurements. These measurements can be performed completely nondestructively on the actual reinforcing steel or on sensors that are embedded in the concrete structure. The corro-

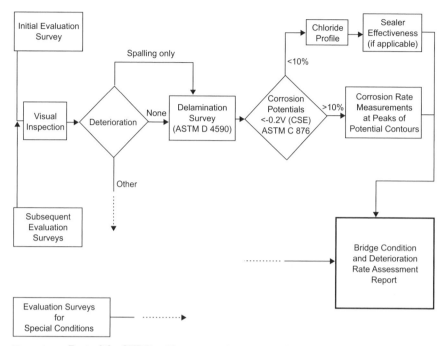

Figure 2.34 Part of the SHRP guide to assessing concrete bridge components (applicable to uncoated steel rebar).

sion sensors are essentially small sections of rebar steel, with shielded electrical leads attached for potential and current measurements. Preferably these sensors are embedded at different depths of cover, and their use must obviously be defined in the design stage of the structure. In general, electrochemical techniques are highly sensitive and therefore can detect corrosion damage at a very early stage. More detail on individual electrochemical techniques has been provided in Sec. 7.2.3.

Chloride content. Samples for determining the chloride level in concrete are collected in the form of powder produced by drilling or by the extraction of cores, sections of which are subsequently crushed. The latter method can provide a more accurate chloride concentration depth profile. The chloride ion concentration, used as a measure of the risk of corrosion damage and degree of chloride penetration, is subsequently determined by potentiometric titration. Two distinctions are made in chloride ion concentration testing: Acid-soluble chloride content (ASTM C 114) refers to the total chloride ion content, while the water-soluble content represents a lower value.

Petrographic examination. Petrographic examination is a microscopic analysis of concrete, performed on core samples removed from the structure. Further details may be found in the ASTM C856 standard. It yields information such as the depth of carbonation, density of the cement paste, air content, freeze-thaw damage, and direct attack of the concrete.

Permeability tests. These tests are based either on ponding core samples in chloride solution (with subsequent chloride content analysis) or on "forced" migration of chloride ions under the influence of an external electric field. The application of the electric field accelerates chloride ion migration and hence reduces the testing time.

2.5.6 Life prediction for corroding reinforced concrete structures

In a sense, inspection of concrete structures provides short-term qualitative life prediction. The essential performance of a structure from one inspection period to the next can probably be anticipated. However, prediction of remaining life over a longer time horizon is obviously important for decision-making, planning, licensing, life-cycle costing, and budgeting purposes.

A popular, fundamental conceptual model of concrete degradation by rebar corrosion involves two separate phases, initiation and propagation (Fig. 2.35). In the initiation phase, no significant corrosion damage takes place, but increasingly corrosive conditions develop that with time will eventually lead to depassivation of the reinforcing steel. The rate of damage in the propagation phase is significantly higher, leading to maintenance requirements and eventually large-scale rehabilitation.

Provided that the concrete is not water-saturated, it may be reasonable to assume that the initiation phase is considerably longer than the propagation period and that the end of the initiation period alone is a useful indicator of service life. Clifton and Pommersheim have reviewed simple models based on this approach.[62] For chloride-induced rebar corrosion, one of these is the use of Fick's second law of diffusion and the concept of a critical chloride concentration. Limitations and simplifying assumptions of this approach have been discussed in previous sections. Actual chloride concentration profiles can be measured on structures, to estimate parameters such as the diffusion coefficient used in the model. For carbonation, it has been proposed that the depth of carbonation is proportional to the square root of the exposure time. Again, the measurement of actual carbonation depth with time can be used to estimate a proportionality constant for a specific structure.

Cady and Weyers described the morphology and chronology of the degradation of bridge decks with chloride-induced corrosion damage.[63] Their work focused on actual bridge decks, and represents a more com-

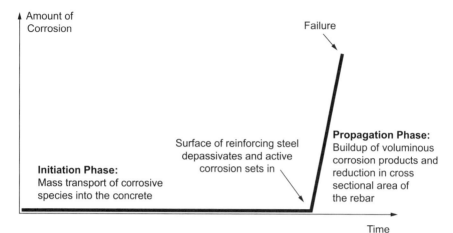

Figure 2.35 Conceptual model of rebar corrosion (schematic).

prehensive methodology than the models described above. A brief summary of their work is as follows: The process of damage often begins during construction, when subsidence cracking provides a direct path by which chlorides can reach some of the reinforcing steel. This cracking phenomenon in fresh concrete was described by a probabilistic function of the slump, rebar diameter, and depth of cover. The probability of cracking increased with increasing rebar size and slump but decreased with increasing cover. Reinforcing steel corrosion and fracture of the concrete could be manifested by this mechanism in a matter of months to a few years.

For the bridge decks investigated, the next phase of concrete damage was noticed after about 7 years. This mechanism of damage involves diffusion of chlorides through uncracked concrete to the reinforcing steel and its subsequent depassivation. The damage was manifested as concrete fracture and spalling under the wedging action of voluminous corrosion products. The requirement for periodic patch-up of small damaged areas thus began after some 7 years. The percentage of concrete surface damaged by this mechanism increases at a steady-state rate of about 2 percent per year. Typically, after 40 percent of the deck is affected, major deck rehabilitation is called for. Once the critical chloride level for rebar corrosion is attained, the time to cracking of concrete is inversely proportional to the corrosion rate. The rebar corrosion rate thus clearly assumes an important role in this model. An approach based on Fick's second law and a critical chloride level was followed to estimate the time between the initial "construction phase" damage and the steady-state damage phase in the concrete. The difficulty in determining values for the chloride diffusion coefficient and the critical chloride level were highlighted.

An interesting approach to life prediction attempted by Buenfeld and Hassanein involves the use of neural networks.[64] They (correctly) argued that deterioration rates should ideally be predicted on the basis of condition surveys on real structures or natural exposure trials, rather than laboratory studies. Clearly, the enormous number of variables involved in such "uncontrolled" tests cannot be tackled with conventional computational approaches. Neural network analysis was directed at large data sets from different sources for predicting chloride profiles and carbonation depth in concrete.

Neural networks are inherently suitable for a combined analysis (using a high number of combined variables) of different individual studies, each with a limited number of variables. For neural network studies, all variables have to be quantified. When no numerical values are available, such as for environmental corrosivity or cement type, a rating scale has to be assigned. The usefulness of a neural network for predicting the time to first cracking in concrete as a result of reinforcement corrosion was highlighted, but the researchers concluded that this is more difficult than predicting chloride profiles or carbonation depths. It was concluded that at present insufficient relevant data for developing a suitable neural network for this purpose are available. A further observation was that available "training data" for neural networks rarely extend to the design life of structures, which is typically more than 50 years. Therefore, while a neural network can determine the time dependence of a concrete degradation process, extrapolation to design life with appropriate safety factors will be required.

2.5.7 Other forms of concrete degradation

Besides corrosion-induced rebar damage, there are three other commonly cited forms of concrete degradation, namely, alkali-aggregate reaction, freeze-thaw damage, and sulfate attack.

Alkali-aggregate reaction. Alkali-aggregate reaction refers to chemical reactions between certain reactive aggregates and the highly alkaline concrete pore solution. Reactive silica is known for such reactions, and in this case the term *alkali-silica reaction* is often used. The damage is associated with an internal volume increase, producing cracking and spalling of the concrete. The expansion of aggregate particles and the formation of hygroscopic gels that swell are thought to produce the internal stresses. The cracking and spalling of alkali-aggregate reaction damage can make the underlying steel more susceptible to corrosion damage.

For this type of damage, the adage "prevention is better than cure" certainly holds true. Screening tests to identify problematic aggregates (such as ASTM C 289 and C 227) are available. Methodologies to improve the reliability and reduce the testing time of existing screening tests are under development. The addition of lithium salts to concrete mixes may arrest the undesirable expansive effects. Drying of concrete and sealing to minimize ingress of moisture have been suggested to limit the damage in existing structures; in practice, these methods are obviously not always easy to implement.

Freeze-thaw damage. Freeze-thaw damage is related to the porous nature of concrete. If the solution trapped in the pores freezes, a volume expansion occurs, which results in tensile stresses. When the hydraulic pressure exceeds the strength of the cement paste, cracking and spalling of the concrete results. Concrete with a high moisture content is most susceptible to this damage mechanism.

In cold climates, where this form of damage is a problem, the use of air-entrained concrete is specified. Such concrete has demonstrated its ability to provide durable long-term service. Essentially, additives create air voids entrained in the concrete. The freezing pore solution can then expand into this interconnected system of air voids. Usually the air content of this type of concrete is between 3 and 8 volume percent. However, the total air content alone is not necessarily adequate to assure resistance to freeze-thaw damage. The distribution and size of the entrained air voids are also of major importance. A tradeoff exists between air content and strength.

Scaling of concrete surfaces is closely related to freeze-thaw damage. Repetitive cycles of freezing and thawing can cause concrete surfaces to scale, leading to a pitted surface morphology. Contact between deicing salts and concrete surfaces plays an important role in scaling damage. By their hygroscopic action, deicing salts can concentrate moisture in the surface layers of concrete. Additional buildup of pressure can be created when dissolved salts recrystallize in the concrete pores.

Sulfate attack. Soluble sulfate species can cause deterioration of concrete as a result of expansive reactions between sulfate and calcium aluminates in the cement paste. Sulfate ions are ubiquitous; they are found in soils, seawater, groundwater, and effluent solutions. Use of cement with a low tricalcium aluminate content is beneficial for reducing the severity of attack.

2.6 Microbes and Biofouling

2.6.1 Basics of microbiology and MIC

Microorganisms pervade our environment and readily "invade" industrial systems wherever conditions permit. These agents flourish in a wide range of habitats and show a surprising ability to colonize water-rich surfaces wherever nutrients and physical conditions allow. Microbial growth occurs over the whole range of temperatures commonly found in water systems, pressure is rarely a deterrent, and limited access to nitrogen and phosphorus is offset by a surprising ability to sequester, concentrate, and retain even trace levels of these essential nutrients. A significant feature of microbial problems is that they can appear suddenly when conditions allow exponential growth of the organisms.[65] Because they are largely invisible, it has taken considerable time for a solid scientific basis for defining their role in materials degradation to be established. Many engineers continue to be surprised that such small organisms can lead to spectacular failures of large engineering systems.

The microorganisms of interest in microbiologically influenced corrosion are mostly bacteria, fungi, algae, and protozoans.[66] Bacteria are generally small, with lengths of typically under 10 μm. Collectively, they tend to live and grow under wide ranges of temperature, pH, and oxygen concentration. Carbon molecules represent an important nutrient source for bacteria. Fungi can be separated into yeasts and molds. Corrosion damage to aircraft fuel tanks is one of the well-known problems associated with fungi. Fungi tend to produce corrosive products as part of their metabolisms; it is these by-products that are responsible for corrosive attack. Furthermore, fungi can trap other materials, leading to fouling and associated corrosion problems. In general, the molds are considered to be of greater importance in corrosion problems than yeasts.[66] Algae also tend to survive under a wide range of environmental conditions, having simple nutritional requirements: light, water, air, and inorganic nutrients. Fouling and the resulting corrosion damage have been linked to algae. Corrosive by-products, such as organic acids, are also associated with these organisms. Furthermore, they produce nutrients that support bacteria and fungi. Protozoans are predators of bacteria and algae, and therefore potentially ameliorate microbial corrosion problems.[66]

MIC is responsible for the degradation of a wide range of materials. An excellent representation of materials degradation by microbes has been provided by Hill in the form of a pipe cross section, as shown in Fig. 2.36.[67] Most metals and their alloys (including stainless steel, aluminum, and copper alloys) are attacked by certain microorganisms. Polymers, hessian, and concrete are also not immune to this form of damage. The synergistic effect of different microbes and degradation mechanisms should be noted in Fig. 2.36.

In order to influence either the initiation or the rate of corrosion in the field, microorganisms usually must become intimately associated with the corroding surface. In most cases, they become attached to the metal surface in the form of either a thin, distributed film or a discrete

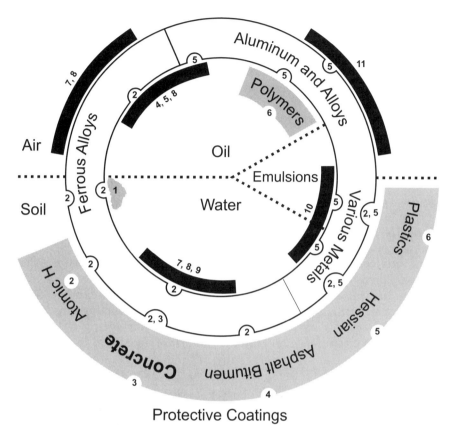

Figure 2.36 Schematic illustration of the principal methods of microbial degradation of metallic alloys and protective coatings. 1. Tubercle leading to differential aeration corrosion cell and providing the environment for 2. 2. Anaerobic sulfate-reducing bacteria (SRB). 3. Sulfur-oxidizing bacteria, which produce sulfates and sulfuric acid. 4. Hydrocarbon utilizers, which break down aliphatic and bitumen coatings and allow access of 2 to underlying metallic structure. 5. Various microbes that produce organic acids as end products of growth, attacking mainly nonferrous metals and alloys and coatings. 6. Bacteria and molds breaking down polymers. 7. Algae forming slimes on aboveground damp surfaces. 8. Slime-forming molds and bacteria (which may produce organic acids or utilize hydrocarbons), which provide differential aeration cells and growth conditions for 2. 9. Mud on river bottoms, etc., provides a matrix for heavy growth of microbes (including anaerobic conditions for 2). 10. Sludge (inorganic debris, scale, corrosion products, etc.) provides a matrix for heavy growth and differential aeration cells, and organic debris provides nutrients for growth. 11. Debris (mainly organic) on metal above ground provides growth conditions for organic acid–producing microbes.

biodeposit. The thin film, or biofilm, is most prevalent in open systems exposed to flowing seawater, although it can also occur in open freshwater systems. Such thin films start to form within the first 2 to 4 h of immersion, but often take weeks to become mature. These films will usually be spotty rather than continuous in nature, but will nevertheless cover a large proportion of the exposed metal surface.[68]

In contrast to the distributed films are discrete biodeposits. These biodeposits may be up to several centimeters in diameter, but will usually cover only a small percentage of the total exposed metal surface, possibly leading to localized corrosion effects. The organisms in these deposits will generally have a large effect on the chemistry of the environment at the metal/film or the metal/deposit interface without having any measurable effect on the bulk electrolyte properties. Occasionally, however, the organisms will be concentrated enough in the environment to influence corrosion by changing the bulk chemistry. This is sometimes the case in anaerobic soil environments, where the organisms do not need to form either a film or a deposit in order to influence corrosion.[68]

The taxonomy of microorganisms is an inexact science, and microbiological assays typically target functional groups of organisms rather than specific strains. Most identification techniques are designed to find only certain types of organisms, while completely missing other types. The tendency is to identify the organisms that are easy to grow in the laboratory rather than the organisms prevalent in the field. This is particularly true of routine microbiological analyses by many chemical service companies, which, although purporting to be very specific, are often based on only the crudest of analytical techniques.

Bacteria can exist in several different metabolic states. Those that are actively respiring, consuming nutrients, and proliferating are said to be in a growth stage. Those that are simply existing, but not growing because of unfavorable conditions, are said to be in a resting state. Some strains, when faced with unacceptable surroundings, form spores that can survive extremes of temperature and long periods without moisture or nutrients, yet produce actively growing cells quickly when conditions again become acceptable. The latter two states may appear, to the casual observer, to be like death, but the organisms are far from dead. Cells that actually die are usually consumed rapidly by other organisms or enzymes. When looking at an environmental sample under a microscope, therefore, it should be assumed that most or all of the cell forms observed were alive or capable of life at the time the sample was taken.

Classification of microorganisms. Microorganisms are first categorized according to oxygen tolerance. There are[68]

- Strict (or obligate) anaerobes, which will not function in the presence of oxygen
- Aerobes, which require oxygen in their metabolism
- Facultative anaerobes, which can function in either the absence or presence of oxygen

- Microaerophiles, which use oxygen but prefer low levels

Strictly anaerobic environments are quite rare in nature, but strict anaerobes are commonly found flourishing within anaerobic microenvironments in highly aerated systems. Another way of classifying organisms is according to their metabolism:

- The compounds or nutrients from which they obtain their carbon for growth and reproduction
- The chemistry by which they obtain energy or perform respiration
- The elements they accumulate as a result of these processes

A third way of classifying bacteria is by shape. These shapes are predictable when organisms are grown under well-defined laboratory conditions. In natural environments, however, shape is often determined by growth conditions rather than by pedigree. Examples of shapes are

- *Vibrio* for comma-shaped cells
- *Bacillus* for rod-shaped cells
- *Coccus* for round cells
- *Myces* for fungilike cells

Bacteria commonly associated with MIC
Sulfate-reducing bacteria. Sulfate-reducing bacteria (SRB) are anaerobes that are sustained by organic nutrients. Generally they require a complete absence of oxygen and a highly reduced environment to function efficiently. Nonetheless, they circulate (probably in a resting state) in aerated waters, including those treated with chlorine and other oxidizers, until they find an "ideal" environment supporting their metabolism and multiplication. There is also a growing body of evidence that some SRB strains can tolerate low levels of oxygen. Ringas and Robinson have described several environments in which these bacteria tend to thrive in an active state.[69] These include canals, harbors, estuaries, stagnant water associated with industrial activity, sand, and soils.

SRB are usually lumped into two nutrient categories: those that can use lactate, and those that cannot. The latter generally use acetate and are difficult to grow in the laboratory on any medium. Lactate, acetate, and other short-chain fatty acids usable by SRB do not occur naturally in the environment. Therefore, these organisms depend on other organisms to produce such compounds. SRB reduce sulfate to sulfide, which usually shows up as hydrogen sulfide or, if iron is available, as black ferrous sulfide. In the absence of sulfate, some strains can function as fermenters and use organic compounds such as pyruvate to produce acetate, hydrogen, and carbon dioxide. Many SRB strains also contain hydrogenase enzymes, which allow them to consume hydrogen.

Most common strains of SRB grow best at temperatures from 25° to 35°C. A few thermophilic strains capable of functioning efficiently at more than 60°C have been reported. It is a general rule of microbiology that a given strain of organism has a narrow temperature band in which it functions well, although different strains may function over widely differing temperatures. However, there is some evidence that certain organisms, especially certain SRB, grow well at high temperatures (around 100°C) under high pressures—e.g., 17 to 31 MPa—but can also grow at temperatures closer to 35°C at atmospheric pressure.[68]

Tests for the presence of SRB have traditionally involved growing the organisms on laboratory media, quite unlike the natural environment in which they were found. These laboratory media will grow only certain strains of SRB, and even then some samples require a long lag time before the organisms will adapt to the new growth conditions. As a result, misleading information regarding the presence or absence of SRB in field samples has been obtained. Newer methods that do not require the SRB to grow to be detected have been developed. These methods are not as sensitive as the old culturing techniques but are useful in monitoring "problem" systems in which numbers are relatively high.

SRB have been implicated in the corrosion of cast iron and steel, ferritic stainless steels, 300 series stainless steels (and also very highly alloyed stainless steels), copper-nickel alloys, and high-nickel molybdenum alloys. Selected forms of SRB damage are illustrated in Fig. 2.37.[70] They are almost always present at corrosion sites because they are in soils, surface-water streams, and waterside deposits in general. Their mere presence, however, does not mean that they are causing corrosion. The key symptom that usually indicates their involvement in the corrosion process of ferrous alloys is localized corrosion filled with black sulfide corrosion products. While significant corrosion by pure SRB strains has been observed in the laboratory, in their natural environment these organisms rely heavily on other organisms to provide not only essential nutrients, but also the necessary microanaerobic sites for their growth. The presence of shielded anaerobic microenvironments can lead to severe corrosion damage by SRB colonies thriving under these local conditions, even if the bulk environment is aerated. The inside of tubercles covering ferrous surfaces corroded by SRB is a classic example of such anaerobic microenvironments.

Sulfur–sulfide-oxidizing bacteria. This broad family of aerobic bacteria derives energy from the oxidation of sulfide or elemental sulfur to sul-

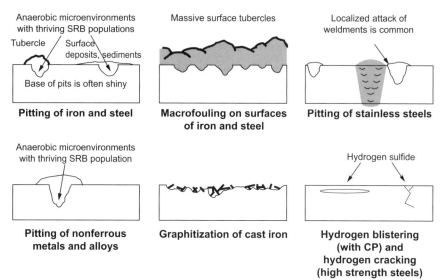

Figure 2.37 Forms of corrosion damage produced by SRB.

fate. Some types of aerobes can oxidize the sulfur to sulfuric acid, with pH values as low as 1.0 reported. These *Thiobacillus* strains are most commonly found in mineral deposits, and are largely responsible for acid mine drainage, which has become an environmental concern. They proliferate inside sewer lines and can cause rapid deterioration of concrete mains and the reinforcing steel therein. They are also found on surfaces of stone buildings and statues and probably account for much of the accelerated damage commonly attributed to acid rain.

Where *Thiobacillus* bacteria are associated with corrosion, they are almost always accompanied by SRB. Thus, both types of organisms are able to draw energy from a synergistic sulfur cycle. The fact that two such different organisms, one a strict anaerobe that prefers neutral pH and the other an aerobe that produces and thrives in an acid environment, can coexist demonstrates that individual organisms are able to form their own microenvironment within an otherwise hostile larger world.

Iron/manganese-oxidizing bacteria. Bacteria that derive energy from the oxidation of Fe^{2+} to Fe^{3+} are commonly reported in deposits associated with MIC. They are almost always observed in tubercles (discrete hemispherical mounds) over pits on steel surfaces. The most common iron oxidizers are found in the environment in long protein sheaths or filaments.[68] While the cells themselves are rather indistinctive in appearance, these long filaments are readily seen under the microscope and are not likely to be confused with other life forms. The observation that filamentous iron bacteria are "omnipresent" in tubercles might, therefore, be more a matter of their easy detection than of their relative abundance.

An intriguing type of iron oxidizers is the *Gallionella* bacterium, which has been blamed for numerous cases of corrosion of stainless steels. It was previously believed that *Gallionella* simply caused bulky deposits that plugged water lines. More recently, however, it has been found in several cases in which high levels of iron, manganese, and chlorides are present in the deposits. The resulting ferric manganic chloride is a potent pitting agent for stainless steels.

Besides the iron-manganese oxidizers, there are organisms that simply accumulate iron or manganese. Such organisms are believed to be responsible for the manganese nodules found on the ocean floor. The accumulation of manganese in biofilms is blamed for several cases of corrosion of stainless steels and other ferrous alloys in water systems treated with chlorine or chlorine–bromine compounds.[71] It is likely that the organisms' only role, in such cases, is to form a biofilm rich in manganese. The hypochlorous ion then reacts with the manganese to form permanganic chloride compounds, which cause distinctive subsurface pitting and tunneling corrosion in stainless steels.

Aerobic slime formers. Aerobic slime formers are a diverse group of aerobic bacteria. They are important to corrosion mainly because they produce extracellular polymers that make up what is commonly referred to as "slime." This polymer is actually a sophisticated network of sticky strands that bind the cells to the surface and control what permeates through the deposit. The stickiness traps all sorts of particulates that might be floating by, which, in dirty water, can result in the impression that the deposit or mound is an inorganic collection of mud and debris. The slime formers and the sticky polymers that they produce make up the bulk of the distributed slime film or primary film that forms on all materials immersed in water.

Slime formers can be efficient "scrubbers" of oxygen, thus preventing oxygen from reaching the underlying surface. This creates an ideal site for SRB growth. Various types of enzymes are often found within the polymer mass, but outside the bacterial cells. Some of these enzymes are capable of intercepting and breaking down toxic substances (such as biocides) and converting them to nutrients for the cells.[68] Tubercles, though attributed to filamentous iron bacteria by some, usually contain far greater numbers of aerobic slime formers. Softer mounds, similar to tubercles but lower in iron content, are also found on stainless steels and other metal surfaces, usually in conjunction with localized MIC. These, too, typically contain high numbers of aerobic bacteria, either *Gallionella* or slime formers.

The term *high numbers* is relative. A microbiologist considers 10^6 cells per cubic centimeter or per gram in an environmental sample to represent high numbers. However, these organisms make up only a minuscule portion of the overall mass. Biomounds, whether crusty tubercles on steel surfaces or the softer mounds on other metals, typically analyze approximately 10 percent by weight organic matter, most of that being extracellular polymers.

Methane producers. Only in recent years have methane-producing bacteria (methanogens) been added to the list of organisms believed responsible for corrosion. Like many SRB, methanogens consume hydrogen and thus are capable of performing cathodic depolarization. While they normally consume hydrogen and carbon dioxide to produce methane, in low-nutrient situations these strict anaerobes will become fermenters and consume acetate instead. In natural environments, methanogens and SRB frequently coexist in a symbiotic relationship: SRB producing hydrogen, CO_2, and acetate by fermentation, and methanogens consuming these compounds, a necessary step if fermentation is to proceed. The case for facilitation of corrosion by methanogens still needs to be strengthened, but methanogens are as common in the environment as SRB and are just as likely to be a problem. The reason they have not been implicated before now is most likely because they do not produce distinctive, solid byproducts.

Organic acid–producing bacteria. Various anaerobic bacteria such as *Clostridium* are capable of producing organic acids. Unlike SRB, these bacteria are not usually found in aerated macroenvironments such as open, recirculating water systems. However, they are a problem in gas transmission lines and could be a problem in closed water systems that become anaerobic.

Acid-producing fungi. Certain fungi are also capable of producing organic acids and have been blamed for corrosion of steel and aluminum, as in the highly publicized corrosion failures of aluminum aircraft fuel tanks. In addition, fungi may produce anaerobic sites for SRB and can produce metabolic byproducts that are useful to various bacteria.

Effect of operating conditions on MIC. Biocorrosion problems occur most often in new systems when they are first wetted. When the problem occurs in older systems, it is almost always a result of changes, such as new sources or quality of water, new materials of construction, new operating procedures (e.g., water now left in system during shutdowns, whereas it used to be drained), or new operating conditions (especially temperature). Some of the operating parameters known to have or suspected of having an effect on MIC are temperature, pressure, flow velocity, pH, oxygen level, and cleanliness.[72]

Temperature. All microorganisms have an optimum temperature range for growth. Observation of the water or surface temperatures at which corrosion mounds or tubercles do or do not grow may offer important clues as to how effective slight temperature changes may be. The normal expectation is that increasing temperature increases corrosion problems. With MIC, this is not necessarily so.

Flow velocity. Flow velocity has little long-term effect on the ability of cells to attach to surfaces. Once attachment takes place, however, flow affects the nature of the biofilm that forms. It has been observed that low-velocity biofilms tend to be very bulky and easily disturbed, while films that form at higher velocities are much denser, thinner, and more tenacious.

As a rule, flow velocities above 1.5 m/s are recommended in water systems to minimize settling out of solids. Such velocities will not prevent surface colonization in systems that are prone to biofouling, however. Stagnant conditions, even for short periods of time, generally result in problems. Increasing velocity to discourage biological attachment is not always feasible, since it can promote erosion corrosion of the particular metal being used. Copper, for instance, suffers erosion corrosion above 1.5 m/s at 20°C.

pH. Bulk water pH can have a significant effect on the vitality of microorganisms. Growth of common strains of SRB, for example, slows above pH 11 and is completely stifled at pH 12.5. Some researchers have speculated that this is why cathodic protection is effective against these microbes, since cathodic protection has a net effect of increasing the pH of the metallic surface being protected.

Oxygen level. Many bacteria require oxygen for growth. There is reason to believe that many biological problems could be partly alleviated if a system were completely deaerated. Many aerobes can function adequately with as little as 50 ppb O_2, and facultative organisms, of course, simply convert to an anaerobic metabolism if oxygen is depleted. Practically speaking, removing dissolved oxygen from the system can affect MIC, but it is not likely to eliminate a severe problem.

Cleanliness. The "cleanliness" of a given water usually refers to the water's turbidity or the amount of suspended solids in that water. Settling of suspended solids enhances corrosion by creating occlusions and surfaces for microbial growth and activity. The organic and dissolved solids content of the water are also important. These factors may be significantly reduced by "cleaning up" the water. Improving water quality is not necessarily a solution to MIC.

With respect to water cleanliness, one rule is that as long as any microorganisms can grow in the water, the potential for MIC exists. On the surfaces of piping and equipment, however, "cleanliness" is much more important. Anything that can be done to clean metal surfaces physically on a regular basis (i.e., to remove biofilms and deposits) will help to prevent or minimize MIC. In summary, any time the operating conditions in a water system are changed, extra attention should be paid to possible biological problems that may result.

Identification of microbial problems

Direct inspection. Direct inspection is best suited to enumeration of planktonic organisms suspended in relatively clean water. In liquid suspensions, cell densities greater than 10^7 cells·cm^{-3} cause the sample to appear turbid. Quantitative enumerations using phase contrast microscopy can be done quickly using a counting chamber which holds a known volume of fluid in a thin layer. Visualization of microorganisms can be enhanced by fluorescent dyes that cause cells to light up under ultraviolet radiation. Using a stain such as acridine orange, cells separated by filtration from large aliquots of water can be visualized and counted on a 0.25-μm filter using the epifluorescent technique. Newer stains such as fluorescein diacetate, 5-cyano-2,3-ditolyltetrazolium chloride, or p-iodonitrotetrazolium violet indicate active metabolism by the formation of fluorescent products.[65]

Identification of organisms can be accomplished by the use of antibodies generated as an immune response to the injection of microbial cells into an animal, typically a rabbit. These antibodies can be harvested and will bind to the target organism selectively in a field sample. A second antibody tagged with a fluorescent dye is then used to light up the rabbit antibody bound to the target cells. In effect, the staining procedure can selectively light up target organisms in a mixed population or in difficult soil, coating, or oily emulsion samples.[73]

Such techniques can provide insight into the location, growth rate, and activity of specific kinds of organisms in mixed populations in biofilms. Antibodies which bind to specific cells can also be linked to enzymes that produce a color reaction in an enzyme-linked immunosorbent assay. The extent of the color produced in solution can then be correlated with the number of target organisms present.[74] While antibody-based stains are excellent research tools, their high specificity means that they identify only the target organisms. Other organisms potentially capable of causing problems are missed.

Growth assays. The most common way to assess microbial populations in industrial samples is through growth tests using commercially available growth media for the groups of organisms that are most commonly associated with industrial problems. These are packaged in a convenient form suitable for use in the field. Serial dilutions of suspended samples are grown on solid agar or liquid media. Based on the growth observed for each dilution, estimates of the most probable number (MPN) of viable cells present in a sample can be obtained.[75] Despite the common use of growth assays, however, only a small fraction of wild organisms actually grow in commonly available artificial media. Estimates of SRB in marine sediments, for example, suggest that as few as one in a thousand of the organisms present actually show up in standard growth tests.[76]

Activity assays.
Whole cell. Approaches based on the conversion of a radioisotopically labeled substrate can be used to assess the potential activity of microbial populations in field samples. The radiorespirometric method allows use of field samples directly, without the need to separate organisms, and is very sensitive. Selection of the radioactively labeled substrate is key to interpretation of the results, but the method can provide insights into factors limiting growth by comparing activity in native samples with supplemented test samples under various conditions. Oil-degrading organisms, for example, can be assessed through the mineralization of ^{14}C-labeled hydrocarbon to carbon dioxide. Radioactive methods are not routinely used by field personnel but have found use in a number of applications, including biocide screening programs, identification of nutrient sources, and assessment of key metabolic processes in corrosion scenarios.[65]

Enzyme-based assays. An increasingly popular approach is the use of commercial kits to assay the presence of enzymes associated with microorganisms that are suspected of causing problems. For example, kits are available for the sulfate reductase enzyme[77] common to SRB associated with corrosion problems and for the hydrogenase enzyme implicated in the acceleration of corrosion through rapid removal of cathodic hydrogen formed on the metal surface.[78] The performance of several of these kits has been assessed by field personnel in round-robin tests. Correlation of activity assays and population estimates is variable. In general, these kits have a narrower range of application than growth-based assays, making it important to select a kit with a range of response appropriate to the problem under consideration.[79]

Metabolites. An overall assessment of microbial activity can be obtained by measuring the amount of adenosine triphosphate (ATP) in field samples. This key metabolite drives many cellular reactions.

Commercial instruments are available which measure the release of light by firefly luciferin/luciferase with ATP. The method is best suited to clean aerobic aqueous samples; particulate and chemical quenching can affect results. Detection of metabolites such as organic acids in deposits or gas compositions including methane or hydrogen sulfide by routine gas chromatography can also indicate biological involvement in industrial problems.[65]

Cell components. Biomass can be generally quantified by assays for protein, lipopolysaccharide, or other common cell constituents, but the information gained is of limited value. An alternative approach is to use cell components to define the composition of microbial populations, with the hope that the insight gained may allow damaging situations to be recognized and managed in the future. Fatty acid analysis and nucleic acid sequencing provide the basis for the most promising methods.

Fatty acid profiles. Analyzing fatty acid methyl esters derived from cellular lipids can fingerprint organisms rapidly. Provided that pertinent profiles are known, organisms in industrial and environmental samples can be identified with confidence. In the short term, the impact of events such as changes in operating conditions or application of biocides can be monitored by such analysis. In the longer term, problem populations may be identified quickly so that an appropriate management response can be implemented in a timely fashion.

Nucleic acid–based methods. Specific DNA probes can be constructed to detect segments of genetic material coding for known enzymes. A gene probe developed to detect the hydrogenase enzyme which occurs broadly in SRB from the genus *Desulfovibrio* was applied to samples from an oilfield waterflood plagued with iron sulfide–related corrosion problems. The enzyme was found in only 12 of 20 samples, suggesting that sulfate reducers which did not have this enzyme were also present.[80] In principle, probes could be developed to detect all possible sulfate reducers, but application of such a battery of probes becomes daunting when large numbers of field samples are to be analyzed.

To overcome this obstacle, the reverse sample genome probe (RSGP) was developed. In this technique, DNA from organisms previously isolated from field problems is spotted on a master filter. DNA isolated from field samples of interest is then labeled with either a radioactive or a fluorescent indicator and exposed to this filter. Where complementary strands of DNA are present, labeled DNA from the field sample sticks to the corresponding spot on the master filter. Organisms represented by the labeled spots are then known to be in the field sample. The technique is quantitative, and early work with oilfield populations suggests that a significant fraction of all the DNA present in a field corrosion site sample can be correlated with known isolates.[80]

Sampling. Samples for analysis can be obtained from industrial systems by scraping accessible surfaces. In open systems or on the outside of pipelines or other underground facilities, this can be done directly. Bull plugs, coupons, or inspection ports can provide surface samples in low-pressure water systems.[81] More sophisticated devices are commercially available for use in pressurized systems.[82] In these devices, coupons are held in an assembly which mounts on a standard pressure fitting. If biofilms are to be representative of a system, it is important that the sampling coupons are of the same material as the system and flush-mounted in the wall of the system so that flow effects match those of the surrounding surface. While pressure fittings allow coupons to be implanted directly in process units, the fittings are expensive, pressure vessel codes and accessibility can restrict their location, and the removal and installation of coupons involves exact technical procedures. For these reasons, sidestream installations are often used instead.

Handling of field samples should be done carefully to avoid contamination with foreign matter, including biological materials. A wide range of sterile sampling tools and containers is readily available. Because many systems are anaerobic, proper sample handling and transport is essential to avoid misleading results brought about by excessive exposure to oxygen in the air. One option is to analyze samples on the spot using commercially available kits, as described above. Where transportation to a laboratory is required, Torbal jars or similar anaerobic containers can be used.[83] In many cases, simply placing samples directly in a large volume of the process water in a completely filled screw-cap container is adequate. Processing in the lab should also be done anaerobically, using special techniques or anaerobic chambers designed for this purpose. Because viable organisms are involved, processing should be done quickly to avoid growth or death of cells that are stimulated or inhibited by changes in temperature, oxygen exposure, or other factors.[65]

2.6.2 Biofouling

For the first 200+ years of microbiology, organisms were studied exclusively in planktonic form (freely floating in water or nutrient broth). In the late 1970s, with the advent of advanced microscopic methods, microbiologists were surprised to find that biofilms are the predominant form of bacterial growth in almost all aquatic systems. Since that time, it has become apparent that organisms living within a biofilm can behave very differently from the same species floating freely. In water treatment, biofilms are undesirable because they harbor pathogenic organisms such as *Legionella*, reduce heat transfer, cause increased friction or complete blockage of pipes, and contribute to corrosion.[84]

Nature of biofilm. A biofilm is said to consist of microbial cells (algal, fungal, or bacterial) and the extracellular biopolymer they produce. Generally, it is bacterial biofilms that are of most concern in industrial water systems, since they are generally responsible for the fouling of heat-transfer equipment. This is due in part to the minimal nutrients that many species require in order to grow.

Biofilm contributes to corrosion in several ways. The simplest is the difference in oxygen concentration depending on the thickness of the biofilm.[85] In addition to this effect, biofilm allows accumulation of frequently acidic metabolic products near the metal surface, which accelerates the cathodic reaction.[86] One particular metabolic product, hydrogen sulfide, will also promote the anodic reaction through the formation of highly insoluble ferrous sulfide. Finally, certain bacteria will oxidize Fe^{2+} produced by these first two effects to form ferric hydroxide in the form of tubercles. The tubercles greatly steepen the oxygen gradient and accelerate the corrosion process. The corrosion products of MIC also interfere with the performance of biocides, resulting in a vicious cycle.[84]

The microorganisms themselves may make up from 5 to 25 percent of the volume of a biofilm. The remaining 75 to 95 percent of the volume, the biofilm matrix, is actually 95 to 99 percent water. The dry weight consists primarily of acidic exopolysaccharides excreted by the organisms. Very close to the bacteria cells, the biofilm matrix is more likely to consist of lipopolysaccharides (fatty carbohydrates), which are more hydrophobic than the exopolysaccharides. The exopolysaccharide/water mixture gels when enough calcium ions replace the acidic protons of the polymers. The chemically very similar alginates are used in water treatment because of this calcium-binding property. The same anionic sites on the polymers will also bind other divalent cations, such as Mg^{2+}, Fe^{2+}, and Mn^{2+}.[87]

The biofilm allows enzymes to accumulate and act on food substrates without being washed away as they would be in the bulk water. The presence of the biofilm causes often acidic metabolic products to accumulate within 0.5 μm or so of the colony. When one species can use the metabolic products of another, colonies of the two species will often be found adjacent to each other within the biofilm. An example of this type of cooperation occurs in MIC, where one can find *Desulfovibrio*, *Thiobacillus*, and *Gallionella* forming a miniature ecosystem within a corrosion pit.[86] The biofilm matrix can also protect organisms within it from the grazing of larger protozoa such as amoeba and from antibodies or leukocytes of a host organism. Because of these many advantages, almost all microorganisms are capable of producing some amount of biofilm. Biofilm is most stable when conditions in the ambient water are stable. Changes in ionic strength, pH, or temperature will all destabilize biofilm.[84]

Biofilm formation. In industrial systems, direct and indirect biomineralization processes can influence scale formation and mineral deposition within the biofilm. Clay particles and other debris become trapped in the extracellular slime, adding to the thickness and heterogeneity of the biofilm. Iron, manganese, and silica are often elevated in biofilms as a result of mineral deposition and ion exchange. In the case of iron-oxidizing bacteria found in aerobic water systems, metal oxides are an important component of the biofilm. In steel systems operating under anaerobic conditions, iron sulfides can be deposited when ferrous ions released by corrosion of steel surfaces precipitate with sulfide generated by bacteria in the biofilm.[65]

A completely clean surface will display an induction period during which colonization occurs. After a previously clean surface has been colonized, a biofilm will grow exponentially at first, until either the thickness of the film interferes with diffusion of nutrients to the organisms within it or the flow of water causes matrix material to slough off at the surface as fast as it is being produced below. Biofilm development is most rapid when consortia of mutually beneficial species are involved. In the absence of antimicrobial agents, biofilms in cooling water typically take 10 to 14 days to reach equilibrium. The equilibrium thickness of biofilms varies widely but can reach the 500- to 1000-µm range in a cooling-water system. The thickness of biofilm is seldom uniform, and patches of exposed metal may even be found in systems with significant biofilm present.

As a biofilm matures, enzymes and other proteins accumulate. These can react with polysaccharides to form complex biopolymers. A selective process occurs in which biopolymers that are most stable under the ambient conditions remain while those that are less stable are sloughed off. Thus a mature biofilm is generally more difficult to remove than a new biofilm. Studies have shown that biofilm growth is due primarily to reproduction within the biofilm rather than to the adherence of planktonic organisms.[88] The shedding of biofilm organisms into the bulk water serves to spread a given species from one region of the system to another, but once species are widespread, the concentration of organisms in the water is merely a symptom of the amount of biofilm activity rather than a cause of biofilm formation. Consequently, planktonic bacteria counts can be misleading. A biocide may kill a large percentage of the planktonic organisms while having little effect on anything but the outer surfaces of the biofilm. In this case, planktonic bacteria counts may rise quickly after the biocide has left the system as shedding of organisms from the biofilm resumes.[84]

In cooling towers and spray ponds, algal biofilms are also a concern. Not only will algal biofilms foul distribution decks and tower fill, but algae will also provide nutrients (organic carbon) that will help support the growth of bacteria and fungi. Algae do not require organic carbon for growth, but instead utilize CO_2 and the energy provided by the sun to manufacture carbohydrate.

In aquatic environments, microorganisms may be suspended freely in the bulk water (planktonic existence) or attached to an immobile substratum or surface (sessile existence). The microorganisms may exist as solitary individuals or in colonies that contain from a few to more than a million individuals. Complex assemblages of various species may occur within both planktonic and sessile microbial populations. The environmental conditions largely dictate whether the microorganisms will exist in a planktonic or sessile state. Sessile microorganisms do not attach directly to the substratum surface, but rather attach to a thin layer of organic matter (the conditioning film) adsorbed on the surface (Fig. 2.38, Stages 1 and 2). As microbes attach to and replicate on the substratum, a biofilm is formed over the surface. The biofilm is composed of immobilized cells and their extracellular polymeric substances.

The characteristics of a biofilm may change with time. During the early stages of development, a biofilm is composed of the pioneering microbial species, which are distributed as individual cells in a heterogeneous manner over the surface. Within a matter of minutes, some of the attached species produce adhesive exopolymers that encapsulate the cells and extend from the cell surface to the substratum and into the bulk fluid (Fig. 2.38, Stage 2). The adhesive exopolymers restrict the dissemination of microbial cells as they replicate on the surface (Fig. 2.38, Stage 3). At this stage of development, the biofilm is less than 10 µm in thickness and exists as a discontinuous matrix of exopolymers interspersed with cells.[72]

As the immobilized cells continue to replicate and excrete more exopolymer material, the biofilm forms a confluent blanket of increasing thickness over the surface (Fig. 2.38, Stage 4). Bacteria attach to surfaces by proteinaceous appendages referred to as fimbriae. Once a number of fimbriae have "glued" the cell to the surface, detachment of the organism becomes very difficult. One reason bacteria prefer to attach to surfaces is the adsorbed organic molecules that can serve as nutrients. Once attached, the organisms begin to produce material

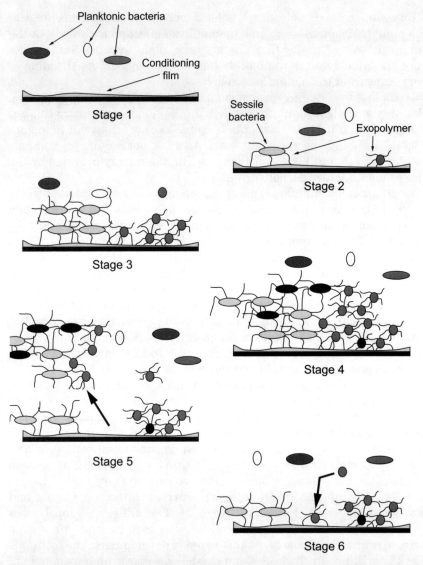

Figure 2.38 Different stages of biofilm formation and growth. Stage 1: Conditioning film accumulates on submerged surface. Stage 2: Planktonic bacteria from the bulk water colonize the surface and begin a sessile existence by excreting exopolymer that anchors the cells to the surface. Stage 3: Different species of sessile bacteria replicate on the metal surface. Stage 4: Microcolonies of different species continue to grow and eventually establish close relationships with one another on the surface. The biofilm increases in thickness. Conditions at the base of the biofilm change. Stage 5: Portions of the biofilm slough away from the surface. Stage 6: The exposed areas of surface are recolonized by planktonic bacteria or sessile bacteria adjacent to the exposed areas.

called extracellular biopolymer or slime. The amount of biopolymer produced can exceed the mass of the bacterial cell by a factor of 100 or more. The extracellular polymer produced may tend to provide a more suitable protective environment for the survival of the organism.

The extracellular biopolymer consists primarily of polysaccharides and water. The polysaccharides produced vary depending on the species but are typically made up of repeating oligosaccharides, such as glucose, mannose, galactose, xylose, and others. An often-cited example of a bacterial-produced biopolymer is xanthan gum, produced by *Xanthomonas campestris*. This biopolymer is used as a thickening agent in a variety of foods and consumer products. Gelation of some biopolymers can occur upon addition of divalent cations, such as calcium and magnesium. The electrostatic interaction between carboxylate functional groups on the polysaccharide and the divalent cations results in a bridging effect between polymer chains. Bridging and cross-linking of the polymers help to stabilize the biofilm, making it more resistant to shear.

Over time, species of planktonic bacteria and nonliving particles become entrained in the biofilm and contribute to a growing community of increasing complexity. At this stage, the mature biofilm may be visibly evident. Its morphology and consistency vary depending on the types of microorganisms present and the conditions in the surrounding bulk liquid. The time it takes to achieve this stage may vary from a few days to several weeks.

As the biofilm increases in thickness, diffusion of dissolved gases and other nutrients from the bulk liquid to the substratum becomes impeded. Conditions become inhospitable to some of the microorganisms at the base of the biofilm, and eventually many of these cells die. As the foundation of the biofilm weakens, shear stress from the flowing liquid causes sloughing of cell aggregations, and localized areas of bare surface are exposed to the bulk liquid (Fig. 2.38, Stage 5). The exposed areas are subsequently recolonized, and new microorganisms and their exopolymers are woven into the fabric of the existing biofilm (Fig. 2.38, Stage 6). This phenomenon of biofilm instability occurs even when the physical conditions in the bulk liquid remain constant. Thus, biofilms are constantly in a state of flux.[72]

Marine biofouling. Marine biofouling is commonplace in open waters, estuaries, and rivers. It is commonly found on marine structures, including pilings, offshore platforms, and boat hulls, and even within piping and condensers. The fouling is usually most widespread in warm conditions and in low-velocity (<1 m/s) seawater. Above 1 m/s, most fouling organisms have difficulty attaching themselves to surfaces. There are various types of fouling organisms, particularly plants (slime algae), sea mosses, sea anemones, barnacles, and mollusks (oysters and mussels). In steel, polymer, and concrete marine construction, biofouling can be detrimental, resulting in unwanted excess drag on structures and marine craft in seawater or causing blockages in pipe systems. Expensive removal by mechanical means is often required. Alternatively, costly prevention methods are often employed, which include chlorination of pipe systems and antifouling coatings on structures.[89]

Marine organisms attach themselves to some metals and alloys more readily than to others. Steels, titanium, and aluminum will foul readily. Copper-based alloys, including copper-nickel, have very good resistance to biofouling, and this property is used to advantage. Copper-nickel is used to minimize biofouling on intake screens, seawater pipe work, water boxes, cladding of pilings, and mesh cages in fish farming.[89]

Problems associated with biofilms. Once bacteria begin to colonize surfaces and produce biofilms, numerous problems begin to arise, including reduction of heat-transfer efficiency, fouling, corrosion, and scale. When biofilms develop in low-flow areas, such as cooling-tower film fill, they may initially go unnoticed, since they will not interfere with flow or evaporative efficiency. Over time, the biofilm becomes more complex, often with filamentous development. The matrix provided will accumulate debris that may impede or completely block flow.

Biofilms may be patchy and highly channelized, allowing nutrient-bearing water to flow through and around the matrix. When excessive algal biofilms develop, portions may break loose and be transported to other parts of the system, causing blockage as well as providing nutrients for accelerated bacterial and fungal growth. Biofilms can cause fouling of filtration and ion-exchange equipment.

Calcium ions are fixed into the biofilm by the attraction of carboxylate functional groups on the polysaccharides. In fact, divalent cations, such as calcium and magnesium, are integral in the formation of gels in some extracellular polysaccharides. A familiar biofilm-induced mineral deposit is the calcium phosphate scale that the dental hygienist removes from teeth. When biofilms grow on tooth surfaces, they are referred to as plaques. If these plaques are not continually removed, they will accumulate calcium salts, mainly calcium phosphate, and form tartar (scale).

When iron- and manganese-oxidizing organisms colonize a surface, they begin to oxidize available reduced forms of these elements and produce a deposit. In the case of iron-oxidizing organisms, ferrous iron is oxidized to the ferric form, with the electron lost in the process being utilized by the bacterium for energy production. As the bacterial colony becomes encrusted with iron (or manganese) oxide, a differential oxygen concentration cell may develop, and the corrosion process will begin. The ferrous iron produced at the anode will then provide even more ferrous iron for the bacteria to oxidize. The porous encrustation (tubercle) may potentially become an autocatalytic corrosion cell or may provide an environment suitable for the growth of sulfate-reducing bacteria.

Friction factor. A fluid flowing through a pipe experiences drag from the pipe surface. This drag reduces flow velocity and increases the pressure required to sustain a given flow rate. Microbial fouling can lead to a sharply increased friction factor with a marked loss of system capacity. Losses up to 55 percent have been reported for water supply systems, with significant effects being seen in large-diameter conduits made of cement and concrete as well as in steel piping.[90] Most of the loss is attributable to increased surface roughness (Table 2.36). Laboratory studies indicate that the friction factor does not increase until the biofilm extends beyond the viscous sublayer of fluid flow normally associated with the pipe wall (typically 30 μm). The friction factor is a function of Reynolds number for different biofilm thickness in turbulent flow.

Unlike hard scale deposits, the biofilm has an irregular surface and spongy (viscoelastic) behavior that exaggerate its drag on fluid flow. Extraordinary increases in friction factor may be related to cells protruding into the bulk water flow and influencing the hydrodynamics at the biofilm–bulk water interface. The extra drag on fluid flow would be analogous to that caused by waving water weeds in a stream. Another common problem encountered in industrial operations is the fouling of screens or pumping systems with debris sloughed off or eroded from fouling deposits. Again, the presence of biological slimes exacerbates such problems by capturing clays and other particulates which might have otherwise remained suspended and passed through the system.[65]

Heat exchange. Bacterial fouling of heat exchangers can occur quickly as a result of a process leak or influx of nutrients. The sudden increase in nutrients in a previously nutrient-limited environment will send

TABLE 2.36 **Roughness of Biofilms Compared to Inorganic Deposits**

Material	Thickness, μm	Relative roughness
Biofilm	40	0.003
	165	0.01
	300	0.06
	500	0.15
Scale, $CaCO_3$	165	0.0001
	224	0.0002
	262	0.0006

bacterial populations into an accelerated logarithmic growth phase, with rapid accumulation of biofilm. The biofilms that develop will then interfere with heat-transfer efficiency.

Sizing of heat exchangers assumes a certain heat-transfer efficiency between the bulk fluid and metal wall. Because biofilms more or less behave like gels on the metal surface, heat transfer can occur only by conduction through the biofilm. The thermal conductivity of biofilms is similar to that of water but much less than that of metals.[87] On the basis of relative thermal conductivities (Table 2.37), a biofilm layer 41 μm thick offers the same resistance to heat transfer as a titanium tube wall 1000 μm thick.

In calculating the impact of biofouling, changes in the advective (convective) heat transfer from the bulk fluid to the biofilm must also be considered because biofilm roughness can influence turbulence at the interface between the biofilm and the bulk fluid. This increase in local turbulence may actually improve the advective heat transfer to the biofilm, partially offsetting the loss in conductive heat transfer. On balance, inorganic deposits give a lower net increase in heat-transfer resistance than biofilms of similar thickness. Case histories in power plant operations have shown that decreases of 30 percent in heat-transfer efficiency can occur in 30 to 60 days as a result of biofouling.

2.6.3 Biofilm control

Introduction. In the natural gas industry, MIC has been estimated to cause 15 to 30 percent of corrosion-related pipeline failures. The growth of bacteria on surfaces in cooling and process-water systems can lead to significant deposits and corrosion problems. Once the severity of these problems is understood, the importance of controlling biofilms becomes quite clear.

Protection from microbial problems can be designed into a system by selection of materials which do not support microbial growth, use of

TABLE 2.37 Thermal Conductivity of Biofilms Compared to Inorganic Deposits and Metals

Material	Thermal conductivity, $W \cdot m^{-1} \cdot K^{-1}$
Biofilm	0.6
Scale, $CaCO_3$	2.6
Iron oxide, Fe_2O_3	2.3
Water	0.6
Carbon steel	52
Stainless steel	16
Copper	384
Titanium	16

cathodic protection, or use of protective coatings. Operating conditions can sometimes be altered to discourage growth, and addition of biocides is common. Avoiding and removing surface deposits is a very effective control procedure. In industrial plant settings, this usually involves physically cleaning production units during shutdowns. Table 2.38 presents some physical methods that have been used to clean fouled surfaces.

In pipelines, cleaning tools called pigs can be pushed through the line by fluid flow without shutdown, often accompanied by slugs of treatment chemicals designed to coat freshly exposed metal surfaces with corrosion inhibitors or to kill microbial communities disturbed by passage of the cleaning tool. In practice, the strategy adopted is an

TABLE 2.38 Some Physical Methods of Cleaning Biofouled Surfaces

Method	Comments
Flushing	Simplest method Limited efficacy Biofilms thinner than viscous sublayer not sheared
Backwashing	Effective for loosely adherent films in tubes, on filters, to a certain extent in ion exchangers
Air bumping	Very limited efficacy
Sponge balls	
Abrasive	Demonstrated efficacy, but possible problems because of the abrasion of protective oxide films
Nonabrasive	Extensively used in industry Problems with thick biofilms and with smearing organics
Sand scouring	Difficult to control abrasive effects
Brushing	Very effective Limited applicability Expensive Can lead to the selection of firmly adhering species
Hot water, steam	Used in high-purity water systems with good results Saves expensive and possibly harmful and toxic chemicals Hot-water systems may select for thermophiles and are reported to carry biofilms including mycobacteria
Irradiation	Very low effectiveness against biofilms Entrapped particles and opaque biofilms may shield bacteria
Ultrasonic energy	Promising method for soft biofilms Application limited to nonsensitive material Some biofilms are extremely stable

exercise in risk management in which capital and operating costs are balanced against the chance and consequence of operating inefficiencies caused by undue fouling or leaks.[65] Biofilms can be controlled through the use of biocides or biodispersants and by limiting nutrients. In the United States, industries spend $1.2 billion annually on biocidal chemicals to fight MIC.[91] Biocides, both oxidizing and nonoxidizing, can be effective in overall biofilm control when applied properly. Table 2.39 lists some of the advantages and disadvantages related to the use of some of the biocides that have been used in the past or are being considered for usage in the future.

The effectiveness of biocides depends on a number of factors, such as the kind of biocide, the biocide concentration, the biocide demand, interference with other dissolved substances, pH, temperature, contact time, types of organisms present, their physiological state, and, most important, the presence of biofilms. As a general rule, the higher the temperature, the longer the contact time needs to be, and the higher the concentration of the disinfectant, the greater should be the degree of disinfecting. A sanitation program will include weakening the biofilm matrix and the strength of the adhesion to the supporting surface by chemicals prior to the application of shear stress by flushing.[92]

The oxidizing biocides, such as chlorine, bromine, chlorine dioxide, and ozone, can be extremely effective in destroying both the extracellular polysaccharides and the bacterial cells. When using oxidizing biocides, one must be sure to obtain a sufficient residual for a long enough duration to effectively oxidize the biofilm. It is generally more effective to maintain a higher residual for several hours than to continuously maintain a low residual. Continuous low-level feed may not achieve an oxidant level sufficient to oxidize the polysaccharides and expose the bacteria to the oxidant.

Too often, microbiological control efforts focus only on planktonic counts, i.e., the number of bacteria in the bulk water. While some useful data may be gathered from monitoring daily bacterial counts, monthly or weekly counts have little meaningful use. Planktonic counts do not necessarily correlate with the amount of biofilm present. In addition, planktonic organisms are not generally responsible for deposit and corrosion problems. There are a few exceptions, such as a closed-loop system, in which planktonic organisms may degrade corrosion inhibitors, produce high levels of H_2S, or reduce pH.

Another misconception involves the use of chlorine at alkaline pH (> 8.0). It is often thought that chlorine is ineffective in controlling microorganisms at elevated pH. This is only partly true. Certainly, the hypohalous acid form of chlorine (HOCl) is more effective at killing cells than the hypohalite form (OCl^-). However, the hypohalite is actually very effective at oxidizing the extracellular polysaccharides and

TABLE 2.39 Advantages and Disadvantages of Industrial Biocides

	Advantages	Disadvantages
Chlorine	Broad spectrum of activity Residual effect Advanced technology available Can be generated on site Active in low concentrations Destroys biofilm matrix and supports detachment	Toxic by-products Degradation of recalcitrant compounds to biodegradable products Development of resistance Corrosive Reacts with extracellular polymer substances (EPS) in biofilms Low penetration characteristic in biofilms Oxidizes to elemental sulfur (extremely difficult to remove from surfaces)
Hypochlorite	Cheap Effective Destabilizes and detaches the biofilm matrix Easy to handle Used for biofilm thickness control	Poor stability Oxidizing Rapid aftergrowth observed Toxic by-products Corrosive Does not control initial adhesion
ClO_2	Can be generated on site Low pH dependency Low sensitivity to hydrocarbons Effective in low concentrations	Explosive gas Safety problems Toxic by-products
Chloramine	Good penetration of biofilms Specific to microorganisms Less toxic by-products High residual effect because of lower reactivity with water ingredients	Less effective than chlorine against suspended bacteria Bacterial resistance observed
Bromine	Very effective against broad microbial spectrum	Toxic by-products Development of bacterial resistance

TABLE 2.39 Advantages and Disadvantages of Industrial Biocides (Continued)

	Advantages	Disadvantages
H_2O_2	Decomposes to water and oxygen Relatively nontoxic Can easily be generated in situ Weakens biofilm matrix and supports detachment and removal	High concentrations (>3%) necessary Frequent resistance Corrosive
Peracetic acid	Very effective in small concentrations Broad spectrum Kills spores Decomposes to acetic acid and water No toxic by-products known Penetrates biofilms	Corrosive Not very stable Increases DOC[†]
Formaldehyde	Low costs Broad antimicrobial spectrum Stability Easy application	Resistance in some organisms Toxicity Suspected of promoting cancer Reacts with protein-fixing biofilms on surfaces Legal restrictions
Glutaraldehyde	Effective in low concentrations Cheap Nonoxidizing Noncorrosive	Does not penetrate biofilms well Degrades to formic acid Raises DOC[+]
Isothiazolones	Effective at low concentrations Broad antibiotic spectrum	Problems with compatibility with other water ingredients Inactivation by primary amines
QUAC*	Effective in low concentrations Surface activity supports biofilm detachment Relatively nontoxic Adsorb to surfaces and prevent biofilm growth	Inactivation at low pH or in the presence of Ca^{2+} or Mg^{2+} Development of resistance

*Quarternarg ammonia compounds.
[†]Dissolved organic carbon.

the proteinaceous attachment structures. Therefore, the use of chlorine in alkaline cooling waters can still be extremely effective.

In order to enter biofilm bacteria cells, chemical species in the water must run the gauntlet of biopolymers that range in properties from strongly anionic and hydrophilic to hydrophobic. This is exacerbated by the fact that many species will accelerate their production of exopolysaccharides in response to stress, including that caused by biocides. The amount of a biocide necessary to achieve a given level of disinfection is often expressed as the product of concentration and time. The same organisms living within a biofilm have been found to require 150 times the C × T factor of hypochlorous acid to achieve a 2 log reduction in activity as they do in planktonic form.[93]

Oxidizers such as chlorine, bromine, and especially peroxide can break down the polymers making up the biofilm; however, this activity is greatest at high pH, where they are in the form of anions. It is in their neutral forms (hypochlorous acid, hypobromous acid, and hydrogen peroxide) at lower pH that they are able to diffuse across the membranes of cells and enter them. Within the cells, each of these oxidizers causes damage by producing free radicals that destroy proteins and nucleic acids. The anions, however, are repelled by the negative charges of the biofilm polymers and act only superficially on the biofilm.

Biodispersants are usually nonionic molecules which adsorb to metal surfaces more readily than biofilm polymers. By reducing the size of the biofilm points of contact with the surface, these materials cause biofilm to detach from the surface. In practice, since these materials do not actually destroy biofilm, the biofilm detaches from high-flow areas and accumulates in low-flow areas. The low solubility of these materials can also lead to fouling by the biodispersants themselves.

Continuous use of nonoxidizing biocides has been avoided in water treatment, partly because of the expense, but also because of the risk of selecting for organisms that are resistant to one particular biocide. Thus, dual alternating slug-fed biocide programs have become more common over the past decade. Continuous use of oxidizing biocides has increased, however, based on the assumption that very few organisms show any resistance to them and that if a nearly sterile system is maintained, biofilm will not develop. While planktonic organisms may not show increased resistance to oxidizers, biofilms do. Studies have shown that with the continued use of chlorine, biofilms will display an increased iron content.[93] The iron acts as a reducing agent, limiting the ability of chlorine to diffuse into the biofilm. In drinking-water lines, biofilm can accumulate at a continuous chlorine concentration of 0.8 mg/L. There has not been found to be a level of continuous halogenation at which biofilm is controlled without significantly increased corrosion. Halogenation also produces carcinogenic halogenated organics and in some applications unpleasant odors.[84]

Hydrogen peroxide and per-salts have been applied for biofilm removal to eliminate the odor and environmental drawbacks of halogens and reduce corrosion; however, the extremely high doses required can still be corrosive and have been regarded as uneconomical for routine use. An alternative route is based on the use of hydroperoxide ion (O_2H^-), a hydrolytic agent far more powerful even than hydroxide, to destroy the polymers making up the biofilm matrix. Because this anion alone would be repelled by these anionic polymers, a phase transfer catalyst is included. The phase transfer catalyst carries the hydroperoxide past the protective biofilm matrix to where it can do its destructive duty. This combination, along with peroxide activators, was originally developed for the detoxification of biological warfare agents. In addition to hydrolysis, the formation of oxygen bubbles from the decomposition of peroxide within the biofilm disrupts the biofilm. The combination is available as a powder or as binary liquids.[93]

Nonoxidizing biocides are also effective in controlling biofilm. Effective control is greatly dependent on frequency of addition, level of feed, and resistance of the incumbent population to the product being fed. A typical application for effective control may include a slug addition of product two to five times a week. As with oxidizing biocides, frequency and dosage will depend on the system conditions. It is generally most effective to alternate nonoxidizing biocides at every addition to ensure broad-spectrum control. Most nonoxidizing biocides will have little effect in destroying the extracellular polysaccharides found in the biofilm. However, many biocides may be able to penetrate and kill bacteria found within the biofilm. Combining the use of nonoxidizing and oxidizing biocides is a very effective means of controlling biofilm.

Biofilm control programs can be made more effective through the utilization of a biopenetrant/dispersant product. Products that penetrate and loosen the biopolymer matrix will not only help to slough the biofilm but also expose the microorganisms to the effects of the biocide. These products are especially effective in systems that have a high organic carbon loading and a tendency to foul. They are typically fed in slug additions prior to biocide feed. A recent development in biodispersant technology is making this approach more effective and popular than ever before. Enzyme technologies that will break down the extracellular polysaccharides and degrade bacterial attachment structures are currently being developed and patented. These technologies, although expensive, may provide biofilm control where biocide use is environmentally restricted or provide a means of quickly restoring fouled cooling-water systems to a clean, efficient operable state.

A practical example: ozone treatment for cooling towers. Ozone is a molecule consisting of three oxygen atoms and is commonly denoted O_3. Under ambient conditions, ozone is very unstable, and as a result it has a relatively short half-life, usually less than 10 min. Ozone is a powerful biocide and virus deactivant and will oxidize many organic and inorganic substances. These properties have made ozone an effective chemical for water treatment for nearly a century. During the last 20 years, technological improvements have made smaller-scale, stand-alone commercial ozone generators both economically feasible and reliable. Using ozone to treat cooling-tower water is a relatively new practice; however, its market share is growing as a result of water and energy savings and environmental benefits relative to traditional chemical treatment processes.

Since the 1970s, when ozonation was first used in cooling towers, a number of cooling-tower operators have switched to it from multichemical treatment and are satisfied with the results. Ozone generation is accomplished by passing a high-voltage alternating current (6 to 20 kV) across a dielectric discharge gap through which air is injected. As air is exposed to the electricity, oxygen molecules disassociate and form single oxygen atoms, some of which combine with other oxygen molecules to form ozone. Different manufacturers have their own variations of components for ozone generators. Two different dielectric configurations exist, flat plates and concentric tubes. Most generators are installed with the tube configuration, since it offers easier maintenance.

Mass transfer of the ozone gas stream to the cooling-tower water is usually accomplished through a venturi in a recirculation line connected to the sump of the cooling tower, where the temperature of the water is the lowest. Since the solubility of ozone is very temperature-dependent, the point of lowest temperature allows the maximum amount of ozone to be introduced in solution to the tower. Mass-transfer equipment can take other forms: column-bubble diffusers, positive-pressure injection (U-tube), turbine mixer tank, and packed tower. The countercurrent column-bubble contactor is the most efficient and cost-effective but is not always useful in a cooling-tower setting because of space constraints.[94]

In a properly installed and operating system, bacterial counts are reduced, with a subsequent minimization of the buildup of biofilm on heat-exchanger surfaces. The resulting reduction in energy use, increase in cooling-tower operating efficiency, and reduction in maintenance effort provide cost savings as well as environmental benefit and compliance with regulations concerning discharge of wastewater from blowdown.

Most cooling-tower ozone treatment systems include the following components: an air dryer, an air compressor, water and oil coalescing filters, a particle filter, ozone injectors, an ozone generator, and a control system. Ambient air is compressed, dried, and then ionized in the generator to produce ozone. Ozone is typically applied to cooling water through a side stream of the circulating tower water.

Field tests have demonstrated that the use of ozone in place of chemical treatment can reduce the need for blowdown, and in some cases, where makeup water and ambient air are relatively clean, can eliminate it. As a result, cost savings accrue from decreased chemical and water use requirements and from a reduction of wastewater volume. There are also environmental benefits, as fewer chlorine or chlorinated compounds and other chemicals are discharged.

References

1. Agarwala, V. S., Bhagat, P. K., and Hardy, G. L., "Corrosion Detection and Monitoring of Aircraft Structures: An Overview," *AGARD Conference Proceedings 565, Corrosion Detection and Management of Advanced Airframe Materials,* Seville, Spain, AGARD, 1994.
2. Money, K. L., Corrosion Testing in the Atmosphere, in *Metals Handbook: Corrosion,* Metals Park, Ohio, ASM International, 1987, pp. 204–206.
3. Steinmayer, R. F., Land Vehicle Management, in *AGARD Lecture Series No. 141,* Neuilly-sur-Seine, France, NATO, 1985.
4. Callaghan, B. G., *Atmospheric Corrosion Testing in Southern Africa,* South Africa, Scientia Pub., 1991.
5. Cooke, G., Koch, G., and Frechan, R., Corrosion Detection & Life Cycle Analysis for Aircraft Structural Integrity, Report ADB-171678, Washington, D.C., Defense Technical Information Center, 1992.
6. Suga, S., and Suga, S., Cyclic Corrosion Tests in Japanese Industries, in Haynes, G. S. and Tellefsen, K., *Cyclic Cabinet Corrosion Testing,* STP 1238, Philadelphia, American Society for Testing and Materials, 1994, pp. 99–112.
7. Oesch, S., and Faller, M., Environmental Effects on Materials: The Effect of the Air Pollutants SO_2, NO_2, NO and O_3 on the Corrosion of Copper, Zinc and Aluminum. A Short Literature Survey and Results of Laboratory Exposures, *Corrosion Science,* **39**(9): 1505–1530 (1997).
8. Barton, K., *Protection against Atmospheric Corrosion,* London, John Wiley and Sons, 1976.
9. Graedel, T. E., GILDES Model Studies of Aqueous Cemistry. I. Formulation and Potential Applications of the Multi-Regime Model, *Corrosion Science,* **38**(12): 2153–2199 (1996).
10. Fyfe, D., *Corrosion,* Oxford, Butterworth-Heinemann, 1994.
11. Sharp, S., Protection of Control Equipment from Atmospheric Corrosion, *Materials Performance,* **29**(12): 43–48 (1990).
12. Dean, S. W., Classifying Atmospheric Corrosivity—A Challenge for ISO, *Materials Performance,* **32**(10): 53–58 (1993).
13. Dean, S. W., Analyses of Four Years of Exposure Data from the USA Contribution to the ISO CORRAG Program, in Kirk, W. W., and Lawson, H. H. (eds.), *Atmospheric Corrosion,* STP 1239, Philadelphia, American Society for Testing and Materials, 1995.
14. Tullmin, M., Roberge, P. R., and Little, M. A., "Aircraft Corrosion Surveillance in the Military," in *Corrosion 97,* Houston, NACE International, 1997, pp. 527-1–527-10.
15. Summitt, R., and Fink, F. T., PACER LIME: An Environmental Corrosion Severity Classification System, AFWAL-TR-80-4102, 1980.
16. Doyle D. P., and Wright, T. E., Rapid Method for Determining Atmospheric Corrosivity and Corrosion Resistance, in Ailor, W. H. (ed.), *Atmospheric Corrosion,* New York, John Wiley and Sons, 1982, pp. 227–243.

17. Pourbaix, M., The Linear Bilogarithmic Law for Atmospheric Corrosion, in Ailor, W. H. (ed.), *Atmospheric Corrosion,* New York, John Wiley and Sons, 1982, pp. 107–121.
18. *BETZ Handbook of Industrial Water Conditioning,* Trevose, Pa., BETZ, 1980.
19. *Manual on Water,* Philadelphia, American Society for Testing and Materials, 1969.
20. Chapter Six, Types of Pollutants,*http://riceinfo.rice.edu/armadillo/Galveston/Chap6/type.pollutant.html,* 1998.
21. Cole, R. H., Frederick, R. E., Healy, R. P., et al., Preliminary Findings of the Priority Pollutant: Monitoring Project of the Nationwide Urban Runoff Program, *Journal of the Water Pollution Control Federation,* **56:**898–908 (1984).
22. Issues of Water Quality: Heavy Metals in Drinking Water, *http://www.everpure.com/issues/iowq003.html,* 1998.
23. DeSilva, F. J., Essentials of Ion Exchange, *http://www.wqa.org/Technical/essentials-of-ion-exchange.html,* 1998.
24. Truesdell, A. H., and Jones, B. F., A Computer Program for Calculating Chemical Equilibria of Natural Waters, *Journal of Research, U.S. Geological Survey,* **2:**233–248 (1974).
25. Ferguson, R. J., Freedman, A. J., Fowler, G., et al., The Practical Application of Ion Association Model Saturation Level Indices to Commercial Water Treatment Problem Solving, in Amjad, Z. (ed.), *Mineral Scale Formation and Inhibition,* New York, Plenum Press, 1995, pp. 323–340.
26. Ferguson, R. J., "Computerized Ion Association Model Profiles Complete Range of Cooling System Parameters," *52nd Annual Meeting, International Water Conference,* Pittsburgh, Pa., IWC-91-47, 1991.
27. Musill, R. R., and Nielsen, H. J., "Computer Modeling of Cooling Water Chemistry," *45th Annual Meeting, International Water Conference,* Pittsburgh, Pa., IWC-84-104, 1984.
28. Pitzer, K. S., Thermodynamics of Electrolytes, *Journal of Physical Chemistry,* **77:**268–277 (1973).
29. Ferguson, R. J., Codina, O., Rule, W., and Baebel, R., "Real Time Control of Scale Inhibitor Feed Rate," *49th Annual Meeting, International Water Conference,* Pittsburgh, Pa., IWC-88-57, 1988.
30. Fowler, V. L., and Perona, J. J., Evaporation Studies on Oak Ridge National Laboratory Liquid Low Level Waste, ORNL/TM-12243, Oak Ridge National Laboratory, 1990.
31. Ferguson, R. J., and Freedman, A. J., "A Comparison of Scale Potential Indices with Treatment Program Results in Ozonated Systems," Paper #279, in *Corrosion 93,* Houston, NACE International, 1993.
32. Watercycle, Kimberton, Pa., French Creek Software, 1997.
33. DownHole SAT, Kimberton, Pa., French Creek Software, 1997.
34. Gill, J. S., Anderson, C. D., and Varsanik, R. G., "Mechanism of Scale Inhibition by Phosphonates," *44th Annual Meeting, International Water Conference,* Pittsburgh, Pa., 1983.
35. Ijseling, F. P., *General Guidelines for Corrosion Testing of Materials for Marine Applications,* London, The Institute of Materials, 1989.
36. Tomczak, T., Properties of Seawater, *http://gaea.es.flinders.edu.au/~mattom/ES1/lecture03.html,* 1998.
37. Hartt, W. H., Culberson, C. H., and Smith, S. W., Calcareous Deposits on Metal Surfaces in Seawater—A Critical Review, *Corrosion,* **40:**609–618 (1984).
38. LaQue, F. L., *Marine Corrosion: Causes and Prevention,* New York, John Wiley and Sons, 1975.
39. Schumacker, M., *Seawater Corrosion Handbook,* Park Ridge, N.J., Noyes Data Corporation, 1979.
40. Todd, B., Materials Selection for High Reliability Seawater Systems, *http://marine.copper.org/,* 1998.
41. Butler, G., and Mercer, A. D., "Corrosion of Steel in Potable Waters," *Nature,* **256:**719–720 (1975).
42. Robinson, W. C., Testing Soil for Corrosiveness, *Materials Performance,* **32:**56–58 (1993).
43. Pope, D. H., and Morris, E. A., III, Some Experiences with Microbiologically Influenced Corrosion of Pipelines, *Materials Performance,* **34:**23–28 (1995).

44. Palmer, J. D., Environmental Characteristics Controlling the Soil Corrosion of Ferrous Piping, in Chaker, V., and Palmer, J. D. (eds.), *Effects of Soil Characteristics on Corrosion,* Philadelphia, American Society for Testing and Materials, 1989, pp. 5–17.
45. Soil, *http://www.metalogic.be/MatWeb/reading/soil/e_soil.htm,* 1998.
46. Wagner, P., and Little, B., Impact of Alloying on Microbiologically Influenced Corrosion—A Review, *Materials Performance,* **32:**65–68 (1993).
47. Bianchetti, R. L., Corrosion and Corrosion Control of Prestressed Concrete Cylinder Pipelines—A Review, *Materials Performance,* **32:**62–66 (1993).
48. Fasullo, E. J., Infrastructure: The Battlefield of Corrosion, in Chaker, V. (ed.), *Forms and Control for Infrastructure,* Philadelphia, American Society for Testing and Materials, 1992.
49. Mehta, P. K., Durability—Critical Issues for the Future, *Concrete International,* **19**(7):27–33 (1997).
50. Nürnberger, U., Chloride Corrosion of Steel in Concrete, Fundamental Relationships—Practical Experience, Part 1 and Part 2, *Betonwek und Fertigteil-Technik,* 601–704 (1984).
51. Glass, G. K., and Buenfeld, N. R., The Presentation of the Chloride Threshold Level for Corrosion of Steel in Concrete, *Corrosion Science,* **39:**1001–1013 (1997).
52. Hansson, C. M., Concrete: The Advanced Industrial Material of the 21st Century, *Metallurgical and Materials Transactions A,* **26A:**1321–1341 (1995).
53. Perchanok, M. S, Manning, D. G., and Armstrong, J. J., Highway Deicers: Standards, Practice and Research in the Province of Ontario, MAMAT-91-13T-91-13, Toronto, Ministry of Transportation, Ontario, 1991.
54. Hartt, W. H., A Critical Evaluation of Cathodic Protection for Prestressing Steel in Concrete, in Page, C. L., Treadaway, K. W. J., and Bamforth, P. B. E. (eds.), *Corrosion of Reinforcement in Concrete,* London, Applied Science, 1990, pp. 515–524.
55. Electrochemical Chloride Extraction from Concrete Bridge Components, Technical Brief #2, Toronto, Canadian Strategic Highway Research Program (C-SHRP), 1995.
56. Sagues, A., Perez-Duran, H. M., and Powers, R.G., Corrosion Performance of Epoxy-Coated Reinforcing Steel in Marine Substructure Service, *Corrosion,* **47:**884–893 (1991).
57. Nürnberger, U., *Stainless Steel in Concrete—State of the Art Report,* London, The Institute of Materials, European Federation of Corrosion, 1996.
58. McDonald, D. D., Sherman, M. R., Pfeifer, D. W., et al., Stainless Steel Reinforcing as Corrosion Protection, *Concrete International,* **17**(5):65–70 (1995).
59. El-Jazairi, B., and Berke, N. S., The Use of Calcium Nitrite as a Corrosion Inhibiting Admixture to Steel Reinforcement in Concrete, in Page, C. L., Treadaway, K. W. J., and Bamforth, P. B. E. (eds.), *Corrosion of Reinforcement in Concrete,* London, Applied Science, 1990, pp. 571–585.
60. Tullmin, M., Mammoliti, L., Sohdi, R., et al., The Passivation of Reinforcing Steel Exposed to Synthetic Pore Solution and the Effect of Calcium Nitrite Inhibitor, *Cement Concrete and Aggregates,* **17:**134–144 (1995).
61. Concrete Bridge Component Evaluation Manual, Technical Brief #2, Toronto, Canadian Strategic Highway Research Program (C-SHRP), (1995).
62. Clifton, J. R., and Pommersheim, J. M., Predicting Remaining Service Life of Concrete, in Swamy, R. N. (ed.), *Corrosion and Corrosion of Steel in Concrete,* London, Sheffield Academic Press, 1994, pp. 619–638.
63. Cady, P. D., and Weyers, R. E., Predicting Service Life of Concrete Bridge Decks Subject to Reinforcement Corrosion, in Chaker, V. (ed.), *Forms and Control for Infrastructure,* Philadelphia, American Society for Testing and Materials, 1992, pp. 328–338.
64. Buenfeld, N. R., and Hassanein, N. M., Predicting the Life of Concrete using Neural Networks, *Proceedings of the Institution of Civil Engineers Structures and Buildings,* **128:**38–48 (1998).
65. Jack, T. R., Monitoring Microbial Fouling and Corrosion Problems in Industrial Systems, *Corrosion Reviews,* **17**(1):1-31 (1999).
66. Pope, D. H., Duquette, D., Wayner, P. C., Jr., et al., *Microbiologically Influenced Corrosion: A State-of-the-Art Review,* Columbus, Ohio, Materials Technology Institute, 1989.

67. Hill, E. C., *Microbial Aspects of Metallurgy,* New York, American Elsevier, 1970.
68. Tatnall, R. E., Introduction Part I, in Kobrin, G. (ed.), *Microbiologically Influenced Corrosion,* Houston, NACE International, 1993.
69. Ringas, C., and Robinson, F. P. A., Microbial Corrosion of Iron-Based Alloys, *Journal of SAIMM,* **87:**425–437 (1987).
70. Sanders, P. F., Biological Aspects of Marine Corrosion, *Metals Society World,* **12:** (1983).
71. Butler, M. A., and Ison, R. C. K., *Corrosion and Its Prevention in Waters,* New York, Van Nostrand Reinhold, 1966.
72. Geesey, G. G., Introduction Part II—Biofilm Formation, in Kobrin, G. (ed.), *Microbiologically Influenced Corrosion,* Houston, NACE International, 1993.
73. Hunik, J. H., van den Hoogen, M. P., de Boer, W, et al., Quantitative Determination of the Spatial Distribution of *Nitrosomonas europaea* and *Nitrobacter agilis* Cells Immobilized in k-Carrageenan Gel Beads by a Specific Fluorescent-Antibody Labelling Technique, *Applied and Environmental Microbiology,* **59:**1951–1954 (1993).
74. Pope, D. H., State of the Art Report on Monitoring, Prevention and Mitigation of Microbiologically Influenced Corrosion in the Natural Gas Industry, GRI-92/0382, Chicago, Gas Research Institute, 1992.
75. Costerton, J. W., and Colwell, R. R., *Native Aquatic Bacteria: Enumeration, Activity and Ecology,* (STP 695), Philadelphia, American Society for Testing and Materials, 1977.
76. Jorgenson, B. B., A Comparison of Methods for the Quantification of Bacterial Sulfate Reduction in Coastal Marine Sediments, *Geomicrobiology Journal,* **1:**49–64 (1978).
77. Odom, J. M., Jessie, K., Knodel, E., et al., Immunological Cross-Reactivities of Adenosine-5'-Phosphosulfate Reductases from Sulfate-Reducing and Sulfide Oxidizing Bacteria, *Applied and Environmental Microbiology,* **57:**727–733 (1991).
78. Bryant, R. D., Jansen, W., Boivin, J., et al., Effect of Hydrogenase and Mixed Sulfate-Reducing Bacterial Populations on the Corrosion of Steel, *Applied and Environmental Microbiology,* **57:**2804–2809 (1991).
79. Scott, P. J. B., and Davies, M., Survey of Field Kits for Sulfate Reducing Bacteria, *Materials Performance,* **31:**64–68 (1992).
80. Voordouw, G., Telang, A. J., Jack, T. R., et al., Identification of Sulfate-Reducing Bacteria by Hydrogenase Gene Probes and Reverse Sample Genome Probing, in Minear, R. A., Ford, A. M., Needham, L. L., et al. (eds.), *Applications of Molecular Biology in Environmental Chemistry,* Boca Raton, Fla., Lewis Publishers, 1995.
81. Sanders, P. F., Monitoring and Control of Sessile Microbes: Cost Effective Ways to Reduce Microbial Corrosion, in Sequeira, C. A. C., and Tiller, A. K. (eds.), *Microbial Corrosion—1,* New York, Elsevier Applied Science, 1988, pp. 191–223.
82. Gilbert, P. D., and Herbert, B. N., *Monitoring Microbial Fouling in Flowing Systems Using Coupons,* in Hopton, J. W., and Hill, E. C. (eds.), *Industrial Microbiological Testing,* London, Blackwell Scientific Publications, 1987, pp. 79–98.
83. Gerhardt, P., Murray, R. G. E., Costilow, R. N., et al., *Manual of Methods for General Bacteriology,* Washington, D.C., American Society of Microbiology, 1981.
84. deLaubenfels, E., Control of Biofilm in Aquatic Systems,*http://biofilms.com/index%20(report).html,* 1998.
85. Little, B., Wagner, P., Gerchakov, S. M., et al., The Involvement of a Thermophilic Bacterium in Corrosion Processes, *Corrosion,* **42:**533–536 (1986).
86. Caldwell, D. E., Korber, D. R., and Lawrence, J. R., Confocal Laser Microscopy and Computer Image Analysis in Microbial Ecology, in Marshall, K. C. (ed.), *Advances in Microbial Ecology,* vol. 12, New York, Plenum Press, 1992.
87. Christensen, B. E., Characklis, W. G., Physical and Chemical Properties of Biofilms, in Characklis, W. G., and Marshall, K. C. (eds.), *Biofilms,* Toronto, Wiley Interscience, 1990, pp. 93–130.
88. Turakhia, M. H., Cooksey, K. E., Characklis, W. G., Influence of a Calcium Specific Chelant on Biofilm Removal, *Applied Environmental Microbiology,* **46:**1236–1238 (1983).

89. Powell, C. A., Copper-Nickel Alloys—Resistance to Corrosion and Biofouling, *http://marine.copper.org/*, 1998.
90. Characklis, W. G., Turakhia, M. H., and Zelver, N., Transport and Interfacial Transfer Phenomena, in Characklis, W. G., and Marshall, K. C. (eds,), *Biofilms,* Toronto, Wiley Interscience, 1990, pp. 265–340.
91. Frank, J. R., and Enzien, M., Environmentally Acceptable Methods Control Pipeline Corrosion at Lower Costs, *http://www.es.anl.gov/htmls/corrosion.html,* 1997.
92. Flemming, H. C., Biofouling in Water Treatment, in Flemming, H. C., and Geesey, G. G. (eds.), *Biofouling and Biocorrosion in Industrial Water Systems,* Berlin, Springer-Verlag, 1991, pp. 47–80.
93. LeChevallier, M. W., Biocides and the Current Status of Biofouling Control in Water Systems, in Flemming, H. C., and Geesey, G. G. (eds.), *Biofouling and Biocorrosion in Industrial Water Systems,* Berlin, Springer-Verlag, 1991, pp. 113–132
94. Lamarre, L., A Fresh Look at Ozone, *The EPRI Journal,* **22:**6–15 (1997).

Chapter

3

High-Temperature Corrosion

3.1 Thermodynamic Principles	222
3.1.1 Standard free energy of formation versus temperature diagrams	222
3.1.2 Vapor species diagrams	223
3.1.3 Two-dimensional isothermal stability diagrams	229
3.2 Kinetic Principles	229
3.2.1 The Pilling-Bedworth relationship	231
3.2.2 Micromechanisms and rate laws	233
3.3 Practical High-Temperature Corrosion Problems	237
3.3.1 Oxidation	238
3.3.2 Sulfidation	245
3.3.3 Carburization	255
3.3.4 Metal dusting	258
3.3.5 Nitridation	260
3.3.6 Gaseous halogen corrosion	260
3.3.7 Fuel ash and salt deposits	262
3.3.8 Corrosion by molten salts	263
3.3.9 Corrosion in liquid metals	263
3.3.10 Compilation and use of corrosion data	264
References	265

High-temperature corrosion is a form of corrosion that does not require the presence of a liquid electrolyte. Sometimes, this type of damage is called *dry corrosion* or *scaling*. The term *oxidation* is ambivalent because it can either refer to the formation of oxides or to the mechanism of oxidation of a metal (i.e., its change to a higher valence than the metallic state). Strictly speaking, high-temperature oxidation is only one type of high-temperature corrosion, but it is the most important high-temperature corrosion reaction. In most industrial environments, oxidation often participates in the high-temperature corrosion reactions, regardless of the predominant mode of corrosion.[1] Alloys often rely upon the oxidation reaction to develop a protective scale to resist corrosion attack such as sulfidation, carburization, and other forms of high-temperature attack. In general, the names of the corrosion mechanisms are determined by the most abundant dominant corrosion products. For example, oxidation implies oxides, sulfidation implies sulfides, sulfidation/oxidation implies sulfides plus oxides, and carburization implies carbides.[2]

Oxidizing environments refer to high-oxygen activities, with excess oxygen. Reducing environments are characterized by low-oxygen activities, with no excess oxygen available. Clearly, oxide scale formation is more limited under such reducing conditions. It is for this reason that reducing industrial environments are generally considered to be more corrosive than the oxidizing variety. However, there are important exceptions to this generalization. At high temperatures, metals can react "directly" with the gaseous atmosphere. Electrochemical reaction sequences remain, however, the underlying mechanism of high-temperature corrosion. The properties of high-temperature oxide films, such as their thermodynamic stability, ionic defect structure, and detailed morphology, play a crucial role in determining the oxidation resistance of a metal or alloy in a specific environment. High-temperature corrosion is a widespread problem in various industries such as

- Power generation (nuclear and fossil fuel)
- Aerospace and gas turbine
- Heat treating
- Mineral and metallurgical processing
- Chemical processing
- Refining and petrochemical
- Automotive
- Pulp and paper
- Waste incineration

3.1 Thermodynamic Principles

3.1.1 Standard free energy of formation versus temperature diagrams

Often determination of the conditions under which a given corrosion product is likely to form is required (e.g., in selective oxidation of alloys). The plots of the standard free energy of reaction (ΔG^0) as a function of temperature, commonly called Ellingham diagrams, can help to visualize the relative stability of metals and their oxidized products. Figure 3.1 shows an Ellingham diagram for many simple oxides.[3] The values of ΔG^0 on an Ellingham diagram are expressed as kilojoules per mole of O_2 to normalize the scale and be able to compare the stability of these oxides directly (i.e., the lower the position of the line on the diagram, the more stable is the oxide).[4] For a given reaction [Eq. (3.1)] and assuming that the activities of M and MO_2 are taken as unity, Eq. (3.2) or its logarithmic form [Eq. (3.3)] may be used to express the oxygen partial pressure at which the metal and oxide coexist (i.e., the dissociation pressure of the oxide).

$$M + O_2 = MO_2 \qquad (3.1)$$

$$p_{O_2}^{M/MO_2} = e^{\Delta G^0/RT} \qquad (3.2)$$

$$\text{Log } p_{O_2}^{M/MO_2} = \frac{\Delta G^0}{RT} \qquad (3.3)$$

The values of $p_{O_2}^{M/MO_2}$ may be obtained directly from the Ellingham diagram by drawing a straight line from the origin marked O through the free-energy line at the temperature of interest and reading the oxygen pressure from its intersection with the scale at the right side labeled $\text{Log}(p_{O_2})$. Values for the pressure ratio H_2/H_2O for equilibrium between a given metal and oxide may be obtained by drawing a similar line from the point marked H to the scale labeled H_2/H_2O ratio, and values for the equilibrium CO/CO_2 ratio may be obtained by drawing a line from point C to the scale CO/CO_2 ratio. See Gaskell, Chap. 10, for a more detailed discussion of the construction and use of Ellingham diagrams for oxides.[3] Table 3.1 lists the coexistence equations, temperature ranges, and standard energy changes that can be used to construct such diagrams.[5] Ellingham diagrams may, of course, be constructed for any class of compounds.

3.1.2 Vapor species diagrams

Vapor species that form in any given high-temperature corrosion situation often have a strong influence on the rate of attack, the rate generally being accelerated when volatile corrosion products form. Gulbransen and Jansson have shown that metal and volatile oxide species are important in the kinetics of high-temperature oxidation of carbon, silicon, molybdenum, and chromium.[6] Six types of oxidation phenomena were identified:

1. At low temperature, diffusion of oxygen and metal species through a compact oxide film
2. At moderate and high temperatures, a combination of oxide film formation and oxide volatility

TABLE 3.1 Thermodynamic Data for Reactions Involving Oxygen

Range, K	Coexistence equation (oxidation reaction)	Standard free energy change, J
900–1154	$Pd + 0.5O_2 = PdO$	$-114{,}200 + 100\,T\ (°K)$
884–1126	$2\,Mn_3O_4 + 0.5O_2 = 3\,Mn_2O_3$	$-113{,}360 + 92.0\,T$
298–1300	$3\,CoO + 0.5O_2 = Co_3O_4$	$-183{,}200 + 148\,T$
892–1302	$Cu_2O + 0.5O_2 = 2\,CuO$	$-130{,}930 + 94.5\,T$
1396–1723	$1.5\,UO_2 + 0.5O_2 = 0.5\,U_3O_8$	$-166{,}900 + 84\,T$
878–1393	$U_4O_9 + 0.5O_2 = 4/3\,U3°8\text{-}z$	$-164{,}400 + 82\,T$
967–1373	$2\,Fe_3O_4 + 0.5O_2 = 3\,Fe_2O_3$	$-246{,}800 + 141.8\,T$
1489–1593	$2\,Cu + 0.5O_2 = Cu_2O$	$-166{,}900 + 43.5\,T$
1356–1489	$2\,Cu + 0.5O_2 = Cu_2O$	$-190{,}300 + 89.5\,T$
924–1328	$2\,Cu + 0.5O_2 = Cu_2O$	$-166{,}900 + 71.1\,T$
992–1393	$3\,MnO + 0.5O_2 = Mn_3O_4$	$-222{,}540 + 111\,T$
1160–1371	$Pb + 0.5O_2 = PbO$	$-190{,}580 + 74.9\,T$
772–1160	$Pb + 0.5O_2 = PbO$	$-215{,}000 + 96.0\,T$
911–1376	$Ni + 0.5O_2 = NiO$	$-233{,}580 + 84.9\,T$
1173–1373	$Co + 0.5O_2 = CoO$	$-235{,}900 + 71.5\,T$
973–1273	$10\,WO_{2.90} + O_2 = 10\,WO_3$	$-279{,}400 + 112\,T$
973–1273	$10\,WO_{2.72} + O_2 = 10\,WO_{2.90}$	$-284{,}000 + 101\,T$
973–1273	$1.39\,WO_2 + 0.5O_2 = 1.30\,WO_{2.72}$	$-249{,}310 + 62.7\,T$
973–1273	$0.5\,W + 0.5O_2 = 0.5\,WO_2$	$-287{,}400 + 84.9\,T$
949–1273	$3\,FeO + 0.5O_2 = Fe_3O_4$	$-311{,}600 + 123\,T$
770–980	$Sn + 0.5O_2 = SnO_2$	$-293{,}230 + 108\,T$
903–1540	$Fe + 0.5O_2 = FeO$	$-263{,}300 + 64.8\,T$
1025–1325	$0.5\,Mo + 0.5O_2 = 0.5\,MoO_2$	$-287{,}600 + 83.7\,T$
1050–1300	$2\,NbO_2 + 0.5O_2 = Nb_2O_5$	$-313{,}520 + 78.2\,T$
693–1181	$Zn + 0.5O_2 = ZnO$	$-355{,}890 + 107.5\,T$
1300–1600	$0.66\,Cr + 0.5O_2 = 0.33\,Cr_2O_3$	$-371{,}870 + 83.7\,T$
1050–1300	$NbO + 0.5O_2 = NbO_2$	$-360{,}160 + 72.4\,T$
923–1273	$Mn + 0.5O_2 = MnO$	$-388{,}770 + 76.3\,T$
1539–1823	$Mn + 0.5O_2 = MnO$	$-409{,}500 + 89.5\,T$
1073–1273	$0.4\,Ta + 0.5O_2 = 0.2\,Ta_2O_5$	$-402{,}400 + 82.4\,T$
1050–1300	$Nb + 0.5O_2 = NbO$	$-420{,}000 + 89.5\,T$
298–1400	$0.5\,U + 0.5O_2 = 0.5\,UO_2$	$-539{,}600 + 83.7\,T$
1380–2500	$Mg_{(v)} + 0.5O_2 = MgO$	$-759{,}600 - 30.83\,T\log T + 317\,T$
923–1380	$Mg_{(l)} + 0.5O_2 = MgO$	$-608{,}200 - 1.00\,T\log T + 105\,T$
1124–1760	$Ca + 0.5\,O_2 = CaO$	$-642{,}500 + 107\,T$
1760–2500	$Ca_{(v)} + 0.5O_2 = CaO$	$-795{,}200 + 195\,T$

3. At moderate and high temperatures, the formation of volatile metal and oxide species at the metal-oxide interface and transport through the oxide lattice and mechanically formed cracks in the oxide layer

4. At moderate and high temperatures, the direct formation of volatile oxide gases

5. At high temperature, the gaseous diffusion of oxygen through a barrier layer of volatilized oxides

6. At high temperature, spalling of metal and oxide particles.

High-Temperature Corrosion 225

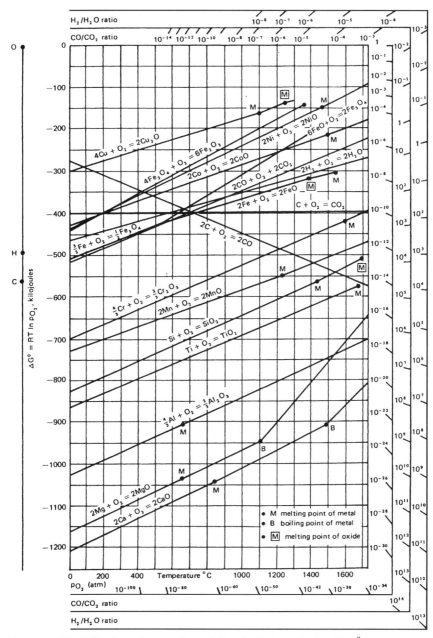

Figure 3.1 The Ellingham diagram for metallurgically important oxides.[3]

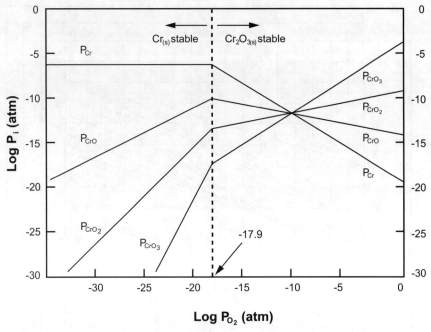

Figure 3.2 Vapor equilibria in the Cr-O system at 1250°C.

The diagrams most suited for presentation of vapor pressure data in oxide systems are $\text{Log}(p_{M_xO_y})$ versus $\text{Log}(p_{O_2})$ at constant temperature and Arrhenius diagrams of $\text{Log}(p_{M_xO_y})$ versus $1/T$ at constant oxygen pressure. The following example will illustrate the construction of the first type of these diagrams by considering the Cr-O system at 1200°C or 1473 K. Only one condensed oxide, Cr_2O_3, is formed under these conditions of high-temperature oxidation.[4] The thermochemical data for producing the vapor equilibria diagram shown in Fig. 3.2 are given in Table 3.2. The thermochemical data for $Cr_2O_{3(s)}$, $Cr_{(s)}$, and the four possible gaseous species $Cr_{(g)}$, $CrO_{(g)}$, $CrO_{2(g)}$, and $CrO_{3(g)}$ were obtained from a commercial database.[7]

The enthalpy (H_T), entropy (S), and heat capacity (C_p) of each species considered were calculated using Eqs. (3.4) to (3.6) in which T was set at 1473 K (Table 3.3). In these equations H_{tr} and T_{tr} represent, respectively, the enthalpy and temperature associated with any phase transition encountered between the reference temperature (298.15 K) and the temperature of interest.

$$H_{(T)} = H_{f(298.15)} + \int_{298.15}^{T} C_p \, dT + H_{tr} \qquad (3.4)$$

TABLE 3.2 Thermochemical Data for the Cr-O System

Species	State	Enthalpy H, kJ·mol⁻¹	Entropy S, J·mol⁻¹·K⁻¹	Heat capacity, Cp				Temperature range	
				A, J·mol⁻¹·K⁻¹	B	C	D	T_1, K	T_2, K
Cr	Gas	397.48	174.305	20.786	0	0	0	298.15	700
	Gas			15.456	2.556	16.828	0.874	700	3000
Cr	Solid	0	23.64	17.715	22.966	−0.377	−9.033	298.15	1000
	Solid			18.067	15.531	−16.698	0	1000	2130
	Liquid	16.933		39.33	0	0	0	2130	2945
Cr_2O_3	Solid	−1139.7	81.199	104.307	24.451	7.59	−3.807	298.15	2603
CrO	Gas	188.28	239.266	24.518	25.33	0.222	−11.201	298.15	400
	Gas			34.292	3.979	−4.351	−1.138	400	1600
CrO_2	Gas	−75.312	269.24	30.878	46.689	0	−15.782	298.15	400
	Gas			45.309	18.552	−7.222	−7.632	400	1100
CrO_3	Gas	−292.88	266.169	74.002	7.686	−18.393	−1.644	298.15	3000
O_2	Gas	0	205.146	31.321	3.895	−3.105	−0.335	298.15	5000

TABLE 3.3 Thermochemical Data for the Cr-O System at 1473 K

Species	State	H, kJ·mol^{-1}	S, J·mol^{-1}·K^{-1}	G, kJ·mol^{-1}
Cr	Gas	422.02	207.58	116.25
Cr	Solid	36.97	70.78	−67.29
Cr$_2$O$_3$	Solid	−993.71	276.68	−1401.27
CrO	Gas	230.37	295.28	−204.57
CrO$_2$	Gas	−12.73	351.72	−530.81
CrO$_3$	Gas	−204.60	381.78	−766.96
O$_2$	Gas	39.67	257.73	−339.97

$$S = S^0_{(298.15)} + \int_{298.15}^{T} \frac{C_p}{T} dT + \sum \frac{H_{tr}}{T_{tr}} \tag{3.5}$$

$$C_p = A + B\,10^{-3}\,T + C\,10^5\,T^{-2} + D\,10^{-6}\,T^2 \tag{3.6}$$

The free energy (G) for each species considered was then calculated with Eq. (3.7) and used to evaluate the stability of these species and the predicted energy of reaction for each equilibrium (Table 3.4).

$$G = H - TS \tag{3.7}$$

Vapor pressures of species at equilibrium with either the metal or its most stable oxide (i.e., Cr$_2$O$_3$) must then be determined. The boundary between these regions is the oxygen pressure for the Cr/Cr$_2$O$_3$ equilibrium expressed in Eq. (3.8).

$$2\text{Cr}_{(s)} + 1.5\text{O}_{2(g)} = \text{Cr}_2\text{O}_{3(s)} \tag{3.8}$$

for which the equilibrium constant (K_p) is evaluated with Eq. (3.9), giving an equilibrium pressure of oxygen calculated with Eq. (3.10).

$$\text{Log } K_p = \frac{-\Delta G^0}{2.303RT} \tag{3.9}$$

$$\text{Log}(p_{O_2}) = -\frac{2}{3} \text{Log } K_p^{\text{Cr}_2\text{O}_3} = -17.90 \tag{3.10}$$

The dotted vertical line in Fig. 3.2 represents this boundary. At low oxygen pressure it can be seen that the presence of Cr$_{(g)}$ is independent of oxygen pressure. For oxygen pressures greater than the Cr/Cr$_2$O$_3$ equilibrium, the Cr$_{(g)}$ vapor pressure may be obtained from the equilibrium expressed in Eq. (3.11).

$$0.5\text{Cr}_2\text{O}_{3(s)} = \text{Cr}_{(g)} + 0.75\text{O}_{2(g)} \tag{3.11}$$

TABLE 3.4 Standard Energy of Reactions for the Cr-O System at 1473 K

Reaction	ΔG^0, kJ·mol^{-1}
$2\ Cr_{(s)} + 1.5\ O_2 = Cr_2O_3$	−756.72
Over $Cr_{(s)}$	
$Cr_{(s)} = Cr_{(g)}$	183.54
$Cr_{(s)} + 0.5\ O_2 = CrO_{(g)}$	32.71
$Cr_{(s)} + O_2 = CrO_{2(g)}$	−123.54
$Cr_{(s)} + 1.5\ O_2 = CrO_{3(g)}$	−189.71
Over Cr_2O_3	
$0.5\ Cr_2O_{3(s)} = Cr_{(g)} + 0.75\ O_2$	561.90
$0.5\ Cr_2O_{3(s)} = CrO_{(g)} + 0.25\ O_2$	411.07
$0.5\ Cr_2O_{3(s)} + 0.25\ O_2 = CrO_{2(g)}$	254.81
$0.5\ Cr_2O_{3(s)} + 0.75\ O_2 = CrO_{3(g)}$	188.65

The other lines in Fig. 3.2 are obtained by using similar equilibrium equations (Table 3.4). The vapor equilibria presented in Fig. 3.2 show that significant $Cr_{(g)}$ vapor pressures are developed at low-oxygen partial pressure (e.g., at the alloy-scale interface of a Cr_2O_3-forming alloy) but that a much larger pressure of $CrO_{3(g)}$ develops at high-oxygen partial pressure. This high $CrO_{3(g)}$ pressure is responsible for the thinning of Cr_2O_3 scales by vapor losses during exposure to oxygen-rich environments.

3.1.3 Two-dimensional isothermal stability diagrams

When a metal reacts with a gas containing more than one oxidant, a number of different phases may form depending on both thermodynamic and kinetic considerations. Isothermal stability diagrams, usually constructed with the logarithmic values of the activities or partial pressures of the two nonmetallic components as the coordinate axes, are useful in interpreting the condensed phases that can form. The metal-sulfur-oxygen stability diagrams for iron, nickel, cobalt, and chromium are shown in Figs. 3.3 to 3.6. One important assumption in these diagrams is that all condensed species are at unit activity. This assumption places important limitations on the use of the diagrams for alloy systems.

3.2 Kinetic Principles

The first step in high-temperature oxidation is the adsorption of oxygen on the surface of the metal, followed by oxide nucleation and the growth

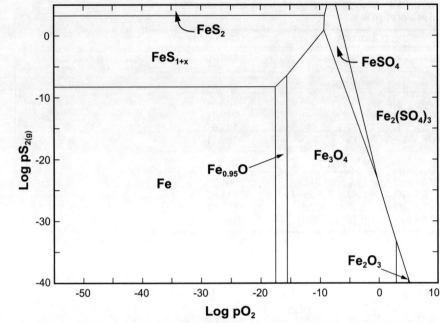

Figure 3.3 Stability diagram of the Fe-S-O system at 870°C.

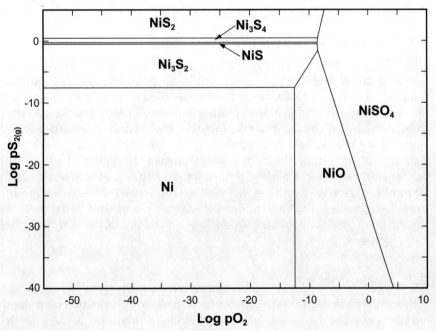

Figure 3.4 Stability diagram of the Ni-S-O system at 870°C.

High-Temperature Corrosion 231

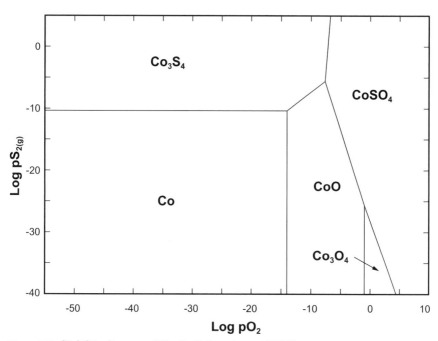

Figure 3.5 Stability diagram of the Co-S-O system at 870°C.

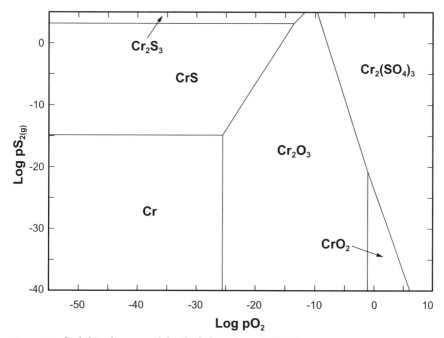

Figure 3.6 Stability diagram of the Cr-S-O system at 870°C.

of the oxide nuclei into a continuous oxide film covering the metal substrate. Defects, such as microcracks, macrocracks, and porosity may develop in the film as it thickens. Such defects tend to render an oxide film nonprotective, because, in their presence, oxygen can easily reach the metal substrate to cause further oxidation.

3.2.1 The Pilling-Bedworth relationship

The volume of the oxide formed, relative to the volume of the metal consumed, is an important parameter in predicting the degree of protection provided by the oxide scale. If the oxide volume is relatively low, tensile stresses can crack the oxide layers. Oxides, essentially representing brittle ceramics, are particularly susceptible to fracture and cracking under such tensile stresses. If the oxide volume is very high, stresses will be set up that can lead to a break in the adhesion between the metal and oxide. For a high degree of protection, it can thus be argued that the volume of the oxide formed should be similar to that of the metal consumed. This argument is the basis for the Pilling-Bedworth ratio:

$$\text{PB} = \frac{\text{volume of oxide produced}}{\text{volume of metal consumed}} = \frac{Wd}{nDw}$$

where W = molecular weight of oxide
D = density of the oxide
n = number of metal atoms in the oxide molecule
d = density of the metal
w = atomic weight of the metal

PB ratios slightly greater than 1 could be expected to indicate "optimal" protection, with modest compressive stresses generated in the oxide layer. Table 3.5 provides the PB ratio of a few metal/oxide systems.[4] In practice, it has been found that PB ratios are generally poor predictors of the actual protective properties of scales. Some of the reasons advanced for deviations from the PB rule include[8]

- Some oxides actually grow at the oxide-air interface, as opposed to the metal-oxide interface.
- Specimen and component geometries can affect the stress distribution in the oxide films.
- Continuous oxide films are observed even if PB < 1.
- Cracks and fissures in oxide layers can be "self-healing" as oxidation progresses.
- Oxide porosity is not accurately predicted by the PB parameter.

TABLE 3.5 Oxide-Metal Volume Ratios of Some Common Metals

Oxide	Oxide/metal volume ratio
K_2O	0.45
MgO	0.81
Na_2O	0.97
Al_2O_3	1.28
ThO_2	1.30
ZrO_2	1.56
Cu_2O	1.64
NiO	1.65
FeO (on α-Fe)	1.68
TiO_2	1.70–1.78
CoO	1.86
Cr_2O_3	2.07
Fe_3O_4 (on α-Fe)	2.10
Fe_2O_3 (on α-Fe)	2.14
Ta_2O_5	2.50
Nb_2O_5	2.68
V_2O_5	3.19
WoO_3	3.30

- Oxides may be highly volatile at high temperatures, leading to nonprotective properties, even if predicted otherwise by the PB parameter.

3.2.2 Micromechanisms and rate laws

Oxide microstructures. On the submolecular level, metal oxides contain defects, in the sense that their composition deviates from their ideal stoichiometric chemical formulas. By nature of the defects found in their ionic lattices, they can be subdivided into three categories:[8]

A *p-type metal-deficit oxide* contains metal cation vacancies. Cations diffuse in the lattice by exchange with these vacancies. Charge neutrality in the lattice is maintained by the presence of electron holes or metal cations of higher than average positive charge. Current is passed by positively charged electron holes.

An *n-type cation interstitial metal-excess oxide* contains interstitial cations, in addition to the cations in the crystal lattice. Charge neutrality is established through an excess of negative conduction electrons, which provide for electrical conductivity.

An *n-type anion vacancy oxide* contains oxygen anion vacancies in the crystal lattice. Current is passed by electrons, which are present in excess to establish charge neutrality.

Electrochemical nature of oxidation reactions. High-temperature oxidation reactions proceed by an electrochemical mechanism, with some similarities to aqueous corrosion. For example, the reaction

$$M + \tfrac{1}{2}O_2 \rightarrow MO$$

proceeds by two basic separate reactions:

$$M \rightarrow M^{2+} + 2e^- \text{ (anodic reaction)}$$

and

$$\tfrac{1}{2}O_2 + 2e^- \rightarrow O^{2-} \text{ (cathodic reaction)}$$

The growth of an n-type cation interstitial oxide at the oxide-gas interface is illustrated in Fig. 3.7. Interstitial metal cations are liberated at the metal-oxide interface and migrate through the interstices of the oxide to the oxide-gas interface. Conduction band electrons also migrate to the oxide-gas interface, where oxide growth takes place. For the n-type anion vacancy oxide, film growth tends to occur at the metal-oxide interface, as shown in Fig. 3.8. Conduction band electrons migrate to the oxide-gas interface, where the cathodic reaction occurs. The oxygen anions produced at this interface migrate through the oxide lattice by exchange with anion vacancies. The metal cations are provided by the anodic reaction at the metal-oxide interface.

In the case of the p-type metal deficit oxides, metal cations produced by the anodic reaction at the metal-oxide interface migrate to the oxide-gas interface by exchange with cation vacancies. Electron charge is effectively transferred to the oxide-gas interface by the movement of electron holes in the opposite direction (toward the metal-oxide interface). The cathodic reaction and oxide growth thus tend to occur at the oxide-gas interface (Fig. 3.9).

The important influence of the diffusion of defects (excess cations, cation vacancies, or anion vacancies) through the oxide film on oxidation rates should be apparent from Figs. 3.7 to 3.9. Conduction electrons (or electron holes) are much more mobile compared to these larger defects and therefore are not important in controlling the reaction rates. For example, if nickel oxide (NiO) is considered as a p-type metal deficient oxide, the oxidation rate of nickel depends on the diffusion rate of cation vacancies. If this oxide is doped with Cr^{3+} impurity ions, the number of cation vacancies increases to maintain charge neutrality. A higher oxidation rate is thus to be expected in the presence of these impurities. By this mechanism, a nickel alloy containing a few percentages of chromium does indeed oxidize more rapidly than pure nickel.[9] From these considerations, a clearer picture of require-

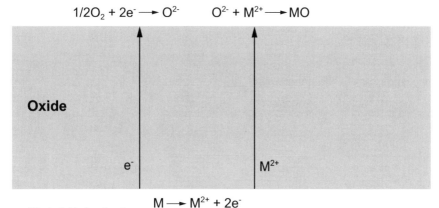

Figure 3.7 Schematic description of the growth of a cation interstitial n-type oxide occurring at an oxide-gas interface.

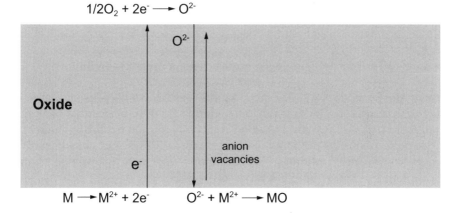

Figure 3.8 Film growth of an n-type anion vacancy oxide occurring at a metal-oxide interface.

ments for protective oxides has emerged. Oxide film properties imparting high degrees of protection include

- Good film adherence to the metal substrate
- High melting point
- Resistance to evaporation (low vapor pressure)

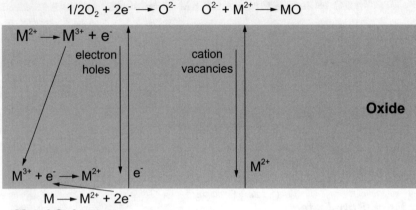

Figure 3.9 Schematic description of a cathodic reaction and oxide growth occurring at the oxide-gas interface.

- Thermal expansion coefficient similar to that of the metal
- High temperature plasticity
- Low electrical conductivity
- Low diffusion coefficients for metal cations and oxygen anions

Basic kinetic models. Three basic kinetic laws have been used to characterize the oxidation rates of pure metals. It is important to bear in mind that these laws are based on relatively simple oxidation models. Practical oxidation problems usually involve alloys and considerably more complicated oxidation mechanisms and scale properties than considered in these simple analyses.

Parabolic rate law. The parabolic rate law [Eq. (3.12)] assumes that the diffusion of metal cations or oxygen anions is the rate controlling step and is derived from Fick's first law of diffusion. The concentrations of diffusing species at the oxide-metal and oxide-gas interfaces are assumed to be constant. The diffusivity of the oxide layer is also assumed to be invariant. This assumption implies that the oxide layer has to be uniform, continuous, and of the single phase type. Strictly speaking, even for pure metals, this assumption is rarely valid. The rate constant, k_p, changes with temperature according to an Arrhenius-type relationship.

$$x^2 = k_p t + x_0 \tag{3.12}$$

where x = oxide film thickness (or mass gain due to oxidation, which is proportional to oxide film thickness)

t = time
k_p = the rate constant (directly proportional to diffusivity of ionic species that is rate controlling)
x_0 = constant

Logarithmic rate law. The logarithmic rate law [Eq. (3.13)] is a following empirical relationship, which has no fundamental underlying mechanism. This law is mainly applicable to thin oxide layers formed at relatively low temperatures and therefore is rarely applicable to high-temperature engineering problems.

$$x = k_e \log(ct + b) \tag{3.13}$$

where k_e = rate constant and c and b are constants.

Linear rate law and catastrophic oxidation. The linear rate law [Eq. (3.14)] is also an empirical relationship that is applicable to the formation and buildup of a nonprotective oxide layer:

$$x = k_L t \tag{3.14}$$

where k_L = rate constant.

It is usually to be expected that the oxidation rate will decrease with time (parabolic behavior), due to an increasing oxide thickness acting as a stronger diffusion barrier with time. In the linear rate law, this effect is not applicable, due to the formation of highly porous, poorly adherent, or cracked nonprotective oxide layers. Clearly, the linear rate law is highly undesirable.

Metals with linear oxidation kinetics at a certain temperature have a tendency to undergo so-called catastrophic oxidation (also referred to as breakaway corrosion) at higher temperatures. In this case, a rapid exothermic reaction occurs on the surface, which increases the surface temperature and the reaction rate even further. Metals that may undergo extremely rapid catastrophic oxidation include molybdenum, tungsten, osmium, rhenium, and vanadium, associated with volatile oxide formation.[9] In the case of magnesium, ignition of the metal may even occur. The formation of low-melting-point oxidation products (eutectics) on the surface has also been associated with catastrophic oxidation. The presence of vanadium and lead oxide contamination in gases deserves special mention because they pose a risk to inducing extremely high oxidation rates.

3.3 Practical High-Temperature Corrosion Problems

The oxidation rate laws described above are simple models derived from the behavior of pure metals. In contrast, practical high-temperature corrosion problems are much more complex and involve the use of alloys. For practical problems, both the corrosive environment and the high-temperature corrosion mechanism(s) have to be understood. In the introduction, it was pointed out that several high-temperature corrosion mechanisms exist. Although considerable data is available from the literature for high-temperature corrosion in air and low-sulfur flue gases and for some other common refinery and petrochemical environments, small variations in the composition of a process stream or in operating conditions can cause markedly different corrosion rates. Therefore, the most reliable basis for material selection is operating experience from similar plants and environments or from pilot plant evaluation.[10]

There are several ways of measuring the extent of high-temperature corrosion attack. Measurement of weight change per unit area in a given time has been a popular procedure. However, the weight change/area information is not directly related to the thickness (penetration) of corroded metal, which is often needed in assessing the strength of equipment components. Corrosion is best reported in penetration units, which indicate the sound metal loss. A metallographic technique to determine with relative precision the extent of damage is illustrated in Fig. 3.10.[11] The parameters shown in Fig. 3.10 relate to cylindrical specimens and provide information about the load-bearing section (metal loss) and on the extent of grain boundary attack that can also affect structural integrity.

When considering specific alloys for high-temperature service, it is imperative to consider other properties besides the corrosion resistance. It would be futile, for example, to select a stainless steel with high-corrosion resistance for an application in which strength requirements could not be met. In general, austenitic stainless steels are substantially stronger than ferritic stainless steels at high temperatures, as indicated by a comparison of stress rupture properties (Fig. 3.11) and creep properties (Fig. 3.12).[11] The various high-temperature corrosion mechanisms introduced earlier are described in more detail in the following sections. The common names for the alloys mentioned in these sections are listed in Table 3.6 with their Unified Numbering System (UNS) alloy number, when available, and their generic type. The composition of these alloys can be found in App. E.

3.3.1 Oxidation

Oxidation is generally described as the most commonly encountered form of high-temperature corrosion. However, the oxidation process itself is not always detrimental. In fact, most corrosion and heat-resistant alloys rely on the formation of an oxide film to provide corrosion resistance. Chromium oxide (Cr_2O_3, chromia) is the most common of such films. In many industrial corrosion problems, oxidation does not occur in isolation; rather a combination of high-temperature corrosion

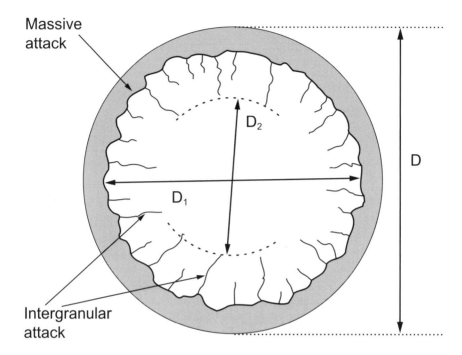

D = original diameter
D_1 = diameter of apparently useful metal
D_2 = diameter of metal unaffected by intergranular attack

Figure 3.10 Metallographic method of measuring hot corrosion attack.

mechanisms causes material degradation when contaminants (sulfur, chlorine, vanadium, etc.) are present in the atmosphere. Strictly speaking, the oxidation process is only applicable to uncontaminated air and clean combustion atmospheres.

For a given material, the operating temperature assumes a critical role in determining the oxidation rate. As temperature is increased, the rate of oxidation also increases. Sedriks has pointed out important differences in temperature limits between intermittent and continuous service.[11] It has been argued that thermal cycling in the former causes cracking and spalling damage in protective oxide scales, resulting in lower allowable operating temperatures. Some alloys' behavior (austenitic stainless steels) follows this argument, whereas others (ferritic stainless steels) actually behave in the opposite manner.[11] Increased chromium content is the most common way of improving oxidation resistance.

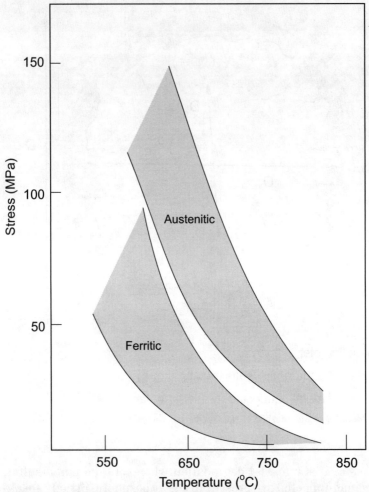

Figure 3.11 Ranges of rupture strength (rupture in 10,000 h) for typical ferritic and austenitic stainless steels.

Apart from chromium, alloying additions used to enhance oxidation resistance include aluminum, silicon, nickel, and some of the rare earth metals. For oxidation resistance above 1200°C, alloys that rely on protective Al_2O_3 (alumina) scale formation are to be preferred over those forming chromia.[12] Increasing the nickel content of the austenitic stainless steels up to about 30%, can have a strong beneficial synergistic effect with chromium.

Fundamental metallurgical considerations impose limits on the amount of alloying additions that can be made in the design of engineering alloys. Apart from oxidation resistance, the mechanical prop-

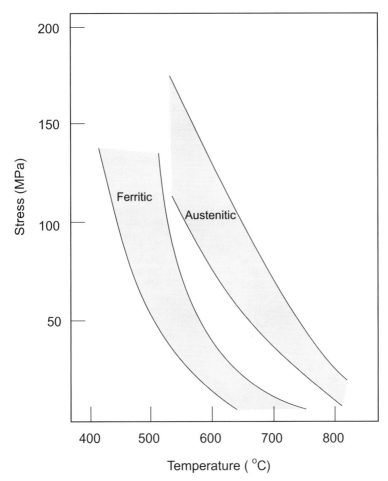

Figure 3.12 Ranges of creep strength (1% in 10,000 h) for typical ferritic and austenitic stainless steels.

erties must be considered together with processing and manufacturing characteristics. Metallurgical phases that can result in severe embrittlement (such as sigma, Laves, and Chi phases) tend to form in highly alloyed materials during high-temperature exposure. In the presence of embrittling metallurgical phases, the ductility and toughness at room temperature are extremely poor. A practical example of such problems involves the collapse of the internal heat-resisting lining of a cement kiln. Few commercial alloys contain more than 30% chromium. Silicon is usually limited to 2% and aluminum to less than 4% in wrought alloys. Yttrium, cerium, and the other rare earth elements are usually added only as a fraction of a percent.[10]

TABLE 3.6 Common Names and UNS Alloy Number of Alloys Used in High-Temperature Applications (Compositions Given in App. E)

Common name	UNS alloy number	Generic family
6	R30016	Ni-, Ni-Fe-, Co-base alloy
25	R30605	Ni-, Ni-Fe-, Co-base alloy
188	R30188	Ni-, Ni-Fe-, Co-base alloy
214	N07214	Ni-, Ni-Fe-, Co-base alloy
230	N06230	Ni-, Ni-Fe-, Co-base alloy
263	N07041	Ni-, Ni-Fe-, Co-base alloy
304	S30400	Austenitic stainless steel
310	S31000	Austenitic stainless steel
316	S31600	Austenitic stainless steel
330	S33000	Austenitic stainless steel
333	N06333	Ni-, Ni-Fe-, Co-base alloy
410	S41000	Martensitic stainless steel
430	S43000	Ferritic stainless steel
446	S44600	Ferritic stainless steel
556	R30556	Ni-, Ni-Fe-, Co-base alloy
600	N06600	Ni-, Ni-Fe-, Co-base alloy
601	N06601	Ni-, Ni-Fe-, Co-base alloy
617	N06617	Ni-, Ni-Fe-, Co-base alloy
625	N06625	Ni-, Ni-Fe-, Co-base alloy
718	N07718	Ni-, Ni-Fe-, Co-base alloy
825	N08825	Ni-, Ni-Fe-, Co-base alloy
2205	S31803	Duplex stainless steellex
1Cr-0.5Mo	K11597	Steel
2.25Cr-1Mo	K21590	Steel
253 MA	S30815	Austenitic stainless steel
5Cr-0.5Mo	K41545	Steel
6B	R30016	Ni-, Ni-Fe-, Co-base alloy
800 H	N08810	Ni-, Ni-Fe-, Co-base alloy
9Cr-1Mo	S50400	Steel
ACI HK	J94224	Cast SS
Alloy 150(UMCo-50)		Ni-, Ni-Fe-, Co-base alloy
Alloy HR-120		Ni-, Ni-Fe-, Co-base alloy
Alloy HR-160		Ni-, Ni-Fe-, Co-base alloy
Carbon Steel	G10200	Steel
Copper	C11000	Copper
Incoloy DS		Ni-, Ni-Fe-, Co-base alloy
Incoloy 801		Ni-, Ni-Fe-, Co-base alloy
Incoloy 803		Ni-, Ni-Fe-, Co-base alloy
Inconel 602		Ni-, Ni-Fe-, Co-base alloy
Inconel 671		Ni-, Ni-Fe-, Co-base alloy
Multimet	R30155	Ni-, Ni-Fe-, Co-base alloy
Nickel	N02270	Ni-, Ni-Fe-, Co-base alloy
René 41		Ni-, Ni-Fe-, Co-base alloy
RA330	N08330	Ni-, Ni-Fe-, Co-base alloy
S	N06635	Ni-, Ni-Fe-, Co-base alloy
Waspaloy		Ni-, Ni-Fe-, Co-base alloy
X	N06002	Ni-, Ni-Fe-, Co-base alloy

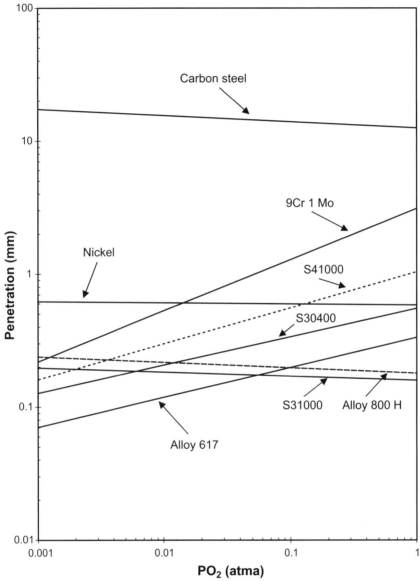

Figure 3.13 Effect of oxygen partial pressure upon metal penetration of some common alloys by oxidation after exposure for 1 year at 930°C.

An interesting approach to circumvent the above problems of bulk alloying is the use of surface alloying. In this approach, a highly alloyed (and highly oxidation resistant) surface layer is produced, whereas the substrate has a conventional composition and metallurgical properties. Bayer has described the formation of a surface alloy

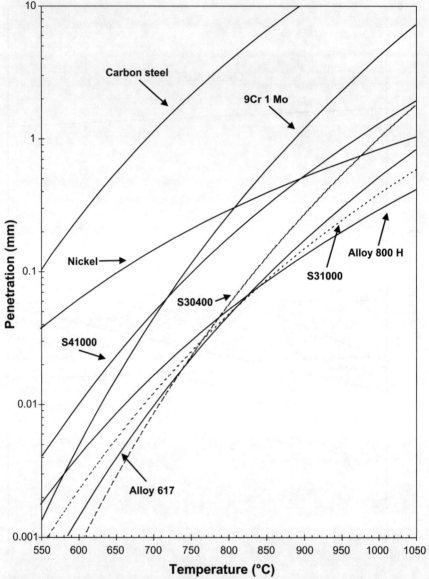

Figure 3.14 Effect of temperature upon metal penetration of some common alloys by oxidation after exposure for 1 year to air.

containing as much as 50% aluminum, by using a pack cementation vapor aluminum diffusion process.[13] The vapor aluminum-diffused surface layer is hard and brittle, but the bulk substrate retains the properties of conventional steels.

Extensive testing of alloys has shown that many alloys establish parabolic time dependence after a minimum time of 1000 h in air at temperatures above 900°C. If the surface corrosion product (scale) is removed or cracked so that the underlying metal is exposed to the gas, the rate of oxidation is faster. The influence of O_2 partial pressure on oxidation above 900°C is specific to each alloy, as illustrated for some common alloys in Fig. 3.13. Most alloys do not show a strong influence of the O_2 concentration upon the total penetration. Alloys such as Alloy HR-120, and Alloy 214 even exhibit slower oxidation rates as the O_2 concentration increases. These alloys are rich in Cr or Al, whose oxides are stabilized by increasing O_2 levels. Alloys, which generally exhibit increased oxidation rates as the O_2 concentration increase, are S30400, S41000, and S44600 stainless steels and 9Cr-1Mo, Incoloy DS, alloys 617, and 253MA. These alloys tend to form poor oxide scales.[2]

Most alloys tend to have increasing penetration rates with increasing temperature for all oxygen concentrations. Some exceptions are alloys with 1 to 4% Al such as alloy 214. These alloys require higher temperatures to form Al_2O_3 as the dominant surface oxide, which grows more slowly than the Cr_2O_3 that dominates at lower temperatures. Figure 3.14 summarizes oxidation after 1 year for some common alloys exposed to air.[2]

The alloy composition can influence metal penetration occurring by subsurface oxidation along grain boundaries and within the alloy grains, as schematically shown in Fig. 3.15.[2] Most of the commercial heat-resistant alloys are based upon combinations of Fe-Ni-Cr. These alloys show about 80 to 95% of the total penetration as subsurface oxidation. Some alloys change in how much of the total penetration occurs by subsurface oxidation as time passes, until long-term behavior is established, even though the corrosion product morphologies may remain constant. Alloys vary greatly in the extent of surface scaling and subsurface oxidation. Tests were conducted in flowing air at 980, 1095, 1150, and 1250°C for 1008 h. The results of these tests, in terms of metal loss and average metal affected (metal loss and internal penetration), are presented in Table 3.7.[1]

3.3.2 Sulfidation

Sulfidation is a common high-temperature corrosion-failure mechanism. As the name implies, it is related to the presence of contamination by sulfur compounds. When examining this form of damage microscopically, a "front" of sulfidation is often seen to penetrate into the affected alloy. Localized pitting-type attack is also possible. A distinction can be made between sulfidation in gaseous environments and corrosion in the presence of salt deposits on corroding surfaces. Only the former is considered in this section; the latter is included in the section on salt and ash deposit corrosion. Lai has divided gaseous environments associated with sulfidation into the following three categories:[12]

TABLE 3.7 Results of 1008-h Static Oxidation Tests on Iron, Nickel, and Cobalt Alloys in Flowing Air at Different Temperatures

Alloy	Temperature, °C							
	980		1095		1150		1250	
	Loss, mm	Affected, mm	Loss, mm	Affected, mm	Loss, mm	Affected, mm	Loss, mm	Affected, mm
214	0.0025	0.005	0.0025	0.0025	0.005	0.0075	0.005	0.018
601	0.013	0.033	0.03	0.067	0.061	0.135	0.11	0.19
600	0.0075	0.023	0.028	0.041	0.043	0.074	0.13	0.21
230	0.0075	0.018	0.013	0.033	0.058	0.086	0.11	0.20
S	0.005	0.013	3.01	0.033	0.025	0.043	> 0.81	> 0.81
617	0.0075	0.033	3.015	0.046	0.028	0.086	0.27	0.32
333	0.0075	0.025	0.025	0.058	0.05	0.1	0.18	0.45
X	0.0075	0.023	0.038	0.069	0.11	0.147	> 0.9	> 0.9
671	0.0229	0.043	0.038	0.061	0.066	0.099	0.086	0.42
625	0.0075	0.018	0.084	0.12	0.41	0.46	> 1.2	> 1.2
Waspaloy	0.0152	0.079	0.036	0.14	0.079	0.33	> 0.40	> 0.40
R-41	0.0178	0.122	0.086	0.30	0.21	0.44	> 0.73	> 0.73
263	0.0178	0.145	0.089	0.36	0.18	0.41	> 0.91	> 0.91
188	0.005	0.015	0.01	0.033	0.18	0.2	> 0.55	> 0.55
25	0.01	0.018	0.23	0.26	0.43	0.49	> 0.96	> 0.96
150	0.01	0.025	0.058	0.097	> 0.68	> 0.68	> 1.17	> 1.17
6B	0.01	0.025	0.35	0.39	> 0.94	> 0.94	> 0.94	> 0.94
556	0.01	0.028	0.025	0.067	0.24	0.29	> 3.8	> 3.8
Multimet	0.01	0.033	0.226	0.29	> 1.2	> 1.2	> 3.7	> 3.7
800H	0.023	0.046	0.14	0.19	0.19	0.23	0.29	0.35
RA330	0.01	0.11	0.02	0.17	0.041	0.22	0.096	0.21
S31000	0.01	0.028	0.025	0.058	0.075	0.11	0.2	0.26
S31600	0.315	0.36	> 1.7	> 1.7	> 2.7	> 2.7	> 3.57	> 3.57
S30400	0.14	0.21	> 0.69	> 0.69	> 0.6	> 0.6	> 1.7	> 1.73
S44600	0.033	0.058	0.33	0.37	> 0.55	> 0.55	> 0.59	> 0.59

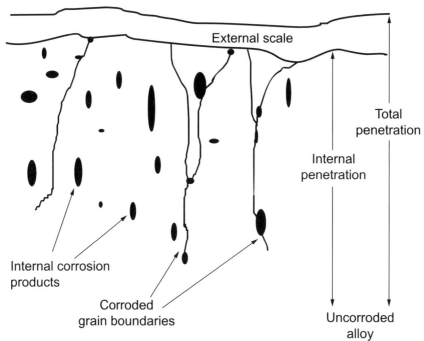

Figure 3.15 Schematic view of total penetration measurement for a typical corrosion product morphology.

- Hydrogen-hydrogen sulfide mixtures or sulfur vapor of a highly reducing nature
- Moderately reducing mixed gas environments that contain mixtures of hydrogen, water, carbon dioxide, carbon monoxide, and hydrogen sulfide
- Sulfur dioxide-containing atmospheres

In the first category, sulfides rather than protective chromia are thermodynamically stable. Hydrogen-hydrogen sulfide mixtures are found in catalytic reformers in oil refining operations. Organic sulfur compounds such as mercaptans, polysulfides, and thiophenes, as well as elemental sulfur, contaminate practically all crude oils in various concentrations and are partially converted to hydrogen sulfide in refining operations. Hydrogen sulfide in the presence of hydrogen becomes extremely corrosive above 260 to 288°C. Sulfidation problems may also be encountered at lower temperatures. Increased temperatures and higher hydrogen sulfide contents generally lead to higher degradation rates.

For catalytic reforming, the 18Cr-8Ni austenitic stainless steels grades are considered to be adequately resistant to sulfidation. The use of stabilized grades is advisable. Some sensitization is unavoidable if exposure in the sensitizing temperature range is continuous or long term. Stainless equipment subjected to such exposure and to sulfidation corrosion should be treated with a 2% soda ash solution or an ammonia solution immediately upon shutdown to avoid the formation of polythionic acid, which can cause severe intergranular corrosion and stress cracking.[10] Vessels for high-pressure hydrotreating and other heavy crude fraction upgrading processes (e.g., hydrocracking) are usually constructed of one of the Cr-Mo alloys. To control sulfidation, they are internally clad with one of the 300 series austenitic stainless steels. In contrast, piping, heat exchangers, valves, and other components exposed to high-temperature hydrogen-hydrogen sulfide environments are usually entirely constructed out of these austenitic stainless alloys. Figure 3.16 illustrates the corrosion behavior of austenitic steels as a function of hydrogen concentration and temperature.[11] In some designs alloy 800H has been used for piping and headers. In others, centrifugally cast HF-modified piping has been used.[10]

The effects of temperature and H_2S concentration upon sulfidation of alloys often used in oil refining services are shown in Figs. 3.17 to 3.21, which represent the metal losses expected after 1 year of exposure (note the decreasing corrosion penetration scale in Figs. 3.18 to 3.20). The carbon steel line, in Fig. 3.17, stops for lower concentrations of H_2S because FeS is not stable and the steel does not corrode in such environment.[2] Increasing the temperature and H_2S concentration increases the sulfidation rate. It is typical that a temperature increase of 55°C will double the sulfidation rate, whereas increasing the H_2S concentration by a factor of 10 may be needed to double the sulfidation rate. Therefore, changes of H_2S concentration are generally less significant than temperature variations.

Increasing the Cr content of the alloy greatly slows the sulfidation, as seen in progression from 9Cr-1Mo, S41000, S30400, 800H, 825, and 625 (Fig. 3.21). The ranges of H_2S concentration represented in these figures span the low H_2S range of catalytic reformers to the high H_2S concentrations expected in modern hydrotreaters. A summary of maximum allowable temperatures that will limit the extent of metal loss by sulfidation to less than 0.25 mm is shown in Table 3.8 for several gas compositions of H_2S-H_2 at a pressure of 34 atm, which is similar to hydrotreating in an oil refinery.[2] The maximum allowable temperatures for alloys exposed to different gas pressures and compositions can be evaluated with this information.

In the second category, the presence of oxidizing gases such as H_2O (steam) or CO_2 slow the sulfidation rate below that expected if only the H_2S-H_2 concentrations were considered. This can be important because gases, which are thought to contain only H_2S-H_2, often also contain some H_2O. For example, a gas, which has been well mixed and equili-

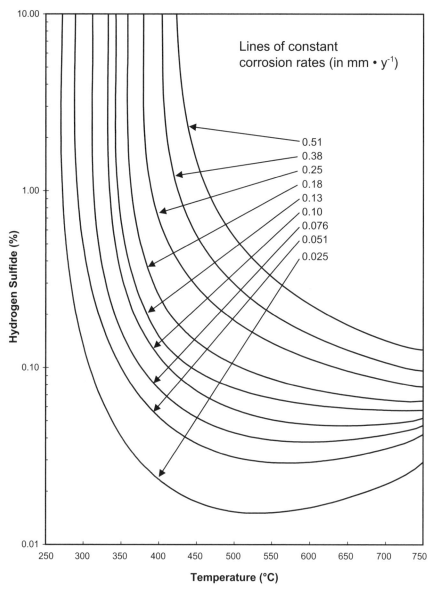

Figure 3.16 Effect of temperature and hydrogen sulfide concentration on corrosion rates of austenitic stainless steels for exposure longer than 150 h.

brated with water at room temperature, may contain up to 2% water vapor in the gas. Sulfidation rate predictions based only upon the H_2S-H_2 concentrations may overestimate the rate of metal loss. The precise mechanism of how H_2O slows sulfidation by H_2S is still unclear, although numerous studies have confirmed this effect. This slowed corrosion rate is sometimes called sulfidation/oxidation because it repre-

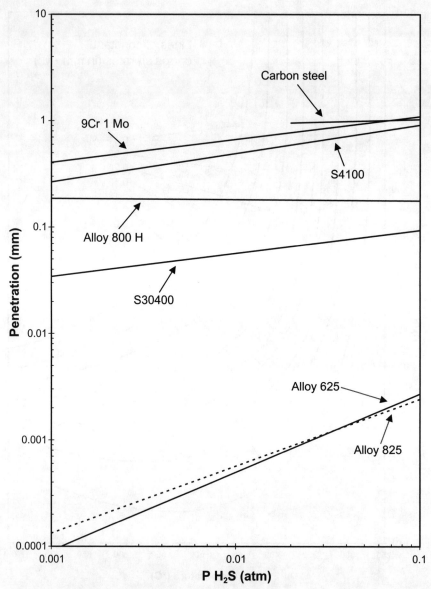

Figure 3.17 Effect of H_2S partial pressure upon sulfidation corrosion after 1 year in H_2-H_2S gases at 34 atm and 540°C.

sents a transition between the rapid corrosion of sulfidation and the slow corrosion of oxidation of alloyed metals containing either Cr or Al.[2]

Atmospheres high in sulfur dioxide are encountered in sulfur furnaces, where sulfur is combusted in air for manufacturing sulfuric acid. Lower levels of sulfur dioxide are encountered in flue gases when fossil fuels contaminated with sulfur species are combusted. It has

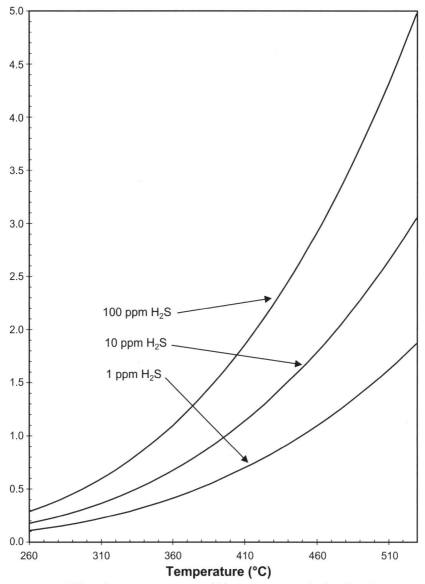

Figure 3.18 Effect of temperature upon sulfidation corrosion of 9Cr-1Mo after 1 year in H_2-H_2S gases at 34 atm.

been pointed out that relatively little corrosion data exist for engineering alloys in these atmospheres.[12] Beneficial effects (retardation of sulfidation) of chromium alloying additions and higher oxygen levels in the atmosphere have been noted.

A tricky situation can arise when designing equipment that requires resistance for variable times of exposure to multiple envi-

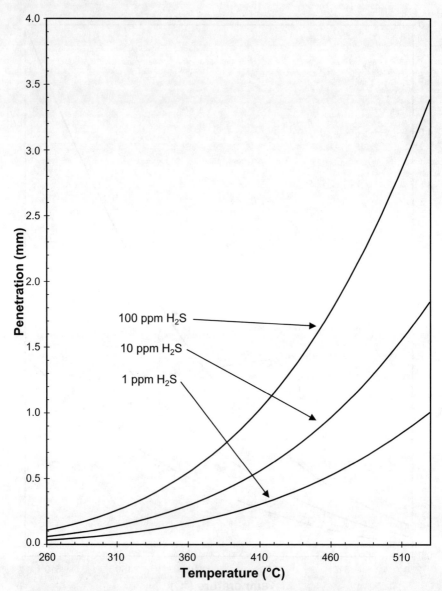

Figure 3.19 Effect of temperature upon sulfidation corrosion of S41000 after 1 year in H_2-H_2S gases at 34 atm.

ronments such as oxidizing and sulfidizing conditions. If oxidation times dominate significantly over sulfidation, it may be prudent to select a high-nickel, high-chromium alloy. Alloys such as HR-120, HR-160, 602CA, or 45TM belong to this category. If sulfidation dominates, low-nickel, high-iron, high-chromium alloys are more appro-

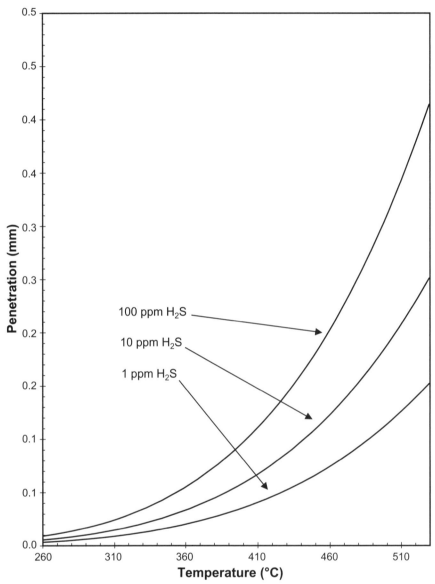

Figure 3.20 Effect of temperature upon sulfidation corrosion of S30400 after 1 year in H_2-H_2S gases at 34 atm.

priate. Increasing the concentration of H_2S tends to increase the sulfidation rate of alloys.[2]

High Ni alloys (greater than 35% Ni) used either as base metals or as welding filler metals are a special concern in sulfidation conditions. Sulfidation of high Ni alloys can be especially rapid and yield corrosion

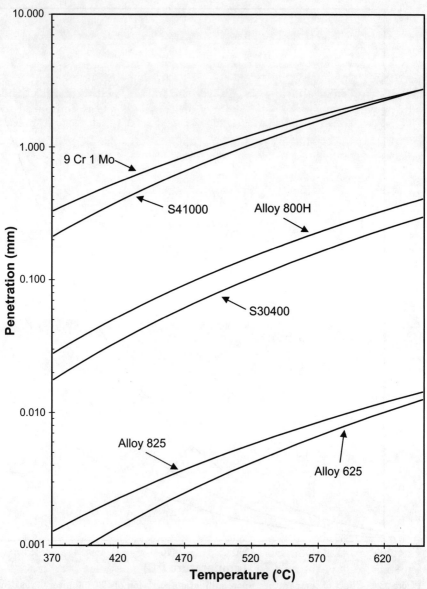

Figure 3.21 Effect of temperature upon sulfidation corrosion after 1 year in H_2-H_2S gases containing 1% H_2S (vol) at 34 atm.

rates greater than 2.5 mm·y^{-1} if the temperature exceeds 630°C, which is the melting point of a potential corrosion product that forms as a mixture of Ni and nickel sulfide. A reasonable approach for high Ni alloys is that they should not be used in sulfidation conditions when metal temperatures will approach or exceed 630°C. High Ni alloys with high Cr levels (such as alloys 625 or 825) can be very suitable with low corrosion rates at lower temperatures.[2]

Alloys that have high concentrations of cobalt are some of the commercially available alloys that are most resistant to sulfidation at temperatures in excess of 630°C. The superior resistance of the cobalt-containing alloys is a result of the higher melting point of the sulfide corrosion products that form on these alloys, as compared to the lower melting points of iron and nickel sulfides. Examples of cobalt-containing alloys, which find application in high-temperature equipment, are alloys 617, HR-160, 6, 188, and Multimet.[2]

3.3.3 Carburization

Carburization can occur when metals are exposed to carbon monoxide, methane, ethane, or other hydrocarbons at elevated temperatures. Carbon from the environment combines primarily with chromium but also with any other carbide formers (Nb, W, Mo, Ti, etc.) present in the alloy to form internal carbides. Carbides formed in the microstructure can be complex in composition and structure and can be found to precipitate on the grain boundaries or inside the grains. The main undesirable effect of carbide formation is embrittlement and reduced ductility at temperatures up to 482 to 538°C. By tying up chromium in the form of stable chromium-rich carbides, carburization also reduces oxidation resistance. Creep strength may also be adversely affected, and internal stresses can arise from the volume increase associated with the carbon uptake and carbide formation. This internal pressure represents additional stress superimposed on operational stresses. Localized bulging, or even cracking, of carburized components is indicative of high internal stress levels that can be generated.

Carburization damage is mainly associated with high-temperature exposure to carbon dioxide, methane, and other hydrocarbons. Heat-treating equipment used for gas carburization (surface hardening) of steels is also vulnerable. An insidious aspect of carburization is its nonuniform nature. Just as for other forms of localized corrosion, it is extremely difficult to predict and model localized carburization damage. As a rule of thumb, carburization problems only occur at temperatures above 815°C, because of unfavorable kinetics at lower temperatures. Carburization is therefore not a common occurrence in most refining operations because of the relatively low tube temperatures of most refinery-fired heaters.

Carburization is more common in the petrochemical industry. A notable problem area has been the radiant and shield sections of ethylene cracking furnaces, due to high tube temperatures up to 1150°C. Apart from temperature, an increase in carbon potential of the gas mix is responsible for a higher severity of damage. High carbon potentials

TABLE 3.8 Sulfidation Corrosion Temperatures Corresponding to a Maximum Metal Loss of 0.25 mm after 1 Year in H_2S-H_2 Gases at 34 atm Gas Pressure

Alloy/H_2S concentration	Maximum allowable temperature, °C				
	0.001%	0.01%	0.1%	1%	10%
Nickel	395	360	340	310	295
Carbon Steel	430	415	405	400	390
9Cr-1Mo	505	445	395	350	310
S41000	570	500	440	390	345
800 H	580	575	575	575	575
430	760	680	615	555	500
S30400	880	790	700	625	565
825	930	630	630	630	630
625	760	630	630	630	630
718	760	630	630	630	630

are associated with the ethane, propane, naphtha, and other hydrocarbons as reactants that are cracked. Carburization has been identified as the most frequent failure mechanism of ethylene furnace tubes. Experience has indicated that the severity of carburization damage in ethylene cracking is process dependent. Some important factors identified include the following:

- Steam dilution, which tends to decrease the rate of damage
- The use of lighter feeds versus heavier feeds, the former having a higher carbon potential
- The frequency and nature of decoking operations; decoking is thought to be a major contributor to carburization damage

Less severe and frequent carburization damage has been reported in reforming operations and in other processes handling hydrocarbon streams or certain ratios of $CO/CO_2/H_2$ gas mixtures at high temperature.[10] As in the case of oxidation and sulfidation, chromium is considered to impart the greatest resistance to carburization.[11] Other beneficial elements include nickel, silicon, columbium, titanium, tungsten, aluminum, and molybdenum. The most important characteristic of a successful alloy is its ability to form and maintain a stable, protective oxide film. Aluminum and silicon alloying additions can contribute positively to this requirement. Unfortunately, the addition of aluminum or silicon to the heat-resistant alloys in quantities to develop full protection involves metallurgical trade-offs in strength, ductility, and/or weldability. Considering fabrication requirements and mechanical properties, viable alloys are generally restricted to about 2 percent of either element. This is helpful but not a total solution.

The tubes of ethylene-cracking furnaces were originally largely manufactured out of the cast HK-40 alloy (Fe-25Cr-20Ni). Since the mid-1980s, more resistant HP alloys have been introduced, but carburization problems have not been eliminated, probably due to more severe operating conditions in the form of higher temperatures. Some operators have implemented a 35Cr-45Ni cast alloy, with various additions, to combat these conditions. For short residence-time furnaces with small tubes, wrought alloys including HK4M and HPM, Alloy 803, and Alloy 800H have been used. Other wrought alloys (e.g., 85H and HR-160, both with high silicon) have been applied to combat carburization of trays, retorts, and other components used in carburizing heat treatments. However, their limited fabricability precludes broad use in the refining or petrochemical industry.[10]

Carburization causes the normally nonmagnetic wrought and cast heat-resistant alloys to become magnetic. The resulting magnetic permeability provides a methodology for monitoring the extent of carburization damage. Measurement devices range from simple hand-held magnets to advanced multifrequency eddy current instruments. Carburization patterns can also reveal uneven temperature distributions that might otherwise have gone undetected. Most alloys tend to have more carburization penetration with increasing temperatures. Figure 3.22 summarizes carburization after 1 year for some common alloys exposed to solid carbon and 200 ppm H_2S.[2]

The time dependence of carburization has been commonly reported to be parabolic. Removal or cracking of any surface carbide scale will tend to increase the rate of carburization. One thousand hours may be required to establish the time dependence expected for long-term service. Carburization data are properly used when the time dependence is considered. Increasing the concentration of H_2S tends to slow the carburization rate of alloys. Figure 3.23 shows the effect for several alloys widely used in petrochemical equipment. The effect of H_2S is to slow decomposition of the CH_4, which adsorbs onto the metal surface, thus slowing the rate of carburization. Increasing concentrations will slow carburization until the concentrations become high enough to cause sulfidation to become the dominant corrosion mechanism. The conditions for the initiation of sulfidation depend upon the alloy and gas compositions.[2]

High Ni alloys used either as base or welding filler metals are often used to resist carburizing conditions. Ni slows the diffusion of carbon in alloys, which is important because carburization is essentially a corrosion mechanism limited by the rate of carbon diffusion in the alloy. However, carburization of high Ni alloys can be especially rapid and yield rates greater than 2.5 mm·y^{-1}, if the temperature exceeds 980°C.

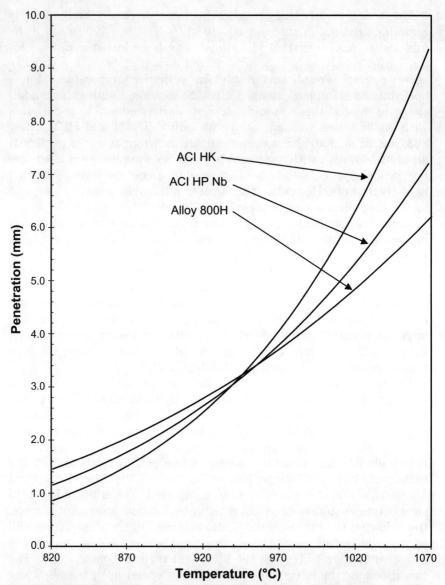

Figure 3.22 Effect of temperature upon carburization of several alloys exposed to solid carbon and 200 ppm H$_2$S at 1 atm.

3.3.4 Metal dusting

Metal dusting is related to carburization and has been reported in similar industries. In this form of degradation, the corrosion products appear as fine powders (hence the term *dusting*) consisting of carbides, oxide, and graphite (soot). The morphology of attack can be localized

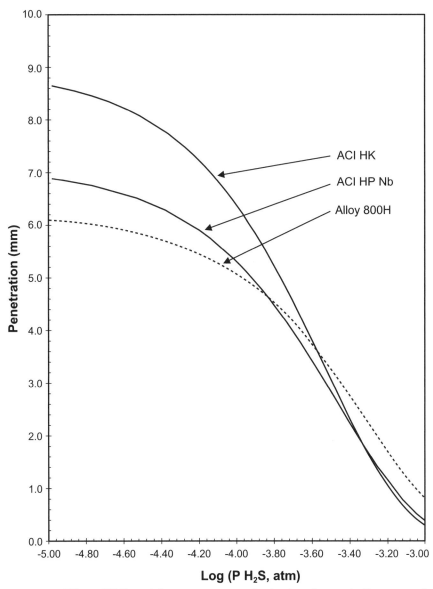

Figure 3.23 Effect of H_2S partial pressure upon carburization of several alloys exposed to solid carbon and 200 ppm H_2S at 982°C.

pitting or relatively uniform damage. The underlying alloy may or may not display evidence of carburization in the microstructure. Metal dusting is manifested at lower temperatures than carburization, typically between 425 and 815°C. Maximum rates of metal dusting damage are considered to occur around 650 to 730°C.

Metal dusting is usually associated with gas streams rich in carbon monoxide and hydrogen. Prediction and modeling of metal dusting are difficult, and little relevant quantitative data is available for engineering alloys to assist designers. It appears that most stainless steels and heat-resistant alloys can be attacked and that the rate of damage can be extremely high. The mechanisms of metal dusting attack are not understood. One remedial measure is adjusting the gas composition by reducing the CO partial pressure.[10]

3.3.5 Nitridation

Nitridation usually occurs when carbon, low-alloy, and stainless steels are exposed to an ammonia-bearing environment at elevated temperatures. The production of ammonia, nitric acid, melamine, and nylon generate such conditions. Nitridation can also result from nitrogen atmospheres, especially under reducing conditions and high temperatures. There are many parallels to carburization; nitridation occurs when chromium and other elements combine with nitrogen to form embrittling nitrides in the microstructure.

Although stainless steels may have adequate resistance, high-nickel alloys tend to be more resistant. Increasing nickel and cobalt contents are also considered to be beneficial. However, pure nickel has shown poor resistance. Alloy 600, with 72% nickel, is often used in the heat-treating industry and occasionally in refining and petrochemical applications involving ammonia at temperatures above 340°C. Economics and its lower strength, compared with Alloy 800H and cast-modified HP, have limited its applications in the latter industries.

3.3.6 Gaseous halogen corrosion

The corrosive effect of halogen on passivating alloys is well known in aqueous media. Chlorides and fluorides also contribute to high-temperature corrosion by interfering with the formation of protective oxides or breaking them down if already formed. The main reason for the reduced corrosion resistance in the presence of halogens is the formation of volatile corrosion products that are nonprotective. The melting points, boiling points, and temperature at which the vapor pressure reaches 10^{-4} atm of selected metal chlorides is presented in Table 3.9.[12] The high volatility and relatively low melting points of these chloride species should be noted. Clearly these properties are not conducive to establishing an effective diffusion barrier on the corroding alloy surfaces.

In refining operations, chlorides most commonly enter the process operations as salt water or brine. Organic chlorides find their way into crude feed. These are not removed in the desalters but are generally removed in the distillation process. Chlorides can enter the down-

TABLE 3.9 Melting Points, Temperatures at Which Chloride Vapor Pressure Reaches 10^{-4} atm and Boiling Points of Various Chlorides

Chlorides	Melting point, °C	Temperature at 10^{-4} atm, °C	Boiling point, °C
$FeCl_2$	676	536	1026
$FeCl_3$	303	167	319
$NiCl_2$	1030	607	987
$CoCl_2$	740	587	1025
$CrCl_2$	820	741	1300
$CrCl_3$	1150	611	945
CrO_2Cl_2	−95		117
$CuCl$	430	387	1690
$MoCl_5$	194	58	268
WCl_5	240	72	
WCl_6	280	11	337
$TiCl_2$	1025	921	
$TiCl_3$	730	454	750
$TiCl_4$	−23	−38	137
$AlCl_3$	193	76	
$SiCl_4$	−70	−87	58
$MnCl_2$	652	607	1190
$ZrCl_4$	483	146	
$NbCl_5$	205		250
$NbCl_4$		239	455
$TaCl_5$	216	80	240
$HfCl_4$	434	132	
CCl_4	−24	−80	77
$NaCl$	801	742	1465
KCl	772	706	1407
$LiCl$	610	665	1382
$MgCl_2$	714	663	1418
$CaCl_2$	772	1039	2000
$BaCl_2$	962		1830
$ZnCl_2$	318	349	732
$PbCl_2$	498	484	954

stream processes. Fluoride contamination is usually the result of blending streams from an alkylation operation. This downstream contamination cannot spill over to petrochemical facilities that take feed from these sources. Halogen contamination during shipment and storage are also of concern to petrochemical facilities.[10] Chlorination processes are used to produce certain metals, as well as in nickel extraction. Calcining operations used in the production of certain rare earth metals and for producing ceramic ferrites for permanent magnets are also associated with high-temperature chlorine-containing environments.

In high-temperature chlorine atmospheres chromium and nickel additions to iron are both regarded as beneficial. Stainless steels are therefore more resistant than the lower-alloyed steels. Austenitic stainless steels tend to outperform the ferritic grades (at equivalent chromium levels). Nickel and nickel-based alloys are widely used under such conditions. The high-nickel alloys are significantly more resistant than the stainless steels to chlorine but not to fluorine, which is more soluble in nickel. When both chlorine and oxygen are present in the environment, essentially a competing situation arises between the formation of stable oxide and volatile chloride species. Therefore, the degradation rate can fluctuate between parabolic, linear, and hybrid behavior. Molybdenum and tungsten alloying additions are considered highly undesirable in such service environments due to the formation of highly volatile oxychlorides (Table 3.9). Aluminum additions are regarded as beneficial, due to the formation of a protective alumina scale at high temperatures.

3.3.7 Fuel ash and salt deposits

In many industrial applications, the surfaces undergoing high-temperature corrosion are not clean; rather, surface deposits of ash and/or salt form on the components. Chemical reactions between these deposits and the protective surface oxide can lead to destruction of the oxide and rapid corrosive attack. In gas turbines, oxidized sulfur contaminants in fuel and sodium chloride from ingested air (marine atmospheres) tend to react to form sulfates that are subsequently deposited on surfaces. The presence of sodium sulfate, potassium sulfate, and calcium sulfate together with magnesium chloride has been reported in such deposits for compressor-stage components.[14] Sodium sulfate is usually regarded as the dominant component of the salt deposits. The detailed mechanisms of hot corrosion have been described by Rapp and Zhang.[15] Hot corrosion is generally considered to occur in the temperature range of 800 to 950°C, although attack at lower temperatures has also been reported.

Testing has indicated that in commercial nickel- and cobalt-based alloys, chromium additions play an important role in limiting this type of damage. Alloys with less than 15% of chromium as alloying addition are considered highly vulnerable to attack.

Refinery heaters and boilers that are fired with low-grade fuels may be vulnerable to corrosion damage, especially if vanadium, sulfur, and sodium contaminants are present at high levels. Vanadium pentoxide and sodium sulfate deposits assume an important role in this type of corrosion damage. The melting point of one of these mixed compound deposits (Na_2SO_4-V_2O_5) can be as low as 630°C, at which point catastrophic corrosion can set in. In these severe operating conditions the use of special high-chromium alloys is required. A 50Ni-50Cr alloy has been recommended over the use of 25Cr-12Ni and 25Cr-20Ni alloys for hangers, tube sheets, and other supports. Ash and salt deposit corrosion is also a problem area in fireside corrosion of waste incinerators, in calcining operations, and in flue gas streams.

3.3.8 Corrosion by molten salts

Corrosion damage from molten salts can occur in a wide variety of materials and by different mechanisms. It has been pointed out that although many studies have been performed, quantitative data for materials selection and performance prediction are rarely available.[16] Molten salt corrosion is usually applicable to materials retaining the molten salt, as used in heat treating, solar and nuclear energy systems, batteries, fuel cells, and extractive metallurgical processes. Some factors that can make molten salts extremely corrosive include the following:

- By acting as fluxes, molten salts destabilize protective oxide layers (on a microscopic scale, this effect contributes toward fuel ash corrosion described above).
- High temperatures are typically involved.
- Molten salts are generally good solvents, preventing the precipitation of protective surface deposits.
- Direct chemical reaction between the containment material and the salt.
- The presence of noble metal ions in the molten salt, more noble than the containment material itself.

3.3.9 Corrosion in liquid metals

Corrosion in liquid metals is applicable to metals and alloy processing, metals production, liquid metal coolants in nuclear and solar power generation, other nuclear breeding applications, heat sinks in automotive and aircraft valves, and brazing operations. Corrosion damage to containment materials is usually the concern. Again, practical design and performance data are extremely limited. In materials selection several possible corrosion mechanisms need to be considered. The most severe problems arise at high temperatures and aggressive melts. Molten steel is typically regarded as a nonaggressive melt, whereas molten lithium is much more corrosive. A brief description of degradation mechanisms follows.[17] Practical problems are complicated by the fact that several of these forms can occur simultaneously. In fact, opposing actions may be required for individual effects that act in combination.

Corrosion reactions can occur by a simple dissolution mechanism, whereby the containment material dissolves in the melt without any impurity effects. Material dissolved in a hot zone may be redeposited in a colder area, possibly compounding the corrosion problem by additional plugging and blockages where deposition has taken place. Dissolution damage may be of a localized nature, for example, by selective dealloying. The second corrosion mechanism is one of reactions involving interstitial (or impurity) elements (carbon, oxygen, etc.) in the melt or containment material. Two further subforms are corrosion product formation and elemental transfer. In the former the liquid metal is directly involved in corrosion product formation. In the latter the liquid metal does not react directly with the containment alloy; rather, interstitial elements are transferred to, from, or across the liquid.

Alloying refers to the formation of reaction products on the containment material, when atoms other than impurities or interstitials of the liquid metal and containment material react. This effect can sometimes be used to produce a corrosion-resistant layer, separating the liquid metal from the containment (for example, aluminum added to molten lithium contained by steel). Lastly, liquid metal can attack ceramics by reduction reactions. Removal of the nonmetallic element from such solids by the melt will clearly destroy their structural integrity. Molten lithium poses a high risk for reducing ceramic materials (oxides).

3.3.10 Compilation and use of corrosion data

A large compilation of corrosion data for metals and alloys in high-temperature gases has been created and is publicly available. The Alloy Selection System for Elevated Temperatures (ASSET) software is based on data compilation representing millions of exposure hours of 70 commercial alloys exposed to industrial environments. The data compilation has been developed and organized to allow prediction of sound metal thickness losses by several corrosion mechanisms at high temperatures as functions of gas composition, temperature, time, and alloy type. Several charts and tables have been prepared as examples of predicted metal losses of alloys corroding in standard conditions for several corrosion mechanisms expected in high-temperature gases.[2] The equations, which correlate the corrosion measurements with exposure conditions and the data, are stored in databases. The corrosion mechanisms for which corrosion predictions can be made are

- Sulfidation
- Sulfidation/oxidation

- Isothermal oxidation
- Carburization

The software uses the composition of the alloy and the corrosive environment information such as gas composition, temperature, and gas pressure to calculate the stable corrosion products and the equilibrium gas composition for a given combination of alloy and exposure conditions. These computations use the Equilib program from F*A*C*T, a Gibbs' free energy minimization program.[18] The calculations can be used to determine the proximity of the corrosive gas to equilibrium by comparing the calculated equilibrium gas composition to the real gas composition in the process equipment. Thermochemical characteristics such as the oxygen and sulfur partial pressure, and carbon activity of the environment, which determine corrosion product stability, are also provided by the calculation and retained for subsequent use. The software also assists the user identify the likely corrosion mechanism, by knowing the stable corrosion products that tend to form at the corrosion product/corrosive gas interface, the alloy in question, and the partial pressures of oxygen and sulfur. Alloys with different compositions in the same exposure conditions may exhibit different stable corrosion products and therefore undergo different corrosion mechanisms.

References

1. Lai, G. Y., *High Temperature Corrosion of Engineering Alloys,* Materials Park, Ohio, American Society for Metals, 1990.
2. John, R. C., *Compilation and Use of Corrosion Data for Alloys in Various High-Temperature Gases,* Corrosion 99 Paper 99073, 1999. Houston, Tex., NACE International, 1999.
3. Gaskell, D. R., *Introduction to Metallurgical Thermodynamics,* New York, McGraw-Hill, 1981.
4. Birks, N., and Meier, G. H., *Introduction to High Temperature Oxidation of Metals,* London, Edward Arnold, 1983.
5. Rapp, R. A., *High Temperature Corrosion,* Washington, D.C., The American Chemical Society, 1980.
6. Gulbransen, E. A., and Jansson, S. A., Thermochemical Considerations of High Temperature Gas-Solid Reactions, in Belton, G. R., and Worrell, W. F. (eds.), *Heterogeneous Kinetics at Elevated Temperatures,* New York, Plenum Press, 1970, pp. 34–46.
7. Roine, A., *Outokumpu HSC Chemistry for Windows (3.0),* Finland, Outokumpu Research Oy, 1997.
8. Jones, D. A., *Principles and Prevention of Corrosion,* Upper Saddle River, N.J., Prentice Hall, 1996.
9. Fontana, M. G., *Corrosion Engineering,* New York, McGraw Hill, 1986.
10. Tillack, D. J., and Guthrie, J. E., *Wrought and Cast Heat-Resistant Stainless Steels and Nickel Alloys for the Refining and Petrochemical Industries,* NiDI Technical Series 10071, Toronto, Canada, Nickel Development Institute, 1992.
11. Sedriks, A. J., *Corrosion of Stainless Steels,* New York, John Wiley, 1979.
12. Lai, G. Y., High-Temperature Corrosion: Issues in Alloy Selection, *Journal of Materials,* **43**:11, 54–60 (1991).

13. Bayer, G. T., Vapor Aluminum Diffused Steels for High-Temperature Corrosion Resistance, *Materials Performance,* **34:**34–38 (1995).
14. Bornstein, N. S., Reviewing Sulfidation Corrosion—Yesterday and Today, *Journal of Materials* 37–39 (1996).
15. Rapp, R. A., and Zhang, Y. S., Hot Corrosion of Materials: Fundamental Studies, *Journal of Materials* 47–55 (1994).
16. Koger, J. W., Fundamentals of Hrigh-Temperature Corrosion in Molten Salts, in *Metals Handbook: Corrosion,* Metals Park, ASM International, 1987, pp. 50–55.
17. Tortorelli, P. F., Fundamentals of High-Temperature Corrosion in Liquid Metals, in *Metals Handbook: Corrosion,* Metals Park, ASM International, 1987, pp. 56–60.
18. Bale, C. W., Pelton, A. D., and Thompson, W. T., *Facility for the Analysis of Chemical Thermodynamics (F*A*C*T) (2.1),* 1996, Montreal, Canada, Ecole Polytechnique/McGill University.

Chapter 4

Modeling, Life Prediction, and Computer Applications

4.1	Introduction	267
4.2	Modeling and Life Prediction	268
	4.2.1 The bottom-up approach	268
	4.2.2 The-top down approach	277
	4.2.3 Toward a universal model of materials failure	291
4.3	Applications of Artificial Intelligence	303
	4.3.1 Expert systems	306
	4.3.2 Neural networks	318
	4.3.3 Case-based reasoning	321
4.4	Computer-Based Training or Learning	322
4.5	The Internet and the Web	324
	References	326

4.1 Introduction

Predictive modeling and statistical process control have become integral components of the modern science and engineering of complex systems. The massive introduction of computers in the workplace has also drastically changed the importance of these machines in daily operations. Computers play important roles in data acquisition in laboratory and field environments, data processing and analysis, data searching, and data presentation in understandable and useful formats. Computers also assist engineers in transforming data into usable and relevant information.

The connectivity of computers to the outside world through the Internet and the Web has opened up tremendous channels of communication that never existed before. This chapter covers a variety of topics related to modeling of corrosion processes, from fundamental expressions to pragmatic models, and to applications of computers such as expert systems and computer-based training.

4.2 Modeling and Life Prediction

The complexity of engineering systems is growing steadily with the introduction of advanced materials and modern protective methods. This increasing technical complexity is paralleled by an increasing awareness of the risks, hazards, and liabilities related to the operation of engineering systems. However, the increasing cost of replacing equipment is forcing people and organizations to extend the useful life of their systems. The prediction of damage caused by environmental factors remains a serious challenge during the handling of real-life problems or the training of adequate personnel. Mechanical forces, which normally have little effect on the general corrosion of metals, can act in synergy with operating environments to provide localized mechanisms that can cause sudden failures.

Models of materials degradation processes have been developed for a multitude of situations using a great variety of methodologies. For scientists and engineers who are developing materials, models have become an essential benchmarking element for the selection and life prediction associated with the introduction of new materials or processes. In fact, models are, in this context, an accepted method of representing current understandings of reality. For systems managers, the corrosion performance or underperformance of materials has a very different meaning. In the context of life-cycle management, corrosion is only one element of the whole picture, and the main difficulty with corrosion knowledge is to bring it to the system management level. This chapter is divided into three main sections that illustrate how corrosion information is produced, managed, and transformed.

4.2.1 The bottom-up approach

Scientific models can take many shapes and forms, but they all seek to characterize response variables through relationships with appropriate factors. Traditional models can be divided into two main categories: mathematical or theoretical models and statistical or empirical models.[1] Mathematical models have the common characteristic that the response and predictor variables are assumed to be free of specification error and measurement uncertainty.[2] Statistical models, on the other hand, are derived from data that are subject to various types of specification, observation, experimental, and/or measurement errors. In general terms, mathematical models can guide investigations, and statistical models are used to represent the results of these investigations.

Mathematical models. Some specific situations lend themselves to the development of useful mechanistic models that can account for the principal features governing corrosion processes. These models are most naturally expressed in terms of differential equations or another nonexplicit form of mathematics. However, modern developments in computing facilities and in mathematical theories of nonlinear and chaotic behaviors have made it possible to cope with relatively complex problems. A mechanistic model has the following advantages:[3]

- It contributes to our understanding of the phenomenon under study.
- It usually provides a better basis for extrapolation.
- It tends to be parsimonious, i.e., frugal, in the use of parameters and to provide better estimates of the response.

The modern progress in understanding corrosion phenomena and controlling the impact of corrosion damage was greatly accelerated when the thermodynamic and kinetic behavior of metallic materials was made explicit in what became known as E-pH or Pourbaix diagrams (thermodynamics) and mixed-potential or Evans diagrams (kinetics). These two models, both established in the 1950s, have become the basis for most of the mechanistic studies carried out since then.

The multidisciplinary nature of corrosion science is reflected in the multitude of approaches to explaining and modeling fundamental corrosion processes that have been proposed. The following list gives some scientific disciplines with examples of modeling efforts that one can find in the literature:

- *Surface science.*   Atomistic model of passive films
- *Physical chemistry.*   Adsorption behavior of corrosion inhibitors
- *Quantum mechanics.*   Design tool for organic inhibitors
- *Solid-state physics.*   Scaling properties associated with hot corrosion
- *Water chemistry.*   Control model of inhibitors and antiscaling agents
- *Boundary-element mathematics.*   Cathodic protection

The following examples illustrate the applications of computational mathematics to modeling some fundamental corrosion behavior that can affect a wide range of design and material conditions.

A numerical model of crevice corrosion. Many mathematical models have been developed to simulate processes such as the initiation and propagation of crevice corrosion as a function of external electrolyte composition and potential. Such models are deemed to be quite important for predicting the behavior of otherwise benign situations that can progress into aggravating corrosion processes. One such model was published recently with a review of earlier efforts to model crevice corrosion.[4] The model presented in that paper was applied to several experimental data sets, including crevice corrosion initiation on stainless steel and active corrosion of iron in several electrolytes. The model was said to break new ground by

- Using equations for moderately concentrated solutions and including individual ion-activity coefficients. Transport by chemical potential gradients was used rather than equations for dilute solutions.
- Being capable of handling passive corrosion, active corrosion, and active/passive transitions in transient systems.
- Being generic and permitting the evaluation of the importance of different species, chemical reactions, metals, and types of kinetics at the metal/solution interface.

Solution of the model for a particular problem requires specification of the chemical species considered, their respective possible reactions, supporting thermodynamic data, grid geometry, and kinetics at the metal/solution interface. The simulation domain is then broken into a set of calculation nodes, as shown in Fig. 4.1; these nodes can be spaced more closely where gradients are highest. Fundamental equations describing the many aspects of chemical interactions and species movement are finally made discrete in readily computable forms.

During the computer simulation, the equations for the chemical reactions occurring at each node are solved separately, on the assumption that the characteristic times of these reactions are much shorter than those of the mass transport or other corrosion processes. At the end of each time step, the resulting aqueous solution composition at each node is solved to equilibrium by a call to an equilibrium solver that searches for minima in Gibbs energy. The model was tested by

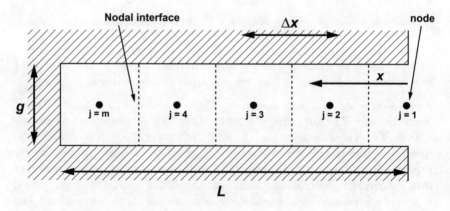

Figure 4.1 Schematic of crevice model geometry.

comparing its output with the results of several experiments with three systems:

- Crevice corrosion of UNS 30400 stainless steel in a pH neutral chloride solution
- Crevice corrosion of iron in various electrolyte solutions
- Crevice corrosion of iron in sulfuric acid

Comparison of modeled and experimental data for these three systems gave agreement ranging from approximate to very good.

A fractal model of corroding surfaces. Surface modifications occurring during the degradation of a metallic material can greatly influence the subsequent behavior of the material. These modifications can also affect the electrochemical response of the material when it is submitted to a voltage or current perturbation during electrochemical testing, for example. Models based on fractal and chaos mathematics have been developed to describe complex shapes and structures and explain many phenomena encountered in science and engineering.[5] These models have been applied to different fields of materials engineering, including corrosion studies. Fractal models have, for example, been used to explain the frequency dependence of a surface response to probing by electrochemical impedance spectroscopy (EIS)[6] and, more recently, to explain some of the features observed in the electrochemical noise generated by corroding surfaces.[7]

In an experiment designed to reveal surface features, a sample of rolled aluminum 2024 sheet (dimensions 100 × 40 × 4 mm) was placed in a 250-mL beaker in such a way that it was immersed in aerated 3% NaCl solution to a level about 30 mm from the top of the specimen.[8] The effect of aeration created a "splash zone" over the portion of the surface that was not immersed. During the course of exposure, a portion of the immersed region in the center of the upward-facing surface became covered with gas bubbles and suffered a higher level of attack than the rest of the immersed surface. After 24 h, the plate was removed from the solution. Figure 4.2 shows the specimen and the areas where the surface profiles were measured in diagrammatic form.

Surface profile measurements were made by means of a Rank Taylor Hobson Form Talysurf with a 0.2-μm diamond-tip probe in all the various planes and directions in these planes, i.e., LT, TL, LS, SL, ST, and TS. The instrument created a line scan of a real surface by pulling the probe across a predefined part of the surface at a fixed scan rate of 1 mm/s. All traces were of length 8 mm, generating 32,000 points with a sampling rate of 0.25 μm per point, except for the SL and ST directions, which, because of the plate thickness, were limited to 2-mm

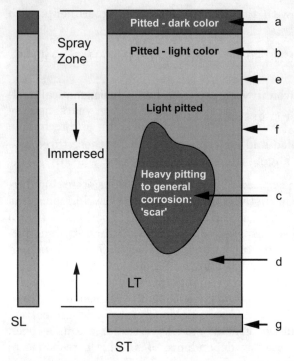

Figure 4.2 Diagram of Al sheet specimen with locations of corroded zones.

traces or 8000 points. The manufacturer's software for the Talysurf instrument was capable of generating more than 20 surface profile parameters. In this study, two parameters, Ra and Rt, were retained. Ra, the roughness average, described the average deviation from a mean line, whereas Rt described the distance from the deepest pit to the highest peak of the profile, an index which was taken as an engineering "worst-case" parameter for pitting severity.

The corrosion found on the plate varied considerably from area to area. The region of the plate beneath the gas bubbles was found to be particularly corroded, with a very high concentration of pits. Across the remainder of the immersed upward-facing surface, pitting was scattered. The splash zone of the surface above the electrolyte was also badly pitted. On the sides, the pits had a geometry and orientation which conformed to the expected grain structure of the rolled material. In all cases, changes noted in traditional Talysurf parameters were consistent with expectations. The severity of the corrosion was indicated by an increase in Ra and Rt, and the profiles obtained gave good general indications of the degree of pitting and the size of pits. There was an approximately tenfold increase in Ra and Rt between the freshly polished surface (reference data in Table 4.1) and the heavily corroded profiles such as a, b, e, and g on Fig. 4.2.

All profiles measured and analyzed with the Talysurf equipment were also analyzed with the rescaled range (R/S) analysis technique. The R/S technique, which can provide a direct evaluation of the fractal dimension of a signal, was derived from one of the most useful mathematical models for analyzing time-series data, proposed a few years ago by Mandelbrot and van Ness.[9] A detailed description of the R/S technique [in which R or R(t,s) stands for the sequential range of the data-point increments for a given lag s and time t, and S or S(t,s) stands for the square root of the sample sequential variance] can be found in Fan et al.[10] Hurst[11] and, later, Mandelbrot and Wallis[12] have proposed that the ratio R(t,s)/S(t,s), also called the rescaled range, was itself a random function with a scaling property described by relation (4.1), in which the scaling behavior of a signal is characterized by the Hurst exponent (H), also called the scaling parameter, which can vary over the range $0 < H < 1$.

$$\frac{R(t,s)}{S(t,s)} \propto s^H \quad (4.1)$$

It has additionally been shown[13] that the local fractal dimension D of a signal is related to H through Eq. (4.2), which makes it possible to characterize the fractal dimension of a given time series by calculating the slope of an R/S plot.

$$D = 2 - H \quad 0 < H < 1 \quad (4.2)$$

Examining the data in Table 4.1, it is apparent that the ground, uncorroded surfaces exhibited behavior close to that of a brownian profile, for which the fractal dimension D equals 1.5. The corroded areas with the biggest reduction in D were those with the most pitting, i.e., traces a, b, and e, all of which occurred in the spray zone above the water. The reduction in fractal dimension at the fine-texture resolution

TABLE 4.1 Calculated Surface Parameters for Regions Identified on Fig. 4.2

Plane	Zone	Ra, μm	Rt, μm	D
	Reference*	0.14	2.95	1.45
Long transverse (LT)	a	1.12	17.6	1.27
	b	1.36	20.0	1.27
	c	0.48	8.82	1.36
	d	0.71	12.8	1.42
Short longitudinal (SL)	e	1.59	15.7	1.23
	f	0.84	14.9	1.30
Short transverse (ST)	g	1.01	17.6	1.35

*Average reference trace measured before corrosion exposure.

of the Talysurf, from about 1.5 to about 1.2, would indicate a "smoothing," which might be explained by a greater loss of mass from the peaks than from the valleys of the profiles.

The correlation coefficients between the fractal dimension and the surface parameters presented in Table 4.1 were calculated to be 0.89 for Ra and 0.76 for Rt. This would indicate that the fractal dimension is slightly better related to a short-range descriptor or an average quantity such as Ra than to a longer-range descriptor or a worst-case distance quantity such as Rt. R/S analysis can provide a direct method for determining the fractal dimension of surface profiles measured with commercial equipment. Such analysis was helpful in shedding a new light on the real nature of the microscopic transformations occurring during the corrosion of aluminum.

Statistical models. Frequently, the mechanism underlying a process is not understood sufficiently well or is simply too complicated to allow an exact model to be formulated from theory. In such circumstances, an empirical model may be useful. The degree of complexity that should be incorporated in an empirical model can seldom be assessed in the first phase of designing the model. The most popular approach is to start by considering the simplest model with a limited set of variables, then increase the complexity of the model as evidence is collected.

Statistical assessment of time to failure is a basic topic in reliability engineering for which many mathematical tools have been developed. Evans, who also pioneered the mixed-potential theory to explain basic corrosion kinetics (see Chap. 1, Aqueous Corrosion), launched the concept of corrosion probability in relation to localized corrosion. According to Evans, an exact knowledge of the corrosion rate was less important than ascertaining the statistical risk of its initiation.[14] Pitting is, of course, only one of the many forms of localized corrosion, and the same argument can be extended to any form of corrosion in which the mechanisms controlling the initiation phase differ from those controlling the propagation phase. The following examples illustrate the applications of empirical modeling in two areas of high criticality.

Pitting corrosion in oil and gas operations. Engineers concerned with soil corrosion of underground steel piping are aware that the maximum pit depth found on a buried structure is somehow related to the percentage of the structure inspected. Finding the deepest actual pit requires a detailed inspection of the whole structure, and as the percentage of the structure inspected decreases, so does the probability of finding the deepest actual pit. A number of statistical transformations to quantify the distributions in pitting variables have been proposed. Gumbel is given the credit for the original development of extreme value statistics (EVS) for the characterization of pit depth distribution.[15]

The EVS procedure is to measure maximum pit depths on several replicate specimens that have pitted, then arrange the pit depth values in order of increasing rank. The Gumbel distribution, expressed in Eq. (4.3), where λ and α are the location and scale parameters, respectively, can then be used to characterize the data set and estimate the extreme pit depth that possibly can affect the system from which the data were initially produced.

$$F(x) = \exp\left[-\exp\left(-\frac{x-\lambda}{\alpha}\right)\right] \qquad (4.3)$$

In reality, there are three types of extreme value distributions:[16]

- *Type 1.* $\exp[-\exp(-x)]$, or the Gumbel distribution
- *Type 2.* $\exp(-x^{-k})$, the Cauchy distribution
- *Type 3.* $\exp[-(\omega - x)^k]$, the Weibull distribution

where x is a random variable and k and ω are constants.

To determine which of these three distributions best fits a specific data set, a goodness-of-fit test is required. The chi-square test or the Kolmogorov-Simirnov test has often been used for this purpose. A simpler graphical procedure using a generalized extreme value distribution with a shape factor dependent on the type of distribution is also possible. There are two expressions for the generalized extreme value distribution, Eq. (4.4) when $kx \leq (\alpha + uk)$ and $k \neq 0$,

$$F(x) = \exp\left(-1 - k\frac{x-u}{\alpha}^{1/k}\right) \qquad (4.4)$$

and Eq. (4.5) when $x \geq u$ and $k = 0$,

$$F(x) = \exp\left(-\exp -\frac{x-u}{\alpha}\right) \qquad (4.5)$$

EVS were put to work on real systems in the oil and gas industries on several occasions for two main reasons. The first reason was the critical nature of many operations associated with the transport of gas and other petroleum products, and the second was the predictability of localized corrosion of steel, the main material used by the oil and gas industry.

Meany has, for example, reported four detailed cases in which extreme value distribution proved to be an adequate representation of corrosion problems:[17]

For underground piping
- In a cathodic protection feasibility study
- For the evaluation of a gas distribution system

For power plant condenser tubing
- During the assessment of stainless steel tube leaks
- During the assessment of Cu-Ni tube pitting performance

In another study, data from water injection pipeline systems and from the published literature were used to simulate the sample functions of pit growth on metal surfaces.[18] This study, by Sheikh et al., concluded that

- Maximum pit depths were adequately characterized by extreme value distribution.
- Corrosion rates for water injection systems could be modeled by a gaussian distribution.
- An exponential pipeline leak growth model was appropriate for all operation regimes.

A more recent publication reported the development of a risk model to identify the probability that unacceptable downhole corrosion could occur as a gas reservoir was depleted.[19] Integration of reservoir simulation data, tubing hydraulics calculations for the downhole wellbore environments, and corrosion pit distribution provided the framework for the risk model. Multiparameter regression showed that the ratio of the volume of liquid water to the volume of liquid hydrocarbon on the tubing walls had a significant influence on corrosion behavior in that field. Using EVS fits for field workover corrosion logging and also laboratory data, a series of extreme value equations with the best fits ($r^2 > 0.95$) was assembled and plotted collectively. It was shown that EVS provided a good representation of the distribution of corrosion pit depths.

A validity analysis of the risk model with a 95 percent corrosion probability indicated at least an 80 percent confidence level for the prediction. Life expectancy calculations using the corrosion risk model provided the basis for the development of an optimized corrosion management strategy to minimize the impact of corrosion on gas deliverability as the reservoir was depleted.

Failure of nuclear waste containers. The regulations pertaining to the geologic disposal of high-level nuclear waste in the United States and Canada require that the radionuclides remain substantially contained within the waste package for 300 to 1000 years after permanent closure of the repository. The current concept of a waste package involves the insertion of spent fuel bundles inside a container, which is then placed in a deep borehole, either vertically or horizontally, with a small air gap between the container and the borehole. For vitrified wastes, a pour canister inside the outer container acts as an additional barrier. Currently, no other barrier is being planned, making the successful performance of the container material crucial to fulfilling the containment requirements over long periods of time.

Provided that no failures occur as a result of mechanical effects, the main factor limiting the survival of these containers is expected to be corrosion caused by the groundwater to which they would be exposed. Two general classes of container materials have been studied internationally: corrosion-allowance and corrosion-resistant materials. Corrosion-allowance materials have a measurable general corrosion rate but are not susceptible to localized corrosion. By contrast, corrosion-resistant materials are expected to have very low general corrosion rates because of the presence of a protective surface oxide film. However, they may be susceptible to localized corrosion damage.

A model developed to predict the failure of Grade 2 titanium was recently published in the open literature.[20] Two major corrosion modes were included in the model: failure by crevice corrosion and failure by hydrogen-induced cracking (HIC). It was assumed that a small number of containers were defective and would fail within 50 years of emplacement. The model was probabilistic in nature, and each modeling parameter was assigned a range of values, resulting in a distribution of corrosion rates and failure times. The crevice corrosion rate was assumed to be dependent only on the properties of the material and the temperature of the vault. Crevice corrosion was also assumed to initiate rapidly on all containers and subsequently propagate without repassivation. Failure by HIC was assumed to be inevitable once a container temperature fell below 30°C. However, the concentration of atomic hydrogen needed to render a container susceptible to HIC would be achieved only very slowly, and the risk might even be negligible if that container had never been subject to crevice corrosion.

Figure 4.3 illustrates the thin-shell packed-particulate design chosen as a reference container for this study. The mathematical procedure to combine various probability functions and arrive at a probability of failure of a hot container as a result of crevice corrosion at a certain temperature is illustrated in Fig. 4.4. The failure rate due to HIC was arbitrarily assumed to have a triangular distribution in order to simplify the calculations, given that HIC is predicted to be only a marginal failure mode under the burial conditions considered.

On the basis of these assumptions and the calculations described in the full paper, it was predicted that 96.7 percent of all containers would fail by crevice corrosion and the remainder by HIC. However, only 0.137 percent of the total number of containers were predicted to fail before 1000 years (0.1 percent by crevice corrosion and 0.037 percent by HIC), with the earliest failure after 300 years.

4.2.2 The top-down approach

The transformation of laboratory results into usable real-life functions for service applications is almost impossible. In the best cases, laboratory

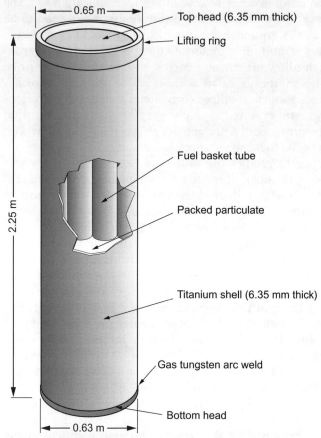

Figure 4.3 Packed-particulate supported-shell container for waste nuclear fuel bundles.

tests can provide a relative scale of merit in support of the selection of materials to be exposed to specific conditions and environments. From an engineering management standpoint, mapping of the parameters defining an operational envelope can reduce the need for exhaustive mechanistic models, since any potential problem should be avoidable by controlling the conditions of its occurrence.

Some of the issues involved in deciding on a cost-effective method for combating corrosion are generic to sound management of engineering systems. Others are specifically related to the impact of corrosion damage on system integrity and operating costs. In process operations, where corrosion risks can be extremely high, costs are often categorized by equipment type and managed as an asset loss risk (Fig. 4.5).[21] The quantification or ranking of risk, defined as the

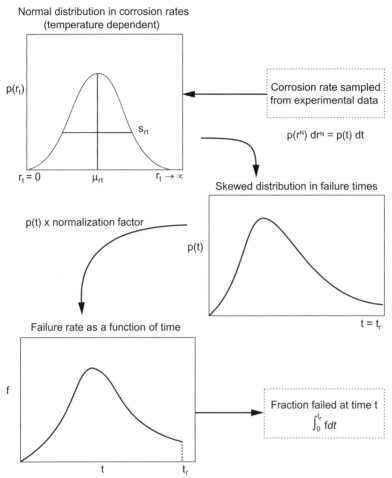

Figure 4.4 Procedure used to determine the failure rate of hot containers as a function of time.

product of the probability and consequences of specific events, should dictate the preferential order in which inspection and maintenance are performed. By referring to Fig. 4.5, the operations department of a process plant should adjust the maintenance schedule, considering the decreasing attention given to piping, reactors, tanks, and process towers. Similar logic applies to all industries. The following examples will illustrate how these considerations are manifested in practice and how corrosion information is integrated into efficient management systems.

A fault tree for the risk assessment of gas pipeline. Fault tree analysis (FTA) is the process of reviewing and analytically examining a system

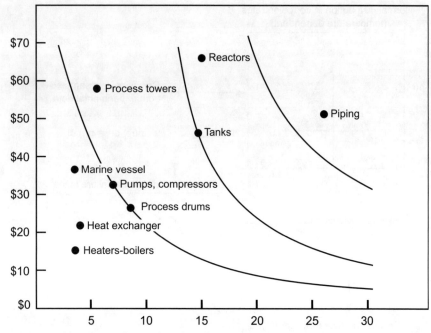

Figure 4.5 Asset loss risk as a function of equipment type.

or equipment in such a way as to emphasize the lower-level fault occurrences which directly or indirectly contribute to a major fault or undesired event. The value of performing FTA is that by developing the lower-level failure mechanisms necessary to produce higher-level occurrences, a total overview of the system is achieved. Once completed, the fault tree allows an engineer to fully evaluate a system's safety or reliability by altering the various lower-level attributes of the tree. Through this type of modeling, a number of variables may be visualized in a cost-effective manner.

A fault tree is a diagrammatic representation of the relationship between component-level failures and a system-level undesired event. A fault tree depicts how component-level failures propagate through the system to cause a system-level failure. The component-level failures are called the terminal events, primary events, or basic events of the fault tree. The system-level undesired event is called the top event of the fault tree. Figure 4.6 presents, in graphical form, the tree and gate symbols most commonly used in the construction of fault trees.[22] A brief description of these symbols is given in the following list:

- *Fault event (rectangle).* A system-level fault or undesired event.
- *Conditional event (ellipse).* A specific condition or restriction applied to a logic gate (mostly used with an inhibit gate).

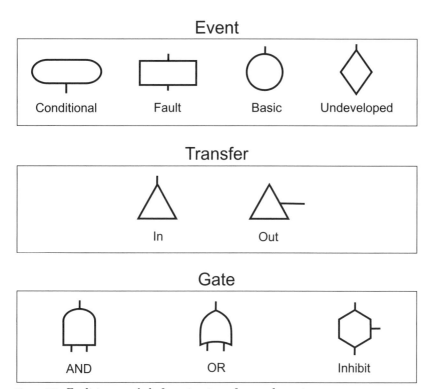

Figure 4.6 Fault tree symbols for gates, transfers, and events.

- *Basic event (circle).* The lowest event examined which has the capability of causing a fault to occur.
- *Undeveloped event (diamond).* A failure which is at the lowest level of examination in the fault tree, but which can be further expanded.
- *Transfer (triangle).* The transfer function is used to signify a connection between two or more sections of the fault tree.
- *AND gate.* The output occurs only if all inputs exist. (Probabilities of the inputs are multiplied, decreasing the resulting probability.)
- *OR gate.* The output is true only if one or more of the input events occur. (Probabilities of the inputs are added, increasing the resulting probability.)
- *Inhibit gate (hexagon).* One input is a lower fault event and the other input is a conditional qualifier or accelerator [direct effect as a decreasing (<1) or increasing factor (>1)].

The FTA methodology was adopted by Nova Corp., a major natural gas transport and processing company in Canada, for the risk assessment of its 18,000-km gas pipeline network.[23] FTA is normally performed for the review and analytical examination of systems or equipment to emphasize the lower-level fault occurrences, and the results of the FTA calculations are regularly validated with inspection results. These results are also used to schedule maintenance operations, conduct surveys, and plan research and development efforts.

Figures 4.7 and 4.8 illustrate respectively the SCC branch and the uniform corrosion branch of the Nova Corp. pipeline outage FTA system. Each element of the branches in Figs. 4.7 and 4.8, which are part of a larger tree that estimates the overall probability of pipeline failure, contains numeric probability information related to technical and historical data for each segment of the 18,000-km pipeline.

The Maintenance Steering Group (MSG) system. The aircraft industry and its controlling agencies have developed another top-down approach to represent potential failures of aircraft components. The Maintenance Steering Group (MSG) system has evolved from many years of corporate knowledge. The first generation of formal air carrier maintenance programs was based on the belief that each part on an aircraft required periodic overhaul. As experience was gained, it became apparent that some components did not require as much attention as others, and new methods of maintenance control were developed. Condition monitoring was thus introduced into the decision logic of the initial Maintenance Steering Group document (MSG-1) and applied to Boeing 747 aircraft.

The MSG system has now evolved considerably. The experience gained with MSG-1 was used to update the decision logic and create a more universal document that is applicable to other aircraft and powerplants.[24] When applied to a particular aircraft type, the MSG-2 logic would produce a list of maintenance significant items (MSIs), to each of which one or more process categories would be applied, such as "hard time," "on-condition," and/or "reliability control."

The most recent update to the system was initiated in 1980. The resultant MSG-3 system has the same basic philosophy as MSG-1 and MSG-2, but prescribes a different approach to the assignment of maintenance requirements. Instead of the process categories typical of MSG-1 and MSG-2, the MSG-3 logic identifies maintenance requirements. The processes, tasks, and intervals arrived at with MSG can be used by operators as the basis for their initial maintenance program. In 1991, industry and regulatory authorities began working together to provide additional enhancements to MSG-3. As a result of these efforts, Revision 2 was submitted to the Federal Aviation Administration (FAA) in September 1993 and accepted a few weeks later. Major enhancements include

Modeling, Life Prediction, and Computer Applications 283

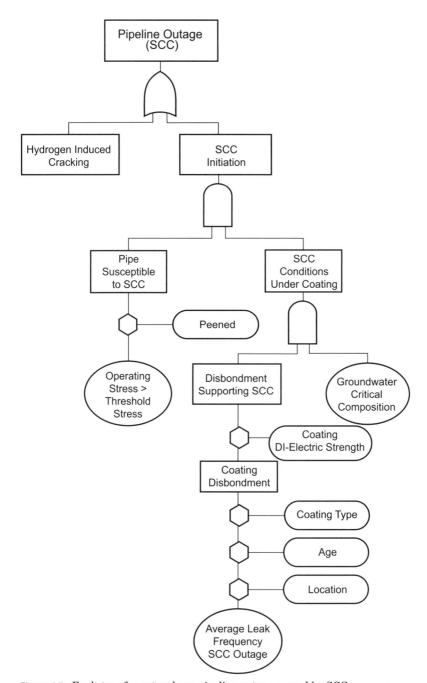

Figure 4.7 Fault tree for natural gas pipeline outage caused by SCC.

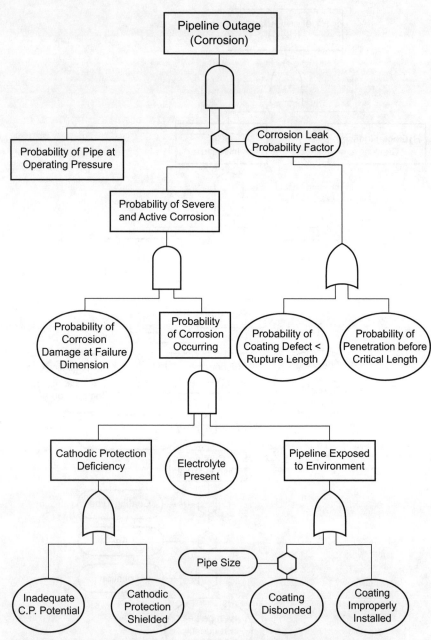

Figure 4.8 Fault tree for natural gas pipeline outage caused by general corrosion.

- Expansion of the systems/powerplant definition of inspection
- Guidelines for the development of a corrosion prevention and control program (CPCP)
- Increased awareness of the requirements of aging aircraft
- Extensive revision of the structure logic

The MSG-3 structure analysis begins with the development of a complete breakdown of the aircraft systems, down to the component level. All structural items are then classified as either structure significant items (SSIs) or other structure. An item is classified as an SSI on the basis of consideration of the consequences of failure and the likelihood of failure, along with material, protection, and probable exposure to corrosive environments. All SSIs are then listed and categorized as damage-tolerant or safe life items to which life limits are assigned.[25] For all SSIs, accidental damage, environmental deterioration, corrosion prevention and control, and fatigue damage evaluations are performed following the logic diagram illustrated in Fig. 4.9.

Once the MSG-3 structure analysis is completed, each element of the structural analysis diagram (Fig. 4.9) can be expanded right to the individual components and associated inspection and maintenance tasks.

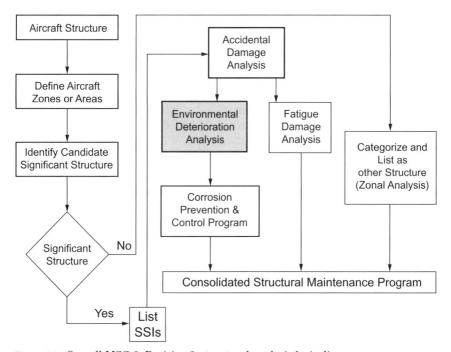

Figure 4.9 Overall MSG-3, Revision 2, structural analysis logic diagram.

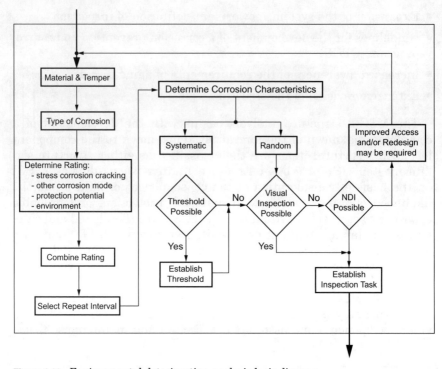

Figure 4.10 Environmental deterioration analysis logic diagram.

The procedure for MSG-3 environmental deterioration analysis (EDA), for example, involves the evaluation of the structure in terms of probable exposure to adverse environments. The evaluation of deterioration is based on a series of steps supported by reference materials containing baseline data expressing the susceptibility of structural materials to various types of environmental damage. While the end product of the MSG-3 is very component-specific, its information contains much of what is required to create a more generic system based on materials instead of part numbers. The logic of the EDA, illustrated in Fig. 4.10, requires the input of a multitude of parameters, given in the following list, guided by the use of a template, shown in Fig. 4.11.

- Item location/accessibility/visibility
- Item material/temper/manufacturing specification
- Material of adjacent items
- Finish protection
- Accidental damage impact
- Area/zone

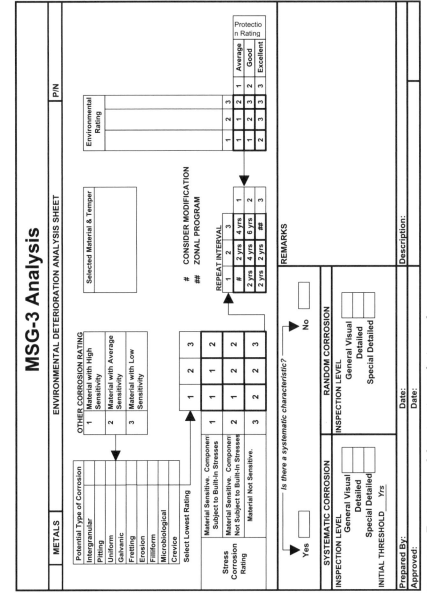

Figure 4.11 Environmental deterioration analysis template.

Modern aircraft are built from a great variety of materials with state-of-the art protective coatings and exemplary design and maintenance constraints. Table 4.2 contains a list of materials that are commonly used in the construction of aircraft with some of the associated problems and solutions. Once data are entered in the MSG system, the predefined relations in the logic permit detailed information to be obtained on the following:

- Likelihood of exposure to corrosive products
- Random/systematic corrosion characteristics
- Required inspection level
- Inspection threshold/repeat cycle intervals
- Corrosion-inhibiting compound application requirements

The corrosion ratings supporting the calculations identified in the EDA sheet (Fig. 4.11) have been adapted from various sources of information. As can be seen in this figure, the impact of SCC on the opera-

TABLE 4.2 Materials Used for the Construction of Modern Aircraft with Associated Problems and Solutions

Alloy	Problems	Solutions
Aluminum		
Wrought 2000 and 7000 series sheets, extrusions, forgings	Galvanic corrosion Pitting Intergranular corrosion Exfoliation Stress corrosion cracking (SCC)	Cladding Anodizing Conversion coatings Ion vapor deposited (IVD) Al Paint
Cast, i.e., Al-Si-(Mg-Cu)	Usually corrosion resistant	
Low-alloy steels		
4000 and 8000 series, 300M fasteners, forgings	Uniform corrosion Pitting SCC Hydrogen embrittlement	Cadmium plating Phosphating Ion vapor deposited (IVD) Al Paint
Stainless steels		
300 series austenitic	Intergranular corrosion Pitting	
400 series martensitic and precipitation hardening (PH) series	Pitting SCC Hydrogen embrittlement	
Magnesium alloys	Uniform corrosion Pitting SCC	Anodizing Conversion coating Painting

tion of aircraft is given special consideration by separating it from the other types of corrosion, which are otherwise considered equally important. The information itself is stored in six tables relating specific materials used in aircraft to the other factors affecting environmental deterioration:[25]

EDA Table 1.   Materials and temper vs. SCC and intergranular, pitting, and uniform corrosion

EDA Table 2.   Combinations of materials vs. galvanic corrosion

EDA Table 3.   Circumstantial conditions vs. fretting, filiform, microbiological, and crevice corrosion

EDA Table 4.   Finish protection vs. added resistance to corrosion

EDA Table 5.   Probable exposure to corrosive environments

EDA Table 6.   Rules to classify corrosion problems as systematic, when they develop gradually with time, or random, when they result from accidental causes

But while the information in these tables appears to reflect the overall knowledge of materials degradation correctly, there is no provision for validating the sources or integrating more detailed mechanisms, even if the information were available. The whole system is built on implicit expertise without the possibility of critically verifying some of its calculated predictions against maintenance observations. Only some vague information concerning the probable exposure to corrosive environments can be found in EDA Table 5, for example, thus opening a finite door to subjectivity in the overall task assessment.

A corrosion index for pipeline risk evaluation.   A risk assessment technique is described in much detail in the second edition of a popular book on pipeline risk management.[26] The technique proposed in that book is based on subjective risk assessment, a method that is particularly well adapted to situations in which knowledge is perceived to be incomplete and judgment is often based on opinion, experience, intuition, and other nonquantifiable resources. A detailed schema relating an extensive description of all the elements involved in creating risk compensates for the fuzziness associated with the manipulation of nonquantifiable data. Figure 4.12 illustrates the basic pipeline risk assessment model or tool proposed in that book.

The technique used for quantifying risk factors is described as a hybrid of several methods, allowing the user to combine scores obtained from statistical failure data with operator experience. The subjective scoring system permits examination of the pipeline risk picture in two general parts. The first part is a detailed itemization and relative weighting of all reasonably foreseeable events that may lead to the failure of a pipeline, and the second part is an analysis of the potential consequences of each failure. The itemization is further broken down into four indexes, illustrated in Fig. 4.12, corresponding to typical categories of pipeline failures. By considering each item in each index, an expert evaluator arrives at a numerical value for that index. The four index values are then summed to obtain the total index value. In the second part, a detailed analysis is made of the potential consequences of a pipeline failure, taking into consideration product characteristics, pipeline operating conditions, and the line location. Building the risk assessment tool requires four steps:

1. *Sectioning.* Dividing a system into smaller sections. The size of each section should reflect practical considerations of operation, maintenance, and cost of data gathering vs. the benefit of increased accuracy.
2. *Customizing.* Deciding on a list of risk contributors and risk reducers and their relative importance.

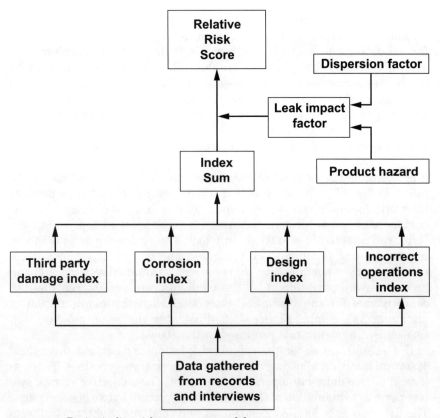

Figure 4.12 Basic pipeline risk assessment model.

3. *Data gathering.* Building the database by completing an expert evaluation of each section of the system.

4. *Maintenance.* Identifying when and how risk factors can change and updating these factors accordingly.

The potential for pipeline failure caused either directly or indirectly by corrosion is probably the most common hazard associated with steel pipelines. The corrosion index was organized in three categories to reflect three types of environment to which pipelines are exposed, i.e., atmospheric corrosion, soil corrosion, and internal corrosion. Table 4.3 contains the elements contributing to each type of environment and the suggested weighting factors.

The basic risk assessment model can be expanded to incorporate additional features that may be of concern in specific situations, as illustrated in Fig. 4.13. Since these features do not necessarily apply to all pipelines, this permits the use of distinct modules that can be activated by an operator to modify the risk analysis.

4.2.3 Toward a universal model of materials failure

One of the principal goals of scientific discovery is the development of a theory, i.e., a coherent body of knowledge that can be used to provide

TABLE 4.3 Corrosion Risk Subjective Assessment

Problem	Weight
Atmospheric corrosion	
1. Facilities	0–5 pts
2. Atmospheric type	0–10 pts
3. Coating/inspection	0–5 pts
	0–20 pts
Internal corrosion	
1. Product corrosivity	0–10 pts
2. Internal protection	0–10 pts
	0–20 pts
Soil corrosion	
1. Cathodic protection	0–8 pts
2. Coating condition	0–10 pts
3. Soil corrosivity	0–4 pts
4. Age of system	0–3 pts
5. Other metals	0–4 pts
6. AC induced currents	0–4 pts
7. SCC and HIC	0–5 pts
8. Test leads	0–6 pts
9. Close internal surveys	0–8 pts
10. Inspection tool	0–8 pts
	0–60 pts
Total	0–100 pts

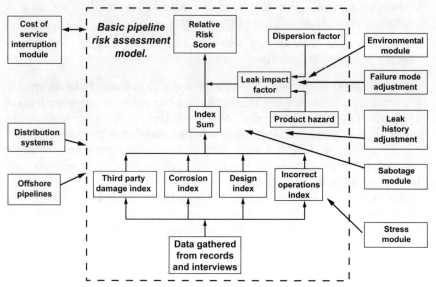

Figure 4.13 Optional modules to customize the basic pipeline risk assessment model.

explanations and predictions for a specific domain of knowledge. Theory development is a complex process involving three principal activities: theory formation, theory revision, and paradigm shift. A theory is first developed from a collection of known observations. It then goes through a series of revisions aimed at reducing the shortcomings of the initial model. The initial theory can thus evolve into one that can provide sophisticated predictions. But a theory can also become much more complex and difficult to use. In such cases, the problems can be partly eliminated by a paradigm shift, i.e., a revolutionary change that involves a conceptual reorganization of the theory.[27] The Venn diagrams of Fig. 4.14 illustrate the three stages of a theory revision.[28] In the first stage of theory revision, (a), an anomaly is noted, a new observation that is not explained by the current model. In a subsequent stage, (b), the old theory is reduced to its most basic or fundamental expression before it finally serves as the basis of a new theory formulation, (c).

A sound corrosion failure model should thus be based on core principles with extensions into real-world applications through adaptive revision mechanisms. A universal representation describing the interactions among defects, faults, and failures of a system is shown in Fig. 4.15. The arrows in this figure imply that quantifiable relations, characteristic of a specific system, exist between a defect, a fault, and a failure. The nature of various corrosion defects is introduced in Chap. 5, Corrosion Failures, in the section on forms of corrosion. Also in Chap. 5, the factors causing these defects have been related to the fun-

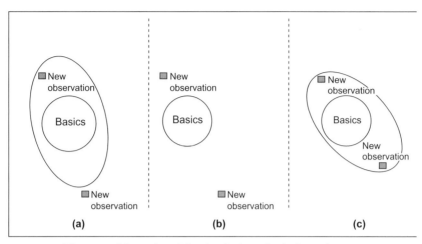

Figure 4.14 Theory revision using abduction for hypothesis formation.

damental work published by Staehle in his review on the progress in science and engineering of SCC problems.[29] The following sections describe how the framework proposed by Staehle was reengineered into a robust and flexible model for different engineering applications.

An object-oriented template. Object orientation (OO) belongs to a series of paradigms that have been generalized far beyond the goal of their initial development. The OO paradigm was created in the 1960s to represent knowledge in artificial intelligence (AI) and expert systems (ES) research[30] and is a fundamentally different way of approaching the organization and processing of information. OO programming tools were specifically designed to fit reality as perceived by humans, yet the OO tools of today cover a much broader realm of technologies than just software programming, and the OO methodology has now been applied to almost every information technology–related activity.[31]

OO in a programming language, system design, or software system is characterized by two key features, (1) abstraction or encapsulation and (2) extensibility.[32] The same features are typical of most memory-based human thoughts. The notion of encapsulation has proved to be a natural paradigm for various applications and environments, such as graphical user interface systems. The extensibility concept refers to the ability to extend an existing system without introducing changes to its fundamental structure. This was an exciting approach to software engineers, who, until the development of OO tools, seemed to require a clean sheet of paper with every new project.

The framework described in Staehle's work was generalized in a structure analogous to the OO paradigm, which was found to be a flexible

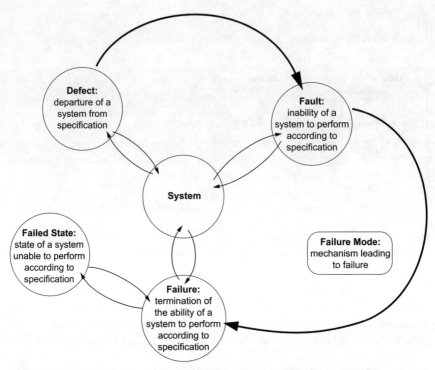

Figure 4.15 Interrelation among defects, failures, and faults.

method of representing such a complex engineering situation. Figure 4.16 illustrates the OO version of the main factors controlling the probability of corrosion problems. In this figure, the focus is on the material factor; the other five factors identified by Staehle are related to the material factor and the overall probability of a corrosion failure by concentric rings representing their influence on the overall probability of failure.

According to the basic materials degradation model, the principal features underlying the environment factor consist of a long list of elements describing the chemical makeup of the environment and the aggravating contributors that can be part of operating conditions, as schematically illustrated in the OO representation of Fig. 4.17. A testing program that investigates only the nominal condition without consideration of effects such as flow, pH cells, deposits, and other galvanic effects is useless for lifetime prediction. An exact and complete environmental definition must include a description of the microenvironment actually in contact with a metallic surface. However, the circumstances producing this microenvironment are also important. Processes such as wetting and drying, buildup of deposits, and changes in flow patterns greatly influence the chemistry of a surface.

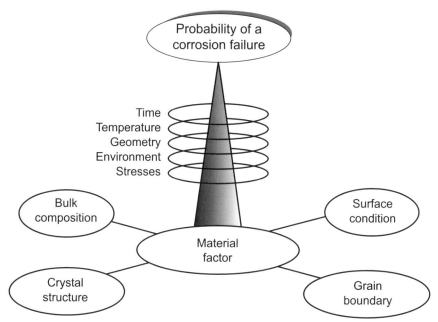

Figure 4.16 An object-oriented representation of the material factor controlling the probability of a corrosion failure.

Once the elements controlling a situation have been explicitly organized, minimal effort is required to translate the information into another probability representation. This point of view is illustrated in the fault tree of Fig. 4.18, where the nominal and circumstantial (nonnominal) environments are separated into two parallel branches. The OR gates in this fault tree would allow evaluation of the top event probability, i.e., the probability that the environment P_{En} will influence a situation, provided that the functions of the two branches P_{No} and P_{Ci} are known.

In Fig. 4.18, the nominal definition of the environment is divided into three components; these are also related to the top event by an OR gate, which indicates that a situation is affected equally by these three components, i.e., major nominal, accidental nominal, and minor nominal. From a probability point of view, the effect of this branch would be quantified by evaluating the probability that a specific nominal composition P_{No} would lead to a corrosion failure as a function of the influence of the major P_{Major}, minor P_{Minor}, and accidental $P_{Accidental}$ components of the environment [Eq. (4.6)].

$$P_{No} = P_{Major} + P_{Accidental} + P_{Minor} \quad (4.6)$$

Figure 4.18 also expresses how the probability of the circumstantial factor is influenced by three gradients expressing temperature

296 Chapter Four

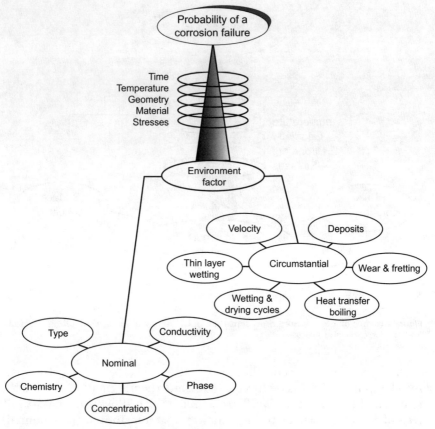

Figure 4.17 An object-oriented representation of the environment factor controlling the probability of a corrosion failure.

differences ΔT, chemical differences ΔChem, and movement v. Because these subfactors are also linked by an OR gate, their impact would be described by Eq. (4.7).

$$P_{Ci} = P_{\Delta T} + P_{\Delta \text{Chem}} + P_v \qquad (4.7)$$

Aluminum failure predictor. Human experts sort evidence by experience. Based on their assessment of a given piece of information, experts will form an initial hypothesis and determine what additional information or tests are required to prove or disprove this initial hypothesis. Further information normally raises more questions, and experts may go through several iterations before feeling confident that the causes and mechanisms of a failure have really been determined with an acceptable degree of confidence. The high-strength aluminum alloy Failure Predictor mimics human analysis from the general to the specific.

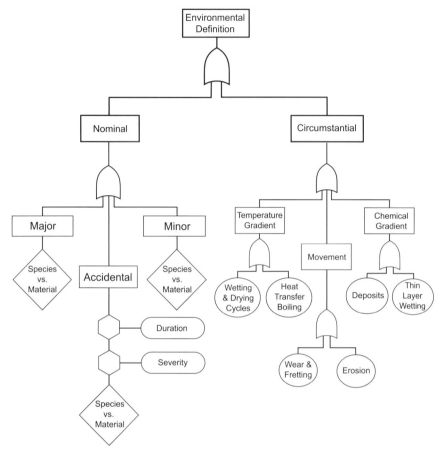

Figure 4.18 A fault tree description of the environment factor controlling the probability of a corrosion failure.

During the last decades, massive efforts and studies of all kinds have paralleled the development of aluminum alloys. These efforts have resulted in the production of an impressive number of reviews and standards that can serve as a starting point for the construction of a knowledge-based system (KBS). The main elements affecting the SCC situation of aluminum alloys are given in Table 4.4 as a function of the six factors proposed in Staehle's framework. Besides the obvious complexity of environmental cracking (EC) problems visible in Table 4.4, there are some important limitations on depending on published data in the development of a KBS for predicting EC problems. The first is that it is almost impossible to separate the individual parameters of the metallurgy of a system, since they tend to be interdependent.

Another serious limitation with most published mechanistic models of the environmental cracking behavior of aluminum alloys is that almost

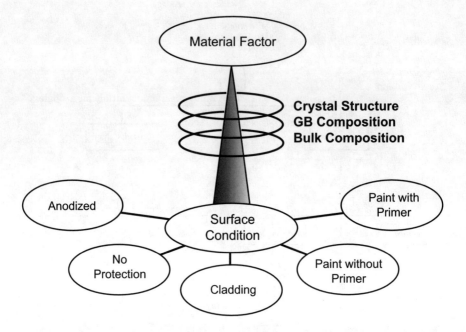

Figure 4.19 An object-oriented representation of the surface condition subfactor of the material factor controlling the probability of a corrosion failure.

all of them refer to either laboratory cast and processed alloys or commercial alloys that were subsequently subjected to laboratory-based heat treatments.[33] It is therefore very important to be very selective when choosing data to support lifetime predictions. The construction of Failure Predictor was based on the elicitation of a total of 12 critical factors and subfactors controlling the probability of EC failures in high-strength aluminum alloys.

In order to actuate the OO framework, each element was linked to the structure and thus to all other parameters in a semiquantitative way. Some accepted methods exist that deal with such a situation and allow different sources of knowledge to be combined with a quantifiable degree of confidence. There are basically three ways, in order of increasing complexity, to represent uncertain information in knowledge engineering: subjective probabilities, certainty factors (CF), and fuzzy logic.[34] The CF approach gives a good approximation of the measure of belief without being too complex to manage during the knowledge elicitation. The two main principles for applying CF to a particular knowledge engineering situation are as follows: (1) The CF must be a quantity that describes the credibility of a given conclusion, and (2) the rules must be structured in such a way that any particular rule will add to either the belief or disbelief in a given conclusion. In

TABLE 4.4 Specific Considerations for the Life Prediction of Aluminum Components as a Function of the Six Factors Controlling the Framework of SCC Information.

Framework factor/observations

Environment
- Measured crack velocities can differ by over nine orders of magnitude for a single alloy as a result of a change in the environment.
- Aluminum will not corrode without the presence of water.
- Environments as low as 0.8% relative humidity will promote SCC.
- Environments the most favorable to crack growth are those containing Cl^-, I^-, and Br^-.

Geometry
- The geometry factor can promote the localization of even mildly corrosive environments.
- There is a sharp drop in the time to failure when the pH of the bulk solution falls below 11.
- Crack tips can have a pH of approximately 3.5.

Service temperature
- Temperature excursions in the thermal aging range affect the strength of alloys and their susceptibility to SCC depending on their position relative to the peak aged condition and their sensitivity to aging.
- For 7XXX series alloys, the beneficial effects that can result from overaging are strongly influenced by the presence of copper.

Time
- No model can predict the occurrence of SCC with satisfaction.
- A semiempirical model was developed for 7079 aluminum alloys to predict crack growth from double cantilever beam test results, but no attempt was made to validate the model with actual service data.

Stress
- Two concepts are necessary to describe the stress factor: (1) the stress definition with all its components and (2) the origin of the stresses, which can be external or internal, such as the wedge action of corrosion products.
- The time to failure vs. applied stress diagrams are often used for empirical determination of the design life expectancy and stress level where SCC crack growth initiates (K_{ISCC}).

Material
- The dominant theme for defining the reactivity of materials is the internal composition of the grain and the grain boundary (GB).
- The composition of the GB can be dominated either by the formation and concentration of precipitates or by the adsorption and concentration of species collected from the environment.
- EC fracture is intergranular (IG) unless loading conditions are severe.
- The influence of quench rate upon IG cracking for 2XXX series alloys is relatively well understood.
- For 7XXX series alloys, it is generally believed that the influence of quench rate upon SCC is dependent upon an alloy's copper content.
- The risk of SCC prevents exploitation of the maximum strength of aluminum because SCC intensity increases with precipitation hardening, reaching a maximum before peak strength.

Eq. (4.8), which summarizes this second principle, MB is a measure of belief in the occurrence of event P given the occurrence of event E, and MD is a measure of disbelief. And the measure of belief that results from considering two sources of evidence, rule 1 (R_1) and rule 2 (R_2), can be calculated by using Eq. (4.9) when these sources have MBs > 0 or using Eq. (4.10) for MDs > 0. The data and CF values would have to reflect the knowledge of a given component in a given situation and at a particular time.

$$CF(P,E) = MB(P,E) - MD(P,E) \tag{4.8}$$

$$CF(P,E) = MB(R_1) + MB(R_2)[1 - MB(R_1)] \tag{4.9}$$

$$CF(P,E) = MD(R_1) + MD(R_2)[1 - MD(R_1)] \tag{4.10}$$

While some certainty factor values were derived from interviewing experts, others were adapted from the literature. In the first case, three experts were consulted and asked to assign a value between −100 and +100 to each subfactor based on its contribution to aluminum SCC, and their answers were averaged for the computation of certainty factors. Table 4.5 details the average values obtained from the three experts for the surface condition subfactor of Fig. 4.19.

An example of the second case, i.e., using a literature source to assign CF values, is the way the bulk composition and crystal structure subfactors of Fig. 4.16 were given probability values. These values were obtained by translating a system for rating the resistance to SCC of various aluminum alloys and their tempers into a linear scale (Table 4.6). This system had been developed by a joint task group of ASTM and the Aluminum Association to assist in alloy and temper selection.[35] The information contained in ASTM G 64-85, Standard Classification of the Resistance to Stress-Corrosion Cracking of High-Strength Aluminum Alloys, was collected from at least 10 random lots which were tested in accordance with the practice recommended in ASTM G 44, Practice for Evaluating Stress Corrosion Cracking Resistance of Metals and Alloys by Alternate Immersion in 3.5% Sodium Chloride Solutions. The highest rating was assigned for results that showed 90 percent conformance at the 95 percent confidence level when tested at the following stresses:

TABLE 4.5 Average CF Values Gathered from Three Experts on the Impact of Surface Conditions on the Probability of an SCC Failure with High-Strength Aluminum Alloys

Surface condition	Paint with primer	Without primer	Cladding	Anodized	None
Good	−0.30	−0.10	−0.50	−0.50	0
Poor	−0.10	0	−0.30	−0.30	0.10
Very poor	0.10	0.10	0.10	0.10	0.15
Localized defects	0.10	0.10	0.10	0.10	0.20

TABLE 4.6 Some CF Values Adapted from ASTM G 64-85 for the Alloy and Temper Subfactors Contributing to an SCC Failure of Aluminum Alloys

Alloy	Temper	Direction of rolling*	Plate†	Rod/bar	Extrusion	Forging
7005	T63	L	x	x	0	0
		LT	x	x	0	0
		ST	x	x	0.6	0.6
7039	T63/T64	L	0	x	0	x
		LT	0	x	0	x
		ST	0.6	x	0.6	x
7049	T73	L	0	x	0	0
		LT	0	x	0	0
		ST	0	x	0.2	0
7075	T6	L	0	0	0	0
		LT	0.2	0.2	0.2	0.2
		ST	0.4	0.2	0.4	0.4
7075	T73	L	0	0	0	0
		LT	0	0	0	0
		ST	0	0	0	0
7075	T76	L	0	x	0	x
		LT	0	x	0	x
		ST	0.4	x	0.4	x
7079	T6	L	0	x	0	0
		LT	0.2	x	0.2	0.2
		ST	0.6	x	0.6	0.6
7175	T736	L	x	x	x	0
		LT	x	x	x	0
		ST	x	x	x	0.2
7475	T6	L	0	x	x	x
		LT	0.2	x	x	x
		ST	0.6	x	x	x
7475	T73	L	0	x	x	x
		LT	0	x	x	x
		ST	0	x	x	x

*L = longitudinal, LT = long transverse, ST = short transverse.
†x means that product not commercially offered.

A. Equal to or greater than 75 percent of the specified minimum yield strength.

B. Equal to or greater than 50 percent of the specified minimum yield strength.

C. Equal to or greater than 25 percent of the specified minimum yield strength or 100 MPa, whichever is higher.

D. Fails to meet the criterion for rating C.

Once the subfactors had been assigned acceptable values, the performance of Failure Predictor was verified with a series of test cases, and the results obtained with the KBS were compared to diagnoses given by human experts. Failure Predictor passed, without difficulty, the Turing test, which states that a KBS is acceptable when its user

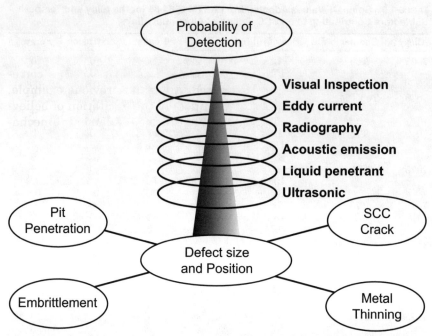

Figure 4.20 An object-oriented representation of the probability of detection of a corrosion defect.

cannot differentiate between diagnoses produced by the software and those produced by credible human experts.[36]

Nondestructive testing. Failure analysis and nondestructive evaluation (NDE) are two complementary aspects of materials engineering. The probability of detection of a defect is another multidimensional parameter that could be appropriately represented in an object-oriented architecture (Fig. 4.20). The integration of NDE into a maintenance program to extend the life of complex structures has to be based on the probability of defect detection by various NDE methods compared with damage tolerance allowances. The probability of detection itself depends on a multitude of parameters pertaining to each technique, to operator familiarity with the technique, and to all other factors describing the materials, flaw sizes and shapes, etc. The knowledge base required to decide which technique to use, when, and by whom can be quite extensive if it is completely based on classical probability mathematics, since this theory assumes that all possible events are known and that each is as likely to occur as any other.

The object-oriented representation of the probability of detection of a corrosion defect in Fig. 4.20 illustrates the flexibility of such a representation. In a global system, the defect size module would only be one version of the overall corrosion factor module, and the attributes specific to defect size and position would be actuated when necessary. With the help of such a tool, the overall probability of detection can be computed by loading a database of CFs corresponding to each system investigated. As in the previous example, the values of the CFs can be determined by the elicitation of believable expertise or, alternatively, by going to the fundamental mechanisms describing the failure modes, which could be computed using more traditional procedural routines.

Table 4.7 contains an example of CF values adapted from textbook information[37] describing the sensitivity of NDE techniques to SCC defects as a function of material composition. These CF values do not take into account defect size, position, and morphology. However, such values can be used as initial default values during the activation of the OO module. Subsequent information can then be used to improve and refine the pertinence of the CF values to the specific context and expertise. The following example illustrates how the rule propagation would be made with even a limited information base such as that contained in Table 4.7. In this example, an operator would ask the system if there would be any advantage in combining two techniques for the inspection of a component made of austenitic stainless steel. According to the data in Table 4.7, one would always obtain an increased confidence if two techniques with positive CFs were used. Table 4.8 illustrates some of the combinations envisaged in this example and the estimated gain in probability of detection from using two techniques instead of the better of the two techniques considered.

4.3 Applications of Artificial Intelligence

The modern world has produced an unprecedented quantity of technical information that merits being preserved and managed. From a societal point of view, knowledge and information are synonymous with energy if one considers the effort required to produce either. This is illustrated in Fig. 4.21, where knowledge is shown at the top of the value scale representing all aspects of materials processes, from extraction to maintenance. In an age of conservation and recycling, it is important to recognize the fact that the most valuable commodities are information as a vehicle and knowledge as the essence. It is surely sensible to preserve and recycle these most valuable commodities.

The rapid development of accessible computing power in the 1980s has led to the use of computers and direct or indirect applications of machine intelligence in every sphere of engineering. As a modern science philosopher said, "The emergence of machine intelligence during

TABLE 4.7 Sensitivity of NDE Techniques to SCC Defects in Various Materials

Technique/ material	UT*			Penetrant		Radiography		Eddy current	Acoustic emission	Visual
	ShW	LoW	SuW	V	F	γ ray	X ray			
SS										
Austenitic	20	20	40	60	80	40	60	80	40	40
Martensitic	20	20	40	60	80	40	60	0	40	20
Ferritic	20	20	40	60	80	40	60	0	40	20
Ni alloys	20	20	40	60	80	40	60	80	40	20
Cu alloys	20	20	40	60	80	40	60	80	40	20
Al alloys	20	20	40	60	80	20	60	80	40	20
Ti alloys	20	20	40	60	80	20	60	80	40	20
Steels	40	20	40	60	80	20	60	0	40	40

*ShW = ultrasonic shear, LoW = longitudinal waves, SuW = surface waves.
†V = visible penetrants, F = fluorescent penetrants.

TABLE 4.8 Examples of Sensitivities Achievable by Using a Combination of Two NDE Techniques

Combination	MB(1)	MB(2)	MB(1) + MB(2)(1 − MB(1))	Gain (%)
LoW and X ray	20	60	0.2 + 0.6 (1 − 0.2)	8
LoW and AE	20	40	0.2 + 0.4 (1 − 0.2)	12
SuW and X ray	40	60	0.4 + 0.6 (1 − 0.4)	16
SuW and P (F)	40	80	0.4 + 0.8 (1 − 0.4)	8
P (F) and X ray	80	60	0.8 + 0.6 (1 − 0.8)	12
P (F) and EC	80	80	0.8 + 0.8 (1 − 0.8)	16
EC and X ray	80	60	0.8 + 0.6 (1 − 0.8)	12
EC and AE	80	40	0.8 + 0.4 (1 − 0.8)	8

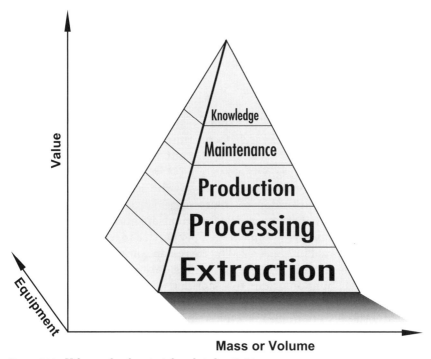

Figure 4.21 Value scale of materials-related activities.

the second half of the twentieth century is the most important development in the evolution of this planet since the origin of life two to three thousand million years ago."[38] However, efficient methodologies have to be developed in order to make use of so much new power in support of human intelligence.

The application of artificial intelligence in performing expert functions has opened new communication channels between various strata of corrosion knowledge holders. The adequate transfer and reuse of information covering corrosion problems and solutions involves the development of information-processing strategies that can become very complex. A typical corrosion engineering task involves different types of knowledge and disciplines such as metallurgy, chemistry, cost engineering, safety, and risk analysis. The expected corrosion behavior of engineering materials is thus only one component of the multifaceted life-cycle management of systems. The increasing availability of computerized information is also making the software accessibility and portability increasingly important. While it has become possible to consult either shallow or very deep information systems at the touch of a few buttons, it remains difficult to move horizontally between these systems without going through a series of menus and introductory screens. Some of the artificial intelligence tools that have been recently developed in support of corrosion control and protection will be reviewed in the following sections.

4.3.1 Expert systems

During the 1970s, research in expert systems (ESs) was mostly a laboratory curiosity. The research focus then was really on developing ways of representing and reasoning about knowledge in a computer rather than on designing actual systems.[39] In 1985 only about 50 systems had been deployed and reported, but the success of some of these had captured the attention of many organizations and individuals. One of the main attractions of ESs for scientists and engineers was the possibility of transferring some level of expertise to a less skilled workforce, as illustrated in Fig. 4.22. The corrosion community reacted with interest to the advent of these new information-processing technologies by establishing programs to foster and encourage the introduction of ESs in the workplace. While some of these programs were relatively modest, others were quite ambitious and important both in scope and in funding.

The main argument in support of these efforts was that many of the common failures caused by corrosion could have been avoided simply by implementing proper measures based on existing information. The evident gap that exists between corrosion science and the real world, where a heavy toll is continuously paid to corrosion, was probably the single main argument for proposing the ES route as a viable alternative for information processing of corrosion data. But there are several problems associated with knowledge engineering methodologies that can contribute to what has been called "the knowledge transformation bottleneck."[40] The availability of cost-effective tools and knowledge

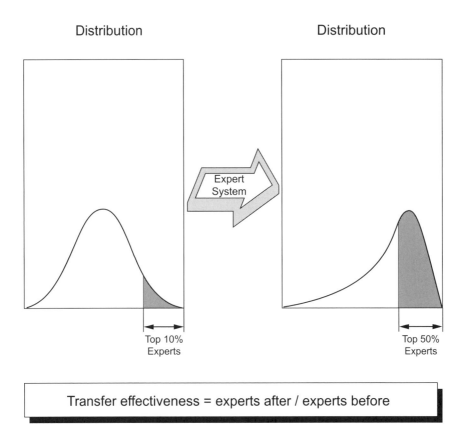

Figure 4.22 Representation of the transfer-of-expertise process possible with an expert system.

elicitation techniques is only part of the picture. The eventual integration of an ES prototype in a user community requires the tacit approval of all parties involved during the various phases of ES development. It also requires the fundamental acceptance of the expertise being computerized.

The advantages and limitations of using ES technology were analyzed in great detail in one of the first reported efforts on combating corrosion with ESs.[41] The Stress Corrosion Cracking ES (SCCES) had been created to calculate the risk of various factors involved in SCC, such as crack initiation, when evidence was supplied by the user. The main goal of this effort was to support the decision process of "general" materials engineers. The system would be initially called on to play the role of a consultant, but it was anticipated that SCCES had the potential to become

- An intelligent checklist
- A trainer
- An expert sharpener
- A communication medium
- A demonstration vehicle

The author of this review has written a more recent paper on a methodology for assessing the general benefits of ESs in the workplace.[42] In this paper, a simple three-level model of benefits is proposed: feature benefits, task benefits, and role benefits. This model is said to illustrate how technological features like expert knowledge and explanation facilities can contribute to the eventual success or failure of a system.

ESs in corrosion. On the European continent, work on ESs for Corrosion Technology (ESCORT), conceived in 1984, served to seed the establishment of a link to the European Strategic Programme on Information Technology (ESPRIT) and the creation of a series of specialized modules.[43] While ESCORT was to deal with the integration of corrosion-related issues such as troubleshooting and selection of preventive measures (materials, coatings, or inhibitors), each module was to be specialized. PRIME, which stood for Process Industries Materials Expert, was the first of these modules. PRIME specifically dealt with the selection of materials typically encountered in the chemical process industries (CPI). PRIME could consider complex chemical processes equipment in contact with a wide range of environments. The materials descriptors were complete with generic information and specialized corrosion behavior.

In the United Kingdom, the experience gained at Harwell in collecting and structuring corrosion knowledge for a computer-based ES served as the foundation for the development of two systems: ACHILLES and MENTOR.[44] ACHILLES dealt broadly with localized corrosion and provided general advice on the problems likely to be encountered in process plants and other similar environments. On the other hand, MENTOR was said to be a faithful adviser of marine engineers. The experience gained during these projects was summarized as follows:

- The front-end interface to the user has to be friendly.
- Transparency of the system is essential.
- A good knowledge base should contain a mixture of heuristics and factual information.

ACHILLES later became the cornerstone of the ACHILLES Club Project, which was given a mandate to develop a series of ES modules that would incorporate a substantial digest of expertise in particular areas of corrosion and corrosion control. The first two modules dealt with cathodic protection and microbial corrosion. The intention was to integrate a number of these modules into a global structure that could access individual modules during the course of a user consultation. This pioneering work also led to the creation of SPICES, an inference engine based on PROLOG, which was said to be particularly adapted to the multidisciplinary nature of corrosion phenomena.[45]

During the same period, the National Association of Corrosion Engineers (NACE) and the National Bureau of Standards [NBS, now called the National Institute of Standards and Technology (NTIS)] were establishing a collaborative program to collect, analyze, evaluate, and disseminate corrosion data.[46] In April 1986, the Materials Technology Institute (MTI) of the Chemical Process Industries decided to sponsor the development of an ES for material selection. During the following year, MTI initiated a project within the NACE-NIST Corrosion Data Program to develop a series of knowledge-based ESs concerning materials for handling hazardous chemicals. These systems became commercially available and are known as the ChemCor series.

Since the mid 1980s, a multitude of other projects have attempted to transfer corrosion expertise into ESs. The NACE conference proceedings, for example, regularly contain papers that illustrate the continuous interest in the application of knowledge engineering to corrosion. Unfortunately, many systems reported in the literature have never been commercialized. This has resulted in a lack of impartial and practical information concerning the performance and accuracy of these systems. It is indeed very difficult to believe everything that is said in a paper, even when the information is apparently there. To remedy this situation, the European Federation of Corrosion (EFC) and MTI have performed two surveys, between 1988 and l990, requesting recognized developers of ESs in corrosion-related areas to provide very specific information concerning the availability, scope, and performance of their systems.[47]

The EFC survey. In the EFC survey, developers of ESs were asked to elaborate on the following salient features of their systems:

- Shell used
- Area of application
- Language (user language? programming language?)
- Hardware (platform and peripherals)
- Development expenditure
- Field evaluation status

Table 4.9 summarizes the results of the 1989 EFC survey, which covered 30 systems developed in 6 countries. A summary of the survey itself indicated that the development effort reported on 22 systems averaged 4.1 person-years (PY), with a median of 2 PY; two of the systems reported efforts exceeding 10 PY. The expenditures for development reported for 11 systems averaged $490,000/year, with a median of $127,000/year. Only 4 systems were available at the time of the survey, but some were expected to be put on the market later. A total of 17 different software shells were used by the developers, with each developer tending to stay with a specific shell once a project had started.

The MTI survey. In the MTI survey, developers of ESs were asked to provide, in a well-defined grid, answers to some slightly more specific questions than those in the EFC survey, such as.

- Availability outside own organization (price, terms)
- Primary objective of the system
- Description of development team
- Application: diagnostics, prescriptive, monitor/control, design/planning, training
- Development effort and expenditure
- Hardware (development, delivery)
- Audience (targeted users)

The MTI survey, summarized in Table 4.10, encompassed descriptions of 36 systems developed in 9 countries, with only 9 systems overlapping those in the EFC survey. Most systems reported were focused on prescription, diagnosis, and training for corrosion prevention. Only a few systems dealt with the monitoring and planning aspects of corrosion prevention and control. The median development time, for the 26 systems for which values were given, was 1 to 3 PY, with two systems again exceeding 10 PY. The average budget for the 16 systems for which this information was given was $126,000/year, with a median of $100,000/year.

The survey also revealed that a total of 18 different software shells were used by the developers, with each developer again tending to stay with a specific shell once a project had started. Most systems were developed and distributed on personal computers (PCs), which is very different from the practice reported during the early days of ES development. Seven systems were available for purchase at the time of the MTI survey, but the survey failed to request information on the validation of the products themselves.

Survey of the literature before 1992. A survey of the open literature also revealed the existence of many ESs dealing with various aspects of corrosion prevention and control.[48] The following list indicates the major areas for which some systems have been reported in support of corrosion prevention and control:

- Cathodic protection
- Cooling waters
- Diagnostics
- Inhibitors
- Materials selection
- Petroleum industries
- Reinforced concrete
- Risk analysis

A compilation of the ESs reported in the EFC and MTI surveys was compared to the literature survey published in 1992.[49] Table 4.11 lists a few of these systems—approximately half of the total number surveyed by EFC and MTI—which overlapped with the literature survey. A rapid examination of the 49 literature references not related to any of the systems cited in the surveys of developers indicated that many of the articles in the literature were published after these surveys had been initiated (1988). In fact, the average date of publication of the references not related to the systems described in the surveys of developers was 1988.8 ($\sigma = 1.5$ year).

Survey of the literature between 1992 and 1995. The period following the first literature survey has seen an extremely rapid evolution of available information-processing tools and a constant progress in the introduction of personal computers in the workplace. Only a few years ago, the tremendous amount of energy required to produce and maintain software systems was responsible for a good part of the high price of development of ESs. It was thus deemed interesting to redo the literature search for applications of ESs or knowledge-based systems to prevent and protect against corrosion. The titles of papers gathered in a search of the recent literature abstracted in the Compendex*Plus system are presented in Table 4.12. The breakdown of the 37 papers identified during that search is as follows:

- 1992: 9 papers
- 1993: 5 papers
- 1994: 13 papers
- 1995: 10 papers

TABLE 4.9 Results of the EFC Survey on Expert Systems in Corrosion

Name	Country	Shell	Rules	Applications	P*	B†	Evaluation	A‡
ACHILLES	UK	SPICES		Diagnosis, prediction, prevention	3	660		
ALUSELECT	Sweden	ORACLE FOCUS	200	Selection of aluminum alloys	2.5	127.5		
AURORA	Finland	LEVEL5	830	Prediction, failure analysis, materials selection	3.3	312	+ Feedback	Buy
AURORA-STACOR	Finland	LEVEL5	126	Prediction (stainless steels)	1.5	120		
AUSCOR	UK	SAVOIR		Prediction (austenitic stainless steels)	6	825		
BANDMAT	Italy	DB CLIPPER		Materials selection, maintenance, monitoring			ENI Consult	
BENTEN	UK	ADVISOR	200	Selection inhibitor	2			
CAMS4	UK			Knowledge-based system (?)				
COMETA	Italy			Database (?)			By experts	
COREX	France	GENESTA II	80	Prevention (low-alloy steel, atmospheric)			In use (EDF)	
CORRBAS	Sweden	FOCUS	20	Diagnosis	0.5	22.5		
CORREAU	France	SPECIAL	150	Copper tubing	1			Buy
CORSER	France	SPECIAL	5000	Materials selection, diagnosis, prevention				3

Name	Country	Shell	Application	Effort*	Budget†	Availability‡	
CRAI	Belgium	KEE	Training, materials selection	1.25			
DB-CTW			Water treatment	4	100		
DOCES	Italy	PC+	Boilers		120		
ERICE	Italy	PC+	Monitoring, diagnosis (power plant)		200	2	
EXPRESS	UK	XI+	Pipeline, risk	2	1000	115.5	
GRADIENT	Belgium	KEE	CAD (heat exchangers)	2		1 company	Loan
H2 DATA	France		Database (?)				Buy
MATEDS	Sweden	FOCUS	Selection of aluminum alloys	2	300	90	
PETROCRUDE	Belgium	KEE	Prediction (refinery)	2		Demo only	
PRIME	Belgium	KEE	Materials selection	25		2500	3 companies
PROP	Italy		Monitoring, diagnosis, pollution (thermal power plant)	12	700		Used (87)
RIACE	Italy	IBM ISE	Materials selection (seawater, exchangers)	3	500		
SECOND	Belgium	KAPPA	Control (cooling tower)	7		50	Used 3 plants
SMI	Sweden		Materials selection				
STM/H2OMON	Italy	ART	Operator support power plant	4	200		
VASMIT	Finland	DBASE	Fatigue	0.8			
VULCAIN-BDM	France	(Minitel)	Database (?)				Loan

*Development effort in person-years.
†Development budget ($000 U.S.).
‡Availability.

TABLE 4.10   Results of the MTI Survey on Expert Systems in Corrosion

Name	Country	Shell	Applications
ACHILLES	UK	SPICES	Prevention
ACORD	Japan	OPS83	Prediction (seawater)
ADVICE	USA		Prediction (high temperature)
AURORA-STACOR	Finland	LEVEL5	Prediction (SSs)
AUSCOR	UK	SAVOIR	Prediction (austenitic SSs)
BENTEN	UK	ADVISOR	Selection inhibitor
BLEACH	USA	EXXYS	Materials selection (beach plant)
BLEACHER	Finland	KEE	Materials selection (beach plant)
BWR	Japan	OPS5	Prediction (IGSCC)
CHEM*COR	USA	KES	Materials selection (hazardous chemicals)
CL2	USA	LEVEL5	Materials selection (Cl2 service)
CORRCON	Israel	OPS5	Design diagnosis
CORREAU	France	NOVYS	Copper tubing
CORRES	Japan	SOHGEN	Prediction
CORSER	France		Materials selection, diagnosis, prevention
CRAI	Belgium	KEE	Training, materials selection
DESAD	USA	PC+	Prevention (desalter unit)
DIASCC	Japan	OPS83	Risk of SCC (SSs)
ECHOS	Japan	ESHELL	Prediction, maintenance, shutdowns
FERPRED	USA	PC+	Ferrite in welds
GENERAL	UK		Materials selection, prediction
JUNIPER	UK		Authoring tools
KISS	Germany	NEXPERT	Materials selection (CPI)
MATGEO	New Zealand	KES	Materials selection (geothermal plants)
OILSTO	Japan		Prediction, inspection
PBCORR	UK	CAMS4	Corrosion of lead
PC6493	UK	CAMS4	Defect assessment (PC6493)
PETRO-COR1	New Zealand	KES	Materials selection (sucker rod pumps)
POURBAIX	Belgium		
PRIME	Belgium	KEE	Materials selection
REFMAIN	Japan		On-line prediction (refinery)
SECOND	Belgium	KAPPA	Control (cooling tower)
SSCP-PH1	USA	PC+	Materials selection (H2S)
WELDPLAN	Japan	OPS83	Advise (weld parameter)
WELDSEL	USA	PC+	Advise (weld rod)
WELDSYM	USA	PC+	Advise (symbol)

*Diagnose (Di), prescribe (Ps), predict (Pd), monitor (M), train (Tr).
†Expert (E), Professional (P), Novice (N).
‡Development effort in person-years.
§Development budget ($000 U.S.).
¶Availability.

TABLE 4.10 Results of the MTI Survey on Expert Systems in Corrosion (*Continued*)

Name	Roles*						Target†			P‡	B§	A
	Di	Ps	Pd	M	Pl	Tr	E	P	N			
ACHILLES	*	*	*	*	*	*		*	*	10	200	
ACORD	*	*								1	140	
ADVICE		*		*						3	27	
AURORA-STACOR	*	*	*				*	*	*	5	94	
AUSCOR			*		*	*	*			10	100	
BENTEN			*			*		*		3		
BLEACH		*	*					*		1	100	
BLEACHER		*	*				*	*	*	5		
BWR			*									
CHEM*COR		*						*	*	3	177	Buy
CL2		*	*		*				*	0.5		
CORRCON	*	*	*				*	*		0.5		
CORREAU	*	*				*		*	*	3	20	Buy
CORRES												
CORSER	*	*	*			*	*		*	50	236	
CRAI						*				1	75	
DESAD	*	*	*			*			*	0.5		
DIASCC	*		*						*	0.5		Buy
ECHOS												
FERPRED	*	*					*					Buy
GENERAL	*	*	*	*	*	*	*			1		
JUNIPER					*	*	*					
KISS	*	*	*	*	*	*		*	*	3	355	
MATGEO												
OILSTO	*	*			*		*			0.5	58	
PBCORR	*	*			*	*	*			3		Buy
PC6493					*	*	*			3	99	
PETRO-COR1												
POURBAIX	*			*	*	*	*			3	75	Buy
PRIME	*	*		*	*		*					
REFMAIN							*	*			150	
SECOND			*			*		*			50	
SSCP-PH1	*				*		*			1		
WELDPLAN												
WELDSEL		*				*						Buy
WELDSYM					*			*				Buy

TABLE 4.11 Cross Compilation of the ESs Identified in the Literature Survey with the EFC and MTI Surveys of Developers

Name	Country	EFC	MTI
ACHILLES	UK	*	*
ADVICE	USA		*
AURORA	Finland	*	
AURORA-STACOR	Finland	*	*
AUSCOR	UK	*	*
BENTEN	UK	*	*
CAMS4	UK	*	
CHEM*COR	USA		*
COMETA	Italy	*	
COREX	France	*	
CORREAU	France	*	*
CRAI	Belgium	*	*
DOCES	Italy	*	
ERICE	Italy	*	
EXPRESS	UK	*	
GRADIENT	Belgium	*	
JUNIPER	UK		*
MATGEO	New Zealand		*
PETRO-COR1	New Zealand		*
PETROCRUDE	Belgium	*	
PRIME	Belgium	*	*
PROP	Italy	*	
RIACE	Italy	*	
SECOND	Belgium	*	*
SSCP-PH1	USA		*
STM/H2OMON	Italy	*	

The list of papers in Table 4.12 is far from a complete inventory of those published during that period. It does not include, for example, any of the papers published in NACE International Proceedings or any thesis work or industrial reports. But, as it stands, this survey can provide a relatively good indication of the recent trends in the efforts to develop ESs or KBSs to combat corrosion problems. As can be observed in Table 4.12, the progress in software technologies has opened new avenues for developers of intelligent systems in corrosion. While most of the tools developed up to the early 1990s were primarily constructed using rules and database management principles, the later systems include object orientation and other paradigms such as artificial neural networks and case-based reasoning.

While visions of systems that would answer all questions and solve all corrosion-related problems have faded away, the broad acceptance of the computer in the workplace has facilitated the introduction of new concepts and methods to manage corrosion information. Very

TABLE 4.12 Titles of References Related to KBSs and ESs Dealing with Corrosion Published between 1992 and 1995

1995
- Knowledge-Based Shell for Selecting a Nondestructive Evaluation Technique
- Knowledge-Based Concrete Bridge Inspection System
- Generalized Half-Split Search for Model-Based Diagnosis
- Development of Expert System for Fractography of Environmentally Assisted Cracking
- Lifetime Prediction in Engineering Systems: The Influence of People
- Discovering Expert System Rules in Data Sets
- Modeling Contact Erosion Using Object-Oriented Technology
- Object-Oriented Representation of Environmental Cracking
- Systems Approach to Completing Hostile Environment Reservoirs
- Storage and Retrieval of Corrosion Data of Desalination Plant Owners

1994
- Bridging the Gap between the World of Knowledge and the World that Knows
- ESs for Material Selection and Analysis for the Oil Industry: An Application-Oriented Perspective
- ANN Predictions of Degradation of Nonmetallic Lining Materials from Laboratory Tests
- Reliability Based Inspection Scheduling for Fixed Offshore Structures
- Automated Corrective Action Selection Assistant
- Fracture Mechanics Limit States for Reassessment and Maintenance of Fixed Offshore Structures
- Corrosion Consultant Expert System
- Computer Knowledge-Based System for Surface Coating and Material Selection
- Databases and Expert Systems for High Temperature Corrosion and Coatings
- CORIS: A Knowledge Based System for Pitting Corrosion
- CORIS: An Expert System for the Selection of Materials Used in Sulfuric Acid
- Expert System to Choose Coatings for Flue Gas Desulphurisation Plant
- ACHILLES Expert System on Corrosion and Protection: Consultations on Aspects of SCC

1993
- Ways to Improve Computerizing of Cathodic Protective Systems for Pipelines
- Investigation of Corrosion Prevention Method for Determination of Steel Structure Condition
- Corrosion Control in Electric Power Plants
- Reliability-Based Expert Systems for Optimal Maintenance of Concrete Bridges
- SEM: Un Sistema Esperto per la Scelta dei Materiali nella Progettazione

1992
- Reliability Assessment of Wet H_2S Refinery and Pipeline Equipment: A KBSs Approach
- Exacor: An ES for Evaluating Corrosion Risks and Selecting Precoated Steel Sheets for Auto Bodies
- Expert Systems in Corrosion Engineering
- Expert Electronic System for Ranking Developments in Sphere of Corrosion Protection
- Research Needs Related to Forensic Engineering of Constructed Facilities
- Expert Computer System for Evaluating Scientific-Research Studies on the Development of Methods of Anticorrosion Protection
- DEX: An Expert System for the Design of Durable Concrete
- Automated System for Selection of a Constructional Material
- Informational Component in Systems of Corrosion Diagnostics for Engineering Equipment

focused ESs have been integrated into large systems as controllers or decision support systems to prevent corrosion damage. Other very focused ESs are also being built and tested to simplify the requirements for multidisciplinary expertise associated with corrosion engineering practices. A few of these computerized methodologies have reached mainstream applications and are readily available. It is expected that the continuous evolution of information-processing technologies will greatly facilitate the development of increasingly sophisticated computer tools and their introduction in the corrosion prevention workplace.

4.3.2 Neural networks

An artificial neural network (ANN) is a network of many very simple processors, or neurons (Fig. 4.23), each having a small amount of local memory. The interaction of the neurons in the network is roughly based on the principles of neural science. The neurons are connected by unidirectional channels that carry numeric data based on the weights of connections. The neurons operate only on their local data and on the inputs they receive via the connections. Most neural networks have some sort of training rule. The training algorithm adjusts the weights on the basis of the patterns presented. In other words, neural networks "learn" from examples. ANNs excel particularly at problems where pattern recognition is important and precise computational answers are not required. When ANNs' inputs and/or outputs contain evolved parameters, their computational precision and extrapolation ability significantly increase, and they can even outperform more traditional modeling techniques. Only a few applications of ANN to solving corrosion problems have been reported so far. Some of these systems are briefly described here:

- *Predicting the SCC risk of stainless steels.* The risk of encountering a stress corrosion cracking situation was functionalized in terms of the main environment variables.[50] Case histories reflecting the influence of temperature, chloride concentration, and oxygen concentration were analyzed by means of a back-propagation network. Three neural networks were developed. One was created to reveal the temperature and chloride concentration dependency (Fig. 4.24), and another to expose the combined effect of oxygen and chloride content in the environment. The third ANN was trained to explore the combined effect of all three parameters. During this project, ANNs were found to outperform traditional mathematical regression techniques, in which the functions have to be specified before performing the analysis.

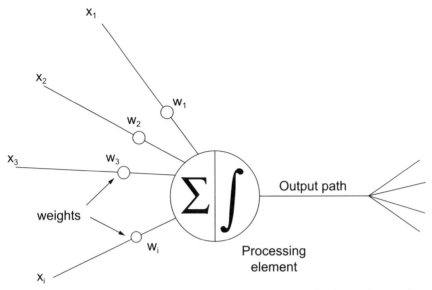

Figure 4.23 Schematic of a single processor or neuron in an artificial neural network.

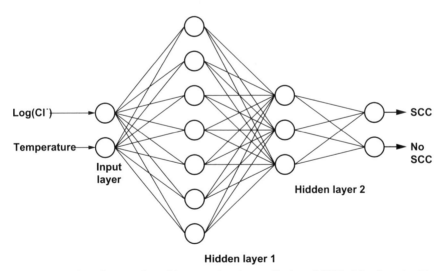

Figure 4.24 Neural network architecture for the prediction of SCC risk of austenitic stainless steels in industrial processes.

- *Corrosion prediction from polarization scans.* An ANN was put to the task of recognizing certain relationships in potentiodynamic polarization scans in order to predict the occurrence of general or localized corrosion, such as pitting and crevice corrosion.[51] The initial data inputs were derived by carefully examining a number of polarization scans for a number of systems and recording those features that were used for the predictions. Table 4.13 lists the initial inputs used and how the features were digitized for computer input. The variables shown were chosen because they were thought to be the most significant in relation to the predictions (Table 4.13). The final ANN proved to be able to make appropriate predictions using scans outside the initial training set. This ANN was embedded in an ES to facilitate the input of data and the interpretation of the numerical output of the ANN.

- *Modeling CO_2 corrosion.* A CO_2 corrosion "worst-case" model based on an ANN approach was developed and validated against a large experimental database.[52] An experimental database was used to train and test the ANN. It consisted initially of six elemental descriptors (temperature, partial CO_2 pressure, ferrous and bicarbonate ion concentrations, pH, and flow velocity) and one output, i.e., the corrosion rate. The system demonstrated superior interpolation performance compared to two other well-known semiempirical models. The ANN model also demonstrated extrapolation capabilities comparable to those of a purely mechanistic electrochemical CO_2 corrosion model.

- *Predicting the degradation of nonmetallic lining materials.* An ANN was trained to recognize the relationship between results of a sequential immersion test for nonmetallic materials and the behavior of the same materials in field applications.[53] In this project, 89 cases were used for the supervised training of the network. Another 17 cases were held back for testing of the trained network. An effort was made to ensure that both sets had experimental data taken from the same test but using different samples. Appropriate choice of features enabled the ANN to mimic the expert with reasonable accuracy. The successful development of this ANN was another indication that ANNs could seriously aid in projecting laboratory results into field predictions.

- *Validation and extrapolation of electrochemical impedance data.* The ANN developed in this project had three independent input vectors: frequency, pH, and applied potential.[54] The ANN was designed to learn from the invisible or hidden information at high and low frequencies and to predict in a lower frequency range than that used for training. Eight sets of impedance data acquired on nickel electrodes in phosphate solutions were used for this project. Five sets were used for training the ANN, and three for its testing. The ANN proved to be a powerful technique for generating diagnostics in these conditions.

TABLE 4.13 Data Inputs and Outputs for Predicting Corrosion Out of Polarization Scans with an Artificial Neural Network

Input parameter	Value of feature
Prepassivation potential	$E_{prot} - E_{corr}$
Pitting potential	$E_{pit} - E_{corr}$
Hysteresis	+1 = positive
	0 = none
	−1 = negative
Current density at scan reversal	$\mu A \cdot cm^{-2}$
Anodic nose	+1 = yes
	0 = no
Passive current density	$\mu A \cdot cm^{-2}$
Potential at anodic-cathodic transition	$E_{A\,to\,C} - E_{corr}$

Output parameter	Value of feature
Crevice corrosion predicted	+1 = yes
	0 = no
Pitting predicted	+1 = yes
	0 = no
Should general corrosion be considered?	+1 = yes
	0 = no

4.3.3 Case-based reasoning

Much of human reasoning is case-based rather than rule-based. When people solve problems, they frequently are reminded of previous problems they have faced. For many years, both law and business schools have used cases as the foundation of knowledge in their respective disciplines. Within AI, when one talks of learning, it usually means the learning of generalizations, either through inductive methods or through explanation-based methods. Case-based reasoning (CBR) is unique in that it makes the learning little more than a by-product of reasoning.[55] CBR has met with tangible success in such diverse human decision-making applications as banking, autoclave loading, tactical decision making, and foreign trade negotiations. The CBR approach is particularly valuable in cases containing ill-structured problems, uncertainty, ambiguity, and missing data. Dynamic environments can also be tackled, as can situations in which there are shifting, ill-defined, and competing objectives. Cases in which there are action feedback loops, involvement of many people, and multiple and potentially changing organizational goals and norms can also be tackled.

A critical issue for the successful development of such systems is the creation of a solid indexing system, since the success of a diagnosis depends heavily on the selection of the best stored case. Any misdirection can lead a query down a path of secondary symptoms and factors. It is therefore very important to establish an indexing system that will effectively indicate or contraindicate the applicability of a stored case. Three issues are particularly important in deciding on the indices:[56]

- Indices must be truly relevant.
- Indices must be generalized; otherwise, only an exact match will be the criterion for case applicability.
- But indices shall not be overgeneralized.

Failure analysts and corrosion engineers also reason by analogy when faced with new situations or problems. Two CBR systems have been recently developed in support of corrosion engineering decisions. Both systems derived their reasoning from a combination of two industrial alloy performance databases. The general architecture of these two CBR systems is presented in Fig. 4.25. The first, M-BASE, facilitates the process of determining materials that have a given set of desired properties and/or specifications. The second, C-BASE, helps the materials engineer in the difficult task of selecting materials for corrosion resistance in complex chemical environments.

4.4 Computer-Based Training or Learning

Potential advantages of the computer-based learning approach over a conventional course offering include access to a larger target population and optimization of the shrinking expert instructor pool. However, experience has shown that, despite advances in software applications, an enormous investment in professional time for planning and developing course material is required. Course modules have been created initially in paper-based format, to place the scientific/technical course content on a sound footing. Selected case studies and assignments have subsequently been designed in electronic format to develop skills in applying the knowledge and understanding gained from the paper-based course notes.

The advantages and disadvantages of computer-based learning (CBL) and more conventional education techniques have been described as follows:[57,58]

Advantages

- Access to a large student and professional "market"
- Potential for achieving higher student cognition
- Student interaction with course material
- Direct linkages to Internet resources
- Higher student attention levels through stimulating multimedia presentations
- Rapid updating of information and course materials

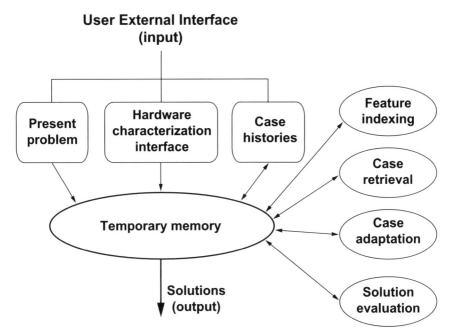

Figure 4.25 Case-based reasoning architecture for the prediction of materials behavior.

- Tracking user interaction with the course material
- Efficient retrieval of specific information using electronic text processing
- Optimization of a steadily shrinking expert instructor pool
- Wider choice of course offerings for students
- Freedom for students to follow individual pace and learning styles
- Achievement of special learning objectives through computer simulations (for example, key technical concepts, role playing, decision-making processes and their consequences)

Disadvantages

- Lack of face-to-face interaction and engagement
- Low inspiration factor, especially when working in isolation
- Lack of teamwork
- Limited communication skills development
- Production of CBL material is (extremely) time-consuming and costly
- Need for special computing and software skills, mainly on the part of the developers

- Requirement for expensive hardware
- Nonuniformity of hardware affecting product quality
- Need for support staff

A venture undertaken by a consortium based at the Corrosion & Protection Centre, UMIST, Manchester, UK, and incorporating the universities of Nottingham, Aston, Leeds, and Glasgow has resulted in CBL course materials, called Ecorr, to support the teaching of corrosion principles and corrosion control methods to engineering students. Ecorr takes a case study approach, with the student learning about corrosion through specific examples of initially simple corrosion phenomena and then real-world corrosion engineering problems. Version 1.0 includes seven case study modules:

Introductory modules:
- Introduction to Corrosion
- Corrosion of Zinc
- Corrosion Kinetics
- Potential Measurements

Advanced modules:
- Pipeline Corrosion
- Drill Pipe Failures
- Cathodic Protection

4.5 The Internet and the Web

The Internet has revolutionized both the computer and communication worlds like nothing before. The invention of the telegraph, telephone, radio, and computer set the stage for this unprecedented integration of capabilities. The Internet is at once a worldwide broadcasting capability, a mechanism for information dissemination, and a medium for collaboration and interaction between individuals and their computers without regard to geographic location. The Internet represents one of the most successful examples of the benefits of sustained investment and commitment to research and development of an information infrastructure. Beginning with the early research in packet switching, the government, industry, and academia have been partners in evolving and deploying this exciting new technology.

The first recorded description of the social interactions that could be enabled through networking was a series of memos written by J. C. R. Licklider of MIT in August 1962, in which he discussed his "galactic network" concept.[59] He envisioned a globally interconnected network through which everyone could quickly access data and programs from any site. In spirit, the concept was very much like the Internet of today. The combination of the powerful communication medium with other advances in computer interfaces and hypertext linkages set the stage for the creation of a global environment that has revolutionized modern computing. The timeline of important milestones in the history of Internet is presented in Fig. 4.26.

The World Wide Web was set up in 1990 by the European Laboratory for Particle Physics (or CERN) as a way for physicists to track one another's progress. The idea was that people working in different places could learn what others were doing by looking at a hypertextual document set up on a computer which could be accessed through the Internet. This idea grew into the much bigger and large-scale operation that we now know as the Web. There are currently well over 10,000 Web servers, the computers which store and handle requests for Web pages, and a great number of people all over the world access the Web for various reasons every day. The Web is continually being enhanced and developed, as a result of rapid technological changes and the addressing of various questions and problems raised by the current state of the Web.

A *Web browser* is a software application used to locate and display Web pages. Three of the most popular browsers are *Netscape Navigator,* Microsoft *Internet Explorer,* and *Spyglass Mosaic.* All of these are *graphical browsers,* which means that they can display graphics as well as text. In addition, most modern browsers can present multimedia information, including sound and video. A full gamut of tools has also been developed to navigate the Web and search for specific information. The speed and functionality of these tools increase at a very fast rate. The following is only a short list of some of these Web exploratory aids:

- Metacrawler
- YAHOO
- LYCOS
- Open Text
- Infoseek
- Excite
- Webcrawler
- Galaxy
- WWWW—the WORLD WIDE WEB WORM
- The Whole Internet Catalog
- World Virtual Tourist (World Map of the Web)
- WebWorld
- Sprawl

Date		Operational Networks
1968	Formation of ARPANET working group (ARPA = Advanced Research Project Agency)	
1969	First 'packets' sent by Charley Kline at UCLA	
1970	ARPANET hosts start using Network Control Protocol (NCP)	
1971	BBN develops a terminal Interface Message Processor (IMP) or TIP that supports up to 64 hosts	15 nodes (23 hosts)
1972	The @ sign is chosen for its 'at' meaning	
1974	Design of a Transmission Control Program (TCP)	
1982	TCP and Internet Protocol (IP) are established as the protocol suite that is known as TCP/IP	
1983	First desktop workstations	113 nodes
1984	Moderated newsgroups introduced on USENET (mod.*)	> 1,000 hosts
1987	Email link established between Germany and China	> 10,000 hosts
1989	Creation of 'Archie' (archiver for ftp sites) by Peter Deutsch, McGill U., the first effort to index the Internet	> 100,000 hosts
	Development of a new protocol for information distribution (Tim Berners-Lee, CERN)	
1991	Development of 'gopher' the first friendly interface to the Internet	
	World-Wide-Web (WWW) released by CERN	
1992	The term 'surfing the net' is coined by Jean Armour Polly	> 1,000,000 hosts
1993	Mosaic, the graphical browser developed by Marc Andreessen (Netscape), takes the Internet by storm	600+ web sites
1995	A number of Net related companies go public, with Netscape leading the pack	> 25,000 web sites
1998	Release of Windows 98 with web browser integrated into desktop operating system	> 3,500,000 web sites

Figure 4.26 Timeline of important Internet milestones.

References

1. Box, G. E., Hunter, W. G., and Hunter, J. S., *Statistics for Experiments,* New York, John Wiley and Sons, 1978.
2. Mason, R. L., Gunst, R. F., and Hess, J. L., *Statistical Design and Analysis of Experiments,* New York, John Wiley and Sons, 1989.
3. Box G. E. P., and Draper, N. R., *Empirical Model-Building and Response Surfaces,* New York, John Wiley and Sons, 1987.

4. Walton, J. C., Cragnolino, G., and Kalandros, S. K., A Numerical Model of Crevice Corrosion for Passive and Active Metals, *Corrosion,* **38:**1–18 (1996).
5. Feigenbaum, M., Quantitative Universality for a Class of Nonlinear Transformations, *Journal of Statistical Physics,* **19:**25–33 (1978).
6. Pajkossy, T. L., and Nyikos, L., *Journal of the Electrochemical Society,* **133:**2063 (1986).
7. Roberge, P. R., The Analysis of Spontaneous Electrochemical Noise for Corrosion Studies, *Journal of Applied Electrochemistry,* **23:**1223–1231 (1993).
8. Roberge, P. R., and Trethewey, K. R., The Fractal Dimension of Corroded Aluminum Surfaces, *Journal of Applied Electrochemistry,* **25:**962–966 (1995).
9. Mandelbrot, B. B., and van Ness, J. W., *SIAM Review,* **10:**421 (1968).
10. Fan, L. T., Neogi, D., and Yashima, M., *Elementary Introduction to Spatial and Temporal Fractals,* Berlin, Germany, Springer-Verlag, 1991.
11. Hurst, E. H., Methods of Using Long-Term Storage in Reservoirs, *Proceedings Institute of Civil Engineering,* 5:519 (1956).
12. Mandelbrot, B. B., and Wallis, J. R., *Water Resources Research,* 5:321 (1969).
13. Feder, J., *Fractals,* New York, Plenum, 1988.
14. Evans, U. R., Mears, R. B., and Queneau, P. E., Corrosion Probability and Corrosion Velocity, *Engineering,* 136:689 (1933).
15. Gumbel, E. J., *Statistical Theory of Extreme Values and Some Practical Applications,* Mathematics Series 33, Washington, D.C., National Bureau of Standards, 1954.
16. Shibata, T., Statistical and Stochastic Approaches to Localized Corrosion, *Corrosion,* 52:813–830 (1996).
17. Meany, J. J., and Ault, J. P., Extreme Value Analysis of Pitting Corrosion, in Parkins, R. N. *Life Prediction of Corrodible Structures,* Houston, Tex., NACE International, 1991, pp. 2-1–2-14.
18. Sheikh, A. K., Boah, J. K., and Hansen, D. A., Statistical Modeling of Pitting Corrosion and Pipeline Reliability, *Corrosion,* 46:190–197 (1990).
19. Riekels, L. M., Seetharam, R. V., Krishnamurthy, R. M., et al., Management of Corrosion in the Arun Field, *Corrosion,* 53:72–81 (1997).
20. Shoesmith, D. W., Ikeda, B. M., and LeNeveu, D. M., Modeling the Failure of Nuclear Waste Containers, *Corrosion,* 53:820–829 (1997).
21. Timmins, P. F., *Predictive Corrosion and Failure Control in Process Operations,* Metals Park, Ohio, ASM International, 1996.
22. RAC, *Fault Tree Analysis Application Guide,* Rome, N.Y., Reliability Analysis Center, 1990.
23. Roberge, P. R., Eliciting Corrosion Knowledge through the Fault-Tree Eyeglass, in Trethewey, K. R., and Roberge, P. R. (eds.), *Modelling Aqueous Corrosion: From Individual Pits to Corrosion Management,* The Netherlands, Kluwer Academic Publishers, 1994, pp. 399–416.
24. Maintenance of Aeronautical Products, in *Airworthiness Manual,* TP6197E, Ottawa, Transport Canada Aviation, 1987, chap. 571.
25. *BBAD Maintenance Program: Policy and Procedures Handbook,* Montreal, Bombardier Inc., 1994.
26. Muhlbauer, W. K., *Pipeline Risk Management Manual,* Houston, Tex., Gulf Publishing Co., 1996.
27. Kuhn, T., *The Structure of Scientific Revolutions,* Chicago, University of Chicago Press, 1970.
28. O'Rorke, P., Morris, S., and Schulenburg, D., Theory Formation by Abduction: A Case Study Based on the Chemical Revolution, in Shrager, J., and Langley, P. (eds.), *Computational Models of Scientific Discovery and Theory Formation,* San Mateo, Calif., Morgan Kaufmann Publishers, 1990.
29. Staehle, R. W., Understanding "Situation-Dependent Strength": A Fundamental Objective in Assessing the History of Stress Corrosion Cracking, in *Environment-Induced Cracking of Metals,* Houston, Tex., NACE International, 1989, pp. 561–612.
30. Dahl, O. J., and Nygaard, K., SIMULA, An Algol Based Simulation Language, *Communications of the ACM,* **9:**671–678 (1966).

31. Ek, T., Working with Objects, *Oracle Magazine,* 40–44 (1994).
32. Kim, W., *Introduction to Object Oriented Databases,* Boston, The MIT Press, 1990.
33. Holroyd, N. J. H., Vasudevan, A. K., and Christodoulou, L., Stress Corrosion of High-Strength Aluminum Alloys, in *Treatise on Materials Science and Technology,* Vol. 31, New York, Academic Press, 1989, pp. 463–483.
34. Turban, E., *Decision Support and Expert Systems: Management Support Systems,* New York, Macmillan, 1990.
35. Hollingsworth, E. H., and Hunsicker, H. Y., Corrosion of Aluminum and Aluminum Alloys, in *Metals Handbook: Corrosion,* Vol. 13, Metals Park, Ohio, American Society for Metals, 1987, pp. 583–609.
36. Bryson, G. D., and Roberge, P. R., Development of an Empirical Model for the Evaluation of Susceptibility to Stress Corrosion Cracking, in Roberge, P. R., Szklarz, K., and Sastri, S. (eds.), *Materials Performance: Sulphur and Energy,* Montreal, The Canadian Institute of Mining, Metallurgy and Petroleum, 1992, pp. 247–257.
37. Stafford, S. W., and Mueller, W. H., Failure Analysis of Stress-Corrosion Cracking, in Jones, R. H. (ed.), *Stress Corrosion Cracking,* Metals Park, Ohio, American Society for Metals, 1992, pp. 417–436.
38. Stonier, T., *Beyond Information,* London, Springer-Verlag, 1992.
39. Durkin, J., *Expert Systems: Design and Development,* New York, Macmillan, 1994.
40. Williams, C., Expert Systems, Knowledge Engineering, and AI Tools: An Overview, *IEEE Expert,* 66–70 (1986).
41. Basden, A., On the Applications of Expert Systems, *International Journal of Man Machine Studies,* **19:**461–472 (1983).
42. Basden, A., Three Levels of Benefits in Expert Systems, *Expert Systems,* **11:**99–107 (1994).
43. Jadot, A., and Lanclus, L., ESCORT: Expert Software for Corrosion Technology, K.V. Leuven University, 1985.
44. Wanklyn, J. N., and Wilkins, N. J. M., Development of an Expert System for Design Consultation on Marine Corrosion, *British Corrosion Journal,* **20:**161–166 (1985).
45. Westcott, C., Williams, D. E., Croall, I. F., Patel, S., and Bernie, J. A., "The Development and Application of Integrated Expert Systems and Databases for Corrosion Consultancy," in *CORROSION 86,* Houston, Tex., NACE International, 1986, paper 54.
46. Ugiansky, G. M., Van Orden, A. C., and Clausen, D. E., The NACE-NBS Corrosion Data Program, in Fu, J., Heidersbach, R., and Erbar, R. (eds.), *Computers in Corrosion Control,* Houston, Tex., NACE International, 1986, pp. 15–20.
47. MTI, Report of Task Group 1 of the Working Party on Expert Systems in Materials Engineering, St. Louis, Materials Technology Institute, 1990.
48. Roberge, P. R., Bridging the Gap between the World of Knowledge and the World that Knows, *Materials Performance,* **33:**52–56 (1994).
49. Roberge, P. R., Expert Systems for Corrosion Prevention and Control, *Corrosion Reviews,* **15:**1–14 (1997).
50. Smets, H. M. G., and Bogaerts, W. F. L., SCC Analysis of Austenitic Stainless Steels in Chloride-Bearing Water by Neutral Network Techniques, *Corrosion,* **48:**618–623 (1992).
51. Rosen, E. M., and Silverman, D. C., Corrosion Prediction from Polarization Scans Using Artificial Neural Network Integrated with an Expert System, *Corrosion,* **48:**734–745 (1992).
52. Nesic, S., and Vrhovac, M., A Neural Network Model for CO_2 Corrosion of Carbon Steel, *Journal of Corrosion Science and Engineering,* **1:**1–11 (1998).
53. Silverman, D. C., Artificial Neural Network Predictions of Degradation of Nonmetallic Lining Materials from Laboratory Tests, *Corrosion,* **50:**411–418 (1994).
54. Urquidi-Macdonald, M., and Egan, P. C., Validation and Extrapolation of Electrochemical Impedance Spectroscopy Data, *Corrosion Reviews,* **15:**169–194 (1997).
55. Kolodner, J., *Case-Based Reasoning,* San Mateo, Calif., Morgan Kaufmann Publishers, 1993.

56. Barletta, R., and Mark, W., "Explanation-Based Indexing of Cases," in *Proceedings of AAAI-88,* Cambridge, Mass., MIT Press, 1988, pp. 50–60.
57. Basu, P., De, D. S., Basu, A., et al., Development of a Multimedia-Based Instructional Program, *Chemical Engineering Education,* 272–277 (1996).
58. Nobar, P. M., Crilly, A. J., and Lynkaran, K., The Increasing Influence of Computers in Engineering Education: Teaching Vibration via Multimedia Programs, *Engineering Education,* **12:** (1996).
59. Leiner, B. M., *A Brief History of the Internet,* 1998.

Chapter 5

Corrosion Failures

5.1	Introduction	332
5.2	Mechanisms, Forms, and Modes of Corrosion Failures	332
	5.2.1 Forms of corrosion	332
	Uniform (or general) corrosion	333
	Pitting	335
	Crevice corrosion	336
	Galvanic corrosion	339
	Selective leaching	344
	Erosion corrosion	345
	Environmental cracking	346
	Intergranular corrosion	349
	5.2.2 Modes and submodes of corrosion	352
	5.2.3 Corrosion factors	354
	5.2.4 The distinction between corrosion-failure mechanisms and causes	357
5.3	Guidelines for Investigating Corrosion Failures	359
5.4	Prevention of Corrosion Damage	360
	5.4.1 Uniform corrosion	362
	5.4.2 Galvanic corrosion	363
	5.4.3 Pitting	364
	5.4.4 Crevice corrosion	365
	5.4.5 Intergranular corrosion	365
	5.4.6 Selective leaching	366
	5.4.7 Erosion corrosion	366
	5.4.8 Stress corrosion cracking	366
5.5	Case Histories in Corrosion Failure Analysis	368
References		369

5.1 Introduction

System failures and subsequent failure investigations have become increasingly important in modern societies. Besides liability issues, an important reason for conducting a failure investigation is to identify the mechanisms and causes of a problem to prevent its reoccurrence. The recommendation of remedial action is indeed an important aspect of the failure analysis process. Neglecting to get to the underlying causes of corrosion failures and to take corrective action can expose an organization to litigation, liability, and loss of customer and public confidence in its product(s). Such risks are unacceptable in the modern global competitive business environment.

Conducting a failure analysis is not an easy or straightforward task. Early recognition of corrosion as a factor in a failure is critical because much important corrosion information can be lost if a failure scene is altered or changed before appropriate observations and tests are made. To avoid these pitfalls, certain systematic procedures have been proposed to guide an investigator through the failure analysis process. But failure analysis is ultimately best learned by experience, and a failure analyst must earn proper credentials by living through actual investigations and having successfully solved a variety of problems.

The use of correct and consistent terminology in failure analyses is vital. The value of information in reports is greatly diminished by inattention to this detail, especially if the information is to be subsequently stored, retrieved, and processed by computers. Unfortunately, critical terminology is often used too loosely in practice. This chapter starts with key concepts for identifying corrosion damage. This is followed by a review of guides for conducting corrosion-failure investigations and a discussion on the usefulness of case histories in failure analysis. The chapter concludes with fundamental remedial measures.

5.2 Mechanisms, Forms, and Modes of Corrosion Failures

In practice, the terms *mechanisms, forms,* and *modes of corrosion failures* are often used interchangeably, leading to some confusion. Such loose usage does not do justice to the significant serious work that has been invested in defining these separate concepts. A clearer picture should emerge from the material that follows.

5.2.1 Forms of corrosion

Form of corrosion is generally well known from one of the most enduring books on corrosion engineering.[1] The different forms of corrosion represent corrosion phenomena categorized according to their appearance. Dillon[2] considered Fontana's basic forms of corrosion and divided them into three groups, based on their ease of identification. The three categories used are

- *Group 1.* Readily identifiable by ordinary visual examination.
- *Group 2.* May require supplementary means of examination.
- *Group 3.* Verification is usually required by microscopy (optical, electron microscopy, etc.).

The main forms of corrosion are shown in Fig. 5.1, together with the respective group categories. In this figure, the number of forms has been expanded somewhat from Fontana's original grouping of eight basic forms. A description and an example of each basic form of corrosion follows.

Uniform (or general) corrosion. Uniform corrosion is characterized by corrosive attack proceeding evenly over the entire surface area or a large fraction of the total area. General thinning takes place until failure. On the basis of tonnage wasted, this is the most important form of corrosion. However, uniform corrosion is relatively easily measured and predicted, making disastrous failures relatively rare. In many cases, it is objectionable only from an appearance standpoint. The breakdown of protective coating systems on structures often leads to this form of corrosion. Dulling of a bright or polished surface, etching by acid cleaners, or oxidation (discoloration) of steel are examples of surface corrosion. Corrosion-resistant alloys and stainless steels can become tarnished or oxidized in corrosive environments. Surface corrosion can indicate a breakdown in the protective coating system, however, and should be examined closely for more advanced attack. If surface corrosion is permitted to continue, the surface may become rough, and surface corrosion can lead to more serious types of corrosion.

An example of uniform corrosion damage on a rocket-assisted artillery projectile is illustrated in Fig. 5.2.[3] The cause of failure was poor manufacturing practices, which included acid pickling prior to the phosphatizing surface protection treatment, inadequate rinsing after the phosphate coating process, excessive drying temperatures, and a poor-quality top coat applied without a primer. Undesirable effects of the corrosion damage include loss of troop confidence (who would like to handle a rusty container packed with TNT!) and a possible impairment of accuracy. Specifying an abrasive blast as phosphate pretreatment, adequate spray rinsing, lower drying temperatures, and the use of a primer-top coat combination were recommended to overcome this problem.

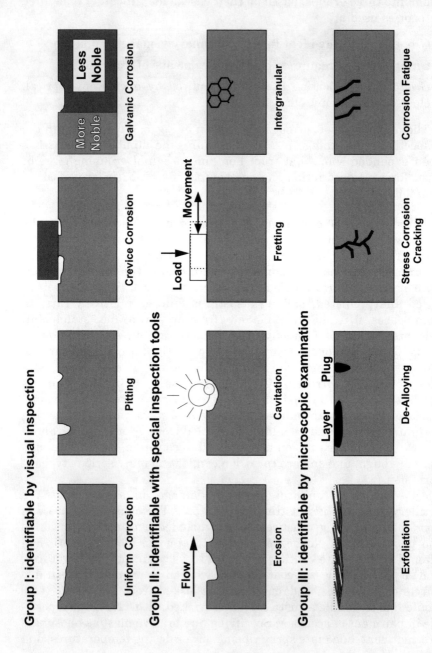

Figure 5.1 Main forms of corrosion regrouped by their ease of recognition.

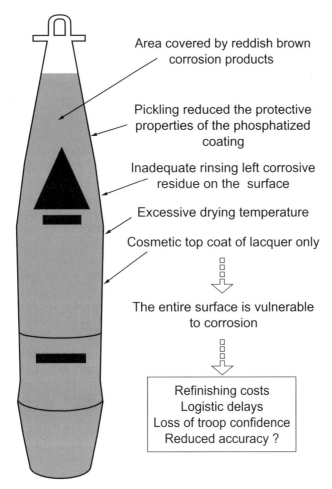

Figure 5.2 Uniform corrosion damage on a rocket-assisted artillery projectile.

Pitting. Pitting corrosion is a localized form of corrosion by which cavities, or "holes," are produced in the material. Pitting is considered to be more dangerous than uniform corrosion damage because it is more difficult to detect, predict, and design against. Corrosion products often cover the pits. A small, narrow pit with minimal overall metal loss can lead to the failure of an entire engineering system. Pitting corrosion, which, for example, is almost a common denominator of all types of localized corrosion attack, may assume different shapes, as illustrated in Fig. 5.3. Pitting corrosion can produce pits with their mouth open (uncovered) or covered with a semipermeable membrane of corrosion products. Pits can be either hemispherical or cup-shaped.

In some cases they are flat-walled, revealing the crystal structure of the metal, or they may have a completely irregular shape.[4]

Pitting corrosion occurs when discrete areas of a material undergo rapid attack while most of the adjacent surface remains virtually unaffected. Such localization of the anodic and cathodic corrosion processes is characterized by the surface area ratio (S_a/S_c) for these two processes, where S_a and S_c are the areas supporting, respectively, the anodic and cathodic reactions. The S_a/S_c ratio, or degree of localization, can be an important driving force of all localized corrosion problems because a corrosion situation corresponds to equal anodic and cathodic absolutes currents (see the section on Mixed Potential Diagrams in Chap. 1, Aqueous Corrosion). Corrosive microenvironments, which tend to be very different from the bulk environment, often play a role in the initiation and propagation of corrosion pits. This greatly complicates the prediction task. Apart from the localized loss of thickness, corrosion pits can also be harmful by acting as stress risers. Fatigue and stress corrosion cracking may initiate at the base of corrosion pits.

Crevice corrosion. Crevice corrosion is a localized form of corrosion usually associated with a stagnant solution on the microenvironmental level. Such stagnant microenvironments tend to occur in crevices (shielded areas) such as those formed under gaskets, washers, insulation materi-

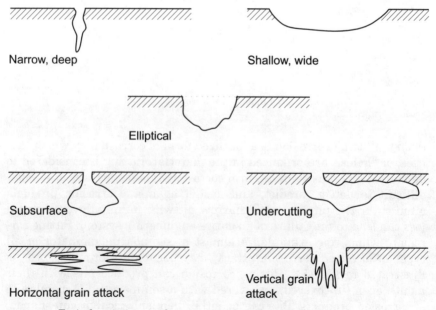

Figure 5.3 Typical variations in the cross-sectional shape of pits.

al, fastener heads, surface deposits, disbonded coatings, threads, lap joints, and clamps. Because oxygen diffusion into the crevice is restricted, a differential aeration cell tends to be set up between crevice (microenvironment) and the external surface (bulk environment). The cathodic oxygen reduction reaction cannot be sustained in the crevice area, giving it an anodic character in the concentration cell. This anodic imbalance can lead to the creation of highly corrosive microenvironmental conditions in the crevice, conducive to further metal dissolution. The formation of an acidic microenvironment, together with a high chloride ion concentration, is illustrated in Fig. 5.4. Filiform corrosion is closely related to crevice attack. It occurs under protective films such as lacquers and is characterized by an interconnected trail of corrosion

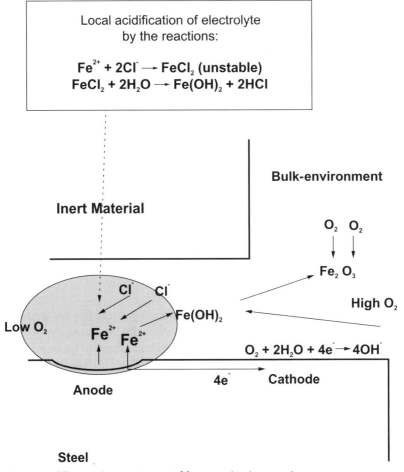

Figure 5.4 Microenvironment created by corrosion in a crevice.

products. Active corrosion occurs at the head of the filament, where a corrosive microenvironment is established, whereas the trailing tail is largely inactive (see Figs. 1.10 and 1.11 in Chap. 1, Aqueous Corrosion). Even though filiform corrosion is often largely a "cosmetic" problem, the impeccable appearance of a product can be very important, as, for example, in the food packaging industry.

Crevice corrosion damage in the lap joints of aircraft skins has become a major safety concern, particularly after the Aloha airline incident. Corrosion damage to aircraft fuselages is an example of atmospheric corrosion, a topic that is described more fully in a separate chapter. On April 28, 1988, a 19-year-old Boeing 737 aircraft, operated by Aloha, lost a major portion of the upper fuselage near the front of the plane, in full flight at 24,000 ft.[5] The extent of the damage is shown schematically in Fig. 5.5. Miraculously, the pilot managed to land the plane on the island of Maui, Hawaii, but one flight attendant died, and several passengers sustained serious injuries. In the Aloha Boeing 737 aircraft, evidence was found of multiple site fatigue damage leading to structural failure. The resulting National Transportation Safety Board investigation report issued in 1989 attributed the incident to the failure of the operator's maintenance program to detect corrosion damage.[6] Earlier, in 1981, a similar aircraft had suffered an in-flight break-up with more than 100 fatalities. Investigations pointed to corrosion-accelerated fatigue of the fuselage skin panels as the failure mechanism.[7]

The three basic types of aircraft fuselage lap splices are shown in Fig. 5.6. A particular aircraft design normally incorporates two or three different types of splices in the fuselage. The fuselages of commercial aircraft are typically constructed from 2024 T3 aluminum alloy. The lap joints are riveted and sealed by some manufacturers, whereas others employ a combination of riveting and adhesive bonding.[8] Corrosion damage in the crevice geometry of the lap joints is highly undesirable. Fatigue cracking in the Aloha case was not anticipated to be a problem, provided the overlapping fuselage panels remained firmly bonded together.[9]

Corrosion processes in this crevice geometry and the subsequent buildup of voluminous corrosion products inside the lap joints lead to *pillowing,* a dangerous condition whereby the overlapping surfaces are separated (Fig. 5.7). The prevalent corrosion product identified in corroded fuselage joints is aluminum oxide trihydrate, with a particularly high-volume expansion relative to aluminum, as shown in Fig. 5.8. The buildup of voluminous corrosion products also leads to an undesirable increase in stress levels near critical fastener holes; rivets have been known to fracture due to the high tensile stresses resulting from pillowing.[10]

Corrosion damage on commercial and military aircraft, such as the pillowing in lap splices described above, is becoming a major concern

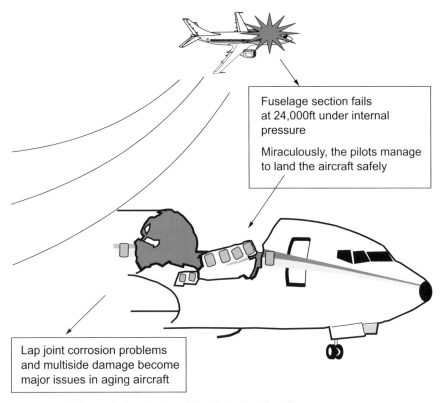

Figure 5.5 A schematic description of the Aloha "incident."

in the context of the global aging aircraft problem. By the turn of the century, 64 percent of the U.S. commercial carrier fleet will be at least 20 years old. In 1970, the average age of this fleet was under 5 years.[7] It is well known that the costs and safety risks associated with aircraft corrosion damage are highest in aging fleets. Lengthy and detailed inspection and maintenance procedures, as part of periodic checks and overhauls, represent a substantial portion of the corrosion costs. A pertinent example involving a recently inspected 28-year-old Boeing 747 has been documented.[6] This aging aircraft, placed on a more stringent inspection program, required 65 days and 90,000 person hours of work in a major overhaul. Apart from replacement parts and maintenance personnel costs, the lost revenue resulting from this lengthy grounding must be factored into the operator's corrosion costs.

Galvanic corrosion. Galvanic corrosion occurs when dissimilar metallic materials are brought into contact in the presence of an electrolyte. Such damage can also occur between metals and alloys and other conducting

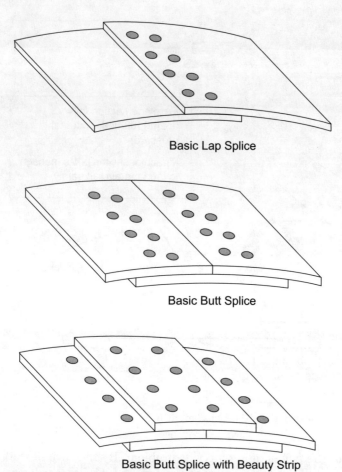

Figure 5.6 The three basic types of lap splices used for construction of aircraft fuselage.

materials such as carbon or graphite. An electrochemical corrosion cell is set up due to differences in the corrosion potentials of the dissimilar materials. The material with the more noble corrosion potential then becomes the cathode of the corrosion cell, whereas the less noble material is consumed by anodic dissolution. The area ratio of the two dissimilar materials is extremely important. If the anode-to-cathode surface area ratio is small, or S_a/S_c high, the galvanic current can be concentrated on a small anodic area. The corrosion rate, visible as thickness loss over time, can then become very high for the anode. For example, if aluminum rivets were used on steel plates, the rivets would corrode extremely rapidly (Fig. 5.9).

The galvanic series (Fig. 5.10) shows the relative nobility of a range of materials in seawater. In general terms, the further two materials are

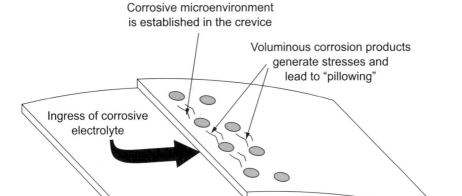

Figure 5.7 Pillowing of lap splices.

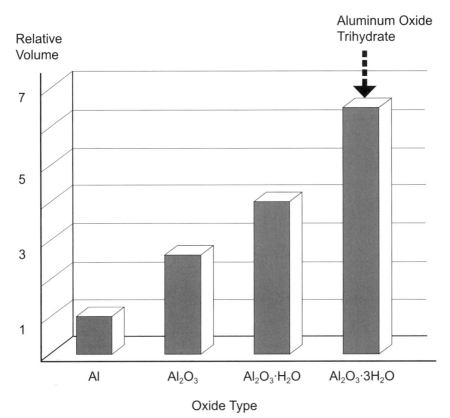

Figure 5.8 Relative volume of aluminum corrosion products.

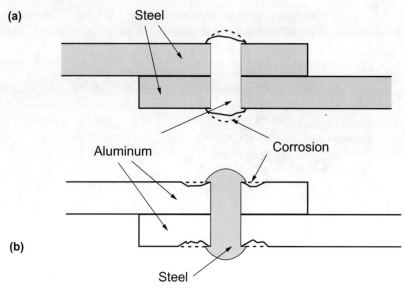

Figure 5.9 Galvanic coupling caused by riveting plates.

apart in the galvanic series, the greater the risk of galvanic corrosion. However, this series does not give any information on the rate of corrosion attack. Furthermore, the relative position of the materials can change in other environments. It is important to realize that galvanic corrosion effects can be manifested not only on the macroscopic level but also within the microstructure of a material. Certain phases or precipitates will undergo anodic dissolution under microgalvanic effects. Because the principle of galvanic corrosion is widely known, it is remarkable that it still features prominently in numerous corrosion failures. Figure 5.11 illustrates the main factors affecting a galvanic corrosion situation.[11]

One well-known landmark that has undergone severe galvanic corrosion in its history is the Statue of Liberty. An excellent publication edited by Baboian, Bellante, and Cliver[12] details corrosion damage to this structure and remedial measures undertaken. It also provides a fascinating historical engineering perspective. The Statue of Liberty was officially inaugurated on October 28, 1866, on Bedloe's Island, in the New York harbor. The design of the statue, which rises more than 91 m into the air, essentially involves a rigid central pylon and a secondary frame, to which further framework, the armature, and the skin are attached. The entire skeleton was manufactured out of wrought iron (more specifically, puddled iron), a common construction material of that era. Copper was selected as skin material for ease of shaping the artistic detail, durability, and good strength to weight ratio for materials commercially available at the time.[13]

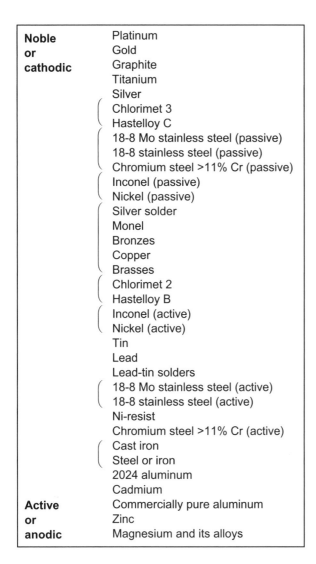

Figure 5.10 Galvanic series of some commercial metals and alloys in seawater.

Crucial components in attaching the copper sheets to the armature were the copper saddles, depicted in Fig. 5.12. These U-shaped components wrapped around the armature iron bars and were riveted to the copper skin. Some 1500 of these copper saddles were used. During construction in the United States, shellac-impregnated asbestos was placed between the armature and the skin, but this disintegrated with time and, through wicking action, acted as an undesirable trap

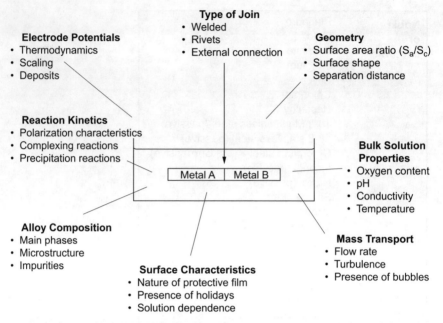

Figure 5.11 Factors affecting galvanic corrosion.

for corrosive electrolyte. The statue was never "water-tight." For example, significant leakage took place through the torch where additional lights had been retrofitted. From the galvanic series and the unfavorable area ratio, it is apparent that the skin attachment arrangement is particularly vulnerable to severe galvanic corrosion, as depicted in Fig. 5.12. The expansive force of the iron corrosion products disfigured the copper skin extensively. It was this galvanic corrosion problem that most necessitated a massive restoration project,[12] at a cost exceeding $200 million in private funding.

After testing for galvanic compatibility and considering the need for mechanical requirements close to those of the original wrought iron, it was decided to replace the iron armature with one of AISI type 316L stainless steel. This stainless alloy was selected on the basis that electrical contact with the skin was unavoidable. Inspection of the galvanic series reveals that assuming the stainless steel remains passive, the larger copper surfaces will tend to act as the anodes. As a further, secondary precaution, a Teflon barrier was applied between the new stainless armature and the copper skin and saddles.

Selective leaching. Selective leaching refers to the selective removal of one element from an alloy by corrosion processes. A common example is the dezincification of unstabilized brass, whereby a weakened,

porous copper structure is produced. The selective removal of
proceed in a uniform manner or on a localized (plug-type) sc
difficult to rationalize dezincification in terms of preferential
solution out of the brass lattice structure. Rather, it is believed that
brass dissolves with Zn remaining in solution and Cu replating out of
the solution.[1] Graphitization of gray cast iron, whereby a brittle
graphite skeleton remains following preferential iron dissolution, is a
further example of selective leaching.

Erosion corrosion. Erosion corrosion is the cumulative damage
induced by electrochemical corrosion reactions and mechanical effects
from relative motion between the electrolyte and the corroding surface. Erosion corrosion is defined as accelerated degradation in the
presence of this relative motion. The motion is usually one of high
velocity, with mechanical wear and abrasion effects. Grooves, gullies,
rounded edges, and waves on the surface usually indicating directionality characterize this form of damage. Erosion corrosion is found in
systems such as piping (especially bends, elbows, and joints), valves,
pumps, nozzles, heat exchangers, turbine blades, baffles, and mills.
Impingement and cavitation are special forms of erosion corrosion. In
the former, moving liquid particles cause the damage, whereas

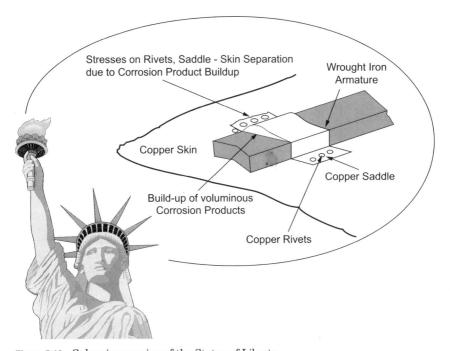

Figure 5.12 Galvanic corrosion of the Statue of Liberty.

collapsing (unstable) vapor bubbles induce surface damage in the latter. Fretting corrosion, which refers to corrosion damage at the asperities of contact surfaces, may also be included in this category. This damage is induced under load and in the presence of repeated relative surface motion, as induced, for example, by vibration. Pits or grooves and oxide debris characterize this damage, typically found in machinery, bolted assemblies, and ball or roller bearings. Contact surfaces exposed to vibration during transportation are exposed to the risk of fretting corrosion.

Environmental cracking. Environmental cracking (EC) is a very acute form of localized corrosion. Because of the intrinsic complexity of the situations leading to different forms of EC, the parameters leading to this class of problems have often been described in qualitative terms such as those in Table 5.1. Table 5.1 lists the factors contributing to one of three forms of EC, that is, stress corrosion cracking (SCC), fatigue corrosion, and hydrogen embrittlement.

SCC is the cracking induced from the combined influence of tensile stress and a corrosive medium. The impact of SCC on a material seems

TABLE 5.1 Characteristics of Environmental Cracking

Factor	SCC	Fatigue corrosion	Hydrogen induced cracking
Stress	Static tensile	Cyclic + tensile	Static tensile
Aqueous corrosive	Specific to the alloy	Any	Any
Temperature increase	Accelerates	Accelerates	< Ambient: increases < Ambient: increases
Pure metal	Resistant	Susceptible	Susceptible
Crack morphology	Transgranular Intergranular Branched	Transgranular Unbranched Blunt tip	Transgranular Intergranular Unbranched Sharp tip
Corrosion products in cracks	Absent	Present	Absent
Crack surface appearance	Cleavagelike	Beach marks and/or striations	Cleavagelike
Cathodic polarization	Suppresses	Suppresses	Accelerates
Near maximum strength	Susceptible but minor	Accelerates	Accelerates

to fall between dry cracking and the fatigue threshold of that material (Fig. 5.13). The required tensile stresses may be in the form of directly applied stresses or residual stresses. Cold deformation and forming, welding, heat treatment, machining, and grinding can introduce residual stresses. The magnitude and importance of such stresses is often underestimated. The residual stresses set up as a result of welding operations tend to approach the yield strength. The buildup of corrosion products in confined spaces can also generate significant stresses and should not be overlooked. SCC usually occurs in certain specific alloy-environment-stress combinations (Fig. 5.14) and has been shown to be, at least for aluminum alloys, very dependent on grain orientation (Fig. 5.15).

Usually, most of the surface remains unattacked, but with fine cracks penetrating into the material. In the microstructure, these cracks can have an intergranular or a transgranular morphology. Macroscopically, SCC fractures have a brittle appearance. SCC is classified as a catastrophic form of corrosion because the detection of such fine cracks can be very difficult and the damage not easily predicted. Experimental SCC data is notorious for a wide range of scatter. A disastrous failure may occur unexpectedly, with minimal overall material loss.

Hydrogen embrittlement is sometimes classified separately from SCC. It refers to the embrittlement and resulting increased cracking

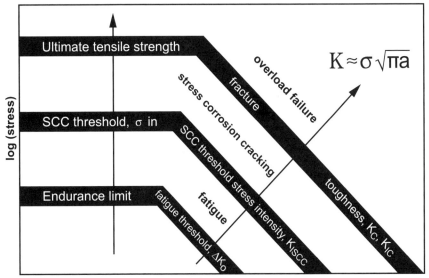

Figure 5.13 Continuum between failure modes.

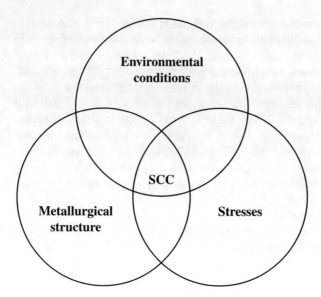

Figure 5.14 The three main factors contributing to the occurrence of SCC.

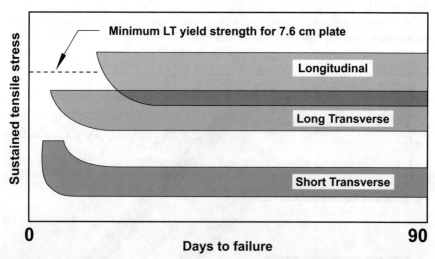

Figure 5.15 SCC susceptibility of 7075-T651 aluminum alloy immersed in 3.5% NaCl solution.

risk due to uptake of hydrogen into the material's structure. The cathodic reduction of water to form hydrogen is a potential source of embrittlement. *Hydrogen stress cracking* and *sulfide stress cracking* are terms used for hydrogen embrittlement from interactions with hydrogen gas and hydrogen sulfide, respectively.

In practice, materials used for their strength are the most susceptible to suffer from SCC problems when some environmental elements render them vulnerable. Such vulnerability exists for stainless steels when chloride ions are present in the environment, even at very low concentrations. Unfortunately, the term *stainless steel* is sometimes interpreted too literally. Structural engineers need to be aware that stainless steels are certainly not immune to corrosion damage and can be particularly susceptible to localized corrosion damage and SCC. The austenitic stainless steels, mainly UNS S30400 and UNS S31600, are used extensively in the construction industry. The development of SCC in S30400 bars, on which a concrete ceiling was suspended in a swimming pool building, had disastrous consequences.

In May 1985, the heavy ceiling in a swimming pool located in Uster, Switzerland, collapsed with fatal consequences[14] after 13 years of service. The failure mechanism was established to be transgranular SCC, as illustrated in Fig. 5.16. The presence of a tensile stress was clearly created in the stainless rods by the weight of the ceiling. Chloride species dispersed into the atmosphere, together with thin moisture films, in all likelihood represented the corrosive environment. A characteristic macroscopic feature of the failed stainless steel rods was the brittle nature of the SCC fractures, with essentially no ductility displayed by the material in this failure mode.

Subsequent to this failure, further similar incidents (fortunately without fatalities) have been reported in the United Kingdom, Germany, Denmark, and Sweden. Although chloride-induced SCC damage is recognized as a common failure mechanism in stainless steels, a somewhat surprising element of these failures is that they occurred at room temperature. As a general rule of thumb, it has often been assumed that chloride-induced SCC in these alloys is not a practical concern at temperatures below 60°C.

Under the assumption that a low-pH–high-chloride microenvironmental combination is responsible for the SCC failures, several factors were identified in UK pool operations that could exacerbate the damage. Notable operational changes included higher pool usage and pool features such as fountains and wave machines, resulting in more dispersal of pool water (and chloride species) into the atmosphere. The importance of eliminating the use of the S30400 and S31600 alloys for stressed components exposed to swimming pool atmospheres should be apparent from this example.

Intergranular corrosion. The microstructure of metals and alloys is made up of grains, separated by grain boundaries. Intergranular corrosion is localized attack along the grain boundaries, or immediately adjacent to grain boundaries, while the bulk of the grains remain

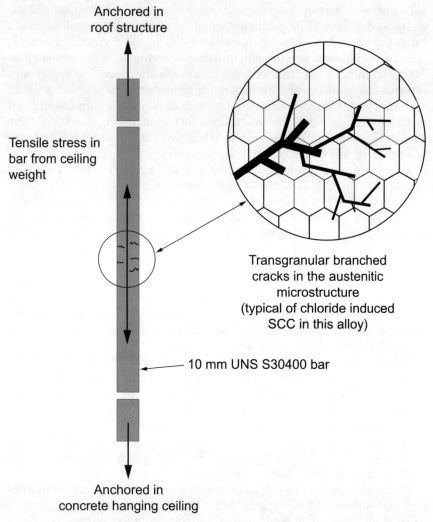

Figure 5.16 Transgranular SCC on stainless steel supporting rods.

largely unaffected. This form of corrosion is usually associated with chemical segregation effects (impurities have a tendency to be enriched at grain boundaries) or specific phases precipitated on the grain boundaries. Such precipitation can produce zones of reduced corrosion resistance in the immediate vicinity. A classic example is the sensitization of stainless steels. Chromium-rich grain boundary precipitates lead to a local depletion of chromium immediately adjacent to these precipitates, leaving these areas vulnerable to corrosive attack in certain electrolytes (Fig. 5.17). This problem is often manifested in the heat-affected zones of welds, where the thermal cycle of welding has produced a sensitized structure.

Knife-line attack, immediately adjacent to the weld metal, is a special form of sensitization in stabilized austenitic stainless steels. Stabilizing elements (notably Ti and Nb) are added to stainless steels to prevent intergranular corrosion by restricting the formation of Cr-rich grain boundary precipitates. Basically, these elements form carbides in preference to Cr in the austenitic alloys. However, at the high temperatures experienced immediately adjacent to the weld fusion zone, the stabilizer carbides dissolve and remain in solution during the subsequent rapid

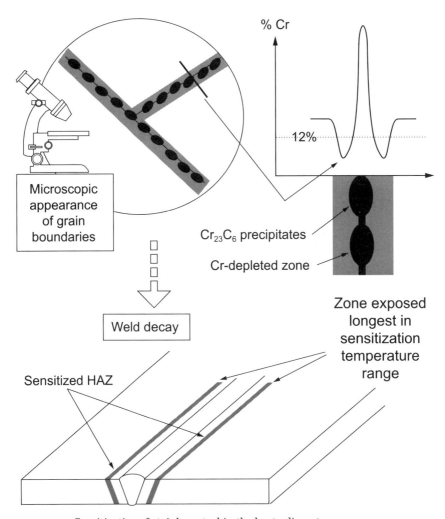

Figure 5.17 Sensitization of stainless steel in the heat-adjacent zone.

cooling cycle. Thereby this zone is left prone to sensitization if the alloy is subsequently reheated in a temperature range where grain boundary chromium carbides are formed. Reheating a welded component for stress relieving is a common cause of this problem. In the absence of the reheating step, the alloy would not be prone to intergranular attack.

Exfoliation corrosion is a further form of intergranular corrosion associated with high-strength aluminum alloys. Alloys that have been extruded or otherwise worked heavily, with a microstructure of elongated, flattened grains, are particularly prone to this damage. Figure 5.18 illustrates the anisotropic grain structure typical of wrought aluminum alloys, and Fig. 5.19 shows how a fraction of material is often sacrificed to alleviate the impact on the susceptibility to SCC of the short transverse sections of a component. Corrosion products building up along these grain boundaries exert pressure between the grains, and the end result is a lifting or leafing effect. The damage often initiates at end grains encountered in machined edges, holes, or grooves and can subsequently progress through an entire section.

5.2.2 Modes and submodes of corrosion

As part of a framework for predicting and assuring corrosion performance of materials, Staehle introduced the concept of modes and sub-

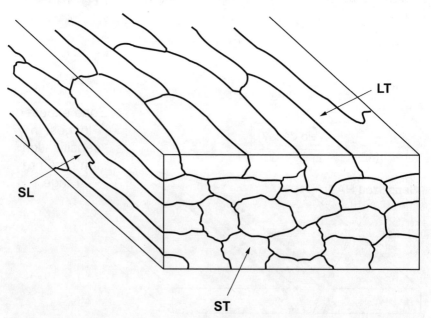

Figure 5.18 Schematic representation of the anisotropic grain structure of wrought aluminum alloys.

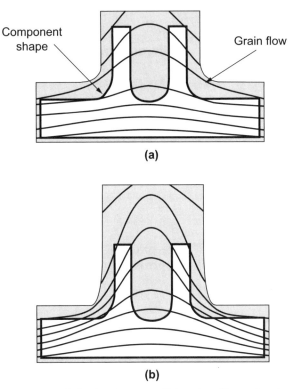

Figure 5.19 Machining for neutralizing the effects of grain flow on corrosion resistance: (a) saving on material and loosing on lifetime and (b) loosing on material for increased lifetime.

modes of corrosion.[15] In this context, a corrosion mode was to be defined by the morphology of corrosion damage, as shown for the four intrinsic modes in Fig. 5.20. Submode categories were also proposed to differentiate between several manifestations of the same mode, for a given material-environment system. For example, Staehle illustrated two submodes of SCC in stainless steel exposed to a boiling caustic solution. A transgranular SCC submode prevailed at low corrosion potentials, whereas an intergranular submode occurred at higher potentials. The identification and distinction of submodes is very important for performance prediction because different submodes respond differently to corrosion variables. Controlling one submode of corrosion successfully does not imply that other submodes will be contained.

A useful analogy to differentiating corrosion submodes is the distinction between different failure mechanisms in the mechanical world. For example, nickel may fracture by intergranular creep or by transgranular creep, depending on the loading and temperature conditions.

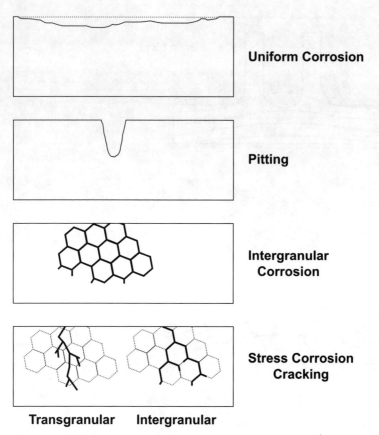

Figure 5.20 The four intrinsic modes of corrosion damage.

The organization of corrosion damage into modes and submodes is important for rationalizing and predicting corrosion damage, in a manner comparable to mechanical damage assessment.

5.2.3 Corrosion factors

Six important corrosion factors were identified in a review of scientific and engineering work on SCC damage,[16] generally regarded as the most complex corrosion mode. According to Staehle's materials degradation model, all engineering materials are reactive and their strength is quantifiable, provided that all the variables involved in a given situation are properly diagnosed and their interactions understood. For characterizing the intensity of SCC the factors were material, environment, stress, geometry, temperature, and time. These factors represent independent variables affecting the intensity of stress corrosion cracking. Furthermore, a number of subfactors were identified for each of the six main factors, as shown in Table 5.2.

The value of this scheme, extended to other corrosion modes and forms, should be apparent. It is considered to be extremely useful for analyzing corrosion failures and for reporting and storing information and data in a complete and systematic manner. An empirical correlation was established between the factors listed in Table 5.2 and the forms of corrosion described earlier (Fig. 5.1). Several recognized corrosion experts were asked to complete an opinion poll listing the main subfactors and the common forms of corrosion as illustrated in the example shown in Fig. 5.21. Background information on the factors and forms of corrosion was attached to the survey. The responses were then analyzed and represented in the graphical way illustrated in Fig. 5.22.

TABLE 5.2 Factors and Contributing Elements Controlling the Incidence of a Corrosion Situation According to Staehle[16]

Factor	Subfactors and contributing elements
Material	Chemical composition of alloy Crystal structure Grain boundary (GB) composition Surface condition
Environment	
Chemical definition	Type, chemistry, concentration, phase, conductivity
Circumstance	Velocity, thin layer in equilibrium with relative humidity, wetting and drying, heat-transfer boiling, wear and fretting, deposits
Stress	
Stress definition	Mean stress, maximum stress, minimum stress, constant load/constant strain, strain rate, plane stress/plane strain, modes I, II, III, biaxial, cyclic frequency, wave shape
Sources of stress	Intentional, residual, produced by reacted products, thermal cycling
Geometry	Discontinuities as stress intensifiers Creation of galvanic potentials Chemical crevices Gravitational settling of solids Restricted geometry with heat transfer leading to concentration effects Orientation vs. environment
Temperature	At metal surface exposed to environment Change with time
Time	Change in GB chemistry Change in structure Change in surface deposits, chemistry, or heat-transfer resistance Development of surface defects, pitting, or erosion Development of occluded geometry Relaxation of stress

Factor	Forms I			
	Uniform	Pitting	Crevice	Galvanic
Material				
Composition				
Crystal structure				
GB composition				
Surface condition				
Environment				
nominal				
circumstantial		▼		
Stress				
applied				
residual				
product built-up				
cyclic				
Geometry				
galvanic potentials				
restricted geometries				
settling of solids				
Temperature				
changing T	▼			
T of surface				
Time				
changes over time				

Figure 5.21 Opinion poll sheet for the most recognizable forms of corrosion problems.

The usefulness of this empirical correlation between the visible aspect of a corrosion problem and its intrinsic root causes has not been fully exploited yet. It is believed that such a tool could be used to

1. *Guide novice investigators.* The identification of the most important factors associated with different forms of corrosion could serve to provide guidance and assistance for inexperienced corrosion-failure investigators. Many investigators and troubleshooters are not corrosion specialists and will find such a professional guide useful. Such guidelines could be created in the form of computer application. A listing of the most important factors would ensure that engineers with little or no corrosion training were made aware of the complexity and multitude of variables involved in corrosion damage. Inexperienced investigators would be reminded of critical variables that may otherwise be overlooked.

2. *Serve as a reporting template.* Once all relevant corrosion data has been collected or derived, the framework of factors and forms could be used for storing the data in an orderly manner in digital databases as illustrated in Fig. 5.23. The value of such databases is greatly diminished if the information is not stored in a consistent manner, making retrieval of pertinent information a nightmarish experience. Analysis of numerous corrosion failure analysis reports has revealed that information on important variables is often lacking.[17] The omission of important information from corrosion reports is obviously not always an oversight by the professional author. In many cases, the desirable information is simply not (readily) available. Another application of the template or framework thus lies in highlighting data deficiencies and

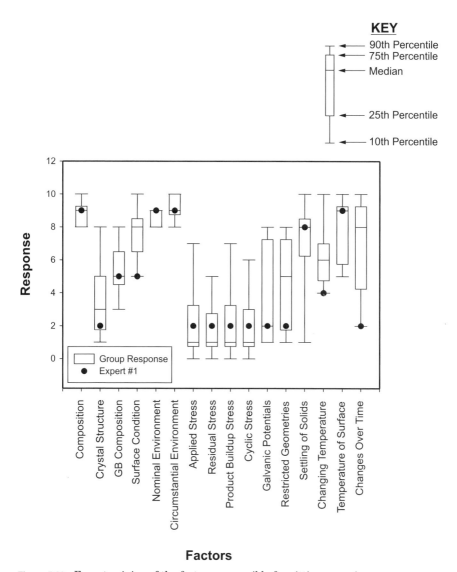

Figure 5.22 Expert opinion of the factors responsible for pitting corrosion.

the need of rectifying such situations. As such, the factors represent a systematic and comprehensive information-gathering scheme.

5.2.4 The distinction between corrosion-failure mechanisms and causes

One thesis is that the scientific approach to failure analysis is a detailed mechanistic "bottom-up" study. Many corrosion-failure analyses are

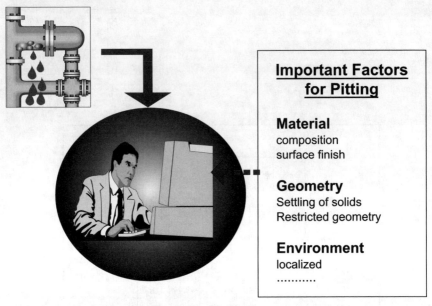

Figure 5.23 The factor/form correlation used as a reporting template.

approached in this manner. A failed component is analyzed in the laboratory using established analytical techniques and instrumentation. Chemical analysis, hardness testing, metallography, optical and electron microscopy, fractography, x-ray diffraction, and surface analysis are all elements of this approach. On conclusion of all these analytical procedures the mechanism of failure, for example "chloride induced transgranular stress corrosion cracking," can usually be established with a high degree of confidence by an expert investigator.

However, this approach alone provides little or no insight into the real causes of failure. Underlying causes of serious corrosion damage that can often be cited include human factors such as lack of corrosion awareness, inadequate training, and poor communication. Further underlying causes may include weak maintenance management systems, insufficient repairs due to short-term profit motives, a poor organizational "safety culture," defective supplier's products, incorrect material selection, and so forth. It is thus apparent that there can be multiple causes associated with a single corrosion mechanism. Clearly, a comprehensive failure investigation providing information on the cause of failure is much more valuable than one merely establishing the corrosion mechanism(s). Establishing the real causes of corrosion failures (often related to human behavior) is a much harder task than merely identifying the failure mechanisms. It is disconcerting that in many instances of technical reporting, causes and mechanisms of corrosion damage are used almost interchangeably. Direct evidence of this problem was obtained when searching a commercial engineering database.[18]

In contrast to the traditional scientific mechanistic approach, systems engineers prefer the "top-down" approach that broadens the definition of the system (see Chap. 4, Corrosion Information Management) and is more likely to include causes of corrosion failures such as human behavior. This is more consistent with the lessons to be learned from the UK Hoar Report, which stated that corrosion control of even small components could result in major cost savings because of the effect on systems rather than just the components.[19]

5.3 Guidelines for Investigating Corrosion Failures

Several guides to corrosion-failure analysis have been published. These are valuable for complementing the expertise of an organization's senior, experienced investigators. These investigators are rarely in a position to transfer their knowledge effectively under day to day work pressures. The guides have been found to be particularly useful in filling this knowledge "gap."

The Materials Technology Institute of the Chemical Process Industries' *Atlas of Corrosion and Related Failures*[20] maps out the process of a failure investigation from the request for the analysis to the submission of a report. It is a comprehensive document and is recommended for any serious failure investigator who has to deal with corrosion damage. The step-by-step procedure section, for example, contains two flow charts, one for the on-site investigation and the other for the laboratory component. The procedural steps and decision elements are linked to tables describing specific findings and deductions, supported by micrographs and actions. Some of the elements of information contained in Sec. 4.5 of the MTI Atlas (the section that relates the origin(s) of failure to plant or component geometry) are illustrated in Figs. 5.24 and 5.25.

In the NACE guidelines,[2] failures are classified into the eight forms of corrosion popularized by Fontana, with minor modifications. The eight forms of corrosion are subdivided into three further categories to reflect the ease of visual identification (Fig. 5.1). Each form of corrosion is described in a separate chapter, together with a number of case histories from diverse branches of industry. An attempt was made to treat each case study in a consistent manner with information on the corrosion mechanism, material, equipment, environment, time to failure, comments, and importantly, remedial actions. It is interesting to note that if stress, geometry, and temperature factors had also been

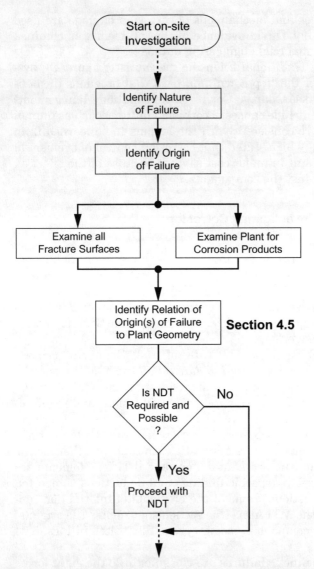

Figure 5.24 Decision tree to guide on-site investigations dealing with corrosion damage.

described for each case history, the complete set of corrosion factors proposed by Staehle would have been documented.

5.4 Prevention of Corrosion Damage

Recognizing the symptoms and mechanism of a corrosion problem is an important preliminary step on the road to finding a convenient solution. There are basically five methods of corrosion control:

- Change to a more suitable material
- Modifications to the environment
- Use of protective coatings
- The application of cathodic or anodic protection
- Design modifications to the system or component

Some preventive measures are generic to most forms of corrosion. These are most applicable at the design stage, probably the most important phase in corrosion control. It cannot be overemphasized that corrosion control must start at the "drawing board" and that design details are critical for ensuring adequate long-term corrosion protection. It is generally good practice to

- Provide adequate ventilation and drainage to minimize the accumulation of condensation (Figs. 5.26 and 5.27)

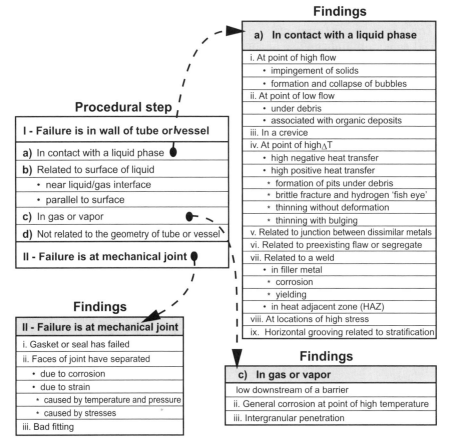

Figure 5.25 Recommendations for relating the origin(s) of failure to plant geometry.

- Avoid depressed areas where drainage is inadequate (Fig. 5.27)
- Avoid the use of absorptive materials (such as felt, asbestos, and fabrics) in contact with metallic surfaces
- Prepare surfaces adequately prior to the application of any protective coating system
- Use wet assembly techniques to create an effective sealant barrier against the ingress of moisture or fluids (widely used effectively in the aerospace industry)
- Provide easy access for corrosion inspection and maintenance work

Additionally, a number of basic technical measures can be taken to minimize corrosion damage in its various forms. A brief summary of generally accepted methods for controlling the various forms of corrosion follows.

5.4.1 Uniform corrosion

The application of protective coatings, cathodic protection, and material selection and the use of corrosion inhibitors usually serves to con-

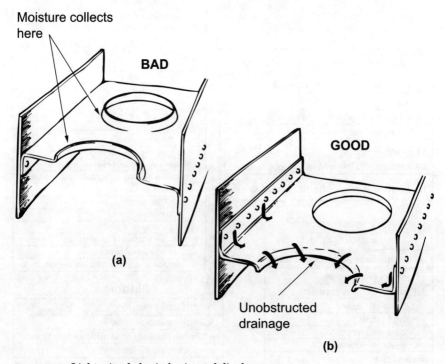

Figure 5.26 Lightening holes in horizontal diaphragms.

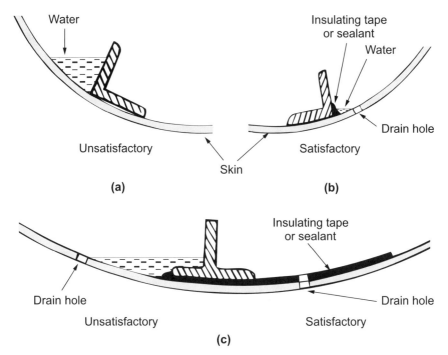

Figure 5.27 Water traps and faying surfaces.

trol uniform corrosion. Some of these methods are used in combination. For example, on buried oil and gas pipelines the primary corrosion protection is provided by organic coatings, with the cathodic protection system playing a secondary role to provide additional protection at coating defects or weaknesses.

5.4.2 Galvanic corrosion

For controlling galvanic corrosion, materials with similar corrosion potential values in a given environment should be used. Unfavorable area ratios (S_a/S_c) should be avoided. Insulation can be employed to physically separate galvanically incompatible materials (Fig. 5.28), but this is not always practical. Protective barrier coatings can be used with an important provision (i.e., coating the anodic material only is not recommended) because it can have disastrous consequences in practice. At defects (which are invariably present) in such coatings, extremely rapid corrosion penetration will occur under a very unfavorable area ratio. It is much better practice to coat the cathodic surface in the galvanic couple. An example of rapid tank failures that resulted from a tank design with coated steel side walls (the anode) and stainless clad tank bottoms (the cathode) is described by Fontana.[1]

If dissimilar materials junctions cannot be avoided at all, it is sensible to design for increased anodic sections and easily replaceable anodic parts. Corrosion inhibitors may be utilized, bearing in mind that their effects on different materials will tend to be variable.

5.4.3 Pitting

Material selection plays an important role in minimizing the risk of pitting corrosion. For example, the resistance to chloride-induced pitting in austenitic stainless steels is improved in alloys with higher molybdenum contents. Thus AISI type 317 stainless steel has a higher resistance than the 316 alloy, which in turn is more resistant than the 304 grade. The following pitting index (PI) [Eq. (5.1)] has been proposed to predict the pitting resistance of austenitic and duplex stainless steels (it is not applicable to ferritic grades):

$$PI = Cr + 3.3Mo + xN \tag{5.1}$$

where Cr, Mo, and N = the chromium, molybdenum, and nitrogen contents, $x = 16$ for duplex stainless steel, and $x = 30$ for austenitic alloys.

Generally speaking, the risk of pitting corrosion is increased under stagnant conditions, where corrosive microenvironments are established on the surface. Drying and ventilation can prevent this accumulation of stagnant electrolyte at the bottom of pipes, tubes, tanks, and so forth. Agitation can also prevent the buildup of local highly corrosive conditions. The use of cathodic protection can be considered for pitting corrosion, but anodic protection is generally unsuitable.

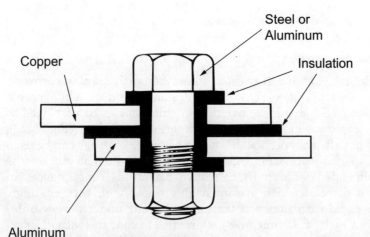

Figure 5.28 Insulating two dissimilar metals for protection against galvanic corrosion.

Environmental modifications such as deaeration, chloride ion removal, and the addition of corrosion inhibitors can reduce the risk of pitting. However, the beneficial effects on existing pits with established highly corrosive microenvironments may be minimal. Furthermore, if the pitting attack is not eliminated completely through the use of corrosion inhibitors, penetration can actually be accelerated due to the concentration of metal dissolution onto a smaller area.

5.4.4 Crevice corrosion

Whenever possible, crevice conditions should be avoided altogether. Welded joints offer alternatives to riveted or bolted joints. In heat exchangers, welded tube sheets are to be preferred over the rolled variety. Harmful surface deposits can be removed by cleaning. Filtration can eliminate suspended solids that could otherwise settle out and form harmful crevice conditions; agitation can also be beneficial in this sense. Where gaskets have to be used, nonabsorbent gasket materials (such as Teflon) are recommended. Cathodic protection can be effective in preventing crevice corrosion, but anodic protection is generally unsuitable. Environmental modifications are not usually effective once crevice corrosion has initiated because the corrosive microenvironment inside the crevice is not easily modified.

5.4.5 Intergranular corrosion

The susceptibility of alloys to intergranular corrosion can often be reduced through heat treatment. For example, in sensitized austenitic stainless steels, high-temperature solution annealing at around 1100°C followed by rapid cooling can restore resistance to intergranular corrosion resistance. In general, alloys should be used in heat-treated conditions associated with least susceptibility to intergranular corrosion. Composition is also an important factor. Grades of stainless steels with sufficiently low interstitial element levels (carbon and nitrogen) are immune to this form of corrosion. The stabilized stainless alloys with titanium and/or niobium additions rarely suffer from this form of corrosion, with the exception of knife-line attack. The L grades of austenitic stainless steels, such as 304L and 316L with carbon levels below 0.03 percent, are widely used in industry and are recommended whenever welding of relatively thick sections is required.

For aluminum alloys it is advisable to avoid exposure of the short transverse grain structure. Protective films such as anodizing, plating, and cladding can reduce the intergranular corrosion risk. Shot peening to induce cold working in the surface grains can also be beneficial.

Selective leaching

Selective leaching is usually controlled by material selection. For example, brass is resistant to dezincification if traces of arsenic, phosphorous, or antimony are added to the alloy. Modern brass plumbing fixtures are made exclusively from these stabilized alloys. Brass with a low Zn content generally tends to be less susceptible. In more corrosive environments the use of cupro-nickel alloys has been advocated.

5.4.7 Erosion corrosion

Materials selection plays an important role in minimizing erosion corrosion damage. Caution is in order when predicting erosion corrosion behavior on the basis of hardness. High hardness in a material does not necessarily guarantee a high degree of resistance to erosion corrosion.

Design features are also particularly important. It is generally desirable to reduce the fluid velocity and promote laminar flow; increased pipe diameters are useful in this context. Rough surfaces are generally undesirable. Designs creating turbulence, flow restrictions, and obstructions are undesirable. Abrupt changes in flow direction should be avoided. Tank inlet pipes should be directed away from the tank walls and toward the center. Welded and flanged pipe sections should always be carefully aligned. Impingement plates of baffles designed to bear the brunt of the damage should be easily replaceable. The thickness of vulnerable areas should be increased. Replaceable ferrules, with a tapered end, can be inserted into the inlet side of heat-exchanger tubes to prevent damage to the actual tubes.

Several environmental modifications can be implemented to minimize the risk of erosion corrosion. Abrasive particles in fluids can be removed by filtration or settling, and water traps can be used in steam and compressed air systems to decrease the risk of impingement by droplets. Deaeration and corrosion inhibitors are additional measures that can be taken. Cathodic protection and the application of protective coatings may also reduce the rate of attack.

For minimizing cavitation damage specifically, steps that can be taken include the minimization of hydrodynamic pressure gradients, designing to avoid pressure drops below the vapor pressure of the liquid, the prevention of air ingress, the use of resilient coatings, and cathodic protection.

5.4.8 Stress corrosion cracking

The use of materials exhibiting a high degree of resistance to SCC is a fundamental measure. Modification of the environment (removal of the critical species, corrosion inhibitor additions) is a further important means of control. In principle, reduced tensile stress levels is a means of controlling SCC. In practice, maintaining tensile stress levels below a critical stress intensity level is difficult because residual stresses often play an important role. These are difficult to quantify. Stress-relieving heat treatments usually do not eliminate residual stresses completely. Furthermore the wedging action of corrosion products can lead to unexpected increases in tensile stress levels. Stress raisers should obviously be avoided. The introduction of residual compressive surface stresses by shot peening is a further remedial possibility. Fit-up stresses should be minimized by close control over tolerances.

Serious attempts are still being made to elucidate and quantify the parameters controlling the incidence of cracking. For this purpose empirical equations have often been derived from laboratory tests. Equation (5.2), for example, summarizes the effects of different alloying elements on the resistance of ferritic steels exposed to a boiling 8.75N-NaOH solution during slow strain tests.[21] The stress corrosion index in that environment (SCI_{OH}) integrates the beneficial (−) or deleterious (+) effect of the alloying elements (in %) when the steels are in contact with such a caustic environment.

$$SCI_{OH} = 105 - 45C - 40Mn - 13.7Ni - 12.3Cr - 11Ti + 2.5Al + 87Si + 413Mo \qquad (5.2)$$

The optimum choice of a steel for a particular application should be made in the light of expressions such as Eq. (5.2), which reflects the corrosivity of the environment as a function of the metallurgical composition and structure. But other practical considerations such as availability of the materials, maintainability, and economical requirements inevitably dictate the use of an alloy out of its safe envelope, in which case the application of coatings, cathodic protection, and/or some other protection scheme, appropriate for the operating conditions, have to be considered. Another important consideration is the accidental damage that can locally modify the pattern of stresses imposed on a metallic component or can destroy some of the protective barriers.

Microstructural anisotropy is an important variable in SCC, especially for aluminum alloys. Tensile stresses in the transverse and short transverse plane should be minimized. Components should be designed with grain orientation in mind (Fig. 5.19). The use of cathodic protection for SCC control is restricted to situations where hydrogen embrittlement effects do not play any role, because hydrogen embrittlement-related SCC damage will be accelerated by the impressed current.

5.5 Case Histories in Corrosion Failure Analysis

Most corrosion failures are not unique in nature. For any given failure, it is likely that a similar problem has been encountered and solved previously. Practicing failure analysis experts rely heavily on their experience from previous cases; it is the extensive experience gained in previous cases that makes them highly effective and successful in their profession. A number of excellent paper-based resources document corrosion case histories.[2,22] Investigators of all experience levels frequently consult such collections of case histories. By learning as much as possible from previous cases, the laboratory work and testing effort of the investigation can be minimized.

A collection of documented corrosion-failure case histories represents a valuable corporate asset. However, information retrieval from a paper-based system can be laborious and time consuming. Typically, hundreds of failure analysis reports are generated each year by an active team of investigators and thousands of such reports are stored in filing cabinets, with no convenient mechanism available to reuse this valuable information. Searching for patterns in accumulated documents and databases is a process regularly performed in large organizations. The weakness in managing large volumes of paper-based information tends to be sporadically compensated by in-depth surveys of available information. For example, a survey of failure analysis reports of landing gear failures in the Canadian Forces revealed that 200 case histories had been investigated over the past 25 years.[23] The survey was successful in determining the dominant failure mechanisms and ranking the importance of root causes as shown in Table 5.3.[24] However, the fundamental need for more efficient methodologies for improving knowledge reuse is not addressed by surveys of this nature. Some new promising options are emerging from the field of computerized knowledge discovery (see Chap. 4, Modeling, Life Prediction, and Computer Applications).

TABLE 5.3 Breakdown of Causes of Landing Failures as a Function of the Failure Mechanism

Mechanisms-Causes	Design	Material selection	Manufacturing	Field maintenance
Overload		8	4	13
Fatigue	59	22	65	24
Cosmetic pitting	3	6	2	6
SCC	7	34	7	6
Structural pitting	22	17	6	41
Wear	9			10
False call		13	16	

References

1. Fontana, M. G., *Corrosion Engineering,* New York, McGraw Hill, 1986.
2. Dillon, C. P., *Forms of Corrosion: Recognition and Prevention,* Houston, Tex., NACE International, 1982.
3. Gilbert, L. O., *Materiel Deterioration Problems in the Army,* unpub., 1979.
4. Szklarska-Smialowska, Z., *Pitting Corrosion,* Houston, Tex., NACE International, 1986.
5. Miller, D., Corrosion Control on Aging Aircraft: What Is Being Done? *Materials Performance,* **29:**10–11 (1990).
6. Hoffman, C., 20,000-Hour Tuneup, *Air & Space,* **12:**39–45 (1997).
7. Seher, C. and Broz, A. L., National Research Program for Nondestructive Inspection of Aging Aircraft, *Materials Evaluation,* **49:**1547–1550 (1991).
8. Komorowski, J. P., Krishnakumar, S., Gould, R. W., et al., Double Pass Retroreflection for Corrosion Detection in Aircraft Structures, *Materials Evaluation,* **54:**80–86 (1996).
9. Wildey, II, J. F., Aging Aircraft, *Materials Performance,* **29:**80–85 (1990).
10. Komorowski, J. P., Bellinger, N. C., Gould, R. W., et al., Quantification of Corrosion in Aircraft Structures with Double Pass Retroreflection, *Canadian Aeronautics and Space Journal,* **42:**76–82 (1996).
11. Oldfield, J. W., Electrochemical Theory of Galvanic Corrosion, in Hack, H. P. (ed.), *Galvanic Corrosion,* Philadelphia, Penn., American Society for Testing of Materials, 1988, pp. 5–22.
12. Baboian, R., Bellante, E. L., and Cliver, E. B., *The Statue of Liberty Restauration,* Houston, Tex., NACE International, 1990.
13. Perrault, C. L., Liberty: To Build and Maintain Her for a Century, in Baboian, R., Bellante, E. L., and Cliver, E. B. (eds.), *The Statue of Liberty Restauration,* Houston, Tex., NACE International, 1990, pp. 15–30.
14. Page, C. L., and Anchor, R. D., Stress Corrosion Cracking in Swimming Pools, *Materials Performance,* **29:**57–58 (1990).
15. Staehle, R. W., Predicting the Performance of Pipelines, Revie, R. W. and Wang, K. C. *International Conference on Pipeline Reliability,* VII-1-1-VII-1-13. 1992. Ottawa, Ont., CANMET.
16. Staehle, R. W., Understanding "Situation-Dependent Strength:" A Fundamental Objective, in *Assessing the History of Stress Corrosion Cracking. Environment-Induced Cracking of Metals,* Houston, Tex., NACE International, 1989, pp. 561–612.
17. Roberge, P. R., An Object-Oriented Model of Materials Degradation, in Adey, R. A., Rzevski, G., and Tasso, C. (eds.), *Applications of Artificial Intelligence,* in *Engineering X,* Southampton, UK, Computational Mechanics Pub., 1995, pp. 315–322.
18. Roberge, P. R., Tullmin, M. A. A., and Trethewey, K., "Knowledge Discovery from Case Histories of Corrosion Problems," *CORROSION 97,* Paper 319. 1997. Houston, Tex., NACE International.
19. Hoar, T. P., *Report of the Committee on Corrosion and Protection,* London, UK, Her Majesty's Stationary Office, 1971.
20. Wyatt, L. M., Bagley, D. S., Moore, M. A., et al., *An Atlas of Corrosion and Related Failures,* St. Louis, Mo., Materials Technology Institute, 1987.
21. Parkins, R. N., *Materials Performance,* **24:**9–20 (1985).
22. EFC, *Illustrated Case Histories of Marine Corrosion,* Brookfield, UK, The Institute of Metals, 1990.
23. Beaudet, P., and Roth, M., Failure Analysis Case Histories of Canadian Forces Aircraft Landing Gear Components, *Landing Gear Design Loads,* Neuilly-sur Seine, France, NATO, 1990, pp. 1.1–1.23.
24. Roberge, P. R., and Grenier, L., "Developing a Knowledge Framework for the Organization of Aircraft Inspection Information," *CORROSION 97,* Paper 382. Houston, Tex., NACE International, 1997.

Chapter 6

Corrosion Maintenance through Inspection and Monitoring

6.1	Introduction	372
6.2	Inspection	374
	6.2.1 Selection of inspection points	375
	6.2.2 Process piping	375
	6.2.3 Risk-based inspection	377
6.3	The Maintenance Revolution	383
	6.3.1 Maintenance strategies	384
	6.3.2 Life-cycle asset management	387
	6.3.3 Maintenance and reliability in the field	394
6.4	Monitoring and Managing Corrosion Damage	406
	6.4.1 The role of corrosion monitoring	406
	6.4.2 Elements of corrosion monitoring systems	409
	6.4.3 Essential considerations for launching a corrosion monitoring program	410
	6.4.4 Corrosion monitoring techniques	416
	6.4.5 From corrosion monitoring to corrosion management	428
6.5	Smart Sensing of Corrosion with Fiber Optics	448
	6.5.1 Introduction	448
	6.5.2 Optical fiber basics	451
	6.5.3 Emerging corrosion monitoring applications	452
	6.5.4 Summary	460
6.6	Nondestructive Evaluation (NDE)	461
	6.6.1 Introduction	461
	6.6.2 Principles and practices	462
	6.6.3 Data analysis	478
References		481

6.1 Introduction

In the modern business environment, successful enterprises cannot tolerate major corrosion failures, especially those involving unscheduled shutdowns, environmental contamination, personal injuries, and fatalities. For this reason, considerable effort must be expended on corrosion control at the design stage and also in the operational phase. Typically, once a system, a plant, or any piece of equipment is put into service, maintenance is required in order to keep it operating safely and efficiently. This is particularly true for aging systems and structures, many of which may operate beyond their original design life.

The required level of maintenance will vary greatly with the severity of the operating environment and the criticality of the engineering system. Some buildings require only regular repainting and occasional inspection of electrical and plumbing lines, but chemical processing plants, power generation plants, aircraft, and marine equipment have extensive maintenance schedules. Even the best of designs cannot be expected to anticipate all conditions that may arise during the life of a system. Corrosion inspection and monitoring are used to determine the condition of a system and, importantly, to determine how well corrosion control and maintenance programs are performing. Corrosion monitoring embraces a host of techniques, from simple exposure of coupons to smart structure computerized sensing systems.

The dividing line between corrosion inspection and corrosion monitoring is not always clear. Usually inspection refers to short-term "one-off" measurements taken in accordance with maintenance and inspection schedules. Corrosion monitoring describes the measurement of corrosion damage over a longer time period and often involves an attempt to gain a deeper understanding of how and why the corrosion rate fluctuates over time. Corrosion inspection and monitoring are most beneficial and cost-effective when they are utilized in an integrated manner. They are complementary and should not be viewed as substitutes for each other. Figure 6.1 illustrates how data from various sources should be combined to ultimately produce management information for decision making.

Inspection techniques for the detection and measurement of corrosion range from simple visual examination to nondestructive evaluation. Significant technological advances have been made in the last decade. For example, the combined use of acoustic emission (AE) and ultrasonics (UT) can, in principle, allow an entire structure to be inspected and growing defects to be quantified in terms of length and depth. Advanced corrosion monitoring methods have been developed that have both on-line capability and the ability to detect problems at an early stage. The oil and gas production and petrochemical industries have assumed a

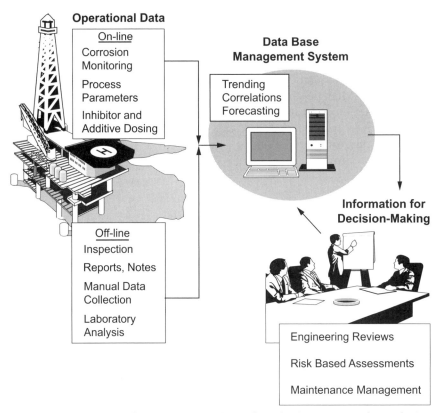

Figure 6.1 Integration of corrosion inspection and monitoring programs for producing management information.

leading role in the advancement of corrosion monitoring. Many techniques that have been accepted in these industries for years are only beginning to be applied in other industries, such as transportation, mining, and construction.

A considerable catalyst for the advances in corrosion inspection and monitoring technology has been the exploitation of oil and gas resources in extreme environmental conditions, such as the North Sea offshore fields. Operation under such extreme conditions has necessitated enhanced instrument reliability and the automation of many tasks, including inspection. The development of powerful user-friendly software has allowed some techniques that were once perceived as mere laboratory curiosities to be brought to the field. In addition to the usual uncertainty concerning the onset or progression of corrosion of equipment, the oil industry has to face everchanging corrosivity of processing streams. During the life of an exploitation system, the corrosivity at a wellhead can oscillate many times between being benign and becoming extremely corrosive.[1] Such changes require more corrosion vigilance in terms of inspection and monitoring.

Considering the many complex forms and mechanisms of corrosion damage, the limitations of individual plant inspection and monitoring techniques are considerable. The large number of variables involved also implies that no single method can be expected to be satisfactory in all possible conditions and environments. Many of these variables are described in other parts of the book. The factors involved in the corrosive environment have, for example, been introduced in Chap. 5, Corrosion Failures.

6.2 Inspection

Inspection normally refers to the evaluation of the quality of some characteristic in relation to a standard or a specification. As products and their manufacturing processes have grown more complex and been divided among many departments, the job of inspection has also become complex and distributed. A flow diagram is useful for showing the various materials, components, and processes that collectively or sequentially make up the system. The main purpose of inspection is to determine whether components, systems, or products conform to specifications. Inspection consists of the following series of actions:

- Interpretation of the specifications
- Measurement and comparison with a specification
- Judging conformance
- Classification of conforming cases
- Classification of nonconforming cases
- Recording and reporting the data obtained

In practice, inspectors may experience difficulty in interpreting a particular specification. Assistance for this task can be provided in several ways:

- Clearing up the meaning of terminology used. Descriptions of sensory qualities, such as "cosmetic corrosion defects," are often confusing. While such defects may not affect the functionality of a component, customers may find them objectionable for aesthetic reasons.
- Eliminating vague or incomplete information in specifications.
- Classifying the importance (seriousness) of product characteristics, to emphasize the most important features of the product.

- Provision of samples, photographs, or other references to assist in interpretation of the meaning of a specification. The importance of visual standards cannot be overemphasized.
- Periodic review and revision of specifications for clarity to eliminate recurring, chronic problems of interpretation.

6.2.1 Selection of inspection points

The selection of inspection points is of paramount importance, as corrosion factors to be considered are often related to the geometry of systems and components. Selection of inspection points should be based on a thorough knowledge of process conditions, materials of construction, the geometry of the system, external factors, and historical records.

An example of an industrial concern being translated into an inspection program is the loss of plant profitability due to production losses associated with boiler tube failures. Underdeposit corrosion in steam-generating systems is caused or enhanced by the breakdown of a protective magnetite film and/or the inability to form such a film. Production losses resulting from reduced steam capacity are far greater than the actual repair and maintenance costs incurred during shutdowns. The major contributors to the formation of deposits and scale are

- Accumulation of corrosion products, mostly consisting of iron oxides introduced into the boiler from the feedwater and condensate systems
- Contaminants present in the makeup water
- Contaminants introduced to the condensate return from process equipment
- Solids introduced from leakages

The methods used to detect and monitor underdeposit corrosion involve an investigation of the water treatment practices accompanied by an evaluation of the amount of feedwater corrosion product deposited on the boiler heat-transfer surfaces. The detection and monitoring methods used in such an environment are presented in Table 6.1, highlighting the main characteristics of the techniques and their applications.

6.2.2 Process piping

Probably the most important inspection function related to plant reliability involves process piping systems. Piping systems not only connect all other equipment within the unit, but also interconnect units within the operation. Thus, they can be considered to be an accurate barometer of conditions occurring within the process. It has been

TABLE 6.1 Inspection Techniques Useful for the Detection of Underdeposit in Boiler Systems

Inspection methods	Application
On-line	
Hydrogen analysis in saturated background	General and steam localized corrosion
Tube temperature monitoring	Deposit buildup
Chemistry (phosphate and pH)	Buffering potential
Off-line	
Visual examination (fiberscope, videoprobe)	Steam blanketing
	Gouging and tubercles
Tube sampling	Deposit amount
	Deposit constituents

demonstrated repetitively that if an inspection department has control over the condition of piping within a unit, the condition of the remaining equipment will also be known with a relatively high degree of confidence.[2] It is rare that corrosion or other forms of deterioration found in major components of process equipment are not found in the interconnecting piping. The latter is generally more vulnerable to corrosion and subject to initial failure because

- The corrosion allowance on piping generally is only one-half that provided for other pieces of refinery equipment.
- Fluid velocities are often higher in piping, leading to accelerated corrosion rates. (This is not always the case for certain localized corrosion processes.)
- Piping design stresses normally are higher, and the piping system may be subject to external loading, vibration, and thermal stresses that are more severe than those encountered in other pieces of equipment.
- The larger number of inspection points in a piping system makes the task of controlling and monitoring the system bigger.

Leaks in pressurized piping systems are extremely hazardous and have led to several catastrophes. Components requiring close attention include

- Lines operating at temperatures below the dew point
- Lines operating in an industrial marine atmosphere
- Points of entry and exit from a building, culvert, etc., where a break in insulation could occur
- Pipe support condition and fireproofing

- Piping alignment, provision for thermal expansion, and position of pipe shoes on supports
- Welded joints, because they can have elevated stress levels (from residual stress effects and stress concentrations), geometrical discontinuities, complex metallurgical structures, and possible galvanic cells (preferential weld corrosion)
- Flanged or screwed joints for evidence of leakage
- Geometrical changes that affect fluid flow characteristics (bends, elbows, section changes, etc.), with a resulting risk of erosion/corrosion

Lines handling corrosive materials such as saltwater ballast, acids, bases, and brine are subject to internal corrosion and require frequent inspection until a satisfactory service history is developed. Frequency and degree of inspection must be individually developed, taking into consideration rate of deterioration and seriousness of an unpredicted leak. Testing of piping systems is done using various techniques, including pressure testing; radiography; and dye-penetrant, magnetic-particle, and ultrasonic testing. Table 6.2 describes many of the problems or materials damage commonly encountered in process-industry piping systems.[2]

The inspection of a new, unfired pressure vessel should begin at the time of manufacture and continue through field construction. Thickness readings and other information obtained during fabrication and construction should be incorporated into the inspection records and should constitute the "baseline" to which subsequent readings are compared. During the life of a vessel, various metallurgical changes can occur in the pressure-holding components that could significantly affect their physical properties. These changes are not apparent with the use of normal inspection techniques. However, an inspector should be aware of these possibilities.

6.2.3 Risk-based inspection

Risk analysis refers to techniques for identifying, characterizing, and evaluating hazards. Risk-based inspection (RBI) is the application of risk analysis principles to the management of inspection programs for plant equipment. RBI has been used in the nuclear power generation industry for some time and is also employed in refineries and petrochemical plants. The ultimate goal of RBI is to develop a cost-effective inspection and maintenance program that provides assurance of acceptable mechanical integrity and reliability. Clearly, it has an important role in today's competitive business environment, where limited technical and financial resources have to be optimized. An

TABLE 6.2 Problems or Materials Damage Commonly Encountered in Piping Systems of Process Industries

Carbon steels	At temperatures above 400 to 430°C, pearlite will convert to a spheroidal form of carbide and eventually, under suitable conditions, to graphite. Spheroidization and graphitization lower the yield stress and ultimate tensile strength, while increasing the ductility. The effect is significant in the heat-affected zone of a welded joint, where graphite tends to form chains in a form known as "eyebrow" graphitization. This condition can lead to severe embrittlement. Some weld failures caused by this type of deterioration have been reported in the literature. In-place metallography and removal of samples can be used to check for this condition. C steels operating above 430°C should be evaluated for possible graphitization after the first 30,000 h of operation, and every 50,000 h thereafter.
Carbon-Mo steels	Three types of damage to 0.5 Mo C steels are elevated-temperature hydrogen attack, graphitization, and temper embrittlement. Where 0.5 Mo C steels are used in hydrogen service above the limits of the C-steel line, pressure vessels (and heat exchangers) should be monitored using ultrasonic attenuation measurements during unit downtime. Each plate in the vessel should be examined at each turnaround or at a maximum interval of 2 years. The readings should be in the plate material immediately adjacent to a main seam weld, which represents an area of maximum residual stress. In addition, any defects identified by other inspection practices should be investigated by metallographic examination for hydrogen attack.
Low Cr-Mo steels	While C steels tend to soften and become more ductile when exposed to temperatures around 400°C, low Cr-Mo steels tend to undergo temper embrittlement. Embrittlement increases the strength of the material but markedly decreases toughness by inhibiting plastic deformation. The 2.25 Cr 0.5 Mo steels are more susceptible to temper embrittlement in the 370 to 480°C range. Not all the factors that affect temper embrittlement in Cr-Mo steels are fully defined, but some estimate of fracture toughness after service can be made from the chemical composition. The amount of shift in transition temperature for a 2.25 Cr-Mo material is commonly expressed by the J factor: J factor = $(Si + Mn)(P + Sn) \times 10^4$. Steel containing 1.25 Cr 0.5 Mo may temper-embrittle at a temperature around 400°C if P + Sn exceeds 0.03%. Steels containing 1.0 Cr 0.5 Mo do not undergo a serious loss of room-temperature ductility when used at this temperature.

Medium Cr-Mo steels	While 5 to 9% Cr materials are often used for pipe and tubing, pressure vessels of this composition are seldom encountered, because the required corrosion resistance is imparted to the base material by stainless cladding or weld overlays. Cr-Mo steels with this range of chromium do not markedly temper-embrittle and retain reasonable room-temperature toughness.
11 to 13% Cr steels	The 12% Cr steels are often used in pressure vessel service, both as a corrosion-resistant cladding and as a material for trays and other components. All the 12% Cr steels will embrittle in a temperature range of 430 to 540°C (800 to 1000°F). Where room-temperature ductility is an important consideration, the low-carbon Type 410S should be selected, because it has the least tendency toward elevated-temperature embrittlement.
Austenitic stainless steels	While austenitic stainless steels do not lose ductility when heated in the 400 to 510°C temperature range, the unstabilized grades are subject to carbide precipitation that may affect their corrosion resistance. Weld overlays normally use a stabilized Type E347 as the last pass for enhanced corrosion resistance. Selection of stainless materials and/or weldments should be made with regard for such phenomena as sigma-phase formation, underbead cracking, fissuring, differential thermal expansion, stress corrosion cracking, etc.
Grain size	Fine-grain steels improve both strength and toughness. Fine grain size promotes a more uniform distribution of plastic deformation, thus preventing the local buildup of stress, particularly in the area of defects. Use of coarse-grained materials such as ASTM A515 and use of heat treatments that lead to grain coarsening should be avoided. Should vessels be involved in unusual conditions, such as unit fires, that may cause changes in grain size, the affected areas should be checked by field metallography on removed samples.
Hardening	Low Cr-Mo steels, when welded or when cooled rapidly from elevated temperatures, form hard, inherently brittle microstructures consisting of martensite and bainite. These structures have limited capacity for plastic deformation and, thus, have low fracture toughness in the as-welded form. It is of prime importance in repair welding that the original ductility be restored to these air-hardenable materials by maintaining the proper preheat and postweld heat treatment requirements of the welding procedure.

TABLE 6.2 Problems or Materials Damage Commonly Encountered in Piping Systems of Process Industries (*Continued*)

Dissolved hydrogen	Hydrogen picked up in the steel during operation will diffuse at atmospheric temperature, and this is primarily a function of time, temperature, and thickness. Care should be exercised on heavy-wall equipment to ensure that the cooldown from operating temperature is as slow as practical so that maximum out gassing can occur, and precautions should be taken during shutdown to avoid unnecessary impact loading. Hydrogen gas can readily dissolve in the molten weld metal during the welding operation. The source of the hydrogen is generally moisture from the surrounding atmosphere or from damp electrodes. Because hydrogen trapped in the weldment can seriously reduce the ductility, cracking can result. Hydrogen exists in steel as an interstitial atom in the solid lattice. Therefore, detection by normal NDE methods is not possible, and the embrittling effect will remain undetected. It is mandatory that all low-hydrogen electrodes be kept warm prior to use, and that sufficient preheat be applied to the base material to ensure a dry weld joint.
Reheat cracking	Cracking can occur in weldments as a result of the heat treatment used to relieve stresses after welding. Cracking is generally confined to the HAZ, and is normally intergranular in nature. Reheat cracking is of particular concern in low Cr-Mo steels that are prone to cracking in the as-welded condition, or in heavy carbon-steel sections that are highly restrained.
Lamellar tearing	Lamellar tearing is generally found in the HAZ of weldments of tee and corner joints. The cause of cracking appears to be inclusions that are parallel to the rolling direction of the plate section being welded. The restraining forces in the welded joint cause the inclusions to open up and run together to form a crack. Set on-type connections welded to heavy sections are particularly susceptible to this type of defect. Where set on-type connections are used, shear-wave ultrasonic inspection of the plate in the area of the attachment and of the completed weld is recommended.

excellent review of RBI, including the relevance of corrosion engineering, was recently published.[3]

Risk-based inspection is a methodology for using risk as a basis for ranking or prioritizing equipment for inspection purposes. Risk is defined as the combination of probability and consequence. Probability is the likelihood of a failure occurring, and consequence is a measure of the damage that could occur as a result of the failure (in terms of injury, fatalities, and property damage). Increased risk resulting from increased probability and higher degree of consequence is illustrated in Fig. 6.2. The highest risk is generally associated with a small percentage of plant items.

Risk-based inspection procedures can use either qualitative or quantitative methodologies. Qualitative procedures provide a ranking of equipment, based largely on experience and engineering judgment. Quantitative risk-based methods use several engineering disciplines to set priorities and develop programs for equipment inspection. Some of the engineering disciplines include nondestructive examination, system and component design and analysis, fracture mechanics, probabilistic analysis, failure analysis, and operation of facilities. Quantitative analysis methods can be expensive, time-consuming, and tedious, and are therefore not commonly used. Often, the information

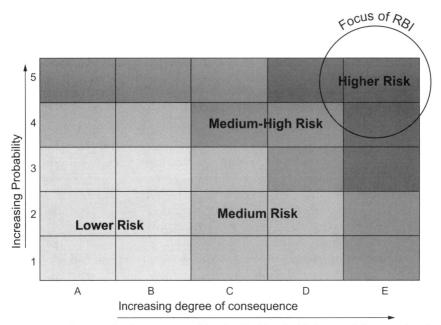

Figure 6.2 Degree of risk as measured by the likelihood of failure and the severity of failure. In RBI, attention is focused on high-risk items.

available is insufficient for conducting a quantitative risk analysis. Two organizations that are currently working on quantitative risk-based analysis procedures for use by the chemical industry are the American Society of Mechanical Engineers (ASME) and the American Petroleum Institute (API).

Probability of failure. To determine the probability of a failure, two fundamental issues must be considered: first, the different forms of corrosion and their rate, and second, the effectiveness of inspection. Clearly the input of corrosion experts is required in order to identify the relevant forms of corrosion in a given situation and to determine the key variables affecting the propagation rate. It is also important to realize that full consensus and supporting data on the variables involved are highly unlikely in real-life complex systems and that simplification will invariably be necessary.

One semiquantitative approach for ranking process equipment is based on internal probability of failure (POF). The procedure involves an analysis of equipment process and inspection parameters, and ranks equipment on a scale of 1 to 3 with 1 being the highest priority. It requires a fair degree of engineering judgment and experience, and therefore is dependent on the background and expertise of the analyst. The procedure is designed to be both practical and efficient. The POF is intended to be a convenient and reproducible means for establishing equipment inspection priorities. As such, it facilitates the most efficient use of finite inspection resources when and where 100 percent inspection is not practical.

The POF approach is based on a set of rules that are heavily dependent on detailed inspection histories, knowledge of corrosion processes, and knowledge of normal and upset conditions. The equipment rankings may have to be changed and can require updating as additional knowledge is gained, process conditions change, and equipment ages. Maximum benefits of the procedure depend on fixed equipment inspection programs that permit the capture, documentation, and retrieval of inspection, maintenance, and corrosion/failure mechanism information. However, the POF procedure is only one-half of a risk-based inspection procedure. The POF ranking has to be combined with a consequence ranking to provide a true risk-based ranking.

Consequences of failure. To assess the consequences of failure, input from experts in process engineering, safety, health, environmental engineering, etc., is obviously important. Three dominant factors are considered in consequence analysis: the types of species that could be released into the environment and their associated hazards, the amount available for release, and the rate of release.[3] Corrosion and materials engineering expertise is required to estimate the amount and nature of damage that could result. Different corrosion mechanisms can produce different morphologies of damage. The difference in impact of the release rate created at a pinhole leak and that at a large rupture is a good example of this aspect of consequence sensitivity. Another important field covered by corrosion engineering is that of materials properties. For example, the risk of a catastrophic explosion as a result of cracks is obviously greater in a brittle material (where they are associated with high release rates) than in a material with higher fracture toughness. The toughness of a material is a key parameter in determining so-called leak-before-break safety criteria and the general tolerance toward defects. An understanding of how toughness can be reduced in service over time is thus obviously important.

Application of RBI. Horvath has outlined three approaches to risk reduction that are incorporated in the API RBI program:[3]

- Optimization of the inspection and monitoring strategy
- Changes in materials of construction
- Control of key process parameters

The inspection and monitoring plan can be reviewed and modified, essentially to shift inspection from overinspected low-risk to underinspected high-risk equipment. Furthermore, inspection techniques should be selected to address all relevant damage mechanisms identified in the RBI program. Inspection points should correspond to the most likely areas of corrosion damage. Inspection intervals need to reflect the rate of corrosion damage and how it may change over time.

Since RBI analysis includes corrosion rates for different materials, such data can be used for alloy selection on a risk reduction vs. cost basis. The RBI methodology will assist in selecting materials rationally, rather than merely succumbing to the temptation of minimizing initial construction costs. Such a short-term view obviously does not take future inspection requirements and future materials upgrades into account.

The benefits of identifying key process parameters affecting corrosion (and other) damage should be apparent, especially if these parameters are subsequently monitored to ensure that they remain within safe operating windows. The impact on risk of any process changes can also be rationally assessed on the basis of RBI.

6.3 The Maintenance Revolution

Maintenance costs represent a significant portion of operating budgets in most industrial sectors, particularly where aging structures or plant is involved. Modern approaches to maintenance management (sometimes referred to as profit-centered maintenance) are designed to minimize these costs and to improve the reliability and availability of plant and equipment. In this context, maintenance activities are treated as an investment, not as an organizational liability. However, as part of overall rationalization, the maintenance function often has to be performed with shrinking technical and financial resources, making focus on the most critical items a logical development. In many cases, "old" corrective maintenance and time-based preventive maintenance practices are inadequate to meet modern demands. The consequences of poor maintenance practices and/or inadequate investment in the maintenance function are the following:

- *Reduced production capacity.* Not only will there be an increase in downtime, but, importantly, assets will underperform during uptime.
- *Increased production costs.* Whenever assets are not performing at optimal level, real cost and opportunity cost penalties are incurred.
- *Lower-quality products and services.* The ultimate consequence will be customer dissatisfaction and probably lost sales.
- *Safety hazards.* Failures can lead to loss of life, injuries, and major financial losses.

6.3.1 Maintenance strategies

Four general types of maintenance philosophies or strategies can be identified, namely, corrective, preventive, predictive, and reliability-centered maintenance. Predictive maintenance is the most recent development. In practice, all these types are used in maintaining engineering systems. The challenge is to optimize the balance among them for maximum profitability. In general, corrective maintenance is the least cost-effective option when maintenance requirements are high.

Corrective maintenance. Corrective maintenance refers to action taken only after a system or component failure has occurred. It is thus a retroactive strategy. The task of the maintenance team in this scenario is usually to effect repairs as soon as possible. Costs associated with corrective maintenance include repair costs (replacement components, labor, and consumables), lost production, and lost sales. To minimize the effects of lost production and speed up repairs, actions such as increasing the size of maintenance teams, using backup systems, and implementation of emergency procedures can be considered. Unfortunately, such measures are relatively costly and/or effective only in the short term. For example, if heat-exchanger tubes have leaked as a result of pitting corrosion and it is urgent that production proceed, it may be possible to plug the leaking tubes on a short-term basis. Obviously, such measures do not assure the longer-term performance of a heat exchanger.

Preventive maintenance. In preventive maintenance, equipment is repaired and serviced before failures occur. The frequency of maintenance activities is predetermined by schedules. The greater the consequences of failure, the greater the level of preventive maintenance that is justified. This ultimately implies a tradeoff between the cost of performing preventive maintenance and the cost of running the equipment until failure occurs. Of course, preventive maintenance tasks can also be dictated by safety, environmental, insurance, or other regulatory considerations.

Inspection assumes a crucial role in preventive maintenance strategies. Components are essentially inspected for corrosion and other damage at planned intervals, in order to enable corrective action before failures actually occur. Performing preventive maintenance at regular intervals will usually result in reduced failure rates. As significant costs are involved in performing preventive maintenance, especially in terms of scheduled downtime, good planning is vital. To maximize asset value and performance, the basic aim is to perform preventive maintenance just before serious damage would set in otherwise.

Furthermore, the level of preventive maintenance activity needs to be driven by the importance of the equipment to the process and the desired level of reliability. In modern complex systems, computerized preventive maintenance systems are used to accomplish these objectives in plants of most sizes. A preventive maintenance system also needs to be dynamic; for example, there should be some mechanism for review of preventive tasking to ensure that the tasks are still valid and to see if any task can be replaced with a predictive task.[4]

Predictive maintenance. Predictive maintenance refers to maintenance based on the actual condition of a component. Maintenance is performed not according to fixed preventive schedules, but rather when a certain change in characteristics is noted. Corrosion sensors that supply diagnostic information on the condition of a system or component play an important role in this maintenance strategy. Preventive maintenance aims to eliminate unnecessary inspection and maintenance tasks, to implement additional maintenance tasks when and where needed, and to focus efforts on the most critical items.

A useful analogy to automobile oil changes can be made. Changing the oil every 5000 km to prolong engine life, irrespective of whether the oil change is really needed or not, is a preventive maintenance strategy. Predictive maintenance would entail changing the oil based on changes in its properties, such as the buildup of wear debris. When a car is used exclusively for long-distance highway travel and is driven in a very responsible manner, oil analysis may indicate a longer critical service interval.

Some of the resources required to perform predictive maintenance will be available from the reduction in breakdown maintenance and the increased utilization that results from proactive planning and scheduling. Good record keeping is very important in identifying repetitive problems and the problem areas with the highest potential impact.

Reliability-centered maintenance. Reliability-centered maintenance (RCM) involves the establishment or improvement of a maintenance program in the most cost-effective and technically feasible manner. It utilizes a systematic, structured approach that is based on the consequences of failure. As such, it represents a shift away from time-based maintenance tasks and emphasizes the functional importance of system components and their failure and maintenance history. RCM is not a particular maintenance strategy, such as preventive maintenance; rather, it can be employed to determine whether preventive maintenance is the most effective approach for a particular system component.

The concept of RCM has its roots in the early 1960s. RCM strategies for commercial aircraft were developed in the late 1960s, when widebody jets were introduced into commercial airline service.[5] A major concern of airlines was that existing time-based preventive maintenance programs would threaten the economic viability of larger, more complex aircraft. With the time-based maintenance approach, components are routinely overhauled after a certain amount of flying time. In contrast, as pointed out above, RCM determines maintenance intervals based on the criticality of a component and its performance history. The experience of airlines with the RCM approach was that maintenance costs remained roughly constant, but that the availability and reliability of their planes improved.[5] RCM is now standard practice for most of the world's airlines.

The initial development work was done by the North American civil aviation industry through "maintenance steering groups," or MSGs. The MSGs were established to reexamine everything that was being done to keep aircraft airborne. These groups consisted of representatives of the aircraft manufacturers, the airlines, and the FAA. The first attempt at a rational, zero-based process for formulating maintenance strategies was promulgated by the Air Transport Association in Washington, D.C., in 1968. This first attempt is now known as MSG 1 (from the first letters of maintenance steering group). A refinement, now known as MSG 2, was promulgated in 1970.

In the mid 1970s, the U.S. Department of Defense wanted to know more about the then state of the art in aviation maintenance thinking. It commissioned a report on the subject from the aviation industry. This report was written by Stanley Nowlan and Howard Heap of United Airlines. They gave it the title "Reliability Centered Maintenance." The report was published in 1978, and it is still one of the most important documents in the history of physical asset management.[6,7] Nowlan and Heap's report represented a considerable advance on MSG 2 thinking. It was used as a basis for MSG 3, which was promulgated in 1980. MSG 3 has since been revised twice. Revision 1 was issued in 1988, and revision 2 in 1993. (See Sec. 4.2.2.) It is used to this day to develop prior-to-service maintenance programs for new types of aircraft (recently including the Boeing 777 and Airbus 330/340).

Following the application of RCM in commercial aviation and defense, these methodologies have also been applied to maintenance programs in the nuclear power, chemical processing, fossil fuel power generation, and other industries. Potential benefits of RCM include

- Maintaining high levels of system reliability and availability
- Minimizing "unnecessary" maintenance tasks
- Providing a documented basis for maintenance decision making
- Identifying the most cost-effective inspection, testing, and maintenance methods

6.3.2 Life-cycle asset management

There are significant improvements, in terms of both costs and efficiency, that can be made through the implementation of asset management and maintenance systems and practices. It is critical that assets be fit for their purpose, perform safely and with respect for environmental integrity, and, most of all, deliver what the users want, when and where they want it.[8] Asset management refers to the effective management of assets from the time of planning for their acquisition until their eventual disposal.

In life-cycle asset management, the aim is to maximize the return on the investment in assets by providing comprehensive information about their condition and value throughout their life. The emphasis is not on the short-term costs of an asset, but rather on the total value (performance) through its entire life. The optimum value of an asset is dependent upon an optimum level of investment. Both the asset value and the available investment levels are a function of time, a variable that assumes major importance in life-cycle asset management (Fig. 6.3).

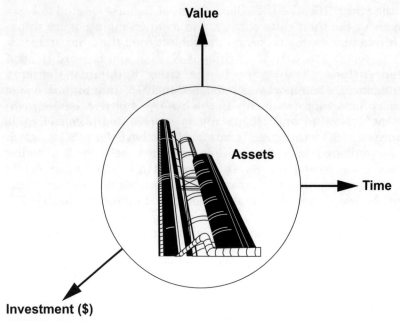

Figure 6.3 The three key variables in life-cycle asset management: the value obtained from an asset and the investment made in the asset, both considered as a function of time.

The pressure to make sound decisions with respect to construction, inspection and maintenance methods and priorities is never greater than during difficult economic times. The main challenge for maintenance managers is to ensure that (scarce) available resources are applied optimally to the (extensive) maintenance requirements. The most critical requirements should be addressed first, followed by prioritization of the remaining maintenance needs. Life-cycle asset management can go a long way toward providing solutions to this challenge. It can be used to justify maintenance budgets, prioritize maintenance expenditures, and predict the need to acquire new assets. Life-cycle asset management focuses on the application of three basic facility management tools: life-cycle costing, condition assessment, and prioritization.

Life-cycle costing. Life-cycle costing utilizes universally accepted accounting practices for determining the total cost of asset ownership or projects over the service life. The basics of corrosion economics are detailed in Appendix C. The economic analysis is usually performed in order to compare competing alternatives. Since the initial capital outlay, support and maintenance costs over the service life, and disposal costs are considered, the time value of money is of major importance in life-cycle costing. Discounting future cash flows to present values essentially reduces all associated costs to a common point in time for objective comparison.

In practice, defining and controlling life-cycle costs is difficult. The future behavior of materials is often uncertain, as are the future uses of most systems, the environmental conditions to which they may be exposed, and the financial and economic conditions that influence relationships between present and future costs. An effective life-cycle cost analysis depends on having a reasonable range of possible alternatives that are likely to deliver equally satisfactory service over a given service life. Substantial obstacles to implementing life-cycle cost control in practice include[9]

- Failure of designers to include life-cycle cost goals in their design criteria
- Failure of owners or managers with short-term responsibility to consider effectively the longer-term impact of their decisions on operations and maintenance requirements
- General desires to minimize the initial expenditures to create short-term "gains" that will increase return on investment, meet budgetary restrictions, or both
- Lack of data and accepted industry standards for describing the maintenance effect and operational performance of components
- Procurement procedures that limit design specificity to enhance competition
- Administrative separation of responsibilities for design, construction, and maintenance

Several decades of experience suggest that improved life-cycle cost management can be achieved through development and application of systematically structured and comprehensive life-cycle cost management. Over the longer term, there is a broad range of actions that managers should consider:[9]

- Formally recognize control of life-cycle cost as an essential and effective element of success.
- Include explicit assessment of design alternatives that influence life-cycle cost as an element of designers' scope of work and fees.
- Assure that value engineering programs and production contract incentives and other procurement mechanisms demonstrate savings in expected life-cycle cost.
- Direct designers to document clearly their design decisions made to control life-cycle cost and the subsequently expected operating consequences.

- Implement cross training and exchange of design and operations and maintenance management personnel to assure that life-cycle cost is controlled at all stages of service life.
- Establish a life-cycle cost management system to maintain operations and maintenance (O&M) data and design decisions in a form that supports operations and maintenance.
- Assign accountability for maintenance and repair at the highest levels in the organization. Responsibilities should include effective use of maintenance and repair funds and other actions required to validate prior facility life-cycle cost management decisions.

Condition assessment. A second major component of life-cycle asset management is systematic condition assessment surveys (CAS). The objective of CAS is to provide comprehensive information about the condition of an asset. This information is imperative for predicting medium- and long-term maintenance requirements, projecting remaining service life, developing long-term maintenance and replacement strategies, planning future usage, determining the available reaction time to damage, etc. Therefore, CAS is in direct contrast to a short-term strategy of "fixing" serious defects as they are found. As mentioned previously, such short-sighted strategies often are ultimately not cost-effective and will not provide optimum asset value and usage in the longer term. CAS includes three basic steps:[9]

- The facility is divided into its systems, components, and subcomponents, forming a work breakdown structure (WBS).
- Standards are developed to identify deficiencies that affect each component in the WBS and the extent of the deficiencies.
- Each component in a WBS is evaluated against the standard.

CAS allows maintenance managers to have the solid analytical information needed to optimize the allocation of financial resources for repair, maintenance, and replacement of assets. Through a well-executed CAS program, information will be available on the specific deficiencies of a facility system or component, the extent and coverage of those deficiencies, and the urgency of repair. The following scenarios, many of which will be all too familiar to readers, indicate a need for CAS as part of corrosion control strategies:

- Assets are aging, with increasing corrosion risks.
- Assets are complex engineering systems, although they may not always appear to be (for example, "ordinary" concrete is actually a highly complex material).

- Assets fulfilling a similar purpose have variations in design and operational histories.
- Existing asset information is incomplete and/or unreliable.
- Previous corrosion maintenance or repair work was performed but poorly documented.
- Information on the condition of assets is not transferred effectively from the field to management, leaving the decision makers ill informed.
- Maintenance costs are increasing, yet asset utilization is decreasing.
- There is great variability in the condition of similar assets, from poor to excellent. The condition appears to depend on local operating microenvironments, but no one is sure where the next major problem will appear.
- The information for long-term planning is very limited or nonexistent.
- An organization's commitment to long-term strategies and plans for corrosion control is limited or lacking.

A requirement of modern condition assessment surveys is that the data and information ultimately be stored and processed using computer database systems. As descriptive terms are unsuitable for these purposes, some form of numerical coding to describe the condition of engineering components is required. An example of assigning such condition codes to galvanized steel electricity transmission towers is shown in Table 6.3.[10] Such numbers will tend to decrease as the system ages, while maintenance work will have the effect of upgrading them. The overall trend in condition code behavior will thus indicate whether maintenance is keeping up with environmental deterioration.

Prioritization. Prioritizing maintenance activities is central to a methodical, structured maintenance approach, in contrast to merely addressing maintenance issues in a reactive, short-term manner. From the preceding sections, it should be apparent that life-cycle asset management can be used to develop a prioritization scheme that can be employed in a wide set of funding decisions, not just maintenance go–no-go decisions. This entails the methodical evaluation of an action against preestablished values and attributes. Prioritization methodologies usually involve a numerical rating system, to ensure that the most important work receives the most urgent attention. The criticality of equipment is an important element of some rating systems. Such an unbiased, "unemotional" rating will ensure that the decisions made will lead to the best overall performance of an engineering system, rather than overemphasizing one of its parts. Preventive maintenance work generally receives a high priority rating.

TABLE 6.3 Selected Condition Coding Criteria Described by Marshall (1998)[10] for Galvanized Electricity Transmission Towers

Condition code, %	Equivalent field assessment
100	New steel; bright, smooth spangled surface. Dark patches on some thicker members.
90	Surface dulled to a matte gray finish.
60	Threads and heads on nuts and bolts start to develop speckled rust. Some darkening red-brown on the undersides of light bracing in cleaner areas, thick crusting in coastal areas.
30	Many bracing members now rusty or turning brown. Large numbers of bolts need to be replaced to retain structural integrity.
10	Holes through many light bracing members, some falling off structure. Severe metal loss on medium-thickness members; flaking rust on legs.

Computerized asset management and maintenance system. In view of the potential increase in efficiency, it is not surprising that computerized asset management and maintenance systems (CAMMS) are becoming increasingly important. Their acquisition alone, however, does not guarantee success in solving problems and increasing profitability. In fact, in the short term, considerable resources may have to be invested before longer-term benefits can be realized. Once a decision has been made to launch a CAMMS initiative, there are six basic issues that deserve special consideration: planning, integration, technology, ease of use, asset management functionality, and maintenance functionality.

Planning. A decision to introduce CAMMS in an organization is a major one, representing a fundamental shift in business culture. The lack of proper planning for CAMMS has been identified as one of the biggest obstacles to success. The planning phase needs to be tackled before the purchasing phase, and significantly more time and effort should be spent in planning than in purchasing. The formulation of detailed goals and objectives is obviously important, together with developing a game plan for companywide commitment to the implementation process.

Integration. The vast number of capabilities and features of modern CAMMS can be overwhelming and confusing. Furthermore, an enormous amount of data will typically have to be collected and entered into the computer system. A sensible approach, therefore, is to gradually integrate CAMMS into the existing system. Implementation in an incremental manner is assisted by software that has a modular architecture. Planning this incremental integration has been shown to be a keystone for success. In this strategy, CAMMS is initially complementary to the existing system while providing long-term capabilities for full integration with other company divisions, such as human resources, finance, scheduling, regulation, condition monitoring, etc. The compatibility of computerized data and information used across different departments with CAMMS is an important requirement in the longer run.

Technology. The investment in computerization is obviously a considerable one in terms of both software and hardware. While the technology should obviously be up to date and leading edge, it is also important to consider how adaptable it is for future use and how easily it can be upgraded, to avoid having to make major reinvestments. At present, a good example of positioning products for future use is a focus on network (intranet and Internet) applications. The nature of the hardware platforms and software development tools used is important in this respect. If these are of a "mainstream" nature, they are more likely to be flexible and adaptable to future requirements. Furthermore, compatibility across different departments is more likely to be achieved with mainstream software development tools and operating systems.

Ease of use. User-friendliness is obviously a key element for the successful implementation of CAMMS. If PC software is based on a dominant operating system, user confidence in it will be greater. After-sale support and service will invariably be required in order to make optimal use of the product, unless a sizable information management department is available in-house to give comprehensive support. In selecting a CAMMS vendor, therefore, the ability to provide support service should be factored in. Multilingual capabilities may be required for corporations with multilanguage needs. Several countries, such as Canada, have more than one official language. In such cases, government departments/agencies and their suppliers typically have multilanguage needs. User-friendliness is also most important to the (major) task of inputting data/information and doing so accurately. Spelling and typing mistakes in data entry can prove to be a major headache in subsequent information retrieval. Modern database software tools can make provision for validating data entries in a user-friendly manner.

Asset management functionality. The key function of CAMMS is to track and measure the output and contribution of the company's maintenance operation relative to overall operations. When comparing one computerized maintenance management solution to another, the ability to measure the impact of maintenance on producing quality goods and services through the use of the organization's assets is ultimately the most important factor. If this requirement is satisfied, maintenance managers will ultimately benefit because they can justify the human and financial resources used for maintenance tasks to senior management.

Maintenance functionality. The maintenance functionality of the system represents the core operations that need to be carried out by the maintenance department. Desired features include the capabilities of managing the maintenance budgets, purchasing functions, and work order scheduling, as well as project and materials management. For example, daily work orders can be uploaded from CAMMS by middle management for use by shop-floor maintenance supervisors. At the end of the day, these processed orders can be downloaded back into CAMMS. Modern computing networks and software can facilitate the seamless transfer of such information. Thus, using CAMMS, this information can be processed, stored, and retrieved in a highly efficient manner. In an alternative "conventional" system, a work order would have to be drawn up on paper; it would then change hands several times and ultimately be filed manually. If, say, 50 paper-based work orders are processed daily in this manner, the risk of losing information and the human effort of storing, retrieving, and reporting information are considerably greater than with the CAMMS alternative.

6.3.3 Maintenance and reliability in the field

The minimization or elimination of corrective maintenance is important from the perspective of introducing statistical process control, identifying bottlenecks in integrated processes, and planning an effective maintenance strategy. Process data are obviously of vital importance for these aspects, but processes operating in a breakdown mode are not stable and yield data of very little, if any, value.

The shift from reactive corrective maintenance toward proactive predictive maintenance represents a significant move toward enhanced reliability. However, efforts designed to identify problems before failure are not sufficient to optimize reliability levels. Ultimately, for enhanced reliability, the root causes of maintenance problems have to be determined, in order to eliminate them. The highest-priority use of root cause analysis (RCA) should be for chronic, recurring problems (often in the form of "small" events), since these usually consume the majority of maintenance resources. Isolated problems can also be analyzed by RCA.

RCA is a structured, disciplined approach to investigating, rectifying, and eliminating equipment failures and malfunctions. RCA procedures are designed to analyze problems to much greater depth (the "roots") than merely the mechanisms and human errors associated with a failure. The root causes lie in the domain of weaknesses in management systems. For example, a pump component may repeatedly require maintenance because it is being damaged by a general corrosion mechanism. The root cause of the problem may have been incorrect purchasing procedures.

The maintenance revolution at electric utilities. Douglas has described the changing maintenance philosophy at electric utilities. The maintenance revolution in electric utility operations has been driven by several factors. A brief summary of these follows:[5,11]

- Markets are becoming more open and competitive, leading to emphasis on cost issues.
- Operating and maintenance costs can be directly controlled by a utility.
- The relative importance of operating and maintenance costs has been rising for more than a decade.
- Assets are aging, leading to increasing maintenance requirements, especially on the fossil fuel generation side.
- At the turn of the century, nearly 70 percent of U.S. fossil fuel plants (43 percent of fossil fuel generation capacity in the United States) will be more than 30 years old, with many critical plants approaching the end of their nominal design life. Utilities are often planning to extend the service life of these plants even further, possibly even under more severe operating conditions.

To meet the above challenges, two fundamental initiatives are under way, namely, shifts to reliability-centered maintenance and predictive maintenance. Broadly speaking, prior to the maintenance revolution, the utilities' maintenance approach had essentially been one of preventive maintenance on "all" components after "fixed" time intervals, irrespective of the components' criticality and actual condition. The shortcomings of this approach included the following: (1) overly conservative maintenance requirements, (2) limited gains in reliability from investments in maintenance, (3) inadequate preventive maintenance on key components, and (4) added risk of worker exposure to radiation through unnecessary maintenance. Anticipated benefits of the revised approach are related not only to reduced maintenance costs but also to improved overall operational reliability.

The nuclear power generating industry followed the aviation sector in RCM initiatives, with an emphasis on preventing failures in the most critical systems and components (those with the most severe consequences of failure). The following three tasks dominated the implementation of RCM in nuclear power generation:

- Failure modes and effects analysis (FMEA) to identify the components that were most vital to overall system functionality
- Logic tree analysis to identify the most effective maintenance procedures for preventing failure in the most critical parts
- Integration of RCM into the existing maintenance programs

The introduction of RCM procedures into fossil fuel plants and power delivery systems can be streamlined because of less restrictive regulations. For example, the FMEA and logic tree analyses were combined into a process called criticality analysis. The main difference in implementing RCM in power generation compared with the aviation industry is that for power plants, RCM has to be implemented in existing plants with existing "established" maintenance practices. The airline industry had the benefit of creating new RCM programs for new aircraft, in collaboration with suppliers of the new airliners. Successes cited by Douglas from the implementation of RCM programs include the following:[5]

- Savings in annual maintenance costs (excluding benefits from improved plant availability), with a payback period of about four and a half years
- Reduced outage rate at a nuclear plant and an estimated direct annual maintenance cost saving of half a million dollars
- A 30 percent reduction in annual maintenance tasks in the ash transport system of a fossil fuel plant
- A fivefold reduction in annual maintenance tasks in a wastewater treatment system
- Maintenance cost savings and increased plant availability at fossil fuel generating units
- In the long term, improved design changes for improved plant reliability

The predictive maintenance component involves the use of a variety of modern diagnostic systems and is viewed as a natural outcome of RCM studies. Such "smart" systems diagnose equipment condition (often in real time) and provide warning of imminent problems. Hence, timely maintenance can be performed, while avoiding unnecessary maintenance and overhauls.

Two types of diagnostic technologies are available. Permanent, online systems provide continuous coverage of critical plant items. The initial costs tend to be high, but high levels of automation are possible. Systems that are designed for periodic condition monitoring are less costly in the short term but more labor-intensive in the long run.

Developments in advanced sensor technologies, some of them spin-offs from military and space programs, are expected to expand predictive maintenance capabilities considerably. Ultimately, the information obtained from such sensors is to be integrated into RCM programs.

Even with automated and effective diagnostic systems in place, plant personnel have experienced some difficulties with data evaluation. These problems arose when diagnostic systems provided more data than maintenance personnel had time to evaluate, or when the systems provided inaccurate or conflicting data. Efforts to correct such counterproductive situations have required additional corporate resources for evaluating, demonstrating, and implementing diagnostic systems, together with increased focus on automation and computerization of analysis and reporting tasks.

The use of corrosion sensors in flue gas desulfurization (FGD) systems falls into the predictive maintenance domain. This application, initiated by the Electric Power Research Institute (EPRI), was related to corrosion of outlet ducts and stacks, a major cause of FGD system unavailability.[12] If condensation occurs within the stack and ducting, rapid corrosion damage will occur in carbon steel as a result of the formation of sulfuric acid. Options for corrosion control include maintaining the temperature of the discharged flue gas above the dew point and the introduction of a corrosion-resistant lining material. Both these options have major cost implications. The corrosion sensors were of the electrochemical type and were designed specifically to perform corrosion measurements under thin-film condensation conditions and to provide continuous information on the corrosion activity. Major benefits obtained from this information included a delay in relining the outlet ducts and stack (estimated cost saving of $3.2 million) and more efficient operations with reduced outlet gas temperatures.

PWR corrosion issues. The significance of corrosion damage in electric utility operations, in terms of its major economic and enormous public safety implications, is well illustrated in the technical history of nuclear pressurized water reactors (PWRs). The majority of operational nuclear power reactors in the United States are of this reactor design. The principle of operation of such a reactor is shown schematically in Fig. 6.4. In the so-called reactor vessel, water is heated by nuclear reactions in the reactor core. This water is radioactive and is pressurized to keep it from boiling, thereby maintaining effective heat transfer. This hot, radioactive water is then fed to a steam generator through U-shaped tubes. A reactor typically has thousands of such tubes, with a total length of several kilometers. In the steam generator, water in contact with the outside surfaces of the tubes is converted to steam. The steam produced drives turbines, which are connected to electricity generators. After passing over the turbine blades, the steam is condensed in a heat exchanger and returned to the steam generator.

Steam generator problems, notably deterioration of the steam generator tubes, have been responsible for forced shutdowns and capacity losses. These tubes are obviously a major concern, as they represent a fundamental reactor coolant pressure boundary. The wall thickness of these tubes has been compared to that of a dime. The safety issues concerning tube failures are related to overheating of the reactor core (multiple tube ruptures) and also release of radioactivity from a rupture in the pressurized radioactive water loop. The cost implications of repairing and replacing steam generators are enormous: replacement costs are $100 to $300 million, depending on the reactor size. Costs of forced shutdowns of a 500-MW power plant may exceed $500,000 per day. Costs of decommissioning a plant because of steam generator problems run into hundreds of millions of dollars.

Corrosion damage in steam generator tubes. The history of corrosion damage in steam generator tubes has been described in detail elsewhere.[11,13] The problems have mainly been related to Alloy 600 (a Ni, Cr, Fe alloy) and have contributed to seven steam generator tube ruptures, numerous forced reactor shutdowns, extensive repair and maintenance work, steam generator replacements, and also radiation exposure of plant personnel. A brief summary follows.

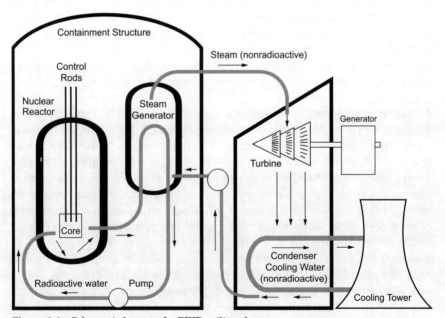

Figure 6.4 Schematic layout of a PWR utility plant.

In the early to mid-1970s, problems of wall thinning were identified. Tube degradation resulted in a need for steam generator replacement in several plants after only 10 to 13 years of operation, a small fraction of the design life and licensing period. Initially, water treatment practices were based on experience from fossil fuel plants. While the water chemistry was obviously closely controlled and monitored to minimize corrosion damage, a fundamental phenomenon tended to lead to more corrosive conditions than had been anticipated from the bulk water chemistry. The formation of steam on the external tube surfaces implied that boiling and drying out could occur in numerous crevices between the tubes and the support structures. Clearly, this could lead to a concentration of corrosive species and the formation of highly corrosive microenvironments. Furthermore, corrosion products tended to accumulate at the bottom of steam generators, again creating crevice corrosion conditions together with surface drying, and producing highly corrosive microenvironments. This effect proved to be very severe at the tube sheet, where the tubes enter the reactor. Not surprisingly, excessive local tube thinning was found to occur at such crevice sites.

The early corrosion problems were partly addressed by replacing sodium phosphate water treatment with an all-volatile treatment (AVT), whereby water was highly purified and ammonia additions were made. The addition of volatile chemicals essentially does not add to the total dissolved solids in the water, and hence concentration of species is ameliorated. However, with AVT, a new corrosion problem was manifested, namely, excessive corrosion of carbon steel support plates. The buildup of voluminous corrosion products at the tube–support plate interface led to forces high enough to dent the tubes. These problems were overcome by modifications to the water treatment programs.

A more recent corrosion problem identified is intergranular corrosion, again in the crevices between tubes and tube sheets, where deposits tend to accumulate. In the presence of stresses, either residual or operational, the problem can be classified as intergranular stress corrosion cracking (IGSCC). This form of cracking has been common in the U-bend region of tubes and also where tubes have been expanded at the top of tube sheets, where residual fabrication stresses prevail. Most recently, localized intergranular corrosion damage has been observed in older steam generators in the vicinity of support plates.

Inspection and maintenance for steam generator tubes. The scope and frequency of steam generator tube inspections depends on the operating history of the individual plant. In cases where operating records show extensive tube degradation, all the tubes are inspected at each shutdown. Modern inspection techniques are listed in Table 6.4, and Table 6.5 shows what

TABLE 6.4 Advanced Inspection Techniques for the Characterization of Equipment Integrity

Inspection method	Special advantage
X-ray	Interior of opaque parts
Gamma radiography	Heavy material sections
Magnetic particle	Discontinuities near the surface
Contact ultrasonic	Simple geometries—all materials
Visible and fluorescent liquid penetrant	Surface discontinuities
Eddy-current/electromagnetic	Discontinuities
Infrared inspection	Temperature differentials
Metallographic/replication	Grain growth–life expectancy
Acoustic emission	Active/growing defects

TABLE 6.5 Summary of Corrosion Mechanisms Detected by In-Service Inspection Methods in LWR, BWR, and PWR systems

Uniform corrosion
   Visual, leakage testing
Service corrosion
   Leakage testing
Microbiologically influenced corrosion
   Visual, leakage testing
Pitting corrosion
   Visual, leakage testing
   Eddy-current, optical scanner
   Sonic leak detector
Intergranular stress corrosion cracking
   Surface examination
   Visual, leakage testing
   Weld inspection, ultrasonic
   Moisture-sensitive tape
Transgranular stress corrosion cracking
   Visual, leakage testing
Differential aeration
   Visual, leakage testing
Galvanic corrosion
   Visual, leakage testing
Erosion corrosion
   Wall thickness, eddy-current
   Surface examination
   Ultrasonic
   Radiography
Fatigue/corrosion
   Surface examination
Thinning
   Eddy-current
Stress corrosion cracking
   Visual
   Surface examination

corrosion mechanisms have been detected with certain inspection techniques in the nuclear power generation industry.

If severe damage is detected, two basic choices are available: The tube can be either plugged (provided that the fraction of plugged tubes is only 10 to 20 percent) or covered with a metallic sleeve. Initial guidelines established by the Nuclear Regulatory Commission (NRC) called for such actions when the defect size reached 40 percent of wall thickness. Efforts are under way to refine this approach by considering allowable flaw sizes in relation to the mechanism of degradation, the material type, the tube dimensions, and the expected stress levels. New experimental initiatives in tube repair include laser welding of sleeves, direct laser melting of damaged tubes to cover damaged areas, and laser repairs using additional alloy wire.

Corrosion prevention measures have included even more stringent water treatment and removal of problematic corrosion product deposits. Chemical cleaning guidelines have been established for critical areas, and a robotic device for inspection and high-water-pressure cleaning of crevice geometries has been developed.

Replacement generators feature more corrosion-resistant materials, such as Alloy 690 tubes and stainless steel support plates, and new fabrication methods designed to minimize residual stresses in the tubes. The methodologies for removal and replacement of steam generators have also been improved, especially the design of the containment structures, which originally did not consider a need for replacement.

Aircraft maintenance. Despite the intense media coverage of air tragedies, flying remains the safest mode of transportation by far. The reliability and safety record of aircraft operators is indeed enviable by most industrial standards. This success is directly attributable to the fact that modern aircraft maintenance practices are far removed from reliance on retroactive corrective procedures. Other industries can learn several valuable lessons from current aircraft maintenance methodologies.

In the design of modern aircraft, ease of maintenance is a critical item. Manufacturers elicit feedback from operators on maintenance issues as part of the design process. As discussed earlier, RCM is fundamental to maintenance programs in modern aircraft operations. Importantly, RCM principles are already invoked at the design stage.

Preventive maintenance is particularly important on a short-term day-to-day basis. Strict scheduling and adherence to regulations are rigorously employed. Documentation is also an essential part of aircraft maintenance; essentially, all maintenance procedures have to be fully documented. The extent of preventive maintenance procedures increases with increasing flying time. A so-called D check represents a major maintenance overhaul, with major parts of the aircraft dismantled, inspected, and rebuilt. Hoffman has provided a fascinating insight into such inspection and maintenance procedures, including the issue of finding and repairing aircraft corrosion damage.[14] For example, on a Boeing 747, one-quarter of a D check involved 38,000 planned hours of labor, tens of thousands of unplanned hours, completion of a 5000-page checklist, and some 1600 nonroutine discrepancies. A North American airline performs these preventive maintenance procedures after every 6200 hours of flight. As aircraft get older, the time between maintenance checks is decreased.

The galley and washroom areas on aircraft are notorious for their high risk of corrosion, particularly because of the corrosive effects of beverage (e.g., coffee) and human excrement spills. An aircraft operator reported to one of the authors a reduction in corrosion maintenance tasks following the replacement of notoriously awkward stand-up washroom facilities in military transport planes!

Predictive maintenance efforts are directed at ensuring long-term aircraft reliability. The nature of these programs is evolving as a result of technology innovations and improvements. While several forms of diagnostic procedures are available for on-line condition assessment, such as advanced engine diagnostic telemetry, the aircraft industry still lags behind in this area, as discussed in a separate section.

There are several organizational and human factors that contribute to the success of aircraft maintenance programs. Technical maintenance information flows freely across organizations, even among business competitors. Procedures are documented, and a clear chain of responsibility exists, with special emphasis on good, open communication channels. Airline mechanics receive intense training and rigorous testing before certification. Ongoing training and skills upgrading is standard for the industry. Efforts are made to feed maintenance information back to aircraft design teams. Computer technology is used extensively by the larger airlines to track and manage aircraft maintenance activities. This is further supported by the provision of computerized technical drawings, parts lists, and maintenance to aircraft maintenance personnel. Figures 6.5 to 6.8 illustrate how advances in information technology have made the collection and presentation of historical data quite straightforward for maintenance personnel.[15]

Measuring reliability—downtime. One of the most visible effects of improvements in maintenance is a reduction in downtime, with higher equipment availability. In most industries, a reduction in downtime is vital to commercial success. The aircraft industry provides an excellent example of the direct major economic implications that arise from downtime caused by corrosion or other damage. The

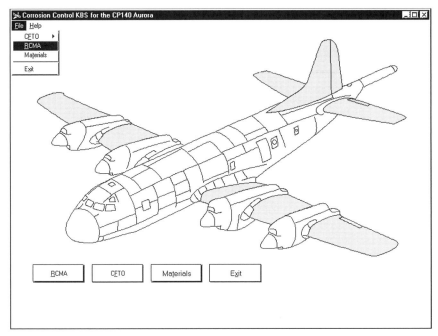

Figure 6.5 Main screen of a knowledge-based system (KBS), showing the areas of a patrol aircraft covered by an aircraft structural integrity program (ASIP).

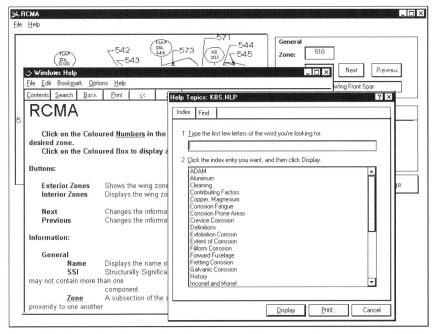

Figure 6.6 Example of integration of graphics and database information into a KBS for an ASIP.

404   Chapter Six

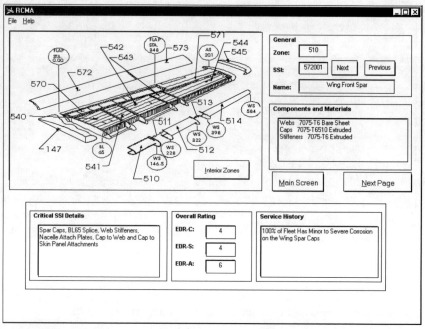

Figure 6.7 Example of context-sensitive help in a KBS for an ASIP.

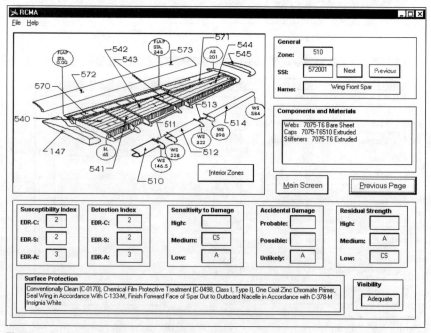

Figure 6.8 Display of some critical component information resident in a KBS for an ASIP.

obvious starting objective is a reduction in unscheduled downtime. The shift away from purely corrective maintenance is at the core of this task. To show progress in maintenance programs and maintain momentum in improvement initiatives, cost savings resulting from reduced unscheduled downtime and the prevention of component failures should be recorded and communicated effectively. Scheduled shutdowns are usually of significantly shorter duration than an unscheduled shutdown resulting from corrosion (or some other) failure. A sensible initial maintenance goal would therefore be a shift from unplanned, unscheduled downtime to planned, scheduled downtime.

In several industries, scheduled shutdowns are an integral part of preventive maintenance. Valid concerns about losing production during such scheduled interruptions can be raised, and there is an obvious incentive to increase the time between such scheduled shutdowns and to minimize their duration by implementing predictive maintenance. Following the minimization of unscheduled downtime, a reduction in scheduled downtime is the next essential challenge.[4]

To maximize the use of scheduled downtime, good planning of all maintenance work is essential. Critical path analysis can be used for such purposes. The ultimate goal is to run the equipment at its maximum sustainable rate, at the desired level of quality and with maximum availability. To initiate such predictive maintenance efforts, the following methodologies have been suggested for industrial plants:[4]

- Categorizing the importance of equipment and how the equipment in each category will be monitored
- Identifying database architectures, including point identification, analysis parameter sets, alarm limits, etc.
- Defining the frequency and quantity of data points collected for each unit
- Performing planning and walk-through inspections
- Defining data review and problem prioritization
- Identifying means of communicating the equipment's condition
- Determining methods of identifying repetitive problems and dealing with them
- Defining repair follow-up procedures

The development of these methodologies represents a starting point; they can be refined further as data and information are analyzed.

6.4 Monitoring and Managing Corrosion Damage

Corrosion monitoring refers to corrosion measurements performed under industrial operating conditions. In its simplest form, corrosion monitoring may be described as acquiring data on the rate of material degradation. However, such data are generally of limited use. They have to be converted to information for effective decision making in the management of corrosion control. This requirement has led to the expansion of corrosion monitoring into the domains of real-time data acquisition, process control, knowledge-based systems, smart structures, and condition-based maintenance. Additional terminology, such as "corrosion surveillance" and "integrated asset management," has been applied to these advanced forms of corrosion monitoring, which are included in this section.

An extensive range of corrosion monitoring techniques and systems for detecting, measuring, and predicting corrosion damage has evolved, particularly in the last two decades. Developments in monitoring techniques coupled with the development of user-friendly software have permitted new techniques that were once perceived as mere laboratory curiosities to be brought to the field. Noteworthy catalysts to the growth of the corrosion monitoring market have been the expansion of oil and gas production under extremely challenging operating conditions (such as the North Sea), cost pressures brought about by global competition, and the public demand for higher safety standards. A listing of corrosion monitoring applications in several important industrial sectors is presented in Table 6.6. In several sectors, such as oil and gas production, sophisticated corrosion monitoring systems have achieved successful track records and credibility, while in other sectors their application is only beginning.

6.4.1 The role of corrosion monitoring

Fundamentally, four strategies for dealing with corrosion are available to an organization. Corrosion can be addressed by

- Ignoring it until a failure occurs
- Inspection, repairs, and maintenance at scheduled intervals
- Using corrosion prevention systems (inhibitors, coatings, resistant materials, etc.)
- Applying corrosion control selectively, when and where it is actually needed

The first strategy represents corrective maintenance practices, whereby repairs and component replacement are initiated only after a

TABLE 6.6 Examples of Industrial Corrosion Monitoring Activities

Industrial sector	Corrosion monitoring applications
Oil and gas production	Seawater injection systems, crude piping systems, gas piping systems, produced water systems, offshore platforms
Refining	Distillation columns, overhead systems, heat exchangers, storage tanks
Power generation	Cooling-water heat exchangers, flue gas desulfurization systems, fossil fuel boilers, steam generator tubes (nuclear), air heaters, steam turbine systems, vaults, atmospheric corrosion, gasification systems, mothballing
Petrochemical	Gas pipelines, heat exchangers, cooling-water systems, atmospheric corrosion, storage tanks
Chemical processing	Chemical process streams, cooling-water circuits and heat exchangers, storage tanks, ducting, atmospheric corrosion
Mining	Mine shaft corrosivity, refrigeration plants, water piping, ore processing plants, slurry pipelines, tanks
Manufacturing	Cooling-water systems and heat exchangers, ducting
Aerospace	On-board and ground level, storage and mothballing
Shipping	Wastewater tanks, shipboard exposure programs
Construction	Reinforced concrete structures, pretensioned concrete structures, steel bridges, hot and cold domestic water systems
Gas and water distribution	Internal and external corrosion of piping systems (including stray current effects)
Paper and pulp	Cooling water, process liquors, clarifiers
Agriculture	Crop spraying systems, fencing systems

failure has occurred. In this reactive philosophy, corrosion monitoring is completely ignored. Obviously this practice is unsuitable for safety-critical systems, and in general it is inefficient in terms of maintenance cost considerations, especially in extending the life of aging engineering systems.

The second strategy is one of preventive maintenance. The inspection and maintenance intervals and methodologies are designed to prevent corrosion failures while achieving "reasonable" system usage. Corrosion monitoring can assist in optimizing these maintenance and inspection schedules. In the absence of information from a corrosion monitoring program, such schedules may be set too conservatively, with excessive downtime and associated cost penalties. Alternatively, if inspections are too infrequent, the corrosion risk is excessive, with associated safety hazards and cost penalties. Furthermore, without input from corrosion monitoring information, preventive inspection and maintenance intervals will be of the routine variety, without accounting for the time dependence of critical corrosion variables. In the oil and gas industry, for example, the corrosivity at a wellhead can fluctuate significantly between being benign and being highly corrosive over the lifetime of the production system. In oil-refining plants, the corrosivity can vary with time, depending on the grade (hydrogen sulfide content) of crude that is processed.

The application of corrosion prevention systems is obviously crucial in most corrosion control programs. However, without corrosion monitoring information, the application of these systems may be excessive and overly costly. For example, a particular inhibitor dosage level on a pipeline may successfully combat corrosion damage, but real-time corrosion monitoring may reveal that a lower dosage would actually suffice. Ideally, the inhibitor feed rate would be continuously adjusted based on real-time corrosion monitoring information. Performance evaluation of in-service materials by corrosion monitoring is highly relevant, as laboratory data may not be applicable to actual operating conditions.

In an idealized corrosion control program, inspection and maintenance would be applied only where and when they are actually needed, as reflected by the "maintenance on demand" (MOD) concept. In principle, the information obtained from corrosion monitoring systems can be of great assistance in reaching this goal. Conceptually, the application of a monitoring system essentially creates a smart structure, which ideally reveals when and where corrective action is required.

The importance of corrosion monitoring in industrial plants and in other engineering systems should be apparent from the above. However, in practice it can be difficult for a corrosion engineer to get management's commitment to investing funds for such initiatives. Significant benefits that can be obtained from such investments include

- Improved safety
- Reduced downtime
- Early warning before costly serious damage sets in
- Reduced maintenance costs
- Reduced pollution and contamination risks
- Longer intervals between scheduled maintenance
- Reduced operating costs
- Life extension

6.4.2 Elements of corrosion monitoring systems

Corrosion monitoring systems vary significantly in complexity, from simple coupon exposures or hand-held data loggers to fully integrated plant process surveillance units with remote data access and data management capabilities. Experience has shown that the potential cost savings resulting from the implementation of corrosion monitoring programs generally increase with the sophistication level (and cost) of the monitoring system. However, even with simple monitoring devices, substantial financial benefits are achievable.

Corrosion sensors (probes) are an essential element of all corrosion monitoring systems. The nature of the sensors depends on the specific techniques used for monitoring (refer to Sec. 6.4.4, Corrosion Monitoring Techniques), but often a corrosion sensor can be viewed as an instrumented coupon. A single high-pressure access fitting for insertion of a retrievable corrosion probe (Fig. 6.9) can accommodate most types of retrievable probes (Fig. 6.10). With specialized tools (and brave specialist operating crews!), sensor insertion and withdrawal under pressurized operating conditions can be possible (Fig. 6.11).

The signal emanating from a corrosion sensor usually has to be processed in some way. Examples of signal processing include filtering, averaging, and unit conversions. Furthermore, in some corrosion sensing techniques, the sensor surface has to be perturbed by an input signal to generate a corrosion signal output. In older systems, electronic sensor leads were usually employed for these purposes and to relay the sensor signals to a signal-processing unit. Advances in microelectronics are facilitating sensor signal conditioning and processing by microchips, which can essentially be considered to be integral to the sensor units. The development of reinforcing steel and aircraft corrosion sensors on these principles has been described.[16,17] Wireless data communication with such sensing units is also a product of the microelectronic revolution.

Irrespective of the sensor details, a data acquisition system is required for on-line and real-time corrosion monitoring. For several plants, the data acquisition system is housed in mobile laboratories, which can be made intrinsically safe. Real-time corrosion measurements are highly sensitive measurements, with a signal response taking place essentially instantaneously as the corrosion rate changes. Numerous real-time corrosion monitoring programs in diverse branches of industry have revealed that the severity of corrosion damage is rarely (if ever) uniform with time. Rather, serious corrosion damage is usually sustained in time frames in which operational parameters have deviated "abnormally." These undesirable operating windows can be identified only with the real-time monitoring approach.

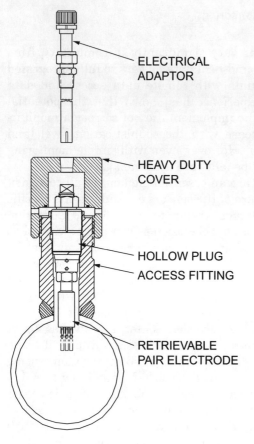

Figure 6.9 High-pressure access fitting for insertion of a retrievable corrosion probe. (*Courtesy of Metal Samples.*)

A computer system often performs a combined role as a data acquisition, data processing, and information management system. In data processing, a process is initiated to transform corrosion monitoring data (low intrinsic value) into information (higher intrinsic value). Complementary data from other relevant sources, such as process parameter logging and inspection reports, can be acquired along with the data from corrosion sensors, for use as input to the management information system. In such a system, more extensive database management and data presentation applications are employed to transform the basic corrosion data into management information for decision-making purposes (Fig. 6.1).

6.4.3 Essential considerations for launching a corrosion monitoring program

One of the most important decisions that have to be made is the selection of the monitoring points or sensor locations. As only a finite number

Corrosion Maintenance through Inspection and Monitoring   411

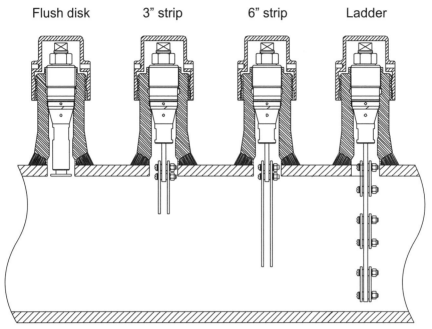

Figure 6.10 A single high-pressure access fitting can be fitted with different types of retrievable corrosion probes. (*Courtesy of Metal Samples.*)

Figure 6.11 Retrieval tool for removing corrosion probes under pressure. (*Courtesy of Metal Samples.*)

of points can be considered, it is usually desirable to monitor the worst-case conditions, the points where corrosion damage is expected to be most severe. Often, such locations can be identified by reasoning from basic corrosion principles, analysis of in-service failure records, and consultation with operational personnel. For example, the most corrosive conditions in water tanks are usually found at the water/air interface. In order to monitor corrosion under these conditions, corrosion sensors could be attached to a floating platform so that the location of the sensor would change as the water level changes.

Dean has presented an example of identifying critical sensor locations in a distillation column.[18] The feed point, overhead product receiver, and bottom product line represent locations of temperature extremes and also points where products with different degrees of volatility concentrate. In many cases, however, the highest corrosivity is encountered at an intermediate height in the column, where the most corrosive species concentrate. Initially, therefore, several monitoring points would be required in such a column, as shown in Fig. 6.12. As monitoring progresses and data from these points become available, the number of monitoring points could be narrowed down.

In practice, the choice of monitoring points is also dictated by the existence of suitable access points, especially in pressurized systems. It is usually preferable to use existing access points, such as flanges, for sensor installations. If it is difficult to install a suitable sensor in a given location, additional bypass lines with customized sensors and access fittings may be a practical alternative. One advantage of a bypass is that it provides the opportunity to manipulate local conditions to highly corrosive regimes in a controlled manner, without affecting the actual operating plant.

It is imperative that the corrosion sensors be representative of the actual component being monitored. If this requirement is not met, all subsequent signal processing and data analysis will be negatively affected and the value of the information will be greatly diminished or even rendered worthless. For example, if turbulence is induced locally around a protruding corrosion sensor mounted in a pipeline, the sensor will in all likelihood give a very poor indication of the risk of localized corrosion damage to the pipeline wall. A flush-mounted sensor should be used instead (Fig. 6.13).

The surface condition of the sensor elements is also very important. Surface roughness, residual stresses, corrosion products, surface deposits, preexisting corrosion damage, and temperature can all have an important influence on corrosion damage and need to be taken into account in making representative probes. Considering these factors, it can be desirable to manufacture corrosion sensors from precorroded material that has experienced actual operational conditions. Corrosion

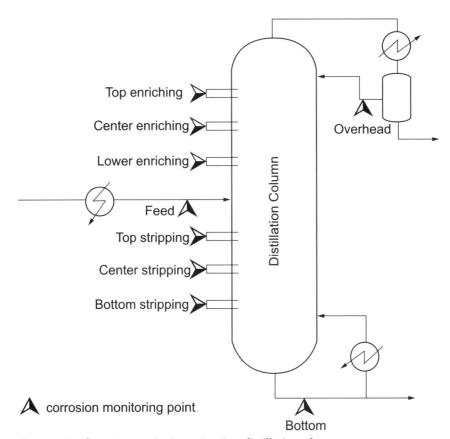

Figure 6.12 Corrosion monitoring points in a distillation column.

sensors may also be heated and cooled, using special devices, so that their surface conditions reflect certain plant operating domains. Sensor designs such as spool pieces in pipes and heat-exchanger tubes, flanged sections of candidate materials, or test paddles bolted to agitators also represent efforts to make the sensors' environment represent actual operational conditions.

Numerous corrosion monitoring techniques and associated sensors are available. All of these techniques have certain advantages and disadvantages, which are discussed in detail in Sec. 6.4.4. There are many pitfalls in selecting suitable techniques, and the advice of a corrosion monitoring expert is usually required. An algorithm, described by Cooper,[19] for evaluating the suitability of two commonly utilized techniques, LPR (one of the electrochemical techniques) and ER (electrical resistance), is shown in Fig. 6.14.

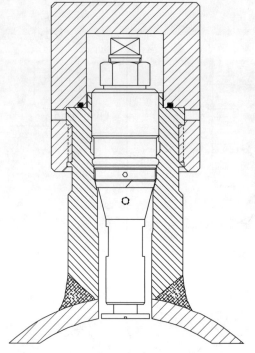

Figure 6.13 Flush-mounted corrosion sensor in an access fitting. (*Courtesy of Metal Samples.*)

In general, it can be said that no individual technique alone is suitable for monitoring corrosion under complex industrial conditions. Therefore, a multitechnique approach is advocated. In many cases, this approach does not require a higher number of sensors, but rather only an increased number of sensor elements for a given probe and access fitting. Considering the overall costs of supporting a corrosion monitoring program such as the one shown in Fig. 6.1, the additional costs associated with a multitechnique philosophy are usually insignificant. Furthermore, greater confidence can be placed in the sensor data if several techniques provide the same response.

Another important consideration is that, irrespective of the technique, instrumented sensors usually provide semiquantitative corrosion damage information at best. It is thus sensible to correlate monitoring data from these sensors with long-term coupon exposure programs and actual plant damage. Unfortunately, nonspecialists may put too much faith in the numerical corrosion rate displayed by a commercial corrosion monitoring device. A suitable example is the LPR technique used in many commercial monitoring systems to derive a certain corrosion rate, commonly displayed as mm/year or milli-inches/year (mpy). Such systems are used extensively in industry for monitoring the effectiveness of

Corrosion Maintenance through Inspection and Monitoring

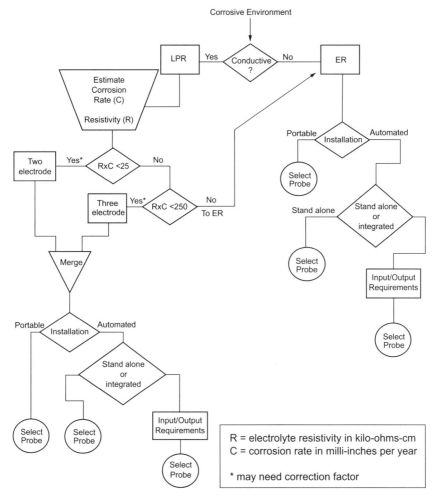

Figure 6.14 Algorithm for suitability of ER and LPR corrosion monitoring techniques. (*Adapted from Cooper.*[19])

water treatment additives and various other applications. From fundamental theoretical considerations, the derived LPR corrosion rate is subject to the following assumptions, which, strictly speaking, rarely apply under actual operating conditions:

- There is only one simple anodic reaction.
- There is only one simple cathodic reaction.
- The anodic and cathodic Tafel constants are known and invariant with time.

- The corrosion reactions proceed by a simple charge transfer mechanism under activation control, which essentially implies that the corroding surface is clean, without corrosion product buildup, scale deposits, or solids settled out of solution.
- Corrosion proceeds in a uniform manner (whereas the vast majority of industrial corrosion problems are related to localized attack).
- The solution resistance is negligible. (Some instruments make a solution resistance compensation, but this is not necessarily accurate.)
- The corrosion potential has reached a steady-state value.

Following the selection of the sensors and monitoring techniques, the type and location of the ancillary monitoring hardware need to be considered. Many industrial plants have intrinsic safety requirements that impose important restrictions on corrosion monitoring systems. To ensure flexibility in large plants, some organizations have adopted the strategy of using a "mobile" corrosion monitoring laboratory that meets their safety regulations. Such a laboratory housing the corrosion monitoring instrumentation can be conveniently moved to different locations as required, to overcome the problems associated with excessive lengths of sensor leads. Furthermore, this arrangement provides a protective environment for measuring and data storage hardware, which could otherwise be damaged in corrosive atmospheres. Mobile laboratories have also been used for corrosion measurements on treated-water circuits. In this case, the corrosion sensors can be "lab-based" along with the instrumentation, through the use of a water bypass flowing through the mobile laboratory.

6.4.4 Corrosion monitoring techniques

To the uninitiated engineer, the plethora of available corrosion monitoring techniques can be overwhelming in the absence of a categorization scheme. The first classification can be to separate *direct* from *indirect* techniques. Direct techniques measure parameters that are directly associated with corrosion processes. Indirect techniques measure parameters that are only indirectly related to corrosion damage. For example, measurements of potentials and current flow directly associated with corrosion reactions in the linear polarization resistance technique represent a direct corrosion rate measurement. The measurement of the corrosion potential only is an indirect method, as there is at best an indirect relationship between this potential and the severity of corrosion damage.

A second categorization scheme is into *intrusive* and *nonintrusive* forms. Intrusive techniques require direct access to the corrosive environment through a structure (pipe wall, aircraft skin, etc.). Sensors and test specimens typify this approach. Nonintrusive methods require no additional hardware to perform a corrosion measurement. A further distinction is possible between *on-line* and *off-line* techniques. On-line techniques are those with continuous monitoring capabilities during operation, whereas off-line methods require periodic sampling and separate analysis. The basic principles for selecting important corrosion monitoring techniques are described below, and the advantages and limitations of these techniques are listed in Table 6.7.

Direct techniques

Corrosion coupons (intrusive). In what is perhaps the simplest form of corrosion monitoring, small specimens are exposed to an environment for a specific period of time and subsequently removed for weight loss measurement and more detailed examination. Even though the principle is very simple, there are numerous potential pitfalls, which can be avoided by following the recommendations of a comprehensive ASTM guide (ASTM G4 standard).

Electrical resistance (intrusive). The underlying principle of the widely used electrical resistance (ER) probes is the simple concept that there is an increase in electrical resistance as the cross-sectional area of a sensing element is reduced by corrosion damage. Since temperature has a strong influence on electrical resistance, ER sensors usually measure the resistance of a corroding sensor element relative to that of an identical shielded element. Commercial sensor elements are in the form of plates, tubes, or wires (Fig. 6.15). Reducing the thickness of the sensor elements can increase the sensitivity of these sensors. However, improved sensitivity involves a tradeoff with reduced sensor lifetime. ER probe manufacturers provide guidelines showing this tradeoff for different sensor geometries (Fig. 6.16). The useful life of ER probes other than wire sensors is usually up to the point where their original thickness has been halved. For ER wire sensors the lifetime is lower, corresponding to loss of a quarter of the original thickness. It is obvious that erroneous results will be obtained if conductive corrosion products or surface deposits form on the sensing element. Iron sulfide formed in sour oil/gas systems or in microbial corrosion and carbonaceous deposits in atmospheric corrosion are relevant examples.

Inductive resistance probes (intrusive). This recently developed technology is a derivative of ER corrosion sensing.[20] The reduction in the thickness of a sensing element is measured by changes in the inductive resistance

TABLE 6.7 Advantages and Disadvantages of Corrosion Monitoring Techniques

Advantages	Disadvantages
Corrosion Coupons	
Coupon exposures are simple and usually of low cost. Many forms of corrosion can be monitored if detailed analysis is performed subsequent to the exposure, but erosion and heat-transfer effects are not easily simulated with coupons.	Long exposure periods may be required to obtain meaningful and measurable weight loss data. Coupons have to be removed from plant or equipment for analysis and corrosion rate determination. (*Note:* Sample removal and cleaning affects the corrosion rate if the coupons are subsequently reexposed.) These devices provide cumulative retrospective information only. For example, if a stress corrosion crack is found in a coupon after a 12-month exposure period, it is not possible to say when the crack initiated and what specific conditions led to the initiation or propagation of this crack. Importantly, the crack growth rate also cannot be established with confidence, as the time of its initiation is unknown. The cleaning, weighing, and microscopic examination of coupons is usually labor-intensive.
Electrical Resistance (ER)	
ER results are easily interpreted, and the technology is well supported by several commercial suppliers. Continuous corrosion monitoring and correlation with operational parameters are possible, provided sufficiently sensitive sensor elements are selected. ER probes are more convenient than coupons in the sense that results can be obtained without retrieval and weight loss measurements. A combined thickness loss due to corrosion and erosion can be measured.	ER probes are more convenient than coupons in the sense that results can be obtained without retrieval and weight loss measurements. A combined thickness loss due to corrosion and erosion can be measured. ER probes are essentially suitable for monitoring only uniform corrosion damage, whereas localized corrosion is usually of more concern to industry. Generally, the sensitivity of ER probes is insufficient to qualify for real-time corrosion measurements, with transients of short duration going undetected. The probes are unsuitable in the presence of conductive corrosion products or deposits.
Inductive Resistance	
The measurement principle of detecting a thickness change in the sensor element is relatively simple, and sensitivity is improved over that of ER probes. The sensor signals are affected by temperature changes to a lesser degree than electrical resistivity signals are.	The technique has been introduced only recently. In its present commercial form, it would appear to be largely applicable to uniform corrosion measurements only.

Linear Polarization Resistance (LPR)	
Interpretation of the measurements is straightforward. Continuous on-line monitoring is possible, as the measurements take only take a few minutes. The high sensitivity of this technique facilitates real-time monitoring in appropriate environments.	The technique is based on uniform corrosion principles only. An environment with relatively high ionic conductivity is required for accurate measurements. Unstable corrosion potentials will produce erroneous results. Even though the applied sensor perturbation is small, repeated application over long times can lead to "artificial" surface damage. In long-term exposures, surface colors visibly different from those on freely corroding sensor elements have been noted on LPR sensors. Idealized theoretical polarization conditions are assumed, which is not necessarily the case in practice (see Sec. 7.3). The short-circuiting of electrodes by conductive species will preclude valid measurements.

Electrochemical Impedance Spectroscopy (EIS)	
This technique is more suited to low-conductivity environments than DC polarization and can also provide information on the state of organic coatings. Recently, EIS-based systems for practical coating integrity assessment have been introduced. Detailed characterization of the corroding surface is theoretically possible.	The instrumentation and interpretation required to obtain full results are typically complex. Consequently, full-frequency spectrum analysis is very rarely applied in the field. Limited-frequency units, which are comparable to LPR devices, have been developed specifically for field use. The corrosion potential has to be very stable to permit meaningful measurements at low frequencies. The technique is essentially limited to uniform corrosion damage only, although it may be possible to detect pitting damage in certain systems. The applied potential perturbation may influence the condition of the corroding sensor element, especially in repeated application over long time periods.

Harmonic Analysis	
Theoretically, a rapid determination of all important kinetic parameters may be possible.	At present, the reliability and application of this technique remain essentially unproven. The instrumentation and theoretical basis of the technique are complex and require specialized electrochemical knowledge.

TABLE 6.7 Advantages and Disadvantages of Corrosion Monitoring Techniques (*Continued*)

Advantages	Disadvantages
Electrochemical Noise (EN)	
The technique is highly sensitive and performs well under conditions of limited conductivity, such as thin-film corrosion. It is one of the very few techniques with the ability to detect localized corrosion damage, such as pitting damage to otherwise passive surfaces and certain submodes of stress corrosion cracking.	Although the number of applications has grown considerably, the technique remains somewhat controversial. The data analysis requirements are complex, and substantial experience is required for interpreting "raw" noise records.
Zero Resistance Ammetry (ZRA)	
These measurements represent a simple method of monitoring galvanic corrosion and the effect of treatments to prevent it.	The measured currents may not represent actual galvanic corrosion rates, as this form of corrosion is highly dependent on the anode:cathode area ratio. An increase in current readings is not always directly associated with an actual increase in corrosion rates.
Potentiodynamic Polarization	
Kinetic information and an overall picture of the material's corrosion behavior can be obtained relatively quickly (compared to, say, coupon exposures).	These techniques are usually limited to laboratory studies, as specialized skills are required to interpret the data. The applied polarization levels may change the sensor surface irreversibly, especially if pitting damage is induced in the anodic cycle. These measurements are generally applicable only to fully immersed probes in conducting solutions.

Thin-Layer Activation (TLA)

The measurement principle of the technique is relatively simple, and the direct measurement on actual components is a desirable feature. Selected small areas can be irradiated, to monitor the degradation of a particular weld zone, for example. The technique can be applied to study erosion effects.	The instrumentation used to irradiate the surfaces can accommodate only small components. The technological infrastructure required to perform the surface activation is substantial and is not readily accessible in all countries. From fundamental principles, the measurements are meaningful only if the radioactive isotopes are removed from the surface undergoing corrosion damage. If they remain in the corrosion products building up on surfaces, the thickness reduction will not be detected. The sensitivity of the technique is generally relatively low. The technique is not yet widely utilized.

Electrical Field Signature Method (EFSM)

Corrosion damage is monitored over large sections of actual structures. Once the instruments are installed, corrosion monitoring can be performed over many years with minimal maintenance.	The technique does not distinguish between internal and external flaws. The interpretation of the voltage signals for localized corrosion damage is not straightforward. The resolution for studying corrosion damage over small specific areas is limited. In general, the technique is not regarded as highly sensitive.

Acoustic Emission (AE)

In principle, this technique is applicable to a wide range of materials, even nonconducting ones. It can be applied to vessels without the need for draining them. Monitoring can be performed over relatively large areas of structures, rather than the specific measuring points of other techniques.	This technique is limited to defects that are actively growing at the time of measurement. Defects that are present but not growing will not be detected. The technique does not provide a quantitative measure of defect size and requires a high level of specialized skill for application and for interpretation of results.

Corrosion Potential

The measurement technique and required instrumentation are relatively simple.	While the technique may indicate changes in corrosion behavior over time, it does not provide any indication of corrosion rates.

TABLE 6.7 Advantages and Disadvantages of Corrosion Monitoring Techniques (*Continued*)

Advantages	Disadvantages
Hydrogen Probes	
The technique represents a useful monitoring tool when the measurements are correlated with actual observed damage under specific operational conditions. The attachment of the probes on external surfaces is convenient, as the probe locations can be easily changed (in principle).	The measurements are restricted to a limited area. The technique may not reflect the corrosion rate, as it detects only the fraction of hydrogen passing into the metallic substrate. The ratio of hydrogen taken up by the substrate to that taken up by the environment may vary with time. The technique is obviously restricted to systems where hydrogen is produced in the cathodic reaction. Generic guidelines relating the measured hydrogen flux and actual damage (thickness loss, cracking, blistering) have not been established.
Chemical Analyses	
In specific, well-characterized systems, it may be possible to perform corrosion monitoring cost-effectively with such techniques. The techniques provide useful supplementary information to direct corrosion measurement techniques, for identifying causes of corrosion damage and solutions to corrosion problems.	No direct information on corrosion rates is obtained; correlation with actual damage or directly measured corrosion data is required. Measuring the chemistry of the bulk environment does not provide information on microenvironments established on corroding surfaces, with the latter often governing the actual rate of damage. Some of these techniques require laboratory measurements, and the results may thus not be immediately available. Fouling of on-line sensor surfaces and interference effects from other chemical species can lead to inaccurate results.

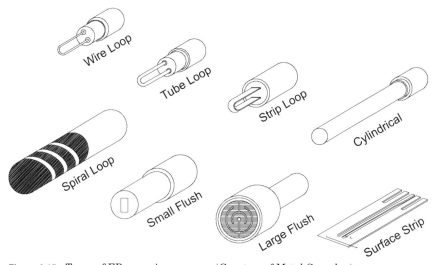

Figure 6.15 Types of ER corrosion sensors. (*Courtesy of Metal Samples.*)

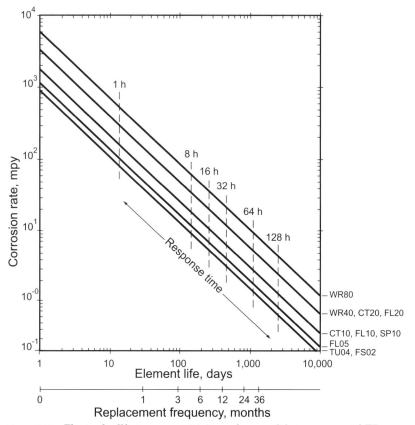

Figure 6.16 The tradeoff between sensitivity and sensor life in commercial ER sensors. (*Note:* 1 mpy (milli-inch/year)=0.0254 mm/year.) (*Courtesy of Metal Samples.*)

of a coil embedded in the sensor. Sensing elements with high magnetic permeability intensify the magnetic field around the coil; therefore, thickness changes affect the inductive resistance of the coil. Sensitivity (decrease in response time) has been claimed to be several orders of magnitude higher than with comparable ER probes.

Linear polarization resistance (intrusive). The linear polarization resistance (LPR) technique is an electrochemical method that uses either three or two sensor electrodes. In this technique, a small potential perturbation (typically of the order of 20 mV) is applied to the sensor electrode of interest, and the resulting direct current is measured. The ratio of the potential to current perturbations, known as the polarization resistance, is inversely proportional to the uniform corrosion rate. The accuracy of the technique can be improved by measuring the solution resistance independently and subtracting it from the apparent polarization resistance value. The technique is well known (its theoretical basis had already been developed in the 1950s), and it is widely used under full immersion aqueous conditions.

Electrochemical impedance spectroscopy (intrusive). This electrochemical technique is related to LPR in the sense that it also relies on the polarization of sensor elements for obtaining corrosion information. In electrical impedance spectroscopy (EIS), the sensor perturbation and phase-shifted response are of the ac type. The application of an alternating potential will produce an alternating-current response which is out of phase with the potential. The frequency range of the applied perturbation signal is typically 0.1 Hz to 100 kHz. To characterize the corrosion behavior in detail, measurements throughout the entire frequency range are required. Some simplified commercial systems rely on measurements at two frequencies only, to extract kinetic information. Full-frequency scans provide phase shift information that can be utilized in equivalent-circuit models. With such models, relatively complex corrosion phenomena can be described and kinetic information derived. Commonly used equivalent circuits for modeling corrosion processes are presented in Chap. 7, Acceleration and Amplification of Corrosion Damage.

Harmonic analysis (intrusive). This technique is related to EIS in that an alternating potential perturbation is applied to one sensor element in a three-element probe, with a resultant current response. Not only the primary frequency but higher-order harmonic oscillations are analyzed in this technique. Theory has been formulated whereby all kinetic parameters (including the Tafel slopes) can be calculated explicitly. No other technique offers this facility. At present, the technique remains largely unproven and rooted in the laboratory domain.

Electrochemical noise (intrusive). Contrary to common perceptions, the underlying measurement principles of this electrochemical technique are extremely simple. The technique is not related to acoustic noise in any way; rather, fluctuations in potential and current between freely corroding electrodes are measured. Because of the small scale of these fluctuations of interest (in many cases <1 µV and <1 nA), sensitive instrumentation is required. Many high-precision digital multimeters facilitate these measurements. A measurement frequency of 1 Hz usually suffices. For simultaneous measurement of electrochemical potential and current noise, a three-electrode sensor is required. The current noise is measured between two of these sensor elements. The potential noise is measured between the third element and the two coupled (for current measurements) elements. For industrial corrosion monitoring purposes, all three sensor elements are usually constructed from the same material.

Higher levels of electrochemical corrosion activity are generally associated with higher noise levels. Certain electrochemical phenomena, such as the breakdown of passivity during pit initiation, have distinct noise "signatures" which can be exploited for corrosion monitoring purposes. Pit initiation and growth can be detected with electrochemical noise measurements long before it becomes evident by visual examination.

While the measurement of electrochemical noise is straightforward, the data analysis can be complex. Such analysis is usually directed at distinguishing among different forms of corrosion, quantifying the noise signals, and processing the vast number of accumulated data points into a summarized format. Data processing approaches have included frequency spectral analysis [fast Fourier transforms (FFT) and the maximum entropy method (MEM)] and chaotic dynamics.

The number of industrial applications of electrochemical noise monitoring has grown significantly in recent years, yet skepticism about the universal usefulness of this technique is still encountered. It has been shown to be well suited to monitoring under thin moisture film conditions, such as those encountered in flue gas condensation and atmospheric corrosion. Present concerns and controversy are mainly related to the validity of corrosion rates derived from noise records.

Zero-resistance ammetry (intrusive). In zero-resistance ammetry (ZRA), galvanic currents are usually measured between dissimilar materials. The dissimilarities between the sensor elements may be related to different compositions, heat treatments, stress levels, or surface conditions. The technique may also be applied to nominally identical electrodes, to indicate changes in the corrosive environment and serve as a broad indicator of changes in corrosion rate.

Potentiodynamic polarization (intrusive). This method is best known for its fundamental role in electrochemistry in the measurement of Evans diagrams. A three-electrode corrosion probe is used to polarize the electrode of interest. The current response is measured as the potential is shifted away from the free corrosion potential. The basic difference from the LPR technique is that the applied potentials for polarization are normally stepped up to levels of several hundred millivolts. These polarization levels facilitate the determination of kinetic parameters, such as the general corrosion rate and the Tafel constants. The formation of passive films and the onset of pitting corrosion can also be identified at characteristic potentials, which can assist in assessing the overall corrosion risk.

Thin-layer activation and gamma radiography (intrusive or nonintrusive). In this technique, developed from the field of nuclear science, a small section of material is exposed to a high-energy beam of charged particles, producing a radioactive surface layer. For example, a proton beam may be used to produce the radioactive isotope ^{56}Co within a steel surface. This isotope decays to ^{56}Fe, with the emission of gamma radiation. The concentration of radioactive species is sufficiently low that the metallurgical properties of the monitored component are essentially unchanged. The radioactive effects utilized are at very low levels and should not be compared to those of conventional radiography. The change in gamma radiation emitted from the surface layer is measured with a separate detector to study the rate of material removed from the surface. The radioactive surfaces can be produced directly on components (nonintrusive) or on separate sensors (intrusive).

Electrical field signature method (nonintrusive). The original development of this technique was largely directed at oil and gas production. The technique measures corrosion damage over several meters of an actual structure, clearly distinguishing it from other smaller sensor systems. An induced current is fed into the monitored section of interest, and the resulting voltage distribution is measured to detect corrosion damage. An array of pins is attached strategically over the structure for measuring purposes. Increased pin spacing implies lower resolution for localized corrosion. Typical applications involve attaching pins to the external surface of a pipeline to monitor corrosion damage to the inside of the pipe walls.

Acoustic emission (nonintrusive). This technique involves measuring acoustic sound waves that are emitted during the growth of microscopic defects, such as stress corrosion cracks. The sensors can thus essentially be viewed as microphones which are strategically positioned on structures. The sound waves are generated from mechanical stresses during pressure or temperature changes. Background noise effects have to be taken into consideration and can be particularly troublesome in on-line measurements.

Indirect techniques. A plethora of indirect corrosion monitoring techniques for different modes and submodes of corrosion damage is available. A multidisciplinary science and engineering team approach is often required for implementation. For example, the indirect monitoring of microbiologically induced corrosion in water systems typically requires expertise and laboratory infrastructure from the fields of corrosion engineering, microbiology, and chemistry.

Corrosion potential (intrusive). Measurement of the corrosion potential is a relatively simple concept, and the underlying principle is widely used in industry for monitoring the corrosion of reinforcing steel in concrete and structures such as buried pipelines under cathodic protection. Monitoring of anodic protection systems is a further application area. Changes in corrosion potential can also give an indication of active/passive behavior in stainless steel. Furthermore, when viewed in the context of Pourbaix diagrams, the corrosion potential can give a fundamental indication of the thermodynamic corrosion risk. The corrosion potential is measured relative to a reference electrode, which is characterized by a stable half-cell potential. The electrochemical details of important reference electrodes are presented in Sec. D-2, Chemical Thermodynamics. Corrosion-potential measurements are usually classified as an intrusive indirect method. Either a reference electrode (and possibly a separate sensor of the material to be monitored) has to be introduced into the corrosive medium for these measurements, or an electrical connection to a structure in conjunction with an external reference electrode has to be established.

Hydrogen monitoring (nonintrusive). The generation of atomic hydrogen as part of the cathodic half-cell reaction in acidic environments can be used for both intrusive or nonintrusive forms of corrosion monitoring. In the latter, hydrogen monitoring sensors are often attached to the outside walls of vessels and piping. It is the diffusion of atomic hydrogen into the metallic substrate that is of most concern, as this can lead to problems such as hydrogen-induced cracking. This "uptake" of hydrogen occurs when the recombination of hydrogen atoms and their subsequent release into the environment as molecular hydrogen are impeded. Hydrogen monitoring is highly applicable to the oil refining and petrochemical industries, where there are hydrocarbon process streams. The presence of hydrogen sulfide in these industries promotes the uptake of hydrogen into plant items.

Hydrogen monitoring probes can be based on any of the following three principles:

- Pressure increase with time in a controlled chamber, as hydrogen passes through the material into the probe chamber
- An electrochemical current resulting from the oxidation of hydrogen under an applied potential
- Current flow in an external circuit, based on a fuel cell principle whereby hydrogen entering the miniature fuel cell causes the current flow

Chemical analyses. Different types of chemical analyses can provide valuable information in corrosion monitoring programs. Measurements of pH, conductivity, dissolved oxygen, metallic and other ion concentrations, water alkalinity, concentration of suspended solids, inhibitor concentrations, and scaling indices all fall within this domain. Several of these measurements can be made on-line using appropriate sensors.

6.4.5 From corrosion monitoring to corrosion management

Several factors have been identified that are essential if corrosion monitoring programs are to make a useful contribution toward the management of corrosion control, that is, to have a real impact on safety and profitability. If these conditions do not exist, data collected with corrosion monitoring systems will not be used for more informed decision making and are likely to gather dust in a filing cabinet. Britton and Tofield identified several technical and personnel requirements for success; additional factors are pertinent in the age of information technology:[21]

- "Correct" location of monitoring points
- Selection of suitable techniques
- A commitment to meaningful corrosion measurements with the highest possible "accuracy"
- Reliability and safety of monitoring hardware
- Establishment of a confidence factor through experience and correlation with other data/information, such as NDE, failure analysis, and plant operational parameters
- Appreciation of the qualitative nature of most monitoring data
- Designing for corrosion monitoring
- Data analysis, interpretation, and presentation in a clear, unequivocal format (computing systems and database software can be highly effective for these tasks)

- The transformation of data into management information
- Availability of specialist corrosion personnel and provision of supporting infrastructure (laboratories, computing systems, test rigs, etc.)
- The patience to learn by experience
- Adequate budgets to support monitoring programs and a positive management attitude

Milliams and Van Gelder have presented a model for corrosion management, illustrating information flow and the multidisciplinary nature of this process.[22] Their basic model, shown in Fig. 6.17, serves as a useful basis for discussing corrosion monitoring in corrosion management. Corrosion control and management starts at the design stage. In this phase, it is important that consideration be given to feedback from existing corrosion monitoring programs and to design requirements for future corrosion monitoring activities. The feedback of information from previous monitoring campaigns to designers is obviously important for design improvements and essentially closes the loop of technical information flow.

Design engineers should be considered to be among the most important users of corrosion information, as numerous corrosion problems can be traced to design inadequacies. Design engineers also have to take future corrosion monitoring requirements into consideration. Monitoring points and access fittings should be defined and incorporated into the design. Corrosion management manuals prepared during the design phase should document anticipated forms of corrosion and means of controlling them (including corrosion monitoring programs) in a formal manner.

Corrosion monitoring obviously assumes an important role in the operations phase. The performance of any operational engineering system has to be characterized and evaluated in order to manage the system so that it functions at an optimal level. Corrosion monitoring can, for example, be used in operations to optimize the dosage of corrosion inhibitors and other additives. Importantly, in an industrial plant, it can be employed as a form of process control, to ensure that operating conditions or practices do not lead to excessive corrosion damage. Arguably, corrosion monitoring applications have historically been generally lacking in this respect. Two fundamental requirements have to be met to achieve this form of process control:

- A link between the measured corrosion data and operational parameters
- The transformation of corrosion, operational, and production data into integrated management information for effective decision making

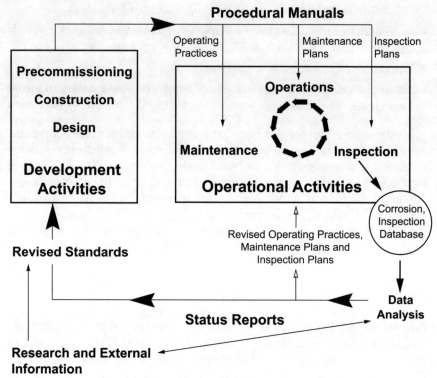

Figure 6.17 Information flow in corrosion management. (*Adapted from Milliams and Van Gelder.*[22])

The first requirement can be met with real-time corrosion monitoring systems, provided that the monitoring techniques selected are sufficiently sensitive to respond rapidly to changes in the process conditions. Corrosion monitoring techniques (such as coupons) that yield only retrospective, cumulative corrosion damage data are not suitable for this purpose.

Modern industrial facilities usually are equipped with systems that form the foundation for the second requirement. Historical inspection data, failure analysis reports, analytical chemistry records, databases of operational parameters, and maintenance management systems are usually in place. The main task, therefore, is one of combining and integrating corrosion data into these existing (computerized) systems. In many organizations, much of the technical infrastructure required for achieving "corrosion process control" is already in place. Only the addition of certain corrosion-specific elements to existing systems may be needed.

As discussed earlier, corrosion monitoring plays a pivotal part in moving away from corrective corrosion maintenance practices to more effective preventive and predictive strategies. As confidence in monitoring data is established over time, through experience and correlation with other data/information such as that found through nondestructive evaluation and failure analysis, these data can assist in defining suitable maintenance schedules. If the rate of corrosion can be estimated from corrosion monitoring data (precise measurements are rarely achieved in practice) and the existing degree of corrosion damage is known from inspection, an estimate of corrosion damage as a function of time is available for maintenance scheduling purposes. Furthermore, sensitive corrosion monitoring techniques can provide early warning of imminent serious corrosion damage so that maintenance action can be taken before costly damage or failure occurs.

In practice, corrosion monitoring is generally considered to be a supplement to conventional inspection techniques, not a replacement. Once a serious corrosion problem has been identified through inspection, a corrosion monitoring program is usually launched to investigate the problem in greater depth. Corrosion monitoring and inspection are thus usually utilized in tandem. In the case of the smart structures monitoring concept, corrosion monitoring can essentially be considered to be a real-time ("live") inspection technique. The combination of corrosion monitoring and inspection data/information is a major organizational asset with the following uses:[22]

- Verifying design assumptions and confirming the design approach
- Identifying possible threats to an installation's integrity
- Planning operation, maintenance, and inspection requirements in the longer term
- Confirming and modifying standards and guides for future designs

Modern computerized database tools can be used to great advantage in the above tasks. The cause of many corrosion failures can be traced to underutilization of inspection and corrosion monitoring data and information.

From the above model, it is apparent that any leader of a corrosion monitoring program has to be comfortable with functioning in a multidisciplinary environment. Furthermore, corrosion monitoring information should be communicated to a wide range of functions, including design, operations, inspection, and maintenance. To facilitate effective communication and involvement of management in corrosion issues, corrosion monitoring data have to be processed into information suitable for management and nonspecialist "consumption." Enormous advances in computing technology can be exploited to meet the above requirements.

Corrosion monitoring examples

Monitoring reinforcing steel corrosion in concrete. In view of the large-scale environmental degradation of the concrete infrastructure in North America and many other regions, the ability to assess the severity of corrosion in existing structures for maintenance and inspection scheduling and the use of corrosion data to predict the remaining service life are becoming increasingly important. Several electrochemical techniques have been used for these purposes, with either embedded probes or the actual structural reinforcing steel (rebar) serving as sensing elements. A few indirect methods of assessing the risk of corrosion are also available.

In the civil engineering and construction industry, corrosion measurements are usually "one-off" periodic inspections. While such measurements can be misleading, it is at times difficult to make a persuasive argument for continuous measurements, in view of the fact that rebar corrosion is often manifested only after decades of service life. As a result of advances in corrosion monitoring technology and selected on-line monitoring studies that have demonstrated the highly time-dependent nature of rebar corrosion damage, continuous measurements may gradually find increasing application. Furthermore, the concept of smart reinforced concrete structures is gaining momentum through the utilization of a variety of diagnostic sensing systems. The integration of corrosion monitoring technology into such systems to provide early warning of costly corrosion damage and information on where the damage is taking place appears to be a logical evolution.

Rebar potential measurements. The simplest electrochemical rebar corrosion monitoring technique is measurement of the corrosion potential. A measurement procedure and data interpretation procedure are described in the ASTM C876 standard. The basis of this technique is that the corrosion potential of the rebar will shift in the negative direction if the surface changes from the passive to the actively corroding state. A simplified interpretation of the potential readings is presented in Table 6.8.

Apart from its simplicity, a major advantage of this technique is that large areas of concrete can be mapped with the use of mechanized devices. This approach is typically followed on civil engineering structures such as bridge decks, for which potential "contour" maps are produced to highlight problem areas. The potential measurements are usually performed with the reference electrode at the concrete surface and an electrical connection to the rebar.

TABLE 6.8 Significance of Rebar Corrosion Potential Values (ASTM C876)

Potential (volts vs. CSE)	Significance
> -0.20	Greater than 90% probability that no corrosion is occurring
≤ -0.20 and ≥ -0.35	Uncertainty over corrosion activity
< -0.35	Greater than 90% probability that corrosion is occurring

In a more recent derivative of this technique, a reference electrode has been embedded as a permanent fixture, in the form of a thin "wire."[23] With this technique, the corrosion potential can be monitored over the entire length of a rebar section, rather than relying on point measurements above the surface. However, this method will not reveal the location of corroding areas along the length of the rebar. A proposed hybrid of this technique is the measurement of potential gradients between two surface reference electrodes, eliminating the need for direct electrical contact with the rebar.

The results obtained with this technique are only qualitative, without any information on actual rebar corrosion rates. Highly negative rebar corrosion values are not always indicative of high corrosion rates, as the unavailability of oxygen may stifle the cathodic reaction.

LPR technique. This technique is widely used to monitor rebar corrosion. It has been used with embedded sensors, which may be positioned at different depths from the surface to monitor the ingress of corrosive species. Caution needs to be exercised in the sensor design in view of the relatively low conductivity of the concrete medium. Furthermore, the current response to the applied perturbation does not stabilize quickly in concrete, typically necessitating a polarization time of several minutes for these readings.

Efforts have also been directed at applying the LPR technique directly to structural rebars, with the reference electrode and counterelectrode positioned above the rebar on the surface. It was realized that the applied potential perturbation and the resulting current response may not be confined to a well-defined rebar area. The development of guard ring devices, which attempt to confine the LPR signals to a certain measurement area, resulted from this fundamental shortcoming. The guard ring device shown schematically in Fig. 6.18 can be conveniently placed directly over the rebar of interest and requires only one lead attachment to the rebar, as for the simple potential measurements. The guard ring is maintained at the same potential as the counterelectrode to minimize the current from the counterelectrode flowing beyond the confinement of the guard ring. An evaluation of several LPR-based rebar corrosion measuring systems has been published.[24]

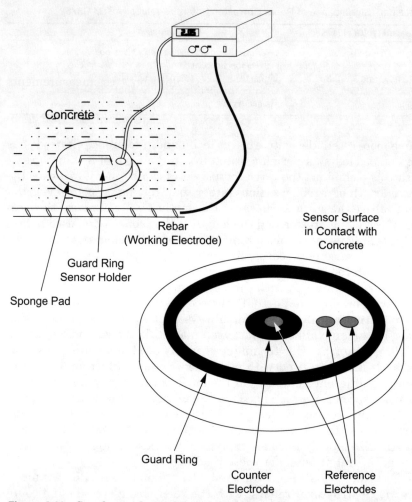

Figure 6.18 Guard ring device for electrochemical rebar corrosion monitoring (schematic).

Corrosion rates (expressed as thickness loss/time) can be derived from guard ring devices following the polarization cycle, but there are many simplifying assumptions in these derivations, and so they should be treated as semiquantitative at best. Important limitations include the following:

- Corrosion damage is assumed to be uniform over the measurement area, whereas chloride-induced rebar corrosion is localized.
- *IR* drop errors are problematic in rebar corrosion measurements, and "compensation" for them by commercial instruments is not necessarily accurate.

- Even if the guard ring confines the measurement signals perfectly, the exact rebar area of the measurement is not known. (How far does the polarization applied from above the rebar actually spread around the circumference of the rebar?)
- The influence of cracks and concrete spalling on these measurements remains unclear at present
- There are fundamental theoretical considerations in the LPR technique (described earlier).

Galvanostatic pulse technique. This technique also uses an electrochemical perturbation applied from the surface of the concrete to the rebar. A current pulse is imposed on the rebar, and the resultant rebar potential change ΔE is recorded by means of a reference electrode. Typical current pulse duration Δt and amplitude have been reported to be 3 s and 0.1 mA, respectively.[25]

The slope $\Delta E/\Delta t$, measured during the current pulse, has been used to provide information on rebar corrosion. High slopes have been linked to passive rebar, whereas localized corrosion damage was associated with a very low slope. This behavior can be rationalized on the basis of potentiodynamic polarization curves for systems displaying pitting corrosion.

Electrochemical impedance spectroscopy. Like those made by dc polarization techniques, EIS measurements can be applied to separate, small, embedded corrosion probes or directly to structural rebars. Efforts to accomplish the latter have involved guard ring devices and the modeling of signal transmission along the length of the rebar. Using a so-called transmission-line model, it has been shown that the penetration depth of the perturbation signal along the length of the rebar is dependent on the perturbation frequency.[26]

A number of different equivalent-circuit models have been proposed for the steel-in-concrete system; one relatively complex example is shown in Fig. 6.19.[27] By accounting for the concrete "solution" resistance and the use of more sophisticated models, a more accurate corrosion rate value than that provided by the more simplistic LPR analysis should theoretically be obtained. The main drawbacks of EIS rebar measurements over a wide frequency range are their lengthy nature and the requirement for specialized electrochemistry knowledge.

Zero-resistance ammetry. The macrocell current measured between embedded rebar probes has been used for monitoring the severity of corrosion. This principle has been widely used, as part of the ASTM G102-92 laboratory corrosion test procedure, with current flow between probes located at different depths of cover. For the monitoring of actual structures, a similar approach has been adopted.[28] Here, current flow has been measured between carbon steel probe elements strategically positioned at

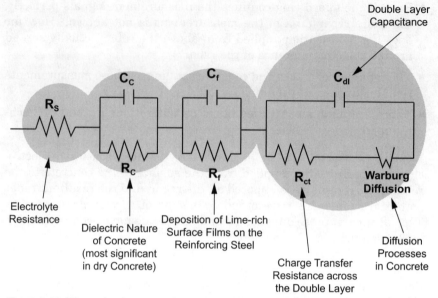

Figure 6.19 Example of an equivalent circuit for the steel-in-concrete system. (*Adapted from Jafar et al.*[27])

different levels within the concrete and an inert material such as stainless steel. Current flows between the carbon steel and stainless steel sensing elements are insignificant when the former alloy remains in the passive condition. Initiation of corrosion attack on the carbon steel is detected by a sudden increase in the measured current. Positioning the carbon steel elements at different depths from the concrete surface reveals the progressive ingress of corrosive species such as chlorides and provides a methodology for providing early warning of damage to the actual structural rebar, located at a certain depth of cover.

The current flowing between identical probe elements can also be used for corrosion monitoring purposes, even if the elements are located at similar depths. It can be argued that such measurements are mainly relevant to detecting the breakdown of passivity and the early stages of corrosion damage, before extensive corrosion damage is manifested on both of the probe elements.

Electrochemical noise measurements. There may be skepticism about the application of electrochemical noise measurements to industrial rebar corrosion monitoring. Concerns about the perceived "oversensitivity" of the technique and fears of external signal interference have been raised. While such concerns may be justified in certain cases, electrochemical noise measurements have been performed with probes embedded in large concrete prisms (up to 4 m long). These prisms were exposed in the Vancouver harbor and in clarifier tanks of the paper and pulp industry.[29] Initial results from this long-term monitoring program suggested that the noise signals did provide a sensible indication of rebar corrosion activity, and no major signal interference problems were encountered. In a more fundamental analysis of the application of electrochemical noise to rebar corrosion, Bertocci[30] concluded that this technique had considerable limitations and that further studies were required before the method could be used with confidence. Much work remains to be done in the signal analysis field, to automate data analysis procedures.

Monitoring aircraft corrosion. In the present economic climate, both commercial and military aircraft operators are faced with the problem of aging fleets. Some aircraft in the U.S. Air Force (USAF) currently have projected life spans of up to 60 to 80 years, compared with design lives of only 20 to 30 years. It is no secret that corrosion problems and the associated maintenance costs are highest in these aging aircraft. Aircraft corrosion falls into the atmospheric corrosion category, details of which are provided in Sec. 2.1, Atmospheric Corrosion.

While corrosion inspection and nondestructive testing of aircraft are obviously widely practiced, corrosion monitoring activity is only beginning to emerge, led by efforts in the military aircraft domain. In recent years, prototype corrosion monitoring systems have been installed on operational aircraft in the United States, Canada, Australia, the United Kingdom, and South Africa. Several systems are in the laboratory and ground-level research and testing phases, particularly those involving the emerging corrosion monitoring techniques described earlier. The "bigger picture" role of corrosion monitoring in a research program on corrosion control for military aircraft is illustrated in Fig. 6.20. The interest in aircraft corrosion monitoring activities is related to three potential application areas:

- Reducing unnecessary inspections
- Optimizing certain preventive maintenance schedules
- Evaluating materials performance under actual operating conditions

The first application area arises from the fact that many corrosion-prone areas of aircraft are difficult to access and costly to inspect. Typically, these areas are inspected on fixed schedules, regardless of whether corrosion has taken place or not on a particular aircraft. Unnecessary physical inspections could be eliminated and substantial cost savings could be realized if the severity of corrosion damage in inaccessible areas could be determined by corrosion sensors. Several prototype on-board corrosion monitoring systems have already been

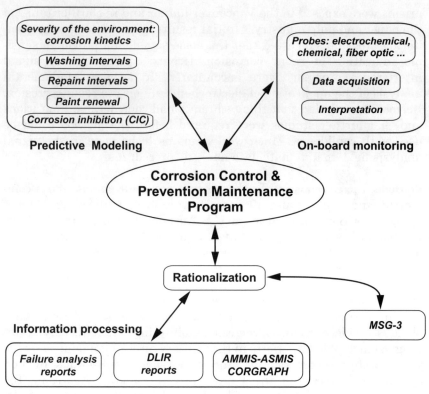

Figure 6.20 Research program for military aircraft, including the role of corrosion monitoring.

installed, to demonstrate the ability of corrosion sensors to detect different levels of corrosive attack in different parts of an aircraft.

One such corrosion surveillance system was installed on an unpressurized transport aircraft. Electrochemical probes in the form of closely spaced probe elements were manufactured from an uncoated aluminum alloy (Fig. 6.21). All but one of the probes were located inside the aircraft, in the areas that were most prone to corrosion attack and difficult to access. Another probe was located outside the aircraft, in its wheel bay.[31] In flights from inland to marine atmospheres, a distinct increase in corrosiveness was recorded by potential noise surveillance signals during the landing phase in the marine environment (Fig. 6.22). However, the strongest localized corrosion signals were recorded at ground level in a humid environment (Fig. 6.23).

A different system based on ER sensors was installed on a CP-140 maritime patrol aircraft, as illustrated in Fig. 6.24. In this case, high corrosion rates were measured in the wheel bay, relative to corrosion

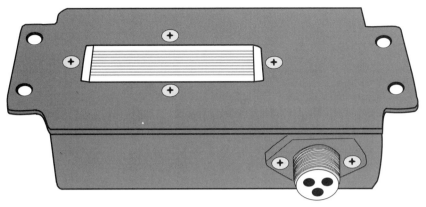

Figure 6.21 Electrochemical probe in the form of closely spaced elements manufactured from an uncoated aluminum alloy.

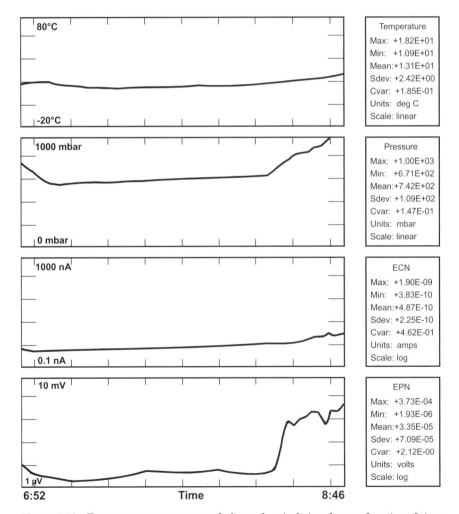

Figure 6.22 Temperature, pressure, and electrochemical signals as a function of time during a flight to a marine environment in South Africa.

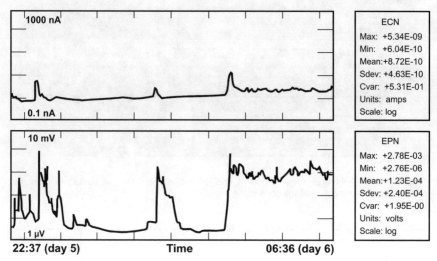

Figure 6.23 Electrochemical signals as a function of time in a marine environment in South Africa.

rates in other locations. More recent developments in this field include the use of thin-film electrochemical corrosion sensors (including wireless communication with these sensors) and the development of customized electrochemical sensors for monitoring corrosion in lap joints. Some new corrosion monitoring techniques for measuring aircraft corrosion in a more distributed manner are under development. Practical criticism has been directed at electrochemical sensors because they are restricted to measuring corrosion over a small surface area only.

One of the primary forms of preventive maintenance in maritime military aircraft is washing. The corrosiveness of the environment in which an aircraft operates usually is not a factor in the washing schedule. The unsatisfactory nature of this approach with respect to controlling corrosion damage has been highlighted. Corrosion monitoring systems installed at ground level and on board flying aircraft have demonstrated that the environmental corrosivity changes significantly over time and also varies for different parts of an aircraft. Arguably, therefore, selected inspection and maintenance schedules could be optimized based on the severity of the environmental corrosivity to which a particular aircraft has been exposed, as measured by corrosion monitoring systems.

On-board corrosion monitoring systems can facilitate the testing and evaluation of aircraft materials and corrosion control methods under actual operating conditions. Sensitive techniques make such evaluations possible in short time frames.

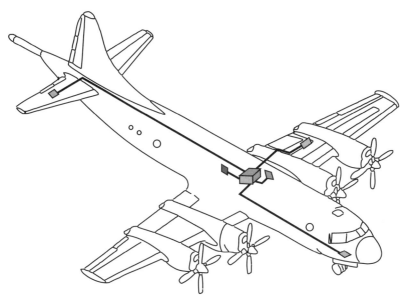

Figure 6.24 On-board ER corrosion sensors installed on a CP-140 maritime patrol aircraft.

Monitoring corrosion under thin-film condensate conditions. Highly corrosive thin-film electrolytes can be formed in several industrial processes. These conditions arise when gas streams are cooled to below the dew point. The resulting thin electrolyte layer (moisture) often contains highly concentrated corrosive species. Probe design and establishment of suitable measuring techniques for corrosion monitoring under such conditions are relatively difficult. One technique, electrochemical noise, has shown considerable promise; it is extremely sensitive and can be used in environments of low conductivity. Since the surface coverage of thin-film electrolytes is discontinuous at times, the latter aspect is important.

A corrosion probe used for electrochemical noise measurements in a gas scrubbing tower of a metal production plant is illustrated in Figs. 6.25 and 6.26. A retractable probe was selected so that the sensor surface could be mounted flush with the internal scrubber wall surface. The close spacing of the carbon steel sensor elements, designed specifically for (discontinuous) thin surface electrolyte films, should be noted. This corrosion sensor was connected to a computer-controlled miniaturized multichannel corrosion monitoring system by shielded multistrand cabling. As the ducting of the gas scrubbing tower was heavily insulated, no special measures were taken to cool the corrosion sensor surfaces. Cooling of probes in such applications is usually necessary if the corrosion sensor surfaces are to attain the same temperature as the internal duct surfaces. In general, the sensor surfaces of an electrochemical corrosion probe positioned in an access fitting will reach higher steady-state temperatures than the actual ducting surface—hence the requirement for cooling.

Potential noise and current records recorded at a conical section at the base of the gas scrubbing tower are presented in Fig. 6.27. At this location, condensate tended to accumulate, and highly corrosive conditions were noted from the operational history of the plant. The high levels of potential noise and current noise in Fig. 6.27 are entirely consistent with the operational experience. It should be noted that the current noise is actually off-scale, in excess of 10 mA, for most of the monitoring period. The high corrosivity indicated by the electrochemical noise data from this sensor location was confirmed by direct evidence of severe pitting attack on the sensor elements, revealed by scanning electron microscopy (Fig. 6.28). In contrast, at a position higher up in the tower, where the sensor surfaces remained dry, the electrochemical noise remained at completely negligible levels (refer to Fig. 6.27).

Corrosion monitoring studies of this nature have proved useful for identifying process conditions that lead to the formation of highly corrosive thin-film electrolytes, revealing the most corrosive areas, and evaluating materials designed to resist such attack in the most cost-

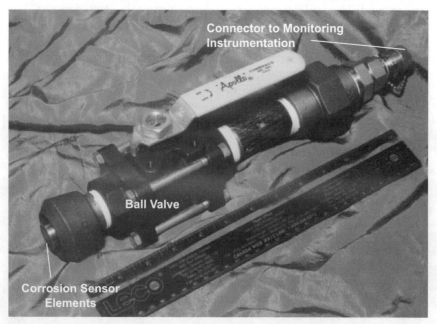

Figure 6.25 Corrosion sensor and access fitting used for thin-film corrosion monitoring.

Figure 6.26 Close-up of corrosion sensing elements used for thin-film corrosion monitoring.

effective manner. Such monitoring programs have been performed in gas ducting, gas stacks, and also gas piping.

Monitoring corrosion in heat-exchanger tubes of cooling-water circuits. Tube-and-shell heat exchangers are widely used in the cooling-water circuits of diverse branches of industry. Corrosion damage is usually a major concern in such units, and water treatment is commonly used as a means of corrosion control. Despite water treatment additives, however, corrosion failures continue to occur, and numerous corrosion failure modes have been documented. Localized corrosion damage can include pitting, crevice corrosion, and stress corrosion cracking. Such localized failures are typically related to fouling or scaling of the tube surfaces, chloride ions in the water, or microbial activity. Uniform corrosion damage may be sustained during acid descaling operations, if these are not closely controlled. Corrosion monitoring of heat-exchanger tube surfaces is technically extremely challenging for the following reasons:

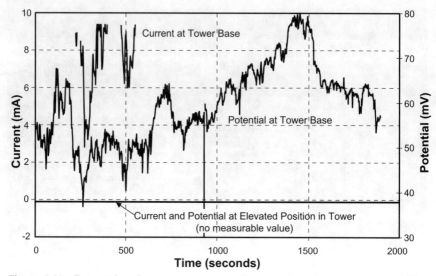

Figure 6.27 Potential and current noise records at two locations in a gas scrubbing tower.

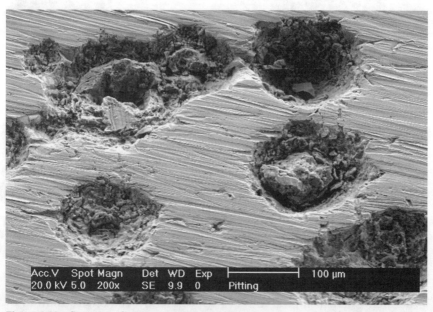

Figure 6.28 Scanning electron microscope image of a sensor element surface after exposure at the base of the scrubbing tower. Microscopic corrosion pits are clearly evident.

- A multitude of corrosion modes can lead to damage.
- Monitoring localized corrosion damage, a common problem, is difficult.
- Corrosion damage occurs under heat-transfer conditions.
- Access to the tightly packed tubes is extremely limited.

In order to overcome the access problems of fitting corrosion sensors into the heat exchanger, a bypass strategy can be followed. Water flowing through the actual heat exchanger is deviated to a side stream, which then flows through a model heat exchanger. The model heat exchanger can be instrumented with corrosion sensors relatively easily. If electrochemical corrosion sensors are used, these can be made representative of an actual heat-exchanger tube by using electrically isolated spool pieces as sensing electrodes.

In order to simulate actual operating conditions, the corrosion sensors in the model heat exchanger need to be subjected to heat flux and scale formation. The use of unheated sensor surfaces would not reflect the operational scaling characteristics accurately, and hence the corrosion damage on the sensors would not be representative of that on the operating unit. Heating elements, temperature sensors, and heat-transfer calculations can be used to mimic the heat flux of the actual heat-exchanger tubes in the model heat exchanger. The use of multiple corrosion monitoring techniques applied to multiple corrosion sensing elements in a model heat exchanger can address the issue of detecting various forms of corrosion damage.

A corrosion monitoring system based on the above principles has been described.[32] It uses a single heat-exchanger tube in the bypass model heat-exchanger loop, with multiple electrochemical corrosion sensing techniques applied to segmented corrosion sensing elements. The principle of this monitoring system is illustrated in Fig. 6.29. Flow controls and varying degrees of heat flux conveniently facilitate the simulation of varying operational conditions, an important capability for "what-if" analysis. A more detailed schematic of this model heat exchanger is given in Fig. 6.30, showing five segmented corrosion sensing elements, each with an individual heater block for heat flux simulations. With these five sensing elements, it was possible to measure both localized and general corrosion damage. The corrosion monitoring techniques utilized in this particular device were electrochemical noise (potential and current), zero-resistance ammetry, and linear polarization resistance.

Monitoring preferential weld corrosion with ZRA. Any weldment is a complex metallurgical structure. The weld metal is essentially a miniature casting, with a composition and microstructure that may differ sub-

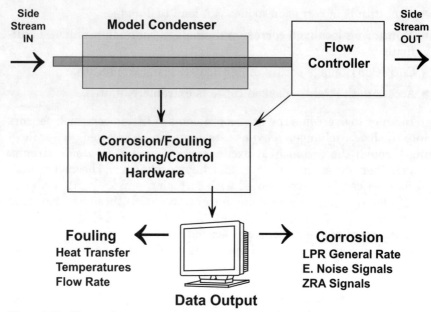

Figure 6.29 Heat-exchanger monitoring systems using the bypass approach (schematic). (*Adapted from Winters et al.*[32])

stantially from those of the parent plate. On a microstructural scale, the weld metal itself is not homogeneous. Typically the weld centerline has a higher impurity content, and the microstructure changes at different stages in the weld solidification cycle. The microstructure of the heat-affected zone (HAZ) also tends to vary from that of the parent plate, as it is subjected to the weld thermal cycles, which change with distance from the fusion line. Consequently, the microstructure of the HAZ is also not uniform (refer to intergranular corrosion in Sec. 5.2.1). It should thus be apparent that the different zones of a weldment can be susceptible to galvanic corrosion as a result of their compositional and microstructural differences.

Differential weld corrosion has been found to be particularly problematic in oil and gas flow lines. Even minor differences in composition and microstructure have been found to result in severe preferential galvanic dissolution of pipeline weldments. The selection of welding consumables and welding procedures to minimize this risk is critical. However, even with these precautions, operating conditions can induce severe preferential weld corrosion. On-line corrosion monitoring programs have been conducted in oil and gas pipelines to identify these operating conditions and to optimize the application of corrosion inhibitors to control the problem.

The ZRA technique lends itself ideally to these monitoring purposes, as outlined by Walsh.[33] Suitable corrosion sensors can be manufactured

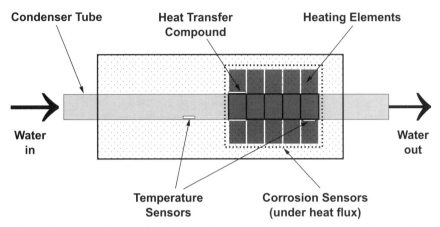

Figure 6.30 Corrosion sensing elements in model heat exchanger for multitechnique electrochemical monitoring (schematic). (*Adapted from Winters et al.*[32])

from representative pipeline weldments, as shown schematically in Fig. 6.31. It should be noted that the internal weld surfaces are used as the exposed sensor elements for monitoring purposes. Essentially, selected strips from the different weld zones are sectioned from the weld and incorporated in a "standard" probe body designed for high-temperature and high-pressure service. A larger number of sensor elements than are depicted in Fig. 6.31 can be incorporated into a single sensor, to investigate different weld compositions and structures. The so-called 2-inch access fittings widely used in the oil and gas industry can be used to mount the sensor surfaces flush with the internal pipeline wall.

ZRA readings can be accomplished with relatively simple instrumentation, and with a sufficiently high sampling frequency, a real-time weld corrosion profile can be obtained for correlation with the operating parameters and process control. Provided that all the sensor elements are connected to the monitoring instrumentation in a consistent manner, the sign and magnitude of the ZRA responses monitored between the elements indicate the severity of galvanic attack and which part(s) of the weldment are dissolving preferentially.

Examples of contrasting highly undesirable and favorable ZRA monitoring profiles are presented schematically in Fig. 6.32. In case A, the ZRA sensor response indicates that the HAZ is subject to intense preferential anodic dissolution. Both the weld metal and the parent plate are more noble (cathodic) than the HAZ. The narrow HAZ surrounded by the weld metal and the large parent plate produces an extremely unfavorable galvanic area effect. These conditions lead to weld failure by extremely rapid preferential penetration of the weldment along the HAZ. Actual HAZ corrosion rates could well exceed the values measured with the sensor, as the most severe area effect cannot be repro-

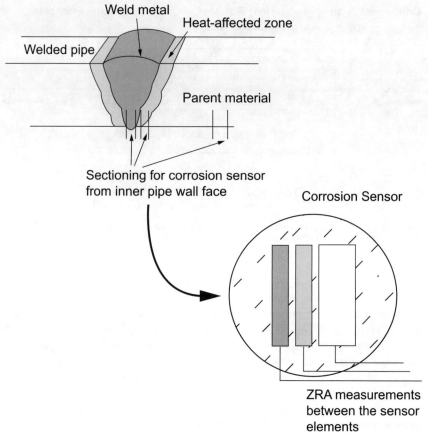

Figure 6.31 Manufacture of preferential weld corrosion sensor (schematic).

duced in the probe. Case B shows a desirable ZRA profile. Essentially, all three weld zones are galvanically compatible, with very low galvanic current levels. The weld metal is only marginally more noble than the HAZ and the parent plate. In practice, addition of inhibitors can be used to achieve this type of situation.

6.5 Smart Sensing of Corrosion with Fiber Optics

6.5.1 Introduction

The techniques described so far have all progressed to industrial applications. A number of less well-known techniques are currently emerging from research and development efforts. There can be little doubt that several of these will find increasing commercial application. Some

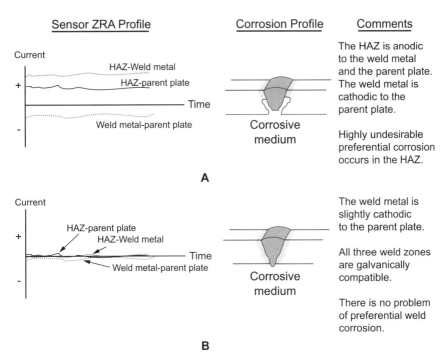

Figure 6.32 Undesirable and favorable weld corrosion profiles from ZRA monitoring (schematic).

promising emerging techniques based on fiber optics are described here. The development of fiber optic technologies for communication applications has sparked interest in creating new sensors by modifying a section of the fiber itself. The range of physical and chemical parameters that can be detected so far is remarkable. Physical and mechanical parameters that can be measured include temperature, strain, pressure, displacement, vibration, magnetic fields, and electric fields.

Chemical parameters that can be measured include pH; some organic compounds; moisture; chloride ions; dissolved gases such as oxygen and carbon monoxide; gases such as oxygen, steam, and ammonia; and compounds that fluoresce as a result of specific interactions, such as enzyme-substrate and antibody-antigen complexes. Some of these parameters have been recognized in the last few years as being potentially useful for monitoring either the effects of corrosion on a structure or some of the factors that induce corrosion. Emerging applications for monitoring the corrosion of structures include

- Detection of moisture and increasing pH in aircraft lap joints
- Measurement of the shift in the light spectrum reflected off rebar as a result of corrosion

- Detection of chloride ions near rebar
- Detection of rebar strain in a bridge due to corrosion

Generic advantages of fiber optic sensing systems include their passive nature, immunity to electromagnetic interference, light weight, small size (an analogy to a human hair may be cited), large bandwidth, mechanical ruggedness, high sensitivity, and ease of multiplexing. A fiber optic sensing system consists of a light source, a detector, a sensing element, and the optical fiber for transmitting the light from the source to the detector. An important concept is the use of the fiber optic sensor itself as a corrosion sensing element, the so-called intrinsic sensor. Corrosion sensing elements in fiber optic sensing systems have been based on the following principles:

- A change in the reflectivity of light from highly polished surfaces, induced by formation of corrosion products
- The detection of chemical species and pH changes associated with corrosion processes
- Changes in strain as the thickness of the corroding material is reduced

Another important corrosion monitoring concept in which fiber optics can play an important role is that of smart coatings. The basic idea is for a coating to reveal where it has been damaged and corrosion attack has been initiated. This form of corrosion sensing has the major advantage that it can be applied over extensive surface areas; the sensing is not restricted to a local measuring point. Fundamental principles that have been proposed for smart coatings include

- The incorporation into coatings of chemicals that induce a color change when corrosion or coating damage occurs[34]
- A fluorescent response to corrosion damage or coating discontinuities[35]

There is a trend toward utilizing the versatility of fiber optic sensors to monitor atmospheric corrosivity and the effects of corrosion on a structure. Emerging techniques for monitoring air corrosivity include

- An optically thin metal that reflects less light as it corrodes
- A thin metal wire that can be configured to function as a corrosion fuse
- A metal coating that undergoes strain relaxation as it corrodes
- Gas sensors that measure the concentration of species that promote corrosion

6.5.2 Optical fiber basics

Optical fibers typically consist of four layers, as shown in Fig. 6.33: (1) an inner core, (2) cladding, (3) a protective buffer, and (4) a jacket. Light is launched into the end of an optical fiber by a light source and is guided down the inner core. Most inner cores are made of silica glass, but some are made of sapphire, fluoride glasses, or neodymium-doped silica. Glass fibers have very low light-loss characteristics and therefore are capable of transmitting a light signal hundreds of miles. The cladding is usually made of a silica glass that has an index of refraction lower than that of the core, so that light is refracted back into the inner core. Protective buffers are usually made of plastic. The function of the plastic buffer layer and jacket is to provide mechanical protection and thus allow optical fibers to be flexible and robust, and also to provide a moisture seal. A typical diameter for a jacket is 125 μm, and that for an inner core is 10 μm.

An environmental parameter can be measured by its influence on one or more of the following characteristics of light through a sensor: (1) intensity, (2) phase, (3) wavelength, or (4) polarization. Changes in the refractive index of the cladding by an environmental parameter can affect both the intensity and the phase of the light. Any fluorescence in the cladding caused by a specific chemical interaction with the environment causes wavelength changes in the light that is refracted back into the inner core. In a common sensor design, an environmental parameter affects the intensity and phase of the light that is reflected back from the sensor toward the light source.

The signal from a fiber optic sensor is analog, not digital as in fiber optic communications, and therefore needs a reference signal. A typical

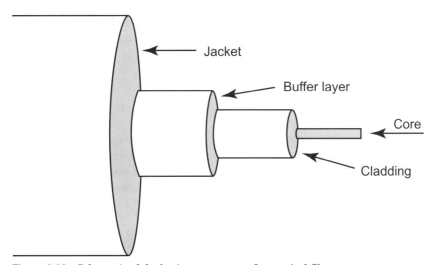

Figure 6.33 Schematic of the basic components of an optical fiber.

method of providing a reference for sensors that modify intensity is to use two wavelengths of light, with the sensing element having a larger effect on the light at one wavelength than at the other. Unwanted environmental effects can be eliminated by taking a ratio of the intensity of the two wavelengths from the sensor.

System requirements for a fiber optic sensor involve light source and signal detection components as well as the optical fiber. Distributed sensors provide continuous spatial resolution of the parameter along the length of the fiber. A quasi-distributed sensor is an optical fiber with a series of sensors at discrete locations along its length, therefore providing discrete spatial resolution. The small diameter of optical fibers limits the amount of light power that can be launched into and detected leaving the fiber. This usually means that fiber optic sensors have relatively low signal-to-noise ratios, which limits the methods of light detection and multiplexing that are feasible in a cost-effective manner.

6.5.3 Emerging corrosion monitoring applications

Atmospheric corrosivity monitoring

Micro-mirror. A method of measuring the corrosivity of an atmosphere that was developed at Sandia National Laboratories involves measuring the reflectivity of an optically thin metal mirror. A thin layer of metal (i.e., from 10 to 30 nm) is applied to the end of a fiber by thermal or vacuum evaporation to form a micro-mirror. A schematic of a micro-mirror system is shown in Fig. 6.34. Light passes through the optical fiber to the metal at the end of the fiber and is partially reflected. The main signal output is either the ratio of the intensity of the reflected light to that of the incident light or the ratio of the reflectivity to the initial reflectivity with a clean micro-mirror. Species from the atmosphere that chemisorb and/or react with the metal reduce the reflectivity.

Butler and Ricco reported that the reflectivity of silver micro-mirrors decreased as species such as H_2S, CO, O_2, SO_2, and H_2 chemisorbed onto the external surface of the metal.[36] The change in reflectivity caused by chemisorption ranged from 0.7 to 0.1 percent. However, the change in reflectivity caused by the reaction of H_2S and Ag to form Ag_2S and H_2 was an order of magnitude larger. These results indicate that corrosive influences that change the composition of a metal can be measured in this manner.

Ammonium sulfate particles have been implicated in the corrosion of microelectronics in humid air. Smyrl and Butler placed a copper micro-mirror on the end of a fiber into an aerated solution of ammo-

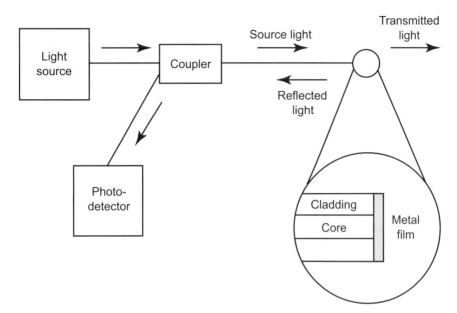

Figure 6.34 Schematic of a light reflection system with a micro-mirror at the end of the optical sensor.

nium sulfate.[37] The thickness of the copper was related to the reflectivity, and therefore the reaction and dissolution of the copper film were measured by the degree of reflection. Corrosion occurred only in the presence of dissolved oxygen. The copper micro-mirror, which was initially 30.5 nm, was dissolved by the aerated ammonium sulfate solution in less than 1 h.

Hydrogen is often a by-product of corrosion. A sensor was formed with a micro-mirror of palladium that was responsive to hydrogen concentration in air up to approximately 5 percent. The interaction of hydrogen and palladium reversibly forms a hydride, PdH_x, which has a lower reflectivity than pure palladium. Smyrl and Butler illustrated that this sensor is responsive to hydrogen that is dissolved in water.[37] Thus, monitoring dissolved hydrogen in small areas such as crevices is a potential application for fiber optic micro-mirrors.

Corrosion fuse. Bennett and McLaughlin described a method for monitoring the corrosion of a metal called a "corrosion fuse."[38] A schematic of a prototype is shown in Fig. 6.35. Attenuation of light through an optical fiber becomes significant when the fiber is bent into a loop smaller than about 3 mm. A thin metal rod maintains the fiber in a microbend with slight tension from a spring. When the metal rod corrodes to the point that it breaks, the fiber straightens because of the spring, and the intensity of the light downstream of the fuse increases. Obviously the composition and thickness of the metal fuse may be readily designed. Decreasing the thickness of the metal fuse increases the sensitivity to corrosion. Bennett and McLaughlin demonstrated that three fuses in series could be monitored, in a quasi-distributed fashion, on the same fiber. The corrosivity of an atmosphere is expected to be inversely related to the time required for a given fuse to break.

The design was tested by placing three units on a single fiber above a salt solution within an enclosed chamber. When very little corrosion was observed after 30 days at 30°C, the bath temperature was raised to 44°C. The sensors broke after another 34, 41, and 44 days, and these events were readily monitored by the light signal.

Strain relaxation. A fiber optic technique for measuring corrosivity by the degree of strain relaxation of a plastically deformed metal coating has been developed. The degree of residual strain in the sensor jacket depends on (1) the coating material, (2) the coating thickness, and

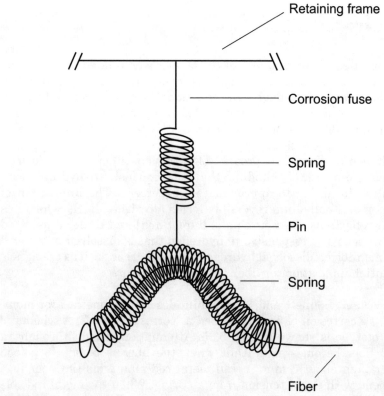

Figure 6.35 Schematic of a corrosion-fuse arrangement.

(3) the load history of the sensor. As the metal coating of the sensor corrodes, the wall weakens, which causes shrinkage as a result of a relaxation of the residual strain.[39]

The arrangement of light source, photodetector, and coupler is the same as in Fig. 6.34. The sensor, which back-reflects light, was an external Fabry-Perot interferometer. Light is reflected back to a photodetector from two semireflective mirrors that are separated to form a cavity within the inner core of an optical fiber. The degree of phase difference between the reflected waves is a linear function of the cavity length. A metal cylinder was fixed to the outside of the inner core by an adhesive. Changes in the length of the metal cylinder cause changes in the cavity length, which produce a linear change in the phase difference of the reflected light waves.

The sensor was tested in an accelerated corrosion experiment by immersion in nitric acid. Most of the strain relaxation occurred within 40 s, and the balance within 200 s. The strain relaxation due to corrosion matched the measured residual strain within an average of 4.5 percent. A disadvantage of this technique is that the response to uniform corrosion conditions is nonlinear. The explanation is that the residual strains are not uniformly distributed, in a cross-sectional sense, across the metal shell. This technique is currently being developed to monitor atmospheric corrosivity on aircraft.

Gas sensors. Systems for measuring process gases with fiber optic sensors have been commercially available for about 10 years from Altoptronic AB of Sweden. Detectors that measure the concentrations of the following gases are currently available: NH_3, O_2, H_2O, HCl, HCN, HF, H_2S, CH_4, and C_2H_4.[40] Some of these gases obviously promote corrosion.

The system configuration is a modification of the reflected-light arrangement shown in Fig. 6.34. The light source is a semiconductor laser diode that operates in the 1.3–1.9-μm or the 0.7–0.9-μm region, depending on what gas species the system is configured for. Light from the laser diode is split into two beams—a measurement path and a reference path. The measurement path continues to the measurement probe, which provides a path length of about 1 m through the gas being sensed. The distance between the light source and the sensor may be up to several kilometers. Almost 100 percent of the measurement light is reflected back by a reflector to the control unit, where its spectrum is compared to the light from the reference path. The signal compensates in real time for transmission degradation in the sensor head and losses caused by dust and chemical salt deposits on the optics in the measurement path.

Monitoring of structures. There is a trend toward making such things as aircraft and concrete structures smarter by embedding or attaching sensors in order to aid in monitoring the structures' health. The corrosion of rebar in bridges is a serious problem, especially in climates where salt is used to deice the road surface.

Lap joints in aircraft. A method for monitoring the presence of moisture and the pH level in aircraft lap joints was described by Mendoza.[41] A lap joint consists of overlapping metal sheets, such as the skin of an aircraft, held together by rows of rivets. Common locations for cracks to form are on the metal sheets close to the rivets. A sealant is applied between the sheets to prevent moisture intrusion. The small diameter of optical fibers makes it possible to place them within the sealant layer along a row of rivets. Each sensor for this project was designed to provide continuous spatial resolution of the response over a 20-m section of optical fiber. Incorporation of the sensor into the lap joint under pressure caused a slight increase in light absorption.

The moisture sensor was made by replacing the cladding with a polymer/solvatochromatic dye compound that absorbed light in the presence of moisture. The degree of attenuation increased with wetness over most of the band between 0.4 and 1.4 μm. At an input power of 2 mW, the dry and wet output power at two wavelengths is shown in Table 6.9.

The difference in the response at the two wavelengths suggested that the main output should be the ratio of the output powers at two wavelengths. The response to moisture was found to be completely reversible.

A prototype pH sensor was constructed by incorporating a pH indicator in a polymer to form the fiber cladding. The pH indicator responded to moisture and increasing pH by increased fluorescence. The sensor was tested in two solutions—a pH 7 buffer and a pH 10 buffer. The light output from the sensor increased as the sensor environment changed from dry to pH 7 to pH 10. Virtually no response at 0.77 μm and a peak response at 0.539 μm to the same environmental change suggested a dual-wavelength design similar to that of the moisture sensor.

Figure 6.36 is a schematic of a detection system for the moisture sensor that permits continuous spatial resolution based on time-division multiplexing. The photodetectors, PD_x, also function as wavelength filters. PD1 and PD1' produce electrical signals that are proportional to the light intensity at 0.67 μm before and after the fiber sensor. The ratio of these outputs is calculated by R1 in order to nullify the effects of any

TABLE 6.9 Dry and Wet Output Power at Two Wavelengths

Wavelength, μm	Optical power (dry), μW	Optical power (wet)
0.67	0.22	0.7
1.30	43.7	0.8

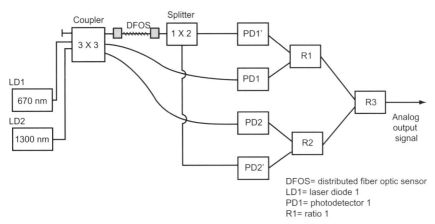

Figure 6.36 Schematic of a detection system for the moisture sensor that permits continuous spatial resolution based on time-division multiplexing.

source fluctuations. Similarly, the reference wavelength at 1.30 μm is detected before and after the fiber sensor by PD2 and PD2′, and the ratio of these responses is the output of R2. The ratio of the output from R1 and R2 is calculated by R3 and is the main output response.

Rebar appearance. A fiber optic sensor was designed to monitor the corrosion of rebar based on the change in color of the surface of rebar as a result of corrosion. A "twin-fiber" approach and a "windowed" approach have been reported so far. Both approaches are currently being tested in bridges in Vermont.[42]

In the twin-fiber design, the transmit fiber illuminates with broadband light a section of rebar, which modifies the spectrum through absorption and reflects light into the receive fiber. The gap between the fiber and the surface to be monitored is less than 10 mm. The signal travels back down the receive fiber, and the spectrum is measured. A spectrum shift indicates that corrosion is present. There is a clear difference in the spectrum between 0.5 and 0.8 μm of uncorroded rebar and corroded rebar.

The windowed approach has the potential to provide spatial resolution of the response at discrete points along the fiber. The spectrum of light that reflects from the rear surface may be detected by measuring the backscattered light at the end of the fiber near the light source. A low signal-to-noise ratio for this design limits the number of discrete sensors along one fiber to about nine.

Chloride detection. Fiber optic sensors that measure chloride ion concentration have been developed.[43] Potential applications include monitoring chloride penetration into steel-reinforced concrete structures and monitoring the leakage of contaminated water from a landfill. Several configurations were tested by Cosentino et al.:[43] (1) absorption mode with silver nitrate, (2) reflection mode with silver chromate, and (3) transmission mode with silver chromate. In the presence of chloride ions, a silver nitrate solution changes from a clear solution to a white silver chloride suspension, and silver chromate changes from red-brown to white silver chloride.

In sensor 1, there was a 1-mm gap between the ends of two fibers that were bathed in a solution of silver nitrate. A semipermeable membrane permitted chloride ions to exchange with nitrate ions. The change from clear solution to whitish suspension increased the scattering of light in the gap and therefore increased light absorption. One drawback was that silver chloride eventually settles as a precipitate, which reverses the response.

In sensor 2, silver chromate was applied to the end of a fiber with a thin layer of glue and the entire end was surrounded with a porous plastic casing. The amount of light reflected back from the end of the fiber was the primary signal. The increase in reflected light caused by the conversion of silver chromate to silver chloride was modest.

With sensor 3, silver chromate was deposited within a hollow porous disk and the optical fiber was passed through its center. As chloride ions seeped into the porous disk, some silver chromate was converted to silver chloride, which increased the amount of light that reflected back into the fiber. Thus the light output from the sensor increased in the presence of chloride ions. The range of detection for this sensor was rather large—from 100 mg/L to greater than 3000 mg/L. There was, however, a nonlinear response to different chloride concentrations as well as a significant time effect.

A disadvantage of all three designs was that the detection of chloride is irreversible—the silver chloride does not revert back to either silver nitrate or silver chromate when the concentration of chloride ions decreases. It has been suggested that a reversible chloride sensor may be based on the absorption peak at 0.360 μm for salt water.

Fuhr and Huston adapted from analytical chemistry a method for measuring chloride concentration known as Fajan's method.[44] A chloride sample is added to a solution of $AgNO_3$ and a dye until the solution turns from milky white to pink. The adaptation of this method to a fiber optic sensor involved measuring the change in the transmission spectrum of a sample of $AgNO_3$ and dye. Light from the input fiber entered one end of a tubing tee and was directed toward the entrance of the output fiber at the opposite end of the tee. A porous membrane over the other opening of the tee permitted the exchange of chloride and nitrate ions. Two wavelengths were selected, 0.60 μm and 0.725 μm. The former wavelength indicated the degree of milkiness, and the latter indicated the degree of pinkness. These sensors are currently under test within three bridges in Vermont.

Embedded strain. Another method for monitoring the deterioration of a bridge as a result of corrosion is to monitor the internal strain of support girders. One of the difficulties with this approach is that the degree to which a tendon stretches during initial stressing (over 8000 με) causes sensors to be under a high degree of tensile strain during most of their operation. This problem was largely solved by embedding the fiber optic sensors in concrete when the tendons were under tensile stress. This, however, complicated the construction of the girders. Of the 18 Bragg grating sensors that were embedded, 3 did not survive.

A Bragg sensor consists of a section of fiber from 1 to 20 mm in length in which the inner core is modified to form a grating of pitch ρ, as shown in Fig. 6.37. The Bragg wavelength λ_B is reflected back toward the light source and is directly proportional to the grating pitch:

$$\lambda_B = 2mn_c\rho$$

where n_c is the index of refraction, ρ is the grating pitch, and m is an integer from 1 to infinity that determines the order of response (e.g., $m = 1$ for the first-order response). Changes in the grating pitch caused by stretching or compressing of the sensor change the Bragg wavelength linearly.

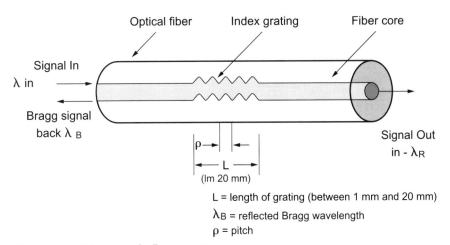

Figure 6.37 Schematic of a Bragg sensor.

The method of monitoring the Bragg wavelength is shown in Fig. 6.38.[45] The reflected light from the Bragg grating is split into two beams. One beam is filtered in proportion to its wavelength, while the other beam is used as a reference to compensate for intensity fluctuations. The ratio of the proportionally filtered and reference signals is a linear function of the Bragg wavelength. Therefore, there is a linear proportionality between the grating pitch ρ and the electrical signal from R1.

6.5.4 Summary

Each of the fiber optic sensors reviewed either already has found or may in the future find a niche application for monitoring corrosion. Advantages of the micro-mirror sensor are its sensitivity, small size, and geometrical flexibility and the possibility of using different metals as the semireflective mirror. For instance, if one were concerned with the corrosivity of an atmosphere toward aluminum, then the micro-mirror could be constructed of aluminum. Although the corrosion fuse seems to be relatively easy to construct, it provides a limited amount of data, e.g., it takes 44 days for a rod of given thickness and composition to corrode through. The strain relaxation sensor has the advantage of small size, but it has a nonlinear response to a corrosive medium, which complicates interpretation of the data. A limitation of the gas sensors is that they require a path length of about 1 m. The moisture sensor in aircraft lap joints seems to be without technical difficulties. One limitation of the pH sensor in lap joints is that corrosion usually causes the pH to decrease near the corroding metal, whereas the pH sensor, as configured, has an increasing response with increas-

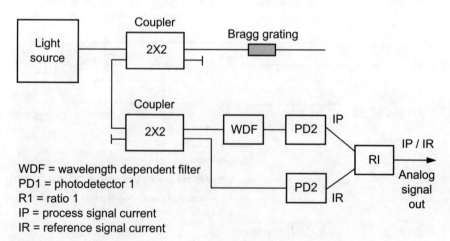

Figure 6.38 Schematic of a system for measuring the Bragg wavelength.

ing pH. One difficulty with both the rebar appearance and embedded strain sensors is the process of incorporating them into the concrete structure. A limitation of all the chloride sensors tested so far is that there is an irreversible response to the presence of chloride.

The advantages of fiber optic sensors have to be weighed against the disadvantages and compared with other techniques for a given corrosion monitoring application. The cost per sensor depends mostly on the extent of multiplexing that is possible. A technical cost of increasing the number of sensors per multiplexer, however, is an eventual decrease in the signal-to-noise ratio per sensor. Table 6.10 presents a summary of the published applications of fiber optic sensors in monitoring corrosion. Also included is an assessment of the degree of commercialization of each technique. Four methods are nearly ready, two are available currently, and two require more design and testing work.

6.6 Nondestructive Evaluation (NDE)

6.6.1 Introduction

Recent advances in nondestructive evaluation (NDE) technologies have led to improved methods for quality control and in-service inspection, and the development of new options for material diagnostics.

TABLE 6.10 Summary of Published Applications of Monitoring Corrosion with Fiber Optic Sensors

Method	Potential application	Degree of commercialization
Micro-mirror	Atmospheric monitoring	Available for licensing
Corrosion fuse	Atmospheric monitoring	Needs custom design for specific applications
Strain relaxation	Atmospheric monitoring	Currently being tested as an on-board corrosion monitoring system on an aircraft
Gas sensors	Atmospheric monitoring	Sensors specific for several gases currently available
Moisture and pH	Aircraft lap joints	In final design and testing stage
Rebar appearance	Rebar	Under test within three bridges in Vermont
Chloride detection	Rebar/landfill monitoring Rebar	At laboratory stage Under test within bridges in Vermont
Embedded Bragg strain sensors	Rebar/composite bridge tendon	Under test in a bridge in Calgary—systems are currently available

Detailed defect sizing and characterization has become the major objective of much NDE work under way today. To address this challenge, the NDE community has turned to a combination of multiple-mode inspections and computer-aided data analyses. Success in this activity has generated quantitative NDE capabilities that can be used both as improved quality assurance tools and as new options for material diagnostics.

NDE is the discipline used to assess the integrity of a system or component without compromising its performance. NDE uses sensors to acquire information about these objects and perform modeling and analysis to convert the information into materials and defect parameters for performance and in-service life prediction. Figure 6.39 illustrates the specific knowledge domains involved in NDE. The inspection of in-service systems can also be complicated by the fact that these systems often operate at relatively high temperature in a closed mode.

- Traditional inspection techniques can be used on high-temperature process equipment only when the equipment is out of service.
- Inspection is generally limited to suspected problem areas as a result of time, accessibility, and cost factors.

When several inspection techniques can be used, the choice of a specific schedule will depend on the accuracy and cost of the inspection, balancing the money spent on safety measures with the business return of the system being maintained (Fig. 6.40). The accuracy of a given technique must also be sufficient to detect defects considerably smaller than those which could result in failure because these defects can grow in size between inspections. A cheaper and less accurate technique used frequently could be equivalent costwise to a more expensive accurate technique used less frequently (Fig. 6.41).[46] However, regardless of which technique has been chosen, the critical decision remains the frequency of application. This decision depends on three factors:

- The extent of damage that might remain invisible to the technique
- The rate of damage occurrence with time
- The extent of damage that the structure can tolerate

6.6.2 Principles and practices

NDE technology denotes application of a diverse array of nondestructive processes to monitor, probe, and measure material response. The measured response is related to a desired material property or test object attribute by interpretation. The main NDE methods are

Corrosion Maintenance through Inspection and Monitoring

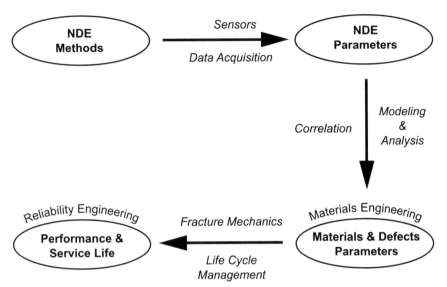

Figure 6.39 Disciplines involved in NDE.

Figure 6.40 The effects of increasing expenditure on safety and inspection.

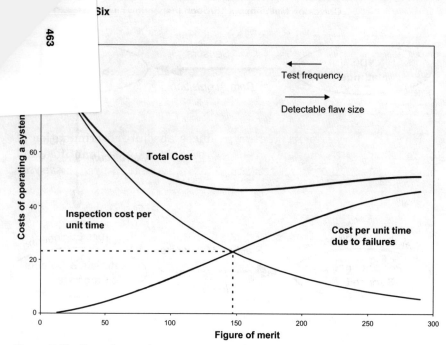

Figure 6.41 Cost of operating a system as a function of expected cost per unit time associated with failures and inspection.

- Visual inspection
- Liquid penetrant inspection
- Magnetic particle inspection
- Radiographic inspection (x-ray and gamma ray)
- Electromagnetic inspection
- Ultrasonic inspection
- Thermographic inspection

Although each method is dependent on different basic principles in both application and output, repeatable and reproducible NDE results depend on specific understanding and control of the

- Material composition (magnetic or nonmagnetic, metallic or nonmetallic, etc.)
- Part thickness, size, and geometry
- Material condition (heat treatment, grain size, etc.)
- Inspection scanning rate
- Fabrication method (casting, forging, weldment, adhesive or brazing bonded, etc.)

- Surface condition (rough, plated, bright, scaled, etc.)
- Nature or use of the part (critical or noncritical, high or low stress, etc.)
- Human factors

No NDE process or procedure produces absolute discrimination of anomalies, but the end output of a procedure may be quantified and the anomaly or flaw detection capability may be measured, analyzed, quantified, and documented.

The diverse nature of different NDE processes results in different sources of variance and possible impact on detection output capabilities. For example, a manually applied liquid penetrant process is dominated by the skill of the operator in process application and interpretation. An automated eddy current process is dominated by calibration, instrument, and procedure variances. It is important to recognize the source of variance in each NDE process and to take the nature of the variance and process control (Table 6.11) into account in applying margins to the NDE processes.[47] NDE methods and procedures are selected using a variety of practical implementation criteria, such as the relative ratings presented in Table 6.12. The lowest-cost method that produces the required result is usually the method of choice. Table 6.13 presents a general overview of the procedural steps required for the main NDE techniques considered here.

Visual inspection. Visual inspection is the oldest and most common form of NDE used in inspecting for corrosion. The physical principle behind visual inspection is that visible light is reflected from a surface, revealing some of its features. By observing the appearance of the part, an inspector can infer its condition. Surface corrosion, exfoliation, pitting, and intergranular corrosion can be detected visually when proper access to the inspection area is obtained. Obviously, visual inspection can detect only surface anomalies. However, some internal corrosion processes do produce surface indications, such as pillowing or flaking.

Visual inspection is a quick and economical method of detecting various types of defects before they cause failure. Its reliability depends upon the ability and experience of the inspector. The inspector must know how to search for critical flaws and how to recognize areas where failure could occur. The human eye is a very discerning instrument, and, with training, the brain can interpret images much better than any automated device can. Visual inspection can be done in many ways, either directly or remotely, by using borescopes, fiberscopes, or video cameras. Visual methods can provide a simple and speedy way to

TABLE 6.11 Dominant Sources of Variance in NDE Procedure Application

	Materials	Equipment	Procedure	Calibration	Criteria	Human factors
Liquid penetrant	X					X
Magnetic particle	X	X	X			X
Radiography	X	X	X			X
Manual eddy current		X	X	X	X	X
Automatic eddy current		X	X	X	X	
Manual ultrasonic		X	X	X	X	X
Automatic ultrasonic		X	X	X	X	
Manual thermographic		X	X	X		X
Automatic thermographic		X	X	X	X	

TABLE 6.12 Relative Cost and Requirement Ratings for the Main NDE Techniques

	Cost		Requirements		
	Inspection	Equipment	Skill	Process control	Process variance
Liquid penetrant	Low	Low	High	High	High
Magnetic particle	Low	Moderate	High	High	High
Radiography	Moderate	High	High	High	High
Manual eddy current	Low	Moderate	High	Moderate	Moderate
Automatic eddy current	Moderate	High	Moderate	High	Low
Manual ultrasonic	Low	Moderate	High	Moderate	Moderate
Automatic ultrasonic	Moderate	High	Moderate	High	Low
Manual thermographic	Low	High	High	High	Moderate
Automatic thermographic	Low	High	Moderate	High	Low

TABLE 6.13 General Process Steps for the Main NDE Techniques

Liquid penetrant inspection
   The test object is cleaned to remove both surface and materials in the capillary opening.
   A penetrant fluid is applied and allowed to penetrate into the capillary opening.
   The surface penetrant fluid is removed without removing fluid from the capillary.
   A developer is applied to provide a visible contrast to the penetrant fluid material.
   The test object is visually inspected to detect, classify, and interpret the presence, type, and size of the penetrant indications.

Magnetic particle inspection
   The test object is cleaned to remove surface contaminants.
   A magnetic field is induced in the object.
   A fluid or powder containing finely divided particles that are attracted by the presence of a discontinuity in a magnetic field is applied.
   The test object is visually inspected to detect, classify, and interpret the presence, type, and size of indications.

X-radiographic inspection
   A sheet of X-ray-sensitive film is located on one side of the test object.
   An X-ray source is located on the opposite side of the test object.
   The X-ray source is activated to "expose the film" in a through-transmission mode.
   The film is developed.
   The resultant film image is visually inspected to detect, classify, and interpret the presence, type, and size (magnitude) of included indications.

Eddy-current inspection
   The eddy-current probe is placed in contact with the test object.
   An alternating magnetic field is induced in the probe by an alternating current in the probe coil.
   Eddy-current flow is induced in the test object.
   The magnitude and phase of the induced current flow is sensed by a secondary coil in the probe or by a change of inductance in the probe.
   A localized change in induced current flow indicates the presence of a discontinuity in the test object.
   The size of the discontinuity is indicated by the extent of the response change as the probe is scanned along the test object.

Ultrasonic inspection
   An ultrasonic transducer is located in contact with or in close proximity to the test object.
   The transducer is energized in a pulsed mode to direct and propagate acoustic energy into the test object.
   Acoustic energy is transmitted, reflected, and scattered within the test object.
   Energy within the test object is transmitted or redirected by internal interfaces caused by test object geometry features or internal anomalies.
   Transmitted or redirected energy from the test object is detected by a transducer located on or near the test object.
   The transmitted or redirected energy is analyzed in the time or frequency domain and the internal condition of the test object is determined by interpretation of the pattern and amplitude features.

TABLE 6.13 General Process Steps for the Main NDE Techniques *(Continued)*

Thermographic inspection
  A pulse of thermal energy is introduced into the test object.
  Energy is diffused within the test object according to the thermal conductivity, the thermal mass, inherent temperature differentials, and the time of observation.
  The temperature of the test object surface is monitored by a thermographic camera with a capability for detection in the infrared energy spectrum.
  Interpretation is completed by visually monitoring the relative surface temperature as a function of time and relating temperature differences in the time domain to the internal condition and/or structure of the test object.
  A relative change in surface temperature is indicative of a change in continuity or disbondment in a bonded structure.
  The size of an unbond is indicated by the location of the temperature gradient on the surface at a specific time and is modified by comparison with responses from similar test objects with similar geometry and thermal mass.

assess questionable components and can help in deciding what to do next. Where necessary, permanent records can be obtained by photography or digital imaging and storage.

Visual inspection is often conducted using a strong flashlight, a mirror mounted on a ball joint, and a magnifying aid. Magnifying aids range in power from 1.5× to 2000×. Fields of view typically range from 90 to 0.2 mm, with resolutions ranging from 50 to 0.2 μm. A 10× magnifying glass is recommended for positive identification of suspected cracks or corrosion. The disadvantage of visual inspection is that the surface to be inspected must be relatively clean and accessible either to the naked eye or to an optical aid such as a borescope. Typically, visual inspection lacks the sensitivity of other surface NDE methods. Further, visual methods are qualitative and do not provide quantitative assessments of either material loss or residual strength. In addition, visual inspection techniques can be labor-intensive and monotonous, leading to errors.

Borescopes. A borescope is a long, thin, rigid rodlike optical device that allows an inspector to see into inaccessible areas by transmitting an image from one end of the scope to the other. Certain structures, such as engines, are designed to accept the insertion of borescopes for the inspection of critical areas. A borescope works by forming an image of the viewing area with an objective lens. That image is transferred along the rod by a system of intermediate lenses. The image arrives at the ocular lens, which creates a viewable virtual image. The ocular lens can be focused for comfortable viewing. Borescopes typically range from 6 to 13 mm in diameter and can be as long as 2 m. Borescopes often incorporate a light near the objective lens to illuminate the viewing area. Different borescopes are designed to provide direct, forward oblique, right angle, and retrospective viewing of the area in question.

Fiberscopes. Fiberscopes are bundles of fiber optic cables that transmit light from end to end. They are similar to borescopes, but they are flexible. They can be inserted into openings and curled into otherwise inaccessible areas. They also incorporate light sources for illumination of the subject area and devices for bending the tip in the desired direction. Like borescope images, fiberscope images are formed at an ocular or eyepiece.

Video imaging systems. Video imaging systems (or "videoscopes") consist of tiny charge-coupled device (CCD) cameras at the end of a flexible probe. Borescopes, fiberscopes, and even microscopes can be attached to video imaging systems. These systems consist of a camera to receive the image, processors, and a monitor to view the image. The image on the monitor can be enlarged or overlaid with measurement scales. Images can also be printed on paper or stored digitally to obtain a permanent record. Video images can be processed for enhancing and analyzing video images for flaw detection. Specialized processing algorithms may be applied which can identify, measure, and classify defects or objects of interest.

Advanced methods. Moiré interferometry is a family of techniques that visualize surface irregularities. Many variations are possible, but the technique most applicable to corrosion detection is shadow moiré (sometimes called projection moiré) for surface height determination. The structured light technique is geometrically similar to projected or shadow moiré methods, and can be thought of as an optical straight-edge. Instead of fringe contours, the resultant observation is the departure from straightness of a projected line. The surface profile can be calculated using image processing techniques.

D-Sight has the potential to map areas of surface waviness as well as to identify cracks, depressions, evidence of corrosion, and other surface anomalies. D-Sight is a method by which slope departures from an otherwise smooth surface are visualized as shadows. It can be used in direct visual inspection or combined with photographic or video cameras and computer-aided image processing. The concept of D-Sight is related to the schlieren method for visualizing index of refraction gradients or slopes in an optical system. One possible problem with D-Sight is that the technique shows virtually every deviation on the surface, regardless of whether it is a defect or a normal result of manufacture.

Liquid penetrant inspection. The liquid penetrant NDE method is applied to detection of faults that have a capillary opening to the test object surface. The nature of this NDE method demands that attention be given to material type, surface condition, and rigor of cleaning. Liquid penetrant inspection can be performed with little capital expenditure, and the materials used are low in cost per use. This technique is applicable to complex shapes and is widely used for general product assurance.

This technique is easy, completely portable, and highly accurate if performed properly. It detects open-to-the-surface crack indications. Rigorous surface cleaning is required. This technique is applicable only to cleaned surfaces; unclean ones will give unsatisfactory results. It is readily used on external and accessible surfaces that have been subjected to minimal corrosion deterioration and can be cleaned. It readily detects any open-to-the-surface cracks, surface defects, and pitting.

Magnetic particle inspection. Magnetic particle inspection is applied to the detection of surface-connected or near-surface anomalies in test objects that are made from materials that sustain a magnetic field. Special equipment is required in order to induce the required magnetic field. Procedure development and process control are required in order to use the proper voltage, amperage, and mode of induction. Test object materials must be capable of sustaining an induced magnetic field during the period of inspection. The concentration and mode of application of the magnetic particles must be controlled. Material characteristics or surface treatments which result in variable magnetic properties will decrease detection capabilities. Magnetic particle inspection can be performed with little capital expenditure and, as with the liquid penetrant technique, the materials used are low in cost per use, the technique is applicable to complex shapes, and it is widely used for general product assurance.

Magnetic inspection can be portable. It requires only a magnetization power source, such as that provided by an electrical outlet. It is most frequently used in evaluating the quality of weld deposits and subsurface weld indications such as cracks. This is the preferred method for detecting cracks in deaerators, for example.

Radiographic inspection. Radiographic inspection is a nondestructive method of inspecting materials for surface and subsurface discontinuities. This method utilizes radiation in the form of either x-rays or gamma rays, both of which are electromagnetic waves of very short wavelength. The waves penetrate the material and are absorbed, depending on the thickness or the density of the material being examined. By recording the differences in absorption of the transmitted waves, variations in the material can be detected. The variations in transmitted waves may be recorded by either film or electronic devices, providing a two-dimensional image that requires interpretation. The method is sensitive to any discontinuities that affect the absorption characteristics of the material.

The techniques and technologies of x-ray radiography have most to do with the design of the x-ray tube itself. There are many different types of tubes used for special applications. The most common is the directional tube, which emits radiation perpendicular to the long axis of the tube in a cone of approximately 40°. Another type is the panoramic tube, which emits x-rays in a complete 360° circle. This type of tube would be used, for example, to examine the girth welds in a jet engine with a single exposure.

- *Real-time radiography.* This is the new form; it presents an instant image, much like a video camera. It is mostly used for examining the surfaces of piping beneath insulation with the insulation in place. It is completely portable, and its operators are required to be licensed. This technique allows the instant viewing of a radiographic image on a cathode-ray tube. The image may be captured on any electronic medium in use today. This electronic/digital imaging technique is the only data retention system available.

- *Classical radiography.* This is similar to a medical radiograph that generates a film record. It is a completely portable inspection procedure, and extensive training and licensing of personnel are required. This technique is used to examine piping for interior corrosion and deposits, weld quality, and conditions of internal valving or components. A limitation is that it cannot be used on piping systems filled with water or other liquids, since the radiation cannot penetrate water. Extensive calibration and destructive verification of actual conditions allow achievement of a high level of confidence in the radiographic technique.

Advances in the use of radiography are being made that involve using computers and high-powered algorithms to manipulate the data. This is termed computed tomography, or CT scanning. By scanning a part from many directions in the same plane, a cross-sectional view of the part can be generated, and a two-dimensional view of the internal structure may be displayed. The tremendous advantage of this method is that internal dimensions can be measured very accurately to determine such conditions as wall thinning in tubes, size of internal discontinuities, relative shapes, and contours. More advanced systems can generate three-dimensional scans when more than one plane is scanned. CT scanning is costly and time-consuming. Radiography in general and CT scanning in particular are extremely useful in validating and calibrating other, less complex and less costly methods.

Radioisotope sources can be used in place of x-ray tubes. Radioisotope equipment has inherent hazards, and great care must be taken with its use. Only fully trained and licensed personnel should work with this equipment. As with x-rays, the most common method of measuring gamma ray transmission is with film.

Compton backscatter imaging (CBI) is emerging as a near-surface NDE measurement and imaging technique. CBI can detect critical embedded flaws such as cracks, corrosion, and delaminations in metal and composite aircraft structures. In CBI, a tomographic image of the inspection layer is obtained by raster scanning the collimated source-detector assembly over the object and storing the measured signal as a function of position. Rather than measuring the x-rays that pass through the object, CBI measures the backscattered beam to generate the image. This enables single-sided measurement.

Eddy-current inspection. When an electrically conductive material is exposed to an alternating magnetic field that is generated by a coil of wire carrying an alternating current, eddy currents are induced on and below the surface of the material. These eddy currents, in turn, generate their own magnetic field, which opposes the magnetic field of the test coil. This magnetic field interaction causes a resistance to current flow, or impedance, in the test coil. By measuring this change in impedance, the test coil or a separate sensing coil can be used to detect any condition that would affect the current-carrying properties of the test material. Eddy currents are sensitive to changes in electrical conductivity, changes in magnetic permeability (the ability of a material to be magnetized), the geometry or shape of the part being analyzed, and defects. Among these defects are cracks, inclusions, porosity, and corrosion.

Eddy-current methods are used to measure a variety of material characteristics and conditions. They are applied in the flaw detection mode for the detection of surface-connected or near-surface anomalies. The test objects must be electrically conductive and be capable of uniform contact by an eddy-current probe. Special equipment and specialized probes are required to perform the inspection. Procedure development, calibration artifacts, and process control are required to assure reproducibility of response in the selected test object.

Initially, eddy-current devices utilized a meter to display changes of voltage in the test coil. Currently, phase analysis instruments provide both impedance and phase information. This information is displayed on an oscilloscope or an integrated LCD display on the instrument. Results of eddy-current inspections are obtained immediately. The other type of eddy-current instrument displays its results on planar form on a screen. This format allows both coil impedance components to be viewed. One component consists of the electrical resistance due to the metal path of the coil wire and the conductive test part. The other component consists of the resistance developed by the inducted magnetic field on the coil's magnetic field. The combination of these two components on a single display is known as an impedance plane.

Automated scanning is performed using an instrumented scanner that keeps track of probe position and automated signal detection so that a response map of the test object surface can be generated. Resolution of the inspection system is somewhat dependent on the fidelity of the scan index and on the filtering and signal processing that are applied in signal detection. A scan map can be generated by automated eddy-current scanning and instrumentation systems.

The results of eddy-current inspection are extremely accurate if the instrument is properly calibrated. Most modern eddy-current instruments are relatively small and battery-powered. In general, surface detection is accomplished with probes containing small coils (3 mm diameter) operating at a high frequency, generally 100 kHz and above. Low-frequency eddy current (LFEC) is used to penetrate deeper into a part to detect subsurface defects or cracks in the underlying structure. The lower the frequency, the deeper the penetration. LFEC is generally considered to be between 100 Hz and 50 kHz.

A major advantage of eddy-current NDE is that it requires only minimal part preparation. Reliable inspections can be performed through normal paint or nonconductive materials up to a thickness of approximately 0.4 mm. Eddy-current technology can be used to detect surface and subsurface flaws on single- and multiple-layered materials.

Advanced methods

Scanned pulsed eddy current. This technique for application of eddy-current technology uses analysis of the peak amplitude and zero crossover of the response to an input pulse to characterize the loss of material. This technology has been shown to measure material loss on the bottom of a top layer, the top of a bottom layer, and the bottom of a bottom layer in two-layer samples. Material loss is displayed according to a color scheme to an accuracy of about 5 percent. A mechanical bond is not necessary, as it is with ultrasonic testing. The instrument and scanner are rugged and portable, using conventional coils and commercial probes. The technique is sensitive to hidden corrosion and provides a quantitative determination of metal loss.

Magneto-optic eddy-current imaging. Magneto-optic eddy-current (MOI) images result from the response of the Faraday magneto-optic sensor to the weak magnetic fields that are generated when eddy currents induced by the MOI interact with defects in the inspected material. Images appear directly at the sensor and can be viewed directly or imaged by a small CCD camera located inside the imaging unit. The operator views the image on the video monitor while moving the imaging head continuously along the area to be inspected. In contrast to conventional eddy-current methods, the MOI images resemble the defects that produce them, making the interpretation of the results more intuitive than the interpretation of traces on a screen. Rivet holes, cracks, and subsurface corrosion are readily visible. The image is in video format and therefore is easily recorded for documentation.

Ultrasonic inspection. Ultrasonic inspection, one of the most widely used NDE techniques, is applied to measure a variety of material characteristics and conditions. Ultrasonic examination is performed using a device which generates a sound wave through a piezoelectric crystal at a frequency between 0.1 and 25 MHz into the piece being examined and analyzes the return signal. The device measures the time it takes for the signal to return and the amount and shape of that signal. It is a completely portable device that requires only that the probe be in direct contact with a clean surface in order to obtain accurate information.

Test objects must support propagation of acoustic energy and have a geometric configuration that allows the introduction and detection of acoustic energy in the reflection, transmission, or scattered energy configurations. The frequencies of the transducer and the probe diameter have a direct effect on what is detected. Lowering the testing frequency increases depth of penetration, while increasing the probe diameter reduces the beam spread. Increasing the frequency also increases the beam spread for a given diameter.

Manual scanning is performed using instruments that have an oscilloscope-type readout. Operator interpretation uses pattern recognition, signal magnitude, timing, and respective hand-scan position. Variations in instrument readout and variations in scanning can be significant. Automated scanning is performed using an instrumented scanner that keeps track of probe position and automated signal detection (time, phase, and amplitude), so that a response map of the internal structure of the test object can be generated. The resolution of the system is somewhat dependent on the fidelity of the scan index and on the filtering and signal processing that are applied in signal detection. A scan map may be generated by automated ultrasonic scanning and instrumentation systems.

The most fundamental technique used is that of thickness testing. In this case, the ultrasonic pulse is a compression or longitudinal wave that is sent in a perpendicular direction into the metal being measured. The signal reflects off the back wall of the product being analyzed, and the time of flight is used to establish the thickness. There are instruments that allow the testing to be conducted through paint coatings. This is done by looking at the waveform and selecting the area that represents the actual material, not the signal developed by the coatings.

Techniques have been developed that employ different types of waves, depending on the type of inspection desired. Compression waves are the type most widely used. They occur when the beam enters the surface at an angle near 90°. These waves travel through materials as a series of alternating compressions and dilations in which the vibrations of the particles are parallel to the direction of the wave travel. This wave is easily generated and easily detected, and has a high velocity of travel in most materials. Longitudinal waves are used for the detection and location of defects that present a reasonably large frontal area parallel to the surface from which the test is being made, such as corrosion loss and delaminations. They are not very effective, however, for the detection of cracks which are perpendicular to the surface.

Shear or transverse waves are also used extensively in ultrasonic inspection; these are generated when the beam enters the surface at a moderate angle. Shear-wave motion is similar to the vibrations of a rope that is being shaken rhythmically: Particle vibration is perpendicular to the direction of propagation. Unlike longitudinal waves, shear waves do not travel far in liquids. Shear waves have a velocity that is about 50 percent of that of longitudinal waves in the same material. They also have a shorter wavelength than longitudinal waves, which makes them more sensitive to small inclusions. This also makes them more easily scattered and reduces penetration.

Surface waves (Rayleigh waves) occur when the beam enters the material at a shallow angle. They travel with little attenuation in the direction of the propagation, but their energy decreases rapidly as the wave penetrates below the surface. They are affected by variations in hardness, plated coatings, shot peening, and surface cracks, and are easily dampened by dirt or grease on the specimen.

Lamb waves, also known as plate waves and guided waves, occur when ultrasonic vibrations are introduced at an angle into a relatively thin sheet. A lamb wave consists of a complex vibration that occurs throughout the thickness of the material, somewhat like the motion of surface waves. The propagation characteristics of lamb waves depend on the density, elastic properties, and structure of the material as well as the thickness of the test piece and the frequency of the vibrations. There are two basic forms of lamb waves: symmetrical (dilational) and asymmetrical (bending). Each form is further subdivided into several modes, which have different velocities that can be controlled by the angle at which the waves enter the test piece. Lamb waves can be used for detecting voids in laminated structures, such as sandwich panels and other thin, bonded laminated structures.

Advanced methods

Dripless bubbler. One of the most promising improvements in ultrasonic testing technology is the dripless bubbler. This is a development not in the ultrasonic probe itself but in the mechanism for employing it consistently on curved, irregular, vertical, and inverted surfaces. The dripless bubbler itself is a pneumatically powered device that holds a water column between the ultrasonic probe and the inspected surface. With software control of the movement of the probe, a fast and accurate map of the inspected surface can be obtained.

Laser ultrasound. There is also emerging interest in the area of laser ultrasonics, or laser-based ultrasound (LUS). The innovation is the use of laser energy to generate sound waves in a solid. This obviates the need for a couplant between the transducer and the surface of the inspected material. The initial application of this new technology seems to be directed toward process control. However, the technology can also be applied for thickness measurement, inspection of welds and joints, surface and bulk flaw detection on a variety of materials, and characterization of corrosion and porosity on metals.

Thermographic inspection. Thermographic inspection methods are applied to measure a variety of material characteristics and conditions. They are generally applied in the flaw detection mode for the detection of interfaces and variation of the properties at interfaces within layered test objects. Test objects must be thermally conductive, and the test object surface must be reasonably uniform in color and texture. This technique uses the infrared energy associated with the part or system being examined. It is noninvasive and gives a photographic image of the thermal conditions present on the surface being examined. It can be used to accurately measure metal temperatures to establish whether brittle or overheated conditions exist. The method is a volume inspection process and therefore loses resolution near edges and at locations of nonuniform geometry change.

Manual inspection is performed using manual control of the thermal pulse process and human observation and interpretation of the thermal images produced as a function of time. A false-color thermal map presentation may be used to aid in discrimination of fine image features and pattern recognition. The thermal map may be recorded on videotape as a function of time. Automated scanning is performed using an instrumented scanner which reproducibly introduces a pulse of thermal energy into the test object and synchronizes pulse introduction with the "start time" for use in automated image readout. Automated readout is effected via preprogrammed digital image processing and is test object– and inspection procedure–specific. Several techniques have been developed that use this temperature information to characterize the thermal properties of the sample being tested.

Many defects affect the thermal properties of materials. Examples are corrosion, debonds, cracks, impact damage, and panel thinning. With judicious application of external heat sources, these defects can be detected by an appropriate infrared survey. Uses of thermography techniques currently range from laboratory investigations to field equipment. Thermography, in its basic form, has the limitation that it measures only the surface temperature of the inspected structure or assembly. Therefore, it does not provide detailed insight into defects or material loss located more deeply in the structure. Because it is an area-type technique, it is most useful for identifying areas that should be inspected more carefully using more precise techniques, such as eddy-current and ultrasonic methods.

Thermal wave imaging overcomes some of these limitations by measuring the time response of a thermal pulse rather than the temperature response. The thermal pulse penetrates multiple layers when there is a good mechanical bond between the layers. The benefits of thermal wave imaging technology include the ability to scan a wide area quickly and to provide fast, quantitatively defined feedback with minimal operator interpretation required.

Advanced methods. The raw image displayed by an IR camera conveys only information about the temperature and emissivity of the surface of the target it views. To gain information about the internal structure of the target, it is necessary to observe the target either as it is being heated or as it cools. Since it takes heat from the surface longer to reach a deeper obstruction than to reach a shallow one, the effect of a shallow obstruction appears at the surface earlier than that of a deep one. The thermal response to a pulse over time, color-coded by time of arrival, is displayed as a two-dimensional, C-scan image for interpretation by the operator.

Dual-band infrared computed tomography uses flash lamps to excite the material with thermal pulses and detectors in both the 3–5- and the 8–12-μm ranges to obtain the results. This technique gives three-dimensional, pulsed-IR thermal images in which the thermal excitation provides depth information, while the use of tomographic mapping techniques eliminates deep clutter.

6.6.3 Data analysis

When an NDE process is applied to a test object, the output response to an anomaly within the test object will depend on the form of detection, the magnitude of the feature that is used in detection, and the relative response magnitude of the material surrounding the anomaly. In an ultrasonic inspection procedure, for example, the amplitude of the response from an anomaly within a structure may be used to differentiate the response from the grain structure (noise) surrounding the anomaly. If the ultrasonic procedure (measurement) is applied repetitively to the same anomaly, a distribution of responses to both the anomaly and the surrounding material will be obtained.

The measured response distribution reflects the variance in the NDE measurement process and is typical of that obtained for any measurement process. The response from the surrounding material constitutes the baseline level for use in discrimination of responses from internal anomalies. The baseline response may be termed *noise,* and both the discrimination capability and anomaly sizing capability of the NDE procedure are dependent on the relative amplitudes and the rate of change of the anomaly response with increasing anomaly size (slope). The considerable flaw-to-flaw variance and the variance in signal response to flaws of equal size cause increased spread in the probability density distribution of the signal response. If a threshold decision (amplitude) level is applied to the responses, clear flaw discrimination (detection) can be achieved, as shown in Fig. 6.42. If the same threshold decision level (acceptance criterion) is applied to a set of flaws of a smaller size (as shown in Fig. 6.43), clear discrimination cannot be accomplished.

In this example, the threshold decision level could be adjusted to a lower signal magnitude to produce detection. As the signal magnitude is adjusted downward to achieve detection, a slight increase in the noise level will result in a "false call." As the flaw size decreases, the noise and signal plus noise responses will overlap. In such cases, a downward adjustment in the threshold decision level (to detect all flaws) will result in an increase in false calls. Figure 6.44 shows an example in which the threshold decision level (acceptance criterion) has been adjusted to a level where a significant number of false calls will occur. In this example, a slight change in flaw signal distribution will also result in failure to detect a flaw. The NDE procedure is not robust and is not subject to qualification or certification for purposes of primary discrimination. The procedure may, however, be useful as a prescreening tool, if it is followed by another procedure that provides discrimination of the residuals. For example, a neural network detection process structured to provide discrimination at a high false call rate may be a useful in-line tool if other features are used for purposes of discrimination after the anomaly or variance is identified.

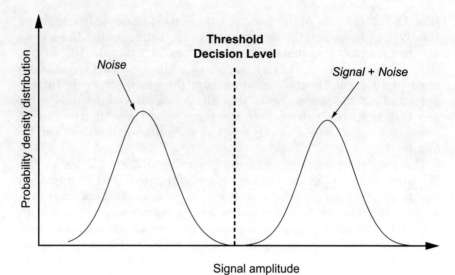

Figure 6.42 Flaw detection at a threshold signal level.

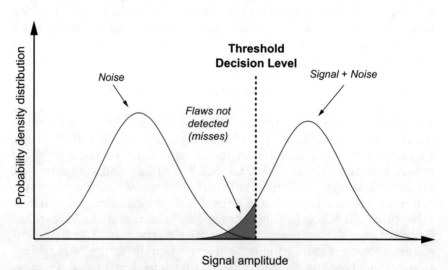

Figure 6.43 Failure to detect smaller flaws at the same threshold signal level.

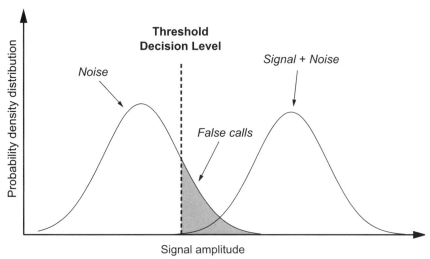

Figure 6.44 Threshold decision level results in false calls.

References

1. Moreland, P. J., and Hines, J. G., *Hydrocarbon Processing,* **17**(1):1–31 (1978).
2. Timmins, P. F., *Predictive Corrosion and Failure Control in Process Operations,* Materials Park, Ohio, ASM International, 1996.
3. Horvath, R. J., The Role of the Corrosion Engineer in the Development and Application of Risk-Based Inspection for Plant Equipment, *Materials Performance,* 39:70–75 (1998).
4. Wheaton, R., The Components of Reliability, *http://www.afe.org/19wheaton.html,* 1998.
5. Douglas, J., The Maintenance Revolution, *EPRI Journal,* **20**(3):6–15 (1995).
6. Nowlan, F. S., and Heap, H. F., *Reliability Centered Maintenance,* AD/AAo66-579, Washington, D.C., National Technical Information Service, 1978.
7. Nowlan F. S., and Heap, H. F., *Reliability Centered Maintenance,* Dolby Access Press, 1979.
8. Griffiths, D., Evolution of Computerized Asset Management and Maintenance, *http://www.afe.org/25cmms.html,* 1998.
9. Coullahan R., and Siegfried, C., Life Cycle Asset Management, *Facilities Engineering Journal* (1996).
10. Marshall, W., Condition Assessment Linked to Predictive Maintenance Modeling: A Valuable Management Tool, *http://www.plsolutions.co.nz,* 1998.
11. Douglas, J., Solutions for Steam Generators, *EPRI Journal,* **20**(3):28–34 (1995).
12. *Corrosion Monitoring System Enables Utility to Avoid Relining Stack and Ducts,* EPRI RP1871-17, Palo Alto, Calif., Electric Power Research Institute, 1990.
13. NRC, Steam Generator Tube Issues, *http://www.nrc.gov,* 1998.
14. Hoffman, C., 20,000-Hour Tuneup, *Air & Space,* **12**:39–45 (1997).
15. Townley, N. J., Roberge, P. R., and Little, M. A., Corrosion Maintenance through Knowledge Re-Use: The CP140 Aurora, *Canadian Aeronautics and Space Journal,* **43**:189–194 (1997).
16. Zollars, B., Salazar, N., Gilbert, J., et al., *Remote Datalogger for Thin Film Sensors,* Houston, Tex., NACE International, 1997.
17. Kelly, R. G., Yuan, J., Jones, S. H., Blanke, W., Aylor, J. H., Wang, W., and Batson, A. P., "Embeddable Microinstruments for Corrosion Monitoring," in *Corrosion 97,* Houston, Tex., NACE International, 1997, pp. 1–12.

18. Dean, S. W., Overview of Corrosion Monitoring in Modern Industrial Plants, in Moran, G.C., and Labine, P. (eds.), *Corrosion Monitoring in Industrial Plants Using Nondestructive Testing and Electrochemical Methods*, ASTM STP 908, Philadelphia, American Society for Testing and Materials, 1986, pp. 197–220.
19. Cooper, G. L., Sensing Probes and Instruments for Electrochemical and Electrical Resistance Corrosion Monitoring, in Moran, G. C., and Labine, P. (eds.), *Corrosion Monitoring in Industrial Plants Using Nondestructive Testing and Electrochemical Methods*, ASTM STP 908, Philadelphia, American Society for Testing and Materials, 1986, pp. 237–250.
20. Denzine, A. F., and Reading, M. S., An Improved, Rapid Corrosion Rate Measurement Technique for All Process Environments, *Materials Performance,* **37:**35–41 (1998).
21. Britton, C. F., and Tofield, B. C., Effective Corrosion Monitoring, *Materials Performance,* **4:**41–44 (1988).
22. Milliams, D. E., and Van Gelder, K., Corrosion Management, *Materials Performance,* **35:**13–15 (1996).
23. Wietek, B., "Monitoring the Corrosion of Steel in Concrete," *F.I.P. Symposium on Post-Tensioned Concrete Structures,* Slough, U.K., Concrete Society, 1996.
24. Flis, J., Pickering, H. W., and Osseo-Asare, K., Assessment of Data from Three Electrochemical Instruments for Evaluation of Reinforcement Corrosion Rates in Concrete Bridge Components, *Corrosion,* **51:**602–609 (1995).
25. Elsener, B., and Bohni, H., Potential Mapping and Corrosion of Steel in Concrete, in Berke, N. S., Chaker, V., and Whiting, D. (eds.), *Corrosion Rates of Steel in Concrete,* ASTM STP 1065, Philadelphia, American Society for Testing and Materials, 1990, pp. 143–156.
26. Macdonald, D. D., Urquidi-Macdonald, M., Rocha-Filho, R. C., et al., Determination of the Polarization Resistance of Rebar in Reinforced Concrete, *Corrosion,* **47:**330–335 (1991).
27. Jafar, M. I., Dawson, J. L., and John, D. G., Electrochemical Impedance and Harmonic Analysis Measurements on Steel in Concrete, in Scully, J. R., Silverman, D. C., and Kendig, M. W. (eds.), *Electrochemical Impedance: Analysis and Interpretation,* ASTM STP 1188, Philadelphia, American Society for Testing and Materials, 1993, pp. 384–403.
28. Schiessl, P., and Raupach, M., "Macrocell Steel Corrosion in Concrete Caused by Chlorides," *Second CANMET/ACI International Conference on Durability of Concrete,* Montreal, Canada, CANMET, 1991, pp. 565–583.
29. Weiermair, R., Hansson, C. M., Seabrook, P. T., and Tullmin, M., "Corrosion Measurements on Steel Embedded in High Performance Concrete," 1996. *Third CANMET/ACI Conference on Performance of Concrete in Marine Environment,* CANMET, St. Andrews by the Sea, 1996, pp. 293–308.
30. Bertocci, U., A Comparison of Electrochemical Noise and Impedance Spectroscopy for the Detection of Corrosion in Reinforced Concrete, in Kearns, J. R., Scully, J. R., Roberge, P. R., et al. (eds.), *Electrochemical Noise Measurement for Corrosion Applications,* ASTM 1277, Montreal, American Society for Testing and Materials, 1996, pp. 39–58.
31. Roberge, P. R., Tullmin, M. A. A., Grenier, L., et al., Corrosion Surveillance for Aircraft, *Materials Performance,* **35:**50–54 (1996).
32. Winters, M. A., Stokes, P. S. N., and Nichols, H. F., Simultaneous Corrosion and Fouling Monitoring under Heat Transfer in Cooling Water Systems, in Kearns, J. R., Scully, J. R., Roberge, P. R., Reichert, D. L., and Dawson, J., *Electrochemical Noise Measurements for Corrosion Applications,* STP 1277, Philadelphia, American Society for Testing and Materials, 1996, pp. 230–246.
33. Walsh, T. G., "Continuous On-Line Weld Corrosion Monitoring for the Oil and Gas Industry," in Revie, R. W., and Wang, K. C., *International Conference on Pipeline Reliability,* Ottawa, Canada, CANMET, 1992, pp. 17-1–17-7.
34. Agarwala, V. S., "Chemical Sensors for Integrity of Coatings," *Tri-Service Conference on Corrosion,* 1992, pp. 315–325.

35. Johnson, R. E., and Agarwala, V. S., "Fluorescence Based Chemical Sensors for Corrosion Detection," in *Corrosion 97,* Houston, Tex., NACE International, 1997, pp. 1–5.
36. Butler, M. A., and Ricco, A. J., Chemisorption-induced Reflectivity Changes in Optically Thin Silver Films, *Applied Physics Letters,* **53:**1471–1473 (1988).
37. Smyrl, W. H., and Butler, M. A., Corrosion Sensors, *The Electrochemical Society Interface,* **2:**35–39 (1993).
38. Bennett, K. D., and McLaughlin, L. R., "Monitoring of Corrosion in Steel Structures using Optical Fiber Sensors," in *Proceedings of SPIE—The International Society for Optical Engineering,* 1995, 2446:48–59.
39. Poland, S. H., Duncan, P. G., Alcock, M. A., Zeakes, J., Sherrer, D., Murphy, K. A., and Claus, R. O., "Corrosion Sensing Technique Using Metal Coated Fiber Optic," *Strain Gages—40th International Symposium,* Anaheim, Calif., 1995.
40. Ahlberg, H., Lundquist, S., Tell, R., et al., Laser Spectroscopy for In Situ Ammonia Monitoring, *Spectroscopy Europe,* **6:**22 (1994).
41. Mendoza, E. A., Khalil, A. N., Sun, Z., et al., "Embeddable Distributed Moisture and pH Sensors for Non-Destructive Inspection of Aircraft Lap Joints," in *Proceedings of SPIE—The International Society for Optical Engineering,* 1995, 2455:102–112.
42. Fuhr, P. L., Ambrose, T. P., Huston, D. R., et al., Fiber Optic Corrosion Sensing for Bridges and Roadway Surfaces, in *Proceedings of SPIE—The International Society for Optical Engineering,* 1995, 2446:2–8.
43. Cosentino, P., Grossman, B., Shieh, C., et al., Fiber-Optic Chloride Sensor Development, *Journal of Geotechnical Engineering,* **121**(8): 610–617 (1995).
44. Fuhr, P. L., and Huston, D. R., Corrosion Detection in Reinforced Concrete Roadways and Bridges via Embedded Fiber Optic Sensors, *Smart Materials and Structures,* **7**(2):217–228 (1998).
45. Melle, S. M., Liu, K., and Measures, R. M., A Passive Wavelength Demodulation System for Guided-Wave Bragg Grating Sensors, *IEEE Photonics Technology Letters,* **4:**515–518 (1992).
46. Bray, D. E., and Stanley, R. K., *Nondestructive Evaluation,* New York, McGraw-Hill, 1989.
47. Rummel, W. D., and Matzkanin, G. A., *Nondestructive Evaluation (NDE) Capabilities Data Book.* Austin, Tex., Nondestructive Testing Information Analysis Center (NTIAC), 1997.

Chapter 7

Acceleration and Amplification of Corrosion Damage

7.1	Introduction	486
7.2	Corrosion Testing	488
	7.2.1 Corrosion tests and standards	491
	7.2.2 Examples of corrosion acceleration	500
	The anodic breakthrough method for testing anodized aluminum	500
	Intergranular anodic test for heat-treatable aluminum alloys	505
	The corrosion resistance of aluminum and aluminum-lithium alloys in marine environments	507
	7.2.3 Laboratory tests	512
	Cabinet tests	513
	Immersion testing	516
	High-temperature/high-pressure (HT/HP) testing	517
	Electrochemical test methods	522
	7.2.4 Field and service tests	555
	Selecting a test facility	557
	Types of exposure testing	557
	Optimizing test programs	559
7.3	Surface Characterization	562
	7.3.1 General sensitivity problems	566
	7.3.2 Auger electron spectroscopy	566
	7.3.3 Photoelectron spectroscopy	567
	7.3.3 Rutherford backscattering	568
	7.3.5 Scanning probe microscopy (STM/AFM)	569
	7.3.6 Secondary electron microscopy and scanning Auger microscopy	571
	SEM	571
	SAM	572
	7.3.7 Secondary ion mass spectroscopy	572
References		574

7.1 Introduction

The corrosion resistance of metals is one of many quality assurance parameters that determine the operational lifetime of systems and components. Beyond the basic properties of the metal related to chemical composition, structure, and surface finish, an investigator needs to consider the requirements for the metal in terms of achieving a necessary level of corrosion resistance. This is particularly true for metals and alloys that are generally used in their passive state, such as stainless steels. Other fundamental characteristics of a metallic material also have to be considered in planning a testing program. These characteristics are determined by the production history of the material and the final forming, machining, welding, and heat-treating steps.

- *Forming.* Forming can affect the structure of the metal profoundly. For example, forming can create internal stresses that may lead to such detrimental manifestations as stress corrosion cracking seen in brasses and stainless steels. In other cases, structural changes from forming can cause stress-induced intergranular corrosion in aluminum alloys. During forming, the metal surface is often contaminated with aggressive substances or substances that hinder subsequent coatings, for example, fatty acid esters.
- *Machining.* Machining, including grinding, grit blasting, and mechanical polishing, affects the surface structure of the metal and its profile. Machining processes can chemically change the surface of the metal by adsorption or inclusion of components from coolants, grinding compounds, and blasting media. In addition, local high temperatures often occur during machining operations, resulting in substantially changed chemical or microstructural properties. The interaction of the mechanical action and the presence of coolants with these high temperatures tends to alter the surface properties of the metal during machining. For example, the contamination of stainless steel surfaces with carbon steel particles from cutting tools, grinding media, and grit blasting can result in severe corrosion damage to the stainless steel. Residues of active anions from electrochemical machining processes, especially electrochemical deburring, can also cause damage. Furthermore, some heat-treatment methods produce similar results.
- *Welding.* Welding is another processing step that changes the metal structure and can have significant effects on the corrosion behavior of the metal. Galvanic corrosion cells can arise between the weld metal, the heat-affected zone (HAZ), and the parent plate. Weld filler metals can vary in composition, and flux residues can alter the metal surface. In addition, the heat-affected zone of the base metal

sometimes receives a damaging heat treatment, and the heating and cooling causes residual stresses in the structure. Weld spatter and weld oxides tend to drastically reduce the corrosion resistance of stainless steels, for example.

- *Heat treatment.* A large group of iron-based alloys has been found to be susceptible to rapid intergranular attack in a wide range of plant environments when the compositions at the grain boundaries have been changed by equilibrium segregation of alloying elements, especially the precipitation of carbides, nitrides, and other intermetallics. These changes are a result of exposure of the alloys during production of mill forms (rods, sheet, plates, and tubes) to temperatures at which solid-state reactions occur preferentially at grain boundaries. Because welding operations are used in the production of tubes from sheet material and during shop fabrication and field erection, there are further opportunities for the exposure of alloys to the range of temperatures that may result in the depletion of essential chromium.

Figure 7.1 illustrates the weld decay zone as a function of the welding temperature of a stainless steel containing what was a common carbon content only a decade ago. The extent of sensitization for a given temperature and time was found to depend very much on the carbon content. An 18-8 stainless steel containing more than 0.1% C may be severely sensitized after heating for 5 min at 600°C, whereas a similar alloy containing 0.06% C is affected less. The physical properties of stainless steels do not change greatly after sensitization. Because precipitation of chromium carbide accompanies sensitization, the alloy becomes slightly stronger and slightly less ductile. Damage occurs only upon exposure to a corrosive environment, with the alloy corroding along grain boundaries at a rate depending on the severity of the environment and the extent of sensitization.

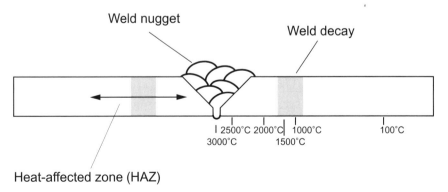

Figure 7.1 Weld decay zone as a function of the welding temperature of stainless steel.

In order to quantify the corrosion resistance of a material, it is common practice to submit the material to harsher environments than are normally encountered in service, hoping to accelerate the damage. Alternatively, a corroded surface and the corrosion products formed during normal exposure can be studied with very sensitive surface analysis techniques, with the aim of amplifying the visibility and characteristics of the damage. Since most corrosion processes occur at the metal/environment interface, much progress in the study of corrosion mechanisms can be related to the gigantic advances made in surface analysis techniques. In fact, scientists involved in the study of fundamental processes of corrosion have often been the first to explore the application of new surface analysis techniques to materials engineering problems.

There are, however, a number of limitations that have prevented many of those techniques from becoming more widely used. The first of these limitations is that corrosion processes are inherently dynamic, while most techniques are essentially static in nature. In this chapter, the reader will be introduced to corrosion testing and to some of the modern techniques that are available for surface analysis in corrosion studies.

7.2 Corrosion Testing

Test programs can provide useful information for a variety of tasks, such as the development of new materials and coatings and the choice of protective schemes for new and old equipment. Test methods for determining corrosion resistance are specific and must be based on the conditions prevailing in certain environments and applications. A large number of factors affect corrosion behavior, and therefore there is no universal corrosion test. The most reliable indication of corrosion behavior is service history. However, that information is rarely available exactly as needed, and therefore other tests are required, ranging from simple field trials to highly accelerated laboratory tests. It is the need to obtain information beyond the service history that introduces ambiguity into corrosion testing.

In most practical applications of materials, knowledge of the corrosion properties of the candidate materials is essential for selection purposes. Such knowledge can be derived from service experience, which usually involves long-term exposure under frequently ill-defined and ill-documented conditions, and corrosion testing.[1] However, because the corrosion mechanism of a system depends on many variables, corrosion testing itself has many pitfalls. Often tests are designed to investigate specific forms of corrosion or specific applications. In this respect, uniform corrosion is the least disturbing. However, tests relating to localized corrosion are far from ideal.

Corrosion tests are an important tool for a variety of industrial tasks that can vary greatly over the life of a system. A decision that makes economic sense at design time may not make any sense by the time the same system is in its 20th year of operation. In some process applications, the materials selected may have been the optimum choice for the initial operating conditions. However, unintended minor changes in the operating conditions can easily increase the corrosivity of a process. For tests to yield meaningful results, knowledge of the environment that exists under actual service conditions is necessary. Quite often the water quality within a plant, under normal operating conditions, differs significantly from that at the intake to the plant. In order to conduct realistic corrosion tests, these variations must be taken into account. The bulk environmental conditions can be clean seawater, e.g., around offshore structures and some power stations. In other instances the water is polluted or brackish, while in still other cases, e.g., ships, a variety of water qualities will be encountered during service.[1]

Some of the factors leading to corrosion damage can be reproduced relatively easily by creating a situation favorable to their occurrence. However, other factors depend entirely on the development of local defects that often become visible only after long and highly variable periods of exposure, such as the effects caused by the neutral salt spray test commonly known as ASTM B 117, Method for Salt Spray (Fog) Testing. When an experiment or test is planned, many factors have to be considered. The following list enumerates some of the most standard considerations for the design of a test program:[2]

- What are the objectives of the test?
- How should the results be interpreted?
- How can the information be integrated with earlier or other tests?
- How many specimens are available, and what is their production schedule (batch, sequential)?
- How many factors control the specimen's behavior?
- How many factors are to be included in the tests?
- Which of these factors interact and which have negligible interaction?
- What type of data are to be measured?
- Is the sample homogeneous?
- How representative is the sample?
- Are the tests destructive?
- How expensive are the tests and/or specimens?

- How much control is there over testing?
- How difficult would it be to include human errors of different kinds in the planning?

With such a long list of questions and the continuously increasing number of testing methods, it is important to simplify the design of test plans by adopting a testing strategy that relates requirements to the main test parameters. The decision tree presented in Fig. 7.2 has been developed to facilitate the selection of tests designed to verify the susceptibility of steels to various forms of stress corrosion cracking (SCC). The strategy would be to start with the most severe and least expensive SCC test, i.e., the slow strain-rate test, in which a bar made from the relevant material is exposed to the environment of interest and slowly monotonically strained to fracture.[3] When cracks are found, the susceptibility of the material should then be further evaluated by performing a battery of other tests designed to differentiate among the various mechanisms leading to SCC and hydrogen embrittlement.

Statistical methods are essential for determining the significance levels of results and corresponding material specifications. Corrosion resistance is only one of many characteristics of a material. Together with the physical, mechanical, and fabrication properties, the corrosion resistance determines the applicability of a material for a specific purpose. These properties may be measured or verified by tests. However, unlike physical and mechanical results, which can be used immediately, corrosion resistance results are often presented in a descriptive or qualitative manner and therefore are difficult to utilize. In order to use the results of these tests for life prediction, consideration of the methodologies presented in Chap. 4, Modeling, Life Prediction, and Computer Applications, is recommended.

Test methods for determining corrosion resistance are specific and must be based on the conditions prevailing and the materials to be used, including coatings and other protective measures planned for the specific application. All these details, including the specification ranges for significant variables, must be determined from individually formulated tests based on the desired service life and other requirements of an application. The emphasis placed on the individual characteristics of a test program and the evaluation methods for each metal-environment combination does not preclude the possibility of standardizing the testing and evaluation methods because many applications are identical or similar and the information gathered from one system is applicable to others.

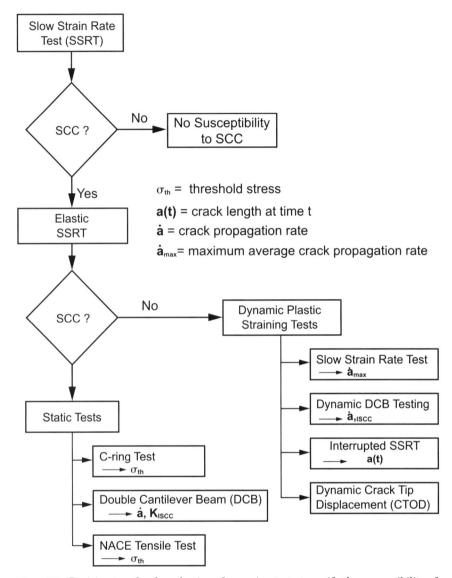

Figure 7.2 Decision tree for the selection of corrosion tests to verify the susceptibility of steels to SCC.

7.2.1 Corrosion tests and standards

There is a multitude of organizations around the world dealing with the production of test methods and standards related to the overall behavior and performance of materials. Organized in 1898, the American Society for Testing and Materials (ASTM) has grown into one of the largest institutions of this kind. ASTM is a not-for-profit organization that provides a forum for producers, users, ultimate consumers, and those having a general interest to meet on common ground and write standards for materials, products, systems, and services. There are 132 ASTM main technical committees, and each is divided into subcommittees. The subcommittee is the primary unit in this organization, as it comprises the highest degree of expertise in any given area. Subcommittees are further subdivided into task groups. Committee G-1, Corrosion of Metals, is thus subdivided into the following subcommittees:

G01.02	Terminology
G01.03	Computers in Corrosion
G01.04	Atmospheric Corrosion
G01.05	Laboratory Corrosion Tests
G01.06	Stress Corrosion Cracking and Corrosion Fatigue
G01.07	Galvanic Corrosion
G01.08	Corrosion of Nuclear Materials
G01.09	Corrosion in Natural Waters
G01.10	Corrosion in Soils
G01.11	Electrochemical Measurements in Corrosion Testing
G01.12	In-Plant Corrosion Tests
G01.14	Corrosion of Reinforcing Steel
G01.91	Standing Committee on Editorial Review
G01.93	Standing Committee on Long Range Planning
G01.95	Standing Advisory Committee for ISO/TC 156
G01.96	Standing Committee on Awards
G01.97	Publicity, Symposia and Workshops
G01.99	Standing Committee on Liaison
G01.99.01	Corrosion of Implant Materials

Besides its regular standard-development meetings, the G-1 committee has sponsored an impressive series of highly focused technical symposia that have led to the publication of over 1300 special technical publications (STP). Committee G-1 has also produced some generic reference documents summarizing state-of-the-art information related to corrosion testing. One such publication, *Corrosion Tests and Standards,* is a very valuable source of information for planning corrosion tests.[4] The information contained in that publication summarizes the efforts of over 400 experts in the field of corrosion testing and evaluation. The ASTM corrosion test handbook is highly redundant by design, and its users will find considerable overlap of subject matter (Fig. 7.3). For example, a specific type of corrosion can be thoroughly discussed in the section Testing for Corrosion Types and in the section Testing in Environments. If a specific metal or alloy is susceptible to that type of corrosion, the subject would also be discussed in the appropriate chapter in Materials Testing. And when a specific industry is involved, the appropriate chapter under Testing in Industries would include a discussion on testing for that type of corrosion in that industry. The test handbook is divided into the following five main sections:

1. *Types of tests.* Each chapter includes basic principles, describes test techniques and important variables, discusses testing considerations such as specimen preparation and evaluation, and includes pertinent standards used.

2. *Testing for Corrosion Types.* Each chapter provides an overview and includes a description of the basic principles and factors controlling the type of corrosion.

3. *Testing in Environments.* The chapters in this section provide a description of each environment, including factors and variables affecting corrosion rates and mechanisms, and the unique characteristics of testing in the specific environment.

4. *By Materials.* This section includes a discussion of the nature of each material, such as the effects of composition, alloying, metallurgical treatments, microstructure, surface effects, and natural protective films on the corrosion behavior.

5. *Testing in Industries.* The chapters in this section provide an overview of the unique situations encountered by various industries, and how corrosion tests are used to combat the corrosion problems faced in these industries.

The development of laboratory corrosion tests should be based on a previous determination of the dominant corrosion factors. Even if the preferred practice is to design such tests so that they represent the most severe conditions for the type of corrosion involved, it is still important to investigate the kinetic components involved in corrosion problems in order to understand the mechanisms and causes for failure. With these points in mind, it is useful to consider how realistic corrosion acceleration may be achieved. Raising the temperature can be useful but may cause changes in the form and nature of hydrous gels, which are often important in the initial stages of corrosion.[5] Increasing the concentration or corrosiveness of salt spray, for example, may not necessarily be appropriate during cyclic testing, since

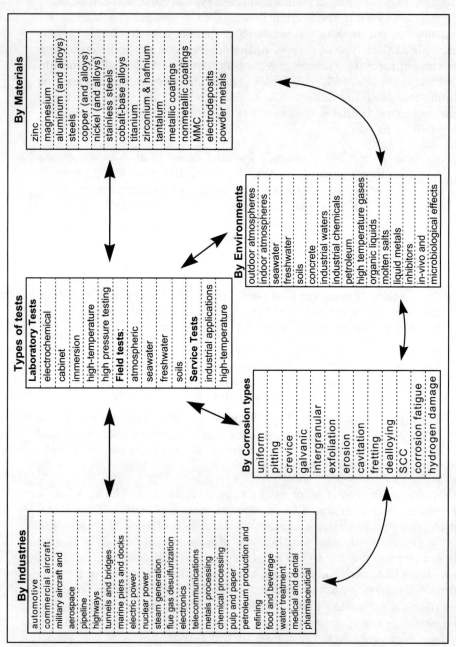

Figure 7.3 A graphical representation of the highly redundant index of the *Corrosion Tests and Standards* handbook.

even an initially dilute spray will, after a sufficient number of cycles, result in the solubility of ionic species being exceeded.[6]

Generally, corrosion products developed in synthetic environments such as those produced in the ASTM B 117 test are substantially different from those produced during natural weathering or even during wet-dry mixed salt spray tests.[5] For example, corrosion of aluminum or zinc specimens in B 117 primarily produces soluble species such as $AlCl_3$ or $ZnCl_2$, and so little corrosion product remains on surfaces. Exposures in a wet-dry test, in contrast, cause the formation of corrosion products on those metals that are more representative of those formed during natural exposure. On aluminum, for example, hydrated aluminas containing chloride and amorphous material are produced in both the high-sulfate and high-chloride cyclic salt spray tests.[5] The reality can be even more complex, as illustrated in Table 7.1, where it can be seen that the products found on specimens exposed to real environments often consist of corrosion products mixed with various foreign materials.[7]

A good example of an element that can be reproduced and accelerated in a laboratory environment is the formation of occluded cells; this can be achieved with multiple crevice assemblies, as described in ASTM G 78, Standard Guide for Crevice Corrosion Testing of Iron-Base and Nickel-Base Stainless Alloys in Seawater and Other Chloride-Containing Aqueous Environments. In this test, washers make a number of contact sites on either side of the specimens (Fig. 7.4). The number of sites showing attack in a given time can be related to the resistance of a material to initiation of localized corrosion, and the average or maximum depth of attack can be related to the rate of propagation. The large number of sites in duplicate or triplicate specimens is amenable to probabilistic evaluation. The same test can be extended to other alloy systems or situations as illustrated in Figs. 7.5 to 7.8, which show the results of four-month exposure of four aircraft aluminum materials partly submerged in a circulating seawater tank.[8]

In other cases, the effect of a test on one of the elements contributing to the corrosion damage can be quantified and recorded for the evaluation of the materials being tested. Amplification of the impact that corrosion has on materials is particularly attractive when the results of a test cannot be easily evaluated. Evaluating test results can be difficult either because these results depend on a slow, solid state transformation of the materials or because they are produced by tests that run for a prespecified time and end with a pass/fail assessment, thus generating censored data. A good example of monitoring specific signals provoked by a particularly aggressive environment is the automated stress corrosion testing method called the ASCOR (automated stress corrosion ring) test, which was specifically developed to evaluate the performance of

TABLE 7.1 Results of X-Ray Diffraction of Products Found on Specimens Exposed to Real Environments

Sample description	Chemical or mineral name*	Chemical formula
Product formed on magnesium during 3-month immersion in tap water	Nesquehonite *Calcium fluosilicate* *Beta silicon carbide* *Sodium sulfide* *Sodium fluoride* Magnesium carbonate chloride hydroxide hydrate Magnesium pyrophosphate *Anorthoclase* *Alpha cristobalite* *Sodium hydroxide* Calcium aluminum oxide sulfate	$MgCO_3 \cdot 3H_2O$ $CaSiF_6$ $\beta\text{-}SiC$ Na_2S NaF $MgCl_2 \cdot MgCO_3 \cdot Mg(OH)_2 \cdot 6H_2O$ $Mg_2P_2O_7$ $(Na,K)AlSi_3O_8$ SiO_2 $NaOH$ $Ca_4Al_6O_{12}SO_4$
Substance found on heat exchanger	Halite	$NaCl$
Substance found beneath paint on metal surface	*Alpha quartz*	SiO_2
Product formed on automobile bumper support during 3-year service	Lepidocrocite Goethite	$\gamma\text{-}Fe_2O_3 \cdot H_2O$ $Fe_2O_3 \cdot 2H_2O$
Product from conversion unit in marine environment	Zinc ferrite Cobalt ferrite *Halite* Chromic oxide Nickel, zinc ferrospinel *Sodium fluothorate* *Embolite* Magnesioferrite *Beryllium palladium* Magnetite *Nickel titanium*	$ZnO \cdot Fe_2O_3$ $CoO \cdot Fe_2O_3$ $NaCl$ Cr_2O_3 $(Ni,Zn)O \cdot Fe_2O_3$ $Na_3Th_2F_{11}$ $Ag(Cl,Br)$ $MgFe_2O_4$ $BePd$ Fe_3O_4 $NiTi$
Product formed on copper during 3-month immersion in tap water	Botallackite *Ilvaite*	$CuCl_2 \cdot 3Cu(OH)_2 \cdot 3H_2O$ $Ca(Fe,Mn,Mg)_2(Fe,Al)(SiO_4)_2OH$
Product from Al-Cu alloy exposed to deep-sea environment	Ammonium copper fluoride dihydrate *Potassium cyanide* Chi alumina Calcium aluminate *Alpha cadmium iodide*	$(NH_4)_2 \cdot CuF_4 \cdot 2H_2O$ KCN Al_2O_3 $3CaO \cdot Al_2O_3$ CdI_2

TABLE 7.1 Results of X-Ray Diffraction of Products Found on Specimens Exposed to Real Environments (*Continued*)

Sample description	Chemical or mineral name*	Chemical formula
Product from Al-Zn-Mg-Cu alloy exposed to deep-sea environment	Chi alumina *Alpha cadmium iodide*	Al_2O_3 CdI_2
Product from Al-Mn alloy exposed to deep-sea environment	Ammonium copper fluoride dihydrate *Nobleite*	$(NH_4)_2CuF_4 \cdot 2H_2O$ $CaB_6O_{10} \cdot 4H_2O$

*Substances shown in italics are not corrosion products of the primary metals or alloys involved in the system.

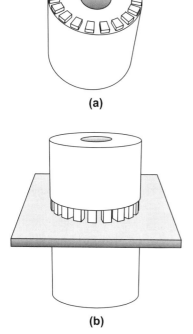

Figure 7.4 A schematic representation of (*a*) the washer and (*b*) a washer assembly for conducting an ASTM G 78 crevice susceptibility test.

aluminum alloys submitted to ASTM G 44, Standard Practice for Evaluating Stress Corrosion Cracking Resistance of Metals and Alloys by Alternate Immersion in 3.5% NaCl Solution. This method involves testing cylindrical and sheet specimens in a loading ring instrumented with strain gauges to measure the load.[9] Initiation of a stress corrosion

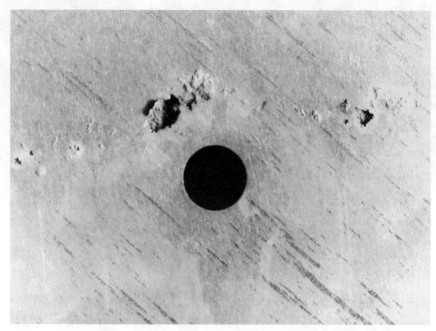

Figure 7.5 Appearance of 8090-T851 aluminum panels with crevice washers after partial immersion in seawater for 4 months. The air/water interface was near the top of the central hole.

Figure 7.6 Appearance of 7075-T6 aluminum panels with crevice washers after partial immersion in seawater for 4 months. The air/water interface was near the top of the central hole.

Acceleration and Amplification of Corrosion Damage 499

Figure 7.7 Appearance of 2090-T8 aluminum panels with crevice washers after partial immersion in seawater for 4 months. The air/water interface was near the top of the central hole.

Figure 7.8 Appearance of 2024-T3 aluminum panels with crevice washers after partial immersion in seawater for 4 months. The air/water interface was near the top of the central hole.

crack results in a small load decrease that can be recorded at regular intervals.

In coupon immersion tests, the accurate determination of low corrosion rates may not be achievable, given the uncertainty of weight loss measurements, when changes in the environment do not permit long exposure. In such cases, the use of additional techniques, such as solution analysis by inductively coupled plasma (ICP) or, alternatively, electrochemical techniques such as electrochemical impedance spectroscopy (EIS),[10] to monitor and record the progress of corrosion by means of amplification has been proposed. The same reasoning can be applied to monitor the progress of a corrosion situation in environmental chambers. The sensitivity of electrochemical techniques to changes in a metallic interface during the first moments of localized corrosion can be used to predict long-term exposure results. While polarization test techniques cannot easily monitor the progress of a localized corrosion situation, they can reliably detect the onset of such problems.[11]

7.2.2 Examples of corrosion acceleration

The following examples illustrate efforts to quantify the results of corrosion tests by acceleration of the processes and amplification of the resulting effects.

The anodic breakthrough method for testing anodized aluminum. Quality control for aluminum anodizing presents a difficult analytical problem. The use of conventional coating thickness measurements is not satisfactory, since the anodizing agents will have dissolved some of the substrate material as the oxide layer was formed. Techniques that determine the coating weight per unit area could be useful in evaluating the extent of anodized film coverage, but these techniques are not sensitive to flaws such as porosity or the presence of contaminants in the coating. Such factors can strongly influence the corrosion protection afforded by the anodized layer. The Standard Method for Measurement of Anodic Coatings on Aluminum (ASTM B 457) describes a procedure to determine the impedance at 1000 Hz as a measure of the quality of sealing anodized aluminum. While this method is applicable to the rapid, nondestructive testing of anodic coatings, its results were never related to the actual corrosion resistance provided by these coatings.

In the absence of a better testing method, industrial and military specifications often require the use of a 300+-h salt spray exposure method to test anodized aluminum coupons and evaluate the quality of the anodizing processes. Not only is the salt spray (fog) exposure test time-consuming, but its results are highly interpretive and therefore relatively imprecise. The long duration of the test itself makes it a poor tool for monitoring daily plant operation of anodizing baths. The anodic breakthrough test method is based on electrochemical activation of the material under test by accelerating the breakdown of the anodized barrier. In this activation step, anodized layers are partially degraded by applying a potential across the metal/electrolyte interface. The total charge passed between the test coupon and the counterelectrode can then be related to the resistance of the coating to corrosion.[12,13]

A cell consisting of a hollow plastic cylinder is quite adequate for this test. The test coupon can be inserted between the hollow cylinder and the solid base, which also serves as the bottom of the cell (Fig. 7.9). A

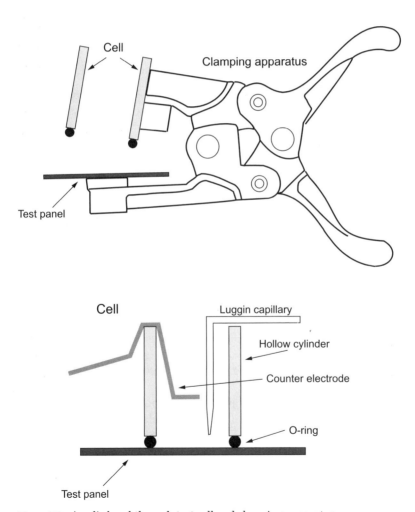

Figure 7.9 Anodic breakthrough test cell and clamping apparatus.

S30400 stainless steel counterelectrode is typically placed above the test coupon. The reference electrode can be contained in a Luggin capillary attached to the main cylinder, and a rubber O-ring was used along with a clamp to create a watertight seal. An applied potential of 600 mV vs. SCE (25°C) for 7 min has been found suitable for the characterization of the breakdown of anodized aluminum. Shorter anodic polarization periods would be more appropriate for less protective coatings, such as those produced with chemical passivation. The current is monitored during the application of the potential, and its integration is used to quantify the breakdown of the anodized barrier. The specimen is exposed in part to a solution containing boric acid and NaCl, with the pH adjusted to 10.5 using a NaOH solution. The current can be monitored during the application of the potential and integrated to quantify the breakdown of the anodized barrier. The anodic current can also be considered to be equivalent to a metal dissolution or corrosion current and converted into corrosion penetration rates using Faraday's law expressed in Eq. (7.1):

$$\text{Corrosion rate} = \frac{3.3\, i_{corr} \cdot \text{E.W.}}{d} \qquad (7.1)$$

where corrosion rate is in $\mu m \cdot year^{-1}$, i_{corr} is in $\mu A \cdot cm^{-2}$, E.W. is the equivalent weight of the element being oxidized (8.99 $g \cdot eq^{-1}$ in the case of Al alloys), and d is the density of the element being oxidized (2.699 $g \cdot cm^{-3}$ for aluminum).

Table 7.2 illustrates typical results obtained with the anodic breakthrough method on eight types of panels, and Table 7.3 compares the observations made after exposing the same types of panels to salt fog testing.[14] Table 7.4 contains corrosion rates determined using the anodic breakthrough method for 2024-T3 specimens that were all anodized in chromic acid with different procedural variants. A wide range of corrosion rates was observed. The worst case was obviously bare aluminum, while lower corrosion rates were observed for specimens that were anodized but not sealed. The lowest corrosion rate was found for coupons anodized and sealed in freshly deionized water at pH 5.5.

Four years of tests with panels processed under different conditions also helped to establish the following correlation among results obtained with an earlier version of the anodic breakthrough technique, results obtained with conventional salt spray testing, and process control parameters:[15]

- Coupons with measured corrosion rates lower than 2.5 $\mu m \cdot year^{-1}$ will not fail salt spray testing.

TABLE 7.2 Results Obtained by Testing, with the Anodic Breakthrough Method, Different Types of Anodized Aluminum Panels on 10 Nonoverlapping Positions

Process:	Boric-sulfuric		Sulfuric			Chromic			
Alloy:	7075	2024	7075	2024	7075	2024^A	2024^B	2024^C	2024^D
	3.3	4.6	−0.2	1.3	2.7	5.7	7.0	6.2	3.4
	3.5	4.6	−0.1	1.7	2.8	5.7	7.0	6.3	3.5
	3.5	4.6	0.1	1.8	2.9	6.0	7.0	6.3	3.5
	3.5	4.7	0.2	2.0	3.1	6.0	7.1	6.7	3.7
	3.8	4.8	0.4	2.1	3.4	6.1	7.2	6.8	3.8
	3.9	4.9	0.5	2.5	3.4	6.2	7.3	6.9	5.1
	4.3	5.0	1.1	2.9	3.5	6.2	7.3	7.0	5.4
	4.5	5.2	1.9	2.9	3.8	6.3	7.3	7.0	5.5
	5.1	5.3	2.4	2.9	4.1	6.3	7.3	7.0	5.6
	5.5	5.6	2.5	3.5	4.7	6.7	7.4	7.2	5.7
Mean:	4.1	4.9	0.9	2.3	3.4	6.1	7.2	6.7	4.5
StDev:	0.7	0.3	1.0	0.7	0.6	0.3	0.1	0.3	1.0

*The results are expressed as $\log_{10}$ (charge density), and the charge density is in $\mu C \cdot cm^{-2}$.

TABLE 7.3 Observations Made on the Same Types of Panels Exposed to Salt Spray Testing for 336 h

Process	Alloy	Observations
Boric-sulfuric	7075	Passed—no change in appearance
	2024	Passed—no change in appearance
Sulfuric	7075	Passed—slight fading of color
	2024	Passed—slight fading of color
Chromic	7075	Passed—overall fading of color
	2024[A]	Failed severely at 48 h
	2024[B]	Failed severely at 120 h
	2024[C]	Failed severely at 48 h
	2024[D]	Failed at 336 h—cloudy white corrosion products

TABLE 7.4 Comparison of Anodizing (Chromic Acid) Operating Conditions with Corrosion Rates Obtained with the Anodic Breakthrough Method

Chromic acid anodizing conditions	Corrosion rate, μm/year
Bare aluminum	1200
Anodized + no sealant	130
Anodized + sealed in tap water* (pH = 3)	110
Anodized + sealed in tap water (pH = 4)	50
Anodized + sealed in tap water (pH = 4.5)	0.8
Anodized + sealed in tap water (pH = 7.0)	5.0
Anodized + sealed in tap water (pH = 7.5)	8.9
Anodized + sealed in deionized water (pH = 5.5)	0.025

*All sealing solutions were maintained at 91°C.

- If the corrosion rates are between 2.5 and 15 μm·year^{-1}, a warning is raised that the anodizing process is deteriorating, and corrective measures are taken. Panels processed in such conditions would pass the salt spray test 90 percent of the time.

- When the corrosion rates exceed 15 μm·year^{-1}, immediate corrective action is required, and parts are reprocessed if corrosion rates exceed 25 μm·year^{-1}. Between 15 and 25 μm·year^{-1}, a judgment call is made depending on the applications.

- The seal time was optimized. An optimum seal time would be 8 min for freshly deionized water, whereas it could be up to 15 min for a one-month-old seal solution.

- The seal solution temperature was also optimized. Modifications to previous specifications were made when it was discovered that cooler seal solutions produced coupons with higher corrosion rates. The minimum seal solution temperature was raised to 90°C (95 ± 5°C in present specifications).

Intergranular anodic test for heat-treatable aluminum alloys. This electrochemical method was designed to determine the susceptibility to intergranular corrosion of solution-heat-treatable aluminum alloys—that is, 2xxx, 6xxx, 7xxx, and 8xxx alloys—without protective coatings and under various aging conditions.[16] The sensitivity of solution-heat-treatable aluminum alloys to intergranular corrosion is a function of the alloy composition, method of manufacturing, solution heat treatment, quench treatment, and artificial precipitation-hardening (aging) treatment. In the naturally aged condition, the sensitivity of solution-heat-treatable aluminum alloys to intergranular corrosion is a function primarily of the rate of cooling during quenching over a critical temperature range. Test results using this technique cannot be regarded as absolute, because they are not applicable to all environments that can be met in service. They are best used in a relative manner, to compare the intergranular corrosion resistance of various heats of solution-heat-treatable aluminum alloys. This method is applicable to cast and wrought heat-treatable aluminum alloys in the form of castings, forging, plates, sheets, extrusions, and semifinished or finished parts. The test results provide information to help determine the intergranular corrosion resistance and thermal processing quality of tested materials.

In this technique, the electrochemical signal serves to both amplify and accelerate a specific corrosion mechanism, i.e., intergranular corrosion (IC). Since that mechanism is often a precursor for the more catastrophic SCC, the same testing technique has been applied to the characterization of the susceptibility of aluminum alloys to SCC. For many industries, indirect testing for IC has replaced carrying out expensive tests for the characterization of aluminum alloys' susceptibility to SCC. The following list summarizes the empirical correlation between IC and SCC reflecting industrial experience:

- *2xxx series,* e.g., 2024. There is very good overlap between susceptibility to IC and to SCC.
- *5xxx series.* This series behaves similarly to 2xxx.
- *6xxx series.* With no Cu, susceptibility to IC is high and that to SCC is low.
- *7xxx series.* With low Cu, i.e., < 0.5%, and weldable—e.g., 7020, 7039, or 7005—there is susceptibility to SCC and SCC by hydrogen embrittlement, but little susceptibility to IC. With higher Cu content—e.g., 7075, 7010, 7050, or 7055—there is susceptibility to SCC by mixed modes and no correlation between SCC and IC susceptibility.

The method is based on the principle that if an aluminum alloy is susceptible to intergranular corrosion, this susceptibility will show as a breakdown during the anodic polarization of the alloy when it is exposed to solutions containing chloride ions. Historically, acceleration of intergranular corrosion tests have tended to be arbitrary and quite drastic. Such testing can be improved by taking the relative anodic characteristics of the material, the relative cathodic phases of the material, and the chemical composition of the test medium into consideration when selecting the electrochemical force.[16]

This test method starts with the anodic polarization of a specimen to determine the subsequent impressed potential. As with any other accelerated test, the test results must be correlated with the service performance of the materials being tested. Before testing, specimens should be degreased with an organic solvent (hydrocarbon, with a boiling point between 60 and 120°C), using a clean soft brush or an ultrasonic cleaning device and carrying out the cleaning in a vessel full of solvent. After cleaning, the specimens should be rinsed with fresh solvent and then dried. Naturally aged alloys are tested not earlier than 24 h after quenching. Artificially aged alloys may be tested by this method at any time. The principle of the technique involves the anodic polarization of specimens in aqueous sodium chloride solution up to the potential at which the alloy shows intergranular corrosion susceptibility, and exposure at this potential E_{ic} (Fig. 7.10).

The tests are typically carried out in a thermostatically controlled electrochemical cell at a temperature between 18 and 25°C with a test solution containing 0.3% sodium chloride. A potentiostat is required to polarize the test electrode at controlled scan rates.[16] The auxiliary electrode can be any inert electrode, and the reference electrode can be either calomel or silver/silver chloride. The anodic polarization curve is plotted for one specimen by scanning the potential from a cathodic value of $E = -1.16$ V vs. SHE at a scan rate of 0.15 mV·s^{-1} to the pitting potential E_{pf}, i.e., the potential at which the density of current is increased by at least one order of magnitude in the anodic polarization process. Another specimen made of the same alloy and temper is then immersed in the cell and allowed to rest for 5 min, and the potential is then moved to

$$E_{ic} = E_{pf} + 20 \text{ mV}$$

Exposure at this potential is continued as follows:

- For copper-containing alloys (range 0.25 to 6.5% Cu): 15 min
- For copper-free alloys (maximum 0.25% Cu): 90 min

After the electrochemical polarization, the specimens are taken out of the cell, washed in distilled water, dried, and metallographically examined. Interpretation of the sensitivity of solution-heat-treatable

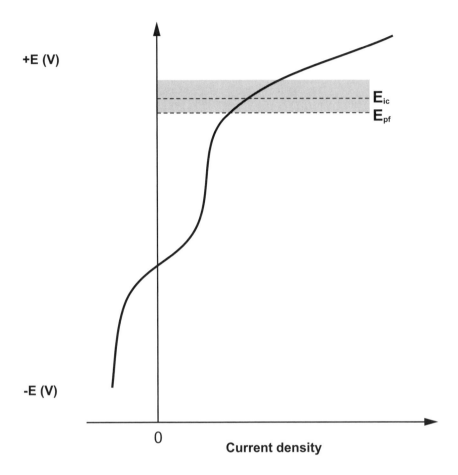

E_{pf} = pitting formation potential
$E_{ic} = E_{pf} + 20$ mV

Figure 7.10 Typical aluminum alloy anodic polarization diagram.

aluminum alloys to intergranular corrosion is based on the type, depth, and relative extent of the attack (the length along the surface of the metallographic cross section) and is expressed as a percentage. Table 7.5 suggests a scale of relative corrosion resistance as a function of observed pit depths.

The corrosion resistance of aluminum and aluminum-lithium alloys in marine environments. In this example, electrochemical impedance spectroscopy (EIS) was used to characterize the corrosion resistance of the three orthogonal faces of aluminum and aluminum-lithium sheet material exposed to a synthetic marine environment, and the results

TABLE 7.5 Intergranular Corrosion Ratings for Aluminum Alloys

Corrosion resistance	Rating	Maximum pit depth, μm
Excellent	1	0
	2	100
	3	100–200
	4	200–400
	5	400–700
Very poor	6	> 700

were compared with those obtained during different long-term exposure of the same alloys to seawater.[8] Figures 7.5 to 7.8 already showed the results of 4-month exposure of these aluminum materials when clamped with crevice washers and partly submerged in circulating seawater.

A summary of the results of the long-term exposures is shown in Table 7.6. After the exposure to the seawater fog, crevice corrosion had initiated under one site on the 8090 alloy. Some minor corrosion was also observed along rolling marks, which became visible after the exposure. Deep pits were observed on the surface of the 2024 alloy, which also suffered crevice corrosion under 6 of the 24 initiation sites, with 3 of these sites showing substantial penetration. Deep pits were also observed on the surface of the 2090 alloy, but crevice corrosion had not been initiated. Pitting had occurred on the surface of the 7075 alloy as well, but the depth of attack was not as severe as that which occurred on either the 2090 or 2024 alloy. This observation was consistent with the higher copper concentration present in the 2000-series alloys. Crevice corrosion was observed under one initiation site on the 7075 alloy.[8]

In all cases, the specimens that were completely immersed in seawater showed a reduction in the intensity (depth of penetration) of corrosion, if any, in the crevices from that observed in the seawater fog exposures. This suggests that the corrosion that occurred in the crevices in the fog exposures was more a result of moisture retention in the crevices than of mechanisms such as differential aeration that are normally associated with crevice corrosion. The rolled surface of the 8090 sheet showed selective corrosion, without deep pits, that etched the surface in a manner which served to highlight the rolling direction of the sheet without causing much metal loss. Corrosion of the surface of the 2024 alloy was characterized by a large number of deep pits. Corrosion of the 2090 alloy took the form of a smaller number of broad pits that were scattered over the surface. The number and severity of these pits increased near the edges of the panel. The surface of the 7075 alloy was characterized by a larger number of pits than that of the 2090, with the average pit on the 7075 having a smaller surface area.

The partial immersion test presented the toughest challenge to the alloys because the air/water interface created the possibility of differential aeration cells resulting from changing oxygen concentrations as a function of distance from the interface. In all cases, the interface was near the central hole. The extent and type of corrosion on the edges of the exposed panels were often quite different from those found on the rolled surfaces. The 8090 panel exhibited one large pit on one of the edges but was essentially unattacked elsewhere. The edges of the 2024 panel exhibited two pits, one of which had initiated on the edge while the other was associated with a pit on the rolled surface. Large cracks were observed on the edges of the 7075 panel. These cracks are consistent with the development of intergranular cracking, to which 7075 is known to be susceptible. The edges of the 2090 panels were essentially unaffected by the fog exposure.

The 8090 panel that was completely immersed in seawater exhibited only superficial corrosion on the rolled surfaces, while being severely attacked on the edges in a manner consistent with the selective corrosion often associated with intergranular cracking. This observation was confirmed by metallographic examination, which showed extensive intergranular cracking associated with the corrosion that had been initiated on the edge. The 2024 panel was severely corroded along the edges, with a combination of pitting of the edge and selective attack. No evidence of intergranular cracks associated with this corrosion could be found.

TABLE 7.6 Results of Long-Term Exposure Tests

Alloy		Salt fog	Total immersion	Partial immersion (air/water interface)
8090-T851	Rolled surface Edges Crevices	Minor 1 pit 1 site	Minor but selective Severe selective (cracking) 3 sites	A few deep pits
2024-T3	Rolled surface Edges Crevices	Many deep pits 2 pits 6 sites	Many deep pits Severe pitting 24 sites	No special corrosion
2090-T3	Rolled surface Edges Crevices	Deep pits No attack None	Broad pits Pits coming from rolled surface 5 sites	Extensive "poultice" corrosion
7075-T6	Rolled surface Edges Crevices	Shallow pits Large cracks 1 site	Many small pits Severe selective attack 24 sites	Extensive "poultice" corrosion

The electrochemical testing results indicated that short-term EIS measurements can provide good predictions for longer exposure of this material to seawater. According to the EIS polarization resistance data, which are summarized in Table 7.7, the 8090 alloy showed roughly equal corrosion rates for all three faces. Except for the rolled face, the corrosion rate of the 8090 alloy was substantially lower than that for the corresponding faces of the 2024 alloy. One interesting and omnipresent characteristic of EIS spectra seemed to be closely related to the long-term localized corrosion behavior of aluminum alloys. This intriguing feature is often described as a constant phase element (CPE) and introduced as an empirical factor in fitting procedures to account for the angle of tilt often visible in complex-plane plots. The empirical factor would typically appear as an exponent β, with a value between 0 and 1, which would be added to the imaginary component of the response $Z(\omega)$ to an impedance frequency (ω) [Eq. (7.2)].

$$Z(\omega) = R_s + \frac{R_p}{1 + (j\omega R_p C_{dl})^\beta} \qquad (7.2)$$

where R_s = solution resistance
R_p = polarization resistance
ω = frequency
C_{dl} = double-layer capacitance

With the assumption that the CPE, which is directly proportional to $(1 - \beta)$, increases in some manner with increased pitting, the EIS data indicated that the rolled surface of the 8090 had the lowest susceptibility to pitting, followed by the long transverse edge and the short transverse edge, which had the highest rate (Table 7.7). Examination of these

TABLE 7.7 Analyzed EIS Results Obtained with Aluminum Alloys

Alloy	Face	Corrosion rate, mm·year^{-1}	Angle of depression, °
8090-T8	Rolled surface	0.05	6
	Long transverse	0.04	12
	Short transverse	0.03	17
2024-T3	Rolled surface	0.05	17
	Long transverse	0.16	21
	Short transverse	0.22	23
2090-T3	Rolled surface	0.06	10
	Long transverse	0.08	16
	Short transverse	0.09	19
7075-T6	Rolled surface	0.14	12
	Long transverse	0.11	33
	Short transverse	0.12	29

surfaces with optical and scanning electron microscopy suggested that the correlation between the CPE and the pitting rate involved the number of pits formed in any given area (pit density) rather than the pit depth. The low pitting rate suggested by EIS for the rolled surface was consistent with visual observation of the long-term-exposure panels. However, the approximate equivalence for all three faces was not. If the interpretation of EIS data is correct, the corrosion of the rolled surface must occur initially at this high rate. However, the corrosion rate would then fall to a much lower value over the longer term. The R_p values for the 2024-T3 alloy showed a pronounced difference in overall corrosion rate between the rolled surface and the edges, with the edges having consistently higher rates. After about 50 h, a similar trend was observed for the CPE. These results were consistent with observations made on the long-term-exposure panels, which were characterized by a higher density of localized corrosion sites on the edges.[17]

On the basis of the EIS data, the conclusion would be reached that the edges of the 8090-T8 alloy had lower overall corrosion rates and were less prone to pitting than their 2024-T3 counterparts. The edges of the 8090 long-term-exposure panels had substantial areas where no visible corrosion had occurred. This could be consistent with the lower overall corrosion rates and lower pitting density in comparison with the 2024. However, the depth of attack within each pit (Fig. 7.11) was as large as or larger than that of a corresponding pit on 2024. Thus the rate of corrosion within a pit was at least as severe for 8090 as for 2024.

As was the case for the 8090 alloy, the corrosion rate determined with EIS for the rolled surface of the 7075 was approximately equal to that measured for the edges. This was not consistent with the appearance of the long-term panels, which suffered more metal loss along the edges

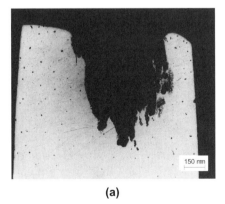
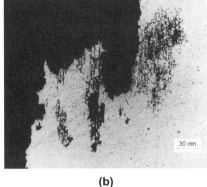

(a)   (b)

Figure 7.11 Photomicrograph of a section through an edge of the 8090-T851 panel immersed in seawater during 4 months (a) at 64× and (b) at 320× to illustrate the intergranular nature of the corrosion attack.

than on the rolled surface. The CPE values obtained for these experiments indicated that the rolled surface of the 7075 alloy had the lowest pitting density, while the long and short edges had higher rates. The higher rates reached similar and essentially constant values after 200 h. These results correlated very well with the long-term-exposure tests, in which the edges did indeed suffer much worse localized attack.

According to the EIS results, the rolled surface of the 2090 alloy had a consistently lower general corrosion rate than the same surface of the 7075. This did not appear to be consistent with the long-term-exposure tests, in which corrosion damage seemed to be more extensive on the surface of the 2090 alloy. In addition, the EIS data suggested that the edges of the 2090 were only slightly more corrosion-resistant than the 7075 edges. Once again this did not appear to be consistent with visual observation of the long-term-exposure panels. In this case, the edges of the 2090 panels suffered noticeably less corrosion than their 7075 counterparts. The CPE data indicated that the pit density should be lower on the rolled surface of the 2090 than on that of the 7075 and that the pit density should be much lower on the edges of the 2090 than on the edges of the 7075. These results are completely consistent with the appearance of the long-term-exposure panels.

The long-term-exposure tests indicated that the rolled surfaces of the 8090-T851 sheet were more resistant to corrosion than those of the conventional 2024-T3 sheet. Except for some pits that developed at an air/water interface, these surfaces suffered only minor corrosion. The same tests indicated that the rolled surfaces of the 2090-T8 sheet suffered at least as much corrosion damage as their counterparts on the 7075-T6 sheet. Some fairly deep pits occurred on the rolled surfaces of the 2090, even during the exposure to seawater fog.

The results obtained during the electrochemical testing of various faces of aluminum sheet material indicated that short-term EIS measurements could provide good predictions of the general and localized corrosion behavior of this material when exposed to seawater. In fact, the prediction of the localized corrosion behavior with the CPE calculated from the EIS data seemed to agree more closely to the long-term test results than the general corrosion estimation.[17]

7.2.3 Laboratory tests

In well-designed chemical processing plants, materials selection is based on a number of factors, such as service history, field in-plant corrosion tests, and pilot plant and laboratory corrosion tests. But, over time, laboratory tests have proven to be the most reliable and simple mean to generate information for the selection of process materials. Many of these tests are routinely performed to provide information on

- Fundamental corrosion evaluation
- Failure analysis
- Corrosion prevention and control
- Acceptance of quality assurance
- Environmental issues involving corrosion
- New alloy/nonmetallic or product process development

The *Corrosion Tests and Standards* handbook subdivides laboratory corrosion tests into four categories: cabinet tests, immersion tests, high-pressure/high-temperature tests, and electrochemical tests. While these four categories represent different sets of conditions accelerating corrosion processes, only electrochemical tests can directly amplify the impact of corrosion processes. The main reason why this is possible is that all electrochemical tests use some fundamental model of the electrode kinetics associated with corrosion processes to quantify corrosion rates. The amplification of the electrical signals generated during these tests has permitted very precise and sensitive measurements to be carried out.

In order to understand how environmental conditions can be accelerated, one has to first recognize the complexity of this factor. An important point for the description of the environment is the distinction between nominal and local (or near-surface) environments. Generally, components are designed to resist nominal environments specified by the applications and service conditions. The planning of testing programs is based on these specifications. Modern testing practices reflect this complexity by building variations into the tests or by focusing on the worst-case aspect of a situation.

Cabinet tests. Cabinet testing refers to tests conducted in closed cabinets where the conditions of exposure are controlled and mostly designed to accelerate specific corrosion situations while trying to emulate as closely as possible the corrosion mechanisms at play. Cabinet tests are generally used to determine the corrosion performance of materials intended for use in natural atmospheres. In order to correlate test results with service performance, it is necessary to establish acceleration factors and to verify that the corrosion mechanisms are indeed following the same paths. Modern surface analysis techniques can be quite useful to ascertain that the corrosion products have the same morphologies and crystallographic structures as those typically found on equipment used in service. There are basically three types of cabinet tests:

Controlled-humidity tests. There are 15 ASTM standards covering different variations on creating and controlling fog and humidity in cabinets for corrosion testing of a broad spectrum of products, from decorative electrodeposited coatings to solder fluxes for copper tubing systems. The basic humidity test is most commonly used to evaluate the corrosion resistance of materials or the effects of residual contaminants. Cyclic humidity tests are conducted to simulate exposure to the high humidity and heat typical of tropical environments. The cabinet in which such tests are performed should be equipped with a solid-state humidity sensor reading the current humidity condition and a feedback controller. The mechanism used to control the humidity moves chamber air via a blower motor and passes it over a heater coil in the bottom of the chamber with an atomizer nozzle fogging into this air stream (Fig. 7.12).

Corrosive gas tests. In these tests, controlled amounts of corrosive gases are added to humidity to replicate more severe environments. Some of these tests are designed to reveal and amplify certain characteristics of a material. ASTM B 775, Test Method for Porosity in Gold Coatings on Metal Substrates by Nitric Acid Vapor, and B 799, Test Method for Porosity in Gold or Palladium Coatings by Sulfurous Acid/Sulfur-Dioxide Vapor, employ very high concentrations of corrosive gases to amplify the presence of pores in gold or palladium coatings. The moist SO_2 test (ASTM G 87) is intended to produce corrosion in a form resembling that in industrial environments. A very sophisticated variation of these tests is the flowing of mixed gas test (ASTM B 827), in which parts per billion levels of pollutants such as chlorine, hydrogen sulfide, and nitrogen dioxide are introduced into a chamber at controlled temperature and humidity.

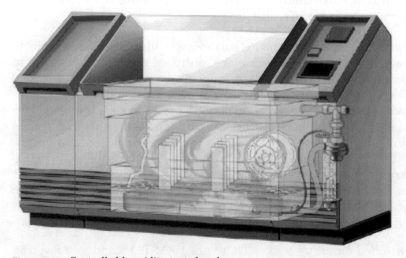

Figure 7.12 Controlled-humidity test chamber.

This test is particularly adapted to the needs of the electronics industry.

Salt spray testing. The oldest and most widely used cabinet test is ASTM B 117, Method for Salt Spray (Fog) Testing, a test that introduces a spray into a closed chamber where some specimens are exposed at specific locations and angles. The concentration of the NaCl solution has ranged from 3.5 to 20%. There is a wide range of chamber designs and sizes including walk-in rooms that are capable of performing this test. Although used extensively for specification purposes, results from salt spray testing seldom correlate well with service performance. Hot, humid air is created by bubbling compressed air through a bubble (humidifying) tower containing hot deionized water. Salt solution is typically moved from a reservoir through a filter to the nozzle by a gravity-feed system (Fig. 7.13). When the hot, humid air and the salt solution mix at the nozzle, the solution is atomized into a corrosive fog. This creates a 100 percent relative humidity condition in the exposure zone. For a low-humidity state in the exposure zone of the chamber, air is forced into the exposure zone via a blower motor that directs air over the energized chamber heaters (Fig. 7.14).

The inspection of specimens exposed to cabinet testing is often done visually or with the use of a microscope when localized corrosion is

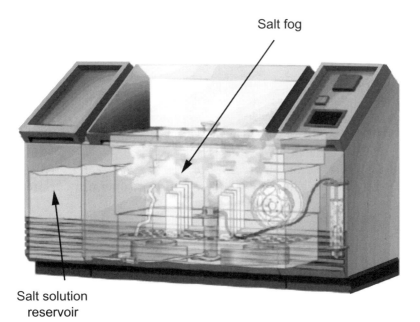

Figure 7.13 Controlled salt fog test chamber during a humid cycle.

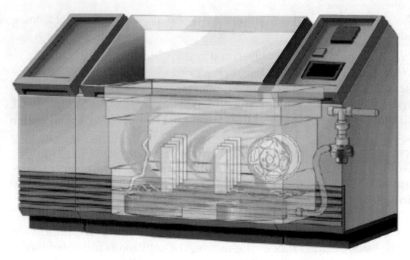

Figure 7.14 Controlled salt fog test chamber during a dry cycle.

suspected. The literature on the results and validity of these tests is abundant. After visual examination, more destructive procedures can be used to quantify test results. Measurement of physical properties or other functional properties often provides valuable information about corrosion damage.

Immersion testing. The environmental conditions that must be simulated and the degree of acceleration that is required often determine the choice of a laboratory test. In immersion testing, acceleration is achieved principally by

- Lengthening the exposure to the critical conditions that are suspected of causing corrosion damage. For example, if a vessel is to be batch-processed with a chemical for 24 h, then laboratory corrosion exposure of 240 h should be considered.
- Intensifying the conditions in order to increase corrosion rates, i.e., increasing solution acidity, salt concentration, temperature or pressure, etc.

Once the environmental conditions have been determined and the test designed, the test should be repeated a sufficient number of times to determine whether it meets the desired standard for reproducibility. Immersion tests can be divided into two categories:

Simple immersion tests. Basically, small sections of the candidate material are exposed to the test medium for a period of time and the loss of weight of the material is measured. Immersion testing remains the best method of screening and eliminating from further consideration those materials that should not be considered for specific applications. But while these tests are the quickest and most economical means for providing a preliminary selection of best-suited materials, there is no simple way to extrapolate the results obtained from these simple tests to the prediction of system lifetime.

Alternative immersion tests. Another variation of the immersion test is the cyclic test procedure, in which a test specimen is immersed for a period of time in a test environment, then removed and dried before being reimmersed to continue the cycle. Normally hundreds of these cycles are completed during the course of a test program.

High-temperature/high-pressure (HT/HP) testing. Autoclave corrosion tests are a convenient means for laboratory simulation of many service environments. The reason for such tests is to recreate the high temperatures and pressures commonly occurring in commercial or industrial processes. Factors affecting corrosion behavior are often intimately linked to the conditions of total system pressure, partial pressures of various soluble gaseous constituents, and temperature. There are many HT/HP environments of commercial interest, including those in industries such as petroleum, nuclear power, chemicals, aerospace, and transportation, where reliability, serviceability, and corrosion concerns are paramount.[18]

Corrosion coupons can be placed in the aqueous phase, in vapor space, or at phase interfaces, depending on the specific conditions that are of interest. Additionally, it is also possible to conduct electrochemical tests in HT/HP vessels. If multiple liquid phases are present, it can be necessary to stir or agitate the media or test vessel to produce mixing and create conditions in which the corrosion test specimens are contacted by all of the phases present. Special magnetic and mechanical stirrers are available that can be used to produce movement of the fluid, leading to a mixing of the phases. In some cases, where contact of the specimens with both liquid and gaseous phases is important in the corrosion process, it may be necessary to slowly rotate or rock the test vessel to produce the intended results.[18] HT/HP corrosion tests have special requirements not common to conventional corrosion experiments conducted in laboratory glassware.

Four variations of common HT/HP test methods that have been found to be useful in materials evaluation involving corrosion phenomena will be briefly described. However, these types of evaluations can be accomplished through careful planning and test vessel design. These include:[18]

Windowed test vessels. Special transparent windows and other fixtures such as fiber optics have been used to permit visual measurements or observations within the confines of test vessels. Besides being able to withstand the pressures, temperatures, and corrosion environments, these windows may have to perform other functions related to the introduction of light or other radiation if these are among the test variables.

Electrochemical measurements. Most conventional electrochemical techniques have been used for experiments conducted inside HT/HP vessels. The most critical electrochemical component in these experiments has always been the reference electrode. The design and construction of the reference electrode are particularly important, as it must provide a stable and standard reference potential. In many applications, test vessels have been modified to accommodate an external reference electrode to minimize the effects of temperature, pressure, contamination, or a combination thereof.

Hydrogen permeation. Hydrogen charging is often a problem that affects materials submitted to HT/HP test conditions. In such cases, it may be necessary to measure hydrogen permeation rates and diffusion constants in order to estimate the potential hazard of hydrogen attack. For hydrogen permeation measurements at high temperatures, it may be imperative to use solid-state devices.

Mechanical property testing. HT/HP vessels have been designed to conduct a variety of mechanical tests, such as slow strain rate (SSR), fracture, or fatigue testing. The main problem is always one of selecting fixtures that can withstand the corrosive environments generated in HT/HP tests.

Static tests. The simplest type of HT/HP corrosion test is conducted in a sealed and static pressurized test vessel. The test vessel typically contains a solution and a vapor space above the solution. In static corrosion tests, the only form of agitation of the test environment is convection produced by heating of the solution. The solution itself can be anything from a single liquid to water-based solutions containing various dissolved salts, such as chlorides, carbonates, bicarbonates, alkali salts, and other constituents or mixtures. The aim of these tests is to reproduce service environments as closely as possible. The liquid and gas phases will be determined by the amounts and vapor pressures of the constituents in the test vessel and by the test temperature. In general, the degree of difficulty of these tests and the amount of expense required for them increase with increasing test pressure and temperature.

Refreshed and recirculating tests. The depletion of volume of the corrosive environment in HT/HP tests is a serious limitation that often has to be overcome by the introduction of fresh environment, either continuously or by periodic replenishment of the gaseous and liquid phases being depleted by the corrosion processes. The limitation of the volume of the corrosive environment in most HT/HP tests makes issues such as the ratio of solution volume to specimen surface area a critical factor. In most cases, it is advantageous to limit this ratio to no less than 30 $cm^3 \cdot cm^{-2}$. In any event, care should be taken to prevent depletion of critical corrosive species or contamination of the test solution with unacceptably high levels of corrosion-produced metal ions.

Such conditions may require changes in the test constituents after a certain period of testing time, depending on their rate of consumption or contamination by corroding specimens. In particularly critical situations, it is possible to minimize such concerns by using constant or periodic replenishment of either the gaseous or the liquid phase in the autoclave under pressurized conditions. The need for agitation is particularly required when multiple liquid phases are present. Special magnetic and mechanical stirrers are available that can be used to produce movement of the fluid. Magnetic or mechanical stirring can also be employed to spin the specimens in the test environment, or alternatively a high-velocity flow system can be employed to induce cavitation or erosion damage on the specimens.

Factors affecting HT/HP test environments. For simple HT/HP exposure tests involving either aqueous or nonaqueous phases, the total pressure is usually determined by the sum of the pressures of the constituents of the test environment, which will vary with temperature. Where liquid constituents are being used for the test environment, the partial pressure is usually taken to be the vapor pressure of the liquid at the intended test temperature. Vapor pressures for several other volatile compounds used in HT/HP corrosion testing can be found in the technical literature. In some cases, higher test pressures can be obtained by pumping additional gas into the test vessel using a special gas pump. Alternatively, hydrostatic pressurization may be employed, in which there is no gas phase in the test vessel and the pressure is increased by pumping additional liquid into the test vessel in a controlled manner.[18] The importance of partial pressure in HT/HP corrosion testing is that the solubility of the gaseous constituents in the liquid phase is usually determined by its partial pressure, which explains why the effect of some gaseous corrosives is often magnified at high pressure.

Special HT/HP corrosion test conditions. A chemical species whose chemical behavior affects corrosion resistance and materials performance is hydrogen. It has been known for decades that atomic hydrogen can produce embrittlement in many metallic materials. Under high hydrogen environment pressure, electrochemical reaction, or both, atomic hydrogen can penetrate structural materials, where it can react by one of the following mechanisms:[18]

- Recombination to form pressurized molecular hydrogen blisters at internal sites in the metal
- Chemical reaction with metal atoms to form brittle metallic hydrides
- Solid-state interaction with metal atoms to produce a loss of ductility and cracks

There has been much interest in conducting hydrogen-induced cracking (HIC) tests in aqueous media that can produce atomic hydrogen on the surface of materials as a result of corrosion or cathodic charging. In most cases, these tests can be conducted at ambient pressure and at temperatures from ambient to elevated, depending on the application. When aqueous hydrogen charging is involved, pressure is usually not a major factor. However, as in the case of steels exposed to aqueous hydrogen sulfide–containing environments, the atomic hydrogen is produced as a result of sulfide corrosion. The severity of the mass-loss corrosion and hydrogen charging is directly dependent on the amount of hydrogen sulfide dissolved in the aqueous solution. In applications involving petroleum production and refining, compressed natural gas storage, chemical processing, and heavy-water production, such effects are compounded by exposure to HT and/or HP conditions. Additionally, variations in pH which control the type and amount of dissolved sulfide species and the severity of corrosion and hydrogen charging can be affected by hydrogen sulfide pressure.

Special considerations for testing in high-purity water. There is a growing awareness that differences in testing procedures in high-temperature high-purity water, such as that used in the nuclear industry, can produce very large scatter in the SCC growth rate data. For example, data from single or multiple laboratories often show scatter of a thousand or even more, which is too high to establish reliable quantitative dependencies unless very large data sets are generated. Environmental cracking is influenced by dozens of interdependent material, environment, and stressing parameters. While there are numerous factors that need to be controlled for optimal experiments, an even bigger challenge revolves around interpreting existing data in which critical measurements were not made and other measurements may be misleading. In general, there is some concern with regard to almost all existing SCC data, partly because the optimal measurements and techniques are not fully known, much less agreed upon or standardized.[19]

Extensive, careful studies show that the scatter in SCC growth-rate data can be collapsed substantially from, e.g., the 1000X range that is observed in some data sets to perhaps a factor of 2 to 5X. Accomplishing this requires very stable loading and tight control on temperature and water chemistry, as well as uniform metallurgical characteristics. While these optimized conditions often yield reproducible crack growth-rate data, it is not uncommon to find no growth or retarded growth rates in some specimens.

Some distinction must be made among phenomena that involve stochastic processes, like discrete birth and death processes in pit nucleation. These are still subject to errors in measurement and experimental technique, but are known to possess well-defined, inherent "scatter." The discrete nature and characteristics of pit nucleation processes generally justify their being treated separately from a macroscopically continuous process like SCC. The types of problems that commonly appear in SCC crack growth data obtained in high-temperature high-purity water can be broken down into the following categories:[19]

- *Stress intensity.* "Constant" active-K testing (vs. wedge loading) is preferred, although use of constant displacement is acceptable if it meets other criteria and less than 15 percent K relaxation has occurred during the test.[19]

- *Test preliminaries.* Careful control and documentation of machining, surface condition, precracking procedures, and preoxidation are important. Final precracking conditions and SCC loading procedure are also particularly important.

- *Test temperature.* The temperature that is most relevant to boiling water reactors (BWRs) is between 274 and 288°C.[19]

- *Inlet and outlet solution conductivity.* Given modern BWR operation, tests in "high-purity" water require that outlet conductivity <0.1 μS·cm^{-1} be achieved, and <0.07 μS·cm^{-1} at the outlet is both desirable and achievable for oxygen concentrations <2 ppm. In most tests in "high-purity water," the actual outlet conductivity is dramatically higher than that of the inlet, as a result of
  1. Chromate release by the autoclave chromium-rich materials
  2. Decomposition of organic species
  3. Release of fluorine from fluorinated polymers or chloride from reference electrodes
  4. In-leakage of carbon dioxide from the air

- *Inlet and outlet dissolved oxygen and hydrogen.* These should generally be measured, unless there is a very strong basis for accepting nominal values of oxygen for the inlet and outlet. Dissolved hydrogen levels are important because (1) hydrogen affects the corrosion potential whether oxygen is present or not, and (2) hydrogen levels even below 100 ppb may have a significant effect on SCC of high-nickel alloys below 300°C.

- *Corrosion potentials.* These should be measured on the test specimen, since it is widely accepted that corrosion potential is a more fundamental measure of SCC effect than the dissolved oxygen level, although it is not a truly fundamental parameter in SCC crack growth.[8] The effect on corrosion potential of acidic/basic impurities or flow rate may be reported but misunderstood. Since the effect of corrosion potential is primarily to create a potential gradient in the crack, the effects of such changes must be carefully interpreted. The same is true of effects of flow rate on corrosion potential.[19]

- *The autoclave refresh rate.* This should be high enough to control intentional (dissolved gases and ionic impurities) and unintentional contributions (usually ionic impurities) to water chemistry. This usually requires that the autoclave volume be refreshed 2 to 4 times per hour.

- *Flow rate.* The flow rate should never be a compromising element of a test program. Since there are few cases in which flow rate is expected to play a large role in SCC in plant components, laboratory data under high-flow-rate conditions should automatically be viewed with caution and concern because the crack tip chemistry can be readily flushed under these conditions.

- *Continuous crack monitoring.* This is essential. Reversed DC potential drop is most commonly used, and good data require a well-behaved crack extension. Good crack length resolution in modern test facilities is a few micrometers. The minimum acceptable crack increments need to be based partly on microstructural considerations. While a wide variety of microstructures are "sampled" across the width of the specimen, there are some concerns that small increments might do a poor job of sampling and exhibit anomalous behavior.[19]

- *Material characteristics.* Typical material characteristics should be known, such as composition, crack orientation, yield strength/hardness, heat-treatment conditions, carbide/phase distribution, and derived parameters. Composition and welding conditions are also valuable in discerning whether weld metal is likely to have experienced hot cracking, since distinguishing hot cracking from SCC is essential even though both may contribute to through-wall penetration.

- **Electrochemical test methods.** In view of the electrochemical nature of corrosion, it is not surprising that measurements of the electrical properties of the metal/solution interface are extensively used across the whole spectrum of corrosion science and engineering, from fundamental studies to monitoring and control in service. Electrochemical testing methods involve the determination of specific interface properties that can be divided into three broad categories:

1. *Potential difference across the interface.* The potential at a corroding interface arises from the mutual polarization of the anodic and cathodic half-reactions constituting the overall corrosion reaction. Potential is intrinsically the most readily observable parameter and, with proper modeling of its value in relation to the thermodynamics of a system, can provide the most useful information on the state of a system. The following examples illustrate various applications of potential measurements to the study of corrosion processes:
   - Determination of the steady-state corrosion potential E_{corr}
   - Determination of E_{corr} trends over time
   - Electrochemical noise (EN) as fluctuations of E_{corr}

2. *Reaction rate as current density.* Partial anodic and cathodic current densities cannot be measured directly unless they are purposefully separated into a bimetallic couple. By polarizing a metal immersed in a solution, it is possible to estimate a net current for the anodic polarization and for the cathodic polarization, from which a corrosion current density i_{corr} can be deduced. Two broad categories summarize the great number of techniques that have been developed around these concepts:
   - Determination of E-i relationships by changing the applied potential, i.e., potentiostatic methods
   - Determination of E-i relationships by changing the applied current, i.e. galvanostatic methods

3. *Surface impedance.* A corroding interface can also be modeled for all its impedance characteristics, therefore revealing subtle mechanisms not visible by other means. Electrochemical impedance spectroscopy is now well established as a powerful technique for investigating corrosion processes and other electrochemical systems.

Types of polarization test methods. Polarization methods such as potentiodynamic polarization, potentiostaircase, and cyclic voltammetry are often used for laboratory corrosion testing. These techniques can provide significant useful information regarding the corrosion mechanisms, corrosion rate, and susceptibility to corrosion of specific materials in designated environments. Although these methods are well established, the results they provide are not always clear and occasionally can be misleading.[20]

Polarization methods involve changing the potential of the working electrode and monitoring the current which is produced as a function of time or potential. For anodic polarization, the potential is changed in the anodic (or more positive) direction, causing the working electrode to become the anode and causing electrons to be withdrawn from it. For cathodic polarization, the working electrode becomes more negative and electrons are added to the surface, in some cases causing electrodeposition. For cyclic polarization, both anodic and cathodic polarization are performed in a cyclic manner.[20] The instrumentation for carrying polarization testing consists of

- A potentiostat which will maintain the potential of the working electrode close to a preset value.
- A current-measuring device for monitoring the current produced by an applied potential. Some potentiostats output the logarithm of the current directly, which will allow plotting of the current vs. potential curves. The ability of the current-measuring device to autorange or to change the scale automatically is also important.
- Ability to store the data directly in a computer or plot them out directly. This is also important.
- Polarization cells. Several test cells for making polarization measurements are available commercially. Polarization cells can have various configurations specific to the testing requirements, whether testing small coupons or testing sheet materials or testing inside autoclaves. In a plant environment, the electrodes may be inserted directly into a process stream. Some of the features of a cell include[20]
    1. The working electrode, i.e., the sample for testing or analysis, which may be accompanied by one or more auxiliary or counterelectrodes.
    2. The reference electrode, which is often separated from the solution by a solution bridge and Luggin probe. This combination eliminates solution interchange with the reference electrode but allows it to be moved very close to the surface of the working electrode to minimize the effect of the solution resistance.
    3. A thermometer to determine temperature.
    4. An inlet and outlet for gas to allow deaeration, aeration, or introduction of specific gases into the solution.
    5. Ability to make an electrical connection directly with the working electrode, which will not be affected by the solution.
    6. Introduction of the working electrode into the solution completely so as to eliminate any crevice at the solution interface, unless this is a desired effect.

7. The test cell itself, composed of a material that will not corrode or deteriorate during the test, and that will not contaminate the test solution. The volume of the cell must be large enough to allow removal of the corroding ions from the surface of the working electrode without affecting the solution potential.
8. If necessary, a mechanism for stirring the solution, such as a stirring bar or bubbling gas, to ensure uniformity of the solution chemistry.

In ASTM G 3, Standard Practice for Conventions Applicable to Electrochemical Measurements in Corrosion Testing, there are several examples of polarization curves. Figure 7.15 illustrates the ideal polarization behavior one could obtain, for example, using the linear polarization method briefly described below. Figures 7.16 and 7.17 show hypothetical curves for, respectively, active and active-passive behavior, while Fig. 7.18 was plotted from actual polarization data obtained with a S43000 steel specimen immersed in a 0.05 M H_2SO_4 solution.

Several methods may be used in polarization of specimens for corrosion testing. Potentiodynamic polarization is a technique in which the potential of the electrode is varied at a selected rate by application of a current through the electrolyte. It is probably the most commonly

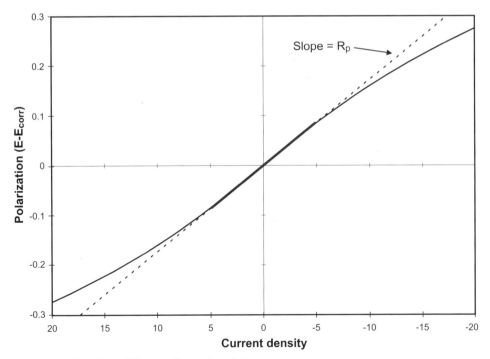

Figure 7.15 Hypothetical linear polarization plot.

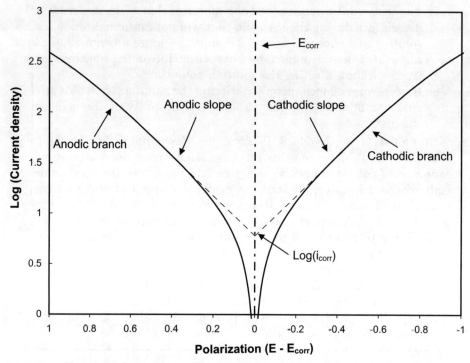

Figure 7.16 Hypothetical polarization diagram for an active system with anodic and cathodic branches.

used polarization testing method for measuring corrosion resistance and is used for a wide variety of functions.[20]

An important variant of potentiodynamic polarization is the cyclic polarization test. This test is often used to evaluate pitting susceptibility. The potential is swept in a single cycle (or slightly less than one cycle), and the size of the hysteresis is examined along with the differences between the values of the starting open-circuit corrosion potential and the return passivation potential. The existence of the hysteresis is usually indicative of pitting, while the size of the loop is often related to the amount of pitting.

Another variant of potentiodynamic polarization is cyclic voltammetry, which involves sweeping the potential in a positive direction until a predetermined value of current or potential is reached, then immediately reversing the scan toward more negative values until the original value of potential is reached. In some cases, this scan is done repeatedly to determine changes in the current-potential curve produced with scanning.

Another variation of potentiodynamic polarization is the potentiostaircase method. This refers to a technique for polarizing an electrode

Acceleration and Amplification of Corrosion Damage 527

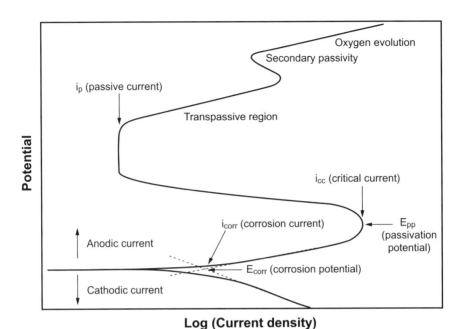

Figure 7.17 Hypothetical polarization diagram for a passivable system with anodic and cathodic branches.

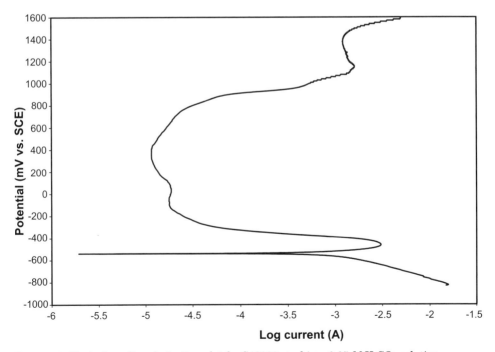

Figure 7.18 Typical anodic polarization plot for S43000 steel in a 0.05 M H_2SO_4 solution.

in a series of potential steps in which the time spent at each potential is constant and the current is often allowed to stabilize prior to changing the potential to the next step. The step increase may be small, in which case the technique resembles a potentiodynamic curve, or it may be large.[20] Another polarization method is electrochemical potentiodynamic reactivation (EPR), which measures the degree of sensitization of stainless steels such as S30400 and S30403 steels. This method uses a potentiodynamic sweep over a range of potentials from passive to active (called reactivation).

Another widely used polarization method is linear polarization resistance (LPR). The polarization resistance of a material is defined as the slope of the potential–current density ($\Delta E/\Delta i$) curve at the free corrosion potential (Fig. 7.15), yielding the polarization resistance R_p, which can be itself related to the corrosion current with the help of Eq. (7.3).[21]

$$R_p = \frac{B}{i_{\text{corr}}} = \frac{(\Delta E)}{(\Delta i)_{\Delta E \to 0}} \tag{7.3}$$

where R_p = polarization resistance
i_{corr} = corrosion current
B = empirical polarization resistance constant that can be related to the anodic (b_a) and cathodic (b_c) Tafel slopes with Eq. (7.4)

$$B = \frac{b_a b_c}{2.3 (b_a + b_c)} \tag{7.4}$$

The Tafel slopes themselves can be evaluated experimentally using real polarization plots similar to those presented in Figs. 7.16 and 7.17 or obtained from the literature.[21] The corrosion currents estimated using these techniques can be converted into penetration rates using Faraday's law, expressed earlier in Eq. (7.1). Alternatively, corrosion currents can be transformed using a generic conversion chart such as that found in Table 7.8 or an alloy-specific conversion table like the one for converting steel corrosion data in Table 7.9.

The study of uniform corrosion and studies assuming corrosion uniformity are probably the most widespread application of electrochemical measurements both in the laboratory and in the field. The widespread use of these electrochemical techniques does not mean that they are without complications. Both linear polarization and Tafel extrapolation need special precautions for their results to be valid. The main complications or obstacles in performing polarization measurements can be summarized in the following categories:

TABLE 7.8 Conversion between Current, Mass Loss, and Penetration Rates for All Metals

	mA·cm^{-2}	mm·year^{-1}	mpy	g·m^{-2}·day^{-1}
mA·cm^{-2}	1	3.28 M/nd	129 M/nd	8.95 M/n
mm·year^{-1}	0.306 nd/M	1	39.4	2.74 d
mpy	0.00777 nd/M	0.0254	1	0.0694 d
g·m^{-2}·day^{-1}	0.112 n/M	0.365/d	14.4/d	1

mpy = milli-inches per year; n = number of electrons freed by the corrosion reaction; M = atomic mass; d = density. As an example, if the metal is iron (Fe), n = 2, M = 55.85 g, and d = 7.88 g·cm^{-3}.

TABLE 7.9 Conversion between Current, Mass Loss, and Penetration Rates for Steel

	mA·cm^{-2}	mm·year^{-1}	mpy	g·m^{-2}·day^{-1}
mA·cm^{-2}	1	11.6	456	249
mm·year^{-1}	0.0863	1	39.4	21.6
mpy	0.00219	0.0254	1	0.547
g·m^{-2}·day^{-1}	0.00401	0.0463	1.83	1

mpy = milli-inches per year.

- *Effect of scan rate.* The rate at which the potential is scanned may have a significant effect on the amount of current produced at all values of potential.[20] The rate at which the potential is changed, the scan rate, is an experimental parameter over which the user has control. If not chosen properly, the scan rate can alter the scan and cause a misinterpretation of the features. The problem is best understood by picturing the surface as a simple resistor in parallel with a capacitor. In such a model, the capacitor would represent the double-layer capacitance and the resistor the polarization resistance, which is inversely proportional to the corrosion rate [Eq. (7.3)]. The goal is for the polarization scan rate to be slow enough so that this capacitance remains fully charged and the current-voltage relationship reflects only the interfacial corrosion process at every potential. If this is not achieved, some of the current being generated would reflect charging of the surface capacitance in addition to the corrosion process, with the result being that the measured current would be greater than the current actually generated by the corrosion reactions. When this happens, the polarization measurement does not represent the corrosion process, often leading to an erroneous prediction.[22]

  The question is, what is that proper scan rate? A relatively valid method would be to use the lower breakpoint frequency of the impedance spectrum as the starting point, provided such EIS measurement is available. The method is based on the premise that the scan rate (voltage rate of change) is analogous to a frequency at every applied potential. That frequency must be low enough so that the impedance magnitude is independent of frequency. Then the polarization or charge transfer resistance is being measured with no interference from the capacitance.

The frequency below which there is no capacitive contribution is about an order of magnitude lower than the breakpoint frequency. The assumption is that this lower frequency is analogous to a scan rate. The conversion to a scan rate is made by assuming that over some small voltage amplitude, e.g., 5 mV, the voltage-current relationship is linear and the linear range corresponds to half of a sinusoidal wave. Table 7.10 shows estimated maximum scan rates for several polarization resistance, solution resistance, and capacitance values typically encountered in practice.

- *Effect of solution resistance.* The distance between the Luggin probe (of the salt bridge to the reference electrode) and the working electrode is purposely minimized in most measurements to limit the effect of the solution resistance. In solutions that have extremely high resistivity, this can be an extremely significant effect. Many materials of importance to corrosion measurements, such as concrete, soil, organic solutions, and many others, have high resistivity, but can also be strongly corrosive to some metals. It is important to be able to make polarization measurements in these high-resistivity environments. A method of interrupting the current and monitoring

TABLE 7.10 Examples of Maximum Scan Rates for Performing Valid Polarization Plots

Solution resistance, $\Omega \cdot cm^2$	Polarization resistance, $k\Omega \cdot cm^2$	Capacitance, $\mu F \cdot cm^{-2}$	Maximum scan rate, $mV \cdot s^{-1}$
10	1	100	5.1
10	10	100	0.51
10	100	100	0.05
10	1000	100	0.005
100	1	100	6.3
100	10	100	0.51
100	100	100	0.05
100	1000	100	0.005
10	1	20	25
10	10	20	2.5
10	100	20	0.25
10	1000	20	0.025
100	1	20	50
100	10	20	2.6
100	100	20	0.25
100	1000	20	0.025

the decay of the potential as a function of time can be used to measure the solution resistance and to determine the actual resistance between the reference and working electrodes.

- *Changing surface conditions.* Since corrosion reactions take place at the surface of materials, when the surface is changed as a result of processing conditions, active corrosion, or other reasons, the potential is usually also changed. This can have a strong effect on the polarization curves.[20]

- *Determination of pitting potential.* In analyzing polarization curves, the appearance of a hysteresis (or loop) between the forward and reverse scans is often thought to denote the presence of localized corrosion (pitting or crevice corrosion). This observation is particularly valid when the corrosion potential is higher or more noble than the pitting potential.

The need for further testing in the face of ambiguous or conflicting polarization results is one of the most important things that can be learned from a single test. The additional steps required when the results of a single test or type of test are ambiguous include[20]

1. Rerun the test under equivalent conditions. This will minimize test-to-test variations.
2. Identify conflicting or ambiguous results. Careful identification of the areas of conflict can provide a starting point for further analysis or testing.
3. Evaluate alternative answers to the conflict or ambiguity. Is there another possible explanation for the results (such as changes in the sample, surface, solution, or stirring rate; possible contamination; or electronic hardware problems)?
4. Run another type of test. Many tests give complementary information which may uncover the difficulty with the initial result. Sometimes a simple examination of the sample visually will locate crevice attack, oxide buildup, or surface changes that have occurred and have led to the ambiguous or conflicting data obtained initially.

Cyclic potentiodynamic polarization. The electrochemical technique that has gained the most widespread acceptance as a general tool for assessing the possibility of an alloy suffering localized corrosion is probably the cyclic potentiodynamic polarization technique. This technique has been especially useful in assessing localized corrosion for passivating alloys such as S31600 stainless steel, nickel-based alloys containing chromium, and other alloys such as titanium and zirconium.[22]

The cyclic potentiodynamic polarization technique for corrosion studies was introduced in the 1960s and refined during the 1970s into a fairly simple technique for routine use. In this technique, the voltage applied to an electrode under study is ramped at a continuous rate relative to a reference electrode using a potentiostat. The voltage is first increased in the anodic or noble direction (forward scan). At some chosen current or voltage, the voltage scan direction is reversed toward the cathodic or active direction (backward or reverse scan). The scan is terminated at another chosen voltage, usually either the corrosion potential or some active potential. The potential at which the scan is started is usually the corrosion potential. The corrosion behavior is predicted from the structure of the polarization scan. Though the generation of the polarization scan is simple, its interpretation can be difficult.[22]

Features useful in interpretation. Figures 7.19 through 7.22 show typical polarization scans that might be observed in practice. The figures are drawn assuming an arbitrary minimum recorded current (e.g., 100 nA·cm^{-2}) that would lie above the actually measured minimum current (e.g., 1 nA·cm^{-2}) sometimes observed in an experiment. Hence, the scan may sometimes cross the potential axis, set at some arbitrary current.

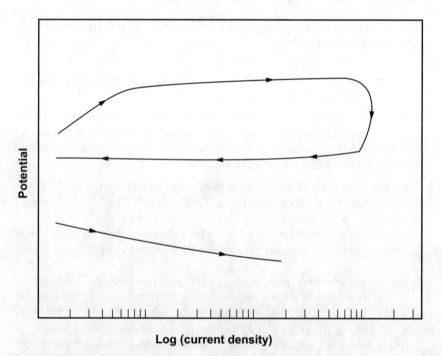

Figure 7.19 Typical polarization scan for an alloy suggesting a significant risk of localized corrosion in the form of crevice corrosion or pitting (the arrow indicates scanning direction).

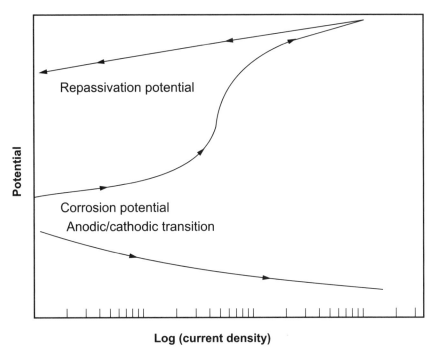

Figure 7.20 Typical polarization scan for a completely passive alloy suggesting little risk of crevice corrosion, pitting, or general corrosion (the arrow indicates scanning direction).

Pitting and repassivation potentials. Two potentials that are often thought to characterize an alloy in terms of localized corrosion are the *repassivation potential* and the *pitting potential* and their values relative to the corrosion potential. A common interpretation is that pitting would occur if the hysteresis between the forward and reverse scans appeared as in Fig. 7.19 and the corrosion potential were equal to or anodic with respect to the pitting potential. The specimen under test would be expected to resist localized corrosion if the corrosion potential lay cathodic with respect to the repassivation potential or if the polarization scan appeared as in Fig. 7.20.[22]

There are several ways to choose the repassivation potential. It can be chosen as the potential at which the anodic forward and reverse scans cross each other. Alternatively, it can be chosen as that potential at which the current density reaches its lowest readable value on the reverse portion of the polarization scan. One reason to choose the latter is that for some polarization scans, such as that in Fig. 7.20, the forward and reverse portions of the polarization scan do not cross each other. In any case, the choice should be consistent for all scans in any particular study.

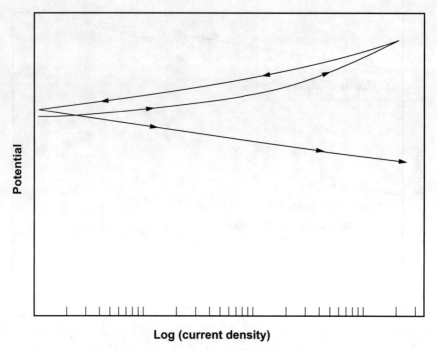

Figure 7.21 Typical polarization scan for an alloy possibly suffering from general high corrosion (the arrow indicates scanning direction).

The pitting potential is that potential at which the forward or ascending portion of the scan shows a rapid rise in current, followed by a negative hysteresis between the forward and reverse portions of the scan, as in Fig. 7.19. Often, the electrode surface exhibits small pits after the experiment. Controversy still surrounds the meaning of these potentials. The values measured are not intrinsic properties of the alloy and are influenced by a variety of experimental variables. The pitting potential as determined by the potentiodynamic scan has been shown to be related qualitatively to the resistance of a material to a loss of passivity by pit initiation. If a crevice develops in a portion of the specimen—between the electrode and its holder, for example—the pitting potential will probably reflect the breakdown of passivity in that crevice.[22]

Hysteresis. The hysteresis refers to a feature of the polarization scan in which the forward and reverse portions of the scan do not overlay each other. The hysteresis shown in both Figs. 7.19 and 7.20 is the result of the disruption of the passivation chemistry of the surface by the increase in potential and reflects the ease with which that passivation is restored as the potential is decreased back toward the corrosion potential. For a given experimental procedure, the larger the hysteresis, the greater the disruption of surface passivity, the greater the difficulty in restoring passivity, and, usually, the greater the risk of localized corrosion.

Approaching a potential from more active potentials at a certain scan rate will create a surface structure different from that created when approaching the potential from more noble potentials. The "positive" hysteresis shown in Fig. 7.20 is caused by the polarization to more noble potentials making the surface more passive. The "negative" hysteresis in Fig. 7.19 is caused by a decrease in passivity, often produced by the initiation of localized corrosion. This latter phenomenon is usually a reflection of a propensity for localized corrosion in the form of either pitting or crevice corrosion. From a practical standpoint, a positive hysteresis usually signifies that the alloy will be more resistant to localized corrosion than does a negative hysteresis.[22]

Active-passive transition or anodic nose. The anodic nose reflects the characteristic in which the current increases rapidly with increasing potential in the anodic direction near the corrosion potential, goes

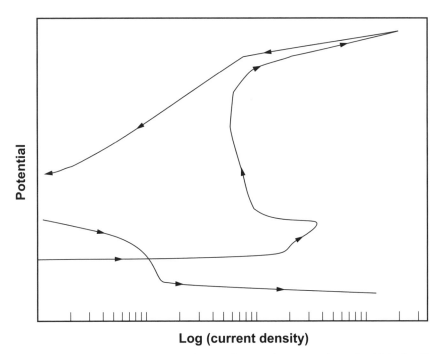

Figure 7.22 Typical polarization scan for an alloy that has an easily oxidizable/reducible surface species without being passive at the corrosion potential (the arrow indicates scanning direction).

through a maximum value, and then decreases to a low value. Iron and some austenitic alloys may demonstrate this type of behavior in acidic environments, for example. The decrease in current may suggest an alloy surface undergoing some type of passivation process or valence change (Fe^{2+} to Fe^{3+}) as the potential is increased. Figure 7.22 shows such an example. The presence of this feature typically means that the alloy has a finite corrosion rate at the corrosion potential.

Anodic-to-cathodic transition potential. The potential at which the current changes from anodic to cathodic during the reverse portion of the scan is assumed to be the potential of the anodic-to-cathodic transition. The difference between this potential and the corrosion potential is another useful feature. If the polarization scan appears as in Figs. 7.19 and 7.20, this potential still exists, but the current at the transition is lower than the lowest recorded value of the current density. Under these circumstances, this potential might be assumed to be the potential at which the cathodic current rises above the lowest recorded value. The difference between this potential and the corrosion potential can provide an additional indication of the persistence of passivity.[22]

Point of scan reversal. The current density (potential) at which the polarization scan is reversed can play a significant role in the appearance of the polarization scan and the value of the repassivation or protection potentials. The reason is that the value of the repassivation potential is dictated by the amount of prior damage to the passive surface. The farther the polarization scan is generated in the anodic direction, the greater tends to be the degree of upset of the surface region. The effect of the point of reversal on the repassivation potential is especially pronounced if the pitting potential is exceeded or some other electrochemical transformation is precipitated, especially if it does not reflect behavior at the corrosion potential. The result can be an erroneous prediction of corrosion behavior. No specific potential can be recommended, since the amount of upset of the surface required for a prediction is somewhat related to the information desired. Maintaining a constant reversal point can be most important if alloys are being compared in a specific environment or if a single alloy is being evaluated across a number of environments.[22]

Potentiodynamic polarization in service. In the following examples, the polarization scans were generated after 1 and 4 days of exposure to a chemical product maintained at 49°C. The potential scan rate was 0.5 mV·s^{-1}, and the scan direction was reversed at 0.1 mA·cm^{-2}. Coupon immersion tests were run in the same environment for 840 h. The goal of these tests was to examine whether S31600 steel could be used for short-term storage of a 50% commercial solution of aminotrimethylene phosphonic acid in water. A small amount of chloride ion (1%) could be present in this acidic chemical. The S31600 steel specimens were exposed to the liquid, at the vapor/liquid interface, and in the vapor. The reason for the three exposures was that in most storage situations, the containment vessel would be exposed to a vapor/liquid interface and a vapor phase at least part of the time. Corrosion in these regions can be very different from that resulting from exposures to liquid. The specimens were fitted with artificial crevice formers.[22]

Figure 7.23 shows the polarization scan generated after 1 day, and Fig. 7.24 shows the polarization scan generated after 4 days of exposure. Considering the parameters mentioned above and how they changed between the polarization scans was the basis for the interpretation results presented in Table 7.11. The important parameters considered were the position of the "anodic-to-cathodic" transition relative to the corrosion potential, the existence of the repassivation potential and its value relative to the corrosion potential, the existence of the pitting potential and its value relative to the corrosion potential, and the hysteresis (positive or negative).

The presence of the negative hysteresis would typically suggest that localized corrosion is possible, depending on the value of the corrosion

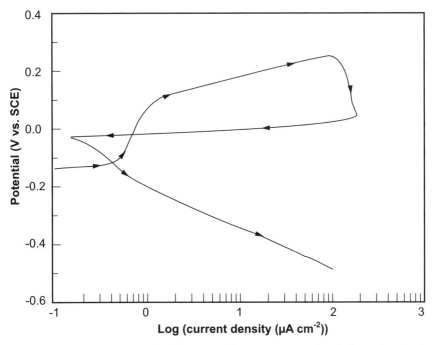

Figure 7.23 Polarization scan for S31600 steel in 50% aminotrimethylene phosphonic acid after 1 day of exposure (the arrow indicates scanning direction).

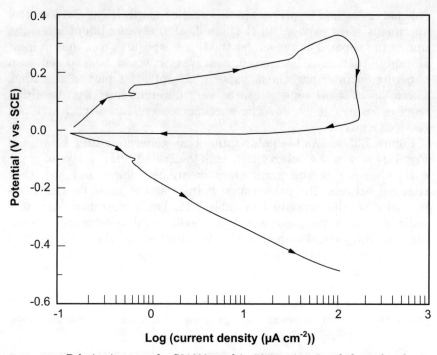

Figure 7.24 Polarization scan for S31600 steel in 50% aminotrimethylene phosphonic acid after 4 days of exposure (the arrow indicates scanning direction).

TABLE 7.11 Features and Values Used to Interpret Figs. 7.23 and 7.24

Feature	Value in Fig. 7.23	Value in Fig. 7.24
Repassivation potential − corrosion potential	0.12 V	0.0 V
Pitting potential − corrosion potential	0.22 V	0.12 V
Potential of anodic-to-cathodic transition − corrosion potential	0.12 V	0.0 V
Hysteresis	Negative	Negative
Active-to-passive transition	No	No

potential relative to the characteristic potentials discussed above. After the first day of exposure, pitting was not expected to be a problem because the pitting potential was well removed from the corrosion potential. One concern was that the repassivation potential and the potential at which the polarization scan shifted from anodic to cathodic current were identical and the cathodic current rose almost linearly with potential. The currents generated were much higher than those normally associated with S31600 steel in a passive state. This observation would suggest that at the corrosion potential, S31600 steel could show a possibly slight corrosion rate, not a surprise in pH environment. If one extrapolates the anodic and cathodic c the corrosion rate might be estimated to be less than 0.12 mm·year. In total, these observations suggested that there was a risk of initiation of corrosion, particularly in localized areas where the pH can decrease drastically.[22]

After 4 days, the risk of localized corrosion increased. At this time, the repassivation potential and the potential of the change from anodic to cathodic current were equal to the corrosion potential. The pitting potential was only about 0.1 V more noble than the corrosion potential, and the hysteresis was still negative. The risk of pitting had increased to become a concern. The corrosion rate, while low, remained greater than zero, suggesting that the alloy would not resist changes in the environment. These results would suggest that corrosion in occluded areas, such as under deposits, was more likely.

Coupon immersion tests confirmed the long-term predictions. Slight attack was found under the special artificial crevice formers in the complete liquid exposure. Pits were found on the coupons mounted at the vapor/liquid interface, especially under deposits. Some pits were found on the coupons mounted in the vapor region. The practical conclusion of this in-service study was that, since localized corrosion often takes time to develop, exposure to this chemical product for several days could be acceptable. However, it was recommended that long-term exposure be avoided because both pitting and crevice corrosion would be expected.

Electrochemical impedance spectroscopy. Electrochemical impedance spectroscopy has been successfully applied to the study of corrosion systems for almost 30 years[23] and has been proven to be a powerful and accurate method for measuring corrosion rates. But in order to access the charge transfer resistance or polarization resistance R_p, which is proportional to the corrosion rate at the monitored interface [Eq. (7.1)], EIS results have to be interpreted with the help of a model of the interface. Since the early work published by Epelboin and coworkers, EIS has gained tremendous momentum and popularity in corrosion laboratories around the world.[24] An important advantage of EIS over other laboratory techniques is the possibility of using very small-amplitude signals without significantly disturbing the properties being measured.

To make an EIS measurement, a small-amplitude signal, usually a voltage between 5 and 50 mV, is applied to a specimen over a range of frequencies from 0.001 to 100,000 Hz. The EIS instrument records the real (resistance) and imaginary (capacitance) components of the impedance response of the system. Depending upon the shape of the EIS spectrum, a circuit model or circuit description code and initial circuit parameters are assumed and input by the operator. The program then fits the best frequency response to the given EIS spectrum to obtain parameters. The quality of the fitting is judged by how well the fitting curve overlaps the original spectrum. By fitting the EIS data, it is possible to obtain a set of parameters which can be correlated with the coating condition and the corrosion of the steel substrate.

Of the numerous equivalent circuits that have been proposed to describe electrochemical interfaces, only a few really apply in the context of a freely corroding interface at or close to kinetic equilibrium. The first circuit (Fig. 7.25a) corresponds to Eq. (7.2) and to the simplest equivalent circuit that can describe a metal/electrolyte interface. Following Boukamp,[25] the term Q has been adopted here to describe the "leaky capacitor" behavior corresponding to the presence of a constant phase element explained by a fundamental dispersion effect. The admittance representation Y^* of the CPE behavior with frequency ω can be described by Eq. (7.5). For $n = (1 - \beta)$, Eq. (7.5) describes the behavior of a resistor with $R = Y_0^{-1}$ and for $n = \beta$, that of a capacitor with $C = Y_0$. For $n = 0.5$, Eq. (7.5) becomes the expression of a Warburg (W) component, and when $n = -\beta$, it emulates an inductance with $L = Y_0^{-1}$.[25]

$$Y^*(\omega) = Y_0 \omega^n \cos(n\pi/2) + jY_0 \omega^n \sin(n\pi/2) \qquad (7.5)$$

Figure 7.26 illustrates the complex-plane presentation of EIS simulated data corresponding to the model circuit in Fig. 7.25a when $R_s = 10\ \Omega$, $R_p = 100\ \text{k}\Omega$, and Q decomposes into $C_{dl} = 40\ \mu\text{F}$ and $n = 0.8$, and Fig. 7.27 shows how the same data would appear in a Bode plot format.

The second circuit (Fig. 7.25b) was proposed by Hladky et al.[26] to take into account a diffusion-limited behavior corresponding to a Warburg component which can be described by Eq. (7.6). The exponent n in Eq. (7.6) can vary between 0.5 and 0.25 depending on the smoothness of the metallic surface, i.e., 0.5 for highly polished surfaces and 0.25 for porous or very rough materials.[27] R and C in Eq. (7.6) are the resistance and capacitance associated with the distributed R-C line of infinite length.

$$Z(\omega) = (0.5R/C)^{0.5} \omega^{-n} \qquad (7.6)$$

Figure 7.28 illustrates the complex-plane presentation of simulated data corresponding to the model circuit in Fig. 7.25b when $R_s = 10\ \Omega$, $R_p = 100\ \text{k}\Omega$, $C_{dl} = 40\ \mu\text{F}$, and the exponent n of the Warburg component $= 0.4$. Figure 7.29 shows the same data in a Bode representation.

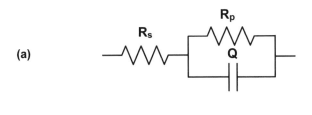

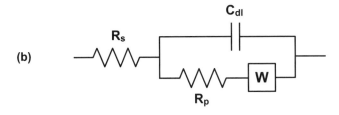

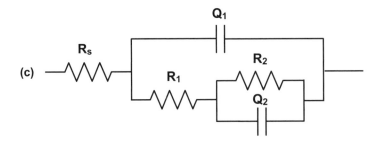

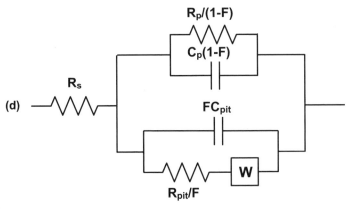

Figure 7.25 Equivalent circuit models proposed for the interpretation of EIS results measured in corroding systems: (a) simplest representation of an electrochemical interface; (b) one relaxation time constant with extended diffusion; (c) two relaxation time constants; and (d) the impedance of pitting processes of Al-based materials.

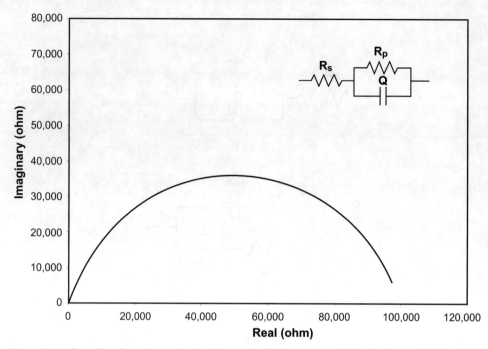

Figure 7.26 Complex-plane presentation of simulated data corresponding to the model circuit in Fig. 7.25a when $R_s = 10\ \Omega$, $R_p = 100\ \text{k}\Omega$, and Q decomposes into $C_{dl} = 40\ \mu\text{F}$ and $n = 0.8$.

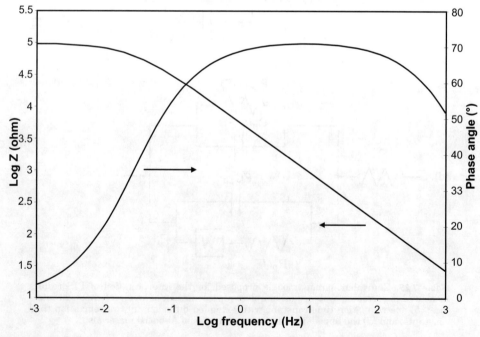

Figure 7.27 Bode representation of the same data illustrated in Fig. 7.26 in complex-plane format.

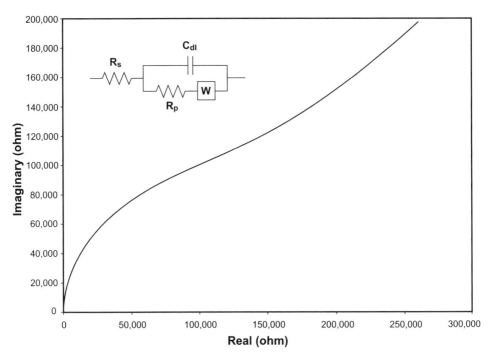

Figure 7.28 Complex-plane presentation of simulated data corresponding to the model circuit in Fig. 7.25b when $R_s = 10\ \Omega$, $R_p = 100\ \mathrm{k}\Omega$, $C_{dl} = 40\ \mu\mathrm{F}$, and the exponent n of the Warburg component $= 0.4$.

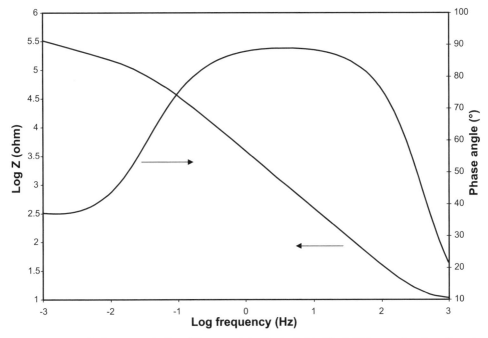

Figure 7.29 Bode representation of the same data illustrated in Fig. 7.28 in complex-plane format.

The third circuit (Fig. 7.25c) has been proposed to describe EIS results containing two relaxation time constants. Such behavior is commonly encountered for corrosion under coatings or under scale, for corrosion-inhibited systems, or even for localized corrosion.[28] The meaning of the circuit elements in Fig. 7.25c will vary with the physical systems represented, but their significance has been validated through additional measurements and calculations. Figure 7.30 illustrates the model circuit in Fig. 7.25c with simulated data obtained with $R_s = 10$ Ω, $R_1 = 40$ kΩ, and $Q_1 = 40$ μF with exponent $n = 1$, $R_2 = 20$ kΩ, and $Q_2 = 20$ μF with exponent $n = 1$. Figure 7.31 is a Bode representation of the same data illustrated in Fig. 7.30 in complex-plane format.

The fourth and last circuit (Fig. 7.25d) was proposed to describe the events which occur on a metallic corroding surface before and after localized corrosion has been observed. This model has been said to be in agreement with a large number of EIS data collected during the study of aluminum and aluminum-based metal matrix composites.[29] The factor in this model attempts to represent the surface ratio

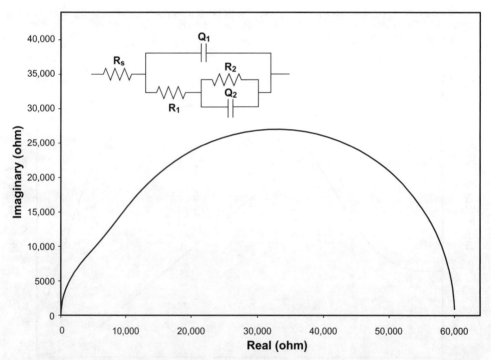

Figure 7.30 Complex-plane presentation of simulated data corresponding to the model circuit in Fig. 7.25c when $R_s = 10$ Ω, $R_1 = 40$ kΩ, and $Q_1 = 40$ μF with exponent $n = 1$, $R_2 = 20$ kΩ, and $Q_2 = 20$ μF with exponent $n = 1$.

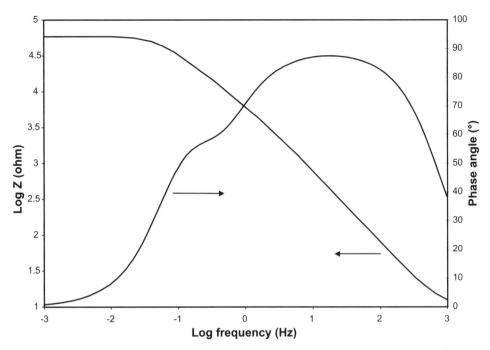

Figure 7.31 Bode representation of the same data illustrated in Fig. 7.30 in complex-plane format.

between the pitted surface and the remaining surface of a specimen. Figure 7.32 is a complex-plane presentation of simulated data corresponding to the model circuit in Fig. 7.25d when $R_s = 10\ \Omega$, $R_p = 20$ kΩ, and $C_p = 40\ \mu$F, a pit surface ratio factor $F = 10^{-3}$, and a Warburg exponent $n = 0.8$. Figure 7.33 is a Bode representation of the same data illustrated in Fig. 7.32 in complex-plane format.

A critical problem in EIS (and in other electrochemical techniques) is the validation of the experimental data. This problem is more obvious in EIS than in time-domain techniques because of the manner in which the experimental data are displayed. For example, it is not uncommon to observe negative resistance (second quadrant) and inductive (fourth quad-rant) behavior when the experimental impedance data are plotted in the complex plane. Also, the impedance loci frequently take the form of depressed and/or distorted semicircles, and these may contain multiple loops. These features are not readily accounted for by using simple electric equivalent circuits. However, the inability to represent electrochemical impedance data by simple equivalent electric circuits is not in itself a problem, since there is no a priori reason why an interfacial impedance could be represented by such electrical analogs.[30] Most companies selling impedance equipment are providing software

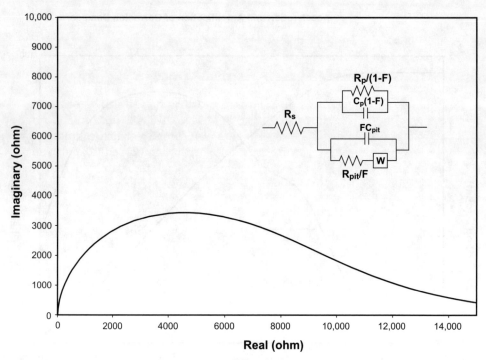

Figure 7.32 Complex-plane presentation of simulated data corresponding to the model circuit in Fig. 7.25d when $R_s = 10\ \Omega$, $R_p = 20$ kΩ, $C_p = 40$ μF, a pit surface ratio factor $F = 10^{-3}$, and a Warburg exponent $n = 0.8$.

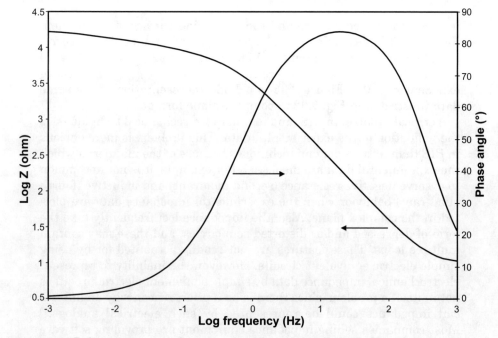

Figure 7.33 Bode representation of the same data illustrated in Fig. 7.32 in complex-plane format.

to create complex circuits with an easy graphical user interface, as illustrated in Fig. 7.34.

To address the validation of impedance data, the use of Kramers-Kronig (KK) transforms has been proposed. More than 60 years ago, Kramers and Kronig developed a number of integral transforms between the real and the imaginary components of a complex transfer function. However, only recently was a practical algorithm developed to apply the KK transforms to validate electrochemical impedance data. Currently, the KK algorithm is commercially available, and it is used routinely in some laboratories to assess the quality of the measured impedance data.[30] These integral transforms were derived assuming four basic conditions, which are discussed prior to stating the transforms themselves.

- *Linearity.* A system is said to be linear if the response to a sum of individual inputs is equal to the sum of the responses to the

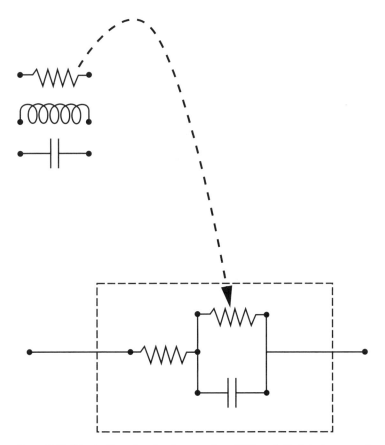

Figure 7.34 Illustration showing how modern EIS analytical tools support the construction of complex equivalent circuits with single circuit elements.

individual inputs. Practically, this also implies that the impedance (or admittance) is not a function of the magnitude of the perturbation.

- *Causality.* The temporal response of a system to an arbitrary excitation must be real and not complex. If the system is at rest and a perturbation is applied at time $t = 0$, the response must be zero for $t < 0$. Physically, this means that the system does not generate noise independent of the applied signal. This is an important consideration in electrochemical systems, because charge transfer interfaces are often active and do in fact generate noise in the absence of an external stimulus.[30]

- *Stability.* A system is said to be stable if it comes back to its original state after a perturbation is removed. Otherwise the system would supply power independently of the input. For a passive system with loss, the transient response must approach zero for a sufficiently long time. This condition ensures that there is no negative resistance in the system. This apparent restriction on the presence of negative resistance is also of great interest in electrochemistry, is frequently observed experimentally, and is predicted theoretically from mechanisms describing active-to-passive transitions.[30]

- *Finite value.* The real and imaginary components of a complex impedance must be finite over the entire frequency range sampled. This condition is also of great interest in electrochemistry because the real and imaginary components of a CPE vary with frequency.

If a system satisfies the conditions of linearity, stability, and causality, it will a priori satisfy the KK transforms, provided that the frequency range is sufficiently broad for the integrals to be evaluated. Accordingly, passive electric circuits provide an absolute measure against which any numerical algorithm for evaluating the integrals can be assessed.[30]

Electrochemical noise. The use of electrochemical noise (EN) for corrosion monitoring is very attractive, as was illustrated in a few examples presented in Chap. 6, Corrosion Maintenance through Inspection and Monitoring. Fluctuations of potential or current of a corroding metallic specimen are a well-known and easily observable phenomenon, and the evaluation of EN as a corrosion tool has increased steadily since Iverson's paper in 1968.[31] The extensive development in the sensitivity of the equipment for studying electrochemical systems has rendered the study of oscillations in electrochemical processes that translate into measurable EN increasingly accessible. The study of corrosion potential fluctuations was applied, for example, to monitor the onset of events characterizing localized corrosion such as pitting or SCC, exfoliation, and erosion-corrosion in either laboratory or diverse and complex industrial environments.[32]

The study of EN has repeatedly been found uniquely appropriate for monitoring the onset of events leading to localized corrosion and understanding the chronology of the initial events typical of this type of corrosion. No other technique, electrochemical or otherwise, is even remotely as sensitive as EN to system changes and upsets. During localized corrosion, EN is believed to be generated by a combination of stochastic processes, such as passivation breakdown and repassivation events, and deterministic processes which can be caused by film formation or pit propagation processes.

The most traditional way to analyze electrochemical noise data has been to transform time records in the frequency domain in order to obtain power spectra. Spectral or power density plots would thus be computed, utilizing fast Fourier transforms (FFT) or other algorithms such as the maximum entropy method (MEM). Some studies have indicated that the roll-off of the voltage noise amplitude from corroding electrodes could be a useful characteristic of corrosion processes.[33,34] In these studies, a roll-off of -20 dB/decade was associated with pitting attack, whereas one of -40 dB/decade was found to be characteristic of general corrosion processes. When converted into a spectral density plot, a roll-off of -20 dB/decade would correspond to a spectral exponent of 1, and a roll-off of -40 dB/decade, to an exponent of 2.

For stochastic signals, the spectral exponents β of spectral density plots can be related to the fractal dimension D of the signals with Eq. (7.7).[35] But since the noise signals often contain deterministic features that can induce variations in the slope of a spectral density plot, such an analytical method is not the most reliable way to evaluate the fractal dimension of a signal.

$$D = \frac{5 - \beta}{2} \qquad (7.7)$$

Another very useful mathematical model has been proposed to specifically reveal the fractal characteristics of signals.[36] A detailed description of this technique, also called rescaled range analysis or the R/S technique [where R or R(t,s) stands for the sequential range of the data point increments for a given lag s and time t, and S or S(t,s) stands for the square root of the sample sequential variance], can be found in Fan et al.[37] Hurst[38] and later Mandelbrot and Wallis[39] have proposed that the ratio R(t,s)/S(t,s) is itself a random function with a scaling property described by relation (7.8), where the scaling behavior of a signal is characterized by the Hurst exponent (H), which can vary over the range $0 < H < 1$.

$$\frac{R_{(t,s)}}{S_{(t,s)}} \propto S^H \qquad (7.8)$$

It has additionally been shown that the local fractal dimension of a noise trace is related to H through Eq. (7.9), which makes it possible to characterize the fractal dimension of a given time series by simply calculating the slope of an R/S plot.[40]

$$D = 2 - H \qquad 0 < H < 1 \qquad (7.9)$$

In contrast with any other signal analysis technique, the stochastic process detector (SPD) technique attempts to quantify the stochasticity of a noise record.[41] The SPD technique involves two levels of transformation. First, the noise records are transformed into series of singular events, i.e., each point of a time series is examined for its appurtenance to either positive or negative noise peaks. In the second level of transformation, the distribution of peak lengths $f(t)$ is examined and compared to the theoretical exponential decay distribution, represented by Eq. (7.10), where λ_t is the mean value and t is the peak length (time) that would characterize a series of stochastic events. A flowchart describing the logic of the SPD technique is presented in Fig. 7.35. The goodness of fit (GF) of real data to the exponential function is then calculated to serve as a measure of the stochasticity of the time records.

$$f(t) = +\lambda_t e^{-\lambda t} \qquad (7.10)$$

The main idea behind applying the SPD technique to the analysis of electrochemical noise is the belief that localized corrosion should induce deterministic features in the overall noise signatures. And since the SPD technique is particularly sensitive to any digression from purely stochastic signals, such an analysis could serve as an advance warning method to detect the onset of a localized corrosion situation. The following experimental results and their analysis illustrate the information one could get with both the SPD and R/S analytical methods.

Cylindrical specimens of S30400 stainless steel containing a crevice in the form of a close-fitting PTFE ring have been exposed to acidified $FeCl_3$ solutions maintained at 60°C, and the electrochemical noise has been monitored. Fractal and stochasticity analysis techniques were used on the time records to characterize the processes leading to

Acceleration and Amplification of Corrosion Damage 551

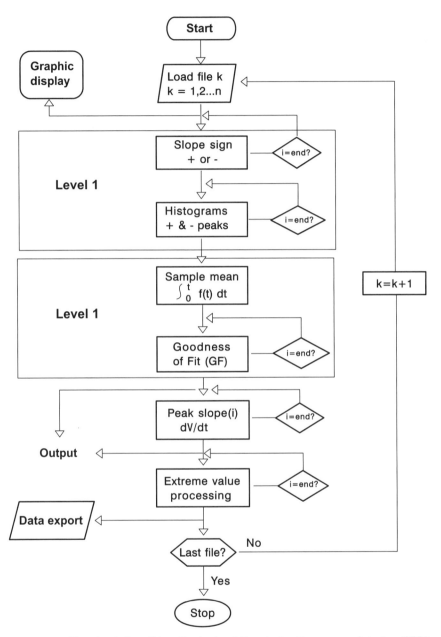

Figure 7.35 Flowchart describing the logic of the stochastic process detector (SPD) technique.

crevice initiation. The stochasticity technique was found to be particularly sensitive to the onset of crevice attack. By using a combination of noise analysis techniques, it was possible to identify three distinct corrosion modes during these experiments: pitting, massive pitting, and crevice attack.[42]

Figures 7.36 to 7.39 contain the E_{corr} measurements obtained during four consecutive experiments made with these S30400 steel cylindrical specimens equipped with the crevice collar and the results obtained by analyzing the voltage fluctuations by the SPD and R/S techniques. At the end of these tests, the specimens were removed from the electrolyte, the PTFE collar was removed, and the severity of the corrosion attack was assessed. In all four cases, severe crevice attack was observed beneath the collar around the majority of the circumference. Knowing that a Brownian motion behavior is equivalent to a fractal dimension of 1.5, as can be verified by the R/S technique, while the presence of persistence causes an increase in D, it is possible to divide the results presented in Figs. 7.36 to 7.39 into two zones: those with $D < 1.5$, and those where $D \geq 1.5$. The transition between these two zones is quite evident in all four experiments carried out during this study. In the first experiment (Fig. 7.36), it occurred at approximately 4.5 h in the test, whereas it occurred at 3.1 h for the second experiment (Fig. 7.37), 3.2 h for the third (Fig. 7.38), and 4.1 h during the fourth (Fig. 7.39).

The switch from antipersistence, i.e., $D < 1.5$, to persistence, i.e., $D > 1.5$, was accompanied, in all four cases, by a permanent transition of E_{corr} toward values that were more cathodic by approximately 80 to 100 mV. It was also accompanied by a sudden burst of electrochemical energy that could be picked up by a scanning platinum probe with a commercial instrument, a Unican Instruments SRET. The combination of a permanent cathodic shift of E_{corr} and a prolonged period of persistence in the EN records have thus come to signify that a stable crevice situation had formed. The results obtained with the SPD technique revealed another aspect of the EN that could be useful for monitoring purposes: The results indicate that the transition from antipersistence to persistence was itself preceded by a change in the level of stochasticity of the EN. In the cases of experiments 2 and 3, the loss of stochasticity, i.e., when GF < 95 percent or (1 − GF) > 5 percent, was quite focused, whereas it was much more diffuse in experiments 1 and 4. This temporary loss of stochasticity was interpreted as being indicative of the presence of chaotic features caused by the presence of two relatively stable states, general pitting and crevice corrosion. The chaotic nature of the voltage fluctuations between these two states, as revealed by the SPD technique, would give an early indication of the tendency to form a crevice.

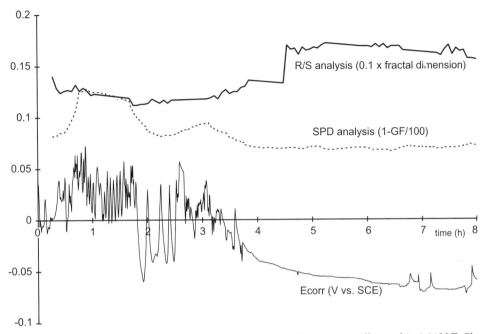

Figure 7.36 First experiment with S30400 steel specimen with a crevice collar and in $0.01M$ FeCl$_3$ acidified to pH 2 and maintained at 60°C.

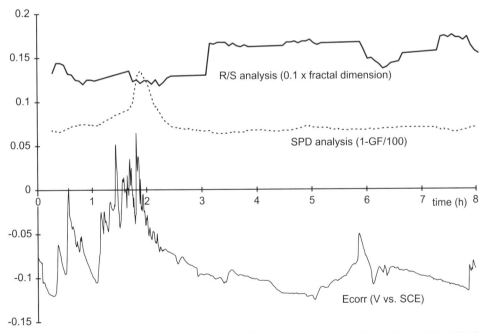

Figure 7.37 Second experiment with S30400 steel specimen with a crevice collar and in $0.01M$ FeCl$_3$ acidified to pH 2 and maintained at 60°C.

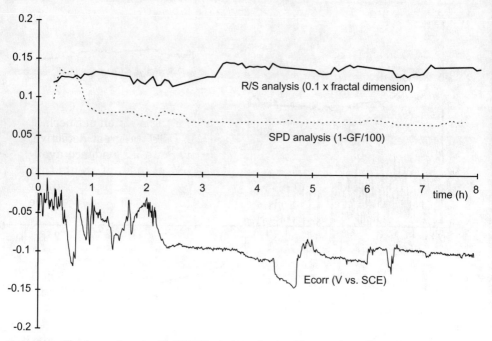

Figure 7.38 Third experiment with S30400 steel specimen with a crevice collar and in $0.01M$ FeCl$_3$ acidified to pH 2 and maintained at 60°C.

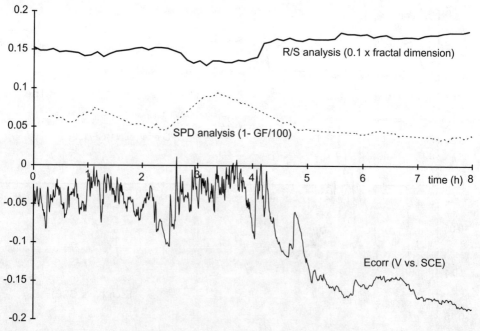

Figure 7.39 Fourth experiment with S30400 steel specimen with a crevice collar and in $0.01M$ FeCl$_3$ acidified to pH 2 and maintained at 60°C.

7.2.4 Field and service tests

In investigating an in-service failure, the analyst must consider a broad spectrum of possibilities or reasons for its occurrence. Often a large number of factors must be understood in order to determine the cause of the original failure. The analyst is in the position of Sherlock Holmes attempting to solve a baffling case. Like the great detective, the analyst must carefully examine and evaluate all evidence available and prepare a hypothesis or a model of the chain of events that could have caused the "crime." If the failure can be duplicated under controlled simulated service conditions in the laboratory, much can be learned about how the failure actually occurred.

The salt spray test, for example, which was originally designed to test coatings on metals, has been widely used to evaluate the resistance of metals to corrosion in marine service or on exposed shore locations.[43,44] Extensive experience has shown that, although salt spray tests yield results that are somewhat similar to those of exposure to marine environments, they do not reproduce all the factors causing corrosion in marine service. Salt spray tests should thus be considered to be arbitrary performance tests and their validity dependent on the extent to which a correlation has been established between the results of the test and the behavior under expected conditions of service. Despite the current widespread use of continuous salt spray methods, their unrealistic simulation of outdoor environments is a serious shortcoming.

The reviews made by F. L. LaQue on this subject indicate that the salt spray test cannot realistically be used, for example, for parts with complicated shapes. This deficiency is principally due to the fact that the salt spray particles fall in vertical patterns, creating a strong orientation dependency.[45,46] Another major inadequacy of the test is the variable sensitivity of different metallic materials to the ions present in various service environments. Since different metals also are affected differently by changes in the concentrations of salt solutions, the salt spray test is not really appropriate for ranking different materials in an order of relative resistance to salt water or salt air. The variability of the environments, even for seagoing equipment, is another factor that is extremely difficult to reproduce in a laboratory. Before attempting to simulate such natural environments, it is thus recommended that the chemistry of the environment and all other parameters controlling the corrosion mechanisms be monitored over time, in a serious attempt to characterize the worst exposure conditions.

Further developments in accelerated testing should be based on modern scientific principles and incorporate an appreciation of the mechanisms of natural atmospheric degradation of the metal being studied. The development of laboratory corrosion tests should be based on a previous determination of the dominant corrosion factors. Even if the preferred practice is to design such tests to represent the most severe conditions for the corrosion involved, it is still important to investigate the kinetic component involved in environmental corrosion in order to understand the causes and reasons for failure. With these points in mind, it is useful to consider how the corrosion acceleration may realistically be achieved. Increasing the concentration or corrosiveness of the salt spray may not necessarily be appropriate during cyclic testing, since even an initially dilute spray will, after a sufficient number of cycles, result in the solubility of ionic species being exceeded. Since the development of an accelerated testing program should focus on the parameters which govern the lifetime behavior of the materials being tested, it is important to establish a general framework of the factors behind corrosion damage and, hence, behind continuous and cyclic cabinet testing.

The lack of correlation between corrosion rates measured during conventional salt spray testing and during outdoor exposure to marine environments and the drastic differences in the nature of the corrosion products formed by these two types of tests have created a general feeling that ASTM B 117 is not an appropriate test environment for anything other than products intended for continuous immersion in seawater environments. The mass loss results presented in Table 7.12 were obtained by Harper over 30 years ago on untreated and anodized aluminum casting alloys exposed to a marine environment for 10 years and in a salt spray test for 1500 h.[47] On some untreated specimens (LM1M, LM4M, and LM5M), mass loss in the marine atmosphere was approximately half of the mass loss measured with salt spray, while for others (LM6M, LM14WP, and LM23P), very different results were obtained. The results on anodized coatings did not correlate much better, although the anodized specimens resisted the salt spray tests consistently better than they did the marine environment.

TABLE 7.12 Mass Loss Comparison between Salt Spray Tests and Marine Atmosphere Exposure Results

Alloy*	Untreated		Anodized	
	Salt spray, g/1500 h	Marine atmosphere, g/10 years	Salt spray, g/1500 h	Marine atmosphere, g/10 years
LM1M†	0.87	0.43	0.06	0.09
LM4M	0.34	0.18	0.02	0.09
LM5M	0.19	0.06	0	0.10
LM6M	0.05	0.12	0	0.04
LM14WP	0.14	0.25	0	0.06
LM23P	0.26	0.23	0.02	0.07

*British Standard aluminum casting alloy (BS 1490).
†M = as cast, W = solution, P = precipitation heat treatment.

Selecting a test facility. There are many factors to consider when selecting a weathering test station to conduct a test program. These can be divided into two categories:

- *Location.* An ideal test site should be located in a clean, pollution-free area, if pollution is not deemed to be a parameter, within the geoclimatic region to be used. This is important for the prevention of unnatural effects on the specimens. Within the local area chosen, there must be no isolated sources of pollution or deleterious atmospheric contamination. This could result from construction, emissions from a manufacturing plant, or chemical spraying in farming areas. The layout of the test field itself is very important. The characteristics of the test field will be determined by its location. For example, if trees enclose the field, the test area will be affected by mildew spores, will have lower sunlight levels, and possibly will have lower temperatures. If the field is on low land and poorly drained, it will flood in times of heavy rainfall, humidity will be higher, and algae growth and dirt attachment will increase.

- *Maintenance.* The exposure maintenance program followed by the test site will also play a major role in determining the accuracy of testing. It is important that the specimens on the test racks be correctly maintained. This involves ensuring that the mounting method is correct and giving constant follow-up attention to maintain the quality. The racks themselves are in contact with the specimens. The racks must be cleaned regularly to remove any dirt, mildew, or algae which would otherwise contaminate the specimens.

Types of exposure testing. As a general principle, the type of exposure is selected to represent usage. Some of the possible types are as follows:

Direct weathering. For direct exposure, the specimen is mounted on the exposure frame, open-backed or solid-backed, and subject to all atmospheric effects. This type can be used at a number of exposure angles. The standard angles used are 45°, 5°, and 90°, these angles being referenced from a horizontal angle of 0°. The angle chosen should be one that matches as closely as possible the position of the end use of the material.[48] The racks should be cleaned on a regular basis to remove mildew and algae if these contaminant producers are present on the test site. Figure 7.40 is an aerial view of the Kennedy Space Center beach corrosion test site, and Fig. 7.41 is a ground view with a background view of the Shuttle.[58]

Black box weathering. Black box exposure is used primarily to recreate the exposure conditions of the horizontal surfaces of an automobile.

Figure 7.40 Aerial view of the Kennedy Space Center beach corrosion test site.

The "box" creates an enclosed air space beneath the panels that form the top surface of the box. The modified environment is similar to that of a parked automobile. The black box can be used at a number of exposure angles. However, for automotive testing, the black box is usually placed at 5°. The box is typically made of aluminum painted black, with the test panels forming the top surface. The black box also serves to lower the panel temperature overnight to below that of the surrounding air, creating a longer condensation period.

Under-glass weathering. This exposure technique places the specimen behind a glass-covered frame, protecting it from any direct rainfall. The solar transmittance properties of the glass filter out a significant amount of the harmful ultraviolet. This method is used to test interior materials.

Tropical weathering. Tropical weathering involves a naturally humid environment that accelerates fungal and algae growth at a significantly faster rate than standard outdoor weathering. Since microbial resistance is a very important characteristic of paints and paint films, considerable attention has been given to developing a field test that provides the optimal conditions for the accelerated growth of mildew and algae. In turn, companies that need to test their algicides and fungicides in paint and paint films can do so in a much shorter period of time. In these tests, specimens are exposed on a standard aluminum frame with a vertical north orientation. Specimens should ideally have a wood or Styrofoam substrate that will also allow for water capillary action from the sample sitting on the test rack.[48]

Optimizing test programs. The iterative process described as "experimental design" consists of planning both the test variables and their subsequent logical analysis. Applied to a corrosion problem, such a process can combine modern scientific principles with an appreciation of the mechanisms of degradation of the material being studied. The role of experimental design in acquiring the knowledge of a process is illustrated in Fig. 7.42, where the loop emphasizes the iterative aspect of the process, leading to increased knowledge of a system behavior. The main idea behind experimental design is to minimize the number of steps before an acceptable understanding becomes possible. One of the first descriptions of an experimental design application to a corrosion situation estimated that such statistics could[49]

Figure 7.41 Ground view of the Kennedy Space Center beach corrosion test site with a background view of the Shuttle.

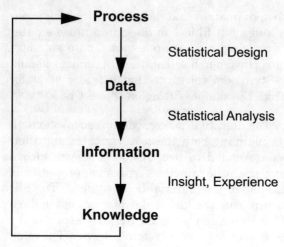

Figure 7.42 The experimental design loop for acquiring knowledge of a process.

- Save time and money: Fewer experiments are required per firm conclusion.
- Simplify data handling: Data are digested in a readily reusable form.
- Establish better correlations: Variables and their effects are isolated.
- Provide greater accuracy: The estimation of errors is the cornerstone of statistical design.

As expressed in Fig. 7.42, the selection of an experimental strategy should precede and influence data acquisition. It is indeed difficult, if not impossible, to retrofit experiments to satisfy the statistical considerations necessary for the construction of valid models. Any time spent in preparing a test program is a good investment. The most important consideration, at the initial planning stages, should be to integrate the available information in order to limit future setbacks. For complex situations, a good compromise is to employ what is called a screening design technique. The purpose of running screening experiments is to identify a small number of dominant factors, often with the intent of conducting a more extensive and systematic investigation. An important application of screening experiments is to perform ruggedness tests that, once completed, will permit the control or limitation of environmental factors or test conditions that can easily influence a test program. There are, of course, many subtleties in designing experiments that may intimidate a person who has limited familiarity with statistics. But fortunately there are a growing number of software

packages that can guide and support a user in a friendly
through the process of designing experiments. The following exa..
illustrate the application of such methodology to practical and comp..
corrosion testing situations.

The selection of a cast superalloy. In this study, a series of cast Ni-base
γ/γ' superalloys were systematically varied at selected levels of Co, Cr,
Mo, Ta, and Al, and the alloys' weight change performance was monitored.[50] A full factorial central experimental design was used, with five sets of star points to completely map the response of these alloys. For five elemental variables, 43 alloy compositions were required, and regression analysis of the results permitted the production of a complete second-degree equation describing the total variability of the alloy performance.

Improved cosmetic corrosion test. As part of the efforts of the American Iron and Steel Institute's (AISI) Task Force on Automotive Corrosion, a design of experiments (DOE) program was initiated. The aim of this program was to study the effects of a number of carefully selected test parameters on the performance of automotive steel sheet products subjected to a cyclic corrosion test and to "on-vehicle" tests.[51] A review of the literature guided the initial selection of seven test variables considered to be of major importance to the corrosion performance of automotive steel. Triplicate 100-mm by 150-mm panels were exposed to eight test runs designed according to a Plackett-Burman partial factorial design. The results of these tests were to be used as a guide to develop an improved test procedure.

Container material for nuclear waste disposal. In this project, a two-level factorial design was adopted to map the effects of five factors (Cl^-, SO_4^{2-}, NO_3^-, F^-, and temperature) on a candidate material for containers in the Yucca Mountain repository site.[52] The trial order was randomized before testing, and a five-factor interaction was used to block the experiments in terms of two different potentiostats used in the study, in order to verify the instrumental variability. A complete response surface of alloy 825 resistance to localized corrosion, estimated by cyclic polarization, was produced as a function of environmental variables.

Managing water chemistry. Experimental design was also used to study. mild steel under water conditions containing various scaling agents and the addition of an organophosphorus inhibitor.[53] The corrosion rates were measured in laboratory experiments using linear polarization, and mathematical models were generated relating the corrosion behavior to solution flow rate and the concentrations of Ca^{2+}, Cl^-, HCO_3^-, and the organic inhibitor. Two models were developed in this bed the corrosion rates of mild steel, and the secncies of the water. These models were then validated in tower experiments.

environments. Experimental design techniques to develop models for understanding the effects of complex materials considered for the operation of different processes such as wood pyrolysis, gas desulfurization, and continuous digestion.[54] In these studies, it was demonstrated that to reduce the complexity of the environments (solution variables, constraints, etc.) to a manageable level, designed experiments are essential. When such studies are properly done, the results can be used to predict the corrosion performance of alloys as a function of solution composition. For the interested reader, reference 54 gives additional details on the actual statistical procedures used for a few typical designs for complex corrosion.

7.3 Surface Characterization

From an engineering materials viewpoint, the impact of corrosion on a system is mostly a surface phenomenon, and the scientists and engineers interested in fundamental corrosion processes have always been among the first to explore the utility of surface analysis techniques. Surface analysis is the use of microscopic chemical and physical probes that give information about the surface region of a sample. The probed region may be the extreme top layer of atoms, or it may extend up to several microns beneath the sample surface, depending on the technique used. These techniques have been increasingly successful in shedding light on many facets of corrosion mechanisms. Surface analysis techniques are fundamentally destructive, since they generally require that the sample be placed in an ultrahigh vacuum to prevent contamination from residual gases in the analysis chamber. A rule of thumb is that up to an atomic layer per second can be formed at pressures of 10^{-4} Pa if each collision of a gas molecule results in its sticking to the surface.[55] Since surface analysis is an extremely specialized field, it has its own nomenclature; the reader is referred to ASTM E 673, Terminology Relating to Surface Analysis. Table 7.13 presents a representative list of various techniques with their fundamental principles, and Table 7.14 identifies the types of information and resolution that they can produce.[56] The following list gives some of the common acronyms used to describe some of the surface analysis techniques.

- Auger electron spectroscopy (AES)
- Electron spectroscopy for chemical analysis (ESCA)

TABLE 7.13 Surface Analytical Techniques with Typical Applications and Signal Detected

Analytical technique	Typical applications	Signal detected
Auger	Surface analysis, high-resolution depth profiling	Auger electrons from near-surface atoms
FE Auger	Surface analysis, microanalysis, microarea depth profiling	Auger electrons from near-surface atoms
AFM/STM	Surface imaging with near-atomic resolution	Atomic-scale roughness
Micro-FTIR	Identification of polymers, organic films, liquids	Infrared absorption
XPS/ESCA	Surface analysis of organic and inorganic molecules	Photoelectrons
HFS	Hydrogen in thin films (quantitative)	Forward-scattered hydrogen atoms
RBS	Quantitative thin-film composition and thickness	Backscattered He atoms
SEM/EDS	Imaging and elemental microanalysis	Secondary and backscattered electrons and X-rays
FE SEM	High-resolution imaging of polished precision cross sections	Secondary and backscattered electrons
FE SEM (in lens)	Ultra-high-resolution imaging with unique contract mechanism	Secondary and backscattered electrons
SIMS	Dopant and impurity depth profiling, surface microanalysis	Secondary ions
Quad SIMS	Dopant and impurity depth profiling, surface microanalysis, insulators	Secondary ions
TOF SIMS	Surface microanalysis of polymers, organics	Secondary ions, atoms, molecules

- Field-emission auger electron spectroscopy (FE Auger)
- Scanning auger microscopy (SAM)
- Scanning probe microscopy (SPM)
- Scanning tunnelling microscopy (STM)
- Secondary electron microscopy (SEM)
- Secondary ion mass spectrometry (SIMS)
- Time-of-flight (TOF)
- Ultraviolet photoelectron spectroscopy (UPS)
- X-ray photoelectron spectroscopy (XPS)

TABLE 7.14 Surface Analytical Techniques with Detection Characteristics

Analytical technique	Elements detected	Organic information	Detection limits	Depth resolution	Imaging/ mapping	Lateral resolution (probe size)
Auger	Li–U	—	0.1–1 at%	<2 nm	Yes	100 nm
FE Auger	Li–U	—	0.01–1 at%	2–6 nm	Yes	<15 nm
AFM/STM	—	—	—	0.01 nm	Yes	1.5–5 nm
Micro-FTIR	—	Molecular groups	0.1–100 ppm	—	No	5 μm
XPS/ESCA	Li–U	Chemical bonding	0.01–1 at%	1–10 nm	Yes	10 μm–2 mm
HFS	H, D	—	0.01 at%	50 nm	No	2 mm × 10 mm
RBS	Li–U	—	1–10 at% ($Z < 20$) 0.01–1 at% ($20 < Z < 70$) 0.001–0.01 at% ($Z > 70$)	2 mm	Yes	
SEM/EDS	B–U	—	0.1–1 at%	1–5 μm (EDS)	Yes	4.5 nm (SEM)
FE SEM	—	—	—	—	Yes	1.5 nm
FE SEM (in lens)	—	—	—	—	Yes	0.7 nm
SIMS	H–U	—	ppb–ppm	5–30 nm	Yes	1 μm (imaging), 30 μm (depth profiling)
Quad SIMS	H–U	—	10^{14}–10^{17} at/cm^3	<5 nm	Yes	<5 mm (imaging), 30 μm (depth profiling)
TOF SIMS	H–U	Molecular ions to mass 10,000	<1 ppma, 10^8 at cm^{-2}	1 monolayer	Yes	0.10 μm

Some of these techniques require ultrahigh vacuum for the analysis of interfaces and others do not. They all involve irradiating the interface with a beam of photons, electrons, or ions and analyzing the reflected beam to determine the chemical nature of the interface. For any of these techniques, the conditions of the film must not change drastically during the measurement. Many of the techniques used to probe surfaces use a beam of ions to strike the surface and knock off atoms of the sample material. These atoms are ionized and are identified and measured using a mass spectrometry technique. Other techniques strike the surface with electrons (AES, EDS) or x-rays (ESCA) and measure the resulting electron or photon emissions to probe the sample. Measurements of the way high-energy helium nuclei bounce off a sample can be used as a sensitive measure of surface-layer composition and thickness (RBS). Surface structure on a microscopic scale is observed by using electron microscopes (SEM), optical microscopes, and atomic force or scanning probe microscopes (AFM/SPM). One common way of characterizing surface analysis techniques is by tabulating the incoming and outgoing particles. These techniques can be classified according to whether they utilize photons, electrons, or ions.[57]

Surface analysis is mainly used in two separate modes. One is in surface science, where the goal is to fundamentally understand the causes of the problem and the mechanisms that are occurring in a system. Usually a model system is picked to eliminate as many confounding variables as possible in order to get a system about which firm conclusions can be drawn. Often, many different techniques will be used on the same problem in order to illuminate as many facets as possible of the problem. The other mode is failure analysis. The goal here is to determine which of the failure modes is the most important one for a particular failure. The samples are real and, hence, nonideal. This analysis mode is often used to identify the elements present, their distribution pattern, and their oxidation state.[55]

Auger electron spectroscopy and x-ray photoelectron spectroscopy are probably the two surface analysis techniques that have found the greatest use in corrosion-related work. One of the first applications of surface analysis techniques to corrosion was an examination of the composition of the passive film on stainless steel. This investigation was undertaken to rationalize the substantial improvement in resistance to pitting and acid solutions that is found when Mo and/or Si are present in stainless steels. The AES results obtained in these early studies challenged the generally accepted explanation of the mid-1970s that the beneficial effects of Mo and Si were due to their enrichment of the passive film. In fact, the AES results indicated that Mo and Si were depleted in the film. There are basically two approaches in using surface-sensitive techniques to elucidate the mechanistic details of the interfacial processes and determine the molecular nature of the surface products, i.e., ex situ and in situ techniques. To investigate the interfacial processes at the metal/liquid interface, one has to resort to in situ techniques; however, one can use both in situ and ex situ techniques for the characterization of the interphase.

7.3.1 General sensitivity problems

The problems of sensitivity and detection limits are common to all forms of spectroscopy. In its simplest form, the question of sensitivity boils down to whether it is possible to detect the desired signal above the noise level. In virtually all surface studies, sensitivity is a major problem. Consider the case of a sample with a surface of size 1 cm^2 with typically 10^{15} atoms in the surface layer. In order to detect the presence of impurity atoms present at the 1 percent level, a technique must be sensitive to 10^{13} atoms.[56] Contrast this with a spectroscopic technique used to analyze a 1 cm^3 bulk liquid sample, typically containing 10^{22} molecules. The detection of 10^{13} molecules in this sample would require 1 part per billion (ppb) sensitivity, a level provided by only a few techniques.

Assuming that a technique of sufficient sensitivity can be found, another major problem that needs to be addressed in surface spectroscopy is distinguishing between signals from the surface and signals from the bulk of the sample. To ensure that the surface signal is distinguishable (shifted) from the comparable bulk signal, either the detection system must have sufficient dynamic range to detect very small signals in the presence of neighboring large signals or the bulk signal must be small compared to the surface signal, i.e., the vast majority of the detected signal must come from the surface region of the sample. It is the latter approach that is used by the majority of surface spectroscopic techniques; such techniques can then be said to be surface-sensitive.

7.3.2 Auger electron spectroscopy

AES is the most commonly used surface technique on metal samples because of the following advantages:[55]

- High surface sensitivity
- Acceptable detectability for many corrosion problems
- Simultaneous detection of all elements (except hydrogen and helium)
- Very good small-area analysis (mapping)
- Ability to probe deeper into the surface by sputter profiling

- Analysis time not excessively long
- Readily available instrumentation

The Auger process gives electrons of characteristic energy for each element, which are determined by the differences in energy of the orbitals involved. In addition to the Auger electrons, there are also much more plentiful secondary electrons with a broad energy distribution that overlies the characteristic peaks. To highlight the characteristic peaks, differentiation is performed on a plot of the number of electrons emitted by the sample versus the energy of those electrons. This results in a spectrum that ignores the more plentiful background (secondary) electrons and emphasizes the characteristic electrons that are used to identify the elements present. In some cases, the exact peak shape and energy can be used to identify the oxidation state of the elements present.[55]

One of the attractions of Auger analysis is that it is quite surface-sensitive, since an Auger spectrum typically represents information about the composition of the top 0.5 to 2 nm of the surface, depending upon the sample analyzed and the analysis conditions. Although Auger electrons can be generated at depths of several micrometers into the sample, the Auger electrons must be able to escape to the surface without undergoing an inelastic collision in order to be detected. Compilations of elemental spectra and charts of atomic number versus electron energy are available to help assign peaks. Modern data processing (background subtraction, peak fitting to standard spectra) has made it possible to correctly resolve many peak-identification problems caused by peak overlap.[55]

7.3.3 Photoelectron spectroscopy

Photoelectron spectroscopy utilizes photoionization and energy-dispersive analysis of the emitted photoelectrons to study the composition and electronic state of the surface region of a sample. Traditionally, when the technique has been used for surface studies, it has been subdivided according to the source of exciting radiation into

- X-ray photoelectron spectroscopy, which uses soft (200 to 2000 eV) x-ray excitation to examine core levels
- Ultraviolet photoelectron spectroscopy, which uses vacuum UV (10 to 45 eV) radiation from discharge lamps to examine valence levels

Photoelectron spectroscopy is based upon a single photon-in/electron-out process, and from many viewpoints this underlying process is much simpler than the Auger process. In XPS the photon is absorbed by an atom in a molecule or solid, leading to ionization and the emission of a core (inner-shell) electron. By contrast, in UPS the photon interacts with valence levels of the molecule or solid, leading to ionization by removal of one of these valence electrons. The kinetic energy distribution of the emitted photoelectrons can be measured using any appropriate electron energy analyzer, and a photoelectron spectrum can thus be recorded. An alternative approach is to consider a one-electron model along the following lines.

The realization that the energy of the ejected photon could be used to determine the chemical state of an atom gave rise to the name ESCA (electron spectroscopy for chemical analysis). Because x-ray photons are necessary to generate the appropriate electrons, the technique is also called x-ray photoelectron spectroscopy. XPS shares the Auger characteristic of good surface sensitivity, since this is driven by the same need for the electrons to be able to reach the detector unscathed. It is possible to vary the depth of analysis in both techniques by varying the tilt angle with regard to the detector. This technique is used more extensively in XPS, where it is often called angle-resolved depth profiling. XPS also has the very important advantage that it can obtain chemical state information on most atoms.[55]

For each and every element, there will be a characteristic binding energy associated with each core atomic orbital, i.e., each element will give rise to a characteristic set of peaks in the photoelectron spectrum at kinetic energies determined by the photon energy and the respective binding energies. The presence of peaks at particular energies therefore indicates the presence of a specific element in the sample under study, and the intensity of the peaks is related to the concentration of the element within the sampled region.[56]

7.3.4 Rutherford backscattering

Rutherford backscattering (RBS) is based on collisions between atomic nuclei and derives its name from Lord Ernest Rutherford, who in 1911 was the first to present the concept of atoms having nuclei. It involves measuring the number and energy of ions in a beam that scatters back after colliding with atoms in the near-surface region of a sample at which the beam has been targeted. With this information, it is possible to determine atomic mass and elemental concentrations versus depth below the surface. RBS is ideally suited for determining the concentration of trace elements that are heavier than the major constituents of the substrate. Its sensitivity for light molecular masses and for the makeup of samples well below the surface is poor.[56]

When a sample is bombarded with a beam of high-energy particles, the vast majority of the particles are implanted into the material and do not escape. This is because the diameter of an atomic nucleus is on the order of 10^{-6} nm while the spacing between nuclei is on the order of 0.2 nm. A small fraction of the incident particles do undergo a direct collision with a nucleus of one of the atoms in the upper few micrometers of the sample. This collision does not actually involve direct contact between the projectile ion and the target atom. Energy exchange occurs because of Coulomb forces between nuclei in close proximity to each other. However, the interaction can be modeled accurately as an elastic collision using classical physics.[56]

The energy measured for a particle backscattering at a given angle depends upon two processes. Particles lose energy as they pass through the sample, both before and after a collision. The amount of energy lost is dependent on the material's stopping power. A particle will also lose energy as the result of the collision itself. The collisional loss depends on the masses of the projectile and the target atoms. The ratio of the energy of the projectile before and after collision is called the kinematic factor.[56] The number of backscattering events that occur from a given element in a sample depend upon two factors: the concentration of the element and the effective size of its nucleus. The probability that a material will cause a collision is called its scattering cross section.

7.3.5 Scanning probe microscopy (STM/AFM)

In the early 1980s, two IBM scientists, Binnig and Rohrer, developed a new technique for studying surface structure, scanning tunneling microscopy (STM). This invention was quickly followed by the development of a whole family of related techniques which, together with STM, may be classified in the general category of scanning probe microscopy (SPM) techniques. Of these later techniques, the most important is undoubtedly atomic force microscopy (AFM).[56] The development of these techniques has without doubt been the most important event in the surface science field in recent times, and has opened up many new areas of science and engineering at the atomic and molecular level. All of the SPM techniques are based upon scanning a probe (typically called the tip in STM, since it literally is a sharp metallic tip) just above a surface while monitoring some interaction between the probe and the surface. The interaction that is monitored is:

> In STM, the tunnelling current between the metallic tip and a conducting substrate which are in very close proximity but not actually in physical contact.

In AFM, the van der Waals force between the tip and the surface; this may be either the short-range repulsive force (in contact mode) or the longer-range attractive force (in noncontact mode).

For the techniques to provide information on the surface structure at the atomic level, the position of the tip with respect to the surface must be very accurately controlled (to within about 0.1 Å) by moving either the surface or the tip.

If the tip is biased with respect to the surface by the application of a voltage between them, then electrons can tunnel between the two, provided that the separation of the tip and the surface is sufficiently small; this gives rise to a tunneling current. The direction of current flow is determined by the polarity of the bias. If the sample is biased $-ve$ with respect to the tip, then electrons will flow from the surface to the tip, while if the sample is biased $+ve$ with respect to the tip, then electrons will flow from the tip to the surface.

The name of the technique arises from the quantum mechanical tunneling-type mechanism by which the electrons can move between the tip and the substrate. Quantum mechanical tunneling permits particles to tunnel through a potential barrier which they could not surmount according to the classical laws of physics; in this case, electrons are able to traverse the classically forbidden region between the two solids. In this model, the probability of tunneling is exponentially dependent upon the distance of separation between the tip and the surface; the tunneling current is therefore a very sensitive probe of this separation. Imaging of the surface topology may then be carried out in one of two ways:

- In constant-height mode, the tunneling current is monitored as the tip is scanned parallel to the surface.
- In constant-current mode, the tunneling current is maintained constant as the tip is scanned across the surface.

If the tip is scanned at what is nominally a constant height above the surface, there is actually a periodic variation in the separation distance between the tip and the surface atoms. At one point the tip will be directly above a surface atom and the tunneling current will be large, whereas at other points the tip will be above hollow sites on the surface and the tunneling current will be much smaller. In practice, however, the normal way of imaging the surface is to maintain the tunneling current constant while the tip is scanned across the surface. This is achieved by adjusting the tip's height above the surface so that the tunneling current does not vary with the lateral tip position. In this mode, the tip will move slightly upward as it passes over a surface atom and, conversely, move slightly in toward the surface as it passes over a hollow.

7.3.6 Secondary electron microscopy and scanning Auger microscopy

The two forms of electron microscopy which are commonly used to provide surface information are secondary electron microscopy (SEM), which provides a direct image of the topographical nature of the surface from all the emitted secondary electrons, and scanning Auger microscopy (SAM), which provides compositional maps of a surface by forming an image from the Auger electrons emitted by a particular element.[56]

SEM. As the primary electron beam is scanned across the surface, electrons with a wide range of energies will be emitted from the surface in the region where the beam is incident. These electrons will include backscattered primary electrons and Auger electrons, but the vast majority will be secondary electrons formed in multiple inelastic scattering processes (these are the electrons that contribute to the background and are completely ignored in Auger spectroscopy). The secondary electron current reaching the detector is recorded, and the microscope image consists of a "plot" of this current I against probe position on the surface. The contrast in the micrograph arises from several mechanisms, but first and foremost from variations in the surface topography. Consequently, the secondary electron micrograph is virtually a direct image of the real surface structure.[55]

The attainable resolution of the technique is limited by the minimum spot size that can be obtained with the incident electron beam, and ultimately by the scattering of this beam as it interacts with the substrate. With modern instruments, a resolution of better than 5 nm is achievable. This is more than adequate for imaging semiconductor device structures, for example, but is insufficient to enable many supported metal catalysts to be studied in any detail.

Although not a true surface technique, SEM-EDS (energy dispersive spectrometer) often provides useful information concerning surface corrosion mechanisms. The ubiquitous nature, low cost, and ease of use of this technique cause it to be used as a tool in many failure analyses involving corrosion. Because its analysis depth is much larger (approximately a micrometer) than that of the true surface techniques, it is not necessary to analyze samples that are high-vacuum-compatible. As a result, almost no sample preparation is needed for many different kinds of samples.

The sample is scanned with a high-energy (typically 5 to 30 keV) electron beam in a raster pattern which causes the ejection of a number of particles, including secondary electrons, backscattered electrons, and x-rays. Secondary electrons (with energies less than 50 eV) are detectable only if they are generated in the top surface of a sample; this causes the secondary electron output to be responsive to topographical detail and therefore gives an image that is remarkably similar to that seen with an optical microscope. Added advantages are greater magnification and depth of field. The contrast in backscattered electron images is mainly dependent on atomic number, so these images provide rough elemental distribution information.[55]

Element identification is provided by analysis of the characteristic x-rays that are emitted with an energy dispersive spectrometer. Quantification can be quite good if appropriate standards are used. The x-ray detector can be set to detect and count only x-rays that have energies within a narrow range. This output can then be used to generate elemental distribution maps, or line scans. Newer detectors with ultrathin windows can easily detect all elements with an atomic number of 5 (boron) or greater.

SAM. With this technique, the incident primary electrons cause ionization of atoms within the region illuminated by the focused beam. Subsequent relaxation of the ionized atoms leads to the emission of Auger electrons characteristic of the elements present in this part of the sample surface. As with SEM, the attainable resolution is ultimately limited by the incident beam characteristics. More significantly, however, the resolution is also limited by the need to acquire sufficient Auger signal to form a respectable image within a reasonable time period, and for this reason the instrumental resolution achievable rarely approaches 20 nm.

7.3.7 Secondary ion mass spectroscopy

Secondary ion mass spectroscopy is the third of the three most common surface analysis techniques. In SIMS, the sample is irradiated with a primary ion beam (normally argon), the impact of which sputters away the surface atoms, some as neutrals and others as ions. Those atoms which become ionized are then detected in a mass spectrometer, where their masses are measured.

SIMS is the most sensitive of all the commonly employed surface analytical techniques. This is because of the inherent sensitivity associated with mass spectrometric–based techniques. There are a number of different variants of the technique:[56]

- Static SIMS, used for submonolayer elemental analysis
- Dynamic SIMS, used for obtaining compositional information as a function of depth below the surface
- Imaging SIMS, used for spatially resolved elemental analysis

These variations are all based on the same basic physical process, and it is this process which is discussed here, together with a brief introduction to the field of static SIMS. In SIMS, the surface of the sample is subjected to bombardment by high-energy ions; this leads to the ejection (or sputtering) of both neutral and charged ($+/-$) species from the surface. The ejected species may include atoms, clusters of atoms, and molecular fragments.

In traditional SIMS, it is only the positive ions that are mass analyzed. This is primarily for practical ease, but it does lead to problems with quantifying the compositional data, since the positive ions are but a small, nonrepresentative fraction of the total sputtered species. It should be further noted that the displaced ions have to be energy filtered before they are mass analyzed (i.e., only ions with kinetic energies within a limited range are mass analyzed). The most commonly employed incident ions used for bombarding the sample are argon ions (Ar^+), but other ions have been used in some applications.[56] The mass analyzer is typically a quadrupole mass spectroscopy analyzer with unit mass resolution, but high-specification time-of-flight (TOF) analyzers are also used and provide substantially higher sensitivity and a much greater mass range (albeit at a higher cost).

In static SIMS (SSIMS), the aim is to obtain sufficient signal to provide compositional analysis of the surface layer without actually removing a significant fraction of a monolayer, i.e., to be able to analyze less than 10^{14} atoms or molecules, or approximately 10 percent of a monolayer, for a 1-cm^2 sample. The technique is then capable of providing information about the topmost single atomic layer of the surface. In dynamic SIMS, which is more common, a high-energy ion beam removes layers of the surface. The beam is so energetic that little chemical information is retained, since the vast majority of any molecular species is fragmented. Although destroying the surface obviously prevents its reexamination, it is not a total disadvantage, since it allows depth profiling to occur naturally.

Some of the advantages of SIMS are that it has a very low detection limit and that it can detect all elements. These advantages make it able to address many problems that neither AES nor XPS is suitable for. Reasonably small (micrometer or smaller) spot sizes allow elemental mapping. A major disadvantage of SIMS is that there is a very great range of ionization rates for different elements. Furthermore, the rates will vary depending upon the other species present (matrix effects). A beam of either positive or negative ions can be used as the exciting beam, with very different response factors. The biggest differences are found with the very electronegative halogens and the electropositive alkali metals.[55]

References

1. Ijseling, F. P., *General Guidelines for Corrosion Testing of Materials for Marine Applications,* London, The Institute of Materials, 1989.
2. Lloyd, D. K., and Lipow, M., *Reliability: Management, Methods and Mathematics,* Milwaukee, Wisc., The American Society for Quality Control, 1984.
3. Erlings, J. G., de Groot, H. W., and Nauta J., The Effect of Slow Plastic and Elastic Straining on Sulphide Stress Cracking and Hydrogen Embrittlement of 3.5% Ni Steel and API 5L X60 Pipeline Steel, *Corrosion Science,* **27:**1153–1167 (1987).
4. Baboian, R., *Corrosion Tests and Standards,* Philadelphia, American Society for Testing and Materials, 1995.
5. Lyon, S. B., Thompson, G. E., and Johnson, J. B., in Agarwala, V. S., and Ugiansky, G. M. (eds.), *New Methods for Corrosion Testing of Aluminum Alloys,* Philadelphia, American Society for Testing and Materials, 1992, pp. 20–31.
6. Treseder, R. S., Haynes, G. S., and Baboian, R. (eds.), *Laboratory Corrosion Tests and Standards,* Philadelphia, American Society for Testing and Materials, 1985, pp. 5–23.
7. Mapes, R. S., and Berkey, W. W., X-Ray Diffraction Methods for the Analysis of Corrosion Products, in Ailor, W. H. (ed.), *Handbook on Corrosion Testing and Evaluation,* New York, John Wiley and Sons, 1971, pp. 697–730.
8. Lenard, D. R., Moores, J. G., Roberge, P. R., and Halliop, E., "The Use of Electrochemical Impedance Spectroscopy to Predict the Corrosion of Aluminum-Lithium Alloys in Marine Environments," AGARD CP-565, in *AGARD Conference Proceedings: Corrosion Detection and Management of Advanced Airframe Materials,* Hull, Canada, Canada Publication Group, 1995, pp. 8-1–8-12.
9. Schra, L. and Groep, F. F., *The ASCOR Test: A Simple Automated Method for Stress Corrosion Testing of Aluminum Alloys,* Report NLR-TP-91438-U, 22, Amsterdam, Netherlands, National Aerospace Laboratory, 1991.
10. Freeman, R. A., and Silverman, D. C., Error Propagation in Coupon Immersion Tests, *Corrosion,* **48:**463–466 (1992).
11. Baker, S. V., Lyon, S. B., Thompson, G. E., et al., in Agarwala, V. S., and Ugiansky, G. M. (eds.), *New Methods for Corrosion Testing of Aluminum Alloys,* Philadelphia, American Society for Testing and Materials, 1992, pp. 32–49.
12. Roberge, P. R., Yousri, S., and Halliop, E., Potentiodynamic Polarization and Impedance Spectroscopy for the Statistical Process Control of Aluminum Anodizing, in Silverman, D. C., Kendig, M. W., and Scully, J., (eds.), *Electrochemical Impedance: Analysis and Interpretation,* STP 1188, Philadelphia, American Society for Testing and Materials, 1993, pp. 313–329.
13. Roberge, P. R., and Halliop, E., An Alternate Electrochemical Procedure for the Testing of Anodized Aluminum, in Haynes, G. S., and Tellefsen, K. (eds.), *Cyclic Cabinet Corrosion Testing,* STP 1238, Philadelphia, American Society for Testing and Materials, 1995, pp. 49–58.
14. Roberge, P. R., and Ash, P., The Anodic Breakthrough Method for Testing Anodized Aluminum, *Metal Finishing,* **93:**22–25 (1995).
15. Yousri, S., and Tempel, P., *Plating and Surface Finishing,* **74:**36–43 (1987).
16. Jirnov, A. D., and Karimova, S. N., "Some Peculiarities of Al-Li Alloys Corrosion Behavior," 2(Sixth), *International Aluminium-Lithium Conference,* Garmisch-Partenkirchen, 1991, pp. 825–829.
17. Roberge, P. R., and Lenard, D. R., The Evaluation of Marine Corrosion Resistance of Aluminum and Aluminum-Lithium Alloys, *Corrosion Reviews,* **15:**631–645 (1997).
18. Kane, R. D., High-Temperature and High Pressure, in Baboian, R. (ed.), *Corrosion Tests and Standards,* Philadelphia, American Society for Testing and Materials, 1995, pp. 106–115.
19. Andresen, P. L., "Effects of Testing Characteristics on Observed SCC Behavior in BWRs," in *Corrosion 98,* Houston, Tex., NACE International, 1998, Paper # 137.
20. Van Orden, A. C., "Applications and Problem Solving Using the Polarization Technique," in *Corrosion 98,* Houston, Tex., NACE International, 1998, Paper # 301.

21. Grauer R., Moreland, P. J., and Pini, G., *A Literature Review of Polarisation Resistance Constant (B) Values for the Measurement of Corrosion Rate,* Houston, Tex., NACE International, 1982.
22. Silverman, D. C., "Tutorial on Cyclic Potentiodynamic Polarization Technique," in *Corrosion 98,* Houston, Tex., NACE International, Paper # 299.
23. Zeller, R. L., III, and Savinell, R. F., *Corrosion Science,* **26:**591 (1986).
24. Epelboin, I., Keddam, M., and Takenouti, H., *Journal of Applied Electrochemistry,* **2:**71 (1972)
25. Boukamp, B. A., Equivalent Circuit (Equivcrt.PAS) Users Manual, Report CT89/214/128, The Netherlands, University of Twente, 1989.
26. Hladky, K., Callow, L. M., and Dawson, J. L., *British Corrosion Journal,* **15:**20 (1980).
27. de Levie, R., *Advances in Electrochemistry and Electrochemical Engineering,* 1969.
28. Silverman, D. C., *Corrosion,* **47:**87 (1991)
29. Mansfeld, F., and Shih, H., *Journal of the Electrochemical Society,* **135:**1171 (1988).
30. Urquidi-MacDonald, M., and Egan, P. C., Validation and Extrapolation of Electrochemical Impedance Spectroscopy Data Analysis, *Corrosion Reviews,* **15:** (1997).
31. Iverson, W. P., *Journal of the Electrochemical Society,* **115:**617 (1968).
32. Kearns, J. R., Scully, J. R., Roberge, P. R., et al., *Electrochemical Noise Measurement for Corrosion Measurements,* Philadelphia, American Society for Testing and Materials, 1996.
33. Hladky, K., and Dawson, J. L., *Corrosion Science,* **21:**317
(1981).
34. Searson, P. C., and Dawson, J. L., *Journal of the Electrochemical Society,* **135:**1908–1915 (1988).
35. Peitzen, H. O., and Saupe, D., *The Science of Fractal Images,* New York, Springer-Verlag, 1988.
36. Mandelbrot, B. B., and Van Ness, J. W., *SIAM Review,* **10:**422 (1968).
37. Fan, L. T., Neogi, D., and Yashima, M., *Elementary Introduction to Spatial and Temporal Fractals,* Berlin, Springer-Verlag, 1991.
38. Hurst, E. H., Methods of Using Long-term Storage in Reservoirs, *Proceedings of the Institute of Civil Engineering,* **5** (Part I):519 (1956).
39. Mandelbrot, B. B., and Wallis, J. R., *Water Resources Research,* **5:**321 (1969).
40. Feder, J., *Fractals,* New York, Plenum, 1988.
41. Roberge, P. R., The Analysis of Spontaneous Electrochemical Noise by the Stochastic Process Detector Method, *Corrosion,* **50:**502 (1994).
42. Roberge, P. R., Trethewey, K. R., Marsh, D. J., et al., Application of Fractals to the Analysis of Electrochemical Noise in 304 Stainless Steel, in LeMay, I., Mayer, P., Roberge, P. R., et al. (eds.), *Materials Performance, Maintenance and Plant Life Assessment,* Montréal, The Canadian Institute of Mining, Metallurgy and Petroleum, 1994, pp. 47–56.
43. Capp, J. A., A Rational Test for Metallic Protective Coatings, *Proceedings of the American Society for Testing and Materials,* **14:**474–481 (1914).
44. Finn, A. N., Method of Making the Salt-Spray Corrosion Test, *Proceedings of the American Society for Testing and Materials,* **18:**237–238 (1918).
45. LaQue, F. L., *Marine Corrosion: Causes and Prevention,* New York, John Wiley and Sons, 1975.
46. LaQue, F. L., *Materials & Methods,* **35:**77–81 (1952).
47. Harper, R., *Metal Industry,* **99:**454–458 (1961).
48. Sub-Tropical Testing Service, *http://www.Sub-Tropical.com/exptest.html,* 1999.
49. Twitchell, S. B., and Lackmeyer, P. J., Experimental Design in Corrosion Control, *Materials Performance,* **14:**14 (1975).
50. Barrett, C. A., "The Effects of Cr, Co, Al, Mo and Ta on the Cyclic Oxidation Behavior of a Prototype Cast Ni-Base Superalloy Based on a 25 Composite Statistically Designed Experiment," in *Conference on High Temperature Corrosion Energy Systems,* Detroit, The Metallurgical Society/AIME, 1984, pp. 667–680.
51. Roudabush, L. A., Towsend, H. E., and McCune, D. C., "Update on the Development of an Improved Cosmetic Corrosion Test by the Automotive and Steel Industries,"

Proceedings of the 6th Automotive and Prevention Conference, P-268, Warrendale, Pa., SAE International, 1993, pp. 53–63.
52. Gragnolino, G. A., and Sridhar, N., Localized Corrosion of a Candidate Container Material for High-Level Nuclear Waste Disposal, *Corrosion,* **47:**464–472 (1991).
53. Davis, R. V., "Investigation of Factors Influencing Mild Steel Corrosion Using Experimental Design," in *Corrosion/93, Houston, Tex.,* NACE International, 1993, *Paper # 280.*
54. Koch, G. H., Spangler, J. M., and Thompson, N. G., Corrosion Studies in Complex Environments, in Francis, P. E., and Lee, T. S. (eds.), *The Use of Synthetic Environments for Corrosion Testing,* Philadelphia, American Society for Testing and Materials, 1988, pp. 3–17.
55. Hopkins, A. G., Surface Analysis, in Baboian, R. (ed.), *Corrosion Tests and Standards,* Philadelphia, American Society for Testing and Materials, 1995, pp. 55–61.
56. The CEA Online Tutorial, *http://www.cea.com/tutorial.htm,* 1998.
57. Shaw, S., *Surface and Interface Characterization in Corrosion,* Houston, Tex., NACE International, 1994.
58. http://ftp-msd.ksc.nasa.gov/msd/ftp/rust/PICTURES/BCH-SITE/

Chapter 8

Materials Selection

8.1	Introduction	578
	8.1.1 Mechanical properties	579
	8.1.2 Fabricability	581
	8.1.3 Availability	581
	8.1.4 Cost	581
	8.1.5 Corrosion resistance	582
8.2	Aluminum Alloys	584
	8.2.1 Introduction	584
	8.2.2 Applications of different types of aluminum	595
	8.2.3 Weldability of aluminum alloys	598
	8.2.4 Corrosion resistance	601
8.3	Cast Irons	612
	8.3.1 Introduction	612
	8.3.2 Carbon presence classification	613
	8.3.3 Weldability	616
	8.3.4 Corrosion resistance	617
8.4	Copper Alloys	622
	8.4.1 Introduction	622
	8.4.2 Weldability	627
	8.4.3 Corrosion resistance	630
	8.4.4 Marine application of copper-nickel alloys	650
	8.4.5 Decorative corrosion products	659
8.5	High-Performance Alloys	664
	8.5.1 Ni- and Fe-Ni-base alloys	666
	8.5.2 Co-base alloys	670
	8.5.3 Welding and heat treatments	671
	8.5.4 Corrosion resistance	676
	8.5.5 Use of high-performance alloys	691
8.6	Refractory Metals	692
	8.6.1 Introduction	692
	8.6.2 Molybdenum	694

8.6.3	Niobium	697
8.6.4	Tantalum	705
8.6.5	Tungsten	708
8.7	Stainless Steels	710
8.7.1	Introduction	710
8.7.2	Welding, heat treatments, and surface finishes	716
8.7.3	Corrosion resistance	723
8.8	Steels	736
8.8.1	Introduction	736
8.8.2	Carbon steels	737
8.8.3	Weathering steels	738
8.8.4	Weldability	739
8.8.5	Corrosion resistance	741
8.9	Titanium	748
8.9.1	Introduction	748
8.9.2	Titanium alloys	750
8.9.3	Weldability	752
8.9.4	Applications	754
8.9.5	Corrosion resistance	755
8.10	Zirconium	769
8.10.1	Applications	773
8.10.2	Corrosion resistance	774
References		777

8.1 Introduction

From a purely technical standpoint, an obvious answer to corrosion problems would be to use more-resistant materials. In many cases, this approach is an economical alternative to other corrosion control methods. Corrosion resistance is not the only property to be considered in making material selections, but it is of major importance in the chemical process industries. Table 8.1 lists the questions that should be answered to estimate the corrosion behavior of materials either in service or considered for such usage.[1]

The choice of a material is the result of several compromises. For example, the technical appraisal of an alloy will generally be a compromise between corrosion resistance and some other properties such as strength and weldability. And the final selection will be a compromise between technical competence and economic factors. In specifying a material, the task usually requires three stages:

1. Listing the requirements
2. Selecting and evaluating the candidate materials
3. Choosing the most economical material

TABLE 8.1   Information Necessary for Estimating Corrosion Performance

Corrodent variables

Main constituents (identity and amount)
Impurities (identity and amount)
Temperature
pH
Degree of aeration
Velocity or agitation
Pressure
Estimated range of each variable

Type of application

What is the function of part or equipment?
What effect will uniform corrosion have on serviceability?
Are size change, appearance, or corrosion product a problem?
What effect will localized corrosion have on usefulness?
Will there be stresses present?
Is SCC a possibility?
Is design compatible with the corrosion characteristics of the material?
What is the desired service life?

Experience

Has the material been used in identical situation?
  With what specific results?
  If equipment is still in operation, has it been inspected?
Has the material been used in similar situations?
What are the differences in performance between the old and new situations?
Any pilot-plant experience?
Any plant corrosion-test data?
Have laboratory corrosion tests been run?
Are there any available reports?

Some particular requirements and typical selection considerations are presented in Table 8.2. The materials selection process is also influenced by the fact that the materials are either considered for the construction of a new system or for the modification or repairs in an existing facility. For the construction of new equipment, the selection procedure should begin as soon as possible and before the design is finalized. The optimum design for corrosion resistance will often vary with the material used. In a repair application, there is usually less opportunity for redesign, and the principal decision factors will be centered on delivery time and ease of fabrication in the field. It is also advisable to estimate the remaining life of the equipment so that the repair is not overdesigned in terms of the corrosion allowance.

8.1.1   Mechanical properties

The selection criteria used by materials engineers in choosing from a group of materials includes a list of qualities that are either desirable

TABLE 8.2 Checklist for Materials Selection

Requirements to be met
Properties (corrosion, mechanical, physical, appearance)
Fabrication (ability to be formed, welded, machined, etc.)
Comparability with existing equipment
Maintainability
Specification coverage
Availability of design data
Selection considerations
Expected total life of plant or process
Estimated service life of material
Reliability (safety and economic consequences of failure)
Availability and delivery time
Need for further testing
Material costs
Fabrication costs
Maintenance and inspection costs
Return on investment analysis
Comparison with other corrosion-control methods

or necessary. Unfortunately, the optimum properties associated with each selection criteria can seldom all be found in a single material, especially when the operating conditions become aggressive. Thus, compromises must frequently be made to realize the best performance of the material selected.[2]

A wide variety of iron- and nickel-based materials are used for pressure vessels, piping, fittings, valves, and other equipment in process industries. The most common of these is plain carbon steel. Although it is often used in applications up to 482 to 516°C, most of its use is limited to 316 to 343°C due to loss of strength and susceptibility to oxidation and other forms of corrosion at higher temperature. Ferritic alloys, with additions consisting primarily of chromium (0.5 to 9%) and molybdenum (0.5 to 1%), are most commonly used at temperatures up to 650°C. Their comparative cost, higher strength, oxidation and sulfidation resistance, and particular resistance to hydrogen, for example, result in their being the material of choice. However, these low-alloy steels have inadequate corrosion resistance to many other elevated temperature environments for which more highly alloyed Ni-Cr-Fe alloys are required.[2]

For applications for which carbon or low-alloy steels are not suitable, the most common choice of material is from within the 18Cr-8Ni austenitic group of stainless steels. These alloys and the 18Cr-12Ni steels are favored for their corrosion resistance in many environments and their oxidation resistance at temperatures up to 816°C. Above 650°C their decreasing strength becomes a consideration and more heat-resistant alloys must often be used.

Most chemical process equipment is designed and fabricated to the requirements of specific pressure vessel and piping codes. These codes include only approved materials and establish the basis for and the setting of allowable stresses. Thus, the mechanical properties of a material are usually the first criteria that materials engineers apply in the selection process. This is especially important for applications at temperatures in the creep range where a minor difference in operating temperature can significantly affect the load-carrying ability of the material.[2]

8.1.2 Fabricability

There are many outstanding materials with highly desirable mechanical properties and corrosion resistance that are seldom used because they cannot be fabricated. There are some materials that have excellent properties that can be fabricated as produced but, because of aging, cannot be modified or repaired after exposure to operating conditions. Materials should therefore be selected on the basis of their maintainability as well as their original fabricability. In general, the wrought heat-resistant alloys have greater fabricability than the cast materials. Cast alloy steels, for example, can typically tolerate significantly higher concentrations of carbon, silicon, tungsten, molybdenum, and so forth, which are added to enhance mechanical properties, corrosion resistance, or both. But, these elements also can adversely affect the original, as-produced fabricability and make maintainability, particularly weldability, difficult, if not impossible.[2]

8.1.3 Availability

Materials engineers and purchasing agents become frustrated when trying to obtain materials that have a limited number of producers or a limited production volume. Such frustration can be particularly high when a small amount of material is needed to finish a job or replace a failed piece. Prior to the original specification of a material, consideration should be given to its future availability for repairs or replacement in the form or forms that it will be used. In those cases where it might not be available, alternative replacement materials should be identified.[2]

8.1.4 Cost

Economics enter into every business decision. However, the important criterion should not be the initial cost of a material, but its life-cycle cost or cost effectiveness. It usually is much more cost effective to specify a material that will provide an extended life, particularly in areas that are difficult to repair or in components that would cause major shutdowns in case of failure. In these situations, the original cost of the material can be insignificant compared to the loss of production caused by the use of a lower-cost, but less-effective, material. The following two extreme alternatives describe the consideration given to economic factors when selecting materials for specific service:

1. A low initial cost system largely based on carbon steel and cast iron that will require considerable maintenance over the life of the plant. Such a system is a reasonable choice in areas where labor costs are low and material is readily available.

2. A system based mainly on alloy materials that, if correctly designed and fabricated, will require minimum maintenance and will function reliably. Rising labor costs in most industries, together with the need for high reliability in capital-intensive plants has produced a trend to this type of system.

In practice many systems are a mixture of these extreme options, resulting in the high initial costs of one and the high maintenance costs of the other. For example, a plant that has experienced costly replacement to galvanized steel piping may replace it with copper alloy piping, leaving valve fittings and other equipment in carbon steel and cast iron. The resulting galvanic corrosion effects result in reduced life for these parts. Thus, avoidance of higher initial costs has resulted in reduced reliability and high maintenance costs.[3] Unfortunately, competitive bidding and corporate bottom lines frequently create barriers that inhibit realization of long equipment life. The enlightened company will recognize the value of the life-cycle cost approach on long-term financial health and not embrace only the low initial cost option.[2]

8.1.5 Corrosion resistance

The additional cost usually associated with choosing increased corrosion resistance during the selection process is invariably less than that due to product contamination or lost production and high maintenance costs due to premature failure. Without adequate corrosion resistance, or corrosion allowance, components often fall short of the expected design life. Unlike mechanical properties, there are no codes governing corrosion properties. For some applications or services, recommended practices have been published by NACE International or other societies.

Many extensive sources of information concerning specific corrosion resistance or corrosivity assessment data have been published in the form of handbooks. Some advanced information systems and modeling tools, specially adapted to corrosion damage prediction, are reviewed in Chap. 4, Modeling, Life Prediction, and Computer Applications. Data upon which to base material selection are also available in inexpensive literature and manufacturers' publications that are available upon request or can be downloaded from the Internet. These sources are referenced in the next sections dealing with specific alloy systems.

A common form of representing the corrosion resistance of materials is what is known as *iso-corrosion diagrams* or charts. These diagrams are two-dimensional representations of three-dimensional corrosion data. Iso-corrosion diagrams present corrosion behavior as a function of corrosive concentration (usually the abscissa) and temperature. The use of the prefix *iso* refers to lines (or regions) of constant corrosion behavior across variations in concentration and temperature. *Corrosion Data Survey—Metals*[4] and *Corrosion Data Survey—Nonmetals*[5] are the most popular publications using such a scheme for representing corrosion information. In these publications the corrosion behavior of metals is expressed in units of penetration rates [i.e., $mm \cdot y^{-1}$ or milli-inch per year (mpy)], and the corrosion behavior of nonmetals is expressed in qualitative terms such as *recommended, questionable,* and *not recommended*. Figure 8.1 illustrates the elemental matrix used to express corrosion resistance in *Corrosion Data Survey—Metals,* and Fig. 8.2 presents iso-corrosion diagrams for S30400 and S31600 stainless steels exposed to aerated acetic acid. Another interesting visual representation of corrosion information can also found in *Corrosion Data Survey—Metals.*[4] Figure 8.3 is a rendition of the hydrochloric acid graph, and Fig. 8.4 shows a mixed acid graph.

In 1984 the National Association of Corrosion Engineers (NACE, now NACE International) and National Institute of Standards and Technology (NIST) agreed to process these two books for use on a personal computer as the first major project of the NACE-NIST Corrosion Data Program. The resultant DOS-based programs were released as the personal computer databases COR·SUR (Metals) and COR·SUR2 (Nonmetals). Over the past decade, they have become the most widely distributed corrosion databases, with over 2000 copies in use.[6] The information can be searched and sorted for tabular or graphical presentations. Color coding is employed to facilitate the interpretation of the iso-corrosion diagrams.

The COR·SUR databases have since been updated. The information was revised, adding new materials and different environments as well as additional information on the existing materials and corrosives. Corrosion data for 37 alloys in over 900 chemical environments and 52 nonmetals in over 700 chemical environments are available in this database. The specific metals and nonmetals covered in the latest version are listed in Table 8.3. A significant enhancement to the latest

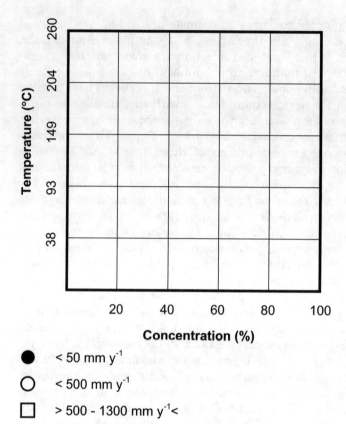

Figure 8.1 Iso-corrosion matrix and key to data points.

version is that the user can add, edit, or delete proprietary records. Searching through the data requires defining the environment and the conditions of interest. The environment is defined by selecting the appropriate corrosive from an alphabetical listing. After the corrosive has been determined, the program steps through the procedure for defining the limits of the search. The user can obtain a printed copy of the iso-corrosion diagrams and a report that includes associated information such as footnotes, references, and personal notes, if available.

8.2 Aluminum Alloys

8.2.1 Introduction

Aluminum is the second most plentiful metallic element on earth. It has been estimated that 8 percent of the earth crust is composed of alu-

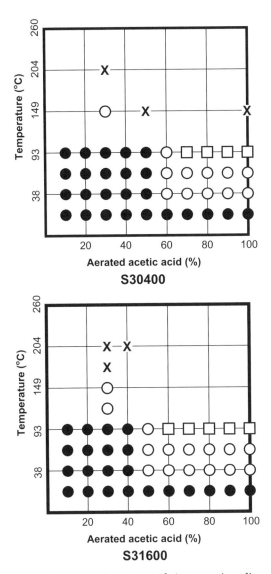

Figure 8.2 Aerated acetic acid iso-corrosion diagrams for S30400 and S31600 stainless steels.

minum, usually found in the oxide form known as bauxite. Aluminum has become the most widely used nonferrous metal on a volume basis. Although more expensive on a tonnage basis, it is the least expensive of metals other than steel on the basis of volume or area. Aluminum and its alloys are divided into two broad classes, castings and wrought, or mechanically worked, products. The latter is subdivided into heat-

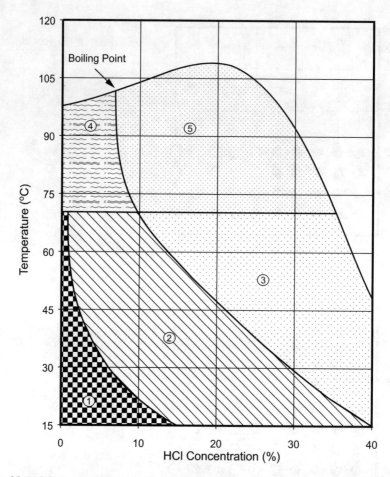

Materials in shaded zones have reported corrosion rates of < 0.5 mm·y^{-1}

Zone 1	Zone 2	Zone 3	Zone 4	Zone 5
20Cr 30Ni[a]	Zirconium	62Ni 28Mo[e]	66Ni 32Cu[b, f]	62Ni 28Mo[e]
66Ni 32Cu[b]	62Ni 32Cu	Molybdenum	62Ni 28Mo[e]	Platinum
62Ni 28Mo	Molybdenum	Platinum	Platinum	Silver
Copper[b]	Platinum	Silver	Silver	Tantalum
Nickel[b]	Silicon bronze[b]	Tantalum		Zirconium
Platinum	Silicon cast iron[c]	Zirconium		
Silicon bronze[b]	Silver			
Silicon cast iron[c]	Tantalum			
Silver	Zirconium			
Tantalum				
Titanium[d]			[a] < 2% at 25°C	
Tungsten			[b] No air	
Zirconium			[c] No FeCl$_3$	
Tantalum			[d] < 10% at 25°C	
Tungsten			[e] No chlorine	
			[f] < 0.05% concentration	

Figure 8.3 Hydrochloric acid graph.

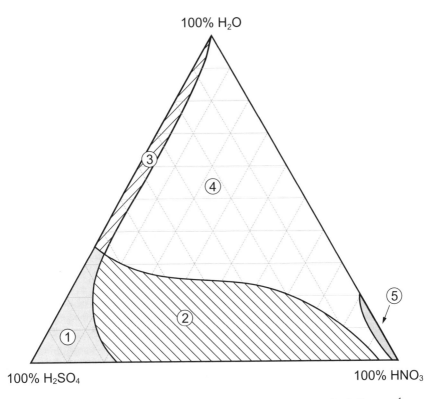

Figure 8.4 Mixed acid graph.

Materials in shaded zones have reported corrosion rates of < 0.5 mm·y^{-1}

Zone 1	Zone 2	Zone 3	Zone 4	Zone 5
20Cr 30Ni	18Cr 8Ni	20Cr 30Ni	18Cr 8Ni	18Cr 8Ni
Gold	20Cr 30Ni	Gold	20Cr 30Ni	20Cr 30Ni
Lead	Cast Iron	Platinum	Gold	Aluminum
Platinum	Gold	Silicon Iron	Platinum	Gold
Silicon Iron	Lead	Tantalum	Silicon iron	Platinum
Steel	Platinum		Tantalum	Silicon Iron
Tantalum	Silicon Iron			Tantalum
	Tantalum			

treatable and non-heat-treatable alloys and into various forms produced by mechanical working.

Production of aluminum. All aluminum production is based on the Hass-Heroult process. Alumina refined from bauxite is dissolved in a cryolite bath with various fluoride salt additions made to control bath temperature, density, resistivity, and alumina solubility. An electrical current is then passed through the bath to electrolyze the dissolved alumina, with oxygen forming at and reacting with the carbon anode

TABLE 8.3   Metals and Nonmetals Included in Cor·Sur

Metals	Nonmetals
1. Alloy 20-25-4Mo (904L)	1. Acetal (polyoxymethylene)
2. Alloy 20-25-6Mo (254SMO/6XN)	2. Acrylonitrile-butadiene-styrene (ABS)
3. Alloy 20-38-3-3Cu (20Cb3/825)	3. Bisphenol A-fumarate
4. Alloy 30-44-5-3W (G-30)	4. Carbon-graphite
5. Aluminum (3003/5154)	5. Cellulose acetate
6. Austenitic (17-12-3) stainless steel (316L/317L)	6. Cellulose acetate butyrate (316L/317L)
7. Austenitic (18-8) stainless steel (304/304L/347)	7. Ceramics
8. Brass (> 15Zn)	8. Chlorendic fiberglass (190 + 47)
9. Cast iron, gray/ductile	9. Chlorendic fiberglass (869 + 44.5)
10. Copper/bronze/Low brass	10. Chlorinated polyvinylchloride
11. Copper-nickel	11. Chlorine sulfonyl polyethylene
12. Duplex (25-6-3)	12. Concrete
13. Gold/platinum	13. Epoxy cements
14. Lead	14. Epoxy-asbestos-glass
15. Ni-16 Cr-16 Mo (C276)	15. Epoxy-fiberglass
16. Ni-20 Cr-16 Mo-4 W (686)	16. Ethylene-tetrafluoroethylene
17. Ni-22 Cr-16 Mo (C22/59)	17. Fluorinated ethylene propylene
18. Ni-23 Cr-16 Mo-1 Cu (C20000)	18. Fluorocarbons FEP and TFE
19. Ni-30 Mo (B-2)	19. Furan laminates
20. Nickel (200)	20. Furans
21. Nickel cast iron (15-35Ni)	21. Furfuryl alcohol-asbestos
22. Nickel-chromium-iron (600/690)	22. Furfuryl alcohol-glass
23. Nickel-copper (400)	23. Glass chemical
24. Ni-Cr-Fe-9 Mo (625/725)	24. Glassed steel
25. Niobium (columbium)	25. Magnesia partially stabilized zirconia
26. PH grade 15-7Mo (S15700)	26. Nylon
27. PH grade 17-4 (S17400)	27. Perfluoroalkoxy
28. PH grade 17-7 (S17700)	28. Phenol-formaldehyde-asbestos
29. Silicon cast iron (14Si)	29. Phenolic, asbestos
30. Silver	30. Phenolic, nonasbestos
31. Stainless steel (12Cr)	31. Phenolics
32. Stainless steel (17Cr)	32. Polychloroprene
33. Steels, carbon/low alloy	33. Polyester-fiberglass
34. Superferritic (26-1) stainless steel	34. Superferritic (26-1) stainless steel
35. Tantalum	34. Polyethylene
36. Titanium, unalloyed	35. Polymethyl methacrylate
37. Zirconium	36. Polyphenylene sulfide (40% glass-filled)
	37. Polyphenylene sulfide + fiberglass
	38. Polypropylene
	39. Polysulfone
	40. Polyvinylchloride
	41. Polyvinylidene chloride
	42. Polyvinylidene fluoride
	43. Rubber, butyl
	44. Rubber, fluorine
	45. Rubber, natural
	46. Silicates
	47. Soapstone
	48. Styrene acrylonitrile
	49. Vinyl ester cements
	50. Vinyl ester resin
	51. Vinyl polybutadiene
	52. Wood

and aluminum collecting as a metal pad at the cathode. The separated metal is periodically removed by siphon or vacuum methods into crucibles, which are then transferred to casting facilities where remelt or fabricating ingots are produced. The major impurities of smelted aluminum are iron and silicon, but zinc, gallium, titanium, and vanadium are typically present as minor contaminants. Refining steps are available to obtain high levels of purity. Purity of 99.99 percent is achieved through fractional crystallization or Hoopes cell operation.

Some aluminum alloys can be solution treated to increase their strength. This consists of heating the alloy to a temperature between 460 and 530°C, at which all the alloying elements are in solid solution (solution treated). The alloy is then rapidly cooled, usually by quenching in water. The metal is then in an unstable state, forcing the alloying elements to precipitate from solution as hard, intermetallic particles, a process known as natural aging, which takes about 5 days to complete. However, some alloys naturally age too slowly and incompletely, so the process needs to be accelerated by a precipitation treatment. This consists of raising the temperature of the alloy to a temperature lower than 200°C for a specified time. Under these conditions artificial aging is promoted.

Annealing, most commonly used with non-heat-treatable alloys to soften the metal so that it can be more easily formed, consists of heating it to a temperature between 350 and 425°C and allowing it to cool. The rate of cooling is not important except with heat-treatable alloys; they must be cooled slowly to prevent age hardening. Although aluminum alloy is approximately a third of the weight of steel, it is about three times the cost.

Mechanical properties. The mechanical properties of aluminum may be improved by alloying, by strain hardening, by thermal treatment, or by combinations of all three techniques. Copper, magnesium, manganese, silicon, and zinc are used as the major constituents in aluminum alloys. Chromium, lead, nickel, and other elements are used for special purposes as minor alloy constituents. Impurities such as iron affect the performance of aluminum alloys and must be considered. Pure aluminum can be strengthened by alloying with small amounts of manganese (up to 1.25%) and magnesium (up to 3.5%). The addition of larger percentages of magnesium produces still higher strengths, but precautions are needed for satisfactory performance. These alloys and pure aluminum can be further hardened by cold work up to tensile strengths of 200 or even 300 MPa. Higher strengths are achieved in alloys that are heat treatable.

Cast aluminum. Three processes, sand, permanent mold, and die-casting, are commonly used for aluminum alloys. As a general rule, heat-treatable alloys are either sand or permanent mold cast. The following combinations of elements constitute the main families of cast aluminum alloys:

Pure aluminum. Rarely used in the cast condition, but special castings may be required for electrical applications.

Al-Si. Silicon is the principal element for conferring good castibility on aluminum alloys.

Al-Si-Mg. Good casting alloys, which can be solution treated and age hardened to give good mechanical properties.

Al-Mg. Best combination of strength and toughness of all the aluminum casting alloys.

Al-Cu. Moderately high strength, medium or poor impact resistance, and fast casting properties; poorest corrosion resistance of aluminum alloys.

Al-Mn. A cheap, non-heat-treatable alloy of poor mechanical properties, fair castibility, but exceptional for non-load-bearing applications at temperatures of up to 500°C (for instance, gas burners of domestic cookers).

The following are the main casting methods for casting aluminum alloys:

High-pressure vacuum die casting. A casting formed when metal is injected into a cavity containing no air under pressures ranging from 14 to 70 MPa. Such castings may be subjected to temperatures of up to 450°C without blisters.

Ultra-high-pressure vacuum die-casting. A casting formed when metal is injected into a cavity containing no air under pressures ranging from 14 to 140 MPa.

Solution heat-treated vacuum die casting. A die casting that has been solution heat treated to enhance its performance properties.

Permanent mold alloy high-pressure vacuum die casting. A casting formed when permanent mold alloy is injected into a cavity containing no air under high pressure.

Casting compositions are described by a three-digit system followed by a decimal value. The decimal .0 in all cases pertains to casting alloy limits. Decimals .1 and .2 concern ingot compositions.

1xx.x. Controlled unalloyed (pure) compositions, especially for rotor manufacture

2xx.x. Alloys in which copper is the principal alloying element

3xx.x. Alloys in which silicon is the principal alloying element, but other alloying elements such as copper and magnesium are specified

4xx.x. Alloys in which silicon is the principal alloying element

5xx.x. Alloys in which magnesium is the principal alloying element

6xx.x. Unused

7xx.x. Alloys in which zinc is the principal alloying element, but other alloying elements such as copper and magnesium may be specified

8xx.x. Alloys in which tin is the principal alloying element

9xx.x. Unused

Wrought aluminum. Superpurity aluminum (99.99+%) is limited to certain chemical plant items, flashing for buildings, and other applications requiring maximum resistance to corrosion and/or high ductility, justifying high cost. Other alloys are Al-Mn, Al-Mg, Al-Mg-Si, Al-Cu-Mg, Al-Zn-Mg, Al-Li, and Al-Sn (used as bearing materials, particularly clad onto steel shells for automobile engines and similar applications).

For wrought alloys, a four-digit system is used to produce a list of wrought composition families as follows:

1xxx. Controlled unalloyed compositions of 99% or higher purity are characterized by generally excellent resistance to attack by a wide range of chemical agents, high thermal and electrical conductivity, and low mechanical properties. For example, 1100-O has a room-temperature minimum tensile strength of 75 MPa and a yield strength of 25 MPa. Iron and silicon are the major impurities. Commercial purity metal (99.00 to 99.80%) is available in three purities and a range of work-hardened grades, for a wide variety of general applications plus a special composition for electrical purposes. High-purity aluminum is used for many electrical and process equipment applications. The higher-purity members of the *1xxx* group are used in equipment handling such products as hydrogen peroxide and fuming nitric acid.

2xxx. Alloys in which copper is the principal alloying element, although other elements, notably magnesium, may be specified. This group involves the first age-hardening alloys and covers a range of compositions. The *2xxx* alloys are high-strength materials, but their copper content reduces their corrosion resistance. Rolled plate and sheet are often clad with a layer of pure aluminum approximately 5% of the sheet thickness on each side. Alclad is a well-known trade name for this coating process.

3xxx. Alloys in which manganese is the principal alloying element. The addition of about 1.25% Mn increases strength without impairing ductility. Alternative alloys with not only Mn but also small additions of Mg have slightly higher strength while retaining good ductility. In general, these alloys are characterized by fairly good corrosion resistance and moderate strength. For example, 3003-O has a room-temperature minimum tensile strength of 125 MPa and a yield strength of 35 MPa. It is formable, readily weldable, can be clad to provide excellent resistance to pitting attack, and is one of the more widely used aluminum alloys for tanks, heat-exchanger components, and process piping.

4xxx. Alloys in which silicon is the principal alloying element. Silicon added to aluminum substantially lowers the melting point without causing the resulting alloys to become brittle.

5xxx. Alloys in which magnesium is the principal alloying element. These alloys are characterized by corrosion resistance and moderate strength. For example, 5858-O has a room-temperature minimum tensile strength of 215 MPa and a yield strength of 80 MPa. There are five standard compositions with Mg contents up to 4.9%, with Mn or Cr in small amounts. There are work-hardening alloys with high to moderated strength and ductility, and high resistance to seawater corrosion, but alloys with > 3.5% Mg require care because corrosion resistance may be impaired. They are widely used for cryogenic equipment and large storage tanks for ammonium nitrate solutions and jet fuel. Alloys of the 5*xxx* group can be readily welded using filler metal of slightly higher Mg content than the parent metal. They anodize well. Certain limitations must be observed regarding cold working during fabrication. In the case of 5*xxx* alloys containing over 3.0% Mg, operating temperatures are limited to 66°C to avoid establishing susceptibility to SCC.

6xxx. Alloys in which magnesium and silicon are the principal alloying element. They can be readily extruded, possess good formability, and can be readily welded and anodized. The 6*xxx* alloys offer moderate strength with good ductility in the heat-treated and aged condition. The popular 6061-T6 has 260 MPa minimum tensile strength and a 240 MPa minimum yield strength. Alloy 6063 has good resistance to atmospheric corrosion and is the most commonly used aluminum alloy for extruded shapes such as windows, doors, store fronts, and curtain walls. Alloys such as 6061 and 6063 contain balanced proportions of magnesium and silicon to form a stoichiometric second-phase intermetallic constituent, magnesium silicide (Mg_2Si). Alloys such as 6351 contain an excess of silicon over magnesium and are termed unbalanced.

7xxx. Alloys in which zinc is the principal alloying element, but other alloying elements such as copper, magnesium, chromium, and zirconium may be specified. A lower range of Zn/Mg additions provides reasonable levels of strength and good weldability. Rolled flat products may be clad with Al-1% Zn alloy.

8xxx. Alloys including tin and some lithium compositions characterizing miscellaneous compositions. Most of the *8xxx* alloys are non-heat-treatable, but when used on heat-treatable alloys, they may pick up the alloy constituents and acquire a limited response to heat treatment.

9xxx. Unused

Special aluminum products. In recent years, a number of new aluminum alloys have been developed. For example, the powder metallurgy route can be a cost-effective method for manufacturing components with conventional aluminum alloys, especially for small parts requiring close dimensional tolerances (e.g., connecting rods for refrigeration compressors). But this process is still relatively expensive. Rapid solidification and vapor deposition processes permit production of aluminum alloys with compositions and microstructures that are not possible by conventional cast or wrought methods.

Reinforcing aluminum alloys with ceramic fibers can provide a useful increase in elastic modulus (especially at elevated temperatures) and improve creep strength and heat erosion resistance. The disadvantages are decreased elongation to fracture and more difficult machining characteristics.

Temper designation system for aluminum alloys. The following lists the temper designations for aluminum alloys:

F. As fabricated. Applies to products shaped by cold working, hot working, or casting processes in which no special control over thermal conditions or strain hardening is employed.

O. Annealed. Applies to wrought products that are annealed to obtain lowest-strength temper, and to cast products that are annealed to improve ductility and dimensional stability. The O may be followed by a digit other than zero. Such a digit indicates special characteristics. For example, for heat-treatable alloys, O1 indicates a product that has been heat treated at approximately the same time and temperature required for solution heat treatment and then air cooled to room temperature.

H. Strain hardened (wrought products only). Applies to products that have been strengthened by strain hardening, with or without supplementary heat treatment to produce some reduction in strength. The H is always followed by two or more digits. The digit following the designation Hl, H2, and H3, which indicates the degree of strain hardening, is a numeral from 1 through 8. An 8 indicates tempers with ultimate tensile strength equivalent to that achieved by about 75 percent cold reduction (temperature during reduction not to exceed 50°C) following full annealing.

- *H1.* Strain hardened only. Applies to products that are strain hardened to obtain the desired strength without supplementary thermal treatment. The digit following the H1 indicates the degree of strain hardening.
- *H2.* Strain hardened and partially annealed. Applies to products that are strain hardened more than the desired final amount and then reduced in strength to the desired level by partial annealing. The digit following the H2 indicates the degree of strain hardening remaining after the product has been partially annealed.
- *H3.* Strain hardened and stabilized. Applies to products that are strain hardened and whose mechanical properties are stabilized by a low-temperature thermal treatment that slightly decreases tensile strength and improves ductility. This designation is applicable only to those alloys that, unless stabilized, gradually age soften at room temperature. The digit following the H3 indicates the degree of strain hardening after stabilization.

W. Solution heat treated. An unstable temper applicable only to alloys that naturally age after solution heat treatment. This designation is specific only when the period of natural aging is indicated.

T. Heat treated to produce stable tempers other than F, O, or H. Applies to products that are thermally treated, with or without supplementary strain hardening, to produce stable tempers. The T is always followed by one or more digits:

- *T1.* Cooled from an elevated temperature-shaping process and naturally aged to a substantially stable condition. Applies to products that are not cold worked after an elevated temperature-shaping process such as casting or extrusion and for which mechanical properties have been stabilized by room-temperature aging.
- *T2.* Cooled from an elevated temperature-shaping process, cold worked, and naturally aged to a substantially stable condition. Applies to products that are cold worked specifically to improve strength after cooling from a hot working process such as rolling or extrusion and for which mechanical properties have been stabilized by room-temperature aging.

- *T3.* Solution heat treated, cold worked, and naturally aged to a substantially stable condition. Applies to products that are cold worked specifically to improve strength after solution heat treatment and for which mechanical properties have been stabilized by room-temperature aging.
- *T4.* Solution heat treated and naturally aged to a substantially stable condition. Applies to products that are not cold worked after solution heat treatment and for which mechanical properties have been stabilized by room-temperature aging.
- *T5.* Cooled from an elevated temperature-shaping process and artificially aged. Applies to products that are not cold worked after an elevated temperature-shaping process such as casting or extrusion and for which mechanical properties, dimensional stability, or both have been substantially improved by precipitation heat treatment.
- *T6.* Solution heat treated and artificially aged. Applies to products that are not cold worked after solution heat treatment and for which mechanical properties, dimensional stability, or both have been substantially improved by precipitation heat treatment.
- *T7.* Solution heat treated and stabilized. Applies to products that have been precipitation heat treated to the extent that they are overaged. Stabilization heat treatment carries the mechanical properties beyond the point of maximum strength to provide some special characteristic, such as enhanced resistance to stress corrosion cracking or exfoliation P corrosion.
- *T8.* Solution heat treated, cold worked, and artificially aged. Applies to products that are cold worked specifically to improve strength after solution heat treatment and for which mechanical properties, dimensional stability, or both have been substantially improved by precipitation heat treatment.
- *T9.* Solution heat treated, artificially aged, and cold worked. Applies to products that are cold worked specifically to improve strength after they have been precipitation heat treated.
- *T10.* Cooled from an elevated temperature-shaping process, cold worked, and artificially aged. Applies to products that are cold worked specifically to improve strength after cooling from a hot working process such as rolling or extrusion and for which mechanical properties, dimensional stability, or both have been substantially improved by precipitation heat treatment.

8.2.2 Applications of different types of aluminum

Building and construction applications. Aluminum is used extensively in buildings of all kinds, bridges, towers, and storage tanks. Because structural steel shapes and plate are usually lower in initial cost, aluminum is used when engineering advantages, construction features, unique architectural designs, light weight, and/or corrosion resistance are considerations. Corrugated or otherwise stiffened sheet products are used in roofing and siding for industrial and agricultural building construction. Ventilators, drainage slats, storage bins, window and door frames, and other components are additional applications for sheet, plate, castings, and extrusions.

Aluminum products such as roofing, flashing, gutters, and downspouts are used in homes, hospitals, schools, and commercial and office buildings. Exterior walls, curtain walls, and interior applications such as wiring, conduit, piping, duct-work, hardware, and railings utilize aluminum in many forms and finishes. Construction of portable military bridges and superhighway overpass bridges has increasingly relied on aluminum elements. Scaffolding, ladders, electrical substation structures, and other utility structures utilize aluminum, chiefly in the form of structural and special extruded shapes. Water storage tanks are often constructed of aluminum alloys to improve resistance to corrosion and to provide an attractive appearance.

Containers and packaging. Low-volumetric-specific heat results in economies when containers or conveyers must be moved in and out of heated or refrigerated areas. The nonsparking property of aluminum is valuable in flour mills and other plants that are subject to fire and explosion hazards. Corrosion resistance is important in shipping fragile merchandise, valuable chemicals, and cosmetics. Sealed aluminum containers designed for air, shipboard, rail, or truck shipments are used for chemicals not suited for bulk shipment. Packaging has been one of the fastest-growing markets for aluminum. Products include household wrap, flexible packaging and food containers, bottle caps, collapsible tubes, and beverage and food cans. Beverage cans have been the aluminum industry's greatest success story, and market penetrations by the food can are accelerating. Soft drinks, beer, coffee, snack foods, meat, and even wine are packaged in aluminum cans. Draft beer is shipped in Alclad aluminum barrels. Aluminum is used extensively in collapsible tubes for toothpaste, ointments, food, and paints.

Transportation. Both wrought and cast aluminum have found wide use in automobile construction. Aluminum sand, die, and permanent mold castings are critically important in engine construction. Cast aluminum wheels are growing in importance. Aluminum sheet is used for hoods, trunk decks, bright finish trim, air intakes, and bumpers. Because of weight limitations and desire to increase effective payloads, manufacturers have intensively employed aluminum cab, trailer, and truck designs. Sheet alloys are used in truck cab bodies, and dead weight is also reduced using extruded stringers, frame rails, and cross members. Extruded or formed sheet bumpers and forged wheels are usual.

Aluminum is also used in truck trailers, mobile homes, and travel trailers and buses, mainly to minimize dead weight. Other uses are in railroad cars, bearings, marine, and aerospace applications. Aluminum is used in virtually all segments of the aircraft, missile, and spacecraft industry. Aluminum is widely used in these applications because of its high strength-to-density ratio, corrosion resistance, and weight efficiency, especially in compressive designs.

Process industries. In the chemical industries aluminum is used for the manufacture of hydrogen peroxide and the production and distribution of nitric acid. It is also used in the manufacture and distribution of liquefied gases, because it retains its strength and ductility at low temperatures, and its lower density is also an advantage over nickel steels.

Aluminum cannot be used with strong caustic solutions, although mildly alkaline solutions—when inhibited—will not attack aluminum. Aluminum may also be used to handle NH_4OH (hot and cold). It does not, however, resist the effects of most other strong alkalis. Salts of strong acids and weak bases, except salts of halogens, have little effect. Aluminum may also be used to handle sulfur and its compounds. It will also be attacked by mercury and its salts.

Its use for handling chlorinated solvents requires careful consideration. Under most conditions, particularly at room temperatures, aluminum alloys resist halogenated organic compounds, but under some conditions they may react rapidly or violent with some of these chemicals. If water is present, these chemicals may hydrolyze to yield mineral acids that destroy the protective oxide film on the aluminum surface. Such corrosion by mineral acids may in turn promote reaction with the chemicals themselves, because the aluminum halides formed by this corrosion are catalysts for some such reactions. To ensure safety, service conditions should be ascertained before aluminum alloys are used with these chemicals.

Electrical applications. Aluminum is used in conductor applications, because of its combination of low cost, high conductivity, adequate mechanical strength, low specific gravity, and excellent resistance to corrosion. It is used in motors and generators (stator frames and end shields, field coils for direct current machines, stator windings in motors, transformer windings and large turbogenerator field coils). It is also used in dry-type power transformers and has been adapted to secondary coil windings in magnetic-suspension-type constant current transformers. Aluminum is used in lighting and capacitors.

Machinery and equipment. Aluminum is used in processing equipment in the petroleum industry such as aluminum tops for steel storage tanks and aluminum pipelines for carrying petroleum products. It is also used in the rubber industry because it resists all corrosion that occurs in rubber processing and is nonadhesive. Aluminum alloys are widely used in the manufacture of explosives because of their nonpyrophoric characteristics. Aluminum is used in textile machinery and equipment, paper and printing industries, coal mine machinery, portable irrigation pipe and tools, jigs, fixtures and patterns, and many instruments.

8.2.3 Weldability of aluminum alloys

The oxide film on aluminum surfaces must be removed or broken up during welding to allow coalescence of the base and the filler metal. The molten aluminum in the fusion zone must be shielded from the atmosphere until it has resolidified. There are several techniques for oxide removal and protection of the weld puddle. Aluminum can be welded by gas and coated electrodes where a fluxing agent is used to penetrate the alumina film and shield the molten metal. Unless completely removed following welding, this flux can be corrosive. The two most common commercial techniques used to weld aluminum are gas metal arc welding (GMAW) and gas tungsten arc welding (GTAW). In both cases, the oxide film is decomposed by the high temperature and shock effect of the arc. The weld puddle is protected from the atmosphere by an inert gas, such as argon or helium, flowing from the welding gun tip and around the electrode.[7]

For non-heat-treatable alloys, material strength depends on the effect of work hardening and solid solution hardening of alloy elements such as magnesium and manganese; the alloying elements are mainly found in the 1xxx, 3xxx, and 5xxx series of alloys. When welded, these alloys may lose the effects of work hardening, which results in softening of the heat-affected zone (HAZ) adjacent to the weld.

For heat-treatable alloys, material hardness and strength depend on alloy composition and heat treatment (solution heat treatment and quenching followed by either natural or artificial aging produces a fine dispersion of the alloying constituents). Principal alloying elements are found in the 2xxx, 6xxx, 7xxx, and 8xxx series. Fusion welding redistributes the hardening constituents in the HAZ, which locally reduces material strength.

Most of the wrought grades in the 1xxx, 3xxx, 5xxx, 6xxx, and medium-strength 7xxx (e.g., 7020) series can be fusion welded using tungsten inert gas (TIG), metal inert-gas (MIG), and oxyfuel processes. The 5xxx series alloys, in particular, have excellent weldability. High-strength alloys (e.g., 7010 and 7050) and most of the 2xxx series are not recommended for fusion welding because they are prone to liquation and solidification cracking.

Filler alloys. Filler metal composition is determined by.

- Weldability of the parent metal
- Minimum mechanical properties of the weld metal
- Corrosion resistance
- Anodic coating requirements

Nominally matching filler metals are often employed for non-heat-treatable alloys. However, for alloy-lean materials and heat-treatable alloys, nonmatching fillers are used to prevent solidification cracking.

Imperfections in welds. Aluminum and its alloys can be readily welded providing appropriate precautions are taken.

Porosity. Porosity is often regarded as an inherent feature of MIG welds. The main cause of porosity is absorption of hydrogen in the weld pool that forms discrete pores in the solidifying weld metal. The most common sources of hydrogen are hydrocarbons and moisture from contaminants on the parent material and filler wire surfaces, and water vapor from the shielding gas atmosphere. Even trace levels of hydrogen may exceed the threshold concentration required to nucleate bubbles in the weld pool, aluminum being one of the metals most susceptible to porosity.[7]

To minimize the risk, the material surface and filler wire should be rigorously cleaned. Three cleaning techniques are suitable: mechanical cleaning, solvent degreasing, and chemical etch cleaning. In gas-shielded welding, air entrainment should be avoided by making sure there is an efficient gas shield and the arc is protected from drafts. Precautions should also be taken to avoid water vapor pickup from gas lines and welding equipment.

Cracking. Cracking occurs in aluminum alloys because of high stresses generated across the weld resulting from high thermal expansion, twice that of steel, and the substantial contraction on solidification, typically 5 percent more than in equivalent steel welds. Solidification cracks form in the center of the weld, usually extending along the centerline during solidification. Solidification cracks also occur in the weld crater at the end of the welding operation. The main causes of solidification cracks are

- Incorrect filler wire/parent metal combination
- Incorrect weld geometry
- Welding under high restraint conditions

The cracking risk can be reduced by using a nonmatching crack-resistant filler, usually from the 4xxx or 5xxx series alloys. The disadvantage is that the resulting weld metal may have a lower strength than the parent metal and not respond to a subsequent heat treatment. The weld bead must be thick enough to withstand contraction stresses. Also, the degree of restraint on the weld can be minimized by using correct edge preparation, accurate joint setup, and correct weld sequence.

Liquation cracking occurs in the HAZ, when low-melting-point films are formed at the grain boundaries. These cannot withstand the contraction stresses generated when the weld metal solidifies and cools. Heat-treatable alloys, 6xxx, 7xxx, and 8xxx series alloys, are more susceptible to this type of cracking. The risk can be reduced by using a filler metal with a lower melting temperature than the parent metal; for example, the 6xxx series alloys are welded with a 4xxx filler metal. However, 4xxx filler metal should not be used to weld high magnesium alloys, such as 5083, because excessive magnesium-silicide may form at the fusion boundary, decreasing ductility and increasing crack sensitivity.[7]

Poor weld bead profile. Incorrect welding parameter settings or poor welder technique can introduce weld profile imperfections such as lack of fusion, lack of penetration, and undercut. The high thermal conductivity of aluminum and the rapidly solidifying weld pool make these alloys particularly susceptible to profile imperfections.

When a filler alloy is used, the weld nugget becomes an aluminum alloy composed of elements of the alloys being joined and the filler alloy. Proper selection of filler alloys is required to minimize the possibility of the weld bead becoming anodic to the adjacent HAZ or to the alloys being welded. The effect of welding on the corrosion resistance of aluminum in a specific environment is determined by the alloy or alloys being joined, the welding filler alloy, and the welding procedure employed. The following factors may influence the corrosion behavior of a welded aluminum assembly in a specific environment:

- Differences in composition of the weld bead and the alloys being welded
- The cast structure of the weld bead as compared to the structure of the welded alloys
- Segregation of constituents of the welded alloys as the welded metal solidifies
- Segregation of constituents of the welded alloys due to precipitation caused by overaging in the HAZ
- Crevice effects due to porosity exposed at the weld bead surface, cold folds in the weld bead, and microcracks

8.2.4 Corrosion resistance

Corrosion resistance of aluminum is dependent upon a protective oxide film. This film is stable in aqueous media when the pH is between about 4.0 and 8.5. The oxide film is naturally self-renewing and accidental abrasion or other mechanical damage of the surface film is rapidly repaired. The conditions that promote corrosion of aluminum and its alloys, therefore, must be those that continuously abrade the film mechanically or promote conditions that locally degrade the protective oxide film and minimize the availability of oxygen to rebuild it.[8]

The acidity or alkalinity of the environment significantly affects the corrosion behavior of aluminum alloys. At lower and higher pH, aluminum is more likely to corrode but by no means always does so. For example, aluminum is quite resistant to concentrated nitric acid. When aluminum is exposed to alkaline conditions, corrosion may occur, and when the oxide film is perforated locally, accelerated attack occurs because aluminum is attacked more rapidly than its oxide under alkaline conditions. The result is pitting. In acidic conditions, the oxide is more rapidly attacked than aluminum, and more general attack should result.

As a general rule, aluminum alloys, particularly the 2*xxx* series, are less corrosion resistant than the commercial purity metal. Some aluminum alloys, for example, are susceptible to intergranular corrosion as a result of low-temperature aging reactions and the subsequent precipitation in the grain boundaries. Susceptibility to intergranular attack in these alloys shows up as exfoliation and stress-corrosion cracking (SCC).

Aluminum is used in high-purity-water systems and to hold and transfer a variety of organic solutions. Lower alcohol may give problems in storage, and organic halides and completely anhydrous organic acids should be avoided. Mercury and heavy metal salt solutions will also give problems. Exfoliation and SCC are not commercial problems with the 1*xxx*, 3*xxx*, 4*xxx*, and 6*xxx* series, or the 5*xxx* alloys containing less than 3% magnesium. The susceptible alloys (2*xxx*, 5*xxx* with higher magnesium, and 7*xxx*) have not been used in major amounts in the chemical process industries. Heat treatments, such as overaging, can be used to improve systems that are susceptible. Historically, the Al-Zn-Mg alloys have been the most susceptible to cracking.

Galvanic corrosion is a potential problem when aluminum is used in complex structures. It is anodic to most of the common construction materials such as iron, stainless steel, titanium, copper, and nickel alloys. If a galvanic situation arises, the aluminum will preferentially corrode. This may cause unsatisfactory service. Aluminum can be used in a wide range of environmental conditions without surface protection and with minimum maintenance. It is often used for its good resistance to atmospheric conditions, as well as industrial fumes and vapors. It is also widely used in cryogenic applications because of its favorable mechanical properties at low temperature (it can be used down to −250°C). Table 8.4 presents the results of atmospheric exposure of different aluminum materials in a wide variety of testing sites around the world.[9]

Effect of alloying. The additions of alloying elements to aluminum change the electrochemical potential of the alloy, which affects corrosion resistance even when the elements are in solid solution. Zinc and magnesium tend to shift the potential markedly in the anodic direction, whereas silicon has a minor anodic effect. Copper additions cause marked cathodic shifts. This results in local anodic and cathodic sites in the metal that affect the type and rate of corrosion.

Very high-purity aluminum, 99.99% or purer, is highly resistant to pitting. Any alloying addition will reduce this resistance. The 5xxx Al-Mg alloys and the 3xxx Al-Mn alloys resist pitting corrosion almost as well. The pure metal and the 3xxx, 5xxx, and 6xxx series alloys are resistant to the more damaging forms of localized corrosion, exfoliation, and SCC. However, cold-worked 5xxx alloys containing magnesium in excess of the solid solubility limit (above 3% magnesium) can become susceptible to exfoliation and SCC when heated for long times at temperatures of about 80 to 175°C.[10]

Effect of metallurgical and mechanical treatments. Metallurgical and mechanical treatments often act in synergy to produce desired or undesired microstructural features in aluminum alloys. Variations in thermal treatments can have marked effects on the local chemistry and hence the local corrosion resistance of high-strength, heat-treatable aluminum alloys. Ideally, all the alloying elements should be fully dissolved, and the quench cooling rate should be rapid enough to keep them in solid solution.

Generally, practices that result in a nonuniform microstructure will lower corrosion resistance, especially if the microstructural effect is localized. Precipitation treatment or aging is conducted primarily to increase strength. Some precipitation treatments purposely overage the aluminum beyond the maximum strength condition (T6 temper) to improve its resistance to IGC, exfoliation, and SCC through the formation of randomly distributed, noncoherent precipitates (T7 tempers). This diminishes the adverse effect of highly localized precipitation at grain boundaries resulting from slow quenching, underaging, or aging to peak strengths.

TABLE 8.4 Results of Atmospheric Exposure of Different Aluminum Materials in a Wide Variety of Testing Sites Around the World

Alloy	City	State/province, country	Exposure, y	Atmosphere	Rate, $\mu m \cdot y^{-1}$
1094.88	Key West	FL, USA	10	Marine	0.1
1095	Cristobal	Panama	10	Marine	0.2
1095	Sandy Hook	NJ, USA	10	Marine	0.1
1098.25	La Jolla	CA, USA	10	Marine	0.7
1100	Panama inland	Panama	16	Inland	12.7
1100	Panama marine	Panama	16	Marine	17.3
1100	Cape Beale	BC, Canada	10	Marine	0
1100	Durban	South Africa	10	Marine	0.6
1100	Halifax	NS, Canada	10	Industrial marine	1.1
1100	Kingston	ON, Canada	10	Rural	0.1
1100	Kure Beach-800	NC, USA	10	Marine (800 ft)	0.1
1100	Kure Beach-80	NC, USA	10	Marine (80 ft)	0.3
1100	Montreal	QC, Canada	10	Severe industrial	0.8
1100	Newark	NJ, USA	10	Industrial	0.4
1100	Point Reyes	CA, USA	10	Marine	0.1
1100	Toronto	ON, Canada	10	Industrial	0.6
1100	University Park	PA, USA	10	Rural	0.1
1100	Vancouver	BC, Canada	10	Urban	0.5
1199	Chicago	IL, USA	7	Industrial	0.6
1199	Richmond	VA, USA	7	Mild industrial	0
1199	Widnes	UK	7	Severe industrial	1.2
3003	Cape Beale	BC, Canada	10	Marine	0
3003	Durban	South Africa	10	Marine	0.7
3003	Esquimalt	BC, Canada	10	Marine	0
3003	Halifax	NS, Canada	10	Industrial marine	1.2
3003	Kingston	ON, Canada	10	Rural	0.1
3003	Kure Beach-800	NC, USA	10	Marine (800 ft)	0.1
3003	Kure Beach-80	NC, USA	10	Marine (80 ft)	0.1
3003	Montreal	QC, Canada	10	Severe industrial	0.7
3003	Newark	NJ, USA	10	Industrial	0.6
3003	Point Reyes	CA, USA	10	Marine	0.1
3003	Saskatoon	SA, Canada	10	Rural	0
3003	Toronto	ON, Canada	10	Industrial	0.9
3003	Trail	BC, Canada	10	Semirural	0.4
3003	University Park	PA, USA	10	Rural	0.1
3003	Vancouver	BC, Canada	10	Urban	0.5
3005	Aruba	Dutch Antilles	7	Marine	0.2
3005	Denge Marsh	UK	7	Marine	0.7
3005	Kure Beach-80	NC, USA	7	Marine	0.2
3005	Manila	Philippines	7	Marine	0.1

TABLE 8.4 Results of Atmospheric Exposure of Different Aluminum Materials in a Wide Variety of Testing Sites Around the World (*Continued*)

Alloy	City	State/province, country	Exposure, y	Atmosphere	Rate, $\mu m \cdot y^{-1}$
5052	Cape Beale	BC, Canada	10	Marine	0
5052	Durban	South Africa	10	Marine	0.6
5052	Esquimalt	BC, Canada	10	Marine	0.1
5052	Halifax	NS, Canada	10	Industrial marine	1
5052	Kingston	ON, Canada	10	Rural	0.1
5052	Kure Beach-800	NC, USA	10	Marine (800 ft)	0.1
5052	Kure Beach-80	NC, USA	10	Marine (80 ft)	0.2
5052	Montreal	QC, Canada	10	Severe industrial	0.7
5052	Newark	NJ, USA	10	Industrial	0.5
5052	Point Reyes	CA, USA	10	Marine	0.1
5052	Saskatoon	SA, Canada	10	Rural	0.1
5052	Toronto	ON, Canada	10	Industrial	0.6
5052	Trail	BC, Canada	10	Semirural	0.3
5052	University Park	PA, USA	10	Rural	0.1
5052	Vancouver	BC, Canada	10	Urban	0.5
6061	Cape Beale	BC, Canada	10	Marine	0
6061	Durban	South Africa	10	Marine	0.9
6061	Esquimalt	BC, Canada	10	Marine	0.1
6061	Halifax	NS, Canada	10	Industrial marine	1.1
6061	Kingston	ON, Canada	10	Rural	0.2
6061	Kure Beach-800	NC, USA	10	Marine (800 ft)	0.1
6061	Kure Beach-80	NC, USA	10	Marine (80 ft)	0.3
6061	Montreal	QC, Canada	10	Severe industrial	0.8
6061	Newark	NJ, USA	10	Industrial	0.5
6061	Point Reyes	CA, USA	10	Marine	0.1
6061	Saskatoon	SA, Canada	10	Rural	0
6061	Toronto	ON, Canada	10	Industrial	0.6
6061	Trail	BC, Canada	10	Semirural	0.2
6061	University Park	PA, USA	10	Rural	0.2
6061	Vancouver	BC, Canada	10	Urban	0.6
6063	Kure Beach-80	NC, USA	10	Marine	0.2
6063	Montreal	QC, Canada	10	Severe industrial	0.7
6063	Toronto	ON, Canada	10	Industrial	0.6
6063	Vancouver	BC, Canada	10	Urban	0.5
1100 H14	Arenzano	Italy	1.75		0.9
1100 H14	Bohus-Malmon	Sweden	5.12		0.3
1100 H14	Kure Beach-800	NC, USA	5	Marine (800 ft)	0.2
1100 H14	Kure Beach-80	NC, USA	5	Marine (80 ft)	0.5
1100 H14	La Jolla	CA, USA	18.15	Marine	12.4

TABLE 8.4 Results of Atmospheric Exposure of Different Aluminum Materials in a Wide Variety of Testing Sites Around the World (*Continued*)

Alloy	City	State/province, country	Exposure, y	Atmosphere	Rate, $\mu m \cdot y^{-1}$
1100 H14	New York	NY, USA	20.55	Industrial	15
1100 H14	Phoenix	AZ, USA	19.15	Rural	1.5
1100 H14	Sandy Hook	NJ, USA	20.37	Marine	5.6
1100 H14	State College	PA, USA	20.15	Rural	1.5
1135 H14	State College	PA, USA	7	Rural	0.1
1180 H14	Arenzano	Italy	1.75		0.6
1180 H14	Bohus-Malmon	Sweden	5.12		0.2
1180 H14	Kure Beach-800	NC, USA	5	Marine (800 ft)	0.2
1180 H14	Kure Beach-80	NC, USA	5	Marine (80 ft)	0.6
1195 H14	Kure Beach-80	NC, USA	7	Marine	0.1
1199 H14	Aruba	Dutch Antilles	7	Marine	0.2
1199 H14	Denge Marsh	UK	7	Marine	0.2
1199 H14	Manila	Philippines	7	Marine	0.1
2014 T3	Aruba	Dutch Antilles	7	Marine	17.8
2014 T3	Denge Marsh	UK	7	Marine	1
2014 T3	Kure Beach-80	NC, USA	7	Marine	0.4
2014 T3	Manila	Philippines	7	Marine	0.3
2017 T3	La Jolla	CA, USA	18.15	Marine	45.2
2017 T3	New York	NY, USA	20.55	Industrial	25.1
2017 T3	Phoenix	AZ, USA	19.15	Rural	1.5
2017 T3	State College	PA, USA	20.15	Rural	2
2024 T3	Panama rain forest	Panama	1	Rain forest	0.4
2024 T3	Panama open field	Panama	1	Open field	1.3
2024 T3	Panama marine	Panama	1	Marine	6.2
3003 H14	Arenzano	Italy	1.75		0.9
3003 H14	Bohus-Malmon	Sweden	5.12		0.2
3003 H14	Chicago	IL, USA	7	Industrial	1.1
3003 H14	Kure Beach-800	NC, USA	5	Marine (800 ft)	0.2
3003 H14	Kure Beach-80	NC, USA	5	Marine (80 ft)	0.5
3003 H14	La Jolla	CA, USA	18.15	Marine	12.2
3003 H14	New York	NY, USA	20.55	Industrial	19.3
3003 H14	Phoenix	AZ, USA	19.15	Rural	0.3
3003 H14	Richmond	VA, USA	7	Mild industrial	0.5
3003 H14	Sandy Hook	NJ, USA	20.37	Marine	7.1
3003 H14	State College	PA, USA	7	Rural	0.1
3003 H14	State College	PA, USA	20.15	Rural	1.8
3003 H14	Widnes	UK	7	Severe industrial	3.8
3004 H34	State College	PA, USA	7	Rural	0.1
3004 H36	Chicago	IL, USA	7	Industrial	1.4
3004 H36	Richmond	VA, USA	7	Mild industrial	0.5

TABLE 8.4 Results of Atmospheric Exposure of Different Aluminum Materials in a Wide Variety of Testing Sites Around the World (*Continued*)

Alloy	City	State/province, country	Exposure, y	Atmosphere	Rate, $\mu m \cdot y^{-1}$
3004 H36	Widnes	UK	7	Severe industrial	2.3
5005 H34	State College	PA, USA	7	Rural	0.1
5050 H34	Arenzano	Italy	1.75		0.6
5050 H34	Bohus-Malmon	Sweden	5.12		0.2
5050 H34	Kure Beach-800	NC, USA	5	Marine (800 ft)	0.2
5050 H34	Kure Beach-80	NC, USA	5	Marine (80 ft)	0.4
5052 H34	Arenzano	Italy	1.75		0.5
5052 H34	Aruba	Dutch Antilles	7	Marine	0.2
5052 H34	Bohus-Malmon	Sweden	5.12		0.2
5052 H34	Denge Marsh	UK	7	Marine	0.3
5052 H34	Kure Beach-800	NC, USA	7	Marine (800 ft)	0.1
5052 H34	Kure Beach-80	NC, USA	5	Marine (80 ft)	0.2
5052 H34	Kure Beach-80	NC, USA	5	Marine (80 ft)	0.3
5052 H34	Manila	Philippines	7	Marine	0.1
5083 H116	Wrightsville Beach	NC, USA	2	Marine	2.5
5083 H116	Wrightsville Beach	NC, USA	1	Marine	2.8
5083 H116	Wrightsville Beach	NC, USA	1	Marine	3.3
5083 H116	Wrightsville Beach	NC, USA	2	Marine	0
5086 H116	Wrightsville Beach	NC, USA	1	Marine	2.3
5086 H116	Wrightsville Beach	NC, USA	2	Marine	2.5
5086 H116	Wrightsville Beach	NC, USA	1	Marine	3
5086 H116	Wrightsville Beach	NC, USA	2	Marine	0
5086 H117	Wrightsville Beach	NC, USA	1	Marine	3.3
5086 H117	Wrightsville Beach	NC, USA	2	Marine	0
5086 H32	Aruba	Dutch Antilles	7	Marine	0.3
5086 H32	Denge Marsh	UK	7	Marine	0.4
5086 H32	Kure Beach-80	NC, USA	7	Marine (80 ft)	0.2
5086 H32	Manila	Philippines	7	Marine	0.1
5086 H34	Arenzano	Italy	1.75		0.6
5086 H34	Bohus-Malmon	Sweden	5.12		0.2
5086 H34	Kure Beach-800	NC, USA	5	Marine (800 ft)	0.3

TABLE 8.4   Results of Atmospheric Exposure of Different Aluminum Materials in a Wide Variety of Testing Sites Around the World (*Continued*)

Alloy	City	State/province, country	Exposure, y	Atmosphere	Rate, $\mu\text{m·y}^{-1}$
5086 H34	Kure Beach-80	NC, USA	5	Marine (80 ft)	0.3
5154 H34	Arenzano	Italy	1.75		0.6
5154 H34	Bohus-Malmon	Sweden	5.12		0.2
5154 H34	Chicago	IL, USA	7	Industrial	1.4
5154 H34	Kure Beach-800	NC, USA	5	Marine (800 ft)	0.2
5154 H34	Kure Beach-80	NC, USA	5	Marine (80 ft)	0.3
5154 H34	Richmond	VA, USA	7	Mild industrial	0.4
5154 H34	Widnes	UK	7	Severe industrial	2.7
5456 H116	Wrightsville Beach	NC, USA	2	Marine	1.3
5456 H116	Wrightsville Beach	NC, USA	2	Marine	2.5
5456 H116	Wrightsville Beach	NC, USA	1	Marine	2.8
5456 H116	Wrightsville Beach	NC, USA	1	Marine	3.3
5456 H116	Wrightsville Beach	NC, USA	1	Marine	3.3
5456 H116	Wrightsville Beach	NC, USA	2	Marine	0
5456 H321	Aruba	Dutch Antilles	7	Marine	0.6
5456 H321	Kure Beach-80	NC, USA	7	Marine	0.2
5456 H321	Manila	Philippines	7	Marine	0.1
6051 T4	Key West	FL, USA	19.67	Marine	1.5
6051 T4	La Jolla	CA, USA	18.15	Marine	15.5
6051 T4	New York	NY, USA	20.55	Industrial	18.3
6051 T4	Phoenix	AZ, USA	19.15	Rural	0.3
6051 T4	Sandy Hook	NJ, USA	20.37	Marine	6.9
6051 T4	State College	PA, USA	20.15	Rural	1.5
6061 T	Panama inland		16	Inland	14.2
6061 T	Panama marine		16	Marine	17.3
6061 T6	Arenzano	Italy	1.75		1
6061 T6	Aruba	Dutch Antilles	7	Marine	0.9
6061 T6	Bohus-Malmon	Sweden	5.12		0.3
6061 T6	Chicago	IL, USA	7	Industrial	1.7
6061 T6	Kure Beach-80	NC, USA	7	Marine (80 ft)	0.2
6061 T6	Kure Beach-800	NC, USA	5	Marine (800 ft)	0.3
6061 T6	Kure Beach-80	NC, USA	5	Marine (80 ft)	0.5
6061 T6	Manila	Philippines	7	Marine	0.1
6061 T6	Richmond	VA, USA	7	Mild industrial	0.4
6061 T6	State College	PA, USA	7	Rural	0.1

TABLE 8.4 Results of Atmospheric Exposure of Different Aluminum Materials in a Wide Variety of Testing Sites Around the World (*Continued*)

Alloy	City	State/province, country	Exposure, y	Atmosphere	Rate, $\mu m \cdot y^{-1}$
6061 T6	Widnes	UK	7	Severe industrial	2.6
6062 T5	Aruba	Dutch Antilles	7	Marine	1.2
6062 T5	Kure Beach-80	NC, USA	7	Marine	0.2
6062 T5	Manila	Philippines	7	Marine	0.1
6063 T6	Chicago	IL, USA	7	Industrial	1.3
6063 T6	Richmond	VA, USA	7	Mild industrial	0.3
6063 T6	Widnes	UK	7	Severe industrial	1.5
7075 T6	Andrews AFB	MD, USA			0.4
7075 T6	Aruba	Dutch Antilles	7	Marine	10.2
7075 T6	Barksdale AFB	LA, USA			0.2
7075 T6	Francis Warren AFB	WY, USA			0.1
7075 T6	Kure Beach-80	NC, USA	7	Marine	0.5
7075 T6	Manila	Philippines	7	Marine	0.3
7075 T6	Tinker AFB	OK, USA			0.1
7079 T6	Andrews AFB	MD, USA			0.5
7079 T6	Davis Monthan AFB	USA			0.5
7079 T6	Francis Warren AFB	WY, USA			0.1
7079 T6	Tinker AFB	OK, USA			0
Al 7 Mg O	Aruba	Dutch Antilles	7	Marine	0.4
Al 7 Mg O	Denge Marsh	UK	7	Marine	1
Al 7 Mg O	Kure Beach-80	NC, USA	7	Marine	0.2
Al 7 Mg O	Manila	Philippines	7	Marine	0.6
Alclad 2017 T3	Key West	FL, USA	19.67	Marine	1
Alclad 2017 T3	La Jolla	CA, USA	18.15	Marine	11.7
Alclad 2017 T3	New York	NY, USA	20.55	Industrial	20.3
Alclad 2017 T3	Phoenix	AZ, USA	19.15	Rural	0.3
Alclad 2017 T3	State College	PA, USA	20.15	Rural	1.5
Alclad 6061 T6	Aruba	Dutch Antilles	7	Marine	2.2
Alclad 6061 T6	Kure Beach-80	NC, USA	7	Marine	0.2
Alclad 6061 T6	Manila	Philippines	7	Marine	0.1

Mechanical working influences the grain morphology and the distribution of alloy constituent particles. Both of these factors can affect the type and rate of localized corrosion. Cast aluminum products normally have an equiaxed grain structure. Special processing routes can be taken to produce fine, equiaxed grains in a thin rolled sheet and certain extruded shapes, but most wrought products (rolled, forged, drawn, or extruded products) normally have a highly directional, anisotrophic grain structure. Rectangular products have a three dimensional (3D) grain structure. Figure 8.5 shows the 3D longitudinal (principal working direction), long transverse, and short transverse grain structures typically present in rolled plate. Almost all forms of corrosion, even pitting, are affected to some degree by this grain directionality. However, highly localized forms of corrosion, such as exfoliation and SCC that proceed along grain boundaries, are highly affected by grain structure. Long, wide, and very thin pancake-shaped grains are virtually a prerequisite for a high degree of susceptibility to exfoliation.

These directional structures markedly affect resistance to SCC and to exfoliation of high-strength alloy products, as evidenced by the SCC susceptibility ratings presented in Table 8.5. The information presented in that table was collected from at least 10 random lots that were then tested in Recommended Practice ASTM G 44 (Practice for Evaluating Stress Corrosion Cracking Resistance of Metals and Alloys by Alternate Immersion in 3.5% Sodium Chloride Solutions). The highest rating was assigned for results that showed 90 percent conformance at the 95 percent confidence level when tested at the following stresses:[8]

A. ≥75 percent of the specified minimum yield strength

B. ≥50 percent of the specified minimum yield strength

C. ≥25 percent of the specified minimum yield strength or 100 MPa, whichever is higher

D. Failure to meet the criterion for rating level C

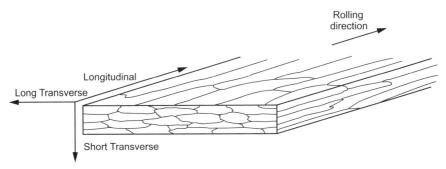

Figure 8.5 Schematic representation of the 3D grain structure typically present in rolled aluminum plates.

TABLE 8.5 Resistance to SCC of Various Aluminum Alloys in Different Temper and Work Conditions

Alloy	Temper	Direction	Plate	Rod/bar	Extrusion	Forging
2011	T3	L		B		
		LT		D		
		ST		D		
2011	T4	L		B		
		LT		D		
		ST		D		
2011	T8	L		A		
		LT		A		
		ST		A		
2014	T6	L	A	A	A	B
		LT	B	D	B	B
		ST	D	D	D	D
2024	T3	L	A	A	A	
		LT	B	D	B	
		ST	D	D	D	
2024	T4	L	A	A	A	
		LT	B	D	B	
		ST	D	D	D	
2024	T6	L		A		A
		LT		B		A
		ST		B		D
2024	T8	L	A	A	A	A
		LT	A	A	A	A
		ST	B	A	B	C
2048	T851	L	A			
		LT	A			
		ST	B			
2124	T851	L	A			
		LT	A			
		ST	B			
2219	T351X	L	A		A	
		LT	B		B	
		ST	D		D	
2219	T37	L	A		A	
		LT	B		B	
		ST	D		D	
2219	T6	L	A	A	A	A
		LT	A	A	A	A
		ST	A	A	A	A
2219	T85XX	L	A		A	A
		LT	A		A	A
		ST	A		A	A
2219	T87	L	A		A	A
		LT	A		A	A
		ST	A		A	A
6061	T6	L	A	A	A	A
		LT	A	A	A	A
		ST	A	A	A	A
7005	T63	L			A	A
		LT			A	A
		ST			D	D

TABLE 8.5 Resistance to SCC of Various Aluminum Alloys in Different Temper and Work Conditions (*Continued*)

Alloy	Temper	Direction	Plate	Rod/bar	Extrusion	Forging
7005	T53	L			A	A
		LT			A	A
		ST			D	D
7039	T64	L	A		A	
		LT	A		A	
		ST	D		D	
7049	T73	L	A		A	A
		LT	A		A	A
		ST	A		B	A
7049	T76	L			A	
		LT			A	
		ST			C	
7149	T73	L			A	A
		LT			A	A
		ST			B	A
7050	T736	L	A		A	A
		LT	A		A	A
		ST	B		B	B
7050	T76	L	A	A	A	
		LT	A	B	A	
		ST	C	B	C	
7075	T6	L	A	A	A	A
		LT	B	D	B	B
		ST	D	D	D	D
7075	T73	L	A	A	A	A
		LT	A	A	A	A
		ST	A	A	A	A
7075	T736	L				A
		LT				A
		ST				B
7075	T76	L	A		A	
		LT	A		A	
		ST	C		C	
7079	T6	L	A		A	A
		LT	B		B	B
		ST	D		D	D
7175	T736	L				A
		LT				A
		ST				B
7178	T6	L	A		A	
		LT	B		B	
		ST	D		D	
7178	T76	L	A		A	
		LT	A		A	
		ST	C		C	
7475	T6	L	A			
		LT	B			
		ST	D			
7475	T73	L	A			
		LT	A			
		ST	A			
7475	T76	L	A			
		LT	A			
		ST	C			

Because SCC in aluminum alloys characteristically is intergranular, susceptible alloys and tempers are most prone to SCC when the tensile stress acts in the short-transverse, or thickness direction, so that the crack propagates along the aligned grain structure. The same material (e.g., 7075-T651 plate) will show a much higher resistance to stress acting in the longitudinal direction, parallel to the principal grain flow. In this case the intergranular crack must follow a very meandering path and usually does not propagate to any major extent. Special agings to various highly resistant T7 tempers have been developed to counteract this adverse effect of directional grain structure. Various artificially aged tempers are available for both 2xxx and 7xxx alloys that provide a range of compromise choices between maximum strength and maximum resistance to exfoliation and SCC.[5]

Role of hydrogen. Hydrogen will dissolve in aluminum alloys in the molten state and during thermal treatments at temperatures close to the melting point in atmospheres containing water vapor or hydrocarbons. Upon solidification, this causes porosity and surface blistering. Recent literature surveys show there is still considerable dispute as to how much, if at all, high-strength aluminum alloys are embrittled by hydrogen. There is some evidence that hydrogen evolving from anodic dissolution at a crack tip can dissolve into the metal at the grain boundary ahead of the crack tip and can thus be a factor in SCC of some 7xxx and possibly 2xxx alloys. Hydrogen embrittlement, however, has not restricted the commercialization of high-strength aluminum alloys.[10]

Protective coatings. As mentioned earlier, pure aluminum, the 3xxx, 5xxx, and most 6xxx series alloys, are sufficiently resistant to be used in industrial atmospheres and waters without any protective coatings. Examples of this are cookware, boats, and building products. Generally coatings are used to enhance an alloy's resistance, and protection is considered necessary for the higher-strength 6xxx alloys and for all 2xxx and 7xxx alloys. Chapter 9, Protective Coatings, describes many of the coatings and coating technologies that have been employed successfully with aluminum alloys for improved service and performance.

8.3 Cast Irons

8.3.1 Introduction

Cast iron is a generic term that identifies a large family of ferrous alloys. Cast irons are primarily alloys of iron that contain more than 2 percent carbon and 1 percent or more silicon. Low raw material costs and relative ease of manufacture make cast irons the least expensive of the engineering metals. Cast irons may often be used in place of steel at considerable cost savings. The design and production advantages of cast iron include

- Low tooling and production cost
- Ready availability
- Good machinability without burring
- Readily cast into complex shapes
- Excellent wear resistance and high hardness (particularly white irons)
- High inherent damping

Cast irons can be cast into intricate shapes because of their excellent fluidity and relatively low melting points and can be alloyed for improvement of corrosion resistance and strength. With proper alloying, the corrosion resistance of cast irons can equal or exceed that of stainless steels and nickel-base alloys.[11] The wide spectrum of properties of cast iron is controlled by three main factors: the chemical composition of the iron, the rate of cooling of the casting in the mold, and the type of carbide or graphite formed.

8.3.2 Carbon presence classification

Cast irons are often classified on the basis of the forms taken by the high level of carbon present.

White cast iron: Iron carbide compound. By reducing the carbon and silicon content and cooling rapidly, much of the carbon is retained in the form of iron carbide without graphite flakes. However, iron carbide, or cementite, is extremely hard and brittle, and these castings are used where high hardness and wear resistance are needed.

Unalloyed white cast iron. Unalloyed white cast iron is a very hard, abrasion-resistant, and low-cost material compared with competitive materials such as carbon steels. The main limitation comes from its brittleness when subjected to compressive loads. White irons are not machinable and are finished by grinding when necessary.

Low-alloy white cast iron. Low-alloy white cast iron has improved toughness and wear resistance. The main limitation is that a better performance or a longer life must justify its extra cost.

Martensitic white cast iron. Martensitic white cast iron has a higher hardness and toughness than other types of white iron. It is stable at high temperatures (480 to 540°C) due to presence of Cr. Low-carbon compositions have higher toughness but lower hardness. The main disadvantage is again higher cost. Stress-relieving heat treatment is also necessary for optimum properties.

High-chromium white cast iron. High-chromium white cast iron has an abrasion resistance similar to martensitic white iron but with higher toughness, strength, and corrosion resistance. Its limitation is high cost.

Malleable cast iron: Irregularly shaped nodules of graphite. Malleable cast iron is produced by heat treatment of closely controlled compositions of white irons that are decomposed to give carbon aggregates dispersed in a ferrite or pearlitic matrix. Because the compact shape of the carbon does not reduce the matrix ductility to the same extent as graphite flakes, a useful level of ductility is obtained. Malleable iron may be divided into the following classes: whiteheart, blackheart, and pearlitic irons.

Malleable iron castings are often selected because the material has excellent machinability in addition to significant ductility. In other applications, malleable iron is chosen because it combines castability with toughness and machinability. Malleable iron is often chosen because of its shock resistance alone. It is used for low-stress parts requiring good machinability such as steering gear, housings, carriers, and mounting brackets. It is used for compresser crankshafts and hubs; for high-strength parts such as connecting rods and universal-joint yokes; in transmission gear, differential cases, and certain gears; and for flanges, pipe fittings, and valve parts for railroad, marine, and other heavy-duty service.

Whiteheart malleable cast iron. Whiteheart malleable castings are produced from high-carbon white cast irons annealed in a decarburizing medium. Carbon is removed at the casting surface, the loss being only compensated for by the diffusion of carbon from the interior. Whiteheart castings are inhomogeneous with a decarburized surface skin and a higher carbon core.

It has a higher-carbon content than other types of malleable iron, which gives better castibility, especially for thin sections. The decarburized layer improves weldability and provides a soft, ductile surface to absorb local-impact blows. Whiteheart malleable cast iron has a marked increase in shock resistance above 100°C and can be used in furnaces up to 450°C. It can also be galvanized and does not suffer galvanizing embrittlement. This iron has very good machinability but is limited by a long heat-treatment time.

Blackheart malleable cast iron. Blackheart malleable irons are produced by annealing low-carbon (2.2 to 2.9%) white iron castings without decarburization. The resulting structure of carbon in a ferrite matrix is homogenous and has better mechanical properties than those of whiteheart irons. It has the best combination of machinability and strength of any ferrous material and a lower cost than nodular cast iron. However, it is not suitable for wear-resistant applications unless it is surface treated. Long heat-treatment cycle times compared with ferritic nodular cast iron are required.

Pearlitic blackheart malleable cast iron. Pearlitic blackheart malleable iron has a pearlitic rather than ferritic matrix, which provides higher strength but lower ductility than ferritic blackheart irons. It has good wear resistance and the highest strength of malleable irons. It can be hardened, and a wide range and combination of properties are possible by control of matrix microstructure. However, it is difficult to weld and requires longer heat-treatment cycle times compared with nodular cast iron.

Gray cast iron: Graphite flakes. Gray cast irons contain 2.0 to 4.5% carbon and 1 to 3% silicon. Their structure consists of branched and interconnected graphite flakes in a matrix of pearlite, ferrite, or a mixture of the two. The graphite flakes form planes of weakness, and so strength and toughness are inferior to those of structural steels. Gray cast iron is used for many different types of parts in a very wide variety of machines and structures. The advantages and limitations of this widely used cast iron are presented in Table 8.6.

Low-alloy gray cast iron enables casting formerly produced in unalloyed gray cast iron to be used in higher-duty applications without redesign or need for costly materials. Alloy additions can cause foundry problems with reuse of scrap. The increase in strength does not bring a corresponding increase in fatigue strength. Cr, Mo, and V are carbide stabilizers that improve strength and heat resistance but impair machinability.

Nodular or ductile cast iron: Spherical graphite nodules. The mechanical properties of gray irons can be greatly improved if the graphite shape is modified to eliminate planes of weakness. Such modification is possible if molten iron, having a composition in the range 3.2 to 4.5% C and 1.8 to 2.8% Si, is treated with magnesium or cerium additions before casting. This produces castings with graphite in spheroidal form instead of flakes, known as nodular, spheroidal graphite, or ductile irons. Nodular irons are available with pearlite, ferrite, or pearlite-ferrite matrixes that offer a combination of greater ductility and higher tensile strength than gray cast irons.

Nodular iron castings are used for many structural applications, particularly those requiring strength and toughness combined with good machinability and low cost. The automotive and agricultural industries are the major users of ductile iron castings. Almost a million tons of duc-

TABLE 8.6 Advantages and Limitations of Gray Cast Iron

Advantages
Most common type of cast iron.
Cheapest material for metal castings, especially for small quantity production.
Very easy to cast—much narrower solidification temperature range than steel.
Low shrinkage in mold due to formation of graphite flakes.
Good machinability, faster material removal rates, but poorer surface finish with ferritic matrix and vice versa for pearlitic matrix.
Graphite acts as a chip breaker and a tool lubricant.
Very high damping capacity.
No difference in notched and unnotched fatigue strength.
Good dry bearing qualities due to graphite.
After formation of protective scales, it resists corrosion in many common engineering environments.

Limitations
Brittle (low impact strength) due to sharp ends of graphite flakes; severely limits use for critical applications.
Graphite acts as a void and reduces strength.
Maximum recommended design stress is one-quarter of ultimate tensile strength.
Maximum fatigue loading limit is one-third of fatigue strength.
Changes in section size will cause variations in machining characteristics (due to variation in microstructure).
Higher-strength irons are more expensive to produce.

tile iron castings were produced in the in the United States in 1988.[11] Because of its economic advantage and high reliability, ductile iron is used for such critical automotive parts as crankshafts, front wheel spindle supports, complex shapes of steering knuckles, disk brake calipers, engine connecting rods, idler arms, wheel hubs, truck axles, suspension system parts, power transmission yokes, high-temperature applications for turbo housings and manifolds, and high-security valves for various applications. Nodular cast iron can be rolled or spun to the desired shape or coined to the exact dimension. The cast iron pipe industry is another major user of ductile iron.

8.3.3 Weldability

The weldability of cast irons depends on their microstructure and mechanical properties. For example, gray cast iron is inherently brittle and often cannot withstand stresses set up by a cooling weld. Because the lack of ductility is caused by the coarse graphite flakes, the graphite clusters in malleable irons and the nodular graphite in irons with spheroidal graphite give significantly higher ductility, which improves the weldability. The weldability may be lessened by the formation of hard and brittle microstructures in the HAZ, consisting of iron carbides and martensite. Because nodular and malleable irons are less likely to form martensite, they can be more readily welded, particularly if the ferrite content is high. White cast iron that is very hard and contains iron carbides normally cannot be unwelded.[12]

Bronze welding is frequently employed to avoid cracking. Because oxides and other impurities are not removed by melting, and mechanical cleaning tends to smear the graphite across the surface, surfaces must be thoroughly cleaned, for example, by means of a salt bath. The potential problem of high-carbon weld metal deposits is avoided by using a consumable nickel or nickel alloy that produces finely divided graphite, lower porosity, and a readily machinable deposit. However, nickel deposits that are high in sulfur and phosphorus from parent metal dilution may result in solidification cracking.

The formation of hard and brittle HAZ structures makes cast irons particularly prone to HAZ cracking during postweld cooling. Preheating and slow postweld cooling reduces HAZ cracking risk. Because preheating will slow the cooling rate both in weld deposit and HAZ, martensitic formation is suppressed and the HAZ hardness is somewhat reduced. Preheating can also dissipate shrinkage stresses and reduce distortion, lessening the likelihood of weld cracking and HAZ. Typical preheat temperatures are given in Table 8.7. Because cracking may also result from unequal expansion, which is especially likely during preheating of complex castings or when preheating is localized on large components, preheat should always be applied gradually. Also, the casting should always be allowed to cool slowly to avoid thermal shock.

8.3.4 Corrosion resistance

Corrosion forms

Selective leaching. Graphitic corrosion is a selective leaching attack observed in gray cast irons in relatively mild environments in which

TABLE 8.7 Typical Preheat Levels for Welding Cast Irons

	Preheat temperature, °C			
Cast iron type	Manual metal arc	Metal inert gas	Gas (fusion)	Gas (powder)
Ferritic flake	300	300	600	300
Ferritic nodular	RT-150	RT-150	600	200
Ferritic whiteheart malleable	RT*	RT*	600	200
Pearlitic flake	300–330	300–330	600	350
Pearlitic nodular	200–330	200–330	600	300
Pearlitic malleable	300–330	300–330	600	300

*RT 5 room temperature; 200°C if high C core is involved.

of iron leaves a structurally very poor graphite network, aching of the iron takes place because the graphite is and the gray iron structure establishes an excellent galvanic form of corrosion generally occurs only when corrosion the metal corrodes more rapidly, the entire surface, including graphite, is removed, and more or less uniform corrosion occurs. Graphitic corrosion is observed only in gray cast irons. The lack of graphite flakes, in both nodular and malleable irons, provides no network to hold the corrosion products together.

Fretting corrosion. The relatively good resistance of cast irons to fretting corrosion is influenced by such variables as lubrication, hardness variations between materials, the presence of gaskets, and coatings.

Pitting and crevice corrosion. The presence of chlorides and crevices or other shielded areas presents conditions that are favorable to the pitting and/or crevice corrosion of cast iron. Pitting has been reported in such environments as dilute alkylaryl sulfonates, antimony trichloride ($SbCl_3$), and calm seawater. Alloying can influence the resistance of cast irons to pitting and crevice corrosion. For example, in calm seawater, nickel additions reduce the susceptibility of cast irons to pitting attack. High-silicon cast irons with chromium or molybdenum offer enhanced resistance to pitting and crevice corrosion.

Intergranular corrosion. The only reference to intergranular attack in cast irons involves ammonium nitrate (NH_4NO_3), in which unalloyed cast irons are reported to be intergranularly attacked.

Erosion corrosion. Fluid flow by itself or in combination with solid particles can cause erosion corrosion attack in cast irons. Two methods are known to enhance the erosion-corrosion resistance of cast irons. First, the hardness of the cast irons can be increased through solid solution hardening or phase transformation–induced hardness increases.

Second, better inherent corrosion resistant can also be used to increase the erosion-corrosion resistance of cast irons. Austenitic nickel cast irons can have hardness similar to unalloyed cast irons but may exhibit better erosion resistance because of the improved inherent resistance of nickel alloyed irons compared to unalloyed irons.

Stress corrosion cracking. SCC is observed in cast irons under certain combinations of environment and stress, and under certain conditions, SCC can be a serious problem. Because unalloyed cast irons are generally similar to ordinary steels in resistance to corrosion, the same environments that cause SCC in steels are likely to cause problems in cast irons. Environments that may cause SCC in unalloyed cast irons include[11]

- Sodium hydroxide (NaOH) solutions
- NaOH-Na$_2$SiO$_2$ solutions
- Calcium nitrate [Ca(NO$_3$)$_2$] solutions
- NH$_4$NO$_3$ solutions
- Sodium nitrate (NaNO$_3$) solutions
- Mercuric nitrate [Hg(NO$_3$)$_2$] solutions
- Mixed acids (H$_2$SO$_4$-HNO$_3$)
- Hydrogen cyanide (HCN) solutions
- Seawater
- Acidic hydrogen sulfide (H$_2$S) solutions
- Molten sodium-lead alloys
- Acid chloride solutions
- Fuming H$_2$SO$_4$

Effect of alloying. Alloying elements can play a dominant role in the susceptibility of cast irons to corrosion attack. Silicon is the most important alloying element used to improve the corrosion resistance of cast irons. Silicon is generally not considered an alloying element in cast irons until levels exceed 3%. Silicon levels between 3 and 14% offer some increase in corrosion resistance to the alloy, but above about 14% Si, the corrosion resistance of the cast iron increases dramatically. Silicon levels up to 17% have been used to enhance the corrosion resistance of the alloy further, but silicon levels over 16% make the alloy extremely brittle and difficult to manufacture. Alloying with silicon promotes the formation of strongly adherent surface films in cast irons. Considerable time may be required to establish these films fully on the castings. Consequently, in some services, corrosion rates may be relatively high for the first few hours or even days of exposure and then may decline to extremely low steady-state rates for the rest of the time the parts are exposed to the corrosive environment.

Nickel increases corrosion resistance by the formation of protective oxide films on the surfaces of the castings. Up to 4% Ni is added in combination with chromium to improve both strength and corrosion resistance in cast iron alloys. The enhanced hardness and corrosion resistance obtained is particularly important for improving the erosion-corrosion resistance of the material. Nickel additions enhance the corrosion resistance of cast irons to reducing acids and alkalies. Nickel additions of 12% or greater are necessary to optimize the corrosion resistance of cast irons.

Chromium is frequently added alone and in combination with nickel and/or silicon to increase the corrosion resistance of cast irons. As with nickel, small additions of chromium are used to refine graphite and matrix microstructures. These refinements enhance the corrosion resistance of cast irons in seawater and weak acids. Chromium additions of 15 to 30% improve the corrosion resistance of cast irons to oxidizing acids. Chromium increases the corrosion resistance of cast iron by the formation of protective oxides on the surfaces of castings. The oxides formed will resist oxidizing acids but will be of little benefit under reducing conditions.

Copper is added to cast irons in special cases. Copper additions of 0.25 to 1% increase the resistance of cast iron to dilute acetic, sulfuric, and hydrochloric (HCl) acids as well as acid mine water. Small additions of copper are also made to cast irons to enhance atmospheric corrosion resistance. Additions of up to 10% are made to some high nickel-chromium cast irons to increase corrosion resistance.

Classification based on corrosion resistance. Cast irons can also be classified on the basis on their corrosion resistance, as in the following section.[11]

Unalloyed gray, ductile, malleable, and white cast irons. Unalloyed gray, ductile, malleable, and white cast irons represent the first and largest category. All of these materials contain carbon and silicon of 3% or less and no deliberate additions of nickel, chromium, copper, or molybdenum. As a group, these materials exhibit corrosion resistance that equals or slightly exceeds that of unalloyed steels, but they show the highest rates of attack for cast irons. These materials are available in a wide variety of configurations and alloys.

Low and moderately alloyed irons. Low and moderately alloyed irons constitute the second major class. These irons contain the iron and silicon of unalloyed cast irons plus up to several percentages of nickel, copper, chromium, or molybdenum. As a group, these materials exhibit 2 to 3 times the service life of unalloyed cast irons.

High-nickel austenitic cast irons. High-nickel austenitic cast irons contain large percentages of nickel and copper and are fairly resistant to acids. When nickel levels exceed 18%, austenitic cast irons are nearly immune to alkalies or caustics, although SCC can occur. High-nickel cast irons can be nodularized to yield ductile irons.

High-chromium cast irons. High-chromium cast irons are basically white cast irons alloyed with 12 to 30% Cr. Other alloying elements may also be added to improve resistance to specific environments. When chromium levels exceed 20%, high-chromium cast irons exhibit good resistance to oxidizing acids. High-chromium irons are not resistant to reducing acids. They are used in saline solutions, organic acids, and marine and industrial atmospheres. These materials display excellent resistance to abrasions, and with proper alloying additions, they can also resist combinations of abrasives and liquids, including some dilute acid solutions.

High-silicon cast irons. The principal alloying element in high-silicon cast irons is 12 to 18% Si, with more than 14.2% Si needed to develop excellent corrosion resistance. Chromium and molybdenum are also used in combination with silicon to develop corrosion resistance to specific environments. High-silicon cast irons represent the most universally corrosion-resistant alloys available at moderate cost. When silicon levels exceed 14.2%, high-silicon cast irons exhibit excellent resistance to most mineral and organic acids. These materials display good resistance in oxidizing and reducing environments and are not appreciably affected by concentration or temperature. Exceptions to universal resistance are hydrofluoric acid (HF), fluoride salts, sulfurous acid (H_2SO_3), sulfite compounds, strong alkalies, and alternating acid-alkali conditions.

The corrosion resistance of high-silicon cast iron is attributed to the development of a thin passive barrier film of hydrated oxides of silicon on the metal surface. This film develops with time due to the dissolution of iron from the metal matrix, which leaves behind silicon that hydrates due to the presence of moisture. Any flaws in the barrier film will reduce its effectiveness.

The passive hydrated silicon film bridges over and forms an impervious barrier layer on a fine-grained high-silicon cast iron with spheroidal graphite areas much more readily than on a high-silicon cast iron with coarse graphite flakes. Thus, a coarse-grained high-silicon cast iron that contains graphite flakes is much more likely to have structural defects and flaws in the passive film than a fine-grained material with spheroidal graphite. Flaws in the passive film are sites for film breakdown. Penetration of the corrosive medium below the film results in localized areas of corrosion and preferential current flow due to lower resistance at graphite flakes, and so on, than on the hydrated silicon film.

Thus, due to the fine grain size with spheroidal graphite and more uniform composition, chill-cast high-silicon cast iron would be expected to have better corrosion resistance than a sand-cast high-silicon cast iron. The shape of the graphite present in an alloy affects the mechanical properties of the material. Flake graphite acts as a severe stress raiser, but the spheroidal graphite does not. A classic example of this effect is the difference between gray cast iron and ductile iron.

8.4 Copper Alloys

8.4.1 Introduction

Copper occurs naturally with elements such as lead, nickel, silver, and zinc. It is widely used in industry both as a pure metal and as an alloying material. The copper industry is composed of two segments: producers (mining, smelting, and refining industries) and fabricators (wire mills, brass mills, foundries, and powder plants). The end products of copper producers, the most important of which are refined cathode copper and wire rod, are sold almost entirely to copper fabricators. The end products of copper fabricators can be generally described as mill and foundry products, and they consist of wire and cable, sheet, strip, plate, rod, bar, mechanical wire, tubing, forgings, extrusions, castings, and powder metallurgy shapes. These products are sold to a wide variety of industrial users.

Mining companies remove vast quantities of low-grade material from open-pit mines to extract copper from the crust of the earth. Approximately 2 tons of overburden must be removed to extract 1 ton of copper. Copper ore is normally crushed, ground, and concentrated, usually by flotation, to produce a beneficiated ore containing about 25% copper. The ore concentrates are then reduced to the metallic state, most often by pyrometallurgical process. The concentrated ore is processed by oxygen/flash smelting to produce a copper sulfide-iron sulfide matte containing up to 60% copper. Sulfuric acid is manufactured from the sulfur dioxide contained in the gases given off and is an important coproduct of copper smelting. The matte is oxidized in a converter to transform the iron sulfides to iron oxides, which separate out in a slag, and to reduce the copper sulfide to blister copper, which contains at least 98.5% copper. Fire refining of blister copper then removes most of the oxygen and other impurities, leaving a product 99.5% pure, which is cast into anodes. Finally most anode copper is electrolytically refined, usually to a purity of at least 99.95%.

Table 8.8 describes briefly some of the advantages and limitations of copper and its alloys. Copper-based alloys are usually classified in terms of one of the main alloying elements. Two main categories of copper alloys are brass and bronze. Brasses are essentially copper-zinc alloys to which other elements may be added. True bronzes are copper-tin alloys.

The Unified Numbering System (UNS) is the accepted alloy designation system in North America for wrought and cast copper and copper alloy products.[13] The three-digit system developed by the U.S. copper and brass industry was expanded to five digits following the prefix letter C and made part of the UNS for metals and alloys. UNS designations are simply expansions of the former designations. For example, Copper Alloy No. 377 (forging brass) in the original three-digit system became C37700 in the UNS system. The UNS is managed jointly by the

TABLE 8.8 Advantages and Limitations of Copper and Its Alloys

Advantages	Limitations
High conductivity of electrical grades superior to all other metals except silver on a volume basis and aluminum on weight basis.	High cost relative to other common metals.
High thermal conductivity.	Conductivity reduced by small quantities of other elements.
Excellent ductility permits easy working.	High casting temperatures of the metal and its alloys.
Wide range of copper-base alloys, most types having good ductility and malleability in the annealed condition and being particularly appropriate for tube forming, hot forming, spinning, deep drawing, etc.	High-temperature properties of the metal impose limitations on its use.
Mechanical properties of copper strength, creep resistance, and fatigue performance are improved by alloying (but conductivity is impaired)	The "gasing" reaction of copper with oxygen requires precautions when temperatures exceed 700°C.
Good corrosion resistance to potable water and to atmospheric and marine environments; can be further improved by alloying.	Toxic; therefore must not be used in contact with foodstuff (e.g., food processing plant).
Useful biocidal properties of the metal and salts.	Some alloys are prone to stress corrosion and other forms of attack (e.g., dezincification of brasses).
Wide range of alloys with special properties (e.g., very high damping capacity).	
Mechanical and electrical properties retained at cryogenic temperatures.	
Weldability of alloys good by appropriate process.	
Nonmagnetic, except some Cu-Ni alloys.	

American Society for Testing and Materials (ASTM) and the Society of Automotive Engineers (SAE).

The designation system is administered by the Copper Development Association (CDA). New designations are assigned as new coppers and copper alloys come into commercial use, and designations are discontinued when an alloy composition ceases to be used commercially. The standard designation composition limits do not preclude the possible presence of other unnamed elements. However, analysis will regularly be made only for the minor elements listed in the table, plus either copper or zinc or plus all major elements except one. The major element that is not analyzed is determined by the difference between the sum of those elements analyzed and 100%. New designations are assigned if a copper or copper alloy meets three criteria:

1. The complete chemical composition is disclosed.
2. The copper or copper alloy is in commercial use or is proposed for commercial use.
3. The composition does not fall within the limits of any designated composition already in the list.

In the designation system, numbers from C10000 through C79999 denote wrought alloys. Cast alloys are numbered from C80000 through C99999. Within these two categories, the compositions are grouped into the following coppers and copper alloys. More detailed families are described in Tables 8.9 and 8.10 for wrought and cast alloys. Detailed compositions of individual alloys can be found in App. E.[14] The main trade names associated with some of these copper alloys are given in Table 8.11.

- *Coppers.* Metals with a designated minimum copper content of 99.3% or higher.
- *Brasses.* These alloys contain zinc as the principal alloying element with or without other designated alloying elements such as iron, aluminum, nickel, and silicon. The wrought alloys comprise three main families of brasses. The cast alloys comprise five main families of brasses. Ingot for remelting for the manufacture of castings may vary slightly from the ranges shown.
- *Bronzes.* Broadly speaking, bronzes are copper alloys in which the major alloying element is not zinc or nickel. Originally *bronze* described alloys with tin as the only or principal alloying element. Today, the term is generally used with a modifying adjective. Bronzes are unquestionably one of the most versatile classes of corrosion- and wear-resistant materials, offering a broad range of properties from a wide selection of alloys and compositions.
- *Copper-nickels.* These are alloys with nickel as the principal alloying element, with or without other elements designated commonly as "nickel silvers" (i.e., alloys containing zinc and nickel as the principal and secondary alloying elements).
- *Leaded coppers.* These comprise a series of cast alloys of copper with 20% or more lead, sometimes with a small amount of silver but without tin or zinc.
- *Special alloys.* Alloys whose chemical compositions do not fall into any of the above categories are combined in "special alloys."

TABLE 8.9 Generic Classification of Wrought Copper Alloys

	UNS Number	Composition
Coppers		
Coppers	C10100-C15760	> 99% Cu
High-copper alloys	C16200-C19600	> 96% Cu
Brasses		
Brasses	C20500-28580	Cu-Zn
Leaded brasses	C31200-C38590	Cu-Zn-Pb
Tin brasses	C40400-C49080	Cu-Zn-Sn-Pb
Bronzes		
Phosphor bronzes	C50100-C52400	Cu-Sn-P
Leaded phosphor bronzes	C53200-C54800	Cu-Sn-Pb-P
Copper-phosphorus and copper-silver-phosphorus alloys	C55180-CS5284	Cu-P-Ag
Aluminum bronzes	C60600-C64400	Cu-Al-Ni-Fe-Si-Sn
Silicon bronzes	C64700-C66100	Cu-Si-Sn
Others		
Other copper-zinc alloys	C66400-C69900	
Copper-nickels	C70000-C72950	Cu-Ni-Fe
Nickel silvers	C73200-C79900	Cu-Ni-Zn

TABLE 8.10 Generic Classification of Cast Copper Alloys

	UNS Number	Composition
Coppers		
Coppers	C80100-C81100	> 99% Cu
High-copper alloys	C81300-C82800	> 94% Cu
Brasses and Bronzes		
Red and leaded red brasses	C83300-C85800	Cu-Zn-Sn-Pb (75-89% Cu)
Yellow and leaded yellow brasses	C85200-C85800	Cu-Zn-Sn-Pb (57-74% Cu)
Manganese and leaded manganese bronzes	C86100-C86800	Cu-Zn-Mn-Fe-Pb
Silicon bronzes, silicon brasses	C87300-C87900	Cu-Zn-Si
Tin bronzes and leaded tin bronzes	C90200-C94500	Cu-Sn-Zn-Pb
Nickel-tin bronzes	C94700-C94900	Cu-Ni-Sn-Zn-Pb
Aluminum bronzes	C95200-C95810	Cu-Al-Fe-Ni
Others		
Copper-nickels	C96200-C96800	Cu-Ni-Fe
Nickel silvers	C97300-C97800	Cu-Ni-Zn-Pb-Sn
Leaded coppers	C98200-C98800	Cu-Pb
Miscellaneous alloys	C99300-C99750	

TABLE 8.11 Trade Names Associated with Some Commonly Used Copper Alloys

Alloy	Trade name	Alloy	Trade name
Coppers		Brasses	
C10100	Oxygen-free, electronic (OFE)	(Cont.)	
		C32000	Leaded red brass
C10200	Oxygen-free (OF)	C33000	Low-leaded brass (tube)
C10300	OFXLP	C33200	High-leaded brass (tube)
C10400	Oxygen-free with Ag (OFS)	C33500	Low-leaded brass
		C34000	Medium-leaded brass, 64.5%
C10500	OFS	C34200	High-leaded brass, 64.5%
C10700	OFS	C35000	Medium-leaded brass, 62%
C10800	OFLP	C35300	High-leaded brass, 62%
C11000	Electrolytic, tough pitch (ETP)	C35600	Extra-high-leaded brass
		C36000	Free-cutting brass
C11010	Remelted high conductivity (RHC)	C36500	Leaded muntz metal, uninhibited
C11020	Fire-refined high conductivity (FRHC)	C37000	Free-cutting muntz metal
		C37700	Forging brass
C11030	Chemically refined tough pitch (CRTP)	C38500	Architectural bronze
		C44300	Admiralty, arsenical
C11100	Electrolytic tough pitch, anneal resistant	C44400	Admiralty, antimonial
		C44500	Admiralty, phosphorized
C11300	Tough pitch with Ag (STP)	C46200	Naval brass, 63.5%
C11400	STP	C46400	Naval brass, uninhibited
C11500	STP	C46500	Naval brass, arsenical
C11600	STP	C47000	Naval brass welding and brazing rod
C12200	Phosphorus-deoxidized, high-residual phosphorus (DHP)		
		C48200	Naval brass, medium leaded
C12900	Fire-refined tough pitch with Ag (FRSTP)	C48500	Naval brass, high leaded
		Bronzes	
C14200	Phosphorus-deoxidized, arsenical (DPA)	C50500	Phosphor bronze, 1.25% E
		C51000	Phosphor bronze, 5% A
C14300	Cadmium copper, deoxidized	C51800	Phosphor bronze
C14500	Tellurium-bearing	C52100	Phosphor bronze, 8% C
C14510	Tellurium-bearing	C52400	Phosphor bronze, 10% D
C14520	Phosphorus-deoxidized, tellurium-bearing (DPTE)	C53400	Phosphor bronze B-1
		C54400	Phosphor bronze B-2
C14700	Sulfur-bearing	C65100	Low-silicon bronze B
C15000	Zirconium copper	C65500	High-silicon bronze A
High coppers		C66700	Manganese brass
C16200	Cadmium copper	C67000	Manganese bronze B
C17000	Beryllium copper	C67500	Manganese bronze A
C17200	Beryllium copper	C68000	Bronze, low fuming (nickel)
C17500	Beryllium copper	C68100	Bronze, low fuming
C18200	Chromium copper	C68700	Aluminum brass, arsenical
C18400	Chromium copper	C69400	Silicon red brass
Brasses		Copper-Nickel alloys	
C21000	Gilding, 95%	C70400	Copper-nickel, 5%
C22000	Commercial bronze, 90%	C70500	Copper-nickel, 7%
C22600	Jewelry bronze, 87.5%	C70600	Copper-nickel, 10%
C23000	Red brass, 85%	C70800	Copper-nickel, 11%
C24000	Low brass, 80%	C71000	Copper-nickel, 20%
C26000	Cartridge brass, 70%	C71500	Copper-nickel, 30%
C26800	Yellow brass, 66%	Nickel-Silvers	
C27000	Yellow brass, 65%	C74500	Nickel silver, 65-10
C27400	Yellow brass, 63%	C75200	Nickel silver, 65-18
C28000	Muntz metal, 60%	C75400	Nickel silver, 65-15
C31400	Leaded commercial bronze	C75700	Nickel silver, 65-12
C31600	Leaded commercial bronze (nickel-bearing)	C76700	Nickel silver, 56.5-15
		C77000	Nickel silver, 55-18

8.4.2 Weldability

In terms of weldability, copper alloys have a wide spectrum of welding characteristics. Copper, because of its high thermal conductivity, needs substantial preheat to counteract its very high heat sink. However, some of the alloys that have a thermal conductivity similar to low-carbon steel, such as cupro-nickel alloys, can normally be fusion welded without a preheat.

Coppers. Tough pitch copper contains stringers of copper oxide ($< 0.1\%$ oxygen as Cu_2O), which does not impair the mechanical properties of wrought material and has high electrical conductivity. Oxygen-free and phosphorus deoxidized copper are more easily welded. TIG and MIG are the preferred welding processes, but oxyacetylene and MMA welding can be used in the repair of tough pitch copper components. To counteract the high thermal conductivity, helium- and nitrogen-based gases, which have higher arc voltages, can be used as an alternative to argon.[15]

High copper alloys. Low alloying additions of sulfur or tellurium can made to improve machining. However, these grades are normally considered to be unweldable. Small additions of chromium, zirconium, or beryllium will produce precipitation hardened alloys that, on heat treatment, have superior mechanical properties. Chromium and beryllium copper may suffer from HAZ cracking unless they are heat treated before welding. When welding beryllium copper, care should be taken to avoid inhaling the welding fumes.

Brasses and nickel silvers. When considering weldability, brasses can be conveniently separated into two groups, low zinc (up to 20% Zn) and high zinc (30 to 40% Zn). Nickel silvers contain 20 to 45% zinc and nickel to improve strength. The main problem in fusion welding these alloys is the volatilization of the zinc, which results in white fumes of zinc oxide and weld metal porosity. Only low-zinc brasses are normally considered suitable for fusion welding using the TIG and MIG processes.

TIG and MIG processes are used with argon or an argon-helium mixture but not nitrogen. A preheat is normally used for low zinc ($< 20\%$ Zn) to avoid fusion defects because of the high thermal conductivity. Although preheat is not needed in higher zinc content alloys, slow cooling reduces cracking risk. Postweld heat treatment also helps reduce the risk of stress corrosion cracking in areas where there is high restraint.

Bronzes. Tin bronzes can contain between 1 and 10% tin. Phosphor bronze contains up to 0.4% phosphorus. Gunmetal is essentially a tin bronze with up to 5% zinc and may additionally have up to 5% lead.

Silicon bronze typically contains 3% silicon and 1% manganese and is probably the easiest of the bronzes to weld.

Bronzes are generally considered to be weldable, apart from phosphor bronze and leaded gunmetal, and a matching filler composition is normally employed. Autogenous welding of phosphor bronzes is not recommended due to porosity, but the risk can be reduced by using a filler wire with a higher level of deoxidants. Gunmetal is not considered weldable due to hot cracking in the weld metal and HAZ.[15]

There are essentially two types of aluminum bronzes: single-phase alloys containing between 5 and 10% aluminum, with a small amount of iron or nickel, and more complex, two-phase alloys containing up to 12% aluminum and about 5% of iron with specific alloys also containing nickel and manganese and silicon. Gas-shielded welding processes are preferred for welding this group of alloys. In TIG welding, the presence of a tenacious, refractory oxide film requires ac (argon) or dc with a helium shielding gas. Because of its low thermal conductivity, a preheat is not normally required except when welding thick-section components.

Rigorous cleaning of the material surface is essential, both before and after each run, to avoid porosity. Single-phase alloys can be susceptible to weld metal cracking, and HAZ cracking can occur under highly restrained conditions. It is often necessary to use matching filler metals to maintain corrosion resistance, but a nonmatching, two-phase filler will reduce the cracking risk. Two-phase alloys are more easily welded. For both types, preheat and interpass temperatures should be restricted to prevent cracking. Table 8.12 gives a brief description of the uses made of some of these alloys.[16]

Copper-nickel alloys. Cupro-nickel alloys contain between 5 and 30% nickel, with specific alloys having additions of iron and manganese; 90/10 and 70/30 (Cu/Ni) alloys are commonly welded grades. These alloys are single phase and are generally considered to be readily weldable using inert gas processes and, to a lesser extent, MMA. A matching filler is normally used, but 70/30 (C18) is often regarded as a universal filler for these alloys. Because the thermal conductivity of cupro-nickel alloys is similar to low-carbon steels, preheating is not required.[15]

These alloys do not contain deoxidants; therefore, autogenous welding is not recommended because of porosity. Filler metal compositions typically contain 0.2 to 0.5% titanium to prevent weld metal porosity. Argon shielding gas is normally used for both TIG and MIG, but in TIG welding, an argon-H_2 mixture, with an appropriate filler, improves weld pool fluidity and produces a cleaner weld bead. Gas backing (usually argon) is recommended, especially in pipe welding, to produce an oxide-free underbead.

TABLE 8.12 Properties and Uses of Main Bronze Bearing Materials

Manganese bronzes: C86300, C86400

Manganese bronzes are modifications of the Muntz metal-type alloys (60% copper 40% zinc brasses) containing small additions of manganese, iron, and aluminum, plus lead for lubricity, antiseizing, and embeddibility. Like the aluminum bronzes, they combine very high strength with excellent corrosion resistance. Manganese bronze bearings can operate at high speeds under heavy loads but require high shaft hardness and nonabrasive operating conditions.

Tin bronzes: C90300, C90500, C90700

The principal function of tin in these bronzes is to strengthen the alloys. (Zinc also adds strength, but more than about 4% zinc reduces the antifrictional properties of the bearings alloy.) The tin bronzes are strong and hard and have very high ductility. This combination of properties gives them a high load-carrying capacity, good wear resistance, and the ability to withstand pounding. The alloys are noted for their corrosion resistance in seawater and brines.

The tin bronzes' hardness inhibits them from conforming easily to rough or misaligned shafts. Similarly, they do not embed dirt particles well and therefore must be used with clean, reliable lubrication systems. They require a shaft hardness between 300 and 400 BHN. Tin bronzes operate better with grease lubrication than other bronzes. They are also well suited to boundary-film operation because of their ability to form polar compounds with small traces of lubricant. Differences in mechanical properties among the tin bronzes are not great. Some contain zinc as a strengthener in partial replacement for more expensive tin.

Leaded tin bronzes: C92200, C92300, C92700

Some tin bronzes contain small amounts of lead. In this group of alloys, lead's main function is to improve machinability. It is not present in sufficient concentration to change the alloys' bearing properties appreciably. A few of the leaded bronzes also contain zinc, which strengthens the alloys at a lower cost than tin. The leaded bronzes in this family otherwise have properties and application that are similar to the tin bronzes.

High-leaded tin bronzes: C93200, C93400, C93500, C93700, C93800, C94300

The family of high-leaded tin bronzes include the workhorses of the bearing bronze alloys. Alloy C3200 has a wider range of applicability and is more often specified than all other bearing materials. It and the other high-leaded tin bronzes are used for general utility applications under medium loads and speeds (i.e., those conditions that constitute the bulk of bearing uses). Strengths and hardnesses are somewhat lower than those of the tin bronzes, but this group of leaded alloys excels in its antifriction and machining properties.

Alloy C93200 utilizes a combination of tin and zinc for cost-effective strengthening, whereas C93700 relies solely on tin to obtain the same strength level. In addition to its good strength, C93700 is known for its corrosion resistance to mildly acidic mine waters and to mineral waters and paper mill sulfite liquors. Wear resistance is good at high speeds and under high-load, shock, and vibration conditions. The alloy has fair casting properties, something to be considered when large or complex bearing shapes must be produced. Alloy C93700 contains enough lead to permit use under doubtful or interruptible lubrication, but it must be used with hardened shafts. The lead addition makes these alloys easy to machine. High strength is sacrificed for superior lubricity in the bronzes containing 15 and 25% lead (C93800 and C94300).

TABLE 8.12 Properties and Uses of Main Bronze Bearing Materials (*Continued*)

As in all leaded bronzes the lead is present as discrete microscopic particles. In alloys C93800 and C94300 there is ample lead available to smear onto the journal to prevent welding and seizing, should the lubricant supply be interrupted. The lead also provides excellent machinability.

Because of their comparatively lower strength and somewhat reduced ductility, alloys C93800 and C94300 should not be specified for use under high loads or in applications where impacts can be anticipated. They operate best at moderate loads and high speeds, especially where lubrication may be unreliable. They conform well and are very tolerant of dirty operating conditions, properties which have found them extensive use in off-highway, earthmoving, and heavy industrial equipment.

Aluminum bronzes: C95300, C95400, C95500, C95510

The aluminum bronzes are the strongest and most complex of the copper-based bearing alloys. Their aluminum content provides most of their high strength and makes them the only bearing bronzes capable of being heat treated. Their high strength, up to 470 MPa yield strength and 820 MPa tensile strength, permits them to be used at unit loads up to 50 percent higher than those for leaded tin bronze alloy C93200. Because of their high strength, however, they have fairly low ductility and do not conform or embed well. They consequently require shafts hardened to 550 to 600 BHN. Surfaces must also be extremely smooth, with both shaft and bearing finished to 1520 μm in RMS.

Careful attention should be given to lubricant cleanliness and reliability, the latter because these alloys do not have the antiseizing properties typical of the leaded and tin bearing bronzes. On the other hand, the aluminum bronzes have excellent corrosion resistance and are ideally suited for such applications as marine propellers and pump impellers. The aluminum bronzes also have superior elevated temperature strength. These bronzes are the only conventional bearing materials able to operate at temperatures exceeding 260°C.

8.4.3 Corrosion resistance

The resistance of all grades of copper to atmospheric corrosion is good, hence their wide usage for roofing and for contact with most waters. The metal develops adherent protective coatings, initially of oxide, but subsequently thickening to give a familiar green patina on roofs and the dark brownish color of bronze statuary. Because copper is largely unaffected by potable water, its is widely used for tubes carrying domestic and industrial water. In the following broad classifications, copper and copper alloys have demonstrated superior corrosion performance:[17]

- Atmospheric exposure such as roofing and other architectural applications
- Plumbing systems, with superior corrosion resistance to both potable waters and soils
- Marine applications involving supply lines, heat exchangers, and hardware where resistance to seawater and biofouling are mandatory
- Industrial and chemical plant process equipment involving exposure to a wide variety of organic and inorganic chemicals

Brasses are the most numerous and the most widely used of the copper alloys because of their low cost, easy or inexpensive fabrication and machining, and relative resistance to aggressive environments. They are, however, generally inferior in strength to bronzes and must not be used in environments that cause dezincification. In these alloys, zinc is added to copper in amounts ranging from about 5 to 45%. As a general rule, corrosion resistance decreases as zinc content increases. It is customary to distinguish between those alloys containing less than 15% zinc (better corrosion resistance) and those with higher amounts. The main problems with the higher zinc alloys are dezincification and SCC. In dezincification, a porous layer of zinc-free material is formed locally or in layers on the surface. Dezincification in the high-zinc alloys can occur in a wide variety of acid, neutral, and alkaline media.[18]

Dezincification can be avoided by maintaining the zinc content below about 15%, and can be minimized by adding 1% tin such as in admiralty (C44300) and naval brass (C46400). Adding less than 0.1% of arsenic, antimony, or phosphorus gives further protection, provided the brass has the single α-phase structure. SCC occurs readily in the high-zinc brasses in the presence of moisture and ammonia. Again, a decrease in the zinc content to less than 15% is beneficial. Brasses containing less than 15% zinc can be used to handle many acid, alkaline, and salt solutions, provided

1. There is a minimum of aeration.
2. Oxidizing materials, such as nitric acid and dichromates, and complexing agents, such as ammonia and cyanides, are absent.
3. There are no elements or compounds that react directly with copper such as sulfur, hydrogen sulfide, mercury, silver salts, or acetylene.

Table 8.13 presents corrosion-resistance ratings for some coppers (C11000, C12200), brasses (C22000, C23000, C26000, 28000), leaded brasses (C36000, C38500), and tin brasses (C42000, C44300, C44500, C46400) in different chemical environments. Table 8.14 presents corrosion ratings for some phosphor-bronzes (C51000, C52100), aluminum-bronzes (C61300, C62700, C63700, C64200), silicon-bronzes (C65100, C65500), copper-nickel alloys (C70600, C71500), aluminum brass (C68700), and one nickel-silver alloy (C75200).[19]

Atmospheric exposure. Copper and copper alloys perform well in industrial, marine, and rural atmospheres except in atmospheres containing ammonia, which have been observed to cause SCC in brasses containing over 20% zinc. Alloy C11000 (ETP copper) is the most widely used,

TABLE 8.13 Corrosion-Resistance Ratings* for Coppers (C11000, C12200), Brasses (C22000, C23000, C26000, 28000), Leaded Brasses (C36000, C38500), and Tin Brasses (C42000, C44300, C44500, C46400) in Different Chemical Environments

Environment/alloy	11000	12200	22000	23000	26000	28000	36000	38500	42000	44300	46400
Alkalies											
Aluminum hydroxide	E	E	E	E	E	E	E	E	NA	E	E
Ammonium hydroxide	P	P	P	P	P	P	P	P	NA	P	P
Barium carbonate	E	E	E	E	E	E	E	E	NA	E	E
Barium hydroxide	E	E	E	E	VG	VG	VG	VG	NA	P	VG
Black liquor-sulfate process	G	G	G	G	P	P	P	P	NA	P	P
Calcium hydroxide	E	E	E	E	VG	VG	VG	VG	NA	P	VG
Lime	E	E	E	E	E	E	E	E	NA	E	E
Lime-sulfur	G	G	G	G	VG	VG	VG	VG	NA	VG	VG
Magnesium hydroxide	E	E	E	E	E	E	E	E	NA	E	E
Potassium carbonate	E	E	E	E	VG	VG	VG	VG	NA	VG	VG
Potassium hydroxide	VG	VG	VG	VG	G	G	G	G	NA	VG	G
Sodium bicarbonate	VG	VG	VG	VG	G	G	G	G	NA	VG	G
Sodium carbonate	E	E	E	E	VG	VG	VG	VG	NA	E	VG
Sodium hydroxide	VG	VG	VG	VG	G	G	G	G		VG	G
Sodium phosphate	E	E	E	E	VG	VG	VG	VG	NA	E	VG
Sodium silicate	E	E	E	E	VG	VG	VG	VG	NA	E	VG
Sodium sulfide	G	G	E	G	VG	VG	VG	VG	NA	VG	VG
Atmosphere											
Industrial	E	E	E	E	VG	VG	VG	VG	NA	E	VG
Marine	E	E	E	E	VG	VG	VG	VG	NA	E	VG
Rural	E	E	E	E	E	E	E	E	NA	E	E
Chlorinated organics											
Carbon tetrachloride, dry	E	E	E	E	E	E	E	E	NA	E	E
Carbon tetrachloride, moist	VG	VG	VG	VG	P	P	P	P	NA	VG	P
Chloroform, dry	E	E	E	E	E	E	E	E	NA	E	E
Ethyl chloride	VG	VG	VG	VG	G	G	G	G	NA	VG	G
Methyl chloride, dry	E	E	E	E	E	E	E	E	NA	E	E

632

	C1	C2	C3	C4	C5	C6	C7	C8	C9	C10	C11
Trichlorethylene, dry	E	E	E	E	E	E	E	E	NA	E	E
Trichlorethylene, moist	VG	VG	VG	VG	VG	VG	G	G	NA	VG	G
Fatty acid											
Oleic acid	E	E	E	E	E	E	G	G	NA	E	G
Palmitic acid	VG	VG	VG	VG	VG	VG	G	G	NA	VG	G
Stearic acid	VG	VG	VG	VG	VG	VG	G	G	NA	VG	G
Food/beverage											
Beer	E	E	E	E	E	E	VG	VG	NA	E	VG
Beet sugar syrups	E	E	E	E	E	E	VG	VG	NA	E	VG
Cane sugar syrups	E	E	E	E	E	E	VG	VG	NA	E	VG
Carbonated beverages	VG	VG	VG	VG	VG	VG	G	G	NA	VG	G
Carbonated water	VG	VG	VG	VG	VG	VG	G	G	NA	VG	G
Cider	E	E	E	E	E	E	G	G	NA	E	G
Coffee	E	E	E	E	E	E	E	E	NA	E	E
Corn oil	E	E	E	E	E	E	VG	VG	NA	E	VG
Cottonseed oil	E	E	E	E	E	E	VG	VG	NA	E	VG
Fruit juices	VG	VG	VG	VG	VG	VG	P	P	NA	G	P
Gelatine	E	E	E	E	E	E	E	E	NA	E	E
Milk	E	E	E	E	E	E	VG	VG	NA	E	VG
Sugar solutions	E	E	E	E	E	E	VG	VG	NA	E	VG
Vinegar	VG	VG	VG	VG	VG	VG	P	P	NA	G	P
Gases											
Ammonia, absolutely dry	E	E	E	E	E	E	E	E	NA	E	E
Ammonia, moist	P	P	P	P	P	P	P	P	NA	P	P
Carbon dioxide, dry	E	E	E	E	E	E	E	E	NA	E	E
Carbon dioxide, moist	VG	VG	VG	VG	VG	VG	G	G	NA	VG	G
Hydrogen	E	E	E	E	E	E	E	E	NA	E	E
Nitrogen	E	E	E	E	E	E	E	E	NA	E	E

TABLE 8.13 Corrosion-Resistance Ratings* for Coppers (C11000, C12200), Brasses (C22000, C23000, C26000, 28000), Leaded Brasses (C36000, C38500), and Tin Brasses (C42000, C44300, C44500, C46400) in Different Chemical Environments (*Continued*)

Environment/alloy	11000	12200	22000	23000	26000	28000	36000	38500	42000	44300	46400
Oxygen	E	E	E	E	E	E	E	E	NA	E	E
Bromine, dry	E	E	E	E	E	E	E	E	NA	E	E
Bromine, moist	VG	VG	VG	VG	VG	VG	VG	P	NA	G	P
Chlorine, dry	E	E	E	E	E	E	E	E	NA	E	E
Chlorine, moist	G	G	G	G	P	P	P	P	NA	G	P
Hydrocarbons											
Acetylene	P	P	P	P	P	P	E	E	NA	P	E
Asphalt	E	E	E	E	E	E	E	E	NA	E	E
Benzene	E	E	E	E	E	E	E	E	NA	E	E
Benzol	E	E	E	E	E	E	E	E	NA	E	E
Butane	E	E	E	E	E	E	E	E	NA	E	E
Creosote	E	E	E	E	VG	VG	VG	VG	NA	E	VG
Crude oil	VG	VG	VG	VG	G	G	G	G	NA	VG	G
Freon, dry	E	E	E	E	E	E	E	E	NA	E	E
Fuel oil, light	E	E	E	E	VG	VG	VG	VG	NA	E	VG
Gasoline	E	E	E	E	E	E	E	E	NA	E	E
Hydrocarbons, pure	E	E	E	E	E	E	E	E	NA	E	E
Kerosene	E	E	E	E	E	E	E	E	NA	E	E
Natural gas	VG	VG	VG	VG	E	E	E	E	NA	E	E
Paraffin	E	E	E	E	E	E	E	E	NA	E	E
Propane	E	E	E	E	E	E	E	E	NA	E	E
Tar	NA	NA	NA	NA	NA	NA	NA	NA	NA	NA	NA
Turpentine	E	E	E	E	VG	VG	VG	VG	NA	E	VG
Inorganic acids											
Boric acid	E	E	E	E	VG	VG	VG	VG	NA	E	VG
Carbolic acid	VG	VG	VG	VG	VG	VG	VG	VG	NA	VG	VG
Hydrobromic acid	G	G	G	G	P	P	P	P	NA	G	P
Hydrochloric acid	G	G	G	G	P	P	P	P	NA	G	P

Material													
Hydrocyanic acid, dry	P	P	P	P	P	P	P	P	P	P	NA	NA	P	P
Hydrofluosilicic acid, anhydrous	VG	VG	VG	VG	VG	VG	P	P	P	P	NA	NA	VG	P
Phosphoric acid	VG	VG	VG	VG	VG	P	P	P	P	P	NA	NA	G	P
Sulfuric acid, 80–95%	VG	VG	VG	VG	VG	VG	P	P	P	P	NA	NA	G	P
Chromic acid	P	P	P	P	P	P	P	P	P	P	NA	NA	P	P
Nitric acid	P	P	P	P	P	P	P	P	P	P	NA	NA	P	P
Sulfurous acid	VG	VG	VG	VG	VG	VG	P	P	P	P	NA	NA	VG	P
Liquid metal														
Mercury	P	P	P	P	P	P	P	P	P	P	NA	NA	P	P
Miscellaneous														
Glue	E	E	E	E	E	E	VG	VG	VG	VG	NA	NA	E	VG
Linseed oil	VG	VG	VG	VG	VG	VG	VG	VG	VG	VG	NA	NA	VG	VG
Rosin	E	E	E	E	E	E	E	E	E	E	NA	NA	E	E
Sewage	E	E	E	E	E	E	G	G	G	G	NA	NA	E	VG
Soap solutions	E	E	E	E	E	E	VG	VG	VG	VG	NA	NA	E	VG
Varnish	E	E	E	E	E	E	E	E	E	E	NA	NA	E	E
Neutral/acid salts														
Alum	VG	VG	VG	VG	VG	VG	P	P	P	P	NA	NA	VG	P
Alumina	E	E	E	E	E	E	E	E	E	E	NA	NA	E	E
Aluminum chloride	VG	VG	VG	VG	VG	VG	P	P	P	P	NA	NA	G	P
Aluminum sulfate	VG	VG	VG	VG	VG	VG	P	P	P	P	NA	NA	VG	P
Ammonium chloride	P	P	P	P	P	P	P	P	P	P	NA	NA	P	P
Ammonium sulfate	G	G	G	G	G	G	P	P	P	P	NA	NA	P	P
Barium chloride	VG	VG	VG	VG	VG	VG	P	P	P	P	NA	NA	G	P
Barium sulfate	E	E	E	E	E	E	E	E	E	E	NA	NA	E	E
Barium sulfide	G	G	G	G	G	G	G	G	G	G	NA	NA	VG	VG
Calcium chloride	VG	VG	VG	VG	VG	VG	P	P	P	P	NA	NA	VG	G

TABLE 8.13 Corrosion-Resistance Ratings* for Coppers (C11000, C12200), Brasses (C22000, C23000, C26000, 28000), Leaded Brasses (C36000, C38500), and Tin Brasses (C42000, C44300, C44500, C46400) in Different Chemical Environments *(Continued)*

Environment/alloy	11000	12200	22000	23000	26000	28000	36000	38500	42000	44300	46400
Carbon disulfide	VG	VG	VG	VG	E	E	E	E	NA	E	E
Magnesium chloride	VG	VG	VG	VG	P	P	P	P	NA	G	P
Magnesium sulfate	E	E	E	E	G	G	G	G	NA	E	G
Potassium chloride	VG	VG	VG	VG	P	P	P	P	NA	VG	G
Potassium cyanide	P	P	P	P	P	P	P	P	NA	P	P
Potassium dichromate acid	P	P	P	P	P	P	P	P	NA	P	P
Potassium sulfate	E	E	E	E	P	P	P	P	NA	E	P
Sodium bisulfate	VG	VG	VG	VG	VG	VG	VG	VG	NA	E	VG
Sodium chloride	VG	VG	VG	VG	P	P	P	P	NA	VG	G
Sodium cyanide	P	P	P	P	P	P	P	P	NA	VG	G
Sodium dichromate, acid	P	P	P	P	P	P	P	P	NA	P	P
Sodium sulfate	E	E	E	E	P	P	P	P	NA	P	P
Sodium sulfite	VG	VG	VG	VG	VG	VG	VG	VG	NA	E	VG
Sodium thiosulfate	VG	VG	VG	VG	P	P	P	P	NA	VG	P
Zinc chloride	G	G	G	G	VG	VG	VG	VG	NA	VG	VG
Zinc sulfate	G	G	G	G	VG	VG	VG	VG	NA	G	P
					P	P	P	P	NA	VG	P
Organic acids											
Acetic acid	VG	VG	VG	VG	P	P	P	P	NA	G	P
Acetic anhydride	VG	VG	VG	VG	P	P	P	P	NA	G	P
Benzoic acid	E	E	E	E	VG	VG	VG	VG	NA	E	E
Butyric acid	E	E	E	E	G	G	G	G	NA	E	VG
Chloracetic acid	VG	VG	VG	VG	P	P	P	P	NA	VG	G
Citric acid	E	E	E	E	G	G	G	G	NA	E	P
Formic acid	E	E	E	E	G	G	G	G	NA	E	G
Lactic acid	E	E	E	E	G	G	G	G	NA	E	G
Oxalic acid	E	E	E	E	G	G	G	G	NA	E	G
Tannic acid	E	E	E	E	VG	VG	VG	VG	NA	E	E
Tartaric acid	E	E	E	E	G	G	G	G	NA	E	G
Trichloracetic acid	VG	VG	VG	VG	P	P	P	P	NA	G	P

Material										
Organic compounds										
Aniline	G	G	G	G	G	G	G	NA	G	G
Aniline dyes	G	G	G	G	G	G	G	NA	G	G
Castor oil	E	E	E	E	E	E	E	NA	E	E
Ethylene glycol	E	E	E	E	E	E	E	NA	E	E
Formaldehyde (aldehydes)	E	E	E	E	VG	VG	VG	NA	VG	VG
Furfural	E	E	E	E	G	G	G	NA	G	G
Glucose	E	E	E	E	E	E	E	NA	E	E
Glycerine	E	E	E	E	E	E	E	NA	E	E
Lacquers	E	E	E	E	E	E	E	NA	E	E
Organic solvents										
Acetone	E	E	E	E	E	E	E	NA	E	E
Alcohols	E	E	E	E	E	E	E	NA	E	E
Amyl acetate	E	E	E	E	VG	VG	VG		VG	VG
Amyl alcohol	E	E	E	E	E	E	E	NA	E	E
Butyl alcohol	E	E	E	E	E	E	E	NA	E	E
Ethers	E	E	E	E	E	E	E		E	E
Ethyl acetate	E	E	E	E	VG	VG	VG	NA	VG	VG
Ethyl alcohol	E	E	E	E	E	E	E	NA	E	E
Lacquer solvents	E	E	E	E	E	E	E	NA	E	E
Methyl alcohol	E	E	E	E	E	E	E		E	E
Toluene	E	E	E	E	E	E	E	NA	E	E
Oxidizing salts										
Ammonium nitrate	P	P	P	P	P	P	P	NA	P	P
Bleaching powder, wet	VG	VG	VG	VG	VG	VG	VG	NA	VG	VG
Borax	E	E	E	E	E	E	E	NA	E	E
Bordeaux mixture	E	E	E	E	E	E	E	NA	E	E
Calcium bisulfite	VG	VG	VG	VG	VG	VG	VG	NA	VG	P

TABLE 8.13 Corrosion-Resistance Ratings* for Coppers (C11000, C12200), Brasses (C22000, C23000, C26000, C28000), Leaded Brasses (C36000, C38500), and Tin Brasses (C42000, C44300, C44500, C46400) in Different Chemical Environments (*Continued*)

Environment/alloy	11000	12200	22000	23000	26000	28000	36000	38500	42000	44300	46400
Calcium hypochlorite	VG	VG	VG	VG	P	P	P	P	NA	VG	P
Copper chloride	G	G	G	G	P	P	P	P	NA	G	P
Copper nitrate	G	G	G	G	P	P	P	P	NA	G	P
Copper sulfate	VG	VG	VG	VG	VG	P	P	P	NA	P	VG
Ferric chloride	P	P	P	P	P	P	P	P	NA	P	P
Ferric sulfate	P	P	P	P	P	P	P	P	NA	P	P
Ferrous chloride	VG	VG	VG	VG	P	P	P	P	NA	VG	P
Ferrous sulfate	VG	VG	VG	VG	P	P	P	P	NA	VG	P
Hydrogen peroxide	VG	VG	VG	VG	G	G	G	G	NA	VG	G
Mercury salts	P	P	P	P	P	P	P	P	NA	P	P
Potassium chromate	E	E	E	E	E	E	E	E	NA	E	E
Silver salts	P	P	P	P	P	P	P	P	NA	P	P
Sodium bisulfite	VG	VG	VG	VG	P	P	P	P	NA	VG	G
Sodium chromate	E	E	E	E	E	E	E	E	NA	E	E
Sodium hypochlorite	G	G	G	G	P	P	E	P	NA	G	P
Sodium nitrate	VG	VG	VG	VG	G	G	G	G	NA	VG	G
Sodium peroxide	G	G	G	G	P	P	P	P	NA	G	P

Sulfur compounds									
Hydrogen sulfide, dry	E	E	E	E	E	E	NA	E	E
Hydrogen sulfide, moist	P	P	P	P	G	G	NA	G	G
Sulfur, dry (solid)	VG	VG	VG	VG	E	E	NA	E	E
Sulfur, molten	P	P	P	P	P	P	NA	P	P
Sulfur chloride, dry	E	E	E	E	E	E	NA	E	E
Sulfur dioxide, dry	E	E	E	E	E	E	NA	E	E
Sulfur dioxide, moist	VG	VG	VG	VG	P	P	NA	VG	P
Sulfur trioxide, dry	E	E	E	E	E	E	NA	E	E
Waters									
Brines	VG	VG	VG	VG	P	P	NA	VG	G
Mine water	G	G	G	G	P	P	NA	G	P
Seawater	VG	VG	VG	VG	G	G	NA	E	VG
Steam	E	E	E	E	G	G	NA	E	E
Water, potable	E	E	E	E	G	G	NA	E	G

*Rating: Excellent (E), very good (VG), good (G), poor (P), not acceptable (NA).

TABLE 8.14 Corrosion Ratings* for Some Phosphor Bronzes (C51000, C52100), Aluminum Bronzes (C61300, C62700, C63700, C64200), Silicon Bronzes (C65100, C65500), Copper-Nickel Alloys (C70600, C71500), Aluminum Brass (C68700), and One Nickel-Silver Alloy (C75200)

Environment/alloy	51000	52100	61300	62700	63700	65100	65500	68700	70600	71500	75200
Alkalies											
Aluminum hydroxide	E	E	E	NA	E	E	E	E	E	E	E
Ammonium hydroxide	P	P	P	NA	P	P	P	P	P	G	P
Barium carbonate	E	E	E	NA	E	E	E	E	E	E	E
Barium hydroxide	E	E	E	NA	E	E	E	E	E	E	E
Black liquor-sulfate process	G	G	P	NA	G	G	G	G	G	VG	G
Calcium hydroxide	E	E	E	NA	E	E	E	E	E	E	E
Lime	E	E	E	NA	E	E	E	E	E	E	E
Lime-sulfur	G	G	VG	NA	G	G	G	VG	G	VG	VG
Magnesium hydroxide	E	E	E	NA	E	E	E	E	E	E	E
Potassium carbonate	E	E	E	NA	E	E	E	E	E	E	E
Potassium hydroxide	VG	VG	E	NA	VG	VG	VG	VG	E	VG	E
Sodium bicarbonate	VG	VG	E	NA	VG	VG	VG	VG	E	E	E
Sodium carbonate	E	E	E	NA	E	E	E	E	E	E	E
Sodium hydroxide	VG	VG	E	NA	VG	VG	VG	VG	E	E	E
Sodium phosphate	E	E	E	NA	E	E	E	E	E	E	E
Sodium silicate	E	E	E		E	E	E	E	E	E	E
Sodium sulfide	G	G	G	NA	G	G	G	VG	G	VG	VG
Atmosphere											
Industrial	E	E	E	NA	E	E	E	E	E	E	E
Marine	E	E	E	NA	E	E	E	E	E	E	E
Rural	E	E	E	NA	E	E	E	E	E	E	E
Chlorinated organics											
Carbon tetrachloride, dry	E	E	E	NA	E	E	E	E	E	E	E
Carbon tetrachloride, moist	VG	VG	G	NA	VG	VG	VG	VG	VG	E	VG
Chloroform, dry	E	E	E	NA	E	E	E	E	E	E	E
Ethyl chloride	VG	VG	VG	NA	VG	VG	VG	VG	VG	VG	VG

	1	2	3	4	5	6	7	8	9	10	11	12
Methyl chloride, dry	E	E	E	E	E	E	NA	E	E	E	E	E
Trichlorethylene, dry	E	E	E	E	E	E	NA	E	E	E	E	E
Trichlorethylene, moist	VG	VG	VG	VG	VG	VG	NA	VG	VG	VG	E	VG
Fatty acid												
Oleic acid	E	E	E	E	E	E	NA	E	E	E	E	E
Palmitic acid	VG	VG	VG	VG	VG	VG	NA	VG	VG	VG	VG	VG
Stearic acid	VG	VG	VG	VG	VG	VG	NA	VG	VG	VG	VG	VG
Food/beverage												
Beer	E	E	E	E	E	E	NA	E	E	E	E	E
Beet sugar syrups	E	E	E	E	E	E	NA	E	E	E	E	E
Cane sugar syrups	E	E	E	E	E	E	NA	E	E	E	E	E
Carbonated beverages	VG	VG	VG	E	E	VG	NA	VG	VG	VG	VG	VG
Carbonated water	VG	VG	VG	E	E	VG	NA	VG	VG	VG	VG	VG
Cider	E	E	E	E	E	E	NA	E	E	E	E	E
Coffee	E	E	E	E	E	E	NA	E	E	E	E	E
Corn oil	E	E	E	E	E	E	NA	E	E	E	E	E
Cottonseed oil	E	E	E	E	E	E	NA	E	E	E	E	E
Fruit juices	VG	VG	VG	VG	VG	VG	NA	VG	VG	VG	VG	VG
Gelatine	E	E	E	E	G	E	NA	E	E	G	E	E
Milk	E	E	E	E	E	E	NA	E	E	E	E	E
Sugar solutions	E	E	E	E	E	E	NA	E	E	E	E	E
Vinegar	VG	VG	VG	VG	VG	VG	NA	VG	VG	VG	VG	VG
Gases												
Ammonia, absolutely dry	E	E	E	E	E	E	NA	E	E	E	E	E
Ammonia, moist	P	P	P	P	P	P	NA	P	P	P	G	P
Carbon dioxide, dry	E	E	E	E	E	E	NA	E	E	E	E	E
Carbon dioxide, moist	VG	VG	VG	VG	VG	VG	NA	VG	VG	VG	VG	VG

TABLE 8.14 Corrosion Ratings* for Some Phosphor Bronzes (C51000, C52100), Aluminum Bronzes (C61300, C62700, C63700, C64200), Silicon Bronzes (C65100, C65500), Copper-Nickel Alloys (C70600, C71500), Aluminum Brass (C68700), and One Nickel-Silver Alloy (C75200) (*Continued*)

Environment/alloy	51000	52100	61300	62700	63700	65100	65500	68700	70600	71500	75200
Hydrogen	E	E	E	NA	E	E	E	E	E	E	E
Nitrogen	E	E	E	NA	E	E	E	E	E	E	E
Oxygen	E	E	E	NA	E	E	E	E	E	E	E
Bromine, dry	E	E	E	NA	E	E	E	E	E	E	E
Bromine, moist	VG	VG	G	NA	VG	VG	VG	G	VG	VG	VG
Chlorine, dry	E	E	E	NA	E	E	E	E	E	E	E
Chlorine, moist	G	G	G	NA	G	G	G	G	G	VG	G
Hydrocarbons											
Acetylene	P	P	P	NA	P	P	P	P	P	P	P
Asphalt	E	E	E	NA	E	E	E	E	E	E	E
Benzene	E	E	E	NA	E	E	E	E	E	E	E
Benzol	E	E	E	NA	E	E	E	E	E	E	E
Butane	E	E	E	NA	E	E	E	E	E	E	E
Creosote	E	E	E	NA	E	E	E	E	E	E	E
Crude oil	VG	VG	VG	NA	VG	VG	VG	VG	VG	VG	VG
Freon, dry	E	E	E	NA	E	E	E	E	E	E	E
Fuel oil, light	E	E	E	NA	E	E	E	E	E	E	E
Gasoline	E	E	E	NA	E	E	E	E	E	E	E
Hydrocarbons, pure	E	E	E	NA	E	E	E	E	E	E	E
Kerosene	E	E	E	NA	E	E	E	E	E	E	E
Natural gas	E	E	E	NA	E	E	E	E	E	E	E
Paraffin	E	E	E	NA	E	E	E	E	E	E	E
Propane	E	E	E	NA	E	E	E	E	E	E	E
Tar	NA	NA	NA	NA		NA	NA		NA	NA	NA
Turpentine	E	E	E	NA	E	E	E	E	E	E	E
Inorganic acids											
Boric acid	E	E	E	NA	E	E	E	E	E	E	E
Carbolic acid	VG	VG	VG	NA	VG	VG	VG	VG	VG	VG	VG

Material											
Hydrobromic acid	G	G	G	G	NA	G	G	G	G	G	G
Hydrochloric acid	G	G	G	G	NA	G	G	G	G	G	G
Hydrocyanic acid, dry	P	P	P	P	NA	P	P	P	P	P	P
Hydrofluosilicic acid, anhydrous	VG	VG	VG	VG	NA	VG	VG	VG	VG	VG	VG
Phosphoric acid	VG	VG	VG	VG	NA	VG	VG	VG	VG	VG	VG
Sulfuric acid, 80–95%	VG	VG	VG	VG	NA	VG	VG	VG	VG	VG	VG
Chromic acid	P	P	P	P	NA	P	P	P	P	P	P
Nitric acid	P	P	P	P	NA	P	P	P	P	P	P
Sulfurous acid	VG	VG	VG	VG	NA	VG	VG	VG	VG	VG	G
Liquid metal											
Mercury	P	P	P	P	NA	P	P	P	P	P	P
Miscellaneous											
Glue	E	E	E	E	NA	E	E	E	E	E	E
Linseed oil	VG	VG	VG	VG	NA	VG	VG	VG	VG	VG	VG
Rosin	E	E	E	E	NA	E	E	E	E	E	E
Sewage	E	E	E	E	NA	E	E	E	E	E	E
Soap solutions	E	E	E	E	NA	E	E	E	E	E	E
Varnish	E	E	E	E	NA	E	E	E	E	E	E
Neutral/acid salts											
Alum	VG	VG	VG	VG	NA	VG	VG	VG	VG	VG	VG
Alumina	E	E	E	E	NA	E	E	E	E	E	E
Aluminum chloride	VG	VG	VG	VG	NA	VG	VG	VG	VG	VG	VG
Aluminum sulfate	VG	VG	VG	VG	NA	VG	VG	VG	VG	VG	VG
Ammonium chloride	P	P	P	P	NA	P	P	P	P	P	P
Ammonium sulfate	G	G	G	G	NA	G	G	G	G	G	G
Barium chloride	VG	VG	VG	VG	NA	VG	VG	VG	VG	VG	VG

TABLE 8.14 Corrosion Ratings* for Some Phosphor Bronzes (C51000, C52100), Aluminum Bronzes (C61300, C62700, C63700, C64200), Silicon Bronzes (C65100, C65500), Copper-Nickel Alloys (C70600, C71500), Aluminum Brass (C68700), and One Nickel-Silver Alloy (C75200) *(Continued)*

Environment/alloy	51000	52100	61300	62700	63700	65100	65500	68700	70600	71500	75200
Barium sulfate	E	E	E	NA	E	E	E	E	E	E	E
Barium sulfide	G	G	VG	NA	G	G	G	VG	G	VG	VG
Calcium chloride	VG	E	E	NA	VG	VG	VG	VG	E	E	E
Carbon disulfide	VG	VG	VG	NA	VG	VG	VG	E	VG	VG	VG
Magnesium chloride	VG	VG	E	NA	VG	G	G	VG	VG	VG	VG
Magnesium sulfate	E	E	E	NA	E	E	E	E	E	E	E
Potassium chloride	VG	E	E	NA	VG	E	E	VG	E	E	E
Potassium cyanide	P	P	P	NA	P	P	P	P	P	P	P
Potassium dichromate acid	P	P	P	NA	P	P	P	P	P	P	P
Potassium sulfate	E	E	E	NA	E	E	E	E	E	E	E
Sodium bisulfate	VG	VG	E	NA	VG	VG	VG	VG	E	E	E
Sodium chloride	VG	E	P	NA	VG	VG	VG	VG	E	E	E
Sodium cyanide	P	P	P	NA	P	P	P	P	P	P	P
Sodium dichromate, acid	P	P	P	NA	P	P	P	P	P	P	P
Sodium sulfate	E	E	E	NA	E	E	E	E	E	E	E
Sodium sulfite	VG	VG	VG	NA	VG	VG	VG	VG	VG	VG	VG
Sodium thiosulfate	G	G	G	NA	G	G	G	P	G	VG	VG
Zinc chloride	G	G	G	NA	G	G	G	G	G	G	G
Zinc sulfate	VG	VG	VG	NA	VG	VG	VG	VG	VG	VG	VG
Organic acids											
Acetic acid	VG	VG	VG	NA	VG	VG	VG	G	VG	VG	VG
Acetic anhydride	VG	VG	VG	NA	VG	VG	VG	G	VG	VG	VG
Benzoic acid	E	E	E	NA	E	E	E	E	E	E	E
Butyric acid	E	E	E	NA	E	E	E	E	E	E	E
Chloracetic acid	VG	VG	VG	NA	VG	VG	VG	G	VG	VG	VG
Citric acid	E	E	E	NA	E	E	E	E	E	E	E
Formic acid	E	E	E	NA	E	E	E	E	E	E	E
Lactic acid	E	E	E	NA	E	E	E	E	E	E	E

Chemical										
Oxalic acid	E	E	E	NA	E	E	E	E	E	E
Tannic acid	E	E	E	NA	E	E	E	E	E	E
Tartaric acid	E	E	E	NA	E	E	E	E	E	E
Trichloracetic acid	VG	VG	VG	NA	VG	VG	G	VG	VG	VG
Organic compounds										
Aniline	G	G	G	NA	G		G			
Aniline dyes	G	G	G	NA	G		G			
Castor oil	E	E	E	NA	E		E			
Ethylene glycol	E	E	E	NA	E		E			
Formaldehyde (aldehydes)	E	E	E	NA	E		E			
Furfural	E	E	E	NA	E		E			
Glucose	E	E	E	NA	E		E			
Glycerine	E	E	E	NA	E		E			
Lacquers	E	E	E	NA	E		E			
Organic solvents										
Acetone	E	E	E	NA	E		E			
Alcohols	E	E	E	NA	E		E			
Amyl acetate	E	E	E		E		E			
Amyl alcohol	E	E	E		E		E			
Butyl alcohol	E	E	E	NA	E		E			
Ethers	E	E	E	NA	E		E			
Ethyl acetate	E	E	E		E		E			
Ethyl Alcohol	E	E	E	NA	E		E			
Lacquer solvents	E	E	E	NA	E		E			
Methyl alcohol	E	E	E		E		E			
Toluene	E	E	E	NA	E		E			

TABLE 8.14 Corrosion Ratings* for Some Phosphor Bronzes (C51000, C52100), Aluminum Bronzes (C61300, C62700, C63700, C64200), Silicon Bronzes (C65100, C65500), Copper-Nickel Alloys (C70600, C71500), Aluminum Brass (C68700), and One Nickel-Silver Alloy (C75200) *(Continued)*

Environment/alloy	51000	52100	61300	62700	63700	65100	65500	68700	70600	71500	75200
Oxidizing salts											
Ammonium nitrate	P	P	P	NA	P			P			
Bleaching powder, wet	VG	VG	G	NA	VG			VG			
Borax	E	E	E	NA	E			E			
Bordeaux mixture	E	E	E	NA	E			E			
Calcium bisulfite	VG	VG	VG	NA	VG			VG			
Calcium hypochlorite	VG	VG	G	NA	VG			VG			
Copper chloride	G	G	G	NA	G			G			
Copper nitrate	G	G	G	NA	G			G			
Copper sulfate	P	VG	VG	NA	P			VG			
Ferric chloride	P	P	P	NA	P			P			
Ferric sulfate	P	P	P	NA	P			P			
Ferrous chloride	VG	VG	VG	NA	VG			VG			
Ferrous sulfate	VG	VG	VG	NA	VG			VG			
Hydrogen peroxide	VG	VG	G	NA	VG			VG			
Mercury salts	P	P	P	NA	P			P			
Potassium chromate	E	E	E	NA	E			E			
Silver salts	P	P	P	NA	P			P			
Sodium bisulfite	VG	VG	VG	NA	VG			VG			
Sodium chromate	E	E	E	NA	E			E			
Sodium hypochlorite	G	G	G	NA	G			G			
Sodium nitrate	VG	VG	VG	NA	VG			VG			
Sodium peroxide	G	G	G	NA	G			G			

Sulfur compounds						
Hydrogen sulfide, dry	E	E	E	NA	E	E
Hydrogen sulfide, moist	P	P	P	NA	P	G
Sulfur, dry (solid)	VG	VG	VG	NA	VG	VG
Sulfur, molten	P	P	P	NA	P	P
Sulfur chloride, dry	E	E	E	NA	E	E
Sulfur dioxide, dry	E	E	E	NA	E	E
Sulfur dioxide, moist	VG	VG	G	NA	VG	VG
Sulfur trioxide, dry	E	E	E	NA	E	E
Waters						
Brines	VG	VG	E	NA	VG	VG
Mine water	G	G	G	NA	G	G
Seawater	VG	VG	E	NA	VG	E
Steam	E	E	E	NA	VG	E
Water, potable	E	E	E	NA	E	E

*Rating: Excellent (E), very good (VG), good (G), poor (P), not acceptable (NA).

particularly for roofing, flashing, gutters, and downspouts, with alloys C22000 (commercial bronze), C23000 (red brass), C38500 (architectural bronze), and C75200 (65-12 nickel silver) accounting for much of the remainder.

Water and soils. The largest single application of copper tube is for hot and cold water distribution lines in building construction, with smaller amounts for heating and drainage lines and fire safety systems. Copper protects itself by forming a protective film, the degree of protection depending on mineral, oxygen, and carbon dioxide contents. The brasses also perform well in unpolluted freshwaters but may experience dezincification in stagnant or slowly moving brackish or slightly acid waters. The copper-nickels, silicon, and aluminum bronzes display excellent resistance to corrosion.[17]

Copper exhibits high resistance to corrosion in most soil types. Studies of samples exposed underground have shown that tough pitch coppers, deoxidized coppers, silicon bronzes, and low-zinc brasses behave essentially alike. Soils containing cinders with high concentrations of sulfides, chlorides, or hydrogen ions corrode these materials. In this type of contaminated soil, alloys containing more than 22% zinc experience dezincification. In soils that contain only sulfides, corrosion rates of the brasses decrease with increasing zinc content and no dezincification occurs. The corrosion rate of copper in quiescent ground water tends to decrease with time, the rate depending on the amount of dissolved oxygen present.

Steam systems. Copper and copper alloys resist attack by pure steam, but if carbon dioxide, oxygen, or ammonia is present, condensates can be quite corrosive to copper alloys. Modern power utility boiler feedwater treatments commonly include the addition of organic amines to inhibit the corrosion of iron components of the system by scavenging oxygen and increasing the pH of the feedwater. These chemicals tend to release ammonia, which can be corrosive to some copper alloys.

Salts. The superior seawater performance of many tin brasses, aluminum bronzes, and copper-nickels over copper is the result of corrosion product insolubility combined with erosion and biofouling resistance. Both alloys C70600 and C71500, for example, display excellent resistance to pitting in seawater. The next section is dedicated to the behavior of these alloys in marine environments. In general, the copper-base alloys are galvanically compatible with one another in seawater. Although the copper-nickel alloys are slightly cathodic (noble) to the nickel-free copper base alloys, the small differences in corrosion potential generally do not lead to serious galvanic effects unless unusually adverse anodic/cathodic area ratios are involved.

Copper metals are widely used in equipment for handling various kinds of salt solutions including the nitrates, sulfates, and chlorides of sodium and potassium. Although alkaline sodium salts such as silicate, phosphate, and carbonate attack copper alloys at low rates, alkaline cyanide is aggressive and attacks copper alloys fairly rapidly because of the formation of soluble complex copper species such as $Cu(CN)$, $Cu(CN)_2^{1-}$ and $Cu(CN)_3^{2-}$.

Polluted cooling waters. The primary causes of accelerated attack of copper alloys by polluted seawater are the action of sulfate-reducing bacteria under anaerobic conditions and the putrefaction of organic sulfur compounds from decaying plant and animal matter within seawater systems during periods of extended shutdown. However, the copper alloys have long been recognized for their inherent resistance to marine fouling, mostly due to the biocidal effect copper ions have on microorganisms in general.

Acids and alkalies. In general, copper alloys are successfully used with nonoxidizing acids as long as the concentration of oxidizing agents, such as dissolved oxygen or air, and ferric (Fe^{3+}) or dichromate ions $(CrO_7)^{2-}$ is low. Successful applications of copper and its alloys are in phosphoric, acetic, tartaric, formic, oxalic, malic, and other organic acids that react in a manner similar to sulfuric. Copper and its alloys resist alkaline solutions, except those containing ammonium hydroxide, or compounds that hydrolyze to ammonium hydroxide or cyanides. Ammonium hydroxide reacts with copper to form the soluble complex copper-ammonium compound $Cu(NH_3)_4^{2+}$.

Liquid metal embrittlement. Although mercury embrittles copper, the severity increases when copper is alloyed with aluminum or zinc. This embrittlement occurs in both tension and fatigue and varies with grain size and strain rate. Other alloying elements such as lithium, sodium, bismuth, gallium, and indium also affect embrittlement.

Organic compounds. Copper and many of its alloys resist corrosive attack by organic compounds such as amines, alkanolamines, esters, glycols, ethers, ketones, alcohols, aldehydes, naphtha, gasoline, and most organic solvents. Corrosion rates of copper and copper alloys in alkanolamines and amines, although low, can be significantly increased if these compounds are contaminated, particularly at high temperatures.

8.4.4 Marine application of copper-nickel alloys

The excellent corrosion and biofouling resistance of copper-nickel alloys in seawater has led to their substantial use in marine service for many years. Development work began in the 1930s in response to a requirement by the British Navy for an improved condenser material. The 70-30 brass used at that time could not adequately withstand prevailing seawater velocities. Based on observations that the properties of 70-30 copper-nickel tended to vary with iron and manganese levels, a composition was sought to optimize resistance to velocity effects, deposit attack, and pitting corrosion. Typical levels of 0.6% iron and 1.0% manganese were finally chosen.[20]

Since the 1950s, the 90-10 alloy has become accepted for condenser service as well as for seawater pipe work in merchant and naval service. In naval vessels, the 90-10 copper-nickel is preferred for surface ships, whereas the 70-30 alloy is used for submarines because its greater strength makes it more acceptable for the higher pressures encountered. These alloys are also used for power station condensers and offshore seawater pipe work on oil and gas platforms. Large quantities are selected for the desalination industry, and they are additionally used for cladding and sheathing of marine structures and hulls.[21]

The two main wrought copper-nickel alloys chosen for seawater service contain 10 and 30% percent nickel, respectively. When comparing international specifications, the compositional ranges of the two alloys vary slightly between specifications, as can be seen in Tables 8.15 and 8.16 for 90-10 and 70-30 copper-nickel alloys. In practice, these variations have little influence on the overall service performance of the alloys. Iron is essential for both alloys because it provides added resistance to corrosion caused by velocity effects called impingement attack.[22] An optimum level is between 1.5 and 2.5% iron, probably as a result of solid solubility. The corrosion resistance improves with increasing iron so long as it remains in solid solution. The specification limits for alloys were set by this observation.

Manganese is necessary as a deoxidant during the melting process, but its effect on corrosion resistance is less well defined than that for iron. Impurity levels must be tightly controlled because elements such as lead, sulfur, carbon, and phosphorus, although having minimal effect on corrosion resistance, can influence hot ductility and, therefore, influence weldability and hot workability.

A comparison of the physical and mechanical properties of the two alloys is given in Table 8.17. Of particular interest for heat exchangers and condensers are the thermal conductivity and expansion characteristics. Although conductivity values for both are good, the 90-10 alloy has the higher value. This partly explains the alloy's greater pop-

TABLE 8.15 Specifications for 90-10 Copper-Nickel Alloy (Maximum Except Where Range Given)

	ISO CuNi10FelMn	BS CN 102	UNS C70600	DIN CuNi10Fe 2.0872
Copper				
Minimum	Rem.	Rem.	Rem.	Rem.
Maximum				
Nickel				
Minimum	9.0	10.0	9.0	9.0
Maximum	11.0	11.0	11.0	11.0
Iron				
Minimum	1.2	1.0	1.0	1.0
Maximum	2.0	2.0	1.8	1.8
Manganese				
Minimum	0.5	0.5		0.5
Maximum	1.0	1.0	1.0	1.0
Tin				
Minimum				
Maximum	0.02			
Carbon	0.05	0.05	0.05*	0.05
Lead	0.03	0.01	0.02*	0.03
Phosphorus			0.02*	
Sulfur	0.05	0.05	0.02*	0.05
Zinc	0.5	0.5	0.5*	0.5
Total other impurities	0.1			0.1
Total impurities		0.3		

*When required for welding.

ularity for heat exchanger and condenser service, where higher strength is not the most important factor.[21] The 70-30 alloy is essentially nonmagnetic and has a magnetic permeability very close to unity. The 90-10 alloy, with higher iron content, is nonmagnetic if the iron can be retained in solid solution during processing. For 90-10 tubing used in minesweepers, air cooling after the final anneal suppresses precipitation sufficiently to provide low permeability.

Both alloys have good mechanical strengths and ductilities, although the higher-nickel alloy does possess the greater inherent strength. Both alloys are single-phase, solid solution alloys and cannot be hardened by heat treatment. The strengths, however, can be increased by work hardening. Although 90-10 copper nickel tubing can have a proof stress of 100 to 160 MPa when supplied in the annealed condition, this could typically be 345 to 485 MPa in the as-drawn condition.

TABLE 8.16 Specifications for 70-30 Copper-Nickel Alloy (Maximum Except Where Range Given)

	ISO CuNi30MnlFe	BS CN 107	UNS C71500	DIN CuNi30Fe 2.0882
Copper				
Minimum	Rem.	Rem.	Rem.	Rem.
Maximum				
Nickel				
Minimum	29.0	30.0	29.0	30.0
Maximum	32.0	32.0	33.0	32.0
Iron				
Minimum	0.4	0.4	0.4	0.4
Maximum	1.0	1.0	1.0	1.0
Manganese				
Minimum	0.5	0.5		0.5
Maximum	1.5	1.5	1.0	1.5
Tin				
Minimum				
Maximum	0.02			
Carbon	0.06	0.06	0.05*	0.06
Lead	0.03	0.01	0.02*	0.03
Phosphorus			0.02*	
Sulfur	0.06	0.08	0.02*	0.05
Zinc	0.5		0.5*	0.5
Total other impurities	0.1			0.1
Total impurities		0.3		

*When required for welding.

Corrosion behavior. General corrosion rates for 90-10 and 70-30 copper-nickel alloys in seawater are low, ranging between 25 and 2.5 $\mu m \cdot y^{-1}$. For the majority of applications, these rates would allow the alloys to last the required lifetime, and there would be little probability of their premature failure in service due to such a corrosion mechanism.[21]

Pitting corrosion. Although copper-nickels have a passive surface film, they have advantages over some other alloy types by having a high resistance to biofouling, thereby decreasing the number of potential sites where corrosion could occur. The copper-nickels also have a high inherent resistance to pitting and crevice corrosion in quiet seawater. Pitting penetration rates can conservatively be expected to be well below 127 μm/y. Sixteen-year tests on 70-30 alloy reported the average depth of the 20 deepest pits to be less than 127 μm.[21] When pits do

TABLE 8.17 Physical and Mechanical Properties of 90-10 (C70600) and 70-30 (C71500) Copper Nickels

Property	90-10	70-30
Specific gravity (g/cm^3)	8.9	8.95
Specific heat (J/kg·K)	377	377
Melting range (°C)	1100–1145	1170–1240
Thermal conductivity (W/mK)	50	29
Coefficient of linear expansion		
-180 to 10°C 10^{-6}/K	13	12
10 to 300°C 10^{-6}/K	17	16
Electrical resistivity ($\mu\Omega$·cm)	19	34
Coefficient of electrical resistivity (10^{-6})	70	50
Modulus of elasticity (GPa)		
Annealed	135	152
Cold worked 50%	127	143
Modulus of rigidity (GPa)		
Annealed	50	56
Cold worked	47	53
Yield strength (0.2%) (MPa)	140	170
Tensile strength (MPa)	320	420
Elongation (%)	40	42

occur, they tend to be shallow and broad in nature and not the undercut type of pitting that can be expected in some other types of alloys.

Stress corrosion cracking. The 90-10 and 70-30 copper-nickels are resistant to chloride- and sulfide-induced SCC. Some copper-based alloys such as aluminum brass are subject to SCC in the presence of ammonia. In practice, this prevents their use in the air-removal section of power plant condensers. Copper-nickel alloys, however, are resistant to SCC and are commonly used in air-removal sections.

Denickelification. Denickelification of 70-30 alloys (i.e., the selective leaching of nickel out of an alloy matrix) has been encountered occasionally in refinery overhead condenser service, where hydrocarbon streams condense at temperatures above 150°C. This appears to be due to thermogalvanic effects resulting from the occurrence of local "hot spots." The solution has been to remove deposits that lead to the hot spots, either by more frequent cleaning or by increasing flow rates. Denickelification was also observed recently in modern warship heat exchangers where some 70-30 copper-nickel tubes suffered severe hot spots corrosion. To prevent this problem from recurring, it is recommended to maintain a continuous flow of seawater and install sacrificial anodes.[23]

Galvanic effects. As a general rule, the copper-base alloys are galvanically compatible with one another in seawater. The copper-nickel alloys are slightly cathodic (noble) to the nickel-free copper-base alloys, but the small differences in corrosion potential generally do not lead to serious galvanic effects between alloys unless unusually adverse anodic/cathodic area ratios are involved. Corrosion rates for galvanic couples of alloys C70600 and C71500 with other materials are shown in Table 8.18. These data demonstrate the increased attack of less noble carbon steel coupled to copper-nickel alloys, the increased attack on the copper-nickel alloys when coupled to more noble titanium, and the general compatibility of copper-nickel alloys with aluminum bronze. It should be noted that coupling the copper-nickel alloys to less noble materials, such as carbon steel, affords protection to the copper-nickel. This effectively reduces its corrosion rate, thereby inhibiting the natural resistance to biofouling of the alloy.[24]

Alloy C70600 is very slightly anodic to C71500, and some advantage has been taken of this fact. Alloy C70600 has been used as cladding on a substrate of C71500 for oil coolers. Any local penetrations by turbulent seawater, such as by erosion corrosion, of the C70600 are arrested when the underlying C71500 alloy is reached, until some significant

TABLE 8.18 Galvanic Couple Data for C70600 and C71500 with Other Materials in 0.6-m/s Flowing Seawater (One-Year Exposures - Equal Area Couples)

Uncoupled	Corrosion rate, μm/y
C70600	31
C71500	20
Aluminum bronze (C61400)	43
Carbon steel	330
Titanium	2

Coupled	Corrosion rate, μm/y
C70600	25
Al bronze (C61400)	43
C70600	3
Carbon steel	787
C70600	208
Titanium	2
C71500	18
Al bronze (C61400)	64
C71500	3
Carbon steel	711
C71500	107
Titanium	2

TABLE 8.19 Galvanic Corrosion Data for C70600 Cast Alloy Couples in Seawater

Alloy	Galvanic effect	
	C70600	Other alloy
C70600	1.0	
Cast 90-10CuNi	0.8	1.6
Cast 70-30CuNi	0.9	1.0
85-5-5-5 (C83600)	0.9	1.5
Monel bronze (C92200)	0.7	1.8
CN7M stainless steel	1.5	0.6
CF8M stainless steel	1.2	0.1
Gray iron	0.1	6.0
Nickel-resist type I	0.4	2.1
Nickel-resist type II	0.3	2.6
Nickel-resist type D2	0.3	2.0

*Seawater velocity: 1.8 m/s; seawater temperature: 10°C (nickel-resist couple tests: 29°C); exposure time: 32 days; equal area couples; ratio of mass loss in couple to control.

area of the anodic cladding has been consumed. This clad construction increased the life of an all C70600 construction in plate-type coolers from about 6 months to more than 5 years of continuous use.

Results of short-term galvanic couple tests between C70600 and several cast copper-base alloys and ferrous alloys are given in Table 8.19. The corrosion rate of cast 70-30 copper-nickel was unaffected by coupling with an equal area of C70600, whereas some increased corrosion of other cast copper-base alloys was noted. Corrosion rates of cast stainless steels were reduced with a resultant increase in corrosion of C70600. Gray iron displayed the largest galvanic effect, and the corrosion rates of nickel-resist alloys nominally doubled.

The contact between the tubes and tube sheet can lead to galvanic corrosion, particularly if proper attention is not given to materials selection. Key problem material combinations in recent years appear to be in the use of titanium or stainless steel tubing (particularly in retubing existing units) where tube sheets of muntz metal (C63500) or aluminum bronze (C61400) exist. Severe galvanic corrosion of these tube sheets has resulted and has led to studies that showed the effective cathodic area was many times larger than had been assumed, approaching a 1000:1 cathode-to-anode ratio. These copper alloy tube sheets coupled to titanium or stainless steels require a carefully designed cathodic protection system.[24]

Microfouling. Copper alloys have good resistance to microfouling, although they are not totally immune to it. Microfouling can be found in heat-exchanger and condenser tubing. A 90-to 100-day interval between cleanings for copper alloys compared favorably with the 10-day interval found necessary for other alloy condenser tubes in the study.[25] The ability of copper-nickel to resist microfouling and remain effective as a heat-transfer surface in seawater for the 3- to 4-month normal intervals between mechanical cleanings, without chlorination, is of clear benefit and one of the reasons why copper-nickel continues as a useful tubing material wherever saline waters are used for cooling.

Because the condenser is the heart of the heat-reject system in operating power or process industry plants, as well as in ships, its reliability and efficiency affect the overall system performance. Deposits and films that accumulate and grow on the tube's internal surface affect heat-transfer capacity and in turn its ability to condense steam. The heat exchanger is simply a device that directs the flow paths in such a way that the two streams are brought into thermal contact through a conducting wall while being kept physically separate. The relatively thin-walled tube, selected primarily for heat-transfer efficiency, becomes the critical component in condensers and other heat exchangers and must perform well over long periods of time under sometimes very difficult operating conditions.[26]

Protective film formation. The good corrosion resistance in seawater offered by copper-nickel alloys results from the formation of a protective oxide film on the metal surface. The film forms naturally and quickly, changing the alloy's initial exposure to seawater. In clean seawater, the film is predominantly cuprous oxide, with the protective value enhanced by the presence of nickel and iron. Cuprous hydroxychloride and cupric oxide are often also present.[25]

The protective film continues to become more protective with time, as indicated by corrosion rate measures made over several years. Studies in quiet seawater show that the time span approaches 4 years before the decrease in corrosion rate becomes negligible. In flowing water, the corrosion rate was found to decrease continually over at least a 14-year period, the effect being similar for both 90-10 and 70-30 alloys. The normal corrosion product film is thin, adherent, and durable. Once fully formed and reasonably mature, the film on copper-nickel alloys will withstand considerable excursions in water velocity, pollution, and other conditions normally adverse to the good performance of copper alloy tubing. Copper-nickel alloys remain resistant to corrosion in deaerated seawater at low pH, as has been experienced in numerous distillation-type desalination plants.[24]

Effect of velocity. The combination of low general corrosion rates and high resistance to pitting and crevice corrosion ensures that the copper-nickel alloys will perform well in quiet, clean, and aerated seawater. As the flow rate of seawater increases, the corrosion rate remains low due to the protective surface film on the alloys. However, once the velocity is such that the film becomes damaged and the active underlying metal is exposed, erosion corrosion (impingement attack) will occur rapidly. The seawater velocity at which this occurs is often called the *breakaway velocity,* and different copper-based alloys show different breakaway velocities.

The 90-10 copper-nickel has better impingement resistance than aluminum brass, which in turn is better than copper.[27] The 70-30 alloy shows better resistance than the 90-10 alloy. This is not a complete representation, however. Rates of attack are not only dependent on seawater velocity but also on pipe work diameter. Table 8.20 compares results of tests on condenser tube alloys using the jet impingement technique in two independent laboratories. General experience has shown that 90-10 copper-nickel can successfully be used in condensers and heat exchangers with water velocities up to 2.5 m/s. For pipeline systems, higher seawater velocities can safely be used in larger-diameter pipes, as indicated by codes of practice. For 70-30 copper-nickel, the maximum design velocity is given as 4 m/s for diameters of 100 mm or greater.[24]

Although much has been written about the effect of high velocity, much less attention has been given to the extremely damaging effect of low velocities. Several failure analyses conducted on C70600 tubing revealed that the original design flow rates were less than 1 m/s. At such low flow rates, there is time for even very light mud and sediment loadings to deposit out in the tubing, leading to underdeposit corrosion and tube failures. Low flow rates can indeed be more damaging than high flow rates and should be guarded against in the original design and operation.[24]

TABLE 8.20 Comparison of Test Results Obtained on Condenser Tube Alloys by Two Independent Laboratories (Testing Conditions: Velocity of Jet, 4.58 m·s^{-1}, Air Added, 3% by Volume, Duration, 28 days)

Material	Average depth of attack, μm	
	BNFMA*	LCCT†
Arsenical admiralty brass	340	270
Arsenical copper	300	
70-30 copper-nickel, 0.04% Fe	110	220
Aluminum brass	40‡	200
70-30 Copper-nickel, 0.8% Fe	20	
70-30 Copper-nickel, 0.45% Fe		100
90-30 Copper-nickel, 2% Fe	0	150

* British Non Ferrous Metal Research Association Laboratories, U.K.
† LaQue Centre for Corrosion Technology, North Carolina.
‡ One specimen out of 20 pitted to a depth of 650 μm. No other specimen greater than 200 μm.

Effect of sulfides. Sulfides are present in polluted water either as industrial effluent or when water conditions support the growth of sulfate-reducing bacteria. They can also occur in stagnant seawater by decomposition of organic matter to sulfides and ammonia. Sulfides form a black corrosion product that is less adherent and protective than the normal oxide film. Under susceptible conditions, unwanted pitting or accelerated general corrosion may occur.

In the complete absence of oxygen, a sulfide film can show an acceptable degree of protection. However, the sulfides become detrimental if dissolved oxygen is also present in the seawater or if exposure to oxygen-free sulfide-polluted waters is followed by exposure to aerated, unpolluted waters. The presence of as little as 0.01 mg/L of sulfides has been shown to accelerate attack of 90-10 copper-nickel in aerated seawater, although the combined influence of velocity and sulfides makes the effect more significant.[21]

Fortunately, a normal oxide film will replace the sulfide film that forms in polluted water once the polluted seawater is replaced by clean, aerated seawater. This occurs when vessels are fitted out in polluted harbors and then operate in the open sea. Higher corrosion rates do continue for some time during the transition period. Experience has shown that as soon as vessels begin regular operation, the normal protective film, once adequately formed, will also persist during subsequent harbor visits. The ideal situation, whether in a ship or power plant, is to recirculate aerated, clean seawater at initial start-up for sufficient time to form a good protective film. When formed, this provides a high degree of corrosion protection to subsequent exposure to sulfides.

The effect of seawater treatments. Ferrous ion additions can be used to reduce the corrosion of copper-nickel alloys either by a direct addition of ferrous sulfate or by a driven iron anode. Ferrous ions in seawater are very unstable and can decay within as little as 3 min.[22] Therefore, the treatment is more effective when additions are continuous rather than intermediate. Ferrous ion treatment has been found to suppress the corrosion rates of copper-nickel in both polluted and unpolluted conditions. However, it is particularly attractive when supplies are present in the seawater. For example, to encourage good initial film formation during fitting out, a system can be filled initially with fresh water containing 5 ppm ferrous sulfate and left in the system for 1 day. After this, the system can be used for normal fitting out purposes, but the ferrous sulfate solution (5 ppm concentration) should be recirculated for 1 h per day throughout the fitting out period.[21] This practice is also useful when systems are retubed or renewed.

Ferrous sulfate is not essential to successful performance but can be viewed as a remedy when trouble has occurred or as a precaution if trouble is likely. Most ships in service have operated successfully without any ferrous sulfate dosing.

Coastal power and process industry plants have used chlorine to control biofouling and slime formation for many years. Chlorine injection is normally provided for heat exchangers in coastal plants that are seldom cleaned more than once a year and for naval ships that must maintain their equipment at maximum efficiency at all times. Chlorine may be added in the gaseous form or developed in situ via electrolytic chlorine generation.

Chlorine is used as an effective biocide when injected continually so that 0.2 to 0.5 ppm residual is maintained at the outlet tube sheet of a power plant condenser.[25] Copper-nickel tubing is resistant to chlorination at concentrations normally required to control biofouling. Excessive chlorination, however, can damage copper alloy tubing. There is some evidence that at high velocities, chlorination has the effect of increasing the impingement rate in the 90-10 alloy, although in the 70-30 alloy, the rate is decreased.[21] An impingement jet velocity of 9 m/s was used. It should be noted, however, that 9-m/s velocity is not normally encountered or recommended for copper-nickel alloys.

Even though alloy C70600 is inherently resistant to attachment of fouling organisms, a few of the larvae that pass through even the smallest screen openings are able to attach to the surface during periods of low flow or shutdown. The few that do attach determine the interval between mechanical cleanings needed to restore full heat-transfer capability. Without chlorine injection, mechanical cleaning to restore heat transfer may be needed in 1 or 2 months. Chlorine injection will extend the interval between mechanical cleanings and maintain original heat-transfer capability for extended periods.

8.4.5 Decorative corrosion products

The spontaneous surface corrosion of copper and its alloys has been used for centuries to create a spectrum of colors and hues controlled by the nature of the alloy and its relation to the environment. Patina is usually a green or brownish film formed naturally on copper and bronze by long exposure or artificially and is often valued aesthetically for its color. Copper and copper alloys are widely used in architectural applications to take advantage of their inherent range of colors. Although these metals may be used in their natural color, as fabricated, it is sometimes desirable to chemically color pure copper, commercial bronze, and architectural bronze.

The most common colors to be produced are referred to as brown statuary finishes for bronze and green patina finishes for copper. The following sections outline procedures and formulations for producing both. Although the chemical solutions described are those generally accepted in the metal-finishing trade, many variations exist. The wide range of colors and shades that may be achieved are largely a mater of craftsmanship and experience. Chemical coloring techniques depend upon time, temperature, surface preparation, mineral content of the water, humidity, and other variables that influence the ultimate result. This section presents the technology that underlies the craftsmanship and art involved in producing these colored finishes.

Green patina finishes. The much admired natural protective coating of a blue-green patina characterizes older copper roofs, including ancient cathedrals, as well as bronze statues and other copper metal surfaces exposed to the weather. Because of the time required to achieve this, much research has been done on artificial patina. The major coloring agent in natural patina is a film of basic copper sulfate. Carbonate and chloride salts of copper may also be present in varying concentrations. In seacoast locations, chloride salts may form an essential part of the patina film. The basic chloride salts of copper are not only fairly soluble but photosensitive as well.[28]

In artificially producing or accelerating the formation of a patina, success seems to depend on the manner in which the solutions are applied, the weather conditions under which the treatment is carried out, and on the climate to which the treated surfaces are exposed. Because of the number of variables involved, chemically induced patinas are prone to lack of adhesion, excessive staining of adjacent materials, and inability to achieve reasonable color uniformity over large surface areas.

Cleaning. The copper surfaces to be colored must be clean, because any dirt, oil, or grease on the surface will interfere with the chemical action of the solution. This involves removal of the residual film of oil left on copper and brass sheets from mill rolling operations and fingerprints and dirt deposited on the surface during handling and installation. A thorough rinsing to remove all traces of the cleaning compound should follow cleaning. If cleaning has been properly done, the rinsing water will spread uniformly without beading or the formation of globular droplets. If necessary, cleaning should be repeated until this condition is obtained.

Oxide film on the copper will cause poor adherence of the patina. Copper roofs that have weathered for 6 months or more should have the oxide film removed before starting the coloring operations. This is done by swabbing the surface with a cold 5 to 10% sulfuric acid solution. The surface should again be thoroughly rinsed with clean water immediately after this swabbing. This should leave a roof surface, whether old or new, in good condition for coloring.

Coloring. Of the three basic processes for accelerated patina formation, one uses a sulfate solution and two utilize chloride salts.

Ammonium sulfate. The concentrated ammonia should have a specific gravity of 0.900 g·cm^{-3}. The ammonium sulfate solution described in Table 8.21 should be prepared in a corrosion-resistant plastic-lined container. Wooden barrels and tubs are also satisfactory if all exposed metal parts are lead covered. First dissolve the ammonium sulfate in the water. When completely dissolved, add the copper sulfate. Then add the concentrated ammonia slowly, while constantly stirring the solution. It is important that the quantity of ammonia be exact, because the correct ratio of ammonia to water must be maintained.

The solution should be applied by spraying. A satisfactory sprayer is an ordinary plastic or galvanized steel, garden-type tank sprayer, with the inside coated with bituminous paint. Spraying should be done rapidly, using a fine spray. Avoid large drops, which tend to run together, causing streaks. It is better to use too little rather than too much solution at a time. Allow the solution to dry after the first spraying. Spraying and drying are repeated five or six times.[28]

The color does not show up immediately. When the spraying has been completed, the copper surface should appear to be covered with a "glassy" coating somewhat resembling a dark, heavy coat of varnish. The development of color depends on suitable weather conditions. Rain within 6 or 8 hours may wash away some of the solution before it has had a chance to act on the copper.

Ideal weather conditions following the treatment are a moderate-to-heavy dew, a light mist or fog, or other condition of high enough atmospheric moisture to give a relative humidity of 80% or more. The atmospheric moisture combines with the deposited solution to react chemically with the copper, and the desired blue-green patina results. The colored layer should be of a satisfactory depth if the action continues undisturbed for at least 6 hours. Where this has occurred, the next rain should wash off the remaining deposit and bring out the blue-green of the patina. At first the color will be somewhat bluer than natural patina, but it should weather into a natural color in time.

TABLE 8.21 Solutions for Accelerating the Formation of Green Patina Finishes on Copper Alloys

Ammonium sulfate, L	Ammonium chloride	Cuprous chloride, L
Cover approximately 15 m^2 Ammonium sulfate 111 g Copper sulfate 3.5 g Conc. ammonia 1.6 mL	Ammonium chloride (Saturation)	Cuprous chloride 164 g Hydrochloric acid 117 mL Glacial acetic acid 69 mL Ammonium chloride 80 g Arsenic trioxide 11 g

Ammonium chloride (salammontac). The saturated ammonium chloride solution can be applied by brush or spray on a thoroughly clean copper surface. Several applications may be required. Frank Lloyd Wright favored this formula. Wright specified that the solution be mixed 24 h prior to its use. Two applications should be made with a lapse of 48 h between the two. Twenty-four hours after the final application, the copper surface should be sprayed with a cold water mist. Wright emphasized that dry weather was required throughout the entire period.[28] The ammonium chloride solution tends to chalk and flake if applied too heavily and is also apt to dissipate in heavy rain.

Cuprous chloride. The acidic cuprous chloride solution described in Table 8.21 can be applied by spray, brush, or stippling. The solution should be stored and used in nonmetallic containers. The solution is both acid and toxic. It can be applied to either bright or weathered copper. If possible, the desired color should be attained in a single application. Reapplication, particularly in direct sunlight, may cause a reaction between the solution and the salts initially deposited, producing a smooth, hard, colorless film similar in appearance to varnish.

Maintenance. No maintenance is required for an existing natural patina or one which is in the process of formation. If a natural statuary finish is desired on copper, weathering can be arrested at the desired point by applying a suitable oil (e.g., raw linseed oil or lemon oil). Depending on the prevailing climatic conditions and the degree of exposure, the frequency of oiling may be at intervals of from 1 to 3 years. Instances have been recorded where the initial oiling applied in two thin coats has preserved the statuary finish in excess of 10 years.

Copper, brass, and bronze are resistant to destructive corrosion. The patina that forms naturally is in fact a protective film. The copper metals are lightweight, easy to work, easy to join, attractive, and extremely durable. This accounts for their use for centuries for roofs, fascias, gutters, downspouts, flashing, storefronts, railings, grilles, and other architectural applications of many descriptions.

Brown statuary finishes. Statuary finishes are conversion coatings. In conversion coatings, the metal surface is either converted into a protective film, usually an oxide or sulfide of the metal involved, or a compound is precipitated that forms a surface film. The use of chemical solutions is generally termed *oxidizing,* although the oldest method and the one that produces the widest range of brown-to-black stages on copper alloys actually produces not an oxide but a metal sulfide finish by the use of alkaline sulfide solutions. Originally liver of sulfur was employed, this being a crude mixture of potassium polysulfides and thiosulfate, also called potassium sulfuret.[29]

Modifications of these formulas called for the use of sodium, potassium, barium, and ammonium sulfides, which were claimed to produce different shades, but almost all sulfide colors are now produced from solutions of polysulfides, which are sold in concentrated form under a number of trade names.

All sulfide films require wet or dry scratch brushing for good appearance and will look better longer if protected by oiling, waxing, or more permanently, by a good top coat of clear lacquer. The desirable contrast in color can be produced by scratch brushing with a pumice paste or by use of a "greaseless" polishing compound on a buffing wheel. In any case, the sulfide solution employed should be fairly dilute, because concentrated solutions can result in a brittle film that may be nonadherent.

Cleaning. The metal surface should be degreased with trichlorethylene or similar solvents. This not only cleans the surface but enhances the cutting quality of abrasives if subsequent mechanical finishing is to be done before applying the color. Clean to a bright satin finish using a mixture of 5% oxalic acid and water together with fine India pumice powder. The cleaning should be done using a fairly stiff short-bristled cleaning brush in the direction of the grain. The metal should be recleaned using the above mixture and a wet, virgin clean white cloth and applied in conformance with the original motion. The work should be cleaned with a virgin cloth, rinsed in clean clear water, and allowed to dry.

Finish the metal with abrasive belts, abrasive pads or wheels, or greaseless abrasive compounds on portable buffing wheels. As the final operation, give the metal a hand rub with a fine abrasive pad and a slurry of pumice and water to ensure complete removal of all surface films of oil and grease. Then remove all traces of pumice by wiping with a clean damp cloth or sponge.[29]

Statuary finishes on bronze. Statuary finishes can be produced in light, medium, and dark brown depending on both the concentration and the number of applications of the coloring solutions. Solutions of 2 to 10% aqueous ammonium sulfide, potassium sulfide, or sodium sulfide are swabbed or brushed on. Oxide pretreatment may be employed to enhance adherence. Final hand toning or blending may be required to achieve acceptable color match and color uniformity.[29]

Statuary finishes on copper. Clean the copper with pumice and water or pumice and solvent to remove all dirt, grease, oil, and tarnish. Brush the entire surface with a 2% solution of liquid ammonium sulfide in water. Once dried, the color can be evened out by rubbing lightly with pumice and water, using a stub or fine brass wire brush.

Maintenance. When a regular maintenance program is followed, most installations can be maintained by oiling or waxing, some by lacquering, and a few by polishing. Oil and wax coatings look best when applied with a well-impregnated, clean soft cloth followed by rubbing with a second, clean soft cloth to remove excess oil or wax. Frequency of oiling or waxing is as important as the oil or wax used. Newly installed metal should be oiled weekly for the first month to build up a protective film. Metals subject to heavy traffic should be oiled or waxed at 1- to 2-week intervals. Where traffic is moderate to light, monthly treatment may suffice. In nontraffic areas, quarterly or semiannual applications are feasible.

Lacquering. Long-term protection can be achieved by applying a clear organic coating. Air-drying formulations are the most convenient to use, and among them the INCRALAC formulation has proven to be the most protective.[29] When sprayed onto a properly cleaned metal surface, this lacquer provides excellent protection indoors or outdoors, even in highly corrosive industrial and marine atmospheres. The use of abrasive pads followed by washing with a cleaning solvent provides a surface for maximum performance. Steel wool should not be used because it sometimes contains a corrosion inhibitor that may cause discoloration later on.

8.5 High-Performance Alloys

A distinction is often made between those alloys that are primarily used for high-temperature strength, commonly referred to as superalloys, and those that are primarily used for corrosion resistance. High-performance alloys are defined, in the present context, as Ni-, Ni-Fe-, and Co-base alloys able to operate at high temperatures, $> 550°C$, and pressures. Fe, Co, and Ni are transition metals with consecutive positions in the periodic table. The relative order of abundance decreases as Fe, Ni, Co.

The Fe-Ni-base high-performance alloys are an extension of stainless steel technology and generally are wrought, whereas Co- and Ni-base high-performance alloys may be wrought or cast depending on the application or composition involved. Appropriate compositions of all high-performance alloy base metals can be forged, rolled to sheet, or otherwise formed into a variety of shapes. The more highly alloyed compositions normally are processed as castings. Typical high-performance alloys have moduli of elasticity around and above 200 GPa.[30] The principal microstructural variables of high performance alloys are:[30]

- The precipitate amount and its morphology
- Grain size and shape
- Carbide distribution

Ni- and Fe-Ni-base high-performance alloy properties are controlled by all three variables. However, the first variable is essentially absent in Co-base high-performance alloys. Structure control is achieved through composition and by processing. For a given nominal composition, there are property advantages and disadvantages for the structures produced by deformation processing or by casting. Cast high-performance alloys generally have coarser grain sizes, more alloy segregation, and improved creep and rupture characteristics. Wrought high-performance alloys generally have more uniform, and usually finer, grain sizes and improved tensile and fatigue properties.[30]

The inherent strength capability of high-performance alloys is controlled by the intragranular distribution. However, the usable strength in polycrystalline alloys is determined by the condition of the grain boundaries, particularly as affected by the carbide phase morphology and distribution. Wrought Ni- and Fe-Ni-base high-performance alloys generally are processed to have optimum tensile and fatigue properties.[30]

For lower-temperature applications where tensile yield or ultimate strength are critical factors, fine γ' often is produced, but a duplex γ' size (uniform coarse and fine) often is preferred because it tends to disperse slip and reduce notch sensitivity. Complex heat treatments have been developed to produce appropriate γ' dispersions along with a suitable carbide distribution in wrought alloys. Although standard heat treatments generally consist of successive steps at decreasing temperatures, some heat treatments incorporate one or more pairs of aging temperatures in which the lower-temperature age precedes the higher-temperature age.[30]

Grain size also affects high-performance alloy strength. A uniform grain size often is preferred but is difficult to achieve in conventional forging operations or in large structural castings. Grain sizes resulting from isothermal forging, particularly of powder billets, are the most uniform. Grains in small cast parts of Ni- and Co-base alloys can be made fairly uniform. Neither coarse nor extremely fine grain sizes are desired, because optimum creep rupture and fatigue properties are not achieved at the extremes of grain size.

However, the nature and extent of grain boundary hardening is not well identified in Co-base high-performance alloys, and the contribution of matrix carbide precipitation to alloy strengthening is not well defined for any high-performance alloy system. Borides and carbides may play a similar role. Carbides at grain boundaries in Co-base high-performance alloys act to inhibit grain boundary sliding and migration. In the highest C-content cast Co-base high-performance alloys, the skeletal carbide network may actually support a portion of the load much as strengthening is achieved in a composite.

Changes in room-temperature strength and ductility of cast Co-base high-performance alloys have been correlated with precipitation of fine $M_{23}C_6$-type carbides. The response of a Co-base high-performance alloy to aging can be quite complex because a number of carbide reactions are possible. Furthermore, the effect of aging may depend on whether the material is in the as-cast or solution-treated condition.[30]

8.5.1 Ni- and Fe-Ni-base alloys

High-performance alloys consist of an austenitic face-centered cubic (fcc) matrix phase gamma (γ) plus a variety of secondary phases. The principal secondary phases are the carbides MC, $M_{23}C_6$, M_6C, and even M_7C_3 in all high-performance alloy types and gamma prime (γ') fcc ordered Ni_3(Al, Ti) intermetallic compound in Ni- and Fe-Ni-base high-performance alloys.[30] High-performance alloys derive their strength from solid solution hardeners and precipitating phases. Carbides may provide limited strengthening directly through dispersion hardening or, more commonly, indirectly by stabilizing grain boundaries against excessive shear. In addition to those elements that produce solid solution hardening and promote carbide and γ' formation, other elements such as B, Zr, Hf, or Ce are added to enhance mechanical or chemical properties.

Some carbide and γ' forming elements may contribute significantly to chemical properties as well. Table 8.22 gives a generalized list of the ranges of alloying elements and their effects in high-performance alloys.[30] Ni-base high-performance alloys are utilized in both cast and wrought forms. Nickel has good resistance to corrosion in the normal atmospheres, natural freshwaters, and deaerated nonoxidizing acids, and it has excellent resistance to corrosion by caustic alkalis. Therefore, nickel offers very useful corrosion resistance itself, and it is an excellent base on which to develop specialized alloys.[31]

Effects of alloying elements

Copper. Additions of copper provide improvement in the resistance of nickel to nonoxidizing acids. In particular, alloys containing 30 to 40% copper offer useful resistance to deaerated sulfuric acid and excellent resistance to all concentrations of deaerated hydrofluoric acid. Additions of 2 to 39% copper to nickel-chromium-molybdenum-iron alloys are also known to improve resistance to hydrochloric, sulfuric, and phosphoric acids.[31]

Chromium. Chromium additions improve the resistance to oxidizing media such as nitric and chromic acids and other highly corrosive environments such as hot H_3PO_4, high-temperature oxidizing gases, and hot sulfur-bearing gases. Alloying additions are usually in the range of 15 to 30% with exceptions containing up to 50% chromium.

TABLE 8.22 Common Ranges of Main Alloying Additions and Their Effects in High-Performance Alloys

Element	Ni-base	Co-base	Effect
Cr	5–25	19–30	Oxidation and hot corrosion resistance
			Formation of carbides
			Solution hardening
Mo, W	0–12	0–11	Formation of carbides
			Solution hardening
Al	0–6	0–4.5	Precipitation hardening
			Oxidation resistance
Ti	0–6	0–4	Precipitation hardening
			Formation of carbides
Co	0–20		Affects amount of precipitate
Ni		0–22	Stabilizes austenite
			Formation of hardening precipitates
Cb	0–5	0–4	Formation of carbides
			Solution hardening
			Precipitation hardening in Ni-, Fe-Ni-base alloys
Ta	0–12	0–9	Formation of carbides
			Solution hardening
			Oxidation resistance

Iron. Iron is typically used in nickel-base alloys to reduce cost, not to promote corrosion resistance. However, iron does provide nickel with improved resistance to H_2SO_4 in concentrations above 50%. Iron also increases the solubility of carbon in nickel, thereby improving the resistance to high-temperature carburizing environments.[31]

Molybdenum. Molybdenum substantially improves resistance to nonoxidizing acids. Commercial alloys containing up to 28% molybdenum have been developed for service in nonoxidizing solutions of HCl, H_3PO_4, and HF, as well as in H_2SO_4 in concentrations below 60%. Molybdenum also markedly improves the localized corrosion resistance of these alloys and imparts strength for high-temperature service.

Tungsten. Tungsten behaves similarly to molybdenum. However, because of its high atomic weight and cost, additions of molybdenum are generally preferred. Additions of tungsten of 3 to 4% in combination with 13 to 16% molybdenum in a nickel-chromium base result in alloys with outstanding resistance to localized corrosion.[31]

Silicon. Silicon is typically present only in minor amounts in most nickel-base alloys. In alloys containing significant amounts of iron, cobalt, molybdenum, tungsten, or other refractory elements, the level of silicon must be carefully controlled because it can stabilize carbides and harmful intermetallic phases. However, the use of silicon as a major alloying element has been found to greatly improve the resistance of nickel to hot, concentrated H_2SO_4. Alloys containing 9 to 11% silicon are produced for such service in the form of castings.[31]

Cobalt. Cobalt is not generally used as a primary alloying element in materials designed for aqueous corrosion resistance. However, cobalt imparts unique strengthening characteristics to alloys designed for high-temperature service. Cobalt, like iron, increases the solubility of carbon in nickel-base alloys, therefore increasing the resistance to carburization.

Niobium and tantalum. Both niobium and tantalum were originally added as stabilizing elements to tie up carbon and prevent intergranular corrosion attack due to grain-boundary carbides. With the advent of argon-oxygen decarburization melting technology such additions are no longer necessary. In high-temperature alloys both elements are used to promote high-temperature strength through solid solution and precipitation hardening mechanisms. Additions of these elements are also considered to be beneficial in reducing the tendency of nickel-base alloys toward hot cracking during welding.[31]

Aluminum and titanium. Aluminum and titanium are often used in minor amounts in corrosion-resistant alloys to deoxidize or tie up carbon and nitrogen. When added together these elements enable the formulation of age-hardenable high-strength alloys for low- or high-temperature service. Additions of aluminum can also be used to promote the formation of a tightly adherent scale at high temperature that resists attack by oxidation, carburization, and chlorination.[31]

Carbon and carbides. There is evidence that nickel forms carbides at high temperatures that are unstable and decompose into nickel and graphite at lower temperatures. Because this phase mixture tends to have low ductility, low-carbon forms of nickel are usually preferred in corrosion-resistant applications. This problem is also alleviated to some extent when nickel is alloyed with copper. In other nickel alloys, the carbides that form depend on the specific alloying elements present and the level of carbon present. In corrosion-resistant alloys, many types of carbides are considered harmful because they can precipitate at grain boundaries during heat treatment or weld fabrication and subsequently promote intergranular corrosion or cracking in service by depletion of matrix elements essential to corrosion resistance. In high-temperature alloys, the presence of carbides is generally desired to control grain size and to enhance elevated-temperature strength and ductility.[31]

There are basically two types of carbides in these alloys. Primary carbides are interdendritic and form during the solidification process. These carbides are typically metastable and would dissolve if given sufficient time at elevated temperatures. However, during metal manufacture, they can persist in the final product as stringers in the direction of predominant metal flow. Some level of carbide stringers usually must be tolerated because they cannot be economically avoided. However, large amounts of such stringers can adversely affect formability, weld fabrication, and service performance characteristics.

Secondary carbides precipitate as the result of thermal exposures during fabrication operations or during component service life. These carbides precipitate preferentially at grain boundaries and internal structural defects such as twin boundaries and dislocations. The quantity of secondary carbides that precipitate depends on the amount of carbon in solutions, the exposure temperature, and the time at such temperature. Therefore, conditions that generate a supersaturated solution of carbon followed by slow cooling or thermal arrests below carbide solvus temperatures will produce heavy secondary carbide precipitation, which generally reduces ductility and toughness, and this adversely affects fabrication and service performance.

Intermetallics phases. The occurrence of intermetallic phases in nickel-base alloys has positive and negative effects. On the positive side, the nickel-base system has been the most widely and successfully exploited of any alloy base in the development of high-strength high-temperature alloys because of the occurrence of unique intermetallic phases. On the negative side, the precipitation of certain intermetallic phases, such as the carbides discussed previously, can seriously degrade ductility and corrosion resistance. In the case of corrosion-resistant alloys, especially the solid solution type, intermetallic precipitation is rather unusual because service temperatures are typically well below those at which precipitation kinetics become important. In such cases, it is only necessary to restrict alloy composition sufficiently to ensure successful manufacturing, fabrication, and use capabilities. For high-temperature alloys, the precipitation of undesired intermetallics can be a major concern, especially for applications requiring a long service life or ease of repair.[31]

Most high-strength nickel-base alloys depend on the precipitation of an AlB-type compound known as gamma prime (γ'). The strength of γ' alloys can be increased by increasing the Al + Ti content to obtain a higher γ' volume fraction. However, alloys with high Al + Ti levels are difficult to manufacture in wrought forms and to fabricate, and they are best exploited as castings. Additions of refractory metals can also be used to increase strength by altering lattice mismatch and antiphase boundary energy. Additions of Co are also effective by increasing the γ' solvus temperature.[31]

Another important intermetallic phase that can be used to strengthen nickel-base alloys is a metastable form of Ni_3Nb known as gamma double prime (γ''). For the most part, it has been exploited in alloys containing significant amounts of iron. Gamma double prime has a body-centered tetragonal crystal structure. At temperatures of 705°C and above, it overages rapidly and transforms into the orthorhombic form of Ni_3Nb. Because of the sluggish nature of the precipitation reaction, alloys strengthened by γ'' can possess excellent weldability.[31]

8.5.2 Co-base alloys

The Co-base high-performance alloys are invariably strengthened by a combination of carbides and solid solution hardeners. Because of their high strength over a wide temperature range and their resistance to many environments, Co-Cr alloys are used to resist wear, particularly in hostile environments and as structural materials at high temperatures. Typically, the alloys used to resist wear contain higher carbon levels (0.25 to 2.5%) for carbide formation, and they are normally cast or applied to critical surfaces by welding in a process known as hardfacing. Alloys used for structural purposes at high temperatures are normally low in C, contain appreciable quantities of Ni, and are available as wrought products. The carbides in the wear-resistant alloys enhance abrasion resistance but reduce ductility.[32]

Chromium has a dual function in cobalt alloys. It is both the predominant carbide former and the most important alloying element in the matrix. The most common carbide in these alloys is a chromium-rich M_7C_3 type, although chromium-rich $M_{23}C_6$ carbides are abundant in low-carbon alloys. Tungsten and molybdenum serve to provide additional strength to the matrix. When present in large quantities, they participate to the formation of carbides during solidification and promote the precipitation of M_6C. The size and shape of the carbide particles within cobalt alloys are strongly influenced by cooling rate and subtle chemistry changes. Such changes markedly affect abrasion resistance, because there is a distinct relationship among the size of abrading species, the size of the structural hard particles, and the abrasive wear rate.[32]

The success of the structural cobalt alloys can be attributed to their inherent strength over a wide temperature range and their resistance to severe environments. The structural alloys generally contain significant quantities of nickel. This serves to stabilize the fcc structure with a view toward improved ductility during service. With sufficient nickel, the structural cobalt alloys tend to exhibit twinning during deformation. Although the structural cobalt alloys are low in carbon as compared to most of the wear-resistant alloys, they nevertheless depend on carbide precipitation for additional strength. The most abundant carbide in the structural cobalt alloys is chromium-rich $M_{23}C_6$, although M_6C and MC carbides are common, depending on the type and level of other alloying additions.[32]

8.5.3 Welding and heat treatments

In terms of their weldability, high-performance alloys can be classified according to the means by which the alloying elements develop the mechanical properties, namely, solid solution alloys and precipitation hardened alloys. A distinguishing feature of precipitation hardened alloys is that mechanical properties are developed by heat treatment to produce a fine distribution of hard particles in a nickel-rich matrix.

Solid solution alloys are readily fusion welded, normally in the annealed condition. Some noteworthy examples of solid solution alloys are Ni 200, the Monel 400 series, the Inconel 600 series, the Incoloy 800 series, Hastelloys and some Nimonic alloys such as 75, and PE13. Because the HAZ does not harden, heat treatment is not usually required after welding. Precipitation hardened alloys may be susceptible to postweld heat-treatment (PWHT) cracking. Some of these alloys are the Monel 500 series, Inconel 700 series, Incoloy 900 series, and most of the Nimonic alloys.

Weldability. Co-base high-performance alloys are readily welded by gas metal arc (GMA) or gas tungsten arc (GTA) techniques. Some cast alloys and wrought alloys, such as Alloy 188, have been extensively welded. Filler metals generally have been less highly alloyed Co-base alloy wire, although parent rod or wire have been used. Co-base high-performance alloy sheet also is successfully welded by resistance techniques. Appropriate preheat techniques are needed in GMA and GTA welding to eliminate tendencies for hot cracking. Electron beam (EB) and plasma arc (PA) welding can be used on Co-base high-performance alloys but usually are not required in most applications because this alloy class is so readily weldable.[30]

Ni- and Fe-Ni-base high-performance alloys are considerably less weldable than the Co-base high-performance alloys. Because of the presence of the strengthening phase, the alloys tend to be susceptible to hot and PWHT cracking. Hot cracking occurs in the weld heat-affected zone, and the extent of cracking varies with alloy composition and weldment restraint. Ni- and Fe-Ni-base high-performance alloys have been welded by GMA, GTA, EB, laser, and PA techniques. Filler metals, when used, usually are weaker, more ductile austenitic alloys so as to minimize hot cracking. Because of their γ' strengthening mechanism and capability, many Ni- and Fe-Ni-base high-performance alloys are welded in the solution heat-treated condition. Special preweld heat treatments have been used for some alloys. Some alloys (e.g., A-286) are inherently difficult to weld despite only moderate levels of γ' hardeners.[30]

Weld techniques for high-performance alloys must address not only hot cracking but PWHT cracking, particularly as it concerns microfissuring (microcracking), because it can be subsurface and therefore difficult to detect. Tensile and stress rupture strengths may be hardly affected by microfissuring, but fatigue strengths can be drastically reduced. In addition to the usual fusion welding techniques above, Ni- and Fe-Ni-base alloys can be resistance welded when in sheet form. Brazing, diffusion bonding, and transient liquid phase bonding also have been employed to join these alloys. Braze joints tend to be more ductility limited than welds.

Most nickel alloys can be fusion welded using gas-shielded processes such as TIG or MIG. Of the flux processes, MMA is frequently used, but the submerged arc welding (SAW) process is restricted to solid solution alloys (Nickel 200, Inconel alloy 600 series, and Monel alloy 400 series) and is less widely used. Solid solution alloys are normally welded in the annealed condition, and precipitation hardened alloys, in the solution treated condition. Preheating is not necessary unless there is a risk of porosity from moisture condensation. It is recommended that material containing residual stresses be solution treated before welding to relieve the stresses.[33]

Postweld heat treatment is not usually needed to restore corrosion resistance, but thermal treatment may be required for precipitation hardening or stress-relieving purposes to avoid stress corrosion cracking. Filler composition normally matches the parent metal. However, most fillers contain a small mount of titanium, aluminum, and/or niobium to help minimize the risk of porosity and cracking. Nickel and its alloys are readily welded, but it is essential to clean the surface immediately before welding. The normal method of cleaning is to degrease the surface, remove all surface oxide by machining, grinding, or scratch brushing, and finally degrease. However, these alloys can suffer from the following weld imperfections and postweld damage:[33]

Porosity. Porosity can be caused by oxygen and nitrogen from air entrainment and surface oxide or by hydrogen from surface contamination. Careful cleaning of component surfaces and using a filler material containing deoxidants such as aluminum and titanium will reduce this risk. When using argon in TIG and MIG welding, attention must be paid to shielding efficiency of the weld pool, including the use of a gas backing system. In TIG welding, argon-H_2 gas mixtures that provide a slightly reducing atmosphere are particularly effective.

Oxide inclusions. Because the oxide on the surface of nickel alloys has a much higher melting temperature than the base metal, it may remain solid during welding. Oxide trapped in the weld pool will form inclusions. In multirun welds, oxide or slag on the surface of the weld bead will not be consumed in the subsequent run and will cause lack of fusion imperfections. Before welding, surface oxide, particularly if it has been formed at a high temperature, must be removed by machining or abrasive grinding; it is not sufficient to wire brush the surface because this serves only to polish the oxide. During welding, surface oxide and slag must be removed between runs.[33]

Weld metal solidification cracking. Weld metal or hot cracking results from contaminants concentrating at the centerline and an unfavorable weld pool profile. Too high a welding speed produces a shallow weld pool, which encourages impurities to concentrate at the centerline and, on solidification, generates sufficiently large transverse stresses to form cracks. This risk can be reduced by carefully cleaning the joint area and avoiding high welding speeds.[33]

Microfissuring. Similar to austenitic stainless steel, nickel alloys are susceptible to formation of liquation cracks in reheated weld metal regions or parent metal HAZ. This type of cracking is controlled by factors outside the control of the welder such as grain size or content impurity. Some alloys are more sensitive than others. For example, the extensively studied Inconel 718 is now less sensitive than some cast superalloys, which cannot be welded without inducing liquation cracks.

Postweld heat-treatment cracking. This is also known as strain-age or reheat cracking. It is likely to occur during postweld aging of precipitation hardening alloys but can be minimized by preweld heat treatment. Solution annealing is commonly used but overaging gives the most resistant condition. Inconel 718 alloy was specifically developed to be resistant to this type of cracking.

Stress corrosion cracking. Welding does not normally make nickel alloys susceptible to weld metal or HAZ corrosion. However, when the material will be in contact with caustic soda, fluosilicates, or HF acid, stress corrosion cracking is possible.

Heat treatment. Solid-solution-strengthened high-temperature alloys are normally supplied in the solution-heat-treated condition unless otherwise specified. In this condition, microstructures generally consist of primary carbides dispersed in a single-phase matrix, with essentially clean grain boundaries. This is usually the optimum condition for the best elevated temperature properties in service and the best room-temperature fabricability. Typical solution heat-treatment temperatures for these alloys are between 1100 and 1200°C.[34]

Heat treatments performed at temperatures below the solution heat-treating temperature range are classified as mill annealing or stress relief treatments. Mill annealing treatments are generally employed to restore formed, partially fabricated, or otherwise as-worked alloy material properties to a point where continued manufacturing operations can be performed. Such treatments may also be used to produce structures in finished raw materials that are optimum for specific forming operations. Minimum recommended mill annealing temperatures for these vary between 900 and 1050°C.[34]

Unlike mill annealing, stress relief treatments for these alloys are not well defined. Depending upon the particular circumstances, stress relief may be achieved with a mill anneal or may require the equivalent of a full solution anneal. Low-temperature treatments, which work for carbon and stainless steels, generally will not be effective. Effective high-temperature treatments will often be a compromise between how much stress is actually relieved and concurrent changes in the structure or dimensional stability of the component.

Annealing during cold or warm forming. The response of high-temperature alloys to heat treatment is very much dependent upon the condition that the material is in when the treatment is applied. When the material is not in a cold- or warm-worked condition, the principal response to heat treatment is usually a change in the amount and morphology of the secondary carbide phases present. Other minor effects may occur, but the grain structure of the material will normally be unaltered by heat treatment when cold or warm work is absent.[34] Care should be exercised in cold forming these alloys to avoid the imposition of less than 10 percent cold work where possible. Small amounts of cold work can lead to exaggerated or abnormal grain growth during annealing. In the everyday fabrication of complex components, it may be impossible to avoid situations where such low levels of cold work or strain are introduced.

Annealing during hot forming. Components manufactured by hot-forming techniques should generally be solution heat treated rather than mill annealed if in-process heat treatment is required. In cases where forming is required to be performed at furnace temperatures below the solution treatment range, intermediate mill annealing may be employed subject to the limits of the forming equipment. Hot-formed components, particularly when formed at high temperatures, will generally undergo recovery, recrystallization, and perhaps even grain growth during the forming operation itself. Similarly, if the hot-forming session involves a small amount of deformation, the piece to be heat treated may exhibit a nonuniform structure, which will respond nonuniformly to the heat treatment.[34]

Final annealing. Solution heat treating is the most common form of finishing operation applied to high-temperature alloys and is often mandated by the applicable specifications for these materials. Where more than about 10 percent cold work is present in the piece, a final anneal is usually mandatory. Putting as-cold-worked material into service can result in recrystallization to a very fine grain size, which in turn can produce a significant reduction in stress rupture strength. A good example of this is vacuum brazing. Often performed as the final step in the fabrication of some components, such a process precludes the possibility of a subsequent solution treatment because of the low melting point of the brazing compound. Consequently, the actual brazing temperatures used are sometimes adjusted to allow for the simultaneous solution heat treating of the component. Because both heating and cooling rates in vacuum furnaces are relatively slow, even with the benefit of advanced gas cooling equipment, it must be recognized that alloy structure and properties produced may be less than optimum.[34]

Stress relieving. A stress relief anneal should be considered only if the treatment does not produce recrystallization in the material. Relief of residual stress in these alloys, arising from thermal strains produced by nonuniform cooling or slight deformations imparted during sizing operations, is often difficult to achieve. In many cases, stress relieving at mill annealing temperatures about 55 to 110°C above the intended use temperature will provide good results. In other cases, a full solution anneal at the low end of the allowable range may be best, although this can make the material subject to abnormal grain growth.[34]

Heating rate and cooling rate. Heating and cooling rates used in the heat treatments of these alloys should be as rapid as possible. Rapid heating to temperature is usually desirable to help minimize carbide precipitation during the heating cycle and to preserve the stored energy from cold or warm work. Slow heating can promote a somewhat finer grain size than might be otherwise desired or required, particularly for thin-section parts given limited time at the annealing temperature. Rapid cooling through the temperature range of about 980 down to 540°C following mill annealing is required to minimize grain boundary carbide precipitation and other possible phase reactions in some alloys. Again, cooling from the solution annealing temperature down to under 540°C should be as rapid as possible considering the constraints of the equipment and the need to minimize component distortion. Water quenching is preferred where feasible.[34]

Use of protective atmosphere. Most of the high-performance alloys may be annealed in oxidizing environments but will form adherent oxide scales that normally must be removed prior to further processing. Some high-temperature alloys contain low chromium. Atmosphere annealing of these materials should be performed in neutral to slightly reducing environments. Protective atmosphere annealing is commonly performed for all of these materials when a bright finish is desired. The best choice for annealing of this type is a low dew point hydrogen environment. Annealing may also be done in argon and helium. Annealing in nitrogen or cracked ammonia is not generally preferred but may be acceptable in some cases. Vacuum annealing is generally acceptable but also may produce some tinting depending on the equipment and temperature. The gas used for forced gas cooling can also influence results. Helium is normally preferred, followed by argon and nitrogen.[34]

8.5.4 Corrosion resistance

High-performance alloys generally react with oxygen, and oxidation is the prime environmental effect on these alloys. At moderate temperatures, about 870°C and below, general uniform oxidation is not a major problem. At higher temperatures, the commercial nickel- and cobalt-base high-performance alloys are attacked by oxygen. The level of oxidation resistance at temperatures below 1200°C is a function of chromium content, Cr_2O_3 forming as a protective oxide film. Above that temperature, chromium and aluminum act in synergy for oxidation protection. The latter element leads to the formation of protective Al_2O_3 surface films. The higher the chromium level, the less aluminum may be required to form a highly protective Al_2O_3 layer.[30]

In operating temperatures lower than 875°C, accelerated oxidation may occur in high-performance alloys through the operation of selective fluxing agents. One of the better documented accelerated oxidation processes is sulfidation. This hot corrosion process is separated into two regimes: low temperature and high temperature. The principal method for combating sulfidation is the use of a high Cr content (>20%) in the base alloy. Although Co-base high-performance alloys and many Fe-Ni-base alloys have Cr levels in this range, most Ni-base high-performance alloys, especially those of the high creep rupture strength type, do not.[30] SCC can occur in Ni- and Fe-Ni-base high-performance alloys at lower temperatures. Hydrogen embrittlement at cryogenic temperatures has also been reported for these alloys.

Nickel and its alloys generally have good resistance to many of the chloride bearing and reducing media that attack stainless steels. The resistance of nickel alloys to reducing media is further enhanced by molybdenum and copper. Alloy B (N10001), with 28% Mo, is resistant to hydrochloric acid. Monel 400 (N04400), with 30% Cu, is widely used in natural waters and in heat-exchanger applications. It also has good resistance to hydrofluoric acid, although SCC is a potential problem. Although Monel 400 is used in similar applications as S31600 stainless steel, it is its opposite in many aspects of its behavior. For example, it has poor resistance to oxidizing media, whereas stainless steels thrive in these conditions. If chromium is added to nickel, alloys resistant to a wide range of oxidizing and reducing media can be obtained. One example is Inconel 600. If molybdenum is further added, the resulting alloys can possess a resistance to an even wider range of reducing and oxidizing media with very good chloride pitting resistance, for example, Hastelloy C (N10002).

These high-nickel alloys are resistant to transgranular SCC in elevated temperature chlorides, whereas the regular austenitic stainless steels are very susceptible to this type of attack. It is interesting to note that S43000 stainless is also resistant to these corrosive environments. The pitting resistance of high-nickel, chromium-containing alloys is generally better than that obtained with stainless steels. However, they can be more susceptible to intergranular corrosion because

1. The solubility of carbon in austenite decreases as nickel increases, which in turn increases the tendency to form chromium carbide.
2. The higher alloys are generally more prone to precipitate intermetallic compounds that can lower corrosion resistance by depleting the matrix in Ni, Mo, and so forth.

Chromium carbides and intermetallic compounds precipitate out at temperatures in the range of about 600 to 1000°C. Therefore, there are restrictions to the use of these alloys as welded materials. Stress -zaccelerated intergranular corrosion has also been observed with Inconel 600 in high-temperature (300°C) water applications.

The corrosion-resistant Hastelloys have become widely used by the chemical processing industries. The attributes of Hastelloys include high resistance to uniform attack, outstanding localized corrosion resistance, excellent SCC resistance, and ease of welding and fabrication. The most versatile of the Hastelloys are the C series. Hastelloy C-22 (N06022) is particularly resistant to pitting and crevice corrosion. This alloy has been used extensively to protect against the most corrosive flue gas desulfurization (FGD) systems and the most sophisticated pharmaceutical reaction vessels.

Ni-base alloys. Nickel and its alloys, like the stainless steels, offer a wide range of corrosion resistance. However, nickel can accommodate larger amounts of alloying elements, chiefly chromium, molybdenum, and tungsten, in solid solution than iron. Therefore, nickel-base alloys, in general, can be used in more severe environments than the stainless steels. In fact, because nickel is used to stabilize the austenite fcc phase of some of the highly alloyed stainless steels, the boundary between these and nickel-base alloys is rather diffuse. The nickel-base alloys range in composition from commercially pure nickel to complex alloys containing many alloying elements.[31]

The types of corrosion of greatest importance in the nickel-base alloy system are uniform corrosion pitting and crevice corrosion, intergranular corrosion, and galvanic corrosion. SCC, corrosion fatigue, and hydrogen embrittlement are also of great importance. To estimate the performance of a set of alloys in any environment, it is of paramount importance to ascertain the composition and, for liquid environments, the electrochemical interaction of the environment with an alloy. A case in point is the nickel-molybdenum Hastelloy B-2 (N10665). This alloy performs exceptionally well in pure deaerated H_2SO_4 and HCl but deteriorates rapidly when oxidizing impurities, such as oxygen and ferric ions, are present.

Ni-base alloys in acid media. Sulfuric acid is the most ubiquitous environment in the chemical industry. The electrochemical nature of the acid varies wildly, depending on the concentration of the acid and the impurity content. Pure acid is considered to be a nonoxidizing acid up to a concentration of about 50 to 60%, beyond which it is generally considered to be oxidizing. The corrosion rates of nickel-base alloys, in general, increase with acid concentration up to 90%. Higher concentrations of the acid are generally less corrosive.[31] The presence of oxidizing impurities can be beneficial to nickel-chromium-molybdenum alloys because these impurities can aid in the formation of passive films that retard corrosion. Another important consideration is the presence of chlorides (Cl^-). Chlorides generally accelerate the corrosion attack, but the degree of acceleration differs for various alloys.

Commercially pure nickel (N02200 and N02201) and Monels have room-temperature corrosion rates below 0.25 mm·y^{-1} in air-free HCl at concentrations up to 10%. In HCl concentrations of less than 0.5%, these alloys have been used at temperatures up to about 200°C. Oxidizing agents, such as cupric, ferric, and chromate ions or aeration, raise the corrosion rate considerably. Under these conditions nickel-chromium-molybdenum alloys such as Inconel 625 (N06625) or Hastelloy C-276 (N10276) offer better corrosion resistance. They can be made passive by the presence of oxidizing agents.

The nickel-chromium-molybdenum alloys also show higher resistance to uncontaminated HCl. For example, alloys C-276, 625, and C-22 show very good resistance to dilute HCl at elevated temperatures and to a wide range of HCl concentrations at ambient temperature. The corrosion resistance of these alloys depends on the molybdenum content. The alloy with the highest molybdenum content (i.e., Hastelloy B-2) shows the highest resistance in HCl of all the nickel-base alloys. Accordingly, this alloy is used in a variety of processes involving hot HCl or nonoxidizing chloride salts hydrolyzing to produce HCl.[31]

Chromium is an essential alloying element for corrosion resistance in HNO_3 environments because it readily forms a passive film in these environments. Thus, the higher chromium alloys show better resistance in HNO_3. In these types of environments, the highest chromium alloys, such as Hastelloy G-30 (N06030), seem to show the highest corrosion resistance. Molybdenum is generally detrimental to corrosion resistance in HNO_3.

Pitting corrosion in chloride environments. The nickel-chromium-molybdenum alloys, such as Hastelloys C-22 and C-276 as well as Inconel 625, exhibit very high resistance to pitting in oxidizing chloride environments. The critical pitting temperatures of various nickel-chromium-molybdenum alloys in an oxidizing chloride solution are shown in Table 8.23. Pitting corrosion is most prevalent in chloride-containing environments, although other halides and sometimes sulfides have been reported to cause pitting. There are several techniques that can be used to evaluate resistance to pitting. Critical pitting potential and pitting protection potential indicate the electrochemical potentials at which pitting can be initiated and at which a propagating pit can be stopped, respectively. These values are functions of the solution concentration, pH, and temperature for a given alloy; the higher the potentials, the better the alloy. The critical pitting temperature (i.e., the potential below which pitting does not initiate), is often used as an indicator of resistance to pitting, especially in the case of highly corrosion-resistant alloys (Table 8.23). Chromium and molybdenum additions have been shown to be extremely beneficial to pitting resistance.[31]

TABLE 8.23 Critical Pitting Temperatures for Nickel Alloys in 6% $FeCl_3$ during 24 h

Alloy	UNS	Critical pitting temperature, °C	
825	N08825	0.0	0.0
904L	N08904	2.5	5.0
317LM	S31725	2.5	2.5
G	N06007	25.0	25.0
G-3	N06985	25.0	25.0
C-4	N06455	37.5	37.5
625	N06625	35.0	40.0
C-276	N10276	60/0	65/0
C-22	N06022	60.0	65.0

TABLE 8.24 Brief Description, Corrosion Resistance, and Applications of High-Performance Alloys and Some Highly Alloyed Stainless Steels

Alloy 20Cb-3 (N08029)

Description and corrosion resistance. The high nickel content combined with chromium, molybdenum, and copper gives the alloy good resistance to pitting and chloride-ion stress-corrosion cracking. The copper content combined with other elements gives the alloy excellent resistance to sulfuric acid corrosion under a wide variety of conditions. The addition of columbium stabilizes the heat-affected zone carbides, so the alloy can be used in the as-welded condition. Alloy 20 has good mechanical properties and exhibits relatively good fabricability.

Applications. Alloy 20 is a highly alloyed iron-base nickel-chromium-molybdenum stainless steel developed primarily for use in the sulfuric acid-related processes. Other typical corrosion-resistant applications for the alloy include chemical, pharmaceutical, food, plastics, synthetic fibers, pickling, and FGD systems.

Alloy 25 (R30605)

Description and corrosion resistance. This is a cobalt-nickel-chromium-tungsten alloy with excellent high-temperature strength and good oxidation resistance up to about 980°C. Alloy 25 also has good resistance to sulfur-bearing environments. It also has good wear resistance and is used in the cold-worked condition for some bearing and valve applications.

Applications. It is principally used in aerospace structural parts, for internals in older, established gas turbine engines, and for a variety of industrial applications.

Alloy 188 (R30188)

Description and corrosion resistance. Alloy 188 is a cobalt-nickel-chromium-tungsten alloy developed as an upgrade to Alloy 25. It combines excellent high-temperature strength with very good oxidation resistance up to about 1095°C. Its thermal stability is better than that for Alloy 25, and it is easier to fabricate. Alloy 188 also has low-cycle fatigue resistance superior to that for most solid-solution-strengthened alloys and has very good resistance to hot corrosion.

Applications. It is widely used in both military and civil gas turbine engines and in a variety of industrial applications.

Alloy 230 (N06230)

Description and corrosion resistance. This is a nickel-chromium-tungsten-molybdenum alloy that combines excellent high-temperature strength, outstanding oxidation resistance up to 1150°C, premier nitriding resistance, and excellent long-term thermal stability. Alloy 230 also has lower expansion characteristics than most high-temperature alloys, very good low-cycle fatigue resistance, and a pronounced resistance to grain coarsening with prolonged exposure at elevated temperatures. Components of Alloy 230 are readily fabricated by conventional techniques, and the alloy can be cast.

Applications. Principal applications for Alloy 230 include
  Wrought and cast gas turbine stationary components
  Aerospace structurals
  Chemical process and power plant internals
  Heat treating facility components and fixtures
  Steam process internals

TABLE 8.24 Brief Description, Corrosion Resistance, and Applications of High-Performance Alloys and Some Highly Alloyed Stainless Steels *(Continued)*

Cobalt Alloy 6B (R30016)

Description and corrosion resistance. Cobalt 6B is a cobalt-based chromium-tungsten alloy for wear environments where seizing, galling, and abrasion are present. 6B is resistant to seizing and galling and with its low coefficient of friction allows sliding contact with other metals without damage by metal pickup in many cases. Seizing and galling can be minimized in applications without lubrication or where lubrication is impractical.

Alloy 6B has outstanding resistance to most types of wear. Its wear resistance is inherent and not the result of cold working, heat treating, or any other method. This inherent property reduces the amount of heat treating and postmachining. 6B has outstanding resistance to cavitation erosion. Steam turbine erosion shields from 6B have protected the blades of turbines for years of continuous service. 6B has good impact and thermal shock resistance, resists heat and oxidation, retains high hardness even at red heat (when cooled, recovers full original hardness), and has resistance to a variety of corrosive media. 6B is useful where both wear and corrosion resistance are needed.

Applications. Applications for Alloy 6B include half sleeves and half bushings in screw conveyors, tile-making machines, rock-crushing rollers, and cement and steel mill equipment. Alloy 6B is well suited for valve parts, pump plungers. Other applications include
Steam turbine erosion shields
Chain saw guide bars
High-temperature bearings
Furnace fan blades
Valve stems
Food processing equipment
Needle valves
Centrifuge liners
Hot extrusion dies
Forming dies
Nozzles
Extruder screws

Cobalt Alloy 6BH (R30016)

Description and corrosion resistance. Cobalt 6BH has the same composition as Cobalt 6B, except the material is hot rolled and then age hardened. The direct age hardening after hot rolling provides the maximum hardness and wear resistance. The advantages this creates are increased wear life, retained edge characteristics, and increased hardness. These properties are in addition to the galling and seizing resistance of the regular Cobalt 6B. Cobalt 6BH is known in the industry as a metal that retains its cutting edge. The economic advantages are in its long wear time, less downtime, and fewer replacements.

Applications. Cobalt 6BH is used for steam turbine erosion shields, chain saw guide bars, high-temperature bearings, furnace fan blades, valve stems, food processing equipment, needle valves, centrifuge liners, hot extrusion dies, forming dies, nozzles, extruder screws, and many other miscellaneous wear surfaces. Applications also include tile-making machines, rock-crushing rollers, and cement and steel mill equipment. Alloy 6BH is well suited for valve parts and pump plungers.

TABLE 8.24 Brief Description, Corrosion Resistance, and Applications of High-Performance Alloys and Some Highly Alloyed Stainless Steels (*Continued*)

Ferralium 255 (S32550)

Description and corrosion resistance. This alloy's high critical pitting crevice temperatures provide more resistance to pitting and crevice corrosion than lesser-alloyed materials. The very high yield strength of this alloy combined with good ductility allows lower wall thickness in process equipment.

Applications. Alloy 255 is finding many cost-effective applications in the chemical, marine, metallurgical, municipal sanitation, plastics, oil and gas, petrochemical, pollution control, wet phosphoric acid, paper-making, and metal-working industries. It is called *super* because it is more alloyed than ordinary stainless steels and has superior corrosion resistance. Alloy 255 is being used in areas where conventional stainless steels are inadequate or, at best, marginal. One good example is in the paper industry, which was hit with an epidemic of corrosion problems when environmental laws forced recycling of process liquids. In closed systems, chemicals such as chlorides can build up to highly corrosive concentrations over time. Paper makers have found that ordinary stainless equipment, which had previously given good service, was no longer adequate for many applications.

Alloy 255 is a cost-effective alternative to materials such as the nickel alloys, 20-type alloys, brass, and bronze. Marine environments have long been the domain of admiralty bronze. Alloy 255 is replacing admiralty bronze and the nickel alloys in offshore platforms, deck hardware, rudders, and shafting. Alloy 255 is also making inroads in "borderline" corrosion applications where the nickel alloys and high-performance alloys have been used but may not have been absolutely necessary. In some instances, it has even been used to replace high-performance Ni-Cr-Mo-F-Cu alloys in the phosphoric acid industry.

Hastelloy C-276 (N10276)

Description and corrosion resistance. This is a nickel-chromium-molybdenum wrought alloy that is considered the most versatile corrosion-resistant alloy available. It is resistant to the formation of grain boundary precipitates in the weld heat-affected zone, thus making it suitable for most chemical process applications in an as-welded condition. Alloy C-276 also has excellent resistance to pitting, stress-corrosion cracking, and oxidizing atmospheres up to 1050°C. It has exceptional resistance to a wide variety of chemical environments and outstanding resistance to a wide variety of chemical process environments including ferric and cupric chlorides, hot contaminated mineral acids, solvents, chlorine and chlorine contamination (both organic and inorganic), dry chlorine, formic and acetic acids, acetic anhydride, seawater and brine solutions, and hypochlorite and chlorine dioxide solutions. It is one of the few alloys resistant to wet chloride gas, hypochlorite, and chlorine dioxide solutions and has exceptional resistance to strong solutions of oxidizing salts, such as ferric and cupric chlorides.

Applications. Some typical applications include equipment components in chemical and petrochemical organic chloride processes and processes utilizing halide or acid catalysts. Other industry applications are pulp and paper digesters and bleach areas, scrubbers and ducting for flue gas desulfurization, pharmaceutical and food processing equipment.

Hastelloy (N10665)

Description and corrosion resistance. Alloy B-2 is a nickel-molybdenum alloy with significant resistance to reducing environments, such as hydrogen chloride gas and sulfuric, acetic, and phosphoric acids. Alloy B-2 provides resistance to pure sulfuric acid

TABLE 8.24 Brief Description, Corrosion Resistance, and Applications of High-Performance Alloys and Some Highly Alloyed Stainless Steels (*Continued*)

and a number of nonoxidizing acids. The alloy should not be used in oxidizing media or where oxidizing contaminants are available in reducing media. Premature failure may occur if B-2 is used where iron or copper is present in a system containing hydrochloric acid. Industry users like the resistance to a wide range of organic acids and the resistance to chloride-induced stress-corrosion cracking.

Alloy B-2 resists the formation of grain boundary carbide precipitates in the weld heat-affected zone, making it suitable for most chemical process applications in the as-welded condition. The heat-affected weld zones have reduced precipitation of carbides and other phases to ensure uniform corrosion resistance. Alloy B-2 also has excellent resistance to pitting and stress corrosion cracking.

Applications. Alloy B-2 has superior resistance to hydrochloric acid, aluminum chloride catalysts, and other strongly reducing chemicals and has excellent high-temperature strength in inert and vacuum atmospheres. Applications in the chemical process industry involve sulfuric, phosphoric, hydrochloric, and acetic acid. Temperature uses vary from ambient temperature to 820°C depending on the environments.

Hastelloy C-22 (N06022)

Description and corrosion resistance. Hastelloy C-22 is a nickel-chromium-molybdenum alloy with enhanced resistance to pitting, crevice corrosion, and stress corrosion cracking. It resists the formation of grain boundary precipitates in the weld heat-affected zone, making it suitable for use in the as-welded condition. C-22 has outstanding resistance to both reducing and oxidizing media and because of its resistibility can be used where "upset" conditions are likely to occur. It possesses excellent weldability and high corrosion resistance as consumable filler wires and electrodes. The alloy has proven results as a filler wire in many applications when other corrosion resistant wires have failed.

It has better overall corrosion resistance in oxidizing corrosives than C-4, C-276, and 625 alloys, outstanding resistance to localized corrosion, and excellent resistance to stress corrosion cracking. It is the best alloy to use as universal weld filler metal to resist corrosion of weldments.

Applications. C-22 can easily be cold worked because of its ductility, and cold forming is the preferred method of forming. More energy is required because the alloy is generally stiffer than austenitic stainless steels.

Hastelloy G-30 (N06030)

Description and corrosion resistance. Hastelloy Alloy G-30 is an improved version of the nickel-chromium-iron molybdenum-copper alloy G-3. With higher chromium, added cobalt, and tungsten the nickel Hastelloy Alloy G-30 shows superior corrosion resistance over most other nickel- and iron-based alloys in commercial phosphoric acids as well as complex environments containing highly oxidizing acids such as nitric/hydrochloric, nitric/hydrofluoric, and sulfuric acids. Hastelloy Alloy G-30 resists the formation of grain boundary precipitates in the heat-affected zone, making it suitable in the as-welded condition.

Applications. Hastelloy Alloy G-30 is basically the same as other high alloys in regard to formability. It is generally stiffer than austenitics. Because of its good ductility, cold working is relatively easy and is the preferred method of forming. The alloy is easily weldable using gas-tungsten arc, gas metal arc, and shielded metal arc. The welding characteristics are similar to those of G-3.

TABLE 8.24 Brief Description, Corrosion Resistance, and Applications of High-Performance Alloys and Some Highly Alloyed Stainless Steels (*Continued*)

Hastelloy X (N06002)

Description and corrosion resistance. This is a nickel-chromium-iron-molybdenum alloy that possesses an exceptional combination of oxidation resistance, fabricability, and high-temperature strength. Alloy X is one of the most widely used nickel-base superalloys for gas turbine engine components. This solid-solution-strengthened grade has good strength and excellent oxidation resistance beyond 2000°F. Alloy X has excellent resistance to reducing or carburizing atmospheres, making it suitable for furnace components. Due to its high molybdenum content, alloy X may be subject to catastrophic oxidation at 1200°C.

It is exceptionally resistant to SCC in petrochemical applications and to carburization and nitriding. All of the product forms are excellent in terms of forming and welding. Although this alloy is primarily noted for heat and oxidation resistance, it also has good resistance to chloride stress corrosion cracking.

Applications. The alloy finds use in petrochemical process equipment and gas turbines in the hot combustor zone sections. It is also used for structural components in industrial furnace applications because of its excellent oxidation resistance. It is recommended especially for use in furnace applications because it has unusual resistance to oxidizing, reducing, and neutral atmospheres. Furnace rolls made of this alloy are still in good condition after operating for 8700 h at 1200°C. Furnace trays, used to support heavy loads, have been exposed to temperatures up to 1250°C in an oxidizing atmosphere without bending or warping. Alloy X is equally suitable for use in jet engine tailpipes, afterburner components, turbine blades, nozzle vanes, cabin heaters, and other aircraft parts. Alloy X has wide use in gas turbine engines for combustion zone components such as transition duct, combustor cans, spray bars, and flame holders. Alloy X is also used in the chemical process industry for retorts, muffles, catalyst support grids, furnace baffles, tubing for pyrolysis operations, and flash drier components.

Incoloy 800 (N08800)

Description and corrosion resistance. Alloy 800 is a nickel-iron-chromium alloy with good strength and excellent resistance to oxidation and carburization in high-temperature atmospheres. It also resists corrosion by many aqueous environments. The alloy maintains a stable, austenitic structure during prolonged exposure to high temperatures.

Applications. Uses for Incoloy 800 include
  Process piping
  Heat exchangers
  Carburizing equipment
  Heating-element sheathing
  Nuclear steam-generator tubing

Incoloy 825 (N08825)

Description and corrosion resistance. Incoloy 825 is a nickel-iron-chromium alloy with additions of molybdenum and copper. It has excellent resistance to both reducing and oxidizing acids, stress-corrosion cracking, and localized attack such as pitting and crevice corrosion. The alloy is especially resistant to sulfuric and phosphoric acids.

Applications. This alloy is used for the following:
  Chemical processing
  Pollution-control equipment

TABLE 8.24   **Brief Description, Corrosion Resistance, and Applications of High-Performance Alloys and Some Highly Alloyed Stainless Steels** (*Continued*)

Oil and gas well piping
Nuclear fuel reprocessing
Acid production
Pickling equipment

Incoloy 925 (N09925)

Description and corrosion resistance. This is a precipitation-hardenable nickel-iron-chromium alloy with additions of molybdenum and copper. It combines the high strength of a precipitation-hardenable alloy with the excellent corrosion resistance of Alloy 825. The alloy has outstanding resistance to general corrosion, pitting, crevice corrosion, and stress corrosion cracking in many aqueous environments including those containing sulfides and chlorides.

Applications. Uses include surface and downhole hardware in sour gas wells and oil-production equipment.

Inconel 600 (N06600)

Description and corrosion resistance. Alloy 600 is a nickel-chromium alloy designed for use from cryogenic to elevated temperatures in the range of 1093°C. The high nickel content of the alloy enables it to retain considerable resistance under reducing conditions and makes it resistant to corrosion by a number of organic and inorganic compounds. The nickel content gives it excellent resistance to chloride-ion stress corrosion cracking and also provides excellent resistance to alkaline solutions.

Its chromium content gives the alloy resistance to sulfur compounds and various oxidizing environments. The chromium content of the alloy makes it superior to commercially pure nickel under oxidizing conditions. In strong oxidizing solutions like hot, concentrated nitric acid, 600 has poor resistance. Alloy 600 is relatively unattacked by the majority of neutral and alkaline salt solutions and is used in some caustic environments. The alloy resists steam and mixtures of steam, air, and carbon dioxide.

Alloy 600 is nonmagnetic, has excellent mechanical properties and a combination of high strength and good workability, and is readily weldable. Alloy 600 exhibits cold-forming characteristics normally associated with chromium-nickel stainless steels. It is resistant to a wide range of corrosive media. The chromium content gives better resistance than Alloys 200 and 201 under oxidizing conditions, and at the same time the high nickel gives good resistance to reducing conditions. Other qualities are as follows:

Virtually immune to chlorine ion stress corrosion cracking.
Demonstrates adequate resistance to organic acids such as acetic, formic, and stearic.
Excellent resistance to high purity water used in primary and secondary circuits of pressurized nuclear reactors.
Little or no attack occurs at room and elevated temperatures in dry gases, such as chlorine or hydrogen chloride. At temperatures up to 550°C in these media, this alloy has been shown to be one of the most resistant of the common alloys.
At elevated temperatures the annealed and solution annealed alloy shows good resistance to scaling and has high strength.
The alloy also resists ammonia-bearing atmospheres, as well as nitrogen and carburizing gases.
Under alternating oxidizing and reducing conditions the alloy may suffer from selective oxidation.

Applications. Typical corrosion applications include titanium dioxide production (chloride route), perchlorethylene syntheses, vinyl chloride monomer (VCM), and

TABLE 8.24 Brief Description, Corrosion Resistance, and Applications of High-Performance Alloys and Some Highly Alloyed Stainless Steels (*Continued*)

magnesium chloride. Alloy 600 is used in chemical and food processing, heat treating, phenol condensers, soap manufacture, vegetable and fatty acid vessels, among other uses. In nuclear reactors uses are for such components as control rod inlet stub tubes, reactor vessel components and seals, steam dryers, and separators in boiling water reactors. In pressurized water reactors it is used for control rod guide tubes and steam generator baffle plates. Other uses include
  Thermocouple sheaths
  Ethylene dichloride (EDC) cracking tubes
  Conversion of uranium dioxide to tetrafluoride in contact with hydrofluoric acid
  Production of caustic alkalis, particularly in the presence of sulfur compounds
  Reactor vessels and heat-exchanger tubing used in the production of vinyl chloride
  Process equipment used in the production of chlorinated and fluorinated hydrocarbons
  Furnace retort seals, fans, and fixtures
  Roller hearths and radiant tubes, in carbonitriding processes especially

Inconel 601 (N06601)

Description and corrosion resistance. The most important property of Alloy 601 is resistance to oxidation at very high temperatures, up to 1250°C, even under severe conditions such as cyclical heating and cooling. This is possible due to Alloy 601 having a tightly adherent oxide layer that is resistant against spalling. Its resistance to carburization is also good, and it is resistant to carbonitriding conditions. Due to its high chromium and some aluminium content, Inconel 601 has good resistance in oxidizing sulfur-bearing atmospheres at elevated temperatures.

Applications. This alloy is used for
  Trays, baskets, and fixtures used in various heat treatments such as carburizing and carbonitriding
  Refractory anchors, strand annealing and radiant tubes, high-velocity gas burners, wire mesh belts, etc.
  Insulating cans in ammonia reformers and catalyst support grids used in nitric acid production
  Thermal reactors in exhaust system of petrol engines
  Fabricated combustion chambers
  Tube supports and ash trays in the power generation industry

Inconel 625 (N06625)

Description and corrosion resistance. This is a material with excellent resistance to pitting, crevice, and corrosion cracking. It is highly resistant in a wide range of organic and mineral acids and has good high-temperature strength. Other features include
  Excellent mechanical properties at both extremely low and extremely high temperatures
  Outstanding resistance to pitting, crevice corrosion, and intercrystalline corrosion
  Almost complete freedom from chloride-induced stress corrosion cracking
  High resistance to oxidation at elevated temperatures up to 1050°C
  Good resistance to acids, such as nitric, phosphoric, sulfuric, and hydrochloric, as well as to alkalis makes possible the construction of thin structural parts of high heat transfer

Applications. Inconel 625 is used for
  Components where exposure to seawater and high mechanical stresses are required
  Oil and gas production where hydrogen sulfide and elementary sulfur exist at temperatures in excess of 150°C

TABLE 8.24 Brief Description, Corrosion Resistance, and Applications of High-Performance Alloys and Some Highly Alloyed Stainless Steels (*Continued*)

Components exposed to flue gas or in flue gas desulfurization plants
Flare stacks on offshore oil platforms
Hydrocarbon processing from tar-sand and oil-shale recovery projects

Inconel 718 (N07718)

Description and corrosion resistance. This is a gamma prime-strengthened alloy with excellent mechanical properties at elevated as well as cryogenic temperatures. It is suitable for temperatures up to around 700°C, can be readily worked and age hardened, and has excellent strength from −250 to 705°C. It can be welded in fully aged condition and has excellent oxidation resistance up to 980°C.

Applications. Uses for this alloy tend to be in the field of gas turbine components and cryogenic storage tanks. Examples are jet engines, pump bodies and parts, rocket motors and thrust reversers, nuclear fuel element spacers, and hot extrusion tooling.

Monel 400 (N04400)

Description and corrosion resistance. Alloy 400 is a nickel-copper alloy with excellent corrosion resistance in a wide variety of media. The alloy is characterized by good general corrosion resistance, good weldability, and moderate-to-high strength. The alloy has been used in a variety of applications. It has excellent resistance to rapidly flowing brackish water and seawater. It is particularly resistant to hydrochloric and hydrofluoric acids when they are deaerated. The alloy is slightly magnetic at room temperature and is widely used in the chemical, oil, and marine industries.

It has a good corrosion resistance in an extensive range of marine and chemical environments, from pure water to nonoxidizing mineral acids, salts, and alkalis. This alloy is more resistant than nickel under reducing conditions and more resistant than copper under oxidizing conditions. It does show, however, better resistance to reducing media than oxidizing ones. It also has

    Good mechanical properties from subzero temperatures up to about 480°C.
    Good resistance to sulfuric and hydrofluoric acids. Aeration, however, will result in increased corrosion rates. It may be used to handle hydrochloric acid, but the presence of oxidizing salts will greatly accelerate corrosive attack.
    Resistance to neutral, alkaline, and acid salts is shown, but poor resistance is found with oxidizing acid salts such as ferric chloride.
    Excellent resistance to chloride ion stress corrosion cracking.

Applications. Uses for Monel 400 include
    Feed water and steam generator tubing
    Brine heaters and seawater scrubbers in tanker inert gas systems
    Sulfuric acid and hydrofluoric acid alkylation plants
    Pickling bat heating coils
    Heat exchangers in a variety of industries
    Transfer piping from oil refinery crude columns
    Plants for the refining of uranium and isotope separation in the production of nuclear fuel
    Pumps and valves used in the manufacture of perchlorethylene, chlorinated plastics
    Monoethanolamine (MEA) reboiling tubes
    Cladding for the upper areas of oil refinery crude columns
    Propeller and pump shafts

Monel 500 (N05500)

Description and corrosion resistance. Alloy K-500 is a nickel-copper alloy, precipitation hardenable through additions of aluminum and titanium. Alloy K-500

TABLE 8.24 Brief Description, Corrosion Resistance, and Applications of High-Performance Alloys and Some Highly Alloyed Stainless Steels (*Continued*)

retains the excellent corrosion-resistant characteristics of 400 and has enhanced strength and hardness after precipitation hardening when compared with 400. Alloy K-500 has approximately 3 times the yield strength and double the tensile strength when compared with 400. K-500 can be further strengthened by cold working before the precipitation hardening.

It has excellent mechanical properties from subzero temperatures up to about 480°C and corrosion resistance in an extensive range of marine and chemical environments from pure water to nonoxidizing mineral acids, salts, and alkalies.

Applications. Typical applications for the alloy that take advantage of high strength and corrosion resistance are pump shafts, impellers, propeller shafts, valve components for ships and offshore drilling towers, bolting, oil well drill collars, and instrumentation components for oil and gas production. It is particularly well suited for centrifugal pumps in the marine industry because of its high strength and low corrosion rates in high-velocity seawater.

Nickel 200 (N02200)

Description and corrosion resistance. This is commercially pure wrought nickel with good mechanical properties over a wide range of temperature and excellent resistance to many corrosives, in particular hydroxides. Nickel 200 can be hot formed to almost any shape. A temperature range of 650 to 1230°C is recommended and should be carefully adhered to because the proper temperature is the most important factor in achieving hot malleability. Full information of the forming process should be sought and understood before proceeding. 200 can be cold formed by all conventional methods, but because nickel alloys have greater stiffness than stainless steels more power is required to perform the operations. Other properties are

- Good resistance to corrosion in acids and alkalies and is most useful under reducing conditions
- Outstanding resistance to caustic alkalis up to and including the molten state
- In acid, alkaline, and neutral salt solutions the material shows good resistance, but in oxidizing salt solutions severe attack will occur
- Resistant to all dry gases at room temperature and in dry chlorine and hydrogen chloride may be used in temperatures up to 550°C
- Resistance to mineral acids varies according to temperature and concentration and whether the solution is aerated or not; corrosion resistance is better in deaerated acid

Applications. It is used in the following:

- Manufacture and handling of sodium hydroxide, particularly at temperature above 300°C
- Production of viscose rayon and manufacture of soap
- Analine hydrochloride production and the chlorination of aliphatic hydrocarbons such as benzene, methane and ethane
- Manufacture of vinyl chloride monomer
- Storage and distribution systems for phenol; immunity from any form of attack ensures absolute product purity
- Reactors and vessels in which fluorine is generated and reacted with hydrocarbons

Nickel 201 (N02201)

Description and corrosion resistance. Nickel 201 can be hot formed to almost any shape. The temperature range 650 to 1230°C is recommended and should be carefully adhered to because the proper temperature is the most important factor in achieving hot malleability. Full information of the forming process should be sought and

TABLE 8.24 Brief Description, Corrosion Resistance, and Applications of High-Performance Alloys and Some Highly Alloyed Stainless Steels (*Continued*)

understood before proceeding. Nickel 201 can be cold formed by all conventional methods, but because nickel alloys have greater stiffness than stainless steels, more power is required to perform the operations. Nickel 201 is the low-carbon version of Nickel 200. It is preferred to Nickel 200 for applications involving exposure to temperatures above 320°C. With low base hardness and lower work-hardening rate, it is particularly suited for cold forming. Other properties are

   Good resistance to corrosion in acids and alkalies; most useful under reducing conditions

   Outstanding resistance to caustic alkalis up to and including the molten stat.

   In acid, alkaline, and neutral salt solutions the material shows good resistance, but in oxidizing salt solutions severe attack will occur

   Resistant to all dry gases at room temperature and in dry chlorine and hydrogen chloride may be used in temperatures up to 550°C

   Resistance to mineral acids varies according to temperature and concentration and whether the solution is aerated or not; corrosion resistance is better in deaerated acid

   Virtually immune to intergranular attack above 315°C; chlorates must be kept to a minimum

Applications. Nickel 201 has the following uses:

   Manufacture and handling of sodium hydroxide, particularly at temperature above 300°C

   Production of viscose rayon; manufacture of soap

   Analine hydrochloride production and the chlorination of aliphatic hydrocarbons such as benzene, methane and ethane

   Manufacture of vinyl chloride monomer

   Storage and distribution systems for phenol; immunity from any form of attack ensures absolute product purity

   Reactors and vessels in which fluorine is generated and reacted with hydrocarbons

Nitronic 60 (S21800)

Description and corrosion resistance. Nitronic 60 is truly an all-purpose metal. This fully austenitic alloy was originally designed as a high-temperature alloy for temperatures around 980°C. The oxidation resistance of Nitronic 60 is similar to S30900 steel and far superior to S30400 steel. The additions of silicon and manganese have given the alloy a matrix to inhibit wear, galling, and fretting even in the annealed condition. Higher strengths are attainable through cold working the material, and it is still fully austenitic after severe cold working. This working does not enhance the antigalling properties as is normal for carbon steels and some stainless steels. The cold or hot work put into the material adds strength and hardness.

The chromium and nickel additions give it comparable corrosion to S30400 and S31600 stainless steels, while having a twice the yield strengths of regular stainless steels. The high mechanical strength in annealed parts permits use of reduced cross sections for weight and cost reductions. Although uniform corrosion resistance of Nitronic 60 is better than S30400 stainless in most environments, its yield strength is nearly twice that of S30400 and S31600 steels. Chloride pitting resistance is superior to that of type S31600 stainless; Nitronic 60 provides excellent high-temperature oxidation resistance and low-temperature impact.

Nitronic 60 is also readily welded using conventional joining processes. It can be handled similarly to S30400 and S31600 steels. No preheat or postweld heat treatments are necessary, other than the normal stress relief used in heavy fabrication. Most applications use Nitronic 60 in the as-welded condition, unless corrosion resistance is a consideration. Fillerless fusion welds (autogenous) have been made using GTA. These

TABLE 8.24 Brief Description, Corrosion Resistance, and Applications of High-Performance Alloys and Some Highly Alloyed Stainless Steels (*Continued*)

welds are free from cracking and have galling and cavitation resistance similar to the unwelded base metal. Heavy weld deposits using this process are sound and exhibit higher strength then the unwelded base metal. The metal-to-metal wear resistance of the GMA welds are slightly lower than the base metal wear resistance.

Applications. Applications using Nitronic 60 are valve stems, seats and trim, fastening systems, screening, pins, bushings and roller bearings, pump shafts, and rings. Other uses include wear plates, rails guides, and bridge pins. This alloy provides a significant lower-cost way to fight wear and galling compared to nickel- or cobalt-based alloys. It is also used for

Automotive valves; it can withstand gas temperatures of up to 820°C for a minimum of 80,000 km

Fastener galling; it is capable of frequent assembly and disassembly, allowing more use of the fastener before the threads are torn up and also helps to eliminate corroded or frozen fasteners

Pins; it is used in roller prosthetics and chains to ensure a better fit of parts (closer tolerance, nonlubricated) and a longer life

Marine shafts; it has better corrosion than types 304 and 316, with double the yield strength

Pin and hanger expansion joints for bridges; it has better corrosion, galling resistance, low-temperature toughness, and high charpy values at subzero temperatures compared to the A36 and A588 carbon steels commonly used.

Nitronic 50 (S20910)

Description and corrosion resistance. Nitronic 50 stainless steel provides a combination of corrosion resistance and strength not found in any other commercial material available in its price range. This austenitic stainless has corrosion resistance greater than that provided by S31600, plus approximately twice the yield strength at room temperature. In addition to the improved corrosion resistance, Nitronic 50 can be welded successfully using conventional welding processes that are normally employed with the austenitic stainless steels.

Its resistance to intergranular attack is excellent even when sensitized at 675°C for 1 h to simulate the heat-affected zone of heavy weldments. Material annealed at 1066°C has very good resistance to intergranular attack for most applications. However, when thick sections are used in the as-welded condition in certain strongly corrosive media, the 1121°C condition gives optimum corrosion resistance.

Applications. Outstanding corrosion resistance gives Armco's Nitronic 50 stainless steel the leading edge for applications where types 316, 316L, 317, and 317L are only marginal. It's an effective alloy for the petroleum, petrochemical, chemical, fertilizer, nuclear fuel recycling, pulp and paper, textile, food processing, and marine industries. Components using the combination of excellent corrosion resistance and high strength currently include pumps, valves and fittings, fasteners, cables, chains, screens and wire cloth, marine hardware, boat and pump shafting, heat exchanger parts, springs, and photographic equipment. Other uses include

Fastener
Marine hardware, mastings and tie downs
Marine and pump shafts
Valves and fittings
Downhole rigging

Cobalt-base alloys. The corrosion behavior of pure cobalt has not been documented as extensively as that of nickel. The behavior of cobalt is similar to that of nickel, although cobalt possesses lower overall corrosion resistance. For example, the passive behavior of cobalt in 0.5 M sulfuric acid has been shown to be similar to that of nickel, but the critical current density necessary to achieve passivity is 14 times higher for the former. Several investigations have been carried out on binary cobalt-chromium alloys. In cobalt-base alloys, it has been found that as little as 10% chromium is sufficient to reduce the anodic current density necessary for passivation from 500 to 1 mA·cm^{-2}. For nickel, about 14% chromium is needed to reduce the passivating anodic current density to the same level.

It should be noted that all of these alloys, regardless of their chromium and molybdenum contents, exhibit similar corrosion resistance in dilute H_2SO_4. Thus, the high-chromium alloys show approximately the same corrosion rates as the lower-chromium alloys. Similar behavior has been observed in the nickel-iron-chromium-molybdenum alloys. In H_2SO_4 and HCl, the nickel and cobalt contents govern the behavior of the alloy as long as minimum amounts of chromium and molybdenum or tungsten are present. The corrosion resistance of wrought cobalt-base alloys in HCl solutions is not good except in very dilute HCl.[32] However, because many of the commercial alloys contain appreciable amounts of chromium, their corrosion resistance to dilute nitric acid is quite good. In highly oxidizing chromic acid, the chromium-containing alloys, whether cobalt- or nickel-base, do not perform well, probably because the passive, chromium oxide film is unstable in this acid.[32]

Environmental embrittlement. Cobalt-base alloys are primarily used in high-temperature applications. In such uses, hydrogen embrittlement and SCC are generally not thought to be important. However, in applications in which cobalt-base alloys are used for aqueous corrosion service, both of these modes of fracture may become important. Cobalt-base alloys can be used to combat hydrogen embrittlement where steels have failed by this mechanism. Annealed cobalt-base alloys do not show significant susceptibility to hydrogen embrittlement, even in the most severe hydrogen-charging conditions. When cold worked to levels exceeding 1380-MPa yield strength, the cobalt-base alloys may not exhibit embrittlement.[32]

8.5.5 Use of high-performance alloys

High-performance alloys have been used in cast, rolled, extruded, forged, and powder processed forms. Sheet, bar, plate, tubing, airfoils, disks, and pressure vessels are but some of the shapes that have been produced. These metals have been used in aircraft, industrial and marine gas turbines, nuclear reactors, aircraft skins, spacecraft structures, petrochemical production, and environmental protection applications. Although developed for high-temperature applications, some are used at cryogenic temperatures.

The Ni-Cr-Fe alloys are also extensively used in refining and petrochemical plant equipment for both liquid and gaseous low-temperature corrosion resistance and for heat-resistant applications. Table 8.24 describes the practical behavior of the main high-performance alloys and highly alloyed stainless steels in some of the very demanding operational situations in which these alloys are expected to perform satisfactorily. The chemical composition of these alloys can be found in App. E.

8.1 Refractory Metals

8.6.1 Introduction

Refractory metals are characterized by their high melting points, exceeding an arbitrary value of 2000°C, and low vapor pressures, two properties exploited by the electronics industry. Only four refractory metals, molybdenum, niobium, tantalum, and tungsten, are available in quantities of industrial significance and have been produced commercially for many years, mainly as additives to steels, nickels, and cobalt alloys and for certain electrical applications. In addition to high-temperature strength, the relatively low thermal expansions and high thermal conductivity of the refractory metals suggest good resistance to thermal shock. Table 8.25 contains additional data on physical and mechanical properties of refractory metals.

There are, however, two characteristics, ready oxidation at high temperatures and, in the case of molybdenum and tungsten, brittleness at low temperatures, which limit their applications. Of the refractory metals, tantalum has the widest use in the chemical process industries. Most applications involve acid solutions that cannot be handled with iron or nickel-base alloys. Tantalum, however, is not suitable for hot alkalis, sulfur trioxide, or fluorine. Hydrogen will readily be absorbed by tantalum to form a brittle hydride. This is also true of titanium and zirconium. Tantalum is often used as a cladding metal.

Corrosion resistance of the refractory metals is second only to that of the noble metals. Unlike the noble metals, however, refractory metals are inherently reactive. It is this very reactivity that can provide corrosion resistance. On contact with air or any other oxidant, refractory metals immediately form an extremely dense, adherent oxide film. This passivating layer prevents access of the oxidant to the underlying metal and renders it resistant to further attack. Unfortunately, these oxides can spall or volatize at elevated temperatures, leaving the metals susceptible to oxidation at a temperature as

TABLE 8.25 Typical Properties of Molybdenum, Niobium, Tantalum, and Tungsten

Property	Unit	Mo	Nb	Ta	W
Atomic number		42	41	73	74
Atomic weight	(g mol^{-1})	95.95	92.91	180.95	183.86
Atomic radius	(nm)	0.1363	0.1426	0.143	0.1371
Lattice type		bcc	bcc	bcc	bcc
Lattice constant, 20°C	(nm)	0.31468	0.3294	0.33026	0.31585
Mass					
Density at 20°C	(g · cm^{-3})	10.2	8.57	16.6	19.3
Thermal properties					
Melting point	(°C)	2610	2468	2996	3410
Boiling point, °C	(°C)	5560	4927	6100	5900
Linear coefficient of expansion	per °C	4.9×10^{-6}	7.1×10^{-6}	6.5×10^{-6}	4.3×10^{-6}
Thermal conductivity, 20°C	W·m^{-1} K^{-1}	147	219	54	167
Specific heat, 20°C	(J·kg^{-1} K^{-1})	255	525	151	134
Electrical properties					
Conductivity	% IACS (Cu)	30	13.2	13	31
Resistivity, 20°C	µΩ·cm	5.7	15	13.5	5.5
Coefficient of resistivity	per °C (0–100°C)	0.0046		0.0038	0.0046
Mechanical properties					
Tensile strength, 20°C	(MPa)	700–1400	195	240–500	700–3500
500°C	(MPa)	240–450		170–310	500–1400
1000°C	(MPa)	140–210		90–120	350–500
Young's modulus-20°C	(GPa)	320	103	190	410
500°C	(GPa)	280		170	380
1000°C	(GPa)	270		150	340
Working temperature	(°C)	1600		Room	1700
Recrystallizing temp	(°C)	900–1200	800–1100	1000–1250	1200–1400
Stress relieving temp	(°C)	800		850	1100

low as 300°C. For high-temperature applications under nonreducing conditions, the refractory metals must be protected by an applied coating, such as a metal silicide.

8.6.2 Molybdenum

Molybdenum provides a corrosion resistance that is slightly better than that of tungsten. It particularly resists nonoxidizing mineral acids. It is obtained from its chief source ore, molybdenite, and has a high Young's modulus. Worked forms (wire, sheet) are ductile at low temperatures, and it is resistant to mineral acids, unless oxidizing agents are present. Limitations are that it has very low oxidation resistance above 450°C. Ductile-brittle transition temperature may be 200°C. Molybdenum has applications in high-temperature parts (but it must be protected form oxidation by atmosphere or coating), especially windings. It is also used in electrodes in glass melting furnaces, for metallizing, and in aerospace structural parts including leading edges and support frames.

Molybdenum is relatively inert to carbon dioxide, hydrogen, ammonia, and nitrogen to 1100°C and also in reducing atmospheres containing hydrogen sulfide. It has excellent resistance to corrosion by iodine vapor, bromine, and chlorine up to certain well-defined temperature limits. Molybdenum also provides good resistance to several liquid metals including bismuth, lithium, potassium, and sodium.[35]

Molybdenum has been used for many years in the lamp industry for mandrels and supports, usually in wire form. Today, several unique properties of molybdenum that satisfy more demanding industry requirements have increased the use of molybdenum as a material in applications requiring other mill forms.

Molybdenum alloys. Molybdenum has several alloys:

- *TZM (titanium, zirconium, molybdenum).* Molybdenum's prime alloy is TZM. This alloy contains 99% Mo, 0.5% Ti, and 0.08% Zr with a trace of carbon for carbide formations. TZM offers twice the strength of pure molybdenum at temperatures over 1300°C. The recrystallization temperature of TZM is approximately 250°C higher than molybdenum, and it offers better weldability.

  The finer grain structure of TZM and the formation of TiC and ZrC in the grain boundaries of the molybdenum inhibit grain growth and the related failure of the base metal as a result of fractures along the grain boundaries. This also gives it better properties for welding. TZM costs approximately 25 percent more than pure molybdenum and costs only about 5 to 10 percent more to machine. For high-strength applications such as rocket nozzles, furnace structural components, and forging dies, it can be well worth the cost differential. TZM is available in sheet and rod form in basically the same size range as molybdenum with the exception of thin foil.

- *Molybdenum/30% tungsten.* This is another molybdenum alloy that offers unique properties. It was developed for the zinc industry. This alloy resists the corrosive effects of molten zinc. Mo/30W has also proved effective in rocket nozzles and has the potential of offering enhanced performance in applications where any erosive effects are a factor.

- *Molybdenum/50% rhenium.* This alloy offers the strength of molybdenum with the ductility and weldability of rhenium. It is a costly alloy and is only available in a very limited size range. It offers significant advantages in thin foil applications for high-temperature delicate parts, especially those that must be welded. Note that although this alloy is nominally 47% rhenium, it is customarily referred to 50/50 molybdenum/rhenium. Other molybdenum/rhenium alloys include molybdenum/rhenium sheet with 47.5 and 41% rhenium. The molybdenum/41% rhenium alloy does not develop sigma phase. This makes the material even more ductile after exposure to high temperatures.

Applications of molybdenum. There is an increasing demand from the electronics and aerospace industries for materials that maintain reliability under ever-increasing temperature conditions. Because its properties meet these requirements, molybdenum also is experiencing an increasing demand. The following characteristics support the demand for molybdenum in many electronics applications:[35]

- Exceptional strength and stiffness at high temperatures
- Good thermal conductivity
- Low thermal expansion
- Low emissivity
- Low vapor pressure
- Electrical resistivity
- Corrosion resistance
- Purity
- Ductility and fabricability
- Machinability

Some combination of these properties and characteristics predicts increased usage of molybdenum in such applications as rocket nozzles, jet tabs, high-temperature dies, electrodes, boring bars, tools, brazing fixtures, electrical contacts, boats, heat shields, and many others as well as high-vacuum applications. Molybdenum can be furnished in many mill forms such as wire, ribbon, foil, plate, sheet, rod, billet, slab, bar, extruded shapes, tubes, and powder.

One of the unique applications of molybdenum is in glass-to-metal seals. Molybdenum has a straight-line expansion. The mean coefficient of expansion is 4.9×10^{-6} measured between 20 and 500°C. Molybdenum is suitable for sealing to hard glass because it has approximately the same coefficient of expansion and a transition temperature below 700°C. Molybdenum oxides dissolve readily in glass. The adhesion between glass and this metal is very satisfactory and gives an absolutely tight seal.

It is essential for the surface of the metal to be correctly oxidized before it comes into contact with the glass. This is easily done, provided that the surface is clean and free from grooves and cracks. The best method of oxidizing the surface is to heat it for a short time in an air-gas or oxygen-gas flame. Excessive oxidation must be avoided because it results in incomplete absorption of the oxide in the glass. This can possibly make the seal porous.

Molybdenum should be oxidized by rapid heating, maintained at high temperature for a short period. The gas flame itself is a guard against excessive oxidation. This is indicated by a slight emission of smoke. Conversely, the reducing part of the flame provides insufficient oxidation and, therefore, must be avoided. The most favorable sealing-in temperature depends upon the viscosity of the hard glass and lies between 1000 and 1200°C. The preoxidized rod, after slight cooling, has a blue color, indicating a low oxide.

Machining characteristics. Pressed and sintered or recrystallized molybdenum machines very much like medium hard cast iron. Wrought molybdenum machines similar to stainless steel. Once molybdenum's few peculiarities are known and respected, it can be machined with conventional tools and equipment. The machining characteristics of molybdenum differ from those of medium hard cast iron and cold-rolled steel in two ways:

1. It has a tendency to break out on the edges when cutting tools become dull.
2. It is very abrasive and causes tools to wear out much faster than steel. Once the expected tool life has been established for a particular operation, establishing a program of scheduled tool replacement will permit maximum machining efficiency at minimum time and investment.

Welding and brazing. Molybdenum can be joined using conventionally accepted welding techniques except for gas. Heli-arc welding is most common and usually provides satisfactory results. Complex welding operations may require more sophisticated or special techniques.

Careful cleaning of the joint surfaces is essential. Controlled weld atmospheres, such as a dry box, are desirable but not necessary. In designing fixtures, all clamping forces should be compressive and should be released immediately after welding to permit unstressed cooling. Copper-base alloys are normally acceptable in creating relatively low-strength joints. Higher-strength joints can be achieved by using gold, platinum, or other more exotic base brazing alloys.[35]

With proper temperature precautions, brazing will normally produce a more ductile joint than welding. Like tungsten, molybdenum has excellent high-temperature properties; however, poor oxidation resistance requires coating protection at higher temperatures. The presence of minute quantities of oxygen, nitrogen, and carbon lower the ductility of molybdenum. Of all the potential contaminants in wrought products, iron is of primary concern. Others, such as aluminum, carbon, calcium, copper, and nickel, may also be present as elements, but they are more frequently present in the form of oxides. Removal of a controlled amount of base metal may be desired to ensure complete removal of contaminants.[35]

Corrosion resistance. Molybdenum provides corrosion resistance that is similar to tungsten. Molybdenum particularly resists nonoxidizing mineral acids. It is relatively inert to carbon dioxide, ammonia, and nitrogen to 1100°C and also in reducing atmospheres containing hydrogen sulfide. Molybdenum offers excellent resistance to corrosion by iodine vapor, bromine, and chlorine up to clearly defined temperature limits. It also provides good resistance to several liquid metals including bismuth, lithium, potassium, and sodium. Table 8.26 gives ratings for the resistance of molybdenum to a wide spectrum of chemical environments.[36]

8.6.3 Niobium

Niobium, sometimes called columbium, can be a less-expensive alternative to tantalum. However, its corrosion resistance is more limited, mostly because of its susceptibility to attack by most alkalies and certain strong oxidants. Even though the mechanical strength of niobium is less than that of tantalum, it can be used economically where the extreme inertness of tantalum is not required. It occurs naturally with tantalum in the minerals columbite and tantalite.

Niobium remains totally resistant to such highly corrosive media as wet or dry chlorine, bromine, saturated brines, ferric chloride, hydrogen

TABLE 8.26 Chemical Reactivity of Molybdenum

Environment	Resistant	Variable	Nonresistant
Al_2O_3, BeO, MgO, ThO_2, ZrO_2 (<1700°C)	X		
Aluminum (molten)			X
Aqua regia (cold)		X	
Aqua regia (hot)			X
Aqueous ammonia		X	
Aqueous caustic soda/potash	X		
Bismuth	X		
Boron (hot) boride fomation			X
Bromine	X		
Carbon (1100°C) carbide formation			X
Carbon dioxide (1200°C) oxidation			X
Carbon monoxide (1400°C) carbide formation			X
Cesium		X	
Chlorine	X		
Cobalt (molten)			X
Fluorine (room temperature)			X
Gallium		X	
Hydrocarbons (1100°C) carbide formation			X
Hydrochloric acid (cold)	X		
Hydroflouric acid	X		
Hydrogen	X		
Inert gases (all)	X		
Iodine	X		
Iron (molten)			X
KNO_2, KNO_3, $KCLO_3$ (molten)			X
Lead		X	
Lithium		X	
Magnesium		X	
Mercury		X	
Molten caustic		X	
Molten caustics in the presence of KNO_2, KNO_3, $KCLO_3$, PbO_2			X
Molten glass	X		
Nickel (molten)			X
Nitric acid (cold)		X	
Nitric acid (hot)			X
Nitric/hydrofluoric mixture (either hot or cold)			X
Nitrogen	X		
Oxygen or air (>400°C)		X	
Oxygen or air (>600°C)			X
Phosphorous	X		
Potassium		X	
Silicon (1000°C) silicide formation			X
Sodium		X	
Sulfide formation (440°C)			X
Sulfuric acid (hot)		X	
Tin (molten)			X
Water	X		
Zinc (molten)		X	

sulfide, sulfur dioxide, nitric and chromic acids, and sulfuric and hydrochloric acids within specific temperature and concentration limits. Niobium is very similar to tantalum, and several alloys are available in the arc-cast and wrought condition. It has the lowest melting point of all the refractory metals covered, the lowest modulus of elasticity and thermal conductivity, and the highest thermal expansion. It also has the lowest strength and density of the refractory metals.

Niobium's ductile-to-brittle transition temperature ranges from -101 to $-157°C$. This metal also has the low thermal neutron capture cross section required for nuclear applications. Its high melting point warrants its use at temperatures above the maximum service temperatures of the iron-, nickel-, and cobalt-base metals. It has excellent ductility and fabricability.

Niobium was used as an alloy for many years. Nb/1%Zr was, and still is, used in nuclear reactors as the tubing for the fuel pellets because of its resistance to neutron bombardment. As C-103 alloy, it has been used for rocket nozzles and exhaust nozzles for jet engines and rockets because of its high strength and oxidation resistance at a low weight. Recently, it has been gaining favor in its pure form for semiconductor equipment components and corrosion resistant parts.[37]

Niobium can be bent, spun, deep drawn, and formed at room temperature up to its maximum work hardening. Machining is somewhat more difficult. High-speed tooling with a proper lubricant will allow machining of niobium. However, tools will wear quickly and high rake angles should be maintained. Tool maintenance must be taken into consideration when costing niobium parts. Nonetheless, this metal is an ideal candidate for a lower-cost alternative when tantalum is being considered.

Applications of niobium. The combination of niobium strength, melting point, resistance to chemical attack, and low neutron absorption cross section favor niobium's use in the nuclear industry. It has been identified as the preferred construction material for the first reactors in the space power systems programs. Niobium mill products are used in the fabrication of corrosion-resistant process equipment including reaction vessels, columns, bayonet heaters, shell and tube heat exchangers, U-tubes, thermowells, spargers, rupture diaphragms, and orifices. Applications are in gas and turbine rocket motors, high-temperature parts, linings and claddings, and containers for reactor fuel.[37]

Working and machining characteristics. The cold-working properties of niobium are excellent. Because of its body-centered cubic (bcc) crystal structure, niobium is a very ductile metal that can undergo cold reductions of more than 95% without failure. The metal can be easily forged, rolled, or swaged directly from ingot at room temperature. Niobium is well suited to deep drawing. The metal may be cupped and drawn to tube, but special care must be taken with lubrication. Sheet metal can also easily be formed by general sheet metal working techniques. The low rate of work hardening reduces springback and facilitates these operations.

Niobium may be machined using standard techniques. However, due to the tendency of the material to gall, special attention needs to be given to tool angles and lubrication. Niobium also has a tendency to stick to tooling during metal-forming operations. To avoid this, specific lubricant and die material combinations are required in high-pressure forming operations.

Welding. Niobium is a highly active metal. It reacts at temperatures well below its melting point with all the common gases, such as nitrogen, oxygen, hydrogen, and carbon dioxide. At its melting point and above, niobium will react with all the known fluxes. This severely restricts the choice of welding methods. Niobium can be welded to several metals, one of which is tantalum. This can be readily accomplished by resistance welding, tungsten-inert gas, plasma welding, and electron beam welding.

Formation of brittle intermetallic phases is likely with many metals and must be avoided. Surfaces to be heated above 300°C should be protected by an inert gas such as argon or helium to prevent embrittlement. It is critical to ensure that the metal is clean prior to welding. An acid pickle wash is recommended. For ambient-temperature pickling, a typical solution is 25 to 35% HF, 25 to 33% HNO_3.

Corrosion resistance. The corrosion resistance of niobium is more limited than tantalum, and this must be taken into consideration. The limitation stems from its sensitivity to most alkalies and certain strong oxidants. However, niobium is totally resistant to such highly corrosive media as wet or dry chlorine, bromine, saturated brines, ferric chloride, hydrogen sulfide, and sulfur dioxide as well as nitric and chromic acids. It is also resistant to sulfuric and hydrochloric acids within specific temperature and concentration limits.

Niobium is also resistant to attack by many liquid metals such as Li < 1000°C, Na, K + NaK < 1000°C, ThMg < 850°C, U < 1400°C, Zn < 450°C, Pb < 850°C, Bi < 500°C, and Hg < 600°C. Niobium has the ability to form stable, passive oxides, and therefore, it can provide unique solutions to many corrosion problems. However, niobium cannot be used in air at temperatures exceeding 200°C. Table 8.27 contains corrosion rates for niobium exposed to various chemical environments.[38]

TABLE 8.27 Corrosion Rates of Commercially Pure Niobium in Various Environments

Environment	Concentration, %	Temperature, °C	Corrosion rate, mm·y^{-1}
\multicolumn{4}{c}{Mineral acids}			
Hydrochloric	1	Boiling	Nil
Hydrochloric (aerated)	15	Room–60	Nil
Hydrochloric (aerated)	15	100	0.025
Hydrochloric (aerated)	30	35	0.025
Hydrochloric (aerated)	30	60	0.05
Hydrochloric (aerated)	30	100	0.125
Hydrochloric	37	Room	0.025
Hydrochloric	37	60	0.25
Hydrochloric	37% with Cl_2	60	0.5
Hydrochloric	10% with 0.1% $FeCl_3$	Boiling	0.025
Hydrochloric	10% with 0.6% $FeCl_3$	Boiling	0.125
Hydrochloric	10% with 35% $FeCl_2$ and 2% $FeCl_3$	Boiling	0.05
Nitric	65	Room	Nil
Nitric	70	250	0.025
Phosphoric	60	Boiling	0.5
Phosphoric	85	Room	0.0025
Phosphoric	85	88	0.05
Phosphoric	85	100	0.125
Phosphoric	85	Boiling	3.75
Phosphoric	85% with 4% HNO_3	88	0.025
Phosphoric	40–50% with 5 ppm F$^-$	Boiling	0.25
Sulfuric	5-40	Room	Nil
Sulfuric	98	Room	Embrittlement
Sulfuric	10	Boiling	0.125
Sulfuric	25	Boiling	0.25
Sulfuric	40	Boiling	0.5
Sulfuric	40% with 2% $FeCl_3$	Boiling	0.25
Sulfuric	60	Boiling	1.25
Sulfuric	60% with 0.1–1% $FeCl_3$	Boiling	0.5
Sulfuric	20% with 7% HC and 100 ppm F$^-$	Boiling	0.25
Sulfuric	50% with 20% HNO_3	50–80	Nil
Sulfuric	50% with 20% HNO_3	Boiling	0.25
Sulfuric	72% + 3% CrO_3	100	0.025
Sulfuric	72% + 3% CrO_3	125	0.125
Sulfuric	72% + 3% CrO_3	Boiling	3.75
\multicolumn{4}{c}{Organic acids}			
Acetic	5–99.7	Boiling	Nil
Citric	10	Boiling	0.025
Formaldehyde	37	Boiling	0.0025
Formic	10	Boiling	Nil
Lactic	10–85	Boiling	0.025
Oxalic	10	Boiling	1.25

TABLE 8.27 Corrosion Rates of Commercially Pure Niobium in Various Environments (*Continued*)

Environment	Concentration, %	Temperature, °C	Corrosion rate, mm·y^{-1}
Tartaric	20	Room–boiling	Nil
Trichloroacetic	50	Boiling	Nil
Trichloroethylene	99	Boiling	Nil
Alkalies			
NaOH	1–40	Room	0.125
NaOH	1–10	98	Embrittlement
KOH	5–40	Room	Embrittlement
KOH	1–5	98	Embrittlement
NH$_4$OH	all	Room	Nil
Salts			
AlCl$_3$	25	Boiling	0.005
Al$_2$(SO$_4$)$_3$	25	Boiling	Nil
AlK(SO$_4$)$_2$	10	Boiling	Nil
CaCl$_2$	70	Boiling	Nil
Cu(NO$_3$)$_2$	40	Boiling	Nil
FeCl$_3$	10	Room–boiling	Nil
HgCl$_2$	Saturated	Boiling	0.0025
K$_2$CO$_3$	1–10	Room	0.025
K$_2$CO$_3$	10–20	98	Embrittlement
K$_3$PO$_4$	10	Room	0.025
MgCl$_2$	47	Boiling	0.025
NaCl	Saturated; pH = 1	Boiling	0.025
Na$_2$CO$_3$	10	Room	0.025
Na$_2$CO$_3$	10	Boiling	0.5
NaHSO$_4$	40	Boiling	0.125
NaOCl	6	50	1.25
Na$_3$PO$_4$	5–10	Room	0.025
Na$_3$PO$_4$	2.5	98	Embrittlement
NH$_2$SO$_3$H	10	Boiling	0.025
NiCl$_3$	30	Boiling	Nil
ZnCl$_2$	40–70	Boiling	Nil
Others			
Bromine	Liquid	20	Nil
Bromine	Vapor	20	0.025
Chromium plating solution	25% CrO$_3$, 12% H$_2$SO$_4$	92	0.125
Chromium plating solution	17% CrO$_3$, 2% Na$_5$SiF$_6$, trace H$_2$SO$_4$	92	0.125
H$_2$O$_2$	30	Room	0.025
H$_2$O$_2$	30	Boiling	0.5

8.6.4 Tantalum

Tantalum is a relatively high-cost heavy metal with a density more than twice that of steel. The physical properties of tantalum are similar to mild steel, except that tantalum has a much higher melting point (3000°C). The tensile strength is about 345 MPa, which can be approximately doubled by cold work. Tantalum is easy to fabricate. It is soft, ductile, and malleable and can be worked into intricate forms. It can be welded by a number of techniques but requires completely inert conditions during welding.

Tantalum provides good thermal conductivity that, combined with its corrosion resistance, has made it the ideal choice for heat exchangers in acid processing equipment. It is superior to the nickel-based alloys in both these categories. Tantalum also develops a stable oxide that is useful in electronics industry applications. It has gained acceptance as a suitable material for mass spectrometer filaments, providing an alternative to rhenium, historically the only suitable material. Refer to Table 8.28 for additional information.[36]

Tantalum alloys. Two tantalum alloys have found particular commercial significance:

97.5% Ta 2.5% W. This alloy is particularly useful in applications where low-temperature strength is important along with high corrosion resistance and good formability. This alloy offers higher strength than pure tantalum while maintaining the fabricability characteristics. It is available in basically the same sizes and shapes as pure tantalum, at a comparable cost.

90% Ta 10% W. This alloy should be considered when high temperatures and high strength in a corrosive environment are required. The alloy has approximately twice the tensile strength of pure tantalum and yet retains tantalum corrosion resistance and a good portion of its ductility. It is not as readily available as pure tantalum or the alloy given above and its cost is somewhat higher.

Applications of tantalum. Tantalum has gained wide acceptance for use in electronic components, chemical equipment, missile technology, and nuclear reactors. The electronics industry consumes a large fraction (60%) of the tantalum produced for capacitors. Other industries concerned with corrosion, especially the chemical processing industry, are accounting for an increasingly large percentage of the market. Tantalum can be used to fabricate valves for corrosive liquids and to manufacture heaters for acids and heat shields for rocket motors.[39]

It is also used as a component of ion implanters in the manufacture of semiconductors. Also, because tantalum does not have a low neutron

TABLE 8.28 Chemical Reactivity of Tantalum

Environment	Resistant	Variable	Nonresistant
Acetic acid	X		
Acetic anhydride	X		
Aluminum chloride	X		
Aluminum sulfate	X		
Ammonia		X	
Ammonium chloride	X		
Ammonium hydroxide		X	
Ammonium nitrate	X		
Ammonium phosphate	X		
Ammonium sulfate	X		
Amyl acetate or chloride	X		
Aqua regia	X		
Arsenic acid	X		
Barium hydroxide	X		
Bromine, dry (< 200°C)	X		
Calcium hydroxide	X		
Calcium hypochlorite	X		
Chlorinated brine	X		
Chlorinated hydrocarbons	X		
Chlorine, dry (< 175°C)	X		
Chlorine, wet	X		
Chlorine oxides	X		
Chloracetic acid	X		
Chromic acid	X		
Chrome plating solutions	X		
Cleaning solution	X		
Copper salts	X		
Ethylene dibromide	X		
Ethyl chloride	X		
Fatty acids	X		
Ferric chloride	X		
Ferric sulfate	X		
Ferrous sulfate	X		
Fluorine			X
Formic	X		
Fuming nitric acid	X		
Fuming sulfuric acid			X
Hydrobromic acid	X		
Hydrochloric acid	X		
Hydrocyanic acid	X		
Hydroflouric acid			X
Hydrogen bromide	X		
Hydrogen chloride	X		
Hydrogen iodide	X		
Hydrogen peroxide	X		
Hydrogen sulfide	X		
Hypochlorous acid	X		
Iodine (<1000°C)	X		
Lactic acid	X		
Magnesium chloride	X		
Magnesium sulfate	X		
Mercuric chloride	X		

TABLE 8.28 Chemical Reactivity of Tantalum (*Continued*)

Environment	Resistant	Variable	Nonresistant
Methyl sulfuric acid	X		
Nickel chloride	X		
Nickel sulfate	X		
Nitric acid	X		
Nitric acid, fuming	X		
Nitric oxides	X		
Nitrous acid	X		
Nitrosyl chloride	X		
Organic chloride	X		
Oxalic acid	X		
Perchloric acid	X		
Phenol	X		
Phosphoric acid < 4 ppm F^-	X		
Pickling acids (except aqua regia)	X		
Phthalic anhydride	X		
Potassium carbonate		X	
Potassium chloride	X		
Potassium dichromate	X		
Potassium hydroxide (dilute)		X	
Potassium hydroxide (concentrated)			X
Potassium iodide-iodine	X		
Silver nitrate	X		
Sodium bisulfate, molten			X
Sodium bisulfate, solution	X		
Sodium bromide	X		
Sodium carbonate		X	
Sodium chlorate	X		
Sodium chloride	X		
Sodium hydroxide (dilute)		X	
Sodium hydroxide (concentrated)			X
Sodium hypochlorite	X		
Sodium nitrate	X		
Sodium sulfate	X		
Sodium sulfide		X	
Sodium sulfite	X		
Stannic chloride	X		
Sulfur (< 500°C)	X		
Sulfur dioxide	X		
Sulfur trioxide			X
Sulfuric acid (> 160°C)	X		
Zinc chloride	X		
Zinc sulfate	X		
Liquid Metals			
Bismuth (< 900°C)	X		
Gallium (< 450°C)	X		
Lead (< 1000°C)	X		
Lithium (< 1000°C)	X		
Magnesium (< 1150°C)	X		
Mercury (< 600°C)	X		
Sodium (1000°C)	X		
Sodium-potassium alloys (< 1000°C)	X		
Zinc (< 500°C)	X		

absorption cross section, it is used for radiation shielding. Tantalum mill products are used in the fabrication of corrosion-resistant process equipment including reaction vessels, columns, bayonet heaters, shell and tube heat exchangers, U-tubes, thermowells, spargers, rupture diaphragms, and orifices.

Tantalum equipment is frequently used in conjunction with glass, glass-lined steel, and other nonmetallic construction materials. Tantalum is also used extensively to repair damage and flaws in glass-lined steel equipment.

Working characteristics. Tantalum is extremely workable. It can be cold worked with standard equipment. Because of its bcc crystal structure, tantalum is a very ductile metal that can undergo cold reductions of more than 95% without failure. It can be rolled, forged, blanked, formed, and drawn. It is also machinable with high-speed carbide tools using a suitable coolant. Annealing tantalum is accomplished by heating the metal in a high vacuum to temperatures above 1100°.

Most procedures used in working and fabricating tantalum are conventional and can be mastered without very much difficulty. However, two important characteristics of tantalum must constantly be kept in mind:[39]

1. Annealed tantalum, like copper, lead, stainless steel, and some other metals, is "sticky." Therefore, it has a strong tendency to seize, tear, and gall. To avoid this, specific lubricant and die material combinations are required in high-pressure forming operations.
2. All forming, bending, stamping, and deep drawing operations are normally performed cold. Heavy sections can be heated for forging to approximately 425°C.

Welding. Tantalum may be welded to several other metals. This can be readily accomplished by resistance welding, tungsten-inert gas, plasma welding, and electron beam welding. Formation of brittle intermetallic phases is likely with many metals and must be avoided. Surfaces to be heated above 300°C should be protected by an inert gas such as argon or helium to prevent embrittlement. Tantalum may also be welded to itself by inert gas arc welding. Acetylene torch welding is destructive to tantalum.

Resistance welding can be performed with conventional equipment. The methods applied are not substantially different from those used in welding other materials. Because its melting point is 1500°C higher than that of SAE 1020 steel and its resistivity is only two-thirds that of SAE 1020 steel, tantalum requires a higher-power input to accomplish a sound weld. The weld duration should be kept as short as possible (i.e., in the range of one to ten cycles at 60 Hz). This is to prevent excessive external heating. Where possible, the work should be flooded with water for cooling and reduction of oxidation.

Strong, ductile welds can be made by the TIG method. Helium, argon, or a mixture of the two gases creates an atmosphere that prevents embrittlement by absorption of oxygen, nitrogen, or hydrogen into the heated metal. Where a pure, inert atmosphere is provided, the fusion and adjacent area will be ductile. Extreme high ductility can be obtained in a welding chamber that can be evacuated and purged with inert gas.[39]

Corrosion resistance. Tantalum is practically inert to most oxidizing and reducing acids, except fuming sulfuric. It is attacked by hot alkalis and hydrofluoric acid. However, it is very susceptible to hydrogen pickup and embrittlement. In the chemical process industries, tantalum is predominantly used in bayonet heaters, heat exchangers, orifice plates, valves, and tantalum-plated tubes. Tantalum patches are applied for the repair of holidays in glass-lined steel vessels. However, these must be electrically isolated from other metallic components in the vessel to avoid hydrogen embrittlement. Other applications include electrodes in thermionic valves, capacitors, surgical implants, and corrosion-resistant linings in chemical industry. Because of its high cost and lack of strength compared to its easy fabricability, tantalum is usually used as a lining over a stronger, less-expensive base material. Most tantalum piping consists of thin-wall tubing inside of carbon steel pipe.

Two main advantages of tantalum are that its anodic film has better dielectric properties than aluminum, and it has a very low ductile-brittle transition temperature. Tantalum also has a versatile aqueous corrosion resistance. In most environments, tantalum is comparable to glass in corrosion resistance, whereas it has physical and mechanical properties similar to mild steel. Tantalum is also resistant to attack by many liquid metals such as Li<1000°C, Na, K + NaK <1000°C, ThMg <850°C, U <1400°C, Zn <450°C, Pb <850°C, Bi <500°C, and Hg <600°C.

Tantalum resists most acids but is attacked by HF and by caustic acids. Unlike glass, however, it is also attacked by fuming sulfuric acid, sulfur dioxide, and chlorosulfonic acid. Due to its very high cost, its use is limited to extremely severe corrosive conditions. Another limitation is that it combines with most gases above 500°C and is susceptible to hydrogen embrittlement.[39] Table 8.28 gives ratings for the resistance of tantalum to a wide spectrum of chemical environments.[36]

8.6.5 Tungsten

Tungsten is a heavy white metal and possesses the highest melting point of all metals. It is widely distributed in small quantities in nature, being about half as abundant as copper. The metal is brittle and difficult to fabricate. Tungsten has a wide usage in alloy steels, magnets, heavy metals, electric contacts, light bulb filaments, rocket nozzles, and electronic applications. Parts, rods, and sheet are made by powder metallurgy using tungsten powder of 99.99% purity, and rolling and forging are performed at high temperatures. The rolled metal and drawn wire have exceptionally high strength and hardness. Tungsten wire for spark plug and wire electronic use is made by powder metallurgy. Tungsten whiskers are used in copper alloys to provide strength. Tungsten has the highest melting point, 3410°C, of the four common refractory metals. In addition, with a density of 19.3 gm·cm^{-3}, it is only surpassed by rhenium and osmium in weight.

Tungsten has a long history of use for filaments in the lamp industry. It offers exceptionally high strength at very high temperatures. In fact, it has the best high-temperature strength of the four common refractory metals. Its high-temperature strength, combined with its good electrical resistivity, have made it a popular choice for other applications in addition to filaments.[40] It is used for heating elements in vacuum furnaces that exceed the temperatures of molybdenum and tantalum as well as other heater applications. Tungsten has also gained wide acceptance as an essential material in electrical contacts, glass-to-metal seals, supports, and electrodes.

Tungsten's properties lend themselves to other metals when alloyed. Tungsten carbide has long been the choice for durable cutting tools. Tungsten's high density is used in conjunction with copper, nickel, iron, and cobalt to form heavy metal. This is an alloy containing 90 to 97% tungsten, and the other metals are used as a binder to keep the tungsten together and to give it machinable properties as well as to temper the brittleness of pure tungsten.

Applications of tungsten. There is an increasing demand from the electronics, nuclear, and aerospace industries for materials that maintain reliability under ever-increasing temperature conditions. Because its properties meet these requirements, tungsten also is experiencing an increasing demand. Characteristics that support the demand for tungsten in a multitude of electronics and high-temperature applications are as follows:[40]

- Strength and stiffness at high temperatures
- Good thermal conductivity

- Low thermal expansion
- Low emissivity

Tungsten has a coefficient of expansion approximating that of hard glass. For this reason, it is used extensively in glass-to-metal seals in hard glass lamp and electronic applications. Under special conditions, it may also be used with quartz. Because tungsten rod has a high degree of strength at elevated temperatures, it is utilized structurally to hold or support high-temperature sources such as filaments and heaters for lamp and electronic uses. Tungsten rod that is specially processed and manufactured for welding rod applications is used extensively in such processes as inert gas–shielded arc welding and atomic hydrogen arc welding.

Other types of tungsten rod are used for electrodes. These types, both regular and thoriated, are used for electrodes in vacuum melting processes, resistance welding, and electro-discharge machining. For tube applications, especially flash and xenon tubes, tungsten is used either pure or thoriated at 1 and 2% for greater emissivity.

Working characteristics. Tungsten is very difficult to machine and fabricate. With experience, it can be turned. Milling is all but impossible. It is only done with great difficulty and at high cost by those most experienced with it. Forming must be done at very high temperatures and with careful stress relieving. Welding is not recommended, and riveting is difficult at best. Extreme care must be exercised when designing a component from tungsten.

Bonding to other metals. Tungsten is best joined to other metals by brazing. Most of the high-temperature brazes can be used. When brazing, an excess of nickel-base filler metals should be avoided because the interaction between tungsten and nickel results in the recrystallizing of tungsten. Contact with graphite should also be avoided to prevent the formation of brittle tungsten carbides. When welded, the weld will be very brittle and the probability of delamination and cracking is high. For all practical purposes, it is extremely difficult to rivet tungsten because of its fragile nature. However, this may be successfully performed in some low-stress situations.

Of all the potential contaminants in wrought products, iron is of primary concern. Others, such as aluminum, carbon, calcium, copper, or nickel, may also be present as elements, but they are more frequently present in the form of oxides. Removal of a controlled amount of base metal may be desired to ensure complete removal of contaminants. There are four main processes used to clean tungsten:

Molten salt. This is one of the most common cleaning processes, requiring simple immersion in a molten bath containing oxidizing agents. This process will not attack the basis metal.

Aqueous alkaline solutions. This process works well on oxidized (yellow tungsten) surfaces. Reduced or intermediate oxides (brown, purple, etc.) will react more slowly to this process, if at all. This process is similar to the use of molten salts in that it will not attack the base metal and it requires an oxidizing agent to work.

Acid solutions. Tungsten is much less reactive to individual acids than most common metals. HCl, HF, and H_2SO_4 have essentially no effect. When tungsten is treated with acid solutions, it frequently is stained by residual oxides even if rapid and thorough rinsing is used.

Electrolytic methods. Electrolytic etching is the removal of basis metal by an applied voltage in a medium capable of dissolving the products of the electrolytic reaction. This may be done in molten salts or aqueous solutions. Electrical current and time determine the amount of metal removal.

For rapid attack of heavy scale, molten salt is far superior to the other methods. In addition, if no oxidizer is present, it can be performed with no fear of basis metal loss. If appreciable sizes or volumes of material are to be processed, particularly with significant basis metal removal, acid solutions present a disposal, as well as an operational, problem. The utility of electroetching is more dependent on geometry than the other methods. It will work well for treating continuous lengths of wire; however, there is a contact problem if the cleaning is to be performed on many small parts.

Corrosion resistance. Table 8.29 gives ratings for the resistance of tungsten to a wide spectrum of chemical environments.[36]

8.7 Stainless Steels

8.7.1 Introduction

Stainless and heat-resisting steels possess unusual resistance to attack by corrosive media at atmospheric and elevated temperatures and are produced to cover a wide range of mechanical and physical properties for particular applications. Along with iron and chromium, all stainless steels contain some carbon. It is difficult to get much less than about 0.03%, and sometimes carbon is deliberately added up to 1.00% or more. The more carbon there is, the more chromium must be used, because carbon can take from the alloy about 17 times its own

TABLE 8.29 Chemical Reactivity of Tungsten

Environment	Resistant	Variable	Nonresistant
Aluminum oxide-oxidation			X
Ammonia	X		
Ammonia (< 700°C)	X		
Ammonia (> 700°C)		X	
Ammonia in presence of H_2O_2		X	
Aqua regia (cold)	X		
Aqua regia (warm/hot)			X
Aqueous caustic soda/potash	X		
Bromine (at red heat)			X
Carbon (> 1400°C) carbide formation			X
Carbon dioxide (> 1200°C) oxidation			X
Carbon disulfide (red heat)			X
Carbon monoxide (< 800°C)	X		
Carbon monoxide (> 800°C)		X	
Chlorine (> 250°C)		X	
Fluorine			X
Hydrochloric acid	X		
Hydrofluoric acid	X		
Hydrogen	X		
Hydrogen sulfide (red heat)		X	
Hydrogen/chloride gas (< 600°C)	X		
In air		X	
In presence of KNO_2, KNO_3, $KCLO_3$, PbO_2			X
Iodine (at red heat)			X
Magnesium oxide-oxidation			X
Mercury (and vapor)	X		
Nitric acid	X		
Nitric oxide (hot) oxidation			X
Nitric/hydrofluoric mixture			X
Nitrogen	X		
Oxygen or air (< 400°C)	X		
Oxygen or air (> 400°C)		X	
Sodium nitrite (molten)			X
Sulfur (molten, boiling)		X	
Sulfur dioxide (red heat)			X
Sulfuric acid		X	
Thorium oxide (> 2220°C) oxidation			X
Water	X		
Water vapor (red heat) oxidation			X

weight of chromium to form carbides. Chromium carbide is of little use for resisting corrosion. The carbon, of course, is added for the same purpose as in ordinary steels, to make the alloy stronger.

Other alloying elements are added for improved corrosion resistance, fabricability, and variations in strength. These elements include appreciable amounts of nickel, molybdenum, copper, titanium, silicon, aluminum, sulfur, and many others that cause pronounced metallurgical changes. The commonly recognized standard types of stainless steels follow. The chemical compositions of stainless steels are given in App. F.

- *Austenitic.* A family of alloys containing chromium and nickel, generally built around the type 302 chemistry of 18% Cr, 8% Ni. Austenitic grades are those alloys that are commonly in use for stainless applications. The austenitic grades are not magnetic. The most common austenitic alloys are iron-chromium-nickel steels and are widely known as the 300 series. The austenitic stainless steels, because of their high chromium and nickel content, are the most corrosion resistant of the stainless group, providing unusually fine mechanical properties. They cannot be hardened by heat treatment but can be hardened significantly by cold working. The straight grades of austenitic stainless steel contain a maximum of .08% carbon. Table 8.30 describes basic mechanical properties for many commercial austenitic stainless steels.

    The "L" grades are used to provide extra corrosion resistance after welding. The letter L after a stainless steel type indicates low carbon (as in 304L). The carbon content is kept to .03% or less to avoid grain boundary precipitation of chromium carbide in the critical range (430 to 900°C). This deprives the steel of the chromium in solution and promotes corrosion adjacent to the grain boundaries. By controlling the amount of carbon, this is minimized. For weldability, the L grades are used.

    The H grades contain a minimum of .04% and a maximum of .10% carbon and are primarily used for higher-temperature applications.

- *Ferritic.* Ferritic alloys generally contain only chromium and are based upon the type 430 composition of 17% Cr. These alloys are somewhat less ductile than the austenitic types and again are not hardenable by heat treatment. Ferritic grades have been developed to provide a group of stainless steels to resist corrosion and oxidation, while being highly resistant to SCC. These steels are magnetic but cannot be hardened or strengthened by heat treatment. They can be cold worked and softened by annealing. As a group, they are more corrosive resistant than the martensitic grades but are generally inferior to the austenitic grades. Like martensitic grades, these are straight chromium steels with no nickel. They are used for decorative trim, sinks, and automotive applications, particularly exhaust systems. Table 8.31 describes basic mechanical properties for many commercial ferritic stainless steels.

- *Martensitic.* These stainless steels may be hardened and tempered just like alloy steels. Their basic building block is type 410, which consists of 12% Cr, 0.12% C. Martensitic grades were developed to provide a group of corrosion-resistant stainless alloys that can be hardened by heat treating. The martensitic grades are straight chromium steels containing no nickel and they are magnetic. The martensitic grades are mainly used where hardness, strength, and wear resistance are required. Table 8.32 describes basic mechanical properties for many commercial austenitic stainless steels.

TABLE 8.30 Nominal Mechanical Properties of Austenitic Stainless Steels

UNS	Type	Tensile, MPa	Yield (0.2%), MPa	Elongation, %	Hardness (Rockwell)	Product form
S20100	201	655	310	40	B90	
S20200	202	612	310	40	B90	
S20500	205	831	476	58	B98	Plate
S30100	301	758	276	60	B85	
S30200	302	612	276	50	B85	
S30215	302B	655	276	55	B85	
S30300	303	621	241	50		Bar
S30323	303Se	621	241	50		Bar
S30400	304	579	290	55	B80	
S30403	304L	558	269	55	B79	
S30430	S30430	503	214	70	B70	Wire
S30451	304N	621	331	50	B85	
S30500	305	586	262	50	B80	
S30800	308	793	552	40		Wire
S30900	309	621	310	45	B85	
S30908	309S	621	310	45	B85	
S31000	310	655	310	45	B85	
S31008	310S	655	310	45	B85	
S31400	314	689	345	40	B85	
S31600	316	579	290	50	B79	
S31620	316F	586	262	60	B85	
S31603	316L	558	290	50	B79	
S31651	316N	621	331	48	B85	
S31700	317	621	276	45	B85	
S31703	317L	593	262	55	B85	
	317LMN	662	373	49	B88	
S32100	321	621	241	45	B80	
N08830	330	552	262	40	B80	
S34700	347	655	276	45	B85	
S34800	348	655	276	45	B85	
S38400	384	517	241	55	B70	Wire
N08020	20Cb-3	550	240	30		

TABLE 8.31 Mechanical Properties of Ferritic Stainless Steels (Annealed Sheet Unless Noted Otherwise)

UNS	Type	Tensile strength, MPa	Yield strength (0.2%), MPa	Elongation (50 mm), %	Hardness (Rockwell)	Product form
S40500	405	448	276	25	B75	
S40900	409	446	241	25	B75	
S42900	429	483	276	30	B80	Plate
S43000	430	517	345	25	B85	
S43020	430F	655	586	10	B92	
S43023	430FSe	655	586	10	B92	Wire
S43400	434	531	365	23	B83	
S43600	436	531	365	23	B83	
S44200	442	552	310	20	B90	Bar
S44600	446	552	345	20	B83	

TABLE 8.32 Mechanical Properties of Martensitic Stainless Steels (Annealed Sheet Unless Noted Otherwise)

UNS	Type	Tensile strength, MPa	Yield strength (0.2%), MPa	Elongation (50 mm), %	Hardness (Rockwell)	Product form
S40300	403	483	310	25	B80	
S41000	410	483	310	25	B80	
S41400	414	827	724	15	B98	
S43000	416	517	276	30	B82	Bar
S42000	416Se	517	276	30	B82	Bar
S42200	420	655	345	25	B92	Bar
S43100	420F	655	379	22	220 (Brinell)	Bar
S41623	422	1000	862	18		Bar
S42020	431	862	655	20	C24	Bar
S44002	440A	724	414	20	B95	Bar
S44004	440B	738	427	18	B96	Bar
S44004	440C	758	448	14	B97	Bar

*Hardened and tempered.

- *Precipitation-hardening (PH).* These alloys generally contain Cr and less than 8% Ni, with other elements in small amounts. As the name implies, they can be hardened by heat treatment. Precipitation hardening grades, as a class, offer the designer a unique combination of fabricability, strength, ease of heat treatment, and corrosion resistance not found in any other class of material. These grades include 17Cr-4Ni (17-4PH) and 15Cr-5Ni (15-5PH). The austenitic precipitation hardenable alloys have, to a large extent, been replaced by the more sophisticated and higher-strength superalloys. The martensitic precipitation hardenable stainless steels are really the workhorses of the family. Although designed primarily as a material to be used for bar, rods, wire, forgings, and so forth, martensitic precipitation hardenable alloys are beginning to find more use in the flat rolled form. The semi-austenitic precipitation hardenable stainless steels were primarily designed as a sheet and strip product, but they have found many applications in other product forms. Developed primarily as aerospace materials, many of these steels are gaining commercial acceptance as truly cost-effective materials in many applications.
- *Duplex.* This is a stainless steel alloy group, with two distinct microstructure phases—ferrite and austenite. The duplex alloys have greater resistance to chloride SCC and higher strength than the other austenitic or ferritic grades. Duplex grades are the newest of the stainless steels. These materials are a combination of austenitic and ferritic material. Modern duplex stainless steels have been developed to take advantage of the high strength and hardness,

Materials Selection

TABLE 8.33 Minimum Mechanical Properties of Duplex Stainless Steels

UNS	Type	Yield strength (0.2%), MPa	Tensile strength, MPa	Elongation, %
S32900	329	485	620	15
S31200	44LN	450	690	25
S31260	DP-3	450	690	25
S31500	3RE60	440	630	30
S31803	2205	450	620	25
S32550	Ferralium 255	550	760	15
S32950	7-Mo PLUS.	485	690	15

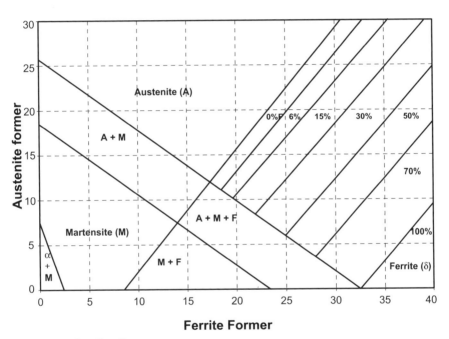

Figure 8.6 Schaeffler diagram.

erosion, fatigue and SCC resistance, high thermal conductivity, and low thermal expansion produced by the ferrite-austenite microstructure. These steels have a high chromium content (18 to 26%), low amounts of nickel (4 to 8%), and generally contain molybdenum. They are moderately magnetic, cannot be hardened by heat treatment, and can readily be welded in all section thicknesses. Duplex stainless steels are less notch sensitive than ferritic types but suffer loss of impact strength if held for extended periods of high temperature above (300°C). Duplex stainless steels thus combine some of the features of the two major classes. They are resistant to SCC, albeit not quite as resistant as the ferritic steels, and their toughness is superior to that of the ferritic steels but inferior to that of the austenitic steels. Duplex steel's yield strength is appreciably greater than that of the annealed austenitic steels by a factor of about two. Table 8.33 describes basic mechanical properties for many commercial austenitic stainless steels.

- *Cast.* The cast stainless steels are similar to the equivalent wrought alloys. Most of the cast alloys are direct derivatives of one of the wrought grades, as C-8 is the cast equivalent of wrought type 304. The C preceding a designation means that the alloy is primarily used for resistance to liquid corrosion. An H designation indicates high-temperature applications.

8.7.2 Welding, heat treatments, and surface finishes

Weldability. An aid in determining which structural constituents can occur in a weld metal is the Schaeffler-de-Long diagram. With knowledge of the properties of different phases, it is possible to judge the extent to which they affect the service life of the weldment. The diagram indicates the structure obtained after rapid cooling to room temperature from 1050°C and is not an equilibrium diagram. It was originally established to provide a rough estimate of the weldability of different austenitic steels. In creating the diagram, the alloying elements commonly used for making stainless steels are categorized as either austenite or ferrite stabilizers.[41] In this diagram the ferrite number (FN) is an international measure of the delta or solidification ferrite content of the weld metal at room temperature. The Cr(ferrite former) and Ni(austenite former) equivalents that form the two axes of the Schaeffler diagram in Fig. 8.6 can be estimated with the following relations:[42]

$$\%\text{Cr equivalent} = 1.5\ \text{Si} + \text{Cr} + \text{Mo} + 2\ \text{Ti} + 0.5\ \text{Nb}$$

$$\%\text{Ni equivalent} = 30\ (\text{C} + \text{N}) + 0.5\ \text{Mn} + \text{Ni} + 0.5\ (\text{Cu} + \text{Co})$$

Austenitic steels. Steels S30400, S31600, S30403, and S31603 have very good weldability. The old problem of intergranular corrosion after welding is very seldom encountered today. The steels suitable for wet corrosion either have carbon contents below 0.05% or are niobium or titanium stabilized. They are also very unsusceptible to hot cracking, mainly because they solidify with a high ferrite content. The higher-alloy steels such as S31008 and N08904 solidify with a fully austenitic structure when welded. They should therefore be welded using a controlled heat input. Steel and weld metal with high chromium and molybdenum contents may undergo precipitation of brittle sigma phase in their microstructure if they are exposed to high temperatures for a certain length of time. The transformation from ferrite to sigma or directly from austenite to sigma proceeds most rapidly within the temperature range 750 to 850°C. Welding with a high heat input leads to slow cooling, especially in light-gage weldments. The weld's holding time between 750 and 850°C then increases, and along with it the risk of sigma phase formation.

Ferritic steels. Ferritic steels are generally more difficult to weld than austenitic steels. This is the main reason they are not used to the same extent as austenitic steels. The older types, such as AISI 430 (S43000), had greatly reduced ductility in the weld. This was mainly due to strong grain growth in the HAZ but also to precipitation of martensite in the HAZ. They were also susceptible to intergranular corrosion after welding. These steels are therefore often welded with preheating and postweld annealing. Modern ferritic steels of type S44400 and S44635 have considerably better weldability due to low carbon and nitrogen contents and stabilization with titanium/niobium. However, there is always a risk of unfavorable grain enlargement if they are not welded under controlled conditions using a low heat input. They do not normally have to be annealed after welding. These steels are welded with matching or austenitic superalloyed filler.[43]

Duplex steels. Modern duplex steels have considerably better weldability than earlier grades. They can be welded more or less as common austenitic steels. Besides being susceptible to intergranular corrosion, the old steels were also susceptible to ferrite grain growth in the HAZ and poor ferrite to austenite transformation, resulting in reduced ductility. Modern steels, which have a higher nickel content and are alloyed with nitrogen, exhibit austenite transformation in the HAZ that is sufficient in most cases. However, extremely rapid cooling after welding, for example, in a tack or in a strike mark, can lead to an unfavorably high ferrite content. Extremely high heat input, as defined subsequently, can also lead to heavy ferrite grain growth in the HAZ.[43]

$$\text{Heat input} = \eta \frac{UI}{1000v}$$

where η = constant dependent on welding method (0.7 to 1.0)
 U = voltage (V)
 I = current (A)
 v = welding speed (mm · s^{-1})

When welding S31803 (alloy 2205) in a conventional way (0.6 to 2.0 kJ·mm^{-1}) and using filler metals at the same time, a satisfactory ferrite-austenite balance can be obtained. For the new superduplex stainless steel S32750 (alloy 2507) a different heat input is recommended (0.2 to 1.5 kJ·mm^{-1}). The reason for lowering the minimum value is that this steel has a much higher nitrogen content than S31803. The nitrogen favors a fast reformation of austenite, which is important when welding with a low heat input. The maximum level is lowered to minimize the risk of secondary phases.

These steels are welded with duplex or austenitic filler metals. Welding without filler metal is not recommended without subsequent quench annealing. Nitrogen affects not only the microstructure but also the weld pool penetration. Increased nitrogen content reduces the penetration into the parent metal. To avoid porosity in TIG welding it is recommended to produce thin beads. To achieve the highest possible pitting corrosion resistance at the root side in ordinary S31803 weld metals, the root gas should be Ar + N^2 or Ar + N^2 + H$_2$. The use of H$_2$ in the shielding gas is not recommended when welding superduplex steels. When welding S31803 with plasma, a shielding gas containing Ar + 5% H$_2$ is sometimes used in combination with filler metal and followed by quench annealing.

Martensitic and martensitic-austenitic steels. The quantity of martensite and its hardness are the main causes of the weldability problems encountered with these steels. The fully martensitic steels are air hardening. The steels are therefore very susceptible to hydrogen embrittlement. By welding at an elevated temperature, the HAZ can be kept austenitic and tough throughout the welding process. After cooling, the formed martensite must always be tempered at about 650 to 850°C, preferably as a concluding heat treatment. However, the weld must first have been allowed to cool to below about 150°C.

Martensitic-austenitic steels, such as 13Cr/6Ni and 16Cr/5Ni/2Mo, can often be welded without preheating and without postweld annealing. Steels of the 13Cr/4Ni type with a low austenite content must, however, be preheated to a working temperature of about 100°C. If optimal strength properties are desired, they can be heat treated at 600°C after welding. The steels are welded with matching or austenitic filler metals.

Filler metals for stainless steels

Austenitic filler metals. Most common stainless steels are welded with filler metals that produce weld metal with 2–12% FN at room temperature. The risk of hot cracking can be greatly reduced with a small percentage of ferrite in the metal because ferrite has much better solubility for impurities than austenite. These filler metals have very good weldability. Heat treatment is generally not required.

High-alloy filler metals with chromium equivalents of more than about 20 can, if the weld metal is heat treated at 550 to 950°C, give rise to embrittling sigma phase. High molybdenum contents in the filler metal, in combination with ferrite, can cause sigma phase during welding if a high heat input is used. Multipass welding has the same effect. Sigma phase reduces ductility and can promote hot cracking. Heat input should be limited for these filler metals. Nitrogen-alloyed filler metals produce weld metals that do not precipitate sigma phase as readily.

Nonstabilized filler metals, with carbon contents higher than 0.05%, can give rise to chromium carbides in the weld metal, resulting in poorer wet corrosion properties. Modern nonstabilized filler metals, however, generally have no more than 0.04% carbon unless they are intended for high-temperature applications.

Superalloyed filler metals with high ferrite numbers (15 to 40%) are often used in mixed weld connections between low-alloy filler metals and stainless steel. Weldability is very good. By using such filler metals, mixed weld metals of the austenitic type can be obtained. The use of filler metals of the ordinary austenitic type for welding low-alloy filler metals to stainless steel can, owing to dilution, result in a brittle martensitic-austenitic weld metal.

Other applications for superalloyed filler metals are in the welding of ferritic and ferritic-austenitic steels. The most highly alloyed, with 29Cr-9Ni, are often used where the weld is exposed to heavy wear or for welding of difficult-to-weld steels, such as 14% Mn steel, tool steel, and spring steel.

Fully austenitic weld metals. Sometimes ferrite-free metals are required because there is usually a risk of selective corrosion of the ferrite. Fully austenitic weld metals are naturally more susceptible to hot cracking than weld metals with a small percentage of ferrite. To reduce the risk, they are often alloyed with manganese, and the level of trace elements is minimized. Large weld pools also increase the risk of hot cracks.

A large fully austenitic weld pool solidifies slowly with a coarse structure and a small effective grain boundary area. A small weld pool solidifies quickly, resulting in a finer-grained structure. Because trace elements are often precipitated at the grain boundaries, the precipitates are larger in a coarse structure, which increases the risk that the precipitates will weaken the grain boundaries to such an extent that microfissures form. Many microfissures can combine to form visible hot cracks.

Fully austenitic filler metals should therefore be welded with low heat input. Because the filler metal generally has lower trace element contents than the parent metal, the risk of hot cracking will be reduced if a large quantity of filler metal is fed down into the weld pool. Because the weld metal contains no ferrite, its impact strength at low temperature is very good. This is important to manufacturers of, for example, welded tanks used to transport cryogenic liquids.

Ferritic filler metals. Fully ferritic filler metals have previously been regarded as very difficult to weld. They also required heat treatment of the weld metal after welding. Those that are used today have very low carbon and nitrogen contents and are often stabilized with titanium. Modern filler metals therefore produce weld metals that are less sensitive to intergranular corrosion. Nor is any postweld heat treatment necessary. Another very important phenomenon that applies to all fully ferritic metals is that they tend to give rise to a coarse crystalline structure in the weld metal. Ductility decreases greatly with increasing grain size. These filler metals must therefore be welded using low heat input.

Weld imperfections

Austenitic stainless steel. Although austenitic stainless steel is readily welded, weld metal and HAZ cracking can occur. Weld metal solidification cracking is more likely in fully austenitic structures, which are more crack sensitive than those containing a small amount of ferrite. The beneficial effect of ferrite has been attributed largely to its capacity to dissolve harmful impurities that would otherwise form low melting-point segregates and interdendritic cracks.

Because the presence of 5 to 10% ferrite in the microstructure is extremely beneficial, the choice of filler material composition is crucial in suppressing the risk of cracking. An indication of the ferrite-austenite balance for different compositions is provided by the Schaeffler diagram. For example, when welding Type 304 stainless steel, a Type 308 filler material that has a slightly different alloy content is used.

Ferritic stainless steel. The main problem when welding ferritic stainless steel is poor HAZ toughness. Excessive grain coarsening can lead to cracking in highly restrained joints and thick-section material. When welding thin-section material (less than 6 mm), no special precautions are necessary.

In thicker material, it is necessary to employ a low heat input to minimize the width of the grain coarsened zone and an austenitic filler to produce a tougher weld metal. Although preheating will not reduce the grain size, it will reduce the HAZ cooling rate, maintain the weld metal above the ductile-brittle transition temperature, and may reduce residual stresses. Preheat temperature should be within the range 50 to 250°C, depending on material composition.

Martensitic stainless steel. The material can be successfully welded, providing precautions are taken to avoid cracking in the HAZ, especially in thick-section components and highly restrained joints. High hardness in the HAZ makes this type of stainless steel very prone to hydrogen cracking. The risk of cracking generally increases with the carbon content. Precautions that must be taken to minimize the risk include

- Using a low-hydrogen process (TIG or MIG) and ensuring that the flux or flux-coated consumable are dried (MMA and SAW) according to the manufacturer's instructions.
- Preheating to around 200 to 300°C. The actual temperature will depend on welding procedure, chemical composition (especially Cr and C content), section thickness, and the amount of hydrogen entering the weld metal.
- Maintaining the recommended minimum interpass temperature.
- Carrying out postweld heat treatment (e.g., at 650 to 750°C). The time and temperature will be determined by chemical composition.

Thin-section, low-carbon material, typically less than 3 mm, can often be welded without preheat, providing that a low-hydrogen process is used, the joints have low restraint, and attention is paid to cleaning the joint area. Thicker-section and higher-carbon (>0.1%) material will probably need preheat and postweld heat treatment. The postweld heat treatment should be carried out immediately after welding not only to temper (toughen) the structure but also to enable the hydrogen to diffuse away from the weld metal and HAZ.

Duplex stainless steels. Modern duplex steels can be readily welded, but the procedure, especially maintaining the heat input range, must be strictly followed to obtain the correct weld metal structure. Although most welding processes can be used, low heat input welding procedures are usually avoided. Preheat is not normally required, and the maximum interpass temperature must be controlled. Choice of filler is important because it is designed to produce a weld metal structure with a ferrite-austenite balance to match the parent metal. To compensate for nitrogen loss, the filler may be overalloyed with nitrogen, or the shielding gas itself may contain a small amount of nitrogen.

Heat treating stainless steels. Wrought stainless steels are solution annealed after processing and hot worked to dissolve carbides and the sigma phase. Carbides may form during heating in the 425 to 900°C range or during slow cooling through this range. Sigma tends to form at temperatures below 925°C. Specifications normally require solution annealing to be done at 1035°C with a rapid quench. The molybdenum-containing grades are frequently solution annealed at somewhat higher temperatures in the 1095 to 1120°C range to better homogenize the molybdenum.

Stainless steels may be stress relieved. There are several stress relief treatments. When stainless steel sheet and bar are cold reduced greater than about 30% and subsequently heated to 290 to 425°C, there is a significant redistribution of peak stresses and an increase in both tensile and yield strength. Stress redistribution heat treatments at 290 to 425°C will reduce movement in later machining operations and are occasionally used to increase strength. Because stress redistribution treatments are made at temperatures below 425°C, carbide precipitation and sensitization to intergranular attack (IGA) are not a problem for the higher carbon grades.

Stress relief at 425 to 595°C is normally adequate to minimize distortion that would otherwise exceed dimensional tolerances after machining. Only the low-carbon L grades or the stabilized S32100 and S34700 grades should be used in weldments to be stress relieved above 425°C because the higher carbon grades are sensitized to IGA when heated above about 25°C.

Stress relief at 815 to 870°C is occasionally needed when a fully stress relieved assembly is required. Only the low-carbon L grades, S32100 and S34700, should be used in assemblies to be heat treated in this range. Even though the low-carbon and stabilized grades are used, it is best to test for susceptibility to IGA per ASTM A262 to be certain there was no sensitization during stress relief treating in this temperature range. Thermal stabilization treatments at 900°C minimum for 1 to 10 h are occasionally employed for assemblies that are to be used in the 400 to 900°C temperature range. Thermal stabilization is intended to agglomerate the carbides, thereby preventing further precipitation and IGA.[44]

Surface finishes. After degreasing, metallic surface contaminants such as iron embedded in fabrication shop forming and handling, weld splatter, heat tint, inclusions, and other metallic particles must be removed to restore the inherent corrosion resistance of the stainless steel surface. Nitric-HF pickling (10% HNO_3, 2% HF at 49 to 60°C) is the most widely used and effective method for removing metallic surface contamination. Pickling may be done by immersion or locally using a pickling paste. Electropolishing, using oxalic or phosphoric acid for the electrolyte and a copper bar or plate for the cathode, can be equally effective. Electropolishing may be done locally to remove heat tint alongside of welds or over the whole surface. Both pickling and electropolishing remove a layer several atoms deep from the surface. Removal of the surface layer has the further benefit of removing surface layers that may have become somewhat impoverished in chromium during the final heat-treatment operation.

Glass bead and walnut shell blasting are very effective in removing metallic surface contamination without damaging the surface. It is sometimes necessary to resort to blasting with clean sand to restore heavily contaminated surfaces such as tank bottoms, but care must be taken to be certain the sand is truly clean, is not recycled, and does not roughen the surface. Steel shot blasting should not be used because it will contaminate the stainless steel with an iron deposit.

Stainless steel wire brushing or light grinding with clean aluminum oxide abrasive disks or flapper wheels are helpful. Grinding or polishing with grinding wheels or continuous belt sanders tend to overheat the surface layers to the point where resistance cannot be fully restored even with subsequent pickling. Brief descriptions of hot-rolled, cold-rolled, and mechanical finishes are presented in Table 8.34.

8.7.3 Corrosion resistance

Stainless steels are mainly used in wet environments. With increasing chromium and molybdenum contents, the steels become increasingly resistant to aggressive solutions. The higher nickel content reduces the risk of SCC. Austenitic steels are more or less resistant to general corrosion, crevice corrosion, and pitting, depending on the quantity of alloying elements. Resistance to pitting and crevice corrosion are very important if the steel is to be used in chloride-containing environments. Resistance to pitting and crevice corrosion typically increases with increasing contents of chromium, molybdenum, and nitrogen. The distribution of stainless steel's failure modes in chemical process industries is illustrated in Fig. 8.7.[45]

Chloride-rich seawater is a particularly harsh environment that can attack stainless steel by causing pitting and crevice corrosion. However, some unique stainless steel grades have been designed to cope with this environment. Alloy 254 SMO (S31254), for example, has a long record of successful installations for seawater handling within offshore, desalination, and coastal process industries. But even with a generally good track record, some crevice corrosion problems have been reported, and for critically severe crevice and temperature situations a better alloy would be 654 SMO (S32654).

Most molybdenum-free steels can be used at high temperatures in contact with hot gases. An adhesive oxide layer then forms on the surface of the steel. At very high temperatures, the oxide begins to scale. The corresponding scaling temperature increases with increasing chromium content. A common high-temperature steel, such as S31008, is Mo free and contains 24 to 26% Cr. Due to a balanced composition and the addition of cerium, among other elements, alloy 253 MA (S30815) can be even used at temperatures of up to 1150 to 1200°C in air.[43]

The influence of alloying elements. Corrosion resistance of stainless steels is a function not only of composition but also of heat treatment, surface condition, and fabrication procedures, all of which may change the thermodynamic activity of the surface and thus dramatically affect the cor-

TABLE 8.34 Descriptions of Common Stainless Steels Finishes

Hot-rolled finishes

No. 0 finish. Also referred to as hot-rolled annealed (HRA). In that process, plates are hot rolled to required thickness and then annealed. No pickling or passivation operations are effected, resulting in a scaled black finish. This does not develop the fully corrosion-resistant film on the stainless steel, and except for certain high-temperature heat-resisting applications, this finish is unsuitable for general use.

No. 1 finish. Plate is hot rolled, annealed, pickled, and passivated. This results in a dull, slightly rough surface, suitable for industrial applications that generally involve the range of plate thicknesses.

Cold-rolled finishes.

No. 2D finish. Material with a No. 1 finish is cold rolled, annealed, pickled, and passivated. This results in a uniform dull matte finish, superior to a No. 1 finish. Suitable for industrial application and eminently suitable for severe deep drawing because the dull surface (which may be polished after fabrication) retains the lubricant during the drawing operation.

No. 2B finish. Material with a 2D finish is given a subsequent light skin pass cold-rolling operation between polished rolls. A No. 2B finish is the most common finish produced and is called for on sheet material. It is brighter than 2D and is semireflective. It is commonly used for most deep drawing operations and is more easily polished to the final finishes required than is a 2D finish.

No. 2BA finish. This is more commonly referred to as a bright annealed (BA) finish. Material with a No. 1 finish is cold rolled using highly polished rolls in contact with the steel surface. This smooths and brightens the surface. The smoothness and reflectivity of the surface improves as the material is rolled to thinner and thinner sizes. Any annealing that needs to be done to effect the required reduction in gage, and the final anneal, is effected in a very closely controlled inert atmosphere. No oxidation or scaling of the surface therefore occurs, and there is no need for additional pickling and passivating. The final surface developed can have a mirror-type finish, similar in appearance to the highly polished No. 7 and No. 8 finishes.

Mechanically polished finishes

No. 3 finish. This is a ground unidirectional uniform finish obtained with 80–100 grit abrasive. It is a good intermediate or starting surface finish for use in such instances where the surface will require further polishing operations to a finer finish after subsequent fabrication or forming.

No. 4 finish. This is a ground unidirectional finish obtained with 150 grit abrasive. It is not highly reflective, but is a good general purpose finish on components that will suffer from fairly rough handling in service.

No. 6 finish. These finishes are produced using rotating cloth mops (tampico fiber, muslin, or linen) that are loaded with abrasive paste. The finish depends on how fine an abrasive is used and the uniformity and finish of the original surface. The finish has a nondirectional texture of varying reflectivity. Satin blend is an example of such a finish.

No. 7 finish. This is a buffed finish and has a high degree of reflectivity. It is produced by progressively using finer and finer abrasives and finishing with buffing compounds. Some fine scratches may remain from the original starting surface.

No. 8 finish. This is produced in an equivalent manner to a No. 7 finish, the final operation being done with extremely fine buffing compounds. The final surface is blemish free with a high degree of image clarity and is the true mirror finish.

Materials Selection 725

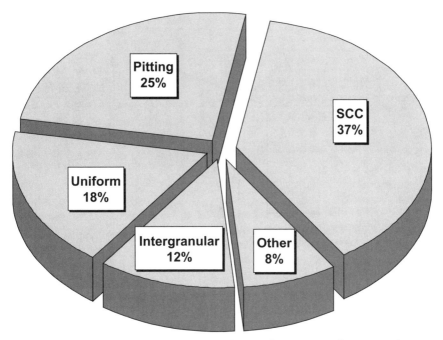

Figure 8.7 Distribution of stainless steel's failure modes in chemical process industries.

rosion resistance. It is not necessary to chemically treat stainless steels to achieve passivity. The passive film forms spontaneously in the presence of oxygen. Most frequently, when steels are treated to improve passivity (passivation treatment), surface contaminants are removed by pickling to allow the passive film to reform in air, which it does almost immediately. The principal alloying elements that affect the corrosion resistance of stainless are discussed below[46] and a schematic summary of the effects of alloying elements on the anodic polarization curve of typical stainless steels, initially presented by Sedriks, is shown in Fig. 8.8.[47]

Chromium. Chromium is, of course, the primary element for forming the passive film or high-temperature, corrosion-resistant chromium oxide. Other elements can influence the effectiveness of chromium in forming or maintaining the film, but no other element can, by itself, create the stainless characteristics of stainless steel. The passive film is observed at about 10.5% chromium, but it affords only limited atmospheric protection at this point. As chromium content is increased, the corrosion protection increases. When the chromium level reaches the 25 to 30% level, the passivity of the protective film is very high, and the high-temperature oxidation resistance is maximized.

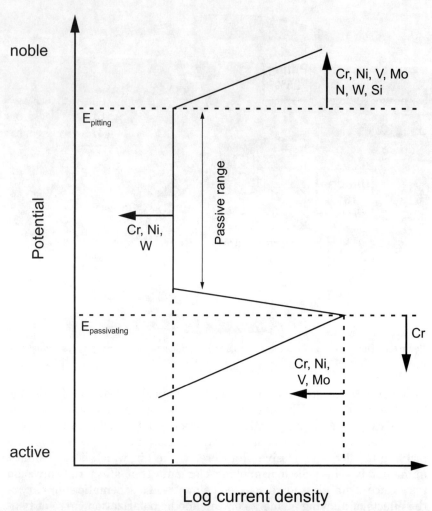

Figure 8.8 Schematic summary of the effects of alloying elements on the anodic polarization curve.

Nickel. In sufficient quantities, nickel is used to stabilize the austenitic phase and to produce austenitic stainless steels. A corrosion benefit is obtained as well, because nickel is effective in promoting repassivation, especially in reducing environments. Nickel is particularly useful in promoting increased resistance to mineral acids. When nickel is increased to about 8 to 10% (a level required to ensure austenitic structures in a stainless that has about 18% chromium), resistance to SCC is decreased. However, when nickel is increased beyond that level, resistance to SCC increases with increasing nickel content.

Manganese. An alternative austenite stabilizer is sometimes present in the form of manganese, which in combination with lower amounts of nickel than otherwise required will perform many of the same functions of nickel in solution. The effects of manganese on corrosion are not well documented. Manganese is known to combine with sulfur to form sulfides. The morphology and composition of these sulfides can have substantial effects on the corrosion resistance of stainless steels, especially their resistance to pitting corrosion.

Other elements. Molybdenum in moderate amounts in combination with chromium is very effective in terms of stabilizing the passive film in the presence of chlorides. Molybdenum is especially effective in enhancing the resistance to pitting and crevice corrosion. Carbon does not seem to play an intrinsic role in the corrosion characteristics of stainless, but it has an important role by virtue of the tendency of carbide formation to cause matrix or grain boundary composition changes that may lead to reduced corrosion resistance. Nitrogen is beneficial to austenitic stainless in that it enhances pitting resistance, retards formation of sigma phase, and may help to reduce the segregation of chromium and molybdenum in duplex stainless steels.

Ferritic steels. Ferritic steels with high chromium contents have good high-temperature properties. However, these steels readily form brittle sigma phase within the temperature range 550 to 950°C. The S44600 steel, with 27% chromium, has a scaling temperature in air of about 1070°C. The modern molybdenum-alloyed ferritic steels have largely the same corrosion resistance as S31600 but are superior to most austenitic steels in terms of their resistance to SCC. A typical application example for these steels is hot water heaters. For chlorine-containing environments, where there is a particular risk of pitting (e.g., in seawater), the high-alloy steel S44635 (25Cr-4Ni-4Mo) can be used. In general the corrosion resistance of ferritic stainless steels is substantially lower than that of the austenitic steels but higher than most of the martensitics. They can withstand only mildly corrosive conditions. As such they find application in the automotive industry and in architectural work as decorative members. They have good oxidation resistance in fresh water but are prone to pitting in brackish and seawater. They can be used for handling dilute alkalis at room temperature and hydrocarbons at moderate temperature.[48]

Ferritic stainless steels cannot be used for any reducing or organic acids such as oxalic, formic, and lactic, but they are used for handling nitric acid and many organic chemicals. S43000 is less costly and most popular for such purposes. Some modifications of S43000 have been developed. S43023 contains selenium, for free-machining use. Various other alloys in the S43000 series, with 1.0 to 2.0% Mo, are also available, such as type S43400, which contains 1.0 to 1.3% Mo. This improves corrosion resistance under reducing conditions and decreases pitting tendencies as well. Because oxidation and scaling tendencies at high temperatures can be reduced by increasing chromium content, two well-known ferritic stainless steels contain 21% Cr (S44200) and 26% Cr (S44600), which increases their service temperature limits to 980 and 1090°C, respectively.[48]

S43000 and S43600 stainless steels are more resistant to SCC than austenitic stainless steels in the presence of small amounts of chloride. Because welding reduces their ductility and resistance to SCC and IGC, they are sometimes alloyed with molybdenum, nickel, and one of the six metals of the platinum group.[48]

Until recently, poor weldability and a lack of toughness and ductility were severe limitations for using ferritic stainless steels. These problems have been addressed by the advent of argon-oxygen decarburization (AOD) and vacuum oxygen decarburization (VOD) processes for stainless steel production. VOD, although more costly, is superior because it reduces interstitial carbon and nitrogen to below 0.025%, compared with 0.035% for AOD. Thus, it is now possible to produce low-carbon, low-nitrogen ferritic stainless steels, with the full benefit of a combination of high chromium and molybdenum (1.5 to 4%) and excellent corrosion resistance, especially to stress corrosion, at a competitive cost.

The corrosion resistance of ferritic steels has been extensively studied. The following expressions summarize the effects of different alloying elements on the resistance of ferritic steels exposed to boiling corrosive solutions during slow strain tests.[49] The stress corrosion indices (SCIs) in each environment integrate the beneficial ($-$) or deleterious ($+$) effect of the alloying elements (in %) when the steels are in contact with such a caustic environment. In boiling 4M $NaNO_3$ at pH 2 the stress corrosion index is

$$SCI_{NO_3} = 1777 - 996C - 390Ti - 343Al - 111Cr - 90Mo$$
$$- 62Ni + 292Si$$

In 8.75 M NaOH it is

$$SCI_{OH} = 105 - 45C - 40Mn - 13.7Ni - 12.3Cr - 11Ti + 2.5Al$$
$$+ 87Si + 413Mo$$

And in 0.5M $NaCO_3$ + 0.5M $NaHCO_3$ at 75°C it is

$$SCI_{CO_3} = 41 - 17.3Ti - 7.8Mo - 5.6Cr - 4.6Ni$$

Austenitic steels. S30400 steel is a great stainless success story. It accounts for more than 50% of all stainless steel produced and finds applications in almost every industry. The S30403 steel is a low-carbon S30400 and is often used to avoid possible sensitization corrosion in welded components. S30409 has a higher carbon content than S30403, which increases its strength (particularly at temperatures above 500°C). This grade is not designed for applications where sensitization corrosion could be expected.

The S30400 steel has excellent corrosion resistance in a wide range of media. It resists ordinary rusting in most architectural applications. It is also resistant to most food processing environments, can be readily cleaned, and resists organic chemicals, dye stuffs, and a wide variety of inorganic chemicals. In warm chloride environments, S30400 is subject to pitting and crevice corrosion and to SCC when subjected to tensile stresses beyond about 50°C. However, it can be successful in warm chloride environments where exposure is intermittent and cleaning is a regular event.

S30400 has good oxidation resistance in intermittent service to 870°C and in continuous service to 925°C. Continuous use of S30400 in the 425 to 860°C range is not recommended if subsequent exposure to room-temperature aqueous environments is anticipated. However, it often performs well in temperatures fluctuating above and below this range. S30403 is more resistant to carbide precipitation and can be used in the above temperature range. Where high-temperature strength is important, higher carbon values are required. S30400 has excellent toughness down to temperatures of liquefied gases and finds application at these temperatures. Like other austenitic grades, S30400 in the annealed condition has very low magnetic permeability.

Austenitic stainless steels are susceptible to SCC in chloride environments. The standard S30400, S30403, S31600, and S31603 stainless steels are the most susceptible. Increasing nickel content above 18 to 20% or the use of duplex or ferritic stainless steels improves resistance to SCC. High residual or applied stresses, temperatures above 65 to 71°C, and chlorides increase the likelihood of SCC. Crevices and wet/dry locations such as liquid vapor interfaces and wet insulation are particularly likely to initiate SCC in susceptible alloys. Initiation may occur in several weeks, in 1 to 2 years, or after 7 to 10 years in service.[2]

Martensitic steels. The corrosion resistance of martensitic stainless steels is moderate (i.e., better than carbon steels and low-alloy steels but inferior to that of austenitic steels). They are typically used under mild corrosion conditions for handling water, steam, gas, and oil. The 17% Cr steels resist scaling up to 800°C and have low susceptibility to corrosion by sulfur compounds at high temperatures.

S41000 is a low-cost, general-purpose, heat-treatable stainless steel. It is used widely where corrosion is not severe (air, water, some chemicals, and food acids). Typical applications include highly stressed parts needing the combination of strength and corrosion resistance such as fasteners. S41008 contains less carbon than S41000 and offers improved weldability but lower hardenability. The S41008 steel is a general-purpose corrosion and heat-resisting chromium steel recommended for corrosion-resisting applications.

S41400 has nickel added (2%) for improved corrosion resistance. Typical applications include springs and cutlery. S41600 contains added phosphorus and sulfur for improved machinability. Typical applications include screw machine parts. S42000 contains increased carbon to improve mechanical properties. Typical applications include surgical instruments. S43100 contains increased chromium for greater corrosion resistance and good mechanical properties. Typical applications include high-strength parts such as valves and pumps. S44000 contains even more chromium and carbon to improve toughness and corrosion resistance. Typical applications include instruments.

Duplex steels. Duplex stainless steels comprise a family of grades with a wide range of corrosion resistance. They are typically higher in chromium than the corrosion-resistant austenitic stainless steels and have molybdenum contents as high as 4.5%. The higher chromium and molybdenum combination is a cost-efficient way to achieve good chloride pitting and crevice corrosion resistance. Many duplex stainless steels exceed the chloride resistance of the common austenitic stainless steels. The constraints of achieving the desired balance of phases define the amount of nickel in duplex stainless steel. The resulting nickel contents, however, are sufficient to provide significant benefit in many chemical environments.[50] Table 8.35 describes the influence of different alloying additions and microstructure on the pitting and crevice corrosion resistance of duplex stainless steels.

Duplex stainless steels have been available since the 1930s. The first-generation duplex stainless steels, such as S32900, have good localized corrosion resistance because of their high chromium and molybdenum contents. When welded, however, these grades lose the optimal balance of austenite and ferrite and, consequently, corrosion resistance and toughness are reduced. Although these properties can be restored by a postweld heat treatment, most of the applications of the first-generation duplexes use fully annealed material without further welding.[50]

In the 1970s, this problem became manageable through the use of nitrogen as an alloy addition. The introduction of AOD technology permitted the precise and economical control of nitrogen in stainless steel.

TABLE 8.35 Influence of Different Alloying Additions and Microstructure on the Pitting and Crevice Corrosion Resistance of Duplex Stainless Steels

Alloying	Effect	Reason	Practical limitation
C	Negative	Causes precipitation of chromium carbides with accompanying chromium-depleted zones.	About 0.03% maximum.
Si	Positive	Si stabilizes the passive film.	About 2% maximum, due to its effect on structural stability and on nitrogen solubility.
Mn	Negative	Mn-rich sulfides act as initiation sites for pitting. Mn may also destabilize the passive film.	About 2%. Higher levels might also increase the risk of intermetallic precipitation.
S	Negative	Sulfides, if not Cr-, Ti-, or Ce-rich, tend to initiate pitting attack.	About 0.003%, if maximum pitting resistance is required. For reasonable machinability, up to 0.02% is allowed.
Cr	Positive	Cr stabilizes the passive film.	Between 25 and 28% maximum depending on the Mo content. Higher Cr content increases the risk of intermetallic precipitation.
Ni	Negative	Increased Ni, other elements constant, dilutes the γ-phase with regard to N, which in turn decreases the PRE of the γ-phase. If the alloy is very sensitive to precipitation of chromium nitrides, Ni can have a positive effect.	Ni should primarily be used to give the alloy the desired austenite content.
Mo	Positive	Mo stabilizes the passive film, either directly or through enrichment beneath the film.	About 4 to 5% maximum depending on the Cr content. Mo enhances the risk of intermetallic precipitation.
N	Positive	N increases the PRE-number of the γ-phase dramatically, not only by increasing the N content of that phase but also by increasing the Cr and Mo contents through their partitioning coefficients.	About 0.15% in Mo-free grades. About 0.3% in superduplex grades and some 0.4% in 25% Cr, high Mo, high Mn alloys.
Cu	Disputed	Marginal positive or negative effects.	About 2.5% maximum. Higher levels reduce hot workability and undesirable hardenability.
W	Positive	Probably the same as for Mo.	Increases the tendency for intermetallic precipitation.

TABLE 8.35 Influence of Different Alloying Additions and Microstructure on the Pitting and Crevice Corrosion Resistance of Duplex Stainless Steels (Continued)

Alloying	Effect	Reason	Practical limitation
Ferrite	Positive	Increased ferrite content increases the N, Cr. and Mo contents of the γ-phase.	Too high ferrite can enhance chromium carbide/nitride precipitation in a coarse microstructure.
Intermetallic phases	Negative	Precipitates with accompanying depletion of alloying elements (Cr, Mo).	If steel manufacturers' recommendations are followed, intermetallic precipitation should not occur during heat treatment or welding.
Chromium carbides and nitrides	Negative	Precipitation of carbides/nitrides causes Cr-depleted zones that are selectively attacked in certain corrosive media.	In older generations of duplex alloys, nitrides were frequently present in welded joints and in base metal with coarse microstructure. This has rarely been the reason for a corrosion failure.

Although nitrogen was first used because it was an inexpensive austenite former, it was quickly found that it had other benefits. These include improved tensile properties and pitting and crevice corrosion resistance.[50] Nitrogen also causes the formation of austenite at a higher temperature, allowing for restoration of an acceptable balance of austenite and ferrite after a rapid thermal cycle in the HAZ after welding. This nitrogen advantage enables the use of duplex grades in the as-welded condition and has created the second generation of duplex stainless steels.

Alloying with nitrogen has stimulated the introduction of many duplex grades, most of them being marketed as proprietary products. Some of these grades are not readily available in product forms other than those produced. However, the S31803 alloy is an exception; it is offered by many producers and is available on an increasingly regular and reliable basis through metal service centers. It has become the most widely used second-generation duplex stainless steel.[50] The latest developed duplex stainless steels with very high Cr, Mo, and N contents, such as alloy 2507 (S32750), have better corrosion resistance than S31803 steel and are in many cases comparable to the 6 Mo steels, that is, 254 SMO (S31254).

One of the primary reasons for using duplex stainless steels is their excellent resistance to chloride SCC. They are quite superior to common austenitic steels in this respect. Modern steels with correctly balanced compositions, such as alloy 2205 (UNS S31803), also possess good pitting properties and are not sensitive to intergranular corrosion after welding, as were the first-generation duplex steels. All duplex stainless steels are susceptible to SCC in the boiling 42% magnesium chloride. Fortunately, this test is so overly severe that its results are not meaningfully related to the SCC that occurs with most austenitic stainless steels in typical applications with less-concentrated chlorides. In boiling 25% sodium chloride and in the sodium chloride "wick test," which have been shown to correlate well with field experience in SCC, the duplex grades are resistant to SCC.[50]

Pitting and crevice. The pitting and crevice behavior of stainless steels in chloride-bearing waters has been studied by a number of investigators. There is considerable variation in the percentage of apparently identical sites where attack occurs, when it occurs. It is useful to describe results in terms of the percentage of apparently identical sites where attack occurs at a given chloride concentration. Very tight crevices increase the likelihood of attack. Rough surfaces, sheared edges, scratches, and similar imperfections also tend to increase the incidence of attack. Crevice or pitting attack also occurs under deposits and under biofouling growths attached to the metal surface. Table 8.36 describes the measured critical crevice corrosion temperatures for many corrosion-resistant austenitic and duplex stainless steels, and Table 8.37 gives the corrosion rates of some of these alloys in selected chemical environments.

Relative resistance can also be described by the chloride concentration below which there is little likelihood of crevice attack occurring. Pitting, particularly at or near welds and in crevices, has often resulted in perforation within a few months. It is necessary, therefore, to chose an alloy with high resistance to localized attack, which is often defined as an alloy with a high pitting-resistance equivalent number (PRE_N). PRE_N is derived from an empirical relationship and can take several forms. The most widely used form to predict the pitting resistance of austenitic and duplex stainless steels is expressed as:[47]

$$PRE_N = Cr + 3.3 (Mo + 0.5 W) + xN$$

where Cr, Mo, W, and N are the chromium, molybdenum, tungsten, and nitrogen contents (%), and $x = 16$ for duplex stainless steel, and 30 for austenitic alloys.

Elevated temperature. The properties of stainless steels at elevated temperatures may degrade from a variety of causes. The consequences of this degradation depend on the process and the expectations of the material.

TABLE 8.36 Critical Crevice Corrosion Temperatures

UNS	Type	Temperature, °C
S32900	329	5
S31200	44LN	5
S31260	DP-3	10
S32950	7-Mo PLUS	15
S31803	2205	17.5
S32250	Ferralium 255	22.5
S30400	304	<−2.5
S31600	316	−2.5
S31703	317L	0
N08020	20Cb-3	0
N08904	904L	0
N08367	AL-6XN	32.5
S31254	254 SMO	32.5

Sigma phase. In ferritic stainless steels the sigma phase is composed only of iron and chromium. In austenitic stainless alloys, it is much more complex and will include nickel, manganese, silicon, and niobium in addition to iron and chromium. The sigma phase forms in ferritic and austenitic stainless steels from ferrite or metastable austenite during exposure at 593 to 927°C. It causes loss of ductility and toughness at temperatures under 120 to 150°C but has little effect on properties in the temperature range where it forms unless the material has been put into service with considerable residual cold work. In this case, creep strength can be adversely affected.[2]

Over time, sigma phase formation is unavoidable in many of the commercial alloys used within the temperature range where it forms. Fortunately, few failures have been directly attributed to it. However, if a component is to be exposed in the critical temperature range and subsequently subjected to extensive cyclic conditions or to shock loading, an immune or more stable material should be used. Increased resistance or immunity is achieved by selecting a composition that is balanced with respect to austenite versus ferrite-forming elements so that no free ferrite is present. This can be determined using the Schaeffler diagram, discussed previously.

Sensitization. Another form of elevated temperature degradation of austenitic stainless steels is sensitization. This is caused by the precipitation of chromium carbides preferentially at grain boundaries. The adjacent chromium-depleted zone then becomes susceptible to accelerated corrosion in some corrosive environments. Sensitization can occur during fabrication from the heat of welding or improper heat treatment or through service exposure in the temperature range of 480 to 815°C. Sensitization has little or no effect on mechanical properties but can lead to severe intergranular corrosion in aggressive aqueous

TABLE 8.37 Corrosion Rates (mm·y^{-1}) in Selected Chemical Environments

Environment	Temperature, °C	S30400	S31600	S31703	N08020	S31803	S32550
1% HCl	Boiling			0.0025		0.0025	0.0025
10% sulfuric	66			0.226		0.030	0.0051
10% sulfuric	Boiling	42.0	21.7	12.4	1.09	5.23	1.01
30% phosphoric	Boiling			0.170		0.0406	0.0051
85% phosphoric	66			0.0051		0.010	0.0025
65% nitric	Boiling	8	0.28	0.533	0.203	0.534	0.13
10% acetic	Boiling			0.0051		0.0025	0.0051
20% acetic	Boiling	7.6	2		0.051		
20% formic	Boiling			0.2159		0.033	0.010
45% formic	Boiling	43.6	520		0.18	0.124	
3% NaCl	Boiling			0.0254		0.0025	0.010

environments such as polythionic acid. Polythionic acid can form during downtime on equipment that has been even mildly corroded by hydrogen sulfide at an elevated temperature. The iron sulfide corrosion product combines with air and moisture to form the acid and induces intergranular corrosion and cracking.[2]

To minimize the chance of sensitization during fabrication, carbide-forming stabilizers are added. The most common are titanium (S32100) and niobium (S34700). As long as their lower strengths are taken into account, another alternative is to use low carbon grades (S30403, S31603) with carbon <0.03%. To minimize the effects of frequent or continuous exposure within the susceptible temperature range, a thermal stabilization treatment of S34700 at 870 to 900°C for 4 h is recommended. S32100 steel does not respond acceptably to this treatment.

The higher carbon content of heat-resistant alloys and the presence of other elements cause these alloys to "age" during exposure to elevated temperatures. Aging results from the formation of secondary carbides and other precipitates. This usually results in higher strength but also causes loss of ductility at ambient temperature, leading to potential fabrication problems. This is more of a problem with cast than wrought heat-resistant alloys because of the typically higher original carbon content.

Recovery from all of the above forms of degradation is possible by solution annealing the material at temperatures appropriate for the alloy grade followed by rapid cooling. For the 300 series stainless steels, annealing can be done at 1066°C, whereas the high-carbon heat-resistant alloys may require treatment as high as 1177°C. Recovery is not permanent. Reexposure to the causative conditions will result in redegradation.[2]

8.8 Steels

8.8.1 Introduction

Iron and steel, the most commonly used metals, corrode in many media, including most outdoor atmospheres. Usually they are selected not for their corrosion resistance but for such properties as strength, ease of fabrication, and cost. These differences show up in the rate of metal lost due to rusting. All steels and low-alloy steels rust in moist atmospheres. In some circumstances, the addition of 0.3% copper to carbon steel can reduce the rate of rusting by one-quarter or even by one-half. The elements copper, phosphorus, chromium, and nickel have all been shown to improve resistance to atmospheric corrosion. Formation of a dense, tightly adhering rust scale is a factor in lowering the rate of attack. The improvement may be sufficient to encourage use without protection and can also extend paint life by decreasing the amount of corrosion underneath the paint. The rate of rusting will usually be higher in the first year of atmospheric exposure than in subsequent years and will increase significantly with the degree of pollution and moisture in the air.

Steel has quite good resistance to alkalies, many organics, and strong oxidizing acids. As a general rule, acids should be avoided. Mild steel can be susceptible to SCC in media that contain nitrates, hydroxides, ammonia, and hydrogen sulfide. Any evolved hydrogen may cause embrittlement and blistering in the steel. Adding copper can offset the harmful effects of phosphorus and sulfur inclusions in the steel in dilute acids. In water, oxygen is detrimental. Like other metals that form passive oxide films, iron benefits in situations in which there is essentially no oxygen to depolarize the cathodic reaction or sufficient oxidizing power to form a stable oxide film.[1]

Low-alloy steels are defined as steels containing up to 5% of the major alloying element. These steels are designed for higher strength and are similar in corrosion resistance to unalloyed steel except for improvements attainable in the rate of atmospheric attack. For example, an alloy steel might rust at one-third the rate of a plain carbon steel without copper. About 10 to 12% chromium is usually needed to avoid rusting in the atmosphere. The silicon irons, particularly those containing about 15% silicon, are more corrosion resistant than steel. Unfortunately, they are available only as castings and are quite brittle. They have good resistance to oxidizing and reducing acids with the exception of hydrofluoric acid and perform particularly well in slurries because of their good erosion-corrosion resistance.

8.8.2 Carbon steels

Ordinary steels are essentially alloys of iron and carbon with small additions of elements such as manganese and silicon added to provide the requisite mechanical properties. The steels are manufactured from a mixture of pig iron and scrap, which is treated in the molten state to remove excess carbon and other impurities. The steel may be continuously cast into strands or cast into individual ingots. The final product is then produced by rolling, drawing, or forging. During hot rolling and forging the steel surface is oxidized by air, and the scale produced is usually termed *millscale*. In air, the presence of millscale on the steel may reduce the corrosion rate over comparatively short periods, but over longer periods the rate tends to rise. In water, severe pitting of the steel may occur if large amounts of millscale are present on the surface.

The addition of about 0.2% copper results in a two- to threefold reduction in the corrosion rate in air compared with a copper-free steel. Variations in the other elements in ordinary steels affect the corrosion rate to a marginal degree, the tendency being for the rate to decrease with increasing content of carbon, manganese, and silicon. For example, in the open air a steel containing 0.2% Si rusts about 10% less rapidly than an otherwise similar steel containing 0.02% Si.[51]

8.8.3 Weathering steels

The mechanical properties of low- or medium-carbon structural steels can be improved considerably by small alloy additions. For example, 1% Cr will raise a steel yield strength (0.2% offset) from 280 to 390 MPa. This has led to the development of a range of so-called high-strength, low-alloy (HSLA) steels with high tensile properties. Although, originally at least, the main purpose was to increase the strength of the steel, improvements in the mechanical properties of unalloyed steels have resulted in a considerable overlap in properties between the two classes. In some cases low-alloy additions, besides making further improvements in properties possible, may even enhance resistance to corrosion. As a class they are by no means corrosion-free but under favorable conditions, such as when they are exposed outdoors, they can rust several times less rapidly than unalloyed mild steel. The low-alloy steels specifically designed to be slow rusting are commonly called weathering steels, and to optimize this corrosion resistance the alloying elements most commonly used are chromium, nickel, and copper.[52]

Uncoated weathering-grade steels have been available for many years. The cost effectiveness of use of this material has been demonstrated in both short- and long-term savings. The additional cost of this grade of steel is offset by the elimination of the need for initially painting structures. Where enhanced atmospheric corrosion resistance is desired, the letter W follows the grade.[53] Environmental benefits also result from the use of this material. The reduction in initial painting reduces emissions of volatile organic compounds (VOCs) when oil-based coatings are used. The elimination of removal of the coating and disposal of contaminated blast cleaning debris over the life span of the structure is another significant environmental benefit.

There are documented cases where the estimated cost of the collection and disposal of materials from a structure repainting project were so great that the structure was either abandoned or replaced. At the same time, there are documented cases where application of this material in improper locations or under improper conditions has resulted in less than desirable performance. In most cases, this poor performance was the result of a lack of understanding of the limitations of weathering-grade steels. The following situations represent conditions where uncoated weathering steel cannot be expected to perform as intended, and continuing corrosion could result in significant damage:[53]

Marine coastal areas. Salt-laden air that is generated along the sea coast may be transported inland by the prevailing winds. The level of chloride concentration caused by the salt-laden air and its effect on the performance of uncoated weathering steel structures depends on the direction of the prevailing winds, the distance from the shore line, and the topographical and environmental characteristics of the area. Thus, the weathering behavior of uncoated weathering steel structures can vary significantly from one location to another.

Areas of frequent high rainfall, high humidity or persistent fog. These climatic conditions can result in excessive condensation and prolonged periods of wetness of the steel. Selection of uncoated steel for use in areas where these conditions persist should not be made without an evaluation of the expected time of wetness of the steel at the particular bridge site.

Industrial areas. In heavy industrial areas with chemical and other manufacturing plants, the air may contain chemical impurities that can be deposited on and decompose the steel surfaces.

8.8.4 Weldability

Commonly used steels can be readily welded. Steels have been grouped in terms of their metallurgical and welding characteristics. The main risks in welding these groups are described below, followed by the main welding imperfections encountered:[54]

Low-carbon unalloyed steels and/or low-alloyed steels. For thin-section, unalloyed materials, these materials can normally be readily welded. However, when welding thicker-section material with a flux process (MMA), there is a risk of HAZ cracking and low-hydrogen electrodes need to be used. The more highly alloyed materials also require preheat or a low-hydrogen welding process to avoid HAZ cracking.

2-5 Ni Steels, CrMo, and CrMoV creep-resisting steel. Thin-section material may be welded without preheat, using a gas-shielded process (TIG and MIG); for thicker-section material, and when using a flux process, preheat with low-hydrogen electrodes (MMA) is needed to avoid HAZ and weld metal cracking. Postweld heat treatment is used to improve HAZ toughness.

Ferritic or martensitic stainless steel, with chromium (12 to 20%). When using filler to produce matching weld metal strength, preheat is needed to avoid HAZ cracking. Postweld heat treatment is essential to restore HAZ toughness. An austenitic stainless steel filler can be used where it is not possible to apply a preheat and postweld treatment.

Porosity. Porosity is formed by entrapment of discrete pockets of gas in the solidifying weld pool. The gas may originate from poor gas shielding, surface contaminants such as rust or grease, or insufficient deoxidants in the parent metal, electrode, or filler wire. A particularly severe form of porosity is "worm holes," caused by gross surface contamination or welding with damp electrodes. The presence of manganese and silicon in the parent metal, electrode, and filler wire is beneficial because they act as deoxidants, combining with entrapped air in the weld pool to form slag. Rimming steels with a high oxygen content can only be welded satisfactorily with a consumable that adds aluminum to the weld pool.[54] To obtain sound porosity-free welds, the joint area should be cleaned and degreased before welding. When using gas-shielded processes, the material surface demands more rigorous cleaning, such as by degreasing, grinding, or machining, followed by final degreasing, and the arc must be protected from draughts.

Solidification cracking. Solidification cracks occur longitudinally as a result of the weld bead having insufficient strength to withstand the contraction stresses within the weld metal. Sulfur, phosphorus, and carbon pickup from the parent metal at high dilution increase the risk of weld metal (solidification) cracking, especially in thick-section and highly restrained joints. When welding high carbon and sulfur content steels, thin weld beads will be more susceptible to solidification cracking. However, a weld with a large depth-to-width ratio can also be susceptible. In this case, the center of the weld, the last part to solidify, will have a high concentration of impurities, increasing the risk of cracking.[54] Solidification cracking is best avoided by careful attention to the choice of consumable, welding parameters and welder technique. To minimize the risk, consumables with low carbon and impurity levels and relatively high manganese and silicon contents are preferred. High–current density processes, such as submerged-arc and CO_2, are more likely to induce cracking.

Hydrogen cracking. A characteristic feature of high-carbon and low-alloy steels is that the HAZ immediately adjacent to the weld hardens on welding with an attendant risk of cold (hydrogen) cracking. The amount of hydrogen generated is determined by the electrode type and the process. Basic electrodes generate less hydrogen than rutile electrodes, and the gas-shielded processes produce only a small amount of hydrogen in the weld pool. Steel composition and cooling rate determine the HAZ hardness. Chemical composition determines material hardenability, and the higher the carbon and alloy content of the material, the greater the HAZ hardness. Section thickness and arc energy influence the cooling rate and, hence, the hardness of the HAZ.

Because cracking only occurs at temperatures slightly above ambient, maintaining the temperature of the weld area above the recommended level during fabrication is especially important. If the material is allowed to cool too quickly, cracking can occur up to several hours after welding, often termed *delayed hydrogen cracking*. After welding, therefore, it is beneficial to maintain the heating for a given period (hold time), depending on the steel thickness, to enable the hydrogen to diffuse from the weld area.[54] When welding C-Mn structural and pressure vessel steels, the measures that are taken to prevent HAZ cracking will also be adequate to avoid hydrogen cracking in the weld metal. However, with increasing alloying of the weld metal (e.g., when welding alloyed or quenched and tempered steels), more stringent precautions may be necessary. The risk of HAZ cracking is reduced by using a low-hydrogen process, low-hydrogen electrodes, and high arc energy and by reducing the level of restraint.

Reheat cracking. Reheat or stress relaxation cracking may occur in the HAZ of thick-section components, usually of greater than 50-mm thickness. The more likely cause of cracking is embrittlement of the HAZ during high-temperature service or stress relief heat treatment. Because a coarse-grained HAZ is more susceptible to cracking, low arc energy input welding procedures reduce the risk. Although reheat cracking occurs in sensitive materials, avoidance of high stresses during welding and elimination of local points of stress concentration (e.g., by dressing the weld toes) can reduce the risk.[54]

8.8.5 Corrosion resistance

Carbon steel. The corrosion rates of wrought iron and mild steel when immersed in seawater or buried in soil are not significantly different when the copper contents are similar. Steel has number of phases and inhomogeneities at the surface, which can cause local cells. The corrosion resistance of iron is low, because cathodic reduction can easily take place on its surface, and moreover, its corrosion product is porous and nonadherent. By contrast, aluminum and other light metals form a compact adhering film that retards corrosion. Steel finds extensive application primarily because of its low cost, reasonably good mechanical properties, and ease of fabrication.[48]

Ambient conditions in an industrial environment are relatively more corrosive because of the presence of moisture and chemical pollutants in the air. Chlorides in coastal areas and sulfur dioxide are highly aggressive, and they lower the critical humidity level for the onset of corrosion.

Sulfur dioxide facilitates depolarization and easily oxidizes to sulfur trioxide on metal surfaces, which in turn forms sulfuric acid. Similarly, acid vapors, H_2S, and organic vapors even in small quantities greatly increase the aggressiveness of the atmosphere.[48] Despite these shortcomings, plain carbon steels, with or without minor alloying elements, are widely used as the most economic materials of construction under ambient, aggressive conditions, and with various combinations of protective coatings and other corrosion prevention or control methods. Conditions permitting the satisfactory use of mild steel are described in Table 8.38.

Aqueous media corrosion. Natural water is widely distributed and stored in steel pipe, galvanized steel pipe, and steel tanks. Natural waters, so long as they are reasonably free from aggressive ions, such as chloride and acidic species, are noncorrosive and have been handled satisfactorily by mild steel pipes and tanks for many years. The primary impurities in these waters are calcium and magnesium salts. These salts can form a hard carbonate protective scale on the surface of steel exposed to hard water. Chemically pure, distilled water is, in fact, corrosive, and when the concentration of these salts is low, the corrosion of steel must be controlled by reducing the oxygen present in the water by chemical treatment or by cathodic protection.

The protective carbonate scale is not just a function of the concentration of calcium and magnesium salts. It is also affected by the alkalinity of the water and concentrations of other salts. Saturation indexes have been developed for monitoring such concentrations. A popular saturation index is the Langelier index, which provides a simple method for determining the conditions and concentrations under which water will form this kind of protective film.[48] Section 2.2 in Chap. 2, Environments, describes in detail the Langelier index and a few other indexes and methods to monitor scaling tendencies of waters.

Brackish waters containing less than 1% NaCl have been handled successfully in steel pipes. Seawater under quiescent conditions can also be stored in steel vessels. The pitting tendency in such cases should be controlled by removing dissolved oxygen. Steel pipes can have a life expectancy of 2 to 5 years in mine waters, depending on their composition. The main factors controlling the corrosion of steel in natural waters follow:

Dissolved gases. Corrosion induced by dissolved oxygen is proportional to its concentration, up to 25 to 30 ppm. Above this level, corrosion decreases at higher concentrations. Higher temperatures and pressures and lower pH increase its corrosivity. Carbon dioxide, although only approximately 10% as corrosive as oxygen, is nearly 100 times more soluble than oxygen. Dissolved H_2S attacks steel

TABLE 8.38 Conditions Permitting the Satisfactory Use of Mild Steel

Service	Pressure, kPa	Temperature, °C
Acetone	1030	370
Acetylene*	1030	150
Air (compressed)	1030	360
Air (compressed)	2070	amb
Alcohol	2070	200
Ammonia (anhydrous gas)	4140	500
Ammonia (anhydrous liquid)	4140	500
Ammonia (aqueous)	4140	500
Benzene*	3240	450
Brine (calcium chloride)†	340	100
Butanol*	1030	385
Carbon dioxide*	3100	150
Carbon disulfide (anhydrous)*	2070	500
Carbon tetrachloride*	2070	500
Caustic (concentration under 5%)*	2760	120
Caustic (concentration 0 to l04 Re)	1030	180
Caustic (concentration 11 to 50%)‡	1030	120–150
Chlorine (anhydrous gas)	340	150
Chloroform*	2070	500
Dowtherm "A"	1030	750
Gas (city)*	140	100
Gas (inert)	1030	350
Gas (natural)*	3401	140
Gas (natural)	690	315
Gas (natural)	4140	80
Hydrogen	1030	450
Hydrogen*	4140	500
Hydrogen chloride (anhydrous gas)*	1030	500
Kerosene*	860	350
Methanol*	1030	390
Nitrogen	4830	500
Sodium cyanide (26% solution)*	170	100
Sodium polysulfide solution	1030	500
Sulfuric acid (commercial grade)§		
60° Be	690	105
66° Be	690	120
109° (40% oleum)	690	160
Xylene*	520	150

*Copper-free steel.
†Economical life of steel, normal maintenance, minimum temperature −26°C.
‡Stress-relieved welds and cold bands, if steam traced.
§Nonflowing or low velocity; 6- to 8-year life at temperatures given.

even in the absence of oxygen.[48]

Hydrogen ion concentration (pH). Very little general corrosion occurs between 4.5 and 9.5 pH. In this range, the corrosion product maintains a pH of approximately 9.5 at the surface of the steel. But in weak acids such as H_2CO_3, hydrogen evolution and just below pH 6 and become rapid at pH 5.0. With ... hydrogen evolution is rapid, and the steel is ...idly.

...Most dissolved salts reduce the solubility of dissolved ...en, and therefore the rate of corrosion in concentrated ...ions is usually lower. Some salts buffer the pH. Some, such as halides and sulfides, are corrosive themselves. Therefore, neutral and acid salt solutions, such as NaCl or Na_2SO_4, which tend to increase the corrosion rate of iron, are not normally handled in steel equipment. Steel can be used for handling alkaline salts that hydrolyze to yield a solution of approximately 9.5 pH. These salts also act as inhibitors for some aggressive solutions. For example, $Ca(OH)_2$ is used to control corrosion caused by $CaCl_2$ in solution in refrigeration systems.[48]

Temperature and velocity effects. Water flowing at velocity higher than 3 m·s^{-1} can cause turbulence and impingement attack. The temperature of the water can be another factor. A rise of 18 to 20°C can double the corrosivity of some waters.

Alkalies. Mild steel is traditionally used for transporting and storing alkalies and alkaline solutions at room temperature and for alkaline salts such as sodium carbonate and phosphate. Iron passivates at a pH higher than 10. However, passivity decreases in concentrated solutions in which iron has a tendency to dissolve as ferrous ions ($HFeO_2^-$). Fortunately, the rate of corrosion is low at room temperature, so concentrated solutions and solid caustic are handled in steel drums. At high temperatures, concentrated alkalies are relatively aggressive, but pots for handling and fusion of caustic soda and potash are made from thick sections of cast iron or steel. Alkaline solutions employed in many chemical processing industries are handled in steel equipment.[48]

However, when steel is exposed to alkaline solutions at high temperatures while stressed in tension, it can crack along intergranular boundaries, in a fashion that has been called caustic embrittlement. This particularly vicious form of attack was first noticed in riveted steam boilers where alkali became concentrated in crevices underneath rivets. Welded construction of boilers has reduced the incidence of such failures.

Acids. Hydrogen evolution is easier on iron than on most other metals, and steel is severely attacked by acidic solutions at levels below pH 4. The presence of oxygen has a depolarizing effect, and corrosion becomes even more severe when oxygen is present. However, strong oxidizing acids can passivate steel, so it can be used for handling, storing, and transporting them. Sulfuric acid below 60% concentration is highly corrosive to steel. However, steel can be used for handling sulfuric acid of

90% concentration and above. Hydrochloric and phosphoric acids in all concentrations attack steel rapidly. Killed steel can be used to handle hydrofluoric acid of above 80% concentration. However, steel in any form is not normally used for handling organic acids.[48]

Nonaqueous organic solvents. In the absence of water, organic solvents do not attack steel, and it is therefore used for handling alcohols and glycols. The addition of small amounts of moisture, on the order of 0.1%, has a detrimental effect, particularly in the case of chlorinated organic solvents. Steel is widely used in petroleum refineries for pressure vessels, crude distillation towers, pipe stills, heat exchangers, piping, valves, all service lines, and storage tanks. The presence of sulfur in crude oil corrodes steel. In cases where sulfur compounds are present, it is economical to line or clad the steel with a corrosion-resistant material or use a low-alloy chromium steel.[48]

Gases and vapors. Most gases and vapors can be handled in steel equipment when they are completely dry. Problems are sometime encountered in subzero temperatures, because some steels lack impact resistance at these low temperatures. Corrosion of steel in the presence of moisture is particularly severe when acidic gases or vapors are involved, such as oxides of nitrogen and sulfur or chlorine. At high temperature, water vapor does not contribute to corrosion, but once the dew point is reached, condensation takes place, and the corrosion rate rises drastically. Steel is used for handling hot dry chlorine, liquid chlorine, and sulfur gases. For steam boilers, steels are the usual materials of construction, and dissolved oxygen in their feed water is their worst enemy, being most corrosive. To remedy this problem, its concentration should be reduced below 0.01 ppm by chemical and mechanical treatment of the feed water. The addition of chromium to steel prevents such attacks.[48]

When hydrogen is handled in steel above 400°C, it reacts with the carbon present in steel to form methane. As a consequence of this decarburization, fissures can form along the grain boundaries. Called hydrogen embrittlement, this phenomenon is controlled by alloying steel with carbide stabilizers such as chromium, molybdenum, and titanium. During ammonia synthesis and petroleum cracking, similar hydrogen and nitrogen embrittlement can take place above 500°C, particularly where high-carbon steels are employed. Alloying with 2% chromium reduces this problem. Low-alloy nickel-chromium-molybdenum steel has been used in the past to resist hydrogen attack at 400 to 450°C and moderate pressures. However, modern high-temperature high-pressure reactors use stainless steel. Despite some of these shortcomings, plain carbon steels are widely used as the most economical materials of construction under various conditions. Additionally, various protective schemes using organic and inorganic coatings, linings, and claddings have been developed to increase the life expectancy of steel under aggressive conditions.[48]

Low-alloy steels. The improvement in rust resistance achieved through low-alloy additions obviously depends on the nature and amounts of the alloying elements. Incidentally, the effects of these additions are not additive. Weathering steels generally perform best when they are freely exposed to the open air in industrial environments.[52] Copper and chromium additions influence the rate of rusting by raising the potential of the surface to more noble values, encouraging passivation. However, HSLA steels in their maximum hardness condition can be very susceptible to SCC in high-humidity environments.

Initially, weathering steels appear to rust like mild steels and quickly assume a fine, sandy appearance. However, unlike mild steel, whose oxide repeatedly spalls off, the surface rust layer stabilizes with time, provided that the exposure conditions allow the steel to dry out periodically. The rust then becomes darker, granular, and tightly adherent, and any pores or cracks become filled with insoluble salts. Because of the need for intermittent drying to stabilize the oxide film, it is doubtful, from the corrosion aspect, whether the use of weathering steels is worthwhile where immersion in natural waters or burial in soil is involved.[52]

When low-alloy steels are exposed outdoors, the rust formed on them is generally darker in color and much finer in grain than that formed on ordinary steel. Moreover, the slowing down in rusting rate with time seems to be more marked for low-alloy steels than for ordinary steels. This can be illustrated by the data presented in Table 8.39.[52] The distinguishing feature of the behavior of the slow-rusting low-alloy steels is the formation of this protective rust layer. Corrosion in conditions where it cannot form is little different from that of unalloyed steel. In particular, the beneficial effects observed in open air do not generally extend to conditions where the steel is enclosed and sheltered from the rain.

Corrosion in natural environments. As mentioned earlier, the effects of the various alloying elements are not additive. Bearing this in mind, the practical effect of individual elements can be summarized as follows:

1. Copper additions up to about 0.4% give a marked improvement, but further additions make little difference.

2. Phosphorus, at least when combined with copper, is also highly beneficial. However, in practice, levels above about 0.10% adversely affect mechanical properties.

3. Chromium, in fractional percentages, has a significant influence on

TABLE 8.39 Variation of Rate of Rusting with Time

Steel	Rate of rusting, (mm·y^{-1})		Ratio (B/A)
	A	B	
	1st and 2d years	6th to 15th year	
Ordinary mild steel (0.02 Cu)	0.129	0.094	0.73
Low-alloy steel (1.0 Cr, 0.6 Cu)	0.077	0.025	0.33

corrosion rates. Although it appears to be beneficial, some conflicting results have been reported, and its contribution to the reduced corrosion of complex low-alloy steels containing copper and phosphorus is not large.

4. Nickel, although reducing corrosion rates a little, is not as important in its effect as the aforementioned elements.

5. Manganese may have a particular value in chloride-contaminated environments, but its contribution is little understood.

6. Silicon is in a similar position to manganese, with conflicting evidence as to its value.

7. Molybdenum has been little used in low-alloy steels but may be as effective as copper and is worthy of further study.

Applications in industry. Most structural steelwork that is exposed to the atmosphere is given a protective coating of some kind. If this coating is continuously maintained in perfect condition, so that no rusting of the steel takes place, there is no advantage from the corrosion aspect in using a low-alloy steel instead of mild steel. If, on the other hand, it is probable that the protective coating will be damaged or allowed to deteriorate, the use of a low-alloy steel should be considered. The more compact rust film formed on these steels will be less likely to cause the coating adjacent to the corroded areas to spall off, and the rate at which breakdown of the coating spreads will be reduced. Several investigators have reported better performance and durability of protective coatings on low-alloy steels than on ordinary steel. Any rust that forms at breaks or holidays or underneath the paint film is less voluminous on the low-alloy steels. Owing to the smaller volume of rust, there is less rupturing of the paint film and, hence, less moisture reaches the steel to promote further corrosion.[52]

However, the most widespread use of weathering steels has been for buildings and bridges, especially where maintenance painting is particularly difficult, dangerous, inconvenient, or expensive. Bridges over land, rivers, railways, roads, and estuaries fall into this category, although in the last two cases care should be taken with respect to airborne salinity. Road bridges can be affected by salt-laden atmospheres or water, produced as a consequence of winter ice and snow clearing with deicing salt and grit. The chloride can be in the form of an airborne spray thrown up by passing vehicles or as a result of leaks in the bridge deck. The important criterion is design. Many bridges have been built successfully from weathering steels but at the design stage it is important to consider the possible effects of road salt to obtain the maximum maintenance-free life.[52]

To obtain a uniform color, it is essential to remove all millscale and residual grease or oil stains, preferably by blasting. The detailing of all sections should be such as to avoid pockets, crevices, and any location that will collect and retain moisture and dirt for long periods. Any such locations, as well as faying surfaces, should be painted for corrosion protection. The paint requirements for weathering steels are exactly the same as for carbon steel, and the slow rusting nature of the weathering steel will result in all paint systems having an extended life before maintenance is required.[52]

An important aspect of design is to predict the lines of runoff of surface water. This is because the water will contain minute particles of brown rust, especially in the prestabilization period, that will stain some surfaces. Matte, porous surfaces stain particularly easily, and runoff should not be over concrete, stucco, galvanized steel, unglazed brick, or stone.

8.9 Titanium

8.9.1 Introduction

Titanium is the fourth most abundant metallic element in the earth's crust. It occurs chiefly as an oxide ore. The commercially important forms are rutile (titanium dioxide) and ilmeite (titanium-iron oxide), the former being richest in titanium content. Metallic titanium was first isolated in impure form in 1887, and with higher purity in 1910. However, it was not until the 1950s that it began to come into use as a structural material. This was initially stimulated by aircraft applications. Although the aerospace industry still provides the major market, titanium and titanium alloys are finding increasingly widespread use in other industries due to their many desirable properties. Titanium is a unique material, as strongas steel with less than 60% of its density but with excellent corrosion resistance. Traditional applications are in the aerospace and chemical industries. More recently, especially as the cost of titanium has fallen significantly, the alloys are finding greater use in other industry sectors, such as offshore.

Titanium commercial extraction process involves treatment of the ore with chlorine gas to produce titanium tetrachloride, which is purified and reduced to a metallic titanium sponge by reaction with magnesium or sodium. The sponge, blended with alloying elements as desired, is then vacuum melted. Several meltings may be necessary to achieve a homogeneous ingot that is ready for processing into useful shapes, typically by forging followed by rolling. For many applications the cost of titanium alloys can be justified on the basis of desirable properties.[55]

Titanium has become increasingly important as a construction material. It is strong and of medium weight. It is very corrosion resistant in many environments. Nevertheless, there are a number of disadvantages to titanium, which have limited its use. One is the high cost relative to the more noble austenitic stainless steels such as S31600. Although it is available in all conventional forms, titanium is, in addition, not easy to shape and form.

Depending on the alloy, titanium alloys may be produced by vacuum arc remelting, electron-beam, or plasma melting. Ingots are commonly 60 to 120 cm in diameter, weighing 2300 to 18,000 kg. Conventional metallurgical processing in air are used to produce wrought alloys. All standard mill products are available. Casting may also be produced using investment casting and rammed graphite molding technologies.

The properties of titanium and its alloys depend on their basic metallurgical structure and the way in which this is manipulated during their mechanical and thermal treatment during manufacture. When heated, titanium atomic structure undergoes a transformation from a close-packed hexagonal arrangement (alpha-titanium) to a body-centered cubic arrangement (beta titanium) at 882°C. This transformation can be considerably modified by the addition of alloying elements to produce four main types of titanium alloys. The chemical compositions of commercial titanium alloys is presented in App. F and some basic mechanical properties of these alloys can be found in Table 8.40. Titanium alloys are also used because of the following properties:

Low coefficient of expansion. Titanium possesses a coefficient of expansion that is significantly less than ferrous alloys. This property also allows titanium to be much more compatible with ceramic or glass materials than most metals, particularly when metal-ceramic/glass seals are involved.

Nonmagnetic. Titanium is virtually nonmagnetic, making it ideal for applications where electromagnetic interference must be minimized. Desirable applications include electronic equipment housing, medical devices, and downhole well logging tools.

Excellent fire resistance. Even at very high temperatures titanium is fire resistant. This is important for applications such as petrochemical plant and firewater systems for offshore platforms, where its ability to survive a hydrocarbon fire is an essential factor.

TABLE 8.40 Mechanical Properties of Titanium Alloys

UNS	ASTM	Trade name	Tensile strength, MPa	Yield strength, MPa	Elastic modulus, GPa
R50250	1	Unalloyed Ti	241	172	103
R50400	2	Unalloyed Ti	345	276	103
R50550	3	Unalloyed Ti	448	379	103
R60700	4	Unalloyed Ti	552	483	103
R56400	5	Ti-6Al-4V	896	827	113
	6	Ti-5Al-2.5Sn	827	793	110
R52400	7	Ti-0.15Pd	345	276	103
R56320	9	Ti-3Al-2.5V	620	483	90
	10	Ti-11.5Mo-6Zr-4.5Sn	689	620	103
R52250	11	Ti-0.15Pd	241	172	103
R53400	12	Ti-0.3-Mo-0.8Ni	483	345	103
	13	Ti-0.5Ni-0.05Ru	276	172	103
	14	Ti-0.5Ni-0.05Ru	414	276	103
	15	Ti-0.5Ni-0.05Ru	483	379	103
R52402	16	Ti-0.05Pd	345	276	103
R52252	17	Ti-0.05Pd	241	172	103
R56322	18	Ti-3Al-2.5V-0.05Pd	620	483	105
R58640	19	Ti-3Al-8V-6Cr-4Zr-4Mo	793	758	103
R58645	20	Ti-3Al-8V-6Cr-4Zr-4Mo-0.05Pd	793	758	103
R58210	21	Ti-15Mo-2.7Nb-3Al-0.25Si	793	758	103
	23	Ti-6Al-4V ELI*	793	758	112
	24	Ti-6Al-4V-0.05Pd	896	827	113
	25	Ti-6Al-4V-0.5Ni-0.05Pd	896	827	113
	26	Ti-0.1Ru	345	276	103
	27	Ti-0.1Ru	241	172	103
	28	Ti-3Al-2.5V-0.1Ru	620	483	90
	29	Ti-6Al-4V-0.1Ru	827	758	112

*Extra low interstitial.

8.9.2 Titanium alloys

Alpha titanium alloys. Alpha titanium alloys are largely single-phase alloys containing up to 7% aluminum (alpha stabilizer) and a small amount (<0.3%) of oxygen, nitrogen, and carbon. Alpha titanium alloys have the lowest strengths of titanium alloys. However, they can be formed and welded. Some contain beta stabilizers to improve strength. Alpha titanium alloys are generally in the annealed or stress-relieved condition. They are considered fully annealed after heating to 675 to 790°C for 1 or 2 h. Alpha alloys range in yield strength from 170 to 480 MPa. Variations are generally achieved by alloy selection. Alpha alloys are generally fabricated in the annealed condition. All fabrication techniques used for stainless steels are generally applicable. Weldability is considered good, given proper gas shielding. Some examples of alpha structure are R50400 and R53400.

Alpha/beta alloys. Alpha plus beta alloys are widely used for high-strength applications and have moderate creep resistance. Alpha/beta titanium alloys are generally used in the annealed or solution-treated and aged condition. Annealing is generally performed in a temperature range 705 to 845°C for $\frac{1}{2}$ to 4 h. Solution treating is generally performed in a temperature range of 900 to 955°C, followed by a water quench. Aging is performed between 480 to 593°C for 2 to 24 h. The precise temperature and time is chosen to achieve the desired mechanical properties. Alpha/beta alloys range in yield strength from 800 MPa to more than 1.2 GPa. Strength can be varied both by alloy selection and heat treatment. Water quenching is required to attain higher strength levels. Section thickness requirements should be considered when selecting these alloys. Generally, alpha/beta alloys are fabricated at elevated temperatures, followed by heat treatment. Cold forming is limited in these alloys. Examples of alpha/beta alloys are R58640 and R56400.

Near alpha alloys. Near alpha alloys have medium strength but better creep resistance than alpha alloys. They can be heat treated from the beta phase to optimize creep resistance and low cycle fatigue resistance. Some can be welded.

Beta phase alloys. Beta phase alloys are usually metastable, formable as quenched, and can be aged to the highest strengths but then lack ductility. Fully stable beta alloys need large amounts of beta stabilizers (vanadium, chromium and molybdenum) and are therefore too dense. In addition, the modulus is low (<100 GPa) unless the beta phase structure is decomposed to precipitate the alpha phase. They have poor stability at 200 to 300°C, have low creep resistance, and are difficult to weld without embrittlement. Metastable beta alloys have some application as high-strength fasteners.

Beta titanium alloys are generally used in the solution-treated and aged condition. High yield strengths (>1.2 GPa) are attainable through cold work and direct age treatments. The annealed condition may also be employed for service temperatures less than 205°C. Annealing and solution treating are performed in a temperature range of 730 to 980°C, with temperatures around 815°C most common. Aging between 482 to 593°C for 2 to 48 h is chosen to obtain the desired mechanical properties. Duplex aging is often employed to improve age response; the first age cycle is performed between 315 and 455°C for 2 to 8 h, followed by the second age cycle between 480 and 595°C for 8 to 16 h. Beta alloys range in yield strength from 780 MPa to more than

1.4 GPa. Current hardness limitations for sour service restrict the use of these alloys to less than the maximum strength.

Beta alloys may be fabricated using any of the techniques employed for alpha alloys, including cold forming in the solution-treated condition. Forming pressure will increase because the yield strength is high compared to alpha alloys. The beta alloys can be welded and may be aged to increase strength after welding. The welding process will produce an annealed condition, exhibiting strength at the low end of the beta alloy range. An example of beta alloys is R56260.

Commercial grades. The strength of titanium can be increased by alloying, some alloys reaching 1.3 GPa, although at a small reduction in corrosion resistance. The commercial types are more commonly known by their ASTM grades than by their UNS numbers. Table 8.41 lists general ASTM specifications for various titanium alloy applications. Titanium grades 1, 2, 3, and 4 are essentially unalloyed Ti. Grades 7 and 11 contain 0.15% palladium to improve resistance to crevice corrosion and to reducing acids, the palladium additions enhancing the passivation behavior of titanium alloys. Titanium grade 12 contains 0.3% Mo and 0.8% Ni and is known for its improved resistance to crevice corrosion and its higher design allowances than unalloyed grades. It is available in many product forms. Other alloying elements (e.g., vanadium, aluminum) are used to increase strength (grades 5 and 9).

8.9.3 Weldability

Commercially pure titanium (98 to 99.5% Ti) or alloys strengthened by small additions of oxygen, nitrogen, carbon, and iron can be readily fusion welded. Alpha alloys can be fusion welded in the annealed condition and alpha/beta alloys can be readily welded in the annealed condition. However, alloys containing a large amount of the beta phase are not easily welded. In industry, the most widely welded titanium alloys are the commercially pure grades and variants of the 6% Al and 4% V alloy, which is regarded as the standard aircraft alloy. Titanium and its alloys can be welded using a matching filler composition; compositions are given in The American Welding Society specification AWS A5.16-90.[56]

Titanium and its alloys are readily fusion welded providing suitable precautions are taken. TIG and plasma processes, with argon or argon-helium shielding gas, are used for welding thin-section components, typically < 10 mm. Autogenous welding can be used for a section thickness of < 3 mm with TIG or < 6 mm with plasma. Pulsed MIG is preferred to dip transfer MIG because of the lower spatter level.

TABLE 8.41 General ASTM Specifications for Titanium Alloys

ASTM B265	Plate and sheet
ASTM B299	Sponge
ASTM B337	Pipe (annealed, seamless, and welded)
ASTM B338	Welded tube
ASTM B348	Bar and billet
ASTM B363	Fittings
ASTM B367	Castings
ASTM B381	Forgings
ASTM B862	Pipe (as welded, no anneal)
ASTM B863	Wire (titanium and titanium alloy)
ASTM F1108	6Al-4V castings for surgical implants
ASTM F1295	6Al-4V niobium alloy for surgical implant applications
ASTM F1341	Unalloyed titanium wire for surgical implant applications
ASTM F136	6Al-4V ELI alloy for surgical implant applications
ASTM F1472	6Al-4V for surgical implant applications
ASTM F620	6Al-4V ELI forgings for surgical implants
ASTM F67	Unalloyed titanium for surgical implant applications

Weld metal porosity. Weld metal porosity is the most frequent weld defect. Because gas solubility is significantly less in the solid phase, porosity arises when the gas is trapped between dendrites during solidification. In titanium, hydrogen from moisture in the arc environment or contamination on the filler and parent metal surface is the most likely cause of porosity. It is essential that the joint and surrounding surface areas are cleaned by first degreasing either by steam, solvent, alkaline, or vapor degreasing. Any surface oxide should then be removed by pickling (HF-HNO$_3$ solution), light grinding, or scratch brushing with a clean, stainless steel wire brush. When TIG welding thin-section components, the joint area should be dry machined to produce a smooth surface finish.

Embrittlement. Embrittlement can be caused by weld metal contamination by either gas absorption or by dissolving contaminants such as dust (iron particles) on the surface. At temperatures above 5000°C, titanium has a very high affinity for oxygen, nitrogen, and hydrogen. The weld pool, HAZ, and cooling weld bead must be protected from oxidation by an inert gas shield (argon or helium). When oxidation occurs, the thin-layer surface oxide generates an interference color. The color can indicate whether the shielding was adequate or an unacceptable degree of contamination has occurred.

Contamination cracking. If iron particles are present on the component surface, they dissolve in the weld metal, reducing corrosion resistance and, at a sufficiently high iron content, causing embrittlement. Iron particles are equally detrimental in the HAZ where local melting of the particles forms pockets of titanium-iron eutectic. Microcracking may occur, but it is more likely that the iron-rich pockets will become preferential sites for corrosion. To avoid corrosion cracking, and minimize the risk of embrittlement through iron contamination, it is a recommended practice to weld titanium in an especially clean area.[56]

8.9.4 Applications

Aircraft. The aircraft industry is the single largest market for titanium products primarily due to its exceptional strength-to-weight ratio, elevated temperature performance, and corrosion resistance. The largest single aircraft use of titanium is in the gas turbine engine. In most modern jet engines, titanium-based alloy parts make up 20 to 30% of the dry weight, primarily in the compressor. Applications include blades, disks or hubs, inlet guide vanes, and cases. Titanium is most commonly the material of choice for engine parts that operate up to 593°C. Titanium alloys effectively compete with aluminum, nickel, and ferrous alloys in both commercial and military airframes. For example, the all-titanium SR-71 still holds all speed and altitude records.

The selection of titanium in both airframes and engines is based upon titanium basic attributes (i.e., weight reduction due to high strength-to-weight ratios coupled with exemplary reliability in service, attributable to outstanding corrosion resistance compared to alternate structural metals). Starting with the extensive use of titanium in the early Mercury and Apollo spacecraft, titanium alloys continue to be widely used in military and space applications. In addition to manned spacecraft, titanium alloys are extensively employed by NASA in solid rocket booster cases, guidance control pressure vessels, and a wide variety of applications demanding light weight and reliability.

Titanium in industry. Industrial applications in which titanium-based alloys are currently utilized include.

- *Gas turbine engines.* Highly efficient gas turbine engines are possible only through the use of titanium-based alloys in components like fan blades, compressor blades, disks, hubs, and numerous non-rotor parts. The key advantages of titanium-based alloys in this application include a high strength-to-weight ratio, strength at moderate temperatures, and good resistance to creep and fatigue. The development of titanium aluminides will allow the use of titanium in hotter sections of a new generation of engines.

- *Heat transfer.* A major industrial application for titanium remains in heat-transfer applications in which the cooling medium is seawater, brackish water, or polluted water. Titanium condensers, shell and tube heat exchangers, and plate and frame heat exchangers are used extensively in power plants, refineries, air conditioning systems, chemical plants, offshore platforms, surface ships, and submarines.

- *Dimensional stable anodes (DSAs)*. The unique electrochemical properties of the titanium DSA make it the most energy efficient unit for the production of chlorine, chlorate, and hypochlorite.
- *Extraction and electrowinning of metals*. Hydrometallurgical extraction of metals from ores in titanium reactors is an environmentally safe alternative to smelting processes. Extended life span, increased energy efficiency, and greater product purity are factors promoting the usage of titanium electrodes in electrowinning and electrorefining of metals like copper, gold, manganese, and manganese dioxide.
- *Medical applications*. Titanium is widely used for implants, surgical devices, pacemaker cases, and centrifuges. Titanium is the most biocompatible of all metals due to its total resistance to attack by body fluids, high strength, and low modulus.
- *Marine applications*. Because of high toughness, high strength, and exceptional erosion-corrosion resistance, titanium is currently being used for submarine ball valves, fire pumps, heat exchangers, castings, hull material for deep sea submersibles, water jet propulsion systems, shipboard cooling, and piping systems.
- *Chemical processing*. Titanium vessels, heat exchangers, tanks, agitators, coolers, and piping systems are utilized in the processing of aggressive compounds, like nitric acid, organic acids, chlorine dioxide, inhibited reducing acids, and hydrogen sulfide.
- *Pulp and paper*. Due to recycling of waste fluids and the need for greater equipment reliability and life span, titanium has become the standard material for drum washers, diffusion bleach washers, pumps, piping systems, and heat exchangers in the bleaching section of pulp and paper plants. This is particularly true for the equipment developed for chlorine dioxide bleaching systems.[57]

8.9.5 Corrosion resistance

Titanium is a very reactive metal that shows remarkable corrosion resistance in oxidizing acid environments by virtue of a passive oxide film. Following its commercial introduction in the 1950s, titanium has become an established corrosion-resistant material. In the chemical industry, the grade most used is commercial-purity titanium. Like stainless steels, it is dependent upon an oxide film for its corrosion resistance. Therefore, it performs best in oxidizing media such as hot nitric acid. The oxide film formed on titanium is more protective than that on stainless steel, and it often performs well in media that cause pitting and crevice corrosion in the latter (e.g., seawater, wet chlorine, organic chlorides). Although titanium is resistant to these media, it is not immune and can be susceptible to pitting and crevice attack at elevated temperatures. It is, for example, not immune to seawater corrosion if the temperature is greater than about 110°C.[1]

Titanium is not a cure-all for every corrosion problem, but increased production and improved fabrication techniques have brought the material cost to a point where it can compete economically with some of the nickel-base alloys and even some stainless steels. Its low density offsets the relatively high materials costs, and its good corrosion resistance allows thinner heat-exchanger tubes. Table 8.42 presents the corrosion rates observed on commercially pure titanium grades in a multitude of chemical environments.[58]

Acid resistance. Titanium alloys resist an extensive range of acidic conditions. Many industrial acid streams contain contaminants that are oxidizing in nature, thereby passivating titanium alloys in normally aggressive acid media. Metal ion concentration levels as low as 20 to 100 ppm can inhibit corrosion extremely effectively. Potent inhibitors for titanium in reducing acid media are common in typical process operations. Titanium inhibition can be provided by dissolved oxygen, chlorine, bromine, nitrate, chromate, permanganate, molybdate, or other cationic metallic ions, such as ferric (Fe^{3+}), cupric (Cu^{2+}), nickel (Ni^{2+}), and many precious metal ions. Figure 8.9 shows the inhibiting effect of ferric chloride on grade 2 titanium exposed to hydrochloric acid at various concentrations and temperatures. Figures 8.10 and 8.11 show similar behavior for, respectively, grade 7 and grade 12 titanium alloys. It is this potent metal ion inhibition that permits titanium to be successfully used for equipment handling hot HCl and H_2SO_4 acid solutions in metallic ore leaching processes.

Oxidizing acids. In general, titanium has excellent resistance to oxidizing acids such as nitric and chromic acid over a wide range of temperatures and concentrations. Titanium is used extensively for handling nitric acid in commercial applications. Titanium exhibits low corrosion rates in nitric acid over a wide range of conditions. At boiling temperatures and above, titanium's corrosion resistance is very sensitive to nitric acid purity. Generally, the higher the contamination and the higher the metallic ion content of the acid, the better titanium will perform. This is in contrast to stainless steels, which is often adversely affected by acid contaminants. Because the titanium corrosion product (Ti^{4+}) is highly inhibitive, titanium often exhibits superb performance

TABLE 8.42 Corrosion Rates of Commercially Pure Titanium Grades

Environment	Concentration, %	Temperature, °C	Corrosion rate, $\mu m \cdot y^{-1}$
Acetaldehyde	75	149	1
	100	149	Nil
Acetic acid	5 to 99.7	124	Nil
Acetic anhydride	99.5	Boiling	13
Acidic gases containing CO_2, H_2O, Cl_2, SO_2, SO_3, H_2S, O_4, NH_3		38–260	< 0.025
Adipic acid	67	232	Nil
Aluminum chloride, aerated	10	100	2*
Aluminum chloride, aerated	25	100	3150*
Aluminum fluoride	Saturated	25	Nil
Aluminum nitrate	Saturated	25	Nil
Aluminum sulfate	Saturated	25	Nil
Ammonium acid phosphate	10	25	Nil
Ammonia anhydrous	100	40	< 125
Ammonia steam, water		222	11,000
Ammonium acetate	10	25	Nil
Ammonium bicarbonate	50	100	Nil
Ammonium bisulfite, pH 2.05	Spent pulping liquor	71	15
Ammonium chloride	Saturated	100	< 13
Ammonium hydroxide	28	25	3
Ammonium nitrate + 1% nitric acid	28	Boiling	Nil
Ammonium oxalate	Saturated	25	Nil
Ammonium sulfate	10	100	Nil
Ammonium sulfate + 12% H_2SO_4	Saturated	25	10
Aqua regia	3:1	25	Nil
Aqua regia	3:1	79	890
Barium chloride	25	100	Nil
Barium hydroxide	Saturated	25	Nil
Barium hydroxide	27	Boiling	Some small pits
Barium nitrate	10	25	Nil
Barium fluoride	Saturated	25	Nil
Benzoic acid	Saturated	25	Nil
Boric acid	Saturated	25	Nil
Boric acid	10	Boiling	Nil
Bromine	Liquid	30	Rapid
Bromine moist	Vapor	30	3
N-butyric acid	Undiluted	25	Nil
Calcium bisulfite	Cooking liquor	26	10
Calcium carbonate	Saturated	Boiling	Nil
Calcium chloride	5	100	5*
Calcium chloride	10	100	7*
Calcium chloride	20	100	15*

TABLE 8.42 Corrosion Rates of Commercially Pure Titanium Grades (*Continued*)

Environment	Concentration, %	Temperature, °C	Corrosion rate, $\mu m \cdot y^{-1}$
Calcium chloride	55	104	1*
Calcium chloride	60	149	< 3*
Calcium hydroxide	Saturated	Boiling	Nil
Calcium hypochlorite	6	100	1
Calcium hypochlorite	18	21	Nil
Calcium hypochlorite	Saturated slurry		Nil
Carbon dioxide	100		Excellent
Carbon tetrachloride	Liquid	Boiling	Nil
Carbon tetrachloride	Vapor	Boiling	Nil
Chlorine gas, wet	> 0.7 H_2O	25	Nil
Chlorine gas, wet	> 1.5 H_2O	200	Nil
Chlorine header sludge and wet chlorine		97	1
Chlorine gas dry	< 0.5H_2O	25	May react
Chlorine dioxide + H_2O and air	5 in steam gas 82		< 3
Chloride dioxide in steam	5	99	Nil
Chlorine trifluoride	100	30	Vigorous reaction
Chloracetic acid	30	82	< 0.125
Chloracetic acid	100	Boiling	< 0.125
Chlorosulfonic acid	100	25	190–310
Chloroform	Vapor & liquid	Boiling	0
Chromic acid	10	Boiling	3
Chromic acid	15	82	15
Chromic acid	50	82	28
Chromium plating bath containing fluoride	240 g/L plating salt	77	1500
Chromic acid + 5% Nitric acid	5	21	3
Citric acid	50	60	0
Citric acid	50 aerated	100	127
Citric acid	50	Boiling	127–1300
Citric acid	62	149	Corroded
Cupric chloride	20	Boiling	Nil
Cupric chloride	40	Boiling	5
Cupric choride	55	119 (boiling)	3
Cupric cyanide	Saturated	25	Nil
Cuprous chloride	50	90	< 3
Cyclohexane (plus traces of formic acid)		150	3
Dichloroacetic acid	100	Boiling	7
Dichlorobenzene + 4–5% HCl		179	102
Diethylene triamine	100	25	Nil
Ethyl alcohol	95	Boiling	130
Ethylene dichloride	100	Boiling	5–125
Ethylene diamine	100	25	Nil
Ferric chloride	10–20	25	Nil

TABLE 8.42 Corrosion Rates of Commercially Pure Titanium Grades (*Continued*)

Environment	Concentration, %	Temperature, °C	Corrosion rate, $\mu\text{m}\cdot\text{y}^{-1}$
Ferric chloride	10–30	100	< 130
Ferric chloride	10–40	Boiling	Nil
Ferric chloride	50	113 (boiling)	Nil
Ferric chloride	50	150	3
Ferric sulfate $9H_2O$	10	25	Nil
Flubonic acid	5–20	Elevated	Rapid
Fluorsilicic	10	25	48,000
Food products		Ambient	No attack
Fomaldehyde	37	Boiling	Nil
Formamide vapor		300	Nil
Formic acid aerated	25	100	1†
Formic acid aerated	90	100	1†
Formic acid nonaerated	25	100	2400†
	90	100	3000†
Furfural	100	25	Nil
Gluconic acid	50	25	Nil
Glycerin		25	Nil
Hydrogen chloride, gas	Air mixture	Ambient	Nil
Hydrochloric acid	1	Boiling	> 2500
Hydrochloric acid	3	Boiling	14,000
Hydrochloric acid chlorine saturated	5	Boiling	10,000
	5	190	< 25
	10	190	> 28,000
200ppm Cl_2	36	25	432
+ 1% HNO_3	5	93	91
+ 5% HNO_3	5	93	30
+ 5% HNO_3	1	Boiling	70
+ 5% HNO_3 + 1.7 g/L $TiCl_4$	1	Boiling	Nil
+ 0.5% CrO_3	5	93	30
+ 1% CrO_3	5	38	18
+ 1% CrO_3	5	93	30
+ 0.05% $CuSO_4$	5	93	90
+ 0.5% $CuSO_4$	5	93	60
+ 0.05% $CuSO_4$	5	Boiling	60
+ 0.5% $CuSO_4$	5	Boiling	80
Hydrofluonic acid	1.48	25	Rapid
Hydrogen peroxide	3	25	< 120
Hydrogen peroxide	6	25	< 120
Hydrogen peroxide	30	25	< 300
Hydrogen sulfide, steam and 0.077% mercaptans	7.65	93–110	Nil
Hypochlorous acid + Cl_2O and Cl_2 gases	17	38	0
Iodine in water + potassium iodide		25	Nil

TABLE 8.42 Corrosion Rates of Commercially Pure Titanium Grades (*Continued*)

Environment	Concentration, %	Temperature, °C	Corrosion rate, $\mu m \cdot y^{-1}$
Lactic acid	10–85	100	< 120
Lactic acid	10	Boiling	< 120
Lead acetate	Saturated	25	Nil
Linseed oil, boiled		25	Nil
Lithium chloride	50	149	Nil
Magnesium chloride	5–40	Boiling	Nil
Magnesium hydroxide	Saturated	25	Nil
Magnesium sulfate	Saturated	25	Nil
Manganous chloride	5–20	100	Nil
Maleic acid	18–20	35	2
Mercuric chloride	10	100	1
Mercuric chloride	Saturated	100	< 120
Mercuric cyanide	Saturated	25	Nil
Methyl alcohol	91	35	Nil
Nickel chloride	5	100	4
Nickel chloride	20	100	3
Nitric acid	17	Boiling	70–100
Nitric acid, aerated	10	25	5
Nitric acid, aerated	50	25	2
Nitric acid, aerated	70	25	5
Nitric acid, aerated	10	40	3
Nitric acid, aerated	50	60	30
Nitric acid, aerated	70	70	40
Nitric acid, aerated	40	200	600
Nitric acid, aerated	70	270	1200
Nitric acid, aerated	20	290	300
Nitric acid, nonaerated	70	80	25–70
Nitric acid white fuming	35	Boiling	120–500
		82	150
		160	< 120
Nitric acid, red fuming	< about 2% H_2O	25	Ignition sensitive
	> about 2% H_2O	25	Not ignition sensitive
Nitric acid + 0.1% $K_2Cr_2O_7$	40	Boiling	Nil–15
Nitric acid + 10% $NaClO_3$	40	Boiling	3–30
Phosphoric acid	10–30	25	20–50
Photographic emulsions			< 120
Potassium bromide	Saturated	25	Nil
Potassium chloride	Saturated	25	Nil
	Saturated	60	< 0.3
Potassium dichromate			Nil
Potassium hydroxide	50	27	10
Potassium permanganate	Saturated	25	Nil
Potassium sulfate	10	25	Nil
seawater, 4 to ½-year test			

TABLE 8.42 Corrosion Rates of Commercially Pure Titanium Grades (*Continued*)

Environment	Concentration, %	Temperature, °C	Corrosion rate, $\mu m \cdot y^{-1}$
Silver nitrate sodium	50	25	Nil
	100	To 590	Nil
Sodium acetate	Saturated	25	Good
Sodium carbonate	25	Boiling	Nil
Sodium chloride	Saturated	25	Nil
Sodium chloride, pH 1.5	23	Boiling	Nil
Sodium chloride, titanium in contact with Teflon	23	Boiling	Attack in crevice
Stannic chloride, molten	100	66	Nil
Stannous chloride	Saturated	25	Nil
Sulfur, molten	100	240	Nil
Sulfur dioxide, water saturated	Near 100	25	< 2
Sulfuric acid, aerated with air	1	60	7
	3	60	12
	5	60	4.8
Sulfuric acid +0.25% $CuSO_4$	30	100	60
0.25% $CuSO_4$	30	93	80
Sulfuric acid +10% nitric acid	90	25	450
30% nitric acid	70	25	630
50% nitric acid	50	25	630
Tannic acid	25	100	< 120
Tartaric acid	10–50	100	< 120
	10	60	2
	25	60	2
Terepthalic	77	218	Nil
Tin, molten	100	498	Resist
Trichloroethylene	99	Boiling	2–120
Uranium chloride	Saturated	21–90	Nil
Urea-ammonia reaction mass		Elevated temperature and pressure	No attack
Urea + 32% ammonia, 20.5% water, 19% carbon dioxide	28	82	80
Water, degassed		315	Nil
X-ray developer solution		25	Nil
Zinc chloride	20	104	Nil
	50	150	Nil
Zinc sulfate	Saturated	25	Nil

*May corrode in crevices.
†Grades 7 and 12 are immune.

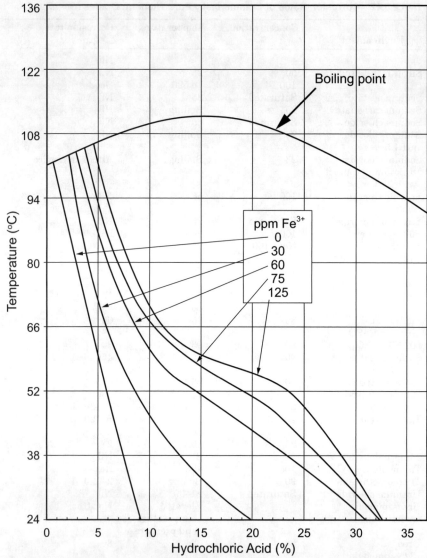

Figure 8.9 Iso-corrosion lines (1 mm·y^{-1}) showing the effect of minute ferric ion concentrations on the corrosion resistance of grade 2 titanium in naturally aerated HCl solutions.

in recycled nitric acid streams such as reboiler loops. One user cites an example of a titanium heat exchanger handling 60% HNO$_3$ at 193°C and 2.0 MPa that showed no signs of corrosion after more than 2 years of operation. Titanium reactors, reboilers, condensers, heaters, and thermowells have been used with solutions containing 10 to 70% HNO$_3$ at temperatures from boiling to 600°C.[57] Although titanium has

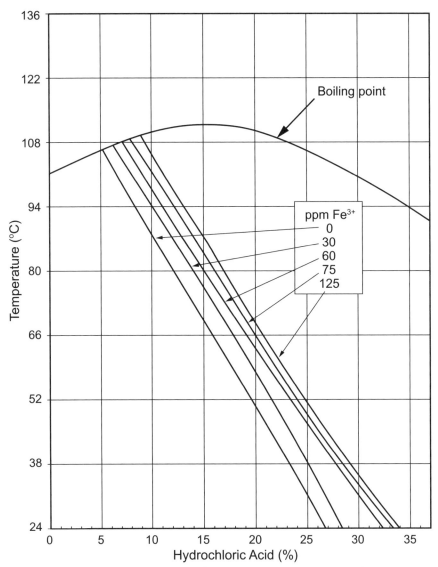

Figure 8.10 Iso-corrosion lines ($1\ mm \cdot y^{-1}$) showing the effect of minute ferric ion concentrations on the corrosion resistance of grade 7 titanium in naturally aerated HCl solutions.

excellent resistance to nitric acid over a wide range of concentrations and temperatures, it should not be used with red fuming nitric acid because of the danger of pyrophoric reactions.

Reducing acids. Titanium alloys are generally very resistant to mildly reducing acids but can display severe limitations in strongly reducing

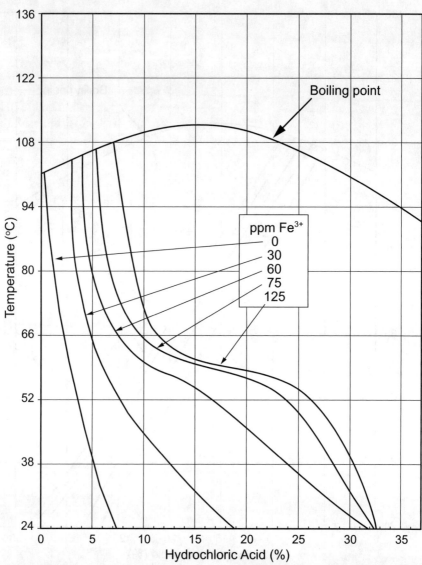

Figure 8.11 Iso-corrosion lines (1 mm·y^{-1}) showing the effect of minute ferric ion concentrations on the corrosion resistance of grade 12 titanium in naturally aerated HCl solutions.

acids. Mildly reducing acids such as sulfurous, acetic, terephthalic, adipic, lactic, and many organic acids generally represent no problem for titanium over the full concentration range. However, relatively pure, strong reducing acids, such as hydrochloric, hydrobromic, sulphuric, phosphoric, oxalic, and sulfamic acids can accelerate general corrosion of titanium depending on acid temperature, concentration, and purity. Titanium-palladium alloys offer dramatically improved corrosion resistance under these severe conditions. In fact, they often compare quite favorably to nickel alloys in dilute reducing acids. Titanium is rapidly attacked by hydrofluoric acid of even very dilute concentrations. Therefore, titanium is not recommended for use with hydrofluoric acid solutions or in fluoride containing solutions below pH 7. Certain complexing metal ions (e.g., aluminum) may effectively inhibit corrosion in dilute fluoride solutions.[57]

Organic acids. Titanium alloys generally exhibit excellent resistance to organic media. Mere traces of moisture, even in the absence of air, normally present in organic process streams assure the development of a stable protective oxide film of titanium. Titanium is highly resistant to hydrocarbons, chloro-hydrocarbons, fluorocarbons, ketones, aldehydes, ethers, esters, amines, alcohols, and most organic acids. Titanium equipment has traditionally been used for production of terephthalic acid, adipic acid, and acetaldehyde. Acetic, tartaric, stearic, lactic, tannic, and many other organic acids represent fairly benign environments for titanium. However, proper titanium alloy selection is necessary for the stronger organic acids such as oxalic, formic, sulfamic, and trichloroacetic acids. Performance in these acids depends on acid concentration, temperature, degree of aeration, and possible inhibitors present. Grades 7 and 12 titanium alloys are often preferred materials in these more aggressive acids.[57]

Titanium and methanol. Anhydrous methanol is unique in its ability to cause SCC of titanium and titanium alloys. Industrial methanol normally contains sufficient water to provide immunity to titanium. In the past the specification of a minimum of 2% water content has proved adequate to protect commercially pure titanium equipment for all but the most severe conditions. In such conditions, due to temperature and pressure, titanium alloys would more than likely be required. A more conservative margin of safety was established by the offshore industry at 5% minimum water content.

Alkaline media. Titanium is generally highly resistant to alkaline media including solutions of sodium hydroxide, potassium hydroxide, calcium hydroxide, magnesium hydroxide, and ammonium hydroxide. In the high basic sodium or potassium hydroxide solutions, however, useful application of titanium may be limited to temperatures below 80°C. This is due to possible excessive hydrogen uptake and eventual embrittlement of titanium alloys in hot, strongly alkaline media. Titanium often becomes the material of choice for alkaline media containing chlorides and/or oxidizing chloride species. Even at higher temperatures, titanium resists pitting, SCC, or the conventional caustic embrittlement observed on many stainless steels in these situations.[57]

Chlorine gas, chlorine chemicals, and chlorine solutions. Titanium is widely used to handle moist or wet chlorine gas and has earned a reputation for outstanding performance in this service. The strongly oxidizing nature of moist chlorine passivates titanium, resulting in low corrosion rates. The selection of a resistant titanium alloy offers a solution to the possibility of crevice corrosion when wet chlorine surface temperatures exceed 70°C (Table 8.42). Dry chlorine can cause rapid attack of titanium and may even cause ignition if moisture content is sufficiently low. However, as little as 1% water is generally sufficient for passivation or repassivation after mechanical damage to titanium in chlorine gas under static conditions at room temperature.

Titanium is fully resistant to solutions of chlorites, hypochlorites, chlorates, perchlorates, and chlorine dioxide. It has been used to handle these chemicals in the pulp and paper industry for many years with no evidence of corrosion. Titanium is used in chloride salt solutions and other brines over the full concentration range, especially as temperatures increase. Near nil corrosion rates can be expected in brine media over the pH range of 3 to 11. Oxidizing metallic chlorides, such as $FeCl_3$, $NiCl_2$ or $CuCl_2$, extend titanium's passivity to much lower pH levels.[57] Localized pitting or corrosion, occurring in tight crevices and under scale or other deposits, is a controlling factor in the application of unalloyed titanium. Attack will normally not occur on commercially pure titanium or industrial alloys below 70°C regardless of solution pH.

Steam and natural waters. Titanium alloys are highly resistant to water, natural waters, and steam to temperatures in excess of 300°C. Excellent performance can be expected in high-purity water and fresh water. Titanium is relatively immune to microbiologically influenced corrosion (MIC). Typical contaminants found in natural water streams, such as iron and manganese oxides, sulfides, sulfates, carbonates, and chlorides do not compromise titanium's performance. Titanium remains totally unaffected by chlorination treatments used to control biofouling.

Seawater and salt solutions. Titanium alloys exhibit excellent resistance to most salt solutions over a wide range of pH and temperatures. Good performance can be expected in sulfates, sulfites, borates, phosphates, cyanides, carbonates, and bicarbonates. Similar results can be expected with oxidizing anionic salts such as nitrates, molybdates, chromates, permanganates, and vanadates and also with oxidizing cationic salts including ferric, cupric, and nickel compounds.

TABLE 8.43 Erosion of Unalloyed Titanium in Seawater Containing Suspended Solids

Flow rate, m·s^{-1}	Suspended matter	Duration, h	Erosion corrosion, μm·y^{-1}		
			Ti Grade 2	Cu/Ni 70/30*	Al brass
7.2	None	10,000	Nil	Pitted	Pitted
2	40 g/L 60 mesh sand	2,000	2.5	99.0	50.8
2	40 g/L 10 mesh sand	2,000	12.7	Severe erosion	Severe erosion

*High iron, high manganese 70/30 copper nickel.

Seawater and neutral brines above the boiling point will develop localized reducing acidic conditions, and pitting may occur. Enhanced resistance to reducing acid chlorides and crevice corrosion is available from alloy grades 7, 11, and 12. Attention to design of flanged joints using heavy flanges and high clamping pressure and to the specification of gaskets may serve to prevent crevices from developing. An alternative strategy is to incorporate a source of nickel, copper, molybdenum, or palladium into the gasket.

Titanium is fully resistant to natural seawater regardless of chemistry variations and pollution effects (i.e., sulfides). Twenty-year corrosion rates well below 0.0003 mm·y^{-1} have been measured on titanium exposed beneath the sea and in splash or tidal zones. In the sea, titanium alloys are immune to all forms of localized corrosion and withstand seawater impingement and flow velocities in excess of 30 m·s^{-1}. Table 8.43 compares the erosion-corrosion resistance of unalloyed titanium with two commonly used seawater materials.[57] In addition, the fatigue strength and toughness of most titanium alloys are unaffected in seawater, and many titanium alloys are immune to seawater stress corrosion.

When in contact with other metals, titanium alloys are not subject to galvanic corrosion in seawater. However titanium may accelerate attack on active metals such as steel, aluminum, and copper alloys. The extent of galvanic corrosion will depend on many factors such as anode-to-cathode ratio, seawater velocity, and seawater chemistry. The most successful strategies eliminate this galvanic couple by using more resistant, compatible, and passive metals with titanium, all-titanium construction, or dielectric (insulating) joints.

Resistance to gases

Oxygen and air. Titanium alloys are totally resistant to all forms of atmospheric corrosion regardless of pollutants present in either marine, rural, or industrial locations. Titanium has excellent resistance to gaseous oxygen and air at temperatures up to 370°C. Above this temperature and below 450°C titanium forms colored surface oxide films that thicken slowly with time. Above 650°C or so titanium alloys suffer from lack of long-term oxidation resistance and will become brittle due to the increased diffusion of oxygen in the metal. In oxygen, the combustion is not spontaneous and occurs with oxygen concentration above 35% at pressures over 2.5 MPa when a fresh surface is created.

Nitrogen and ammonia. Nitrogen reacts much more slowly with titanium than oxygen. However, above 800°C, excessive diffusion of the nitride may cause metal embrittlement. Titanium is not corroded by liquid anhydrous ammonia at ambient temperatures. Moist or dry ammonia gas or ammonia water (NH_4OH) solutions will not corrode titanium to their boiling-point and above.

Hydrogen. The surface oxide film on titanium acts as a highly effective barrier to hydrogen. Penetration can only occur when this protective film is disrupted mechanically or broken down chemically or electrochemically. The presence of moisture effectively maintains the oxide film, inhibiting hydrogen absorption up to fairly high temperatures and pressures. On the other hand, pure, anhydrous hydrogen exposures should be avoided, particularly as pressures and/or temperatures increase. The few cases of hydrogen embrittlement of titanium observed in industrial service have generally been limited to situations involving:

- High temperatures, high alkaline media
- Titanium coupled to active steel in hot aqueous sulfide streams
- Where titanium has experienced severe prolonged cathodic charging in seawater

Sulfur-bearing gases. Titanium is highly corrosion resistant to sulfur-bearing gases, resisting sulfide stress corrosion cracking and sulfidation at typical operating temperatures. Sulfur dioxide and hydrogen sulfide, either wet or dry, have no effect on titanium. Extremely good performance can be expected in sulfurous acid even at the boiling point. Field exposures in flue gas desulfurization (FGD) scrubber systems of coal-fired power plants have similarly indicated outstanding performance of titanium. Wet SO_3 environments may be a problem for titanium in cases where pure, strong, uninhibited sulfuric acid solutions may form, leading to metal attack. In these situations, the background chemistry of the process environment is critical for successful use of titanium.

Reducing atmospheres. Titanium generally resists mildly reducing, neutral, and highly oxidizing environments up to reasonably high temperatures. The presence of oxidizing species including air, oxygen, and ferrous alloy corrosion products often extends the performance limits of titanium in many highly aggressive environments. However, under highly reducing conditions the oxide film may break down, and corrosion may occur.

8.10 Zirconium

Zirconium is generally alloyed with niobium or tin, with hafnium present as a natural impurity, and oxygen content controlled to give specific strength levels. Controlled quantities of the beta stabilizers (i.e., iron, chromium, and nickel) and the strong alpha stabilizers tin and oxygen are the main alloying elements in zirconium alloys.[48] Nuclear engineering, with its specialized demands for materials having a low neutron absorption with adequate strength and corrosion resistance at elevated temperatures, has necessitated the production of zirconium in relatively large commercial quantities. This specific demand has resulted in the development of specially purified zirconium and certain zirconium alloys, for use as cladding material in nuclear reactors.[59]

As it occurs in nature, zirconium is always found in association with hafnium, in the ratio of 1 part hafnium to 50 parts zirconium, and commercial-grade zirconium contains approximately 2% hafnium. Because hafnium has a high absorption capacity for thermal neutrons, nuclear reactor–grade zirconium is not permitted to contain more than 0.025% Hf, and usually it contains closer to 0.01%.

This situation gave rise to bulk production of two families of zirconium alloys, as can be seen in Table 8.44, which describes the composition of these alloys. Both R60804 and R60802 are used in water-cooled nuclear reactors. Generally, for the chemical engineer not particularly associated with atomic energy, unalloyed zirconium containing hafnium is an appropriate choice for those occasions that require the special corrosion-resistant properties exhibited by the metal. The relative costs of some corrosion-resistant alloys, in different manufacturing product forms, are compared to R600802 in Table 8.45.

Mechanical properties of these grades of zirconium depend to a large extent upon the purity of the zirconium sponge used for melting. Hardness and tensile strength increase rapidly with rise in impurity content, notably oxygen, nitrogen, and iron. Typical mechanical properties of chemical grades of zirconium are listed in Table 8.46. Table 8.47 provides additional physical and mechanical properties for alloys R69702 and R69705. Zirconium, specific gravity 6.574, is lighter than most conventional structural materials such as steel copper, brass, and stainless steels. Its melting point of 1850°C

TABLE 8.44 Mechanical Properties of Zirconium Alloys

Alloy	Trade name	Tensile, MPa	Yield (0.2% offset), MPa	Elongation, %
Industrial grades				
R69702	702	379	207	16
R69704	704	413	241	14
R69705	705	552	379	16
R69706	706	510	345	20
Nuclear grades				
R60001 (annealed)	Unalloyed	296	207	18
R60802 (annealed)	Zircalloy-2	386	303	25
R60804 (annealed)	Zircalloy-4	386	303	25
R60901 (annealed)	Zr-2.5Nb	448	344	20
R60901 (cold worked)		510	385	15

TABLE 8.45 Costs Relative to S31600 of Some Commercial Metals in Different Product Forms

UNS	Metal or alloy	Plate	Tubing	Vessel	Heat exchanger
S31600	316	1	1	1	1
R50400	Ti, grade 2	2.0	2.25	2.0	1.5
R53400	Ti, grade 12	3.1	9.6	2.2	1.7
N06600	Inconel 600	3.6	4.0	3.0	1.8
R52400	Ti, grade 7	6.5	8.8	2.0	2.0
R60802	Zircalloy-2	8.0	9.0	3.5	2.2
N10276	Hastelloy C-276	7.0	7.5	4.0	3.0
N10665	Hastelloy B-2	9.7	11.0	4.5	3.0
	Tantalum		24.8		

TABLE 8.46 Compositions of Zirconium Alloys

UNS	Alloy	Hf, %	Fe, %	Cr, %	Sn, %	O, %	Ni, %	Nb, %
Industrial grades								
R69702	702	4.5	0.2	With Fe		0.16		
R69704	704	4.5	0.3	With Fe	1.5	0.18	1.5	1.5
R69705	705	4.5	0.2	With Fe		0.18		
R69706	706	4.5	0.2	With Fe		0.16		
Nuclear grades								
R60001	Unalloyed					0.8		
R60802	Zircalloy-2		0.1	0.1	1.4	0.12	0.05	
R60804	Zircalloy-4		0.2	0.1	1.4	0.12		
R60901	Zr-2.5Nb					0.14		2.6

TABLE 8.47 Physical and Mechanical Properties of R69702 and R69705

Physical properties	Units	R69702	R69705
Density	g·cm^{-3}	6.510	6.640
Crystal structure			
Alpha phase		hcp (< 865°C)	
Beta phase		bcc (> 865°C)	bcc (> 854°C)
Alpha + beta phase			hcp + bcc (< 854°C)
Melting point	°C	1852	1840
Boiling point	°C	4377	4380
Linear coefficient of expansion	per °C	5.89 × 10^{-6}	6.3 × 10^{-6}
Thermal conductivity (300–800 K)	W·m^{-1} K^{-1}	22	17.1
Specific heat (20°C)	J·kg^{-1}·K^{-1}	285	281
Electrical properties (20°C)			
Resistivity	μΩ·cm	39.7	55.0
Coefficient of resistivity	per °C	0.0044	
Mechanical properties			
Modulus of elasticity	GPa	98.5	95.8
Shear modulus	GPa	35.9	34.2
Poisson's ratio (20°C)		0.35	0.33

gives it reasonable temperature resistance and good creep properties. It has a hcp lattice structure (alpha phase) at room temperature that undergoes allotropic transformation to bcc structure (beta phase) at approximately 870°C. This makes zirconium and most of its alloys strongly anisotropic, which has a great effect on their engineering properties.

Small amounts of impurities, especially oxygen, strongly affect its transformation temperature. Oxygen content plays an important role in the strength of zirconium, and therefore it must be carefully controlled. Reducing it to less than 1000 ppm lowers the strength of zirconium alloys to less than acceptable limits. The alpha-stabilizing elements (e.g., aluminum, antimony, beryllium, cadmium, hafnium, lead, nitrogen, oxygen, and tin) raise the alpha-to-beta transformation temperature, whereas the beta-stabilizing elements (e.g., cobalt, chromium, copper, iron, manganese, molybdenum, nickel, niobium, silver, tantalum, thorium, titanium, tungsten, uranium, and vanadium) lower it. Carbon, silicon, and phosphorus have very low solubility in zirconium even at temperatures above 1000°C. They readily form intermetallic compounds and are relatively insensitive to heat treatment. Most elements and impurities are soluble in beta zirconium but relatively insoluble in alpha zirconium, where they exist as secondary-phase intermetallic compounds.

Ingots of zirconium and its alloys are most commonly 40 to 760 mm in diameter and weigh 1100 to 4500 kg. Wrought products are available in a variety of forms and sizes, such as sheet and strip, plate, foil, bar and rod, wire, tube and pipe, and tube shell. Cast parts such as valve bodies and pump castings and impellers are also available.[60] The fabrication characteristics of zirconium are similar to those of titanium, and they impose similar precautions and conditions on forming, machining, and welding it. Because it is even more costly than titanium, zirconium is often used in the form of linings and claddings on lower-cost structural substrates.[48]

Zirconium alloys are generally used in the annealed or stress-relieved condition. They can be fully annealed at a temperature range of 675 to 800°C for 2 to 4 h at temperature. When R69705 is heat treated at temperatures in excess of 675°C, the subsequent cooling rate should be controlled. The cooling rate should not exceed 110°C/h until the temperature of the material is less than 480°C. Stress relieving of zirconium alloys is done at 540 to 595°C for 0.5 to 1 h at temperature.

Zirconium alloys are most commonly welded by gas tungsten arc welding (GTAW) technique. Other welding methods include metal arc gas welding, plasma arc welding, electron beam welding, and resistance welding. All welding of zirconium must be done under an inert atmosphere. It is very important that the welding done with proper shielding because of zirconium's reactivity to gases at welding temperatures.

8.10.1 Applications

Zirconium and its alloys are used in nuclear applications that require good resistance to high-temperature water and steam, as well as a low thermal neutron cross section and good elevated temperature strength. Another major application for zirconium alloys is as a structural material in the chemical processing industry. Zirconium alloys exhibit excellent resistance to corrosive attack in most organic and inorganic acids, salt solutions, strong alkalies, and some molten salts. In certain applications, the unique corrosion resistance of zirconium alloys can extend its useful life beyond that of the remainder of the plant.

Although zirconium and its alloys are costly compared with other common corrosion-resistant materials, their extremely low corrosion rates, resulting in long service life and reduced maintenance and downtime cost, make zirconium and its alloys quite cost effective. Table 8.45, which compares costs between S31600 stainless steel and various corrosion-resistant metals and alloys, shows that although R69702 is more costly than stainless steel, Inconel, and titanium alloys, it costs roughly the same as or less than some of the Hastelloys and considerably less than tantalum.

These costly exotic metals and alloys are often used for heat exchangers. If alternative corrosion-resistant materials such as plastics, ceramics, and composites were used instead, their low thermal conductivity would necessitate greatly increasing their size. Despite its high cost, the excellent corrosion resistance of zirconium and its alloys, because it promises long maintenance-free service life for the equipment, proves to be cost effective in many chemical processing and other applications where corrosion is an important problem.

The material is employed in the form of heat exchangers, stripper columns, reactor vessels, pumps, valves, and piping for a wide variety of chemical processes. These include hydrogen peroxide production, rayon manufacture, and the handling of phosphoric and sulfuric acids and ethyl benzene. Gas scrubbers, pickling tanks, resin plants, and coal gasification reactors are some of the applications in which the good corrosion resistance of zirconium toward organic acids is used. A particularly useful attribute is the ability of the material to withstand environments with alternating acidity and alkalinity.[59]

Heat exchangers. In those areas where zirconium alloys exhibit exceptional corrosion resistance, scaling or scale formation is virtually nonexistent. As a consequence, fouling allowance factors may be markedly reduced or eliminated. Heat exchangers can then be designed and operated on the basis of the calculated overall heat transfer coefficient rather than a design coefficient. The higher design coefficients are the result of noncorroding, nonfouling, high-film-coefficient surfaces. Periodic cleaning is not required on a frequent basis, so the effective on-stream time is dramatically increased.

Columns. Zirconium alloys are frequently used as a structural material in the construction of stripper or drying columns. The choice of zirconium alloy grades depends on the corrosive media involved. R60702 is used for the most severe applications, such as sulfuric acid at concentrations above 55%. With its higher strength, zirconium alloy R60705 can allow significant cost savings over R60702 when the corrosivity of the media permits its use. Zirconium alloys R60702 and R60705 are both qualified for use in the construction of pressure vessels. One of the world's largest zirconium alloy columns, constructed by Nooter Corporation, is 40 m tall and approximately 3.5 m in diameter.[61]

Reactor vessels. Steel shells lined with zirconium alloys solve the most difficult corrosion problems in reactor vessels and tanks. Zirconium alloys' plates can be welded to form vessels of any size. When used as a liner in steel vessels, the strength is enhanced. This can be accomplished as a loose lining, as a resistance welded lining, or as an explosively bonded lining. Large assemblies can be made with minimal weld joints. Zirconium alloys resistance to organic acids led to their acceptance as a construction material for reactors, tanks, and piping in ethylbenzene reactors. Gas scrubbers and pickling tanks, resin plants, chlorination systems, batch reactors, and coal degasification reactors are but a few of the applications in which zirconium alloys will function with superior efficiency compared to many other common metals.

8.10.2 Corrosion resistance

Zirconium resembles titanium from a fabrication point of view. It also resembles titanium in corrosion resistance. However, in hydrochloric acid, zirconium is more resistant. It also resists all chlorides except ferric and cupric chloride. Their excellent corrosion resistance to many chemical corrodants at high concentrations and elevated temperatures and pressures cause zirconium and its alloys to be used in a wide range of chemical processing and industrial applications despite their high cost. Table 8.48 presents the corrosion rates and estimated lives for some zirconium equipment exposed to some corrosive environments.[48]

Like titanium and some of the other nonferrous metals and alloys, the corrosion resistance of zirconium is attributable to the natural formation of a dense, stable, self-healing oxide film on its surface, which protects the base metal from chemical and mechanical attack up to 300°C. Zirconium is highly corrosion-resistant to strong alkalies, most organic and mineral acids, and some molten salts. It is an excellent

TABLE 8.48 Corrosion Rates and Estimated Zirconium Equipment Lives Exposed to Some Corrosive Environments

Environment	Concentration, %	Temperature, °C	Corrosion, $mm \cdot y^{-1}$	Estimated life, y
Acetic acid	100	200	< 0.025	> 20
Hydrochloric acid	32	82	< 0.025	> 20
Hydrochloric acid + 100 ppm $FeCl_3$	20	105	< 0.125	2
Hydrochloric acid	2	225	< 0.025	> 20
Nitric acid	10–70	Room, 200	< 0.025	> 20
Nitric acid + 1% $FeCl_3$	70	120	(Nil)	> 20
Seawater	Natural	200	< 0.025	> 20
NaOH solution	50	57	< 0.025	> 20
NaOH solution	73	129	< 0.05	10
NaOH solution	73	212	< 0.5–1.25	1 or less
NaOH solution +16% ammonia	52	138	< 0.125	2
Sulfuric acid	70	100	< 0.05	10
Sulfuric acid	65	130	< 0.025	> 20
Sulfuric acid +1000 ppm $FeCl_3$	60	Boiling	< 0.025	> 20
Sulfuric acid +10,000 ppm $FeCl_3$	60	Boiling	< 0.125	2
Urea reactor		193	< 0.025	> 20

construction material for processing equipment that will experience alternating contact with strong acids and alkalies. Its alloys are not readily corroded by oxidizing media such as air, carbon dioxide, nitrogen, oxygen, and steam at temperatures through 400°C, except in the presence of halides. It is attacked by fluoride ions, wet chlorine, aqua regia, concentrated sulfuric acid above 80% concentration, and ferric or cupric chlorides. It does not require anodic protection systems.

Both zirconium and titanium are excellent for seawater service, but there are differences in corrosion-resistance properties. In nonacidic chloride corrosion resistance, such as in seawater or chloride solutions where titanium and zirconium are both corrosion resistant over a wide range of conditions, zirconium is better than titanium for resisting crevice corrosion, because crevice environments tend to become reducing with time. Zirconium is also much more reliable than titanium in withstanding organic acids, such as acetic, citric, and formic acids, where zirconium resists corrosion in the entire concentration range and at elevated temperatures. The ability of titanium to resist these acids is affected by aeration and water content. In handling chlorine, although zirconium is resistant to dry chlorine below 200°C, it is susceptible to localized corrosion by wet chlorine.

Acid corrosion. Unalloyed zirconium has excellent resistance to sulfuric acid up to 80% concentration at room temperature and to 60% concentration at the boiling point. The transition from low to high corrosion rate occurs over a very narrow range of acid concentrations. Weld and heat-affected zones corrode at lower acid concentrations than the recrystallized base metal. When such an attack occurs, it is rapid and intergranular, creating a highly pyrophoric surface layer that ignites easily. The effects of corrosion are marginally different for the different zirconium alloys.[48]

Oxidizing impurities such as ferric, cupric, and nitrate ions in concentrations of approximately 200 ppm in sulfuric acid adversely affect corrosion resistance, reducing by approximately 5% the concentration of acid it can withstand, for a corrosion rate of less than 0.125 mm·y^{-1}. R69702 and R69704 are not affected by these oxidizing impurity levels at acid concentrations less than 65%, and R69705, at concentration levels less than 60%. Below 65% sulfuric acid, R69702 does not experience accelerated attack even at cupric and ferric ion contents up to 1% in sulfuric acid. Zirconium has a very low tolerance for fluoride impurities in sulfuric acid even at low concentrations of the acid. At concentrations higher than 50%, even 1 ppm of fluoride ions in the acid will increase the corrosion rate appreciably. Therefore, when zirconium equipment must be used to handle sulfuric acid contaminated with fluoride ions, these ions must be complexed by using inhibitors such as zirconium sponge and phosphorous pentoxide.

Zirconium shows excellent corrosion resistance to hydrochloric acid and is superior to any other engineering metal for this application, with a corrosion rate of less than 0.125 mm·y^{-1} at all concentrations and temperatures well in excess of the boiling point. Aeration does not affect its corrosion resistance, but the presence of oxidizing impurities such as cupric or ferric chlorides in relatively small amounts will decrease it. Therefore, either these ions should be avoided, or suitable electrochemical protection should be provided. Zirconium also shows excellent corrosion resistance to nitric acid in all concentrations up to 90% and temperatures up to 200°C, with only platinum being equal to it for this service. Welded zirconium and its alloys retain this high corrosion resistance. In concentrated nitric acid, zirconium may exhibit SCC at nitric acid concentrations above 70%, if under high tensile stress.[48]

Zirconium is resistant to corrosion by phosphoric acid at concentrations up to 55% and temperatures exceeding the boiling point. Above 55% concentration, the corrosion rate increases with concentration and temperature, but it remains below 0.125 mm·y^{-1} for concentrations up to 85% at 60°C. Fluoride ion impurities in phosphoric acid, originating from the feedstock, can increase the corrosion rate. Zirconium does not withstand hydrofluoric acid even at concentrations as low as 0.001%.

Alkaline corrosion. Zirconium is resistant to corrosion by almost all alkalies, both in solution and in the fused condition, up to the boiling point. It resists sodium and potassium hydroxide solutions even under anhydrous conditions and resists molten potassium hydroxide and molten sodium hydroxide, the latter at temperatures greater than 1000°C. It resists calcium and ammonium hydroxides at concentrations up to 28% up to boiling. Because it is resistant to both alkalies and acids, it is the preferred material of construction for processes that cycle between acid and alkaline solutions.

Aqueous media and marine corrosion. Zirconium has excellent corrosion resistance to seawater, fresh water, brackish water, and other polluted water streams and is a material of choice for heat exchangers, condensers, and other equipment handling these media, where it can replace titanium-palladium alloys. Unlike titanium and its alloys, zirconium is highly resistant to crevice corrosion. With their high corrosion resistance to pressurized water and steam, low neutron absorption (with low hafnium content), good mechanical strength, and ductility, at nuclear reactor service temperatures, and their ability to remain stable even after extensive radiation, zirconium alloys are used extensively in fuel cladding, fuel channels, and pressure tubes for boiling water and pressurized water nuclear power plants. Zirconium is alloyed with tin, iron, chromium, and nickel to improve its strength in these applications.

Corrosion from molten metals and salts. Zirconium is resistant to corrosion in some molten salts. It also withstands molten metals such as sodium, potassium and the sodium-potassium eutectic used in nuclear reactors. Its corrosion rate is less than 0.025 mm·y^{-1} in liquid lead up to 600°C, in liquid lithium up to 800°C, in mercury up to 100°C, and in molten sodium up to 600°C. The corrosion rate is affected by trace impurities such as hydrogen, nitrogen, or oxygen in specific molten metals. Zirconium is severely attacked by molten bismuth, magnesium, and zinc.

Corrosion from organic compounds. Zirconium is very resistant to corrosion by organic compounds, particularly most organic acids. In acetic acid and acetic anhydride, its corrosion rate is less than 0.05 mm·y^{-1} at all concentrations and temperatures. It also has high resistance to citric, formic, lactic, oxalic, tannic, and tartaric acids, as well as to chlorinated organic acids. Corrosion rates for nuclear reactor–grade zirconium alloys used in fuel cladding at temperatures up to 465°C, in contact with organic coolants such as polyphenyls, are similar to those in low-pressure steam. Hydriding because of hydrogen pickup from the coolant, which can cause stress-corrosion cracking and hydrogen embrittlement, is held to a minimum by keeping small amounts of moisture in the coolant and holding dissolved hydrogen and chlorine content of the coolant to a minimum. SCC of zirconium has been found in concentrated methanol solutions containing heavy metal chlorides, gaseous iodine, or fused salts containing iodine, even though zirconium and its alloys are free from stress-corrosion cracking in seawater and most aqueous chemical media.

References

1. Henthorne, M., Materials Selection for Corrosion Control, *Chemical Engineering*, 1139–1146, 1971.
2. Tillack, D. J., and Guthrie, J. E., *Wrought and Cast Heat-Resistant Stainless Steels and Nickel Alloys for the Refining and Petrochemical Industries.* NiDI Technical Series 10071, 1992. Toronto, Canada, Nickel Development Institute.
3. Todd, B., Materials Selection for High Reliability Seawater Systems, *http://marine.copper.org/*, 1998.
4. *Corrosion Data Survey—Metals.* Houston, Tex., National Association of Corrosion Engineers, 1985.
5. *Corrosion Data Survey—Non-Metals.* Houston, Tex., National Association of Corrosion Engineers, 1975.
6. Mashayekhi, B., Sturrock, C. P., and Flanigan, C. D., *Corrosion Data Survey: The Next Generation.* Paper 604, 1997. Houston, Tex., NACE International.
7. Weldability of Materials: Aluminum Alloys, *www.twi.co.uk/bestprac/jobknol/jk21.htm*, 1998.

8. Hollingsworth, E. H., and Hunsicker, H. Y., Corrosion of Aluminum and Aluminum Alloys, in *Metals Handbook: Corrosion,* vol. 13. Metals Park, Ohio, American Society for Metals, 1987, pp. 583–609.
9. Cooke, G., Koch, G., and Frechan, R., *Corrosion Detection & Life Cycle Analysis for Aircraft Structural Integrity,* Report ADB-171678, 1-12-1992. Washington, D.C., Defense Technical Information Center.
10. Lifka, B. W., Aluminum (and Alloys), in Baboian, R. (ed.), *Corrosion Tests and Standards,* Philadelphia, American Society for Testing of Materials, 1995, pp. 447–457.
11. Stickle, D. R., Corrosion of Cast Irons, in *Corrosion.* Metals Park, Ohio, ASM International, 1988, pp. 566-572.
12. Weldability of Materials: Cast Irons, *www.twi.co.uk/bestprac/jobknol/jk25.htm,* 1998.
13. *Source Book on Copper and Copper Alloys,* Metals Park, Ohio, American Society for Metals, 1979.
14. Copper ã Brass ã Bronze: Standard Designations for Wrought and Cast Copper and Copper Alloys, *http://properties.copper.org/standard-designations/homepage.htm,* 1998.
15. Weldability of Materials: Copper and Copper Alloys, *www.twi.co.uk/bestprac/jobknol/jk23.htm,* 1998.
16. Selecting Bronze Bearing Materials,*http://www.copper.org/industrial/bronze_bearing.htm,* 1998.
17. Cohen, A., Copper (and Alloys), in Baboian, R. (ed.), *Corrosion Tests and Standards,* Philadelphia, American Society for Testing of Materials, 1995, pp. 466–475.
18. Polan, N. W., Corrosion of Copper and Copper Alloys, in *Metals Handbook: Corrosion,* Metals Park, Ohio, ASM International, 1987, pp. 610–640.
19. Copper & Copper Alloy: Corrosion Resistance Database, *http://protection.copper.org/database.htm,* 1998.
20. Bailey, G. L., Copper Nickel Iron Alloys Resistant to Seawater Corrosion. *Journal of the Institute of Metals,* **79:** (1951).
21. Powell, C. A., Copper-Nickel Alloys—Resistance to Corrosion and Biofouling, *http://marine.copper.org/,* 1998.
22. Parvizi, M. S., Aladjem, A., and Castle, J. E., Behaviour of 90-10 Cupronickel in Seawater, *International Material Reviews* **33:** (1988).
23. Lenard, D. R., and Welland, R. R., Corrosion Problems with Copper-Nickel Components in Sea Water Systems, *CORROSION/98,* Paper 599. 1998. Houston, Tex., NACE International.
24. Kirk, W. W., and Tuthill, A. H., Copper-Nickel Condenser and Heat Exchanger Systems, *http://marine.copper.org/3-toc.html,* 1998. (GENERIC) Ref Type: Electronic Citation
25. Tuthill, A. H., Guidelines for the Use of Copper Alloys in Seawater, *Materials Performance,* **26** (1987).
26. Gilbert, P. T., A Review of Recent Work on Corrosion Behavior of Copper Alloys in Seawater, *Materials Performance,* **21:**47–53 (1982).
27. Gilbert, P. T., Corrosion Resisting Properties of 90/10 Copper Nickel Iron Alloy with Particular Reference to Offshore Oil and Gas Applications, *British Corrosion Journal,* (1979).
28. Green Patina Finishes, *http://protection.copper.org/green.htm,* 1998.
29. Brown Statuary Finishes, *http://protection.copper.org/brown.htm,* 1998.
30. Donachie, Jr., M. J., Introduction to Superalloys, in *Superalloys Source Book.* Materials Park, Ohio, American Society for Metals, 1984, pp. 3–19.
31. Asphahani, A. I., Corrosion of Nickel-Base Alloys, in *Metals Handbook: Corrosion.* Metals Park, Ohio, ASM International, 1987, pp. 641–657.
32. Asphahani, A. I., Corrosion of Cobalt-Base Alloys, in *Metals Handbook: Corrosion.* Metals Park, Ohio, ASM International, 1987, pp. 658–668.
33. Weldability of Materials: Nickel and Nickel Alloys, *www.twi.co.uk/bestprac/jobknol/jk22.htm,* 1998.
34. High-Temperature Alloys Fabrication Guide—Heat Treatment, *http://www.haynesintl.com/Fabric/FBht.html,* 1998.

35. Molybdenum, *www.rembar.com / moly.htm,* 1998.
36. Rembar/Technical Data, *www.rembar.com / tech2.htm,* 1998.
37. Niobium, *www.rembar.com / niobium.htm,* 1998.
38. Yau, T. L., and Webster, R. T., Corrosion of Niobium and Niobium Alloys, in *Metals Handbook: Corrosion.* Metals Park, Ohio, ASM International, 1987, pp. 722–724.
39. Tantalum, *www.rembar.com / tant.htm,* 1998.
40. Tungsten, *www.rembar.com / tung.htm,* 1998.
41. Sedriks, A. J., *Corrosion of Stainless Steels,* New York, John Wiley, 1979.
42. Stainless Steel from Avesta Sheffield—Steel Grades. 2d ed. Avesta, Sweden, Avesta Sheffield, 1997.
43. Holmberg, B., *Stainless Steels: Their Properties and Their Suitability for Welding.* Avesta, Sweden, Avesta Welding, 1994.
44. *Metals Handbook: Heat Treating, Cleaning and Finishing.* Metals Park, Ohio, American Society for Metals, 1991.
45. Congleton, J., Stress Corrosion Cracking of Stainless Steels, in Shreir, L. L., Jarman, R. A., and Burstein, G. T. (eds.), *Corrosion Control.* Oxford, UK, Butterworths Heinemann, 1994, pp. 8:52–8:83
46. Craig, B. D., and Anderson, D. S., *Handbook of Corrosion Data.* Materials Park, Ohio, ASM International, 1995.
47. Gunn, R. N., *Duplex Stainless Steels.* Cambridge, UK, Abington Publishing, 1997.
48. Chawla, S. L., and Gupta, R. K., *Materials Selection for Corrosion Control.* Materials Park, Ohio, ASM International, 1993.
49. Parkins, R. N., An Overview-Prevention and Control of Stress Corrosion Cracking, *Materials Performance* **24**(8):9–20 (1985).
50. Davidson, R. M., and Redmond, J. D., *Practical Guide to Using Duplex Stainless Steel.* NiDI Technical Series 10044, 1990. Toronto, Canada, Nickel Development Institute.
51. Chandler, K. A., and Hudson, J. C., Iron and Steel, in Shreir, L. L., Jarman, R. A., and Burstein, G.T. (eds.), *Corrosion Control.* Oxford, UK, Butterworths Heinemann, 1994, pp. 3:3–3:22.
52. Hudson, J. C., Stanners, J. F., and Hooper, R. A. E., Low-alloy Steels, in Shreir, L. L., Jarman, R. A., and Burstein, G. T. (eds.), *Corrosion Control.* Oxford, UK, Butterworths Heinemann, 1994, pp. 3:23–3:33
53. Willett, T. O., *Technical Advisory: Uncoated Weathering Steel in Structures.* T5140.22, 10-3-1989. U.S. Department of Transportation, Federal Highway Administration.
54. Weldability of Materials: Steels, *www.twi.co.uk / bestprac / jobknol / jk19.htm,* 1998.
55. *Titanium Industries Data and Reference Guide.* Morristown, N.J., Titanium Industries Inc., 1998.
56. Weldability of Materials: Titanium and Titanium Alloys. *www.twi.co.uk/bestprac/ jobknol/jk24.htm,* 1998.
57. Titanium and Its Alloys, *http:/ /www.titanium.org/alloychartext.html,* 10-24-1996.
58. *Corrosion Resistance of Titanium.* Denver, Colo., Titanium Metals Corporation (TIMET), 1997.
59. Cotton, J. B., and Hanson, B. H., Titanium and Zirconium, in Shreir, L. L., Jarman, R. A., and Burstein, G. T. (eds.), *Corrosion Control.* Oxford, UK, Butterworths Heinemann, 1994, pp. 5:36–5:59.
60. Zirconium & Zr-alloys, *http:/ /www.metalogic.be/MatWeb/reading/m_zr.htm,* 1998.
61. *Zircadyne Properties and Applications,* Albany, Ore., Teledyne Wah Chang Albany, 1991.

Chapter

9

Protective Coatings

9.1	Introduction	781
9.2	Coatings and Coating Processes	782
	9.2.1 Metallic coatings	782
	9.2.2 Inorganic coatings	805
	9.2.3 Organic coatings	810
9.3	Supplementary Protection Systems	829
	9.3.1 Jointing compounds and sealants	830
	9.3.2 Water displacing compounds	830
9.4	Surface Preparation	831
References		831

9.1 Introduction

Protective coatings are probably the most widely used products for corrosion control. They are used to provide long-term protection under a broad range of corrosive conditions, extending from atmospheric exposure to the most demanding chemical processing conditions. Protective coatings in themselves provide little or no structural strength, yet they protect other materials to preserve their strength and integrity. The main function of a protective coating is to isolate structural reactive elements from environmental corrosives. The fact that protective coatings occupy only a very small fraction of the total volume of a system is quite telling of the heavy requirements imposed on these materials. A coating must provide a continuous barrier to a substrate, and any imperfection can become the focal point for degradation and corrosion of the substrate.

Metal finishing comprises a wide range of processes that are practiced by most industries engaged in manufacturing operations using metal parts. Typically, metal finishing is performed on manufactured parts after they have been shaped, formed, forged, drilled, turned, wrought, cast, and so forth. A "finish" can be defined as any final operation applied to the surface of a metal article to alter its surface properties and achieve various goals. The quality of a coating depends on many factors besides the nature of the materials involved. Metal finishing operations are intended to increase corrosion or abrasion resistance, alter appearance, serve as an improved base for the adhesion of other materials, enhance frictional characteristics, add hardness, improve solderability, add specific electrical properties, or improve the utility of the product in some other way.

9.2 Coatings and Coating Processes

Coating fundamentals makes reference to a multitude of concepts and properties. A critical property of antifouling paint is, for example, the inhibition of living organism growth on the coating. A fire-resistant coating, on the other hand, should resist or retard the burning of the substrate. From a corrosion point of view a coating is rated on the resistance it provides against corrosion in a specific environment, and because there are many variations in environment corrosivity, there is also a great variety of corrosion protective coatings. These can be broadly divided into metallic, inorganic, and organic coatings. A general description of how the main elements are used in metallic and inorganic coatings is given in Table 9.1.

9.2.1 Metallic coatings

Metallic coatings provide a layer that changes the surface properties of the workpiece to those of the metal being applied. The workpiece becomes a composite material exhibiting properties generally not achievable by either material if used alone. The coatings provide a durable, corrosion-resistant layer, and the core material provides the load-bearing capability.

The deposition of metal coatings, such as chromium, nickel, copper, and cadmium, is usually achieved by wet chemical processes that have inherent pollution control problems. Alternative metal deposition methods have replaced some of the wet processes and may play a greater role in metal coating in the future. Metallic coatings are deposited by electroplating, electroless plating, spraying, hot dipping, chemical vapor deposition, and ion vapor deposition. Some important coatings are cadmium, chromium, nickel, aluminum, and zinc. Copper, gold, and silver are also used in electrical equipment and occasionally for specialty fastener applications. Copper is used as a base layer in multiple-plate electroplating, silver is used for antifretting purposes, and both silver and gold are sometimes used to provide electrical conductivity in waveguides and at contacts.

TABLE 9.1 Properties and Applications of the Main Metallic Elements Used for Protective Coatings

Aluminum

Aluminum coatings can be applied to steel by hot dipping, cementation, ion vapor deposition, and spraying. Ion vapor deposition is a relatively new process, and spraying is the only process that has been used extensively over a long period of time. Pack cementation is widely used for gas turbine components. In soft waters aluminum is cathodic with respect to steel; however, in seawater or some fresh waters containing chloride ions or sulfate ions, aluminum may become anodic to steel, and aluminum coatings should therefore corrode sacrificially and provide cathodic protection to steel. However, as noted below, this may not always be the case.

Sprayed aluminum coatings provide an adherent, somewhat absorbent film about 100 to 150 μm thick. They provide very good protection to steel, and they may be sealed with organic lacquers or paints to provide further protection and delay the formation of visible surface rust. The surface of the steel must first be grit-blasted to provide a rough surface to aid adhesion. Unfortunately the thickness and relative roughness of the coatings make them unsuitable for close tolerance parts.

Ion vapor-deposited aluminum coatings have been used on a variety of parts including steel and titanium fasteners, electrical connectors, engine mounts and stator vanes, landing gear components, integrally machined wing skins, and a large number of miscellaneous components. These coatings are soft and ductile and are prepared using commercially available aluminum (1100 alloy) feed wire that is melted, vaporized, and ionized in a glow discharge created by an inert gas. The process is applied in a batch mode, where parts to be coated are held at a high negative potential relative to the evaporation source. The positively charged gas ions bombard the surface of the part and perform a final cleaning action. When this is done the aluminum is vaporized and ionized, and the ionized aluminum is accelerated toward the part surface where it plates as a dense, tightly adherent coating.

Minimum coating thickness are in the range 8 to 25 μm, and coatings may be used as prepared or with a supplementary chromate treatment. The thinner coatings are used when close tolerances are required such as on threads, intermediate thickness coatings (> 13 μm) are used on interior parts or where only mildly corrosive environments are expected, and the thicker coatings (> 25 μm) are used for exterior parts operating in highly corrosive environments and for engine parts. Ion vapor-deposited aluminum has been considered as a replacement for diffused nickel cadmium and aluminum pigmented paints for use in the cooler sections of gas turbines, where temperatures are less than 454°C. The process has also been considered as an alternative to pack cementation for the preparation of aluminide coatings on hot-section components. In this case the ion vapor-deposited aluminum is diffused into the nickel-based superalloy substrates to form the nickel aluminide coating.

Ion vapor deposition of aluminum is attractive because it avoids the environmental and toxicological problems associated with cadmium. It does not cause hydrogen embrittlement of steel or solid metal embrittlement of steel or titanium, and it should be more galvanically compatible with aluminum alloy structure and avoid the exfoliation corrosion of sensitive aluminum alloy structure. However, views on the ability of aluminum to protect steel fasteners appear to vary. A view is that the presently available pure aluminum coatings are not able to provide adequate sacrificial protection to steel in a chloride ion environment, and for this reason a recommendation is made for the development of aluminum coatings containing small amounts of zinc or other elements for improved protection.

TABLE 9.1 Properties and Applications of the Main Metallic Elements Used for Protective Coatings (*Continued*)

Cadmium

Cadmium is widely used by the aircraft industry for electroplating steel fasteners and bearing assemblies because it provides a galvanically acceptable couple with aluminum. Cadmium is also anodic to steel and will cathodically protect the substrate at scratches or gaps in the coating and at cut edges. It also exhibits surface lubricity and conductivity and resists fretting and fatigue, and its corrosion products do not cause binding. Platings are usually deposited from cyanide baths, but baths containing fluoroborates or sulfamates are also used. The baths may contain special additives to reduce hydrogen penetration, and the coatings are usually 5 to 25 μm thick.

The use of cadmium introduces four problems. The first problem is that it is highly toxic, and environmental protection agencies have been concerned about its release into the environment. Disposal of wastes from cyanide baths is therefore a problem, as is the eventual disposal of the finished coated part. The second problem is that the electroplating process also exposes parts to cathodically produced hydrogen, and because many of the high-strength steels involved are highly susceptible to hydrogen embrittlement, stringent requirements exist to bake parts immediately after plating to remove this hydrogen. Most process specifications for cadmium plating include requirements for baking and subsequent testing of coupons to demonstrate absence of embrittlement. The problem of hydrogen embrittlement can be avoided by applying cadmium coatings by an ion vapor deposition process, which does not produce hydrogen. This process is sometimes used on very high strength steels where hydrogen would be difficult to remove by baking. Once deposited, these coatings are essentially similar to electrodeposited coatings and should receive the same type of additional paint protection. The third problem is that cadmium has been reported to cause solid metal embrittlement of steel and titanium alloys. Finally, cadmium has also been reported to cause exfoliation corrosion of susceptible aluminum alloys when used on fasteners in contact with these alloys in a riveted or bolted structure.

Many alternatives to cadmium plating exist, with no single universal substitute available. Some cadmium plating alternatives are zinc plating, tin or tin alloy plating, cobalt-zinc plating, zinc-nickel plating, zinc-iron plating, zinc-flake dispersion coating, metallic ceramic coating, and ion vapor deposition of aluminum. The most successful of these alternatives has been zinc-nickel plating, which has a long history in the electroplating industry. Generally, for alternatives to be successful, they must provide sufficient corrosion resistance, as measured by standard tests. For certain military and aerospace applications, the alternative deposits must also provide other desired characteristics, such as lubricity. Many electroplating job shops have eliminated cadmium plating because of a reduced market and the enforcement of local discharge standards.

Chromium

Chromium is used as a protective coating, providing resistance to wear, abrasion, and corrosion. It has hardness in the range 900 to 1100 HV, low-friction characteristics, and high reflectivity. It is used as a thin coating, usually in the range 0.2 to 1 μm thick, as the final layer in a multipleplate copper-nickel-chromium electroplating or as a thick coating up to 300 μm to provide wear resistance. When used as a constituent of a multiple-plate coating, chromium provides hardness, reflectivity, and tarnish resistance. The corrosion resistance is derived primarily from the barrier effect of the thick nickel plate under the chromium. However, copper, nickel, and chromium are all cathodic with respect to steel, and corrosion can be accelerated once the coating is breached and the underlying steel is exposed. For this reason these coatings are not chosen where corrosion protection is the primary concern.

TABLE 9.1 Properties and Applications of the Main Metallic Elements Used for Protective Coatings (*Continued*)

Hard chromium plating is usually applied directly to steel parts in thickness up to about 300 μm to provide resistance to wear, abrasion, and corrosion. It is also used to build up worn or undersized parts. In the thicker applications it may be impervious but is subject to microcracks. Chromium is a metal with low cathode efficiency, and substantial amounts of hydrogen are deposited on the part along with the metal being plated. Because of this, parts must be baked as soon as possible after plating to drive off the hydrogen and prevent embrittlement.

Chromium plating is traditionally performed with a hexavalent chromium bath, but trivalent chromium plating has increased in use, especially during the past 10 years. With either process, an undercoat of nickel/copper or nickel is usually applied. Trivalent chromium plating is an economically attractive alternative to hexavalent plating for some applications. However, its use has been limited due to a difference in appearance from the standard hexavalent bath. The trivalent bath chemistry is more expensive to purchase than the hexavalent bath. The cost savings are a result of reduced metal loading on the treatment system (the trivalent bath contains less total chromium) and the avoidance of the hexavalent chromium reduction step during treatment. The total cost of trivalent chromium plating is about one-third of the costs for hexavalent solution.

Hard chromium plating is applied to tools, hydraulic cylinders, and other metal surfaces that require wear resistance. It is widely used in the mining industry. The major difference between the hard chromium and decorative deposits is their thickness. The hard chromium deposit is typically hundreds of times thicker than decorative ones. Although research efforts have aimed at a trivalent chromium substitute for hard chromium plating, no solutions are available commercially. Input material changes for hard chromium have focused on alternative deposits. Alternative processes have also been used. The most successful alternative input material is electroless nickel. Other alternative input materials under investigation are electroplated nickel alloys and nickel alloy composites. Alternative processes to hard chromium plating include brush plating, vacuum coating, and metal sprays.

Chromium use with aluminum finishing is perhaps most common in the aerospace industry. Chromium combines with aluminum on the surface of parts to provide corrosion and wear resistance and a chemically active surface for painting or coloring. The two most common processes are chromic acid anodizing and chromate conversion coating. Both processes are presently performed in hexavalent chromium baths. The anodizing process is electrolytically performed, and the conversion coating process involves simple immersion. Significant research efforts have been undertaken during the past 10 years to find alternatives to these processes. For many applications, alternatives have been identified and implemented. For example, chromic acid anodizing has been partially replaced by common sulfuric acid anodizing and sulfuric/boric acid anodizing, and chromium baths have been replaced to a lesser extent by nonchromium conversion coatings (e.g., permanganate, rare earth metals, and zirconium oxide).

Another use of chromium during aluminum finishing is for deoxidizing/desmutting. These preliminary processes (sometimes a combined single step) remove oxides and other inorganics that would interfere with aluminum processing (e.g., anodizing). Alternatives to the chromium-based products include iron and ammonium salts and amines mixed with various oxidizers and/or etchants. Owing to the extent of research for nonchromium aluminum finishing and the success rate of these efforts, it is possible that chromium use will eventually be eliminated from the aluminum finishing area. One would expect to see large-scale substitutions during the next 10 years. However, total elimination will take considerable longer because of small residual uses of chromium for which no satisfactory substitute exists.

TABLE 9.1 Properties and Applications of the Main Metallic Elements Used for Protective Coatings (*Continued*)

Nickel
By far the greatest use of nickel plating is on steel in conjunction with copper and chromium as described above. However, nickel can also be deposited, both on metals and nonmetals by an electroless or nonelectrolytic process. The metal is deposited spontaneously on the surface of a catalytic substrate immersed in an aqueous solution containing the metal ion and a reducing agent together with a compound (frequently the salt of an organic acid) that acts as a buffer and a complexing agent for the metallic ion.
The baths often contain phosphorous or boron, and they provide coatings of uniform thickness even over sharp corners and into deep recesses. The coatings have low internal stress and are less magnetic than electrodeposited nickel platings, and they have hardness values of about 500 HV. The coatings can be heat treated to higher hardness of about 1000 HV, which provides wear and abrasion resistance. This increase in hardness is achieved by a precipitation hardening process involving the phosphorous, which is usually present in amounts of 5 to 10%. The heat treatment is carried out at temperatures of about 400°C.
Electroless nickel coatings about 25 μm thick are often used after baking to remove hydrogen and to provide protection against stress corrosion cracking of precipitation hardenable stainless steels.

Zinc
Zinc coatings may be applied either by electroplating or spraying. Electroplatings are normally less than 25 μm thick and may be as thin as 5 μm on threaded parts. However, although they provide good protection to steel in rural atmospheres, they do not perform as well in marine or industrial environments. Zinc coatings 30 μm thick last about 11 years or longer in rural or suburban locations, about 8 years in marine locations, and only 4 years in industrial atmospheres. The short life in industrial atmospheres was attributed to attack by sulfuric acid in polluted atmospheres. Zinc plating does not perform as well as cadmium in tropical and marine atmospheres, and therefore cadmium is preferred for aircraft use. Where thicker coatings are permissible, zinc may be deposited by spraying but must compete with aluminum, which is usually the preferred material.

Plating and surface treatment processes are typically batch operations, in which metal objects are dipped into and then removed from baths containing various reagents to achieve the desired surface condition. The processes involve moving the object being coated through a series of baths designed to produce the desired end product. These processes can be manual or highly automated operations, depending on the level of sophistication and modernization of the facility and the application. Plating operations can generally be categorized as electroplating and electroless plating processes.

Electroplating. Electroplating is achieved by passing an electrical current through a solution containing dissolved metal ions and the metal object to be plated. The metal object serves as the cathode in an electrochemical cell, attracting metal ions from the solution. Ferrous and nonferrous metal objects are plated with a variety of metals, including aluminum, brass, bronze, cadmium, copper, chromium, iron, lead, nickel, tin, and zinc, as well as precious metals, such as gold, platinum, and silver. The process is regulated by controlling a variety of parameters, including the voltage and amperage, temperature, residence times, and the purity of bath solutions. Plating baths are almost always aqueous solutions; therefore, only those metals that can be reduced from aqueous solutions of their salts can be electrodeposited. The only major exception is aluminum, which can be plated from organic electrolytes. The sequence of unit operations in an electroplating operation typically involves various cleaning steps, stripping of old plating or paint, electroplating steps, and rinsing between and after each of these operations. Electroless plating uses similar steps but involves the deposition of metal on a substrate without the use of external electrical energy.

Electroless plating. Electroless nickel (EN) plating is a chemical reduction process that depends upon the catalytic reduction process of nickel ions in an aqueous solution (containing a chemical reducing agent) and the subsequent deposition of nickel metal without the use of electrical energy. Thus in the EN plating process, the driving force for the reduction of nickel metal ions and their deposition is supplied by a chemical reducing agent in solution. This driving potential is essentially constant at all points of the surface of the component, provided the agitation is sufficient to ensure a uniform concentration of metal ions and reducing agents. The electroless deposits are therefore very uniform in thickness all over the part's shape and size. The process thus offers distinct advantages when plating irregularly shaped objects, holes, recesses, internal surfaces, valves, threaded parts, and so forth.

Electroless (autocatalytic) nickel coating provides a hard, uniform, corrosion-, abrasion-, and wear-resistant surface to protect machine components in many industrial environments. EN is chemically deposited, making the coating exceptionally uniform in thickness. Careful process control can faithfully reproduce the surface finish, eliminating the need for costly machining after plating.

In a true electroless plating process, reduction of metal ions occurs only on the surface of a catalytic substrate in contact with the plating solution. Once the catalytic substrate is covered by the deposited metal, the plating continues because the deposited metal is also catalytic. As a result, electroless plating processes are widely used in industry to meet the end-use functional requirements and are only rarely used for decorative purposes. Distinct advantages of EN plating are

- Uniformity of the deposits, even on complex shapes. The electroplated coatings are thinner in recessed areas and thicker on projecting areas.

- Deposits are often less porous and thus provide barrier corrosion protection to steel substrates that is much superior to that of electroplated nickel and hard chrome.
- Deposits cause about one-fifth as much hydrogen absorption as electrolytic nickel and about one-tenth as much as hard chrome.
- Deposits can be plated with zero or compressive stress. Fatigue strength debit on the substrate material is thus zero or positive.
- Deposits have inherent lubricity and nongalling characteristics, unlike electrolytic nickel.
- Deposits have good wetability for oils.
- Deposits have good solderability, braze weldability, and weldability.
- Deposits are much harder with as-plated microhardness of 450 to 600 HV, which can be increased to 1000 to 1100 HV by a suitable heat treatment, the increase being due to the precipitation of Ni3P, which causes general hardening of the alloy.

Deposits have unique magnetic properties. EN deposits containing more than 8% P are generally considered to be essentially nonmagnetic in the as-plated condition. In Ni-P coatings, phosphorus is present as supersaturated solution in fine microcrystalline solid solution, bordering on amorphous or liquidlike (glasslike) metastable structure, and is responsible for nonferromagnetic behavior of as-plated Ni-P deposits (with P > 8%).

A second generation of EN plating has been developed by codepositing micrometer-sized particles of silicon carbide with the nickel, thereby creating an extremely wear- and corrosion-resistant coating. The nickel alloy matrix provides corrosion resistance, and the silicon carbide particles, which are actually the contacting surface, add wear resistance.

Hot dip galvanizing. Hot dip galvanizing is the process of applying a zinc coating to fabricated iron or steel material by immersing the material in a bath consisting primarily of molten zinc. The simplicity of the galvanizing process is a distinct advantage over other methods of providing corrosion protection. The automotive industry depends heavily on this process for the production of many components used in car manufacturing, as illustrated in Table 9.2.

The recorded history of galvanizing goes back to 1742 when a French chemist named Melouin described, in a presentation to the French Royal Academy, a method of coating iron by dipping it in molten zinc. In 1836 another French chemist, Sorel, obtained a patent for a means of coating iron with zinc after first cleaning it with 9% sulfuric acid and fluxing it with ammonium chloride. A British patent for a similar process was granted in 1837. By 1850, the British galvanizing industry was using 10,000 tons of zinc a year for the protection of steel.[1]

Galvanizing can be found in almost every major application and industry where iron or mild steel is used. The utilities, chemical process, pulp and paper, automotive, and transportation industries, to name just a few, have historically made extensive use of galvanizing for corrosion control. They continue to do so today. For over 140 years, galvanizing has had a proven history of commercial success as a method of corrosion protection in a myriad of applications worldwide.

The electrochemical protection provided to steel by zinc coatings is a vital element in the effectiveness of galvanized coatings in protecting steel from corrosion. All pregalvanized products rely on the cathodic protection provided by zinc to prevent corrosion of exposed steel at cut edges. While the potential difference between metals is the prime driving force providing the corrosion current, it is not a reliable guide to the rate and type of corrosion occurring at a particular point. The severity of galvanic corrosion also depends on the ratio of the areas of metals in contact, the duration of wetness (galvanic corrosion can only occur in the presence of a conductive solution), and the conductivity of the electrolyte. The presence of oxide films on the surface of one or both of the metals can greatly inhibit galvanic corrosion.

In any situation where zinc is corroded sacrificially to protect exposed steel, the mass of available zinc will determine the corrosion protection performance. Corrosion rates of zinc coatings required to cathodically protect uncoated steel in aggressive environments (saltwater/marine) may be 25 times higher than the normal zinc corrosion rate.

Pack cementation. Diffusion coatings are formed by depositing a layer of aluminum on the metal surface and then heating the component in a furnace for a period of time. During this heat treatment, the aluminum and metal atoms migrate, or diffuse, into each other, which is the reason these coatings are called diffusion coatings. This processing is usually performed by a pack cementation process in which the aluminum deposition and the heat treatment occur simultaneously.

Pack cementation is widely used to confer oxidation resistance on ferrous alloys. Usually relatively expensive aluminum or binary alloys grade reagent is used during the pack process with aluminum as a source. Pack cementation processes include aluminizing, chromizing, and siliconizing. Components are packed in metal powders in sealed heat-resistant retorts and heated inside a furnace to precisely controlled temperature-time profiles. In the aluminizing process, a source

TABLE 9.2 Coatings for Automotive Sheet Steels

Steel coating	Description	Typical applications
Hot dipped, zinc coated (regular and minimized spangle)	Made on hot-dipped galvanizing lines and supplied in coils and cut lengths. Includes regular and minimized spangle in a wide range of coating designations and is available in extra smooth finish.	Rocker panels, wheelhouse and inner and outer panels, luggage compartment floor pans, bumper reinforcement, body structure inner reinforcements, floor pans
Hot dipped, zinc coated (fully alloyed zinc-iron coated)	Hot-dipped, zinc-coated product that is heat treated or wiped to produce a fully alloyed zinc-iron coating.	Body rails, cross members, light truck box beds
Hot dipped, zinc coated (differentially zinc coated)	Hot-dipped, zinc-coated product that is produced with different specified coating weights on opposite sheet surfaces. Both surfaces are zinc.	Cross members, hoods, fenders, door outer panels, quarter panels, wheelhouses, various underbody components
Hot dipped, zinc coated (differentially zinc-iron coated)	Same as above, except that the coating on the lighter surface is heat treated or wiped to produce a fully alloyed zinc-iron coating.	Fenders, doors, outer body panels, quarter panels, hoods, floor pans, door inner panels, dash panels
Hot dipped, zinc coated, one side	Produced with continuous hot-dipped, zinc coating on one side and a zinc-free, cold-rolled steel surface on the other for superior paint adhesion.	Fenders, door outer panels, quarter panels, deck lids, lower back panels, roofs, hoods
Electrolytic zinc, flash coated	Produced by continuously flash electroplating with zinc—30 to 60 g/m^2 total on both sides. It is used when minimal corrosion resistance is required.	Window guides, wiper blade frames, radio speaker baskets, head rest supports
Electroplated zinc coated	Produced by continuously electroplating zinc. Two side coatings can be produced on an equal basis or differentially. One side of the product has a standard cold-rolled surface.	Exposed and unexposed body panels
Electroplated iron-zinc alloy coated	Produced by the simultaneous electroplating of zinc and iron to form an alloy coating. One and two side coatings can be produced on an equal basis or differentially coated.	Exposed and unexposed body panels

Electroplated zinc-nickel alloy coated	Produced by the simultaneous electroplating of zinc and nickel to form an alloy coating. One and two side coatings can be produced on an equal basis or differentially coated.	Exposed and unexposed body panels
Aluminum coated	Produced by hot-dip coating cold-rolled sheet steel on continuous lines. It provides a material with the superior strength of steel and the surface properties of aluminum.	Exhaust systems, chassis components
Aluminum-zinc coated	Produced by hot-dip coating cold-rolled sheet steel on continuous lines. It has the superior strength of steel and excellent corrosion resistance.	Exhaust systems, air cleaner covers, core plugs, brake shields, floor pan covers
Zinc-aluminum mischmetal coated	Produced by hot-dip coating cold-rolled sheet steel on continuous lines. It provides maximum formability and excellent corrosion resistance.	Fuel-tank shields, fuel oil-filter shields, motor housings, shock towers and other deep-drawn underbody parts
Long terne	Cold-rolled sheet steel coated on both sides with a lead-tin alloy by a continuous hot-dip process.	Fuel tanks, fuel lines, brake lines, radiator and heater parts, air cleaners
Nickel terne	Cold-rolled sheet steel electrolytically nickel flash-plated and then coated on both sides with a lead-tin alloy by a continuous hot-dip process. Corrosion resistance is superior to standard long terne.	Fuel tanks, fuel lines, brake lines, radiator and heater parts, air cleaners
Tin coated	Cold-rolled sheet steel coated with tin by a continuous electrolytic process.	Oil filter and heater components
Zincrometal	A cold-rolled steel product with a base coat containing primarily chromium and zinc, top coated by a weldable zinc-rich primer for corrosion resistance—generally only to one side. The other side is typically a standard cold-rolled surface for superior paint adhesion.	Door inner and outer panels, fenders, quarter panels, hoods, deck lids, lift gate outers, lower back panels

of aluminum reacts with a chemical activator on heating to form a gaseous compound (e.g., pure Al with NaF to form AlF). This gas is the transfer medium that carries aluminum to the component surface. The gas decomposes at the substrate surface, depositing aluminum and releasing the halogen activator. The halogen activator returns to the pack and reacts with the aluminum again. Thus, the transfer process continues until all of the aluminum in the pack is used or until the process is stopped by cooling. The coating forms at temperatures ranging from 700 to 1100°C over a period of several hours.[2]

Pack cementation is the most widely used process for making diffusion aluminide coatings. Diffusion coatings are primarily aluminide coatings composed of aluminum and the base metal. A nickel-based superalloy forms a nickel-aluminide, which is a chemical compound with the formula NiAl. A cobalt-based superalloy forms a cobalt-aluminide, which is a chemical compound with the formula CoAl. It is common to incorporate platinum into the coating to improve the corrosion and oxidation resistance. This is called a platinum-aluminide coating. Diffusion chrome coatings are also available.

Diffusion aluminide coatings protect the base metal by forming a continuous, aluminum oxide layer, Al_2O_3, which prevents further oxidation of the coating. (Actually, oxidation continues but at much slower rates than without a continuous aluminum oxide scale.) When part of the Al_2O_3 scale spalls off, the underlying aluminide layer is exposed to form a new Al_2O_3 scale. Thus, the coating is self-healing.

Pack cementation can also be used to produce chromium-modified aluminide coatings. The addition of chromium is known to improve the hot corrosion resistance of nickel-based alloys. Although chromium can be codeposited with aluminum in a single-step process, a duplex process is frequently used to form the chromium-modified aluminide. The component is first chromized using either pack cementation or a gas phase process, and this is then followed by a standard aluminizing treatment. The final distribution of the chromium in the coating will depend on whether a low- or high-activity aluminizing process is employed.

For a platinum-aluminide coating, a thin (typically 8-μm) layer of platinum is first deposited onto the substrate, usually by a plating process. The second step involves aluminizing for several hours using the conventional packed cementation process to form the platinum-aluminide coating.

Conventional pack cementation processes are unable to effectively coat internal surfaces such as cooling holes. The coating thickness on these internal surfaces is usually less than on the surface due to limited access by the carrier gas. Access can be improved by pulsing the carrier gas,[3] or by use of a vapor phase coating process.

Another method of coating both the internal and external surfaces involves generating the coating gases in a reactor that is separate from the vessel the parts are in. The coating gases are pumped around the outside and through the inside of the parts by two different distribution networks. Internal passages can be coated by filling them with the powder used in the pack (actually a variation of this powder).[4]

Slurry processes can also be used to deposit the aluminum or the aluminum and other alloying elements. The slurry is usually sprayed on the component. The component is then given a heat treatment, which burns off the binder in the slurry and melts the remaining slurry, which reacts with the base metal to form the diffusion coating. After coating, it is usually necessary to heat treat the coated component to restore the mechanical properties of the base metal.

Cladding. Corrosion resistance can be improved by metallurgically bonding to the susceptible core alloy a surface layer of a metal or an alloy with good corrosion resistance. The cladding is selected not only to have good corrosion resistance but also to be anodic to the core alloy by about 80 to 100 mV. Thus if the cladding becomes damaged by scratches, or if the core alloy is exposed at drilled fastener holes, the cladding will provide cathodic protection by corroding sacrificially.

Cladding is usually applied at the mill stage by the manufacturers of sheet, plate, or tubing. Cladding by pressing, rolling, or extrusion can produce a coating in which the thickness and distribution can be controlled over wide ranges, and the coatings produced are free of porosity. Although there is almost no practical limit to the thickness of coatings that can be produced by cladding, the application of the process is limited to simple-shaped articles that do not require much subsequent mechanical deformation. Among the principal uses are lead and cadmium sheathing for cables, lead-sheathed sheets for architectural applications, and composite extruded tubes for heat exchangers. Because of the cathodic protection provided by the cladding, corrosion progresses only to the core/cladding interface and then spreads laterally, thus helping to prevent perforations in thin sheet. The cut edges of the clad product should be protected by the normal finish or by jointing-compound squeezed out during wet assembly.

For aluminum-copper alloys (2000 series) dilute aluminum alloys such as 1230, 6003, or 6053, containing small amounts of manganese, chromium, or magnesium, may be used as cladding material. These have low-copper contents, less than 0.02%, and low-iron content, less than 0.2%. However these alloys are not sufficiently anodic with respect to the Al-Zn-Mg-Cu alloys of the 7000 series, and they do not provide cathodic protection in these cases. The 7000 series alloys are therefore usually clad with aluminum alloys containing about 1% zinc, such as 7072, or aluminum-zinc-magnesium alloys such as 7008 and 7011, which have higher zinc contents.

The thickness of the cladding is usually between 2 and 5% of the total sheet or plate thickness, and because the cladding is usually a softer and lower-strength alloy, the presence of the cladding can lower the fatigue strength and abrasion resistance of the product. In the case of thick plate where substantial amounts of material may be removed from one side by machining so that the cladding becomes a larger fraction of the total thickness, the decrease in strength of the product may be substantial. In these cases the use of the higher-strength claddings such as 7008 and 7011 is preferred.

Thermal spraying. Energy surface treatment involves adding energy into the surface of the work piece for adhesion to take place. Conventional surface finishing methods involve heating an entire part. The methods described in this section usually add energy and material into the surface, keeping the bulk of the object relatively cool and unchanged. This allows surface properties to be modified with minimal effect on the structure and properties of the underlying material.[5] Plasmas are used to reduce process temperatures by adding energy to the surface in the form of kinetic energy of ions rather than thermal energy. Table 9.3 shows the main metallic materials that have been used for the production of spray coatings and Table 9.4 contains a brief description of the main advanced techniques. Similarly, Table 9.5 describes briefly the applications and costs of these advanced techniques, and Table 9.6 summarizes the limits and applicability of each technique.

Advanced surface treatments often require the use of vacuum chambers to ensure proper cleanliness and control. Vacuum processes are generally more expensive and difficult to use than liquid or air processes. Facilities can expect to see less-complicated vacuum systems appearing on the market in the future. In general, use of the advanced surface treatments is more appropriate for treating small components (e.g., ion beam implantation, thermal spray) because the treatment time for these processes is proportional to the surface areas being covered. Facilities will also have to address the following issues when considering the new techniques:[5]

- *Quality control methods.* Appropriate quality assurance tests need to be developed for evaluating the performance of the newer treatment techniques.
- *Performance testing.* New tribological tests must be developed for measuring the performance of surface engineered materials.

TABLE 9.3 Spray-Coating Materials

Type coating	General qualities
Aluminum	Highly resistant to heat, hot water, and corrosive gases; excellent heat distribution and reflection
Babbitt	Excellent bearing wearability
Brass	Machines well, takes a good finish
Bronze	Excellent wear resistance; exceptional machinability; dense coatings (especially Al, bronze)
Copper	High heat and electrical conductivity
Iron	Excellent machining qualities
Lead	Good corrosion protection, fast, deposits and dense coatings
Molybdenum (molybond)	Self-bonding for steel surface preparation
Monel	Excellent machining qualities; highly resistant to corrosion
Nickel	Good machine finishing; excellent corrosion protection
Nickel-chrome	High-temperature applications
Steel	Hard finishes, good machinability
Chrome steel (tufton)	Bright, hard finish, highly resistant to wear
Stainless	Excellent corrosion protection and superior wearability
Tin	High purity for food applications
Zinc	Superior corrosion resistance and bonding qualities

- *Substitute cleaning and coating removal.* The advanced coatings provide excellent adhesion between the substrate and the coating; as a result, these coatings are much more difficult to strip than conventional coatings. Many coating companies have had to develop proprietary stripping techniques, most of which have adverse environmental or health risks.

- *Process control and sensing.* The use of advanced processes requires improvements in the level of control over day-to-day production operations, such as enhanced computer-based control systems.

Coatings can be sprayed from rod or wire stock or from powdered materials. The material (e.g., wire) is fed into a flame, where it is melted. The molten stock is then stripped from the end of the wire and atomized by a high-velocity stream of compressed air or other gas, which propels the material onto a prepared substrate or workpiece. Depending on the substrate, bonding occurs either due to mechanical interlock with a roughened surface, due to localized diffusion and alloying, and/or by means of Van der Waals forces (i.e., mutual attraction and cohesion between two surfaces).

TABLE 9.4 Description of the Main Advanced Techniques for Producing Metallic Coatings

Combustion torch/flame spraying
Flame spraying involves the use of a combustion flame spray torch in which a fuel gas and oxygen are fed through the torch and burned with the coating material in a powder or wire form and fed into the flame. The coating is heated to near or above its melting point and accelerated to speeds of 30 to 90 m/s. The molten droplets impinge on the surface, where they flow together to form the coating.
Combustion torch/high-velocity oxy-fuel (HVOF)
With HVOF, the coating is heated to near or above its melting point and accelerated in a high-velocity combustion gas stream. Continuous combustion of oxygen fuels typically occurs in a combustion chamber, which enables higher gas velocities (550 to 800 m/s). Typical fuels include propane, propylene, or hydrogen.
Combustion torch/detonation gun
Using a detonation gun, a mixture of oxygen and acetylene with a pulse of powder is introduced into a water-cooled barrel about 1 m long and 25 mm in diameter. A spark initiates detonation, resulting in hot, expanding gas that heats and accelerates the powder materials (containing carbides, metal binders, oxides) so that they are converted into a plasticlike state at temperatures ranging from 1100 to 19,000°C. A complete coating is built up through repeated, controlled detonations.
Electric arc spraying
During electric arc spraying, an electric arc between the ends of two wires continuously melts the ends while a jet of gas (air, nitrogen, etc.) blows the molten droplets toward the substrate at speeds of 30 to 150 m/s.
Plasma spraying
A flow of gas (usually based on argon) is introduced between a water-cooled copper anode and a tungsten cathode. A direct current arc passes through the body of the gun and the cathode. As the gas passes through the arc, it is ionized and forms plasma. The plasma (at temperatures exceeding 30,000°C) heats the powder coating to a molten state, and compressed gas propels the material to the workpiece at very high speeds that may exceed 550 m/s.
Ion plating/plasma based
Plasma-based plating is the most common form of ion plating. The substrate is in proximity to a plasma, and ions are accelerated from the plasma by a negative bias on the substrate. The accelerated ions and high-energy neutrals from charge exchange processes in the plasma arrive at the surface with a spectrum of energies. In addition, the surface is exposed to chemically activated species from the plasma, and adsorption of gaseous species form the plasma environment.
Ion plating/ion beam enhanced deposition (IBED)
During IBED, both the deposition and bombardment occur in a vacuum. The bombarding species are ions either from an ion gun or other sources. While ions are bombarding the substrate, neutral species of the coating material are delivered to the substrate via a physical vapor deposition technique such as evaporation or sputtering. Because the secondary ion beam is independently controllable, the energy particles in the beam can be varied over a wide range and chosen with a very narrow window. This

TABLE 9.4 Description of the Main Advanced Techniques for Producing Metallic Coatings (*Continued*)

allows the energies of deposition to be varied to enhance coating properties such as interfacial adhesion, density, morphology, and internal stresses. The ions form nucleation sites for the neutral species, resulting in islands of coating that grow together to form the coating.

Ion implantation

Ion implantation does not produce a discrete coating; the process alters the elemental chemical composition of the surface of the substrate by forming an alloy with energetic ions (10 to 200 keV in energy). A beam of charged ions of the desired element (gas) is formed by feeding the gas into the ion source where electrons, emitted from a hot filament, ionize the gas and form a plasma. The ions are focused into a beam using an electrically biased extraction electrode. If the energy is high enough, the ions will go into the surface, not onto the surface, changing the surface composition. Three variations have been developed that differ in methods of plasma formation and ion acceleration: beamline implantation, direct ion implantation, and plasma source implantation. Pretreatment (degreasing, rinse, ultrasonic cleaner) is required to remove any surface contaminants prior to implantation. The process is performed at room temperature, and time depends on the temperature resistance of the workpiece and the required dose.

Sputtering and sputter deposition

Sputtering is an etching process for altering the physical properties of the surface. The substrate is eroded by the bombardment of energetic particles, exposing the underlying layers of the material. The incident particles dislodge atoms from the surface or near-surface region of the solid by momentum transfer form the fast, incident particle to the surface atoms. The substrate is contained in a vacuum and placed directly in the path of the neutral atoms. The neutral species collides with gas atoms, causing the material to strike the substrate from different directions with a variety of energies. As atoms adhere to the substrate, a film is formed. The deposits are thin, ranging from 0.00005 to 0.01 mm. The most commonly applied materials are chromium, titanium, aluminum, copper, molybdenum, tungsten, gold, silver, and tantalum. Three techniques for generating the plasma needed for sputtering are available: diode plasmas, RF diodes, and magnetron enhanced sputtering.

Laser surface alloying

The industrial use of lasers for surface modifications is increasingly widespread. Surface alloying is one of many kinds of alteration processes achieved through the use of lasers. It is similar to surface melting, but it promotes alloying by injecting another material into the melt pool so that the new material alloys into the melt layer. Laser cladding is one of several surface alloying techniques performed by lasers. The overall goal is to selectively coat a defined area. In laser cladding, a thin layer of metal (or powder metal) is bonded with a base metal by a combination of heat and pressure. Specifically, ceramic or metal powder is fed into a carbon dioxide laser beam above a surface, melts in the beam, and transfers heat to the surface. The beam welds the material directly into the surface region, providing a strong metallurgical bond. Powder feeding is performed by using a carrier gas in a manner similar to that used for thermal spray systems. Large areas are covered by moving the substrate under the beam and overlapping disposition tracks. Shafts and other circular objects are coated by rotating the beam. Depending on the powder and substrate metallurgy, the microstructure of the surface layer can be controlled, using the interaction time and laser parameters. Pretreatment is not as vital to successful performance of laser

TABLE 9.4 Description of the Main Advanced Techniques for Producing Metallic Coatings (*Continued*)

cladding processes as it is for other physical deposition methods. The surface may require roughening prior to deposition. Grinding and polishing are generally required posttreatments.

Chemical vapor deposition (CVD)
Substrate pretreatment is important in vapor deposition processes, particularly in the case of CVD. Pretreatment of the surface involves minimizing contamination mechanically and chemically before mounting the substrate in the deposition reactor. Substrates must be cleaned just prior to deposition, and the deposition reactor chamber itself must be clean, leak-tight, and free from dust and moisture. During coating, surface cleanliness is maintained to prevent particulates from accumulating in the deposit. Cleaning is usually performed using ultrasonic cleaning and/or vapor degreasing. Vapor honing may follow to improve adhesion. Mild acids or gases are used to remove oxide layers formed during heat-up. Posttreatment may include a heat treatment to facilitate diffusion of the coating material into the material.

The basic steps involved in any thermal coating process are substrate preparation, masking and fixturing, coating, finishing, inspection, and stripping (when necessary). Substrate preparation usually involves scale and oil and grease removal, as well as surface roughening. Roughening is necessary for most of the thermal spray processes to ensure adequate bonding of the coating to the substrate. The most common method is grit blasting, usually with alumina. Masking and fixturing limit the amount of coating applied to the workpiece to remove overspray through time-consuming grinding and stripping after deposition. The basic parameters in thermal spray deposition are particle temperature, velocity, angle of impact, and extent of reaction with gases during the deposition process. The geometry of the part being coated affects the surface coating because the specific properties vary from point to point on each piece. In many applications, workpieces must be finished after the deposition process, the most common technique being grinding followed by lapping. The final inspection of thermal spray coatings involves verification of dimensions, a visual examination for pits, cracks, and so forth. Nondestructive testing has largely proven unsuccessful.

There are three basic categories of thermal spray technologies: combustion torch (flame spray, high velocity oxy-fuel, and detonation gun), electric (wire) arc, and plasma arc. Thermal spray processes are maturing, and the technology is readily available.

Environmental concerns with thermal spraying techniques include the generation of dust, fumes, overspray, noise, and intense light. The metal spray process is usually performed in front of a "water curtain" or dry filter exhaust hood, which captures the overspray and fumes.

TABLE 9.5 Applications and Costs of the Main Advanced Techniques for Producing Metallic Coatings

Combustion torch/flame spraying
This technique can be used to deposit ferrous-, nickel-, and cobalt-based alloys and some ceramics. It is used in the repair of machine bearing surfaces, piston and shaft bearing or seal areas, and corrosion and wear resistance for boilers and structures (e.g., bridges).
Combustion torch/high velocity oxy-fuel (HVOF)
This technique may be an effective substitute for hard chromium plating for certain jet engine components. Typical applications include reclamation of worn parts and machine element buildup, abradable seals, and ceramic hard facings.
Combustion torch/detonation gun
This can only be used for a narrow range of materials, both for the choice of coating materials and as substrates. Oxides and carbides are commonly deposited. The high-velocity impact of materials such as tungsten carbide and chromium carbide restricts application to metal surfaces.
Electric arc spraying
Industrial applications include coating paper, plastics, and other heat-sensitive materials for the production of electromagnetic shielding devices and mold making.
Plasma spraying
This techniques can be used to deposit molybdenum and chromium on piston rings, cobalt alloys on jet-engine combustion chambers, tungsten carbide on blades of electric knives, and wear coatings for computer parts.
Ion plating/plasma based
Coating materials include alloys of titanium, aluminum, copper, gold, and palladium. Plasma-based ion plating is used in the production of x-ray tubes; space applications; threads for piping used in chemical environments; aircraft engine turbine blades; tool steel drill bits; gear teeth; high-tolerance injection molds; aluminum vacuum sealing flanges; decorative coatings; corrosion protection in nuclear reactors; metallizing of semiconductors, ferrites, glass, and ceramics; and body implants. In addition, it is widely used for applying corrosion-resistant aluminum coatings as an alternative to cadmium. Capital costs are high for this technology, creating the biggest barrier for ion plating use. It is used where high value-added equipment is being coated such as expensive injection molds instead of inexpensive drill bits.
Ion plating/ion beam enhanced deposition (IBED)
Although still an emerging technology, IBED is used for depositing dense optically transparent coatings for specialized optical applications, such as infrared optics. Capital costs are high for this technology, creating the biggest barrier for ion plating use. Equipment for IBED processing could be improved by the development of low-cost, high-current, large-area reactive ion beam sources.
Ion implantation
Nitrogen is commonly implanted to increase the wear resistance of metals because ion beams are produced easily. In addition, metallic elements, such as titanium, yttrium, chromium, and nickel, may be implanted into a variety of materials to produce a wider

TABLE 9.5 Applications and Costs of the Main Advanced Techniques for Producing Metallic Coatings (*Continued*)

range of surface modifications. Implantation is primarily used as an antiwear treatment for components of high value such as biomedical devices (prostheses), tools (molds, dies, punches, cutting tools, inserts), and gears and ball bearings used in the aerospace industry. Other industrial applications include the semiconductor industry for depositing gold, ceramics, and other materials into plastic, ceramic, and silicon and gallium arsenide substrates. The U.S. Navy has demonstrated that chromium ion implantation could increase the life of ball bearings for jet engines with a benefit-to-cost ratio of 20:1. A treated forming die resulted in the production of nearly 5000 automobile parts compared to the normal 2000 part life from a similar tool hard faced with tank plated chromium. The initial capital cost is relatively high, although large-scale systems have proven cost effective. An analysis of six systems manufactured by three companies found that coating costs range from \$0.04 to \$0.28/cm^2. Depending on throughput, capital cost ranges from \$400,000 to \$1,400,000, and operating costs were estimated to range from \$125,000 to \$250,000.

Sputtering and sputter deposition

Sputter-deposited films are routinely used simply as decorative coatings on watchbands, eyeglasses, and jewelry. The electronics industry relies heavily on sputtered coatings and films (e.g., thin film wiring on chips and recording heads, magnetic and magneto-optic recording media). Other current applications for the electronics industry are wear-resistant surfaces, corrosion-resistant layers, diffusion barriers, and adhesion layers. Sputtered coatings are also used to produce reflective films on large pieces of architectural glass and for the coating of decorative films on plastic in the automotive industry. The food packaging industry uses sputtering for coating thin plastic films for packaging pretzels, potato chips, and other products. Compared to other deposition processes, sputter deposition is relatively inexpensive.

Laser surface alloying

Although laser processing technologies have been in existence for many years, industrial applications are relatively limited. Uses of laser cladding include changing the surface composition to produce a required structure for better wear, or high-temperature performance; build up a worn part; provide better corrosion resistance; impart better mechanical properties; and enhance the appearance of metal parts. The high capital investment required for using laser cladding has been a barrier for its widespread adoption by industry.

Chemical vapor deposition (CVD)

CVD processes are used to deposit coatings and to form foils, powders, composite materials, free-standing bodies, spherical particles, filaments, and whiskers. CVD applications are expanding both in number and sophistication. The U.S. market in 1998 for CVD applications was \$1.2 billion, 77.6 percent of which was for electronics and other large users, including structural applications, optical, optoelectronics, photovoltaic, and chemical. Analysts anticipate that future growth for CVD technologies will continue to be in the area of electronics. CVD will also continue to be an important method for solving difficult materials problems. CVD processes are commercial realities for only a few materials and applications. Start-up costs are typically very expensive.

TABLE 9.6   Limits and Applicability of the Main Advanced Techniques for Producing Metallic Coatings

Combustion torch/flame spraying
Flame spraying is noted for its relatively high as-deposited porosity, significant oxidation of the metallic components, low resistance to impact or point loading, and limited thickness (typically 0.5 to 3.5 mm). Advantages include the low capital cost of the equipment, its simplicity, and the relative ease of training the operators. In addition, the technique uses materials efficiently and has low associated maintenance costs.

Combustion torch/high velocity oxy-fuel (HVOF)
This technique has very high velocity impact, and coatings exhibit little or no porosity. Deposition rates are relatively high, and the coatings have acceptable bond strength. Coating thickness range from 0.000013 to 3 mm. Some oxidation of metallics or reduction of some oxides may occur, altering the coating's properties.

Combustion torch/detonation gun
This technique produces some of the densest of the thermal coatings. Almost any metallic, ceramic, or cement materials that melt without decomposing can be used to produce a coating. Typical coating thickness range from 0.05 to 0.5 mm, but both thinner and thicker coatings are used. Because of the high velocities, the properties of the coatings are much less sensitive to the angle of deposition than most other thermal spray coatings.

Electric arc spraying
Coating thickness can range from a few hundredths of a millimeter to almost unlimited thickness, depending on the end use. Electric arc spraying can be used for simple metallic coatings, such as copper and zinc, and for some ferrous alloys. The coatings have high porosity and low bond strength.

Plasma spraying
Plasma spraying can be used to achieve thickness from 0.3 to 6 mm, depending on the coating and the substrate materials. Sprayed materials include aluminum, zinc, copper alloys, tin, molybdenum, some steels, and numerous ceramic materials. With proper process controls, this technique can produce coatings with a wide range of selected physical properties, such as coatings with porosity ranging from essentially zero to high porosity.

Ion plating/plasma based
This technique produces coatings that typically range from 0.008 to 0.025 mm. Advantages include a wide variety of processes as sources of the depositing material; in situ cleaning of the substrate prior to film deposition; excellent surface covering ability; good adhesion; flexibility in tailoring film properties such as morphology, density, and residual film stress; and equipment requirements and costs equivalent to sputter deposition. Disadvantages include many processing parameters that must be controlled; contamination may be released and activated in the plasma; and bombarding gas species may be incorporated in the substrate and coating.

Ion plating/ion beam enhanced deposition (IBED)
Advantages include increased adhesion; increased coating density; decreased coating porosity and prevalence of pinholes; and increased control of internal stress, morphology, density, and composition. Disadvantages include high equipment and

TABLE 9.6 Limits and Applicability of the Main Advanced Techniques for Producing Metallic Coatings (*Continued*)

processing costs; limited coating thickness; part geometry and size limit; and gas precursors used for some implantation species that are toxic. This technique can produce a chromium deposit 10 μm thick with greater thickness attained by layering. Such thickness is too thin for most hard chrome requirements (25 to 75 μm with some dimensional restoration work requiring 750 μm) and layering would significantly add to the cost of the process. IBED provides some surface cleaning when the surface is initially illuminated with a flux of high-energy inert gas ions; however, the process will still require precleaning (e.g., degreasing).

Ion implantation

Ion implantation can be used for any element that can be vaporized and ionized in a vacuum chamber. Because material is added to the surface, rather than onto the surface, there is no significant dimensional change or problems with adhesion. The process is easily controlled, offers high reliability and reproducibility, requires no posttreatment, and generates minimal waste. If exposed to high temperatures, however, implanted ions may diffuse away from the surface due to limited depth of penetration, and penetration does not always withstand severe abrasive wear. Implantation is used to alter surface properties, such as hardness, friction, wear resistance, conductance, optical properties, corrosion resistance, and catalysis. Commercial availability is limited by general unfamiliarity with the technology, scarcity of equipment, lack of quality control and assurance, and competition with other surface modification techniques. Areas of research include ion implantation of ceramic materials for high-temperature internal combustion engines, glass to reduce infrared radiation transmission and reduce corrosion, as well as automotive parts (piston rings, cylinder liners) to reduce wear.

Sputtering and sputter deposition

This technique is a versatile process for depositing coatings of metals, alloys, compounds, and dielectrics on surfaces. The process has been applied in hard and protective industrial coatings. Primarily TiN, as well as other nitrides and carbides, has demonstrated high hardness, low porosity, good chemical inertness, good conductivity, and attractive appearance. Sputtering is capable of producing dense films, often with near-bulk quantities. Areas requiring future research and development include better methods for in situ process control; methods for removing deposited TiN and other hard, ceramiclike coatings from poorly coated or worn components without damage to the product; and improved understanding of the factors that affect film properties.

Laser surface alloying

This technique can be used to apply most of the same materials that can be applied via thermal spray techniques; the powders used for both methods are generally the same. Materials that are easily oxidized, however, will prove difficult to deposit without recourse to inert gas streams and envelopes. Deposition rates depend on laser power, powder feed rates, and traverse speed. The rates are typically in the region of 2×10^{-4} cm^3 for a 500-W beam. Thickness of several hundred micrometers can be laid down on each pass of the laser beam, allowing thickness of several millimeters to accumulate. If the powder density is too high, this thermal cycling causes cracking and delamination of earlier layers, severely limiting the attainable buildup. Research has found that easily oxidized materials, such as aluminum, cannot be laser clad because the brittle oxide causes cracking and delamination. Some steels may be difficult to coat effectively.

TABLE 9.6 Limits and Applicability of the Main Advanced Techniques for Producing Metallic Coatings (*Continued*)

The small size of the laser's beam limits the size of the workpieces that can be treated cost effectively. Shapes are restricted to those that prevent line-of-sight access to the region to be coated.

Chemical vapor deposition (CVD)
CVD is used mainly for corrosion and wear resistance. CVD processes are also usually applied in cases where specific properties of materials of interest are difficult to obtain by other means. CVD is unique because it controls the microstructure and/or chemistry of the deposited material. The microstructure of CVD deposits depends on chemical makeup and energy of atoms, ions, or molecular fragments impinging on the substrate; chemical composition and surface properties of the substrate; substrate temperature; and presence or absence of a substrate bias voltage. The most useful CVD coatings are nickel, tungsten, chromium, and titanium carbide. Titanium carbide is used for coating punching and embossing tools to impart wear resistance.

Water curtain systems periodically discharge contaminated wastewaters. Noise generated can vary from approximately 80 dB to more than 140 dB. With the higher noise-level processes, robotics are usually required for spray application. The use of metal spray processes may eliminate some of the pollution associated with conventional tank plating. In most cases, however, wet processes, such as cleaning, are necessary in addition to the metal coating process. Therefore, complete elimination of tanks may not be possible. Waste streams resulting from flame spray techniques may include overspray, wastewaters, spent exhaust filters, rejected parts, spent gas cylinders, air emissions (dust, fumes), and wastes associated with the grinding and finishing phases.

Physical vapor deposition. Vapor deposition refers to any process in which materials in a vapor state are condensed through condensation, chemical reaction, or conversion to form a solid material. These processes are used to form coatings to alter the mechanical, electrical, thermal, optical, corrosion-resistance, and wear properties of the substrates. They are also used to form free-standing bodies, films, and fibers and to infiltrate fabric to form composite materials.[5] Vapor deposition processes usually take place within a vacuum chamber.

There are two categories of vapor deposition processes: physical vapor deposition (PVD) and chemical vapor deposition (CVD). In PVD processes, the workpiece is subjected to plasma bombardment. In CVD processes, thermal energy heats the gases in the coating chamber and drives the deposition reaction.

Physical vapor deposition methods are clean, dry vacuum deposition methods in which the coating is deposited over the entire object simultaneously, rather than in localized areas. All reactive PVD hard coating processes combine:

- A method for depositing the metal
- Combination with an active gas, such as nitrogen, oxygen, or methane
- Plasma bombardment of the substrate to ensure a dense, hard coating[6]

PVD methods differ in the means for producing the metal vapor and the details of plasma creation. The primary PVD methods are ion plating, ion implantation, sputtering, and laser surface alloying.

Waste streams resulting from laser cladding are similar to those resulting from high-velocity oxy-fuels and other physical deposition techniques: blasting media and solvents, bounce and overspray particles, and grinding particles. Generally speaking, none of these waste streams are toxic.[6]

CVD is a subset of the general surface treatment process, vapor deposition. Over time, the distinction between the terms *physical vapor deposition* and *chemical vapor deposition* has blurred as new technologies have been developed and the two terms overlap. CVD includes sputtering, ion plating, plasma-enhanced chemical vapor deposition, low-pressure chemical vapor deposition, laser-enhanced chemical vapor deposition, active-reactive evaporation, ion beam, laser evaporation, and many other variations. These variants are distinguished by the manner in which precursor gases are converted into the reactive gas mixtures. In CVD processes, a reactant gas mixture impinges on the substrate upon which the deposit is to be made. Gas precursors are heated to form a reactive gas mixture. The coating species is delivered by a precursor material, otherwise known as a reactive vapor. It is usually in the form of a metal halide, metal carbonyl, a hydride, or an organometallic compound. The precursor may be in gas, liquid, or solid form. Gases are delivered to the chamber under normal temperatures and pressures, whereas solids and liquids require high temperatures and/or low pressures in conjunction with a carrier gas. Once in the chamber, energy is applied to the substrate to facilitate the reaction of the precursor material upon impact. The ligand species is liberated from the metal species to be deposited upon the substrate to form the coating. Because most CVD reactions are endothermic, the reaction may be controlled by regulating the amount of energy input.[7] The steps in the generic CVD process are

- Formation of the reactive gas mixture
- Mass transport of the reactant gases through a boundary layer to the substrate

- Adsorption of the reactants on the substrate
- Reaction of the adsorbents to form the deposit
- Description of the gaseous decomposition products of the deposition process

The precursor chemicals should be selected with care because potentially hazardous or toxic vapors may result. The exhaust system should be designed to handle any reacted and unreacted vapors that remain after the coating process is complete. Other waste effluents from the process must be managed appropriately. Retrieval, recycle, and disposal methods are dictated by the nature of the chemical. For example, auxiliary chemical reactions must be performed to render toxic or corrosive materials harmless, condensates must be collected, and flammable materials must be either combusted, absorbed, or dissolved. The extent of these efforts is determined by the efficiency of the process.[7]

9.2.2 Inorganic coatings

Inorganic coatings can be produced by chemical action, with or without electrical assistance. The treatments change the immediate surface layer of metal into a film of metallic oxide or compound that has better corrosion resistance than the natural oxide film and provides an effective base or key for supplementary protection such as paints. In some instances, these treatments can also be a preparatory step prior to painting.

Anodizing. Anodizing involves the electrolytic oxidation of a surface to produce a tightly adherent oxide scale that is thicker than the naturally occurring film. Anodizing is an electrochemical process during which aluminum is the anode. The electric current passing through an electrolyte converts the metal surface to a durable aluminum oxide. The difference between plating and anodizing is that the oxide coating is integral with the metal substrate as opposed to being a metallic coating deposition. The oxidized surface is hard and abrasion resistant, and it provides some degree of corrosion resistance.

However, anodizing cannot be relied upon to provide corrosion resistance to corrosion-prone alloys, and further protection by painting is usually required. Fortunately the anodic coating provides an excellent surface both for painting and for adhesive bonding. Anodic coatings break down chemically in highly alkaline solutions (pH $>$ 8.5) and highly acid solutions (pH $<$4.0). They are also relatively brittle and may crack under stress, and therefore supplementary protection, such as painting, is particularly important with stress corrosion-prone alloys.

Anodic coatings can be formed in chromic, sulfuric, phosphoric, or oxalic acid solutions. Chromic acid anodizing is widely used with 7000 series alloys to improve corrosion resistance and paint adhesion, and unsealed coatings provide a good base for structural adhesives. However these coatings are often discolored, and where cosmetic appearance is important, sulfuric acid anodizing may be preferred. Table 9.7 shows the alloys suitable for anodizing and describes some of the coating properties obtained with typical usage and finishing advice.

The Al_2O_3 coating produced by anodizing is typically 2 to 25 μm thick and consists of a thin nonporous barrier layer next to the metal

TABLE 9.7 Aluminum Alloys Suitable for Anodizing

Series	Coating properties	Uses	Finishing advice
1xxx	Clear bright	Cans, architectural	Care should be taken when racking this soft material; good for bright coatings susceptible to etch, staining.
2xxx	Yellow poor protection	Aircraft mechanical	Because copper content is >2%, these produce yellow, poor weather-resistant coatings.
3xxx	Grayish-brown	Cans, architectural, lighting	Difficult to match sheet to sheet (varying degrees of gray/brown). Used extensively for architectural painted products
4xxx	Dark gray	Architectural, lighting	Produces heavy black smut, which is hard to remove; 4043 and 4343 used for architectural dark gray finishes in past years.
5xxx	Clear good protection	Architectural, welding, wire lighting	For 5005, keep silicon < 0.1% and magnesium between 0.7 and 0.9%; maximum of ±20% for job; watch for oxide streaks
6xxx	Clear good protection	Architectural, structural	Matte: iron > 0.2%. Bright: iron < 0.1%. 6063 best match for 5005. 6463 best for chemical brightening.
7xxx	Clear good protection	Automotive	Zinc over 5% will produce brown-tinted coatings; watch zinc in effluent stream; good for bright coatings.

SOURCE: Aluminum Anodizers Council (AAC) Technical Bulletin 2-94, *Aluminum Alloy Reference for Anodizing,* March 1994.

with a porous outer layer that can be sealed by hydrothermal treatment in steam or hot water for several minutes. This produces a hydrated oxide layer with improved protective properties. Figure 9.1 illustrates a porous anodic film and its evolution during the sealing process. Improved corrosion resistance is obtained if the sealing is done in a hot metal salt solution such as a chromate or dichromate solution. The oxide coatings may also be dyed to provide surface coloration for decorative purposes, and this can be performed either in the anodizing bath or afterward. International standards for anodic treatment of aluminum alloys have been published by the International Standards Organization and cover dyed and undyed coatings. There are many reasons to anodize a part. Following are a few considerations and the industries that employ them

- *Appearance.* Products look finished, cleaner, and better, and this appearance lasts longer. Color enhances metal and promotes a solid, well-built appearance while removing the harsh metal look. Any aluminum product can be color anodized.
- *Corrosion resistance.* A smooth surface is retained and weathering is retarded. Useful for food handling and marine products.
- *Ease in cleaning.* Any anodized product will stay cleaner longer and is easier to clean when it does get dirty.
- *Abrasion resistance.* The treated metal is tough, harder than many abrasives, and is ideal for caul plates, tooling, and air cylinder applications.
- *Nongalling.* Screws and other moving parts will not seize, drag, or jam, and wear in these areas is diminished. Gun sights, instruments, and screw threads are typical applications.
- *Heat absorption.* This can provide uniform or selective heat-absorption properties to aluminum for the food processing industry.
- *Heat radiation.* This is used as a method to finish electronic heat sinks and radiators. Further, anodizing will not rub off, is an excellent paint base, removes minor scuffs, and is sanitary and tasteless.

There are many variations in the anodization process. The following examples are given to illustrate some of the processes used in the industry:

1. *Hardcoat anodizing.* As the name implies, a hardcoat finish is tough and durable and is used where abrasion and corrosion resistance, as well as surface hardness, are critical factors. Essentially, hardcoating is a sulfuric acid anodizing process, with the electrolyte concentration, temperature, and electric current parameters altered to

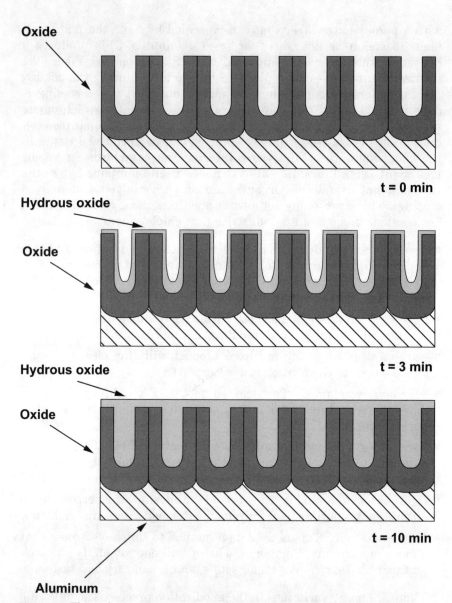

Figure 9.1 The evolution of a porous anodic film on aluminum as a function of the sealing time at 85°C.

produce the hardened surface. Wearing qualities have actually proven to be superior to those of case hardened steel or hard chrome plate.

2. *Bulk anodizing.* Bulk anodizing is an electrochemical process for anodizing small, irregularly shaped parts, which are processed in perforated aluminum, plastic, or titanium baskets. The tremendous

quantity of parts that can be finished in a relatively short time makes this technique highly economical. Another advantage in processing such large volumes at one time is the resulting consistency in color and quality. Finishing items such as rivets, ferrules, medical hubs, and so forth, using the bulk process make production economically feasible.

3. *Sulfuric acid anodizing.* This is the most common method of anodizing. The part is subjected to a specified electric current through a sulfuric acid electrolyte, converting the surface to an aluminum oxide coating capable of absorbing dyes in a wide range of colors. Abrasion and/or corrosion resistance is enhanced, and the surface may also be used as a base for applied coatings, such as paint, Teflon, and adhesives. Custom coloring is available to meet any specification, and through prefinish techniques, matte, satin, or highly reflective surfaces can be furnished.

Anodizing treatments are also available for magnesium and titanium alloys. The treatments commonly used with magnesium alloys involve several processing options to produce either thin coatings of about 5-μm thickness for flexibility and surfaces suitable for paint adhesion, or thick coatings, up to about 30 μm for maximum corrosion and abrasion resistance. When anodizing is used for the treatment of titanium and titanium alloys, it can provide limited protection to the less noble metals against galvanic corrosion, and when used together with solid film lubricants, it helps to prevent galling. The process produces a smooth coating with a uniform texture and appearance and a uniform blue-to-violet color.

Chromate filming. A number of proprietary chromate filming treatments are available for aluminum, magnesium, cadmium, and zinc alloys. The treatments usually involve short-time immersion in strongly acid chromate solutions, but spraying or application by brushing or swabbing can also be used for touchup of parts. The resulting films are usually about 5 μm thick and are colored depending on the base alloy, being golden yellow on aluminum, dull gold on cadmium and zinc, and brown or black on magnesium. The films contain soluble chromates that act as corrosion inhibitors, and they provide a modest improvement in corrosion resistance of the base metal. However, their main purpose is to provide a suitable surface for sealing resins or paints. Epoxy primer, for example, which does not adhere well to bare aluminum, adheres very well to chemical conversion coatings. Among the best-known coatings used with aluminum alloys are those produced by the Alodine 1200 and Alocrom 1200 processes.

A process for zinc alloys has been described to consist of immersion for a few seconds in a sodium dichromate solution at a concentration of 200 g/L and acidified with sulfuric acid at 8 ml/L. The treatment is performed at room temperature and is followed by rinsing and drying to produce a dull yellow zinc chromate coating.

Phosphate coatings. A number of proprietary treatments such as Parkerizing and Bonderizing are available for use on steel. They are applied by brushing, spraying, or prolonged immersion in an acid orthophosphate solution containing iron, zinc, or manganese. For example a solution might contain $Zn(H_2PO_4)_2 \cdot 2H_2O$ with added H_3PO_4. The coatings consist of a thick porous layer of fine phosphate crystals, tightly bonded to the steel. The coatings do not provide significant corrosion resistance when used alone, but they provide an excellent base for oils, waxes, or paints, and they help to prevent the spreading of rust under layers of paint. Phosphating should not be applied to nitrided or finish-machined steel, and steel parts containing aluminum, magnesium, or zinc are subject to pitting in the bath. Some restrictions apply also to heat-treated stainless and high-strength steels.

Nitriding. Steels containing nitride-forming elements such as chromium, molybdenum, aluminum, and vanadium can be treated to produce hard surface layers, providing improved wear resistance. Many of the processes employed are proprietary, but typically they involve exposure of cleaned surfaces to anhydrous ammonia at elevated temperatures. The nitrides formed are not only hard but also more voluminous than the original steel, and therefore they create compressive residual surface stresses. Therefore, nitrided steels usually exhibit improved fatigue and corrosion fatigue resistance. Similar beneficial effects can be achieved by shot peening.

Passive films. Austenitic stainless steels and hardenable stainless steels such as martensitic, precipitation hardening, and maraging stainless steels are seldom coated, but their corrosion resistance depends on the formation of naturally occurring transparent oxide films. These films may be impaired by surface contaminants such as organic compounds or metallic or inorganic materials. Treatments are available for these materials to clean and degrease surfaces and produce uniform protective oxide films under controlled conditions. These usually involve immersion in an aqueous solution of nitric acid and a dichromate solution.

9.2.3 Organic coatings

Paints, coatings, and high-performance organic coatings were developed to protect equipment from environmental damage. Of prime importance in the development of protective coatings was the petroleum industry, which produced most of the basic ingredients from which most synthetic resins were developed. The cracking of petroleum produced a multitude of unsaturated workable compounds that are important in the building of large resin polymers such as vinyls and acrylics. The solvents necessary for the solution of the resins were also derived from petroleum or natural gas. The building blocks for epoxies and modern polyurethane coatings are other derivatives produced by refining petroleum products.[8]

The Steel Structures Painting Council (SSPC) is the world's acknowledged resource and authority for protective coatings technology. SSPC's mission is to advance the technology and promote the use of protective coatings to preserve industrial marine and commercial structure components and substrates. Table 9.8 describes briefly most of the numerous standards and guides currently maintained by SSPC.

Some other concepts important for designing corrosion-resistant coatings include those of coating protection, component design, component function, and coating formulation. Many coatings contain as many as 15 to 20 ingredients with their own range of functionality. Some of the main variables used to design corrosion protective coatings are

- *Impermeability.* The ideal impermeable coating should be completely unaffected by the specific environment it is designed to block, be it most commonly humidity, water, or any other corrosive agent such as gases, ions, or electrons. This ideal impermeable coating should have a high dielectric constant and also have perfect adhesion to the underlying surface to avoid any entrapment of corrosive agents. Good impermeability has been the successful ingredient of many anticorrosion coatings.

- *Inhibition.* In contrast with coatings developed on the basis of impermeability, inhibitive coatings function by reacting with a certain environment to provide a protective film or barrier on the metallic surface. The concept of adding an inhibitor to a primer has been applied to coatings of steel vessels since these vessels were first constructed. Such coatings were originally oil based and heavily loaded with red lead.

- *Cathodically protective pigments.* As with inhibition, cathodic protection in coatings is mostly provided by additives in the primer. The main function of these additives is to shift the potential of the environment to a less-corrosive cathodic potential. Inorganic zinc-based primers are good examples of this concept.

The coating system approach. For serious corrosion situations, the coating system approach (primer, intermediate coat, and topcoat) provides all the ingredients for a long-lasting solution.[8]

TABLE 9.8 Reference, Purpose, and Brief Description of Painting Standards and Specifications

Guide to SSPC-VIS 1-89: Visual Standard for Abrasive Blast Cleaned Steel (Standard Reference Photographs)
This guide describes the use of standard reference photographs depicting the appearance of previously unpainted hot-rolled carbon steel prior to and after abrasive blast cleaning. These photographs are intended to be used to supplement the written SSPC blast cleaning surface preparation specifications. Because the written specifications are the primary means to determine conformance with blast cleaning requirements, the photographs shall not be used as a substitute for these specifications.
Guide to Visual Standard No. 2: Guide to Standard Method of Evaluating Degree of Rusting on Painted Steel Surfaces
This guide describes only the pictorial standard and does not constitute the standard. It is to be used for comparative purposes and is not intended to have a direct relationship to a decision regarding painting requirements.
Guide to SSPC-VIS 3: Visual Standard for Power-and Hand-Tool Cleaned Steel (Standard Reference Photographs)
This guide describes the use of standard reference photographs depicting the appearance of unpainted, painted, and welded hot-rolled carbon steel prior to and after power and hand tool cleaning. These photographs are intended to be used to supplement the written SSPC power and hand tool surface preparation specifications. Because the written specifications are the primary means to determine conformance with cleaning requirements, the photographs shall not be used as a substitute for the written specifications.
Surface Preparation Specification No. 1 (SSPC-SP 1): Solvent Cleaning
This specification covers the requirements for the solvent cleaning of steel surfaces—removal of all detrimental foreign matter such as oil, grease, dirt, soil, salts, drawing and cutting compounds, and other contaminants from steel surfaces by the use of solvents, emulsions, cleaning compounds, steam, or other similar materials and methods that involve a solvent or cleaning action.
Surface Preparation Specification No. 2 (SSPC-SP 2): Hand Tool Cleaning
This specification covers the requirements for the hand tool cleaning of steel surfaces—removal of all rust scale, mill scale, loose rust, and loose paint to the degree specified by hand wire brushing, hand sanding, hand scraping, hand chipping, or other hand impact tools or by a combination of these methods. The substrate should have a faint metallic sheen and also be free of oil, grease, dust, soil, salts, and other contaminants.
Surface Preparation Specification No. 3 (SSPC-SP3): Power Tool Cleaning
This specification covers the requirements for the power tool cleaning of steel surfaces—removal of all rust scale, mill scale, loose paint, and loose rust to the degree specified by power wire brushes, power impact tools, power grinders, power sanders, or by a combination of these methods. The substrate should have a pronounced metallic sheen and also be free of oil, grease, dirt, soil, salts, and other contaminants. Surface should not be buffed or polished smooth.

TABLE 9.8 Reference, Purpose, and Brief Description of Painting Standards and Specifications (*Continued*)

Joint Surface Preparation Standard (SSPC-SP 5/NACE No. 1):
White Metal Blast Cleaning

This standard covers the requirements for white metal blast cleaning of steel surfaces by the use of abrasives—removal of all mill scale, rust, rust scale, paint, or foreign matter by the use of abrasives propelled through nozzles or by centrifugal wheels. A white metal blast cleaned surface finish is defined as a surface with a gray-white, uniform metallic color, slightly roughened to form a suitable anchor pattern for coatings. The surface, when viewed without magnification, shall be free of all oil, grease, dirt, visible mill scale, rust, corrosion products, oxides, paint, and any other foreign matter.

Joint Surface Preparation Standard (SSPC-SP 6/NACE No. 3):
Commercial Blast Cleaning

This standard covers the requirements for commercial blast cleaning of steel surfaces by the use of abrasives—removal of mill scale, rust, rust scale, paint, and foreign matter by the use of abrasives propelled through nozzles or by centrifugal wheels, to the degree specified. A commercial blast cleaned surface finish is defined as one from which all oil, grease, dirt, rust scale, and foreign matter have been completely removed from the surface and all rust, mill scale, and old paint have been completely removed except for slight shadows, streaks, or discolorations caused by rust stain, mill scale oxides, or slight, tight residues of paint or coating that may remain; if the surface is pitted, slight residues of rust or paint may by found in the bottom of pits; at least two-thirds of each square inch of surface area shall be free of all visible residues and the remainder shall be limited to the light discoloration, slight staining, or tight residues mentioned above.

Joint Surface Preparation Standard (SSPC-SP 7/NACE No. 4):
Brush-Off Blast Cleaning

This standard covers the requirements for brush-off blast cleaning of steel surfaces by the use of abrasives—removal of loose mill scale, loose rust, and loose paint, to the degree hereafter specified, by the impact of abrasives propelled through nozzles or by centrifugal wheels. It is not intended that the surface shall be free of all mill scale, rust, and paint. The remaining mill scale, rust, and paint should be tight and the surface should be sufficiently abraded to provide good adhesion and bonding of paint. A brush-off blast cleaned surface finish is defined as one from which all oil, grease, dirt, rust scale, loose mill scale, loose rust, and loose paint or coatings are removed completely, but tight mill scale and tightly adhered rust, paint, and coatings are permitted to remain provided that all mill scale and rust have been exposed to the abrasive blast pattern sufficiently to expose numerous flecks of the underlying metal fairly uniformly distributed over the entire surface.

Surface Preparation Specification No. 8 (SSPC-SP 8): Pickling

This specification covers the requirements for the pickling of steel surfaces—removal of all mill scale, rust, and rust scale by chemical reaction, or by electrolysis, or by both. It is intended that the pickled surface shall be completely free of all scale, rust, and foreign matter. Furthermore, the surface shall be free of unreacted or harmful acid or alkali or smut.

Joint Surface Preparation Standard (SSPC-SP 10/NACE No. 2):
Near-White Blast Cleaning

This standard covers the requirements for near-white metal blast cleaning of steel surfaces by the use of abrasives—removal of nearly all mill scale, rust, rust scale, paint,

TABLE 9.8 Reference, Purpose, and Brief Description of Painting Standards and Specifications (*Continued*)

or foreign matter by the use of abrasives propelled through nozzles or by centrifugal wheels, to the degree hereafter specified. A near-white blast cleaned surface finish is defined as one from which all oil, grease, dirt, mill scale, rust, corrosion products, oxides, paint, and other foreign matter have been completely removed from the surface except for very light shadows, very slight streaks or slight discolorations caused by rust stain, mill scale oxides, or light, tight residues of paint or coating that may remain. At least 95 percent of each square inch of surface area shall be free of all visible residues, and the remainder shall be limited to the light discoloration mentioned above.

Surface Preparation Specification No. 11 (SSPC-SP 11): Power Tool Cleaning to Bare Metal

This specification covers the requirements for the power tool cleaning to produce a bare metal surface and to retain or produce a surface profile. This specification is suitable where a roughened, clean, bare metal surface is required, but where abrasive blasting is not feasible or permissible.

Joint Surface Preparation Standard (SSPC-SP 12/NACE No. 5): Surface Preparation and Cleaning of Steel and Other Hard Materials by High- and Ultrahigh-Pressure Water Jetting Prior to Recoating

This standard provides requirements for the use of high- and ultrahigh-pressure water jetting to achieve various degrees of surface cleanliness. This standard is limited in scope to the use of water only without the addition of solid particles in the stream.

Abrasive Specification No. 1 (SSPC-AB 1): Mineral and Slag Abrasives

This specification defines the requirements for selecting and evaluating mineral and slag abrasives used for blast cleaning steel and other surfaces for painting and other purposes.

Abrasive Specification No. 2 (SSPC-AB 2): Specification for Cleanliness of Recycled Ferrous Metallic Abrasives

This specification covers the requirements for cleanliness of recycled ferrous metallic blast cleaning abrasives used for the removal of coatings, paints, scales, rust, and other foreign matter from steel or other surfaces. Requirements are given for lab and field testing of recycled ferrous metallic abrasives work mix. Recycled ferrous metallic abrasives are intended for use in field or shop abrasive blast cleaning of steel or other surfaces.

Thermal Precleaning (NACE 6G194/SSPC-SP-TR 1): Specifications for Thermal Precleaning

This state-of-the-art report addresses the use of thermal precleaning for tanks, vessels, rail tank cars and hopper cars, and process equipment, when preparing surfaces for the application of high-performance or high-bake coating and lining systems.

Painting System Guide No. 1.00: Guide for Selecting Oil Base Painting Systems

These specifications cover oil base painting systems for steel cleaned with hand or power tools.

TABLE 9.8 Reference, Purpose, and Brief Description of Painting Standards and Specifications (*Continued*)

Painting System Specification No. 1.04: Three-Coat Oil-Alkyd (Lead- and Chromate-Free) Painting System for Galvanized or Non-Galvanized Steel (with Zinc Dust-Zinc Oxide Linseed Oil Primer)

This specification covers an oil-base, lead- and chromate-free painting system for new or weathered (white or red rusted) galvanized steel. It is also effective on nongalvanized steel cleaned with hand or power tools. This system is suitable for use on parts or structures exposed in Environmental Zone 1A (interior, normally dry) and Zone 1B (exterior, normally dry). The finish paint allows for a choice of durable, fade-resistant colors.

Painting System Specification No. 1.09: Three-Coat Oil Base Zinc Oxide Painting System (without Lead or Chromate Pigment)

This specification covers an oil-base, lead- and chromate-free painting system for steel cleaned with hand or power tools. This system is suitable for use on parts or structures exposed in Environmental Zones 1A (interior, normally dry) and 1B (exterior, normally dry). The finish paint allows for a choice of durable, fade-resistant colors.

Painting System Specification No. 1.10: Four-Coat Oil Base Zinc Oxide Paintin System (without Lead or Chromate Pigment)

This specification covers an oil-base, lead- and chromate-free painting system for steel cleaned with hand or power tools. This system is suitable for use on parts or structures exposed in Environmental Zones 1A (interior, normally dry) and 1B (exterior, normally dry). The finish paint allows for a choice of durable, fade-resistant colors.

Painting System Specification No. 1.12: Three-Coat Oil Base Zinc Chromate Painting System

---

This specification covers an oil-base, zinc-chromate painting system for steel cleaned with hand or power tools. This system is suitable for use on parts or structures exposed in Environmental Zones 1A (interior, normally dry) and 1B (exterior, normally dry). The finish paint allows for a choice of durable, fade-resistant colors.

Painting System Specification No. 1.13: One-Coat Oil Base Slow Drying Maintenance Painting System (without Lead or Chromate Pigments)

This specification covers a one-coat oil-base, lead- and chromate-free painting system for steel cleaned with hand or power tools. This system is suitable for use on parts or structures exposed in Environmental Zones 1A (interior, normally dry) and 1B (exterior, normally dry). This system is never used as a shopcoat because of its very long drying time. It is unsuitable for use where the slow drying, slippery paint film would be dangerous to workers when walking or climbing on painted surfaces.

Painting System Specification No. 2.00: Guide for Selecting Alkyd Painting Systems

---

These specifications cover alkyd painting systems for commercial blast cleaned or pickled steel. These systems are suitable for use on parts or structures exposed in Environmental Zones 1A (interior, normally dry) and 1B (exterior, normally dry). The color of the finish paint must be specified.

TABLE 9.8 Reference, Purpose, and Brief Description of Painting Standards and Specifications (*Continued*)

Painting System Specification No. 2.05: Three-Coat Alkyd Painting System for Unrusted Galvanized Steel (for Weather Exposure)
This specification covers an alkyd painting system for new, unrusted, untreated, galvanized steel. This system is suitable for use on parts or structures exposed in Environmental Zones 1A (interior, normally dry) and 1B (exterior, normally dry). The primer has good adhesion to clean galvanized steel but does not adhere properly to rusted galvanized steel. Painting System No. 1.04 should be specified for this condition. The finish paint allows for a choice of durable, fade-resistant colors.
Painting System Specification No. 3.00: Guide for Selecting Phenolic Painting Systems
These specifications cover phenolic painting systems for blast cleaned steel. These systems are suitable for use on parts or structures exposed in Environmental Zones 1A (interior, normally dry), and 1B (exterior, normally dry), and 2A (frequently wet by fresh water). Phenolic paints will normally dry in about 12 h. For optimum intercoat adhesion recoating should take place in less than 24 h. The color of the finish paint must be specified.
Painting System Specification No. 4.00: Guide for Selecting Vinyl Painting Systems
The guide covers vinyl painting system for blast cleaned or pickled steel. These systems are suitable for use on parts or structures exposed in Environmental Zones 1A (interior, normally dry), 2A (frequently wet by fresh water), 2B (frequently wet by salt water), 2C (fresh water immersion), 2D (salt water immersion), 3A (chemical, acidic), and 3B (chemical neutral). The color of the finish paint must be specified.
Painting System Specification No. 9.01: Cold-Applied Asphalt Mastic Painting System with Extra-Thick Film
This specification covers a cold-applied asphalt mastic painting system for above-ground steel structures. This system is suitable for use on parts or structures exposed in Environmental Zones 2A (frequently wet by fresh water), 2B (frequently wet by salt water), 3B (chemical, neutral), and 3C (chemical, alkaline). It should not be used in contact with oils, solvents, or other reagents which tend to soften or attack the coating.
Painting System Specification No. 10.01: Hot-Applied Coal Tar Enamel Painting System
This system is suitable for use on parts or structures exposed in Environmental Zones 2C (fresh water immersion), 3B (chemical, neutral), and 3C (chemical, alkaline). It has good abrasion resistance. It is also suitable for underground use. It must be used with discretion for immersion in corrosive chemicals because the coating is dissolved by some organic solvents and attacked by oxidating solutions. The coal tar enamel must be topcoated with coal tar emulsion when exposed to sunlight to prevent checking and alligatoring.
Painting System Specification No. 10.02: Cold-Applied Coal Tar Mastic Painting System
This specification covers a cold-applied coal tar painting system for underground and underwater steel structures, consisting of two cold-applied coats. This system is suitable for use on parts or structures exposed in Environmental Zones 2C (fresh water

TABLE 9.8 Reference, Purpose, and Brief Description of Painting Standards and Specifications (*Continued*)

immersion), 3B (chemical, neutral), and 3C (chemical, alkaline). It has fairly good abrasion resistance and is suitable for underground use. It must be used with discretion for immersion in corrosive chemicals because the coating is dissolved by some organic solvents and attacked by oxidating solutions. The coal tar mastic must be topcoated with coal tar emulsion when exposed to sunlight to prevent checking and alligatoring.

Painting System Specification No. 11.01: Black (or Dark Red) Coal Tar Epo Polyamide Painting System

This specification covers a complete coal tar epoxy-polyamide black (or dark red) painting system for the protection of steel surfaces that will be exposed to severely corrosive conditions. This system is suitable for use on parts or structures exposed in Environmental Zones 2A (frequently wet by fresh water), 2B (frequently wet by salt water), 2C (fresh water immersion), 2D (salt water immersion), 3A (chemical, acidic), 3B (chemical, neutral), and 3C (chemical, alkaline). Its resistance to chemical fumes, mists, and splashings is generally considered to be good, but its suitability for prolonged immersion in specific chemicals should be confirmed by trial tests in the absence of applicable case histories. It is also suitable for underground exposure and as a protective coating for sound concrete surfaces in marine and some chemical environments. Its good weathering properties can be improved by applying a finish coat of a compatible aluminum pigmented paint. Although it is self-priming and exhibits good adhesion to clean structural steel surfaces, it may also be used over suitable inhibitive primers. The color of paint is black unless red is specified.

Painting System Guide No. 12.00: Guide to Zinc-Rich Coating Systems

This guide provides general information on the description, selection, and applications of zinc-rich coatings and the selection of top coats. Zinc-rich coatings are highly pigmented primer coatings that are uniquely defined by their capability of galvanically protecting steel exposed at discontinuities such as narrow scratches and holidays. Although the major pigment component in a zinc-rich coating is zinc dust, the vehicle may be inorganic or organic. Zinc-rich coatings are classified as follows: Type IA—Inorganic: postcured, water-borne, alkali-silicates; Type IB—Inorganic: self-cured, water-borne, alkali-silicates; Type IC—Inorganic: self-cured, solvent-borne, alkyl-silicates; Type IIA—Organic: thermoplastic binders; Type IIB—Organic: thermoset binders. Certain zinc-rich coating systems are suitable for use in protecting steel surfaces either topcoated or untopcoated. Zinc-rich systems are not suitable for certain exposure conditions.

Painting System Specification No. 12.01: One-Coat Zinc-Rich Painting System

This specification covers a one-coat zinc-rich painting system to be used on steel in mild-to-moderately severe environments. This system is suitable for use on parts or structures exposed in Environmental Zone 3B (chemical, neutral). It is not recommended for environments where corrosive contaminants will have a pH below 5 or above 9 or in severely corrosive environments. The system is recommended as a durable shop primer or as a protective one-coat system for normal atmospheric weathering environments and certain immersion services. This specification does not pertain to weldable prefabrication zinc-rich primers that are applied at lower thicknesses [1 mil (25 μm) or less]. Further information regarding these and other zinc-rich primers can be found in SSPC-PS Guide 12.00, Guide for Selecting Zinc-Rich Painting Systems.

TABLE 9.8 Reference, Purpose, and Brief Description of Painting Standards and Specifications (*Continued*)

Painting System Specification No. 4.02: Four-Coat Vinyl Painting System (for Fresh Water, Chemical, and Corrosive Atmospheres)

This specification covers a complete vinyl painting system for structural steel. This system is suitable for use on parts or structures exposed in Environmental Zones 2C (fresh water immersion), 3A (chemical exposure, acidic), and 3B (chemical exposure, neutral). The finish paint allows for choice of colors.

Painting System Specification No. 4.04: Four-Coat White or Colored Vinyl Painting System (for Fresh Water, Chemical, and Corrosive Atmospheres)

This specification covers a complete vinyl painting system for structural steel. This system is suitable for use on parts or structures exposed in Environmental Zones 2B (frequently wet by salt water), 2C (fresh water immersion), 3A (chemical exposure, acidic), and 3B (chemical exposure, neutral). The finish paint allows for choice of colors.

Painting System Specification No. 7.00: Guide for Selecting One-Coat Shop Painting Systems

This guide covers one-coat shop painting systems for steel that will not be exposed to corrosive conditions for long periods. They are also suitable for steel encased in concrete in those cases where bonding of steel to concrete is not required. They can be used under fireproofing. These systems are suitable for use on parts or structures exposed in Environmental Zones 0 (encased in concrete or masonry, normally dry) and 1A (interior, normally dry). The paints covered by this guide are primers, and if a color other than the standard color is required, the color must be specified.

Painting System Specification No. 8.00: Guide to Topcoating Zinc-Rich Primers

This guide covers the selection and application (including surface preparation) of top coats to surfaces coated with a zinc-rich primer. Both organic and inorganic zinc-rich primers are included. The guide does not cover the selection and application of the zinc-rich primer.

Painting System Specification No. 13.01: Epoxy-Polyamide Painting System

This specification outlines a three-coat epoxy-polyamide painting system for the protection of steel surfaces subject to industrial exposure, marine environments, and areas subject to chemical exposure such as acid and alkali. This system, when properly applied and cured, is capable of giving excellent protection to steel surfaces in Environmental Zones 2A (frequently wet by fresh water), 2B (frequently wet by salt water), 3A (chemical, acidic), 3B (chemical, neutral), and 3C (chemical, alkaline) but not in potable water tanks. Although the coating herein specified has exhibited good chemical protection, its resistance against specific chemicals should, in the absence of applicable case histories, be appropriately tested.

Painting System Specification No. 14.01: Steel Joist Shop Painting System

This specification covers a one-coat shop joist primer that will provide temporary protection to the steel joists during delivery and erection. This system is intended as a one-coat shop paint for open web and long-span steel joists that may be either enclosed or exposed in the interiors of buildings (Environmental Zone 1A, interior, normally dry) where the temperature rarely falls below the dew point, the humidity rarely exceeds 85%, and corrosive protection is not necessary.

TABLE 9.8 Reference, Purpose, and Brief Description of Painting Standards and Specifications (*Continued*)

Painting System Specification No. 15.00: Guide for Selecting Chlorinated Rubber Painting Systems

These specifications cover chlorinated rubber painting systems for blast cleaned or pickled steel. These coatings are not recommended for areas exposed to strong organic solvents, oxidating acids, or the areas where the surface temperature exceeds 74°C). Straight chain unsaturated acids and fats and oils of animal or vegetable origin will cause softening and swelling of these coatings. These systems are suitable for use on parts or structures exposed in Environmental Zones 1A (interior, normally dry), 1B (exterior, normally dry), 2A (frequently wet by fresh water), 2B (frequently wet by salt water), 2C (fresh water immersion), 2D (salt water immersion), 3A (chemical, acidic), 3B (chemical, neutral), and 3C (chemical, alkaline). Chlorinated rubber paints are single-package systems that dry to solvent evaporation and have low permeability to water vapor and oxygen. After drying, they are nonflammable and resistant to mildew growth. The color of the finish must be specified.

Painting System Specification No. 15.01: Chlorinated Rubber Painting System for Salt Water Immersion

This specification covers a complete chlorinated rubber painting system for structural steel. This system is suitable for use on parts or structures exposed in Environmental Zones 2B (frequently wet by salt water) and 2D (salt water immersion). The finish paint allows for a choice of colors.

Painting System Specification No. 15.02: Chlorinated Rubber Painting System for Fresh Water Immersion

This specification covers a complete chlorinated rubber painting system for structural steel. This system is suitable for use on parts or structures exposed in Environmental Zones 2A (frequently wet by fresh water) and 2C (fresh water immersion). It may also be used in nonsolvent chemical atmospheres. The finish paint allows for a choice of colors.

Painting System Specification No. 15.03: Chlorinated Rubber Painting System for Marine and Industrial Atmospheres

This specification covers a complete chlorinated rubber painting system for structural steel. This system is suitable for use on parts or structures exposed in Environmental Zones 1A (interior, normally dry), 1B (exterior, normally dry), 2A (frequently wet by fresh water), 2B (frequently wet by salt water), 3A (chemical, acidic), 3B (chemical, neutral), and 3C (chemical, alkaline). The finish paint allows for a choice of colors.

Painting System Specification No. 15.04: Chlorinated Rubber Painting System for Field Application over a Shop Applied Solvent Base Inorganic Zinc-Rich Primer

This specification covers a field-applied chlorinated rubber painting system for structural steel shop-primed with a solvent base inorganic zinc-rich primer. This system is suitable for use on parts or structures exposed in Environmental Zones 1A (interior, normally dry), 1B (exterior, normally dry), 2A (frequently wet by fresh water), 2B (frequently wet by salt water), 3A (chemical exposure, acidic), 3B (chemical exposure, neutral), and 3C (chemical exposure, alkaline). The finish paint allows for a choice of colors.

TABLE 9.8 Reference, Purpose, and Brief Description of Painting Standards and Specifications (*Continued*)

Painting System Specification No. 16.01: Silicone Alkyd Painting System for New Steel
This specification covers a complete silicone alkyd painting system for structural steel. This system is suitable for use on parts or structures exposed in Environmental Zone 2A (frequently wet by fresh water), including high humidity, infrequent immersion, and mild chemical atmospheres. The primary virtue of this system is the exterior durability and minimum deterioration of the silicone alkyd finish as shown by chalk resistance, gloss retention, and color retention. In addition, the finish paint allows for a choice of colors.
Painting System Guide No. 17.00: Guide for Selecting Urethane Painting Systems
This guide outlines urethane painting systems for structural steel surfaces. There are three types of urethane coatings covered by the guide. They are Types II, IV, and V, as classified by ASTM Standard D 16. These painting systems are suitable for use on parts or structures exposed in varied types of environments ranging from severely corrosive environments to mild atmospheric conditions. These painting systems are intended principally for structural steel where excellent weathering, color retention, and chemical resistance is desired. The color of the finish must be specified.
Painting System Specification No. 18.01: Three-Coat Latex Painting System
This specification covers a complete latex painting system for structural steel. This system is suitable for use on parts or structures exposed in Environmental Zones 1A (interior, normally dry) and 1B (exterior, normally dry) and high-humidity or mild chemical atmospheres. The finish paint is semigloss and chalk resistant and allows for a choice of colors.
Painting System Guide No. 19.00: Guide for Selecting Painting Systems for Ship Bottoms
This guide covers painting systems for ship bottoms from the keel to the light load line on steel ships. The area from the light load line to the deep load line, more commonly called the boot-top area, may also be coated with these systems; however, SSPC-PS Guide 20.00 covers painting systems for this area. It should be noted that boot tops are rarely used with today's commercial ships, and bottom systems may extend up to the deep load line. These coating systems may also be used for other floating or stationary structures exposed to or submerged in salt or brackish water. This would include barges, buoys, oceanographic installations, and so forth.
Painting System Guide No. 20.00: Guide for Selecting Painting Systems for Boottoppings
This guide covers painting systems for the protection of the exterior boot-top areas (the area from the light load line to the deep load line) of steel ships. It should be noted that boottops are rarely used with today's commercial ships, and bottom systems may extend up to the deep load line. In general, the anticorrosive and antifouling paints covered in SSPC-PS Guide 19.00 are applicable to boot-top areas.
Painting System Guide No. 21.00: Guide for Selecting Painting Systems for Topsides
This guide covers painting systems for the protection of the topside or exterior area of steel ships. This includes the area from the deep load line to the rail, more commonly

TABLE 9.8 Reference, Purpose, and Brief Description of Painting Standards and Specifications (*Continued*)

called the freeboard, decks, and superstructure. These systems can also be used for above-water parts of floating structures exposed to salt or fresh water and the normal marine environment. They also cover all above-water areas on ships such as deck equipment or machinery, booms, mast, and bulwarks.

Painting System Guide No. 22.00: Guide for Selecting One-Coat Preconstruction or Prefabrication Painting Systems

This guide covers those shop primers used in today's modern commercial shipyards for preconstruction and prefabrication priming of abrasive blast cleaned structural steel and steel plates. To maximize efficiency in new construction, all ships' steel plates, shapes, and angles are abrasive blast cleaned, shop primed, and stored for future use in preparation of sections of ships, called modules or units. Shop primers are covered by generic classification.

Coating System Guide No. 23.00: Guide for Thermal Spray Metallic Coating Systems

This guide covers the requirements for thermal spray metallic coatings, with and without sealers and topcoats, as a means to prevent corrosion of steel surfaces. Types of metallic coatings included are pure zinc, pure aluminum, and zinc/aluminum alloy, 85% zinc/15% aluminum by weight. This system is suitable for use on structures or parts thereof exposed in SSPC Environmental Zones 1A (interior, normally dry), 1B (exterior, normally dry), 2A (frequently wet by fresh water), and 2C (fresh water immersion). It may be used in Environmental Zones 2B (frequently wet by salt water), 2D (salt water immersion), 3A (chemical exposure, acidic), 3B (chemical exposure, neutral), and 3C (chemical exposure, alkaline) with proper sealing/topcoating (see Section 6 and Note 11.2 of the Guide. This document is intended to serve as a guide for preparing specifications for thermal spray applications.

Painting System Specification No. 24.00: Latex Painting System for Industrial and Marine Atmospheres, Performance-Based

This specification covers a painting system for steel surfaces based on multiple coats of air-drying, single-component latex paints having a total dry film thickness of a minimum of 6 mil (152 µm). The painting system is categorized according to its performance level over blast-cleaned steel, the intended substrate. The painting system is also categorized according to the volatile organic compound (VOC) classes of the latex paints comprising it. The system is suitable for exposure in Environmental Zones 1A (interior, normally dry), 1B (exterior, normally dry), 2A (frequently wet by fresh water), 2B (frequently wet by salt water), 3A (chemical exposure, acidic), 3B (chemical exposure, neutral), 3C (chemical exposure, alkaline), and 3D (chemical exposure, mild solvent). The system is not intended for immersion service.

Paint Specification No.5: Zinc Dust, Zinc Oxide, and Phenolic Varnish Paint

This specification covers a quick-drying zinc dust, zinc oxide, and phenolic varnish paint for steel or galvanized surfaces. It has very good rust inhibitive characteristics but only fair wetting ability for rusting, greasy, or oily surfaces. It has a drying time of about 12 h, good durability even when weathered before finish coating, and may be used for intermediate and finish coats. This paint is supplied in two-package kits: one package contains the liquid vehicle (component A) and the other contains the zinc dust zinc oxide pigment (component B). This paint is suitable for exposure in Environmental Zones 1A (interior, normally dry) and 1B (exterior, normally dry) and is particularly

TABLE 9.8 Reference, Purpose, and Brief Description of Painting Standards and Specifications (*Continued*)

suited for exposure in Environmental Zone 2A (frequently wet by fresh water). It is intended for brush or spray application over steel surfaces prepared in accordance with SSPC-SP 6, Commercial Blast Cleaning; SSPC-SP 10, Near-White Blast Cleaning; SSPC-SP 5, White Metal Blast Cleaning; and SSPC-SP 8, Pickling; or over clean galvanized steel. This paint is suitable as a shop primer, field primer, maintenance primer, or intermediate coat and is to be applied in accordance with SSPC-PA 1, Shop, Field, and Maintenance Painting. This paint will dry in about 12 h and should be recoated within 24 h for optimum intercoat adhesion.

Surface Preparation Specification No. 4 (SSPC-SP 4): Flame Cleaning

Removal of all loose scale, rust, and other detrimental foreign matter by passing high-temperature, high-velocity oxy-acetylene flames over the entire surface, followed by wire brushing. Surface should also be free of oil, grease, dirt, soil, salts, and other contaminants.

Primers. The primer is a universal component of all anticorrosive coatings and is considered to be one of the most important element of a protective system. A good primer generally provides the ability to stifle or retard the spread of corrosion discontinuities such as pinholes, holidays, or breaks in the film. To perform satisfactorily they must themselves adhere well to the base metal or any surface conversion coating that might be present. They should also contain an adequate concentration of a leachable inhibitor, where this is considered an important feature of the protection system, and this is usually a chromate pigment. The primary functions of a primer are

- Adhesion or strong bond to the substrate
- Cohesion or internal strength
- Inertness to the environment
- High bond to intermediate coat
- Appropriate flexibility

Intermediate or body coats. Intermediate or body coats are usually used in coating systems designed for specific applications where coating thickness and structure are advantageous. Intermediate coats with red lead and inert pigments have been giving coatings a heavy body that is impervious to the most demanding applications. The primary purposes of an intermediate coat are to provide

- Thickness to a protective coating
- Strong chemical resistance
- Resistance to moisture vapor transfer

- Increased coating electrical resistance
- Strong cohesion
- Strong bond to primer and topcoat

Topcoats. In the coating system approach, the topcoat provides a resinous seal over the intermediate coats and the primer. The first topcoat may in fact penetrate into the intermediate coat, thus providing the coating system with an impervious top surface. The topcoat is the first line of defense of many coatings against aggressive chemicals, water, or the environment. It is generally more dense than intermediate coats because topcoats are formulated with a lower pigment-to-vehicle (solvent) ratio. The topcoats commonly used include air-drying paints and oil-based varnishes which harden by oxidation; acrylics and other lacquers, which dry by solvent evaporation; and polyurethane and epoxy paints, which dry by cold curing chemical reactions. High-temperature curing or stoving can also be used with certain types of epoxy to produce a harder finish, but this also makes them more difficult to remove.

Polyurethane paints have been widely used in marine applications worldwide. However, these paints are quite brittle and tend to chip and crack, and for these reasons many prefer solvent-drying acrylic paints for the exterior finish. These paints can be removed locally by chemical solvents down to the primer and are reported to be easier to touch up. Whichever paint system is selected for exterior use, it is usual to qualify the system on the basis of its ability to prevent filiform corrosion. Topcoats main functions are to provide

- A resistant seal for the coating system
- An initial barrier to the environment
- Resistance to chemicals, water, and weather
- Toughness and wear resistance to the surface
- A pleasant appearance

However, there are a number of situations where the intermediate coats provide the primary barrier to the environment, and the finish coat is applied for entirely different purposes. The topcoat can be used, for example, to provide a nonskid surface, and the intermediate coat and the primer provide the barrier to the environment, as in a marine environment.

Basic coating components

Binders. To perform in a practical environment, a coating must convert, after its application, into a dense, solid, and adherent membrane that has all the properties discussed previously. The binder is the material that makes this possible. It provides uniformity and coherence to the coating system. Not all binders are corrosion resistant, so only a few serve in the formulation of protective coatings. The binder's ability to form a dense, tight film is directly related to its molecular size and complexity. Binders that have the highest molecular weight will form films by the evaporation of the vehicle, whereas binders with smaller molecular weight will generally be reacted in situ. Binders can be classified according to their essential chemical reactions.

Oxygen-reactive binders. Oxygen-reactive binders are generally low molecular weight resins that are only capable of producing coatings through an intermolecular reaction with oxygen. This reaction is often catalyzed by metallic salts of cobalt or lead. Examples are

- *Alkyds.* Alkyds are produced by chemically reacting natural drying oils to form a synthetic resin with good film curability, chemical resistance, and weather resistance.
- *Epoxy esters.* Epoxy resins react chemically with drying oils to form epoxy esters. The drying oils' part of the molecule determines the basic properties of the epoxy ester coatings. The coating dries by oxidation in the same manner as an alkyd.
- *Urethane alkyds.* Epoxy resins are also chemically combined with drying oils as part of the molecule that further reacts with isocyanates to produce urethane alkyds. Upon application as a liquid coating, the resin-oil combination converts by oxidation to a solid.
- *Silicone alkyds.* Alkyd resins are combined with silicone molecules to form an excellent weather-resistant combination known as silicone alkyds.

Lacquers. Lacquers are coatings that are converted from a liquid material to a solid film by the evaporation of solvents alone. Lacquers have generally a low volume of solids. Examples are

- *Polyvinyl chloride polymers.* This principal corrosion-resistant lacquer is made from polyvinyl chloride copolymers. The vinyl molecule is relatively large and will effectively dissolve in solvent in the 20% range.
- *Chlorinated rubbers.* To be effective, chlorinated rubbers have to be modified by other resistant resins to obtain higher solids, decreased brittleness, and increased adhesion.
- *Acrylics.* Acrylics are also of high molecular weight and may be combined with vinyls to improve exterior weatherability and color retention.

- *Bituminous materials.* Bituminous asphalts and coal tars are often combined with solvents to form lacquer-type films. They can provide good corrosion resistance but can only be applied where appearance is not a factor.

  Heat-conversion binders. Examples of heat-conversion binders are

- *Hot melts.* Hot melts normally involve asphalt or coal tar and are melted and applied as 100% solids in the hot-liquid condition.
- *Organisols and plastisols.* These are high molecular resins (organisols) or vinyl materials (plastisols) that are dispersed in a solvent or plasticizer to solvate them into a filming material upon heating.
- *Powder coatings.* Powder coatings are high molecular weight thermoplastic resins or semithermoset resins applied to a substrate as a very fine powder that is melted to form a coating. Powder coatings can be applied by using an electrostatic field with the coating and substrate charged with opposing polarities. Such an application method is very efficient because the coated section of a substrate becomes insulating, therefore making the uncoated section the only one electrostatically attractive to the powder being applied.

  Coreactive binders. Coreactive binders are formed from two low molecular weight resins that are combined prior to the application to the substrate, where they react to form a very adherent and solid film. Examples are

- *Epoxies.* Epoxy binders are made of relatively low molecular weight resins in which the epoxy group is at the end of each molecule. The epoxy resins are then reacted with amines of various molecular weight and cured to form high molecular weight binders with good solvent and chemical resistance.
- *Polyurethanes.* Polyurethanes are coreactive binders in which low molecular weight resins containing alcohol or amine groups are reacted with di-isocynates into an intermediate resin prepolymer that is then capable of reacting with other groups containing amines, alcohol, or even water.

  Condensation binders. Condensation binders are based primarily on resins that interact to form cross-linked polymers when subject to sufficient thermal energy. These binders are also called high-baked materials and are commonly used as tank and pipe linings. Condensation is essentially the release of water during the polymerization process.

  Coalescent binders. Coalescent binders are coatings where binders of various resin types are emulsified to form a liquid binder. They are primarily emulsified with water or less commonly with some other solvent dispersions. When applied to the surface, the medium evaporates, leaving the coating in such a way that the binder resin gradually flows into itself, or coalesces, to form a continuous film.

Inorganic binders. Inorganic binders are mostly inorganic silicates dissolved in water or solvent that react with moisture in the air after their application to a surface. The type of inorganic binder depends on the form of the silicate during the curing period. Examples are

- *Postcured silicates.* Soluble silicates are combined with zinc dust to form very hard rocklike films that are further stabilized by reacting them with an acidic curing agent.
- *Self-curing water silicates.* In this case, the soluble alkali silicates are combined with colloidal silica to improve the curing speed. Once applied to a surface they develop water insolubility by reacting with carbon dioxide and moisture from the air.
- *Self-curing solvent-based silicates.* These binders are organic esters of silica that are converted from a liquid form to a solid by reaction with moisture from the air, forming a very hard and corrosion-resistant binder. A major advantage of these materials is their conversion to rain- or moisture-resistant form shortly after their application.

Pigments. Pigments are essentially dry powders that are insoluble in the paint medium and that consequently need to be mixed in it by a dispersion technique. They range from naturally occurring minerals to artificial organic compounds. Pigments contribute several properties essential to the effective use of protective coatings. Several different pigments may be used within the same coating, all of them contributing to the coating's general characteristics to perform important functions such as providing

- Color
- Protection to resin binder
- Corrosion inhibition
- Corrosion resistance
- Film reinforcement
- Nonskid properties
- Sag control
- Increased coverage
- Hide and gloss control
- Adhesion

Zinc phosphates are now probably the most important pigments in anticorrosive paints. The selection of the correct binder for use with these pigments is very important and can dramatically affect their performance. Red lead is likely to accelerate the corrosion of nonferrous metals, but calcium plumbate is unique in providing adhesion to newly galvanized surfaces in the absence of any pretreatment and is claimed to behave similarly on other metals.

Solvents. Most coatings are made with multiple solvents and rarely with a single solvent. The choice of solvents influences viscosity, flow properties, drying speed, spraying and brushing characteristics, and gloss. There is no universal solvent for protective coatings, the best solvent in one system being often impractical for another. Asphalts, for example, can be readily dissolved by hydrocarbons but are insoluble in alcohols. One of the most serious problems associated with coatings is the wrong choice of solvent because it can severely affect the curing and adhesion characteristics of the final coating. One convenient way to describe solvents is to regroup them into the following categories:

- *Aliphatic hydrocarbons.* Aliphatic hydrocarbons or paraffins such as naphtha or mineral spirits are typically used with asphalt-, oil-, and vinyl-based coatings.
- *Aromatic hydrocarbons.* Aromatic hydrocarbons, such as toluene, xylene, or some of the higher-boiling homologs, are typically used with chlorinated rubbers, coal tars, and certain alkyds.
- *Ketones.* Ketones such as acetone, methyl ethyl ketone, methyl isobutyl or amyl ketone, and many others, are very effectively used with vinyls, some epoxies, and other resin formulations.
- *Esters.* Esters such as ethyl, n-propyl, n-butyl, and amyl acetates are used commonly as latent solvents (a type of solvent that just swells the binder at room temperature) with epoxy and polyurethane formulations.
- *Alcohols.* Alcohols such as methyl, propyl, iso-propyl or butyl alcohols, and cyclo-hexanol are good solvents for highly polar binders such as phenolics. Some alcohols are used in connection with epoxies.
- *Ethers and alcohol ethers.* Ethers such as ethyl ether are excellent solvents for some of the natural resins, oils, and fats. The usual forms of ether used in protective coatings are alcohol ethers such as ethylene glycol mono methyl ether, known commonly as cellosolve. Cellosolve is a good solvent for many oils, gums, natural resins, and synthetic resins such as alkyds, ethyl-cellulose, nitro-cellulose, polyvinyl acctate, polyvinyl butyryl, and phenolics. Cellosolve is a slow solvent that is used in many lacquers to improve flow-out and gloss.

- *Water.* The recent regulations to reduce the emission of volatile organic compounds (VOCs) produced by organic solvents are forcing the coating industry to reconsider the applicability of water as a solvent. The most common water-borne coatings used for application to metals are air dried or force dried at temperatures below 90°C. A wide range of coating formulations falls into this category. The most commonly available technologies are water-reducible alkyds and modified alkyds, acrylic latexes, and acrylic epoxy hybrids.

Nonstick coatings. Nonstick coatings for industrial, architectural, automotive, and marine use are widespread. Hull coatings that resist the formation of strong bonds to marine organisms keep ships free from marine growth without needing heavy metal toxins that accumulate in the environment. Maintaining sanitation in health-care facilities and food processing plants is eased by surfaces that resist microbial attachment.

Investigations of nonstick surfaces have usually focused on the surface and overlooked the adherent. The free energy of a surface or its critical surface tension has long been believed to be the dominant factor in adhesion. Surface free energy is the excess energy of the groups, atoms, or molecules on the surface compared with their counterparts in the bulk material. The size of the free energy represents the capability of the surface to interact spontaneously with other materials. Organic polymers possess surface free energies typically between 11 and 80 mJ·m^{-2}. Many commercial polymers with surface energies at the lower end of this range (Table 9.9) have been studied in the search for nonstick coatings.

Hydrocarbons such as polyethylene and polypropylene are readily available and inexpensive, but they are not sufficiently soluble to cast as films. Unsubstituted hydrocarbons are easily oxidized and their nonstick properties rapidly deteriorate in exterior usage. Halogenated

TABLE 9.9 Surface Free Energy of Some Polymers

Polymer	Surface energy, mJ·m^{-2}
Polyethylene	34
Polychlorotrifluoroethylene	31
Polypropylene	30
Polyvinyl fluoride	28
Polyethylene-co-tetrafluoroethylene	27
Polyvinylidene fluoride	25
Polydimethylsiloxane	22
Polytetrafluoroethylene	19
Polytetrafluoroethylene-co-hexafluoropropylene	18
Poly[3,3,3-trifluoropropylmethylsiloxane]	18
Polyethylene-co-chlorotrifluoroethylene	15

polymers containing chlorine and especially fluorine have also received a great deal of attention. There are at least half a dozen commodity homopolymers and copolymers containing vinyl fluoride, vinylidene fluoride, tetra-fluoroethylene, hexa-fluoropropylene or chlorotrifluoroethylene. All of these have surface energies between 15 and 31 mJ·m^{-2} and show excellent resistance to chemicals.[9]

Fluorinated precursors are now commercially available to help overcome the obstacles of solubility and adhesion to the substrate. These oligomers have molecular weights of 2000 to 7000 and contain fluorine to impart low surface energy and hydroxyl groups to confer reactivity and adhesion. They are formulated with polyisocyanates to produce urethane coatings. The dominant fluorinated polyols used today are copolymers of chlorotrifluoroethylene and assorted nonhalogenated vinyl ethers. The latter are functionalized to provide reactivity, adhesion, and solubility, and their structure and proportions vary widely. The materials are known generically as *fluorinated ethylene vinyl ether* (FEVE) resins.[10]

Fluorinated polyols derived from hexafluoroacetone (HFA) are also produced for surface coatings. The surface energy of these polyols is close to that of poly(tetrafluoroethylene) (PTFE), and PTFE can be dispersed in the resin as any conventional paint pigment. Flakes of PTFE overlap in the dry film, improving the barrier properties of the coating. These coatings are used as interior linings in large fuel storage tanks and have been used as anticorrosion coatings for ship bilges and tanks and as nontoxic fouling release coatings on small boats.

Industrial and marine coatings containing either FEVE- or HFA-based fluoropolyurethanes are applied as a topcoat over a urethane or epoxy primer. When topcoating is done before the basecoat has fully cured, chemical reaction between the two coats takes place and ensures good adhesion and durability. Fluorinated groups preferentially migrate to the upper surface, where they demonstrate their nonstick behavior.

However, some drawbacks to fluorocarbon surfaces persist. For example, pure PTFE is quite porous and accumulates marine fouling rapidly, in spite of its low surface free energy, because marine organisms inject their adhesive and achieve a mechanical interlock. In addition, fluorine atoms impart stiffness to fluoropolymer chains by raising the barrier to rotation about the backbone bond. In addition, fluorourethane coatings are highly cross-linked thermosets with little or no significant molecular mobility.[9]

9.3 Supplementary Protection Systems

Supplementary protection is provided to surfaces that already have some form of permanent or semipermanent protection such as cladding or conversion coating. The supplementary protection may be in the form of a material that can be easily applied and removed and that will be replaced periodically during the life of the system. Jointing compounds and sealants are examples of this type.

9.3.1 Jointing compounds and sealants

Jointing compounds are used for protection at joints where they act by excluding dirt and moisture and by providing a reservoir of soluble passivators that act as inhibitors. Sealants are applied to joints to prevent the escape of fluids, such as fuel, but they also exclude moisture. Jointing compounds must remain flexible to allow easy disassembly of parts. Various synthetic resins are used for this purpose. The compounds harden sufficiently at edges to take paint, but they remain tacky within the joint so that flexure does not cause cracking. Sealants of the type now being specified are also elastomeric, and the most popular are polysulphide sealants containing corrosion inhibitors. The inhibitive sealants are very effective when used in faying surfaces and butt joints, for wet installation of fasteners, and over fastener patterns. They are also effective in insulating dissimilar metals.

9.3.2 Water-displacing compounds

Water-displacing compounds may be useful in providing supplementary protection for paint systems that have deteriorated or become damaged in service. They are applied as fluids by wiping, brushing, spraying, or dipping, and they are usually immiscible with water and displace water from surfaces and crevices. A number of fluids used are based on lanolin and contain various solvents and inhibitors. The evaporation of the solvents leaves either thin soft films, semihard films, or hard resin films that provide varying degrees of protection. Some of these fluids may be used to provide short-term protection. They should then exhibit excellent water-displacing characteristics and leave a thin oily film, providing short-term corrosion protection.

Two typical water displacement products used in North America by aircraft maintainers are AML-350 and AMLGUARD. AML350 is a petroleum sulfonate in a mineral spirit solvent. When applied to a metal surface, it spreads over the surface and under water droplets, and as the solvent evaporates, it leaves a soft oillike film of sulfonate, which isolates the metal from the environment and acts as a corrosion inhibitor. The film is built up to a thickness of 2 to 5 μm.

AMLGUARD is a water-displacing compound containing solvents, silicone and silicone alkyd resins, barium petroleum sulfonate, and several other additives. It dries to the touch in about 18 h, but continues to cure for 1 to 3 months to form a hard, dry, but flexible finish between 25 and 50 μm thick. It not only displaces water, but it also leaves a protective barrier coating containing barium petroleum sulfonate and alkyl ammonium organic phosphate as inhibitors. AML350 is intended for use on internal metallic parts and electrical connectors. AMLGUARD is intended for temporary use on external aircraft parts, such as wheels, wheel wells, cables, landing gear parts, wing leading edges, and helicopter blades.

9.4 Surface Preparation

It is well recognized that you can make a poor coating perform with excellent pretreatment, but you cannot make an excellent coating perform with poor pretreatment. Surface pretreatment by chemical or mechanical means is also important in painting, and the methods used are designed to ensure good adhesion of the paint to the alloy surface. Surface engineering for increased material performance is one important element in the world of metal finishing. Refer to Table 9.8 for the main specifications concerning surface preperation. Most metal surface treatment and plating operations have three basic steps:

1. Surface cleaning or preparation, which involves the use of solvents, alkaline cleaners, acid cleaners, abrasive materials, and/or water
2. Surface modification, which involves some change in surface properties, such as application of a metal layer or hardening
3. Rinsing or other workpiece finishing operations to produce the final product

References

1. AAA Galvanizing Inc., What Is Galvanizing? http://www.aaagalvanizing.com/gal.htm, 1998.
2. Mevrel, R., Duret, C., and Pichoir, R., Pack Cementation Processes, *Materials Science and Technology,* **2:**201 (1986).
3. Restall, J. E., and Hayman, C., Coatings for Heat Engines, Clarke, R. L. (ed.), pp. 347-357. 1984. Washington, D.C., U.S. Dept. of Energy.
4. Rose, B. R., Simultaneous Internal and External Coating of a First Stage Turbine Bucket with a Chromium Reinforced Aluminide, *Eighth National Research Council of Canada Symposium on Industrial Application of Gas Turbines,* 1989, Ottawa, Canada, National Research Council of Canada.
5. Tsai, EC-E, and Nixon, R., Simple Techniques for Source Reduction of Wastes from Metal Plating Operations, *Hazardous Waste & Hazardous Materials,* **6**(1):67-78 (1989).
6. Jeanmenne, R. A., EN for Hard Chromium, *Products Finishing,* **54**(4):84-93 (1990).
7. Graves, B., *Industrial Toxics Project: The 33/50, Products Finishing,* **56**(9):132-135 (1992).
8. Munger, C. G., *Corrosion Prevention by Protective Coatings,* Houston, Tex., NACE International, 1984.
9. Brady, Jr. R. F., In Search of Non-Stick Coatings, *Chemistry in Britain,* 219–222 (1997).
10. Munekata, S., Flouropolymers as Coating Material, *Progress in Organic Coatings,* **16:**113–134 (1988).

Chapter 10

Corrosion Inhibitors

10.1	Introduction	833
10.2	Classification of Inhibitors	834
	10.2.1 Passivating (anodic)	836
	10.2.2 Cathodic	837
	10.2.3 Organic	837
	10.2.4 Precipitation inhibitors	837
	10.2.5 Volatile corrosion inhibitors	838
10.3	Corrosion Inhibition Mechanism	838
	10.3.1 Inhibitors for acid solutions	839
	10.3.2 Inhibitors in near-neutral solutions	845
	10.3.3 Inhibitors for oil and gas systems	851
	10.3.4 Atmospheric and gaseous corrosion	857
10.4	Selection of an Inhibitor System	860
References		861

10.1 Introduction

The use of chemical inhibitors to decrease the rate of corrosion processes is quite varied. In the oil extraction and processing industries, inhibitors have always been considered to be the first line of defense against corrosion. A great number of scientific studies have been devoted to the subject of corrosion inhibitors. However, most of what is known has grown from trial and error experiments, both in the laboratories and in the field. Rules, equations, and theories to guide inhibitor development or use are very limited.

By definition, a corrosion inhibitor is a chemical substance that, when added in small concentration to an environment, effectively decreases the corrosion rate. The efficiency of an inhibitor can be expressed by a measure of this improvement:

$$\text{Inhibitor efficiency (\%)} = 100\, \frac{(CR_{uninhibited} - CR_{inhibited})}{CR_{uninhibited}} \qquad (10.1)$$

where $CR_{uninhibited}$ = corrosion rate of the uninhibited system
       $CR_{inhibited}$ = corrosion rate of the inhibited system

In general, the efficiency of an inhibitor increases with an increase in inhibitor concentration (e.g., a typically good inhibitor would give 95% inhibition at a concentration of 0.008% and 90% at a concentration of 0.004%). A synergism, or cooperation, is often present between different inhibitors and the environment being controlled, and mixtures are the usual choice in commercial formulations. The scientific and technical corrosion literature has descriptions and lists of numerous chemical compounds that exhibit inhibitive properties. Of these, only very few are actually used in practice. This is partly because the desirable properties of an inhibitor usually extend beyond those simply related to metal protection. Considerations of cost, toxicity, availability, and environmental friendliness are of considerable importance.

Table 10.1 presents some inhibitors that have been used with success in typical corrosive environments to protect the metallic elements of industrial systems. Commercial inhibitors are available under various trade names and labels that usually provide little or no information about their chemical composition. It is sometimes very difficult to distinguish between products from different sources because they may contain the same basic anticorrosion agent. Commercial formulations generally consist of one or more inhibitor compounds with other additives such as surfactants, film enhancers, de-emulsifiers, oxygen scavengers, and so forth. The inhibitor solvent package used can be critical in respect to the solubility/dispersibility characteristics and hence the application and performance of the products.

10.2 Classification of Inhibitors

Inhibitors are chemicals that react with a metallic surface, or the environment this surface is exposed to, giving the surface a certain level of protection. Inhibitors often work by adsorbing themselves on the metallic surface, protecting the metallic surface by forming a film. Inhibitors are normally distributed from a solution or dispersion. Some are included in a protective coating formulation. Inhibitors slow corrosion processes by

- Increasing the anodic or cathodic polarization behavior (Tafel slopes)
- Reducing the movement or diffusion of ions to the metallic surface
- Increasing the electrical resistance of the metallic surface

TABLE 10.1 Some Corrosive Systems and the Inhibitors Used to Protect Them

System	Inhibitor	Metals	Concentration
Acids			
HCl	Ethylaniline	Fe	0.5%
	MBT*	..	1%
	Pyridine + phenylhydrazine	..	0.5% + 0.5%
	Rosin amine + ethylene oxide	..	0.2%
H_2SO_4	Phenylacridine	..	0.5%
H_3PO_4	NaI	..	200 ppm
Others	Thiourea	..	1%
	Sulfonated castor oil	..	0.5–1.0%
	As_2O_3	..	0.5%
	Na_3AsO_4	..	0.5%
Water			
Potable	$Ca(HCO_3)_2$	Steel, cast iron	10 ppm
	Polyphosphate	Fe, Zn, Cu, Al	5–10 ppm
	$Ca(OH)_2$	Fe, Zn, Cu	10 ppm
	Na_2SiO_3	..	10–20 ppm
Cooling	$Ca(HCO_3)_2$	Steel, cast iron	10 ppm
	Na_2CrO_4	Fe, Zn, Cu	0.1%
	$NaNO_2$	Fe	0.05%
	NaH_2PO_4	..	1%
	Morpholine	..	0.2%
Boilers	NaH_2PO_4	Fe, Zn, Cu	10 ppm
	Polyphosphate	..	10 ppm
	Morpholine	Fe	Variable
	Hydrazine	..	O_2 scavenger
	Ammonia	..	Neutralizer
	Octadecylamine	..	Variable
Engine coolants	Na_2CrO_4	Fe, Pb, Cu, Zn	0.1–1%
	$NaNO_2$	Fe	0.1–1%
	Borax	..	1%
Glycol/water	Borax + MBT*	All	1% + 0.1%
Oil field brines	Na_2SiO_3	Fe	0.01%
	Quaternaries	..	10–25 ppm
	Imidazoline	..	10–25 ppm
Seawater	Na_2SiO_3	Zn	10 ppm
	$NaNO_2$	Fe	0.5%
	$Ca(HCO_3)_2$	All	pH dependent
	NaH_2PO_4 + $NaNO_2$	Fe	10 ppm + 0.5%

*MBT = mercaptobenzotriazole.

Inhibitors have been classified differently by various authors. Some authors prefer to group inhibitors by their chemical functionality, as follows:[1]

- *Inorganic inhibitors.* Usually crystalline salts such as sodium chromate, phosphate, or molybdate. Only the negative anions of these compounds are involved in reducing metal corrosion. When zinc is used instead of sodium, the zinc cation can add some beneficial effect. These zinc-added compounds are called mixed-charge inhibitors.
- *Organic anionic.* Sodium sulfonates, phosphonates, or mercaptobenzotriazole (MBT) are used commonly in cooling waters and antifreeze solutions.
- *Organic cationic.* In their concentrated forms, these are either liquids or waxlike solids. Their active portions are generally large aliphatic or aromatic compounds with positively charged amine groups.

However, by far the most popular organization scheme consists of regrouping corrosion inhibitors in a functionality scheme as follows.[2]

10.2.1 Passivating (anodic)

Passivating inhibitors cause a large anodic shift of the corrosion potential, forcing the metallic surface into the passivation range. There are two types of passivating inhibitors: oxidizing anions, such as chromate, nitrite, and nitrate, that can passivate steel in the absence of oxygen and the nonoxidizing ions, such as phosphate, tungstate, and molybdate, that require the presence of oxygen to passivate steel.

These inhibitors are the most effective and consequently the most widely used. Chromate-based inhibitors are the least-expensive inhibitors and were used until recently in a variety of application (e.g., recirculation-cooling systems of internal combustion engines, rectifiers, refrigeration units, and cooling towers). Sodium chromate, typically in concentrations of 0.04 to 0.1%, was used for these applications. At higher temperatures or in fresh water with chloride concentrations above 10 ppm higher concentrations are required. If necessary, sodium hydroxide is added to adjust the pH to a range of 7.5 to 9.5. If the concentration of chromate falls below a concentration of 0.016%, corrosion will be accelerated. Therefore, it is essential that periodic colorimetric analysis be conducted to prevent this from occurring. In general, passivation inhibitors can actually cause pitting and accelerate corrosion when concentrations fall below minimum limits. For this reason it is essential that monitoring of the inhibitor concentration be performed.

10.2.2 Cathodic

Cathodic inhibitors either slow the cathodic reaction itself or selectively precipitate on cathodic areas to increase the surface impedance and limit the diffusion of reducible species to these areas. Cathodic inhibitors can provide inhibition by three different mechanisms: (1) as cathodic poisons, (2) as cathodic precipitates, and (3) as oxygen scavengers. Some cathodic inhibitors, such as compounds of arsenic and antimony, work by making the recombination and discharge of hydrogen more difficult. Other cathodic inhibitors, ions such as calcium, zinc, or magnesium, may be precipitated as oxides to form a protective layer on the metal. Oxygen scavengers help to inhibit corrosion by preventing the cathodic depolarization caused by oxygen. The most commonly used oxygen scavenger at ambient temperature is probably sodium sulfite (Na_2SO_3).

10.2.3 Organic

Both anodic and cathodic effects are sometimes observed in the presence of organic inhibitors, but as a general rule, organic inhibitors affect the entire surface of a corroding metal when present in sufficient concentration. Organic inhibitors, usually designated as *film-forming*, protect the metal by forming a hydrophobic film on the metal surface. Their effectiveness depends on the chemical composition, their molecular structure, and their affinities for the metal surface. Because film formation is an adsorption process, the temperature and pressure in the system are important factors. Organic inhibitors will be adsorbed according to the ionic charge of the inhibitor and the charge on the surface. Cationic inhibitors, such as amines, or anionic inhibitors, such as sulfonates, will be adsorbed preferentially depending on whether the metal is charged negatively or positively. The strength of the adsorption bond is the dominant factor for soluble organic inhibitors.

These materials build up a protective film of adsorbed molecules on the metal surface, which provides a barrier to the dissolution of the metal in the electrolyte. Because the metal surface covered is proportional to the inhibitor concentrates, the concentration of the inhibitor in the medium is critical. For any specific inhibitor in any given medium there is an optimal concentration. For example, a concentration of 0.05% sodium benzoate or 0.2% sodium cinnamate is effective in water with a pH of 7.5 and containing either 17 ppm sodium chloride or 0.5% by weight of ethyl octanol. The corrosion due to ethylene glycol cooling water systems can be controlled by the use of ethanolamine as an inhibitor.

10.2.4 Precipitation inhibitors

Precipitation-inducing inhibitors are film-forming compounds that have a general action over the metal surface, blocking both anodic and cathodic sites indirectly. Precipitation inhibitors are compounds that cause the formation of precipitates on the surface of the metal, thereby providing a protective film. Hard water that is high in calcium and magnesium is less corrosive than soft water because of the tendency of the salts in the hard water to precipitate on the surface of the metal and form a protective film.

The most common inhibitors of this category are the silicates and the phosphates. Sodium silicate, for example, is used in many domestic water softeners to prevent the occurrence of *rust water*. In aerated hot water systems, sodium silicate protects steel, copper, and brass. However, protection is not always reliable and depends heavily on pH and a saturation index that depends on water composition and temperature. Phosphates also require oxygen for effective inhibition. Silicates and phosphates do not afford the degree of protection provided by chromates and nitrites; however, they are very useful in situations where nontoxic additives are required.

10.2.5 Volatile corrosion inhibitors

Volatile corrosion inhibitors (VCIs), also called vapor phase inhibitors (VPIs), are compounds transported in a closed environment to the site of corrosion by volatilization from a source. In boilers, volatile basic compounds, such as morpholine or hydrazine, are transported with steam to prevent corrosion in condenser tubes by neutralizing acidic carbon dioxide or by shifting surface pH toward less acidic and corrosive values. In closed vapor spaces, such as shipping containers, volatile solids such as salts of dicyclohexylamine, cyclohexylamine, and hexamethylene-amine are used. On contact with the metal surface, the vapor of these salts condenses and is hydrolyzed by any moisture to liberate protective ions. It is desirable, for an efficient VCI, to provide inhibition rapidly and to last for long periods. Both qualities depend on the volatility of these compounds, fast action wanting high volatility, whereas enduring protection requires low volatility.

10.3 Corrosion Inhibition Mechanism

The majority of inhibitor applications for aqueous, or partly aqueous, systems are concerned with four main types of environment:

1. Aqueous solutions of acids as used in metal-cleaning processes such as pickling for the removal of rust or mill scale during the production and fabrication of metals or in the postservice cleaning of metal surfaces

2. Natural waters, supply waters, and industrial cooling waters in the near-neutral pH range (5 to 9)

3. Primary and secondary production of oil and subsequent refining and transport processes

4. Atmospheric or gaseous corrosion in confined environments, during transport, storage, or any other confined operation

The following sections describe corrosion mechanisms in terms of these four main environments.

10.3.1 Inhibitors for acid solutions

The corrosion of metals in acid solutions can be inhibited by a wide range of substances, such as halide ions, carbon monoxide, and many organic compounds, particularly those containing elements of Groups V and VI of the Periodic Table (i.e., nitrogen, phosphorus, arsenic, oxygen, sulfur, and selenium). Organic compounds containing multiple bonds, especially triple bonds, are effective inhibitors. The primary step in the action of inhibitors in acid solutions is generally agreed to be adsorption onto the metal surface, which is usually oxide-free in acid solutions. The adsorbed inhibitor then acts to retard the cathodic and/or anodic electrochemical corrosion processes.

Inhibitors of corrosion in acid solution can interact with metals and affect the corrosion reaction in a number of ways, some of which may occur simultaneously. It is often not possible to assign a single general mechanism of action to an inhibitor because the mechanism may change with experimental conditions. Thus, the predominant mechanism of action of an inhibitor may vary with factors such as its concentration, the pH of the acid, the nature of the anion of the acid, the presence of other species in the solution, the extent of reaction to form secondary inhibitors, and the nature of the metal. The mechanism of action of inhibitors with the same functional group may additionally vary with factors such as the effect of the molecular structure on the electron density of the functional group and the size of the hydrocarbon portion of the molecule.

Adsorption of corrosion inhibitors onto metals. The inhibitive efficiency is usually proportional to the fraction of the surface θ covered with adsorbed inhibitor. However, at low surface coverage ($\theta<0.1$), the effectiveness of adsorbed inhibitor species in retarding the corrosion reactions may be greater than at high surface coverage. In other cases, adsorption of inhibitors, such as thiourea and amines, from diluted solutions, may stimulate corrosion.

The information on inhibitor adsorption, derived from direct measurements and from inhibitive efficiency measurements, considered in conjunction with general knowledge of adsorption from solution, indicates that inhibitor adsorption on metals is influenced by the following main features.

Surface charge on the metal. Adsorption may be due to electrostatic attractive forces between ionic charges or dipoles on the adsorbed species and the electric charge on the metal at the metal-solution interface. In solution, the charge on a metal can be expressed by its potential with respect to the zero-charge potential. This potential relative to the zero-charge potential, often referred to as the (φ-potential, is more important with respect to adsorption than the potential on the hydrogen scale, and indeed the signs of these two potentials may be different. As the potential of a metallic surface becomes more positive, the adsorption of anions is favored, and as the φ-potential becomes more negative, the adsorption of cations is favored.

The functional group and structure of the inhibitor. Inhibitors can also bond to metal surfaces by electron transfer to the metal to form a coordinate type of link. This process is favored by the presence in the metal of vacant electron orbitals of low energy, such as occurs in the transition metals. Electron transfer from the adsorbed species is favored by the presence of relatively loosely bound electrons, such as may be found in anions, and neutral organic molecules containing lone pair electrons or π-electron systems associated with multiple, especially triple, bonds or aromatic rings. The electron density at the functional group increases as the inhibitive efficiency increases in a series of related compounds. This is consistent with increasing strength of coordinate bonding due to easier electron transfer and hence greater adsorption.

Interaction of the inhibitor with water molecules. Adsorption of inhibitor molecules is often a displacement reaction involving removal of adsorbed water molecules from the surface. During adsorption of a molecule, the change in interaction energy with water molecules in passing from the dissolved to the adsorbed state forms an important part of the free energy change on adsorption. This has been shown to increase with the energy of solvation of the adsorbing species, which in turn increases with increasing size of the hydrocarbon portion of an organic molecule. Thus increasing size leads to decreasing solubility and increasing adsorbability. This is consistent with the increasing inhibitive efficiency observed at constant concentrations with increasing molecular size in a series of related compounds.

Interaction of adsorbed inhibitor species. Lateral interactions between adsorbed inhibitor species may become significant as the surface coverage, and hence the proximity, of the adsorbed species increases. These lateral interactions may be either attractive or repulsive. Attractive interactions occur between molecules containing large hydrocarbon components (e.g., *n*-alkyl chains). As the chain length increases, the increasing Van der Waals attractive force between adjacent molecules leads to stronger adsorption at high coverage. Repulsive interactions occur between ions or molecules containing dipoles and lead to weaker adsorption at high coverage.

In the case of ions, the repulsive interaction can be altered to an attractive interaction if an ion of opposite charge is simultaneously adsorbed. In a solution containing inhibitive anions and cations the adsorption of both ions may be enhanced and the inhibitive efficiency greatly increased compared to solutions of the individual ions. Thus, synergistic inhibitive effects occur in such mixtures of anionic and cationic inhibitors.

Reaction of adsorbed inhibitors. In some cases, the adsorbed corrosion inhibitor may react, usually by electrochemical reduction, to form a product that may also be inhibitive. Inhibition due to the added substance has been termed *primary inhibition* and that due to the reaction product, *secondary inhibition*. In such cases, the inhibitive efficiency may increase or decrease with time according to whether the secondary inhibition is more or less effective than the primary inhibition. Sulfoxides, for example, can be reduced to sulfides, which are more efficient inhibitors.

Effects of inhibitors on corrosion processes. In acid solutions the anodic process of corrosion is the passage of metal ions from the oxide-free metal surface into the solution, and the principal cathodic process is the discharge of hydrogen ions to produce hydrogen gas. In air-saturated acid solutions, cathodic reduction of dissolved oxygen also occurs, but for iron the rate does not become significant compared to the rate of hydrogen ion discharge until the pH exceeds a value of 3. An inhibitor may decrease the rate of the anodic process, the cathodic process, or both processes. The change in the corrosion potential on addition of the inhibitor is often a useful indication of which process is retarded. Displacement of the corrosion potential in the positive direction indicates mainly retardation of the anodic process (anodic control), whereas displacement in the negative direction indicates mainly retardation of the cathodic process (cathodic control). Little change in the corrosion potential suggests that both anodic and cathodic processes are retarded.

The following discussion illustrates the usage of anodic and cathodic inhibitors for acid cleaning of industrial equipment. The combined action of film growth and deposition from solution results in fouling that has to be removed to restore the efficiency of heat exchangers, boilers, and steam generators. E-pH diagrams indicate that the fouling of iron-based boiler tubes, by Fe_3O_4 and Fe_2O_3, can be dissolved in either the acidic or alkaline corrosion regions. In practice, inhibited hydrochloric acid has been repeatedly proven to be the most efficient method to remove fouling. Four equations are basically needed to explain the chemistry involved in fouling removal. Three of those equations represent cathodic processes [Eqs. (10.2) and (10.3); A, A' and A" in Figs. 10.1 and 10.2; and Eq. (10.4); B in Figs. 10.1 and 10.2] and one anodic process [i.e., the dissolution of tubular material [Eq. (10.5); C in Figs. 10.1 and 10.2]:[3]

$$Fe_2O_3 + 4\ Cl^- + 6\ H^+ + 2\ e^- \rightarrow 2\ FeCl_{2(aq)} + 3\ H_2O \quad (10.2)$$

$$Fe_3O_4 + 6\ Cl^- + 8\ H^+ + 2\ e^- \rightarrow 3\ FeCl_{2(aq)} + 4\ H_2O \quad (10.3)$$

$$2\ H^+ + 2\ e^- \rightarrow H_2 \quad (10.4)$$

$$Fe + 2\ Cl^- \rightarrow FeCl_{2(aq)} + 2\ e^- \quad (10.5)$$

These equations indicate that the base iron functions as a reducer to accelerate the dissolution of iron oxides. Because it is difficult to determine the endpoint for the dissolution of fouling oxides, an inhibitor is generally added for safety purpose. Both anodic and cathodic inhibitors could be added to retard the corrosion of the bare metal after dissolution of the fouling oxides. Figures 10.1 and 10.2 illustrate the action that could be played by either an anodic inhibitor (Fig. 10.1) or a cathodic inhibitor (Fig. 10.2). It can be seen that although the anodic inhibitor retards the anodic dissolution of iron at the endpoint, it concurrently decreases the rate of oxide dissolution permitted by the chemical system.

On the other hand, the cathodic inhibitor retards both the reduction of protons into hydrogen and the dissolution of the base, whereas the reduction of the fouling oxides is left unaffected. The E-pH diagrams also indicate that the dissolution of the fouling oxides is also possible in alkaline solutions. But the kinetics of anodic and cathodic reactions in high pH environments are much slower, and therefore these reactions are less useful.

Electrochemical studies have shown that inhibitors in acid solutions may affect the corrosion reactions of metals in the following main ways.

Formation of a diffusion barrier. The absorbed inhibitor may form a surface film that acts as a physical barrier to restrict the diffusion of ions or molecules to or from the metal surface and so retard the rate of corrosion reactions. This effect occurs particularly when the inhibitor species are large molecules (e.g., proteins, such as gelatin or agar agar, polysaccharides, such as dextrin, or compounds containing long hydrocarbon chains). Surface films of these types of inhibitors give rise to resistance polarization and also concentration polarization affecting both anodic and cathodic reactions.

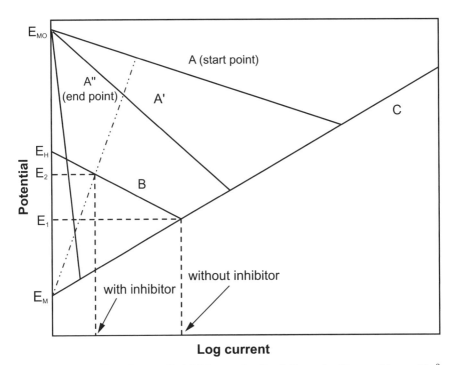

Figure 10.1 The effect of an anodic inhibitor on the dissolution rate of iron and iron oxide.[3]

Blocking of reaction sites. The simple blocking decreases the number of surface metal atoms at which corrosion reactions can occur. The mechanisms of the reactions are not affected, and the Tafel slopes of the polarization curves remain unchanged. It should be noted that the anodic and cathodic processes may be inhibited to different extents. The anodic dissolution process of metal ions is considered to occur at steps or emergent dislocations in the metal surface, where metal atoms are less firmly held to their neighbors than in the plane surface. These favored sites occupy a relatively small proportion of the metal surface. The cathodic process of hydrogen evolution is thought to occur on the plane crystal faces that form most of the metal surface area. Adsorption of inhibitors at low surface coverage tends to occur preferentially at anodic sites, causing retardation of the anodic reaction. At higher surface coverage, adsorption occurs on both anodic and cathodic sites, and both reactions are inhibited.

Participation in the electrode reactions. Corrosion reactions often involve the formation of adsorbed intermediate species with surface metal atoms [e.g., adsorbed hydrogen atoms in the hydrogen evolution reaction and adsorbed (FeOH) in the anodic dissolution of iron].

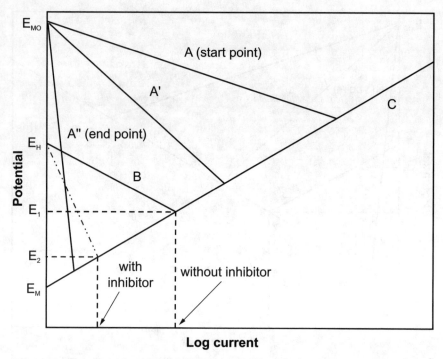

Figure 10.2 The effect of a cathodic inhibitor on the dissolution rate of iron and iron oxide.[3]

The presence of adsorbed inhibitors will interfere with the formation of these adsorbed intermediates, but the electrode processes may then proceed by alternative paths through intermediates containing the inhibitor. In these processes the inhibitor species act in a catalytic manner and remain unchanged. Such participation by the inhibitor is generally characterized by an increase in the Tafel slope of the anodic dissolution of the metal.

Inhibitors may also retard the rate of hydrogen evolution on metals by affecting the mechanism of the reaction, as indicated by increases in the Tafel slopes of cathodic polarization curves. This effect has been observed on iron in the presence of inhibitors such as phenyl-thiourea, acetylenic hydrocarbons, aniline derivatives, benzaldehyde derivatives. and pyrilium salts.

Alteration of the electrical double layer. The adsorption of ions or species that can form ions on metal surfaces will change the electrical double layer at the metal-solution interface, and this in turn will affect the rates of the electrochemical reactions. The adsorption of cations, such as quaternary ammonium ions and protonated amines, makes the potential more positive in the plane of the closest approach to the metal of ions from the solution. This positive potential displacement retards the discharge of the positively charged hydrogen ion.

Conversely, the adsorption of anions makes the potential more negative on the metal side of the electrical double layer, and this will tend to accelerate the rate of discharge of hydrogen ions. This effect has been observed for the sulfosalicylate ion and the benzoate ion.

Measuring the efficiency of a acid inhibitor. The following example illustrates how the corrosion efficiency of an inhibitor can be evaluated with a relatively simple corrosion test. Trans-cinnamaldehyde (TCA) corrosion inhibiting efficiency was evaluated with an electrochemical technique called linear polarization resistance (LPR). TCA can be used to reduce the corrosion of steel during pickling or oil field acidizing treatments. Nearly 40 years ago, Hugel tested a variety of inhibitors for steel in 6 M HCl at 60°C and found that alkenyl and aromatic aldehydes were very effective.[4] Cinnamaldehyde was one of the best, providing almost 99% protection. Numerous patents have been issued since then on the use of aldehydes, and cinnamaldehyde in particular, as steel corrosion inhibitors in acid media.

The LPR polarization resistance (R_p) is typically calculated from the slope of a polarization curve where

$$R_p = \frac{\Delta E}{\Delta I_{app}}$$

and where ΔE is the voltage change for an applied current (ΔI_{app}). R_p itself can be converted in a corrosion current (I_{corr}) using the Stern Geary approximation written as:[5,6]

$$I_{corr} = \frac{\beta_a \beta_c}{2.3\,(\beta_a + \beta_c)\,R_p}$$

where β_a and β_c are, respectively, the anodic and cathodic Tafel slopes.

The polarization curves presented in Figs. 10.3 to 10.5 were obtained with carbon steel exposed to a solution containing, respectively, 250, 1000, and 5000 ppm of TCA in a 6 M HCl solution. Assuming, for this example, that β_a and β_c are both equal to 0.1 V decade^{-1} and that the R_p of uninhibited carbon steel in 6 M HCl is equal to 14 $\Omega \cdot cm^2$, it is possible to obtain the inhibitor efficiency values presented in Table 10.2.

10.3.2 Inhibitors in near-neutral solutions

Corrosion of metals in neutral solutions differs from that in acid solutions in two important respects. In air-saturated solutions, the main

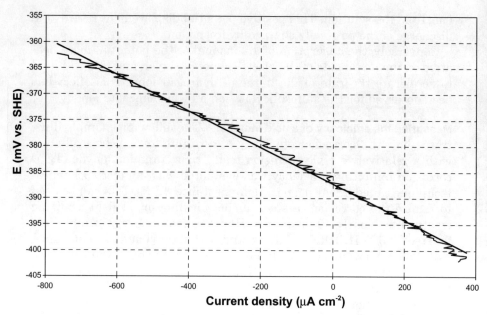

Figure 10.3 Corrosion of AISI 1018 carbon steel in 6 M HCl containing 250 ppm trans-cinnamaldehyde.

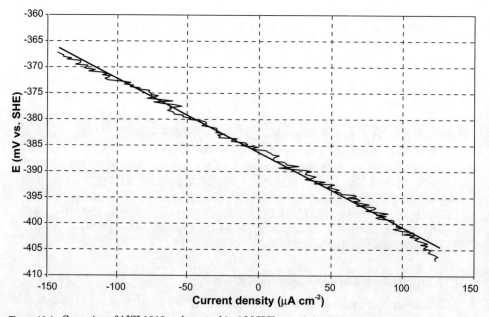

Figure 10.4 Corrosion of AISI 1018 carbon steel in 6 M HCl containing 500 ppm trans-cinnamaldehyde.

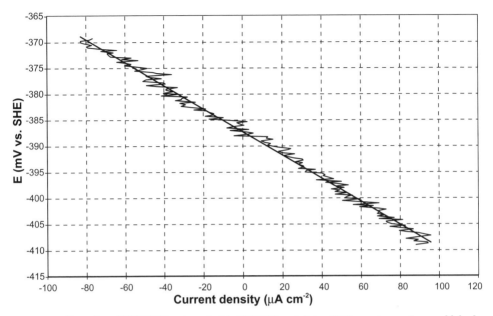

Figure 10.5 Corrosion of AISI 1018 carbon steel in 6 M HCl containing 1000 ppm trans-cinnamaldehyde.

TABLE 10.2 Inhibitor Efficiency of Trans-Cinnamaldehyde (TCA) to the Corrosion of Carbon Steel Exposed to a 6 M HCl Solution

TCA, ppm	R_p, $\Omega \cdot cm^2$	Corrosion current, $mA \cdot cm^{-2}$	Corrosion rate, $mm \cdot y^{-1}$	Efficiency, %
0	14	1.55	18.0	0
250	35	0.62	7.2	60
1000	143	0.152	1.76	90
5000	223	0.097	1.13	94

cathodic reaction in neutral solutions is the reduction of dissolved oxygen, whereas in acid solution it is hydrogen evolution. Corroding metal surfaces in acid solution are oxide-free, whereas in neutral solutions metal surfaces are covered with films of oxides, hydroxides, or salts, owing to the reduced solubility of these species. Because of these differences, substances that inhibit corrosion in acid solution by adsorption on oxide-free surfaces do not generally inhibit corrosion in neutral solution.

Typical inhibitors for near-neutral solutions are the anions of weak acids, some of the most important in practice being chromate, nitrite, benzoate, silicate, phosphate, and borate. Passivating oxide films on metals offer high resistance to the diffusion of metal ions, and the anodic reaction of metal dissolution is inhibited. These inhibitive anions are often referred to as anodic inhibitors, and they are more generally used than cathodic inhibitors to inhibit the corrosion of iron, zinc, aluminum, copper, and their alloys in near-neutral solutions. The action of inhibitive anions on the corrosion of metals in near-neutral solution involves the following important functions:

1. Reduction of the dissolution rate of the passivating oxide film
2. Repair of the oxide film by promotion of the reformation of oxide
3. Repair of the oxide film by plugging pores with insoluble compounds
4. Prevention of the adsorption of aggressive anions

Of these functions, the most important appears to be the stabilization of the passivating oxide film by decreasing its dissolution rate (function 1). Inhibitive anions probably form a surface complex with the metal ion of the oxide (i.e., Fe^{3+}, Zn^{2+}, Al^{3+}), such that the stability of this complex is higher than that of the analogous complexes with water, hydroxyl ions, or aggressive anions.

Stabilization of the oxide films by repassivation is also important (function 2). The plugging of pores by formation of insoluble compounds (function 3) does not appear to be an essential function but is valuable in extending the range of conditions under which inhibition can be achieved. The suppression of the adsorption of aggressive anions (function 4) by participation in a dynamic reversible competitive adsorption equilibrium at the metal surface appears to be related to the general adsorption behavior of anions rather than to a specific property of inhibitive anions.

Inhibition in neutral solutions can also be due to the precipitation of compounds, on a metallic surface, that can form or stabilize protective films. The inhibitor may form a surface film of an insoluble salt by precipitation or reaction. Inhibitors forming films of this type include

- Salts of metals such as zinc, magnesium, manganese, and nickel, which form insoluble hydroxides, especially at cathodic areas, which are more alkaline due to the hydroxyl ions produced by reduction of oxygen
- Soluble calcium salts, which can precipitate as calcium carbonate in waters containing carbon dioxide, again at cathodic areas where the high pH permits a sufficiently high concentration of carbonate ions
- Polyphosphates in the presence of zinc or calcium, which produce a thin amorphous salt film

These salt films, which are often quite thick and may even be visible, restrict diffusion, particularly of dissolved oxygen to the metal surface. They are poor electronic conductors, and so oxygen reduction does not occur on the film surface. These inhibitors are referred to as cathodic inhibitors.

The following sections discuss the mechanism of action of inhibitive anions on iron, zinc, aluminum, and copper.

Iron. Corrosion of iron (or steel) can be inhibited by the anions of most weak acids under suitable conditions. However, other anions, particularly those of strong acids, tend to prevent the action of inhibitive anions and stimulate breakdown of the protective oxide film. Examples of such aggressive anions include the halides, sulfate, and nitrate. The balance between the inhibitive and aggressive properties of a specific anion depends on the following main factors (which are themselves interdependent):

- *Concentration.* Inhibition of iron corrosion in distilled water occurs only when the anion concentration exceeds a critical value. At concentrations below the critical value, inhibitive anions may act aggressively and stimulate breakdown of the oxide films. Effective inhibitive anions have low critical concentrations for inhibition. A number of anions have been classified in order of their inhibitive power toward steel, judged from their critical inhibitive concentrations. The order of decreasing inhibitive efficiency is azide, ferricyanide, nitrite, chromate, benzoate, ferrocyanide, phosphate, tellurate, hydroxide, carbonate, chlorate, *o*-chlorbenzoate, bicarbonates fluoride, nitrate, and formate.
- *pH.* Inhibitive anions are effective in preventing iron corrosion only at pH values more alkaline than a critical value. This critical pH depends on the anion.
- *Dissolved oxygen concentration and supply.* Inhibition of the corrosion of iron by anions requires a critical minimum degree of oxidizing power in the solution. This is normally supplied by the dissolved oxygen present in air-saturated solutions.
- *Aggressive anion concentration.* When aggressive anions are present in the solution, the critical concentrations of inhibitive anions required for protection of iron are increased. It has been shown that the relationship between the maximum concentration of aggressive anion C_{agg} permitting full protection by a given concentration of inhibitive anion C_{inh} is of the form

$$\log C_{inh} = n \log C_{agg} + K$$

where K is a constant dependent on the nature of the inhibitive and aggressive anions, and n is an exponent that is approximately the ratio of the valency of the inhibitive anion to the valency of the aggressive anion

- *Nature of the metal surface.* The critical concentration of an anion required to inhibit the corrosion of iron may increase with increasing surface roughness.
- *Temperature.* In general, the critical concentrations of anions (e.g., benzoate, chromate, and nitrite) required for the protection of steel increase as the temperature increases.

Zinc. The effects of inhibitive and aggressive anions on the corrosion of zinc are broadly similar to the effects observed with iron. Thus with increasing concentration, anions tend to promote corrosion but may give inhibition above a critical concentration. Inhibition of zinc corrosion is somewhat more difficult than that of iron (e.g., nitrite and benzoate are not efficient inhibitors for zinc). However, inhibition of zinc corrosion is observed in the presence of anions such as chromates, borate, and nitrocinnamate, which are also good inhibitors for the corrosion of iron. Anions such as sulfate, chloride, and nitrate are aggressive toward zinc and prevent protection by inhibitive anions. The presence of dissolved oxygen in the solution is essential for protection by inhibitive anions. As in the case of iron, pressures of oxygen greater than atmospheric or an increase in oxygen supply by rapid stirring can lead to the protection of zinc in distilled water. Inhibition of zinc corrosion occurs most readily in the pH range of 9 to 12, which corresponds approximately to the region of minimum solubility of zinc hydroxide.

The ways in which inhibitive anions affect the corrosion of zinc are mainly similar to those described above for iron. In inhibition by chromate, localized uptake of chromium has been shown to occur at low chromate concentrations and in the presence of chloride ions. Inhibitive anions also promote the passivation of zinc (e.g., passivation is much easier in solutions of the inhibitive anion, borate, than in solutions of the noninhibitive anions, carbonate and bicarbonate). A critical inhibition potential, analogous to that on iron, has been observed for zinc in borate solutions. Thus inhibitive anions promote repair of the oxide film on zinc by repassivation with zinc oxide.

Aluminum. When aluminum is immersed in water, the air-formed oxide film of amorphous γ-alumina initially thickens (at a faster rate than in air) and then an outer layer of crystalline hydrated alumina forms, which eventually tends to stifle the reaction. In near-neutral air-saturated solutions, the corrosion of aluminum is generally inhibited by anions that are inhibitive for iron (e.g., chromate, benzoate, phosphate, and acetate). Inhibition also occurs in solutions containing sulfate or nitrate ions, which are aggressive toward iron. Aggressive anions for aluminum include the halide ions, F^-, Cl^-, Br^-, I^-, which cause pitting attack, and anions that form soluble complexes with aluminum (e.g., citrate and tartrate), which cause general attack. Competitive effects, similar to those observed on iron, are observed in the action of mixtures of inhibitive anions and chloride ions on aluminum.

In near-neutral and deaerated solutions, the oxide film on anodized aluminum is stable and protective in distilled water and chloride solutions, as well as in solutions of inhibitive anions. Thus the inhibition of aluminum corrosion by anions differs from that of iron or zinc in that the presence of dissolved oxygen in the solution is not necessary to stabilize the oxide film. In corrosion inhibition by chromate ions, their interaction with the oxide film on aluminum has been shown to result in the formation of an outer layer of the film that is more protective due to its high electronic resistance and low dissolution rate. Chromate ions were also found to prevent the uptake and penetration of chloride ions into the aluminum oxide film.

Copper. Little work has been carried out on the mechanism of inhibition by anions of copper corrosion in neutral solutions. Inhibition occurs in solutions containing chromate, benzoate, or nitrite ions. Chloride and sulfide ions are aggressive, and there is some evidence that chloride ions can be taken up into the cuprous oxide film on copper to replace oxide ions and create cuprous ion vacancies that permit easier diffusion of cuprous ions through the film, thus increasing the corrosion rate.

Copper corrosion can also be effectively inhibited in neutral solution by organic compounds of low molecular weight, such as benzotriazole and 2-mercaptobenzothiazole. Benzotriazole is particularly effective in preventing the tarnishing and dissolution of copper in chloride solutions. In the presence of benzotriazole, the anodic dissolution, oxide film growth, and dissolved oxygen reduction reactions are all inhibited, indicating strong adsorption of the inhibitor on the cuprous oxide surface.

10.3.3 Inhibitors for oil and gas systems

Even in early days, oil producers applied numerous chemical compounds, sometimes with success, to minimize corrosion damage in the oil wells themselves and in surface handling equipment. Once amines and imidazolines came into use, corrosion inhibition in oil wells became dramatically more effective. Modern inhibitors are applied in the field, continuously or by periodic batch, at concentrations of 15 to

50 ppm, based on total liquid production. A much wider variety of inhibitor chemistry is available today for combating oil-field corrosion than existed only a decade ago. In recent years, organic molecules containing sulfur, phosphorus, and nitrogen in various combinations have been developed. These inhibitor types have extended the performance of oil-field inhibitors, particularly in the directions of being tolerant of oxygen contamination and of controlling corrosion associated with high CO_2, low H_2S conditions.[7]

Most of the inhibitors currently used in producing wells are organic nitrogenous compounds. The basic types have long-chain hydrocarbons (usually C_{18}) as a part of the structure. Most inhibitors in successful use today are either based on the long-chain aliphatic diamine, or on long carbon chain imidazolines. Various modifications of these structures have been made to change the physical properties of the material (e.g., ethylene oxide is commonly reacted with these compounds in various molecular percentages to give polyoxy-ethylene derivatives that have varying degrees of brine dispersibility). Many carboxylic acids are used to make salts of these amines or imidazolines. Inhibitors in general petroleum production can be classified as follows:[8]

- Amides/imidazolines
- Salts of nitrogenous molecules with carboxylic acids
- Nitrogen quaternaries
- Polyoxyalkylated amines, amides, and imidazolines
- Nitrogen heterocyclics and compounds containing P, S, O

There are several hypotheses and theories concerning the inhibitive action of the long-chain nitrogenous compounds. One of the classical concepts is the so-called sandwich theory in which the bottom part of the sandwich is the bond between the polar end of the molecule and the metal surface. The strength of the protective action depends on this bond. The center portion of the sandwich is the nonpolar end of the molecule and its contribution toward protection is the degree to which this portion of the molecule can cover or wet the surface. The top portion of the protective sandwich is the hydrophobic layer of oil attached to the long carbon tail of the inhibitor. This oil layer serves as the external protective film, covering the inhibitor film and creating a barrier to both outward diffusion of ferrous ion and inward diffusion of corrosive species.

Water or water solutions of salts alone will not cause damaging corrosion unless they contain specific corrodents, such as CO_2, H_2S, and their products of dissolution. Oil and gas wells are either sweet or sour.

Sweet wells do not contain hydrogen sulfide, whereas sour wells do. The source of CO_2 can be mineral dissolution or a by-product of the petroleum-forming process. The source of H_2S can be dissolution of mineral deposits in the rocks, a by-product of the petroleum-forming process, or bacterial action at any time in the history of the petroleum deposit. Oxygen always originates from air and can only come in contact with petroleum fluids after the recovery process begins. It does not exist in the undisturbed hydrocarbon deposit.

The dissolution products of H_2S in oil-field waters will be dissolved hydrogen sulfide molecules (H_2S) and bisulfide ions (HS^-), and the dissolution products of CO_2 will be dissolved CO_2 molecules (some hydrate to form H_2CO_3) and bicarbonate (HCO_3^-) ions. The pH of these waters is not basic enough to produce appreciable amounts of sulfide or carbonate ions. However, damaging corrosion in the oil field nearly always takes localized forms, often pitting. Corrosion pits in oil-field steels typically penetrate at 10 to 100 times the rate of uniform corrosion. Pit growth in steels exposed to brine, an active corrosion system, occurs because of a galvanic couple between filmed metal and relatively bare metal.

Sweet corrosion. Corrosion in CO_2 gas wells can be divided into three temperature regimes. Below 60°C, the corrosion product is nonprotective and high corrosion rates will occur. Above approximately 150°C, magnetite is formed, and the wells are not corrosive except in the presence of high brine levels. In the middle temperature regime, in which most gas well conditions lie, the iron carbonate corrosion product layer is protective but is affected adversely by chlorides and fluid velocity.[7]

One of the important physical properties of oil-field inhibitors is their volubility or dispersibility characteristic in the oil and the brine being produced. An inhibitor, properly chosen on the basis of the corrosion mechanism, will not be effective if it does not have access to the corroding metal. When it comes to treating oil and gas wells, there are also some important differences. The distinction between an oil well and a gas well is not clear cut. Often the distinction is made on the basis of economics or workload balance within a producing company. The facts that many oil wells produce a considerable volume of gas and many gas wells produce a considerable volume of liquid, plus the fact that wells often experience a shift in production during their lifetime, make a technical distinction difficult. However, there are more important differences. Typical gas wells are much hotter than oil wells, and the hydrocarbon liquids are much lighter. Gas wells are normally much deeper and usually produce lower total dissolved solids (TDS) brines. Oxygen is not a factor to consider in gas well corrosion but can cause major problems in artificial lift oil wells.

Due to the large temperature gradient in many gas wells, corrosion mechanisms can change, resulting in different types of corrosion in the same well, whereas oil wells do not exhibit this behavior. Normally, oil wells produce more liquid than gas wells, resulting in a shorter treatment life when batch treated. Because corrosion in oil wells is electrochemical in nature, an electrolyte must be present for corrosion to occur. In oil wells, the source of the water is nearly always the producing formation, and the water will contain dissolved salts in concentrations ranging from traces to saturation. Water associated with corrosion may be in a thin layer, in droplets, or even the major phase.

Results of the study of corrosion control by inhibitors in producing oil wells in carbon dioxide flooded fields[8] showed imidazolines are successful in protection in CO_2 brines. The inhibitor was found to be incorporated in the carbonate corrosion product layer but was still more effective if the surface film contained sulfide. Also, better results were obtained with inhibitors, such as nitrogen-phosphorus compounds or compounds with sulfur in the organic molecules.

Sour corrosion. In sour wells, hydrogen sulfide is the primary corrosive agent, and frequently carbon dioxide is present as well. The presence of various iron sulfides in the corrosion products at different concentrations of hydrogen sulfide has been identified. Based on this evidence the net corrosion reaction due to hydrogen sulfide can be written as follows:

$$Fe + H_2S \rightarrow FeS + 2H^+ + 2e^- \quad (10.6)$$

The most probable mechanism to explain the accelerating effect of hydrogen sulfide involves the formation of a molecular surface complex that can yield hydrogen atoms according to Eqs. (10.7) to (10.9). Some of the hydrogen produced in the process [Eq. (10.9)] may recombine to form molecular gaseous hydrogen, whereas some can diffuse in the metal and eventually cause blistering or hydrogen induced cracking.[9]

$$Fe + HS^- \rightarrow Fe(HS^-)_{ads} \quad (10.7)$$

$$Fe(HS^-)_{ads} + H^+ \rightarrow Fe(H\text{-}S\text{-}H)_{ads} \quad (10.8)$$

$$Fe(H\text{-}S\text{-}H)_{ads} + e^- \rightarrow Fe(HS^-)_{ads} + H_{ads} \quad (10.9)$$

Corrosion inhibitors used in the past to combat corrosion in sour wells include aldehydes, cyanamide thiourea, and urea derivatives. The most widely used inhibitors are organic amines. Although organic amines are known to be less effective inhibitors in acid solution, inhibition by amines in the presence of hydrogen sulfide is greatly enhanced.[9] Oil-field inhibitors function by incorporating into a thin layer of corrosion product on the metal surface. This surface film may be a sulfide or a carbonate and may be anaerobic or partially oxidized. Some types of inhibitor molecules incorporate better in one type of film than others. For example, amine inhibitors are not effective when oxygen is present. Inhibitor molecules containing nitrogen (e.g., imidazolines) will incorporate into either sulfide or carbonate films but are more effective when the film contains some sulfide.

Acidizing. An important procedure for stimulation of oil and gas well production is acidizing. Because of the very low permeability of certain formations containing hydrocarbons, these are not able to flow readily into the well. Formations composed of limestone or dolomite may be treated with HCl or, if the rock is sandstone, a mixture containing HF. In the acidizing treatment, the acid (e.g., HCl, at a concentration of 7 to 28%) is pumped down the tubing into the well where it enters the perforations and contacts the formation; the acid etches channels that provide a way for oil and gas to enter the well.[8] Many inhibitors are used for well acidizing operations, mainly high molecular weight nitrogenous compounds such as those used in primary production or the reaction products of these compounds with unsaturated alcohols. Many of those commercial inhibitors contain alkyl or alkylaryl nitrogen compounds and acetylenic alcohols, such as 1-octyn-3-ol. These products present serious handling problems because they are very toxic; this can determine which product is actually used by an operator. Furthermore, their effectiveness is limited both in efficiency and time. Acid soaks normally last between 12 and 24 h, after which time inhibitor efficiency can start to fall off alarmingly.

Oxygen-containing inhibitors that are successful in concentrated HCl include cinnamaldehyde and the alkynols containing unsaturated groups conjugated with the oxygen function described as alpha-alkenylphenones.[8] They provide, especially when mixed with small amounts of surfactants, protection similar to that obtained with acetylenic alcohols.

Oxygen-influenced corrosion. Oil-producing formations originally contain no oxygen. During the process of bringing oil to the surface, oxygen from air contamination may dissolve into produced fluids. This oxygen has three consequences:

1. Oxygen can readily accept electrons, so it increases the rate of corrosion.
2. The nature of the surface corrosion product changes, so the chemical properties required for effective inhibitor incorporation change.

3. Oxidation of certain ions in solution leads to increased precipitation of solid phases.

Air may be pulled into the annuli of wells having little gas pressure as a consequence of the artificial lift process or of negative-pressure gas gathering systems. In some cases, in situ combustion stimulation can introduce oxygen into the formation itself. On the surface, small amounts of oxygen can be introduced into production liquids by leaking pump packing or direct contact during storage.[10]

In water flooding, the same types of inhibitors as described for primary production are currently used. The most effective and most frequently used are the quaternary ions of the fatty or the imidazoline types. They are also good bactericides and dispersive agents. Combination of amino-methylene phosphonate and zinc salts have been used successfully in circulating water systems and have provided more effective protection than the inorganic phosphate-zinc salts. Organic sulfonates have recently been introduced into practice.

Oxygen is practically always present in drilling muds. The most effective control of oxygen corrosion would be to keep it out of the system, but this is difficult because the drilling fluid is exposed to the atmosphere as it circulates through the pit. The attack is almost always in the form of pitting, which in a short time can produce irreversible damage to drilling equipment. Oxygen activity in drilling muds is determined by the interplay of a number of factors. For example, phosphorus compounds such as sodium hexametaphosphate, phosphate esters of organic alcohol, and organic phosphonates may act as anodic inhibitors, but a precaution is required in their use because they have a strong tendency to thin nondispersed muds. Tannins and lignins are thinners for high-solid muds, and they also have a certain inhibitive influence.

Application methods. The selection of an inhibitor is of prime importance, but the proper application of an inhibitor is even more important. If an inhibitor does not reach the corrosive areas, it cannot be effective. Maximum corrosion protection can be achieved by continuous injection of inhibitor through a dual tubing string (kill string), a capillary tubing, a side mandrel valve, or even perforated tubing. Any of these methods will supply a continuous residual of inhibitor to maintain corrosion protection. Treating rates or inhibitor concentrations are best based on the volume of fluid produced and can range from near 50 ppm to over 1000 ppm, depending on the severity of the conditions.[7]

Many gas wells are not equipped with facilities for continuous treatment and must be treated by some type of batch or slug treatment. The most commonly used method is the batch or short-batch treatment in which a volume of inhibitor solution (typically 2 to 10%) is injected into a shut-in well and allowed to fall to the bottom. Fall rates are a function of solution viscosity. The common failure of this method is not allowing sufficient time for the inhibitor to reach the hole bottom. A variation on this method is the tubing displacement treatment in which the inhibitor solution is pushed to the bottom by diesel or condensate. This guarantees the inhibitor reaching bottom hole, but it can kill low-pressure wells and is more costly.

Sometimes a short batch is forced down with a nitrogen displacement or compressed gas to speed up the fall rate and reduce shut-in time. An inhibitor squeeze is sometimes used to try to get a longer return time and simulate a continuous treatment. However, there is always the concern of formation damage with squeezes and with tubing displacements.

10.3.4 Atmospheric and gaseous corrosion

VCIs represent a very economical and powerful tool in combating the atmospheric or gaseous corrosion damage done to metals and alloys. Volatile corrosion inhibition is based on conditioning of the environment with trace amounts of inhibitive material to achieve a protective effect. A VCI compound, in addition to being volatile, is required to promote electrochemical effects such as change of the potential in the diffuse part of the double layer that controls the migration of components of the electrode reactions.

The first condition for good efficiency of a vapor phase inhibitor is its capability to reach the metallic surface to be protected. The second is that the rate of transfer of the molecule should not be too slow to prevent an initial attack of the metal surface by the aggressive environment before the inhibitor can act. These two conditions are related partly to the vapor pressure of the inhibitor, partly to the distance between the source(s) of the inhibitor and the metal surfaces, and partly to the accessibility of the surfaces.[11]

The vapor pressure of a chemical compound will depend upon the structure of the crystal lattice and the character of the atomic bonds in the molecule. In this respect, organic components of the molecule will generally ensure its volatility. A convenient volatile inhibitor should not have too high a vapor pressure, because it will be lost as a result of the fact that enclosures are generally not airtight; protection will then drop. A convenient partial vapor pressure for efficient compounds will lie between 10^{-5} and 10^{-1} mm·Hg (i.e., 10^{-3} to 10 Pa).

By definition, only compounds that have an appreciable vapor pressure under atmospheric conditions and can act as electrolyte layer inhibitors by electrochemically changing the kinetics of electrode reactions should be classified as VCIs. Neutralizing amines have an appreciable vapor pressure and are effective inhibitors for ferrous metals, but their mechanism is based on adjusting the pH value of the electrolyte, thus creating conditions that are inhospitable for rust formation. Hence, they should not necessarily be classified as volatile corrosion inhibitors.

Volatile compounds reach the protective vapor concentration rapidly, but in the case of enclosures that are not airtight, the consumption of inhibitor is excessive and the effective protective period is short. Low vapor pressure inhibitors are not rapidly exhausted and can ensure more durable protection. However, more time is required to achieve a protective vapor concentration. Furthermore, there is a possibility of corrosion occurring during the initial period of saturation, and if the space is not hermetically sealed, an effective inhibitor concentration may never be obtained. Therefore, the chemical compound used as a volatile inhibitor must not have too high or too low a vapor pressure, but some optimum vapor pressure.[12]

The comparison between the vapor pressure of a compound and its molecular heat of sublimation shows a marked decrease in vapor pressure values with an increase in heat of sublimation. A plausible explanation is that a decrease in vapor pressure is caused by steric intermolecular actions between functional groups and by an increase in molecular weight of the compound (Table 10.3).[12]

It is significant that the most effective volatile corrosion inhibitors are the products of the reaction of a weak volatile base with a weak volatile acid. Such substances, although ionized in aqueous solutions, undergo substantial hydrolysis, the extent of which is almost indepen-

TABLE 10.3 Saturated Vapor Pressures of Common VCIs

Substance	Temperature, °C	Vapor pressure, mm·Hg	Melting point, °C
Morpholine	20	8.0	
Benzylamine	29	1.0	
Cyclohexylamine carbonate	25.3	0.397	
Diisopropylamine nitrite	21	4.84×10^{-3}	139
Morpholine nitrite	21	3×10^{-3}	
Dicyclohexylamine nitrite	21	1.3×10^{-4}	179
Cyclohexylamine benzoate	21	8×10^{-5}	
Dicyclohexylamine caprylate	21	5.5×10^{-4}	
Guanadine chromate	21	1×10^{-5}	
Hexamethyleneimine benzoate	41	8×10^{-4}	64
Hexamethyleneamine nitrobenzoate	41	1×10^{-6}	136
Dicyclohexylamine benzoate	41	1.2×10^{-6}	210

dent of concentration. In the case of the amine nitrites and amine carboxylates, the net result of those reactions may be expressed as

$$H_2O + R_2NH_2NO_2 \rightarrow (R_2NH_2)^+{:}OH^- + H^+{:}(NO_2^-) \quad (10.10)$$

The nature of the adsorbed film formed at the steel-water interface is an important factor controlling the efficiency of VCIs. Metal surfaces exposed to vapors from VCIs in closed containers give evidence of having been covered by a hydrophobic-adsorbed layer. The contact angle of distilled water on such surfaces increases with time of exposure. Experimental studies on the adsorption of volatile inhibitors from the gas phase confirm the assumption that the VCIs react with the metal surface, thus providing corrosion protection. When a steel electrode is exposed to vapors of a VCI, the steady-state electrode potential shifts considerably into the region of positive values. The higher the vapor pressure, the stronger the shift of the electrode potential in the positive direction. Inhibitor adsorption is not a momentary process and requires much time for completion. This indicates that the adsorption is chemical and not physical in nature, resulting in a chemisorbed layer on the metal surface. In proper conditions, the inhibitor molecule will become dissociated or undissociated from the vapor phase and will dissolve into the water layer, with several possible effects (i.e., on the pH, surface wetting, and electrochemical processes at the metal/aqueous film interface).

It is well known, and shown in potential-pH diagrams, that an alkalization of the corrosive medium has a beneficial effect on the corrosion resistance of some metals, notably ferrous metals. Cyclohexylamine and dicyclohexylamine are moderately strong bases (pK_a = 10.66 and 11.25, respectively). The pH of the solutions of their salts with weak acids will depend on the pK_a of the acid. For example, cyclohexylamine carbonate will have a rather alkaline pH (pK_a for carbonic acid: 6.37), whereas dicyclohexyl ammonium nitrite will have a neutral pH (pK_a = 3.37 for nitrous acid). Guanidine is a strong base (pK_a = 13.54) and is mainly used as an additive in VCI formulations to adjust the alkalinity. Buffers (sodium tetraborate, etc.) may have to be used to maintain the pH of a VCI formulation at a convenient level.

The effect of a volatile inhibitor on the electrochemical processes at the metal surface is first evidenced by the shift in the steady-state electrode potential when an electrode is exposed to vapors of the volatile inhibitor.[11] The positive shift generally observed with most of the VCIs on ferrous metals is indicative of a preferentially anodic effect of the inhibitors. This anodic effect may be related either to a simple blocking effect of the anodic sites by the amine part of the inhibitors or to the contribution of the anionic component (i.e., the weak acid component).

In the case of nitrobenzoates, for example, it has been claimed that an acceleration of the cathodic partial process by reduction of the nitro group may lead, in addition to the effect of oxygen in the thin electrolyte layer, to a complete passivation of iron or ordinary steels. Contributions from the two parts of the dissociated molecule to the inhibitive effect is very likely and explains a synergistic effect of the inhibitor at the cathodic and anodic sites, as was suggested long ago. For example, it was shown by autoradiographic studies that the dissociation products of cyclohexylamine carbonate act separately on anodic and cathodic sites, with the former effect predominating.[11]

10.4 Selection of an Inhibitor System

Proper choices of inhibitors should be made by matching the appropriate inhibitor chemistry with the corrosion conditions and by selection of appropriate physical properties for the application conditions. Method of application and system characteristics must be considered when selecting physical properties of an inhibitor.

Inhibitor selection begins with the choice of physical properties. Must the inhibitor be a solid or liquid? Are melting and freezing points of importance? Is degradation with time and temperature critical? Must it be compatible with other system additives? Are specific solubility characteristics required? This list can be extensive but is important because it defines the domain of possible inhibitors. It must be the first step of the inhibitor evaluation for any new system. The physical measurements are those routinely done as part of minimal quality acceptance testing.

In choosing between possible inhibitors, the simplest corrosion tests should be done first to screen out unsuitable candidates. The philosophy of initial screening tests should be that poorly performing candidates are not carried forward. An inhibitor that does poorly in early screening tests might actually do well in the actual system, but the user seldom has the resources to test all possible inhibitors. The inhibitor user must employ test procedures that rigorously exclude inferior inhibitors even though some good inhibitors may also be excluded.

The challenge in inhibitor evaluation is to design experiments that simulate the conditions of the real-world system. The variables that must be considered include temperature, pressure, and velocity as well as metal properties and corrosive environment chemistry. System corrosion failures are usually localized and attributed to micro conditions at the failure site. Adequate testing must include the most severe conditions that can occur in the system and not be limited to macro or average conditions. Examples of microenvironments are hot spots in heat exchangers and highly turbulent flow at weld beads.

The practice of corrosion inhibition requires that the inhibitive species should have easy access to the metal surface. Ideally, surfaces should therefore be clean and not contaminated by oil, grease, corrosion products, water hardness scales, and so forth. Furthermore, care should be taken to avoid the presence of deposited solid particles. This conditioning is often difficult to achieve, and there are many cases where less than adequate consideration has been given to the preparation of systems to receive inhibitive treatment.

It is also necessary to ensure that the inhibitor reaches all parts of the metal surfaces. Care should be taken, particularly when first filling a system, that all dead ends, pockets, and crevice regions are contacted by the inhibited fluid. This will be encouraged in many systems by movement of the fluid in service, but in nominally static systems it will be desirable to establish a flow regime at intervals to provide renewed supply of inhibitor.[13]

Inhibitors must be chosen after taking into account the nature and combinations of metals present, the nature of the corrosive environment, and the operating conditions in terms of flow, temperature, and heat transfer. Inhibitor concentrations should be checked on a regular basis and losses restored either by appropriate additions of inhibitor or by complete replacement of the whole fluid as recommended, for example, with engine coolants. Where possible, some form of continuous monitoring should be employed, although it must be remembered that the results from monitoring devices, probes, coupons, and so forth, refer to the behavior of that particular component at that particular part of the system. Nevertheless, despite this caution, it must be recognized that corrosion monitoring in an inhibited system is well established and widely used.

References

1. Jones, L. W., *Corrosion and Water Technology for Petroleum Producers,* Tulsa, Okla, Oil and Gas Consultants International, 1988.
2. Hackerman, N., and Snaveley, E. S., Inhibitors, in Brasunas, A. de S. (ed.), *Corrosion Basics,* Houston, Tex., NACE International, 1984, pp. 127–146.
3. Chen, C. M., and Theus, G. J., *Chemistry of Corrosion-Producing Salts in Light Water Reactors,* Report NP-2298, Palo Alto, Cal., Electric Power Research Institute, 1982.
4. Hugel, G., Corrosion Inhibitors—Study of their Activity Mechanism, in *1st European Symposium on Corrosion Inhibitors,* Ferrara, Italy, U. of Ferrara, 1960.
5. Stern, M., Method for Determining Corrosion Rates from Linear Polarization Data, *Corrosion,* **14**(9):440–444 (1958).
6. Stern M., and Geary, A. L., Electrochemical Polarization I: A Theoretical Analysis of the Slope of Polarization Curves, *Journal of the Electrochemical Society,* **104**(1):559–563 (1957).

7. French, E. C., Martin, R. L., and Dougherty, J. A., Corrosion and Its Inhibition in Oil and Gas Wells, in Raman, A., and Labine, P. (eds.), *Reviews on Corrosion Inhibitor Science and Technology,* Houston, Tex., NACE International, 1993, pp. II-1-1–II-1-25.
8. Lahodny-Sarc, O., Corrosion Inhibition in Oil and Gas Drilling and Production Operations, in *A Working Party Report on Corrosion Inhibitors,* London, U.K., The Institute of Materials, 1994, pp. 104–120.
9. Sastri, V. S., Roberge, P. R., and Perumareddi, J. R., Selection of Inhibitors Based on Theoretical Considerations, in Roberge, P. R., Szklarz, K., and Sastri, S. (eds.), *Material Performance: Sulphur and Energy,* Montreal, Canada, Canadian Institute of Mining, Metallurgy and Petroleum, 1992, pp. 45–54.
10. Thomas, J. G. N., The Mechanism of Corrosion, in Shreir, L. L., Jarman, R. A., and Burstein, G. T. (eds.), *Corrosion Control,* Oxford, UK, Butterworths Heinemann, 1994, pp. 17:40–17:65.
11. Fiaud, C., Theory and Practice of Vapour Phase Inhibitors, in *A Working Party Report on Corrosion Inhibitors,* London, U.K., The Institute of Materials, 1994, pp. 1–11.
12. Miksic, B. A., Use of Vapor Phase Inhibitors for Corrosion Protection of Metal Products, in Raman, A., and Labine, P. (eds.), *Reviews on Corrosion Inhibitor Science and Technology,* Houston, Tex., NACE International, 1993, pp. II-16-1–II-16-13.
13. Mercer, A. D., Corrosion Inhibition: Principles and Practice, in Shreir, L. L., Jarman, R. A., and Burstein, G.T. (eds.), *Corrosion Control,* Oxford, UK, Butterworths Heinemann, 1994, pp. 17:11–17:39.

Chapter

11

Cathodic Protection

11.1 Introduction — 863
   11.1.1 Theoretical basis — 864
   11.1.2 Protection criteria — 866
   11.1.3 Measuring potentials for protection criteria — 867
11.2 Sacrificial Anode CP Systems — 871
   11.2.1 Anode requirements — 872
   11.2.2 Anode materials and performance characteristics — 873
   11.2.3 System design and installation — 874
11.3 Impressed Current Systems — 878
   11.3.1 Impressed current anodes — 880
   11.3.2 Impressed current anodes for buried applications — 881
   11.3.3 Ground beds for buried structures — 884
   11.3.4 System design — 885
11.4 Current Distribution and Interference Issues — 886
   11.4.1 Corrosion damage under disbonded coatings — 886
   11.4.2 General current distribution and attenuation — 888
   11.4.3 Stray currents — 892
11.5 Monitoring the Performance of CP Systems for Buried Pipelines — 904
   11.5.1 CP system hardware performance monitoring — 904
   11.5.2 Structure condition monitoring — 905
References — 919

11.1 Introduction

The basic principle of cathodic protection (CP) is a simple one. Through the application of a cathodic current onto a protected structure, anodic dissolution is minimized. Cathodic protection is often applied to coated structures, with the coating providing the primary form of corrosion protection. The CP current requirements tend to be excessive for uncoated systems. The first application of CP dates back to 1824, long before its theoretical foundation was established. This chapter deals mainly with CP related to buried pipelines, an important application field. Other common CP installations include buried tanks, marine structures such as offshore platforms, and reinforcing steel in concrete.

11.1.1 Theoretical basis

The CP principle is illustrated in Fig. 11.1 for a buried pipeline, with the electrons supplied to the pipeline by using a dc source and an ancillary anode. In the case of a coated pipeline, it should be noted that current (using the conventional direction) is flowing to the areas as the coating is defective. The nonuniform current flux arising from the particular geometry in Fig. 11.1 is also noteworthy. Furthermore, it should be noted that an electron current flows along the electric cables connecting the anode to the cathode, and ionic current flows in the soil between the anode and cathode to complete the circuit.

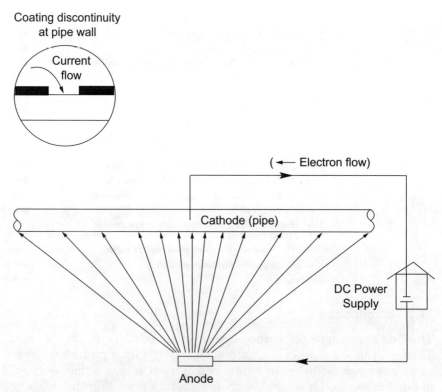

Figure 11.1 Current flow and distribution in cathodic protection of a pipeline (schematic). Note the current flow for a coated pipeline at a coating discontinuity.

An Evans diagram can provide the theoretical basis of CP. Such a diagram is shown schematically in Fig. 11.2, with the anodic metal dissolution reaction under activation control and the cathodic reaction diffusion limited at higher density. As the applied cathodic current density is stepped up, the potential of the metal decreases, and the anodic dissolution rate is reduced accordingly. Considering the logarithmic current scale, for each increment that the potential of the metal is reduced, the current requirements tend to increase exponentially.

In anaerobic, acidic environments the hydrogen evolution reaction tends to occur at the cathodically protected structure, whereas oxygen reduction is a likely cathodic reaction in aerated, near-neutral environments:

$$2H^+ + 2e^- \rightarrow H_2 \quad \text{(anaerobic, acidic environments)}$$

$$O_2 + 2H_2O + 4e^- \rightarrow 4OH^- \quad \text{(near-neutral environments)}$$

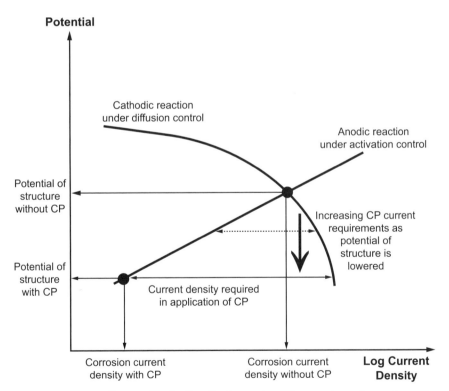

Figure 11.2 Evans diagram illustrating the increasing CP current requirements as the potential of the structure is lowered to reduce the anodic dissolution rate.

The production of hydroxide ions, leading to alkaline surface conditions, should be noted in the oxygen reduction reaction. Pourbaix diagrams are useful to determine the possible cathodic reactions as a function of the potential of the cathodically polarized structure. Combinations of different thermodynamically stable reactions can occur in practice.

The balancing anode reactions depend on the material of the anode and the environment. The following are examples of reactions at the anodes of a CP system:

$$M \rightarrow M^{n+} + ne^- \quad \text{(for a consumable anode)}$$

$$2H_2O \rightarrow O_2 + 4H^+ + 4e^- \quad \text{(inert anode)}$$

$$2Cl^- \rightarrow Cl_2 + 2e^- \quad \text{(inert anode in brackish environment)}$$

11.1.2 Protection criteria

In practical terms, a decision has to be made concerning the level of cathodic protection current that is applied. Too little current will lead to excessive corrosion damage, whereas excessive current (or "overprotection") can lead to disbonding of coatings and hydrogen embrittlement. Furthermore, corroding structures do not have uniform corrosion potentials or protection requirements over their entire surface. Practical protection criteria need to take such variations into account. The following is a list of protection criteria that have been proposed for buried steel structures:

- Potential of structure ≤ −850 mV w.r.t. saturated Cu/CuSO$_4$ reference electrode (under aerobic conditions)

- Potential of structure ≤ −950 mV w.r.t. saturated Cu/CuSO$_4$ reference electrode (under anaerobic conditions where microbial corrosion may be a factor)

- Negative potential shift of ≥ 300 mV when current is applied

- Positive potential shift of ≥ 100 mV when the current is interrupted

The first criterion is probably the best known and widely used in industry due to its ease of application. Using the Nernst equation and a ferrous ion concentration of $10^{-6}\,M$ (a criterion commonly used to define negligible corrosion in thermodynamics), a potential for steel of -930 mV w.r.t. Cu/CuSO$_4$ can be derived, which is somewhat more negative than this criterion. The satisfactory performance under the less stringent potential requirement may be related to the formation of protective ferrous hydroxide on the surface. Strictly speaking, potential protection criteria are based on the potential of the structure at the soil interface. Actual measurements performed by placing the reference electrode some distance away from the structure usually have to be corrected.

The potential shift criteria require measurements with the CP system in the ON and OFF conditions. Ohmic drop errors (see section below) can invalidate the 300-mV shift criterion. The 100-mV shift criterion may be useful for preventing overprotection. Table 11.1[1], provides further information on desirable and undesirable potential ranges for buried steel. The harmful effects of overprotection are included in this table.

Different protection criteria are required for different material-environment combinations. Other construction materials commonly used in buried applications, such as copper, aluminum, and lead, have different potential criteria than those given for steel above. Table 11.2 provides a comprehensive listing of cathodic protection criteria for different materials and environments. It should be noted that excessively negative potentials can be damaging to materials such as lead and aluminum and their alloys, due to the formation of alkaline species at the cathode.

11.1.3 Measuring potentials for protection criteria

Strictly speaking, the potential protection criteria outlined above refer to the structure-to-soil potential. However, in practice, it is clearly difficult to measure this potential of, say, a buried pipeline. In principle, a reference would have to be placed in the soil surrounding the pipeline, at an "infinitely" small distance away from the pipeline surface. When a reference electrode is placed at ground level to measure the potential of the cathodically protected pipeline, this measurement will contain two components: (1) the pipe-to-soil potential and (2) the so-called IR drop (Fig. 11.3). The IR drop error arises from the fact that

TABLE 11.1 Relationship between Potential and Corrosion Risk for Buried Steel

Potential (V vs. $Cu/CuSO_4$)	Condition of steel
-0.5 to -0.6	Intense corrosion
-0.6 to -0.7	Corrosion
-0.7 to -0.8	Some protection
-0.8 to -0.9	Cathodic protection
-0.9 to -1.0	Some overprotection
-1.0 to -1.1	Increased overprotection
-1.1 to -1.4	Increasingly severe overprotection, coating disbondment and blistering, increasing risk of hydrogen embrittlement

TABLE 11.2 Selected Cathodic Protection Criteria for Different Materials

Material	CP criteria	Standard/reference
Buried steel and cast iron (not applicable to applications in concrete)	−850 mV vs. Cu/CuSO$_4$	NACE Standard RP0169-83
	Minimum negative 300-mV shift under application of CP	NACE Standard RP0169-83
	Minimum positive 100-mV shift when depolarizing (after CP current switched off)	NACE Standard RP0169-83
	−850 mV vs. Cu/CuSO$_4$ in aerobic environment	British Standard CP 1021:1973
	−950 mV vs. Cu/CuSO$_4$ in anaerobic environment	British Standard CP 1021:1973
Steel (offshore pipelines)	−850 mV vs. Cu/CuSO$_4$	NACE Standard RP0675-75
	Minimum negative 300-mV shift under application of CP	NACE Standard RP0675-75
	Minimum positive 100-mV shift when depolarizing (after CP current switched off)	NACE Standard RP0675-75
Aluminum	Minimum negative potential shift of 150 mV under application of CP	NACE Standard RP0169-83
	Positive 100-mV shift when depolarizing (after CP current switched off)	NACE Standard RP0169-83
	Positive limit of −950 mV vs. Cu/CuSO$_4$	British Standard CP 1021:1973
	Negative limit of −1200 mV vs. Cu/CuSO$_4$	NACE Standard RP0169-83
	Negative limit of −1200 mV vs. Cu/CuSO$_4$	
Copper	Positive 100-mV shift when depolarizing (after CP current switched off)	NACE Standard RP0169-83
Lead	−650 mV vs. Cu/CuSO$_4$	British Standard CP 1021:1973
Dissimilar metals	Protection potential of most reactive (anodic) material should be reached	NACE Standard RP0169-83

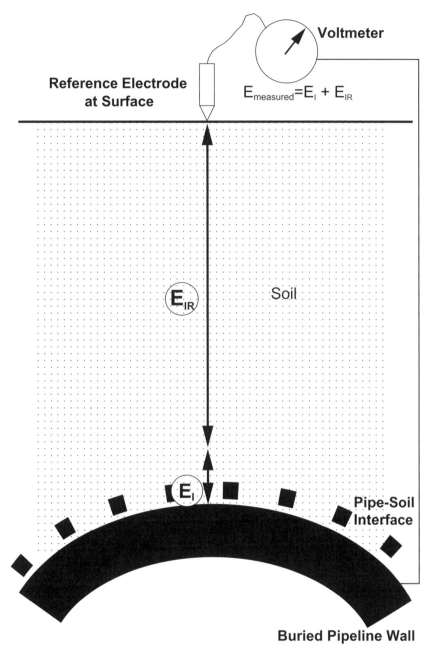

Figure 11.3 Schematic illustration of the IR drop error introduced during pipeline potential measurements at ground level. (E_{IR} = IR drop potential and E_I = pipe-to-soil potential.)

current is flowing through the soil and that the soil between the pipeline and the reference electrode has a certain electrical resistance.

Unfortunately, when a surface potential reading is made, the IR drop error will tend to give a false sense of security. In the presence of the IR drop, the pipeline potential will actually appear to be more negative than the true pipe-to-soil potential. It is thus hardly surprising that regulatory authorities are increasingly demanding that corrections for the IR drop error be made in assessments of buried structures.

To minimize this fundamental error, it has become customary to conduct so-called instant OFF potential readings, mainly in the case of impressed current cathodic protection systems. On the practical level, in systems involving numerous buried sacrificial anodes such readings are usually not possible. In this approach, the impressed CP current is interrupted briefly to theoretically provide a "true" pipe-to-soil potential reading. This momentary interruption of current theoretically produces a reading free from undesirable IR drop effects. The theoretical basis for this methodology is illustrated in Fig. 11.4. In practice, a so-called waveform analysis has to be performed to establish a suitable time interval following the current interruption for defining the OFF potential. As shown in Fig. 11.4, transient potential spikes tend to occur in the transition from the ON to the OFF potential, which should be avoided in establishing the OFF potential. There is thus no incentive to determine the OFF potential as soon as possible after interrupting the current; rather time

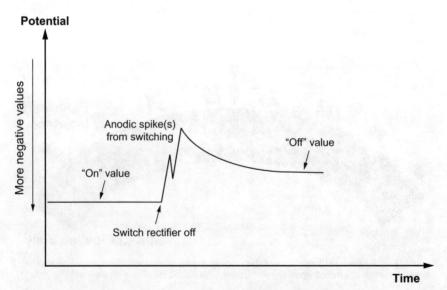

Figure 11.4 Measurement of instant-OFF potentials, by interrupting the CP current supply (schematic).

should be allowed for the spike(s) to dissipate. The total duration of the OFF cycle is only of the order of 1 s or even shorter.

11.2 Sacrificial Anode CP Systems

Cathodic protection can be applied by connecting sacrificial anodes to a structure. Basically, the principle is to create a galvanic cell, with the anode representing the less noble material that is consumed in the galvanic interaction (Fig. 11.5). Ideally, the structure will be protected as a result of the galvanic current flow. In practical applications a number of anodes usually have to be attached to a structure to ensure overall protection levels. The following advantages are associated with sacrificial anode CP systems:

- No external power sources required.
- Ease of installation (and relatively low installation costs).

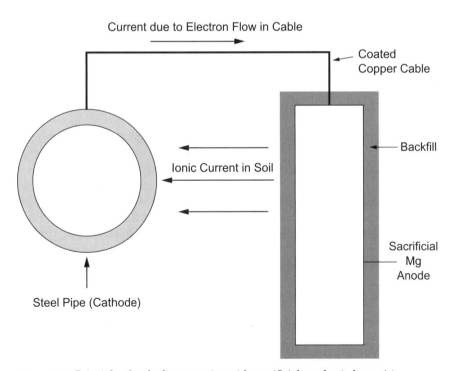

Figure 11.5 Principle of cathodic protection with sacrificial anodes (schematic).

- Unlikely cathodic interference in other structures.
- Low-maintenance systems (assuming low current demand).
- System is essentially self-regulating.
- Relatively low risk of overprotection.
- Relatively uniform potential distributions.

Unfortunately, these relatively simple systems also have some limitations such as

- Limited current and power output.
- High-resistivity environments or large structures may require excessive number of electrodes. Maximum resistivity of 6000 to 10,000 $\Omega \cdot cm$ is generally regarded as the limit, depending on coating quality.
- Anodes may have to be replaced frequently under high current demand.
- Anodes can increase structural weight if directly attached to a structure.

Typical applications include buried tanks, underground pipelines, buried communication and power cables, water and gas distribution systems, internal protection of heat exchangers and hot water tanks, ships, and marine structures.

11.2.1 Anode requirements

The anode material must provide a certain driving voltage to generate sufficient current to adequately protect a structure. The driving voltage is defined as the difference in between the operating voltage of the anode and the potential of the polarized structure it is protecting. A fundamental requirement is for the anode to have a stable operating potential over a range of current outputs. This means that the operating potential should lie very close to the free corrosion potential and that the corrosion potential remains essentially unaffected by current flow. With such characteristics, it is said that sacrificial anode systems are self-regulating in terms of potential. Furthermore, over its lifetime, an anode must consistently have a high capacity to deliver electric current per unit mass of material consumed. The capacity is defined as the total charge (in coulombs) delivered by the dissolution of a unit mass of the anode material. The theoretical capacity can be determined from Faraday's law, and the anode efficiency obtained in practice can be defined as

$$\text{Efficiency} = \frac{\text{actual capacity}}{\text{theoretical capacity}} \times 100\%$$

Ideally an anode will corrode uniformly and approach its theoretical efficiency. Passivation of an anode is obviously undesirable. Ease of manufacturing in bulk quantities and adequate mechanical properties are also important.

11.2.2 Anode materials and performance characteristics

For land-based CP applications of structural steel, anodes based on zinc or magnesium are the most important. Zinc anodes employed underground are high-purity Zn alloys, as specified in ASTM B418-95a. Only the Type II anodes in this standard are applicable to buried soil applications. The magnesium alloys are also high-purity grades and have the advantage of a higher driving voltage. The low driving voltage of zinc electrodes makes them unsuitable for highly resistive soil conditions. The R892-91 guidelines of the Steel Tank Institute give the following driving voltages, assuming a structure potential of -850 mV versus CSE:

High potential magnesium. -0.95 V

High-purity zinc: -0.25 V

Magnesium anodes generally have a low efficiency at 50 percent or even lower. The theoretical capacity is around 2200 Ah/kg. For zinc anodes, the mass-based theoretical capacity is relatively low at 780 Ah/kg, but efficiencies are high at around 90 percent.

Anodes for industrial use are usually conveniently packaged in bags prefilled with suitable backfill material. This material is important because it is designed to maintain low resistivity (once wetted) and a steady anode potential and also to minimize localized corrosion on the anode.

The current output from an anode can be estimated from Dwight's equation (applicable to relatively long and widely spaced anodes) as follows:

$$i = \frac{2\pi EL}{\rho \ln (8L/D - 1)}$$

where i = current output (A)
E = driving voltage of the anode (V)
L = anode length (cm)
ρ = soil resistivity ($\Omega \cdot$cm)
D = anode diameter (cm)

The life expectancy of an anode is inversely proportional to the current flowing and can be estimated with the following expression:

$$\text{Lifetime} = \frac{KUeW}{i}$$

where Lifetime = anode life (years)
- K = anode consumption factor (0.093 for Zn, 0.253 for Mg)
- U = utilization factor, a measure of the allowable anode consumption before it is rendered ineffective (typically 0.85)
- W = mass of the anode (kg)
- e = efficiency of the anode (0.9 for Zn, 0.5 for Mg)
- i = current output (A)

11.2.3 System design and installation

The design of CP systems lies in the domain of experienced specialists. Only the basic steps involved in designing a sacrificial anode system are outlined. Prior to any detailed design work a number of fundamental factors such as the protection criteria, the type and integrity of the coating system, the risk of stray current corrosion, and the presence of neighboring structures that could be affected by the CP system have to be defined.

Buried structures in soils. For structures buried in soil, such as pipelines, the first step in detailed design is usually to determine the resistivity of the soil (or other electrolyte). This variable is essential for determining the anodes' current output and is also a general measure of the environmental corrosiveness. The resistivity essentially represents the electrical resistance of a standardized cube of material. Certain measurement devices thus rely on measuring the resistance of a soil sample placed in a standard box or tube. A common way to make in situ measurement is by the so-called Wenner four-pin method. In this method, four equally spaced pins are driven into the ground along a straight line. The resistivity is derived from an induced current between the outer pin pair and the potential difference established between the inner pair. An additional type of resistivity measurement is based on electromagnetic inductive methods using a transmitter and pickup coils.

The second design step addresses electrical continuity and the use of insulating flanges. These parameters will essentially define the structural area of influence of the CP system. To ensure protection over different structural sections that are joined mechanically, electrical bonding is required. In complex structures, insulated flanges can restrict the spread of the CP influence.

In the third step the total current requirements are estimated. For existing systems, the current that has to be applied to achieve a certain potential distribution can be measured, but this is not possible for new systems. For the latter case, current requirements have to be determined based on experience, with two important variables standing out: First, the type of environment has to be considered for specifying an adequate level of current density. For example, a soil contaminated with active sulfate-reducing bacteria, leading to microbial corrosion effects, typically requires a higher current density for protection. The second important variable is the surface area that requires protection. The total current requirements obviously decrease with increasing quality of the surface coating. Field-coated structures usually have higher current requirements compared with factory-coated structures. The effective exposed area of coated structures used for design purposes should take coating deterioration with time into account.

Following the above, a suitable anode material can be selected, together with the number of anodes and anode size for a suitable output and life combination. The anode spacing also has to be established to obtain a suitable current distribution over the entire structure. Provision also has to be made for test stations to facilitate basic performance monitoring of the CP system. There are two basic types of test station. In one type, a connection to the pipe by means of a shielded lead wire is provided at the surface. Such a connection is useful for monitoring the potential of the pipeline relative to a reference electrode. The reference electrode may be a permanent installation. The second type provides surface access to the anode-structure connection. The current flowing from the anode to the structure can thereby be conveniently monitored at the surface. More details may be found in the publication of Peabody.[2]

In urban centers test stations are usually recessed into the ground with their covers flush with the pavement (Fig. 11.6). In outlying rural areas test stations tend to be above ground in the form of test posts. It is important to record the location of each test station. In urban areas a locating system based on street names and position relative to lot lines is commonly used. Locations relative to landmarks can be used in rural situations. A more recent option is the Global Positioning System (GPS) for finding test stations in the field. The relevant GPS coordinates obviously have to be recorded initially, before GPS positioning units can be used for locating test stations. Affordable handheld GPS systems are now readily available for locating rural test stations with reasonable accuracy.

Professional installation procedures are a key requirement for ensuring adequate performance of sacrificial anode CP systems.

Figure 11.6 Ground-level test station used in urban areas.

Following successful design and installation, the system is essentially self-regulating. Although the operating principles are relatively simple, attention to detail is required, for example, in establishing wire connections to the structure. The R892-91 guidelines of the Steel Tank Institute highlight the importance of an installation information package that should be made available to the system installer. The following are key information elements:

- A site plan drawn to scale, identifying the size, quantity, and location of anodes, location and types of test stations, layout of piping and foundations
- Detailed material specifications related to the anodes, test stations, and coatings, including materials for coating application in the field
- Site-specific installation instructions and/or manufacturer's recommended installation procedures
- Inspection and quality control procedures for the installation phase

Submerged marine structures. Cathodic protection of submerged marine structures such as steel jackets of offshore oil and gas platforms and pipelines is widely provided by sacrificial anode systems. A commonly used protection criterion for such steel structures is −800 mV relative to a silver/silver chloride-reference electrode. In offshore applications, impressed current systems are more vulnerable to mechanical wear and tear of cabling and anodes. Compared to soils, seawater has a low resistivity, and the low driving voltages of sacrificial anodes are thus of lower concern in the sea. The sacrificial anodes in offshore applications are usually based on aluminum or zinc. The chemical composition of an aluminum alloy specified for protecting an offshore gas pipeline is presented in Table 11.3.[3] Close control over impurity elements is crucial to ensure satisfactory electrochemical behavior. Sydberger, Edwards, and Tiller[4] have presented an excellent overview of designing sacrificial anode systems for submerged marine structures, using a conservative approach. A brief summary of this publication follows.

One of the main benefits of adequate design and a conservative design approach is that future monitoring and maintenance requirements will be minimal. Correct design also ensures that the system will essentially be self-regulating. The anodes will "automatically" provide increased current output if the structure potential shifts to more positive values, thereby counteracting this potential drift. Furthermore, a conservative design approach will avoid future costly retrofits. Offshore in situ anode retrofitting tends to be extremely costly and will tend to exceed the initial "savings." Such a design approach has also proven extremely valuable for requalification of pipelines, well beyond their original design life. A conservative design approach is sensible when considering that the cost of CP systems may only be of the order of 0.5 to 1% of the total fabrication and installation costs.

The two main steps involved in the design calculations are (1) calculation of the average current demand and the total anode net mass required to protect the structure over the design life and (2) the initial and final current demands required to polarize the structure to the required potential protection criterion. The first step is associated with

TABLE 11.3 Chemical Composition of Anode Material for an Offshore Pipeline

Element	Maximum, wt. %	Minimum, wt. %
Zinc	5.5	2.5
Indium	0.04	0.015
Iron	0.09	/
Silicon	0.10	/
Copper	0.005	/
Others, each	0.02	/
Aluminum	Balance	/

the anticipated current density once steady-state conditions have been reached. The second step is related to the number and size of individual anodes required under dynamic, unsteady conditions.

The cathodic current density is a complex function of various seawater parameters, for which no "complete" model is available. For design purposes, four climatic zones based on average water temperature and two depth ranges have therefore been defined: tropical, subtropical, temperate, and arctic. For example, in colder waters current densities tend to be higher due to a lower degree of surface protection from calcareous layers.

One major design uncertainty is the quality (surface coverage) of the coating. In subsea pipelines, the coating is regarded as the primary corrosion protection measure, with CP merely as a back-up system. For design purposes, not only do initial defects in the coating have to be considered but also its degradation over time.

In general, because of design uncertainties and simplifications, a conservative design approach is advisable. This policy is normally followed through judicious selection of design parameters rather than using an overall safety factor. Marginal designs will rarely result in underprotection early in the structure's life; rather the overall life of the CP system will be compromised. Essentially, the anode consumption rates will be excessive in underdesigned systems. Further details may be found in design guides such as NACE RP0176-94 and Det Norske Veritas (DNV) Practice RP B401.

11.3 Impressed Current Systems

In impressed current systems cathodic protection is applied by means of an external power current source (Fig. 11.7). In contrast to the sacrificial anode systems, the anode consumption rate is usually much lower. Unless a consumable "scrap" anode is used, a negligible anode consumption rate is actually a key requirement for long system life. Impressed current systems typically are favored under high-current requirements and/or high-resistance electrolytes. The following advantages can be cited for impressed current systems:

- High current and power output range
- Ability to adjust ("tune") the protection levels
- Large areas of protection
- Low number of anodes, even in high-resistivity environments
- May even protect poorly coated structures

The limitations that have been identified for impressed current CP systems are

Cathodic Protection

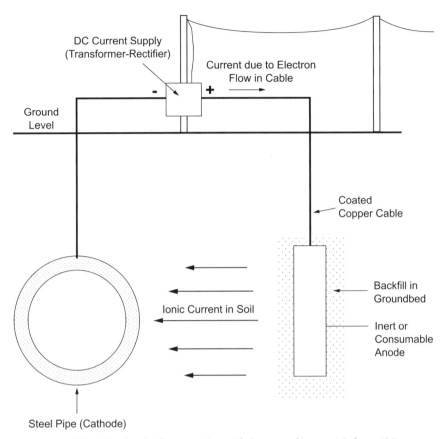

Figure 11.7 Principle of cathodic protection with impressed current (schematic).

- Relatively high risk of causing interference effects.
- Lower reliability and higher maintenance requirements.
- External power has to be supplied.
- Higher risk of overprotection damage.
- Risk of incorrect polarity connections (this has happened on occasion with much embarrassment to the parties concerned).
- Running cost of external power consumption.
- More complex and less robust than sacrificial anode systems in certain applications.

The external current supply is usually derived from a transformer-rectifier (TR), in which the ac power supply is transformed (down) and rectified to give a dc output. Typically, the output current from such units does not have pure dc characteristics; rather considerable "ripple" is inevitable with only half-wave rectification at the extreme end of the spectrum. Other power sources include fuel- or gas-driven generators, thermoelectric generators, and solar and wind generators. Important application areas of impressed current systems include pipelines and other buried structures, marine structures, and reinforcing steel embedded in concrete.

11.3.1 Impressed current anodes

Impressed current anodes do not have to be less noble than the structure that they are protecting. Although scrap steel is occasionally used as anode material, these anodes are typically made from highly corrosion-resistant material to limit their consumption rate. After all, under conditions of anodic polarization, very high dissolution rates can potentially be encountered. Anode consumption rates depend on the level of the applied current density and also on the operating environment (electrolyte). For example, the dissolution rate of platinized titanium anodes is significantly higher when buried in soil compared with their use in seawater. Certain contaminants in seawater may increase the consumption rate of platinized anodes. The relationship between discharge current and anode consumption rate is not of the simple linear variety; the consumption rate can increase by a higher percentage for a certain percentage increase in current.

Under these complex relationships, experience is crucial for selecting suitable materials. For actively corroding (consumable) materials approximate consumption rates are of the order of grams per ampere-hour (Ah), whereas for fully passive (nonconsumable) materials the corresponding consumption is on the scale of micrograms. The consumption rates for partly passive (semiconsumable) anode materials lie somewhere in between these extremes.

The type of anode material has an important effect on the reactions encountered on the anode surface. For consumable metals and alloys such as scrap steel or cast iron, the primary anodic reaction is the anodic metal dissolution reaction. On completely passive anode surfaces, metal dissolution is negligible, and the main reactions are the evolution of gases. Oxygen can be evolved in the presence of water, whereas chlorine gas can be formed if chloride ions are dissolved in the electrolyte. The reactions have already been listed in the theory section of this chapter. The above gas evolution reactions also apply to nonmetallic conducting anodes such as carbon. Carbon dioxide evolution is a further possibility for this material. On partially passive surfaces, both the metal dissolution and gas evolution reactions are important. Corrosion product buildup is obviously associated with the former reaction.

It is apparent that a wide range of materials can be considered for impressed current anodes, ranging from inexpensive scrap steel to high-cost platinum. Shreir and Hayfield[5] identified the following desirable properties of an "ideal" impressed current anode material:

- Low consumption rate, irrespective of environment and reaction products
- Low polarization levels, irrespective of the different anode reactions
- High electrical conductivity and low resistance at the anode-electrolyte interface
- High reliability
- High mechanical integrity to minimize mechanical damage during installation, maintenance, and service use
- High resistance to abrasion and erosion
- Ease of fabrication into different forms
- Low cost, relative to the overall corrosion protection scheme

In practice, important trade-offs between performance properties and material cost obviously have to be made. Table 11.4 shows selected anode materials in general use under different environmental conditions. The materials used for impressed anodes in buried applications are described in more detail below.

11.3.2 Impressed current anodes for buried applications

The NACE International Publication 10A196 represents an excellent detailed description of impressed anode materials for buried

TABLE 11.4 Examples of Impressed Current Anodes Used in Different Environments

Marine environments	Concrete	Potable water	Buried in soil	High-purity liquids
Platinized surfaces	Platinized surfaces	High-Si iron	Graphite	Platinized surfaces
Iron, and steel	Mixed-metal oxides	Iron and steel	High-Si Cr cast iron	
Mixed-metal oxides graphite	Polymeric	Graphite Aluminum	High-Si iron	
Zinc			Mixed-metal oxides	
High-Si Cr cast iron			Platinized surfaces	
			Polymeric, iron and steel	

applications. Further detailed accounts are also given by Shreir and Hayfield[5] and Shreir, Jarman, and Burstein;[6] only a brief summary is provided here.

Graphite anodes have largely replaced the previously employed carbon variety, with the crystalline graphite structure obtained by high-temperature exposure as part of the manufacturing process that includes extrusion into the desired shape. These anodes are highly porous, and it is generally desirable to restrict the anode reactions to the outer surface to limit degradation processes. Impregnation of the graphite with wax, oil, or resins seals the porous structure as far as possible, thereby reducing consumption rates by up to 50 percent. Graphite is extremely chemically stable under conditions of chloride evolution. Oxygen evolution and the concomitant formation of carbon dioxide gas accelerate the consumption of these anodes. Consumption rates in practice have been reported as typically between 0.1 to 1 kg $A^{-1} y^{-1}$ and operating currents in the 2.7 to 32.4 A/m^2 range. Buried graphite anodes are used in different orientations in anode beds that contain carbonaceous backfill.

The following limitations apply to graphite anodes: Operating current densities are restricted to relatively low levels. The material is inherently brittle, with a relatively high risk of fracture during installation and operational shock loading. In nonburied applications, the settling out of disbonded anode material can lead to severe galvanic attack of metallic substrates (most relevant to closed-loop systems) and, being soft material, these anodes can be subject to erosion damage.

Platinized anodes are designed to remain completely passive and utilize a surface coating of platinum (a few micrometers thick) on titanium, niobium, and tantalum substrates for these purposes. Restricting the use of platinum to a thin surface film has important cost advantages. For extended life, the thickness of the platinum surface layer has to be increased. The inherent corrosion resistance of the substrate materials, through the formation of protective passive films, is important in the presence of discontinuities in the platinum surface coating, which invariably arise in practice. The passive films tend to break down at a certain anodic potential, which is dependent on the corrosiveness of the operating environment. It is important that the potential of unplatinized areas on these anodes does not exceed the critical depassivation value for a given substrate material. In chloride environments, tantalum and niobium tend to have higher breakdown potentials than titanium, and the former materials are thus preferred at high system voltages.

These anodes are fabricated in the form of wire, mesh, rods, tubes, and strips. They are usually embedded in a ground bed of carbonaceous material. The carbonaceous backfill provides a high surface area (fine particles are used) and lowers the anode/earth resistance; effective transfer of current between the platinized surfaces and the backfill are therefore important. Reported consumption rates are less than 10 mg A^{-1} y^{-1} under anodic chloride evolution and current densities up to 5400 A/m^2. In oxygen evolution environments reported consumption rates are of the order of 16 mg/A-y at current densities below 110 A/m^2. In the presence of current ripple effects, platinum consumption rates are increased, particularly at relatively low frequencies.

Limitations include current attenuation in long sections of wire. Uneven current distribution results in premature localized anode degradation, especially near the connection to a single current feed point. Multiple feed points improve the current distribution and provide system redundancy in the event of excess local anode dissolution. Current ripple effects, especially at low frequencies, should be avoided. The substrate materials are at risk to hydrogen damage if these anodes assume a cathodic character outside of their normal operational function (for example, if the system is de-energized).

Mixed-metal anodes also utilize titanium, niobium, and tantalum as substrate materials. A film of oxides is formed on these substrates, with protective properties similar to the passive film forming on the substrate materials. The important difference is that whereas the "natural" passive film is an effective electrical insulator, the mixed metal oxide surface film passes anodic current. The product forms are similar to those of the platinized anodes. These anodes are typically used with carbonaceous backfill. Electrode consumption is usually not the critical factor in determining anode life; rather the formation of nonconductive oxides between the substrate and the conductive surface film limits effective functioning. Excessive current densities accelerate the buildup of these insulating oxides to unacceptable levels.

Scrap steel and iron represent consumable anode material and have been used in the form of abandoned pipes, railroad or well casings, as well as any other scrap steel beams or tubes. These anodes found application particularly in the early years of impressed current CP installations. Because the dominant anode reaction is iron dissolution, gas production is restricted at the anode. The use of carbonaceous backfill assists in reducing the electrical resistance to ground associated with the buildup of corrosion products. Periodic flooding with water can also alleviate resistance problems in dry soils.

Theoretical anode consumption rates are at 9 kg A^{-1} y^{-1}. For cast iron (containing graphite) consumption rates may be lower than theoretical due to the formation of carbon-rich surface films. Full utilization of the anode is rarely achieved in practice due to preferential dissolution in certain areas. Fundamentally, these anodes are not prone to failure at a particular level of current density. For long anode lengths, multiple current feed points are recommended to ensure a reasonably even current distribution over the surface and prevent premature failure near the feed point(s).

Limitations include the buildup of corrosion products that will gradually lower the current output. Furthermore, in high-density urban areas, the use of abandoned structures as anodes can have serious consequences if these are shorted to foreign services. An abandoned gas main could, for example, appear to be a suitable anode for a new gas pipeline. However, if water mains are short circuited to the abandoned gas main in certain places, leaking water pipes will be encountered shortly afterward due to excessive anodic dissolution.

High-silicon chromium cast iron anodes rely on the formation of a protective oxide film (mainly hydrated SiO_2) for corrosion resistance. The chromium alloying additions are made for use in chloride-containing environments to reduce the risk of pitting damage. These anodes can be used with or without carbonaceous backfill; in the latter case the resistance to ground is increased (particularly under dry conditions) as are the consumption rates. Consumption rates have been reported to typically range between 0.1 to 1 kg A^{-1} y^{-1}. The castings are relatively brittle and thus susceptible to fracture under shock loading.

Polymeric anodes are flexible wire anodes with a copper core surrounded by a polymeric material that is impregnated with carbon. The impregnated carbon is gradually consumed in the conversion to carbon dioxide, with ultimate subsequent failure by perforation of the copper strand. The anodes are typically used in combination with carbonaceous backfill, which reportedly increases their lifetime substantially. Because these anodes are typically installed over long lengths, premature failures are possible when soil resistivity varies widely.

11.3.3 Ground beds for buried structures

From the above description, the important role played by the ground beds in which the impressed current anodes are located should already be apparent. Carbonaceous material (such as coke breeze and graphite) used as backfill increases the effective anode size and lowers the resistance to soil. It is important to realize that, with such backfill, the anodic reaction is mainly transferred to the backfill. The consumption of the actual anode material is thereby reduced. To ensure low resistivity of the backfill material, its composition, particle size distribution, and degree of compaction (tamping) need to be controlled. The latter two variables also affect the degree to which gases generated at the anode installation can escape. If it is difficult to establish desirable backfill properties consistently in the ground, prepackaged anodes and backfill inside metal canisters can be considered. Obviously these canisters will be consumed under operational conditions.

The anodes may be arranged horizontally or vertically in the ground bed. The commonly used cylindrical anode rods may be the long continuous variety or a set of parallel rods. Some advantageous features of vertical deep anode beds include lower anode bed resistance, lower risk of induced stray currents, lower right-of-way surface area required, and improved current distribution in certain geometries. Limitations that need to be traded off include higher initial cost per unit of current output, repair difficulties, and increased risk of gas blockage.

At very high soil resistivities, a ground bed design with a continuous anode running parallel to a pipeline may be required. In such environments discrete anodes will result in a poor current distribution, and the potential profile of the pipeline will be unsatisfactory. The pipe-to-soil potential may only reach satisfactory levels in close proximity to the anodes if discontinuous anodes are employed in high-resistivity soil.

11.3.4 System design

Just as for sacrificial anode systems, design of impressed current CP systems is a matter for experienced specialists. The first three basic steps are similar to sacrificial anode designs, namely, evaluation of environmental corrosivity (soil resistivity is usually the main factor considered), determining the extent of electrical continuity in the system, and subsequently estimating the total current requirements.

One extremely useful concept to determine current requirements in existing systems is current drain testing. In these tests, a CP current is injected into the structure with a temporary dc power source. Small commercial units supplying up to 10 A of current are available for these purposes. A temporary anode ground bed is also required; grounded fixtures such as fences, fire hydrants, or street lights have been used. Potential loggers have to be installed at selected test stations to monitor the potential response to the injected current. The recorded relationship between potential and current is used to define what current level will be required to reach a certain protection criterion. An example of results from a current drain test performed on a buried, coated steel pipeline is presented in Fig. 11.8. Once the data loggers and current-supply hardware have been installed, these tests usually only require a few minutes of time.

Following the completion of the above three steps, the anode geometry and material have to be specified, together with a ground bed design. The designer needs to consider factors such as uniformity of

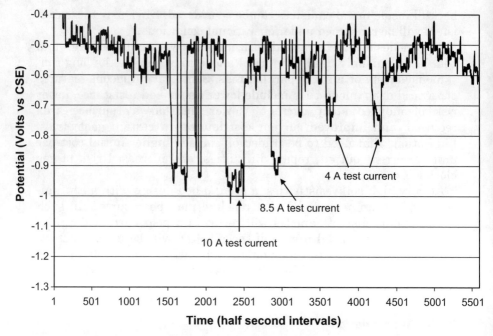

Figure 11.8 Current drain test results for a buried steel pipeline.

current distribution (see separate section below), possible interference effects (see Sec. 11.4.3), the availability of electrical power, and local bylaws and policies with respect to rectifier locations. Once the circuit layout and cabling are defined, the circuit resistance can be calculated and the rectifier can subsequently be sized in terms of current and potential output. Lastly, consideration must be given to the design of ancillary equipment for control purposes and test stations for monitoring purposes. Modern designs include provisions for remote rectifier performance monitoring and remote rectifier output adjustments.

11.4 Current Distribution and Interference Issues

11.4.1 Corrosion damage under disbonded coatings

It has already been stated that in buried cathodically protected structures, a surface coating is in fact the primary form of corrosion protection, with CP as a secondary measure. Users of this double protection methodology are sometimes surprised to find that severe localized corrosion damage has occurred under a coating, despite the two-fold preventive measures. Such localized corrosion damage has been observed in both sacrificial anode and impressed current CP systems. Importantly, it may not be possible to detect such problems in structure-to-soil potential surveys.

The phenomenon of coating disbondment plays a major role in this type of problem. The protective properties of a coating are greatly dependent on its ability to resist disbondment around defects.[7] The protective properties of the coating are compromised when water enters the gap between the (disbonded) coating and the metallic surface. A corrosive microenvironment will tend to develop in such a situation. Depending on the nature of this microenvironment, the CP system may not be able to protect the surface under the disbondment. Only when the trapped water has a high conductivity (e.g., saline conditions) will a protective potential be projected under the disbondment.[8] In the absence of protective CP effects, the surface will corrode under the free corrosion potential of the particular microenvironment that is established.

Jack, Wilmott, and Sutherby[8] identified three primary corrosion scenarios that could be manifested under shielded disbonded coatings on buried steel pipelines, together with secondary transformations of the primary sites (Table 11.5). A brief description follows.

Aerobic sites. Under aerobic conditions, oxygen reduction is the dominant cathodic reaction. Corrosion rates thus depend on the mass transport of oxygen to the steel surface. Under stagnant conditions, oxygen diffusion into the solution under the shielded disbondment is the rate-limiting step. The formation of surface oxides is also important for corrosion kinetics. The main corrosion products expected under aerobic conditions are iron (III) oxides/hydroxides.

Anaerobic sites. Hydrogen evolution is a prime candidate for the cathodic half-cell reaction under anaerobic conditions. Corrosion rates therefore tend to increase with decreasing pH (increasing acidity levels). In the case of ground water saturated with calcium and carbonate, the corrosion product is mainly iron (II) carbonate, a milky white precipitate. On exposure to air this white product will revert rapidly to reddish iron (III) oxides.

TABLE 11.5 Primary Corrosion Scenarios and Transformations at Disbonded Coating Sites for Steel Pipelines Buried in Alberta Soil

Primary corrosion scenario	Secondary transformation
Aerobic	Anaerobic + sulfate reducing bacteria (SRB)
Anaerobic	Aerobic
Anaerobic + SRB	Aerobic

Anaerobic sites with sulfate reducing bacteria (SRB). Highly corrosive microenvironments tend to be created under the influence of SRB; they convert sulfate to sulfide in their metabolism. Likely corrosion products are black iron (II) sulfide (in various mineral forms) and iron (II) carbonate. SRB tend to thrive under anaerobic conditions. These chemical species will again tend to change if the corrosion cell is disturbed and aerated.

Secondary transformations. Changing soil conditions can lead to transformations in the primary corrosion sites. After all, soil conditions are dynamic with variations in humidity, temperature, water table levels, and so forth. For example, mixtures of iron (II) carbonate and iron (III) oxides and the relative position of these species have indicated dominant transformations from anaerobic to aerobic conditions, with the reddish products encapsulating the white species.

The transformation from anaerobic sites to aerobic sites is a drastic one, with high CP current demand and extremely high corrosion rates. Iron (II) sulfides are oxidized to iron (III) oxides and sulfur species. In turn, sulfur is ultimately oxidized to sulfate.

The change of aerobic sites to anaerobic sites with SRB leads to reduction of Fe (III) oxides to iron sulfide species. The conversion kinetics are pH dependent. Increasingly corrosive conditions should be anticipated with the formation of sulfide species.

11.4.2 General current distribution and attenuation

In practice, the current distribution in CP systems tends to be far removed from idealized uniform current profiles. It is the nature of electron current flow in structures and the nature of ionic current flow in the electrolyte between the anode and the structure that influence the overall current distribution. A number of important factors affect the current distribution, as outlined below.

One underlying factor is the anode-to-cathode separation distance. In general, too close a separation distance results in a poor distribution, as depicted in Fig. 11.9. A trade-off that must be made, when increasing this distance, is the increased resistance to current flow. At excessive distances, the overall protection levels of a structure may be compromised for a given level of power supply. Additional anodes can be used to achieve a more homogeneous ionic current flow, where an optimum anode-to-cathode separation distance cannot be achieved.

Resistivity variations in the electrolyte between the anode and cathode also have a strong influence on the current distribution. Areas of low resistivity will "attract" a higher current density, with current flowing preferentially along the path of least resistance. An example of

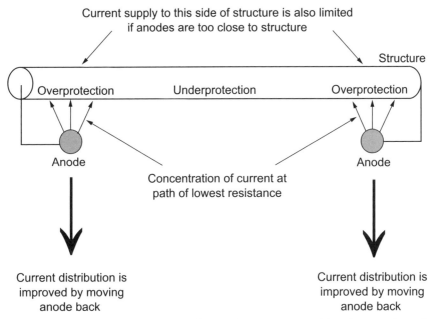

Figure 11.9 Nonuniform distribution of protective current resulting from anode positioning too close to the corroding structure (schematic).

such an unfavorable situation is illustrated in Fig. 11.10. Similar problems may be encountered in deeply buried structures, when different geological formations and moisture contents are encountered with increasing depth from the surface. An indication of resistivity variations across different media is given in Table 11.6.

Another important factor for coated structures is the presence of defects in the protective coating. Not only does the size of a defect affect the current but also the position of the defect relative to the anode. Current tends to be concentrated locally at defects. A fundamental source of nonuniformly distributed CP current over structures results from an effect known as attenuation. In long structures such as pipelines the electrical resistance of the structure itself becomes significant. The resistance of the structure causes the current to decrease nonlinearly as a function of distance from a drain point. A drain point refers to the point on the structure where its electrical connection to the anode is made. This characteristic decrease in current (and also in potential), shown in Fig. 11.11, occurs even under the following idealized conditions:

- The anodes are sufficiently far removed from the structure.
- The electrolyte resistivity is completely uniform between the anode(s) and the structure.

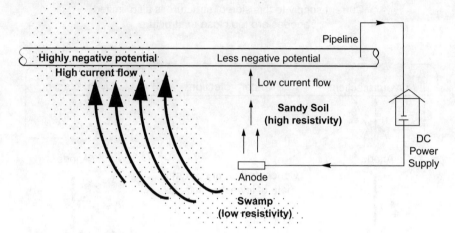

Figure 11.10 Nonuniform current distribution over a pipeline resulting from differences in the electrolyte (soil) resistivity (schematic). The main current flow will be along the path of least resistance.

TABLE 11.6 Resistivities of Different Electrolytes

Soil type	Typical resistivity, $\Omega \cdot cm$
Clay (salt water)	< 1000
Clay (fresh water)	< 2000
Marsh	1000–3000
Humus	1000–4000
Loam	3000–10,000
Sand	> 10,000
Limestone	> 20,000
Gravel	> 40,000

- The coating has a high and uniform ohmic resistance.
- A linear relationship exists between the potential of the structure and the current.

Under these idealized conditions the following attenuation equations apply

$$E_x = E_0 \exp(-\alpha x)$$
$$I_x = I_0 \exp(-\alpha x)$$

where E_0 and I_0 are the potential and current at the drainage point, and x is the distance from the drainage point.

The attenuation coefficient α is defined as

Cathodic Protection 891

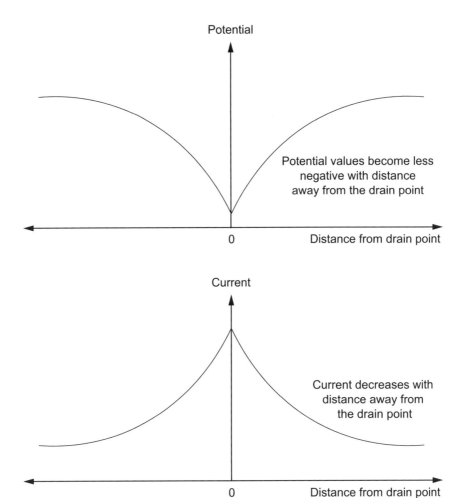

Figure 11.11 Potential and current attenuation as a function of distance from the drain point, due to increasing electrical resistance of the pipeline itself (schematic).

$$\alpha = \frac{R_S}{R_K}$$

where R_S is the ohmic resistance of the structure per unit length and R_K is given by

$$R_K = \sqrt{R_S R_L}$$

where R_L is known as the leakage resistance and refers to the total resistance of the structure-electrolyte interface, including the ohmic resistance of any applied surface coating(s).

To minimize attenuation, the term α should be as small as possible. This implies that for a given material a high R_K value is desirable. Because the ohmic resistance of the structure R_S is fixed for a given material, the leakage resistance R_L needs to be considered. The higher the integrity of the coating, the higher R_L will be. The buildup of calcareous deposits on exposed areas of cathodically protected structures will also tend to increase R_L. The formation of such deposits is therefore desirable for attenuation considerations. For achieving a relatively uniform current distribution in CP systems, the following factors are thus generally regarded as desirable:

- Relatively high electrolyte resistance
- Uniform electrolyte resistance
- Low resistivity of the structure
- High quality of coating (high resistance)
- Relatively high anode to cathode separation distance
- Sufficiently large power supply in the CP system

11.4.3 Stray currents

Stray currents are currents flowing in the electrolyte from external sources, not directly associated with the cathodic protection system. Any metallic structure, for example, a pipeline, buried in soil represents a low-resistance current path and is therefore fundamentally vulnerable to the effects of stray currents. Stray current tends to enter a buried structure in a certain location and leave it in another. It is where the current leaves the structure that severe corrosion can be expected. Corrosion damage induced by stray current effects has also been referred to as *electrolysis* or *interference*. For the study and understanding of stray current effects it is important to bear in mind that current flow in a system will not only be restricted to the lowest-resistance path but will be distributed between paths of varying resistance, as predicted by elementary circuit theory. Naturally, the current levels will tend to be highest in the paths of least resistance.

There are a number of sources of undesirable stray currents, including foreign cathodic protection installations; dc transit systems such as electrified railways, subway systems, and streetcars; welding operations; and electrical power transmission systems. Stray currents can be classified into three categories

1. Direct currents
2. Alternating currents
3. Telluric currents

Direct stray current corrosion. Typically, direct stray currents come from cathodic protection systems, transit systems, and dc high-voltage transmission lines. A distinction can be made between anodic interference, cathodic interference, and combined interference.

Anodic interference is found in relatively close proximity to a buried anode, under the influence of potential gradients surrounding the anode. As shown in Fig. 11.12, a pipeline will pick up current close to the anode. This current will be discharged at a distance farther away from the anode. In the current pickup region, the potential of the pipeline subject to the stray current will shift in the negative direction; in essence it receives a boost of cathodic protection current locally. This local current boost will not necessarily be beneficial, because a state of overprotection could be created. Furthermore, the excess of alkaline species generated can be harmful to aluminum and lead alloys. Conversely, in areas where the stray current is discharged, its potential will rise to more positive values. It is in the areas of current discharge that anodic dissolution is the most severe.

Cathodic interference is produced in relatively close proximity to a polarized cathode. As shown in Fig. 11.13, current will flow away from the structure in the region in close proximity to the cathode. The potential will shift in the positive direction where current leaves this structure, and this area represents the highest corrosion damage risk. Current will flow onto the structure over a larger area, at further distances from the cathode, again with possible damaging overprotection effects.

An example of combined anodic and cathodic interference is presented in Fig. 11.14. In this case current pickup occurs close to an

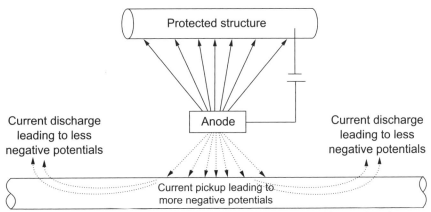

Figure 11.12 Anodic interference example (schematic).

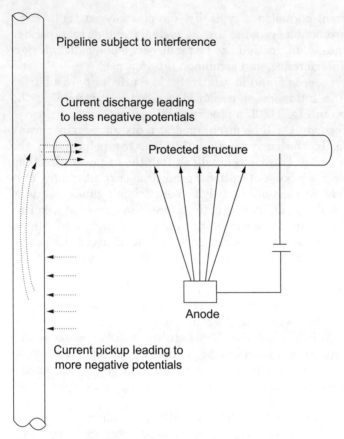

Figure 11.13 Cathodic interference example (schematic).

anode, and current discharge occurs close to a cathodically polarized structure. The degree of damage of the combined stray current effects is greater than in the case of anodic or cathodic interference acting alone. The effects are most pronounced if the current pickup and discharge areas are in close proximity. Correspondingly, the damage in both the current pickup (overprotection effects) and discharge regions (corrosion) will be greater.

Alternating current. There is an increasing trend for pipelines and overhead powerlines to use the same right-of-way. Alternating stray current effects arise from the proximity of buried structures to high-voltage overhead power transmission lines. There are two dominant mechanisms by which these stray currents can be produced in buried pipelines: electromagnetic induction and transmission line faults.

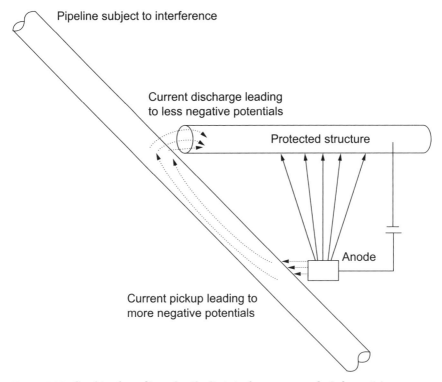

Figure 11.14 Combined anodic and cathodic interference example (schematic).

In the electromagnetic coupling mechanism, a voltage is induced in a buried structure under the influence of the alternating electromagnetic field surrounding the overhead transmission line. This effect is similar to the coupling in a transformer, with the overhead transmission line acting as the primary transformer coil and the buried structure acting as the secondary coil. The magnitude of the induced voltage depends on factors such as the separation distance from the powerline, the relative position of the structure to the powerlines, the proximity to other buried structures, and the coating quality. Such induced voltages can be hazardous to anyone who comes in contact with the pipeline or its accessories.[9]

The second mechanism is one of resistive coupling, whereby ac currents are directly transmitted to earth during transmission line faults. Causes of such faults include grounding of an overhead conductor, lightning strikes, and major load imbalances in the conductors. Usually such faults are of very short duration, but due to the high currents involved, substantial physical damage to coated structures is possible. Ancillary equipment such as motorized valves, sensors, and cathodic protection stations could also be damaged. These faults represent a major threat to human and animal life, even if no contact is made with the pipeline. The example listed in Table 11.7[10] for a pipeline provides an indication of the relative magnitude of these two mechanisms. Further details, including safety issues, may be found in the publication of Kirkpatrick.[9]

Telluric effects. These stray currents are induced by transient geomagnetic activity. The potential and current distribution of buried structures can be influenced by such disturbances in the earth's magnetic field. Such effects, often assumed to be of greatest significance in closer proximity to the poles, have been observed to be most intense during periods of intensified sun spot activity. In general, harmful influences on structures are of limited duration and do not remain highly localized to specific current pickup and discharge areas. Major corrosion problems as a direct result of telluric effects are therefore relatively rare.

Geomagnetic activity for different locations is recorded and reported by organizations such as the Geological Survey of Canada. Activity is classified into quiet, unsettled, and active conditions. Furthermore, charts forecasting magnetic activity are available, similar to short- and long-term weather forecasts. Such forecast data has proven useful to avoid measurements of pipeline "baseline" corrosion parameters during sporadic periods of high geomagnetic transients.

Controlling stray current corrosion. In implementing countermeasures against stray current effects, the nature of the stray currents has to be considered. For mitigating dc interference, the following fundamental steps can be taken:

- Removal of the stray current source or reduction in its output current
- Use of electrical bonding

TABLE 11.7 Example of Fault Effect Calculation

Route length	4.1 km
Overhead supply system voltage	66 kV
Supply system fault current	Three-phase 6350 A
	Single-phase to earth 1600 A
Fault current duration	Three phase: 0.68 s
	Single-phase to earth: 0.12 s
Fault trip operation	Single trip
Maximum induced voltage on pipe under normal current load	−2.5 V
Maximum induced voltage on pipe under fault current	−1050 V

- Cathodic shielding
- Use of sacrificial anodes
- Application of coatings to current pickup areas

To implement the first obvious option in the above listing, cooperation from the owners of the source is a prerequisite. In several cases, so-called electrolysis committees have been formed to serve as forums for cooperation between different organizations.

The establishment of an electrical connection between the interfering and interfered-with structure is a common remedial measure. Figure 11.15 shows how the interference problem presented in Fig. 11.12 is mitigated by an electrical bond created between the two structures. A variable resistance may be used in the bonded connection. A so-called forced drainage bond imposes an additional external potential on the bond to "assist" stray current drainage through the bond. In practice, for complex systems, the design of bonds is not a simple matter. Furthermore, stray currents tend to be dynamic in nature, with the direction of current reversing from time to time. In such cases, simple bonding is insufficient, and additional installation of diodes will be required to protect a critical structure at all times.

In cathodic shielding the aim is to minimize the amount of stray current reaching the structure at risk. A metallic barrier (or "shield") that is polarized cathodically is positioned in the path of the stray current, as shown in Fig. 11.16. The shield represents a low-resistance preferred path for the stray current, thereby minimizing the flow of stray current onto the interfered-with structure.

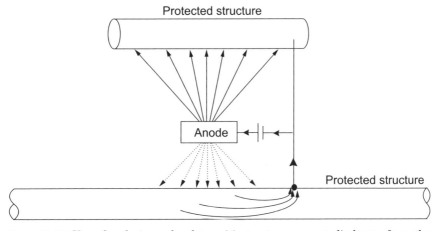

Figure 11.15 Use of a drainage bond to mitigate stray current discharge from the pipeline (schematic).

If the shield is connected to the negative terminal of the power supply of the interfering structure, its effects on the protection levels of the interfering structure have to be considered; these will obviously be reduced for a given rectifier output.

Sacrificial anodes can be installed at the current discharge areas of interfered-with structures to mitigate stray current corrosion. This mitigation method is most applicable to relatively low levels of stray currents. As shown in Fig. 11.17, the current is discharged from these anodes rather than from the structure at risk. The importance of placing the anodes close to the interfering structure is apparent: to minimize the resistance to current flowing from the anodes. The galvanically less noble anodes will generate a cathodic protection current, thereby compensating for small amounts of residual stray currents that continue to be discharged from the interfered-with structure.

The use of protective coatings to reduce stray current damage should be implemented prior to the installation of buried structures. It is usually impractical to apply such coatings after the installation phase. The use of coatings to mitigate the influence of stray currents should only be considered at the current pickup areas. It is not recommended to rely on additional coatings at current discharge areas, because rapid localized corrosive penetration is to be expected at any coating defect. In general, if a macroscopic anode and cathode exists on a structure, coatings should never be applied to the anode alone for corrosion protection. Any discontinuities in the coating covering only the anode represent sites where intense anodic dissolution will occur. It is much better practice to coat the cathode as a corrosion control method.

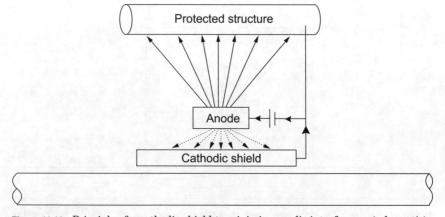

Figure 11.16 Principle of a cathodic shield to minimize anodic interference (schematic).

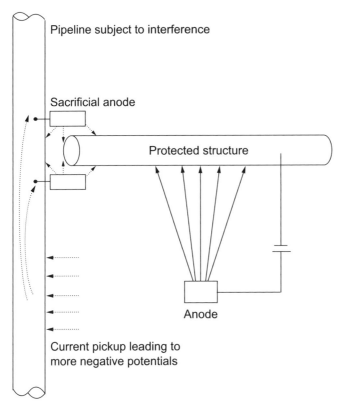

Figure 11.17 Use of sacrificial anodes to mitigate cathodic interference (schematic).

Ac-induced stray current effects can be reduced by locating buried structures sufficiently far away from power lines. Nonmetallic pipeline construction materials such as high-density polyethylene can be used in some cases, where operating pressures are low. The highest potential shifts occur on metallic structures that have high-integrity coatings. With high-quality coatings, grounding of the pipeline is clearly limited. Obviously, it is generally not desirable to sacrifice coating quality to reduce the magnitude of these effects. A similar reduction of the induced potential shifts can be achieved with distributed sacrificial anodes attached to the structure. These anodes provide cathodic protection current and reduce the resistance to ground, which is fundamentally desirable to minimizing the ac-induced voltage shifts. The use of such anodes will also tend to mitigate the influence of telluric effects.

Stray current case study—dc rail transit systems. Stray current-induced corrosion damage has been associated with North American dc rail transit systems for more than a century. In the United States alone, there are more than 20 transit authorities operating electrified rail systems in major urban centers. Stray current corrosion problems continue to plague several North American cities where the transit systems are typically installed in high-density urban areas. Examples of stray current corrosion problems in transit systems, excluding foreign structures, are listed in Table 11.8. Obviously such urban areas are associated with underground cables and piping (water and gas) systems that can also be highly susceptible to this form of corrosion damage.

A recent survey of the cost of corrosion in the United States has estimated that some 5 percent of the total cost is attributable to stray current effects, mostly due to electrified transit systems. This percentage includes the damage to utility structures operated outside the direct activities of the transit authorities. In other parts of the world electrified rail systems can represent the dominant form of rail transportation for passengers and freight. Not surprisingly, major stray current corrosion problems have also been associated with these systems, again with serious economic implications.

These stray current problems stem from the fundamental design of electrified rail transit systems, whereby current is returned to substations via the running rails. The ground surrounding the rails can be viewed as a parallel conductor to the rails. The magnitude of stray current flow in the ground conductor will obviously increase as its resistivity decreases. Any metallic structure buried in ground of this nature will tend to "attract" stray current because it represents a very low-resistance current path. The highest rate of metal dissolution occurs where the current leaves the structure, and undesirable overprotection effects can occur at the points of current pickup.

TABLE 11.8 Examples of Direct Stray Current Damage in Electrified Transit Systems

Type of damage	Comments
Corrosion of steel base plates and anchors in footings of supports	Caused by stray current discharge into the ground at elevated rail sections
Localized thinning of metal spikes in wooden ties	Lifetimes can be drastically reduced
Loss of rail section	
Reinforcing steel corrosion in concrete structures	Applicable to support structures, platforms, subway walls, tunnels, and surrounding buildings
Corrosion of expansion joint bonds	
Corrosion of steel shells in tunnels	Tunnels generally regarded as wet, highly corrosive areas conducive to stray current corrosion

A basic stray current scenario is illustrated in Fig. 11.18, where the negative rail has been grounded. Remote from the substation, due to a voltage drop in the rail itself, the rails will tend to be less negative relative to earth, and stray current flows onto the pipeline. Close to the substation, the rails are highly negative relative to earth, and stray current will tend to leave the pipeline and induce corrosion damage. In essence, the moving electrified vehicle represents a moving stray current source.

Considering that the rails are actually often mounted on wooden ties, some readers may wonder why current flow from the rails to the ground is actually possible. Such ties may after all appear to be insulators separating the rails from ground. First, the presence of water (rain) can obviously negate the insulating properties of the wooden ties by directly acting as an electrolyte. Wooden ties obviously always contain moisture and can retain rain water and therefore will never be perfect insulators, even in nominally dry conditions. Second, accumulation of soil and other debris will tend to bridge the ground and the rails, permitting the transfer of current. Third, metallic fasteners (spikes) holding the rails in place tend to act as short circuits for current flow to the ground. Considering these factors, it is apparent that for all practical purposes grounding of the rail can never be completely eliminated. Lastly, in some countries the outdated practices of deliberately bonding neighboring buried utility structures to the rail return current prevail. This approach is generally unsatisfactory because a large amount of stray current enters the ground that cannot be controlled in complex utility systems.

The stray currents tend to be very dynamic in nature, with the magnitude of stray current varying with usage of the transit system and relative position and degree of acceleration of the electrified vehicles. Fundamentally, the following factors all have an effect on the severity

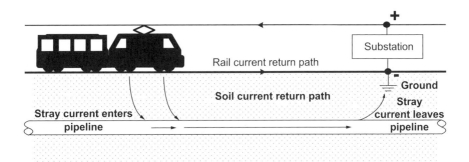

Figure 11.18 Stray current problem associated with an electrified dc transit system (schematic).

of stray currents: magnitude of propulsion current, substation spacing, substation grounding method, resistance of the running rails, usage and location of cross bonds and isolated joints, track-to-earth resistance, and the voltage of the traction power system. At a particular location on an affected pipeline, the presence of stray currents can thus usually be identified when fluctuating pipe-to-soil potentials are recorded with time.

Figure 11.19 illustrates the fluctuations in pipe-to-soil potentials for a pipeline in close proximity to an electrified rail transit system. Typically a number of trains would have passed this monitoring point during the data collection period in Fig. 11.19. Positive potential excursions associated with current discharge at the measuring point and negative potential transients related to current pickup are evident. At greater distances from the stray current source the potential profile is significantly more stable, as indicated by the second potential trace in Fig. 11.19.

Stray current effects are often detected at insulators in pipeline systems, which separate different sections of these systems. An example of stray current effects on two sides of an insulator is presented in Fig. 11.20. The more positive potential profile of pipeline on Side A

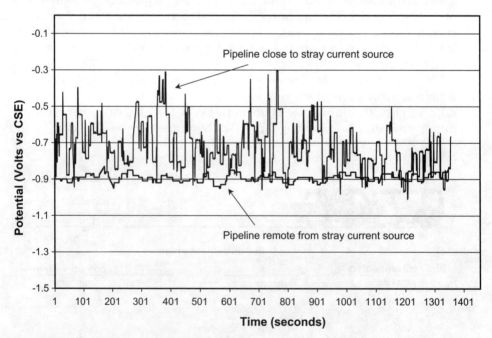

Figure 11.19 Stray current activity on a pipeline revealed by significant potential transients (schematic).

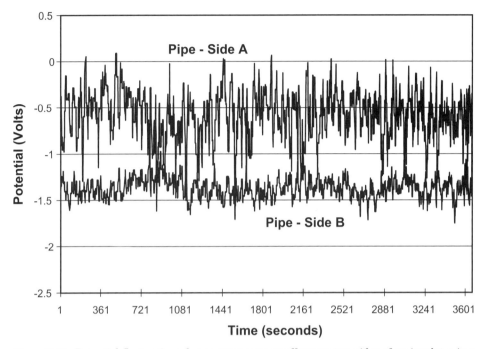

Figure 11.20 Potential fluctuations due to stray current effects on two sides of an insulator in a pipeline. Note the opposite direction of the potential transients on the two sides.

indicates that current discharge is predominantly occurring, representing an increased corrosion risk. Conversely on Side B, current pickup is the overriding effect. However, it should be noted that stray current flow is very dynamic in this situation, with potential transients in the positive and negative direction recorded on both sides of the insulator. Negative potential excursions on Side A correspond to positive excursions on Side B (and vice versa), indicating current pickup and discharge respectively (and vice versa).

The older dc transit systems generally produce the worst stray current problems due to the following factors:

- Relatively high electrical resistance of the running rails (smaller rail cross sections, bolted connections, deterioration of connections over time, etc.)
- Poor isolation from earth of the running rails (intentional grounded negative bus, intimate earth contact, moisture absorbing wood ties, etc.)
- Widely spaced substations leading to a higher voltage drop in the rails

In modern system designs stray current problems are ameliorated with two fundamental measures: (1) decreasing the electrical resistance of the rail return circuit and (2) increasing the electrical resistance between the rails and ground. The first measure makes current return through the ground less likely. Steps taken in this direction include the use of heavier rail sections, continuously welded rails, improved rail bonding, and reduced spacing between substations. It is desirable to combine substations with passenger stations. At passenger stations current flow is highest due to acceleration of trains. This combination ensures that these peak currents have a very short return path. The rail-to-soil resistance can be increased by using insulators placed between the rails and concrete or wooden ties and by using insulated rail fasteners. Stray current concerns are particularly relevant when older rail systems are integrated with newer designs. The higher current demand of modern, high-speed vehicles poses increased stray current risks in the older sections.

Apart from the above design improvements in more modern transit systems, several other remedial measures can be considered. Regular rail inspections are important for detecting problems with electrical continuity and faulty insulators. In addressing stray current problems of this nature, communication and cooperation between different organizations and stakeholders is most beneficial. Corrosion control coordinating committees are usually established for these purposes.

11.5 Monitoring the Performance of CP Systems for Buried Pipelines

Monitoring cathodic protection systems and corrosion damage to pipelines (and other structures) under their influence is a highly specialized subject. This material is therefore presented separately from the general corrosion monitoring chapter, although readers should be able to identify some overlap in basic concepts.

In the discussion of cathodic protection monitoring, two important distinct areas can be identified. The first domain lies in monitoring the condition and performance of the CP system hardware. Monitoring of rectifier output, pipe-to-soil potential and current measurements at buried sacrificial anodes, inspection of bonds, fuses, insulators, test posts, and permanent reference electrodes are relevant to this area. The second domain concerns the condition of the pipeline (or buried structure) itself and largely deals with surveys along the length of the pipeline to assess its condition and to identify high corrosion-risk areas.

11.5.1 CP system hardware performance monitoring

Because CP systems are expected to operate in demanding environmental conditions over long time periods, it should not be surprising that hardware maintenance is a fundamental requirement for reliable performance. Lightning strikes are also a major damage factor. NACE International has produced an excellent guideline,[11] describing procedures on the basis of monthly, quarterly, and annual monitoring. According to these guidelines, selected test stations are visited on a monthly basis. The rectifier units and selected sacrificial anode locations are visited, visually inspected, and their output measured. The quarterly procedures are similar, but the number of monitoring points ideally embraces the entire system. These comprehensive steps also apply to annual monitoring, which additionally focuses on the detailed inspection of system components (bonds, shunts, fuses, surge, divertors, distribution boxes, cable connections, mounting, systems etc.).

Considering the geographic expanse of typical CP systems, it is apparent that these tasks and the associated record keeping can be time consuming and labor intensive. An increasing trend toward selective remote rectifier monitoring, using modern communication systems and computer networks, is emerging to accomplish these tasks with reduced resources. The benefits of this approach were highlighted several years ago.[12] Wireless cell phone and satellite communication systems are available for interrogating rectifiers in remote locations. Importantly, GPS technology represents remote, wireless, low-cost timing devices for performing more specialized recordings with rectifier interruptions. Modern commercial systems presently provide the following remote rectifier monitoring features:

- Rectifier voltage and current output
- Standard on structure-to-soil potential data
- Instant OFF structure-to-soil potential data
- Depolarization data
- Ancillary readings from suitable sensors such as temperature and soil parameters
- Remote adjustment of rectifiers, provided "advanced" control hardware has been installed
- Alarm and alert notification, if preset operating windows are exceeded

11.5.2 Structure condition monitoring

Monitoring the condition of structures protected by CP is a highly specialized subject. In many cases, condition monitoring requirements are specified by regulatory authorities. Further details of techniques used for assessing the condition of buried pipelines are presented below.

Close interval potential surveys. Close Interval Potentials Survey (CIPS) refers to potential measurements along the length of buried pipelines to assess the performance of CP systems and the condition of the cathodically protected pipeline. The potential of a buried pipeline can obviously be measured at the permanent test posts (Fig. 11.21) but, considering that these may be miles apart, only a very small fraction of the overall pipeline surface can be assessed in this manner. The principle of a CIPS is to record the potential profile of a pipeline over its entire length by taking potential readings at intervals of around 1 m.

Figure 11.21 Test station with an electrical connection to the pipeline. (*Courtesy of CSIR North America Inc.*)

Methodology. In principle CIPS measurements are relatively simple. A reference electrode is connected to the pipeline at a test post, and this reference electrode is positioned in the ground over the pipeline at regular intervals (around 1 m) for the measurement of the potential difference between the reference electrode and the pipeline (Figs. 11.22 and 11.23).

In practice, a three-person crew is required to perform these measurements. One person walking ahead locates the pipeline with a pipe locator to ensure that the potential measurements are performed directly overhead the pipeline. This person also carries a tape measure and inserts a distance marker (a small flag) at regular intervals over the pipeline. The markers serve as distance calibration points in the survey. The second person carries a pair of electrodes that are connected to the test post by means of a trailing thin copper wire and the potential measuring instrumentation. This person is also responsible for entering specific features as a function of the measuring distance. Such details (road, creek, permanent distance marker, fence, rectifier,

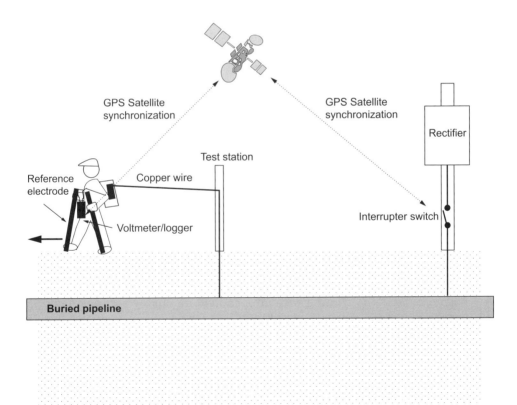

Figure 11.22 CIPS methodology (schematic).

Figure 11.23 CIPS equipment, with the operator starting a survey at a test post. (*Courtesy of CSIR North America Inc.*)

block valve, etc.) serve as useful geographical reference points when corrective actions based on survey results have to be taken. The third person collects the trailing wire after individual survey sections have been completed. (Strictly speaking, the first person may not be required if the distance can be monitored via a counter measuring the length of the unwinding copper wire.)

In practice, CIPS measurements are very demanding on the field crews and require extensive logistical support from both the pipeline operator and the CIPS contractor. Field crews are typically required to move over or around fences (of which there are many in residential areas), roads, highways, and other obstacles and difficult terrain. Breakage of the trailing copper wire is not uncommon, and special strengthened wire has to be taped down onto road surfaces that are crossed. The rate of progress for a survey thus greatly depends on the terrain to be surveyed. Pawson[13] has identified the following responsibilities of a pipeline operator in preparing for a CIPS:

- Preparation of a detailed technical specification for the survey
- Establishing and clearing the right of way (the path of the pipeline)
- Notification of land owners and foreign operators
- Establishing the sphere of influence of existing rectifiers and foreign structures
- Checking the condition and establishing functionality of rectifiers, bonds, and isolation
- Characterizing the effectiveness of the CP systems in difficult terrain (water swamps, etc.)
- Identification of suitable seasonal and weather windows
- Specification of the reporting format
- Ensuring availability of support personnel
- Specification of qualifications and experience required from the CIPS contractor

The CIPS contractor's responsibilities are[13]

- Provision of a proposal detailing specification compliance or, where appropriate, the specification of alternate work methods
- Provision of a sample of the report format(s)
- Provision of project schedules
- Utilization of professional crews and equipment (including materials, spares, support, and back-up equipment)
- Implementation of professional data management and report preparation

A report published by NACE International[14] gives additional guidelines about the nature of information that should be supplied by the pipeline operator and by the contractor. A most important consideration in the potential readings is the IR, or ohmic drop error, that is included in the potential measurements when a CP system is operational. As discussed earlier, the voltage criteria for protection against corrosion are based on the potential of the pipeline at the interface with the soil. The ON potential measurements taken at the surface do not represent this interfacial potential and include, among other effects, an IR drop from the cathodic polarization of the pipeline. In the presence of the IR drop, the measured ON potentials tend to be more negative than the actual interfacial potential, and a false sense of security may be obtained.

For minimization of IR errors, two current interruption criteria used in practice include a 3-s ON, 1-s OFF cycle and a 0.8-s ON, 0.2-s OFF cycle. The ON period is generally selected to be distinctly longer than the OFF period to limit depolarization of the pipeline. A waveform analysis is also important because the transition between the ON and OFF potential readings is not necessarily smooth; spikes of several hundred millivolts may be encountered in practice, and the measuring instrument should obviously be set up to avoid these transients (refer to Fig. 11.4).

To accomplish the ON-OFF switching for the above potential measurements, a current interrupter has to be installed on the rectifiers. Modern interrupters are based on solid-state switches and can be programmed to perform switching only when the survey is performed during the day; this feature minimizes the depolarization of the pipeline that may occur gradually due to the cumulative effects of the OFF periods. Importantly, a particular section of a pipeline will typically be under the influence of several rectifiers, and thus a number of rectifiers will have to be interrupted in a synchronized manner to perform the ON and OFF potential readings. Pipeline operators usually specify that at least two rectifiers ahead of the survey team and two rectifiers behind it have to be interrupted in a fully synchronized manner.

In modern practice, this type of synchronous switching of multiple rectifiers is accomplished by controlling the switch intervals and timing with the GPS. The potential recording device can also be synchronized to the rectifier interruption cycle with GPS technology. The GPS is a satellite-controlled radio navigation system that facilitates the determination of position (in three dimensions), velocity, and time. Users of this system may be based on land, at sea, or in the air, and the system is operational 24 h a day, in any weather conditions and anywhere on earth. The operator of this extremely accurate timing system is the U.S. Department of Defense.

On a single-phase rectifier an interrupter can, in principle, be installed at three locations: the input ac to the rectifier, the transformed ac current, or the dc from the rectifier. The current output of the rectifier to the pipeline should be verified before and after this type of interrupter installation. It will need to be adjusted if the current drawn to power the interrupter unit affects the output to the pipeline.

Data management. Computers are used to process survey data, with the first processing step being the downloading of data from the field measurement units at the end of each survey day. The data is typically transformed further for presentation in graphical form, together with a "client copy" of a database, using data storage devices such as magnetic disk, tape, or CD-ROM.

An example of graphical CIPS data is presented in Fig. 11.24. In the simplest format, the ON and OFF potentials are plotted as a function of distance. The distance is referred to as a station number, with station number 0-00 representing the starting point of a survey. The usual sign convention is for potentials to be plotted as positive values. The difference between the ON and OFF potential values should be noted. As is usually the case, the OFF potentials are less negative than the ON values. If the relative position is reversed, some unusual condition such as stray current interference is likely to have arisen.

NACE International has drawn up a Standard Recommended Practice [RP0792-92] for the format of computerized close interval survey data. These guidelines describe a general organization for CIPS data, applicable to both mainframe and personal computers. According to these recommendations, each data file containing the ON and OFF potential values should be supplemented with the following information, in the form of a separate header file or a header incorporated into the data file:

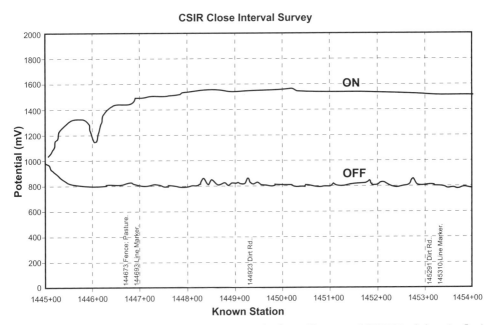

Figure 11.24 CIPS data, showing ON and OFF potential values. (*Courtesy of CSIR North America Inc.*)

- Data file name
- Date
- ON and OFF time cycles
- Environmental conditions
- Survey operator
- Pipeline identification
- Interval used in the survey
- Equipment details
- Location description
- Beginning and end stations
- Direction of survey and type of survey
- Measuring units
- Method of connection to the pipeline
- Additional user-defined information

The provision of comprehensive supporting information is of vital importance because the condition of a cathodically protected pipeline will typically be assessed through a number of surveys, over a number of years. The environmental conditions, measurement instrumentation and technology, and the contractor can therefore easily be different from survey to survey. Meaningful trending and comparison from survey to survey cannot be made if this supplementary information is not available. For example, the potential readings in dry soil may differ from wetter conditions.

Advantages and limitations. The CIPS technique provides a complete pipe-to-soil potential profile, indicating the status of cathodic protection levels. The interpretation of results, including the identification of defects, is relatively straightforward. The rate of progress of the survey team is independent of the coating quality on the pipeline. When the entire pipeline is walked, the condition of the right-of-way and of the cathodic protection equipment can be assessed together with the potential measurements.

Fundamentally, these surveys do not indicate the actual severity of corrosion damage, because the corrosion potential is not a kinetic parameter. The entire length of the pipeline has to be walked by a survey team and significant logistical support is required. The technique is not applicable to certain terrain such as paved areas, roads, rivers, and so forth.

Pearson survey. The Pearson survey, named after its inventor, is used to locate coating defects in buried pipelines. Once these defects have been identified, the protection levels afforded by the CP system can be investigated at these critical locations in more detail.

Methodology. An ac signal of around 1000 Hz is imposed onto the pipeline by means of a transmitter, which is connected to the pipeline and an earth spike, as shown in Fig. 11.25. Two survey operators make earth contact either through metal studded boots or aluminum poles. A distance of several meters typically separates the operators. Essentially, the signal measured by the receiver is the potential gradient over the distance between the two operators. Defects are located by a change in the potential gradient, which translates into a change in signal intensity.

As in the CIPS technique, the measurements are usually recorded by walking directly over the pipeline. As the front operator approaches a defect, increasing signal intensity is recorded. As the front person moves away from the defect, the signal intensity drops and later picks up again as the rear operator approaches the defect. The interpretation of signals can obviously become confusing when several defects are located between the two operators. In this case, only one person walks directly over the pipeline, with the connecting leads at a right angle to the pipeline.

In principle, a Pearson survey can be performed with an impressed cathodic protection system remaining energized. Sacrificial anodes

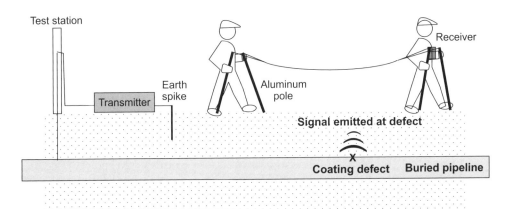

Figure 11.25 Pearson survey methodology (schematic).

should be disconnected because the signal from these may otherwise mask actual coating defects. A three-person team is usually required to locate the pipeline, perform the survey measurements, place defect markers into the ground, and move the transmitters periodically. The operator carrying the receiver should be highly experienced, especially if the survey is based on audible signals and instrument sensitivity settings. Under these conditions, the results are completely dependent on this operator's judgment.

Advantages and limitations. By walking the entire length of the pipeline, an overall inspection of the right-of-way can be made together with the measurements. In principle, all significant defects and metallic conductors causing a potential gradient will be detected. There are no trailing wires and the impressed CP current does not have to be pulsed.

The disadvantages are similar to those of CIPS because the entire pipeline has to be walked and contact established with ground. The technique is therefore unsuitable to roads, paved areas, rivers, and so forth. Fundamentally, no severity of corrosion damage is indicated and no direct measure of the performance of the CP system is obtained. The survey results can be very operator dependent, if no automated signal recording is performed.

Direct current voltage gradient (DCVG) surveys. DCVG surveys are a more recent methodology to locate defects on coated buried pipelines and to make an assessment of their severity. The technique again relies on the fundamental effect of a potential gradient being established in the soil at coating defects under the application of CP current; in general, the greater the size of the defect, the greater the potential gradient. The DCVG data is intricately tied to the overall performance of a CP system, because it gives an indication of current flow and its direction in the soil.

Methodology. The potential gradient is measured by an operator between two reference electrodes (usually of the saturated $Cu/CuSO_4$ type), separated by a distance of say half a meter. The appearance of these electrodes resembles a pair of cross-country ski poles (Fig. 11.26). A pulsed dc signal is imposed on the pipeline for DCVG measurements. The pulsed input signal minimizes interference from other current sources (other CP systems, electrified rail transit lines, telluric effects). This signal can be obtained with an interrupter on an existing rectifier or through a secondary current pulse superimposed on the existing "steady" CP current.

The operator walking the pipeline observes the needle of a millivoltmeter needle to identify defect locations. (More recently devel-

Figure 11.26 DCVG measuring equipment. (*Courtesy of CSIR North America Inc.*)

oped DCVG systems are digital and do not have a needle as such.) It is preferable for the operator to walk directly over the pipeline, but it is not strictly necessary. The presence of a defect is indicated by a increased needle deflection as the defect is approached, no needle deflection when the operator is immediately above the defect, and a decreasing needle deflection as the operator walks away from the defect (Fig. 11.27). It is claimed that defects can be located with an accuracy of 0.1 to 0.2 m, which represents a major advantage in minimizing the work of subsequent digs when corrective action has to be taken.

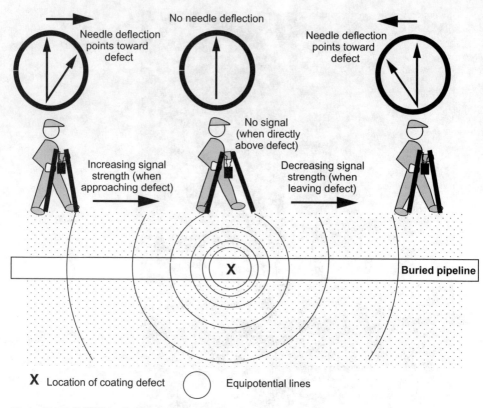

Figure 11.27 DCVG methodology (schematic).

An additional feature of the DCVG technique is that defects can be assigned an approximate size factor. Sizing is most important for identifying the most critical defects and prioritizing repairs. Leeds and Grapiglia[15] have provided details on the sizing procedure. An empirically based rating based on the so-called %IR value has been adopted in general terms as follows:

- 0 to 15%IR ("small"): No repair required usually.
- 16 to 35%IR ("medium"): Repairs may be recommended.
- 36 to 60%IR ("large"): Early repair is recommended.
- 61 to 100%IR ("extra large"): Immediate repair is recommended.

To establish a theoretical basis for the %IR value, the pipeline potential measured relative to remote earth at a test post must be considered. This potential (V_t) is made up of two components:

$$V_t = V_i + V_s$$

where V_i is the voltage across the pipe to soil interface and V_s is the voltage between the soil surrounding the pipe and remote earth. The %IR value is defined as

$$\%\text{IR} = \frac{V_s}{V_t}$$

Essentially the pipe-to-soil interface and the soil between the pipe and remote earth can be viewed as two resistors in series, with a potential difference across each of them. Although V_i cannot be measured easily in practice, V_s can be measured relatively easily with the DCVG instrumentation (one reference electrode is initially placed at the defect epicenter, and the voltage change is then summed as the electrodes are moved away from the epicenter to remote earth). In practice, the V_s value measured at a test post has to be extrapolated to a value at the defect location. Two test post readings bracketing the defect location and simple linear extrapolation are usually employed. For effective protection of the defect by the CP system, the V_s/V_t ratio should be small. The overall shift in pipeline potential due to the application of CP should be manifested by mainly shifting V_i, not V_s. Higher %IR values imply a lower level of cathodic protection.

Because the DCVG technique can be used to determine the direction of current flow in the soil, a further defect severity ranking has been proposed. As indicated in Fig. 11.1, current will tend to flow to a defect under the protective influence of the CP system. Corrosion damage (anodic dissolution) at the defect has an opposite influence; it will tend to make current flow away from the defect. Using an adaptation of the DCVG technique, it has been claimed that it is possible to establish whether current flows to or from a defect, with the CP system switched ON and OFF in a pulsed cycle.

Advantages and limitations. Fundamentally, the DCVG technique is particularly suited to complex CP systems in areas with a relatively high density of buried structures. These are generally the most difficult survey conditions. The DCVG equipment is relatively simple and involves no trailing wires. Although a severity level can be identified for coating defects, the rating system is empirical and does not provide quantitative kinetic corrosion information. The survey team's rate of progress is dependent on the number of coating defects present. Terrain restrictions are similar to the CIPS technique. However, it may be possible to place the electrode tips in asphalt or concrete surface cracks or in between the gaps of paving stones.

Corrosion coupons. Corrosion coupons connected to cathodically protected structures are finding increasing application for performance monitoring of the CP system. Essentially these coupons, installed uncoated, represent a defect simulation on the pipeline under controlled conditions. These coupons can be connected to the pipeline via a test post outlet, facilitating a number of measurements such as potential and current flow.

A publication describing an extensive coupon development and monitoring program on the Trans Alaska Pipeline System[16] serves as an excellent case study. This coupon monitoring program was designed to assess the adequacy of the CP system under conditions where techniques involving CP current interruption on the pipeline were impractical. Although the coupon monitoring methodology is based on relatively simple principles, significant development efforts and attention to detail are typically required in practice, as this case study amply illustrates.

Methodology. Perhaps the most important consideration in the installation of corrosion coupons is that a coupon must be representative of the actual pipeline surface and defect. The exact metallurgical detail and surface finish as found on the actual pipeline are therefore required on the coupon. The influence of corrosion product buildup may also be important. Furthermore the environmental conditions of the coupon and the pipe should also be matched (temperature, soil conditions, soil compaction, oxygen concentration, etc.). Current shielding effects on the bonded coupon should be avoided.

Several measurements can be made after a coupon-type corrosion sensor has been attached to a cathodically protected pipeline.[17] ON potentials measured on the coupon are in principle more accurate than those measured on a buried pipe, if a suitable reference electrode is installed in close proximity to the coupon. The potentials recorded with a coupon sensor may still contain a significant IR drop error, but this error is lower than that of surface ON potential measurements. Instant-OFF potentials can be measured conveniently by interrupting the coupon bond wire at a test post. Similarly, longer-term depolarization measurements can be performed on the coupon without depolarizing the entire buried structure. Measurement of current flow to or from the coupon and its direction can also be determined, for example, by using a shunt resistor in the bond wire. Importantly, it is also possible to determine corrosion rates from the coupon. Electrical resistance sensors provide an option for in situ corrosion rate measurements as an alternative to weight loss coupons.

The surface area of the coupon used for monitoring is an important variable. Both the current density and the potential of the coupon are dependent on the area. In turn, these two parameters have a direct relation to the kinetics of corrosion reactions.

Advantages and limitations. A number of important corrosion parameters can be conveniently monitored under controlled conditions, without any adjustments to the energized CP system of the structure. The measurement principles are relatively simple. It is difficult (virtually impossible) to guarantee that the coupon will be completely representative of an actual defect on a buried structure. The measurements are limited to specific locations. The coupon sensors have to be extremely robust and relatively simple devices to perform satisfactorily under field conditions.

References

1. Ashworth, V., The Theory of Cathodic Protection and Its Relation to the Electrochemical Theory of Corrosion, in Ashworth, V., and Booker, C. J. L. (eds.), *Cathodic Protection,* Chichester, U.K., Ellis Horwood, 1986.
2. Peabody, A. W., *Control of Pipeline Corrosion,* Houston, Tex., NACE International, 1967.
3. Eliassen, S., and Holstad-Pettersen, N., Fabrication and Installation of Anodes for Deep Water Pipelines Cathodic Protection, *Materials Performance,* **36**(6):20–23 (1997).
4. Sydberger, T., Edwards, J. D., and Tiller, I. B., Conservatism in Cathodic Protection Designs, *Materials Performance,* **36**(2):27–32 (1997).
5. Shreir, L. L., and Hayfield, P. C. S., Impressed Current Anodes, in Ashworth, V., and Booker, C. J. L. (eds.) *Cathodic Protection,* Chichester, U.K., Ellis Horwood, 1986.
6. Shreir, L. L., Jarman, R. A., and Burstein, G. T. (eds.), Corrosion, vol. 2, 3d ed., Oxford, Butterworth Heinemann, 1994.
7. Beavers, J. A., and Thompson, N. G., Corrosion Beneath Disbonded Pipeline Coatings, *Materials Performance,* **36**(4):13–19, (1997).
8. Jack, T. R., Wilmott, M. J., and Sutherby, R. L., Indicator Minerals Formed During External Corrosion of Line Pipe, *Materials Performance,* **34**(11):19–22 (1995).
9. Kirkpatrick, E. L., Basic Concepts of Induced AC Voltages on Pipelines, *Materials Performance,* **34**(7):14–18 (1995).
10. Allen, M. D., and Ames, D. W., Interaction and Stray Current Effects on Buried Pipelines: Six Case Histories, in Ashworth, V., and Booker, C. J. L. (eds.), *Cathodic Protection* Chichester, U.K., Ellis Horwood, 1986, pp. 327–343.
11. NACE International and Institute of Corrosion, *Cathode Protection Monitoring for Buried Pipelines,* pub. no. CEA 54276, Houston, Tex, NACE International, 1988.
12. Goloby, M. V., Cathodic Protection on the Information Superhighway, *Materials Performance,* **34**(7):19–21 (1995).
13. Pawson, R. L., Close Interval Potential Surveys—Planning, Execution, Results, *Materials Performance,* **37**(2):16–21 (1998).
14. NACE International, *Specialized Surveys for Buried Pipelines,* pub. no. 54277, Houston, Tex, NACE International, 1990.
15. Leeds, J. M., and Grapiglia, J., The DC Voltage-Gradient Method for Accurate Delineation of Coating Defects on Buried Pipelines, *Corrosion Prevention and Control,* **42**(4):77–86 (1995).
16. Stears, C. D., Moghissi, O. C., and Bone, III, L., Use of Coupons to Monitor Cathodic Protection of an Underground Pipeline, *Materials Performance,* **37**(2):23–31 (1998).
17. Turnipseed, S. P., and Nekoksa, G., Potential Measurement on Cathodically Protected Structures Using an Integrated Salt Bridge and Steel Ring Coupon, *Materials Performance,* **35**(6):21–25 (1996).

Chapter 12

Anodic Protection

12.1	Introduction	921
12.2	Passivity of Metals	923
12.3	Equipment Required for Anodic Protection	927
12.3.1	Cathode	929
12.3.2	Reference electrode	929
12.3.3	Potential control and power supply	930
12.4	Design Concerns	930
12.5	Applications	932
12.6	Practical Example: Anodic Protection in the Pulp and Paper Industry	933
	References	938

12.1 Introduction

In contrast to cathodic protection, anodic protection is relatively new. Edeleanu first demonstrated the feasibility of anodic protection in 1954 and tested it on small-scale stainless steel boilers used for sulfuric acid solutions. This was probably the first industrial application, although other experimental work had been carried out elsewhere.[1] This technique was developed using electrode kinetics principles and is somewhat difficult to describe without introducing advanced concepts of electrochemical theory. Simply, anodic protection is based on the formation of a protective film on metals by externally applied anodic currents. Anodic protection possesses unique advantages. For example, the applied current is usually equal to the corrosion rate of the protected system. Thus, anodic protection not only protects but also offers a direct means for monitoring the corrosion rate of a system. As an enthusiast and famous corrosion engineer claimed, "anodic protection can be classed as one of the most significant advances in the entire history of corrosion science."[2]

Anodic protection can decrease corrosion rate substantially. Table 12.1 lists the corrosion rates of austenitic stainless steel in sulfuric acid solutions containing chloride ions with and without anodic protection. Examination of the table shows that anodic protection causes a 100,000-fold decrease in corrosive attack in some systems. The primary advantages of anodic protection are its applicability in extremely corrosive environments and its low current requirements.[2] Table 12.2 lists several systems where anodic protection has been applied successfully.

Anodic protection has been most extensively applied to protect equipment used to store and handle sulfuric acid. Sales of anodically protected heat exchangers used to cool H_2SO_4 manufacturing plants have represented one of the more successful ventures for this technology.

TABLE 12.1 Anodic Protection of S30400 Stainless Steel Exposed to an Aerated Sulfuric Acid Environment at 30°C with and without Protection at 0.500 V vs. SCE

		Corrosion rate, $\mu m \cdot y^{-1}$	
Acid concentration, M	NaCl, M	Unprotected	Protected
0.5	10^{-5}	360	0.64
0.5	10^{-3}	74	1.1
0.5	10^{-1}	81	5.1
5	10^{-5}	49,000	0.41
5	10^{-3}	29,000	1.0
5	10^{-1}	2,000	5.3

TABLE 12.2 Current Requirements for Anodic Protection

			Current density	
H_2SO	Temperature, °C	Alloy	To passivate, $mA \cdot cm^{-2}$	To maintain, $\mu A \cdot cm^{-2}$
1 M	24	S31600	2.3	12
15%	24	S30400	0.42	72
30%	24	S30400	0.54	24
45%	65	S30400	180	890
67%	24	S30400	5.1	3.9
67%	24	S31600	0.51	0.10
67%	24	N08020	0.43	0.9
93%	24	Mild steel	0.28	23
99.9% (oleum)	24	Mild steel	4.7	12
H_3PO_4				
75%	24	Mild steel	41	20,000
115%	82	S30400	3.2×10^{-5}	1.5×10^{-4}
NaOH				
20%	24	S30400	4.7	10

These heat exchangers are sold complete with the anodic protection systems installed and have a commercial advantage in that less costly materials can be used. Protection of steel in H_2SO_4 (> 78% concentration) storage vessels is perhaps the most common application of anodic protection. There is little activity directed toward developing applications to protect metals from corrosion by other chemicals.[3]

Anodic protection is used to a lesser degree than the other corrosion control techniques, particularly cathodic protection. This is mainly because of the limitations on metal-chemical systems for which anodic protection will reduce corrosion. In addition, it is possible to accelerate corrosion of the equipment if proper controls are not implemented. However, anodic protection has its place in the corrosion control area, provided some important basics are respected.

12.2 Passivity of Metals

The passivation behavior of a metal is typically studied with a basic electrochemical testing setup (App. D, Basic Electrochemical Instrumentation). When the potential of a metallic component is controlled and shifted in the more anodic (positive) direction, the current required to cause that shift will vary. If the current required for the shift has the general polarization behavior illustrated in Fig. 12.1, the metal is active-passive and can be anodically protected. Only a few systems exhibit this behavior in an appreciable and usable way. The corrosion rate of an active-passive metal can be significantly reduced by shifting the potential of the metal so that it is at a value in the passive range shown in Fig. 12.1. The current required to shift the potential in the anodic direction from the corrosion potential E_{corr} can be several orders of magnitude greater than the current necessary to maintain the potential at a passive value. The current will peak at the passivation potential value shown as E_{pp} (Fig. 12.1).

To produce passivation the critical current density (i_{cc}) must be exceeded. The anodic potential must then be maintained in the passive region without allowing it to fall back in the active region or getting into the transpassive region, where the protective anodic film can be damaged and even break down completely. It follows that although a high current density may be required to cause passivation ($> i_{cc}$), only a small current density is required to maintain it, and that in the passive region the corrosion rate corresponds to the passive current density (i_p).

The relative tendency for passivation is strongly dependent on the interactions between a metal and its environment. The passivation behavior can vary extensively with changes in either. Figure 12.2 illustrates how the sensitization of a S30400 stainless steel, for example, can affect its passivation behavior when exposed to sulfuric acid.[4]

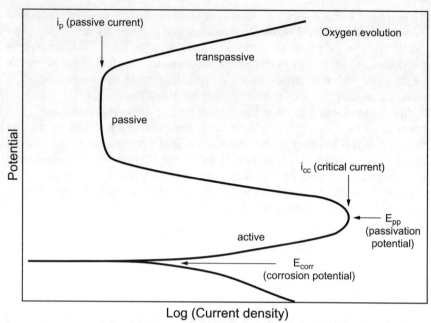

Figure 12.1 Hypothetical polarization diagram for a passivable system with active, passive, and transpassive regions.

Among the parameters that are particularly affected by sensitization are i_p and i_{cc}, as defined in Fig. 12.1. In this example, the ability to sustain passivity increases as the current density to maintain passivity (i_p) decreases and as the total film resistance increases, as indicated from measurements obtained with different metals exposed to 67% sulfuric acid (Table 12.3). The lower or more reducing the potential at which a passive metal becomes active, the greater the stability of passivity. The depassivation potential corresponding to the passive-active transition, called the Flade potential, can differ appreciably from E_{pp} measured by going through the active-passive process of the same system. This technical distinction is important for the control aspect of anodic protection where E_{pp} is the potential to traverse to obtain passivation, and the Flade potential is the potential to avoid traversing back into active corrosion.

Passivity can also be readily produced in the absence of an externally applied passivating potential by using oxidants to control the redox potential of the environment. Very few metals will passivate in nonoxidizing acids or environments, when the redox potential is more cathodic than the potential at which hydrogen can be produced. A good example of that behavior is titanium and some of its alloys, which can be readily passivated by most acids, whereas mild steel requires a strong oxidizing

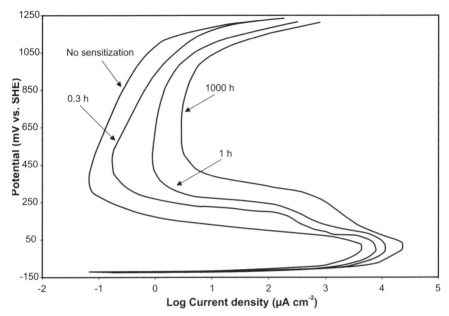

Figure 12.2 Anodic polarization curves of S30400 steel in a 1 M H_2SO_4 at 90°C after sensitization for various times.

agent, such as fuming HNO_3, for its passivation. Alloying with a more easily passivated metal normally increases the ease of passivation and lowers the passivation potential, as in the alloying of iron and chromium in 10% sulfuric acid (Table 12.4). Small additions of copper in carbon steels have been found to reduce i_p in sulfuric acid. Each alloy system has to be evaluated for its own passivating behavior, as illustrated by the case Ni-Cr alloys where both the additions of nickel to chromium and chromium to nickel decrease the critical current density in a mixture of sulfuric acid and 0.25 M K_2SO_4 (Table 12.5).[1]

The parameters defining and controlling the passivation domain of a system are thus directly related to the composition, concentration, purity, temperature, and agitation of the environment. This is illustrated with the current densities required to obtain passivity (i_{cc}), and to maintain passivity (i_p), for a S30400 steel in different electrolytes, as presented in Table 12.6. From the data in this table, it can be seen that it is approximately 100,000 times easier to passivate large areas of this steel in contact with 115% phosphoric acid than in 20% sodium hydroxide. The concentration of the electrolyte is also important, and for a S31600 steel in sulfuric acid, although there is a maximum corrosion rate at about 55%, the critical current density decreases progressively as the concentration of acid increases (Table 12.7).[1]

TABLE 12.3 Current Density to Maintain Passivity and Film Resistance of Some Metals in 67% Sulfuric Acid

Metal or alloy	i_p, µA·cm^{-2}	Film resistance, MΩ·cm
Mild steel	150	0.026
S30400 steel	2.2	0.50
S31000 steel	0.5	2.1
S31600 steel	0.1	17.5
Titanium	0.08	1.75
N08020	0.03	4.6

TABLE 12.4 Effect on Critical Current Density and Passivation Potential of Chromium Content for Iron-Chromium Alloys in 10% Sulfuric Acid

Chromium, %	i_{cc}, mA·cm^{-2}	E_{pp}, V vs. SHE
0	1000	+0.58
2.8	360	+0.58
6.7	340	+0.35
9.5	27	+0.15
14.0	19	−0.03

TABLE 12.5 Effect on Critical Current Density and Passivation Potential on Alloying Nickel with Chromium in 0.5 M and 5 M H$_2$SO$_4$ Containing 0.25 M K$_2$SO$_4$

Ni, %	i_{cc}, mA·cm^{-2}		E_{pp}, V vs. SHE	
	0.5 M	5 M	0.5 M	5 M
100	100	23	+0.36	+0.47
91	0.95	3.9	+0.06	+0.14
77	0.11	0.82	+0.07	+0.08
49	0.020	0.20	+0.03	+0.06
27	0.012	0.041	+0.02	+0.05
10	0.0013	0.011	+0.04	+0.08
1	1.0	5.0	−0.32	−0.20
0	1.5	8.0	−0.30	−0.20

The presence in the environment of impurities that retard the formation of a passive film or accelerate its degradation is often detrimental. In this context, chloride ions can be quite aggressive for many alloys and particularly for steels and stainless steels. As an example, the addition of 3% HCl hydrochloric acid to 67% sulfuric acid raises the critical current density for the passivation of a S31600 stainless steel from 0.7 to 40 mA·cm^{-2} and the current density to maintain passivity from 0.1 to 60 µA·cm^{-2}. Therefore, the use of the calomel electrode in anodic-protection systems is not recommended because of the possible leakage of chloride ions into the electrolyte,

TABLE 12.6 Critical Current Density and Current Density to Maintain Passivity of S30400 Stainless Steel in Various Electrolytes

Electrolyte	i_{cc}, mA·cm^{-2}	i_p, μA·cm^{-2}
20% NaOH	4.65	9.9
67% H_2SO_4 (24°C)	0.51	0.093
LiOH (pH = 9.5)	0.08	0.022
80% HNO_3 (24°C)	0.0025	0.031
115% H_3PO_4 (24°C)	0.000015	0.00015

TABLE 12.7 Effect of Concentration of Sulfuric Acid at 24°C on Corrosion Rate and Critical Current Density of S31600 Stainless Steel

Sulfuric acid, %	Corrosion rate, mm·y^{-1}	i_{cc}, mA·cm^{-2}
0	0	4.7
40	2.2	1.6
45	5.6	1.4
55	8.9	1.0
65	7.8	0.7
75	6.7	0.4
105	0	0.1

and metal/metal oxide and other electrodes are often preferred. Because of this chloride effect the storage of hydrochloric acid requires a more passive metal than mild steel, and titanium anodically protected by an external source of current or galvanic coupling has been reported to be satisfactory although even this oxide film has sometimes been found to be unstable.[1]

An increase in the temperature of an electrolyte may have several effects. An increase in temperature may make passivation more difficult, reduce the potential range in which a metal is passive, and increase the current density or corrosion rate during passivity as indicated in Fig. 12.3 for mild steel in 10% H_2SO_4.[1] Note the magnitude of the critical current density that is slightly higher than 10 mA·cm^{-2}. Such a high current density requirement creates a problem in practical anodic protection systems where the surfaces to be protected can be quite large.

12.3 Equipment Required for Anodic Protection

Figure 12.4 shows a schematic of an anodic protection system for a storage vessel. Some of the basic properties required of the components of an anodic system are described here.

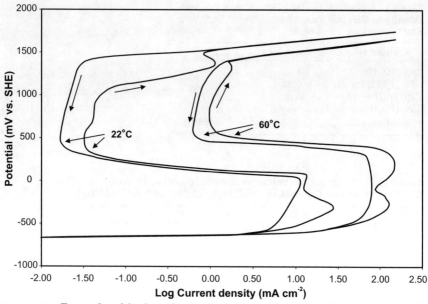

Figure 12.3 Forward and backward potentiostatic anodic polarization curves for mild steel in 10% sulfuric acid at 22 and 60°C.

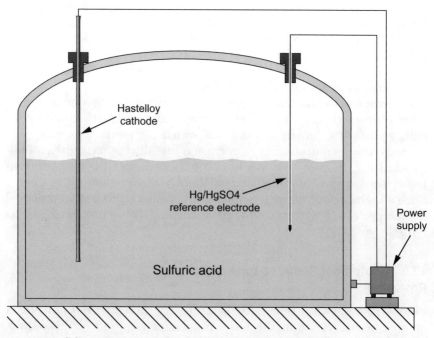

Figure 12.4 Schematic of an anodic protection system for a sulfuric acid storage vessel.

12.3.1 Cathode

The cathode should be a permanent-type electrode that is not dissolved by the solution or the currents impressed between the vessel wall and electrode. The cathodes used in most of the first applications of anodic protection were made of platinum-clad brass. These electrodes were excellent electrochemically but were costly, and the active area contacting the solution was limited by this cost. Because the overall resistance is a direct function of current density, it is advantageous to use large surface area electrodes. Many other, less costly metals have been used for cathodes instead of these costly materials. Some of these metals are listed in Table 12.8 with the chemical environments in which they were used.[3,5] The electrode size is chosen to conform to the geometry of the vessel and to provide as large a surface area as possible. The location of the cathode is not a critical factor in simple geometry, such as storage vessels, but in heat exchangers, it is necessary to extend the electrode around the surface to be protected. Multiple cathodes can be used in parallel to distribute the current and to decrease circuit resistance.

12.3.2 Reference electrode

Reference electrodes must be used in anodic protection systems because the potential of the vessel to be protected has to be carefully controlled. The reference electrode must have an electrochemical potential that is constant with respect to time and that is minimally affected by changes in temperature and solution composition. Several reference electrodes have been used for anodic protection, including those listed in Table 12.9.[3,5] The reference electrode has been a source of many problems in anodic protection installations because of its fragile nature.

TABLE 12.8 Cathode Materials Used in Field Installations

Cathode metal	Environment
Platinum-clad brass	Sulfuric acid of various concentrations
Steel	Kraft pulping liquor
Chromium nickel steel	H_2SO_4 (78–105%)
Silicon cast iron	H_2SO_4 (89–105%)
Copper	Hydroxylamine sulfate
S30400	Liquid fertilizers (nitrate solutions)
	Sulfuric acid
Nickel-plated steel	Chemical nickel plating solutions
Hastelloy C	Liquid fertilizers (nitrate solutions)
	Sulfuric acid of various concentrations
	Kraft digester liquid

TABLE 12.9 Reference Electrodes Used for Anodic Protection Installations

Electrode	Environment
Calomel	Sulfuric acid of various concentrations
	Kraft digester
Ag/AgCl	Sulfuric acid, fresh or spent
	Kraft solutions
	Fertilizer solutions
	Sulfonation plant
Hg/HgSO$_4$	H$_2$SO$_4$
	Hydroxylamine sulfate
Pt/PtO	H$_2$SO$_4$
Au/AuO	Alcohol solution
Mo/MoO$_3$	Sodium carbonate solutions
	Kraft digester
	Green or black liquors
Platinum	H$_2$SO$_4$
Bismuth	NH$_4$OH
S31600 steel	Fertilizer solutions
	H$_2$SO$_4$
Nickel	Fertilizer solutions
	Nickel plating solutions
Silicon	Fertilizer solutions

12.3.3 Potential control and power supply

The dc power supplies used in anodic protection systems have similar design and requirements as the rectifiers for cathodic protection, with one exception. Because of the nature of the active-passive behavior of the vessel, the currents required to maintain the potential of the vessel wall in the passive range can become very small. Some designs of dc power supplies must be specially modified to reduce the minimum amount of current put out of the power supply.[3]

The potential control in anodic protection installations has two functions. First, the potential must be measured and compared to a desired preset value. Second, a control signal must then be sent to the power supply to force the dc current between the cathode and vessel wall. In early systems, this control function was done in an ON-OFF method because of the high costs of electronic circuitry. The recent progress in power electronics has resulted in all systems having a continuous proportional-type control.[3] Packaging these electronic components occasionally involves special requirements because most of the installations are made in chemical plants. Explosion-proof enclosures are sometimes required, and chemically resistant enclosures are necessary in other installations.

12.4 Design Concerns

Designing an anodic protection system requires knowledge of the basic electrochemical behavior of the system and of the geometry of the equipment to be protected, considering any special operational conditions. As described earlier, the electrochemical parameters of concern are the potential at which the vessel must be maintained for corrosion protection, the current required to establish passivity, and the current required to maintain passivity. The electrode potential can be determined directly from polarization curves, and the required currents can be estimated from the polarization data. However, because the current is so strongly time dependent, its variations with respect to time must be carefully estimated. Empirical data available from field installations are the best source for this type of information.[3]

Special care and attention should also be focused on estimating the solution resistivity of a system because it is important in determining the overall circuit resistance. The power requirements for the dc power supply should be as low as possible to reduce operating costs. The solution resistivity should usually be sufficiently low so that the circuit resistance is controlled by the cathode surface area. It is essential for a system to have good throwing power or good ability for the applied current to reach the required value over complex geometry and variable distances. In general, a uniform distribution of potential over a regular-shaped passivated surface can be readily obtained by anodic protection. It is much more difficult to protect surface irregularities, such as the recessions around sharp slots, grooves, or crevices because the required current density will not be obtained in these areas. This incomplete passivation can have catastrophic consequences. This difficulty can be overcome by designing the surface to avoid these irregularities or by using a metal or alloy that is easily passivated with as low a critical current density as possible. In the rayon industry, crevice corrosion in titanium has been overcome by alloying it with 0.1% palladium.[1]

The actual passivation of a surface is very rapid if the applied current density is greater than the critical value. However, because of the high current requirements, it has been found to be neither technically nor economically practical to passivate the whole surface of a large vessel in the same initial period. For a storage vessel with an area of 1000 m², for example, a current of 5000 A could be necessary. It is therefore essential to avoid these very high currents by using one of a few techniques. It may be possible and practical, for example, to lower the temperature of the electrolyte, thereby reducing the critical current density before passivating the metal. If a vessel has a very small floor area, it may be treated in a stepwise manner by passivating the base, then the lower areas of the walls, and finally the upper areas of the walls, but this technique is not practical for very large storage tanks with a considerable floor area.[1]

Another method that has been successful is to passivate the metal by using a solution with a low critical current density (such as phosphoric acid), which is then replaced with the more aggressive acid (such as sulfuric acid) that has to be contained in the vessel (cf. Table 12.6). The critical current density can also be minimized by pretreating the metal surface with a passivating inhibitor.

12.5 Applications

Anodic protection has been used for storage vessels, process reactors, heat exchangers, and transportation vessels that contain various corrosive solutions. The majority of the applications of anodic protection involve the manufacture, storage, and transport of sulfuric acid, more of which is produced worldwide than any other chemicals. Storage of 93% H_2SO_4 and above in low-carbon steel vessels has met with some success in terms of vessel life. Anodic protection has been successful in reducing the amount of iron picked up during storage. Field studies have shown that the iron content of H_2SO_4 in concentrations of 93% and above increases at rates of 5 to 20 ppm per day of storage, depending on acid concentrations vessel size, acid residence time, and storage temperature. Several anodic protection systems have been successful in reducing the rates of iron pickup to 1 ppm per day or less. The level of purity of the acid has been sufficient to meet market demands for low iron content acid. Sulfuric acid will continue to pick up iron during transportation in trucks, railroad cars, and barges. Portable anodic protection has been used for such vessels to maintain the purity of the acid and to extend storage time.[3]

A large market has developed for anodically protected heat exchangers as replacements for cast iron coolers. Shell and tube, spiral, and plate-type exchangers have been sold complete with anodic protection as an integral part of the equipment. Sulfuric acid of 96 to 98% concentration at temperatures up to 110°C has been handled in S31600 stainless steel heat exchangers by the use of anodic protection. Corrosion rates have been reduced from unprotected rates of more than 5 mm·y^{-1} to less than 0.025 mm·y^{-1}, and cost savings have been substantial because of extended equipment life and the higher-purity acid that was produced by using these protected heat exchangers. Several other corrosive systems have also been handled in anodically protected heat exchangers.[3]

One consequence of reducing the rate of corrosion of steel in an acid is also to reduce the production of hydrogen, which has been reported as the cause of explosions in phosphoric acid systems.[1] The presence of hydrogen may also induce the formation of blisters at inclusions in the metal and can also produce grooving on vertical surfaces. Anodic protection has been found to significantly stifle the formation of hydrogen, therefore minimizing such problems.

12.6 Practical Example: Anodic Protection in the Pulp and Paper Industry

Anodic protection is a powerful technique used to mitigate corrosion of liquor tankage. However, the electrochemistry of Kraft liquors is complex due to the multiple oxidation states of sulfur compounds, the number of possible Fe-S-H_2O reactions, and the existence of active-passive behavior. The electrochemical behavior may be further complicated because some Fe-S compounds are semiconductors.[6] The major sulfur species in Kraft liquors are listed in Table 12.10.

Anodic protection of a Kraft liquor tank was first successfully realized at the end of 1984, and the success of this system resulted in many commercial installations. Unfortunately, unexpectedly high corrosion rates were reported at localized areas in several of the tanks even though the remainder of the surfaces corroded at rates less than 0.13 mm·y^{-1}.[1] Most of the problems experienced have been attributed to incomplete understanding of the electrochemistry of carbon steel in these liquors and the coexistence of active and passive areas, which had not been addressed properly in earlier control strategies.

Tromans deduced an elegant yet simple model of passivation in caustic sulfide that explains the role of sulfide in the process.[7] According to Tromans, after the initial nucleation of Fe_3O_4, sulfide is incorporated as substitutional ions into the Fe_3O_4 spinel lattice, forming a nonprotective compound Fe_3O_4-xSx. At the peak of the active-passive transition, Tromans predicted x to be approximately 0.19. Because passivation cannot occur until the sulfide in the film is completely removed by oxidation, high current densities are required to force this reaction. However, once devoid of sulfide, the film can remain stable if the potential is kept more positive than the Flade potential corresponding to the reduction reaction of oxidized sulfur species such as $S_2O_3^{2-}$ and S_2^{2-}.[6]

Figure 12.5 depicts one anodic and four cathodic idealized polarization curves, including the possible intersection points. The number and location of the intersection points creates four types of behavior, namely, monostable (active), bistable, astable, and monostable (pas-

TABLE 12.10 Major Sulfur Species in Kraft Liquors

Species	Symbol	Sulfur valence
Sulfate	(SO_4^{2-})	+6
Sulfite	(SO_3^{2-})	+4
Thiosulfate	($S_2O_3^{2-}$)	+2
Sulfur	S	0
Polysulfide	S_x^{x-}	−1
Sulfide	S^{2-}	−2

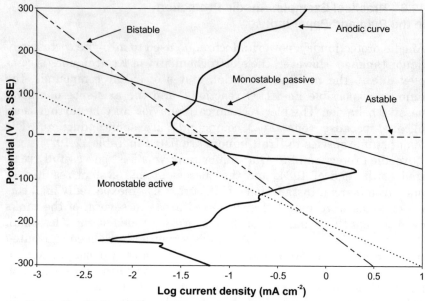

Figure 12.5 Possible combinations of anodic/cathodic intersections in the mixed potential representation of carbon steel exposed to Kraft liquors.

sive). A potentiodynamic curve of each of these types of behavior is shown, respectively, in Figs. 12.6 through 12.9. Astable behavior occurs infrequently because it requires a single anodic-cathodic intersection on the negative resistance portion of the anodic curve. This is an unstable operating condition that results in continuous oscillations between active and passive potentials. Various alloys in elevated temperature sulfuric acid are known to exhibit such behavior.[6]

The four types of mixed potential models presented in Figs. 12.6 to 12.9 are simplistic and do not necessarily reflect the complete behavior of carbon steel in Kraft liquors because the models all assume some sort of steady states. Figure 12.10 depicts typical curves from an in situ test in a white liquor clarifier at different scan rates. The passive state does not exist until after the active-passive transition is traversed. Therefore, unless sufficient anodic current density is discharged from carbon steel by a naturally occurring cathodic reaction or an applied anodic protection current, the carbon steel liquor interface remains monostable (active) because the passive film and its low current density properties do not exist.

Under normal operating chemistries in white and green Kraft liquors, carbon steel exhibits a monostable (active) behavior, and the bistable behavior occurs only after the passivation process has reached some degree of completion, as predicted by Tromans and verified by

Anodic Protection 935

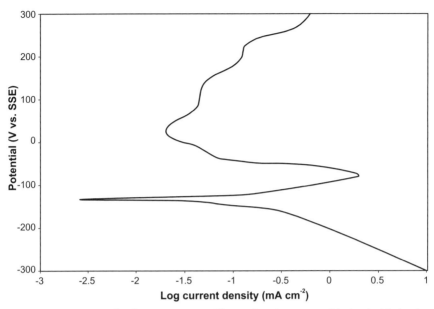

Figure 12.6 Theoretical polarization curve illustrating the monostable (active) behavior of mild steel exposed to Kraft liquors.

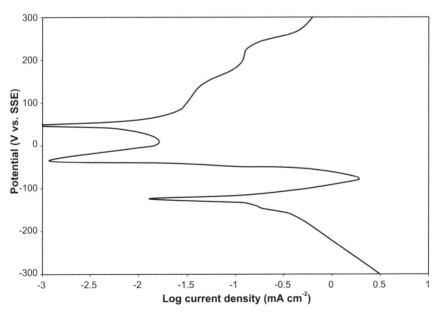

Figure 12.7 Theoretical polarization curve illustrating the bistable behavior of mild steel exposed to Kraft liquors.

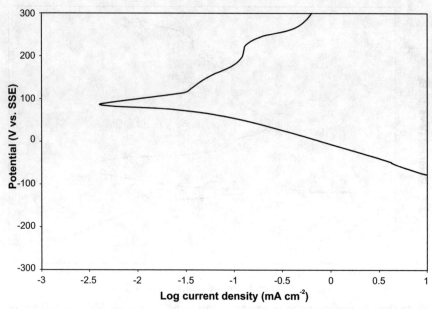

Figure 12.8 Theoretical polarization curve illustrating the monostable (passive) behavior of mild steel exposed to Kraft liquors.

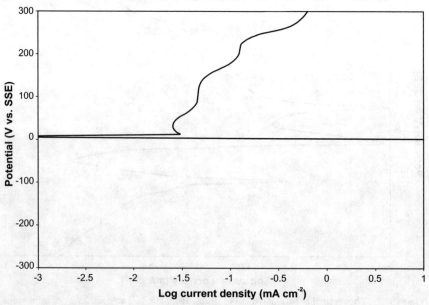

Figure 12.9 Theoretical polarization curve illustrating the astable behavior of mild steel exposed to Kraft liquors.

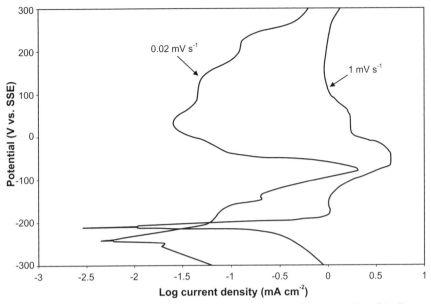

Figure 12.10 Typical in situ polarization curves of carbon steel immersed in white liquor at two scan rates.

typical curves. However, once created, the passive state is not permanently stable. When the direction of the curve is reversed, a second stable equilibrium potential is established. During traverse of the active-passive transition, the corrosion rate has been measured to be only 10 percent of the total Faradaic equivalent; hence 90 percent of the current is consumed in sulfide oxidation. Design of the protection and control systems now incorporates all of the features required to passivate the tank, maintain passivation, detect active areas, and repassivate if required.[6] Some of these features are[6]

1. *The location of cathodes.* Design is based on primary current distribution with the ratio of the minimum to maximum current density around the circumference of the tank greater than 0.9.

2. *Fluctuating liquor level.* This requires higher initial current density and more frequent repassivation cycles to form a tenacious passive layer. When immersed, the wet/dry zone of a tank exhibits a more positive potential than the remainder of tank, which may account for the higher corrosion rates there. However, it has been observed that the wet/dry zone does not get covered with surface buildup or deposits. The constantly immersed zone builds a thick surface deposit on these protected surfaces.

3. *Control scheme.* Conventional control schemes rely on a simple proportional, integral algorithm (PI). This technique is not optimal when active and passive areas exist simultaneously. The use of this

type of control will, in fact, result in accelerated corrosion of active areas. When a surface is entirely passive, the measured potential is uniform to within a few millivolts around the circumference of the tank. An active area of any size significantly distorts this uniformity. The impact is so acute and precise that active areas can be located by triangulation. Once this distortion is detected, the control system initiates an automatic repassivation using high currents for a programmable period of time. Implementation of this scheme has revealed when and where active areas are formed on the vessel. Several repassivations are common during the first month of operation. Activation seldom occurs after this stabilization period.

References

1. Walker, R., Anodic Protection, in Shreir, L. L., Jarman, R. A., and Burstein, G. T. (eds.), *Corrosion Control,* Oxford, U.K., Butterworths Heinemann, 1994, pp. 10:155–10:170.
2. Fontana, M. G., *Corrosion Engineering,* New York, McGraw Hill, 1986.
3. Locke, C. E., Anodic Protection, in *Metals Handbook: Corrosion,* Materials Park, Ohio, ASM International, 1987, pp. 463–465.
4. Sedriks, A. J., *Corrosion of Stainless Steels,* New York, John Wiley, 1979.
5. Riggs, O. L., and Locke, C. E., *Anodic Protection,* New York, Plenum Press, 1981.
6. Munro, J. I., Anodic Protection of White and Green-Liquor Tankage With and Without the Use of Protective Organic Linings, in *Proceedings of the 7th International Symposium on Corrosion in the Pulp and Paper Industry,* Atlanta, Ga., TAPPI, 1992, 117–130.
7. Tromans, D., Anodic Polarization Behavior of Mild Steel in Hot Alkaline Sulfide Solutions, *Journal of the Electrochemical Society,* **127**:1253–1256 (1980).

APPENDIX A

SI Units Conversion Table

How to Read This Table

The table provides conversion factors to SI units. These factors can be considered as unity multipliers. For example,
Length: m/X

$$0.0254 \text{ in}$$
$$0.3048 \text{ ft}$$

means that

$$1 = 0.0254 \text{ (m/in)}$$
$$1 = 0.3048 \text{ m/ft}$$

and similarly,

$$1 = 418.7 \text{ (W/m)} / \text{(cal/s} \cdot \text{cm)}$$

The SI units are listed immediately after the quantity; in this case, length: m/X. The m stands for meter, and the X designates the non-SI units for the same quantity. These non-SI units follow the numerical conversion factors.

Note: In the following table at all locations, *ton* refers to U.S. rather than metric ton.

Acceleration: $(m/s^2)/X$	0.01	cm/s^2
	7.716E-08	m/h^2
	0.3048	ft/s^2
	8.47E-05	ft/min^2
	2.35E-08	ft/h^2
Acceleration, angular: $(rad/s^2)/X$	2.78E-04	rad/min^2
	7.72E+08	rad/h^2
	1.74E-03	rev/min^2

Appendix A

Area: m^2/X	1.0E-04	cm^2
	1.0E-12	μm^2
	0.0929	ft^2
	6.452E-04	in^2
	0.8361	yd^2
	4,047	acre
	2.59E+06	mi^2
Current: A/X	10.0	abampere
	3.3356E-10	statampere
Density: $(kg/m^3)/X$	1000.0	g/cm^3
	16.02	lbm/ft^3
	119.8	lbm/gal
	27,700	lbm/in^3
	2.289E-3	$grain/ft^3$
Diffusion coefficient: $(m^2/s)/X$	1.0E-04	cm^2/s
	2.78E-04	m^2/h
	0.0929	ft^2/s
	2.58E-05	ft^2/h
Electrical capacitance: F/X	1	$A^2 \cdot s^4/kg \cdot m^2$
	1	$A \cdot s/V$
	1.0E+09	abfarad
	1.113E-12	statfarad
	3.28	V/ft
Electric charge: C/X	1	$A \cdot s$
	10	abcoulomb
	3.336E-10	statcoulomb
Electrical conductance: S/X	1	Ω^{-1}
Electric field strength: $(V/m)/X$	1	$kg \cdot m/A \cdot s^3$
	100	V/cm
	1.0E-08	abvolt/m
	299.8	statvolt/m
	39.4	V/in
Electrical resistivity: $(V \cdot m/A)/X$, $(\Omega \cdot m)/X$	1 $kg \cdot m^3/A^2 \cdot s^3$	
	1	$kg \cdot m^5/A^2 \cdot s^3$
	1.0E-09	$abohm \cdot m$
	8.988E+11	$statohm \cdot m$
Energy: J/X	3.6E+06	kWh
	4.187	cal
	4187	kcal
	1.0E-07	erg
	1.356	$ft \cdot lbf$
	1055	Btu
	0.04214	$ft \cdot pdl$
	2.685E+06	$hp \cdot h$
	1.055E+08	therm
	0.113	$in \cdot lbf$
	4.48E+04	$hp \cdot min$
	745.8	$hp \cdot s$

SI Units Conversion Table

Energy density: $(J/m^3)/X$	3.6E + 06	kWh/m^3
	4.187E + 06	cal/cm^3
	4.187E + 09	$kcal/cm^3$
	0.1	erg/cm^3
	47.9	$ft \cdot lbf/ft^3$
	3.73E + 04	Btu/ft^3
	1.271E + 08	kWh/ft^3
	9.48E + 07	$hp \cdot h/ft^3$
Energy, linear: $(J/m)/X$	418.7	cal/cm
	4.187E + 05	$kcal/cm$
	1.0E-05	erg/cm
	4.449	$ft \cdot lbf/ft$
	3461	Btu/ft
	8.81E + 06	$hp \cdot h/ft$
	1.18E + 07	kWh/ft
Energy per area: $(J/m^2)/X$	41,868	cal/cm^2
	4.187E + 07	$kcal/cm^2$
	0.001	erg/cm^2
	14.60	$ft \cdot lbf/ft^2$
	11,360	Btu/ft^2
	2.89E + 07	$hp \cdot h/ft^2$
	3.87E + 07	kWh/ft^2
Flow rate, mass: $(kg/s)/X$	1.0E-03	g/s
	2.78E-04	kg/h
	0.4536	lbm/s
	7.56E-03	lbm/min
	1.26E-04	lbm/h
Flow rate, mass/force: $(kg/N \cdot s)/X$	9.869E-05	$g/cm^2 \cdot atm \cdot s$
	1.339E-08	$lbm/ft^2 \cdot atm \cdot h$
Flow rate, mass/volume: $(kg/m^3 \cdot s)/X$	1000	$g/cm^3 \cdot s$
	16.67	$g/cm^3 \cdot min$
	0.2778	$g/cm^3 \cdot h$
	16.02	$lbm/ft^3 \cdot s$
	0.267	$lbm/ft^3 \cdot min$
	4.45E-03	$lbm/ft^3 \cdot h$
Flow rate, volume: $(m^3/s)/X$	1.0E-06	cm^3/s
	0.02832	ft^3/s (cfs)
	1.639E-05	in^3/s
	4.72E-04	ft^3/min (cfm)
	7.87E-06	ft^3/h (cfh)
	3.785E-03	gal/s
	6.308E-05	gal/min (gpm)
	1.051E-06	gal/h (gph)
Flux, mass: $(kg/m^2 \cdot s)/X$	10	$g/cm^2 \cdot s$
	1.667E-05	$g/m^2 \cdot min$
	2.78E-07	$g/m^2 \cdot h$
	4.883	$lbm/ft^2 \cdot s$
	0.0814	$lbm/ft^2 \cdot min$
	1.356E-03	$lbm/ft^2 \cdot h$

942 Appendix A

Force: N/X	1.0E-05	dyn
	1	kg·m/s
	9.8067	kg(force)
	9.807E-03	g(force)
	0.1383	pdl
	4.448	lbf
	4448	kip
	8896	ton(force)
Force, body: (N/m^3)/X	10	dyn/cm^3
	9.807E + 06	kg(f)/cm^3
	157.1	lbf/ft^3
	2.71E + 05	lbf/in^3
	3.14E + 05	ton(f)/ft^3
Force per mass: (N/kg)/X	0.01	dyn/g
	9.807	kg(f)/kg
	9.807	lbf/lbm
	0.3049	lbf/slug
Heat transfer coefficient: (W/m·K)/X	41,868	cal/s·cm^2·°C
	1.163	kcal/h·m^2·°C
	1.0E-03	erg/s·cm^2·°C
	5.679	Btu/h·ft^2·°F
	12.52	kcal/h·ft^2·°C
Henry's constant: (N/m^2)/X	1.01326E + 05	atm
	133.3	mmHg
	6893	lbf/in^2
	47.89	lbf/ft^2
Inductance: H/X	1	kg·m^2/A^2·s^2
	1	V·s/A
	1.0E-09	abhenry
	8.988E + 11	stathenry
Length: m/X	0.01	cm
	1.0E-06	μm
	1.0E-10	Å
	0.3048	ft
	0.0254	in
	0.9144	yd
	1609.3	mi
Magnetic flux: Wb/X	1	kg·m^2/A·s^2
	1	V·s
Mass: kg/X	1.0E-03	g
	0.4536	lbm
	6.48E-05	grain
	0.2835	oz (avdp)
	907.2	ton (U.S.)
	14.59	slug
Mass per area: (kg/m^2)/X	10	g/cm^2
	4.883	lbm/ft^2
	703.0	lbm/in^2
	3.5E-04	ton/mi^2

SI Units Conversion Table

Quantity	Value	Unit
Moment inertia, area: m^4/X	1.0E-08	cm^4
	4.16E-07	in^4
	8.63E-03	ft^4
Moment inertia, mass: $(kg \cdot m^2)/X$	1.0E-07	$g \cdot cm^2$
	0.04214	$lbm \cdot ft^2$
	1.355	$lbf \cdot ft \cdot s^2$
	2.93E-04	$lbm \cdot in^2$
	0.11	$lbf \cdot in/s$
Momentum: $(kg \cdot m/s)/X$	1.0E-05	$g \cdot cm/s$
	0.1383	$lbm \cdot ft/s$
	2.30E-03	$lbm \cdot ft/min$
Momentum, angular: $(kg \cdot m^2/s)/X$	1.0E-07	$g \cdot cm^2/s$
	0.04215	$lbm \cdot ft^2/s$
	7.02E-04	$lbm \cdot ft^2/min$
Momentum flow rate: $(kg \cdot m/s^2)/X$	1.0E-05	$g \cdot cm/s^2$
	0.1383	$lbm \cdot ft/s^2$
	3.84E-05	$lbm \cdot ft/min^2$
Power: W/X	4.187	cal/s
	4187	$kcal/s$
	1.0E-07	erg/s
	1.356	$ft \cdot lbf/s$
	0.293	Btu/h
	1055	Btu/s
	745.8	hp
	0.04214	$ft \cdot pdl/s$
	0.1130	$in \cdot lbf/s$
	3517	ton refrigeration
	17.6	Btu/min
Power density: $(W/m^3)/X$	4.187E + 06	$cal/s \cdot cm^3$
	4.187E + 09	$kcal/s \cdot cm^3$
	0.1	$erg/s \cdot cm^3$
	47.9	$ft \cdot lbf/s \cdot ft^3$
	3.73E + 04	$Btu/s \cdot ft^3$
	10.36	$Btu/h \cdot ft^3$
	3.53E + 04	kW/ft^3
	2.63E + 04	hp/ft^3
Power flux: $(W/m^2)/X$	41,868	$cal/s \cdot cm^2$
	4.187E + 07	$kcal/s \cdot cm^2$
	0.001	$erg/s \cdot cm^2$
	14.60	$ft \cdot lbf/s \cdot ft^2$
	11,360	$Btu/s \cdot ft^2$
	3.156	$Btu/h \cdot ft^2$
	8028	hp/ft^2
	1.072E + 04	kW/ft^2
Power, linear: $(W/m)/X$	418.7	$cal/s \cdot cm$
	4.187E + 05	$kcal/s \cdot cm$
	1.0E-05	$erg/s \cdot cm$
	4.449	$ft \cdot lbf/s \cdot ft$
	3461	$Btu/s \cdot ft$
	0.961	$Btu/h \cdot ft$
	2447	hp/ft

Appendix A

Pressure, stress: Pa/X	0.1	dyn/cm^2
	1	N/m^2
	9.8067	kg(f)/m^2
	1.0E + 05	bar
	1.0133E + 05	std. atm
	1.489	pdl/ft^2
	47.88	lbf/ft^2
	6894	lbf/in^2 (psi)
	1.38E + 07	ton(f)/in^2
	249.1	in H$_2$O
	2989	ft H$_2$O
	133.3	torr, mmHg
	3386	inHg
Resistance: Ω/X	1	kg·m^2/A^2·s^3
	1	V/A
	1.0E-09	abohm
	8.988E + 11	statohm
Specific energy: (J/kg)/X	1	m^2/s^2
	4187	cal/g
	4.187E + 06	kcal/g
	2.99	ft·lbf/lbm
	2326	Btu/lbm
	5.92E + 06	hp·h/lbm
	7.94E + 06	kWh/lbm
Specific heat, gas constant: (J/kg·K)/X	1	m^2/s^2·K
	4187	cal/g·°C
	1.0E-04	erg/g·°C
	4187	Btu/lbm·°F
	5.38	ft·lbf/lbm·°F
Specific surface: (m^2/kg)/X	0.1	cm^2/g
	2.205E-12	μm^2/lbm
	0.2048	ft^2/lbm
Specific volume: (m^3/kg)/X	1.0E-03	cm^3/g
	1.0E-15	μm^3/g
	0.0624	ft^3/lbm
Specific weight: (N/m^3)/X	10	dyn/cm^3
	157.1	lbf/ft^3
Surface tension: (N/m)/X	1.0E-03	dyn/cm
	14.6	lbf/ft
	175.0	lbf/in
Temperature: K/X (difference)	0.5555	°R
	0.5555	°F
	1.0	°C
Thermal conductivity: (W/m·K)/X	418.7	cal/s·cm·°C
	1.163	kcal/h·m·°C
	1.0E-05	erg/s·cm·°C
	1.731	Btu/h·ft·°F

SI Units Conversion Table

	0.1442	Btu·in/h·ft^2·°F
	2.22E-03	ft·lbf/h·ft·°F
Time: s/X	60.0	min
	3600	h
	86,400	day
	3.156E + 07	year
Torque: N·m/X	1.0E-07	dyn·cm
	1.356	lbf·ft
	0.0421	pdl·ft
	2.989	kg(f)·ft
Velocity: (m/s)/X	0.01	cm/s
	2.78E-04	m/h
	0.278	km/h
	0.3048	ft/s
	5.08E-03	ft/min
	0.477	mi/h
Velocity, angular: (rad/s)/X	0.01667	rad/min
	2.78E-04	rad/h
	0.1047	rev/min
Viscosity, dynamic: (kg/m·s)/X	1	N·s/m^2
	0.1	P
	0.001	cP
	2.78E-04	kg/m·h
	1.488	lbm/ft·s
	4.134E-04	lbm/ft·h
	47.91	lbf·s/ft^2
(g/cm·s)/X	1	P
Viscosity, kinematic: (m^2/s)/X	1.0E-04	St
	2.778E-04	m^2/h
	0.0929	ft^2/s
	2.581E-05	ft^2/h
(cm^2/s)/X	1	St
Volume: m^3/X	1.0E-06	cm^3
	1.0E-03	l
	1.0E-18	μm^3
	0.02832	ft^3
	1.639E-05	in^3
	3.785E-03	gal (U.S.)
Voltage, electrical potential: V/X	1.0	kg·m^2/A·s^3
	1	W/A
	1.0E-08	abvolt
	299.8	statvolt

Using the Table

The quantity in braces {} is selected from the table.

Example 1

To calculate how many meters are in 10 ft, the table provides the conversion factor as 0.3048 m/ft. Hence multiply

$$10 \text{ ft} \times \{ 0.3048 \text{ m/ft} \} = 3.048 \text{ m}$$

Example 2

Convert thermal conductivity of 10 kcal/h·m·°C to SI units. Select the appropriate conversion factor for these units:

$$10 \text{ (kcal/h} \cdot \text{m} \cdot \text{°C)} \times \{ 1.163 \text{ (W/m} \cdot \text{K)} / \text{(kcal/h} \cdot \text{m} \cdot \text{°C)} \}$$
$$= 11.63 \text{ W/m} \cdot \text{K}$$

APPENDIX B

Glossary

abrasive Material such as sand, crushed chilled cast iron, crushed steel grit, aluminum oxide, silicon carbide, flint, garnet, or crushed slag used for cleaning or surface roughening.

absolute pressure Pressure above zero pressure; the sum of the gage and atmospheric pressures.

absorb To take in and engulf wholly.

absorbent A material, usually a porous solid, which takes another material into its interior. When rain soaks into soil, the soil is an absorbent.

absorption A process in which molecules are taken up by a liquid or solid and distributed throughout the body of that liquid or solid; the process in which one substance is taken into the body of an absorbent. Compare with **adsorption.**

ac impedance See **electrochemical impedance spectroscopy.**

acicular ferrite A highly substructured nonequiaxed ferrite formed upon continuous cooling by a mixed diffusion and shear mode of transformation that begins at a temperature slightly higher than the transformation temperature range for upper bainite. It is distinguished from bainite in that it has a limited amount of carbon available; thus, there is only a small amount of carbide present.

acid A substance which releases hydrogen ions when dissolved in water. Most acids will dissolve the common metals and will react with a base to form a neutral salt and water.

acid cleaning The process of cleaning the interior surfaces of steam-generating units by filling the unit with dilute acid accompanied by an inhibitor to prevent corrosion, and subsequently draining and washing the unit and neutralizing the acid by a further wash of alkaline water.

acid embrittlement A form of hydrogen embrittlement that may be induced in some metals by acid.

acid mine drainage Drainage of water from areas that have been mined for coal or other mineral ores; the water has a low pH, sometimes less than 2.0, because of its contact with sulfur-bearing material.

acid rain Atmospheric precipitation with a pH below 3.6 to 5.7. Burning of fossil fuels for heat and power is the major factor in the generation of oxides of nitrogen and sulfur, which are converted into nitric and sulfuric acids washed down in the rain. See also **atmospheric corrosion**.

acidity The quantitative capacity of water or a water solution to neutralize an alkali or base. It is usually measured by titration with a standard solution of sodium hydroxide and expressed in terms of its calcium carbonate equivalent. See also **carbon dioxide**, **mineral acidity**, **total acidity**.

activated carbon A water treatment medium, found in block, granulated, or powdered form, which is produced by heating carbonaceous substances (bituminous coal or cellulose-based substances such as wood or coconut shell) to 700°C or less in the absence of air to form a carbonized char, and then activating (oxidizing) at 800 to 1000°C with oxidizing gases such as steam and carbon dioxide (oxygen is never used as the oxidizing gas because its reaction with the carbon surface is too rapid and violent) to form pores, thus creating a highly porous adsorbent material.

activated silica A material usually formed from the reaction of a dilute silicate solution with a dilute acid. It is used as a coagulant aid.

activation The changing of a passive surface of a metal to a chemically active state. Contrast with **passivation.**

active The negative direction of electrode potential. [Also used to describe corrosion and its associated potential range when an electrode potential is more negative than an adjacent depressed corrosion-rate (passive) range.]

activity A measure of the chemical potential of a substance, where chemical potential is not equal to concentration, that allows mathematical relations equivalent to those for ideal systems to be used to correlate changes in an experimentally measured quantity with changes in chemical potential.

activity (ion) The ion concentration corrected for deviations from ideal behavior. Concentration multiplied by activity coefficient.

activity coefficient A characteristic of a quantity expressing the deviation of a solution from ideal thermodynamic behavior; often used in connection with electrolytes.

adhesion A binding force that holds together molecules of substances whose surfaces are in contact or near proximity.

adhesive strength The magnitude of attractive forces, generally physical in character, between a coating and a substrate. Two principal interactions that contribute to the adhesion are Van der Waals forces and permanent dipole bonds.

adsorb To take in on the surface.

adsorbent A material, usually solid, capable of holding gases, liquids, and/or suspended matter at its surface and in exposed pores. Activated carbon is a common adsorbent used in water treatment.

adsorption The physical process occurring when liquids, gases, or suspended matter adhere to the surfaces of, or in the pores of, an adsorbent medium. Adsorption is a physical process which occurs without chemical reaction.

aeration The process whereby water is brought into intimate contact with air by spraying or cascading, or air is brought into intimate contact with water by an air aspirator or by bubbling compressed air through the body of water.

aeration cell An oxygen concentration cell; an electrolytic cell resulting from differences in dissolved oxygen at two points. See also **differential aeration cell.**

aerobic An action or process conducted in the presence of air, such as aerobic digestion of organic matter by bacteria.

age hardening Hardening by aging, usually after rapid cooling or cold working.

Alclad A composite wrought product made up of an aluminum alloy core that has on one or both surfaces a metallurgically bonded aluminum or aluminum alloy coating that is anodic to the core and thus electrochemically protects the core against corrosion.

algae Small primitive plants containing chlorophyll, commonly found in surface water. Excessive growths may create taste and odor problems and consume dissolved oxygen during decay.

alkali A water-soluble mineral compound, usually considered to have moderate strength as a base (as opposed to the caustic or strongly basic hydroxides, although this differentiation is not always made). In general, the term is applied to bicarbonate and carbonate compounds when they are present in the water or solution. (See **base.**)

alkaline (1) Having properties of an alkali. (2) Having a pH greater than 7.

alkaline cleaner A material blended from alkali hydroxides and such alkaline salts as borates, carbonates, phosphates, or silicates. The cleaning action may be enhanced by the addition of surface-active agents and special solvents.

alkalinity The quantitative capacity of a water or water solution to neutralize an acid. It is usually measured by titration with a standard acid solution of sulfuric acid and is expressed in terms of its calcium carbonate equivalent. If there is not enough buffer, or anything that would reduce the shock of acid rain by neutralization, the water is considered acid-sensitive. (See **alkali, base.**)

alkyl benzene sulfonate A term applied to a family of branched chain chemical compounds formerly used as detergents. Sometimes called "hard" detergents because of their resistance to biological degradation, these compounds have been largely replaced with linear alkyl sulfonate (LAS) compounds, which are more readily degraded to simpler substances.

alpha iron The body-centered cubic form of pure iron, stable below 910°C.

alternate-immersion test A corrosion test in which the specimens are intermittently exposed to a liquid medium at definite time intervals.

alum A common name for aluminum sulfate, used as a coagulant.

aluminizing Forming an aluminum or aluminum alloy coating on a metal by hot dipping, hot spraying, or diffusion.

amoeba A small, single-celled animal or protozoan.

amphoteric A term applied to oxides and hydroxides which can act basic toward strong acids and acidic toward strong alkalis; substances which can dissociate electrolytically to produce hydrogen or hydroxyl ions according to conditions.

anaerobic In the absence of air or unreacted or free oxygen.

angstrom A unit of length equal to one ten-billionth of a meter.

anion A negatively charged ion.

anion exchange An ion-exchange process in which anions in solution are exchanged for other anions from an ion exchanger. In demineralization, for example, bicarbonate, chloride, and sulfate anions are removed from solution in exchange for a chemically equivalent number of hydroxide anions from the anion-exchange resin.

annealing A generic term denoting a treatment consisting of heating to and holding at a suitable temperature, followed by cooling at a suitable rate; used primarily to soften metallic materials, but also to simultaneously produce desired changes in other properties or in microstructure.

anode The electrode of an electrolytic cell at which oxidation is the principal reaction. (Electrons flow away from the anode in the external circuit. It is usually the electrode where corrosion occurs and metal ions enter solution.)

anode corrosion efficiency The ratio of the actual corrosion (weight loss) of an anode to the theoretical corrosion (weight loss) calculated by Faraday's law from the quantity of electricity that has passed.

anodic cleaning Electrolytic cleaning in which the workpiece is done by the anode. Also called *reverse-current cleaning*.

anodic inhibitor An inhibitor that reduces the corrosion rate by acting on the anodic (oxidation) reaction.

anodic polarization The change of the electrode potential in the noble (positive) direction due to current flow. (See **polarization**.)

anodic protection A technique to reduce corrosion of a metal surface under some conditions by passing sufficient current to it to cause its electrode potential to enter and remain in the passive region; imposing an external electrical potential to protect a metal from corrosive attack. (Applicable only to metals that show active-passive behavior.) Contrast with **cathodic protection.**

anodizing Forming a conversion coating on a metal surface by anodic oxidation; most frequently applied to aluminum.

anolyte The electrolyte adjacent to the anode of an electrolytic cell.

antifouling Intended to prevent fouling of underwater structures, such as the bottoms of ships; refers to the prevention of marine organisms' attachment or growth on a submerged metal surface, generally through chemical toxicity caused by the composition of the metal or the coating layer.

antipitting agent An agent added to electroplating solutions to prevent the formation of pits or large pores in the electrodeposit.

aqueous Pertaining to water; an aqueous solution is made by using water as a solvent.

aquifer A layer or zone below the surface of the earth which is capable of yielding a significant volume of water.

arrester A device, usually screening at the top, to impede the flow of large dust particles or sparks from a stack.

artificial aging Aging above room temperature. Compare with **natural aging.**

ash The incombustible inorganic matter in fuel.

ASME The American Society of Mechanical Engineers.

atmospheric air Air under the prevailing atmospheric conditions.

atmospheric corrosion The gradual degradation or alteration of a material by contact with substances present in the atmosphere, such as oxygen, carbon dioxide, water vapor, and sulfur and chlorine compounds.

atmospheric pressure The barometric reading of pressure exerted by the atmosphere. At sea level, 0.101 MPa or 760 mm of mercury.

atom The smallest particle of an element that can exist either alone or in combination.

attrition In water treatment, the process in which solids are worn down or ground down by friction, often between particles of the same material. Filter media and ion-exchange materials are subject to attrition during backwashing, regeneration, and service.

austenite A solid solution of one or more elements in face-centered cubic iron. Unless otherwise designated (such as nickel austenite), the solute is generally assumed to be carbon.

austenitic The name given to the face-centered cubic (FCC) crystal structure of ferrous metals. Ordinary iron and steel has this structure at elevated temperatures; certain stainless steels (300 series) also have this structure at room temperature.

austenitizing Forming austenite by heating a ferrous alloy into the transformation range (partial austenitizing) or above the transformation range (complete austenitizing). When used without qualification, the term implies complete austenitizing.

auxiliary electrode The electrode in an electrochemical cell that is used to transfer current to or from a test electrode, usually made of noncorroding material.

backfill Material placed in a drilled hole to fill space around anodes, vent pipes, and buried components of a cathodic protection system.

backflow Flow of water in a pipe or line in a direction opposite to normal flow. Often associated with back siphonage or the flow of possibly contaminated water into a potable water system.

backflow preventer A device or system installed in a water line to stop backflow.

backwash The process in which beds of filter or ion-exchange media are subjected to flow in the direction opposite to the service flow direction to loosen the bed and to flush suspended matter collected during the service run.

bacteria Single-celled organisms (singular form, *bacterium*) which lack well-defined nuclear membranes and other specialized functional cell parts and reproduce by cell division or spores. Bacteria may be free-living organisms or parasites. Bacteria (along with fungi) are decomposers that break down the wastes and bodies of dead organisms, making their components available for reuse. Bacterial cells range from about 1 to 10 μm in length and from 0.2 to 1 μm in width. They exist almost everywhere on earth. Despite their small size, the total weight of all bacteria in the world likely exceeds that of all other organisms combined.

bactericide Any substance or agent which kills bacteria, both disease-causing and non-disease-causing. Spores and nonbacterial microorganisms (e.g., algae, fungi, and viruses) are not necessarily killed by a bactericide.

baffle A plate or wall for deflecting gases or liquids.

bainite A metastable aggregate of ferrite and cementite resulting from the transformation of austenite at temperatures below the pearlite range but above the martensite start temperature.

barometric pressure Atmospheric pressure as determined by a barometer, usually expressed in mm of mercury or MPa.

base A substance that releases hydroxyl ions when dissolved in water. Bases react with acids to form a neutral salt and water. (See **alkali**.)

base exchange Synonymous with **cation exchange.**

batch A quantity of material treated or produced as a unit.

batch operation A process method in which a quantity of material is processed or treated, usually with a single charge of reactant in a single vessel, and often involving stirring. For example, the neutralization of a specific volume of an acid with a base in a vessel, with stirring or mixing, is a batch operation.

beach marks Macroscopic progression marks on a fatigue fracture or stress corrosion cracking surface that indicate successive positions of the advancing crack front. The classic appearance is of irregular elliptical or semielliptical rings, radiating outward from one or more origins. See also **striation**.

bed The ion exchanger or filter medium in a column or other tank or operational vessel.

bed depth The height of the ion exchanger or filter medium in the vessel after preparation for service.

bed expansion The increase in the volume of a bed of ion-exchange or filter medium during upflow operations, such as backwashing, caused by lifting and separation of the medium. Usually expressed as the percent of increase in bed depth.

bicarbonate alkalinity The alkalinity of a water due to the presence of bicarbonate ions (HCO_3).

bioassay A test which determines the effect of a chemical on a living organism.

biochemical oxygen demand (BOD) The amount of oxygen (measured in mg/L) required in the oxidation of organic matter by biological action under specific standard test conditions. Widely used to measure the amount of organic pollution in wastewater and streams.

biocide A chemical which can kill or inhibit the growth of living organisms such as bacteria, fungi, molds, and slimes. Biocides can be harmful to humans, too. Biocides kill spores of living organisms also, and since spores are the most resistant of all life forms, a biocide may be properly defined as a sterilizing agent.

biodegradable Subject to degradation to simpler substances by biological action, such as the bacterial breakdown of detergents, sewage wastes, and other organic matter.

black oxide A black finish on a metal produced by immersing the metal in hot oxidizing salts or salt solutions.

blasting A method of cleaning or surface roughening by a forcibly projected stream of sharp, angular abrasive.

bleach a strong oxidizing agent and disinfectant formulated to break down organic matter and destroy biological organisms. Commonly refers to a 5.25% nominal solution of sodium hypochlorite (household bleach), which is equivalent to 3 to 5% available free chlorine (strength varies with shelf life). Sodium hypochlorite is also available commercially in concentrations of between 5 and 15% available chlorine. Dry bleach is a dry calcium hypochlorite with 70% available chlorine.

blister A raised area, often dome-shaped, resulting from either loss of adhesion between a coating or deposit and the base metal or delamination under the pressure of expanding gas trapped in a metal in a near-subsurface zone.

blowdown The withdrawal of water containing a high concentration of solids from an evaporating-water system (such as a boiler system) in order to maintain the solids-to-water concentration ratio within specified limits. Blowdown is normally performed in boiler and cooling-water operations. The term may also refer to removal of other solutions of undesirable quality from a system or vessel.

blowdown valve A valve generally used to continuously regulate the concentration of solids in the boiler; not a drain valve.

BOD Abbreviation for **biochemical oxygen demand.**

boiler A closed vessel in which water is heated, steam is generated, steam is superheated, or any combination thereof is done, under pressure or vacuum by the application of heat from combustible fuels, electricity, or nuclear energy.

boiler efficiency A term often substituted for **combustion efficiency** or *thermal efficiency*. True boiler efficiency is the measure of fuel-to-steam efficiency.

boiler water A term construed to mean a representative sample of the circulating water in the boiler after the generated steam has been separated and before the incoming feedwater or added chemical becomes mixed with it so that its composition is affected.

bond coat A preliminary (or prime) coat of material that improves adherence of the subsequent spray deposit.

bonding force The force that holds two atoms together; it results from a decrease in energy as two atoms are brought closer to each other.

brackish water Water having salinity values ranging from approximately 500 to 5000 parts per million (milligrams per liter).

breakdown potential The least noble potential at which pitting or crevice corrosion, or both, will initiate and propagate.

breakpoint chlorination A chlorination procedure in which chlorine is added until the chlorine demand is satisfied and a dip (breakpoint) in the chlorine residual occurs. Further additions of chlorine produce a chlorine residual proportional to the amount added.

breakthrough The appearance in the effluent from a water conditioner of the material being removed by the conditioner, such as hardness in the effluent of a softener or turbidity in the effluent of a mechanical filter; an indication that regeneration, backwashing, or other treatment is necessary for further service.

brine A strong solution of salt(s) (usually sodium chloride and other salts too) with total dissolved solids concentrations in the range of 40,000 to 300,000 or more milligrams per liter. Potassium or sodium chloride brine is used in the regeneration stage of cation- and/or anion-exchange water treatment equipment.

brittle fracture Separation of a solid accompanied by little or no macroscopic plastic deformation. Typically, brittle fracture occurs by rapid crack propagation with less expenditure of energy than for ductile fracture.

buffer A chemical substance which stabilizes pH values in solutions.

buffer capacity A measure of the capacity of a solution or liquid to neutralize acids or bases. This is a measure of the capacity of water to offer a resistance to changes in pH.

bunker oil Residual fuel oil of high viscosity, commonly used in marine and stationary steam power plants (no. 6 fuel oil).

bypass A connection or a valve system that allows untreated water to flow through a water system while a water treatment unit is being regenerated, backwashed, or serviced.

calcareous coating or deposit A layer consisting of a mixture of calcium carbonate and magnesium hydroxide deposited on surfaces being cathodically protected because of the increased pH adjacent to the protected surface.

calcium One of the principal elements in the earth's crust. When dissolved in water, calcium is a factor contributing to the formation of scale and insoluble soap curds, which are a means of clearly identifying hard water.

calcium carbonate equivalent A common basis for expressing the concentration of hardness and other salts in chemically equivalent terms to simplify certain calculations; signifies that the concentration of a dissolved mineral is chemically equivalent to the stated concentration of calcium carbonate.

calcium hypochlorite A chemical compound, [$Ca(ClO)_2 \cdot 4H_2O$], used as a bleach and as a source of chlorine in water treatment; specifically useful because it is stable as a dry powder and can be formed into tablets.

calomel electrode An electrode widely used as a reference electrode of known potential in electrometric measurement of acidity and alkalinity, corrosion studies, voltammetry, and measurement of the potentials of other electrodes. See also **electrode potential, reference electrode.**

calorie The mean calorie is 1/100 of the heat required to raise the temperature of 1 g of water from 0 to 100°C at a constant atmospheric pressure. It is about equal to the quantity of heat required to raise 1 g of water 1°C. Another definition is: A calorie is 4.1860 joules.

capillary action A phenomenon in which water or many other liquids will rise above the normal liquid level in a tiny tube or capillary, as a result of attraction between molecules of the liquid for each other and the walls of the tube.

carbide A chemical compound formed between carbon and a metal or metals; examples are tungsten carbide, tantalum carbide, titanium carbide, chromium carbide.

carbon chloroform extract The matter adsorbed from a stream of water by activated carbon, then extracted from the activated carbon with chloroform, using a specific standardized procedure; a measure of the organic matter in a water.

carbon dioxide A gas present in the atmosphere and formed by the decay of organic matter. The gas in carbonated beverages; in water, it forms carbonic acid.

carbonaceous Materials of or derived from organic substances such as coal, lignite, peat, etc.

carbonaceous exchanger Ion-exchange material produced by the sulfonation of carbonaceous matter.

carbonate alkalinity Alkalinity due to the presence of the carbonate ion (CO_3^{2-}).

carbonate hardness Hardness due to the presence of calcium and magnesium bicarbonates and carbonates in water; the smaller of the total hardness and the total alkalinity. (See **temporary hardness.**)

carboxylic An organic acidic group (COOH) which contributes cation-exchange ability to some resins.

carburizing flame A nonstandard term for **reducing flame.**

CASS test See **copper-accelerated salt-spray test.**

cathode The electrode of an electrolytic cell at which reduction is the principal reaction. Electrons flow toward the cathode in the external circuit.

cathodic corrosion Corrosion of a metal when it is a cathode. (This usually happens to metals because of a rise in pH at the cathode or as a result of the formation of hydrides.)

cathodic disbondment The destruction of adhesion between a coating and its substrate by products of a cathodic reaction.

cathodic inhibitor An inhibitor that reduces the corrosion rate by acting on the cathodic (reduction) reaction.

cathodic pickling Electrolytic pickling in which the work is done by the cathode.

cathodic polarization A change in the electrode potential in the active (negative) direction as a result of current flow. (See **polarization.**)

cathodic protection A corrosion control system in which the metal to be protected is made to serve as a cathode, either by the deliberate establishment of a galvanic cell or by impressed current. (See **anodic protection.**)

cathodic reaction Electrode reaction equivalent to a transfer of negative charge from the electronic to the ionic conductor. A cathodic reaction is a reduction process.

catholyte The electrolyte adjacent to the cathode of an electrolytic cell.

cation A positively charged ion.

cation exchange Ion-exchange process in which cations in solution are exchanged for other cations from an ion exchanger.

caustic Any substance capable of burning or destroying animal flesh or tissue. The term is usually applied to strong bases.

caustic cracking Stress corrosion cracking of metals in caustic solutions. (See also **stress corrosion cracking.**)

caustic dip A strongly alkaline solution into which metal is immersed for etching, for neutralizing acid, or for removing organic materials such as greases or paints.

caustic embrittlement See **caustic cracking.**

caustic soda The common name for sodium hydroxide.

cavitation The formation and rapid collapse within a liquid of cavities or bubbles that contain vapor or gas or both.

cavitation damage The degradation of a solid body resulting from its exposure to cavitation. (This may include loss of material, surface deformation, or changes in properties or appearance.)

cavitation erosion Progressive loss of original material from a solid surface as a result of continuing exposure to cavitation.

cementation coating A coating developed on a metal surface by a high-temperature diffusion process (such as carburization, calorizing, or chromizing).

cementite A compound of iron and carbon known chemically as iron carbide and having the approximate chemical formula Fe_3C. It is characterized by an orthorhombic crystal structure.

cermet A physical mixture of ceramics and metals; examples are alumina plus nickel and zirconia plus nickel.

chalking The development of loose removable powder at the surface of an organic coating, usually caused by weathering.

checking The development of slight breaks in a coating that do not penetrate to the underlying surface.

checks Numerous, very fine cracks in a coating or at the surface of a metal part. Checks may appear during processing or during service and are most often associated with thermal treatment or thermal cycling. Also called *check marks, checking,* or *heat checks.*

chelate A molecular structure in which a heterocyclic ring can be formed by the unshared electrons of neighboring atoms or a coordination compound in which a heterocyclic ring is formed by a metal bound to two atoms of the associated ligand. See also **complexation.**

chelating agent A chemical compound sometimes fed to water to tie up undesirable metal ions, keep them in solution, and eliminate or reduce their normal effects. (See **sequestering agent.**)

chelation The process of forming complex chemical compounds in which certain metal ions are bound into stable ring structures, keeping the ions in solution and eliminating or reducing their normal (and often undesirable) effects. Similar to the process of **sequestration.**

chemical conversion coating A protective or decorative nonmetallic coating produced in situ by chemical reaction of a metal with a chosen environment. (Such a coating is often used to prepare the surface prior to the application of an organic coating.)

chemical oxygen demand The amount of oxygen required for the chemical oxidation of organic matter in a wastewater sample.

chemical potential In a thermodynamic system with several constituents, the rate of change of the Gibbs function of the system with respect to the change in the number of moles of a particular constituent.

chemical stability Resistance to attack by chemical action.

chemical vapor deposition A coating process, similar to gas carburizing and carbonitriding, in which a reactant atmosphere gas is fed into a processing chamber, where it decomposes at the surface of the workpiece, liberating one material for either absorption by or accumulation on the workpiece. A second material is liberated in gas form and is removed from the processing chamber along with excess atmospheric gas.

chemisorption A process related to adsorption in which atoms or molecules of reacting substances are held to the surface atoms of a catalyst by electrostatic forces having about the same strength as chemical bonds. Chemisorption differs from physical adsorption chiefly in the strength of bonding, which is much greater in chemisorption than in adsorption.

chlorides Salts of chloride that are generally soluble. High concentrations contribute to corrosion problems.

chlorination The treatment process in which chlorine gas or a chlorine solution is added to water for disinfection and control of microorganisms. Chlorination is also used in the oxidation of dissolved iron, manganese, and hydrogen sulfide impurities.

chlorinator A device designed to feed chlorine gas or solutions of its compounds, such as hypochlorite, into a water supply.

chlorine A gas, Cl_2, widely used in the disinfection of water and as an oxidizing agent for organic matter, iron, etc.

chlorine demand A measure of the amount of chlorine which will be consumed by organic matter and other oxidizable substances in a water before a chlorine residual will be found. Chlorine demand represents the difference between the total chlorine fed and the chlorine residual.

chlorinity The total halogen ion content as titrated by the addition of silver nitrate, expressed in parts per thousand (o/oo).

chromadizing Improving paint adhesion on aluminum or aluminum alloys, mainly aircraft skins, by treatment with a solution of chromic acid. Also called *chromodizing* or *chromatizing*. Not to be confused with **chromating** or **chromizing**.

chromate treatment A treatment of metal in a solution of a hexavalent chromium compound to produce a conversion coating consisting of trivalent and hexavalent chromium compounds.

chromating Performing a chromate treatment.

chromizing A surface treatment at elevated temperature, generally carried out in pack, vapor, or salt bath, in which an alloy is formed by the inward diffusion of chromium into the base metal.

clad metal A composite metal containing two or more layers that have been bonded together. The bonding may have been accomplished by co-rolling, co-extrusion, welding, diffusion bonding, casting, heavy chemical deposition, or heavy electroplating.

cladding A surfacing variation in which surfacing material is deposited or applied, usually to improve corrosion or heat resistance.

cleavage Splitting (fracture) of a crystal on a crystallographic plane of low index.

cleavage fracture A fracture, usually of polycrystalline metal, in which most of the grains have failed by cleavage, resulting in bright reflecting facets. It is associated with low-energy brittle fracture.

coagulant A material, such as alum, which will form a gelatinous precipitate in water and cause the agglomeration of finely divided particles into larger particles, which can then be removed by settling and/or filtration.

coagulant aid A material which is not a coagulant, but which improves the effectiveness of a coagulant by forming larger or heavier particles, speeding the reactions, or permitting reduced coagulant dosage.

coagulation The clumping together of very fine colloidal (less than 0.1 µm in size) and dispersed (0.1 to 100 µm in size) particles into larger visible agglomerates of these particles (usually between 100 and 1000 µm in size), caused by the use of chemicals (coagulants). The chemicals neutralize the electric charges of the fine particles and cause destabilization of the particles. This clumping together makes it easier to separate the solids from the water by settling, skimming, draining, or filtering.

coalescence The union or growing together of colloidal particles into a group or larger unit as a result of molecular attraction on the surfaces of the particles.

coating strength (1) A measure of the cohesive bond within a coating, as opposed to the coating-to-substrate bond (adhesive strength). (2) The tensile strength of a coating, usually expressed in kPa.

coating stress The stresses in a coating resulting from rapid cooling of molten material or semimolten particles as they come into contact with the substrate. Coating stresses are a combination of body and textural stresses.

COD The abbreviation for **chemical oxygen demand**.

cold cracking A type of weld cracking that usually occurs below 203°C. Cracking may occur during or after cooling to room temperature, sometimes with a considerable time delay. Three factors combine to produce cold cracks: stress (for example, from thermal expansion and contraction), hydrogen (from hydrogen-containing welding consumables), and a susceptible microstructure (plate martensite is most susceptible to cracking, ferritic and bainitic structures are least susceptible).

cold working Deforming metal plastically under conditions of temperature and strain rate that induce strain hardening. Usually, but not necessarily, conducted at room temperature. Contrast with **hot working.**

coliform bacteria A group of microorganisms used as indicators of water contamination and the possible presence of pathogenic (disease-producing) bacteria.

colloid Very finely divided solid particles which do not settle out of a solution; intermediate between a true dissolved particle and a suspended solid, which will settle out of solution. The removal of colloidal particles usually requires coagulation.

combined available chlorine The chlorine present as chloramine or other chlorine derivatives in a water, but still available for disinfection and the oxidation of organic matter. Combined chlorine compounds are more stable than free chlorine forms, but are somewhat slower in disinfection action.

combined carbon   Carbon in iron or steel that is combined chemically with other elements, not in the free state as graphite or temper carbon.

combustion   The rapid chemical combination of oxygen with the combustible elements of a fuel, resulting in the release of heat.

combustion efficiency   The effectiveness of the burner in completely burning the fuel. A well-designed burner will operate with as little as 10 to 20 percent excess air, while converting all combustibles in the fuel to useful energy.

compensated hardness   A calculated value based on the total hardness, the magnesium-to-calcium ratio, and the sodium concentration of a water. It is used to correct for the reductions in hardness removal capacity caused by these factors in zeolite exchange water softeners.

complexation   The formation of complex chemical species by the coordination of groups of atoms, termed *ligands,* to a central ion, commonly a metal ion. Generally, the ligand coordinates by providing a pair of electrons that form an ionic or covalent bond to the central ion.

compressive stress   A stress that causes an elastic body to deform in the direction of the applied load.

concentration cell   An electrolytic cell, the emf of which is caused by a difference in the concentration of some component in the electrolyte. (This difference leads to the formation of discrete cathode and anode regions.)

concentration polarization   That portion of the polarization of a cell produced by concentration changes resulting from passage of current through the electrolyte.

condensate   Condensed water resulting from the removal of latent heat from steam.

conductivity   (1) A material property relating heat flux (heat transferred per unit area per unit time) to a temperature difference. (2) The ability of a water sample to transmit electric current under a set of standard conditions. Usually expressed as microhm conductance.

connate water   Water deposited simultaneously with rock and held with essentially no flow; usually occurs deep in the earth, and usually is high in minerals as a result of long contact.

continuity bond   A metallic connection that provides electrical continuity between metal structures.

continuous blowdown   The uninterrupted removal of concentrated boiler water from a boiler to control total solids concentration in the remaining water.

convection   The transmission of heat by the circulation of a liquid or gas. It may be natural, with the circulation caused by buoyancy effects due to temperature differences, or forced, with circulation caused by a mechanical device such as a fan or pump.

conversion coating   A coating consisting of a compound of the surface metal produced by chemical or electrochemical treatments of the metal. Examples include chromate coatings on zinc, cadmium, magnesium, and aluminum and oxide and phosphate coatings on steel. See also **chromate treatment** and **phosphating**.

copper-accelerated salt-spray (CASS) test An accelerated corrosion test for some electrodeposits for anodic coatings on aluminum.

corrodkote test An accelerated corrosion test for electrodeposits.

corrosion A chemical or electrochemical reaction between a material, usually a metal, and its environment that produces a deterioration of the material and its properties.

corrosion fatigue The process in which a metal fractures prematurely under conditions of simultaneous corrosion and repeated cyclic loading at lower stress levels or after fewer cycles than would be required in the absence of the corrosive environment.

corrosion fatigue strength The maximum repeated stress that can be endured by a metal without failure under definite conditions of corrosion and fatigue and for a specific number of stress cycles and a specified period of time.

corrosion potential The potential of a corroding surface in an electrolyte relative to that of a reference electrode measured under open-circuit conditions.

corrosion product A substance formed as a result of corrosion.

corrosion rate The amount of corrosion occurring per unit time (for example, mass change per unit area per unit time, penetration per unit time).

corrosion resistance The ability of a metal to withstand corrosion in a given corrosion system.

corrosivity The tendency of an environment to cause corrosion in a given corrosion system.

counterelectrode See **auxiliary electrode**.

crazing A network of checks or cracks appearing on a coated surface.

creep Time-dependent strain occurring under stress. The creep strain occurring at a diminishing rate is called *primary creep*; that occurring at a minimum and almost constant rate, *secondary creep*; and that occurring at an accelerating rate, *tertiary creep*.

Crenothrix polyspora A genus of filamentous bacteria which utilize iron in their metabolism and cause staining, plugging, and taste and odor problems in water systems. (See **iron bacteria**.)

crevice corrosion Localized corrosion of a metal surface at or immediately adjacent to an area that is shielded from full exposure to the environment because of close proximity between the metal and the surface of another material.

critical anodic current density The maximum anodic current density observed in the active region for a metal or alloy electrode that exhibits active-passive behavior in an environment.

critical flaw size The size of a flaw (defect) in a structure that will cause failure at a particular stress level.

critical humidity The relative humidity above which the atmospheric corrosion rate of some metals increases sharply.

critical pitting potential The least noble potential where pitting corrosion will initiate and propagate. (See **breakdown potential**.)

cross-sectional area The area of a plane at a right angle to the direction of flow through a tank or vessel; often expressed in square feet and related to the flow rate (for example, 5 gallons per minute per square foot of ion-exchanger bed area).

current density The electric current to or from a unit area of an electrode surface.

current efficiency The ratio of the electrochemical equivalent current density for a specific reaction to the total applied current density.

deactivation The process of prior removal of the active corrosive constituents, usually oxygen, from a corrosive liquid by controlled corrosion of expendable metal or by other chemical means, thereby making the liquid less corrosive.

deaeration Removal of air and gases from boiler feedwater prior to its introduction into a boiler.

dealloying See **parting**.

dechlorination The removal of chlorine residual.

defect A discontinuity or discontinuities that by nature or accumulated effect (for example, total crack length) render a part or product unable to meet minimum applicable acceptance standards or specifications.

degasification Removal of gases from samples of steam taken for purity test. Removal of CO_2 from water, as in the ion-exchange method of softening.

degrease To remove oil or grease from the surface of the workpiece.

deionization The removal of all ionized minerals and salts from a solution by a two-phase ion-exchange process. Positively charged ions are removed by a cation-exchange resin in exchange for a chemically equivalent amount of hydrogen ions. Negatively charged ions are removed by an anion-exchange resin in exchange for a chemically equivalent amount of hydroxide ions.

deliquescent The process of melting or becoming liquid by absorbing moisture from the air.

demineralization The removal of ionized minerals and salts from a solution by a two-phase ion-exchange procedure, similar to deionization (the two terms are often used interchangeably).

dendrite A crystal that has a treelike branching pattern; most evident in cast metals slowly cooled through the solidification range.

denickelification Corrosion in which nickel is selectively leached from nickel-containing alloys. Most commonly observed in copper-nickel alloys after extended service in fresh water.

density The mass of a substance per specified unit of volume; for example, pounds per cubic foot. True density is the mass per unit volume excluding pores; apparent density is the mass per unit volume including pores. (See **specific gravity.**)

deoxidizing (1) The removal of oxygen from molten metals by use of suitable deoxidixers. (2) Sometimes, the removal of undesirable elements other than oxygen by the introduction of elements or compounds that readily react with them. (3) In metal finishing, the removal of oxide films from metal surfaces by chemical or electrochemical reaction.

depolarization Not a preferred term; see **polarization.**

deposit A foreign substance which comes from the environment that adheres to a surface of a material.

deposit corrosion Localized corrosion under or around a deposit or collection of material on a metal surface. (See also **crevice corrosion**).

descaling Removing the thick layer of oxides formed on some metals at elevated temperatures.

desiccant A chemical used to attract and remove moisture from air or gas.

design load The load for which a steam generating unit is designed; considered the maximum load to be carried.

design pressure The pressure used in the design of a boiler for the purpose of calculating the minimum permissible thickness or physical characteristics of the different parts of the boiler.

detergent Any material with cleaning powers, including soaps, synthetic detergents, many alkaline materials and solvents, and abrasives. In popular usage, the term is often used to mean the synthetic detergents such as ABS or LAS. (See **alkyl benzene sulfonate, linear alkyl sulfonate.**)

dew point The temperature at which moisture will condense from humid vapors into a liquid state.

dezincification Corrosion in which zinc is selectively leached from zinc-containing alloys; most commonly found in copper-zinc alloys containing less than 83% copper after extended service in water containing dissolved oxygen. The parting of zinc from an alloy (in some brasses, zinc is lost, leaving a weak, brittle, porous, copper-rich residue behind).

dialysis The separation of components of a solution by diffusion through a semipermeable membrane which is capable of passing certain ions or molecules while rejecting others. (See **electrodialysis, semipermeable membrane.**)

diaphragm pump A type of positive displacement pump in which the reciprocating piston is separated from the solution by a flexible diaphragm, thus protecting the piston from corrosion and erosion, and avoiding problems with packing and seals.

diatomaceous earth A processed natural material, the skeletons of diatoms, used as a filter medium.

diatomite Another name for **diatomaceous earth.**

dielectric fitting A plumbing fitting made of or containing an electrical nonconductor, such as plastic; used to separate dissimilar metals in a plumbing system to control galvanic corrosion.

dielectric shield In a cathodic protection system, an electrically nonconductive material, such as a coating, plastic sheet, or pipe, that is placed between an anode and an adjacent cathode to avoid current wastage and improve current distribution, usually on the cathode.

differential aeration cell (oxygen concentration cell) A concentration cell caused by differences in oxygen concentration along the surface of a metal in an electrolyte. (See **concentration cell.**)

diffusion Spreading of a constituent in a gas, liquid, or solid, tending to make the composition of all parts uniform.

diffusion coating Any process whereby a base metal or alloy is either (1) coated with another metal or alloy and heated to a sufficient temperature in a suitable environment or (2) exposed to a gaseous or liquid medium containing the other metal or alloy, thus causing diffusion of the coating or of the other metal or alloy into the base metal, with resultant changes in the composition and properties of its surface.

diffusion limited current density The current density, often referred to as limiting current density, that corresponds to the maximum transfer rate that a particular species can sustain as a result of the limitation of diffusion.

digestion The process in which complex materials are broken down into simpler substances; may be due to chemical, biological, or a combination of reactions.

disbondment The destruction of adhesion between a coating and the surface coated.

discontinuity Any interruption in the normal physical structure or configuration of a part, such as cracks, laps, seams, inclusions, or porosity. A discontinuity may or may not affect the usefulness of the part.

disinfection A process in which vegetative bacteria are killed. It may involve disinfecting agents such as chlorine or physical processes such as heating.

dislocation A linear imperfection in a crystalline array of atoms. Two basic types are recognized: (1) An edge dislocation corresponds to the row of mismatched atoms along the edge formed by an extra, partial plane of atoms within the body of a crystal; (2) a screw dislocation corresponds to the axis of a spiral structure in a crystal, characterized by a distortion that joins normally parallel planes together to form a continuous helical ramp.

dissociation The process by which a chemical compound breaks down into simpler constituents, such as CO_2 and H_2O, at high temperature.

dissolved solids The weight of matter in true solution in a stated volume of water; includes both inorganic and organic matter; usually determined by weighing the residue after evaporation of the water.

distillate fuels Liquid fuels, usually distilled from crude petroleum.

distillation Vaporization of a substance with subsequent recovery of the vapor by condensation. Often used in a less precise sense to refer to vaporization of volatile constituents of a fuel without subsequent condensation.

distilled water Water with a higher purity, produced by vaporization and condensation.

dolomite A specific form of limestone containing chemically equivalent concentrations of calcium and magnesium carbonates; the term is sometimes applied to limestone with compositions similar to that of true dolomite.

double layer The interface between an electrode or a suspended particle and an electrolyte created by charge-charge interaction (charge separation), leading to an alignment of oppositely charged ions at the surface of the electrode or particle.

downtime The amount of time a piece of equipment is not operational.

drain A pipe or conduit in a piping system which carries liquids to waste by gravity; sometimes the term is limited to liquids other than sewage.

drain line A tube or pipe from a water conditioning unit that carries backwash water, regeneration wastes, and/or rinse water to a drain or waste system.

drainage Conduction of electric current from an underground metallic structure by means of a metallic conductor. Forced drainage is that applied to underground metallic structures by means of an applied electromotive force or sacrificial anode. Natural drainage is that from an underground structure to a more negative (more anodic) structure, such as the negative bus of a trolley substation.

dry corrosion See **gaseous corrosion, hot corrosion.**

dry steam Steam containing no moisture. Commercially, dry steam containing not more than $1/2$ percent moisture.

ductile fracture Fracture characterized by tearing of metal accompanied by appreciable gross plastic deformation and expenditure of considerable energy. Contrast with **brittle fracture.**

ductility The ability of a material to deform plastically without fracturing; measured by elongation or reduction of area in a tensile test, by height of cupping in an Erichsen test, or by other means.

dynamic Active, alive, or tending to produce motion, as opposed to static, resting, or fixed.

dynamic system A system or process in which motion occurs or which includes active forces, as opposed to static conditions with no motion.

economizer A device that utilizes waste heat by transferring heat from flue gases to warm incoming feedwater.

effluent The stream emerging from a unit, system, or process, such as the softened water from an ion-exchange softener.

electrochemical admittance The reciprocal of the electrochemical impedance, I/E.

electrochemical cell An electrochemical system consisting of an anode and a cathode in metallic contact and immersed in an electrolyte. (The anode and cathode may be different metals or dissimilar areas on the same metal surface.)

electrochemical impedance spectroscopy (EIS) The frequency-dependent, complex-valued proportionality factor, E/I, between the applied potential (or current) and the response current (or potential) in an electrochemical cell. This factor becomes the impedance when the perturbation and response are related linearly (the factor value is independent of the perturbation magnitude) and the response is caused only by the perturbation. The value may be related to the corrosion rate when the measurement is made at the corrosion potential.

electrochemical potential (electrochemical tension) The partial derivative of the total electrochemical free energy of the system with respect to the number of moles of the constituent in a solution when all other factors are constant. (It is analogous to the chemical potential of the constituent, except that it includes the electrical as well as the chemical contributions to the free energy.)

electrode potential The potential of an electrode in an electrolyte as measured against a reference electrode. (The electrode potential does not include any resistance losses in potential in either the solution or the external circuit. It represents the reversible work to move a unit charge from the electrode surface through the solution to the reference electrode.)

electrodialysis A process in which a direct current is applied to a cell to draw charged ions through ion-selective semipermeable membranes, thus removing the ions from the solution.

electrolysis Production of chemical changes in the electrolyte by the passage of current through an electrochemical cell.

electrolyte A nonmetallic substance that carries an electric current, or a substance which, when dissolved in water, separates into ions which can carry an electric current.

electrolytic cell An assembly, consisting of a vessel, electrodes, and an electrolyte, in which electrolysis can be carried out.

electrolytic cleaning A process of removing soil, scale, or corrosion products from a metal surface by subjecting it as an electrode to an electric current in an electrolytic bath.

electromotive force series (emf series) A list of elements arranged according to their standard electrode potentials, with "noble" metals such as gold being positive and "active" metals such as zinc being negative.

electron A fundamental particle found in the atom that carries a single negative charge.

electroplating Electrodepositing a metal or alloy in an adherent form on an object serving as a cathode.

electropolishing A technique in which a high polish is produced by making the specimen the anode in an electrolytic cell, where preferential dissolution at high points smooths the surface; commonly used to prepare metallographic specimens.

electrostatic precipitator A device for collecting dust, mist, or fumes from a gas stream by placing an electric charge on the particles and removing those particles onto a collecting electrode.

elution The stripping of ions from an ion-exchange material by other ions, either because of greater affinity or because of much higher concentration.

embrittlement The severe loss of ductility or toughness or both of a material, usually a metal or alloy.

endpoint The point at which a process is stopped because a predetermined value of a measurable variable has been reached.

endurance limit The maximum stress that a material can withstand for an infinitely large number of fatigue cycles; the maximum cyclic stress level that a metal can withstand without fatigue failure. See also **fatigue strength.**

environmental cracking Brittle fracture of a normally ductile material in which the corrosive effect of the environment is a causative factor. Environmental cracking is a general term that includes corrosion fatigue, high-temperature hydrogen attack, hydrogen blistering, hydrogen embrittlement, liquid metal embrittlement, solid metal embrittlement, stress corrosion cracking, and sulfide stress cracking.

equilibrium The state in which the action of multiple forces produces a steady balance.

equilibrium (reversible) potential The potential of an electrode in an electrolytic solution when the forward rate of a given reaction is exactly equal to the reverse rate. (The equilibrium potential can be defined only with respect to a specific electrochemical reaction.)

equilibrium reaction A chemical reaction which proceeds primarily in one direction until the concentrations of reactants and products reach an equilibrium.

equivalent weight The weight in grams of an element, compound, or ion which would react with or replace 1 g of hydrogen; the molecular weight in grams divided by the valence.

erosion The progressive loss of material from a solid surface as a result of mechanical interaction between that surface and a fluid, a multicomponent fluid, or solid particles carried with the fluid.

erosion-corrosion A conjoint action involving corrosion and erosion in the presence of a moving corrosive fluid; it leads to the accelerated loss of material.

evaporation The change of state from a liquid to a vapor.

evaporation rate The number of pounds of water evaporated in a unit of time.

exchange current density The rate of charge transfer per unit area when an electrode reaches dynamic equilibrium (at its reversible potential) in a solution; that is, the rate of anodic charge transfer (oxidation) balances the rate of cathodic charge transfer (reduction).

exfoliation Corrosion that proceeds laterally from the sites of initiation along planes parallel to the surface, generally at grain boundaries; it forms corrosion products that force metal away from the body of the materials, giving rise to a layered appearance.

expansion joint A joint that permits movement to eliminate stress due to expansion.

external circuit The wires, connectors, measuring devices, current sources, etc., that are used to bring about or measure the desired electrical conditions within the test cell.

fatigue strength The maximum stress that can be sustained for a specified number of cycles without failure, with the stress being completely reversed within each cycle unless otherwise staled.

feed pump A pump that supplies water to a boiler.

feedwater Water introduced into a boiler during operation. It includes make-up and return condensate.

feedwater treatment The treatment of boiler feedwater by the addition of chemicals to prevent the formation of scale or to eliminate other objectionable characteristics.

ferrite A solid solution of one or more elements in body-centered cubic iron. Unless otherwise designated (for instance, as chromium ferrite), the solute is generally assumed to be carbon. On some equilibrium diagrams, there are two ferrite regions separated by an austenite area. The lower area is alpha ferrite; the upper, delta ferrite. If there is no designation, alpha ferrite is assumed.

ferritic Pertaining to the body-centered cubic (BCC) crystal structure of many ferrous (iron-based) metals.

filiform corrosion Corrosion that occurs under some coatings in the form of randomly distributed threadlike filaments.

film A thin, not necessarily visible layer of material.

filter Porous material through which fluids or fluid and solid mixtures are passed to separate matter held in suspension.

filtrate The effluent liquid from a filter.

fin An extended surface, a solid, experiencing energy transfer by conduction within its boundaries as well as energy transfer with its surroundings by convection and/or radiation, used to enhance heat transfer by increasing surface area.

fin tube A tube with one or more fins.

fired pressure vessel A vessel containing a fluid under pressure that is exposed to heat from the combustion of fuel.

firetube A type of boiler design in which combustion gases flow inside the tubes and water flows outside the tubes.

fish eyes Areas on a steel fracture surface having a characteristic white, crystalline appearance.

flakes Short, discontinuous internal fissures in wrought metals attributed to stresses produced by localized transformation and decreased solubility of hydrogen during cooling after hot working. In a fracture surface, flakes appear as bright, silvery areas; on an etched surface, they appear as short, discontinuous cracks.

flame spraying A thermal spraying process in which an oxyfuel gas flame is the source of heat for melting the surfacing material. Compressed gas may or may not be used for atomizing and propelling the surfacing material to the substrate.

flammability Susceptibility to combustion.

flash point The lowest temperature at which, under specified conditions, fuel oil gives off enough vapor to flash into a momentary flame when ignited.

flashing The process of producing steam by discharging water into a region with a pressure lower than the saturation pressure that corresponds to the water temperature

floc An agglomeration of finely divided suspended particles in a larger, usually gelatinous particle, the result of physical attraction or adhesion to a coagulant compound.

flocculation The process of causing a floc to form by gentle stirring or mixing after treatment with a coagulant.

flow control A device designed to limit or restrict the flow of water or regenerant; may include a throttling valve, an orifice of fixed diameter, or a pressure-compensating orifice.

flue A passage for products of combustion.

flue gas The gaseous product of combustion in the flue to the stack.

flush tank A tank or chamber in which water is stored for rapid release.

flush valve A self-closing valve designed to release a large volume of water when tripped.

foaming The continuous formation of bubbles which have sufficiently high surface tension to remain as bubbles beyond the disengaging surface.

fogged metal A metal whose luster has been reduced because of a surface film, usually a corrosion product layer.

forced circulation The circulation of water in a boiler by mechanical means external to the boiler.

forced-draft fan A fan supplying air under pressure to the fuel-burning equipment.

foreign structure Any metallic structure that is not intended as part of a cathodic protection system of interest.

fouling The accumulation of refuse in gas passages or on heat-absorbing surfaces, resulting in undesirable restriction to the flow of gas or heat.

fouling organism Any aquatic organism with a sessile adult stage that attaches to and fouls underwater structures of ships.

fractography Descriptive treatment of fracture, especially in metals, with specific reference to photographs of the fracture surface. Macrofractography involves photographs at low magnification ($<25\times$); microfractography, photographs at high magnification ($>25\times$).

fracture mechanics A quantitative analysis for evaluating structural behavior in terms of applied stress, crack length, and specimen or machine component geometry.

fracture toughness A generic term for measures of resistance to extension of a crack. The term is sometimes restricted to results of fracture mechanics tests, which are directly applicable in fracture control; however, it commonly includes results from simple tests of notched or precracked specimens that are not based on fracture mechanics analysis. Results from tests of the latter type are often useful for fracture control, based on either service experience or empirical correlations with fracture mechanics tests. See also **stress-intensity factor.**

free ash Ash which is not included in the fixed ash.

free available chlorine The concentration of residual chlorine present as dissolved gas, hypochlorous acid, or hypochlorite, and not combined with ammonia or in other less readily available forms.

free carbon The part of the total carbon in steel or cast iron that is present in elemental form as graphite or temper carbon. Contrast with **combined carbon.**

free carbon dioxide Carbon dioxide present in water as the gas or as carbonic acid, but not combined in carbonates or bicarbonates.

free corrosion potential Corrosion potential in the absence of net electric current flowing to or from the metal surface.

free ferrite Ferrite that is formed directly from the decomposition of hypoeutectoid austenite during cooling, without the simultaneous formation of cementite, also called *proeutectoid ferrite*.

free machining Pertains to the machining characteristics of an alloy to which one or more ingredients have been introduced to give small broken chips, lower power consumption, better surface finish, and longer tool life; among such additions are sulfur or lead to steel, lead to brass, lead and bismuth to aluminum, and sulfur or selenium to stainless steel.

fretting Surface damage resulting from relative motion between surfaces in contact under pressure.

fretting corrosion The deterioration at the interface between contacting surfaces as the result of corrosion and slight oscillatory slip between the two surfaces.

fuel   A substance containing combustible material used for generating heat.

Gallionella ferruginea   A genus of stalked, ribbonlike bacteria which utilize iron in their metabolism and cause staining, plugging, and odor problems in water systems. (See **iron bacteria.**)

galvanic cell   A cell which generates an electric current, consisting of dissimilar metals in contact with each other and with an electrolyte.

galvanic corrosion   Accelerated corrosion of a metal because of an electrical contact with a more noble metal or nonmetallic conductor in a corrosive electrolyte.

galvanic couple   A pair of dissimilar conductors, commonly metals in electrical contact. (See **galvanic corrosion**).

galvanic current   The electric current between metals or conductive nonmetals in a galvanic couple.

galvanic series   A list of metals and alloys arranged according to their relative corrosion potentials in a given environment.

galvanize   To coat a metal surface with zinc using any of various processes.

galvanneal   To produce a zinc-iron alloy coating on iron or steel by keeping the coating molten after hot-dip galvanizing until the zinc alloys completely with the base metal.

galvanodynamic   Refers to a technique in which current that is continuously varied at a selected rate is applied to an electrode in an electrolyte.

galvanostaircase   A galvanostep technique for polarizing an electrode in a series of constant current steps, with the time duration and current increments or decrements equal for each step.

galvanostatic   An experimental technique in which an electrode is maintained at a constant current in an electrolyte.

galvanostep   A technique in which an electrode is polarized in a series of current increments or decrements.

gamma iron   The face-centered cubic form of pure iron, stable from 910 to 1400°C.

gas pressure regulator   A spring-loaded, dead-weighted, or pressure-balanced device which will maintain the gas pressure to a supply line.

gaseous corrosion   Corrosion with gas as the only corrosive agent and without any aqueous phase on the surface of the metal; also called **dry corrosion.**

gate valve   A valve with a closing element that is a disk which is moved across the stream, often in a groove or slot for support against pressure.

gage pressure   The pressure above atmospheric pressure.

gel zeolite   A synthetic sodium aluminosilicate ion exchanger.

general corrosion   A form of deterioration that is distributed more or less uniformly over a surface; see **uniform corrosion.**

Gibbs free energy Thermodynamic function; also called *free energy, free enthalpy,* or *Gibbs function.*

glass electrode A glass membrane electrode used to measure pH or hydrogen-ion activity.

globe valve A valve in which the closing element is a sphere or a flat or rounded gasket which is moved into or onto a round port.

grab sample A single sample of material collected at one place and one time.

grain An individual crystal in a polycrystalline metal or alloy; it may or may not contain twinned regions and subgrains; a portion of a solid metal (usually a fraction of an inch in size) in which the atoms are arranged in an orderly pattern.

grain boundary A narrow zone in a metal corresponding to the transition from one crystallographic orientation to another, thus separating one grain from another, with the atoms in each grain arranged in an orderly pattern; the irregular junction of two adjacent grains.

grain-boundary corrosion Same as **intergranular corrosion.**

grain dropping The dislodgment and loss of a grain or grains (crystals) from a metal surface as a result of intergranular corrosion.

grains (water) A unit of measure commonly used in water analysis for the measurement of impurities in water [17.1 grains = 1 part per million (ppm)].

grains per cubic foot The term for expressing dust loading in weight per unit of gas volume (7000 grains equals 1 pound).

gram (g) The basic unit of weight (mass) of the metric system, originally intended to be the weight of 1 cubic centimeter of water at 4°C.

gram-milliequivalent The equivalent weight of a substance in grams, divided by one thousand.

graphitic corrosion The deterioration of metallic constituents in gray cast iron, leaving the graphitic particles intact. (The term *graphitization* is commonly used to identify this form of corrosion but is not recommended because of its use in metallurgy for the decomposition of carbide to graphite.) See also **parting, selective leaching.**

graphitization A metallurgical term describing the formation of graphite in iron or steel, usually from decomposition of iron carbide at elevated temperatures. Not recommended as a term to describe graphitic corrosion.

gravimetric Measurement of matter on the basis of weight.

green rot A form of high-temperature corrosion of chromium-bearing alloys in which green chromium oxide (Cr_2O_3) forms, but certain other alloy constituents remain metallic; some simultaneous carburization is sometimes observed.

groundbed A buried item, such as junk steel or graphite rods, that serves as the anode for the cathodic protection of pipelines or other buried structures.

handhole An access opening in a pressure part usually not exceeding 18 cm in its longest dimension.

handhole cover A handhole closure.

hard water Water which contains calcium or magnesium in an amount such that an excessive amount of soap is required in order to form a lather.

hardness A measure of the amount of calcium and magnesium salts in water, usually expressed as grains per gallon or ppm of $CaCO_3$.

head A measure of the pressure at a point in a water system.

head loss The same as **pressure drop.**

heat balance An accounting of the distribution of the heat input, output, and losses.

heat exchanger A vessel in which heat is transferred from one medium to another.

heat release rate A rate that describes the heat available per square foot of heat-absorbing surface in the furnace or per cubic foot of volume.

heating surface A surface which is exposed to products of combustion on one side and water on the other. This surface is measured on the side receiving the heat.

heating value The quantity of heat released by a fuel through complete combustion. It is commonly expressed in Btu per pound, per gallon, or per cubic foot.

hot corrosion An accelerated corrosion of metal surfaces that results from the combined effect of oxidation and reactions with sulfur compounds and other contaminants, such as chlorides, to form a molten salt on a metal surface which fluxes, destroys, or disrupts the normal protective oxide.

hot cracking In weldments, a process caused by the segregation at grain boundaries of low-melting constituents in the weld metal. Hot cracking can be minimized by the use of low-impurity welding materials and proper joint design. Also called *solidification cracking.*

hot dip coating A metallic coating obtained by dipping the base metal into a molten metal.

hot shortness A tendency for some alloys to separate along grain boundaries when stressed or deformed at temperatures near the melting point. Hot shortness is caused by a low-melting constituent, often present in only minute amounts, that is segregated at grain boundaries.

hot working Deforming metal plastically at such a temperature and strain rate that recrystallization takes place simultaneously with the deformation, thus avoiding any strain hardening.

Huey test Corrosion testing in a boiling solution of nitric acid. This test is mainly used to detect the susceptibility to intergranular corrosion of stainless steel.

humidity test A corrosion test involving exposure of specimens at controlled levels of humidity and temperature.

hydration The chemical combination of water into a substance.

hydrocarbon A chemical compound of hydrogen and carbon.

hydrogen blistering The formation of blisters on or below a metal surface as a result of excessive internal hydrogen pressure. (Hydrogen may be formed during cleaning, plating, corrosion, etc.)

hydrogen damage A general term for the embrittlement, cracking, blistering, and hydride formation that can occur when hydrogen is present in some metals.

hydrogen embrittlement Hydrogen-induced cracking or severe loss of ductility caused by the presence of hydrogen in the metal.

hydrogen-induced cracking (HIC) Same as **hydrogen embrittlement.**

hydrogen overvoltage Overvoltage associated with the liberation of hydrogen gas.

hydrolysis (1) Decomposition or alteration of a chemical substance by water. (2) In aqueous solutions of electrolytes, the reactions of cations with water to produce a weak base or of anions to produce a weak acid.

hydrophilic Having an affinity for water.

hydrophobic Lacking an affinity for, repelling, or failing to absorb or adsorb water.

hydrostatic test A strength and tightness test of a closed pressure vessel by water pressure; a pressure test procedure in which a vessel or system is filled with water, purged of air, sealed, subjected to water pressure, and examined for leaks, distortion, and/or mechanical failure.

hydroxide A chemical compound containing hydroxyl (OH) ion. (See **hydroxyl.**)

hydroxyl The OH anion, which has a single negative charge and provides the characteristics common to bases.

hygroscopic Possessing a marked ability to accelerate the condensation of water vapor; applied to condensation nuclei composed of salts that yield aqueous solutions with a very low equilibrium vapor pressure compared with that of pure water at the same temperature.

hypochlorite The OCl anion; calcium and sodium hypochlorite are commonly used as bleaches and disinfecting agents.

ignition The initiation of combustion.

immersion plating Depositing a metallic coating on a metal immersed in a liquid solution, without the aid of an external electric current. Also called *dip plating*.

immunity A state of resistance to corrosion or anodic dissolution of a metal caused by thermodynamic stability of the metal.

impingement corrosion A form of erosion-corrosion generally associated with the local impingement of a high-velocity flowing fluid against a solid surface.

impressed current An electric current supplied by a device employing a power source that is external to the electrode system. (An example is direct current for cathodic protection.)

inclusions Particles of foreign material in a metallic matrix. The particles are usually compounds (such as oxides, sulfides, or silicates), but may be of any substance that is foreign to (and essentially insoluble in) the matrix.

incubation period A period prior to the detection of corrosion during which the metal is in contact with a corrodent.

indicator A material which can be used to show the endpoint of a chemical reaction, usually by a color change, or a chemical concentration, usually by a depth or shade of color.

industrial atmosphere An atmosphere in an area of heavy industry, with soot, fly ash, and sulfur compounds as the principal constituents.

inert anode An anode that is insoluble in the electrolyte under the conditions prevailing in the electrolysis.

influent The stream entering a unit, stream, or process, such as the hard water entering an ion-exchange water softener.

inhibitor A chemical substance or combination of substances that, when present in the proper concentration and forms in the environment, prevents or reduces corrosion.

injector A device utilizing a steam jet to entrain and deliver feedwater into a boiler.

inorganic Being or composed of matter other than hydrocarbons and their derivatives; matter that is not of plant or animal origin.

inorganic matter Matter which is not derived from living organisms and contains no organically produced carbon; includes rocks, minerals, and metals. Contrast with **organic matter.**

inorganic zinc-rich paint A coating containing a zinc powder pigment in an inorganic vehicle.

insulation A material of low thermal conductivity used to reduce heat losses.

intensiostatic See **galvanostatic.**

intercrystalline corrosion See **intergranular corrosion.**

intergranular Between crystals or grains.

intergranular corrosion Preferential corrosion at or adjacent to the grain boundaries of a metal or alloy.

intergranular cracking Cracking or fracturing that occurs between the grains or crystals in a polycrystalline aggregate. Also called *intercrystalline cracking.*

intergranular fracture Brittle fracture of a metal in which the fracture is between the grains, or crystals, that form the metal. Also called *intercrystalline fracture.* Contrast with **transgranular fracture.**

interlock A device to test for the existence of a required condition, and to furnish proof of that condition to the primary safety control circuit.

intermittent blowdown The blowing down of boiler water at intervals.

internal oxidation The formation of isolated particles of corrosion products beneath the metal surface. (This occurs as the result of preferential oxidation of certain alloy constituents by inward diffusion of oxygen, nitrogen, sulfur, etc.)

internal treatment The treatment of boiler water by introducing chemicals directly into the boiler.

intumescence The swelling or bubbling of a coating, usually because of heating (a term currently used in space and fire protection applications).

ion An atom or group of atoms that has gained or lost one or more outer electrons and thus carries an electric charge. Positive ions, or cations, are deficient in outer electrons. Negative ions, or anions, have an excess of outer electrons.

ion exchange (1) The reversible interchange of ions between a liquid and a solid, with no substantial structural changes in the solid. (2) A reversible process in which ions are released from an insoluble permanent material in exchange for other ions in a surrounding solution; the direction of the exchange depends upon the affinities of the ion exchanger for the ions present and the concentrations of the ions in the solution.

ion exchanger A permanent, insoluble material which contains ions that will exchange reversibly with other ions in a surrounding solution. Both cation and anion exchangers are used in water conditioning.

ionization The process in which atoms gain or lose electrons; sometimes used as synonymous with **dissociation,** the separation of molecules into charged ions in solution.

ionization constant A constant specific for each partially ionizable chemical compound that expresses the ratio of the concentration of ions from the compound to the concentration of undissociated compound.

iron bacteria Microorganisms which are capable of utilizing ferrous iron, either from the water or from steel pipe, in their metabolism and precipitating ferric hydroxide in their sheaths and gelatinous deposits. These organisms tend to collect in pipelines and tanks during periods of low flow, and to break loose in slugs of turbid water to create staining, taste, and odor problems.

isocorrosion diagram A graph or chart that shows constant corrosion behavior with changing solution (environment) composition and temperature.

KISCC Abbreviation for the critical value of the plane strain stress-intensity factor that will produce crack propagation by stress corrosion cracking of a given material in a given environment.

knifeline attack Intergranular corrosion of an alloy, usually stabilized stainless steel, along a line adjoining or in contact with a weld after heating into the sensitization temperature range.

Kraft process A wood-pulping process in which sodium sulfate is used in the caustic soda pulp-digestion liquor.

laminar flow The flow of fluid in which the flow paths are in smooth, parallel lines, with essentially no mixing and no turbulence.

Langelier index A calculated number used to predict the calcium carbonate ($CaCO_3$) stability of a water; that is, whether a water will precipitate, dissolve, or be in equilibrium with calcium carbonate. It is sometimes erroneously assumed that any water that tends to dissolve calcium carbonate will always be corrosive.

leakage The uncontrolled quantity of fluid which enters or leaves an enclosure.

lime The common name for calcium oxide (CaO); hydrated lime is calcium hydroxide [$Ca(OH)_2$].

lime scale Hard-water scale containing a high percentage of calcium carbonate.

limestone A sedimentary rock, largely calcium carbonate but usually also containing significant amounts of magnesium carbonate.

linear alkyl sulfonate A term applied to a family of straight chain chemical compounds, widely used as detergents; sometimes called "soft" detergents because they are more readily degraded to simpler substances by biological action than the previously used alkyl benzene sulfonate.

lining The material used on the furnace side of a furnace wall. It is usually of high-grade refractory tile or brick or plastic refractory material.

liquid metal embrittlement Catastrophic brittle failure of a normally ductile metal when it is brought into contact with a liquid metal and is subsequently stressed in tension.

liter The basic metric unit of volume; 3.785 L equal 1 U.S. gal; 1 L of water weighs 1000 g.

load The rate of output required; also the weight carried.

load factor The ratio of the average load in a given period to the maximum load carried during that period.

local action corrosion Corrosion caused by local corrosion cells on a metal surface.

local cell A galvanic cell resulting from inhomogeneities between areas on a metal surface in an electrolyte. The inhomogeneities may be of a physical or chemical nature in either the metal or its environment.

local corrosion cell An electrochemical cell created on a metal surface because of a difference in potential between adjacent areas on that surface.

localized corrosion Corrosion at discrete sites; for example, pitting, crevice corrosion, and stress corrosion cracking.

long-line current Electric current through the earth from an anodic to a cathodic area of a continuous metallic structure. (Usually used only where the areas are separated by considerable distance and where the current results from concentration-cell action.)

low-water cutoff A safety device that shuts off a boiler or burner in the event of low water, preventing pressure vessel failure.

lug Any projection, such as an ear, used for supporting or grasping.

Luggin probe or Luggin-Haber capillary A device used in measuring the potential of an electrode with a significant current density imposed on its surface. (The probe minimizes the IR drop that would otherwise be included in the measurement without significantly disturbing the current distribution on the specimen.)

M alkalinity Methyl orange alkalinity. (See **total alkalinity**.)

MAG The metal active gas welding process; it uses CO_2 and argon-CO_2 mixtures.

makeup The water added to boiler feedwater to compensate for that lost through exhaust, blowdown, leakage, etc.

manganese greensand Greensand which has been processed to incorporate the higher oxides of manganese in its pores and on its surface. The product has a mild oxidizing power, and is often used in the oxidation and precipitation of iron, manganese, and/or hydrogen sulfide, and their removal from water.

manganese zeolite Synthetic gel zeolite which has been processed in the same manner as manganese greensand and is used for similar purposes.

manhole An opening in a pressure vessel of sufficient size to permit a person to enter.

manifold A pipe or header for collection of a fluid from or distribution of a fluid to a number of pipes or tubes.

maximum allowable working pressure The maximum gage pressure permissible in a completed boiler. This pressure is based upon either proof tests or calculations for every pressure part of a vessel using nominal thickness exclusive of allowances for corrosion and thickness required for loadings other than pressure. It is the basis for the settings of the pressure-relieving devices protecting the vessel.

maximum continuous load The maximum load which can be maintained for a specified period.

MCL Abbreviation for maximum contaminant level; the maximum allowable concentration of a contaminant in water as established in the U.S. EPA drinking water regulations.

mechanical filter A filter primarily designed for the removal of suspended solid particles, as opposed to filters with additional capabilities.

media The plural form of *medium*.

medium A material used in a filter bed to form a barrier to the passage of certain suspended solids or dissolved molecules.

metal dusting Accelerated deterioration of metals in carbonaceous gases at elevated temperatures, forming a dustlike corrosion product.

metallizing See **thermal spraying.**

methylene blue active substances Chemical compounds which react with methylene blue to form a blue compound which can be used to estimate the concentration by measurement of the depth of color. Substances measured include ABS and LAS types of detergents; thus the term is commonly used as an expression of detergent concentration.

microbiologically influenced corrosion (MIC) Corrosion that is affected by the action of microorganisms in the environment.

micrometer Formally known as micron, a linear measure equal to one millionth of a meter or 0.00003937 inch. The symbol for the micrometer is μm.

micron See **micrometer.**

microstructure The structure of a prepared surface of a metal as revealed by a microscope at a magnification exceeding 25×.

MIG The metal inert gas (argon or other) welding process.

mil One thousandth of an inch.

mill scale The heavy oxide layer formed during hot fabrication or heat treatment of metals.

mineral A term applied to inorganic substances such as rocks and similar matter found in the earth strata, as opposed to organic substances such as plant and animal matter. Minerals normally have definite chemical composition and crystal structure.

mineral acidity Acidity due to the presence of inorganic acids such as hydrochloric, sulfuric, and nitric acids, as opposed to acidity due to carbonic acid or organic acids.

mixed potential The potential of a specimen (or specimens in a galvanic couple) when two or more electrochemical reactions are occurring simultaneously.

MMA The manual metal arc welding process.

moisture Water in the liquid or vapor phase.

moisture in steam Particles of water carried in steam, expressed as the percentage by weight.

moisture loss The boiler flue gas loss representing the difference between the heat content of the moisture in the exit gases and that at the temperature of the ambient air.

molal solution The concentration of a solution expressed in moles of solute divided by 1000 g of solvent.

molar solution An aqueous solution that contains 1 mole (gram-molecular weight) of solute in 1 L of the solution.

mole The mass (in grams) numerically equal to the relative molecular mass of a substance. It is the amount of substance of a system that contains as many elementary units (6.02×10^{23} atoms of an element or 6.02×10^{23} molecules of a chemical compound) as there are atoms of carbon in 0.012 kg of the pure nuclide C_{12}. The weight of one mole of an element is equal to its atomic weight in grams; the weight of one mole of a compound is equal to its molecular weight in grams.

molecule The simplest combination of atoms that will form a specific chemical compound; the smallest particle of a substance which will still retain the essential composition and properties of that substance, and which can be broken down only into atoms and simpler substances.

Moneypenny-Strauss test Corrosion testing in a copper sulfate solution containing sulfuric acid; used to detect the susceptibility of stainless steel to intergranular corrosion.

monomer A molecule, usually an organic compound, having the ability to join with a number of identical molecules to form a polymer.

most probable number (MPN) The term used to indicate the number of microorganisms which, according to statistical theory, would be most likely to produce the results observed in certain bacteriological tests; usually expressed as a number per 100 mL of water.

MPN The abbreviation for **most probable number.**

nanometer Abbreviated nm, a unit of length equal to one thousandth of a micrometer.

natural aging Spontaneous aging of a supersaturated solid solution at room temperature. Compare with **artificial aging.**

natural circulation The circulation of water in a boiler caused by differences in density.

natural gas Gaseous fuel occurring in nature.

negative charge The electric charge on an electrode or ion in solution resulting from the presence of an excess of electrons. (See **anion, electron**.)

negative head A condition of negative pressure or partial vacuum.

negative pressure A pressure below the surrounding atmospheric pressure at a specific point; a partial vacuum.

Nernst equation An equation that expresses the exact electromotive force of a cell in terms of the activities of the products and reactants of the cell.

neutral In electrical systems, the term used to indicate neither an excess nor a lack of electrons; a condition of balance between positive and negative charges. In chemistry, the term used to indicate a balance between acids and bases; the neutral point on the pH scale is 7.0, indicating the presence of equal numbers of free hydrogen (acidic) and hydroxide (basic) ions.

neutralization The addition of either an acid or a base to a solution as required to produce a neutral solution. The use of alkaline or basic materials to neutralize the acidity of some waters is a common practice in water conditioning.

neutralizer A common designation for alkaline materials such as calcite (calcium carbonate) or magnesia (magnesium oxide) used in the neutralization of acid waters.

neutron embrittlement Embrittlement resulting from bombardment with neutrons, usually encountered in metals that have been exposed to a neutron flux in the core of a reactor. In steels, neutron embrittlement is evidenced by a rise in the ductile-to-brittle transition temperature.

nitriding Introducing nitrogen into the surface layer of a solid ferrous alloy by holding it at a suitable temperature in contact with a nitrogenous material, usually ammonia or molten cyanide of appropriate composition. Quenching is not required to produce a hard case.

nitrocarburizing Any of several processes in which both nitrogen and carbon are absorbed into the surface layers of a ferrous material at temperatures below the lower critical temperature and, by diffusion, create a concentration gradient. Nitrocarburizing is performed primarily to provide an antiscuffling surface layer and to improve fatigue resistance.

noble The positive (increasingly oxidizing) direction of electrode potential.

noble metal A metal with a standard electrode potential that is more noble (positive) than that of hydrogen.

noble potential A potential more cathodic (positive) than the standard hydrogen potential.

noncarbonate hardness Water hardness due to the presence of compounds such as calcium and magnesium chlorides, sulfates, or nitrates; the excess of total hardness over total alkalinity.

normal solution An aqueous solution containing one gram equivalent weight of the active reagent in one liter of the solution.

normal stress The stress component perpendicular to a plane on which forces act.

normalizing Heating a ferrous alloy to a suitable temperature above the transformation range and then cooling it in air to a temperature substantially below the transformation range.

NO$_x$ Abbreviation for all of the family of oxides of nitrogen.

nozzle A short flanged or welded neck connection on a drum or shell for the outlet or inlet of fluids; also, a projecting spout through which a fluid flows.

occluded cell An electrochemical cell created at a localized site on a metal surface which has been partially obstructed from the bulk environment.

oil burner A burner for firing oil.

open-circuit potential The potential of an electrode measured with respect to a reference electrode or another electrode when no current flows to or from it.

operating pressure The pressure at which a boiler or any other vessel is operated.

organic matter Substances of or derived from plant or animal matter. Organic matter is characterized by its carbon-hydrogen structure.

osmosis A process of diffusion of a solvent such as water through a semipermeable membrane which will transmit the solvent but impede most dissolved substances. The normal flow of solvent is from the dilute solution to the concentrated solution in an attempt to bring the solutions on both sides of the membrane to equilibrium.

overpressure The minimum operating pressure of a hot-water boiler that is sufficient to prevent the water from steaming.

overvoltage The change in potential of an electrode from its equilibrium or steady-state value when current is applied.

oxidation Loss of electrons by a constituent of a chemical reaction. Also refers to the corrosion of a metal that is exposed to an oxidizing gas at elevated temperatures.

oxidized surface (on steel) A surface having a thin, tightly adhering, oxidized skin (from straw to blue in color), extending in from the edge of a coil or sheet.

oxidizing agent A compound that causes oxidation, thereby itself being reduced.

oxidizing atmosphere An atmosphere which tends to promote the oxidation of immersed materials.

oxygen attack Corrosion or pitting in a boiler caused by oxygen.

oxygen concentration cell A galvanic cell resulting from differences in oxygen concentration between two locations; see **differential aeration cell.**

ozone A powerfully oxidizing allotropic form of the element oxygen. The ozone molecule contains three atoms (O_3).

P alkalinity Phenolphthalein alkalinity of a water as determined by titration with standard acid solution to the phenolphthalein endpoint (pH approximately 8.3). It includes both carbonate and hydroxide alkalinity.

particle size (1) A measure of dust size, expressed in micrometers or percent passing through a standard mesh screen. (2) The size of a particle suspended in water as determined by its smallest dimension.

parting The selective corrosion of one or more components of a solid solution alloy.

parting limit The minimum concentration of a more noble component in an alloy above which parting does not occur in a specific environment.

parts per billion (ppb) A measure of proportion by weight, equivalent to one unit weight of a material per billion (10^9) unit weights of compound. One part per billion is equivalent to 1 mg/Mg.

parts per million (ppm) A measure of proportion by weight, equivalent to one unit weight of a material per million (10^6) unit weights of compound. One part per million is equivalent to 1 mg/kg.

pass A confined passageway containing a heating surface through which a fluid flows in essentially one direction.

passivation The process in metal corrosion by which metals become passive. (See **passive.**)

passivator A type of inhibitor which appreciably changes the potential of a metal to a more noble (positive) value.

passive The state of a metal surface characterized by low corrosion rates in a potential region that is strongly oxidizing for the metal.

passive-active cell A corrosion cell in which the anode is a metal in the active state and the cathode is the same metal in the passive state.

passivity A condition in which a piece of metal, because of an impervious covering of oxide or other compound, has a potential that is much more positive than that of the metal in the active state.

patina The coating, usually green, that forms on the surface of metals such as copper and copper alloys exposed to the atmosphere; also used to describe the appearance of a weathered surface of any metal.

pearlite A metastable lamellar aggregate of ferrite and cementite resulting from the transformation of austenite at temperatures above the bainite range.

perfect combustion The complete oxidation of all the combustible constituents of a fuel, utilizing all the oxygen supplied.

permanent hardness Water hardness due to the presence of the chlorides and sulfates of calcium and magnesium, which will not precipitate by boiling. This term has been largely replaced by **noncarbonate hardness.**

permanganate Generally refers to potassium permanganate, a chemical compound used in water treatment.

petroleum Naturally occurring mineral oil consisting predominantly of hydrocarbons.

pH A measure of the degree of acidity or alkalinity (basicity) of a solution; the negative logarithm of the hydrogen-ion activity. At 25°C, 7.0 is the neutral value. Decreasing values below 7.0 indicate increasing acidity; increasing values above 7.0, increasing basicity.

phosphating Forming an adherent phosphate coating on a metal by immersion in a suitable aqueous phosphate solution. Also called *phosphatizing*.

physical vapor deposition A coating process in which the cleaned and masked component to be coated is heated and rotated on a spindle above the streaming vapor generated by melting and evaporating a coating material source bar with a focused electron beam in an evacuated chamber.

physisorption The binding of an adsorbate to the surface of a solid by forces whose energy levels approximate those of condensation. Contrast with **chemisorption.**

pickle A solution or process used to loosen or remove corrosion products such as scale or tarnish.

pickling Removing surface oxides from metals by chemical or electrochemical reaction.

pitting Corrosion of a metal surface, confined to a point or small area, that takes the form of cavities.

pitting factor Ratio of the depth of the deepest pit resulting from corrosion to the average penetration as calculated from weight loss.

pitting resistance equivalent number An empirical relationship to predict the pitting resistance of austenitic and duplex stainless steels. It is expressed as $PRE_N = Cr + 3.3 (Mo + 0.5 W) + 16N$.

pK The reciprocal of the logarithm of the ionization constant of a chemical compound.

plane strain The stress condition in linear elastic fracture mechanics in which there is zero strain in a direction normal to both the axis of applied tensile stress and the direction of crack growth (that is, parallel to the crack front); most nearly achieved in loading of thick plates along a direction parallel to the plate surface. Under plane-strain conditions, the plane of fracture instability is normal to the axis of the principal tensile stress.

plasma spraying A thermal spraying process in which the coating material is melted with heat from a plasma torch that generates a nontransferred arc. Molten coating material is propelled against the base metal by the hot, ionized gas issuing from the torch.

plastic deformation The permanent (inelastic) distortion of metals under applied stresses that strain the material beyond its elastic limit.

plasticity The property that enables a material to undergo permanent deformation without rupture.

pOH The negative logarithm of the hydroxyl ion concentration. The pOH is related to the pH by the expression $pH + pOH = 14$.

polarization The change from the open-circuit electrode potential as the result of the passage of current.

polarization admittance The reciprocal of polarization resistance (di/dE).

polarization curve A plot of current density versus electrode potential for a specific electrode-electrolyte combination.

polarization resistance The slope (dE/di) at the corrosion potential of a potential (E) versus current density (i) curve. It is inversely proportional to the corrosion current density when the polarization resistance technique is applicable. The term is also used to describe the method of measuring corrosion rates using this slope.

polymer A chain of organic molecules produced by the joining of primary units called monomers.

polyphosphate A form of phosphate polymer consisting of a series of condensed phosphoric acids containing more than one atom of phosphorus. Polyphosphate is used as a sequestering agent to control iron and hardness, and as a coating agent that forms a thin passivating film on metal surfaces to control corrosion.

porosity A measure of the volume of internal pores, or voids, in ion exchangers and filter media; sometimes expressed as a ratio to the total volume of the medium. (See **void volume**.)

positive charge   The net electric charge on an electrode or ion in solution as a result of the removal of electrons.

postchlorination   The application of chlorine to a water following other water treatment processes.

potential-pH diagram   See **Pourbaix diagram.**

potentiodynamic (potentiokinetic)   A technique wherein the potential of an electrode with respect to a reference electrode is varied at a selected rate by application of a current through the electrolyte.

potentiostaircase   A potentiostep technique for polarizing an electrode in a series of constant-potential steps, with the time duration and potential increments or decrements equal for each step.

potentiostat   An instrument for automatically maintaining an electrode in an electrolyte at a constant potential or controlled potentials with respect to a suitable reference electrode.

potentiostatic   The technique for maintaining a constant electrode potential.

potentiostep   A technique in which an electrode is polarized in a series of potential increments or decrements.

poultice corrosion   A term used in the automotive industry to describe the corrosion of vehicle body parts caused by to the collection of road salts and debris on ledges and in pockets that are kept moist by weather and washing. Also called **deposit corrosion** or *attack*.

Pourbaix diagram (potential-pH diagram)   A graphical representation showing regions of thermodynamic stability of species in metal–water electrolyte systems.

powder metallurgy   The art of producing metal powders and utilizing metal powders for production of massive materials and shaped objects.

prechlorination   The application of chlorine to a water prior to other water treatment processes.

precious metal   One of the relatively scarce and valuable metals: gold, silver, and the platinum-group metals. Also called **noble metal**.

precipitate   To separate materials from a solution by the formation of insoluble matter by chemical reaction and removal of this insoluble matter.

precipitation   The removal of solid or liquid particles from a fluid.

precipitation hardening   Hardening caused by the precipitation of a constituent from a supersaturated solid solution. See also **age hardening.**

precipitation heat treatment   Artificial aging in which a constituent is precipitated from a supersaturated solid solution.

precracked specimen   A specimen that is notched and subjected to alternating stresses until a crack develops at the root of the notch.

preheated air   Air at a temperature exceeding that of the ambient air.

pressure   Force per unit of area.

pressure differential A difference or change in pressure detected between two points in a system as a result of differences in elevation and/or pressure drop caused by flow.

pressure drop The difference in pressure between two points in a system, caused by resistance to flow.

pressure vessel A closed vessel or container designed to confine a fluid at a pressure above atmospheric pressure.

primary current distribution The current distribution in an electrolytic cell that is free of polarization.

primary passive potential (passivation potential) The potential corresponding to the maximum active current density (critical anodic current density) of an electrode that exhibits active-passive corrosion behavior.

primer (prime coat) The first coat of paint applied to a surface, formulated to have good bonding and wetting characteristics. It may or may not contain inhibiting pigments.

priming The discharge of steam containing excessive quantities of water in suspension from a boiler, as a result of violent ebullition.

principal stress (normal) The maximum or minimum value at the normal stress at a point in a plane considered with respect to all possible orientations of the considered plane. On such principal planes, the shear stress is zero. There are three principal stresses on three mutually perpendicular planes. The state of stress at a point may be (1) uniaxial, a state of stress in which two of the three principal stresses are zero, (2) biaxial, a state of stress in which only one of the three principal stresses is zero, or (3) triaxial, a state of stress in which none of the principal stresses is zero. *Multiaxial stress* refers to either biaxial or triaxial stress.

process steam Steam used for industrial purposes other than producing power.

products of combustion The gases, vapors, and solids resulting from the combustion of fuel.

profile Anchor pattern on a surface produced by abrasive blasting or acid treatment.

protection potential The most noble potential where pitting and crevice corrosion will not propagate.

protective potential The threshold value of the corrosion potential that has to be reached to enter a protective potential range. The term is used in cathodic protection to refer to the minimum potential required to control corrosion.

protective potential range A range of corrosion potential values in which an acceptable corrosion resistance is achieved for a particular purpose.

purge To introduce air into a vessel's flue passages in such a volume and manner as to completely replace the air or gas-air mixture contained therein.

radiation damage A general term for the alteration of properties of a material arising from exposure to ionizing radiation (penetrating radiation), such as x-rays, gamma rays, neutrons, heavy-particle radiation, or fission fragments in nuclear fuel material.

rate of blowdown A rate normally expressed as a percentage of the water fed.

raw water Untreated water, or any water before it reaches a specific water treatment device or process.

reaction A chemical transformation or change brought about by the interaction of two substances.

reactive metal A metal that readily combines with oxygen at elevated temperatures to form very stable oxides, for example, titanium, zirconium, and beryllium.

reassociation The recombination of the products of dissociation.

recirculation The reintroduction of part of the flowing fluid to repeat the cycle of circulation.

recrystallization (1) Formation of a new, strain-free grain structure from that existing in cold worked metal, usually accomplished by heating. (2) The change from one crystal structure to another, as occurs on heating or cooling through a critical temperature.

red water Water which has a reddish or brownish appearance as a result of the presence of precipitated iron and/or iron bacteria.

redox potential The potential of a reversible oxidation-reduction electrode measured with respect to a reference electrode, corrected to the hydrogen electrode, in a given electrolyte.

reducing agent A compound that causes reduction, thereby itself becoming oxidized.

reducing atmosphere An atmosphere which tends to (1) promote the removal of oxygen from a chemical compound or (2) promote the reduction of immersed materials.

reducing flame A gas flame resulting from combustion of a mixture containing too much fuel or too little air.

reduction The gain of electrons by a constituent of a chemical reaction.

reference electrode A nonpolarizable electrode with a known and highly reproducible potential.

refractory Brickwork or castable used in boilers to protect metal surfaces and for boiler baffles.

refractory metal A metal having an extremely high melting point, for example, tungsten, molybdenum, tantalum, niobium, chromium, vanadium, and rhenium. In the broad sense, this term refers to metals having melting points above the range for iron, cobalt, and nickel.

regenerant A solution of a chemical compound used to restore the capacity of an ion-exchange system. Sodium chloride brine is used as a regenerant for ion-exchange water softeners, and acids and bases are used as regenerants for the cation and anion resins used in demineralization.

regeneration The process of restoring an ion-exchange medium to a usable state after exhaustion. In general, it includes the backwash, regenerant introduction, and fresh water rinse steps necessary to prepare a water softener exchange bed for service. Specifically, the term may be applied to the step in which the regenerant solution is passed through the exchanger bed (salt brine for softeners, acid and bases for deionizers).

relative humidity The ratio, expressed as a percentage, of the amount of water vapor present in a given volume of air at a given temperature to the amount required to saturate the air at that temperature.

relief valve (safety relief valve) An automatic pressure-relieving device actuated by the pressure upstream of the valve and characterized by an opening pop action with further increase in lift with an increase in pressure over the popping pressure.

residual The amount of a specific material remaining in the water following a water treatment process. The term may refer to material remaining as a result of incomplete removal or to material meant to remain in the treated water.

residual chlorine Chlorine remaining in a treated water after a specified period of contact time to provide continuing protection throughout a distribution system; the difference between the total chlorine added and that consumed by oxidizable matter.

residual stress Stresses that remain within a body as a result of plastic deformation.

resin Synthetic organic ion-exchange material, such as the high-capacity cation-exchange resin widely used in water softeners.

rest potential See **open-circuit potential.**

reverse osmosis A process that reverses, by the application of pressure, the natural process of osmosis so that water passes from the more concentrated to the more dilute solution through a semipermeable membrane, thus producing a stream of water up to 98 percent free of dissolved solids.

ringworm corrosion Localized corrosion frequently observed in oil-well tubing in which a circumferential attack is observed near a region of metal "upset."

riser (1) That section of pipeline extending from the ocean floor up to the platform; also, the vertical tube in a steam generator convection bank that circulates water and steam upward. (2) A reservoir of molten metal connected to a casting to provide additional metal to the casting, required as the result of shrinkage before and during solidification.

rust A corrosion product consisting primarily of hydrated iron oxide. (The term is properly applied only to ferrous alloys.)

safety shut-off valve A manually opened, electrically latched, electrically operated safety valve designed to automatically shut off fuel when deenergized.

safety valve A spring-loaded valve that automatically opens when pressure attains the valve setting, used to prevent excessive pressure from building up in a boiler.

saline water Water containing an excessive amount of dissolved salts, usually over 5000 mg/L.

salinity The total proportion of salts in seawater, often estimated empirically as chlorinity $\times$ 1.80655; also expressed in parts per thousand (o/oo).

salt In chemistry, a class of chemical compounds which can be formed by the neutralization of an acid with a base; the common name for the specific chemical compound sodium chloride, used in the regeneration of ion-exchange water softeners.

salt splitting The process in which neutral salts in water are converted to their corresponding acids or bases by ion-exchange resins containing strongly acidic or strongly basic functional groups.

sampling The removal of a portion of a material for examination or analysis.

saturated air Air which contains the maximum amount of water vapor that it can hold at its temperature and pressure.

saturated steam Steam at the temperature and pressure at which evaporation occurs.

saturated temperature The temperature at which evaporation occurs at a particular pressure.

scale A deposit of mineral solids on the interior surfaces of water lines and containers, often formed when water containing the carbonates or bicarbonates of calcium and magnesium is heated.

season cracking See **stress corrosion cracking.**

secondary treatment Treatment of boiler feedwater or internal treatment of boiler water after primary treatment.

sediment Matter in water which can be removed from suspension by gravity or mechanical means.

sedimentation The process in which solid suspended particles settle out of water, usually when the water has little or no movement. Also called *settling*.

segregation The tendency of refuse of varying compositions to deposit selectively in different parts of the unit.

selective leaching Removal of one element from a solid alloy by corrosion processes.

semipermeable membrane Typically a thin organic film which allows the passage of some ions or materials while preventing the passage of others. Some membranes will allow only the passage of cations.

septic A condition existing during the digestion of organic matter, such as that in sewage, by anaerobic bacteria in the absence of air. A common process for the treatment of household sewage in septic tanks, and in specially designed digesters in municipal sewage treatment systems.

sequestering agent A chemical compound sometimes fed into water to tie up undesirable ions, keep them in solution, and eliminate or reduce the normal effects of the ions. For example, polyphosphates are sequestering agents which sequester hardness and prevent reactions with soap. If the ions involved are metal ions, sequestering agents may also be chelating agents.

sequestration A chemical reaction in which certain ions are bound into a stable, water-soluble compound, thus preventing undesirable action by the ions.

service run That portion of the operating cycle of a water conditioning unit in which treated water is being delivered, as opposed to the period when the unit is being backwashed, recharged, or regenerated.

service water General-purpose water which may or may not have been treated for a special purpose.

shear That type of force that causes or tends to cause two contiguous parts of the same body to slide relative to each other in a direction parallel to their plane of contact.

shear strength The stress required to produce fracture in the plane of cross section, the conditions of loading being such that the directions of force and of resistance are parallel and opposite, although their paths are offset a specified minimum amount. The maximum load divided by the original cross-sectional area of a section separated by shear.

shell The cylindrical portion of a pressure vessel.

shielded The separation of metallic parts by an electrical nonconductor; insulated by other than an air gap.

sigma phase A hard, brittle, nonmagnetic intermediate phase with a tetragonal crystal structure, containing 30 atoms per unit cell, occurring in many binary and ternary alloys of the transition elements.

sigma-phase embrittlement Embrittlement of iron-chromium alloys (most notably austenitic stainless steels) caused by precipitation at grain boundaries of the hard, brittle intermetallic sigma phase during long periods of exposure to temperatures between approximately 560 and 980°C. Sigma-phase embrittlement results in severe loss in toughness and ductility, and can make the embrittled material susceptible to intergranular corrosion.

silica gel or siliceous gel A synthetic hydrated sodium aluminosilicate with ion-exchange properties, once widely used in ion-exchange water softeners.

slip Plastic deformation by the irreversible shear displacement (translation) of one part of a crystal relative to another in a definite crystallographic direction and usually on a specific crystallographic plane.

slow-strain-rate technique An experimental technique for evaluating susceptibility to stress corrosion cracking. It involves pulling the specimen to failure in uniaxial tension at a controlled slow strain rate while the specimen is in the test environment and examining the specimen for evidence of stress corrosion cracking.

sludge The semifluid solid matter collected at the bottom of a system tank or watercourse as a result of the sedimentation or settling of suspended solids or precipitates.

slug A large "dose" of chemical treatment applied internally to a steam boiler intermittently. The term is also sometimes used instead of *priming* to denote a discharge of water through a boiler steam outlet in relatively large intermittent amounts.

slushing compound An obsolete term describing oil or grease coatings used to provide temporary protection against atmospheric corrosion.

smelt Molten slag; in the pulp and paper industry, the cooking chemicals tapped from the recovery boiler as molten material and dissolved in the smelt tank as green liquor.

smoke Small gas-borne particles of carbon or soot, less than 1 μm in size and of sufficient number to be observable, resulting from incomplete combustion of carbonaceous materials.

***s-n* diagram** A plot showing the relationship of stress s and the number of cycles n before fracture in fatigue testing.

soda ash The common name for sodium carbonate, Na_2CO_3, a chemical compound used as an alkalinity builder in some soap and detergent formulations to neutralize acid water, and in the lime-soda water treatment process.

sodium chloride The chemical name for common salt.

sodium cycle The cation-exchange process in which sodium on the ion-exchange resin is exchanged for hardness and other ions in water. Sodium chloride is the common regenerant used in this process.

soft water Water which contains little or no calcium or magnesium salts, or water from which scale-forming impurities have been removed or reduced.

softening The act of reducing scale-forming calcium and magnesium impurities from water.

solid solution A single, solid, homogeneous crystalline phase containing two or more chemical species.

solute (1) A component of either a liquid or a solid solution that is present to a lesser or minor extent; the component that is dissolved in the solution. (2) The substance which is dissolved in and by a solvent. Dissolved solids, such as the minerals found in water, are solutes.

solution (1) A homogeneous dispersion of two or more kinds of molecular or ionic species. Solutions may be composed of any combination of liquids, solids, or gases, but they always consist of a single phase. (2) A liquid, such as boiler water, containing dissolved substances.

solution feeder A device, such as a power-driven pump or an eductor system, designed to feed a solution of a water treatment chemical into the water system, usually in proportion to flow.

solution heat treatment Heating an alloy to a suitable temperature, holding it at that temperature long enough to cause one or more constituents to enter into solid solution, and then cooling it rapidly enough to hold these constituents in solution.

solvent The component of either a liquid or a solid solution that is present to a greater or major extent; the liquid, such as water, in which other materials (solutes) are dissolved.

solvent degreasing The removal of oil, grease, and other soluble contaminants from the surface of the workpiece by immersion in suitable cleaners.

soot Unburned particles of carbon derived from hydrocarbons.

sour gas A gaseous environment containing hydrogen sulfide and carbon dioxide in hydrocarbon reservoirs. Prolonged exposure to sour gas can lead to hydrogen damage, sulfide-stress cracking, and/or stress corrosion cracking in ferrous alloys.

sour water Wastewaters containing fetid materials, usually sulfur compounds.

spalling The flaking or separation of a sprayed coating; the spontaneous chipping, fragmentation, or separation of a surface or surface coating; the breaking off of the surface of refractory material as a result of internal stresses.

specific conductance The measure of the electrical conductance of water or a water solution at a specific temperature, usually 25°C.

specific gravity The ratio of the weight of a specific volume of a substance to the weight of the same volume of pure water at 4°C.

specific humidity The weight of water vapor in a gas–water vapor mixture per unit weight of dry gas.

spheroidite An aggregate of iron or alloy carbides of essentially spherical shape dispersed throughout a matrix of ferrite.

splat A single thin, flattened sprayed particle.

splat cooling An extremely rapid, high rate of cooling, leading to the formation of metastable phases and an amorphous microstructure in thermal spraying deposits.

spray angle The angle included between the sides of the cone formed by liquid discharged from mechanical, rotary atomizers and by some forms of steam or air atomizers.

spray nozzle A nozzle from which a liquid fuel is discharged in the form of a spray.

sputtering A coating process in which thermally emitted electrons collide with inert gas atoms, which accelerate toward and hit a negatively charged electrode that is a target of the coating material. The impacting ions dislodge atoms of the target material, which are in turn projected to and deposited on the substrate to form the coating.

stack A vertical conduit which, because of the difference in density between internal and external gases, develops a draft at its base.

stack draft The magnitude of the draft measured at the inlet to the stack.

stagnation The condition of being free from movement or lacking circulation.

standard electrode potential The reversible potential for an electrode process when all products and reactions are at unit activity on a scale in which the potential for the standard hydrogen half cell is zero.

static pressure The measure of potential energy of a fluid.

static system A system or process in which the reactants are not flowing or moving; the opposite of **dynamic system.**

steam The vapor phase of water, unmixed with other gases.

steam-generating unit A unit to which water, fuel, and air are supplied and in which steam is generated. It consists of a boiler furnace and fuel-burning equipment, and may include as component parts water walls, a superheater, a reheater, an economizer, an air heater, or any combination thereof.

steam separator A device for removing the entrained water from steam.

strain The unit of change in the size or shape of a body as a result of force.

strain hardening An increase in hardness and strength caused by plastic deformation at temperatures below the recrystallization range.

strain rate The time rate of straining for the usual tensile test.

strainer A device, such as a filter, to retain solid particles while allowing a liquid to pass.

stratification Nonhomogeneity existing transversely in a gas stream.

stray current Current flowing through paths other than the intended circuit.

stray current corrosion The corrosion caused by electric current from a source external to the intended electric circuit, for example, extraneous current in the earth.

stress The intensity of the internally distributed forces or components of forces that resist a change in the volume or shape of a material that is or has been subjected to external forces. Stress is expressed in force per unit area and is calculated on the basis of the original dimensions of the cross section of the specimen.

stress concentration factor (K_t) A multiplying factor for applied stress that allows for the presence of a structural discontinuity such as a notch or hole; K_t equals the ratio of the greatest stress in the region of the discontinuity to the nominal stress for the entire section.

stress corrosion cracking A cracking process that requires the simultaneous action of a corrodent and sustained tensile stress. (This excludes corrosion-reduced sections which fail by fast fracture. It also excludes intercrystalline or transcrystalline corrosion, which can cause an alloy to disintegrate without either applied or residual stress.)

stress-intensity factor A scaling factor, usually denoted by the symbol K, used in linear-elastic fracture mechanics to describe the intensification of applied stress at the tip of a crack of known size and shape. At the onset of rapid crack propagation in any structure containing a crack, the factor is called the critical stress-intensity factor, or the fracture toughness.

stress raisers Changes in contour or discontinuities in structure that cause local increases in stress.

stress-relief cracking A cracking process that occurs when susceptible alloys are subjected to thermal stress relief after welding to reduce residual stresses and improve toughness. Stress-relief cracking occurs only in metals that can precipitation-harden during such elevated-temperature exposure; it usually occurs at stress raisers, is intergranular in nature, and is generally observed in the coarse-grained region of the weld heat-affected zone. Also called *postweld heat treatment cracking*.

stress relieving Heat treatment carried out in steel to reduce internal stresses.

striation A fatigue fracture feature, often observed in electron micrographs, that indicates the position of the crack front after each succeeding cycle of stress. The distance between striations indicates the advance of the crack front across the crystal during one stress cycle, and a line normal to the striation indicates the direction of local crack propagation.

stud A projecting pin serving as a support or means of attachment.

subsurface The material, workpiece, or substance on which the coating is deposited.

subsurface corrosion See **internal oxidation.**

sulfate-reducing bacteria (SRB) A group of bacteria which are capable of reducing sulfates in water to hydrogen sulfide gas, thus producing obnoxious tastes and odors. These bacteria have no sanitary significance and are classed as nuisance organisms.

sulfidation The reaction of a metal or alloy with a sulfur-containing species to produce a sulfur compound that forms on or beneath the surface of the metal or alloy.

sulfonic acid A specific acidic group (SO_3H^-) which gives certain cation-exchange resins their ion-exchange capability.

superchlorination The addition of excess amounts of chlorine to a water supply to speed chemical reactions or ensure disinfection with short contact time. The chlorine residual following superchlorination is high enough to be unpalatable, and thus dechlorination is commonly employed before the water is used.

superheated steam Steam with its temperature raised above that of saturation. The temperature in excess of its saturation temperature is referred to as superheat.

supernatant The clear liquid lying above a sediment or precipitate.

surface-active agent The material in a soap or detergent formulation which promotes the penetration of the fabric by water, the loosening of the soil from surfaces, and the suspension of many soils; the actual cleaning agent in soap and detergent formulations.

surface blowoff The removal of water, foam, etc., from the surface at the water level in a boiler; the equipment for such removal.

surface tension The result of attraction between molecules of a liquid which causes the surface of the liquid to act as a thin elastic film under tension. Surface tension causes water to form spherical drops and reduces penetration into fabrics. Soaps, detergents, and wetting agents reduce surface tension and increase penetration by water.

surfactant A contraction of the term *surface-active agent*; usually an organic compound whose molecules contain a hydrophilic group at one end and a lipophilic group at the other.

surge The sudden displacement or movement of water in a closed vessel or drum.

suspended solids Solid particles in water which are not in solution.

synthetic detergent A synthetic cleaning agent, such as linear alkyl sulfonate and alkyl benzene sulfonate. Synthetic detergents react with water hardness, but the products are soluble.

Système International (SI) The system of measurement, otherwise known as the metric system, used in most countries around the world. It is based on factors of 10, and is convenient to use for scientific calculations and with numbers that are very small or very large.

Tafel slope The slope of the straight-line portion of a polarization curve, usually occurring at more than 50 mV from the open-circuit potential, when the curve is presented in a semilogarithmic plot in terms of volts per logarithmic cycle of current density (commonly referred to as volts per decade).

TDS The abbreviation for **total dissolved solids.**

temporary hardness Water hardness due to the presence of calcium and magnesium carbonates and bicarbonates, which can be precipitated by heating the water. Now largely replaced by the term **carbonate hardness.**

tensile strength In tensile testing, the ratio of maximum load to original cross-sectional area; also called *ultimate tensile strength.*

tensile stress A stress that causes two parts of an elastic body on either side of a typical stress plane to pull apart; contrast with **compressive stress.**

tension The force or load that produces elongation.

terne An alloy of lead containing 3 to 15% Sn, used as a hot-dip coating for steel sheet or plate. Terne coatings, which are smooth and dull in appearance, give the steel better corrosion resistance and enhance its ability to be formed, soldered, or painted.

therm A unit of heat applied especially to gas; 1 therm = 100,000 Btu.

thermal shock A cycle of temperature swings that result in failure of metal as a result of expansion and contraction.

thermal spraying A group of coating or welding processes in which finely divided metallic or nonmetallic materials are deposited in a molten or semi-molten condition to form a coating. (The coating material may be in the form of powder, ceramic rod, wire, or molten materials.)

thermocouple A device for measuring temperatures, consisting of lengths of two dissimilar metals or alloys that are electrically joined at one end and connected to a voltage-measuring instrument at the other end. When one junction is hotter than the other, a thermal electromotive force is produced that is roughly proportional to the difference in temperature between the hot and cold junctions.

thermogalvanic corrosion The corrosive effect resulting from the galvanic cell caused by a thermal gradient across the metal surface.

threshold A very low concentration of a substance in water. The term is sometimes used to indicate the concentration which can just be detected.

threshold stress For stress corrosion cracking, the critical cross-sectional stress at the onset of cracking under specified conditions.

throughput volume The amount of solution passed through an ion-exchange bed before the ion exchanger is exhausted.

throwing power The relationship between the current density at a point on a surface and the point's distance from the counterelectrode. The greater the ratio of the surface resistivity shown by the electrode reaction to the volume resistivity of the electrolyte, the better is the throwing power of the process tinning.

TIG The tungsten inert gas welding process.

tile A preformed refractory, usually applied to shapes other than standard brick.

titration An analytical process in which a standard solution in a calibrated vessel is added to a measured volume of sample until an endpoint, such as a color change, is reached. From the volume of the sample and the volume of standard solution used, the concentration of a specific material may be calculated.

total acidity The total of all forms of acidity, including mineral acidity, carbon dioxide, and acid salts. Total acidity is usually determined by titration with a standard base solution to the phenolphthalein endpoint (pH 8.3).

total alkalinity The alkalinity of a water as determined by titration with a standard acid solution to the methyl orange endpoint (pH approximately 4.5); sometimes abbreviated as *M alkalinity*. Total alkalinity includes many alkalinity components, such as hydroxides, carbonates, and bicarbonates.

total carbon The sum of the free carbon and combined carbon (including carbon in solution) in a ferrous alloy.

total chlorine The total concentration of chlorine in a water, including combined and free chlorine.

total dissolved solids (TDS) The weight of solids per unit volume of water which are in true solution, usually determined by the evaporation of a measured volume of filtered water and determination of the residue weight.

total hardness The sum of all hardness constituents in a water, expressed as their equivalent concentration of calcium carbonate. Primarily the result of calcium and magnesium in solution, but may include small amounts of metals such as iron, which can act like calcium and magnesium in certain reactions.

total pressure The sum of the static and velocity pressures.

total solids The weight of all solids, dissolved and suspended, organic and inorganic, per unit volume of water; usually determined by the evaporation of a measured volume of water at 105°C in a preweighted dish.

total solids concentration The weight of dissolved and suspended impurities in a unit weight of boiler water, usually expressed in ppm.

toughness The ability of a metal to absorb energy and deform plastically before fracturing.

trace A very small concentration of a material, high enough to be detected but too low to be measured by standard analytical methods.

transcrystalline See **transgranular.**

transference The movement of ions through the electrolyte associated with the passage of the electric current; also called *transport* or *migration*.

transgranular Through or across crystals or grains.

transgranular cracking Cracking or fracturing that occurs through or across a crystal or grain; also called *transcrystalline cracking*.

transgranular fracture Fracture through or across the crystals or grains of a metal.

transition metal A metal in which the available electron energy levels are occupied in such a way that the d band contains less than its maximum number of 10 electrons per atom, for example, iron, cobalt, nickel, and tungsten. The distinctive properties of the transition metals result from the incompletely filled d levels.

transpassive region The region of an anodic polarization curve, noble to and above the passive potential range, in which there is a significant increase in current density (increased metal dissolution) as the potential becomes more positive (noble).

transpassive state A state of anodically passivated metal characterized by a considerable increase in the corrosion current, in the absence of pitting, when the potential is increased.

trap A receptacle for the collection of undesirable material.

treated water Water which has been chemically treated to make it suitable for boiler feed.

tube A hollow cylinder for conveying fluids.

tube hole A hole in a drum, heater, or tube sheet to accommodate a tube.

tuberculation The formation of localized corrosion products scattered over the surface in the form of knoblike mounds called tubercles. Also, the process in which blisterlike growths of metal oxides develop in pipes as a result of the corrosion of the pipe metal. Iron oxide tubercles often develop over pits in iron or steel pipe, and can seriously restrict the flow of water.

turbidity A measure of the cloudiness of water, the result of finely divided particulate matter suspended in the water; usually reported in arbitrary units determined by measurements of light scattering.

turbulent flow A type of flow characterized by crosscurrents and eddies. Turbulence may be caused by surface roughness or protrusions in pipes, bends and fittings, changes in channel size, or excessive flow rates; turbulence significantly increases pressure drops. Contrast **laminar flow.**

U-bend specimen A horseshoe-shaped test piece used to detect the susceptibility of a material to stress corrosion cracking.

ultimate strength The maximum stress a material can sustain without fracture, determined by dividing maximum load by the original cross-sectional area of the specimen.

undercutting A step in the sequence of surface preparation involving the removal of substrate material.

unfired pressure vessel A vessel designed to withstand internal pressure that is neither subjected to heat from products of combustion nor an integral part of a fired-pressure-vessel system.

uniform corrosion Corrosion that proceeds at about the same rate at all points on a metal surface.

vacuum deposition Condensation of thin metal coatings on the cool surface of work in a vacuum.

valence A whole number (positive or negative) representing the power of one element to combine with another. In general terms, the valence number represents the number of electrons in an atom or combined group of atoms which can be easily given up or accepted in order to react with or bond to another atom or group of atoms to form a molecule.

vapor The gaseous product of evaporation.

vapor deposition See **chemical vapor deposition, physical vapor deposition, sputtering.**

vapor plating Deposition of a metal or compound on a heated surface by reduction or decomposition of a volatile compound at a temperature below the melting points of the deposit and the base material.

vaporization The change from the liquid or solid phase to the vapor phase.

velocity pressure The measure of the kinetic energy of a fluid.

vent An opening in a vessel or other enclosed space for the removal of gas or vapor.

vertical firing An arrangement of a burner such that air and fuel are discharged into the furnace in practically a vertical direction.

viable Alive and capable of continued life.

virus The smallest form of life known to be capable of producing disease or infection, usually considered to be of large molecular size. Viruses multiply by assembly of component fragments in living cells rather than by cell division, like most bacteria.

viscosity The resistance of fluids to flow, as a result of internal forces and friction between molecules, which increases as temperature decreases.

void volume The volume of the spaces between particles of ion exchanger, filter media, or other granular material; often expressed as a percentage of the total volume occupied by the material.

voids A term generally applied to paints to describe holidays, holes, and skips in a film; also used to describe shrinkage in castings and welds.

volatile Capable of vaporization at a relatively low temperature.

volatile matter Those products given off by a material as gas or vapor, determined by definite prescribed methods.

volatile solids Matter which remains as a residue after evaporation at 105 or 180°C, but which is lost after ignition at 600°C. Includes most forms of organic matter.

volumetric Referring to measurement by volume rather than by weight.

wash primer A thin, inhibiting paint, usually chromate pigmented with a polyvinyl butyrate binder.

waste heat Sensible heat in noncombustible gases discharged to the environment.

water A liquid composed of 2 parts of hydrogen and 16 parts oxygen by weight.

water conditioning Virtually any form of water treatment designed to improve the aesthetic quality of water through the neutralization, inhibition, or removal of undesirable substances.

water softener A material that removes hardness ($CaCO_3$) from water through an ion exchange of sodium with calcium and magnesium.

water softening The removal of calcium and magnesium, the ions which are the principal cause of hardness, from water.

water table The level of the top of the zone of saturation, in which free water exists in the pores and crevices of rocks and other earth strata.

water tube A tube in a boiler having water and steam on the inside and heat applied to the outside.

weak base load fraction x The sum of the chloride, sulfate, and nitrate. Also referred to as the *theoretical mineral acidity* (TMA).

weep A term usually applied to a minute leak in a boiler joint at which droplets (or tears) of water form very slowly.

weld cracking Cracking that occurs in the weld metal.

weld decay Intergranular corrosion, usually of stainless steels or certain nickel-base alloys, that occurs as the result of sensitization in the heat-affected zone during the welding operation; not a preferred term.

wet-bulb temperature The lowest temperature which a water-wetted body will attain when exposed to an air current. This is the temperature of adiabatic saturation, and can be used to measure humidity.

wet steam Steam containing moisture.

wetness A term used to designate the percentage of water in steam. also used to describe the presence of a water film on heating surface interiors.

wetting A condition in which the interfacial tension between a liquid and a solid is such that the contact angle is 0 to 90°.

wetting agent A substance that reduces the surface tension of a liquid, thereby causing it to spread more readily on a solid surface.

white rust Zinc oxide, the powdery product of corrosion of zinc or zinc-coated surfaces.

windbox A chamber below the grate or surrounding a burner, through which air under pressure is supplied for combustion of the fuel.

windbox pressure The static pressure in the windbox of a burner or stoker.

work hardening Same as **strain hardening.**

working electrode The test or specimen electrode in an electrochemical cell.

yield Evidence of plastic deformation in structural materials. Also called *creep* or *plastic flow*.

yield point The first stress in a material, usually less than the maximum attainable stress, at which an increase in strain occurs without an increase in stress.

yield strength The stress at which a material exhibits a specified deviation from proportionality of stress and strain. An offset of 0.2 percent is used for many metals.

yield stress The stress level in a material at or above the yield strength but below the ultimate strength, i.e., a stress in the plastic range.

zeolite A group of hydrated sodium aluminosilicates, either natural or synthetic, with ion-exchange properties. (See **gel zeolite.**)

zone of aeration The layer in the ground above an aquifer where the available voids are filled with air. Water falling on the ground percolates through this zone on its way to the aquifer.

zone of saturation The layer in the ground in which all of the available voids are filled with water.

APPENDIX C

Corrosion Economics

C.1	Introduction	1001
C.2	Cash Flows and Capital Budgeting Techniques	1002
C.3	Generalized Equation for Straight-Line Depreciation	1004
C.4	Examples	1006
C.5	Summary	1009
	References	1009

C.1 Introduction

Any engineering project undertaken by a profit-motivated organization has the underlying aim of enhancing the wealth of its owners (shareholders). Management in industry ultimately bases its decisions on this principle, including those related to corrosion control. The selection of optimal projects from the viewpoint of owners' wealth lies in the financial domain of capital budgeting techniques. These techniques determine how capital should be invested in the long term. Four key motives can be identified for making capital investments (expenditures):[1]

1. Expansion for increasing the scope and output of operations
2. Replacement for obsolete or rundown assets
3. Renewal for life extension of assets, as an alternative to replacement
4. Investment in nontangible assets such as advertising, research, information, management consulting, etc.

The formal steps in the capital budgeting process in sequential order are (1) proposal generation, (2) review and analysis of the proposals, (3) decision making, (4) implementation, and (5) monitoring of results, to compare the actual project outcome with the predictions. This section will focus on the second step, namely, how to evaluate the economic viability of corrosion control investments.

C.2 Cash Flows and Capital Budgeting Techniques

Every corrosion engineering project will have a certain cash flow pattern over time. Usually, there is an initial outflow of cash, when a new asset fitted with a certain corrosion control system is apcquired. Subsequently there are inflows of cash, resulting from operations and further cash outflows required for maintenance, corrosion control upgrades, running costs, and so forth. In capital budgeting techniques the different cash flows involved in the project are identified, estimated, and analyzed, with a view to maximizing owners' wealth.

Clearly such cash flows can be complex if all the financial implications of project options are investigated in detail. Invariably in corrosion economics calculations a compromise has to be made between two opposing needs, the need for precision and the need for simplicity. For example, the present costs or investment in two alternative anticorrosion methods may be known with a high degree of certainty. However, the service lives, future maintenance costs, or operating costs may be estimates with only a limited degree of certainty. The need for stringent risk assessment required of many modern engineering systems may also add to the complexity of estimating useful life and cost estimates.

When considering the above cash inflows and outflows over time, the *time value of money* has to be considered. This concept implies that money has a value that varies depending on *when* it is received or disbursed. Readers will have gained first-hand knowledge of this principle from any loans they have taken out with financial institutions. A loan received "now" has to be repaid with interest charges in the future. The following is a generalized formula between the present value and future value of cash flows:

$$\text{PV} = \frac{F_n}{(1 + i)^n} \qquad (\text{C.1})$$

which states that present value (PV) of a future cash flow (F_n) after (n) time periods equals the future amount (C_n) discounted to zero date at some interest rate (i). The value of n is usually specified in years and i as the annual interest rate. Several capital budgeting techniques exist that are based on the time value of money. The five important methods briefly described in Table C.1 vary considerably with regard to their application and complexity.

NACE International, in special report on economics of corrosion, advocates the use of the discounted cash flow method, which provides ready calculation of net present worth.[2] Factors that need to be considered in calculating net present worth include

TABLE C.1 Five Important Capital Budgeting Techniques

Internal Rate of Return (IRR)

The IRR is considerably more difficult to calculate than the NPV without the assistance of a computer, and it represents a sophisticated form of analysis. The IRR is defined as the discount rate that equates the present value of all cash flows with the initial investment made in a project. The IRR consists essentially of the interest cost or borrowed capital plus any existing profit or loss margin. A project is financially more favorable when the positive difference between IRR and the interest rate charged for borrowing increases. Once all the cash flows have been accounted for over the life of a project, the IRR has to be computed by an iterative procedure.

Present Worth of Future Revenue Requirements (PWRR)

The PWRR is particularly applicable to regulated public utilities, for which the rate of return is more or less fixed by regulation. The principal objection to the PWRR method is that it is inadequate where alternatives are competing for a limited amount of capital because it does not identify the alternative that produces the greatest return on invested capital.

Discounted Payback (DBP) and Benefit-Costs Ratios (BCR)

The payback period is a relatively simple concept. It is defined as the amount of time required to recover its initial project expense. DBP takes the time value of money into consideration by adjusting all future cash flows to time zero, before calculating the payback period (in the most simple form of payback analysis, these adjustments are not considered). It is a very basic technique that can be used to screen candidate projects. The BCR method is related to the IROR method.

Present Worth (PW)

The PW, also referred as Net Present Value (NPV), is considered the easiest and most direct of the five methods. It consequently has the broadest application to engineering economy problems. Many industries refer to this method as the discounted cash flow method of analysis.

- Initial cost
- Best estimate of expected life
- Length of typical shutdown for emergency repair
- Cost of planned maintenance during scheduled shutdowns
- Effect of failure on total plant operation

The net present value is a summation of the present value of all cash inflows and outflows minus the initial project cost (C_0). To include the effects of taxation (essentially a business expense), all actual cash flows for tax-paying organizations are reduced by the formula given in Eq. (C.2):

$$\text{Amount after taxes} = C\,(1 - T_x) \qquad (C.2)$$

All expenses allowed to be charged against income for tax purposes, but not representing actual cash flow, are modified by the formula given in Eq. (C.3). Depreciation allowances are an excellent example of where such tax savings are possible; they are treated similarly to income:

$$\text{Cash flow} = \text{noncash expense charge}\,(T_x) \qquad (C.3)$$

The PV of the tax savings cash flow from a depreciation expense series (DES) of an original cost is given in Eq. (C.4):

$$PV_{DES} = C_0\,(Q, i, N_Q)\,T_x \qquad (C.4)$$

where Q is the present value factor for a cash flow stream and N_Q is the time span of the depreciation expense stream.

Combining these definitions, one can obtain the fundamental net present value (NPV) equation that includes the tax effects:

$$NPV = \sum_{n=1}^{N} = \frac{F_n}{(1+i)^n}\,(1 - T_x) + C_0\,(Q, i, N_Q)\,T_x - C_0 \qquad (C.5)$$

Verink has developed a simplified version of this complex equation to fit most engineering systems.[3]

C.3 Generalized Equation for Straight-Line Depreciation

Verink has developed a generalized equation [Eq. (C.6)] that is particularly adapted to corrosion engineering problems. This equation takes into account the influence of taxes, straight-line depreciation, operating expenses, and salvage value in the calculation of present worth and annual cost. Using this equation, a problem can be solved merely by entering data into the equation with the assistance of compound interest data.

$$PW = -P + \left[\frac{t(P-S)}{n}\right]\left(\frac{P}{A, i\%, n}\right)$$
$$- (1-t)(X)\left(\frac{P}{A, i\%, n}\right) + S\left(\frac{P}{F, i\%, n}\right) \quad (C.6)$$

where A = annual end-of-period cash flow
  F = future sum of money
  $i\%$ = interest rate
  n = number of years
  PW = present worth, referred to also as NPV
  P = cost of the system at time 0
  S = salvage value
  t = tax rate expressed as a decimal
  X = operating expenses

First term, $-P$. This term represents the initial project expense, at time zero. As an expense, it is assigned a negative value. There is no need to translate this value to a future value in time because the PW approach discounts all money values to the present (time zero).

Second term, $[t(P-S)/n](P/A, i\%, n)$. The second term in this equation describes the depreciation of a system. The portion enclosed in brackets expresses the annual amount of tax credit permitted by this method of straight-line depreciation. The portion in parentheses translates annual costs in equal amounts back to time zero by converting them to present worth.

Third term, $-(1-t)(X)(P/A, i\%, n)$. The third term in the generalized equation consists of two terms. One is $(X)(P/A, i\%, n)$, which represents the cost of items properly chargeable as expenses, such as the cost of maintenance, insurance, and the cost of inhibitors. Because this term involves expenditure of money, it also comes with a negative sign. The second part, $t(X)(P/A, i\%, n)$, accounts for the tax credit associated with this business expense and because it represents a saving, it is associated with a positive sign.

Fourth term, $S(P/F, i\%, n)$. The fourth term translates the future value of salvage to the present value. This is a one-time event rather than a uniform series, and therefore it involves the single-payment present worth factors. Many corrosion measures, such as coatings and other repetitive maintenance measures, have no salvage value, in which cases this term is zero.

Present worth (PW) can be converted to equivalent annual cost (A) by using the following formula:

$$A = (PW)(A/P, i\%, n)$$

One can calculate different options by referring to interest tables or by simply using the formula describing the various functions. The capital recovery function (P/A), or how to find P once given A, is

$$\left(\frac{P}{A}, i\%, n\right) \quad \text{where} \quad P_n = A\frac{(1+i)^n - 1}{i(1+i)^n}$$

The compound amount factor (P/F), or how to find P once given F, is

$$\left(\frac{P}{A}, i\%, n\right) \quad \text{where} \quad F_n = P(1+i)^{-n}$$

The capital recovery factor (A/P), or how to find A once given P, is

$$\left(\frac{A}{P, i\%, n}\right) \quad \text{where} \quad A_n = P\frac{i(1+i)^n}{(1+i)^n - 1}$$

The following examples serve as illustrations of where these calculations would be useful.

C.4 Examples

C.4.1 Example 1

A new heat exchanger is required in conjunction with a rearrangement of existing facilities. Because of corrosion, the expected life of a carbon steel heat exchanger is 5 years. The installed cost is $9500. An alternative to the heat exchanger is a unit fabricated of AISI type 316 stainless steel, with an installed cost of $26,500 and an estimated life of 15 years, to be written off in 11 years. The minimum acceptable interest rate is 10 percent, the tax rate is 48 percent, and the depreciation method is straight line. Determine which unit would be more economical based on annual costs.

Solution	Option 1, Carbon Steel Heat Exchanger	Option 2, AISI Type 316 Heat Exchanger
$-P$	$-\$9,500$	$-\$26,500$
$\{t(P-S)/n\}(P/A, i\%, n)$	$[0.48(9500-0)/5](3.791) = \$3457$	$[0.48(26,500-0)/11](6.495) = \$7510$
$-(1-t)(X)(P/A, i\%, n)$	0	0
$S(P/F, I\%, n)$	0	0
PW	$-\$6043$	$-\$18,989$
A	$-\$6043(0.2638) = -\$1594$	$-\$18,989(0.15396) = -\$2924$

The carbon steel heat exchanger is thus a cheaper solution.

C.4.2 Example 2

In this case, the carbon steel heat exchanger of Example 1 will require additional protection costing $3000 in yearly maintenance such as painting, use of inhibitors, and so forth. Determine if the choice established in Example 1 would be modified.

Solution	Option 1, Carbon Steel, $3000/Year Maintenance	Option 2, AISI Type 316
Installed cost	−$9500	−$26,500
−P	[0.48(9500 − 0)/5](3.791) = $3457	[0.48(26,500 − 0)/11](6.495) = $7510
$[t(P-S)/n](P/A, i\%, n)$	−(1 − 0.48)(3000)(3.791) = −$5914	0
$-(1-t)(X)(P/A, i\%, n)$	0	0
$S(P/F, I\%, n)$	−$11,957	−$18,989
PW	−$11,957 (0.2638) = −$3154	−$18,989(0.15396) = −$2924
A	−$11,957 (0.2638) = −$3154	−$18,989(0.15396) = −$2924

The choice would be modified in favor of the stainless steel heat exchanger.

C.4.3 Example 3

Given the conditions given in Example 1 but uncertain that a service life of 5 years can be obtained, determined the service life at which the carbon steel heat exchanger is economically equivalent to the type 316 stainless steel unit.

Solution	Option 1, $n = 2$ years	Option 2, $n = 3$ years
−P	−$9500	−$9500
$\{t(P-S)/n\}(P/A, i\%, n)$	[0.48(9500 − 0)/2](1.736) = $3958	[0.48(9500 − 0)/3](2.487) = $3780
$-(1-t)(X)(P/A, i\%, n)$	0	0
$S(P/F, I\%, n)$ 0	0	
PW	−$5542	−$5720
A	−$5542 (0.5762) = −$3193	−$5720(0.40211) = −$2300

Thus a carbon steel heat exchanger must last more than 2 years but will be economically favored in less than 3 years under the conditions given.

C.4.4 Example 4

Under the conditions described in Example 3, it becomes interesting to evaluate how much product loss X can be tolerated after 2 of the 5 years of anticipated life, for example, from roll leaks or a few tube failures, before the selection of type AISI 316 stainless steel could have been justified.

Solution $\quad A_{316} = A_{steel} + A_{product\ loss}$

$-\$2924 = -\$1594 + [\,(1-0.48)\,(X)\,(0.8264)\,]\,[0.2638]$ the third term in the generalized equation

where 0.8264 is the single payment PW factor for 2 years $(P/F)_{10\%,\ 2y}$
Solving for X:

$$-\$1330 = 0.1134\,(X) \qquad X = -\$11,728$$

If production losses exceed $11,728 in year 2, with no losses in any other year, theAISI type 316 stainless steel heat exchanger would be the most cost economical solution.

C.4.5 Example 5

A paint system originally cost $4.88/m² to apply and has totally failed after 4 years. Assume an interest rate of 10%, a tax rate of 48%, and straight-line depreciation to answer the following questions:

1. If the paint system is renewed twice at the same cost for a total life of 12 years, what is the annual cost, assuming the first application is capitalized and those in the fourth and eighth are considered to be expenses?
2. Total maintenance could be avoided by biennial touch-up (wire brush, spot primer, and topcoat). What is the most that can be spent on this preventative maintenance?

Solution

Repaint every four years

$-P$	$-\$4.88 \text{ m}^{-2}$
$[t(P - S)/n](P/A, i\%, n)$	$[0.48(4.88 - 0)/4](3.1699) = \$1.86$
$-(1 - t)(X)(P/A, i\%, n)$, year 4	$-(1 - 0.48)(4.88)(0.6830) = -\$1.73$
$-(1 - t)(X)(P/A, i\%, n)$, year 8	$-(1 - 0.48)(4.88)(0.4665) = -\$1.18$
$S(P/F, I\%, n)$	0
PW	$-\$4.88 + \$1.86 - \$1.73 - \$1.18 =$
	$-\$5.93 \text{ m}^{-2}$
A	$-\$5.93\,(0.1468) = -\$0.871\,m^{-2}$

The annual cost is $0.871/m² of present dollars.

Biennial touch-ups instead of repaints every four years. Biennial touch-up costs (X) at 2, 4, 6, 8, and 10 years can be expressed as an equivalent annual touch-up cost X' with the following equation:

$$X' = X\,[(P/F)_{10\%,\,2y} + (P/F)_{10\%,\,4y} + (P/F)_{10\%,\,6y} + (P/F)_{10\%,\,8y} + (P/F)_{10\%,\,10y}]\,(A/P)_{10\%,\,12\,y}$$

$$X' = X\,(0.8640 + 0.6830 + 0.5645 + 0.4665 + 0.3855)\,(0.1468)$$

$$= 0.4295\,X$$

Repainting costs alone can be expressed similarly with the following formula:

$$A = -P\left[(P/F)_{10\%,\,4y} + (P/F)_{10\%,\,8y}\right](A/P)_{10\%,\,12y}$$
$$A = -4.88\,(0.6830 + 0.4665)\,(0.1468)$$
$$A = -\$0.8235\ \text{m}^{-2}$$

To be equal or less costly, the annual equivalent annual cost of touch-ups should be equal or less that the repaint program, and hence:

$$0.4295X = \$0.8235\ \text{m}^{-2} \quad \text{or} \quad X = -\$1.917\ \text{m}^{-2}$$

is the answer.

C.5 Summary

Capital budgeting techniques represent a powerful methodology for evaluation corrosion engineering projects in terms of financial value to an organization. In their elementary forms, some of these methodologies lend themselves to "back-of-an-envelope" calculations to screen out project proposals. The basis of these techniques is closely related to life-cycle costing, discussed in more detail in other sections. Unfortunately, all too often, selection of materials for corrosion applications is still solely based on a comparison of initial installed costs of alternative materials. The time value of money concept, including important considerations such as ease of repair, costs associated with scheduled and unplanned shutdowns, and the effect of component failure on overall plant operations are thereby inadequately accounted for or completely ignored.

References

1. Gitman, L. J., *Principles of Managerial Finance,* New York, Harper Collins, 1991.
2. *Economics of Corrosion,* NACE 3C194, 1994, Houston, Tex., NACE International.
3. Verink, E. D., Corrosion Economic Calculations, in *Metals Handbook: Corrosion,* Metals Park, Ohio, ASM International, 1987, pp. 369–374.

APPENDIX D

Electrochemistry Basics

D.1	Principles of Electrochemistry	1011
	D.1.1 Introduction	1011
	D.1.2 Electrolyte conductance	1018
	D.1.3 Basic electrochemical instrumentation	1025
D.2	Chemical Thermodynamics	1029
	D.2.1 Free energy and electrochemical cells	1029
	D.2.2 Electrochemical potentials	1030
	D.2.3 Standard electrode potentials	1030
	D.2.4 Introduction to the Nernst Equation	1031
	D.2.5 Advanced thermodynamics	1033
	D.2.6 Potential-pH diagrams	1040
D.3	Kinetic Principles	1047
	D.3.1 Kinetics at equilibrium: The exchange current concept	1047
	D.3.2 Kinetics under polarization	1048
	References	1059

D.1 Principles of Electrochemistry

D.1.1 Introduction

Many significant chemical reactions are electrochemical in nature. To understand electrochemical reactions, it is necessary to understand the terms and concepts of electricity and extend these to apply to electrochemical relationships. Electrochemical reactions are chemical reactions in which electrons are transferred.

The most fundamental quantity used in the study of electricity is electrical current (I). Electrical current is measured in one of the base units of the International System, the ampere (A). There are two types of electrical current: direct current (dc), in which the current flows in only one direction, and alternating current (ac), in which the current flows alternately in opposite directions.

Electrical charge (Q) is the product of electrical current and time (t) (i.e., $Q = It$). Electrical charge is measured in coulombs (C). A charge of 1 C is passed when a current of 1 A flows for a time of 1 s, so the coulomb is the ampere-second.

Energy must be expended to force electrical current to flow through matter. The amount of energy expended in doing so depends upon the electrical resistance (R) of the particular matter through which the current passes. The electrical resistance is measured in ohms (Ω). When 1 joule (J) of energy is expended in driving a current of 1 A through a resistance, the electrical resistance is 1 Ω.

One of the most useful of the laws of electricity is Ohm's law ($E = IR$). This law links the electrical quantities of current I and resistance R with the electrical potential or potential difference (E) that is driving the current through the resistance. The SI unit for the electrical potential difference is the volt (V). A potential difference of 1 V will drive 1 A of current through 1 Ω of resistance, expending as it does so 1 J of energy and passing in the process 1 C of charge. The joule is also the volt-coulomb.

Basic definitions

Oxidation and reduction. Electrons are always transferred from one atom or molecule to another in an electrochemical reaction, even though electrons may not appear explicitly in the global balanced equation for the reaction.

There are three different types of electrochemical reactions that are distinguished by the changes of oxidation state that occur in them. They are called oxidation reactions, reduction reactions, and redox reactions; the term *redox* is an abbreviation for oxidation-reduction reactions, because both oxidation and reduction occur in redox reactions. Electrons appear explicitly in oxidations or reductions but appear only implicitly in redox reactions.

In an oxidation reaction, atoms of the element(s) involved in the reaction lose electrons. The charge on these atoms must then become more positive.

Example

$$Fe^{2+}_{(aq)} \rightarrow Fe^{3+}_{(aq)} + e^-$$

A reduction reaction is the reverse of an oxidation reaction. In a reduction reaction, atoms of the elements involved gain electrons.

Example

$$Zn^{2+}_{(aq)} + 2e^- \rightarrow Zn_{(s)}$$

A redox reaction is an electrochemical reaction in which both reduction and oxidation take place together. The electrons lost in an oxidation component are gained in a reduction component. The stoichiometry of a redox reaction is such that all of the electrons lost in the oxidation are gained in the reduction, so electrons can appear only implicitly in a redox reaction.

Electrolyte and electrodes. An ion is an atom or molecule that has acquired an electrical charge. An ion that carries a positive charge is called a cation, and an ion that carries a negative charge is called an anion. Compounds, molecules, and atoms that are uncharged are referred to as neutral species. A solution that contains ions is called an electrolyte solution, or more simply an electrolyte. Electrolyte solutions conduct electricity because charged ions can move through them. Electrolyte solutions are ionic conductors as distinguished from the electronic conductors, such as metallic wires, in which charge is carried by movement of electrons. An electrolyte solution may be used for this purpose alone. An example of this is the salt bridge that is used to permit the flow of ionic charge between different electrolyte solutions.

Two or more electrodes form an electrochemical cell from which two external wires can lead to an external electrical device. An oxidation or reduction reaction takes place at one electrode. Electrochemical reactions in which electrons appear explicitly are also called electrode reactions, or half-reactions. A half-reaction can be either a reaction in which electrons appear as products (oxidation) or a reaction in which electrons appear as reactants (reduction). A combination of two electrode reactions forms a cell reaction, or because one electrode must be carrying out an oxidation while the other is carrying out a reduction, a redox reaction.

A balanced electrochemical half-reaction is a reaction in which all atoms appearing on one side are balanced, in type and number, by atoms appearing on the other. The total charge on one side of a half-reaction is equal to the total charge on the other, but at least one electron appears on one side not balanced by an electron on the other. Electrons are not normally found in aqueous solutions because they react with water; they move through electronic conductors that are generally metallic wires. Ions, however, are quite stable in aqueous solutions and can carry charge as they move through it. When an oxidation reaction takes place at an electrode, that electrode is called an *anode*; when a reduction reaction takes place at an electrode, that electrode is called a *cathode*.

Electrochemical cells. An electrochemical cell can either drive an external electrical device (load) or be driven by it (power supply), depending upon the relative electromotive forces applied by the cell and the device. The current that flows through a cell will produce an electrochemical reaction that follows the principles of electrochemical stoichiometry (i.e., Faraday's law). The electromotive force (emf) of the cell is then called its reversible potential (E_{rev}). An electrochemical cell can be described as galvanic, reversible, or electrolytic:

- A galvanic cell is a cell in which current flows, power is produced, and the cell reaction is proceeding spontaneously.
- An electrolytic cell is a cell in which current flows, power is consumed, and the cell reaction being driven is the reverse of the spontaneous cell reaction.
- A reversible cell is a cell in which no current flows (and therefore no power is involved, because $P = EI$). The cell reaction in a reversible cell is neither spontaneous nor nonspontaneous; it is called reversible because an infinitesimal change in the cell potential can cause it to proceed in either direction.

When cells are under reversible conditions, the potential difference, or emf, across them can be measured without any loss due to ohmic, or IR, drop. In such a particular case E_{rev} is identical to the open-circuit potential.

Electrode structures and notations. At an electrode, conduction changes from ionic to electronic. A half-cell reaction may not contain any electronic conductor explicitly, but the actual physical half-cell corresponding to that reaction must contain an electronic conductor that makes external electrical contact. When writing a real half-cell, we usually specify the physical state of all components (s, l, g, aq, etc.) and denote phase boundaries by a slash, /. Any electrolytic conductor used to separate two electrolyte solutions while allowing the passage of ionic charge between them separation in the electrolyte, such as salt bridges, separators, membranes and so forth, is designated by //, one slash for the phase boundary at each end. Some examples of half-reactions and half-cells are illustrated in Table D.1.

TABLE D.1   Examples of Half-Reactions vs. Half-Cells

Half reaction	Half cell
$Fe^{2+} + 2e^- \Leftrightarrow Fe$	$//Fe^{2+}_{(aq)}/Fe_{(s)}$
$Cr^{3+} + e^- \Leftrightarrow Cr2+$	$//Cr^{3+}_{(aq)}, Cr^{2+}_{(aq)}/Pt_{(s)}$
$AgCl + e^- \Leftrightarrow Cl^- + Ag$	$//AgCl_{(s)}, Cl^-_{(aq)}/Ag_{(s)}$
$Cl_2 + 2e^- \Leftrightarrow 2Cl^-$	$//Cl_{2(aq, sat.)}, Cl^-_{(aq)}/C_{(s)}$

When writing half-reactions, it is necessary to use a consistent style. In 1958, it was agreed to use the reduction style for both half-cell reactions and for physically real half-cells. According to that convention, the reduced species of half-cell reactions are considered to be products, and the oxidized species are considered to be reactants.

Cell structures and notations. The charge, in an electrochemical cell, is carried by electrons in part of the circuit and by ions in the electrolyte. The cell electrode into which electrons flow is the most positive electrode of the cell, and it is conventionally written on the right.

The cathode, at which reduction is taking place, is the most positive electrode so the cathode is written on the right. Then the electrons in the external circuit must flow from left to right; the electrode on the left must be the cell anode because oxidation is taking place there. Charge in the electrolyte is carried by ions, both by the positive cations and the negative anions. The ions move so as to complete the circuit, anions moving from right to left as electrons in the external circuit move from left to right and cations moving in the opposite direction.

Example An aqueous cell that operates spontaneously using the following reaction:

$$Mg_{(s)} + Fe^{2+} \rightarrow Mg^{2+} + Fe_{(s)}$$

would be written in cell notation as

$$Mg/Mg^{2+}//Fe^{2+}/Fe$$

The Fe^{2+} is being reduced at the cathode, so the iron electrode couple is written on the right. The flow of electrons in the external circuit is from left to right, the flow of anions in the electrolyte and separator is from right to left, and the flow of cations in the electrolyte and separator is from left to right.

Faraday's Laws of electrolysis. When an electric current is made to pass through a cell, the current may cause chemical reactions to occur at its electrodes. This process is called electrolysis, and the cell in which it occurs is called an electrolytic cell. In the 1830s, the English scientist Michael Faraday showed that electrochemical reactions follow all normal chemical stoichiometric relations but in addition follow certain stoichiometric rules related to charge. These are known as Faraday's laws of electrolysis. According to Faraday's observations.

1. The masses of primary product formed by electrolysis at an anode or a cathode are directly proportional to the charge passed (Q).

2. For a given charge, the ratio of the masses of the primary products is the same as the ratio of the chemical equivalents (the formula mass divided by the valence change).

Moles and coulombs. Faraday's empirical laws of electrolysis relate to the number of electrons required to discharge 1 mole of an element. Suppose that the charge required were one electron per molecule, as in the case of a reaction such as the electroplating of silver:

$$Ag^+_{(aq)} + e^- \rightarrow Ag_{(s)}$$

Discharging 1 mole of silver would therefore require 1 mole of electrons, or Avogadro's number of electrons. The charge carried by 1 mole of electrons is known as 1 Faraday (F). The Faraday is related to other electrical units because the charge on a single electron is $1.6027733 \times 10^{-19}$ C/electron. Multiplying the electronic charge by the Avogadro number, $6.0221367 \times 10^{+23}$ electrons/mole, tells us that 1 F equals 96,485 C/(mol of electrons). Combination of the principles of Faraday with an electrochemical reaction of known stoichiometry permits us to write Faraday's laws of electrolysis as a single equation. When only one chemical species i is involved, Faraday's can be expressed as Eq. (D.1):

$$Q = F \cdot \Delta N_i \, n_i \tag{D.1}$$

where $Q = \int_0^t I \, dt$

N_i = number of moles of species i
ΔN_i = the change in that amount
n_i = number of electrons per molecule of species i
I = total current
t = time of electrolysis

When the electrolysis results in more than a single reaction, or when the primary products are changed through secondary chemical reactions, the charge Q is divided into the various reaction paths according to each path current efficiency [Eq. (D.2)]:

$$Q = F \sum_i Q_i = F \sum_i \Delta N_i \, n_i \tag{D.2}$$

where Q_i is the charge involved in each reaction. The overall current efficiency can be defined by Eq. (D.3):

$$\text{Current efficiency (\%)} = \frac{Q_i}{Q} 100 = \frac{Q_i}{\sum_i Q_i} 100 \tag{D.3}$$

Balancing electrode reactions. In any stoichiometric half-cell (electrode) reaction, the charge on both sides is balanced explicitly by electrons. The balanced equation gives the ratio of moles of electrons to moles of other species, and the number of moles of electrons can be converted into coulombs using the Faraday. In aqueous solutions these reactions may be complex because the solvent water can become involved in the reaction. In acidic aqueous solution, an electrode reaction is most easily balanced by carrying out the following steps in order:

1. Balance all elements except hydrogen and oxygen using stoichiometric coefficients.
2. Balance oxygen by adding water as necessary.
3. Balance hydrogen by adding hydrogen ions as necessary.
4. Balance charge by adding electrons as necessary. The number of electrons necessary to balance the charge is the charge number of the electrode reaction, z.

Example. The reduction of permanganate ion to manganese ion in acidic aqueous solution is balanced as follows using the above four steps:

1. $MnO_4^- \rightarrow Mn^{2+}$
2. $MnO_4^- \rightarrow Mn^{2+} + 4H_2O$
3. $8H^+ + MnO_4^- \rightarrow Mn^{2+} + 4H_2O$
4. $5e^- + 8H^+ + MnO_4^- \rightarrow Mn^{2+} + 4H_2O$

If the reaction takes place in a basic aqueous solution rather than an acidic one, protons will not be available. Nevertheless, it is easier to balance a reduction or oxidation reaction in basic solution, as if it were in acidic solution, than to use a formal conversion procedure to give the basic stoichiometry. The formal conversion to basic solution is made by

5. Adding to both sides of the equation the number of hydroxide ions equal to the number of hydrogen (H^+ ions appearing on the only side that has any)
6. Combining hydrogen and hydroxide ions to form water wherever possible
7. Canceling any water that now appears on both sides of the reaction equation.

Example The permanganate reduction in acidic solution shown in the previous example would be converted to basic solution as follows using the three steps above:

5. $5e^- + 8H^+ + 8OH^- + MnO_4^- \rightarrow Mn^{2+} + 4H_2O + 8OH^-$
6. $5e^- + 8H_2O + MnO_4^- \rightarrow Mn^{2+} + 4H_2O + 8OH^-$
7. $5e^- + 4H_2O + IO_3^- \rightarrow I^- + 8OH^-$

D.1.2 Electrolyte conductance

Introduction to conductance of electrolytes. Because a compound or mixture of compounds is electrically neutral, a solution made by dissolving a compound or mixture in any solvent must be neutral also. This means that the total positive charge must equal the total negative charge. This statement is known as the law of electroneutrality for electrolyte solutions. When a substance dissociates into ions in solution and the dissociation is essentially complete, the substance is called a strong electrolyte. Incomplete dissociation is found for weak electrolytes.

When an electrolyte dissociates, the resulting ions interact with surrounding solvent molecules or ions, a process known as solvation, to form charged clusters known as solvated ions. These solvated ions can move through the solution under the influence of an externally applied electric field. Such motion of charge is known as ionic conduction, and the resulting current is ionic current. The ionic current is determined by the nature of the ions, their concentrations, the solvent, and the electric field imposed. Ionic conduction of current in that part of the electrolyte is sufficiently removed from the electrodes that it does not influence, nor is it influenced by, the regions adjacent to the electrodes. This region, called the bulk solution or bulk electrolyte, is uniform in concentration.

In the bulk electrolyte current is carried only by means of ions. If a direct current is imposed upon a chemical cell, chemical reactions will occur at the electrodes in accordance with Faraday's laws. If an alternating rather than a direct current is used, the Faradaic reaction that takes place on one half-cycle is reversed on the following half-cycle. There are still flows of current, however, and such currents, which do not produce chemical changes in materials, are called non-Faradaic current. One of these is the current due to the current-carrying ability, or conductance, of ions. Thus measurements of ionic conduction are normally made by ac techniques to avoid complications due to the Faradaic processes taking place at the electrodes.

Theory of ionic conduction. Conductance, whether ionic or electronic, is the reciprocal of resistance. Ionic conductance, which for the bulk solution is the only conductance present, is the reciprocal of ionic resistance. Removing the dependence upon the size and shape of the conductor requires use of conductivity κ rather than conductance G. Because conductance increases directly with the cross-sectional area of a conductor and decreases with its length, conductivity is defined by Eq. (D.4):

$$\kappa = \frac{\ell/A}{R} = G\left(\frac{\ell}{A}\right) \qquad (D.4)$$

In the above equation, ℓ is the length of the conductor (i.e., the gap separating the electrodes in Fig. (D.1), and A is the cross-sectional area of each electrode, assuming that both electrodes have the same dimensions. The ratio ℓ/A is also called the cell constant or shape factor and has units of m^{-1}.

Molar conductivities of salts. The conductivity of a strong electrolyte solution such as KCl decreases as the solution concentration decreases. For dilute solutions, or solutions sufficiently dilute that the ionic environment does not change significantly upon further dilution, the conductivity should decrease as it does with concentration only because the number of charge carriers per unit volume decreases. It is therefore convenient to factor out the dependence upon concentration by defining the molar conductivity Λ of an electrolyte [Eq. (D.5)]:

$$\kappa = \Lambda C \tag{D.5}$$

If the concentration (C) is expressed in $mol \cdot m^{-3}$, the appropriate SI unit for molar conductivity Λ is $Sm^3 \, m^{-1} mol^{-1}$, or $Sm^2 \cdot mol^{-1}$. Other concentration units include the $mol \cdot dm^{-3}$, or molarity, and the $mol(kg^{-1}$ solvent), or molality. Experimentally the value of Λ is found to be independent of concentration for any electrolyte whenever the solution is sufficiently dilute. Λ_0, the molar conductivity extrapolated to infinite dilution, is characteristic only of the ions and the solvent and is independent of ionic interactions.

The German physicist Friedrich Kohlrausch found that for dilute solutions of strong electrolytes, extrapolation of measured values of Λ to infinite dilution was approximately linear when done against the

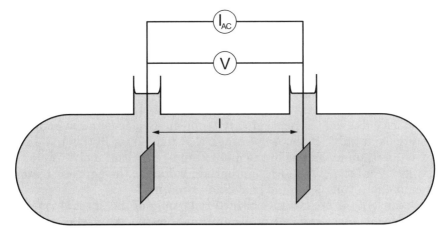

Figure D.1 Schematic of a conductivity cell containing an electrolyte and two inert electrodes of surface A parallel to each other and separated by distance 1.

square root of concentration. This means that the data suggest an equation containing an empirical coefficient B characteristic of the electrolyte [Eq. (D.6)]:

$$\Lambda = \Lambda_0 - B\sqrt{C} \tag{D.6}$$

Molar ionic conductivities. At infinite dilution, each ionic species present contributes a fixed amount to the total ionic conductivity, regardless of the nature of any other ions present. This means that the total conductivity of a sufficiently dilute solution is given by the sum of the individual ionic conductivities of the i different ionic species present [Eq. (D.7)]:

$$\kappa = \sum_i \kappa_i \tag{D.7}$$

It is convenient to define the molar ionic conductivity λ of individual ions (i.e., the conductivity of ions carrying 1 mole of charges), in the same manner as molar conductivity of electrolytes [Eqs. (D.8) and (D.9)]:

$$\kappa_i = \lambda\, C_i \tag{D.8}$$

and

$$\kappa = \sum_i \lambda_i C_i \tag{D.9}$$

The molar conductivity of an electrolyte salt at sufficient dilution is then simply the sum of the molar ionic conductivity of the ions produced by dissociation of the salt [Eq. (D.10)]:

$$\Lambda_0 = \sum_i n_i \lambda_{0i} \tag{D.10}$$

where Λ_0 is the molar conductivity and n_i is the number of moles of ions i produced by the dissociation of 1 mole of the salt.

Values of λ_{0i}, the molar ionic conductivity or, in metric units the equivalent conductance of individual ions, can be obtained from measured values of Λ extrapolated to give Λ_0. Table D.2 contains values of aqueous equivalent ionic conductivity for many ions found in aqueous solutions at 25°C. It should be noted that the values in this table are given in SI units. Values in the metric units of $S^{-1}\cdot cm^2 \cdot mol^{-1}$ would be larger by a factor of 10. An appropriate value for the aqueous hydrogen ion, for example, would be 349.99 $S\cdot cm^2\cdot mol^{-1}$.

Because λ_{0i} is a constant characteristic only of the specific ion i in the solvent, measurement of κ permits following the variation of C_i with time. One of the main applications of the technique is to monitor water quality in modern water purification systems. However, conduc-

TABLE D.2 Values of Limiting Molar Ionic Conductivity at 25°C

Cation	λ_0, mS·m²·mol⁻¹	Anion	λ_0, mS·m²·mol⁻¹
H^+	35.00	OH^-	19.84
$\frac{1}{3}Al^{3+}$	6.30	Br^-	7.82
Ag^+	6.19	$CH_3CO_2^-$	4.09
$\frac{1}{2}Ba^{2+}$	6.36	$C_2H_5CO_2^-$	3.58
$\frac{1}{2}Ca^{2+}$	5.95	$C_6H_5CO_2^-$	3.24
$\frac{1}{2}Cu^{2+}$	5.36	Cl^-	7.64
$\frac{1}{2}Fe^{2+}$	5.40	ClO_3^-	6.46
$\frac{1}{3}Fe^{3+}$	6.84	ClO_4^-	6.74
K^+	7.35	CN^-	4.45
$\frac{1}{3}La^{3+}$	6.97	$\frac{1}{2}CO_3^{2-}$	5.93
Li^+	3.87	F^-	5.54
$\frac{1}{2}Mg^{2+}$	5.30	$\frac{1}{3}Fe(CN)_6^{3-}$	10.1
Na^+	5.01	$\frac{1}{4}Fe(CN)_6^{4-}$	11.1
NH_4^+	7.35	HCO_3^-	4.45
$\frac{1}{2}Ni^{2+}$	5.30	HCO_2^-	5.46
$\frac{1}{2}Pb^{2+}$	6.95	HSO_4^-	5.20
$\frac{1}{2}Zn^{2+}$	5.28	I^-	7.69
		MnO_4^-	6.10
		NO_3^-	7.15
		$\frac{1}{3}PO_4^{3-}$	8.00
		$\frac{1}{2}SO_4^{2-}$	8.00

tance measurements are inherently nonselective because all ions conduct ionic current in an electrolyte solution.

Transport numbers in ionic solutions. When more than one ion is present in an electrolyte solution, it is useful to describe the fraction of the ionic conductance due to each ionic species present. The transport number of the individual ion t_i, sometimes called the migration number or transference number, is therefore defined as the fraction of the conductance due to that ion [Eq. (D.11)]:

$$t_i = \frac{\kappa_i}{\sum_i \kappa_i} \quad \text{(D.11)}$$

and, in sufficiently dilute solutions [Eq. (D.12)],

$$t_i = \frac{\lambda_{0i}}{\sum_i \lambda_{0i}} \quad \text{(D.12)}$$

When the solution contains only a single electrolyte salt, the above equation simplifies to Eq. (D.13):

$$t_i = \frac{\lambda_{0i}}{\Lambda_{0i}} \quad \text{(D.13)}$$

Transport numbers vary with the nature of the dissolved salt and of the solution as well as with concentration of the electrolyte. Transport numbers do change with concentration in a solution of a single salt, but only slightly. However, because the transport number of an ion is the fraction of the total ionic conductance due to that ion, the transport number of any particular ion or ions can be reduced to virtually zero by the addition to the solution of a large concentration of some salt that does not contain them. Electrochemists often make use of this technique.

Note that the units for molar ionic conductivity, $S \cdot m^2 \cdot mol^{-1}$, are units of velocity under a uniform potential gradient. As a consequence these values are also referred to as limiting ionic mobility (u_i), expressed as Eq. (D.14):

$$u_i = \frac{\lambda_i}{|z_i| F} \tag{D.14}$$

leading to the following expression of conductivity [Eq. (D.15)]:

$$\kappa = F \sum_i |z_i| u_i C_i \tag{D.15}$$

and transport number [Eq. (D.16)]:

$$t_i = \frac{|z_i| u_i C_i}{\sum_i |z_i| u_i C_i} \tag{D.16}$$

where $|z_i|$ is the absolute valence of ion species i. Table D.3 contains values of limiting ionic mobility corresponding to the equivalent ionic conductivity values presented in Table D.2.

Example The conductivity of protons (H$^+$), a value of 0.35 $S \cdot m^2 \cdot mol^{-1}$, is converted into a mobility value as follows:

1. 3.5×10^{-2} $S \cdot m^2 \cdot mol^- = 3.5 \times 10^{-2}$ $A \cdot m^2 \cdot mol^{-1} \cdot V^{-1}$
2. 3.5×10^{-2} $A \cdot m \cdot mol^{-1}$ per V/m = 3.5×10^{-2} $C \cdot m \cdot mol^{-1} s^{-1}$ per V/m and dividing by F (C mol^{-1})
3. 3.5×10^{-2} $C \cdot m \cdot mol^{-1} s^{-1}/(96{,}485$ C/mol) per V/m = 3.6×10^{-7} m/s per V/m

Mobility of uncharged species in solution. The mobility of an uncharged particle u_i can be estimated by dividing its diffusion coefficient D_i by the product of the Boltzmann constant and the absolute temperature. Uncharged particles are unaffected by an electric field, and their motion is driven only by diffusion. In molar terms, this is expressed as

$$u_i = \frac{D_i N_A}{RT} \quad \text{because Boltzmann constant} = \frac{R}{N_A}$$

TABLE D.3 Values of Limiting Ionic Mobility at 25°C

Cation	u, 10^{-8} m^2·s^{-1}·V^{-1}	Anion	u, 10^{-8} m^2·s^{-1}·V^{-1}
H^+	36.28	OH^-	20.56
Al^{3+}	6.53	Br^-	8.10
Ag^+	6.42	$CH_3CO_2^-$	4.24
Ba^{2+}	6.59	$C_2H_5CO_2^-$	3.71
Ca^{2+}	6.17	$C_6H_5CO_2^-$	3.36
Cu^{2+}	5.56	Cl^-	7.92
Fe^{2+}	5.60	ClO_3^-	6.70
Fe^{3+}	7.09	ClO_4^-	6.99
K^+	7.62	CN^-	4.61
La^{3+}	7.22	CO_3^{2-}	6.15
Li^+	4.01	F^-	5.74
Mg^{2+}	5.49	$Fe(CN)_6^{3-}$	10.47
Na^+	5.19	$Fe(CN)_6^{4-}$	11.50
NH_4^+	7.62	HCO_3^-	4.61
Ni^{2+}	5.49	HCO_2^-	5.66
Pb^{2+}	7.20	HSO_4^-	5.39
Zn^{2+}	5.47	I^-	7.97
		MnO_4^-	6.32
		NO_3^-	7.41
		PO_4^{3-}	8.29
		SO_4^{2-}	8.29

where R is the gas constant (8.314 J·K^{-1}·mol^{-1}) and N_A is the Avogadro number (6.023 × 10^{23} molecules·mol^{-1}).

For a particle of macroscopic dimensions moving through an ideal hydrodynamic continuum with velocity v, this force will be opposed by the viscous drag of the medium until these two forces are in balance. Stokes treated the ideal case of a spherical particle moving through an ideal hydrodynamic medium. Under these conditions Stokes's law gives the mobility [Eq. (D.17)]:

$$u_i = \frac{1}{6\pi\eta r_i} \qquad (D.17)$$

In this equation η is the viscosity, sometimes called the dynamic viscosity, of the medium, and r_i is the radius of the particle. Equating these two equations of ionic mobility gives the Stokes-Einstein equation [Eq. (D.18)]:

$$D_i = \frac{RT}{6\pi\eta N_A r_i} \qquad (D.18)$$

The Stokes-Einstein equation can be used to calculate the diffusion coefficients of uncharged species. It gives reasonable results if the species diffusing is roughly spherical and much larger than the solvent molecules.

Mobility of ions in solution. When an ion rather than an uncharged species is in motion, the force upon it is determined by the interaction of the electric field and the ionic charge. The conductivity due to a single ion submitted to an electrical field of unity (i.e., 1 V/m), is expressed as Eq. (D.19):

$$\kappa_i = C_i |z_i| u_i F \tag{D.19}$$

or as $\lambda_i = u_i F$ because $\lambda_i = \kappa_i / C_i$

Again, remember that C_i expresses the concentration of species i in terms of moles of charges produced (i.e., 1 mole $CaCl_2$ generates 2 moles of charge). The ionic velocity v_i is the product of ionic mobility and the electrical field (ξ) expressed in V/m [Eq. (D.20)]:

$$v_i = u_i \xi \tag{D.20}$$

The comparison between the mobility of ionic species and Fick's first law of diffusion has permitted Einstein to express the mobility of ionic species as Eq. (D.21):

$$u_i = \frac{z_i F D_i}{RT} \quad \text{or} \quad D_i = \frac{RT u_i}{z_i F} \tag{D.21}$$

or, as expressed in Nernst-Einstein equation [Eq. (D.22)]:

$$\lambda_i = \frac{z_i^2 F^2 D_i}{RT} \tag{D.22}$$

Some care is required in its use because molar conductivity of ions and specific ionic conductance are often given in the form λ_i / Z_i, as indicated by species such as $\tfrac{1}{2} Ca^{2+}$ listed in Table D.2. One redundant z_i must then be dropped from the last equation. Alternatively, the values given in that form may be converted to those of real species, that is $\lambda_0(Ca^{2+})$ is simply $2 \lambda_0(\tfrac{1}{2} Ca^{2+})$.

New expressions of ionic mobility can be obtained [Eqs. (D.23) to (D.25)] by combining Stokes-Einstein equation with Einstein equation:

$$D_i = \frac{RT}{6\pi \eta N_A r_i} = \frac{RT u_i}{z_i F} \tag{D.23}$$

and

$$u_i = \frac{z_i F}{6\pi \eta N_A r_i} \tag{D.24}$$

or

$$\lambda_{0i} = \frac{z_i^2 \cdot F^2}{6\pi \eta N_A r_i} \qquad (D.25)$$

Of the terms on the right-hand side of this equation, only the viscosity η is strongly dependent upon the medium. An approximation for all media in which ions may move a relationship was first suggested empirically by Walden and became known as Walden's rule:

$$\Lambda_0 \eta = \text{constant}$$

D.1.3 Basic electrochemical instrumentation

Most electrochemical work is achieved using what is called a potentiostat. A potentiostat is an electronic device that controls the voltage difference between a working electrode and a reference electrode. Both electrodes are contained in an electrochemical cell. The potentiostat implements this control by injecting current into the cell through an auxiliary electrode. In almost all applications, the potentiostat measures the current flow between the working and auxiliary electrodes. The controlled variable in a potentiostat is the cell potential, and the measured variable is the cell current. A potentiostat typically functions with an electrochemical cell that contains three electrodes, and that is true for both field probes and lab cells. Figure D.2 shows the schematic of a commercial potentiostat connected to an electrochemical cell.

Working electrode. Electrochemical reactions being studied occur at the working electrode. In corrosion testing, the working electrode is a sample of the corroding metal. Generally, the working electrode is not the actual metal structure being studied. Instead a small sample is used to represent the structure. This is analogous to testing using weight loss coupons. The working electrode can be bare or coated metal.

Reference electrode. A reference electrode is used in measuring the working electrode potential. A reference electrode should have a constant electrochemical potential as long as no current flows through it. The most common laboratory reference electrodes are the saturated calomel electrode (SCE) and the silver/silver chloride (Ag/AgCl) electrodes. In field probes, a pseudoreference (a piece of the working electrode material) is often used. A Luggin capillary is often used to position the sensing point of a reference electrode to a desired point in a cell.

1026　Appendix D

The Luggin capillary in a laboratory cell is made from glass or plastic. It is generally filled with the test solution. The Luggin holds the reference electrode, as shown in Fig. D.3. The tip of the Luggin capillary near the working electrode is open to the test solution. The reference electrode senses the solution potential at this open tip. Note that the Luggin tip is significantly smaller than the reference electrode itself. The Luggin capillary allows sensing of the solution potential close to the working electrode without the adverse effects that occur when the large reference electrode is placed near the working electrode.

Auxiliary electrode. The auxiliary electrode is a conductor that completes the cell circuit. The auxiliary (counter) electrode in lab cells is generally an inert conductor like platinum or graphite. In field probes it is generally another piece of the working electrode material. The current that flows into the solution via the working electrode leaves the solution via the auxiliary electrode.

Electrometer. The electrometer circuit measures the voltage difference between the reference and working electrodes. Its output has two major functions: It is the feedback signal in the potentiostat cir-

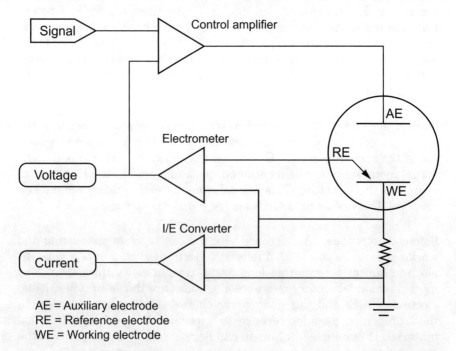

AE = Auxiliary electrode
RE = Reference electrode
WE = Working electrode

Figure D.2 Schematic of a commercial potentiostat connected to an electrochemical cell.

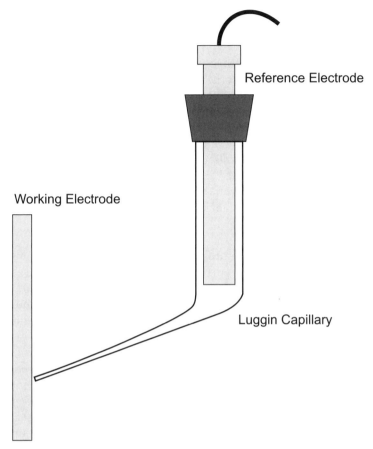

Figure D.3 Schematic of a Luggin capillary positioning a reference electrode in close proximity to an electrochemical cell working electrode.

cuit, and it is the signal that is measured whenever the cell voltage is needed. An ideal electrometer has zero input current and infinite input impedance. Current flow through the reference electrode can change its potential. In practice, all modern electrometers have input currents close enough to zero that this effect can usually be ignored. Two important electrometer characteristics are its bandwidth and its input capacitance.

The electrometer bandwidth characterizes the ac frequencies the electrometer can measure when it is driven from a low-impedance source. The electrometer bandwidth must be higher than the bandwidth of the rest of the potentiostat electronics. The electrometer input capacitance and the reference electrode resistance form an RC filter. If this filter's time constant is too large, it can limit the effective bandwidth of the electrometer and cause system instabilities. Smaller input capacitance translates into more stable operation and greater tolerance for high-impedance reference electrodes.

I/E converter. The current to voltage (I/E) converter in the simplified schematic measures the cell current. It forces the cell current to flow through a current measurement resistor. The voltage drop across that resistor is a measure of the cell current. Cell current in a corrosion experiment can often vary by as much as seven orders of magnitude. Current cannot be measured over such a wide range using a single resistor. A number of different resistors have to be switched into the I/E circuit under computer control. This allows measurement of widely varying currents, with each current measured using an appropriate resistor. An "I/E autoranging" algorithm is often used to select the appropriate resistor values.

Control amplifier. The control amplifier is a servo amplifier. It compares the measured cell voltage with the desired voltage and drives current into the cell to force the voltages to be the same. Under normal conditions, the cell voltage is controlled to be identical to the signal source voltage. The control amplifier has a limited output current capability.

The signal. The signal circuit is a computer-controlled voltage source. It is generally the output of a digital to analog (D/A) converter that converts computer-generated numbers into voltages. Proper choice of number sequences allows the computer to generate constant voltages, voltage ramps, and even sine waves at the signal circuit output. When a D/A converter is used to generate a waveform such as a sine wave or a ramp, the waveform is a digital approximation of the equivalent analog waveform. It contains small voltage steps. The size of these steps is controlled by the resolution of the D/A converter and the rate at which it is being updated with new numbers.

Galvanostats and zero resistance amplifiers (ZRAs). Most laboratory-grade potentiostats can also be operated as a galvanostat or as a ZRA. The potentiostat in the simplified schematic (Fig. D.1) becomes a galvanostat when the feedback is switched from the cell voltage signal to the cell current signal. The instrument then controls the cell current rather than the cell voltage. The electrometer output can still be used to measure the cell voltage. A ZRA allows one to force a potential difference of 0 V between two electrodes. The cell current flowing between the electrodes can be measured. A ZRA is often used to measure galvanic corrosion phenomena and electrochemical noise.

D.2 Chemical Thermodynamics

D.2.1 Free energy and electrochemical cells

Electrical work is the product of charge moved Q times the cell potential (E) through which it is moved. If the work done is that of an electrochemical cell in which the potential difference is E, and the charge is that of 1 mole of reaction in which n moles of electrons are transferred, the electrical work $-w$ done by the cell must be nE. In this relationship, the Faraday constant F is necessary to obtain coulombs from moles of electrons. In an electrochemical cell operating reversibly, no current flows and

$$\Delta G = -nFE \tag{D.26}$$

Under standard conditions, the standard free energy of the cell reaction ΔG^0 is directly related to the standard potential difference across the cell, E^0:

$$\Delta G^0 = -nFE^0 \tag{D.27}$$

Electrode potentials can be combined algebraically to give cell potential. For a galvanic cell, which operates spontaneously, a positive cell voltage will be obtained if the difference is taken in the usual way, as Eq. (D.28):

$$E_{cell} = E_{cathode} - E_{anode} \tag{D.28}$$

The free energy change in a galvanic cell, or in a spontaneous cell reaction, is negative and the cell voltage is positive.

In electrolytic cells, the reaction is driven in the nonspontaneous direction by an external electrical force. The free energy change in an electrolytic cell, or in a nonspontaneous cell reaction, is therefore positive and the cell voltage negative.

Other thermodynamic quantities can be derived from electrochemical measurements. For example, the entropy change (ΔS) in the cell reaction is given by the temperature dependence of ΔG:

$$\Delta S = -\left(\frac{\partial \Delta G}{\partial T}\right)_P \tag{D.29}$$

hence

$$\Delta S = nF\left(\frac{\partial E}{\partial T}\right)_P \tag{D.30}$$

and

$$\Delta H = \Delta G + T\Delta S = nF\left[T\left(\frac{\partial E}{\partial T}\right)_P - E\right] \qquad (D.31)$$

The equilibrium constant (K_{eq}) for the same reaction can be obtained with the following equation:

$$RT \ln K_{eq} = -\Delta G^0 = nFE^0 \qquad (D.32)$$

D.2.2 Electrochemical potentials

The potential difference across an electrochemical cell is the potential difference measured between two electronic conductors. In the external circuit connected to an electrochemical cell, the electrons will flow from the most negative point to the most positive point. Because the potentials of electrodes can be either positive or negative, the electrons in the external circuit can also be said to flow from the least-positive electrode to the most-positive electrode. Voltmeters can be used to measure the potential differences across electrochemical cells but cannot measure directly the actual potential of any single electrode. Nevertheless, it is convenient to assign part of the cell potential to one electrode and part to the other.

D.2.3 Standard electrode potentials

Standard potential differences are the actual cell potential differences measured in reversible cells under standard conditions. For solid or liquid compounds or elements, standard conditions are the pure compound or element; for gases they are 100 kPa pressure, and for solutes they are the ideal 1 molar (mol/liter) concentration.

Tables of standard electrode potentials can be obtained if any one electrode, operated under standard conditions, is designated as the standard electrode or standard reference electrode with which all other electrodes will be compared. This electrode is called the standard hydrogen electrode, abbreviated SHE. The potential difference across a reversible cell made up of any electrode and a SHE is called the reversible potential of that electrode, E. If this other electrode is also being operated under standard conditions of pressure and concentration, the reversible potential difference across the cell is the standard electrode potential E_0 of that electrode.

In many practical potential measurements, the standard hydrogen electrode cannot be used because hydrogen reacts with other substances in the cell or because other substances in the cell react with the catalytic platinum electrode surface upon which the H^+/H_2 potential is established. It is often much more convenient to use alternative electrodes whose potentials are precisely known with respect to the SHE. Two of the electrodes most commonly used for this purpose are the AgCl/Ag electrode, //AgCl(s),Cl$^-$/Ag(c) at $E_0 = +0.2224$ V vs. SHE, and the saturated calomel electrode (SCE) at 0.241 V vs. SHE. The effect of changing the reference electrode is to change the zero of a potential scale while leaving the relative positions of all of the potentials unchanged, as shown in Fig. D.4.

D.2.4 Introduction to the Nernst equation

The Nernst equation, named after the German chemist Walther Nernst, can be derived from the equation linking free energy changes to the reaction quotient:

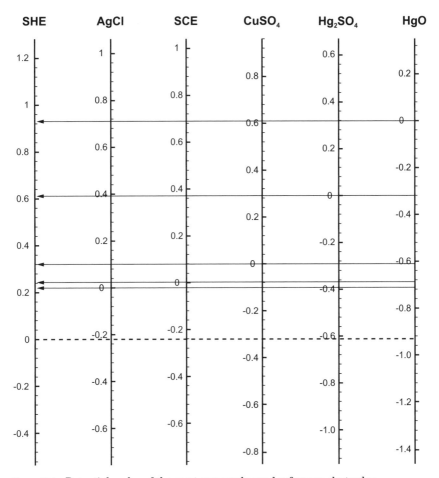

Figure D.4 Potential scales of the most commonly used reference electrodes.

$$\Delta G = \Delta G^0 + RT \ln Q \qquad (D.33)$$

where, for a generalized equation of the form

$$aA + bB + \cdots \rightarrow mM + nN + \cdots \qquad (D.34)$$

$$Q = \frac{a_M^m a_N^n \cdots}{a_A^a a_B^b \cdots} \qquad (D.35)$$

At equilibrium, $\Delta G = 0$ and Q corresponds to K_{eq}, as expressed earlier.

In the case of an electrochemical reaction, substitution of the relationships $\Delta G = -nFE$ and $\Delta G^0 = -nFE^0$ into the expression of a reaction free energy and division of both sides by $-nF$ gives the Nernst equation for an electrode reaction:

$$E = E^0 - \frac{RT}{nF} \ln Q \qquad (D.36)$$

Combining constants at 25°C (298.15 K) gives the simpler form of the Nernst equation for an electrode reaction at this standard temperature:

$$E = E^0 - \frac{0.059}{n} \log_{10} Q \qquad (D.37)$$

In this equation, the electrode potential E is the actual potential difference across a cell in which this electrode and a standard hydrogen electrode are present.

Alternatively, the relationship in Eq. (D.28) can be used to combine two Nernst equations corresponding to two half-cell reactions into the Nernst equation for a cell reaction:

$$E_{cell} = (E^0_{cathode} - E^0_{anode}) - \frac{0.059}{n} \log_{10} Q \qquad (D.38)$$

Some of the species that take part in electrode reactions are pure solid compounds and pure liquid compounds. In dilute aqueous solutions, water can be treated as a pure liquid. For pure solid compounds or pure liquid compounds, activities are constant, and their values are considered to be 1. The activities of gases are usually taken as their partial pressures, and the activities of solutes such as ions are usually taken as their molar concentrations, that is,

$$a_i = \gamma_i [\ ]_i \approx [i] \qquad (D.39)$$

where $[i]$ and γ_i are respectively the molar concentration and the activity coefficient of species i.

D.2.5 Advanced thermodynamics

The present section illustrates in very detailed terms how the calculations from basic thermodynamic data can lead to open circuit cell potential in any condition of temperature and pressure. The aluminum-air corrosion cell was chosen for this example because of the relative simplicity of its chemistry.

The aluminum-oxygen system. The high electrochemical potential and low equivalent weight of aluminum combine to produce a theoretical energy density of 2.6 kWh/kg and make it an attractive candidate as an anode material in metal/air electrochemical cells. The development of aluminum-based cells dates back to 1855 when M. Hulot described a voltaic cell containing aluminum with an acid electrolyte. Since then, many attempts to substitute aluminum for zinc in zinc/carbon and zinc/manganese dioxide cells have been reported. Zaromb first proposed its use in combination with air diffusion electrodes in 1962. Three types of Al-O_2 cells have been developed to date:

1. A solids-free system (SFS) that uses pumps and auxiliary equipment to manage reaction products
2. A solids self-management system (SSMS) that eliminates much of the auxiliary equipment by allowing the reaction products to solidify in the cell enclosure during the discharge cycle
3. A solids management system (SMS) that controls the precipitation of the alumina in a separate tank.

Electrochemistry of the Al-O_2 couple. Figure D.5 shows a general schematic of a typical Al-air system. Tables D.4 and D.5, respectively, contain thermodynamic data for pure species and soluble species involved in the equilibria associated with aluminum, water, and oxygen. Table D.6 contains the chemical and electrochemical reactions possibly occurring in a typical Al-air corrosion cell.

The overall anodic reaction of the aluminum air battery is the corrosion of aluminum into a soluble form stable in a caustic environment, that is, AlO_2 [Eq. (D.40)], that can subsequently precipitate as $Al_2O_3 \cdot H_2O$ [Eq. (D.41)] depending on the concentration of ions in solution, pH, and temperature.

$$Al + 4OH^- \rightarrow AlO_2^- + 2H_2O + 3\,e^- \qquad (D.40)$$

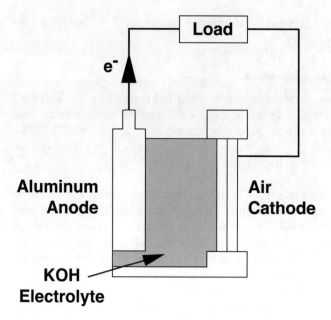

Figure D.5 Schematic of an aluminum-air corrosion cell.

$$2AlO_2^- + 2H_2O \rightarrow Al_2O_3 \cdot H_2O + 2OH^- \quad (D.41)$$

There is also a parasitic reaction at the aluminum anode that has to be considered because it has serious safety implications (i.e., the production of hydrogen gas from the reduction of water):

$$2H_2O + 2e^- \rightarrow H_2 + 2OH^- \quad (D.42)$$

The unique cathodic reaction is the reduction of oxygen [Eq. (D.43)]:

$$O_2 + 2H_2O + 4e^- \rightarrow 4OH^- \quad (D.43)$$

The overall cell voltage can be calculated from thermodynamic data by computing Gibbs free energy for the individual species involved in the global reaction [Eq. (D.44)] and using the coefficients expressed in that equation:

$$4Al + 4OH^- + 3O_2 \rightarrow 4AlO_2^- + 2H_2O \quad (D.44)$$

Detailed calculations

Calculate $G°$ for each species. The free energy of a substance, for which heat capacity data are available, can be calculated as a function of temperature using Eq. (D.45):

TABLE D.4 Pure Species Considered and Their Thermodynamic Data

Species	$G^0_{(298\,K)}$, J·mol^{-1}	$S^0_{(298\,K)}$, J·mol^{-1}	A	$B \times 10^3$	$C \times 10^{-5}$	C_p,* J·mol^{-1}·K^{-1}	$G^0_{(333\,K)}$,† J·mol^{-1}
O$_2$	0	205	29.96	4.184	1.674	29.85	−7,234.04
H$_2$	0	131	27.28	3.263	0.502	28.82	−4,642.01
H$_2$O	−237,000	69.9	10.669	42.284	−6.903	18.54	−239,483
Al	0	28.325	20.67	12.38	0	24.79	−1,040.43
Al(OH)$_3$	−1,136,542					0	−1,136,542
Al$_2$O$_3$·H$_2$O	−1,825,500	96.86	120.8	35.14	0	132.51	−1,829,152

*Calculated with Eq. (D.46).
†Calculated with Eq. (D.48).

TABLE D.5 Soluble Species Considered and Their Thermodynamic Data

Species	$G^0_{(298\ K)}$, $J \cdot mol^{-1}$	$S^0_{(298\ K)}$, $J \cdot mol^{-1}$	$\check{S}^0_{(298\ K)}$, $J \cdot mol^{-1}$	a	b	C_p^*, $J \cdot mol^{-1} \cdot K^{-1}$	$G^0_{(333\ K)}$,† $J \cdot mol^{-1}$
H^+	0	0					
OH^-	$-157{,}277$	41.888	-20.9	0.065	-0.005	118.75	-234.9
Al^{3+}	$-485{,}400$	-321.75	20.968	-0.37	0.0055	-452.03	$-157{,}849$
$Al(OH)^{2+}$	$-694{,}100$	-142.26	-384.45	0.13	-0.00166	372.84	$-474{,}876$
$Al(OH)_2^+$	$-900{,}000$	205.35	-184.06	0.13	-0.00166	267.95	$-689{,}651$
AlO_2^-	$-838{,}968$	96.399	184.43	0.13	-0.00166	75.06	$-907{,}336$
			117.31	-0.37	0.0055	-284.94	$-841{,}778$

*Calculated with Eq. (D.47).
†Calculated with Eq. (D.48).

TABLE D.6 Reactions Considered to Model an Aluminum-Air Corrosion Cell

Water equilibria

$2\,e^- + 2\,H^+ = H_2$
$4\,e^- + O_2 + 4\,H^+ = 2\,H_2O$
$OH^- + H^+ = H_2O$

Equilibria involving aluminum metal

$3\,e^- + Al^{3+} = Al$
$3\,e^- + Al(OH)_3 + 3\,H^+ = Al + 3\,H_2O$
$6\,e^- + Al_2O_3 \cdot H_2O + 6\,H^+ = 2\,Al + 4\,H_2O$
$3\,e^- + AlO_2^- + 4\,H^+ = Al + 2\,H_2O$
$3\,e^- + Al(OH)^{2+} + H^+ = Al + H_2O$
$3\,e^- + Al(OH)_2^+ + 2\,H^+ = Al + 2\,H_2O$

Equilibria involving solid forms of oxidized aluminum

$Al(OH)_3 + H^+ = Al(OH)_2^+ + H_2O$
$Al_2O_3 \cdot H_2O + 2\,H^+ = 2\,Al(OH)_2^+$
$Al(OH)_3 + 2\,H^+ = Al(OH)^{2+} + 2\,H_2O$
$Al_2O_3 \cdot H_2O + 4\,H^+ = 2\,Al(OH)^{2+} + 2\,H_2O$
$Al(OH)_3 + 3\,H^+ = 2\,Al^{3+} + 4\,H_2O$
$Al_2O_3 \cdot H_2O + 6\,H^+ = Al^{3+} + 3\,H_2O$
$Al(OH)_3 = AlO_2^- + H^+ + H_2O$
$Al_2O_3 \cdot H_2O = 2\,AlO_2^- + 2\,H^+$

Equilibria involving only soluble forms of oxidized aluminum

$AlO_2^- + 4\,H^+ = Al^{3+} + 2\,H_2O$

$$G^0_{(T_2)} = G^0_{(T_1)} - S^0_{(T_1)}[T_2 - T_1] - T_2 \int_{T_1}^{T_2} \frac{C_p^0}{T}\,dT + \int_{T_1}^{T_2} C_p^0\,dT \quad (D.45)$$

For pure substances (i.e., solids, liquids, and gases) the heat capacity C_p^0 is often expressed, as in Table D.4, as function of the absolute temperature:

$$C_p^0 = A + BT + CT^{-2} \quad (D.46)$$

For ionic substances, one has to use another method, such as proposed by Criss and Cobble in 1964,[1] to obtain the heat capacity, provided the temperature does not rise above 200°C. The expression of the ionic capacity [Eq. (D.47)] makes use of absolute entropy values and the parameters a and b contained in Table D.4:

$$C_p^0 = (4.186a + bS^0_{(298\,K)})(T_2 - 298.16)/\ln\left(\frac{T_2}{298.16}\right) \quad (D.47)$$

By combining Eq. (D.46) or (D.47) with Eq. (D.45) one can obtain the free energy [Eq. (D.48)] at any given temperature by using the fundamental data contained in Tables D.4 and D.5:

$$G^0_{(T)} = G^0_{(298\ K)} + (C_p^0 - S^0_{(298\ K)})(T_2 - 298.16) - T_2 \ln\left(\frac{T_2}{298.16}\right) C_p^0 \quad (D.48)$$

Although these equations appear slightly overwhelming, they can be computed relatively simply with the use of a modern spreadsheet, where the data in Table D.4 could be imported with the functions in Eqs. (D.46) to (D.48) properly expressed.

Calculate G for each species. For species O, the free energy of 1 mol can be obtained from G^0 with Eq. (D.49):

$$G_{o(T)} = G_{o(T)}^{\ 0} + 2.303\ RT \log_{10} a_O \quad (D.49)$$

For x mol of species O the free energy is expressed by Eq. (D.50):

$$xG_{O(T)} = x\,(G_{O(T)}^{\ 0} + 2.303\ RT \log_{10} a_O) \quad (D.50)$$

For pure substances such as solids, a_O is equal to 1. For a gas, a_O is equal to its partial pressure (p_O), as a fraction of 1 atmosphere. For soluble species, the activity of species O (a_O), is the product of the activity coefficient of that species (γ_O) with its molar concentration ([O]) (i.e., $a_O = \gamma_O[O]$). The activity coefficient of a chemical species in solution is close to 1 at infinite dilution when there is no interference from other chemical species. For most other situations the activity coefficient is a complex function that varies with the concentration of the species and with the concentration of other species in solution. For the sake of simplicity the activity coefficient will be assumed to be of value 1; hence Eq. (D.50) can be written as a function of [O]:

$$xG_{O(T)} = x\,(G_{O(T)}^{\ 0} + 2.303\ RT \log_{10} [O]) \quad (D.51)$$

Taking the global reaction fo the Al-O_2 system expressed in Eq. (D.44) and the G^0 values calculated for 60°C in Tables D.4 and D.5, one can obtain thermodynamic values for the products and reactants, as is done in Table D.7.

Calculate cell ΔG. The DG of a cell can be calculated by subtracting the G values of the reactants from the G values of the products in Table D.7. Keeping the example of the global reaction at 60°C in mind, one would obtain

$$\Delta G = G_{products} - G_{reactants} = -3{,}846{,}087 - (-670{,}615) = -3{,}175{,}472\ J$$

TABLE D.7 Calculated Free Energies for Species Involved in the Global Al-Air Reaction at 60°C

Species	x	$G^0_{(333\ K)}$, J·mol^{-1}	a_o	$2.303RT \log_{10} a_O$	$x\ G^0_{(333\ K)}$	$\Sigma G^0_{(333\ K)}$
Reactants						
Al	4	−1040.43	1	0.00	−4161.72	
OH$^-$	4	−157,849	1	0.00	−631,396.00	
O$_2$	3	−7,234.04	0.2	−4452.02	−35,058.17	−670,615.89
Products						
AlO$_2^-$	4	−841,778	0.1	−6369.39	−3,392,589.58	
H$_2$O	2	−239,483	1	0.00	−478,966.00	−3,846,087.21

Translate ΔG into potential

$$E = \frac{-\Delta G}{nF} = \frac{3{,}188{,}818}{(12 \cdot 96{,}485)} = 2.74\ \text{V}$$

where $n = 12$ because each Al gives off 3 e^- [cf. Eq. (D.40)] and there are four Al in the global Eq. (D.44) representing the cell chemistry.

Calculate the specific capacity (Ah·kg^{-1}). The specific capacity relates the weight of active materials with the charge that can be produced, that is, a number of coulombs or ampere-hours (Ah). Because 1 A = 1 C·s^{-1}, 1 Ah = 3600 C, and because 1 mole of e^- = 96,485 C (Faraday), 1 mole of e^- = 26.80 Ah.

By considering the global expression of the cell chemistry expressed in Eq. (D.44), one can relate the weight of the active materials to a certain energy and power. In the present case 12 moles of e^- are produced by using

4 moles of Al	4 × 26.98 g·mol^{-1}, or 107.92 g
4 moles of OH$^-$ as KOH	4 × 56.11 g·mol^{-1}, or 224.44 g
3 moles of O$_2$ (as air)	0 g
3 moles of O$_2$ (compressed or cryogenic)	3 × 32.00 g mol^{-1}, or 96.00 g

Weight of active materials for the production of 12 moles of e^- is then 332.36 g if running on free air and 428.36 g if running on compressed or cryogenic oxygen. The theoretical specific capacity is thus 26.80 × 12/0.3324 = 967.5 Ah·kg^{-1} if running on air and 26.80 × 12/0.4284 = 750.7 Ah·kg^{-1} if running on compressed or cryogenic oxygen.

Calculate the energy density (Wh·kg^{-1}). The energy density can then be obtained by multiplying the specific capacity obtained from calculating the specific capacity with the thermodynamic voltage calculated when translating ΔG into potentials: $2.74 \times 967.5 = 2651$ Wh·kg^{-1}, or 2.651 kWh·kg^{-1} if running on air and, because the voltage for running on pure oxygen is slightly higher (i.e., 2.78 V), $2.78 \times 750.7 = 2087$ Wh·kg^{-1}, or 2.087 kWh·kg^{-1} if running on compressed or cryogenic oxygen.

Reference electrodes. The thermodynamic equilibrium of any other chemical or electrochemical reaction can be calculated in the same manner, provided the basic information is found. Table D.8 contains the chemical description of most reference electrodes used in laboratories and field units, and Tables D.9 and D.10, respectively, contain the thermodynamic data associated with the solid and soluble chemical species making these electrodes. Table D.11 presents the results of the calculations performed to obtain the potential of each electrode at 60°C (i.e., away from the 25°C standard temperature).

D.2.6 Potential-pH diagrams

Potential-pH (E-pH) diagrams, also called predominance or Pourbaix diagrams, have been adopted universally since their conception in the early 1950s. They have been repetitively proven to be an elegant way to represent the thermodynamic stability of chemical species in given aqueous environments. E-pH diagrams are typically plotted for various equilibria on normal cartesian coordinates with potential (E) as the ordinate (y-axis) and pH as the abscissa (x-axis).[2]

Pourbaix diagrams are a convenient way of summarizing much thermodynamic data, and they provide a useful means of predicting electrochemical and chemical processes that could potentially occur in certain conditions of pressure, temperature, and chemical makeup. These diagrams have been particularly fruitful in contributing to the understanding of corrosion reactions.

Stability of water. Equation (D.52) describes the equilibrium between hydrogen ions and hydrogen gas in an aqueous environment:.

$$2H^+ + 2e^- = H_2 \qquad (D.52)$$

which can be written as Eq. (D.53) in neutral or alkaline solutions:

$$2H_2O + 2e^- = H_2 + 2OH^- \qquad (D.53)$$

Adding sufficient OH^- to both sides of the reaction in Eq. (D.52) results in Eq. (D.53). At higher pH than neutral, Eq. (D.53) is a more appropriate representation. However, both representations signify the same reaction for which the thermodynamic behavior can be expressed by a Nernst Eq. (D.54):

TABLE D.8 Equilibrium Potential of the Main Reference Electrodes Used in Corrosion, at 25°C

Name	Equilibrium reaction	Nernst Equation, V vs. S H E		Potential, V vs. S H E	T coefficient, mV·C^{-1}
Hydrogen	$2H^+ + 2e^- = H_2$ (SHE)	$E^0 - 0.059$ pH		0.00	
Silver chloride	$AgCl + e^- = Ag + Cl^-$	$E^0 - 0.059 \log_{10} a_{Cl^-}$		0.2224	−0.6
			0.1 M KCl	0.2881	
			1.0 M KCl	0.2224	
			Seawater	∼ 0.250	
Calomel	$Hg_2Cl_2 + 2e^- = 2Hg + 2Cl^-$	$E^0 - 0.059 \log_{10} a_{Cl^-}$		0.268	
			0.1 M KCl	0.3337	−0.06
			1.0 M KCl	0.280	−0.24
	(SCE)		Saturated	0.241	−0.65
Mercurous sulfate	$Hg_2SO_4 + 2e^- = 2Hg + SO_4^{-2}$	$E^0 - 0.0295 \log_{10} a_{SO_4^{2-}}$		0.6151	
Mercuric oxide	$HgO + 2e^- + 2H^+ = Hg + H_2O$	$E^0 - 0.059$ pH		0.926	
Copper sulfate	$Cu^{2+} + 2e^- = Cu$ (sulfate solution)	$E^0 + 0.0295 \log_{10} a_{Cu^{2+}}$		0.340	
			Saturated	0.318	

TABLE D.9 Data and Calculations of the Free Energy and Potential of the Main Reference Electodes at 60°C

(TemRef = 25; TemC = 60; TemA = 333.16; $T_2 - T_1 = 35$; $\ln(T_2/T_1) = 0.1109926$)

Species	G^0 (298 K), J·mol^{-1}	S^0 (298 K), J·mol^{-1}	A	B	C	C_p (333K),* J·mol^{-1}·K^{-1}	G^0T (333K),[†] J·mol^{-1}
O$_2$	0	205	29.96	4.184	−1.674	29.85	−7234.04
H$_2$	0	131	27.28	3.263	0.502	28.8	−4642.01
H$_2$O	−237000	69.9	10.669	42.284	−6.903	18.5	−239483.00
Ag	0	42.55	21.297	8.535	1.506	25.5	−1539.69
Cu	0	33.2	22.635	6.276		24.7	−1210.91
Hg	0	76.02	26.94	0	0.795	27.7	−2715.41
AgCl	−109805	96.2	62.258	4.184	−11.297	53.5	−113277.
Hg$_2$Cl$_2$	−210778	192.5	63.932	43.514	0	78.4	−217670.
Hg$_2$SO$_4$	−625880	200.66	131.96			132	−633164.
HgO	−58555	70.29	34.853	30.836	0	45.1	−61104.4

*Calculated with Eq. (D.46).
[†]Calculated with Eq. (D.48).

TABLE D.10 Thermodynamic Data of Soluble Species Associated with the Most Commonly Used Reference Electrodes

Species	G^0 (298K), J·mol^{-1}	S^0 (298K), J·mol^{-1}	$\check{S}^0$ (298K), J·mol^{-1}	a	b	C_p Eq.(D.47)	Eq.(D.48)
H$^+$	0	0	−20.9	0.065	−0.005	118.7525	−234.927
Cu^{2+}	65689	−207.2	−249.04	0.13	−0.00166	301.9618	72343.6
Cl$^-$	−131260	−12.6	8.32	−0.37	0.0055	−473.9694	−129881.
SO$_4^{2-}$	−744600	10.752	52.592	−0.37	0.0055	−397.1863	−744190.

TABLE D.11 Calculations of the Equilibrium Associated with the Most Commonly Used Reference Electrodes at 60°C

Name	$\sum G°$ reactants,* $J \cdot mol^{-1}$	$\sum G°$ products,* $J \cdot mol^{-1}$	$\Delta G°$ reaction, $J \cdot mol^{-1}$	Potential, V
Hydrogen	−470	−46,420	−4,172	0.0216
Silver chloride	−113,277	−131,421	−18,144	0.1880
Calomel	−217,670	−265,193	−47,523	0.2463
Mercurous sulfate	−633,164	−749,621	−116,457	0.6035
Mercuric chloride	−61,574	−242,199	−180,624	0.9360
Copper sulfate	72,344	−1,211	−73,555	0.3812

*Note: all species considered to be of activity = 1

$$E_{H^+/H_2} = E_{H^+/H_2}^0 + \frac{RT}{nF} \ln \frac{[H^+]^2}{p_{H_2}} \quad \text{(D.54)}$$

that becomes Eq. (D.55) at 25°C and p_{H_2} of value unity:

$$E_{H^+/H_2} = E_{H^+/H_2}^0 - 0.059 \, pH \quad \text{(D.55)}$$

Equation (D.52) and its alkaline or basic form, Equation (D.53), delineate the stability of water in a reducing environment and are represented in a graphical form by the sloping line (a) on the Pourbaix diagram in Fig. D.6. Below line (a) in this figure the equilibrium reaction indicates that the decomposition of H_2O into hydrogen is favored, whereas it is thermodynamically stable above that line. As potential becomes more positive or noble, water can be decomposed into its other constituent, oxygen, as illustrated in Eqs. (D.56) and (D.57) for, respectively, the acidic form and neutral or basic form of the same process:

$$O_2 + 4H^+ + 4e^- = 2H_2O \quad \text{(D.56)}$$

$$O_2 + 2H_2O + 4e^- = 4OH^- \quad \text{(D.57)}$$

Again these equivalent equations can be used to develop a Nernst expression of the potential, that is Eq. (D.58) expressed as Eq. (D.59) in standard conditions of temperature and oxygen pressure (i.e., p_{O_2} of value unity):

$$E_{O_2/H_2O} = E_{O_2/H_2O}^0 + \frac{RT}{nF} \ln p_{O_2} [H^+]^4 \quad \text{(D.58)}$$

$$E_{O_2/H_2O} = E_{O_2/H_2O}^0 - 0.059 \, pH \quad \text{(D.59)}$$

The line labeled (b) in Fig. D.6 represents the behavior of E vs. pH for this last equation. Figure D.6 is divided into three regions. In the upper one, water can be oxidized and form oxygen, whereas in the lower one, it can be reduced to form hydrogen gas. In the intermediate region, water is thermodynamically stable. It is common practice to superimpose these two lines (a) and (b) on Pourbaix diagrams to mark the water stability boundaries.

Predominance diagram of aluminum. Aluminum provides one of the simplest cases for demonstrating the construction of E-pH diagrams. In the following discussion, only four species containing the aluminum element will be considered: two solid species (Al and $Al_2O_3 \cdot H_2O$) and two ionic species (Al^{3+} and AlO_2). The first equilibrium

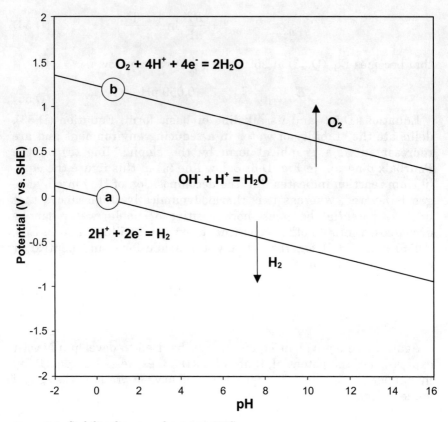

Figure D.6 Stability diagram of water at 25°C.

to consider examines the possible presence of either Al^{3+} or AlO_2^- expressed in Eq. (D.60):.

$$Al^{3+} + 2H_2O = AlO_2^- + 4H^+ \qquad (D.60)$$

Because there is no change in valence of the aluminum present in the two ionic species considered, the associated equilibrium is independent of the potential, and the expression of that equilibrium can be derived to give an expression valid in standard conditions [Eq. (D.61)]:

$$RT \ln K_{eq} = RT \ln Q = -\Delta G^0_{reaction} \qquad (D.61)$$

where

$$Q = \frac{a_{AlO_2^-} \, a^4_{H^+}}{a_{Al^{3+}} \, a^2_{H_2O}}$$

Assuming that the activity of H_2O is unity and that the activities of the two ionic species are equal, one can obtain a simpler expression of the equilibrium based purely on the activity of H^+:

$$4 \log_{10} [H^+] = \frac{-\Delta G^0_{reaction}}{2.303 RT} \quad (D.62)$$

or, if G^0 is expressed in joules,

$$-4 \log_{10} [H^+] = 4\, pH = \Delta G^0_{reaction} \times 1.75 \times 10^{-4} \quad (D.63)$$

By using the thermodynamic data provided in Tables D.4 and D.5 and following the detailed procedure outlined earlier, it is possible to calculate that the free energy of reaction [Eq. (D.60)] is in fact equal to 120.44 kJ·mol^{-1} (for either 1 [Al^{3+}] or 1 [AlO_2^-]). Equation (D.63) then becomes Equation (D.64):

$$pH = 120{,}440 \times 4.38 \times 10^{-5} = 5.27 \quad (D.64)$$

This is represented, in the E-pH diagram shown in Fig. D.7, by a dotted vertical line separating the dominant presence of Al^{3+} at low pH from the dominant presence of AlO_2^- at the higher end of the pH scale.

The next phase for constructing the aluminum E-pH diagram is to consider the equilibria between the four species mentioned earlier. A computer program that would compare all possible interactions and rank them in terms of their thermodynamic stability would typically carry out this work. The steps of this data-crunching process are illustrated in Figs. D.8 to D.10.

D.3 Kinetic Principles

Thermodynamic principles can help explain a situation in terms of the stability of chemical species and reactions associated with corrosion process. However, thermodynamic calculations cannot be used to predict reaction rates. Electrode kinetic principles have to be used to estimate these rates.

D.3.1 Kinetics at equilibrium: The exchange current concept

The exchange current I_o is a fundamental characteristic of electrode behavior that can be defined as the rate of oxidation or reduction at an equilibrium electrode expressed in terms of current. Exchange current, in fact, is a misnomer because there is no net current flow. It is merely a convenient way of representing the rates of oxidation and

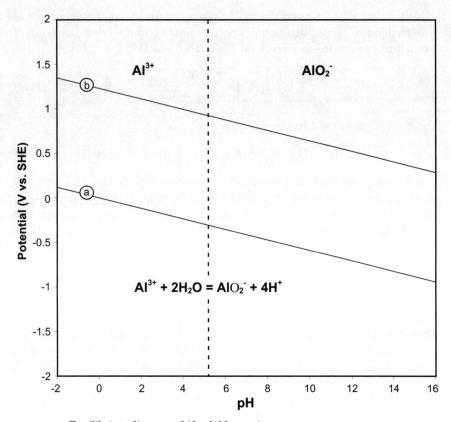

Figure D.7 Equilibrium diagram of Al-soluble species.

reduction of a given electrode at equilibrium, when no loss or gain is experienced by the electrode material. As an example, the exchange current for reducing ferric ions, Eq. (D.65), would be related to the current of each direction of a reversible reaction, that is, a cathodic branch (I_c) representing Eq. (D.65) and an anodic current (I_a) representing Eq. (D.66):

$$Fe^{3+} + 1e^- \rightarrow Fe^{2+} \tag{D.65}$$

$$Fe^{2+} \rightarrow Fe^{3+} + 1e^- \tag{D.66}$$

Because the net current is zero at equilibrium, it implies that the sum of these two currents is zero as in Eq. (D.67). Because I_a is, by convention, always positive, it follows that, when no external voltage or current is applied to the system, the exchange current I_o is equal to I_c or I_a [Eq. (D.68)]:

$$I_a + I_c = 0 \tag{D.67}$$

$$I_a = -I_c = I_o \tag{D.68}$$

There is no theoretical way of accurately determining the exchange current for any given system. This must be determined experimentally. For the characterization of electrochemical processes it is always preferable to normalize the value of the current by the surface area of the electrode and use the current density often expressed as a small i (i.e., $i = I$/surface area).

D.3.2 Kinetics under polarization

Electrodes can be polarized by the application of an external voltage or by the spontaneous production of a voltage away from equilibrium. This deviation from equilibrium potential is called polarization. The

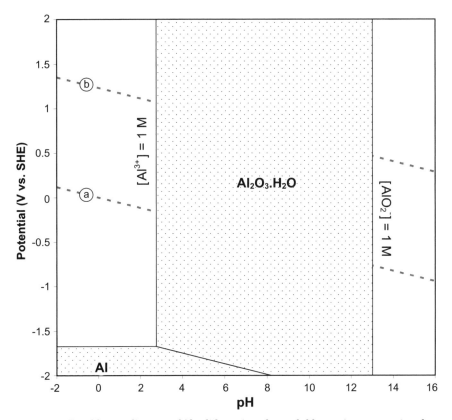

Figure D.8 Equilibrium diagram of Al solid species when soluble species are at a 1-molar concentration.

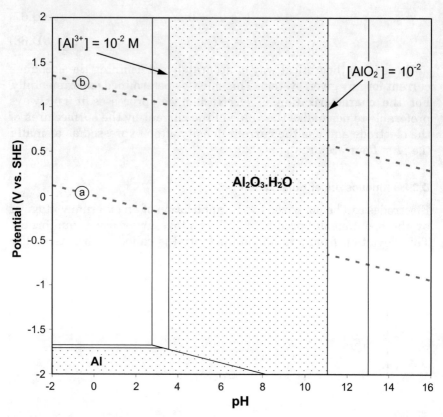

Figure D.9 Equilibrium diagram of Al solid species when soluble species are at a 10^{-2}-molar concentration.

magnitude of polarization is usually described as an overvoltage (η), that is, a measure of polarization with respect to the equilibrium potential (E_{eq}) of an electrode. This polarization is said to be either anodic, when the anodic processes on the electrode are accelerated by changing the specimen potential in the positive (noble) direction, or cathodic, when the cathodic processes are accelerated by moving the potential in the negative (active) direction. There are three distinct types of polarization in any electrochemical cell, the total polarization across an electrochemical cell being the summation of the individual elements as expressed in Eq. (D.69):

$$E_{applied} - E_{eq} = \eta_{total} = \eta_{act} + \eta_{conc} + iR \tag{D.69}$$

where η_{act} = activation overpotential, a complex function describing the charge transfer kinetics of the electrochemical processes. η_{act} is predominant at small polarization currents or voltages.

η_{conc} = concentration overpotential, a function describing the mass transport limitations associated with electrochemical processes. η_{conc} is predominant at large polarization currents or voltages.

iR = is often called the ohmic drop. iR follows Ohm's law and describes the polarization that occurs when a current passes through an electrolyte or through any other interface such as surface film, connectors, and so forth.

Activation polarization. Both the anodic and cathodic sides of a reaction can be studied individually by using some well-established electrochemical methods where the response of a system to an applied polarization, current or voltage, is studied. A general representation of

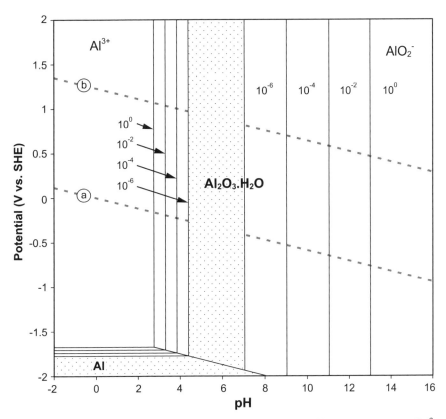

Figure D.10 Equilibrium diagram of Al solid species when soluble species are at a 10^{-6} molar concentration.

the polarization of an electrode supporting one redox system is given in the Butler-Volmer equation:.

$$i = i_0 \left\{ exp\left(\beta \frac{nF}{RT} \eta_{act}\right) - exp\left(-(1-\beta)\frac{nF}{RT}\eta_{act}\right) \right\} \quad (D.70)$$

where: i = anodic or cathodic current.
β = charge transfer barrier or symmetry coefficient for the anodic or cathodic reaction. β values are typically close to 0.5.
η_{act} = $E_{applied} - E_{eq}$ (i.e., positive for anodic polarization and negative for cathodic polarization).
n = number of participating electrons.
R = gas constant.
T = absolute temperature.
F = Faraday.

A polarization plot of the ferric/ferrous oxydo-reduction reaction on palladium (i_o $10^{0.8}$ mA·cm^{-2}), iridium (i_o $10^{0.2}$ mA·cm^{-2}), and rhodium (i_o $10^{-4.8}$ mA·cm^{-2}) is shown in Fig. D.11. The current behavior in Fig. D.11 illustrates the high level of sensitivity of an electrode polarization behavior to even small variations in the exchange current density. The exchange current density reflects the electrocatalytic performance of that electrode toward a specific reaction and can vary over many orders of magnitude. The current density scale in Fig. D.11 had to be changed to much lower values in Fig. D.12 to be able to see the current behavior of the same reaction on rhodium.

The exchange current density for the production of hydrogen on a metallic surface can similarly vary between 10^{-2} A·cm^{-2}, for a good electrocatalytic surface such as platinum, to as low as 10^{-13} A·cm^{-2} for electrode surfaces containing lead or mercury. Added, even in small quantities, to battery electrode materials, mercury will stifle the dangerous production of confined gaseous hydrogen. Mercury and lead were also, for the same hydrogen-inhibiting property, commonly used in many commercial processes as electrode material before their high toxicity was acknowledged a few years ago. It should be noted that the voltage on the polarization plots in Figs. D.11 and D.12 was presented as the overvoltage, with current reversal of its polarity at zero. Figure D.13 shows the data presented in Fig. D.11 with the absolute potential instead of the overvoltage.

The presence of two polarization branches in a single reaction is illustrated in Fig. D.14 for the same Fe^{+3}/Fe^{+2} couple in contact with a palladium electrode. When η_{act} is anodic (i.e., positive), the second term in the Butler-Volmer equation becomes negligible, and

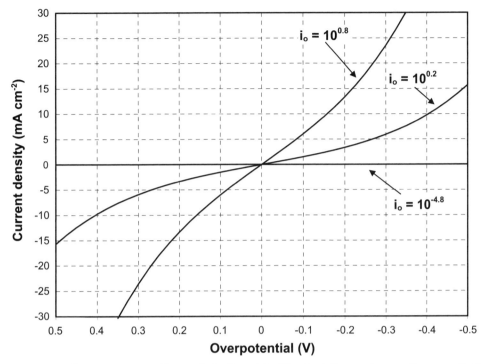

Figure D.11 Current vs. overvoltage polarization plot of the ferric/ferrous ion reaction on palladium ($i_o = 10^{0.8}$ mA·cm^{-2}), iridium ($i_o = 10^{0.2}$ mA·cm^{-2}), and rhodium ($i_o = 10^{-4.8}$ mA·cm^{-2}) on a current scale of 60 mA.

i_a can be more simply expressed by Eq. (D.71) and its logarithm form [Eq. (D.72)]:

$$i_a = i_o \left\{ \exp\left(\beta_a \frac{nF}{RT} \eta_a\right) \right\} \quad \text{(D.71)}$$

$$\eta_a = b_a \log_{10}\left(\frac{i_a}{i_o}\right) \quad \text{(D.72)}$$

where b_a is the Tafel coefficient that can be obtained from the slope [Eq. (D.73)] of a plot of η against $\log i$, with the intercept yielding a value for i_o:

$$b_a = 2.303 \cdot \frac{RT}{\beta nF} \quad \text{(D.73)}$$

Similarly, when η_{reaction} is cathodic (i.e., negative), the first term in the Butler-Volmer equation becomes negligible, and i_c can be more

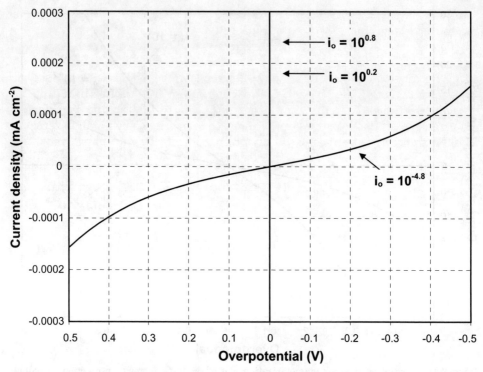

Figure D.12 Current vs. overvoltage polarization plot of the ferric/ferrous ion reaction on palladium ($i_o = 10^{0.8}$ mA·cm^{-2}), iridium ($i_o = 10^{0.2}$ mA·cm^{-2}), and rhodium ($i_o = 10^{-4.8}$·mA cm^{-2}) on a current scale of 0.6 mA.

simply expressed by Eq. (D.74) and its logarithm [Eq. (D.75)], with b_c obtained by plotting η vs. log i [Eq. (D.76)]:

$$i_c = i_o \left\{ -\exp\left(-(1-\beta_c) \frac{nF}{RT} \eta_c \right) \right\} \quad \text{(D.74)}$$

$$\eta_c = b_c \log_{10}\left(\frac{i_c}{i_o}\right) \quad \text{(D.75)}$$

$$b_c = -2.303 \frac{RT}{\beta nF} \quad \text{(D.76)}$$

A Tafel plot for the same data set that was presented in Fig. D.14 is now shown in Fig. D.15 as a log (i)/overpotential plot. It is relatively simple, using such representation, to obtain the exchange current density values and the parameters behind the slopes of the current/voltage behavior, that is, Eq. (D.76).

Concentration polarization. When the cathodic reagent at the corroding surface is in short supply, the mass transport of this reagent could become rate controlling. A frequent case of this type of control occurs when the cathodic processes depend on the reduction of dissolved oxygen.

Because the rate of the cathodic reaction is proportional to the surface concentration of the reagent, the reaction rate will be limited by a drop in the surface concentration. For a sufficiently fast charge transfer (small activation overvoltage), the surface concentration will fall to zero, and the corrosion process will be totally controlled by mass transport. For purely diffusion-controlled mass transport, the flux of a species O to a surface from the bulk is described with Fick's first law [Eq. (D.77)]:

$$J_O = -D_O \left(\frac{\delta C_O}{\delta x} \right) \qquad (D.77)$$

where J_O = flux of species O (mol s^{-1} · cm^{-2})
D_O = diffusion coefficient of species O (cm^2 · s^{-1})
$\delta C_O/\delta x$ = concentration gradient of species O across the interface (mol · cm^{-4})

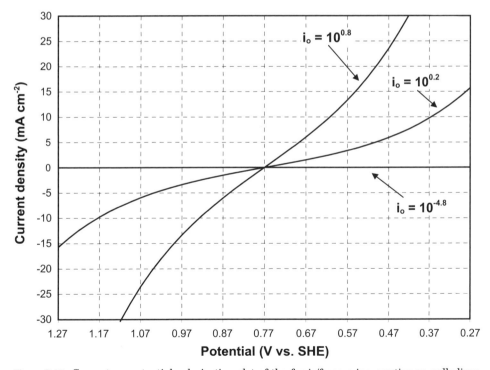

Figure D.13 Current vs. potential polarization plot of the ferric/ferrous ion reaction on palladium ($i_o = 10^{0.8}$ mA·cm^{-2}), iridium ($i_o = 10^{0.2}$ mA·cm^{-2}), and rhodium ($i_o = 10^{-4.8}$ mA·cm^{-2}) current scale of 60 mA.

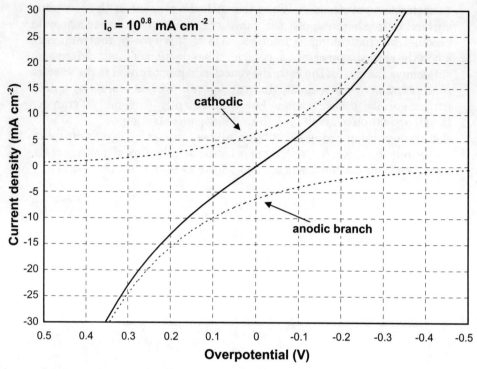

Figure D.14 Current vs. overvoltage polarization plot of the ferric/ferrous ion reaction on palladium showing both the anodic and cathodic branches of the resultant current behavior.

The diffusion coefficient of an ionic species at infinite dilution can be estimated with the help of Nernst-Einstein Eq. (D.78), relating D_O with the conductivity of the species (λ_O):

$$D_O = \frac{RT\,\lambda_O}{|z_O|^2 F^2} \qquad (D.78)$$

where z_O = the valency of species O
R = gas constant (i.e., 8.314 J mol^{-1} · K^{-1})
T = absolute temperature (K)
F = Faraday's constant (i.e., 96487 C · mol^{-1})

Table 1.6 (Aqueous Corrosion) contains values for D_O and λ_O of some common ions. For more practical situations the diffusion coefficient can be approximated with the help of Eq. (D.79), which relates D_O to the viscosity of the solution (μ) and absolute temperature:

$$D_O = \frac{TA}{\mu} \qquad (D.79)$$

where A is a constant for the system.

The region near the metallic surface where the concentration gradient occurs is also called the diffusion layer (δ). Because the concentration gradient $\delta C_O / \delta x$ is greatest when the surface concentration of species O is completely depleted at the surface (i.e., $C_O = 0$), it follows that the cathodic current is limited in that condition, as expressed by Eq. (D.80):

$$i_c = i_L = -nFD_O \frac{C_O^{bulk}}{\delta} \tag{D.80}$$

For intermediate cases, η_{conc} can be evaluated using an expression [Eq. (D.81)] derived from the Nernst equation:

$$\eta_{conc} = \frac{2.303RT}{nF} \log_{10}\left(1 - \frac{i}{i_L}\right) \tag{D.81}$$

where $2.303RT/F = 0.059$ V when $T = 298.16$ K.

When concentration control is added to a process, it simply adds to the polarization as in the following equation:

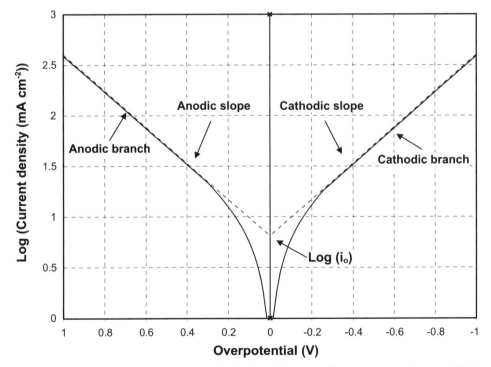

Figure D.15 Log (current) vs. overvoltage Tafel plot of the ferric/ferrous ion reaction on palladium showing how to obtain the exchange current density (intercept) and the slope $b = -2.303$ $(RT/\beta nF)$ of both the anodic and cathodic branches.

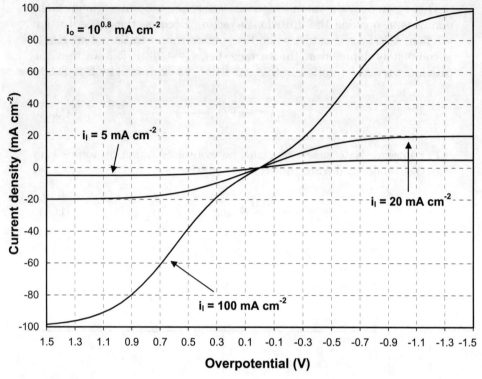

Figure D.16 Current vs. overvoltage epolarization plot of the ferric/ferrous ion reaction on palladium with three levels of concentration overvoltage (100, 20, and 5 mA·cm^{-2}).

$$\eta_{tot} = \eta_{act} + \eta_{conc}$$

We know that, for purely activation controlled systems, the current can be derived from the voltage with the following expression:

$$I = 10^{\,[\,(E - E_{eq})/b \,+\, \log_{10}(I_o)\,]}$$

To simplify the expression of the current in the presence of concentration effects, suppose that

$$A = 10^{\,[\,(E - E_{eq})/b \,+\, \log_{10}(I_o)\,]}$$

$$\eta_{tot} = E - E_{eq} = \eta_{act} + \eta_{conc}$$

and

$$I = \frac{I_l A}{I_l + A}$$

where I_l is the limiting current of the cathodic process.

Figure D.6 illustrates the effect of a limiting current on the polarization of an electrode. For this example three arbitrary limiting current densities were added the activation voltage of the Fe^{3+}/Fe^{2+} reaction on palladium. Figure D.7 presents the same data set on a logarithmic current scale.

Ohmic overpotential. The ohmic drop caused by the electrolytic resistance between two electrodes can be measured by using an alternating current technique (see Sec. D.1.2, Electrolyte Conductance) or minimized by measuring the potential as close as possible to the working electrode. In any case the ohmic overpotential is a simple function described by the product of the effective solution resistance and the cell current, or iR.

References
1. Criss, C. M., and Cobble, J. W., The Thermodynamic Properties of High Temperature Aqueous Solutions, *Journal of the American Chemical Society,* **86:**5385–5393 (1964).
2. Pourbaix, M., *Atlas of Electrochemical Equilibria in Aqueous Solutions,* Houston, Tex., NACE International, 1974.

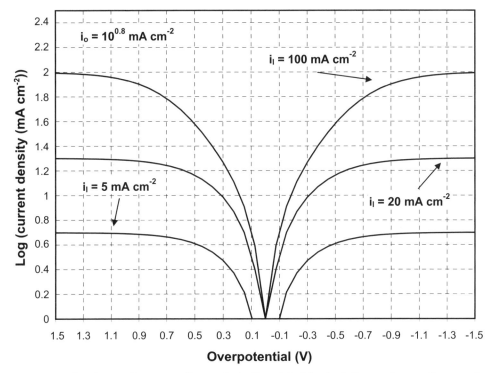

Figure D.17 Log (current) vs. overvoltage polarization plot of the ferric/ferrous ion reaction on palladium with three levels of concentration overvoltage (100, 20, and 5 mA·cm^{-2}).

APPENDIX E

Chemical Compositions of Engineering Alloys

TABLE E.1 Chemical Composition Limits of Wrought Aluminum Alloys

UNS	AA No.	Si	Fe	Cu	Mn	Mg	Cr	Ni	Zn	Ti
A91050	1050	0.25	0.40	0.05	0.05	0.05			0.05	0.03
A91060	1060	0.25	0.35	0.05	0.03	0.03			0.05	0.03
A91100	1100	1.0 Si + Fe		0.05–0.20	0.05				0.10	
A91145	1145	0.55 Si + Fe		0.05	0.05	0.05			0.05	0.03
A91175	1175	0.15 Si + Fe		0.10	0.02	0.02			0.04	0.02
A91200	1200	1.0 Si + Fe		0.05	0.05				0.10	0.05
A91230	1230	0.7 Si + Fe		0.10	0.05	0.05			0.10	0.03
A91235	1235	0.65 Si + Fe		0.05	0.05	0.05			0.10	0.06
A91345	1345	0.30	0.40	0.10	0.05	0.05			0.05	0.03
A91350	1350	0.10	0.40	0.05	0.01		0.01		0.05	
A92011	2011	0.40	0.7	5.0–6.0					0.30	
A92014	2014	0.50–1.2	0.7	3.9–5.0	0.40–1.2	0.20–0.8	0.10		0.25	0.15
A92017	2017	0.20–0.8	0.7	3.5–4.5	0.40–1.0	0.40–0.8	0.10		0.25	0.15
A92018	2018	0.9	1.0	3.5–4.5	0.20	0.45–0.9	0.10	1.7–2.3	0.25	
A92024	2024	0.50	0.50	3.8–4.9	0.30–0.9	1.2–1.8	0.10		0.25	0.15
A92025	2025	0.50–1.2	1.0	3.9–5.0	0.40–1.2	0.05	0.10		0.25	0.15
A92036	2036	0.50	0.50	2.2–3.0	0.10–0.40	0.30–0.6	0.10		0.25	0.15
A92117	2117	0.8	0.7	2.2–3.0	0.20	0.20–0.50	0.10		0.25	
A92124	2124	0.20	0.30	3.8–4.9	0.30–0.9	1.2–1.8	0.10		0.25	0.15
A92218	2218	0.9	1.0	3.5–4.5	0.20	1.2–1.8	0.10	1.7–2.3	0.25	
A92219	2219	0.20	0.30	5.8–6.8	0.20–0.40	0.02			0.10	0.02–0.10
A92319	2319	0.20	0.30	5.8–6.8	0.20–0.40	0.02			0.10	0.10–0.20
A92618	2618	0.10–0.25	0.9–1.3	1.9–2.7		1.3–1.8		0.9–1.2	0.10	0.04–0.10
A93003	3003	0.6	0.7	0.05–0.20	1.0–1.5				0.10	
A93004	3004	0.30	0.7	0.25	1.0–1.5	0.8–1.3			0.25	
A93005	3005	0.6	0.7	0.30	1.0–1.5	0.20–0.6	0.10		0.25	0.10
A93105	3105	0.6	0.7	0.30	0.30–0.8	0.20–0.8	0.20		0.40	0.10
A94032	4032	11.0–13.5	1.0	0.50–1.3		0.8–1.3	0.10	0.50–1.3	0.25	
A94043	4043	4.5–6.0	0.8	0.30	0.05	0.05			0.10	0.20
A94045	4045	9.0–11.0	0.8	0.30	0.05	0.05			0.10	0.20
A94047	4047	11.0–13.0	0.8	0.30	0.15	0.10			0.20	

Designation	No.	Si	Fe	Cu	Mn	Mg	Cr	Zn	Ti	Other each	Other total
A94145	4145	9.3–10.7	0.8	3.3–4.7	0.15		0.15	0.15			0.20
A94343	4343	6.8–8.2	0.8	0.25	0.10		0.10	0.20			0.20
A94643	4643	3.6–4.6	0.8	0.10	0.05	0.10–0.30	0.05	0.10	0.15		0.10	0.15
A95005	5005	0.30	0.7	0.20	0.20	0.50–1.1	0.10	0.25			0.25
A95050	5050	0.40	0.7	0.20	0.10	1.1–1.8	0.10	0.25			0.25
A95052	5052	0.25	0.40	0.10	0.10	2.2–2.8	0.15–0.35	0.10			0.10
A95056	5056	0.30	0.40	0.10	0.05–0.20	4.5–5.6	0.05–0.20	0.10			0.15
A95083	5083	0.40	0.40	0.10	0.40–1.0	4.0–4.9	0.05–0.25	0.25	0.15		0.15
A95086	5086	0.40	0.50	0.10	0.20–0.7	3.5–4.5	0.05–0.25	0.25	0.15		0.15
A95154	5154	0.25	0.40	0.10	0.10	3.1–3.9	0.15–0.35	0.20	0.20		0.20
A95183	5183	0.40	0.40	0.10	0.50–1.0	4.3–5.2	0.05–0.25	0.25	0.15		0.15
A95252	5252	0.08	0.10	0.10	0.10	2.2–2.8		0.05			0.05
A95254	5254	0.45 Si + Fe		0.05	0.01	3.1–3.9	0.15–0.35	0.20	0.05		0.05
A95356	5356	0.25	0.40	0.10	0.05–0.20	4.5–5.5	0.05–0.20	0.10	0.06–0.20		0.06–0.20
A95454	5454	0.25	0.40	0.10	0.50–1.0		0.05–0.20	0.25	0.20		0.20
A95456	5456	0.25	0.40	0.10			0.05–0.20	0.25	0.20		0.20
A95457	5457	0.08	0.10	0.20	0.15–0.45			0.05			0.05
A95554	5554	0.25	0.40	0.10	0.50–1.0		0.05–0.20	0.25	0.05–0.20		0.05–0.20
A95556	5556	0.25	0.40	0.10	0.50–1.0		0.05–0.20	0.25	0.05–0.20		0.05–0.20
A95652	5652	0.40 Si + Fe		0.04	0.01	2.2–2.8	0.15–0.35	0.10			
A95654	5654	0.45 Si + Fe		0.05	0.01	3.1–3.9	0.15–0.35	0.20	0.05–0.15		0.05–0.15
A95657	5657	0.08	0.10	0.10	0.03	0.6–1.0		0.05			
A96003	6003	0.35–1.0	0.6	0.10	0.8	0.8–1.5	0.35	0.20			0.10
A96005	6005	0.6–0.9	0.35	0.10	0.10	0.40–0.6	0.01	0.10			0.10
A96009	6009	0.6–1.0	0.50	0.15–0.6	0.20–0.8	0.40–0.8	0.10	0.25			0.10
A96010	6010	0.8–1.2	0.50	0.15–0.6	0.20–0.8	0.60–1.0	0.10	0.25			0.10
A96053	6053	Mg × 0.5	0.35	0.10		1.1–1.4	0.15–0.35	0.10			0.10
A96061	6061	0.40–0.8	0.7	0.15–0.40	0.15	0.8–1.2	0.04–0.35	0.25	0.15		0.15
A96063	6063	0.20–0.6	0.35	0.10	0.10	0.45–0.9	0.10	0.10	0.10		0.10
A96066	6066	0.9–1.8	0.50	0.7–1.2	0.6–1.1	0.8–1.4	0.40	0.25	0.20		0.20
A96070	6070	1.0–1.7	0.50	0.15–0.40	0.40–1.0	0.50–1.2	0.10	0.25	0.15		0.15
A96101	6101	0.30–0.7	0.50	0.10	0.03	0.35–0.8	0.03	0.10			0.10
A96105	6105	0.6–1.0	0.35	0.10	0.10	0.45–0.8	0.10	0.10	0.10		0.10
A96151	6151	0.6–1.2	1.0	0.35	0.20	0.45–0.8	0.15–0.35	0.25	0.10		0.15

TABLE E.1 Chemical Composition Limits of Wrought Aluminum Alloys (*Continued*)

UNS	AA No.	Si	Fe	Cu	Mn	Mg	Cr	Ni	Zn	Ti
A96162	6162	0.40–0.8	0.50	0.20	0.10	0.7–1.1	0.10		0.25	0.10
A96201	6201	0.50–0.9	0.50	0.10	0.03	0.6–0.9	0.03		0.10	
A96253	6253	Mg × 0.5	0.50	0.10		1.0–1.5	0.04–0.35		1.6–2.4	
A96262	6262	0.40–0.8	0.7	0.15–0.40	0.15	0.8–1.2	0.04–0.14		0.25	0.15
A96351	6351	0.7–1.3	0.50	0.10	0.40–0.8	0.40–0.8			0.20	0.20
A96463	6463	0.20–0.6	0.15	0.20	0.05	0.45–0.9			0.05	
A96951	6951	0.20–0.50	0.8	0.15–0.40	0.10	0.40–0.8			0.20	
A97001	7001	0.35	0.40	1.6–2.6	0.20	2.6–3.4	0.18–0.35		6.8–8.0	0.20
A97005	7005	0.35	0.40	0.10	0.20–0.7	1.0–1.8	0.06–0.20		4.0–5.0	0.01–0.06
A97008	7008	0.10	0.10	0.05	0.05	0.7–1.4	0.12–0.25		4.5–5.5	0.05
A97016	7016	0.10	0.12	0.45–1.0	0.03	0.8–1.4			4.0–5.0	0.03
A97021	7021	0.25	0.40	0.25	0.10	1.2–1.8	0.05		5.0–6.0	0.10
A97029	7029	0.10	0.12	0.50–0.9	0.03	1.3–2.0			4.2–5.2	0.05
A97049	7049	0.25	0.35	1.2–1.9	0.20	2.0–2.9	0.10–0.22		7.2–8.2	0.10
A97050	7050	0.12	0.15	2.0–2.6	0.10	1.9–2.6	0.04		5.7–6.7	0.06
A97072	7072	0.7 Si + Fe		0.10	0.10	0.10			0.8–1.3	
A97075	7075	0.40–0.50		1.2–2.0	0.30	2.1–2.9	0.18–0.28		5.1–6.1	0.20
A97175	7175	0.15 0.20		1.2–2.0	0.10	2.1–2.9	0.18–0.28		5.1–6.1	0.10
A97178	7178	0.40 0.50		1.6–2.4	0.30	2.4–3.1	0.18–0.28		6.3–7.3	0.20
A97475	7475	0.10	0.12	1.2–1.9	0.06	1.9–2.6	0.18–0.25		5.2–6.2	0.06

TABLE E.2 Chemical Composition Limits of Cast Aluminum Alloys

AA No	AA No	Si	Fe	Cu	Mn	Mg	Cr	Ni	Zn	Sn	Ti
A02010	201.0	0.10	0.15	4.0–5.2	0.20–0.50	0.15–0.55					0.15–0.35
A02020	202.0	0.10	0.15	4.0–5.2	0.20–0.8	0.15–0.55	0.20–0.6				0.15–0.35
A02030	203.0	0.30	0.50	4.5–5.5	0.20–0.30	0.10		1.3–1.7	0.10		0.15–0.25
A02040	204.0	0.20	0.35	4.2–5.0	0.10	0.15–0.35		0.05	0.10	0.05	0.15–0.30
A02060	206.0	0.10	0.15	4.2–5.0	0.20–0.50	0.15–0.35		0.05	0.10	0.05	0.15–0.30
A02080	208.0	2.5–3.5	1.2	3.5–4.5	0.50	0.10		0.35	1.0		0.25
A02130	213.0	1.0–30	1.2	6.0–8.0	0.6	0.10		0.35	2.5		0.25
A02220	222.0	2.0	1.5	9.2–10.7	0.50	0.15–0.35		0.50	0.8		0.25
A02240	224.0	0.06	0.10	4.5–5.5	0.20–0.50						0.35
A02380	238.0	3.5–4.5	1.5	9.0–11.0	0.6	0.15–0.35		1.0	1.5		0.25
A02400	240.0	0.50	0.50	7–0–9–0	0.30–0.7	5.5–6.5		0.30–0.7	0.10		0.20
A02420	242.0	1.07	1.0	3.5–4.5	0.35	1.2–1.8	0.25	1.7–2.3	0.35		0.25
A02430	243.0	0.35	0.40	3.5–4.5	0.15–0.45	1.8–2.3	0.20–0.40	1.9–2.3	0.05		0.06–0.20
A02490	249.0	0.05	0.10	3.8–4.6	0.25–0.50	0.25–0.50			2.5–3.5		0.02–0.35
A02950	295.0	0.7–1.5	1.0	4.0–5.0	0.35	0.03			0.35		0.25
A02960	296.0	2.0–3.0	1.2	4.0–5.0	0.35	0.05		0.35	0.50		0.25
A03050	305.0	4.5–5.5	0.6	1.0–1.5	0.50	0.10	0.25		0.35		0.25
A03080	308.0	5.0–6.0	1.0	4.0–5.0	0.50	0.10			1.0		0.25
A03190	319.0	5.5–6.5	1.0	3.0–4.0	0.50	0.10		0.35	1.0		0.25
A03240	324.0	7.0–8.0	1.2	0.40–0.6	0.50	0.40–0.7		0.30	1.0		0.20
A03280	328.0	7.5–8.5	1.0	1.0–2.0	0.20–0.6	0.20–0.6	0.35	0.25	1.5		0.25
A03320	332.0	8.5–10.5	1.2	2.0–4.0	0.50	0.50–1.5		0.50	1.0		0.25
A03330	333.0	8.0–10.0	1.0	3.0–4.0	0.50	0.05–0.50		0.50	1.0		0.25
A03360	336.0	11.0–13.0	1.2	0.50–1.5	0.35	0.7–1.3		2.0–3.0	0.35		0.25

TABLE E.2 Chemical Composition Limits of Cast Aluminum Alloys (*Continued*)

AA No	AA No	Si	Fe	Cu	Mn	Mg	Cr	Ni	Zn	Sn	Ti
A03390	339.0	11.0–13.0	1.2	1.5–3.0	0.50	0.50–1.5		0.50–1.5	1.0		0.25
A03430	343.0	6.7–7.7	1.2	0.50–0.9	0.50	0.10	0.10		1.2–2.0	0.50	
A03540	354.0	8.6–9.4	0.20	1.6–2.0	0.10	0.40–0.6			0.10		0.20
A03550	355.0	4.5–5.5	0.6	1.0–1.5	0.50	0.40–0.6	0.25		0.35		0.25
A03560	356.0	6.5–7.5	0.6	0.25	0.35	0.20–0.45			0.35		0.25
A03570	357.0	6.5–7.5	0.15	0.05	0.03	0.45–0.6			0.05		0.20
A03580	358.0	7.6–8.6	0.30	0.20	0.20	0.40–0.6	0.20		0.20		0.10–0.20
A03590	359.0	8.5–9.5	0.20	0.20	0.10	0.50–0.7			0.10		0.20
A03600	360.0	9.0–10.0	2.0	0.6	0.35	0.40–0.6		0.50	0.50	015	
A03610	361.0	9.5–10.5	1.1	0.50	0.25	0.40–0.6	0.20–0.30	0.20–0.30	0.50	0.10	0.20
A03630	363.0	4.5–6.0	1.1	2.5–3.5	0.25	0.15–0.40	0.20–0.30	0.25	3.0–4.5	0.25	0.20
A03640	364.0	7.5–9.5	1.5	0.20	0.10	0.20–0.40	0.25–0.50	015	0.15	0.15	
A03690	369.0	11.0–12.0	1.3	0.50	0.35	0.25–0.45	0.30–0.40	0.05	1.0	0.10	
A03800	380.0	7.5–9.5	2.0	3.0–4.0	0.50	0.10		0.50	3.0	0.35	
A03830	383.0	9.5–11.5	1.3	2.0–3.0	0.50	0.10		0.30	3.0	0.15	
A03840	384.0	10.5–12.0	1.3	3.0–4.5	0.50	0.10		0.50	3.0	0.35	
A03850	385.0	11.0–13.0	2.0	2.0–4.0	0.50	0.30		0.50	3.0	0.30	
A03900	390.0	16.0–18.0	1.3	4.0–5.0	0.10	0.45–0.65			0.10		0.20
A03920	392.0	18.0–20.0	1.5	0.40–0.8	0.20–0.6	0.8–1.2		0.50	0.50	0.30	0.20
A03930	393.0	21.0–23.0	1.3	0.7–11	0.10	0.7–1.3		2.0–2.5	0.10		0.10–0.20
A04130	413.0	11.0–13.0	2.0	1.0	0.35	0.10		0.50	0.50	0.15	
A04430	443.0	4.5–6.0	0.8	0.6	0.50	0.05	0.25		0.50		0.25
A04440	444.0	6.5–7.5	0.6	0.25	0.35	0.10			0.35		0.25
A05110	511.0	0.30–0.7	0.50	0.15	0.35	3.5–4.5			0.15		0.25

A05120	512.0	1.4–2.2	0.6	0.35	0.8	3.5–4.5			0.35		0.25
A05130	513.0	0.30	0.40	0.10	0.30	3.5–4.5			1.4–2.2		0.20
A05140	514.0	0.35	0.50	0.15	0.35	3.5–4.5			0.15		0.25
A05150	515.0	0.50–1.0	1.3	0.20	0.40–0.6	2.5–4.0			0.10		
A05180	518.0	0.35	1.8	0.25	0.35	7.5–8.5		0.15	0.15	0.15	
A05200	520.0	0.25	0.30	0.25	0.15	9.5–10.6			0.15		0.25
A05350	535.0	0.15	0.15	0.05	0.10–0.25	6.2–7.5					0.10–0.25
A07050	705.0	0.20	0.8	0.20	0.40–0.6	1.4–1.8	0.20–0.40		2.7–3.3		0.25
A07070	707.0	0.20	0.8	0.20	0.40–0.6	1.8–2.4	0.20–0.40		4.0–4.5		0.25
A07100	710.0	0.15	0.50	0.35–0.65	0.05	0.6–0.8			6.0–7.0		0.25
A07110	711.0	0.30	0.7–1.4	0.35–0.65	0.05	0.25–0.45			6.0–7.0		0.20
A07120	712.0	0.30	0.50	0.25	0.10	0.50–0.65	0.40–0.6		5.0–6.5		0.15–0.25
A07130	713.0	0.25	1.1	0.40–1.0	0.6	0.20–0.50	0.35	0.15	7.0–8.0		0.25
A07710	771.0	0.15	0.15	0.10	0.10	0.8–1.0	0.06–0.20		6.5–7.5		0.10–0.20
A07720	772.0	0.15	0.15	0.10	0.10	0.6–0.8	0.06–0.20		6.0–7.0		0.10–0.20
A08500	850.0	0.7	0.7	0.7–1.3	0.10	0.10		0.7–1.3		5.5–7.0	0.20
A08510	851.0	2.0–3.0	0.7	0.7–1.3	0.10	0.10		0.30–0.7		5.5–7.0	0.20
A08520	852.0	0.40	0.7	1.7–2.3	0.10	0.6–0.9		0.9–1.5		5.5–7.0	0.20
A08530	853.0	5.5–6.5	0.7	3.0–40	0.50					5.5–6.0	0.20

TABLE E.3 Wrought-Coppers—Standard Designations for Wrought Coppers (Composition as Maximum % Unless Indicated as Range or Minimum)

Alloy	Cu (+Ag)	Ag	As	Sb	P	Te	Other
C10100	99.99		0.0005	0.0004	0.0003	0.0002	
C10200	99.95						
C10300	99.95				.001–.005		.0010 Oxygen
C10400	99.95	0.027					
C10500	99.95	0.034					
C10700	99.95	0.085					
C10800	99.95				.005–.012		
C10920	99.9						
C10930	99.9	0.044					.02 Oxygen
C10940	99.9	0.085					.02 Oxygen
C11000	99.9						.02 Oxygen
C11010	99.9						
C11020	99.9						
C11030	99.9						
C11040	99.9		0.0005	0.0004		0.0002	
C11100	99.9						
C11300	99.9	0.027					
C11400	99.9	0.034					
C11500	99.9	0.054					
C11600	99.9	0.085					
C11700	99.9				0.04		.004–.02B
C12000	99.9				.004–.012		
C12100	99.9	0.014			.005–.012		
C12200	99.9				.015–.040		
C12210	99.9				.015–.025		
C12220	99.9				.040–.065		
C12300	99.9				.015–.040		
C12500	99.88		0.012	0.003			.025Te + Se, .003Bi, .004Pb, .050Ni
C12510	99.9			0.003	0.03		.025Te + Se, .005Bi, .020Pb, .050Ni, .05Fe, .05Sn, .080Zn

Alloy	Cu	Al	Fe	Pb	O	B	Other
C12900	99.88	0.054	0.012	0.003		0.025	.050Ni, .003Bi, .004Pb
C14180	99.9						.02Pb, .01Al
C14181	99.9						.002Cd, .005C, .002Pb, .002Zn
C14200	99.4		.15–.50		0.075		
C14300	99.9				0.002		.05–.15Cd
C14410	99.9				.015–.040		.05Fe, .05Pb, 0–.20Sn
C14415	99.96				.005–.020		.10–.15Sn
C14420	99.9						.04–.15Sn
C14500	99.9					.005–.05	
C14510	99.85				.004–.012	.40–.7	.05Pb
C14520	99.9				.010–.030	.30–.7	
C14530	99.9				.004–.020	.40–.7	
C14700	99.9				.001–.010	.003–0.023	.003–.023Sn
C15000	99.8				.002–.005		.20–.50S
C15100	99.85						.10–.20Zr
C15500	99.75	.027–.10			.040–.080		.05–.15Zr

Alloy	Cu	Al	Fe	Pb	O	B	
C15715	99.62	.13–.17	0.01	0.01	.12–.19		
C15720	99.52	.18–.22	0.01	0.01	.16–.24		
C15725	99.43	.23–.27	0.01	0.01	.20–.28		
C15760	98.77	.58–.62	0.01	0.01	.52–.59		
C15815	97.82	.13–.17	0.01	0.01	0.19	1.2–1.8	.08–.13Mg

TABLE E.4 Wrought High Coppers Standard Designations for Wrought High Copper Alloys (Composition as Maximum % Unless Indicated as Range or Minimum)

Alloy	Cu (+Ag)	Fe	Sn	Ni	Co	Cr	Si	Be	Pb	Other
C16200	Rem.	0.02								.7–1.2Cd
C16500	Rem.	0.02	.50–.7							.6–1.0Cd
C17000	Rem.						0.2	1.60–1.79		.20Al
C17200	Rem.						0.2	1.80–2.00	0.02	.20Al
C17300	Rem.						0.2	1.80–2.00	.20–.6	.20Al
C17410	Rem.	0.2			.35–.6		0.2	.15–.50		.20Al
C17450	Rem.	0.2	0.25	.50–1.0			0.2	.15–.50		.20Al, .10–.50Zr
C17460	Rem.	0.2	0.25	1.0–1.4			0.2	.15–.50		.20Al, .10–.50Zr
C17500	Rem.	0.1			2.4–2.7		0.2	.40–.7		.20Al
C17510	Rem.	0.1		1.4–2.2	0.3		0.2	.20–.6		.20Al
C17530	Rem.	0.2		1.8–2.5			0.2	.20–.40		.6Al
C18000	Rem.	0.15		1.8–3.0			.40–.8			
C18030	Rem.		.08–.12			.10–.8				.005–.015P
C18040	Rem.		.20–.30			.10–.20				.005–.015P, .05–.15Zn
C18050	Rem.					.25–.35				.005–.015Te
C18070	99					.05–.15				.01–.40Ti
C18090	96.0 min					.15–.40	.02–.07			.15–.8Ti
C18100	98.7 min		.50–1.2	.30–1.2		.20–1.0				.03–.06Mg, .08–.20Zr
C18135	Rem.					.40–1.2				.20–.6Cd
C18140	Rem.					.15–.45	.005–.05			.05–.25Zr
C18150	Rem.					.50–1.5				.05–.25Zr
C18200	Rem.	0.1				.6–1.2	0.1			
C18400	Rem.	0.15				.40–1.2	0.1		0.05	.005As, .005Ca, .05Li, .05P, .7Zn
C18665	99.0 min.									.40–.9Mg, .002–.04P
C18700	99.5 min.								.8–1.5	
C18835	99.0 min	0.1	.15–.55				.15–.40		0.05	.01P, .30Zn
C18980	Rem.		.6–.9				0.5		0.02	.05P, .30Mn, .10Zn
C18990	98		1						0.02	.50Mn, .15P
C18990	Rem.		1.8–2.2							.005–.015P
C19000	Rem.	0.1		.9–1.3		.10–.20			0.05	.8Zn, .15–.35P

Alloy	Cu	Fe	Sn	Zn	Al	Pb	P	Other
C19010	Rem.			.8–1.8				.01–.05P
C19015	Rem.			.50–2.4				.02–.20P, .02–.15Mg
C19020	Rem.		.30–.9	.50–3.0				.01–.20P
C19025	Rem.			.8–1.2				.03–.07P, n + Si
C19030	Rem.	0.1	1.0–1.5	1.5–2.0		0.02		.01–.03P
C19100	Rem.	0.2		.9–1.3		0.1		.50Zn, .35–.6Te, .15–.35P
C19140	Rem.	0.05	0.05	.8–1.2		.40–.8		.50Zn, .15–.35P
C19150	Rem.	0.05	0.05	.8–1.2		.50–1.0		.50Zn, .15–.35P
C19160	Rem.	0.05	0.05	.8–1.2		.8–1.2		.50Zn, .15–.35P

Alloy	Cu	Fe	Sn	Zn	Al	Pb	P	Other
C19200	98.5 min.	.8–1.2					.01–.04	
C19210	Rem.	.05–.15					.025–.040	
C19220	Rem.	.10–.30	.05–.10				.03–.07	.005–.015B, .10–.25Ni
C19260	98.5 min.	.40–.8						.20–.40Ti, .02–.15Mg
C19280	Rem.	.50–1.5	.30–.7	.30–.7			.005–.015	
C19400	97.0 min.	2.1–2.6		.05–.20		0.03	.015–.15	
C19410	Rem.	1.8–2.3	.6–.9	.10–.20			.015–.050	
C19450	Rem.	1.5–3.0	.8–2.5				.005–.05	
C19500	96.0 min.	1.0–2.0	.10–1.0	0.2	0.02	0.02	.01–.35	.30–1.3Co
C19520	96.6 min.	.50–1.5				.01–3.5		
C19700	Rem.	.30–1.2	0.2	0.2		0.05	.10–.40	.01–.20Mg, .05Ni, .05Co, .05Mn
C19710	Rem.	.05–.40	0.2	0.2		0.05	.07–.15	.10 Ni + Co, .05Mn, .03–.06Mg
C19750	Rem.	.35–1.2	.05–.40	0.2		0.05	.10–.40	.01–.20Mg, .05Ni, .05Co, .05Mn
C19900	Rem.							2.9–3.4Ti

TABLE E.5 Wrought-Brasses—Standard Designations for Wrought Brasses (Composition as Maximum % Unless Indicated as Range or Minimum)

Part 1. Copper-zinc alloys (brasses)

Alloy	Cu	Pb	Fe	Zn	Other
C21000	94.0–96.0	0.03	0.05	Rem.	
C22000	89.0–91.0	0.05	0.05	Rem.	
C22600	86.0–89.0	0.05	0.05	Rem.	
C23000	84.0–86.0	0.05	0.05	Rem.	
C23030	83.5–85.5	0.05	0.05	Rem.	.20–.40Si
C23400	81.0–84.0	0.05	0.05	Rem.	
C24000	78.5–81.5	0.05	0.05	Rem.	
C24080	78.0–82.0	0.2		Rem.	.10Al
C26000	68.5–71.5	0.07	0.05	Rem.	
C26130	68.5–71.5	0.05	0.05	Rem.	.02–.08As
C26200	67.0–70.0	0.07	0.05	Rem.	
C26800	64.0–68.5	0.15	0.05	Rem.	
C27000	63.0–68.5	0.1	0.07	Rem.	
C27200	62.0–65.0	0.07	0.07	Rem.	
C27400	61.0–64.0	0.1	0.05	Rem.	
C28000	59.0–63.0	0.3	0.07	Rem.	

Part 2. Copper-zinc-lead alloys (leaded brasses)

Alloy	Cu	Pb	Fe	Zn	Other
C31200	87.5–90.5	.7–1.2	0.1	Rem.	.25Ni
C31400	87.5–90.5	1.3–2.5	0.1	Rem.	.7Ni
C31600	87.5–90.5	1.3–2.5	0.1	Rem.	.7–1.2Ni, .04–.10P, .25Ni
C32000	83.5–86.5	1.5–2.2	0.1	Rem.	
C33000	65.0–68.0	.25–.7	0.07	Rem.	
C33200	65.0–68.0	1.5–2.5	0.07	Rem.	
C33500	62.0–65.0	.25–.7	.15	Rem.	
C34000	62.0–65.0	.8–1.5	.15	Rem.	

Alloy	Cu	Pb	Fe	Sn	Zn			
C34200	62.0–65.0	1.5–2.5	.15					
C34500	62.0–65.0	1.5–2.5	0.15					
C35000	60.0–63.0	.8–2.0	.15					
C35300	60.0–63.0	1.5–2.5	.15					
C35330	59.5–64.0	1.5–3.5			Rem.	.02–.25As		
C35600	60.0–63.0	2.0–3.0	.15		Rem.			
C36000	60.0–63.0	2.5–3.7	0.35		Rem.			
C36500	58.0–61.0	.25–.7	0.15		Rem.	.25Sn		
C37000	59.0–62.0	.8–1.5	0.15		Rem.			
C37100	58.0–62.0	.6–1.2	0.15		Rem.			
C37700	58.0–61.0	1.5–2.5	0.3		Rem.			
C37710	56.5–60.0	1.0–3.0	0.3		Rem.			
C38000	55.0–60.0	1.5–2.5	0.35		Rem.	.50Al, 0.30Sn		
C38500	55.0–59.0	2.5–3.5	0.35		Rem.			

Part 3. Copper-zinc-tin alloys (tin brasses)

Alloy	Cu	Pb	Fe	Sn	Zn	P	Other
C40400				.35–.7	2.0–3.0		
C40500	94.0–96.0	0.05	0.05	.7–1.3	Rem.		
C40500	94.0–96.5	0.05	.08–.12	1.8–2.2	Rem.	.028–.04	.11–.20Ni
C40810	94.5–96.5	0.05	.05–.20	2.6–4.0	Rem.	.02–.04	.05–.20Ni
C40850	94.0–96.0	0.05	.01–.05	1.7–2.3	Rem.	.02–.04	.05–.20Ni
C40860	91.0–93.0	0.05	0.05	2.0–2.8	Rem.		
C41000	89.0–92.0	0.1	0.05	.30–.7	Rem.		
C41100	89.0–93.0	0.1	0.05	.7–1.3	Rem.		
C41300	89.0–93.0	0.1	0.05	1.5–2.2	Rem.		
C41500	88.0–91.0			1.5–2.0	Rem.	0.25	
C42000	86.0–89.0	0.05	0.05	.8–1.4	Rem.	0.35	
C42200	87.0–90.0	0.05	0.05	1.5–3.0	Rem.	0.35	
C42500	88.0–91.0	0.05	.05–.20	1.5–3.0	Rem	.02–.04	.05–.20Ni
C42520	84.0–87.0	0.1	0.05	1.7–2.7	Rem.		
C43000	84.0–87.0	0.05	0.05	.40–1.0	Rem.		

TABLE E.5 Wrought-Brasses—Standard Designations for Wrought Brasses (Composition as Maximum % Unless Indicated as Range or Minimum) (*Continued*)

Part 3. Copper-zinc-tin alloys (tin brasses)

Alloy	Cu	Pb	Fe	Sn	Zn	P	Other
C43500	79.0–83.0	0.1	0.05	.6–1.2	Rem.		
C43600	80.0–83.0	0.05	0.05	.20–.50	Rem.		
C44300	70.0–73.0	0.07	0.06	.8–1.2	Rem.		.02–.06As
C44400	70.0–73.0	0.07	0.06	.8–1.2	Rem.		.02–.10Sb
C44500	70.0–73.0	0.07	0.06	.8–1.2	Rem.	.02–.10	
C46200	62.0–65.0	0.2	0.1	.50–1.0	Rem.		
C46400	59.0–62.0	0.2	0.1	.50–1.0	Rem.		
C46500	59.0–62.0	0.2	0.1	.50–1.0	Rem.		.02–.06As
C47000	57.0–61.0	0.05		.25–1.0	Rem.		.01Al
C47940	63.0–66.0	1.0–2.0	.10–1.0	1.2–2.0	Rem.		.10–.50Ni (+Co)
C48200	59.0–62.0	.40–1.0	0.1	.50–1.0	Rem.		
C48500	59.0–62.0	1.3–2.2	0.1	.50–1.0	Rem.		
C48600	59.0–62.0	1.0–2.5		.30–1.5	Rem.		.02–.25As

TABLE E.6 Wrought-Bronzes—Standard Designations for Wrought Bronzes (Composition as Maximum % Unless Indicated as Range or Minimum)

Part 1. Copper-tin-phosporus alloys (phosphor bronzes)

Alloy	Cu	Pb	Fe	Sn	Zn	P	Elements
C50100	Rem.	0.05	0.05	.50–.8		.01–.05	
C50200	Rem.	0.05	10	1.0–1.5		0.04	
C50500	Rem.	0.05	0.1	1.0–1.7	0.3	.03–.35	
C50510	Rem.			1.0–1.5	.10–.25	.02–.07	.15–.40Ni
C50700	Rem.	0.05	0.1	1.5–2.0		0.3	
C50710	Rem.			1.7–2.3		0.15	.10–.40Ni
C50715	Rem.	0.02	.05–0.15	1.7–2.3		.025–0.04	
C50725	94.0 min.	0.02	.05–.20	1.5–2.5	1.5–3.0	.02–.06	
C50780	Rem.	0.05	.05–.20	1.7–2.3		.02–.10	.05–.20Ni
C50900	Rem.	0.05	0.1	2.5–3.8	0.3	.03–.30	
C51000	Rem.	0.05	0.1	4.2–5.8	0.3	.03–.35	
C51080	Rem.	0.05	.05–.20	4.8–5.8	0.3	.02–.10	.05–.20Ni
C51100	Rem.	0.05	0.1	3.5–4.9	0.3	.03–.35	
C51180	Rem.	0.05	.05–.20	3.5–4.9	0.3	.02–.10	.05–.20Ni
C51800	Rem.	0.02		4.0–6.0		.10–.35	.01Al
C51900	Rem.	0.05	0.1	5.0–7.0	0.3	.03–.35	
C51980	Rem.	0.05	.05–.20	5.5–7.0	0.3	.02–.10	.05–.20Ni
C52100	Rem.	0.05	0.1	7.0–9.0	0.2	.03–.35	
C52180	Rem.	0.05	.05–.20	7.0–9.0	0.3	.02–.10	.05–.20Ni
C52400	Rem.	0.05	0.1	9.0–11.0	0.2	.03–.35	

Part 2. Copper-tin-lead-phosphorus alloys (leaded phosphor bronzes)

Alloy	Cu	Pb	Fe	Sn	Zn	P
C53400	Rem.	.8–1.2	0.1	3.5–5.8	0.3	.03–.35
C54400	Rem.	3.5–4.5	0.1	3.5–4.5	1.5–4.5	.01–.50

TABLE E.6 Wrought-Bronzes—Standard Designations for Wrought Bronzes (Composition as Maximum % Unless Indicated as Range or Minimum) (*Continued*)

Part 3. Copper-phosphorus and copper-silver-phosphorus alloys (brazing alloys)

Alloy	Cu	Ag	P
C55180	Rem.		4.8–5.2
C55181	Rem.		7.0–7.5
C55280	Rem.	1.8–2.2	6.8–7.2
C55281	Rem.	4.8–5.2	5.8–6.2
C55282	Rem.	4.8–5.2	6.5–7.0
C55283	Rem.	5.8–6.2	7.0–7.5
C55284	Rem.	14.5–15.5	4.8–5.2

Part 4. Copper-aluminum alloys (aluminum bronzes)

Alloy	Pb	Fe	Sn	Zn	Al	Mn	Si	Ni (+Co)	Other
C60800	0.1	0.1			5.0–6.5				.02–.35As
C61000	0.02	0.5			6.0–8.5				
C61300	0.01	2.0–3	.20–0.5	0.2	6.0–7.5	0.2	0.1	0.15	.015P
C61400	0.01	1.5–3.5		0.1	6.0–8	1	0.1		.015P
C61500	0.015			0.2	7.7–8.3			1.8–2.2	
C61550	0.05	0.2	0.05	0.8	5.5–6.5			1.5–2.5	.8Zn
C61800	0.02	.50–1.5		0.02	8.5–11		0.1		
C61900	0.02	3.0–4.5	0.6	0.8	8.5–10				
C62200	0.02	3.0–4.2		0.02	11.0–12				
C62300		2.0–4	0.6		8.5–10	0.5	0.1		
C62400		2.0–4.5	0.2		10.0–11.5	0.3	0.25	1	
C62500		3.5–5.5			12.5–13.5	2	0.25		
C62580	0.02	3.0–5		0.02	12.0–13		0.04		
C62581	0.02	3.0–5		0.02	13.0–14		0.04		
C62582	2	3.0–5		0.02	14.0–15		0.04		
C63000		2.0–4	0.2	0.3	9.0–11	1.5	0.25	4.0–5.5	
C63010		2.0–3.5	0.2	0.3	9.7–10.9	1.5		4.5–5.5	

Alloy	Cu(+Ag)	Pb	Fe	Sn	Zn	Mn	Si	Ni(+Co)	Other
C63020	0.03	4.0–5.5	0.25		0.3	10.0–11	1.5		4.2–6	.20Co
C63200	0.02	3.5–4.3				8.7–9.5	1.2–2		4.0–4.8	
C63280	0.02	3.0–5				8.5–9.5	.6–3.5	0.1	4.0–5.5	
C63380	0.02	2.0–4				7.0–8.5	11.0–14		1.5–3	
C63400	0.05	0.15			0.15	2.6–3.2		0.1	0.15	.15As
C63600	0.05	0.15	0.2		0.5	3.0–4		.25–0.45	0.15	.15As
C63800	0.05	0.2	0.2		0.5	2.5–3.1		.7–1.3	0.2	.25–.55Co
C64200	0.05	0.3	0.2		0.5	6.3–7.6	0.1	1.5–2.1	0.25	.15As
C64210	0.05	0.3	0.2		0.5	6.3–7	0.1	1.5–2	0.25	.15As

Part 5. copper-silicon alloys (silicon bronzes)

Alloy	Cu(+Ag)	Pb	Fe	Sn	Zn	Mn	Si	Ni(+Co)	Other
C64700	Rem.	0.1	0.1		0.5			.40–8	1.6–2.2	
C64710	95.0 min.				.20–0.5	0.1		.50–9	2.9–3.5	
C64725	95.0 min.	0.01	0.25	.20–.8	.50–1.5			.20–8	1.3–2.7	.01Ca, .20Mg, .20Cr
C64730	93.5 min.			1.0–1.5	.20–0.5	0.1		.50–9	2.9–3.5	
C64740	95.0 min.	.10max		1.5–2.5	.20			.05–.50	1.0–2	.01Ca, .05Mg
C64900	Rem.	0.05	0.1	1.2–1.6	0.2			.8–1.2	0.1	.10Al
C65100	Rem.	0.05	0.8		1.5			.8–2.0		
C65400	Rem.	0.05		1.2–1.9	0.5	0.7		2.7–3.4		.01–.12Cr
C65500	Rem.	0.05	0.8		1.5	.50–1.3		2.8–3.8	0.6	
C65600	Rem.	0.02	0.5	1.5	1.5	1.5		2.8–4		.01Al
C66100	Rem.	.20–8	0.25		1.5	1.5		2.8–3.5		

TABLE E.6 Wrought-Bronzes—Standard Designations for Wrought Bronzes (Composition as Maximum % Unless Indicated as Range or Minimum) (*Continued*)

Part 6. Other copper-zinc alloys

Alloy	Cu (+Ag)	Pb	Fe	Sn	Zn	Ni (+Co)	Al	Mn	Si	Other
C66300	84.5–87.5	0.05	1.4–2.4	1.5–3	Rem.					.35–.20Co
C66400	Rem.	0.015	1.3–1.7	0.05	11.0–12					.30–.7Co
C66410	Rem.	0.015	1.8–2.3	0.05	11.0–12					
C66420	Rem.		.50–1.5		12.0–17					
C66700	68.5–71.5	0.07	0.1		Rem.			.8–1.5		
C66800	60.0–63	0.5	0.35	0.3	Rem.	0.25	0.25	2.0–3.5	.50	
C66900	62.5–64.5	0.05	0.05		Rem.			11.5–12.5		
C67000	63.0–68	0.2	2.0–4	0.5	Rem.		3.0–6	2.5–5		
C67300	58.0–63	.40–3	0.5	0.3	Rem.	0.25	0.25	2.0–3.5	.50–1.5	
C67400	57.0–60.0	0.5	0.35	0.3	Rem.	0.25	.50–2	2.0–3.5	.50–1.5	
C67420	57.0–58.5	.25–8	.15–55	0.35	Rem.	0.25	1.0–2	1.5–2.5	.25–7	
C67500	57.0–60	0.2	.8–2.0	.50–1.5	Rem.		0.25	.05–0.5		
C67600	57.0–60	.50–1	.40–1.3	.50–1.5	Rem.			.05–.50		
C68000	56.0–60	0.05	.25–1.25	.75–1.1	Rem.	.20–8	0.01	.01–0.5	.04–0.15	
C68100	56.0–60	0.05	.25–1.25	.75–1.1	Rem.		0.01	.01–.50	.04–.15	
C68700	76.0–79	0.07	0.06		Rem.		1.8–2.5			.02–.06As
C68800	Rem.	0.05	0.2		21.3–24.1		3.0–3.8			.25–.55Co
C69050	70.0–75				Rem.	.50–1.5	3.0–4		.10–0.6	.01–.20Zr
C69100	81.0–84	0.05	0.25	0.1	Rem.	.8–1.4	.7–1.2	.10 min.	.8–1.3	
C69400	80.0–83	0.3	0.2		Rem.				3.5–4.5	
C69430	80.0–83	0.3	0.2		Rem.				3.5–4.5	.03–.06As
C69700	75.0–80	.50–1.5	0.2		Rem.			0.4	2.5–3.5	
C69710	75.0–80	.50–1.5	0.2		Rem.			0.4	2.5–3.5	.03–.06As

TABLE E.7 Wrought Copper-Nickel Alloys—Standard Designations for Wrought Copper-Nickel Alloys (Composition as Maximum % Unless Indicated as Range or Minimum)

Alloy	Cu (+Ag)	Pb	Fe	Zn	Ni	Sn	Mn	Other
C70100	Rem.		0.05	0.25	3.0–4.0		0.5	
C70200	Rem.	0.05	0.1		2.0–3.0		0.4	
C70250	Rem.	0.05	0.2	1	2.2–4.2		0.1	.05–.30Mg, .25–1.2Si
C70260	Rem.				1.0–3.0			.20–.7Si, .005P
C70400	Rem.	0.05	1.3–1.7	1	4.8–6.2		.30–.8	
C70500	Rem.	0.05	0.1	0.2	5.8–7.8		0.15	
C70600	Rem.	0.05	1.0–1.8	1	9.0–11.0		1	
C70610	Rem.	0.01	1.0–2.0		10.0–11.0		.50–1.0	.05S, .05C
C70620	86.5 min.	0.02	1.0–1.8	0.5	9.0–11.0		1	.05C, .02P, 02S
C70690	Rem.	0.001	0.005	0.001	9.0–11.0		0.001	
C70700	Rem.		0.05		9.5–10.5		0.5	
C70800	Rem.	0.05	0.1	0.2	10.5–12.5		0.15	
C71000	Rem.	0.05	1	1	19.0–23.0		1	
C71100	Rem.	0.05	0.1	0.2	22.0–24.0		0.15	
C71300	Rem.	0.05	0.2	1	23.5–26.5		1	
C71500	Rem.	0.02	.40–1.0	1	29.0–33.0		1	
C71520	65.0 min.	.40–1.0	0.5	0.5	29.0–33.0		1	.05C, .02P, 02S
C71580	Rem.	0.05	0.5	0.05	29.0–33.0		0.3	
C71581	Rem.	0.02	.40–.7		29.0–32.0		1	
C71590	Rem.	0.001	0.15	0.001	29.0–31.0	0.001	0.5	.03S, .06C
C71640	Rem.	0.01	1.7–2.3		29.0–32.0		1.5–2.5	.30–.7Be
C71700	Rem.		.40–1.0		29.0–33.0			2.2–3.0Cr, .02–.35Zr, .01–.20Ti, .04C, .25Si, .015S, .02P
C71900	Rem.	0.015	0.5	0.05	28.0–33.0		.20–1.0	
C72150	Rem.	0.05	0.1	0.2	43.0–46.0		0.05	.10C, .50Si
C72200	Rem.	0.05	.50–1.0	1	15.0–18.0		1	.30–.7Cr, .03Si, .03Ti
C72420	Rem.	0.02	.7–1.2	0.2	13.5–16.5	0.1	3.5–5.5	1.0–2.0Al, .50Cr, .15Si, .05Mg, .15S, .01P, .05C
C72500	Rem.	0.05	0.6	0.5	8.5–10.5	1.8–2.8	0.2	
C72650	Rem	0.01	0.1	0.1	7.0–8.0	4.5–5.5	0.1	
C72700	Rem.	0.02	0.5	0.5	8.5–9.5	5.5–6.5	.05–.30	.10Nb, .15Mg
C72800	Rem.	0.005	0.5	1	9.5–10.5	7.5–8.5	.05–.30	.10Al, .001B, .001Bi, .10–.30Nb, .005–.15Mg, .005P, .0025S, .02Sb, .05Si, .01Ti
C72900	Rem.	0.02	0.5	0.5	14.5–15.5	7.5–8.5	0.3	.10Nb, .15Mg
C72950	Rem.	0.05	0.6		20.0–22.0	4.5–5.7	0.6	

TABLE E.8 Wrought Nickel-Silvers—Standard Designations for Wrought Nickel-Silver Alloys (Composition as Maximum % Unless Indicated as Range or Minimum)

Alloy	Cu (+Ag)	Pb	Fe	Zn	Ni (+Co)	Mn	Other
C73500	70.5–73.5	0.1	0.25	Rem.	16.5–19.5	0.5	
C74000	69.0–73.5	0.1	0.25	Rem.	9.0–11.0	0.5	
C74300	63.0–66.0	0.1	0.25	Rem.	7.0–9.0	0.5	
C74500	63.5–66.5	0.1	0.25	Rem.	9.0–11.0	0.5	
C75200	63.5–66.5	0.05	0.25	Rem.	16.5–19.5	0.5	
C75400	63.5–66.5	0.1	0.25	Rem.	14.0–16.0	0.5	
C75700	63.5–66.5	0.05	0.25	Rem.	11.0–13.0	0.5	
C76000	60.0–63.0	0.1	0.25	Rem.	7.0–9.0	0.5	
C76200	57.0–61.0	0.1	0.25	Rem.	11.0–13.5	0.5	
C76400	58.5–61.5	0.05	0.25	Rem.	16.5–19.5	0.5	
C76700	55.0–58.0			Rem.	14.0–16.0	0.5	
C77000	53.5–56.5	0.05	0.25	Rem.	16.5–19.5	0.5	
C77300	46.0–50.0	0.05		Rem.	9.0–11.0		.01Al, .25P, .04–.25Si
C77400	43.0–47.0	0.2		Rem.	9.0–11.0		
C78200	63.0–67.0	1.5–2.5	0.35	Rem.	7.0–9.0	0.5	
C79000	63.0–67.0	1.5–2.2	0.35	Rem.	11.0–13.0	0.5	
C79200	59.0–66.5	.8–1.4	0.25	Rem.	11.0–13.0	0.5	
C79800	45.5–48.5	1.5–2.5	0.25	Rem.	9.0–11.0	1.5–2.5	
C79830	45.5–47.0	1.0–2.5	0.45	Rem.	9.0–10.5	.15–.55	

TABLE E.9 Cast Coppers and High Coppers—Standard Designations for Cast Coppers and High Coppers (Composition as Maximum % Unless Indicated as Range or Minimum)

Alloy	Cu (+Ag)	P	Be	Co	Si	Ni	Fe	Al	Sn	Pb	Zn	Cr
C80100	99.95											
C80410	99.9		0.1									
C81100	99.7											
C81200	99.9	.045–.065										
C81400	Rem.		.02–.10									.6–1.0
C81500	Rem.				0.15		0.1	0.1	0.1	0.02	0.1	.40–1.5
C81540	95.1 min.			2.40–2.70	.40–.8	2.0–3.0	0.15	0.1	0.1	0.02	0.1	.10–.6
C82000	Rem.		.45–.8	0.3	0.15	0.2	0.1	0.1	0.1	0.02	0.1	0.1
C82200	Rem.		.35–.8			1.0–2.0						
C82400	Rem.		1.60–1.85	.20–.65		0.2	0.2	0.15	0.1	0.02	0.1	0.1
C82500	Rem.		1.90–2.25	.35–.70	.20–.35	0.2	0.25	0.15	0.1	0.02	0.1	0.1
C82510	Rem.		1.90–2.15	1.0–1.2	.20–.35	0.2	0.25	0.15	0.1	0.02	0.1	0.1
C82600	Rem.		2.25–2.55	.35–.65	.20–.35	0.2	0.25	0.15	0.1	0.02	0.1	0.1
C82700	Rem.		2.35–2.55		0.15	1.0–1.5	0.25	0.15	0.1	0.02	0.1	0.1
C82800	Rem.		2.50–2.85	.35–.70	.20–.35	0.2	0.25	0.15	0.1	0.02	0.1	0.1

TABLE E.10 Cast Brasses

Part 1. Copper-tin-zinc and copper-tin-zinc-lead alloys (red and leaded red brasses)

Alloy	Cu	Sn	Pb	Zn	Fe	Sb	As	Ni	S	P	Al	Si
C83300	92.0–94.0	1.0–2.0	1.0–2.0	2.0–6.0								
C83400	88.0–92.0	0.2	0.5	8.0–12.0	0.25	0.25		1	0.08	0.03	0.005	0.005
C83450	87.0–89.0	2.0–3.5	1.5–3.0	5.5–7.5	0.3	0.25		.8–2.0	0.08	0.03	0.005	0.005
C83500	86.0–88.0	5.5–6.5	3.5–5.5	1.0–2.5	0.25	0.25		.50–1.0	0.08	0.03	0.005	0.005
C83600	84.0–86.0	4.0–6.0	4.0–6.0	4.0–6.0	0.3	0.25		1	0.08	0.05	0.005	0.005
C83800	82.0–83.8	3.3–4.2	5.0–7.0	5.0–8.0	0.3	0.25		1	0.08	0.03	0.005	0.005
C83810	Rem.	2.0–3.5	4.0–6.0	7.5–9.5	0.5			2			0.005	0.1

Part 2. Copper-tin-zinc and copper-tin-zinc-lead alloys (semi-red and leaded semi-red brasses)

Alloy	Cu	Sn	Pb	Zn	Fe	Sb	Ni	S	P	Al	Si	Bi
C84200	78.0–82.0	4.0–6.0	2.0–3.0	10.0–16.0	0.4	0.25	0.8	0.08	0.05	0.005	0.005	
C84400	78.0–82.0	2.3–3.5	6.0–8.0	7.0–10.0	0.4	0.25	1	0.08	0.02	0.005	0.005	
C84410	Rem.	3.0–4.5	7.0–9.0	7.0–11.0			1			0.01	0.2	0.05
C84500	77.0–79.0	2.0–4.0	6.0–7.5	10.0–14.0	0.4	0.25	1	0.08	0.02	0.005	0.005	
C84800	75.0–77.0	2.0–3.0	5.5–7.0	13.0–17.0	0.4	0.25	1	0.08	0.02	0.005	0.005	

Part 3. Copper-zinc and copper-zinc-lead alloys (yellow and leaded yellow brasses)

Alloy	Cu	Sn	Pb	Zn	Fe	Sb	Ni	Mn	As	S	P	Al	Si
C85200	70.0–74.0	.7–2.0	1.5–3.8	20.0–27.0	0.6	0.2	1					0.005	0.05
C85400	65.0–70.0	.50–1.5	1.5–3.8	24.0–32.0	0.7		1					0.35	0.05
C85500	59.0–63.0	0.2	0.2	Rem.	0.2		0.2	0.2			0.02		
C85700	58.0–64.0	.50–1.5	.8–1.5	32.0–40.0	0.7		1			0.05		0.8	0.05
C85800	57.0 min.	1.5	1.5	31.0–41.0	0.5	0.05	0.5	0.25	0.05	0.05	0.01	0.55	0.25

1082

Part 4. Manganese bronze and leaded manganese bronze alloys (high-strength and leaded high-strength yellow brasses)

Alloy	Cu	Sn	Pb	Zn	Fe	Ni	Al	Mn	Si
C86100	66.0–68.0	0.2	0.2	Rem.	2.0–4.0		4.5–5.5	2.5–5.0	
C86200	60.0–66.0	0.2	0.2	22.0–28.0	2.0–4.0	1	3.0–4.9	2.5–5.0	
C86300	60.0–66.0	0.2	0.2	22.0–28.0	2.0–4.0	1	5.0–7.5	2.5–5.0	
C86400	56.0–62.0	.50–1.5	.50–1.5	34.0–42.0	.40–2.0	1	.50–1.5	.10–1.5	
C86500	55.0–60.0	1	0.4	36.0–42.0	.40–2.0	1	.50–1.5	.10–1.5	
C86550	57.0 min.	1	0.5	Rem.	.7–2.0	1	.50–2.5	.10–3.0	0.1
C86700	55.0–60.0	1.5	.50–1.5	30.0–38.0	1.0–3.0	1	1.0–3.0	.10–3.5	
C86800	53.5–57.0	1	0.2	Rem.	1.0–2.5	2.5–4.0	2	2.5–4.0	

Part 5. Copper-silicon alloys (silicon bronzes and silicon brasses)

Alloy	Cu	Sn	Pb	Zn	Fe	Al	Si	Mn	Other
C87300	94.0 min.		0.2	0.25	0.2		3.5–4.5	.8–1.5	
C87400	79.0 min.		1	12.0–16.0		0.8	2.5–4.0		
C87500	79.0 min.		0.5	12.0–16.0		0.5	3.0–5.0		
C87600	88.0 min.		0.5	4.0–7.0	0.2		3.5–5.5	0.25	
C87610	90.0 min.		0.2	3.0–5.0	0.2		3.0–5.0	0.25	
C87800	80.0 min.	0.25	0.15	12.0–16.0	0.15	0.15	3.8–4.2	0.15	.01P

Part 6. Copper-bismuth and copper-bismuth-selenium alloys (high-strength and leaded high-strength yellow brasses)

Alloy	Cu	Sn	Pb	Zn	Fe	Ni	Sb	S	P	Al	Si	Bi	Se	Other
C89320	87.0–91	5.0–7	0.09	1	0.2	1	0.35	0.08	0.3	0.005	0.005	4.0–6		
C89510	86.0–88	4.0–6	0.25	4.0–6	0.3	1	0.25	0.08	0.05	0.005	0.005	.50–1.5	.35–0.7	
C89520	85.0–87	5.0–6	0.25	4.0–6	0.3	1	0.25	0.08	0.05	0.005	0.005	1.9–2.2	.8–1.2	–64
C89550	58.0–64	.50–1.5	0.2	32.0–40	0.7					.30–0.7		.7–2	.07–0.25	
C89844	83.0–86	3.0–5	0.2	7.0–10	0.3	1	0.25	0.08	0.05	0.005	0.005	2.0–4		
C89940	64.0–68.0	3.0–5	0.01	3.0–5	.7–2	20.0–23	0.1	0.05	.10–0.15	0.005	0.15	4.0–5.5		.20Mn

TABLE E.11 Cast Bronzes

Part 1. Copper-tin alloys (tin bronzes)

Alloy	Cu	Sn	Pb	Zn	Fe	Sb	Ni	S	P	Al	Si	Mn
C90200	91.0–94.0	6.0–8.0	0.3	0.5	0.2	0.2	0.5	0.05	0.05	0.005	0.005	
C90300	86.0–89.0	7.5–9.0	0.3	3.0–5.0	0.2	0.2	1	0.05	0.05	0.005	0.005	
C90500	86.0–89.0	9.0–11.0	0.3	1.0–3.0	0.2	0.2	1	0.05	0.05	0.005	0.005	
C90700	88.0–90.0	10.0–12.0	0.5	0.5	0.15	0.2	0.5	0.05	0.3	0.005	0.005	
C90710	Rem.	10.0–12.0	0.25	0.05	0.1	0.2	0.1	0.05	.05–1.2	0.005	0.005	
C90800	85.0–89.0	11.0–13.0	0.25	0.25	0.15	0.2	0.5	0.05	0.3	0.005	0.005	
C90810	Rem.	11.0–13.0	0.25	0.3	0.15	0.2	0.5	0.05	.15–.8	0.005	0.005	
C90900	86.0–89.0	12.0–14.0	0.25	0.25	0.15	0.2	0.5	0.05	0.05	0.005	0.005	
C91000	84.0–86.0	14.0–16.0	0.2	1.5	0.1	0.2	0.8	0.05	0.05	0.005	0.005	
C91100	82.0–85.0	15.0–17.0	0.25	0.25	0.25	0.2	0.5	0.05	1	0.005	0.005	
C91300	79.0–82.0	18.0–20.0	0.25	0.25	0.25	0.2	0.5	0.05	1	0.005	0.005	
C91600	86.0–89.0	9.7–10.8	0.25	0.25	0.2	0.2	12–2.0	0.05	0.3	0.005	0.005	
C91700	84.0–87.0	11.3–12.5	0.25	0.25	0.2	0.2	1.2–2.0	0.05	0.3	0.005	0.005	

Part 2. Copper-tin-lead alloys (leaded tin bronzes)

Alloy	Cu	Sn	Pb	Zn	Fe	Sb	Ni	S	P	Al	Si	Mn
C92200	86.0–90.0	5.5–6.5	1.0–2.0	3.0–5.0	0.25	0.25	1	0.05	0.05	0.005	0.005	
C92210	86.0–89.0	4.5–5.5	1.7–2.5	3.0–4.5	0.25	0.2	.7–1.0	0.05	0.03	0.005	0.005	
C92220	86.0–88.0	5.0–6.0	1.5–2.5	3.0–5.5	0.25		.50–1.0		0.05			
C92300	85.0–89.0	7.5–9.0	.30–1.0	2.5–5.0	0.25	0.25	1	0.05	0.05	0.005	0.005	
C92310	Rem.	7.5–8.5	.30–1.5	3.5–4.5			1			0.005	0.005	
C92400	86.0–89.0	9.0–11.0	1.0–2.5	1.0–3.0	0.25	0.25	1	0.05	0.05	0.005	0.005	0.03
C92410	Rem.	6.0–8.0	2.5–3.5	1.5–3.0	0.2	0.25	0.2			0.005	0.005	
C92500	85.0–88.0	10.0–12.0	1.0–1.5	0.5	0.3	0.25	.8–1.5	0.05	0.3	0.005	0.005	0.05
C92600	86.0–88.5	9.3–10.5	.8–1.5	1.3–2.5	0.2	0.25	0.7	0.05	0.03	0.005	0.005	
C92610	Rem.	9.5–10.5	.30–1.5	1.7–2.8	0.15		1			0.005	0.005	
C92700	86.0–89.0	9.0–11.0	1.0–2.5	0.7	0.2	0.25	1	0.05	0.25	0.005	0.005	0.03
C92710	Rem.	9.0–11.0	4.0–6.0	1	0.2	0.25	2	0.05	0.1	0.005	0.005	

Alloy	Cu	Sn	Pb	Zn	Fe	Sb	Ni	S	P	Al	Si
C92800	78.0–82.0	15.0–17.0	4.0–6.0	0.8	0.2	0.25	0.8	0.05	0.05	0.005	0.005
C92810	78.0–82.0	12.0–14.0	4.0–6.0	0.5	0.5	0.25	0.8–1.2	0.05	0.05	0.005	0.005
C92900	82.0–86.0	9.0–11.0	2.0–3.2	0.25	0.2	0.25	2.8–4.0	0.05	0.5	0.005	0.005

Part 3. Copper-tin-lead alloys (high-leaded tin bronzes)

Alloy	Cu	Sn	Pb	Zn	Fe	Sb	Ni	S	P	Al	Si
C93100	Rem.	6.5–8.5	2.0–5.0	2	0.25	0.25	1	0.05	0.3	0.005	0.005
C93200	81.0–85.0	6.3–7.5	6.0–8.0	1.0–4.0	0.2	0.35	1	0.08	0.15	0.005	0.005
C93400	82.0–85.0	7.0–9.0	7.0–9.0	0.8	0.2	0.5	1	0.08	0.5	0.005	0.005
C93500	83.0–86.0	4.3–6.0	8.0–10.0	2	0.2	0.3	1	0.08	0.05	0.005	0.005
C93600	79.0–83.0	6.0–8.0	11.0–13.0	1	0.2	0.55	1	0.08	0.15	0.005	0.005
C93700	78.0–82.0	9.0–11.0	8.0–11.0	0.8	0.7	0.5	0.5	0.08	0.1	0.005	0.005
C93720	83.0 min.	3.5–4.5	7.0–9.0	4	0.7	0.5	0.5		0.1		
C93800	75.0–79.0	6.3–7.5	13.0–16.0	0.8	0.15	0.8	1	0.08	0.05	0.005	0.005
C93900	76.5–79.5	5.0–7.0	14.0–18.0	1.5	0.4	0.5	0.8	0.08	1.5	0.005	0.005
C94000	69.0–72.0	12.0–14.0	14.0–16.0	0.5	0.25	0.5	.50–1.0	0.08	0.05	0.005	0.005
C94100	72.0–79.0	4.5–6.5	18.0–22.0	1	0.25	0.8	1	0.08	0.05	0.005	0.005
C94300	67.0–72.0	4.5–6.0	23.0–27.0	0.8	0.15	0.8	1	0.08	0.08	0.005	0.005
C94310	Rem.	1.5–3.0	27.0–34.0	0.5	0.5	0.5	.25–1.0		0.05		
C94320	Rem.	4.0–7.0	24.0–32.0		0.35						
C94330	68.5–75.5	3.0–4.0	21.0–25.0	3	0.7	0.5	0.5		0.1		
C94400	Rem.	7.0–9.0	9.0–12.0	0.8	0.15	0.8	1	0.08	0.5	0.005	0.005
C94500	Rem.	6.0–8.0	16.0–22.0	1.2	0.15	0.8	1	0.08	0.05	0.005	0.005

Part 4. Copper-tin-nickel alloys (tin-nickel bronzes)

Alloy	Cu	Sn	Pb	Zn	Fe	Sb	Ni	Mn	S	P	Al	Si
C94700	85.0–90.0	4.5–6.0	0.1	1.0–2.5	0.25	0.15	4.5–6.0	0.2	0.05	0.05	0.005	0.005
C94800	84.0–89.0	4.5–6.0	.30–1.0	1.0–2.5	0.25	0.15	4.5–6.0	0.2	0.05	0.05	0.005	0.005
C94900	79.0–81.0	4.0–6.0	4.0–6.0	4.0–6.0	0.3	0.25	4.0–6.0	0.1	0.08	0.05	0.005	0.005

TABLE E.11 Cast Bronzes (*Continued*)

Part 5. Copper-aluminum-iron and copper-aluminum-iron-nickel alloys (aluminum bronzes)

Alloy	Cu	Pb	Fe	Ni	Al	Mn	Mg	Si	Zn	Sn	Other
C95200	86.0 min.		2.5–4.0		8.5–9.5						
C95210	86.0 min.	0.05	2.5–4.0	1	8.5–9.5	1	0.05	0.25	0.5	0.1	
C95220	Rem.		2.5–4.0	2.5	9.5–10.5	0.5					
C95300	86.0 min.		.8–1.5		9.0–11.0						
C95400	83.0 min.		3.0–5.0	1.5	10.0–11.5	0.5					
C95410	83.0 min.		3.0–5.0	1.5–2.5	10.0–11.5	0.5					
C95420	83.5 min.		3.0–4.3	0.5	10.5–12.0	0.5					
C95500	78.0 min		3.0–5.0	3.0–5.5	10.0–11.5	3.5					
C95510	78.0 min.		2.0–3.5	4.5–5.5	9.7–10.9	1.5			0.3	0.2	
C95520	74.5 min.	0.03	4.0–5.5	4.2–6.0	10.5–11.5	1.5		0.15	0.3	0.25	.20Co–.05Cr
C95600	88.0 min.			0.25	6.0–8.0			1.8–3.2			
C95700	71.0 min.		2.0–4.0	1.5–3.0	7.0–8.5	11.0–14.0		0.1			
C95710	71.0 min.	0.05	2.0–4.0	1.5–3.0	7.0–8.5	11.0–14.0		0.15	0.5	1	.05P
C95720	73.0 min.	0.03	1.5–3.5	3.0–6.0	6.0–8.0	12.0–15.0		0.1	0.1	0.1	.20Cr
C95800	79.0 min.	0.03	3.5–4.5(31)	4.0–5.0(31)	8.5–9.5	.8–1.5		0.1			
C95810	79.0 min.	0.1	3.5–4.5	4.0–5.0	8.5–9.5	.8–1.5	0.05	0.1	0.5		
C95820	77.5 min.		4.0–5.0	4.5–5.8	9.0–10.0	1.5		0.1	0.2	0.2	
C95900	Rem.	0.02	3.0–5.0	0.5	12.0–13.5	1.5					

TABLE E.12 Cast Copper-Nickel-Iron Alloys (Copper-Nickels)

Alloy	Cu	Pb	Fe	Ni	Mn	Si	Nb	C	Be	Other
C96200	Rem.	0.01	1.0–1.8	9.0–11.0	1.5	0.5	1	0.1		.02S, .02P
C96300	Rem.	0.01	.50–1.5	18.0–22.0	.25–1.5	0.5	.50–1.5	0.15		.02S, .02P
C96400	Rem.	0.01	.25–1.5	28.0–32.0	1.5	0.5	.50–1.5	0.15		.02S, .02P
C96600	Rem.	0.01	.8–1.1	29.0–33.0	1	0.15			.40–.7	
C96700	Rem.	0.01	.40–1.0	29.0–33.0	.40–1.0	0.15			1.1–1.2	.15–.35Zr, .15–.35Ti
C96800	Rem.	0.005	0.5	9.5–10.5	.05–.30	0.05	.10–.30			
C96900	Rem.	0.02	0.5	14.5–15.5	.05–.30		0.1			.15Mg, 7.5–8.5 Sn, .50 Zn
C96950	Rem.	0.02	0.5	11.0–15.5	.05–.40	0.3	0.1			5.8–8.5Sn

TABLE E.13 Chemical Compositions of Nickel-, Nickel-Iron-, and Cobalt-Base Alloys

Alloy	UNS	C	Cr	Ni	Co	Fe	Mo	W	Other
263	N07041	0.06	20	Rem.	20		5.8		Al 0.5, Ti 2.2
20Cb-3	N08020	0.02	20	33		Rem.	2.2		Cu 3.3, Cb 0.5
20Mo-4	N08024	0.02	23.5	37		Rem.	3.8		Cu 1.0, Cb 0.25
20Mo-6	N08026	0.02	24	36		Rem.	5.6		Cu 3.0
625 Plus	N07716	0.02	20	Rem.		5	9		Cb 3.1, Al 0.2, Ti 1.3
Alloy 150(UMCo-50)		0.06	27		Rem.	18			
Alloy 188	R30188	0.1	22	22	Rem.	3		14	La 0.04
Alloy 214		0.04	16	Rem.		3			Al 4.5, Y
Alloy 230	N06230	0.1	22	Rem.		3	2	14	La 0.02, B 0.015
Alloy 242		0.03	8	Rem.			25		
Alloy 25 (L-605)	R30605	0.1	20	10	Rem.	3		15	
Alloy 556	R30556	0.1	22	20	18	Rem.	3	2.5	Ta 0.6, La 0.02, N 0.2, Zr 0.02
Alloy 6B	R30016	1.2	30		Rem.	1.5	4.5		Cb 0.7, N 0.2
Alloy HR-120		0.05	25	37		Rem.			Si 2.75
Alloy HR-160		0.05	28	Rem.	29	1.5			Al 3.5, Ta 6.5, Zr 0.15, Y 0.10
AR 213		0.17	19		Rem.			4.5	Al 4.0, Ti 3.5, B 0.03
Astroloy		0.06	15	Rem.	17		5.3		Si 1.5
Chromel D			18.5	36		Rem.			Cu 68.0, Mn 1.1
Cupro 107				Rem.		0.8			Al 1.0, Ti 3.0
D-979	N09979	0.05	15	Rem.		27	4	4	Al 0.25, Ti 1.7
Discalloy	K66220	0.06	14	26		Rem.	3		Al 4.8, Y 0.3
Fecralloy A		0.03	15.8			Rem.			Cu 55.0
Ferry alloy				Rem.					V 0.03
Hastelloy B	N10001	0.05		Rem.	2.5	5	28		
Hastelloy B-2	N10665	0.01		Rem.		2	28		
Hastelloy C	N10002	0.08	15.5	Rem.	2.5	6	17	4	
Hastelloy C-22	N06022	0.01	22	Rem.	2.5	3	13	3	
Hastelloy C-276	N10276	0.01	15.5	Rem.	2.5	5.5	16	4	
Hastelloy C-4	N06455	0.01	16	Rem.	2	3	15.5		
Hastelloy G	N06007	0.05	22	Rem.		19.5	6.5		Cb + Ta 2.0, Cu 2.0
Hastelloy G-3	N06985	0.015	22	Rem.	5	19.5	7	1.5	Cb + Ta 0.3, Cu 2.0

```
Alloy              UNS No.   C      Cr    Ni    Fe    Mo    Co    W     Other
Hastelloy G-30               0.03   29.5  Rem.   5    5           2.5         Cu 2.0
Hastelloy H-9M               0.03   22    Rem.   5    5           9     2
Hastelloy N        N 10003   0.06    7    Rem.  15.5              16.5
Hastelloy S        N06635    0.02   15.5  Rem.   3                14.5        La 0.05, B 0.015
Hastelloy W        N10004    0.12    5    Rem.   6    24
Hastelloy X        N06002    0.1    22    Rem.   2.5  18.5  1.5   9     0.6
IN 100             N13100    0.15   10    Rem.  15    18.5        3           Al 5.5, Ti 4.7, Zr 0.06, V 1.0, B 0.015
IN 100 Gatorize              0.07   12.4  Rem.              3.2               Al 5.0, Ti 4.3, Zr 0.06, B0.02, V 0.8
IN 102             N06102    0.06   15    Rem.        7     3     3           Cb 3.0, Al 0.4, Ti 0.6, Mg0.02, Zr 0.03
IN 587                       0.05   28.5  Rem.  20                3           Cb 0.7, Al 1.2, Ti 2.3, Zr 0.05
IN 597                       0.05   24.5  Rem.  20                            Cb 1.0, Al 1.5, Ti 3.0, Zr 0.05
Incoloy 800        N08800    0.05   21    32.5                                Al 0.3, Ti 0.3
Incoloy 800H       N08810    0.08   21    32.5                                Al 0.4, Ti 0.4
Incoloy 800HT      N08811    0.08   21    32.5                                Al + Ti 1.0
Incoloy 802        N08802    0.4    21    32.5
Incoloy 825        N08825    0.03   21.5  Rem.        30    3     1.5         Cu 2.2
Incoloy 903        N19903                 38    15                            Ti 1.4, Al 0.9, Cb 3.0
Incoloy 904        N19904                 32.5  14.5  Rem.                    Ti 1.6
Incoloy 907        N19907                 38    13                            Ti 1.5, Cb 4.7, Si 0.15
Incoloy 909        N19909                 38    13                            Ti 1.5, Cb 4.7, Si 0.4
Incoloy 925        N09925    0.01   21    Rem.        28                      Cu 1.8, Ti 2.1, Al 0.3
Incoloy DS                   0.06   17    35    Rem.                    3     Si 2.3
Inconel 600        N06600    0.08   15.5  Rem.        8
Inconel 601        N06601    0.1    23    Rem.       14.4                     Al 1.4
Inconel 617        N06617    0.07   22    Rem.  12.5  1.5         9           Al 1.2
Inconel 625        N06625    0.1    21.5  Rem.        2.5         9           Cb 3.6
Inconel 671                  0.05   48    Rem.
Inconel 690        N06690    0.02   29    Rem.        9                       Ti 0.35
Inconel 706        N09706    0.03   16    Rem.       37                       Ti 1.8, Al 0.2, Cb 2.9
Inconel 718        N07718    0.04   18    Rem.       18.5   3                 Cb 5.1
Inconel 751        N07751    0.05   15    Rem.        7                       Ti 2.5, Al 1.1, Cb 1.0
Inconel X-750      N07750    0.04   15.5  Rem.        7                       Ti 2.5, Al 0.7, Cb 1.0
```

TABLE E.13 Chemical Compositions of Nickel-, Nickel-Iron-, and Cobalt-Base Alloys (*Continued*)

Alloy	UNS	C	Cr	Ni	Co	Fe	Mo	W	Other
Kanthal AF			22			Rem.			Al 5.3, Y
Kanthal Al			22			Rem.			Al 5.8
M 252	K92500	0.15	20	Rem.	10		10		Al 1.0, Ti 2.6, B 0.005
MAR-M 918	N07252	0.05	20	20	Rem.				Ta 7.5, Zr 0.10
Monel 400	N04400			Rem.		1.2			Cu 31.5, Mn 1.1
Monel 401	N04401			Rem.		0.3			Cu 55.5, Mn 1.63
Monel 450	C71500			Rem.		0.7			Cu 68.0, Mn 0.7
Monel K-500	N05500			Rem.		1			Cu 29.5, Ti 0.6, Al 2.7
Monel R-405	N04405			Rem.		1.2			Cu 31.5, Mn 1.1
MP 159			19	25.5	Rem.	9	7		Cb 0.6, Al 0.2, Ti 3.0
MP 35N	R30035	0.1	20	35	Rem.		10		
Multimet (N-155)	R30155	0.5	21	20	20	Rem.	3	2.5	Cb + Ta 1.0, N 0.15
NA 224			27	Rem.		18.5		6	
Ni 200	N02200	0.08		99.6					
Ni 201	N02201	0.02		99.6					
Ni 270	N02270	0.01		99.98					
Nichrome 80			20	Rem.					Si 1.0
Nimonic 105		0.08	15	Rem.	20		5		Al 4.7, Ti 1.3, B 0.005
Nimonic 115		0.15	15	Rem.	15.0		4		Al 5.0, Ti 4.0
Nimonic 70			20	Rem.					Al 1.0, Ti 1.25, Cb 1.5
Nimonic 75		0.1	19.5	Rem		25			
Nimonic 80A	N07080	0.06	19.5	Rem.					Al 1.4, Ti 2.4
Nimonic 81		0.03	30	Rem.					Al 0.9, Ti 1.8
Nimonic 86	N07090		25	Rem.			10		Ce 0.03
Nimonic 90		0.07	19.5	Rem.	16.5				Al 1.5, Ti 2.5
Nimonic 901			12.5	Rem.		36	5.8		Ti 2.9
Nimonic 91			28.5	Rem.	20				Al 1.2, Ti 2.3
Nimonic AP 1			15	Rem.	17		5		Al 4.0, Ti 3.5
Nimonic PE 11		0.05	18	Rem.		34	5.2		Al 0.8, Ti 2.3
Nimonic PE 16		0.05	16.5	Rem.		34	3.3		Al 1.2, Ti 1.2
Nimonic PK 31			20	Rem.	14		4.5		Al 0.4, Ti 2.35, Cb 5.0

Alloy	UNS No.	C	Cr	Ni	Co	Fe	Mo	W	Other
Nimonic PK 33		0.04	18	Rem.	14		7		Al 2.1, Ti 2.4
Nimonic PK 37	N07001		19.5	Rem.	16.5				Al 1.5, Ti 2.5
Nimonic PK 50			19.5	Rem.	13.5		4.25		Al 1.4, Ti 3.0
Pyromet 31	N07031	0.04	22.5	Rem.	15		2		Al 1.4, Ti 2.3, Cu 0.9, B 0.005
Pyromet 860		0.05	13	Rem.	4	28.9	6		Al 1.0, Ti 3.0, B 0.01
Pyromet CTX-I		0.03		37.7	16	Rem.			Cb 3.0, Al 1.0, Ti 1.7
RA 330	N08330	0.05	19	35		Rem.			Si 1.2
RA 330HC		0.4	19	35		Rem.			Si 1.2
RA 333	N06333	0.05	25	Rem.	3	18	3	3	Al 0.2, Ti 2.6, B 0.015
Refractory 26		0.03	18	Rem.	20	16	3.2	3	Al 5.5, Ti 4.2, Zr 0.06, B 0.015
René 100		0.16	9.5	Rem.	15		3		Al 1.5, Ti 3.0, B 0.006
René 41		0.09	19	Rem.	11	5	10		Cb 3.5, Al 3.5, Ti 2.5, Zr 0.05
René 95		0.15	14	Rem.	8		3.5	3.5	Cb 4.0
S-816	R30816	0.38	20	20	Rem.	4	4	4	Cu 1.0
Sanicro 28	N08028	0.01	27	31		Rem.	3.5		Cb 0.5, Al 1.5, Ti 2.5, Zr 0.06, B 0.008
Udimet 400		0.06	17.5	Rem.	14		4		Al 2.9, Ti 2.9, Zr 0.05, B 0.006
Udimet 500		0.08	18	Rem.	18.5		4		Al 2.0, Ti 3.0, B 0.005
Udimet 520		0.05	19	Rem.	12		6	1	Cb 6.5, Al 0.5, Ti 1.0
Udimet 630		0.03	18	Rem.		18	3	3	Al 5.3, Ti 3.5, B 0.03
Udimet 700		0.03	15	Rem.	18.5		5.2		Al 2.5, Ti 5.0
Udimet 710		0.07	18	Rem.	15		3	1.5	Al 2.5, Ti 5.0, Zr 0.03, B
Udimet 720		0.03	17.9	Rem.	14.7		3	1.3	N 0.08
Ultimet		0.06	26	9	Rem.	3	5	2	Ta 1.5, Al 4.6, Ti 3.5, Zr 0.10
Unitemp AF2-1DA		0.35	12	Rem.	10		3	6	Ta 1.5, Al 4.0, Ti 2.8, Zr 0.1, B 0.015
Unitemp AF2-1DA6		0.04	12	Rem.	10		2.7	6.5	Al 0.25, Ti 3.0, V 0.5, B 0.01
V-57		0.08	14.8	27		Rem.	1.25		Al 0.2, Ti 2.85, B 0.05
W-545	K66545	0.08	13.5	26		Rem.	1.5		Al 1.5, Ti 3.0, Zr 0.05, B 0.006
Waspaloy		0.08	19	Rem.	14		4.3		

TABLE E.14 Refractory Metals—Typical Analysis of Refractory Metals

Element	Max % Mo	Max % Ta	Max % Nb	Max % W
Aluminum	0.001		0.005	0.002
Calcium	0.003			0.003
Chromium	0.005			0.002
Copper	0.001			0.002
Iron	0.005	0.010	0.01	0.003
Lead	0.002			0.002
Magnesium	0.001			0.002
Molybdenum	99.95 min.	0.010	0.01	
Manganese	0.001			0.002
Nickel	0.001	0.005	0.005	0.003
Silicon	0.003	0.005	0.005	0.002
Tin	0.003			0.002
Titanium	0.002	0.005		0.002
Tantalum		99.90 min.	0.2	
Tungsten		0.030	0.05	99.95 min.
Carbon	0.005	0.0075	0.01	0.005
Oxygen		0.020	0.025	
Nitrogen		0.0075	0.01	
Hydrogen		0.0001	0.0015	
Niobium		0.050	99.9	

TABLE E.15 Austenitic Stainless Steels—Standard Designations for Austenitic Stainless Steels (Composition as Maximum in % Unless Indicated as Range or Minimum)

UNS	Type	C	Mn	P	S	Si	Cr	Ni	Mo	Other
S20100	201	0.15	5.5–7.5	0.060	0.030	1.00	16.00–18.00	3.50–5.50		0.25N
S20200	202	0.15	7.5–10.0	0.060	0.030	1.00	17.00–19.00	4.00–6.00		0.25N
S20500	205	0.2	14.0–15.5	0.030	0.030	0.50	16.50–18.00	1.00–1.75		0.32–0.40N
S30100	301	0.15	2.00	0.045	0.030	1.00	16.00–18.00	6.00–8.00		
S30200	302	0.15	2.00	0.045	0.030	1.00	17.00–19.00	8.00–10.00		
S30215	302B	0.15	2.00	0.045	0.030	2.00–3.00	17.00–19.00	8.00–10.00		
S30300	303	0.15	2.00	0.20	0.15 min.	1.00	17.00–19.00	8.00–10.00	0.60	
S30323	303Se	0.15	2.00	0.20	0.060	1.00	17.00–19.00	8.00–10.00		0.15Se min.
S30400	304	0.08	2.00	0.045	0.030	1.00	18.00–20.00	8.00–10.50		
S30403	304L	0.03	2.00	0.045	0.030	1.00	18.00–20.00	8.00–12.00		
S30430	18-9-LW	0.08	2.00	0.045	0.030	1.00	17.00–19.00	8.00–10.00		3.00–4.00Cu
S30451	304N	0.08	2.00	0.045	0.030	1.00	18.00–20.00	8.00–10.50		0.10–0.16N
S30500	305	0.12	2.00	0.045	0.030	1.00	17.00–19.00	10.50–13.00		
S30800	308	0.08	2.00	0.045	0.030	1.00	19.00–21.00	10.00–12.00		
S30900	309	0.20	2.00	0.045	0.030	1.00	22.00–24.00	12.00–15.00		
S30908	309S	0.08	2.00	0.045	0.030	1.00	22.00–24.00	12.00–15.00		
S31000	310	0.25	2.00	0.045	0.030	1.50	24.00–26.00	19.00–22.00		
S31008	310S	0.08	2.00	0.045	0.030	1.50	24.00–26.00	19.00–22.00		
S31400	314	0.25	2.00	0.045	0.030	.50–3.00	23.00–26.00	19.00–22.00		
S31600	316	0.08	2.00	0.045	0.030	1.00	16.00–18.00	10.00–14.00		
S31620	316F	0.08	2.00	0.20	0.10 min.	1.00	16.00–18.00	10.00–14.00	1.75–2.5	
S31603	316L	0.03	2.00	0.045	0.030	1.00	16.00–18.00	10.00–14.00	2.0–3.0	
S31651	316N	0.08	2.00	0.045	0.030	1.00	16.00–18.00	10.00–14.00	2.0–3.0	0.10–0.16N
S31700	317	0.08	2.00	0.045	0.030	1.00	18.00–20.00	11.00–15.00	3.0–4.0	

TABLE E.15 Austenitic Stainless Steels—Standard Designations for Austenitic Stainless Steels (Composition as Maximum in % Unless Indicated as Range or Minimum) *(Continued)*

UNS	Type	C	Mn	P	S	Si	Cr	Ni	Mo	Other
S31703	317L	0.030	2.00	0.045	0.030	1.00	18.00–20.00	11.00–15.00	3.0–4.0	
	317LMN	0.030	2.00	0.045	0.030	0.75	17.00–20.00	13.50–17.50	4.0–5.0	0.10–0.20N
S32100	321	0.08	2.00	0.045	0.030	1.00	17.00–19.00	9.00–12.00		5 × C Ti min.
N08830	330	0.08	2.00	0.040	0.030	0.75–1.50	17.00–20.00	34.00–37.00		0.10 Ta 0.20Cb
S34700	347	0.08	2.00	0.045	0.030	1.00	17.00–19.00	9.00–13.00		Cb + Ta 10 × C (min.) Ta 0.10 max. Co 0.20 max.
S34800	348	0.08	2.00	0.045	0.030	1.00	17.00–19.00	9.00–13.00		
S38400	384	0.08	2.00	0.045	0.030	1.00	15.00–17.00	17.00–19.00		
S31254	254 MO	0.01	2.00	0.045	0.030	1.00	20	18	6.1	0.20N, Cu
S32654	654 MO	0.01	3.50	0.045	0.030	1.00	24	22	7.3	0.50N, Cu
S30815	253 MA	0.09	2.00	0.045	0.030	1.7	21	11		Ce
N08904	904L	0.01	2.00	0.045	0.030	1.00	20	25	4.5	0.06N, 1.5Cu
N08020	20Cb-3	0.07	2.00	0.045	0.030	1.00	20	34	2.2	Cb
N08367	AL 6XN	0.03	2.00	0.045	0.030	1.00	20	25	6	0.18–25N

TABLE E.16 Ferritic Stainless Steels—Nominal Chemical Composition (%) of Ferritic Stainless Steels (Maximum Unless Noted Otherwise)

UNS	Type	C	Mn	P	S	Si	Cr	Ni	Mo	Other
S40500	405	0.08	1.00	0.040	0.030	1.00	11.50–14.50	0.60		0.10–0.30 Al
S40900	409	0.08	1.00	0.045	0.045	1.00	10.50–11.75	0.50		6 × C–0.75 Ti
S42900	429	0.12	1.00	0.040	0.030	1.00	14.00–16.00	0.75		
S43000	430	0.12	1.00	0.040	0.030	1.00	16.00–18.00	0.75		
S43020	430F	0.12	1.25	0.060	0.15 min.	1.00	16.00–18.00		0.60	
S43023	430FSe	0.12	1.25	0.060	0.060	1.00	16.00–18.00			0.15 Se min.
S43400	434	0.12	1.00	0.040	0.030	1.00	16.00–18.00		0.75–1.25	
S43600	436	0.12	1.00	0.040	0.030	1.00	16.00–18.00		0.75–1.25	5 × C–0.70 Cb + Ta
S44200	442	0.20	1.00	0.040	0.030	1.00	18.00–23.00	0.60		
S44600	446	0.20	1.50	0.040	0.030	1.00	23.00–27.00	0.75		0.25N
S44635	Monit	0.25	1.00	0.040	0.030	0.75	24.50–26.00	4.00	3.5–4.5	0.035N, 4 × (C + N) Ti + Nb

TABLE E.17 Martensitic Stainless Steels—Nominal Chemical Composition (%) of Martensitic Stainless Steels (Maximum Unless Noted Otherwise)

UNS	Type	C	Mn	P	S	Si	Cr	Ni	Mo	Other
S40300	403	0.15	1.00	0.040	0.030	0.50	11.50–13.00			
S41000	410	0.15	1.00	0.040	0.030	1.00	11.50–13.50			
S41400	414	0.15	1.00	0.040	0.030	1.00	11.50–13.50	1.25–2.50		
S41600	416	0.15	1.25	0.060	0.15 min.	1.00	12.00–14.00		0.60	
S41623	416Se	0.15	1.25	0.060	0.060	1.00	12.00–14.00			0.15Se min.
S42000	420	0.15 min.	1.00	0.040	0.030	1.00	12.00–14.00			
S42020	420F	0.15 min.	1.25	0.060	0.15 min.	1.00	12.00–14.00		0.60	
S42200	422	0.20–0.25	1.00	0.025	0.025	0.75	11.00–13.00	0.50–1.00	0.75–1.25	0.15–0.30V, 0.75–1.25W
S43100	431	0.20	1.00	0.040	0.030	1.00	15.00–17.00	1.25–2.50		
S44002	440A	0.60–0.75	1.00	0.040	0.030	1.00	16.00–18.00	0.75		
S44004	440B	0.75–0.95	1.00	0.040	0.030	1.00	16.30–18.00	0.75		
S44004	440C	0.95–1.20	1.00	0.040	0.030	1.00	16.30–18.00	0.75		

TABLE E.18 Nominal Compositions of First- and Second-Generation Duplex Stainless Steels

UNS	Type	Cr	Mo	Ni	Cu	C	N	Other
First generation									
S32900	Type 329	26	1.5	4.5		0.08		
J93370	CD-4MCu	25	2	5	3	0 04		
Second generation									
S32304	SAF 2304	23		4		0.030	0.05–0.20	
S31500	3RE60	18.5	2.7	4.9		0.030	0.05–0.1	1.7 Si
S31803	2205	22	3	5		0.030	0.08–0.20	
S31200	44LN	25	1.7	6		0.030	0.14–0.20	
S32950	7-Mo PLUS	26.5	1.5	4.8		0.03	0.15–0.35	
S32550	Ferralium 255	25	3	6	2	0.04	0.1–0.25	
S31260	DP-3	25	3	7	0.5	0.030	0.10–0.30	0.3 W
S32750	SAF 2507	25	4	7		0.030	0.24–0.32	

TABLE E.19 Compositions of Precipitation-Hardening (PH) Stainless Steels

UNS	Alloy	C	Mn	Si	Cr	Ni	Mo	P	S	Other
					Martensitic					
S13800	PH13-8Mo	0.05	0.10	0.10	12.25–13.25	7.5–8.5	2.0–2.5	0.01	0.008	0.90–1.35 Al, 0.01 N
S15500	15-5PH	0.07	1.00	1.00	14.0–15.5	3.5–5.5		0.04	0.03	2.5–4.5 Cu, 0.15–0.45 Nb
S17400	17-4PH	0.07	1.00	1.00	15.0–17.5	3.0–5.0		0.04	0.03	3.0–5.0 Cu, 0.15–0.45 Nb
S45000	Custom 450	0.05	1.00	1.00	14.0-16.0	5.0–7.0	0.5–1.0	0.03	0.03	1.25–1.75 Cu, 8 × %C min Nb
S45000	Custom 455	0.05	0.50	0.50	11.0–12.5	7.5–9.5	0.50	0.04	0.03	1.5–2.5 Cu, 0.8–1.4 Ti, 0.1–0.5 Nb
					Semiaustenitic					
S15700	PH15-7 Mo	0.09	1.00	1.00	14.5–16.0	6.5–7.75	2.0–3.0	0.04	0.04	0.75–1.50 Al
S17700	17-7PH	0.09	1.00	1.00	16.0–18.0	6.50–7.75		0.04	0.04	0.75–150 Al
S35000	AM-350	0.07–0.11	0.50–1.25	0.50	16.0–17.0	4.5–5.0	2.50–3.25	0.04	0.03	0.07–0.13 N
S35500	AM-355	0.10–0.15	0.50–1.25	0.50	15.0–16.0	4.0–5.0	2.50–3.25	0.04	0.03	0.07–0.13 N
					Austenitic					
S66286	A-286	0.08	2.00	1.00	13.5–16.0	24.0–27.0	1.0–1.5	0.025	0.025	1.90–2.35 Ti, 0.35 max. Al, 0.10–0.50 V, 0.003–0.010 B
	JBK-73	0.015	0.05	0.02	14.5	29.5	1.25	0.006	0.002	2.15 Ti, 0.25 Al, 0.27 V, 0.0015 B

TABLE E.20 Nominal Chemical Composition (%) of Cast Heat-Resistant Stainless Steels

UNS	Type	C	Cr	Ni	Fe
	HA	0.2	8–10		Rem.
	HB	0.3	18–22	2	Rem.
J92605	HC	0.5	26–30	4	Rem.
J93005	HD	0.5	26–30	4–7	Rem.
J93403	HE	0.2–0.5	26–30	8–11	Rem.
J92603	HF	0.2–0.4	19–23	9–12	Rem.
J93503	HH	0.2–0.5	24–28	11–14	Rem.
J94003	HI	0.2–0.5	26–30	14–18	Rem.
J94224	HK	0.2–0.6	24–28	18–22	Rem.
J94604	HL	0.2–0.6	28–32	18–22	Rem.
J94213	HN	0.2–0.5	19–23	23–27	Rem.
J94605	HT	0.35–0.75	13–17	33–37	Rem.
J95405	HU	0.35–0.75	17–21	37–41	Rem.
N08001	HW	0.35–0.75	10–14	58–62	Rem.
N06006	HX	0.35–0.75	15–19	64–68	Rem.
J95705	HP	0.4	25	35	Rem.

TABLE E.21 Titanium—Nominal Chemical Composition of Commercial Titanium Alloys

UNS	ASTM	N	C	H	Fe	O	Al	V	Ni	Mo	Nb	Cr	Zr	Pd
R50250	1	0.03	0.08	0.015	0.2	0.18								
R50400	2	0.03	0.08	0.015	0.3	0.25								
R50550	3	0.05	0.08	0.015	0.3	0.35								
R56400	5	0.05	0.08	0.015	0.4	0.2	5.5–6.75	3.5–4.5						
R52400	7	0.03	0.08	0.015	0.3	0.25								.12–.25
R56320	9	0.03	0.08	0.015	0.25	0.15	2.5–3.5	2.0–3.0						
R52250	11	0.03	0.08	0.015	0.2	0.18								0.1
R53400	12	0.03	0.08	0.015	0.3	0.25			.6–.9	.2–.4				
R52402	16	0.03	0.08	0.015	0.3	0.25								.04–.08
R52252	17	0.03	0.08	0.015	0.2	0.18								.04–.08
R56322	18	0.03	0.08	0.015	0.25	0.15	2.5–3.5	2.0–3.0						.04–.08
R58640	19	0.03	0.05	0.02	0.3	0.12	3.0–4.0	7.5–8.5		3.5–4.5		5.5–6.5	3.5–4.5	
R58645	20	0.03	0.05	0.02	0.3	0.12	3.0–4.0	7.5–8.5		3.5–4.5		5.5–6.5	3.5–4.5	.04–.08
R58210	21	0.03	0.05	0.015	0.4	0.17	2.5–3.5			15–16.0	2.2–3.2			
	23	0.03	0.08	0.015	0.4	0.13	5.5–6.5	3.5–4.5						

APPENDIX

Thermodynamic Data and *E*-pH Diagrams

The tables and graphics in this appendix describe the thermodynamic behavior of the following metals when exposed to pure water at 25 and 60°C:

- Chromium[1,2]
- Copper[3,4]
- Iron[5–8]
- Manganese[9,10]
- Nickel[11–13]
- Zinc[9,14]

Tables F.1 to F.6 contain the basic thermodynamic values for each species, solid or ionic, considered for the construction of the *E*-pH diagrams. The graphics were obtained with a publicly available software system that has been used throughout the book to calculate different equilibrium systems.[15] The basic calculations were detailed in Sec. D.2, Chemical Thermodynamics. The relations between the free energy of the species considered and the associated equations are evaluated with the data presented in Tables F.1 to F.6 and the following equations. The free energy (G^0) of a substance for which heat capacity data are available can be calculated as a function of temperature using Eq. (F.1).

$$G_{(T_2)}^0 = G_{(T_1)}^0 - S_{(T_1)}^0 (T_2 - T_1) - T_2 \int_{T_1}^{T_2} \frac{C_p^0}{T} dT + \int_{T_1}^{T_2} C_p^0 \, dT \quad \text{(F.1)}$$

Appendix F

TABLE F.1 Species Considered for the Cr-H$_2$O System and Their Thermodynamic Data

Species	$G^0_{(298\ K)}$, J·mol^{-1}	$S^0_{(298\ K)}$, J·mol^{-1}	A	$B \times 10^3$	$C \times 10^{-5}$
O$_2$	0	205	29.96	4.184	−1.674
H$_2$	0	131	27.28	3.263	0.502
H$_2$O	−237,000	69.9	75.27	0	0
Cr	0	23.77	17.41	15.15	1.26
CrO	−350,661	44.77	46.48	8.12	−3.68
Cr$_2$O$_3$	−1,058,134	81.17	119.37	9.2	−15.65
CrO$_2$	−539,740	48.12	67.49	12.55	−12.55
CrO$_3$	−502,080	73.22	75.86	16.78	−8.37
Cr(OH)$_3$	−900,815	80.33	0	0	0
CrOOH	−672,955	25.1	0	0	0

		$\check{S}^0_{(298\ K)}$, J·mol^{-1}	a	b
H$^+$	0	0	−20.9	0.065	−0.005
Cr^{2+}	−176,146	−104.6	−146.44	0.13−	0.00166
Cr^{3+}	−215,476	−307.52	−370.28	0.13	−0.00166
Cr(OH)$^{2+}$	−430,950	−68.62	−110.46	0.13	−0.00166
Cr(OH)$_2^+$	−632,663	−144.77	−165.69	0.13	−0.00166
CrO$_4^{2-}$	−727,849	50.21	92.05	−0.37	0.0055
HCrO$_4^-$	−764,835	184.1	205.02	−0.37	0.0055
CrO$_2^-$	−535,929	96.23	117.15	−0.37	0.0055
CrO$_3^{3-}$	−603,416	−238.49	−175.73	−0.37	0.0055

TABLE F.2 Pure Species Considered for the Cu-H$_2$O System and Their Thermodynamic Data

Species	$G^0_{(298\ K)}$, J·mol^{-1}	$S^0_{(298\ K)}$, J·mol^{-1}	A	$B \times 10^3$	$C \times 10^{-5}$
O$_2$	0	205	29.96	4.184	−1.674
H$_2$	0	131	27.28	3.263	0.502
H$_2$O	−237,000	69.9	75.27	0	0
Cu	0	33.2	22.635	6.276	0
Cu$_2$O	−147,904	92.4	62.62	0	0
CuO	−127,905	42.6	42.32	0	0
Cu(OH)$_2$	−358,987	87	87.91	0	0

		$\check{S}^0_{(298\ K)}$, J·mol^{-1}	a	b
H$^+$	0	0	−20.9	0.065	−0.005
Cu$^+$	50,626	−12.6	−33.52	0.13	−0.00166
Cu^{2+}	65,689	−207.2	−249.04	0.13	−0.00166
Cu(OH)$^+$	−129,704	41.89	20.97	0.13	−0.00166
Cu$_2$(OH)$_2^{2+}$	−280,328	−98.22	−140.06	0.13	−0.00166
Cu^{3+}	303,340	−401.8	−464.56	0.13	−0.00166
HCuO$_2^-$	−258,571	96.38	117.3	−0.37	0.0055
CuO$_2^{2-}$	−183,678	−98.22	−56.38	−0.37	0.0055
CuO$_2^-$	−112,550	96.38	117.3	−0.37	0.0055

Thermodynamic Data and E-pH Diagrams 1103

TABLE F.3 Pure Species Considered for the Fe-H$_2$O System and Their Thermodynamic Data

Species	$G^0_{(298\ K)}$, J·mol^{-1}	$S^0_{(298\ K)}$, J·mol^{-1}	A	$B \times 10^3$	$C \times 10^{-5}$
O$_2$	0	205	29.96	4.184	−1.674
H$_2$	0	131	27.28	3.263	0.502
H$_2$O	−237,000	69.9	75.27	0	0
Fe	0	27.1	12.72	31.71	−2.51
Fe$_3$O$_4$	−1,020,000	146	91.55	201.67	0
Fe$_2$O$_3$	−742,000	87.3	98.28	77.82	−14.85
Fe(OH)$_2$	−493,000	92.4	96.3	0	0
Fe(OH)$_3$	−714,000	96.1	105	0	0
			$\check{S}^0_{(298\ K)}$, J·mol^{-1}	a	b
H$^+$	0	0	−20.9	0.065	−0.005
Fe(OH)$_{2(sln)}$	−449,000	38	38	0.13	−0.00166
Fe(OH)$_{3(sln)}$	−661,000	75.2	75.2	0.13	−0.00166
FeOH$^+$	−274,000	−29.3	−50.2	0.13	−0.00166
Fe(OH)$_2^+$	−459,000	−29.3	−50.2	0.13	−0.00166
Fe^{2+}	−92,200	−107	−149	0.13	−0.00166
FeOH^{2+}	−242,000	−105	−147	0.13	−0.00166
Fe^{3+}	−17,800	−279	−342	0.13	−0.00166
Fe(OH)$_3^-$	−621,000	41.8	62.7	−0.37	0.0055
Fe(OH)$_4^-$	−843,000	25.1	46	−0.37	0.0055
FeO$_4^{2-}$	−467,000	37.6	79.5	−0.37	0.0055

TABLE F.4 Pure Species Considered for the Mn-H$_2$O System and Their Thermodynamic Data

Species	$G^0_{(298\ K)}$, J·mol^{-1}	$S^0_{(298\ K)}$, J·mol^{-1}	A	$B \times 10^3$	$C \times 10^{-5}$
O$_2$	0	205	29.96	4.184	−1.674
H$_2$	0	131	27.28	3.263	0.502
H$_2$O	−237,000	69.9	75.27	0	0
Mn	0	32.0076	23.8488	14.14192	−1.54808
MnO	−362,920	59.70568	46.48424	8.11696	−3.68192
Mn$_3$O$_4$	−1,283,233	155.6448	144.9338	45.27088	−9.2048
Mn$_2$O$_3$	−881,150	110.4576	103.4703	35.06192	−13.5143
MnO$_2$	−465,177	53.05312	69.4544	10.20896	−16.2339
			$\check{S}^0_{(298\ K)}$, J·mol^{-1}	a	b
H$^+$	0	0	−20.9	0.065	−0.005
Mn^{2+}	−228,028	−115.478	−157.34	0.13	−0.00166
Mn(OH)$^+$	−405,011	−37.656	−58.576	0.13	−0.00166
Mn^{3+}	−82,006.4	−378.652	−441.41	0.13	−0.00166
HMnO$_2^-$	−507,101	62.76	83.68	−0.37	0.0055
MnO$_4^-$	−447,270	212.1288	233.05	−0.37	0.0055
MnO$_4^{2-}$	−500,825	100.416	142.256	−0.37	0.0055

TABLE F.5 Pure Species Considered for the Ni-H$_2$O System and Their Thermodynamic Data

Species	$G^0_{(298\ K)}$, J·mol^{-1}	$S^0_{(298\ K)}$, J·mol^{-1}	A	B × 10^3	C × 10^{-5}
O$_2$	0	205	29.96	4.184	−1.674
H$_2$	0	131	27.28	3.263	0.502
H$_2$O	−237,000	69.9	75.27	0	0
Ni	0	30.12	16.99	294.55	0
Ni(OH)$_2$	−453,130	79.5	0	0	0
NiO	−215,940	38.58	−20.88	157.23	16.28
Ni$_3$O$_4$	−711,910	146.44	129.03	71.46	−23.93
Ni$_2$O$_3$	−469,740	94.14	98.28	77.82	−14.85
NiO$_2$	−215,140	52.3	69.45	10.21	−16.23

			$\check{S}^0_{(298\ K)}$, J·mol^{-1}	a	b
H$^+$	0	0	−20.9	0.07	−0.01
Ni^{2+}	−46,442	−201.3	−243.14	0.13	0
HNiO$_2^-$	−349,218	62.76	41.84	−0.37	0.01

TABLE F.6 Pure Species Considered for the Ni-H$_2$O System and Their Thermodynamic Data

Species	$G^0_{(298\ K)}$, J·mol^{-1}	$S^0_{(298\ K)}$, J·mol^{-1}	A	B × 10^3	C × 10^{-5}
O$_2$	0	205	29.96	4.184	−1.674
H$_2$	0	131	27.28	3.263	0.502
H$_2$O	−237,000	69.9	75.27	0	0
Zn	0	41.63	25.4	0	0
Zn(OH)$_2$	−559,358	81.6	72.4	0	0

			$\check{S}^0_{(298\ K)}$, J·mol^{-1}	a	b
H$^+$	0	0	−20.9	0.065	−0.005
Zn^{2+}	−147,280	−207.2	−249.04	0.13	−0.00166
Zn(OH)$^+$	−329,438	41.89	20.97	0.13	−0.00166
HZnO$_2^-$	−464,227	96.38	117.3	−0.37	0.0055
ZnO$_2^{2-}$	−389,424	−98.22	−56.38	−0.37	0.0055

For pure substances, i.e., solids, liquids, and gases, the heat capacity C_p^0 is expressed as an empirical function of the absolute temperature [Eq. (F.2)].

$$C_p^0 = A + BT + CT^{-2} \tag{F.2}$$

For ionic substances, one has to use another method, such as that proposed by Criss and Cobble in 1964,[16] to obtain the heat capacity, provided that the temperature does not rise above 200°C. The expression of the ionic capacity [Eq. (F.3)] makes use of absolute entropy values and the parameters a and b contained in Tables F.1 to F.6.

$$C_p^0 = (4.186a + b\check{s}^0{}_{(298\,K)})\,(T_2 - 298.16)\,/\,\ln\left(\frac{T_2}{298.16}\right) \tag{F.3}$$

By combining Eq. (F.2) or (F.3) with Eq. (F.1), one can obtain the free energy [Eq. (F.4)] at a given temperature by using the fundamental data contained in Tables F.1 to F.6.

$$G_t^0 = G^0{}_{(298\,K)} + (C_p^0 - S^0{}_{(298\,K)})\,(T_2 - 298.16)$$

$$- T_2 \ln\left(\frac{T_2}{298.16}\right) C_p^0 \tag{F.4}$$

Table F.7 provides an index for the thermodynamic data of the species considered, the equations possible, and associated E-pH diagrams at two temperatures, 25 and 60°C.

References

1. Silverman, D.C., Absence of Cr(IV) in the EMF-PH Diagram for Chromium, *Corrosion*, **39**:488–491 (1983).
2. Lee, J. B., Elevated Temperature Potential-pH Diagrams for the Cr-H$_2$O, Mo-H$_2$O, and Pt-H$_2$O Systems, *Corrosion*, **37**:467 (1981).
3. Bianchi, G., and Longhi, P., Copper in Sea-Water, Potential-pH Diagrams, *Corrosion Science*, **13**:853–864 (1973).
4. Duby, P., *The Thermodynamic Properties of Aqueous Inorganic Copper Systems*, INCRA Monograph IV, New York, The International Copper Research Association, 1977.
5. Le, H. H., and Ghali, E., Interpretation des diagrammes E-pH du système Fe-H$_2$O en relation avec la fragilisation caustique des aciers, *Journal of Applied Electrochemistry*, **23**:72–77 (1993).
6. Silverman, D. C., Presence of Solid Fe(OH)$_2$ in EMF-pH Diagram for Iron, *Corrosion*, **38**:453–455 (1982).
7. Townsend, H. E., Potential-pH Diagrams at Elevated Temperature for the System Fe-H$_2$O, *Corrosion Science*, **10**:343–358 (1970).
8. Biernat, R. J., and Robins, R. G., High-Temperature Potential/pH Diagrams for the Iron-Water and Iron-Water-Sulphur Systems, *Electrochimica Acta*, **17**:1261–1283 (1972).
9. Pourbaix, M., *Atlas of Electrochemical Equilibria in Aqueous Solutions*, Houston, Tex., NACE International, 1974.

TABLE F.7 Index to Thermodynamic Data, Equilibrium, and Associated E-pH Diagrams for Important Engineering Metals

Element	Equations	Temperature, °C	Figure
Chromium	(Data Table F.1)		
Hydrated state	Table F.8	25	F.1
		60	F.2
Dry state	Table F.9	25	F.3
		60	F.4
Copper	(Data Table F.2)		
Hydrated state	Table F.10	25	F.5
		60	F.6
Dry state	Table F.11	25	F.7
		60	F.8
Iron	(Data Table F.3)		
Hydrated state	Table F.12	25	F.9
		60	F.10
Dry state	Table F.13	25	F.11
		60	F.12
Manganese	(Data Table F.4)		
	Table F.14	25	F.13
		60	F.14
Nickel	(Data Table F.5)		
Hydrated state	Table F.15	25	F.15
		60	F.16
Dry state	Table F.16	25	F.17
		60	F.18
Zinc	(Data Table F.6)		
	Table F.17	25	F.19
		60	F.20

10. Macdonald, D. D., The Thermodynamics and Theoretical Corrosion Behavior of Manganese in Aqueous Systems at Elevated Temperatures, *Corrosion Science*, **16:**482 (1976).
11. Macdonald, D. D., *The Thermodynamics of Metal-Water Systems at Elevated Temperatures,* Part 4, *The Nickel-Water System,* AECL-4139, Pinawa, Canada, Whiteshell Nuclear Research Establishment, 1972.
12. Chen, C. M., and Theus, G. J., *Chemistry of Corrosion-Producing Salts in Light Water Reactors,* NP-2298, Palo Alto, Calif., Electric Power Research Institute, 1982.
13. Cowan, R. L., and Staehle, R. W., The Thermodynamics and Electrode Kinetic Behavior of Nickel in Acid Solution in the Temperature Range 25° to 300°C, *Journal of the Electrochemical Society,* **118:**557–568 (1971).
14. Pan, P., and Tremaine, P. R., Thermodynamics of Aqueous Zinc: Standard Partial Molar Heat Capacities and Volumes of Zn^{2+} (aq) from 10 to 55°C, *Geochimica et Cosmochimica Acta,* **58:**4867–4874 (1994).
15. Roberge, P. R., *KTS-Thermo* (2.01), Kingston, Canada, Kingston Technical Software, 1998.
16. Criss, C. M., and Cobble, J. W., The Thermodynamic Properties of High Temperature Aqueous Solutions, *Journal of the American Chemical Society,* **86:**5385–5393 (1964).

TABLE F.8 Possible Reaction in the Cr-H_2O System between the Species Most Stable in Wet Conditions

	Equilibria
1.	$2e^- + 1CrO + 2H^+ = 1Cr + 1H_2O$
2.	$1e^- + 1Cr(OH)_3 + 1H^+ = 1CrO + 2H_2O$
3.	$3e^- + 1Cr(OH)_3 + 3H^+ = 1Cr + 3H_2O$
4.	$1e^- + 1CrO_2 + 1H_2O + 1H^+ = 1Cr(OH)_3$
5.	$1e^- + 1CrO_2 + 1H^+ = 1Cr(OH)_3$
6.	$3e^- + 1CrO_3 + 3H^+ = 1CrO_2 + 1H_2O$
7.	$2e^- + 1Cr^{2+} = 1Cr$
8.	$3e^- + 1CrO_2^- + 4H^+ = 1Cr + 2H_2O$
9.	$6e^- + 1HCrO_4^- + 7H^+ = 1Cr + 4H_2O$
10.	$6e^- + 1CrO_4^{2-} + 8H^+ = 1Cr + 4H_2O$
11.	$1e^- + 1CrO_2^- + 2H^+ = 1CrO + 1H_2O$
12.	$1CrO_2^- + 1H_2O + 1H^+ = 1Cr(OH)_3$
13.	$1CrO_3 + 1H_2O = 1CrO_4^{2-} + 2H^+$
14.	$1CrO_3 + 1H_2O = 1HCrO_4^- + 1H^+$
15.	$1CrO + 2H^+ = 1Cr^{2+} + 1H_2O$
16.	$3e^- + 1Cr^{3+} = 1Cr$
17.	$3e^- + 1CrO_3^{3-} + 6H^+ = 1Cr + 3H_2O$
18.	$1e^- + 1CrO_3^{3-} + 4H^+ = 1CrO + 2H_2O$
19.	$1e^- + 1CrO_2 + 4H^+ = 1Cr^{3+} + 2H_2O$
20.	$2e^- + 1HCrO_4^- + 3H^+ = 1CrO_2 + 2H_2O$
21.	$2e^- + 1CrO_4^{2-} + 4H^+ = 1CrO_2 + 2H_2O$
22.	$1Cr(OH)_3 + 3H^+ = 1Cr^{3+} + 3H_2O$
23.	$1Cr(OH)_3 = 1CrO_3^{3-} + 3H^+$
24.	$1e^- + 1Cr(OH)_3 + 3H^+ = 1Cr^{2+} + 3H_2O$
25.	$3e^- + 1HCrO_4^- + 4H^+ = 1Cr(OH)_3 + 1H_2O$
26.	$3e^- + 1CrO_4^{2-} + 5H^+ = 1Cr(OH)_3 + 1H_2O$
27.	$3e^- + 1CrO_4^{2-} + 4H^+ = 1CrO_2^- + 2H_2O$
28.	$3e^- + 1CrO_4^{2-} + 2H^+ = 1CrO_3^{3-} + 1H_2O$
29.	$1CrO_2^- + 4H^+ = 1Cr^{3+} + 2H_2O$
30.	$1CrO_3^{3-} + 2H^+ = 1CrO_2^- + 1H_2O$
31.	$1CrO_4^{2-} + 1H^+ = 1HCrO_4^-$
32.	$1e^- + 1Cr^{3+} = 1Cr^{2+}$
33.	$1e^- + 1CrO_2^- + 4H^+ = 1Cr^{2+} + 2H_2O$
34.	$3e^- + 1HCrO_4^- + 7H^+ = 1Cr^{3+} + 4H_2O$
35.	$3e^- + 1CrO_4^{2-} + 8H^+ = 1Cr^{3+} + 4H_2O$
36.	$3e^- + 1HCrO_4^- + 3H^+ = 1CrO_2^- + 2H_2O$
37.	$1Cr(OH)_3 + 2H^+ = 1Cr(OH)^{2+} + 2H_2O$
38.	$1Cr(OH)_3 + 1H^+ = 1Cr(OH)_2^+ + 1H_2O$
39.	$1e^- + 1CrO_2 + 3H^+ = 1Cr(OH)^{2+} + 1H_2O$
40.	$1Cr(OH)^{2+} + 1H^+ = 1Cr^{3+} + 1H_2O$
41.	$1Cr(OH)_2^+ + 1H^+ = 1Cr(OH)^{2+} + 1H_2O$
42.	$1CrO_2^- + 2H^+ = 1Cr(OH)_2^+$
43.	$1e^- + 1Cr(OH)^{2+} + 1H^+ = 1Cr^{2+} + 1H_2O$
44.	$1e^- + 1Cr(OH)_2^+ + 2H^+ = 1Cr^{2+} + 2H_2O$
45.	$3e^- + 1CrO_4^{2-} + 7H^+ = 1Cr(OH)^{2+} + 3H_2O$
46.	$3e^- + 1CrO_4^{2-} + 7H^+ = 1Cr(OH)^{2+} + 3H_2O$
47.	$3e^- + 1HCrO_4^- + 5H^+ = 1Cr(OH)_2^+ + 2H_2O$
48.	$3e^- + 1CrO_4^{2-} + 6H^+ = 1Cr(OH)_2^+ + 2H_2O$
49.	$1CrO_2^- + 3H^+ = 1Cr(OH)^{2+} + 1H_2O$

TABLE F.9 Possible Reactions in the Cr-H_2O System between the Species Most Stable in Dry Conditions

	Equilibria
1.	$2e^- + 1CrO + 2H^+ = 1Cr + 1H_2O$
2.	$2e^- + 1Cr_2O_3 + 2H^+ = 2CrO + 1H_2O$
3.	$6e^- + 1Cr_2O_3 + 6H^+ = 2Cr + 3H_2O$
4.	$2e^- + 2CrO_2 + 2H^+ = 1Cr_2O_3 + 1H_2O$
5.	$6e^- + 2CrO_3 + 6H^+ = 1Cr_2O_3 + 3H_2O$
6.	$2e^- + 1CrO_3 + 2H^+ = 1CrO_2 + 1H_2O$
7.	$2e^- + 1Cr^{2+} = 1Cr$
8.	$3e^- + 1CrO_2^- + 4H^+ = 1Cr + 2H_2O$
9.	$6e^- + 1HCrO_4^- + 7H^+ = 1Cr + 4H_2O$
10.	$6e^- + 1CrO_4^{2-} + 8H^+ = 1Cr + 4H_2O$
11.	$1e^- + 1CrO_2^- + 2H^+ = 1CrO + 1H_2O$
12.	$2CrO_2^- + 2H^+ + 1Cr_2O_3 + 1H_2O$
13.	$1CrO_3 + 1H_2O = 1CrO_4^{2-} + 2H^+$
14.	$1CrO_3 + 1H_2O = 1HCrO_4^- + 1H^+$
15.	$1CrO + 2H^+ = 1Cr^{2+} + 1H_2O$
16.	$3e^- + 1Cr^{3+} = 1Cr$
17.	$3e^- + 1CrO_3^{3-} + r + 3H_2O$
18.	$1e^- + 1CrO_3^{3-} + 4H^+ = 1CrO + 2H_2O$
19.	$1e^- + 1CrO_2 + 4H^+ = 1Cr^{3+} + 2H_2O$
20.	$2e^- + 1HCrO_4^- + 3H^+ = 1CrO_2 + 2H_2O$
21.	$2e^- + 1CrO_4^{2-} + 4H^+ = 1CrO_2 + 2H_2O$
22.	$1Cr_2O_3 + 6H^+ = 2Cr^{3+} + 3H_2O$
23.	$1Cr_2O_3 + 3H_2O = 2CrO_3^{3-} + 6H^+$
24.	$2e^- + 1Cr_2O_3 + 6H^+ = 2Cr^{2+} + 3H_2O$
25.	$6e^- + 2HCrO_4^- + 8H^+ = 1Cr_2O_3 + 5H_2O$
26.	$6e^- + 2CrO_4^{2-} + 10H^+ = 1Cr_2O_3 + 5H_2O$
27.	$3e^- + 1CrO_4^{2-} + 4H^+ = 1CrO_2^- + 2H_2O$
28.	$3e^- + 1CrO_4^{2-} + 2H^+ = 1CrO_3^{3-} + 1H_2O$
29.	$1CrO_2^- + 4H^+ = 1Cr^{3+} + 2H_2O$
30.	$1CrO_3^{3-} + 2H^+ = 1CrO_2^- + 1H_2O$
31.	$1CrO_4^{2-} + 1H^+ = 1HCrO_4^-$
32.	$1e^- + 1Cr^{3+} = 1Cr^{2+}$
33.	$1e^- + 1CrO_2^- + 4H^+ = 1Cr^{2+} + 2H_2O$
34.	$3e^- + 1HCrO_4^- + 7H^+ = 1Cr^{3+} + 4H_2O$
35.	$3e^- + 1CrO_4^{2-} + 8H^+ = 1Cr^{3+} + 4H_2O$
36.	$3e^- + 1HCrO_4^- + 3H^+ = 1CrO_2^- + 2H_2O$
37.	$1Cr_2O_3 + 4H^+ = 2Cr(OH)^{2+} + 1H_2O$
38.	$1Cr_2O_3 + 1H_2O + 2H^+ = 2Cr(OH)_2^+$
39.	$1e^- + 1CrO_2 + 3H^+ = 1Cr(OH)^{2+} + 1H_2O$
40.	$1Cr(OH)^{2+} + 1H^+ = 1Cr^{3+} + 1H_2O$
41.	$1Cr(OH)_2^+ + 1H^+ = 1Cr(OH)^{2+} + 1H_2O$
42.	$1CrO_2^- + 2H^+ = 1Cr(OH)_2^+$
43.	$1e^- + 1Cr(OH)^{2+} + 1H^+ = 1Cr^{2+} + 1H_2O$
44.	$1e^- + 1Cr(OH)_2^+ + 2H^+ = 1Cr^{2+} + 2H_2O$
45.	$3e^- + 1CrO_4^{2-} + 7H^+ = 1Cr(OH)^{2+} + 3H_2O$
46.	$3e^- + 1CrO_4^{2-} + 7H^+ = 1Cr(OH)^{2+} + 3H_2O$
47.	$3e^- + 1HCrO_4^- + 5H^+ = 1Cr(OH)_2^+ + 2H_2O$
48.	$3e^- + 1CrO_4^{2-} + 6H^+ = 1Cr(OH)_2^+ + 2H_2O$
49.	$1CrO_2^- + 3H^+ = 1Cr(OH)^{2+} + 1H_2O$

TABLE F.10 Possible Reactions in the Cu-H$_2$O System between the Species Most Stable in Wet Conditions

	Equilibria
1.	$3H^+ + 1HCuO_2^- = 2H_2O + 1Cu^{2+}$
2.	$4H^+ + 1CuO_2^{2-} = 2H_2O + 1Cu^{2+}$
3.	$1H^+ + 1CuO_2^{2-} = 1HCuO_2^-$
4.	$1e^- + 1Cu^{2+} = 1Cu^+$
5.	$1e^- + 3H^+ + 1HCuO_2^- = 1Cu^+ + 2H_2O$
6.	$1e^- + 4H^+ + 1CuO_2^{2-} = 2H_2O + 1Cu^+$
7.	$2e^- + 2H^+ + 1Cu_2O = 1H_2O + 2Cu$
8.	$2e^- + 2H^+ + 1Cu(OH)_2 = 2H_2O + 1Cu$
9.	$2e^- + 2H^+ + 2Cu(OH)_2 = 3H_2O + 1Cu_2O$
10.	$2H^+ + 1Cu_2O = 1H_2O + 2Cu^+$
11.	$2H^+ + 1Cu(OH)_2 = 2H_2O + 1Cu^{2+}$
12.	$2H^+ + 1CuO_2^{2-} = 1Cu(OH)_2$
13.	$1e^- + 1Cu^+ = 1Cu$
14.	$2e^- + 1Cu^{2+} = 1Cu$
15.	$2e^- + 3H^+ + 1HCuO_2^- = 2H_2O + 1Cu$
16.	$2e^- + 4H^+ + 1CuO_2^{2-} = 2H_2O + 1Cu$
17.	$2e^- + 1H_2O + 2Cu^{2+} = 2H^+ + 1Cu_2O$
18.	$2e^- + 4H^+ + 2HCuO_2^- = 3H_2O + 1Cu_2O$
19.	$2e^- + 6H^+ + 2CuO_2^{2-} = 3H_2O + 1Cu_2O$
20.	$1e^- + 2H^+ + 1Cu(OH)_2 = 2H_2O + 1Cu^+$

TABLE F.11 Possible Reactions in the Cu-H$_2$O System between the Species Most Stable in Dry conditions

	Equilibria
1.	$3H^+ + 1HCuO_2^- = 2H_2O + 1Cu^{2+}$
2.	$4H^+ + 1CuO_2^{2-} = 2H_2O + 1Cu^{2+}$
3.	$1H^+ + 1CuO_2^{2-} = 1HCuO_2^-$
4.	$1e^- + 1Cu^{2+} = 1Cu^+$
5.	$1e^- + 3H^+ + 1HCuO_2^- = 1Cu^+ + 2H_2O$
6.	$1e^- + 4H^+ + 1CuO_2^{2-} = 2H_2O + 1Cu^+$
7.	$2e^- + 2H^+ + 1Cu_2O = 1H_2O + 2Cu$
8.	$2e^- + 2H^+ + 1CuO = 1H_2O + 1Cu$
9.	$2e^- + 2H^+ + 2CuO = 1H_2O + 1Cu_2O$
10.	$2H^+ + 1Cu_2O = 1H_2O + 2Cu^+$
11.	$2H^+ + 1CuO = 1H_2O + 1Cu^{2+}$
12.	$1H^+ + 1HCuO_2^- = 1H_2O + 1CuO$
13.	$1e^- + 1Cu^+ = 1Cu$
14.	$2e^- + 1Cu^{2+} = 1Cu$
15.	$2e^- + 3H^+ + 1HCuO_2^- = 2H_2O + 1Cu$
16.	$2e^- + 4H^+ + 1CuO_2^{2-} = 2H_2O + 1Cu$
17.	$2e^- + 1H_2O + 2Cu^{2+} = 2H^+ + 1Cu_2O$
18.	$2e^- + 4H^+ + 2HCuO_2^- = 3H_2O + 1Cu_2O$
19.	$2e^- + 6H^+ + 2CuO_2^{2-} = 3H_2O + 1Cu_2O$
20.	$1e^- + 2H^+ + 1CuO = 1H_2O + 1Cu^+$

TABLE F.12 Possible Reactions in the Fe-H$_2$O System between the Species Most Stable in Wet Conditions

	Equilibria
1.	$2e^- + 2H^+ = 1H_2$
2.	$4e^- + 1O_2 + 4H^+ = 2H_2O$
3.	$2e^- + 1Fe(OH)_2 + 2H^+ = 1Fe + 2H_2O$
4.	$2e^- + 1Fe^{2+} = 1Fe$
5.	$2e^- + 1Fe(OH)_3^- + 3H^+ = 1Fe + 3H_2O$
6.	$1e^- + 1Fe(OH)_3 + 1H^+ = 1Fe(OH)_2 + 1H_2O$
7.	$1e^- + 1Fe(OH)_3 + 3H^+ = 1Fe^{2+} + 3H_2O$
8.	$1Fe(OH)_3^- + 1H^+ = 1Fe(OH)_2 + 1H_2O$
9.	$1e^- + 1Fe(OH)_3 = 1Fe(OH)_3^-$
10.	$1Fe^{3+} + 3H_2O = 1Fe(OH)_3 + 3H^+$
11.	$1Fe^{2+} + 2H_2O = 1Fe(OH)_2 + 2H^+$
12.	$1e^- + 1Fe^{3+} = 1Fe^{2+}$
13.	$1Fe^{2+} + 1H_2O = 1FeOH^+ + 1H^+$
14.	$1FeOH^+ + 1H_2O = 1Fe(OH)_{2(sln)} + 1H^+$
15.	$1Fe(OH)_{2(sln)} + 1H_2O = 1Fe(OH)_3^- + 1H^+$
16.	$1Fe^{3+} + 1H_2O = 1FeOH^{2+} + 1H^+$
17.	$1FeOH^{2+} + 1H_2O = 1Fe(OH)_2^+ + 1H^+$
18.	$1Fe(OH)_2^+ + 1H_2O = 1Fe(OH)_{3(sln)} + 1H^+$
19.	$1e^- + 1FeOH^{2+} + 1H^+ = 1Fe^{2+} + 1H_2O$
20.	$1e^- + 1Fe(OH)_2^+ + 2H^+ = 1Fe^{2+} + 2H_2O$
21.	$1e^- + 1Fe(OH)_{3(sln)} + 1H^+ = 1Fe(OH)_{2(sln)} + 1H_2O$
22.	$1e^- + 1Fe(OH)_{3(sln)} + 2H^+ = 1FeOH^+ + 2H_2O$
23.	$1e^- + 1Fe(OH)_{3(sln)} + 3H^+ = 1Fe^{2+} + 3H_2O$

TABLE F.13 Possible Reactions in the Fe-H$_2$O System between the Species Most Stable in Dry Conditions

	Equilibria
1.	$2e^- + 2H^+ = 1H_2$
2.	$4e^- + 1O_2 + 4H^+ = 2H_2O$
3.	$8e^- + 1Fe_3O_4 + 8H^+ = 3Fe + 4H_2O$
4.	$2e^- + 1Fe^{2+} = 1Fe$
5.	$2e^- + 1Fe(OH)_3^- + 3H^+ = 1Fe + 3H_2O$
6.	$2e^- + 3Fe_2O_3 + 2H^+ = 2Fe_3O_4 + 1H_2O$
7.	$2e^- + 1Fe_3O_4 + 8H^+ = 3Fe^{2+} + 4H_2O$
8.	$2e^- + 1Fe_2O_3 + 6H^+ = 2Fe^{2+} + 3H_2O$
9.	$2e^- + 1Fe_3O_4 + 5H_2O = 3Fe(OH)_3^- + 1H^+$
10.	$2Fe^{3+} + 3H_2O = 1Fe_2O_3 + 6H^+$
11.	$1e^- + 1Fe^{3+} = 1Fe^{2+}$
12.	$1Fe^{2+} + 1H_2O = 1FeOH^+ + 1H^+$
13.	$1FeOH^+ + 1H_2O = 1Fe(OH)_{2(sln)} + 1H^+$
14.	$1Fe(OH)_{2(sln)} + 1H_2O = 1Fe(OH)_3^- + 1H^+$
15.	$1Fe^{3+} + 1H_2O = 1FeOH^{2+} + 1H^+$
16.	$1FeOH^{2+} + 1H_2O = 1Fe(OH)_2^+ + 1H^+$
17.	$1Fe(OH)_2^+ + 1H_2O = 1Fe(OH)_{3(sln)} + 1H^+$
18.	$1FeOH^{2+} + 1H^+ = 1Fe^{2+} + 1H_2O$
19.	$1e^- + 1Fe(OH)_2^+ + 2H^+ = 1Fe^{2+} + 2H_2O$
20.	$1e^- + 1Fe(OH)_{3(sln)} + 1H^+ = 1Fe(OH)_{2(sln)} + 1H_2O$
21.	$1e^- + 1Fe(OH)_{3(sln)} + 2H^+ = 1FeOH^+ + 2H_2O$
22.	$1e^- + 1Fe(OH)_{3(sln)} + 3H^+ = 1Fe^{2+} + 3H_2O$

TABLE F.14   Possible Reactions in the Mn-H$_2$O System

	Equilibria
1.	$2e^- + 2H^+ = 1H_2$
2.	$4e^- + 1O_2 + 4H^+ = 2H_2O$
3.	$1Mn(OH)^+ + 1H^+ = 1Mn^{2+} + 1H_2O$
4.	$1HMnO_2^- + 3H^+ = 1Mn^{2+} + 2H_2O$
5.	$1HMnO_2^- + 2H^+ = 1Mn(OH)^+ + 1H_2O$
6.	$1MnO + 2H^+ = 1Mn^{2+} + 1H_2O$
7.	$1MnO + 1H^+ = 1Mn(OH)^+$
8.	$1HMnO_2^- + 1H^+ = 1MnO + 1H_2O$
9.	$2e^- + 1Mn_3O_4 + 8H^+ = 3Mn^{2+} + 4H_2O$
10.	$2e^- + 1Mn_3O_4 + 5H^+ = 3Mn(OH)^+ + 1H_2O$
11.	$2e^- + 1Mn_3O_4 + 2H_2O = 3HMnO_2^- + 1H^+$
12.	$2e^- + 1Mn_2O_3 + 6H^+ = 2Mn^{2+} + 3H_2O$
13.	$2e^- + 1MN_2O_3 + 4H^+ = 2Mn(OH)^+ + 1H_2O$
14.	$2e^- + 1Mn_2O_3 + 1H_2O = 2HMnO_2^-$
15.	$2e^- + 1MnO_2 + 4H^+ = 1Mn^{2+} + 2H_2O$
16.	$2e^- + 1MnO_2 + 3H^+ = 1Mn(OH)^+ + 1H_2O$
17.	$2e^- + 1MnO_2 + 1H^+ = 1HMnO_2^-$
18.	$1e^- + 1MnO_2 + 4H^+ = 1Mn^{3+} + 2H_2O$
19.	$3e^- + 1MnO_4^- + 4H^+ = 1MnO_2 + 2H_2O$
20.	$2e^- + 1MnO_4^{2-} + 4H^+ = 1MnO_2 + 2H_2O$
21.	$2e^- + 1MnO + 2H^+ = 1Mn + 1H_2O$
22.	$2e^- + 1Mn_3O_4 + 2H^+ = 3MnO + 1H_2O$
23.	$2e^- + 3Mn_2O_3 + 2H^+ = 2Mn_3O_4 + 1H_2O$
24.	$2e^- + 2MnO_2 + 2H^+ = 1Mn_2O_3 + 1H_2O$
25.	$2e^- + 1Mn^{2+} = 1Mn$
26.	$2e^- + 1Mn(OH)^+ + 1H^+ = 1Mn + 1H_2O$
27.	$2e^- + 1HMnO_2^- + 3H^+ = 1Mn + 2H_2O$
28.	$3e^- + 1Mn^{3+} = 1Mn$
29.	$7e^- + 1MnO_4^- + 8H^+ = 1Mn + 4H_2O$
30.	$6e^- + 1MnO_4^{2-} + 8H^+ = 1Mn + 4H_2O$
31.	$1e^- + 1Mn^{3+} = 1Mn^{2+}$
32.	$4e^- + 1MnO_4^{2-} + 8H^+ = 1Mn^{2+} + 4H_2O$
33.	$4e^- + 1MnO_4^{2-} + 7H^+ = 1Mn(OH)^+ + 3H_2O$
34.	$4e^- + 1MnO_4^{2-} + 5H^+ = 1HMnO_2^- + 2H_2O$
35.	$5e^- + 1MnO_4^- + 8H^+ = 1Mn^{2+} + 4H_2O$
36.	$5e^- + 1MnO_4^- + 7H^+ = 1Mn(OH)^+ + 3H_2O$
37.	$4e^- + 1MnO_4^- + 8H^+ = 1Mn^{3+} + 4H_2O$
38.	$1e^- + 1MnO_4^- = 1MnO_4^{2-}$

TABLE F.15 Possible Reactions in the Ni-H$_2$O System between the Species Most Stable in Wet Conditions

	Equilibria
1.	$1Ni(OH)_2 + 2H^+ = 1Ni^{2+} + 2H_2O$
2.	$2e^- + 8H^+ + 1Ni_3O_4 = 3Ni^{2+} + 4H_2O$
3.	$2e^- + 6H^+ + 1Ni_2O_3 = 3H_2O + 2Ni^{2+}$
4.	$2e^- + 4H^+ + 1NiO_2 = 2H_2O + 1Ni^{2+}$
5.	$2e^- + 1Ni^{2+} = 1Ni$
6.	$2e^- + 3H^+ + 1HNiO_2^- = 2H_2O + 1Ni$
7.	$2e^- + 2H^+ + 1Ni(OH)_2 = 2H_2O + 1Ni$
8.	$2e^- + 2H^+ + 2H_2O + 1Ni_3O_4 = 3Ni(OH)_2$
9.	$1H^+ + 1HNiO_2^- = 1Ni(OH)_2$
10.	$2e^- + 1Ni_3O_4 + 2H_2O = 1H^+ + 3HNiO_2^-$
11.	$2e^- + 2H^+ + 3Ni_2O_3 = 1H_2O + 2Ni_3O_4$
12.	$2e^- + 2H^+ + 2NiO_2 = 1H_2O + 1Ni_2O_3$
13.	$3H^+ + 1HNiO_2^- = 2H_2O + 1Ni^{2+}$
14.	$2e^- + 1H_2O + 1Ni_2O_3 = 2HNiO_2^-$
15.	$2e^- + 1H^+ + 1NiO_2 = 1HNiO_2^-$

TABLE F.16 Possible Reactions in the Ni-H$_2$O System between the Species Most Stable in Dry Conditions

	Equilibria
1.	$2e^- + 8H^+ + 1Ni_3O_4 = 3Ni^{2+} + 4H_2O$
2.	$2e^- + 6H^+ + 1Ni_2O_3 = 3H_2O + 2Ni^{2+}$
3.	$2e^- + 4H^+ + 1NiO_2 = 2H_2O + 1Ni^{2+}$
4.	$2e^- + 1Ni^{2+} = 1Ni$
5.	$2e^- + 2H^+ + 1NiO = 1Ni + 1H_2O$
6.	$2e^- + 3H^+ + 1HNiO_2^- = 2H_2O + 1Ni$
7.	$2e^- + 2H^+ + 1Ni_3O_4 = 1H_2O + 3NiO$
8.	$2H^+ + 1NiO = 1H_2O + 1Ni^{2+}$
9.	$1H^+ + 1HNiO_2^- = 1H_2O + 1NiO$
10.	$2e^- + 1Ni_3O_4 + 2H_2O = 1H^+ + 3HNiO_2^-$
11.	$2e^- + 2H^+ + 3Ni_2O_3 = 1H_2O + 2Ni_3O_4$
12.	$2e^- + 2H^+ + 2NiO_2 = 1H_2O + 1Ni_2O_3$
13.	$3H^+ + 1HNiO_2^- = 2H_2O + 1Ni^{2+}$
14.	$2e^- + 1H_2O + 1Ni_2O_3 = 2HNiO_2^-$
15.	$2e^- + 1H^+ + 1NiO_2 = 1HNiO_2^-$

TABLE F.17 Possible Reactions in the Zn-H$_2$O System

	Equilibria
1.	$2e^- + 2H^+ = 1H_2$
2.	$4e^- + 1O_2 + 4H^+ = 2H_2O$
3.	$3H^+ + 1HZnO_2^- = 2H_2O + 1Zn^{2+}$
4.	$1H^+ + 1Zn(OH)^+ = 1H_2O + 1Zn^{2+}$
5.	$2H^+ + 1HZnO_2^- = 2H_2O + 1Zn(OH)^+$
6.	$4H^+ + 1ZnO_2^{2-} = 2H_2O + 1Zn^{2+}$
7.	$1H^+ + 1ZnO_2^{2-} = 1HZnO_2^-$
8.	$2e^- + 2H^+ + 1Zn(OH)_2 = 2H_2O + 1Zn$
9.	$2H^+ + 1Zn(OH)_2 = 2H_2O + 1Zn^{2+}$
10.	$1H^+ + 1HZnO_2^- = 1Zn(OH)_2$
11.	$2H^+ 1ZnO_2^{2-} = 1Zn(OH)_2$
12.	$2e^- + 1Zn^{2+} = 1Zn$
13.	$2e^- + 3H^+ + 1HZnO_2^- = 2H_2O + 1Zn$
14.	$2e^- + 4H^+ + 1ZnO_2^{2-} = 2H_2O + 1Zn$

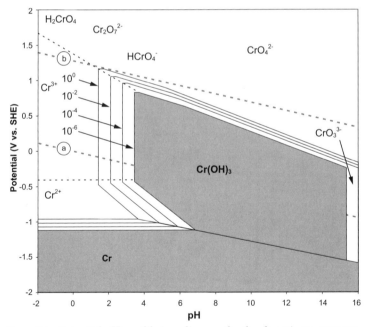

Figure F.1 Potential-pH equilibrium diagram for the chromium-water system at 25°C considering the hydrated oxide forms.

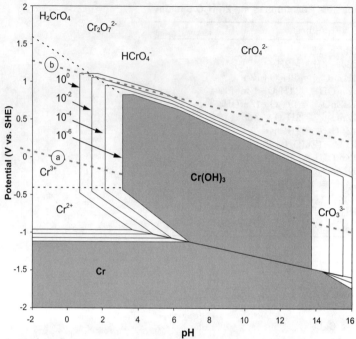

Figure F.2 Potential-pH equilibrium diagram for the chromium-water system at 60°C considering the hydrated oxide forms.

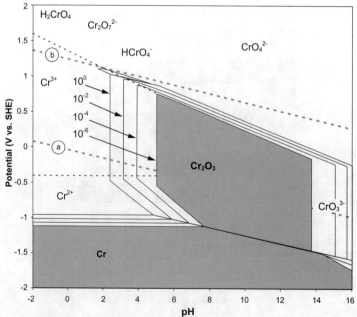

Figure F.3 Potential-pH equilibrium diagram for the chromium-water system at 25°C considering the dry oxide forms.

Thermodynamic Data and E-pH Diagrams 1115

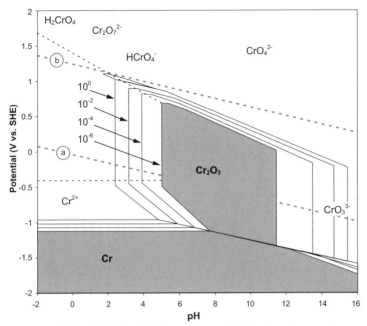

Figure F.4 Potential-pH equilibrium diagram for the chromium-water system at 60°C considering the dry oxide forms.

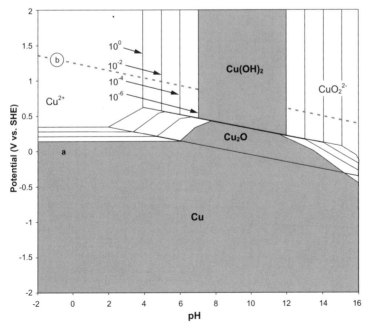

Figure F.5 Potential-pH equilibrium diagram for the copper-water system at 25°C considering the hydrated oxide forms.

1116   Appendix F

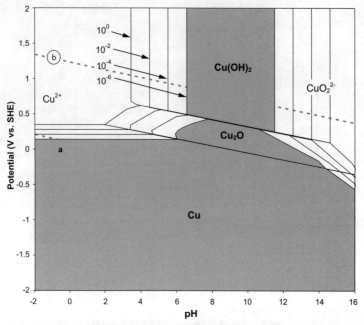

Figure F.6 Potential-pH equilibrium diagram for the copper-water system at 60°C considering the hydrated oxide forms.

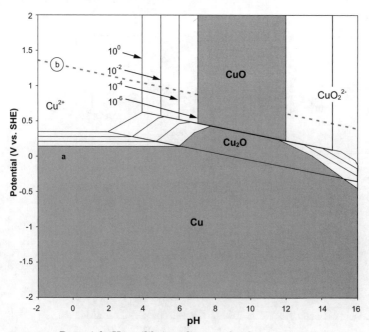

Figure F.7 Potential-pH equilibrium diagram for the copper-water system at 25°C considering the dry oxide forms.

Thermodynamic Data and E-pH Diagrams 1117

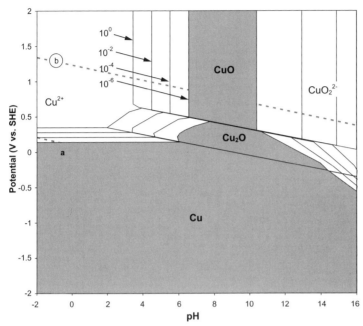

Figure F.8 Potential-pH equilibrium diagram for the copper-water system at 60°C considering the dry oxide forms.

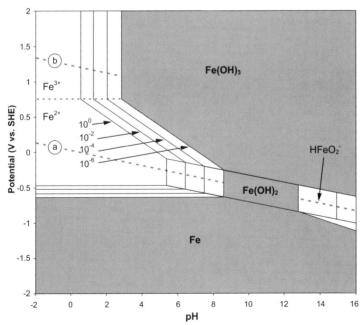

Figure F.9 Potential-pH equilibrium diagram for the iron-water system at 25°C considering the hydrated oxide forms.

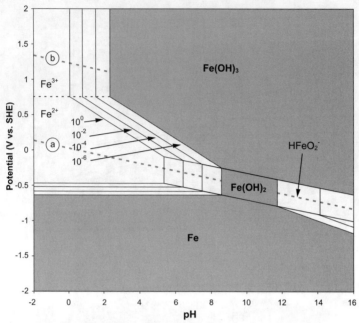

Figure F.10 Potential-pH equilibrium diagram for the iron-water system at 60°C considering the hydrated oxide forms.

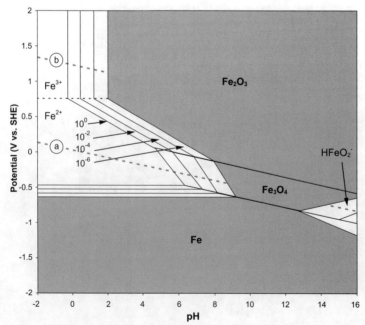

Figure F.11 Potential-pH equilibrium diagram for the iron-water system at 25°C considering the dry oxide forms.

Thermodynamic Data and E-pH Diagrams 1119

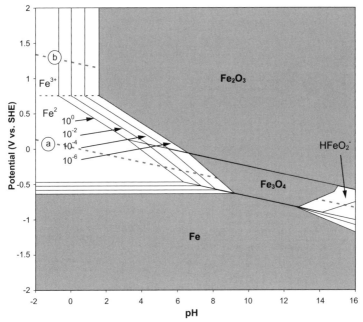

Figure F.12 Potential-pH equilibrium diagram for the iron-water system at 60°C considering the dry oxide forms.

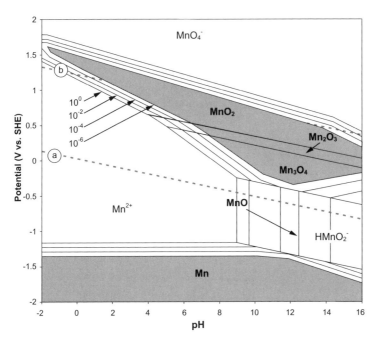

Figure F.13 Potential-pH equilibrium diagram for the manganese-water system at 25°C.

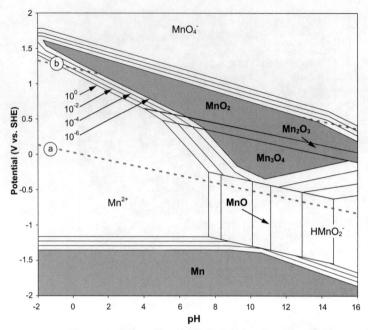

Figure F.14 Potential-pH equilibrium diagram for the manganese-water system at 60°C.

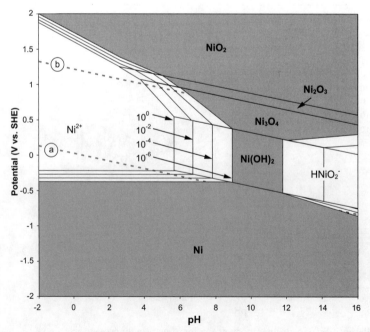

Figure F.15 Potential-pH equilibrium diagram for the nickel-water system at 25°C considering the hydrated oxide forms.

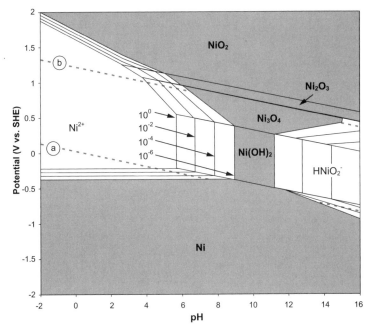

Figure F.16 Potential-pH equilibrium diagram for the nickel-water system at 60°C considering the hydrated oxide forms.

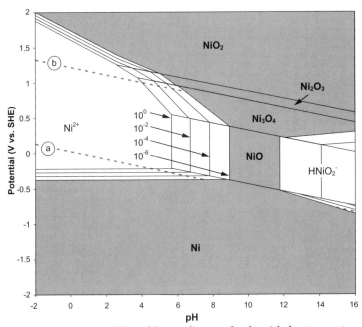

Figure F.17 Potential-pH equilibrium diagram for the nickel-water system at 25°C considering the dry oxide forms.

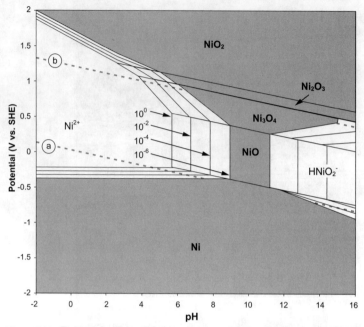

Figure F.18 Potential-pH equilibrium diagram for the nickel-water system at 60°C considering the dry oxide forms.

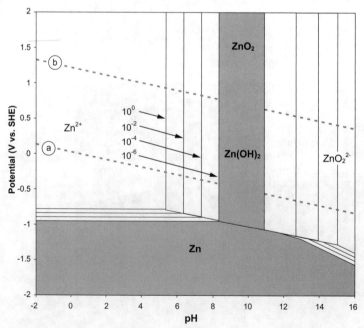

Figure F.19 Potential-pH equilibrium diagram for the zinc-water system at 25°C.

Thermodynamic Data and E-pH Diagrams

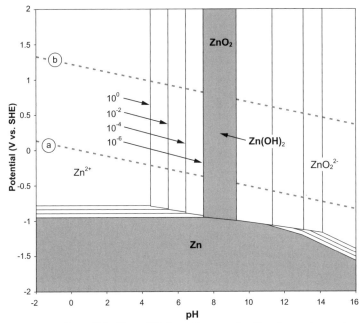

Figure F.20 Potential-pH equilibrium diagram for the zinc-water system at 60°C.

APPENDIX G

Densities and Melting Points of Metals

TABLE G.1 Density of Metals in Decreasing Order of Density

High	g·cm^{-3}	Medium	g·cm^{-3}	Low	g·cm^{-3}
Osmium	22.48	Bismuth	9.90	Gallium	5.97
Iridium	22.42	Erbium	9.16	Arsenic	5.73
Platinum	21.45	Copper	8.96	Germanium	5.32
Rhenium	21.02	Cobalt	8.92	Europium	5.24
Gold	19.30	Nickel	8.90	Selenium	4.81
Tungsten	19.30	Cadmium	8.65	Titanium	4.50
Uranium	19.05	Niobium	8.57	Yttrium	4.34
Tantalum	16.60	Iron	7.87	Barium	3.50
Mercury	13.55	Manganese	7.44	Aluminum	2.70
Hafnium	13.09	Indium	7.31	Strontium	2.60
Rhodium	12.44	Tin	7.30	Boron	2.34
Ruthenium	12.20	Chromium	7.14	Silicon	2.32
Palladium	12.02	Zinc	7.14	Beryllium	1.84
Thallium	11.85	Neodymium	7.00	Magnesium	1.74
Thorium	11.70	Samarium	6.93	Calcium	1.55
Lead	11.34	Cerium	6.78	Sodium	0.97
Silver	10.49	Antimony	6.68	Potassium	0.86
Molybdenum	10.20	Zirconium	6.50	Lithium	0.53
		Tellurium	6.24		
		Lanthanum	6.19		
		Vanadium	6.11		

TABLE G.2 Density of Metals in Alphabetical Order

	g·cm^{-3}		g·cm^{-3}		g·cm^{-3}
Aluminum	2.7	Lanthanum	6.19	Strontium	2.6
Antimony	6.68	Lead	11.34	Tantalum	16.6
Arsenic	5.73	Lithium	0.53	Tellurium	6.24
Barium	3.5	Magnesium	1.74	Thallium	11.85
Beryllium	1.84	Manganese	7.44	Thorium	11.7
Bismuth	9.9	Mercury	13.55	Tin	7.3
Boron	2.34	Molybdenum	10.2	Titanium	4.5
Cadmium	8.65	Neodymium	7	Tungsten	19.3
Calcium	1.55	Nickel	8.9	Uranium	19.05
Cerium	6.78	Niobium	8.57	Vanadium	6.11
Chromium	7.14	Osmium	22.48	Yttrium	4.34
Cobalt	8.92	Palladium	12.02	Zinc	7.14
Copper	8.96	Platinum	21.45	Zirconium	6.5
Erbium	9.16	Potassium	0.86		
Europium	5.24	Rhenium	21.02		
Gallium	5.97	Rhodium	12.44		
Germanium	5.32	Ruthenium	12.2		
Gold	19.3	Samarium	6.93		
Hafnium	13.09	Selenium	4.81		
Indium	7.31	Silicon	2.32		
Iridium	22.42	Silver	10.49		
Iron	7.87	Sodium	0.97		

TABLE G.3 Melting Points of Metals in Decreasing Order of Temperature

High	(°C)	Medium	(°C)	Low	(°C)
Tungsten	3410	Rhodium	1966	Neodymium	1024
Rhenium	3180	Chromium	1930	Silver	961
Tantalum	2996	Zirconium	1857	Germanium	947
Osmium	2700	Thorium	1845	Lanthanum	920
Molybdenum	2610	Platinum	1773	Barium	850
Iridium	2454	Titanium	1725	Calcium	848
Ruthenium	2450	Vanadium	1710	Cerium	815
Niobium	2468	Palladium	1549	Arsenic	814
Boron	2300	Iron	1535	Strontium	774
Hafnium	2230	Cobalt	1495	Aluminum	660
		Yttrium	1490	Magnesium	651
		Nickel	1455	Antimony	630
		Erbium	1450	Tellurium	452
		Beryllium	1278	Zinc	419
		Manganese	1220	Lead	327
		Europium	1150	Cadmium	321
		Uranium	1133	Thallium	302
		Copper	1083	Bismuth	271
		Samarium	1072	Tin	232
		Gold	1063	Selenium	217
		Silicon	1410	Lithium	179
				Indium	156
				Sodium	98
				Potassium	62
				Gallium	30
				Mercury	−38.8

TABLE G.4 Melting Points of Metals in Alphabetical Order

	(°C)		(°C)		(°C)
Aluminum	660	Iridium	2454	Selenium	217
Antimony	630	Iron	1535	Silicon	1410
Arsenic	814	Lanthanum	920	Silver	961
Barium	850	Lead	327	Sodium	98
Beryllium	1278	Lithium	179	Strontium	774
Bismuth	271	Magnesium	651	Tantalum	2996
Boron	2300	Manganese	1220	Tellurium	452
Cadmium	321	Mercury	−38.8	Thallium	302
Calcium	848	Molybdenum	2610	Thorium	1845
Cerium	815	Neodymium	1024	Tin	232
Chromium	1930	Nickel	1455	Titanium	1725
Cobalt	1495	Niobium	2468	Tungsten	3410
Copper	1083	Osmium	2700	Uranium	1133
Erbium	1450	Palladium	1549	Vanadium	1710
Europium	1150	Platinum	1773	Yttrium	1490
Gallium	30	Potassium	62	Zinc	419
Germanium	947	Rhenium	3180	Zirconium	1857
Gold	1063	Rhodium	1966		
Hafnium	2230	Ruthenium	2450		
Indium	156	Samarium	1072		

# Index

Acoustic-emission monitoring, 426
Aircraft, 4, 337–341, 440, 449, 456, 460
    corrosion monitoring, 437, 440
    D check, 402
    maintenance, 401
    structural integrity programs, 403, 404
Aluminum alloys, 584
    aluminum-lithium, 507
    anodic polarization, 507
    anodizing, 806
    applications, 595
    cast, 589
    corrosion resistance, 601
        alloying, 602
        atmospheric corrosion, 602, 603
        grain structure, 507, 609
        mechanical treatments, 602
        metallurgical treatments, 602
    corrosion testing, 498, 499
    exfoliation, 352
    filiform corrosion, 28
    grain structure, 352
    inhibitors, 850
    intergranular corrosion, 508
    mechanical properties, 589
    production, 587
    soil corrosion, 153
    special products, 593
    stress-corrosion cracking, 300, 348, 610
    tempers, 593
    weldability, 598
    welds:
        filler metals, 599
        imperfections, 599
    wrought, 591
American Society for Testing and Materials, 491, 623

Anodic protection, 41, 361, 364, 365, 427, 775, 921–934
    applications, 932
    cathodes, 929
    critical current, 926
    current requirements, 922
    design, 930, 937
    equipment, 927
    film resistance, 926
    heat exchangers, 922, 932
    passivity of metals, 923
    polarization diagrams, 527, 924
    potential control, 930
    power supply, 930
    pulp and paper industry, 933
    reference electrodes, 929, 930
    stainless steels, 922
    steels, 928, 935, 936
    storage vessel, 928
Artificial intelligence, 303
    case-based reasoning, 321
    certainty factors, 298
    expert systems, 306
    knowledge-based systems, 297, 403, 404
    neural networks, 186, 318
Atmospheric corrosion, 58, 62, 81, 84
    aluminum alloys, 602, 603
    atmospheric contaminants, 68
    atmospheric salinity, 67
    copper alloys, 631
    corrosivity maps, 59–61
    evaluation, 81, 84
        CLIMAT test, 82
        microbalance, 84
    ISO methodology, 69, 73, 74, 77
    ISO standards, 70–73
    monitoring, 452
    nature of corrosion products, 65
    PACER LIME algorithm, 78, 80

1130 Index

Atmospheric corrosion (*Cont.*):
  prediction, 84
  sulfur dioxide, 66
  temperature, 68
  theory, 61
    GILDES model, 63
  time of wetness, 66, 69
  weathering steels, 739
Automotive industry:
  coatings, 790
  hot dip galvanizing, 788

Biofouling, 187, 200
  biofilm formation, 201, 202
  control, 208, 214
    biocides, 198, 203, 209, 211–215
    chlorine, 194, 213, 659
    ozone, 215
    physical methods, 210
  enzymes, 202
  formation, 204
  impact, 206
    friction, 207
    thermal conductivity, 207, 773
  marine, 205
  thermal conductivity, 208
Butler-Volmer equation, 36, 37, 1052, 1053

Cast irons, 612
  carbon content, 613
  classification, 613
  corrosion resistance, 617, 618
    alloying, 619
    classification, 620
  gray cast iron, 615, 616
  malleable cast iron, 614
  nodular cast iron, 615
  soil corrosion, 152
  welding, 616
  white cast iron, 613, 617
Cathodic protection, 4, 131, 133, 148, 152, 153, 166–175, 196, 209, 269, 275, 309, 311, 324, 362–367, 427, 655, 742, 789, 793, 811, 863–878, 892–899, 904, 905, 912, 914, 917, 921, 923, 930
  DNV Practice RP B401, 878
  electrical continuity, 171, 874, 885, 904
  electrolytic resistivity, 890
  Evans diagram, 865
  impressed current systems, 878
    anodes, 880, 881
      buried applications, 881
      carbon dioxide evolution, 880
      graphite, 882

Cathodic protection, impressed current systems, anodes (*Cont.*):
      high-silicon chromium cast iron, 884
      materials, 881
      mixed-metal, 883
      platinized, 882
      polymeric materials, 884
      scrap steel and iron, 883
    backfill, 153, 873, 882, 883, 884
    current distribution, 888–891
    design, 885
    disbonded coatings, 337, 886, 887
    ground beds, 884, 885
    interferences, 886
    limitations, 878
    monitoring, 904
      close-interval potential surveys, 906–914, 918
      corrosion coupons, 918
      direct current voltage gradient, 914–917
      hardware performance, 904
      Pearson survey, 913
      structure condition, 905
    potential distribution, 891
    stray currents, 892
      alternating, 894, 895
      controlling, 896–899
      dc rail transit systems, 899, 901
      direct, 893, 894
      gas pipelines, 902
      potential fluctuations, 903
      telluric effects, 896
    theoretical basis, 879
  marine structures, 876
  NACE IP 10A196, 881
  NACE RP0176–94, 878
  protection criteria, 866
    corrosion risk, 867
    IR drop error, 869, 870
    potentials, 867, 870
  reinforced concrete, 168
  sacrificial anodes, 871
    buried structures, 874
    design, 874
    materials, 873
    offshore pipeline, 877
    requirements, 872, 875
  soil resistivity, 874, 885
  test station, 876
  theoretical basis, 864
  Wenner four-pin method, 874
Chlorides:
  detection, 457

Chlorides (*Cont.*):
  high-temperature corrosion, 261
  reinforced concrete, 162, 164, 183
  seawater, 129
  soils, 146
Chlorine, 194, 213, 659
CLIMAT test, 82
Computer-based training, 322
Condition assessment, 180, 390
  surveys, 390, 391
Conductivity, 40
  electrolytic, 890
  resistivity of water, 41
  soil resistivity, 146, 874, 885
Copper alloys, 622
  atmospheric corrosion, 631
  bronzes, 629
  cast, 625
  copper-nickels:
    chlorine, 659
    galvanic corrosion, 655
    marine applications, 650
    microfouling, 655
    passivation, 656
    pitting, 652
    seawater, 656, 658
    sulfides, 658
    UNS C70600 (90–10), 651, 653
    UNS C71500 (70–30), 652, 653
  corrosion resistance, 630, 632
    acids, 649
    alkalis, 649
    organic compounds, 649
    polluted waters, 649
    salts, 648
    soils, 648
    steam, 648
    UNS C70600 (90–10), 652
    UNS C71500 (70–30), 652
  decorative finishes, 659, 663
    brown patina, 662
    green patina, 660, 661
    patina, 659
  dezincification, 152, 344, 345, 366, 631, 648
  inhibitors, 851
  liquid metal embrittlement, 649
  seawater, 140
  soil corrosion, 152
  trade names, 624, 626
  welding, 627
  wrought, 625
Copper Development Association, 623
COR·SUR, 583
Corrosion coupons, 81, 517, 918

Corrosion diagrams:
  iso-corrosion, 584
  mixed acid graph, 587
  stainless steels, 583, 585
  titanium alloys, 762, 763, 764
Corrosion economics, 1001
  capital budgeting techniques, 1003
  cash-flow techniques, 1001
  straight-line depreciation, 1004
  time value of money, 388, 1002, 1009
Corrosion failures, 16, 292, 331, 332, 374
  Aloha incident, 3, 338, 339
  artillery projectile, 333, 335
  consequences, 382
  corrosion of reinforcing steel
    in concrete, 30
    carbonation, 30
    hydrogen embrittlement, 31
    re-alkalization, 30, 171, 172, 173
  evaluation, 414
  factors, 354
  failure analysis, 359, 360, 491
  failure modes, 303, 347, 443, 565
    mode definition, 16, 17
    stainless steels, 723, 725
  filiform corrosion, 26–29
  high-temperature corrosion, 247
  mechanism, 245, 256, 280, 338, 349, 353, 357, 358, 368, 382
  modes, 352
  MV KIRKI, 4
  prevention, 360
  probability, 382
    subjective assessment, 356
  reinforced concrete, 186
  responsibility for corrosion failures, 10
  Statue of Liberty, 58, 342, 345
  steam generator tubes, 397
  system management, 7
Corrosion forms, 193, 332, 334
Corrosion management, 276, 372, 373, 410, 429, 430
  monitoring, 406, 428
  reporting templates, 356, 358
Corrosion monitoring:
  access fittings, 409, 410
  corrosion coupons, 417, 918
  electrical resistance, 423, 441, 919
  electrochemical impedance
    spectroscopy, 424, 435, 539
  electrochemical noise, 425, 436, 548
  harmonic analysis, 424
  inductive resistance, 417
  monitoring points, 413

1132 Index

Corrosion monitoring (*Cont.*):
  sensors, 82, 84, 385, 397, 409–416, 437,
    438, 440, 441, 445–448, 918
  strategies, 406
  systems, 409, 440
  techniques, 182, 406, 413–418, 427,
    430, 431, 437, 440, 445
    direct, 416
    indirect, 416
Corrosion prevention, 360
  control measures, 8, 9, 10
  preventive measures, 308, 361, 401
Corrosion rates:
  conversion, 529
  thermodynamic calculations, 32, 1047
Corrosion surveillance, strategy, 74, 76
Corrosion testing, 148
  accelerating factors, 500
  aluminum alloys, 498, 499, 505, 602
  anodic breakthrough method, 500–504
  ASCOR, 495
  ASTM B 117, 489, 495, 515, 556
  ASTM G 3, 525
  ASTM G 44, 300, 497, 609
  ASTM G 78, 495, 497
  ASTM handbook, 494
  corrosion products, 496
  electrochemical methods, 522
    cyclic potentiodynamic polarization,
      531
    electrochemical impedance
      spectroscopy, 539
    electrochemical noise, 549–551
    polarization, 523
    polarization diagram, 527, 528, 924
    potentiodynamic polarization, 526, 536
    potentiostaircase, 526
  field and service, 555
    potentiodynamic polarization, 536
    test facility, 557
    types of exposure testing, 557
  high-performance alloys, 561
  high-purity water, 520
  intergranular anodic test, 505
  laboratory tests, 512
    cabinet testing, 513
      alternate immersion, 517
      controlled humidity, 513, 514
      corrosive gas, 514
      immersion testing, 516
    high-temperature/high pressure, 517,
      518, 519
    windowed test vessels, 518
  nuclear waste, 561

Corrosion testing (*Cont.*):
  optimization, 559
  protective coatings, 561
  salt-spray testing, 502, 504, 515, 556
  stainless steels, 527, 537, 538
  stress intensity, 521
  welds, 487
Cost of corrosion, 1
  Battelle study, 2
  catastrophic damage, 3
  corrosion of the infrastructure, 4
  Hoar's report, 1, 359
  influence of people, 5
Crevice corrosion, 25, 58, 62, 73, 140, 141,
    157, 269–271, 277, 289, 336, 338, 365,
    399, 443, 508, 531–535, 539, 552, 618,
    652, 656, 677, 678, 723, 727–733, 752,
    756, 766, 767, 775, 776, 931
  aircraft, 4, 337, 338, 440, 449, 456,
    460, 465
  corrosion testing, 552, 553, 554
  critical temperature, 734
  deposits, 401
  duplex stainless steels, 730, 731
  modeling, 269
  prevention, 365
  stainless steels, 733
  underdeposits, 376

DC rail transit systems, 900

Electrochemical impedance spectroscopy,
    320, 545, 547
  constant phase element, 510, 540
  corrosion monitoring, 424, 435, 539
  equivalent circuits, 541
  Kramers-Kronig, 547
  validating, 545
Electrochemical noise, 82, 271, 425, 436,
    437, 441–445, 523, 549, 550, 1028
  corrosion monitoring, 425, 436, 548
Electrochemistry of corrosion:
  diffusion coefficients, 40, 236, 1023
  Evans diagram, 42, 45–54, 269, 426, 865
  exchange current, 32–35, 43, 1047–1049,
    1052, 1054, 1057
    electrode composition, 33
    soluble species concentration, 34
    surface impurities, 35
    surface roughness, 34
  instrumentation, 923, 1025
  kinetic principles, 32, 230
  Luggin capillary, 1025–1027

Electrochemistry of corrosion (*Cont.*):
  overpotential:
    activation overpotential, 35, 42, 1050
    concentration overpotential, 35, 38, 42, 1051, 1055
    hydrogen overpotential, 15
    ohmic drop, 35, 41, 909, 1051, 1059
  polarization, 35, 43, 1049
    Butler-Volmer equation, 36, 37, 1052, 1053
    Tafel slope, 43, 424, 528, 834, 843, 844, 845
  thermodynamics, Nernst potential, 43
Ellingham diagrams, 222, 223, 225
Environmental cracking, 297, 346, 520
  hydrogen embrittlement, 31, 168, 169, 171, 173, 346, 347, 348, 367, 490, 505, 612, 676, 678, 691, 707, 718, 745, 768, 777, 866
  prevention, 367
  stress-corrosion cracking, 82, 282, 283, 288, 289, 293, 297, 299–304, 307, 318, 319, 346–354, 366, 367, 490, 491, 505, 520–522, 549, 592, 601, 602, 609, 610, 612, 618, 620, 631, 653, 676–678, 691, 712, 714, 715, 723, 726–729, 732, 733, 737, 746, 765, 766, 776, 777
Erosion corrosion, 141, 196, 345, 366, 654, 657
Experimental design, 559–561
Extreme value statistics, 274

Failure analysis, 302, 332
Filiform corrosion, 28, 337
Flade potential, 924, 933
Flue gas desulfurization, 397, 677, 768
Fractal geometry, 271
  Hurst exponent, 273, 550

Galvanic corrosion, 32, 58, 174, 289, 339, 342, 344, 345, 363, 364, 446, 486, 582, 601, 655, 678, 767, 789, 809, 1028
  copper-nickel alloys, 655
  galvanic series, 32, 340, 342, 344
  prevention, 363
Gas pipelines, 3, 146, 208, 276, 279, 282, 283, 284, 289–291, 292, 408, 412, 426, 446, 447, 657, 864, 867, 869, 870, 875, 877, 884–893, 895–918
  cathodic protection, 864, 876, 902
  risk assessment, 279
Gas scrubbing, corrosion monitoring, 441, 442, 444

Heat exchanger tubes, corrosion monitoring, 443
Heat exchangers, anodic protection, 922, 932
High performance alloys, 664
  alloying elements, 666
  applications, 680, 691
  carbon and carbides, 668
  cobalt base, 664
  corrosion resistance, 676, 680, 691
    nickel-base alloys, 677
  heat treatments, 673
  intermetallic phases, 669
  iron nickel base, 664–666, 671, 672, 676
  nickel base, 561, 664, 666, 671, 676
  seawater, 140
  specifications, 680
  welding, 671
High-temperature corrosion, 221
  ASSET, 264
  carburization, 255–257, 265
  chlorides, 261
  Co-S-O system, 231
  creep strength, 241
  Cr-O system, 226, 228
  Cr-S-O system, 232
  deposits, 262
  Ellingham diagrams, 222–225
  F*A*C*T, 265
  gaseous halogens, 260
  high-temperature alloys, 242
  hydrogen sulfide, 250, 259
  kinetics, 230, 236
  liquid metals, 263
  mechanism, 233, 234
  metal dusting, 258
  molten salts, 263
  Ni-S-O system, 230
  nitridation, 260
  oxidation, 223, 229, 238, 243
  penetration damage, 239, 247
  Pilling-Bedworth, 231
  rupture strength, 240
  in service, 237
  stability diagrams, 230
  sulfidation, 245, 247, 249, 253, 256, 264
  thermodynamics, 222
  vapor species diagrams, 223
History of corrosion science, 11
Hydrochloric acid, 583, 586, 677, 699, 700, 756, 774, 776, 842, 926, 927
Hydrogen:
  embrittlement, 31, 168, 169, 171, 173, 346–348, 367, 490, 505, 612, 676, 678, 691, 707, 718, 745, 768, 777, 866

1134　Index

Hydrogen, embrittlement, (Cont.):
　　cobalt-base alloys, 691
　　evolution, 15, 744
　　monitoring, 427
　　stress-corrosion cracking, 612

Inhibitors, 178, 833–839, 857
　　acids, 839
　　aluminum alloys, 850
　　cathodic inhibitors, 837
　　classification, 834
　　copper alloys, 851
　　corrosion processes, 841
　　definition, 833
　　efficiency, 834
　　　　evaluation, 845
　　functionality scheme, 836
　　mechanism, 838
　　oil and gas operations, 851
　　　　acidizing, 845
　　　　application methods, 856
　　　　oxygen-influenced corrosion, 855
　　　　sour corrosion, 854
　　　　sweet corrosion, 853
　　organic inhibitors, 269, 837
　　organophosphorus, 561
　　passivating inhibitors, 836
　　precipitation inhibitors, 837, 838
　　reinforced concrete, 175, 178
　　selection, 860
　　steels, 849
　　　　pickling, 845
　　trans-cinnamaldehyde, 846, 847
　　vapor phase, 838
　　volatile, 838, 857, 859
　　zinc, 850
Inspection, 4, 22, 197, 288, 344, 371–374, 376, 383, 385, 399, 406, 462, 464, 548, 876
　　boilers, 376
　　inspection points, 375
　　nuclear reactors, 400
　　process piping, 375, 378
　　risk-based, 377, 381, 383
　　　　probability of failures, 382
　　techniques, 372, 376
Intergranular corrosion, 73, 248, 349, 351, 352, 365, 399, 465, 486, 505, 506, 507, 508, 601, 668, 677, 678, 716, 717, 720, 733–736
　　aluminum alloys, 352
　　prevention, 365
　　stainless steels, 351
Internet, 324–326

Ion exchange, 101, 103
　　selectivity, 103
　　types of resins, 102, 104

Jointing compounds, 830

Kraft liquors, 933–936
　　composition, 933
Kramers-Kronig, 547

Langelier saturation index, 106, 110
Larson-Skold index, 109
Life-cycle costing, 166, 184, 388
　　corrosion economics, 1001
Life prediction, 267
　　aircraft, 282
　　aluminum alloys, 296, 300
　　crevice corrosion, 269
　　extreme value statistics, 274
　　fault tree analysis, 279
　　gas pipelines, 279, 289
　　modeling, 268, 291
　　　　bottom up, 268
　　　　fractal models, 271
　　　　mathematical models, 268
　　　　probability estimation, 295
　　　　statistical models, 268, 274
　　　　subjective assessment, 291
　　　　top down, 277
　　nuclear waste, 276
　　pitting corrosion, 274
　　reinforced concrete, 184
　　stainless steel cracking, 318
　　sweet gas corrosion, 320
Linear polarization, corrosion monitoring, 414–416, 424, 426, 433, 435
Localized corrosion:
　　local environment, 13, 22, 25–31, 81, 82, 147, 192, 336, 337, 349, 364, 365, 391, 399, 887, 888
　　surface area ratio, 336, 340, 519
Luggin capillary, 1025–1027

Maintenance:
　　asset management, 387, 388, 390, 391, 392
　　costs, 383, 391
　　nuclear reactors, 397
　　poor practices, 384
　　predictive, 385, 386, 395–397, 405
　　preventive, 385
　　prioritizing, 391
　　reliability-centered, 386, 387, 395–397, 401

Index   1135

Maintenance (Cont.):
  strategies, 384
Maintenance Steering Group, 282
  environmental deterioration analysis, 286, 287
  structure significant items, 285
Materials selection, 577
  COR SUR, 583
  costs, 770
  pitting index, 364
  seawater, 139, 140
Materials Technology Institute, 309, 310, 311, 314, 316, 359
Maximum entropy method, 425
Methods of corrosion control, 8, 9, 10
Microbiologically influenced corrosion, 144, 147, 152, 187, 188, 191–197, 201, 208, 209, 766
  activity assays, 198
  assays, 190
  classification, 190
    acid-producing fungi, 195
    aerobes, 192, 194
    iron/manganese oxidizing, 193
    methane producers, 195
    organic acid producers, 195
    slime formers, 194
    sulfur/sulfide oxidizing, 192
  direct inspection, 197
  DNA probes, 199
  effect of operating conditions, 195
    cleanliness, 196
    flow velocity, 196
    oxygen, 196
    pH, 196
    temperature, 196
  growth assays, 198
  identification, 197
  most probable number, 198
  planktonic, 200, 213
  sampling, 200
  soils, 148, 149
  sulfate-reducing bacteria, 189, 191–199, 888
Molybdenum alloys:
  applications, 695
  corrosion resistance, 694, 698
  machining, 696
  properties, 693
  specifications, 694
  welding, 697
Monitoring:
  acoustic emission, 426
  aircraft, 437

Monitoring (Cont.):
  corrosion potentials, 416, 427, 432
  electrical field signature method, 426
  fiber optics, 448
  hydrogen, 427
  potentiodynamic polarization, 426, 536
  reinforced concrete, 432
  sensors, 417, 423, 438
  thin layer activation, 426
  zero resistance ammetry, 425, 435

NACE International, 309, 316, 359, 582, 583, 878, 905, 909, 911, 1002
Niobium alloys, 699
  applications, 699
  corrosion resistance, 700, 701
  machining, 699
  properties, 693
  welding, 700
Nondestructive evaluation, 182, 302- 305, 428, 431, 461–479
  eddy current, 473
  flaw size, 480
  liquid penetrant, 471
  magnetic particle, 471
  radiographic inspection, 471
    Compton backscatter imaging, 473
  thermographic, 477
  ultrasonic, 475
  visual inspection, 465
    borescopes, 469, 470
    fiberscopes, 470

Oil and gas operations:
  acidizing, 845
  cathodic protection, 876
  inhibitors, 851
  oxygen-influenced corrosion, 855
  pitting corrosion, 274
  sour corrosion, 854
  sweet corrosion, 853
Oxygen:
  high-temperature corrosion, 223, 229, 243
  microbiologically-influenced corrosion, 196
  oil and gas operations, 855
  seawater, 658
  solubility in seawater, 38, 133
  solubility in soils, 144
  solubility in water, 25, 38
Ozone, 215

Passivation:
  active-passive behavior, 525, 930, 933
  anodic nose, 535
  copper-nickel alloys, 656
  hysteresis, 534
  passivity of metals, 923
  pitting potential, 533
  polarization diagram, 527, 924
  repassivation potential, 533
  stainless steels, 725, 726
Phosphoric acid, 666, 722, 744, 776, 806, 925, 932
Pilling-Bedworth, 231
Pillowing, 4, 338, 465
Pitting corrosion, 22, 274, 335, 336, 364, 385, 426, 435, 602, 650, 679, 718, 727
  copper-nickel alloys, 652
  critical pitting temperature, 679
  duplex stainless steels, 730, 731
  nickel-base alloys, 679
  oil and gas operations, 274
  pitting potential, 531
  pitting resistance, 364, 733
  prevention, 364
  stainless steels, 733
  subjective assessment, 357
Potentiodynamic polarization, 320, 435, 523, 526, 531, 532
  monitoring, 426, 536
  scan rate, 529
  in service, 536
  solution resistance, 530
Pourbaix diagrams, 13–31, 269, 427, 841, 842, 866, 1040–1047
  applications, 16
  aluminum, 29
  corrosion of reinforcing steel in concrete, 29
  filiform corrosion, 26
  hydronic heating of buildings, 22
  iron, 20, 21, 23, 24
  marine boilers, 17
Protective coatings, 362, 612, 781
  inorganic coatings, 805
    anodizing, 805
      suitable alloys, 806
    chromate filming, 809
    nitriding, 810
    passive films, 63, 810
    phosphate coatings, 810
  metallic coatings, 782
    applications, 799
    chemical vapor deposition, 782, 803, 804
    cladding, 793

Protective coatings, metallic coatings, (Cont.):
    costs, 799
    diffusion coatings, 789
    electroless plating, 787
    electroplating, 786
    hot dip galvanizing, 788
    pack cementation, 789, 792
    physical vapor deposition, 803, 804
    properties, 783
    techniques, 796, 801
    thermal spraying, 794
      environmental concerns, 798
      limitations, 801
      materials, 795
  non-stick coatings, 828
    fluorinated polyols, 829
    surface free energy, 828
  organic coatings, 148, 363, 782, 810, 811
    binders, 824
    components, 823
    epoxy, 824, 825
    intermediate coats, 822
    pigments, 826
    polyurethane, 823
    primers, 822
    solvents, 827
    specifications, 812
    standards, 812
    topcoats, 823
    variables, 811
    volatile organic compounds, 738, 828
    water-borne coatings, 828
    zinc phosphates, 827
  processes, 782
  surface preparation, 660, 831
Puckorius scaling index, 108
Pulp and paper industry, 755, 933, 934, 935, 936

Refractory metals, 692
Reinforced concrete, 153, 154, 156
  alkali-aggregate reaction, 186
  carbonation, 30, 157, 165
  chlorides, 162, 163
  condition assessment, 180–183
  corrosion products, 159
  degradation processes, 158, 186
  de-icing salts, 159
  freeze-thaw damage, 187
  high-performance concrete, 179
  hydrogen embrittlement, 31
  life prediction, 184
  mechanism, 162
  monitoring, 182, 432

Index 1137

Reinforced concrete (*Cont.*):
  re-alkalization, 30, 171–173
  rebar corrosion, 30, 155–164, 166, 174, 178, 180, 184–186, 432–437
  remedial measures, 166, 171
    cathodic protection, 168, 170
    concrete design, 178
    de-icing methods, 166, 168
    electrochemical chloride extraction, 170, 171
    epoxy coating, 175
    galvanized rebar, 177
    inhibitors, 175, 178
    stainless steel rebar, 175
  repairs, 173, 174
  sulfate attack, 187
Resistivity, 143, 146, 148, 530, 587, 695, 706, 708, 872–878, 884, 885, 888–892, 900, 931
  electrolytic, 890
  soil resistivity, 146, 874, 885
  water resistivity, 41
Risk assessment:
  evaluation, 402
  failure mode and effects analysis (FMEA), 396
  fault tree analysis, 5, 6, 280, 281, 295, 297
  gas pipelines, 279, 289, 291
  inspection, 383
  safety hazards, 384
Ryznar stability index, 108, 109

Scaling indices, 105
Schaeffler diagram, 715, 716, 720, 734
Sealants, 830
Seawater, 129
  brackish, 137
  calcareous deposits, 131
  calcium carbonate, 132
  copper-nickel alloys, 656
  corrosion resistance, 138
  dissolved materials, 129
  dissolved organic compounds, 135
  materials selection:
    copper alloys, 140
    nickel alloys, 140
    stainless steels, 140
  oxygen, 133
  pH, 132
  polluted, 136, 138
  salinity, 129, 130
  sulfides, 658
  temperature effects, 141

Seawater (*Cont.*):
  treatments, 658
  velocity effects, 140, 656
Selective leaching, 152, 344, 345, 617, 618, 653
  dezincification, 152, 344, 345, 366, 631, 648
  graphitization, 345
SI Units, 939
Soil corrosion, 142, 492
  aluminum alloys, 153
  bacteria, 148, 149
  cast iron, 152
  chloride, 146
  copper alloys, 152, 648
  corrosivity of soils, 143
  ferrous alloys, 151
  ions, 147
  lead, 153
  microbiologically-influenced corrosion, 147
  nonferrous alloys, 152
  oxygen content, 144
  particle size, 143
  pH, 144
  prediction, 150
  redox potential, 146
  reinforced concrete, 153
  resistivity, 146, 150
  soil classification, 142, 145, 147, 148
  soil horizon, 142, 147
  soil profile, 142
  stainless steels, 152
  water, 143
  zinc, 153
Stainless steels, 140, 710, 721
  anodic polarization, 725, 726
  austenitic, 712
    anodic protection, 922, 925
    mechanical properties, 713
  corrosion diagrams, 583, 585
  corrosion resistance, 723
    alloying elements, 723
    austenitic steels, 729
    chemical environments, 735
    crevice corrosion, 733, 734
    duplex steels, 730
    ferritic steels, 727
    martensitic steels, 729
    pitting corrosion, 733
    sensitization, 734
  corrosion testing, 537, 538
  Cr equivalent, 716
  duplex, 714
    mechanical properties, 715

Stainless steels (*Cont.*):
  failure modes, 723, 725
  ferrite number, 716, 719
  ferritic, 712
    mechanical properties, 713
  heat treatments, 716, 721
  high temperatures, 733
  intergranular corrosion, 351
  martensitic, 712
    mechanical properties, 714
  Ni equivalent, 716
  passivation, 63, 810
  pitting corrosion, 364
  precipitation-hardening, 714
  seawater, 140
  sigma phase, 734
  stress-corrosion cracking, 349
  surface finishes, 716, 722, 724
  testing, 525, 528
  welding, 716
    filler metals, 718
    imperfections, 720
Steel Structures Painting Council, 811
Steels, 736
  anodic protection, 928, 934–936
  carbon steels, 737
  corrosion resistance:
    acids, 744
    alkalis, 737, 744
    alloying elements, 746
    carbon steels, 741–743
    organic solvents, 745
    weathering steels, 739
  inhibitors, 849
  pickling, 845
  weathering steels, 738
  welding, 739
Stochastic process detector, 550, 551
Stress-corrosion cracking:
  aluminum alloys, 300, 610
  intergranular, 399
Sulfides:
  copper-nickel alloys, 658
  high-temperature corrosion, 250, 259
  sulfide-stress cracking, 348
Sulfuric acid, 67, 100, 189, 193, 250, 271, 397, 525, 527, 619, 660, 666, 667, 668, 678, 691, 707, 710, 742, 744, 756, 768, 773–775, 806, 807, 809, 810, 921–928, 932, 934
Surface characterization, 562
  ASTM E 673, 562
  Auger electron spectroscopy, 562–566
  detection characteristics, 564

Surface characterization (*Cont.*):
  photoelectron spectroscopy, 567
  Rutherford backscattering, 568
  scanning Auger microscopy, 563, 571
  scanning probe microscopy, 563, 569
  secondary electron microscopy, 563, 571
  secondary ion mass spectroscopy, 572
  sensitivity problems, 566
Surface preparation, 660, 831

Tantalum alloys, 705
  applications, 705
  corrosion resistance, 703, 707
  machining, 706
  properties, 693
  welding, 706
Thermal conductivity, biofouling, 208
Thermodynamics:
  Gibbs energy, 265, 270, 1034
  high-temperature corrosion, 222
  Nernst potential, 43
Titanium alloys, 748
  applications, 754
  commercial grades, 752
  corrosion resistance, 750, 755, 757
    acids, 756, 762–764
    alkalis, 765
    chlorine gas, 766
    gases, 767
    methanol, 765
    natural waters, 766
    organic acids, 765
    oxidizing acids, 756
    seawater and salt solutions, 766
    steam, 766
  properties, 749
  specifications, 753
  welding, 752
Transmission towers, 391, 392
Tungsten alloys, 708
  applications, 708
  corrosion resistance, 711
  machining, 709
  properties, 693
  welding, 709

Uniform corrosion, 333
  prevention, 362

Water chemistry, 99, 108, 110, 117–119, 122–127, 399, 521, 522, 561
  biochemical oxygen demand, 96
  hardness, 94, 95, 96

Water chemistry (*Cont.*):
  ion activity product, 105, 107, 108, 111, 112
  mineral salts, 93
  nuclear reactors, 399
  organic matter, 96
  pH, 96
  priority pollutants, 97
  saturation of water, 105, 106
  scaling, 105–114, 118, 119, 122–128
    carbon dioxide, 92
    ion association model, 112–117, 122, 124
    Langelier saturation index, 106, 110
    Larson-Skold index, 109
    momentary excess index, 110
    Oddo-Tomson index, 110
    prediction, 125
    Puckorius scaling index, 108
    Ryznar stability index, 108, 109
    Stiff-Davis index, 110
  total organic carbon, 97
Water-displacing compounds, 830
Water treatment:
  purification, 87, 88
    ion exchange, 99–103
    methods, 91

Welding;
  aluminum alloys, 598
  cast irons, 616
  copper alloys, 627
  high-performance alloys, 671
  molybdenum alloys, 697
  niobium alloys, 700
  stainless steels, 716
  steels, 739
  tantalum alloys, 706
  tungsten alloys, 709
Welding titanium alloys, 752
Welds:
  corrosion monitoring, 445
  heat adjacent zone, 351, 486

Zero resistance ammetry, monitoring, 425, 435
Zirconium alloys, 769
  applications, 773
  corrosion resistance, 772, 774
    acids, 775
    alkalis, 776
    molten metals and salts, 777
    organic compounds, 777
  mechanical properties, 770
  specifications, 770

ABOUT THE AUTHOR

Pierre R. Roberge, Ph.D., P.Eng., is an internationally recognized expert on corrosion, corrosion prevention, and computer modeling techniques. A professor of chemical and materials engineerng at the Royal Military College of Canada, he has written more than 125 technical and professional articles and book chapters on advances in his field.